J. F. von Raumer F. Neubauer (Eds.)

Pre-Mesozoic Geology in the Alps

With 279 Figures and 32 Tables

Springer-Verlag
Berlin Heidelberg New York London Paris Tokyo
Hong Kong Barcelona Budapest

Prof. Dr. J. F. von RAUMER
Institut für Mineralogie und Petrographie
Universität Pérolles
CH-1700 Fribourg
Switzerland

Prof. Dr. FRANZ NEUBAUER
Institut für Geologie und Paläontologie
Karl-Franzens-Universität
Heinrichstraße 26
A-8010 Graz
Austria

ISBN 978-3-642-84642-7 ISBN 978-3-642-84640-3 (eBook)
DOI 10.1007/978-3-642-84640-3

Library of Congress Cataloging-in-Publication Data
Pre-Mesozoic geology in the Alps / J. F. von Raumer, F. Neubauer (eds.). p. cm. Includes bibliographical references and index.
ISBN-13:978-3-642-84642-7 1. Geology, Stratigraphic–
Paleozoic. 2. Geology–Alps. I. Raumer, J. F. von (Jürgen F.), 1932– . II. Neubauer, Franz, 1951– . QE654.P745 1993' 551.7'2'094947–dc20

Softcover reprint of the hardcover 1st edition 1993

32/3145-5 4 3 2 1 0 Printed on acid-free paper

Preface

The Alps are an arched mountain chain stretching 1500 km between Vienna and Graz in Austria and Genova in Italy. They resulted from the collision of the African and Laurasian plates during Mesozoic and Tertiary times. The high standard of knowledge attained over the last 30 years by the working groups on "Alpine Metamorphism" is well known and helped considerably to recognize pre-Mesozoic elements in the Alps. In Part I of this book the subdivision of the major Alpine units and pre-Mesozoic palinspastic reconstructions are covered before discussion of the pre-Mesozoic geology in Parts II, III and IV. It is understood that the Mesozoic and later events overprinted pre-existing structures veiling the earlier history and the nature of protoliths.

Although the Alpine overprint does not facilitate the recognition of older structures, pre-Mesozoic basement units were recognized during the first beginnings of geological observations in the Alps, about 200 years ago. Fifty percent of the Alpine domain is underlain by basement units that have been unconformably covered since Permian and Mesozoic times. This basement appears today in a complex pattern among the Alpine structures. The history of their discovery and explanation, parallel with a growing sophistication of research methods, are the subject of the introductory chapter of Part II.

Stratigraphic work in the Alpine domains has been undertaken since the middle of the last century by several geological schools. Detailed knowledge of fossil-bearing Ordovician to Permian sedimentary sequences led to very early evolutionary models for the timing of Variscan orogeny in the Alps. However, the relationships of widespread metamorphic units have been a matter of discussion up to the present. As Alpine basement appears in most areas as high-grade metamorphic units, it was not included in the general interpretation of the Variscan framework in Europe. It was the growing knowledge of Alpine metamorphic events and Alpine palaeogeographic framework, and the application of lithostratigraphic comparative work in high-grade metamorphic terrains, that allowed the possibility of comparing pre-Mesozoic structures outside the Alps with those encountered in the Alps. Also, over the last 10 years, the psychological barriers that impeded correlation among the main Variscan structures, including those in the Alps, have been reduced.

Since 1983, three meetings influenced considerably the actual knowledge on basement correlations in the Alps: the conference on "Variscan and Prevariscan elements in the Swiss Alps" (contribution to IGCP No. 5) held at Fribourg in 1983, the conference on "Caledonian-Variscan structures of the Alps" (3rd meeting "Geodynamics in the European Variscides") organized at Fribourg in 1987, and the lectures given during the advanced short course "Comparision of Variscan and pre-Variscan structures inside and outside the Alps", organized at Fribourg, March 1990, and headed by the "3e cycle Romand en Sciences de la Terre". During this last meeting, the lecturers with topics concerning the Alps agreed to contribute their papers to this volume. Encouraged by the positive answers, the editors decided to organize this volume.

Different factors contributed to the rather slow evolution of ideas concerning the basement in the Alps. In the Western Alps, the data collection in the Alpine domain is hindered by the climatic conditions at higher altitudes, the fieldwork seasons being often rather short. The advance of glaciers at the end of the last century (the "small glaciation") slowed down discoveries considerably. In addition to the climatic difficulties, there are also those of accessibility. The steep morphology is the expression of the recent uplift. Although the outcrop conditions are very often excellent, the mapping takes, in consequence, about four times as long as in areas of the Central European Variscan mountains. Also, sampling very often means backpacking heavy loads over many kilometres and a thousand metres change in elevation.

Perhaps the greatest difficulty, however, results from the multiple Alpine overprint. Gathering data in the Alpine domain requires longer experience to enable recognition of the, sometimes subtle, differences between rock units formed by the different Alpine overprints. This holds especially true for the Penninic domain where the general metamorphic overprint accompanied by a multitude of deformations causes most rock units to appear simply to be greenschists! In addition, except for the well-dated fossiliferous Palaeozoic units, most areas are underlain by polymetamorphic lithologies. Keeping such difficulties in mind the reader may understand that even 200 years of data collection may not be sufficient to answer all questions. Nevertheless, the very long tradition of geological research in the Alps has led to a wealth of data found in a vast literature, and presented here in two parts.

Part II, in addition to the introduction to the historical evolution of knowledge of the basement, presents general palaeontological and geochronological data, comparative studies of ultramafic rocks and ore deposits, and the Late Variscan evolution through palaeomagnetic data, the wide evolution of Late Variscan granitoids, and the Late Variscan-Early Mesozoic plate reorganization. These sections give a sound background for the regional descriptions which follow.

In Part III, presentation of the regional data follows the classic separation into the Helvetic, Penninic, Austro-Alpine and South-Alpine domains. The basic idea in planning the book was to cover the entire Alpine basement through independent contributions which can also stand alone. In the regional sections, each basement unit is presented by the working team with personal experience in this area. All contributions reflect specific interests of contributors and only give, therefore, minor importance to other aspects of geology. All significant data, however, are included in the contributions. As independent working groups share specific topics in their research, we included data and interpretations from several groups to cover all recent work. This procedure implies some repetitions but yields more complete models. The same is true for sections treating specific aspects from basically the same area, e.g. the Austro-Alpine basement.

Finally, in the concluding sections of Part IV only do we try to coordinate all the data and to deduce a general, pre-Mesozoic evolution of the Alpine domain, which cannot be separated from the evolution of the Variscan geological framework in Central Europe.

We wish this book to serve as a source of data and as a guide to the corresponding literature. Papers concerning the basement are written in many different languages and are often published in local journals, sometimes not easily obtained, even by people working in the Alps. Although we tried to cover all basement areas of the Alps, we feel that the density of data and discussion vary considerably when comparing the different papers. It becomes evident that the state of knowledge is not equal in the whole area. This should initiate future research in very different domains. Sedimentological records of Permian, Carboniferous and older sediments are needed for documentation of basin dynamics, and more knowledge about regions being eroded. In spite of the Alpine overprint, it should prove possible to gather much more P-T-data from different units and crustal levels. Statistics on zircon morphology, geochemical and geochronological data are still insufficient for many basement units (e.g. the pre-Variscan granites; metabasic rock series of very different ages). New geochemical criteria are needed to obtain better insight into the magmatic and sedimentological evolution through time. And, it is necessary that true geological fieldwork has to be done, to reinterprete well-known structures!

We hope that this book will not only be used by specialists, but that it may serve as an introduction to one aspect of the complex and fascinating Alpine geology for those not familiar with the Alps. We are aware that it is not easy to fulfill this last objective. Only the good will and interest of all participating authors and the patient work of the reviewers has made such an enterprise possible. Our warmest thanks go, therefore, to all authors who participated in this venture, and we point out the special merit of those who, late in the day, were willing to cover subjects that otherwise would not have been included. During the period of publication we lost a very good friend, Georg Troll (München, University) who agreed to collaborate as an author at an early stage.

Besides the reviews presented by the authors of this book, we appreciate those of M. Brown (College Park, Maryland), C. Chopin (Paris), M. Delaloye (Geneva), J. Dostal (Halifax), A. Escher (Lausanne), W. Franke (Giessen), H. Fritz (Graz), B. Hansen (Göttingen), R. D. Hatcher (Knoxville, Tennessee), C. Hawkesworth (Milton Keynes), R. Höll (Munich), H. L. Holzer (Graz), A. Krill (Trondheim), B. Lammerer (Munich), H. Laubscher (Basel), J. P. Lefort (Rennes), E. Marton (Budapest), R. Mason (London), P. Matte (Montpellier), B. Messiga (Pavia), R. D. Nance (Athens, Ohio), O. Oncken (Potsdam), F. Paris (Rennes), G. Pestal (Vienna), G. B. Piccardo (Genoa), W. Pitcher (Liverpool), D. Robinson (Bristol), P. Robinson (Amherst, Massachusetts), J. Rodgers (New Haven, Connecticut), W. Schubert (Würzburg), J. Schumacher (Kiel), J. W. Skehan (Weston, Massachusetts) E. Stein (Darmstadt), B. Stöckhert (Bochum), D. H. Tarling (Newcastle), E. Wallbrecher (Graz), W. Wildi (Geneva). We thank also Massimo Chiaradia (Fribourg) who compiled the subject index and prepared the figures at the beginning of Parts I–IV. Finally, we would like to express our special thanks to Springer-Verlag and the Geosciences Editorial for their pleasant and fruitful cooperation.

Fribourg/Graz, Mai 1993

J. F. von Raumer
F. Neubauer

Contents

Part I: Pre-Mesozoic Basement and Alpine Structures

Part II: Pre-Mesozoic Basement of the Alps

Part III: Specific Descriptions of Pre-Alpine Basement

Chapter 1: The Helvetic Realm

Chapter 2: Penninic Realm

Chapter 3: Austro-Alpine Realm

Contributors

ABRECHT, J.
GEOTEST, Birkenstr. 15
CH-3052 Zollikofen, Switzerland

BIINO, G.
Institut für Mineralogie und Petrographie, Balzerstr. 1
CH-3012 Bern, Switzerland

BONIN, B.
Dept. de Pétrologie – Volcanologie, Bâtiment 504, Université de Paris Sud
F-91405 Orsay Cedex, France

BRÄNDLEIN, P.
Institut für Geologie und Mineralogie Universität Erlangen, Schlossgarten 5
D-8520 Erlangen, Germany

BÜCHI, H.
Institut für Mineralogie und Petrographie, ETH Zentrum
CH-8092 Zürich, Switzerland

BURRI, M.
Institut de Géologie, Université de Lausanne BFSH2
CH-1015 Lausanne, Switzerland

BUSSY, F.
Institut de Minéralogie, UNIL BFSH2
CH-1015 Lausanne, Switzerland

CADOPPI, P.
Dip. di Scienze della Terra, Via Accademia delle Scienze, 5
I-10123 Torino, Italy

CHESSEX, R.
Dépt. de Minéralogie, Rue des Maraîchers 13
CH-1211 Geneve, Switzerland

COLOMBO, A.
Dip. di Scienze della Terra, Sezione Mineralogia, Via Botticelli, 23
I-20133 Milano, Italy

CONTI, P.
Institut für Geologie, Bernoullistr. 32
CH-4056 Basel, Switzerland

CORTESOGNO, L.
Dip. di Scienze della Terra sez. Petrografia, Corso Europa, 30
I-16132 Genova, Italy

DALLAGIOVANNA, G.
Dip. di Scienza della Terra sez. Geologia-Paleontologia, Strada Nuova 65
I-16132 Pavia, Italy

DAL PIAZ, G. V.
Dipartimento di Geologia, Via Giotto, 1
I-35137 Padova, Italy

DE GRACIANSKY, M.
Ecole des Mines, 60, Boulevard Saint-Michel
F-75272 Paris Cedex 06, France

DESMONS, J.
Centre de Recherches Pétrographiques et Géochimiques, CNRS, B. P. 20
F-54501 Vandoeuvre-les-Nancy Cedex, France

DI PISA, A.
Istituto Sc. Geologico-Mineralogiche, Corso Angioi 10
I-07100 Sassari, Italy

EGGENBERGER, U.
Institut für Mineralogie und Petrographie, Balzerstr. 1
CH-3012 Bern, Switzerland

FINGER, F.
Institut für Mineralogie, Universität Salzburg, Hellbrunner Str. 34
A-5020 Salzburg, Austria

FLISCH, M.
Brückfeldstr. 23
CH-3000 Bern, Switzerland

FRASL, G.
Institut für Geologie und Paläontologie, Universität Salzburg, Hellbrunner Str. 34
A-5020 Salzburg, Austria

FRISCH, W.
Institut für Geologie und Paläontologie, Sigwartstr. 10
D-7600 Tübingen, FRG

GAGGERO, L.
Dip. di Scienze della Terra Sez. Petrografia, Corso Europa, 30
I-16132 Genova, Italy

GATTIGLIO, M.
Dipartimento di Scienze della Terra, Via Valperga Caluso 37
I-10125 Torino, Italy

GEBAUER, D.
Institut für Mineralogie und Petrographie, ETH Zentrum
CH-8092 Zürich, Switzerland

GODIZART, G.
KTB-Feldlabor
D-8486 Windischeschenbach, FRG

GOUFFON, Y.
Institut de Géologie, Université de Lausanne BFSH2
CH-1015 Lausanne, Switzerland

Gracianský, P. C. de
Ecole Nationale Supérieure des Mines de Paris
F-75272 Paris Cedex 06, France

Graf, K.
Institut für Mineralogie und Petrographie, Balzerstr. 1
CH-3012 Bern, Switzerland

Haunschmid, B.
Institut für Geologie und Paläontologie, Universität Salzburg, Hellbrunner Str. 34
A-5020 Salzburg, Austria

Heinisch, H.
Geologisches Institut, Albertstr. 23B
D-7800 Freiburg i. Br., FRG

Höck, V.
Institut für Geologie und Paläontologie, Universität Salzburg, Hellbrunner Str. 34
A-5020 Salzburg, Austria

Hoinkes, G.
Institut für Mineralogie und Petrographie, Universitätsplatz 2
A-8010 Graz, Austria

Köppel, V.
Institut für Mineralogie und Petrographie, ETH Zentrum
CH-8092 Zürich, Switzerland

Krainer, K.
Institut für Geologie und Paläontologie, Innrain 52
A-6020 Innsbruck, Austria

Lardeaux, J. M.
Laboratoire des Sciences de la Terre ENS Lyon, 46 Allée d'Italie
F-69364 Lyon Cedex 07, France

Lettner, H.
Institut für Geologie und Paläontologie, Universität Salzburg, Hellbrunner Str. 34
A-5020 Salzburg, Austria

Loeschke, J.
Institut für Geologie und Paläontologie, Sigwartstr. 10
D-7600 Tübingen, FRG

Maggetti, M.
Institut für Mineralogie und Petrographie, Universität-Pérolles
CH-1700 Fribourg, Switzerland

Marro, Chr.
ABA GEOL SA, Rue du Collège 1
CH-1530 Payerne, Switzerland

Mauritsch, H. J.
Institut für Geophysik, Montanuniversität, Franz-Josef-Strasse 18
A-8700 Leoben, Austria

Meccheri, M.
Dipartimento di Scienze della Terra, Via delle Cerchia 3
I-53100 Siena, Italy

Ménot, R. P.
Lab. de Géologie-Pétrologie CNRS, 23 Rue du Docteur P. Michelon
F-42023 Saint Etienne Cedex, France

Mercier, D.
Ecole des Mines de Paris, CGGM, Laboratoire de Sédimentologie,
35, rue Saint-Honoré
F-77305 Fontainebleau Cedex, France

Mercolli, I.
Institut für Mineralogie und Petrographie, Balzerstr. 1
CH-3012 Bern, Switzerland

Neubauer, F.
Institut für Geologie, Heinrichstr. 26
A-8010 Graz, Austria

Nollau, G.
Institut für Geologie und Mineralogie, Schlossgarten 5
D-8520 Erlangen, FRG

Oberhänsli, R.
Institut für Geowissenschaften, Saarstr. 21
D-6500 Mainz, FRG

Paquette, L.
Laboratoire de Géochimie CNRS, Université Paul Sabatier,
38 Rue des trente-six ponts
F-31400 Toulouse, France

Pfeifer, H. R.
Dept. de Minéralogie, UNIL BFSH2
CH-1015 Lausanne, Switzerland

Pfiffner, A.
Institut für Geologie, Balzerstr. 1
CH-3012 Bern, Switzerland

Ploquin, A.
C. R. P. G. B. P. 20
F-54501 Vandoeuvre-les-Nancy Cedex, France

von Quadt, A.
Institut für Kristallographie und Petrographie, ETH Zentrum
CH-8092 Zürich, Switzerland

Ratschbacher, L.
Institut für Geologie und Paläontologie, Sigwartstr. 10
D-7600 Tübingen, FRG

von Raumer, J. F.
Institut für Mineralogie und Petrographie, Universität-Pérolles
CH-1700 Fribourg, Switzerland

Sacchi, R.
Dip. di Scienze della Terra, Via Accademia delle Scienze, 5
I-10123 Torino, Italy

Sandrone, R.
Dipartimento di Georisorse e Territorio, Politecnico, Corso Duca degli Abruzzi 24
I-10129 Torino, Italy

Sartori, M.
Institut de Géologie, Université de Genève
CH-1211 Genève, Switzerland

Sassi, F. P.
Dipartimento di Mineralogia e Petrologia, Corso Garibaldi, 37
I-35137 Padova, Italy

Schaltegger, U.
Université Louis Pasteur, CNRS Centre de Sédimentologie, 1, rue Blessig
F-67084 Strasbourg, France

Schermaier, A.
Institut für Geologie und Paläontologie, Universität Salzburg, Hellbrunner Str. 34
A-5020 Salzburg, Austria

Schindlmayr, A. O.
Institut für Geologie und Paläontologie, Universität Salzburg, Hellbrunner Str. 34
A-5020 Salzburg, Austria

Schmid, St.
Institut für Geologie, Bernoullistr. 32
CH-4056 Basel, Switzerland

Schönlaub, H. P.
Geologische Bundesanstalt, Rasumovskygasse 23
A-1031 Wien, Austria

Schroll, E.
Geotechnisches Institut, Arsenal Objekt 210
A-1030 Wien, Austria

Schulz, B.
Institut für Geologie und Mineralogie, Schlossgarten 5
D-8520 Erlangen, FRG

Siletto, G. B.
Dip. di Scienze della Terra, Sezione Mineralogia, Via Botticelli, 23
I-20133 Milano, Italy

Spalla, M. I.
Laboratoire des Sciences de la Terre ENS Lyon, 46 Allée d'Italie
F-69364 Lyon Cedex 07, France

Spiess, R.
Dipartimento di Mineralogia e Petrologia, Corso Garibaldi, 37
I-35137 Padova, Italy

Spillmann, P.
Institut für Mineralogie und Petrographie, ETH Zentrum
CH-8092 Zürich, Switzerland

Steyrer, H. P.
Institut für Geologie und Paläontologie, Universität Salzburg, Hellbrunner Str. 34
A-5020 Salzburg, Austria

Stille, P.
Université Louis Pasteur, CNRS Centre de Sédimentologie, 1, rue Blessig
F-67084 Strasbourg, France

Thélin, Ph.
Institut de Minéralogie, UNIL BFSH2
CH-1015 Lausanne, Switzerland

Thöni, M.
Institut für Geologie, Universität Wien, Universitätsstr. 7
A-1010 Wien, Austria

Tunesi, A.
Dipartimento di Mineralogia, Via S. Botticelli, 23
I-20133 Milano, Italy

Vanossi, M.
Dip. di Scienza della Terra sez. Geologia-Paleontologia, Strada Nuova 65
I-16132 Pavia, Italy

Vavra, G.
Institut für Mineralogie und Petrographie, ETH Zentrum
CH-8092 Zürich, Switzerland

Vialon, P.
Université J. Fournier, LGIT IRIGM, Domaine Universitaire
B. P. 53X
F-38041 Grenoble, France

Visona, D.
Dip. di Mineralogia e Petrologia, Corso Garibaldi 37
I-35100 Padova, Italy

Vivier, G.
CNRS, 25, Avenue des Martyrs
B. P. 166X
F-38042 Grenoble Cedex, France

Winkler, M.
Institut für Geologie und Paläontologie, Sigwartstr. 10
D-7600 Tübingen, FRG

Ziegler, P.
Kirchweg 41
CH-4102 Binningen, Switzerland

Part I

Pre-Mesozoic Basement and Alpine Structures

Convenor: O. A. PFIFFNER

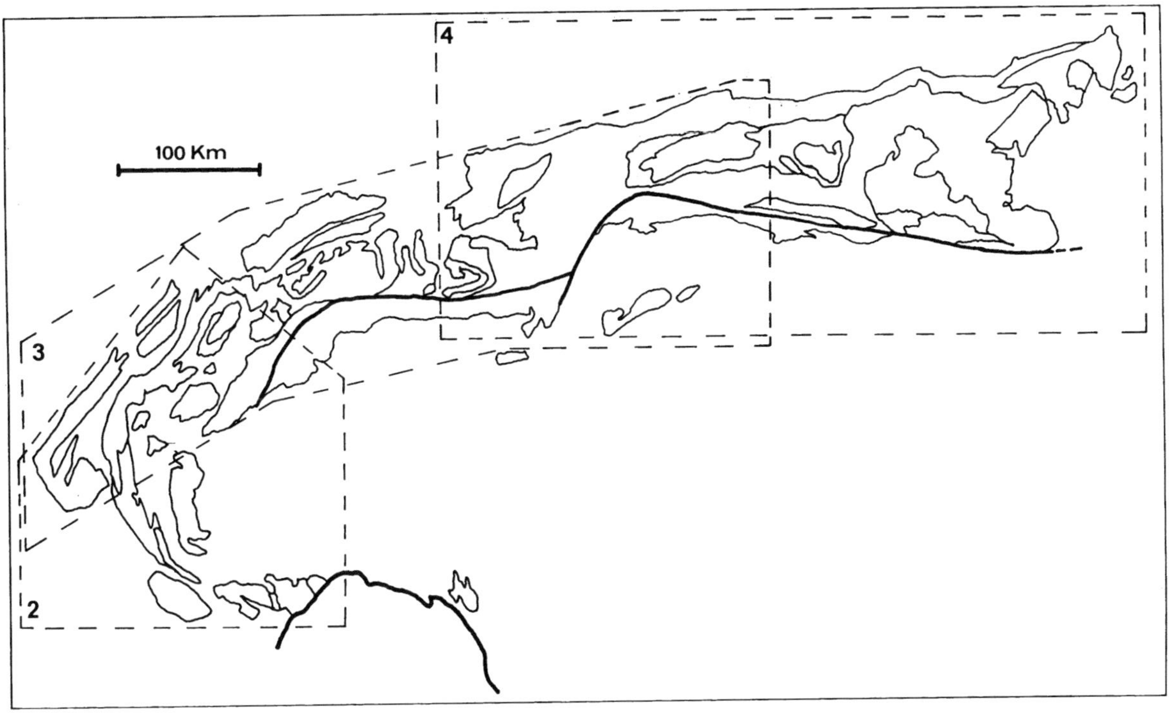

The Structure of the Alps: An Introduction

O. Adrian Pfiffner

The Alps as part of the Alpine or Tethyan belt extend from the Mediterranian to the Vienna basin over a distance of 1000 km. They loop around the Adriatic microplate with the Western Alps at its western and the Central and Eastern Alps on its northern fringe. Much of the intricate architecture of the Alps is related to the complex rotation and motion of the Adriatic microplate nested between the European and African plates.

In Permian times, a phase of post-Variscan stretching and subsidence produced a number of basins filled with red beds on that part of Pangea which was to become the future Alps. The break-up of Pangea from Triassic time onward produced an oceanic basin, which opened by extensional and strike-slip tectonics synchronous with the beginning of the opening of the North Atlantic. Mesozoic sediments associated with this ocean are of quite varied nature. They include e.g. epicontinental shallow marine sediments accumulated on a thermally subsiding platform, as well as fault-scarp breccias and deep-water clastics related to continental stretching and tectonic subsidence. In a general way the future Western Alps were at the site of more or less normal rifting (Piemont ocean), whereas the realm of the future Central and Eastern Alps was characterized by oblique, transform-dominated rifting. The Mesozoic sedimentary column is thus laterally discontinuous making correlations along strike of the Alpine chain often difficult.

The Mesozoic (and Cenozoic) sediments and their pre-Triassic basement were later involved in the Alpine nappe structure in different ways.

1. The sediments were completely detached from their pre-Triassic basement. The latter was either also detached as flakes of upper crust to form nappes of its own, or it remained more or less fixed to its lower crust.
2. The sediments remained attached to their pre-Triassic basement and both were implied in fold-and-thrust structures.
3. There are transitions between the various types.

As a consequence correlation of individual pieces of pre-Triassic basement units outcropping sometimes as isolated patches throughout the Alps is even more difficult than correlation of individual Mesozoic sedimentary sequences.

Convergence between the European plate and the Adriatic (or Apulian) spur of the African continent started some 110 Ma ago. It led to the closure of much of the Tethys basin and the building of a Cretaceous mountain chain. In the Western Alps an east-dipping subduction zone with important stacking and west-directed thrusting of upper crustal flakes (Fig. 1) is indicated by high-pressure assemblages (see also de Graciansky, this Vol., and references therein). Orogenic movements including (north)west-directed thrusting (see Fig. 1) and metamorphism is also documented in the Eastern Alps (see Ratschbacher and Frisch this Vol.), while in the future Central Alps sedimentation continued up into the Tertiary. The later, Tertiary phase of convergence was more north-south oriented. It can be regarded as a pure continent-continent collision. The indentation of the Adriatic microcontinent into the European plate led to the formation of a bivergent fold-and-thrust belt in the Central Alps. In the northern part of the orogen (Helvetic and Penninic units) thrusting was mainly to the north, while the South-Alpine units were thrust to the south (Fig. 1). Important vertical uplift due to a wedge of Adriatic lower crust forced into the European plate (see Pfiffner this Vol.) exposed amphibolite-grade rocks at the present surface. But apart from the vertical escape of material in the collision zone, lateral escape is indicated by important strike slip movements (Ratschbacher and Frisch this Vol., and references therein).

The subdivision of the Alpine chain into Western, Central and Eastern Alps (Fig. 1) follows the general strike of the Alpine structure, which turns from N-S in the Western to E-W in the Eastern Alps. But the three segments also differ in regard to the megatectonic units constituting the various nappe piles. In the Western Alps (see also de Graciansky this Vol.) the **internal units** constitute the eastern part of the oro-

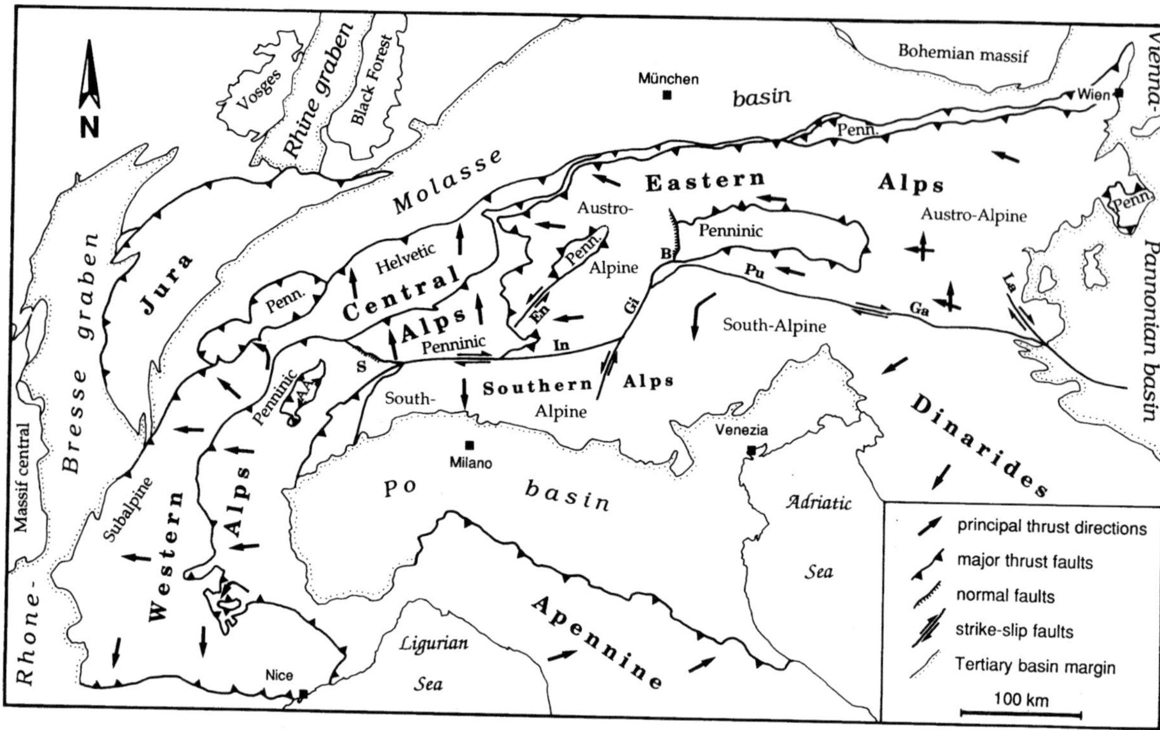

Fig. 1. Schematic tectonic map of the Alps and their surroundings; AA Austro-Alpine. Segments of the Peri-Adriatic Line are: In Insubric; Gi Giudicaria; Pu Pustertal; Ga Gailtal; associated faults are En Engadine; La Lavanttal; S Simplon; B Brenner

gen and tectonically overly the **external units.** The megatectonic units are defined on the basis of the nature of the Mesozoic sediments. Thus, the external or **Subalpine units** represent the former submerged shelf of the European craton, while the internal, **Penninic units** correspond to the stretched and thinned part of the European margin and the newly formed oceanic crust of the Piemont ocean. To the northeast of the Western Alps crustal fragments of the stretched Adriatic margin (Sesia Lanzo zone and Dt. Blanche nappe) overly Penninic nappes and are also referred to as internal units by some geologists. Pre-Triassic basement is involved in the nappe structure in the internal units (e.g. "massifs cristallins internes" Gran Paradiso, Dora-Maira) and, probably to a lesser degree, in the external units ("massifs cristallins externes" Belledonne, Pelvoux, Argentera). To the west the Subalpine units are thrust onto Tertiary sediments of the Molasse basin (to the north) and the Bresse-Rhone graben.

In the case of the Central Alps (see also Pfiffner this Vol. and references therein) the Helvetic zone represents the submerged shelf of the European craton. Here, too, the pre-Triassic basement is involved to some degree in the nappe structure (e.g. Aiguilles Rouges, Mt. Blanc, Aar, Gotthard massifs). The **Penninic nappes** formed from the stretched European crust and newly formed oceanic crust. Basement-involvement is particularly important in the Lepontine area. Both the Helvetic and Penninic nappes formed in response to north-directed thrusting. The **Austro-Alpine** and **South-Alpine nappes** emanated from the stretched margin of the Adriatic microplate. The Austro-Alpine nappes were thrust in a (north)westerly direction onto the Penninic zone and now form the tectonically highest units. Due to extensive Tertiary unroofing, uplift and erosion they outcrop as klippen only in most of the Central Alps (e.g. Dt. Blanche). The South-Alpine units form a southvergent fold-and-thrust belt which is partly buried hidden beneath the Pliocene infill of the Po basin located to the south. Austro-Alpine (and Penninic) nappes are separated from South-Alpine units by the Insubric Line, a major dextral-transpressive fault belonging to the Peri-Adriatic Line. Slices of pre-Triassic basement are involved in both, the Austro-Alpine and South-Alpine units.

In the Eastern Alps (see also Ratschbacher and Frisch, this Vol. and references therein) unroofing was less important so that the surface geology is essentially accounted for by the **Austro-Alpine** and adjacent **South-Alpine** units. As the Engadine, Tauern and Rechnitz windows clearly show, however, the **Penninic nappes** of the Central Alps extend all the way across the Eastern Alps buried underneath the Austro-Alpine nappes. In addition small lenses of **Helvetic nappes** straddle the entire northern fringe of the Austro-Alpine nappes, proving also their lateral continuity. As in the Central Alps pre-Triassic basement is involved in the thrust sheets of the Penninic and Austro-Alpine units.

A foredeep, the Molasse basin, fringes the Central and Eastern Alps to the north. It narrows considerably westwards and pinches out shortly after reaching the Western Alps. This foredeep probably formed in response to downbending and flexing of the European lithosphere in response to loading in the late stage of the continent-continent collision. The southernmost part of the basin (Subalpine Molasse) was overridden by Alpine nappes and incorporated into the Alpine orogen. Towards the craton pre-Triassic basement is exposed, possibly in part due to uplift as a flexural bulge (Bohemian massif, Black Forest, Vosges, Massif central). Eastward the Eastern Alps plunge and disappear under the Tertiary sediments of the (Vienna-)Pannonian basin. This extensional basin is nested within the loop of the Carpathians. On the inner side of the Alpine loop, the Po (and Venetian) basin is a foredeep that the Alps and Dinarides share with the Apennines.

Summarizing, it follows that pre-Triassic basement including sedimentary and magmatic rocks of the Variscan cycle outcrop scattered throughout the Alps, isolated from the Variscan chain of Europe by Tertiary basins, and dismembered and dislocated during the break-up of Pangea and the subsequent convergence between the European plate and the Adriatic spur of the African plate. As a consequence of the varying degree of involvement of the pre-Triassic basement in the Alpine nappe pile and the varying amount of uplift associated with these deformations these basement rocks outcrop as irregular, partly isolated patches within the Alps. In the following three sections rearrangement of these individual basement blocks into their original, pre-Triassic configuration is attempted in order to gain insight into the distribution of the various rock suites of the Variscan cycle throughout the Alpine realm.

Suggested reading

Coward MP, Dietrich D, Park RG (eds) (1989) Alpine tectonics. Geol Soc Lond, Spec Publ 45, 450 pp

Debelmas J (1974) Géologie de la France, vol II. Les chaînes plissées du cycle alpin et leur avant-pays. Doin, Paris

Debelmas J, Lemoine M (1970) The Western Alps: Paleogeography and structure. Earth Sci Rev 6: 221–256

Flügel HW, Faupl P (eds) (1987) Geodynamics of the Eastern Alps. Deuticke, Wien

Oberhauser R (ed) (1980) Der geologische Aufbau Österreichs. Springer, Berlin Heidelberg New York

Roure F, Heitzmann P, Polino R (eds) (1990) Deep structure of the Alps. Mém Soc Géol Fr 156; Mém Soc Géol Suisse 1; Soc Geol Ital Spec vol 1, 367 pp

Tollmann A (1977, 1985, 1986) Geologie von Österreich, Band 1, 2, 3. Deuticke, Wien

Trümpy R (1980) Geology of Switzerland – a guide book. Part A: An outline of the geology of Switzerland. Wepf, Basel

Basement-Cover Relationship in the Western Alps Constraints for Pre-Triassic Reconstructions

P. C. de Graciansky

Abstract

Attempts at pre-Triassic reconstruction in the Western Alps have to take into account the three phases of the development of the deformation from Late Palaeozoic to Cenozoic times. These are (1) extension and block-faulting within the Late Hercynian craton that led to the creation of the Tethyan ocean and margins during Middle Jurassic times; (2) inversion of the strain field that resulted from a change of the movement of Africa with respect to Europe during Middle Cretaceous times; (3) folding and thrusting in a compressional regime that corresponded to the closure of the Tethys and then of the collision of Africa with Europe during Late Cretaceous and Cenozoic times. Another constraint for Late Palaeozoic reconstruction is the role played by the Pelvoux-Argentera line which has been both a major transfer line of the Tethyan rift system and a lateral ramp with respect to early Alpine thrustings. The major part of the structures of the Tethyan European margin in the Central Alps has been lost beneath the huge thrust sheets of internal Alpine origin. Nevertheless, the French subalpine zones escaped severe deformation because of their location at the exterior of the major lateral ramp line. The original shape of the European passive margin is reflected in the present-day structures by the arcuate shape of the Western Alps, the distribution of lateral ramps and thrust planes and lateral asymmetric escape of rock masses during the collision.

1 Introduction

The Alpine realm is classically interpreted as beeing located at the boundary between the African and Eurasian plates. The breakup of Pangea that began during Early Mesozoic time gave birth to the oceanic area known as Tethys, after Suess (1885–1888). The Tethyan ocean was oriented approximately E–W, extending from the Caribbean to the Himalayas. Its complicated shape reflects a series of small pull-apart basins such as the Jurassic Central Atlantic or the Ligurian portion, connected by sinistral transfer lines such as the Gibraltar-Newfoundland one (Fig. 1). Oceanic crust was well developed in the basins but not in the transfer corridors.

Alpine building in Europe basically resulted from the closure of the Tethyan ocean, the inversion of its margins and the continental collision between Africa and western part of Eurasia. The onset of this sequence of events is dated as early Late Cretaceous

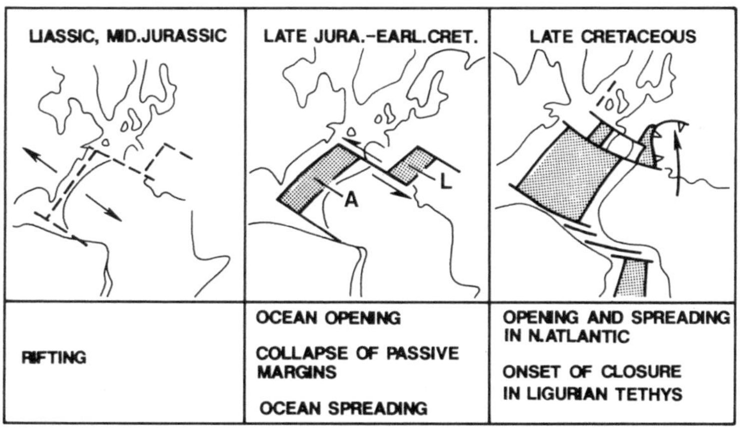

Fig. 1. Three main stages of the development of Tethys and onset of Alpine collision (Lemoine et al. 1988)

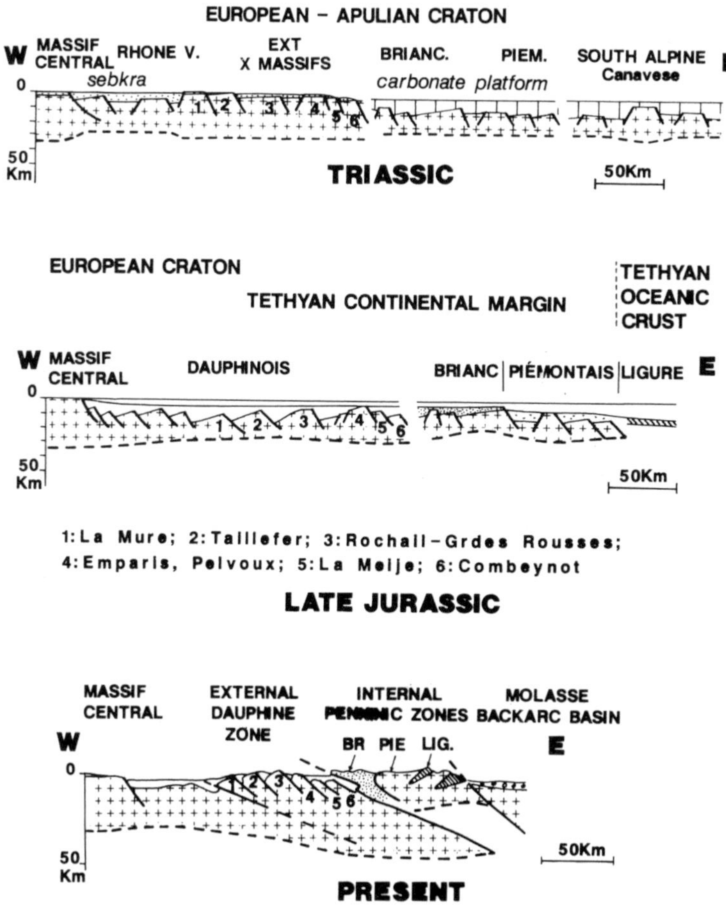

Fig. 2. East-west diagrammatic cross-sections across the French and Italian Alps

(Lemoine and Tricart 1986). Closure resulted from subduction of the oceanic crust and the distal parts of Tethyan margins (Mattauer and Tapponnier 1978) as demonstrated by the existence of high to very high pressure metamorphic rocks (Chopin 1987). This was followed by the obduction of small slabs of the former Tethyan oceanic crust and by the thrusting and stacking of portions of the previously thinned continental crust detached from both the European and African margins (Fig. 2).

A pre-Triassic reconstruction of the Alps therefore has to take into account the threefold history (Fig. 3) that is marked by complementary movements of Africa with respect to Eurasia: (1) divergence during Triassic and Jurassic times with an eastward movement of Africa relative to a fixed Europe, (2) convergence between the two major plates since Cretaceous times. This was a consequence of a change of the direction of movement of Africa during this sequence of events, (3) the collision that occurred during Neogene times. This last phase is marked by left-lateral movement of the African plate along the peri-Adriatic line which resulted of both the westward escape of thrust masses and back-thrusting at the interior of the arc.

The continental edge of the European craton that resulted from the phase of divergence was characteristic of a complicated shape with associated extensional faults and transfer faults. The orientation of such faults with respect to the strain field during the convergence phase strongly influenced the main Alpine structures. Noticeably, the leading edge of the African plate was marked by the Apulo-Adriatic promontory or Adria (Channel et al. 1979) the existence of which was partly responsible for the curvature of the Alpine belt from Nice to Vienna. In consequence the Alps are subdivided structurally and geographically into the Western, Central and Eastern Alps along the strike, each portion being characteristic of its position with respect to Adria.

The area discussed in this paper includes the Western – mainly French and Italian – part that occupies the Western wing of the promontory. It is characteristic of a change in orientation of Alpine trends from

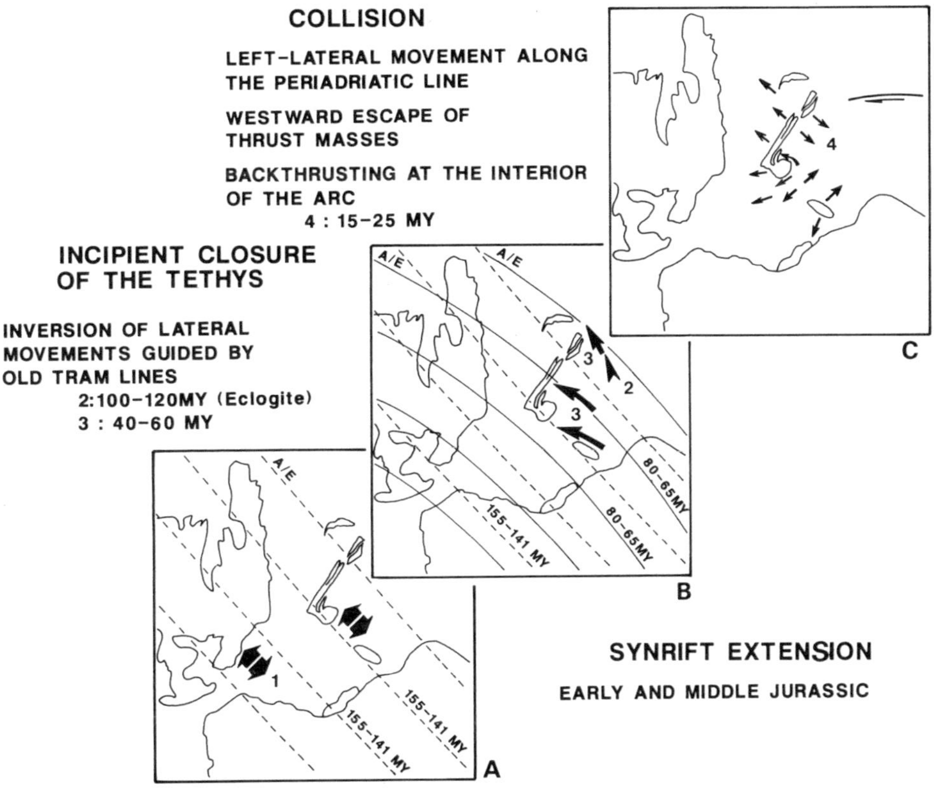

Fig. 3 A–C. Alpine building, inversion of trajectories and continental collision. **A** Extension: the direction of extension *(arrows 1)* during the Tethyan rifting is consistent with the attitude of the small circles related to the movement of Africa with respect to Europe for 155–141 m. y.; **B** inversion: during the first stages of the Tethyan closure the small circles 80–65 m. y. were roughly parallel to the 155-141 m. y. ones. However, both families of circles had different poles of rotation. The trajectories for 100–120 m. y. (eclogite facies: *arrow 2*) and for (50–60 m. y. (*arrow 3*: Choukroune et al. 1986) were parallel to the 80-65 m. y. small circles. This suggests that the same crustal scale transfer lines have guided the movement during both the Tethyan extension and the incipient Alpine compression. **C** Collision: the pattern of the deformation resulting from the collision does not follow regular lines. It has been particularly complex including forethrusting and back thrusting with a radiating pattern *(arrows 4)*. The poles of rotation for 155-144 m. y. is 58°65′ N–56°33′ E and for 80-65 m. y. is 33°41′ N–5°18′ E. Small circles have been kindly drawn by L. Jolivet (Lemoine et al. 1989)

the northwest at the Cote d'Azur to the northeast at Lake Geneva. In this segment, which derives mainly from the European craton basement, rocks of the Variscan cycle have been involved at various degree within the Alpine orogeny (Fig. 4).

The major Alpine structures of the Western Alps comprise classically internal units and external (or subalpine) units that are separated by the frontal Penninic thrust (Gignoux and Moret 1934). This major thrust coincides with a major palaeogeographic boundary (Figs. 2 and 4).

Internal structures comprise *Penninic* and Austro-Alpine units. They are characterized as huge thrust sheets with metamorphic and ductile deformation that increase eastward. The *Penninic* structures comprise, from west to east, the Briançonnais. Piemontese (or Piemontais) and Ligurian units. The Briançonnais and Piemontese units were derived from the stretched, distal, part of the European margin of the Tethys. The Austro-Alpine (and Sud-Alpine) units originated from the Apulo-Adriatic margin. The Ligurian units include mainly ophiolites and attenuated Mesozoic sedimentary cover that have been sliced from the Tethyan-Ligurian oceanic crust (Lemoine and Tricart 1986; Fig. 2).

Slivers of ophiolite material mark the major suture zone between the two cratonic plates that have been dislocated in a complicated manner as a consequence of superimposed alpine phases of deformation (Fig. 2).

Piemontese and Ligurian units comprise a sedimentary cover of Schistes Lustrés (Lustrous shales) whose high degree of metamorphism in places (Chopin 1987; Goffé and Chopin 1986) until recently limited deciphering of the stratigraphy. The Schistes Lustrés are known now to be composed of a post-Triassic sequence corresponding to the cover of the Piemontese basement and a Late Jurassic to Creta-

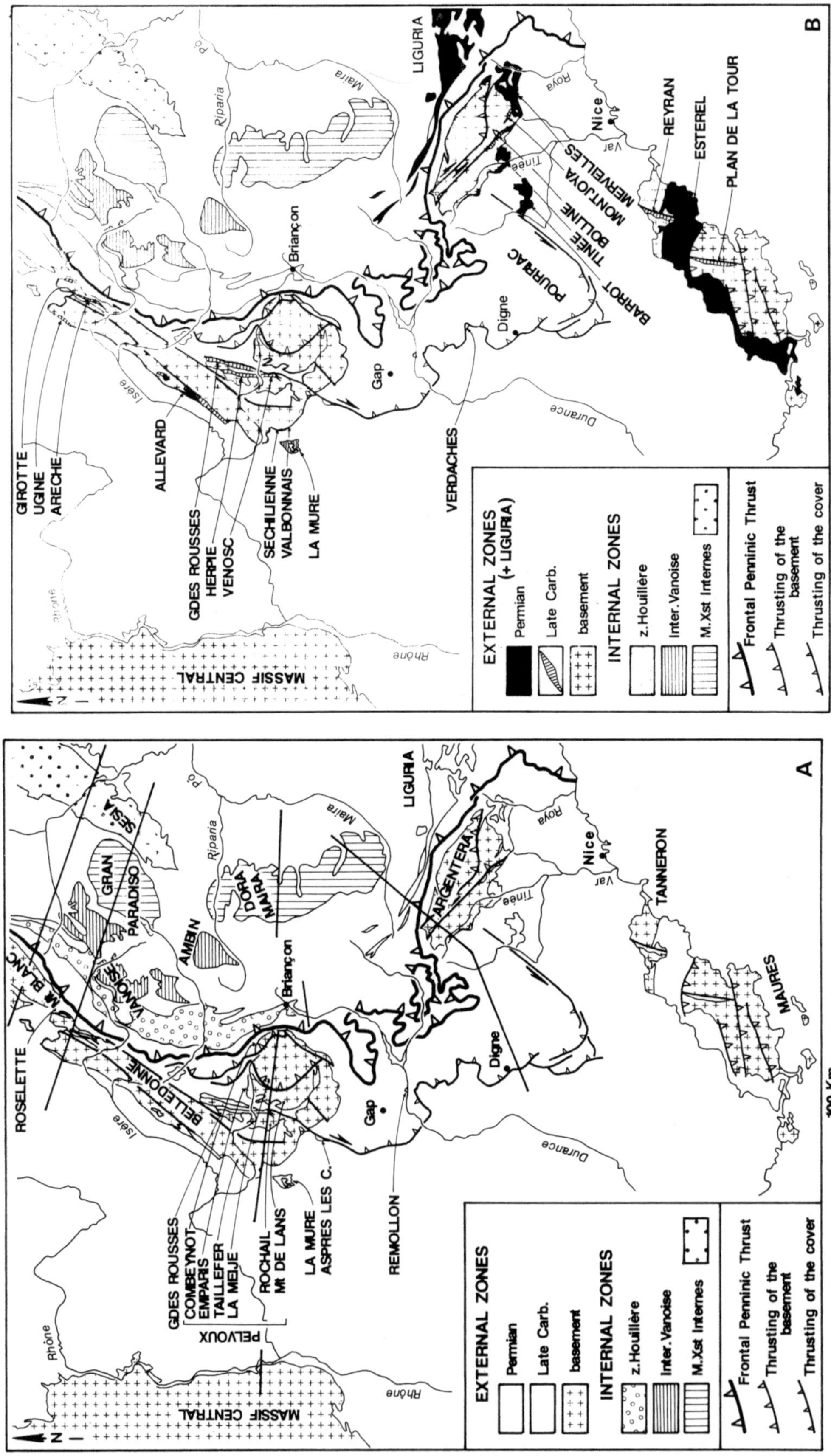
A
B
MASSIF CENTRAL
Rhône
Isère
Durance
Gap
Digne
Briançon
Nice
Var
Tinée
Roya
Maira
Riparia
Po
LIGURIA
100 Km
EXTERNAL ZONES
(+ LIGURIA)
Permian
Late Carb.
basement
INTERNAL ZONES
z.Houillère
Inter.Vanoise
M.Xst Internes
Frontal Penninic Thrust
Thrusting of the basement
Thrusting of the cover
ROSELETTE
Mt BLANC
BELLEDONNE
VANOISE
GRAN PARADISO
SESIA
AMBIN
DORA MAIRA
ARGENTERA
TANNERON
MAURES
PELVOUX
GDES ROUSSES
COMBEYNOT
EMPARIS
TAILLEFER
LA MEIJE
ROCHAIL
Mt DE LANS
LA MURE
ASPRES LES C.
REMOLLON
GIROTTE
UGINE
ARECHE
ALLEVARD
HERPIE
VENOSC
SECHILIENNE
VALBONNAIS
VERDACHES
POURRIAC
BARROT
TINEE
BOLLINE
MONTJOVA
MERVEILLES
REYRAN
ESTEREL
PLAN DE LA TOUR

ceous sequence which rested directly on the Ligurian oceanic crust (the serie de Chabrières; Lemoine and Tricart 1986; Fudral et al. 1987).

Only Briançonnais, Piemontese and Ligurian (ophiolitic) units outcrop in the Western Alps. Austro-Alpine units are widespread in the Central and Eastern Alps but are now buried beneath the Po Valley alluviums at the rear side of the Western Alps along the studied transect. The South-Alpine unit where it outcrops is the Ivrea zone (Fig. 3) which marks a part of the lower crust of the Apulian plate shortened during the Palaeozoic orogeny, later raised to the surface during the Alpine tectonism (Brodie et al. 1989; Dal Piaz this Vol.).

The external units of the Western Alps (Fig. 2) comprise the subalpine chains of Provence and of Dauphiné in the south that pass into the Helvetic nappes to the north of the Arve Valley. This subalpine domain is characterized by mild tectonism and weak to absent metamorphism. Alpine compression led to the detachment of the post-Triassic cover along evaporitic levels dated as Triassic where present. Deformation in the cover is mainly concentric folding with minor thrusting.

Basement response to Alpine shortening is internal thrusting as shown by the ECORS seismic line (Mugnier et al. 1990; Tardy et al. 1990). The intensity of alpine deformation increases from west to east so that extensional structures inherited from the Tethyan rifting are well preserved at the Rhône Valley (western) side of the belt. This area is located mainly at the southwestern border of the domain affected by strong deformation that resulted from convergence and collision.

The external subalpine domain developed only on the proximal part of the former European margin of Tethys (Fig. 2) but has no stratigraphic relationship to the Austro-Alpine domain.

2 Pre-Triassic Reconstruction and Crustal Extension During Late Palaeozoic and Mesozoic Rifting and Spreading

Reconstruction of extensional faults in the Penninic units which have suffered alpine ductile deformation remains a difficult problem. Nevertheless, such reconstructions have been attempted at places in the Briançonnais, Vanoise and external Piemontese Schistes Lustrés (Arnaud et al. 1978; Tricart et Lemoine 1986; Faure and Mégard-Galli 1988; Jaillard 1988; Mégard-Galli and Faure 1988; Graciansky et al. 1989).

In the subalpine zones, the Permian and Mesozoic basins rift systems are reasonably well preserved as a consequence of their position outside of the southwestern wing of the Apulian-African promontory that caused the Alpine collision.

Among the many fault directions two of them are of prime importance (Fig. 5): the Cevenole fault system, which is oriented NNE–SSW to NE–SW, was mainly extensional. In contrast, the Pelvoux-Argentera line oriented NW–SE was mainly a transcurrent direction (Lemoine et al. 1989).

Such an interpretation rests upon two types of arguments.

The available microstructural data on faults dated Liassic (Fig. 5) has shown that the main direction of

Fig. 4 A, B. Outcrops of the Pre-triassic basement in the Western Alps. The pre-Late Carboniferous outcrops **(A)** have been separated from the Late Carboniferous and Permian basins **(B)** of the external zones (plus Liguria) on these two sketch maps. A similar figure cannot be drawn at this scale for the northern internal zones as a consequence of the multiphased Alpine thrustings. Within the external zones, the pre-Triassic basement makes an S-shaped line of outcrops shown by the external crystalline massifs. Main masses are Mont Blanc, Aiguilles Rouges, Belledonne, Taillefer, Rochail, Pelvoux, Argentera and Maures-Esterel-Tanneron. These were originally Mesozoic extensional blocks that were later uplifted during Alpine deformation (Lemoine and Tricart 1986). Independent of these major blocks, small slices of metamorphic Variscan basement such as Roselette, Rocheray (Gillcrist et al. 1987), Aspres les Corps (Gidon and Aprahamian 1980–1981), and Remollon (Gidon 1975; Hayward and Graham 1989) have been detached during the Alpine thrustings. The Aspres les Corps slice is involved within a wrench fault corridor that corresponds to one northern lateral ramp of the Digne thrust (Guérin et al. 1990). Fragments of Late Carboniferous basins have also been carried along by Alpine thrusting. The best-known outcrop is at Verdaches near Barles and there marks the base of the Digne thrust surface. Main masses of the external crystalline massifs comprise Variscan metamorphic basement. Small and isolated Late Carboniferous basins are scattered on the crystalline massifs. The largest one is La Mure (see outline of Late Carboniferous stratigraphy by Barfety 1984) at the southern end of the Belledonne-Taillefer where coal is still mined. Others are located in Belledonne (Ugine, Arèche, Beaufort, Cevins, Petitcœur). In Taillefer (Valbonnais, Séchilienne) in Rochail (Venosc, les Deux Alpes), in the Grandes Rousses (l'Herpie) in Argentera (Mont-Joya-Ferisson, Las Grottas, Pouriac) and in the Maures (Reyran and Plan de la Tour). Permian basins are poorly developed in the Pelvoux and Belledonne (Belledonne: Bisanne at the north and Allevard, northeast from Grenoble). By contrast, wide and thick Permian sequences widely outcrop south of the Argentera in the Tinée valley, the Roya valley and in the Dome de Barrot. Permian outcrops also occur widely around the Massif des Maures in Cuers, and in the Luc and Esterel districts where rhyolites and ignimbrites are especially well known. For a modern synthesis, see data by Toutin-Morin and Vinchon (1989) and Toutin-Morin (1987)

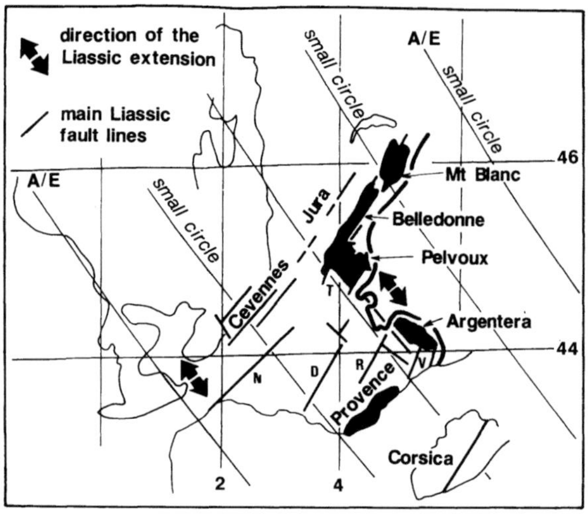

Fig. 5. Directions of extension and tramlines during the Tethyan rifting phase in the Western Alps (from Lemoine et al. 1989). The major Liassic extensional fault lines are those of the Cévennes, and of Nimes *(N)*, Durance *(D)*, Rouaine *(R)*, Var Valley *(V)* and others that bound, to the east, the external crystalline massifs of La Mure, Taillefer, Belledonne, Pelvoux and Grandes Rousses (see Figs. 2 and 4)

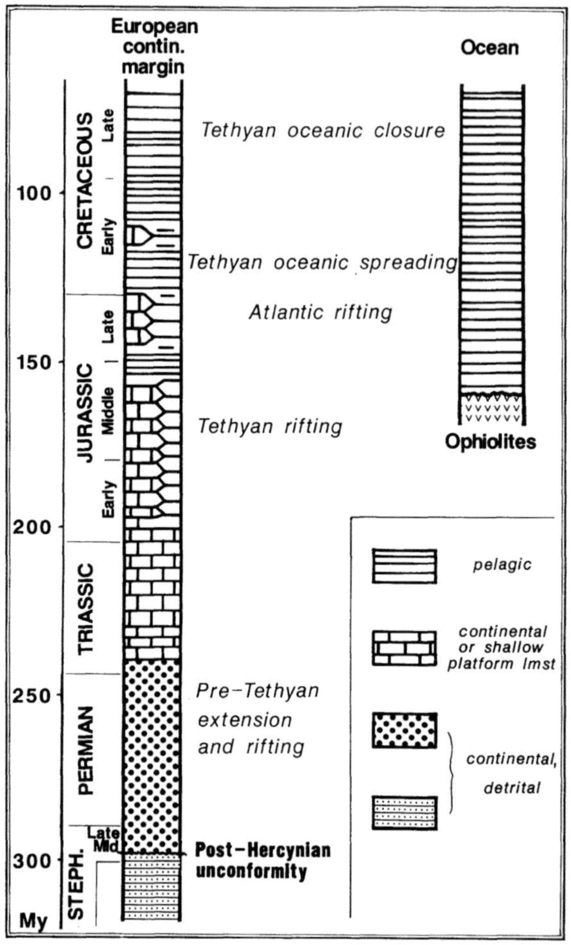

Fig. 6. The main steps of the palaeotectonic evolution of the European margin of the Tethys

extension was NW–SE. This result was acquired independently in the Cevennes (Bonijoly and Delpont 1982; Le Guen and Maurin 1987) and in the Bourg d'Oisans area (Grand 1988) and in Briançonnais (Faure and Mégard-Galli 1988). It has been inferred by several authors such as Coward and Dietrich (1989) but without actual field evidence.

Additional evidence is given by the trajectory of the African-Apulian plate with respect to Europe during the late phase of rifting and Tethyan drifting (155–130 m. y.). The small circles of the rotation of Africa during this period are tangential to the transcurrent fault lines (Fig. 5). This strongly suggests that the Pelvoux Argentera NW–SE direction acted as a major transfer zone during the main rifting phase (Lemoine et al. 1989).

The **transfer fault system** is particularly well exposed at the south–western border of the Pelvoux-Argentera line which acted as a major palaeogeographic boundary during Permian to Jurassic times (Dardeau 1978).

The Nimes, Durance, Rouaine and Var faults in the Southern Subalpine Ranges belong to he **extensional system** like others that bound on the east the external crystalline massifs of La Mure, Taillefer (Ornon fault), Belledonne, Pelvoux, Grandes Rousses (Figs. 2 and 5). The latter are gigantic normal faults whose scarps have been re-eroded and are preserved cliffs on the eastern flank of these basement cores. The early-rift sediment fill of Liassic age is still preserved.

The role that Late Hercynian fracture lines have played on the Mesozoic sediment distribution remains a classical interpretation of Alpine geology (see Debelmas 1974). The close relationship between old fractures in the basement and Mesozoic active faults, such as Belledonne, Aspres and Clamensane, oriented NNE–SSW, have been pointed out earlier by Arnaud et al. (1978). The NW–SE direction of Late Hercynian wrench faults is clear in the crystalline massif of Argentera (Malaroda 1970; Bogdanoff 1986; Fig. 5).

NW–SE structures have a heavy overprint on the west-European Alpine foreland as shown on Ziegler's palaeogeographical maps (1988). Major lineaments of this system such as the Newfoundland, Gibraltar and the Tornquist fracture zones which were prime boundaries in Permo-Triassic palaeogeography. The deep Triassic basins formed southwest of Britain on the hanging walls of Variscan thrusts (Coward and Dietrich 1989) arguably show the influence of the Hercynian grain on the direction of Mesozoic extension (Arthaud and Matte 1977). The coincidence and possibly the influence of major Hercynian lineaments on the principal directions of transfer and of elongation during the Early Tethyan phase gives some coherence to the system.

The main period of Tethyan rifting in the Western Alps is dated Early to Middle Jurassic. Thus, the history of the Western Tethys can be divided into three main phases (Fig. 6): (1) Palaeozoic to Triassic prerift phase; (2) Jurassic rifting (Hettangian to Callovian) including the Central ("Tethyan") part of the Atlantic; (3) Tethyan spreading contemporaneous with rifting in the future Alpine foreland and the North Atlantic (Bay of Biscay) margin (Late Jurassic and Early Cretaceous). This late phase ended with the incipient Alpine orogeny during Late Cretaceous times. It marks the Tethyan oceanic closure.

The prerift phase includes: (a) the Variscan orogeny that ended during Latest Westphalian; then (b) the development of a wrench fault system of Stephanian to Early Permian (Autunian) age and again (c) the development of several rift systems which intersected the Hercynian belt during Triassic times that post-date the "Late Permian tectonic remission" (Ziegler 1988). No oceanic crust was created during the period in the Western Alps (Bechstadt et al. 1978) as it occurred in the Middle East (Channel et al. 1979).

The Rhône Valley graben belongs to this system. The Frontal Penninic thrust which turns nearly at a right angle on the near side of the Pelvoux (Figs. 4, 5) separates two domains significantly distinct in terms of Alpine Triassic palaeogeography. On the east of the L-shaped line, shallow (tidal) platform carbonates and evaporites are found but on the west, i. e. in the Alpine external zones, sebkha (continental) evaporites and detritics were deposited (Mégard-Galli and Faure 1988).

The development of the Jurassic rifting itself was essentially episodic and comprised three major phases dated as Hettangian to Sinemurian, Late Liassic and Bathonian to Early Callovian. These are separated by periods of subsidence and relative tectonic quiescence (Lemoine and Tricart 1986).

The Tethyan spreading during Late Jurassic and Early Cretaceous times was marked by the general post-rift subsidence of the margins partly contemporaneous with renewed extension and block tilting. The latter records the distant effects of North Atlantic rifting. The pre-Cenomanian unconformity in the subalpine ranges (Graciansky and Lemoine 1988) as well as in the Pyrenean (Choukroune 1976) post-date phases of block tilting and crustal extension dated Late Jurassic to Early Cretaceous. During the same period, compression occurred in the Hellenides (Dercourt et al. 1986), in the Northern Calcareous Alps (Frank 1987), and most parts of the European continental margin in the Eastern Alps (Laubscher and Bernoulli 1982). The plate model that has to take into account North Atlantic rifting, Tethyan spreading, change of path of Africa and Adria with respect to Europe at 95–100 Ma (Livermore and Smith 1985) for explaining the evolution of the strain field in Alpine areas is not completely clear yet.

Estimation of the width of Tethys is of prime importance for pre-Triassic palinspastic reconstruction. It can be estimated from the relative motion of the European and African plates with the help of palaeomagnetic data. The Callovian Tethys reconstruction by Dercourt et al. (1986) shows it significantly narrower than another model by Coward and Dietrich (1989) which suggests a width of 800–900 km. Such values must be compared to the estimation by the same authors from Homewood et al.'s (1986) data of a width of 250 km for the European marginal lithosphere involved in the building of the Western Alps. Such data provide a range for the original width of the Tethys and European margin.

Such a figure matches relatively well with the estimation of a 40–50 km shortening for the unaffected European crust to Belledonne eastward added to the 80±15 km given for the apparent extension of the European slab below the internal domain. This value of 80 km, which results from gravity modelling, represents a lower constraint or the amount of shortening by thrust, west of the ophiolite suture (Nicolas et al. 1990).

3 Pre-Triassic Basement and Alpine Structures

The internal Alpine zones comprise thrust sheets that have been imbricated during Late Mesozoic and Cenozoic times in a very complex manner due to the superimposition of several phases of deformation and metamorphism. In the east, they overide the external zones, whose basement was sheared along a deep fault surface (Ménard and Thouvenot 1984; Mugnier et al. 1990) and they are overlain unconformably in the east by the Late Cenozoic alluvium of the Po Valley.

The interpretation of the structures of the internal zones is still controversial in spite of numerous and extensive field studies (Argand 1916; Laubscher 1971; Compagnoni et al. 1977; Mattauer and Tappon-

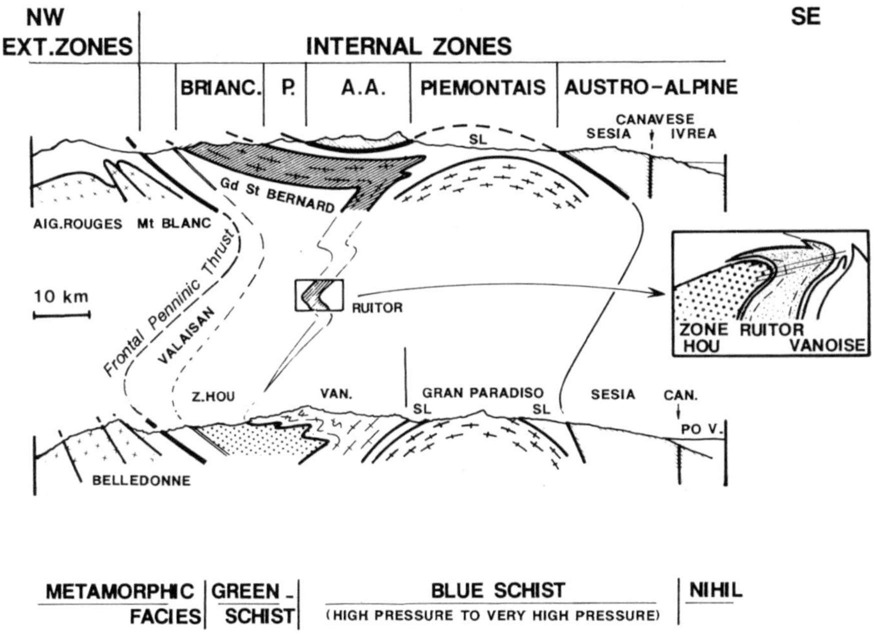

Fig. 7. Structural sketch sections across the internal Alpine units in Savoy (simplified from Debelmas 1974, and unpublished data by Mercier). *Above* Section through the Grand Saint Bernard. *Below* Section through the Belledonne Grand Paradiso line. *Brianc* Briançonnais; *P* Piemontais; *A. A.* Austro-Alpine; *Z. HOU* Zone Houillère; *Van* Vanoise; *Can* Canavese; *PoV.* Po Valley; *S. L.* Schistes Lustrés. These two sections show the superimposition of Austro-Alpine, Piemontais, Briançonnais and Valaisan nappes that are overthrust at their western boundary onto the Massifs Cristallins Externes of Mont Blanc and Belledonne. The section across strike is the well-known "eventail Briançonnais" or fan-shaped structure which derived from a combination of folds of opposite vergence. This results from successive phases of ductile folding dated Late Oligocene and Late Miocene that corresponded to episodes of overthrusting in the western Briançonnais and of simultaneous backthrusting and backfolding in the eastern Briançonnais (Tricart 1984; Platt et al. 1989). Ruitor *(inset)* is considered in classical literature (e.g. Debelmas 1974) as part of the basement of the partly detached Zone Houillère and thus independent of the Vanoise. Recent field observations of prime importance for palinspastic reconstructions (Baudin and Mercier, pers. comm. 1991) have shown that the Ruitor is the southern extension and vertical root zone of the Grand Saint Bernard, a gigantic Pennine-style recumbent fold widely exposed in Switzerland at the core of a main Briançonnais nappe. The Grand Saint Bernard and Ruitor have been thrust westward onto the Zone Houillère during the Early Alpine phases and later backthrust towards Vanoise. This implies that the cores of the Briançonnais nappes have no cylindrical prolongation along the strike

nier 1978; Dal Piaz this Vol.). One recent interpretation proposed by Tardy et al. (1990) relates the slicing of the Ivrea body and the deep fault surfaces that intersects the European continental crust to the Alpine collision.

3.1 Internal Zones

Pre-Triassic masses form three concentric belts within the Pennine units. The easternmost comprises Grand Paradis, Dora-Maira (Fig. 4), the two others are related to the internal and to the external Briançonnais.

Grand Paradis (Dal Piaz this Vol.) and Dora-Maira (Sandrone this Vol.) internal crystalline massifs are surrounded by thrust sheets which are the highly deformed Mesozoic Schistes Lustrés. Both massifs derived from the distal part of the European craton stretched by Tethyan rifting (Figs. 2, 7, 8). Due to severe (high pressure) metamorphism, the Variscan basement can hardly be separated from the Permo-Carboniferous.

Little has been published by French geologists on the alpine structure of Dora Maira since the monographs by Vialon (1966) and Michard (1967). Classical descriptions show the Dora-Maira Massif to consist of three main structural units. The lowest is the Bas-Chisone complex. It comprises the pre-Carboniferous basement and the graphite-bearing Pinerolo group of supposed Carboniferous age, that has been postulated to belong to the same unit as the internal Briançonnais Acceglio zone following Argand (1911 a, b). The middle unit is the Nappe de la Varaita, which involves the Hercynian basement plus a thick Permian sequence with volcanics and detrital rocks. The upper units comprise the Dronero and Sampeyre

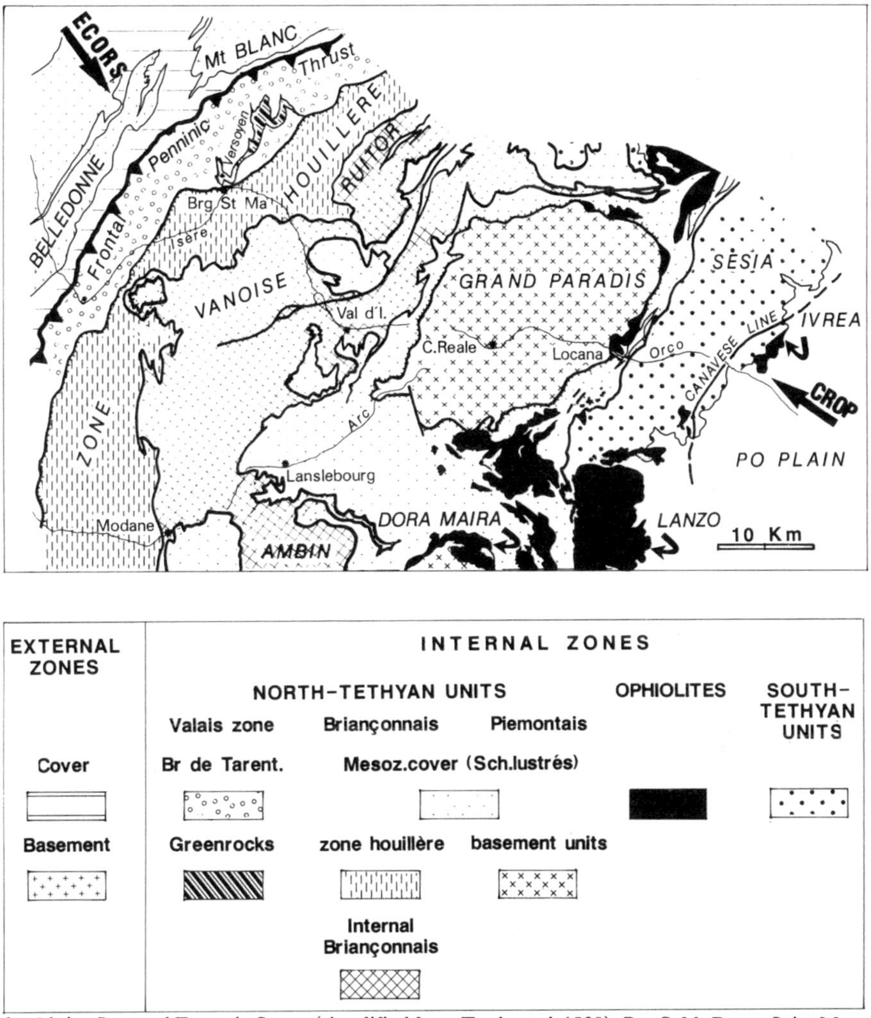

Fig. 8. Structural sketch map of the Alpine Internal Zones in Savoy (simplified from Tardy et al. 1990). *Brg St Ma* Bourg Saint Maurice; *Val d'I* Val d'Isère; *Br de Tarent.* includes here Brèche de Tarentaise, Sub-Briançonnais and External Zone Houillère Briançonnaise. This map shows the location of the *ECORS-CROP* deep seismic line and other structural cross-sections drawn from field data (Fig. 7). Recent field observations have shown that units constituting the internal zones have no cylindrical prolongation along the strike. The Vanoise Mont Pourri and Ambin nappes pinch out both northward at the west of the Ruitor-Grand Saint Bernard and southward between the Piemont units and the Briançonnais Zone Houillère. The Grand Saint Bernard nappe pinches out southward between the Vanoise and the Zone Houillère. The Grand Saint Bernard and Vanoise are not strict longitudinal equivalents. Another point (Mercier, pers. comm. 1990) results from the discovery of floras dated as Namurian to Late Westphalian in the zone du Petit Saint Bernard which was previously considered to be a Mesozoic Valaisan unit (Tardy et al. 1990). This discovery tends to broaden the Briançonnais Zone Houillère

nappes and are composed mainly of Permian detrital rocks resting on a reversed and deeply folded sequence of Schistes Lustrés. The middle and upper units could be correlated tentatively with Vanoise or with Piedmont basement, i. e. deriving from more distal parts of the Tethyan margin that the internal Briançonnais.

According to Sandrone et al. (this Vol.), the stratigraphic succession of the Dora Maira comprises: (1) a polymetamorphic basement; (2) a succession of detrital and graphitic layers (these have been intruded by granitic magmas dated as of the Hercynian cycle; Chopin et al. 1991); (3) detrital, volcanic and intrusive rocks related to the Permian; (4) the Permo-Triassic Sampeyre formation; and (5) metadolomites and marbles of presumed Mesozoic age. The age of the graphite bearing layers of the Pinerolo group can be related to the Ordovician as well as to the Carboniferous as discussed for similar lithologies in Vanoise by Mercier (this Vol.). Terms (2) to (5) are monometamorphic in that sense they have been submitted to Alpine metamorphism only. The age of the pre-Alpine metamorphism is unknown between Variscan or pre-Variscan cycles.

The middle belt, i.e. the internal Briançonnais comprises the nappe of Pontis (Grand Saint Bernard) and other nappe piles in Ruitor, Vanoise, Ambin and presumably small slivers detached from the basement in Liguria (Cortesogno and Vanossi 1982–1985; Cortesogno et al. this Vol.).

The Western belt i.e. the external Briançonnais is the Zone Houillère. It comprises weakly metamorphic, Namurian to Eocene, layers which have been detached from an unknown basement.

The Briançonnais belt trends continuously from the Swiss Valais to near the city of Briançon where it disappears below Mesozoic thrust sheets due to a southward plunge of units. It returns to outcrop southward from the Ubaye Valley near Acceglio to the Ligurian coast west of Genoa (Fig. 4).

The Briançonnais is characterized by: (1) scattered masses of metamorphic pre-Namurian basement (internal Briançonnais only); (2) a thick (3000 m) Namurian to Permian pile of detritics, coal measures and volcanics; (3) a typical sequence of Austro-Alpine-type Triassic platform carbonate; (4) an attenuated sequence of Jurassic to Cretaceous pelagic sediments. The latter sequence is considered to typify the depositional conditions onto a basement outer high of the Tethyan margin (Lemoine and Tricart 1986). The definition of the Briançonnais mainly results from the peculiarities of the Mesozoic sequence rather than from the characteristics of the basement. As a result the Briançonnais is considered classically as a tectonic assemblage with frequent substitution of basement cover rather than as a genuine stratigraphic column (Guillot et al. 1986; Fabre et al. 1987). Therefore, a new interpretation can be expected from better knowledge of its basement.

The middle Penninic belt or internal Briançonnais is characteristic of an old (Precambrian?) polymetamorphic basement and a Palaeozoic to Mesozoic monometamorphic cover (Thelin et al. this Vol.). There is a debate concerning the age of the earliest monometamorphic (Alpine phase) layers: graphite-bearing shales of Vanoise could not derive from Carboniferous coal measures but rather from Ordovician layers (Peruccio-Parison 1984; Mercier this Vol.).

Main internal Briançonnais units are widely exposed in Switzerland. They are, from interior to exterior, the Mont Fort Nappe (southern prolongation is in the inner Vanoise), the Siviez-Mischabel Nappe (southern prolongations are Vanoise and Ambin), the Pontis Nappe (southern prolongations are Ruitor, Aiguilles de Peclet, Sapey and Acceglio units). Detailed study is well advanced in Switzerland (Thelin et al. this Vol.) but still in progress in France. No synthetic map is available for the area south of the Ruitor.

The Zone Houillère (external Briançonnais) is the only Briançonnais Palaeozoic known between the Arc (Maurienne) Valley and the Ubaye Valley. It widely outcrops from the Tarentaise (Isère) Valley in

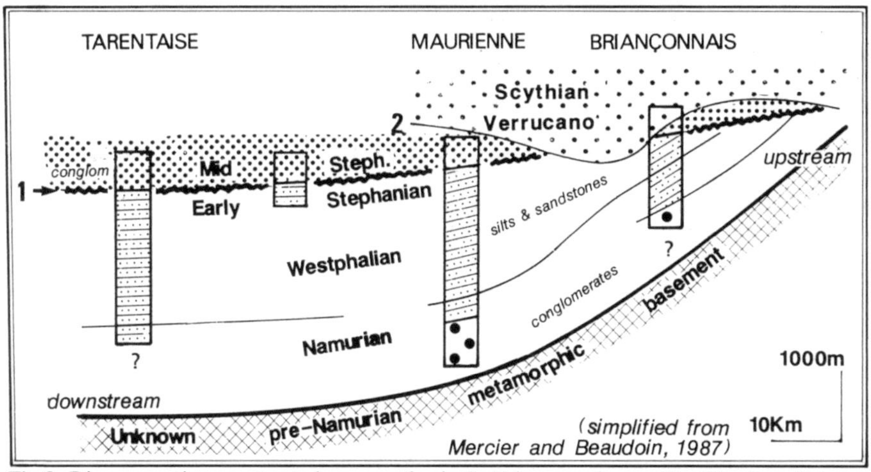

Fig. 9. Diagrammatic reconstructed cross-section in the Briançonnais zone Houillère from Briançon to Isère valley (Tarentaise) (after Mercier and Beaudoin 1987). This section points out the extremely thick Briançonnais Carboniferous sequence and the strong contrast to the external zones where Namurian to Westphalian beds are missing and small Stephanian basins are scattered. This observation suggests that the frontal Pennine thrust is a reactivated Hercynian lineament. *Unconformity (1)* post-dates the latest Hercynian folds and is overlain by Middle Stephanian conglomerates. *Unconformity (2)* records extensional episodes that correspond to the early breakup of Pangea. Four major stratigraphic sequences are bracketed by these unconformities. These are: detritics and coal measures dated Namurian through Early Stephanian; coarse detritics of Middle Stephanian age; "Early" Permian volcanics and detritics equivalent to the Eo-Permian of Ellenberger (1958); and "Late" Permian or Neopermian (Ellenberger 1958), Verrucano facies that grades upwards to the Triassic Scythian sandstones. The four units post-date the Hercynian metamorphism

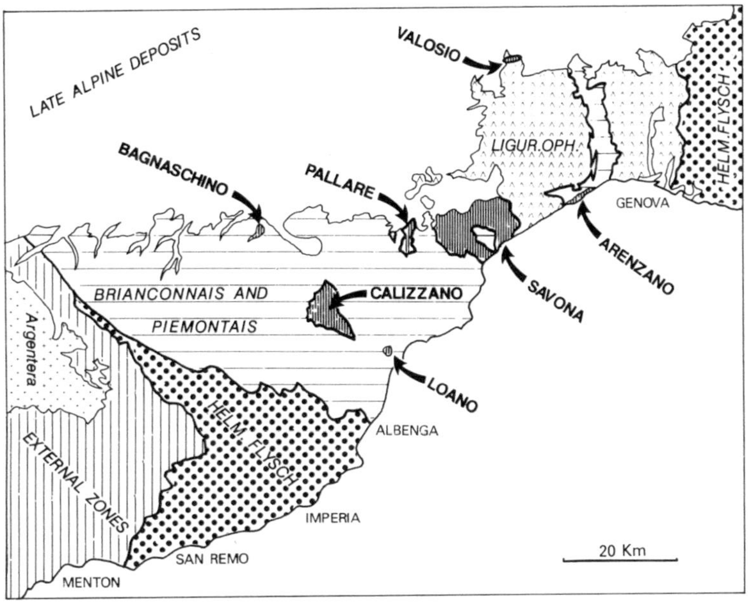

Fig. 10. Structural sketch map of the Ligurian units (after Haccard 1987, unpubl.). Savona, Callizzano, Pallare, Loano, Bagnaschino, Arenzano and Valosio are metamorphic rocks sliced from the pre-Namurian basement during Alpine shortening. *Ligur. oph.* Ligurian ophiolites

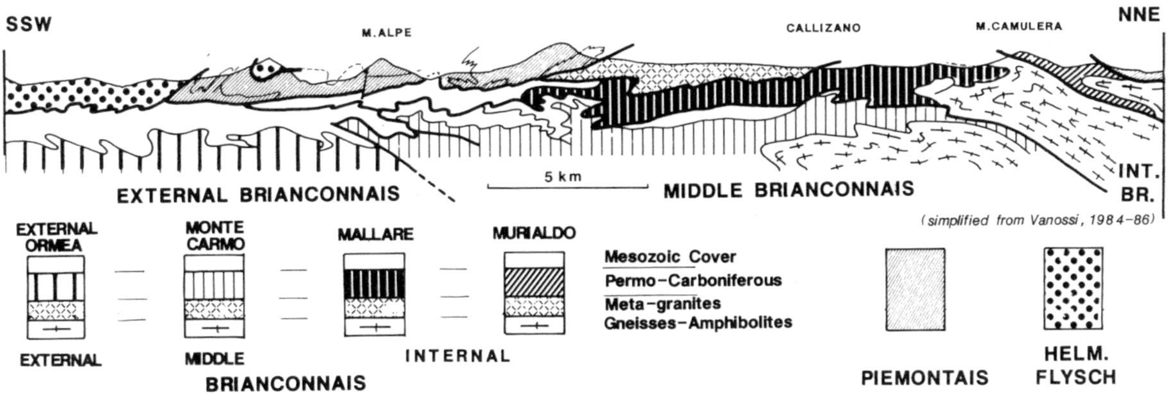

Fig. 11. Structural cross-section in the Ligurian Alps (after Vanossi 1984–1986). The nappe pile comprises units deriving from the Briançonnais at the base from the Piedmont in the Middle and the Helminthoïd flysch at the top. The Ligurian Briançonnais belongs to the three following types depending mainly on their Palaeozoic lithological components:

– External (Ormea) and internal (Mt. Carmo) Briançonnais comprise only a thick sequence of Stephanian to Permian detritics and volcanics metamorphosed in the greenschist facies. Pre-Namurian basement is unknown. The Mesozoic and Cenozoic cover is partly or frankly décollé from its substratum.

– Internal Briançonnais (Mallare, Murialdo) comprise Hercynian metagranitoids overlain by Stephanian to Permian volcanic and detritics. The Mallare unit has been metamorphosed in the pumpellyite-actinolite zone, and the Pamparato-Murialdo in the lawsonite-albite Alpine zone. This latter unit was in a more internal palaeogeographic position than the former. The Mesozoic and Cenozoic cover has been completely detached during Alpine thrustings.

– Isolated basement units comprise only pre-Stephanian gneisses, micaschists and amphibolites with no sedimentary cover preserved. They float as tectonic slivers within the nappe pile. The Calizzano, Savona, Pallare, Loano units are in the Alpine lawsonite-albite facies. There are considered as more external than the Bagnaschino units which belong to the blueschists and quartz-jadeite facies. All are considered to derive from the ultra-internal or innermost Briançonnais.

The overall vergence of Alpine thrust sheets is oriented towards the exterior of the Ligurian belt. The Late Alpine backthrusting seems to have had little effect on the overall amount of tectonic transport. It is therefore accepted that the higher units of the nappe pile were derived from the inner parts of the belt (that is of the distal parts of the inverted European margin), whereas the external Briançonnais units form the base of the nappe stack. From such principles the palaeogeographic reconstruction (Fig. 12) has been proposed from the present-day structures.

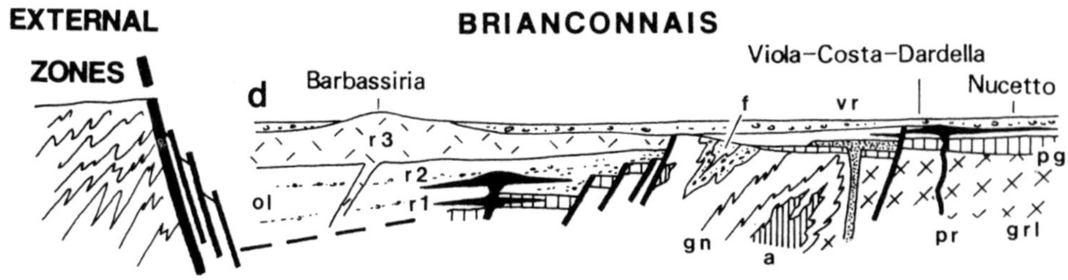

Fig. 12. Reconstructed section across the Ligurian Briançonnais for earliest Triassic times (Cortesogno and Vanossi 1982). Pre-Namurian basement: *a* amphibolites; *gn* gneisses; *gr1* granites; *f* metasedimentary sequence related to the Ordovician to Early Carboniferous. Late Carboniferous to latest Permian cover: *pg* paragneiss di Lisio; *vl* Scisti di Viola (metashales); *Ol* Ollano formation (Late Carboniferous detritics); *pr, r1, r2* associated pyroclastites and rhyolitic volcanics; *r3* Porfiroidi del Melogno (very thick rhyolitic ignimbrites) emplaced during latest Carboniferous to Earliest Permian times; *vr* Verrucano conglomerates (Late Permian to Early Triassic). The Hercynian metamorphic basement comprises gneisses *(gn)* and amphibolites *(a)* and transformed granitic intrusions *(gr 1)*. The Late Palaeozoic cover includes feldspathic sandstones (now paragneisses *(pg)*, several bodies of rhyolites and ignimbrites *(r1, r2)* associated with the Carboniferous Ollano detrital formation *(ol)* that might be equivalent to the Namurian to Early Stephanian layers of Savoy. The Early Permian is characteristic of the development of huge masses of rhyolites *(r3)*. The Late Permian is represented by the Verrucano *(vr)* conglomerates, metashales and metarenites *(f)*. The Late Palaeozoic sediments and volcanics were laid down during several episodes of extensional block faulting. The Late Carboniferous to Permian Briançonnais can be considered as a major tilted block whose raised edge coincides with the internal units (Murialdo, Bagnaschino) of the present-day structures. The thicker post-Namurian sequence is located within the external units (Ormea). This feature was maintained during the Early Jurassic phase of Tethyan extension. Thus the future Alpine frontal Penninic thrust will coincide roughly with and possibly will reactivate the inferred listric fault that probably separated the Dauphinois and Provençal external zone from the Briançonnais during Permian times

the north to the Durance near Briançon in the south (Fig. 4). It shows a decrease in metamorphic gradient from north to south. It is separated from the internal Briançonnais by a major thrust surface associated with a jump in metamorphic grade and change in lithology. It can be divided into two structural units, the internal Zone Houillère showing a more complete stratigraphic sequence than the externalmost one (Mercier pers. comm. 1991). The Zone Houillère, as a whole, is interpreted as a broad intracontinental subsiding basin delineated by extensional faults and block tilting on the post-Hercynian platform dated as Namurian to Permian (Mercier this Vol.; Fig. 9).

In the southern part of the Western Alps, i.e. in Liguria, the Briançonnais outcrops along the Adriatic coast southwest of Genova (Figs. 10, 11).

The stratigraphy (Fig. 12) comprises: (1) the crystalline basement including gneisses, amphibolites and metagranites dated as pre-Namurian in age; (2) a sequence of detritics and volcanics of Permian and Late Carboniferous ages; (3) a cover of Mesozoic to Cenozoic age with typical, Briançonnais, attenuated sequences (Cortesogno and Vanossi 1982–1985; Vanossi 1984–1986). Lithologies and ages are comparable to those of Savoy with the exception of the major development of volcanics in Liguria (Cabella et al. 1987–1988; Cortesogno et al. 1988 a, b; Cortesogno et al. this Vol.).

Basement units related to the Briançonnais – including possibly the Bas Chisone Complex of the Lower Dora Maira – occupied a width of several hundreds of kilometers across the former Tethyan European margin of the Western Alps. But most of the Mesozoic cover has been lost which makes attempts at a reliable palinspastic reconstruction difficult.

It is important to note the divergent pattern of the lineations all around the curvature of the Western Alps from the external zones (Fry 1989) to the internal ones including the Briançonnais (Vanossi 1984–1986; Choukroune et al. 1986; Platt et al. 1989 and others). In detail, the regularity of the distribution of the lineation has been perturbated by the morphology of the foreland at the base of the advancing nappe. For example, clear divergence of stretching lineation has been observed within Briançonnais units at the eastern side of the Pelvoux massif which acted there as a buttress (Tricart 1980). Such structures hamper any attempt at simple palinspastic reconstruction in these areas.

3.2 External Zones

Basement outcrops comprise the "Massifs Cristallins Externes" as named by classical French geologists (Gignoux and Moret 1934). The external crystalline massifs have a dog-leg shape in map view (Fig. 4).

The northern branch consists of two lines of massifs comprising the Mont Blanc-Aiguilles Rouges, the

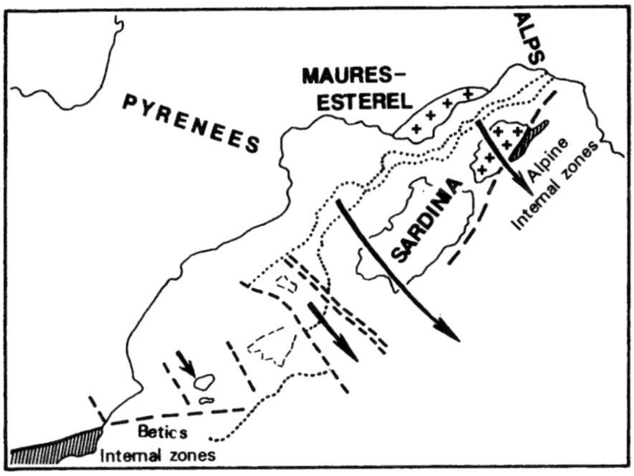

Fig. 13. Reconstructed position of the Corsica-Sardinia small plate prior to the Late Oligocene extension in Provence. Western Alps and Gulf of Lions (see Burrus 1984). This shows that the Maures-Esterel-Tanneron block has, with Corsica and Sardinia, been a nearly continuous unit until as late as Early Cenozoic times. *Arrows* corespond to the trajectory of the block of Corsica and Sardinia during the Tertiary

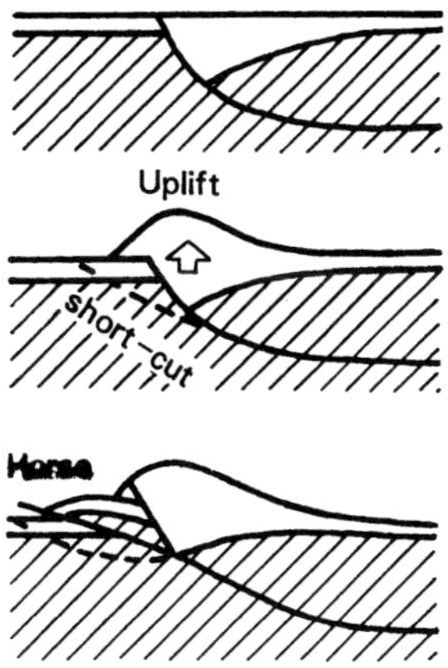

Fig. 14. Model for the development of a footwall short-cut thrust system (Hayward and Graham 1989). Such a mechanism may explain the presence of isolated slabs of the basement within a stack of Mesozoic thrust sheets, such as Rocheray (Gillcrist et al. 1987), Roselette, Remollon or Verdaches (see locations in Fig. 4). Tectonic slabs have been detached from the raised edge of tilted blocks

Belledonne-Grandes Rousses, and the Taillefer. Its orientation is NE–SW, i.e. parallel to the Cévennes direction, which corresponds to an old Hercynian lineament and to an extensional fault trend associated with Tethyan rifting (Fig. 3 A). The middle branch of the dog-leg-shape comprises the Rochail, Mont de Lans, Pelvoux and Combeynot, then the Argentera-Mercantour. The orientation is NW–SE and corresponds to another Hercynian lineament and to a transfer line of the Tethyan rifting.

The southern branch is occupied by the Tanneron-Esterel-Maures along the Côte d'Azur to which Corsica and Sardinia were attached prior to the Oligocene oceanic opening in the Gulf of Lions (Fig. 13). For discussions on pre-opening reconstructions, see Burrus (1984).

In addition to the major crystalline massifs, some pre-Triassic slivers are scattered within the subalpine domains. These are sliced from basement by Alpine thrusts and are generally found inserted along major décollement surfaces within folded Mesozoic sequences (Figs. 4, 14).

The stratigraphical sequence comprises first, the pre-Late Westphalian metamorphics which make up the main part of the External Crystalline Massifs (Ménot 1988; von Raumer, Ménot and others, this Vol.) and second, Late Westphalian to Permian basins scattered on the basement outcrops. The latter seem to just post-date the latermost Variscan compression by extrapolating data by Mercier (this Vol.) in the Briançonnais. They occur within small extensional half-grabens which record the initial phase in the disintegration of Pangea. These half-grabens were compressed during the Alpine diastrophism.

Wide Permian basins are developed in the Alpes Maritimes, Provence and Corsica, i.e. southwestward of the Pelvoux Argentera line (Fig. 4B). They are nearly absent northeast of this line which acted, therefore, as an early palaeogeographic boundary as it did during the Tethyan rifting. The Barrot, high Tinée Valley, Roya Valley and Esterel Permian sequences were known long ago for their classical red beds which were deposited contemporaneously with extension (Toutin-Morin 1987). The Esterel and Corsican Permian comprises thick andesitic to rhyolitic volcanics including ignimbrites (Gondolo and Toutin-Morin 1989), which are closely comparable to those of Liguria.

The style and intensity of deformation in the External Crystalline Massifs depend on their location with respect to the Alpine strain field. Deformation has been very intense in the areas submitted directly to the collision, i.e. at the front of main alpine thrusts, as in Switzerland (Pfiffner, Ratschbacher this Vol.) and further eastward. Such areas correspond to the northern branch of the dog-leg shape of outcrops mentioned, i.e. from Belledonne up to the Aar-Gotthard. The Pelvoux-Argentera branch of the dog-leg shape which is oriented NW–SE acted as a lateral ramp during both the Tethyan extension (Lemoine et al. 1989) and then during the early phases of Alpine shortening (Choukroune et al. 1986). So the overall displacement of Argentera with respect to the European craton certainly has been minor.

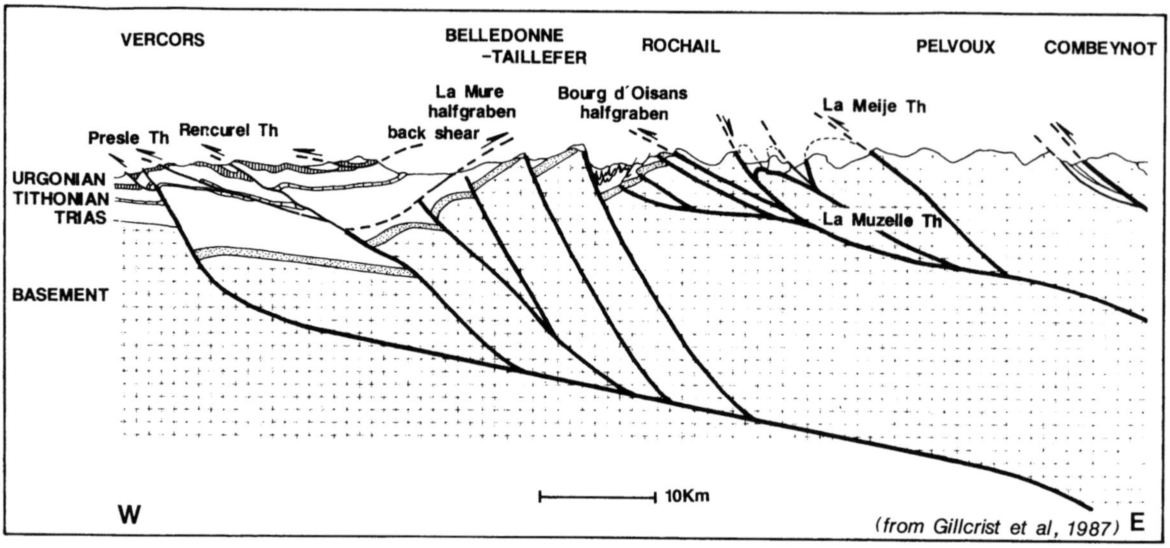

Fig. 15. Detachment surfaces below the Massifs Cristallins Externes of Pelvoux and Belledonne. The existence of detachment surfaces below the Massifs Cristallins Externes can be inferred from the considerable elevation of the basement of the Pelvoux (La Barre des Ecrins: 3774 m) which sharply contrasts with its altitude (below sea level?) below Chartreuse and Vercors. Attempts to balance sections according to the models of Boyer and Elliot (1982) imply a detachment surface through the basement, dying somewhere below the Rhône valley (Vann et al. 1986). The reality of this detachment surface has been recently documented by the discovery of an eastward dipping reflector below the Massif de Belledonne from the ECORS deep seismic line across the Alps (Mugnier et al. 1990) which is consistent with an interpreted section by Gillcrist et al. (1987; and Fig. 16)

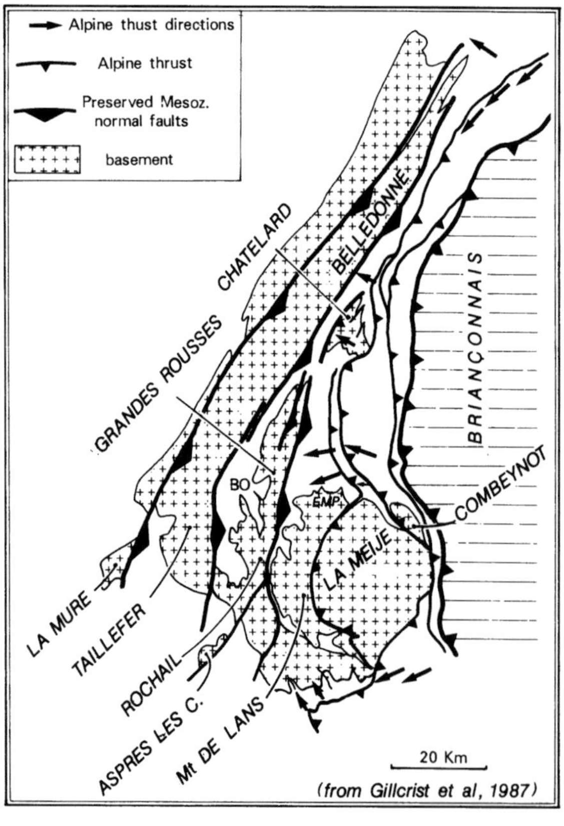

Fig. 16. Alpine thrust surfaces within the Massifs Cristallins Externes of Pelvoux and Belledonne (Gillcrist et al. 1987). The Pelvoux, Belledonne and neighbouring massifs display extended surfaces of basement outcrops that are interrupted by narrow and elongated "synclines" infilled with mainly Liassic shales and limestones. The Belledonne is divided into two wings in this manner by a Liassic "syncline" and is separated from the Grandes Rousses by the Bourg d'Oisans "syncline" (see interpretation by Ménard and Thouvenot 1984). The Combeynot, Pelvoux, Mont de Lans and Taillefer are separated by intercalated highly deformed Liassic units (Graciansky et al. 1989). All these were infact old half grabens resulting from the pre-Tethyan extensional tectonics partly inverted during the Alpine shortening (Lemoine and Tricart 1986). The resulting "synclines" area are generally over-ridden on their eastern flank by westward-verging reverse faults that involve the basement as well as its cover. Note the change of transport directions along thrust planes from northwest at the north of Pelvoux to southeast at the south. This radiating pattern of the directions of the final thrustings around the arc is a consequence of the continental collision

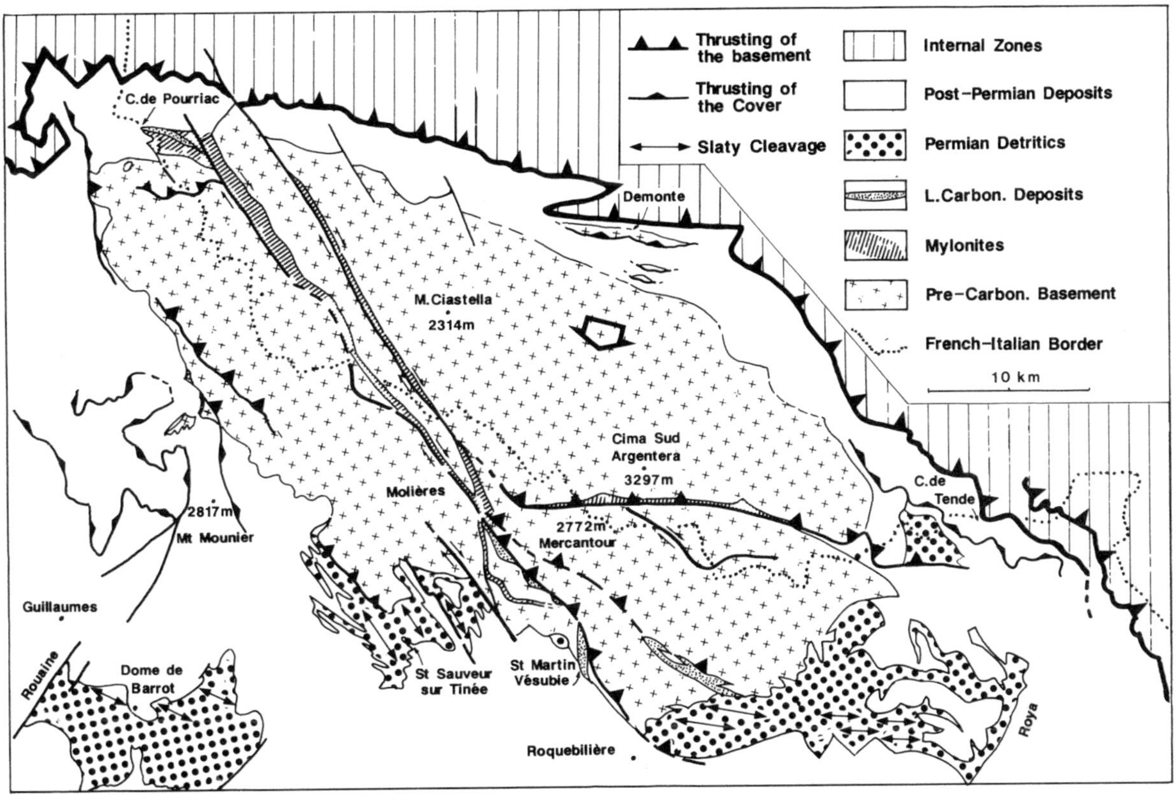

Fig. 17. Structural sketch map of the Argentera massif. Alpine deformation within the Argentera gave birth to independent elongated basement massifs with an overall NW–SE trend. These are separated by reverse faults or strike-slip fault depending on their orientation with respect to the regional trend of the strain. The main direction of shortening was mainly north–south (Horrenberger et al. 1978; Guardia and Ivaldi 1985). As a result, the old Hercynian NW–SE lineaments (Malaroda, 1970) underwent dextral strike-slip and the east–west became southward overthrusts. The deformation was concentrated along the main fault corridors where mylonites developed in the greenschist metamorphic facies corresponding to a retrograde metamorphism of the surrounding mesozonal to catazonal (Malaroda 1966) Variscan metamorphic rocks (Horrenberger et al. 1978; Bogdanoff 1986). Pervasive deformation and Alpine cleavage affected the whole mass of the basement between the mylonitic zones and partially re-oriented the Hercynian structures but in a relatively modest manner

The External Crystalline Massifs of Pelvoux and Belledonne can be considered as detached parts of the European craton. These have been transported along rampflats coinciding with deep shear surfaces cutting across the fragile part of the lithosphere as a result of Alpine shortening (Fig. 15).

The geometry of the Alpine deformation was guided by old listric faults inherited from the Tethyan rifting: it can be shown that the old listric (extensional) faults caused heterogeneities that controlled the emplacement of the thrust lines during the Alpine compressions (Fig. 16) in the various geometrical styles discussed by Gillcrist et al. (1987), or by Hayward and Graham (1989).

Pronounced local variations in thrust directions, often within individual thrust sheets have been shown by Gillchrist et al. (1987), Platt et al. (1989) and Vialon et al. (1989). Fry (1989) stated that the radiating pattern results from the superimposition with time of several directions of transport which seem to be constant over long distances along major individual thrust planes. The radiating pattern of the directions of the final thrustings and of lineations around the arc is a consequence of the continental collision (Fig. 3C).

The Pelvoux-Argentera line marks the southwestern limit of major Alpine nappes of internal origin (including the ophiolites) with respect to the autochthonous Provence. Merle and Brun (1984) have clearly shown that the particle paths in the internal zones during the early phases of the Alpine shortening were roughly parallel to this line. The Pelvoux-Argentera line can be considered therefore as a strike-slip corridor of prime importance within the Alpine mountain building.

As a result the Alpine features of the Argentera (Fig. 17) can be interpreted as a large-scale "pop up" with SW-vergent structures at its SW flank and NE-

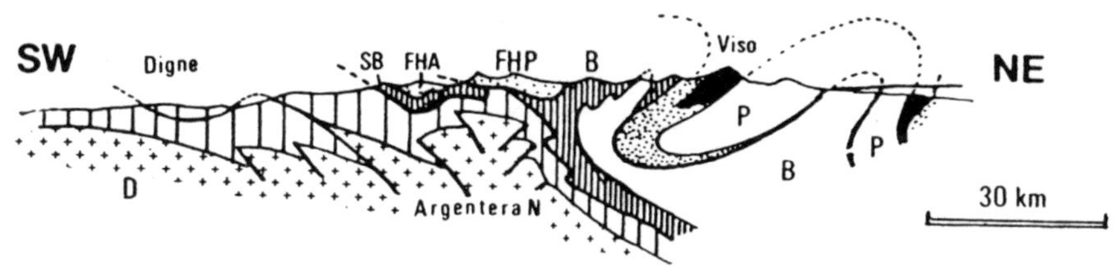

Fig. 18. Structural cross-section across the northern part of the Argentera massif (Sturani 1963; Debelmas 1983). The Argentera Massif is shown as bounded by reverse faults of opposite vergence. This situation could result in a major strike-slip corridor. *D* Dauphinois; *SB* sub-Briançonnais; *B* Briançonnais; *SL* Schistes Lustrés; *B* basement of the Schistes Lustrés; *FH* helminthoïd Flysch; *OPH* ophiolites

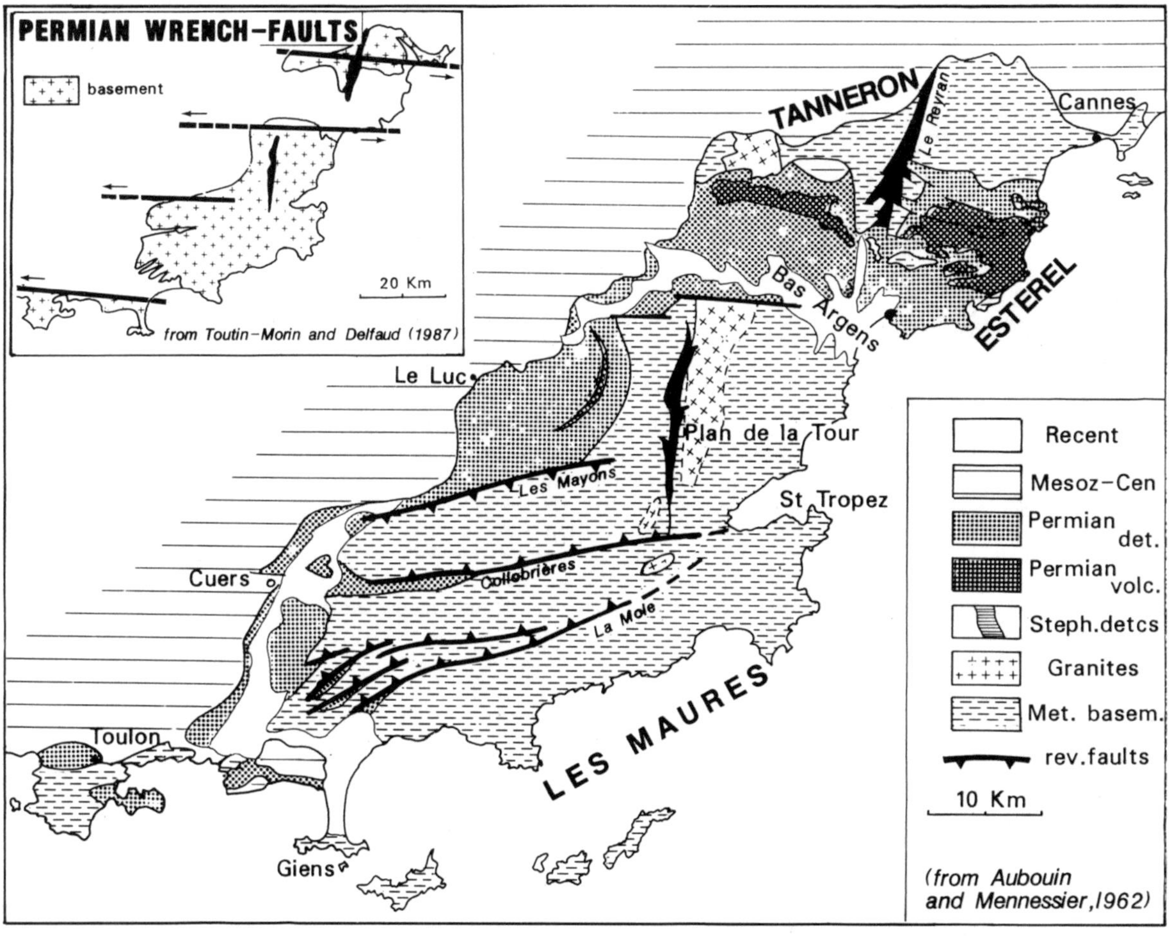

Fig. 19. Structural sketch map of the Maures-Esterel-Tanneron (Aubouin and Mennessier 1962). Permian left-lateral wrench faulting was accompanied by the deposition of thick detrital and volcanic sequences (Toutin-Morin and Delfaud 1987). The Alpine shortening in the Massif des Maures is oriented north/south and resulted from the Pyreneo-Provençal orogeny. It amounts to 5 to 10% only. There is no suspected major thrust surface at the outcrop between the Massif des Maures and the Massif Central but an overall detachment of the Mesozoic cover which is characteristic of Alpine tectonism in Provence. The massif can be considered as still being linked with the Western European craton

vergent structures at its NE flank as pointed out by Fry (1989) and represented already (Fig. 18) by Debelmas (1983) and Sturani (1963). This was accompanied by a strong north-south shortening (Horrenberger et al. 1978). However, the amount of displacement of the Argentera with respect to the Western European craton was certainly very small in comparison to Belledonne and Mont Blanc.

The dating of the vertical movement of the Argentera is compatible with these views. This area was most probably emergent during Late Cretaceous times (Guardia and Ivaldi 1985) as were many external crystalline massifs. The Argentera began to subside at the beginning of the Tertiary because of the general flexure of the European craton in response to tectonic loading by the advancing Alpine internal nappes. This is shown by the burial of the Alpes Maritimes and adjacent areas below the deep marine Grès d'Annot (Stanley 1961; Ravenne et al. 1987) dated Late Eocene to Early Oligocene. The idea of a rotation of the Argentera has been excluded from both microstructural and palaeomagnetic data (Bogdanoff and Schott 1977).

The Maures-Esterel-Tanneron Massif (Fig. 19) is located outside of the domain affected by Alpine thrustings. It can be considered as part of the Alpine foreland and is still linked with the western European craton.

4 Summary, Discussion and Conclusions

Identifying problems of Alpine geology within Hercynian basement units has been considered of little reward by both Hercynian petrographers and Alpine structural geologists. In consequence, the available information on the pre-Triassic basement is sparse with respect to the large amount of data on the Mesozoic and Cenozoic. In fact, our overviews on the geology of the basement in the western Alps are extrapolated partly from geological knowledge on the cover.

The development of the deformation in the Alps from Late Palaeozoic to Late Cenozoic times is characteristic of, firstly block-faulting and extension that led to the creation of the Tethyan ocean and margins (Fig. 1), secondly, folding and thrusting in a compressional regime that corresponded to the closure of the Tethys and to the collision of Africa with Europe (Fig. 3).

4.1 Hercynian Crustal Stretching and Tethyan Opening

The extension and crustal stretching that commenced during the first half of the Stephanian recorded the Late Hercynian post-orogenic collapse. The Triassic rift system in the Western Alps is linked with the development of the Neo-tethys spreading axis in the Middle East (Channel et al. 1979). Volcanic activity of that age was intense in the Italian dolomites and rare in the Pelvoux area but no Triassic oceanic crust was created in the Western Alps. Episodic rifting phases dated as Liassic resulted in a complicated pattern of submarine troughs and swells on the European, Apulian and North African cratons. The Late Bathonian to Callovian extension is the main Tethyan rifting phase preceding continental break-up and the onset of spreading in the Tethyan ocean in our Western Alps.

The African and European blocks began to be separated clearly by Middle Jurassic times as a consequence of the eastward movement of Africa with respect to Europe (Livermore and Smith 1985, and others). A width of 250 km across the strike has been estimated for the European lithosphere involved in the Western Alps building. Such an estimation is of prime importance to attempts at viable pre-Triassic reconstructions. The major part of the structures of the Tethyan European margin in the Central Alps has been lost beneath the huge thrust sheets of internal Alpine origin (Fig. 3). Nevertheless, the French subalpine zones escaped severe deformation because of their location at the exterior of one of the major lateral ramp systems that bounded the main thrust sheets (Figs. 4, 5).

They therefore preserved Jurassic palaeo-faults and transfer lines whose geometry suggests that synrift extension was oriented NW to SE (present-day orientation). The Pelvoux-Argentera line (Dardeau 1978) was one these principal transfer lines well preserved today as a major Jurassic palaeogeographic boundary.

The network of synrift extensional and transfer faults has probably been influenced by main Hercynian lineaments. This does not imply, however, that the old Hercynian faults and structural discontinuities were reactivated themselves during later phases of extension but rather that faults of Late Palaeozoic to Mesozoic age are roughly coincident in space.

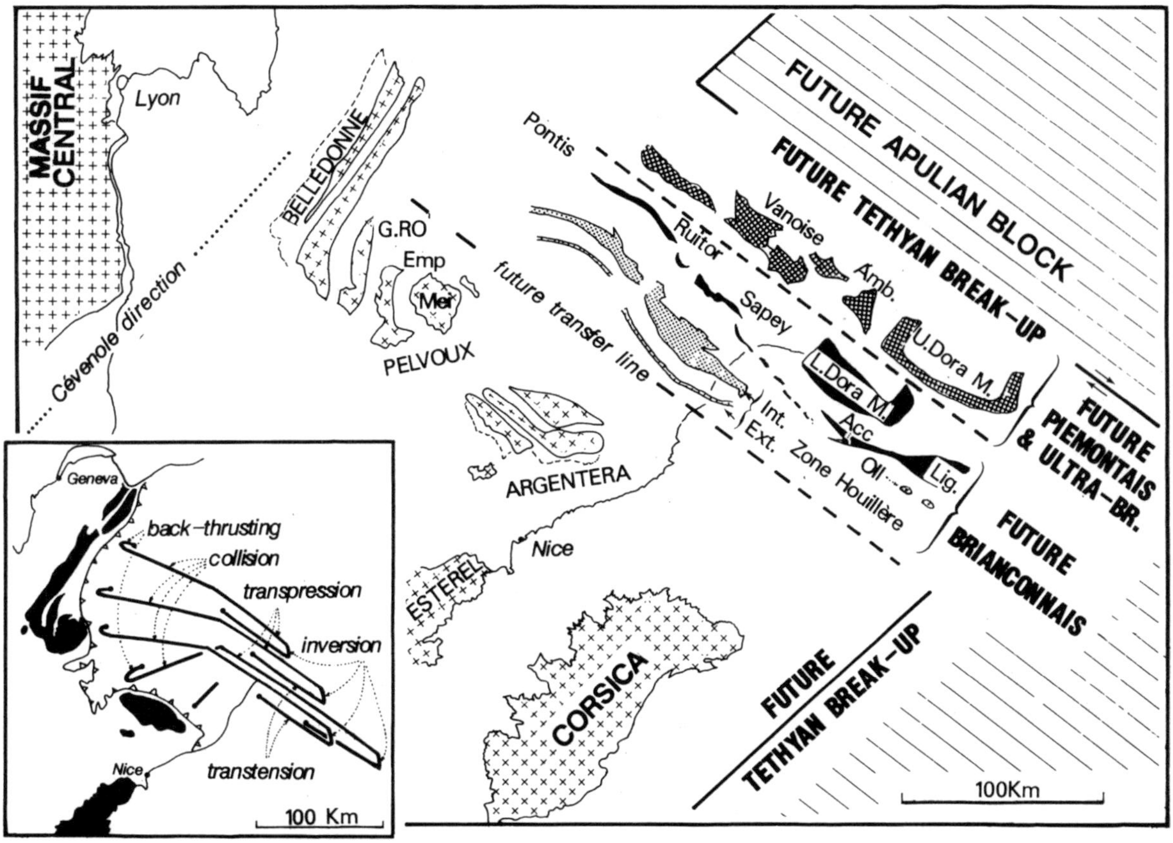

Fig. 20. An attempt at pre-Triassic reconstruction in the Western Alps: an unsolvable problem. External zones: Belledonne, Grandes Rousses *(G. Ro)*, Pelvoux (*Emp* Emparis-Mont de Lans; *Mei* Meije), Argentera, Esterel-Maures. Future Briançonnais: External and Internal Zone Houillère with possible equivalent in Liguria (*Oll* Ollano). Internal Briançonnais or Pontis – St Bernard, Ruitor, Sapey, Acceglio *(Acc)* and possible equivalents in the Lower Dora Maira (*L. Dora M.* or Bas Chisone) and in Liguria *(Lig.)*. Future Ultra-Briançonnais and Piemontais: nappes in Switzerland (Siviez-Mischabel and Montfort: Thélin this Vol.) with probable extension towards Vanoise, Ambin, Upper Dora Maira *(U. Dora M.)* and Liguria (Savona, Callizzano: Cortesogno this Vol.). These southern prolongations of the Swiss nappes have not yet been identified clearly. This fancy artwork pretends to take into account the multiphased development of the deformation since Late Palaeozoic times *(insert)*, including – southeastward transtensional dextral movement along transfer lines shown by the Pelvoux-Argentera lineament (Early and Middle Jurassic); – inversion of the transport direction (Mid-Cretaceous); – northwestward transpressional sinistral movement along the previous tramlines (Late Cretaceous); – collision with right-lateral movement along the Insubric line (Cenozoic; Fig. 3 C); – backthrusting at the interior of the arc. But there is no way to estimate the amplitude of the horizontal displacement of units. The successive right-lateral and left-lateral movements along the same tramlines probably allowed some parts of the European margin to return close to their Late Palaeozoic position at the end of the Mesozoic. Therefore an approximate pre-Triassic reconstruction might be drawn, by taking into account mainly the directions of the thrusts. But the radial pattern of the lineations that converge towards the centre of the Western Alpine Arc (in the area of Turin) raise an unsolvable space problem for any pre-collisional reconstruction. If the idea of the transtensional, NW–SE trending, Tethyan margin is accepted, one way to solve the space problem is to admit that the Briançonnais and Piemontais units, as early as Late Palaeozoic times, were located in a more southeastern position than at present. Such a position may result simply from unbending the arc. But the sketchmap obtained cannot show the downward plunge of several tens of kilometres that marked the subduction of the distal part of the European craton (Dora-Maira and Gran Paradiso, Chopin et al. 1991) as a consequence of the collision

4.2 Alpine Building and Continental Collision

The change in the movement of Africa with respect to Europe since Mid-Cretaceous times resulted in: (1) the closure of Tethys; then (2) the European-African collision (Dercourt et al. 1986; Dewey et al. 1989).

During the early phases of the closure, the transfer faults of the Tethyan extension were re-used as tram lines during the northward movements of Adria. This is documented by the direction of particle paths during Late Cretaceous time (Choukroune et al. 1986). Support is also given by the small circles corresponding to the movement of Africa with respect to Europe for the same period which are tangential in the Western Alps to the small circles corresponding to the period Tethyan rifting during the Middle Jurassic (Fig. 3).

The Pelvoux-Argentera lineament now appears as a palaeogeographic limit both during extension and compression. It has been both a major transfer line of the Tethyan rift system and a lateral ramp with respect to Alpine early thrustings. Ricou (1981) and Ricou and Siddans (1986) have considered the frontal Penninic thrust at the front of the Briançonnais as a corridor of wrench faults. This idea is consistent with these views for the section which runs from the rear of the Argentera to the south of Pelvoux during the early phases of Alpine thrusting (Merle and Brun 1984; Choukroune et al. 1986). However, elsewhere and for later periods, Fry (1989) and Platt et al. (1989) have shown from their field observations that the gross geometry is dominated by thrusts and by patterns of later normal faults rather than by strike-slip faults.

Vialon et al. (1989) have invoked the idea of the counterclockwise rotation of a block belonging to the European margin that would comprise Mont Blanc, Belledonne and Pelvoux. The whole Alpine edifice would have rotated over a deep, continent-wide, detachment surface. However, from these ideas the hypothesis of a major rotated block is not a requisite for interpreting western Alpine structures. The constancy of displacement directions suggests that there has been no significant relative rotation across thrusts (Fry 1989). The re-orientation with time of the directions of extension within thrust sheets (as observed by Merle and Brun 1984) does not imply the rotation of the whole mass during its transportation.

The deformation pattern that resulted from the collision has been particularly complex (Fig. 3) compared to many other mountain belts. This is due to both the long duration, approx. 100 Ma, and to the complicated shape of the edges of the colliding continents and to the poorly understood behaviour at depth of the deep structure. Notably, the Adria (Adriatic promontory: Channel et al. 1979) at the African front and the original shape of the European passive margin are accurately reflected by the arcuate shape of the Western Alps, the distribution of lateral ramps and thrust planes and the pattern of mass displacement. The result of the indentation of the two passive margins is the lateral asymmetric escape of rock masses during the collision as well shown by Schmid et al. (1989) on the Lepontine dome or by Choukroune et al. (1986) in the Western Alps. The radiating pattern of the directions of thrusting and of lineations around the arc makes coaxial restoration across the Western Alps impossible (Coward and Dietrich 1989).

We hope that new data will emerge from improved knowledge of the basement which has been partly neglected by many western Alpine geologists. Such progress would bring new constraints for our reconstructions (Fig. 20) and give some enlightenment on problems still the subject of debate.

Acknowledgements
Most of the ideas in this chapter resulted from several years of teamwork in the Alps which were carried out with Gérard Dardeau, Rod Graham, Marcel Lemoine, Daniel Mercier, Pierre Tricart and with many young colleagues and students whom we have been pleased to bring to the field. The manuscript was carefully revised by D. Roberts, G. Dardeau, A. Pfiffner and W. Wildi. Figures were completed thanks to G. Dardeau. Fieldwork was supported by grants from Institut Français du Pétrole, Société Nationale des Pétroles d'Aquitaine and British Petroleum Cy.

References

Argand E (1911 a) Sur la tectonique de la grande zone permo-carbonifère du Valais à la Mediterranée. Actes Soc Helv Sci Nat 94ème session, vol 1

Argand E (1911 b) Les Nappes de recouvrement des Alpes penniques et leurs prolongements structuraux. Beitr Geol Karte Schw XXXI: 1–25

Argand E (1916) Sur l'arc des Alpes occidentales. Eclogae Geol Helv 14: 145–191

Arnaud H, Barfety JC, Gidon M, Pairis JL (1978) A propos du rhegmatisme des zones externes alpines au sud de Grenoble. C R Acad Sci Paris 286: 1335–1338

Arthaud F, Matte Ph (1977) Late Paleozoic strike slip faulting in southern Europe and northern Africa. Geol Soc Am Bull 88: 1305–1320

Aubouin J, Mennessier G (1962) Essai sur la structure de la Provence. Livre Mém Prof Paul Fallot, Soc Géol Fr 2: 45–98

Barféty JC (1984) Socle. In: Feys R (ed) Synthèse géologique du Sud-Est de la France. Mém Bur Rech Géol Min 125: 1–45

Baudrimont AF, Dubois P (1977) Un bassin mésogéen du domaine périalpin: le Sud-Est de la France. Bull Cent Rech Explor Prod Elf-Aquitaine I, 1: 261–308

Bechstadt T, Brandner R, Mostler H, Schmidt K (1978) Aborted rifting in the Triassic of the eastern and southern Alps. Neues Jahrb Geol Palaeontol Abh 156 (2): 157–178

Blondel T, Charollais J, Sambeth U, Pavoni N (1988) La faille du Vuache (Jura Méridional): un exemple de failles à caractère polyphasé. Bull Soc Vaudoise Sci Nat 79 (2): 65–91

Bogdanoff S (1980) Analyse structurale dans la partie occidentale de l'Argentera-Mercantour (Alpes Maritimes). Thèse, Fac Sci Univ Orsay, 316 pp

Bogdanoff S (1986) Evolution de la partie occidentale du massif cristallin externe de l'Argentera placé dans l'arc alpin. Géol Fr 4: 433–453

Bogdanoff S, Schott JJ (1977) Etude paléomagnétique et analyse tectonique dans les Schistes Rouges permiens du Sud de l'Argentera. Bull Soc Géol Fr (7) XIX, No 4: 909–916

Bonijoly R, Delpont G (1982) Etude du Bassin des Causses et de la bordure cévenole par la télédétection et la géologie structurale. Doc Bur Rech Geol Min, Orléans 46: 1–40

Boyer SE, Elliott D (1982) Thrust systems. Bull Am Assoc Pet Geol 66: 1196–1230

Brodie KH, Rex D, Rutter EH (1989) On the age of deep crustal extensional faulting in the Ivrea zone, Northern Italy. In: Coward MP, Dietrich D, Park RG (eds) Alpine tectonics. Geol Soc Lond Spec Publ 45: 203–210

Burrus J (1984) Contribution to a Geodynamic synthesis of the Provençal basin (North-Western Mediterranean). Mar Geol 55: 247–269

Cabella R, Cortesogno L, Dalla Giovanna G, Vannucci R, Vanossi M (1987–1988) Vulcanismo, sedimentazione e tettonica nel Brianzonese Ligure esterno durante il Permo-Carbonifero. Atti Ticinensi di Scienze della Terra, Pavia 31: 269–326

Channel JET, D'Argenio B, Horwath F (1979) Adria, the African promontory. In Mesozoic Mediterranean paleogeography. Earth Sci Rev 15: 213–292

Charollais J, Clavel B, Amato E, Escher A, Busnardo R, Steinhauser N, Mac Sotay O, Donze P (1983) Etude préliminaire sur la faille du Vuache (Jura méridional). Bull Soc Vaudoise Sci Nat 303, 76 (3): 217–256

Chopin C (1987) Very-high-pressure metamorphism in the Western Alps: implications for subduction of continental crust. Phil Trans R Soc Lond A 321: 183–197

Chopin C, Henry C, Michard A (1991) Geology and petrology of the coesite bearing terrain, Dora Maira massif, Western Alps High pressure metamorphism in the Western Alps. Eur J Mineral 3: 263–29

Choukroune P (1976) Structure et évolution tectonique de la zone nordpyrénéenne. Analyse de la déformation dans une portion de la chaîne à schistosité verticale. Mém Soc Géol Fr Paris NS 40 (127): 116 pp

Choukroune P, Ballèvre M, Cobbold P, Gautier V, Merle O, Michard JP (1986) Deformation and motion in the Western Alpine arc. Tectonics 5: 215–226

Compagnoni R, Dal Piaz GV, Hunziker JC, Gosso G, Lombardo B, Williams PF (1977) The Sesia-Lanzo zone, a slice of continental crust with alpine high pressure low temperature assemblages in the western Italian Alps. Rend Soc Ital Min Pet 33 (1): 281–334

Cortesogno L, Vanossi M (1982–1985) Introduzione alla Geologia delle Alpi Liguri. Lav Soc Ital Biogeogr N S IX: 1–48

Cortesogno L, Dalla Giovanna G, Vannucci R, Vanossi M (1988 a) Volcanisme, sédimentation et tectonique pendant le Permo-Carbonifère en Briançonnais ligure: une revue. Eclogae Geol Helv 81 (2): 487–510

Cortesogno L, Dalla Giovanna G, Vanossi M (1988b) Paleogéographie du Briançonnais ligure d'après les lithoclastes des formations permocarbonifères. Géol Alp Mém 14: 9–28

Coward M, Dietrich D (1989) Alpine tectonics – an overview. In: Coward MP, Dietrich D, Park RG (eds) Alpine tectonics. Geol Soc Lond Spec Publ 45: 1–29

Curnelle R, Dubois P (1986) Evolution mésozoïque des grands bassins sédimentaires français: bassins de Paris, d'Aquitaine et du Sud-Est. Bull Soc Géol Fr 8, II: 529–546

Dardeau G (1978) Un dispositif seuil-bassin dans le Lias et le Jurassique des Alpes Maritimes, prolongement vers le sud-est des structures du bassin de Gap-Digne. C R Somm Soc Geol Fr 8: 173–177

Debelmas J (1955) Les zones subbriançonnaise et briançonnaise occidentales entre Vallouise et Guillestre (Hautes Alpes). Mem Expl Carte Géol Fr, Paris, 172 pp

Debelmas J (1974) Géologie de la France. Doin, Paris, 544 pp

Debelmas J (1983) Alpes du Dauphiné, Coll Guides géologiques régionaux. Masson, Paris, 198 pp

Debrand-Passard S, Courbouleix S, Lienhardt MJ (1984) Synthèse géologique du Sud-Est de la France. Mém Bur Rech Géol Min Fr 125: 1–615

Dercourt J, Zonenshain LP, Ricou LE, Kazmin VG, Le Pichon X, Knipper AL, Grandjacquet C, Sborshchikov IM, Boulin J, Sorokhtin O, Geyssant J, Lepvrier C, Biju Duval B, Sibuet JC, Savostin LA, Westphal M, Laner JP (1986) Geological evolution of the Tethys belt from Atlantic to Pamir since Liassic. Tectonophysics 123, 1–4: 241–315

Dewey JF, Helman ML, Turco E, Hutton DHW, Knott SD (1989) Kinematics of the western Mediterranean. In: Coward MP, Dietrich D, Park RG (eds) Alpine tectonics. Geol Soc Lond Spec Publ 45: 265–283

Durand Delga M (1980) La Méditerranée occidentale: étapes de sa genèse et problèmes structuraux liés à celle-ci. Mém H S Soc Géol Fr 10: 205–224

Ellenberger F (1958) Etude géologique du pays de Vanoise. Mém Expl Carte Géol Fr, Paris, 560 pp

Fabre J, Schade J, Baudin T, Desmons J, Mercier D, Peruccio-Parison MD (1987) Relics of pre-Mesozoic events in the Briançon zone (northern French Alps). Pre-Variscan and Variscan events in the Alpine-Mediterranean mountain belts. In: Flügel HW, Sassi FP, Grecula P (eds) Mineralia Slovaca Monography. Alfa, Bratislava, pp 183–208

Faure JL, Mégard-Galli J (1988) L'évolution jurassique en Briançonnais: sédimentation continentale et fracturation distensive. Bull Soc Géol Fr Paris 8 (4): 681–692

Faure-Muret A (1955) Etudes géologiques sur le massif de l'Argentera-Mercantour et ses enveloppes sédimentaires. Mém Serv Carte Géol Fr, 336 pp

Frank W (1987) Evolution of Austro-Alpine elements in the Cretaceous. In: Flügel HW, Faupl P (eds) Geodynamics of the Eastern Alps. Deuticke, Vienna

Fry N (1989) Southwestward thrusting and tectonics of the western Alps. In: Coward MP, Dietrich D, Park RG (eds) Alpine Tectonics. Geol Soc Lond Spec Publ 45: 83–109

Fudral S, Deville E, Marthaler M (1987) Distinction de trois ensembles d'unités dans les "Schistes Lustrés" compris entre la Vanoise et le Val de Suse (Alpes franco-italiennes septentrionales): aspects lithostratigraphiques, paléogéographiques et géodynamiques. C R Acad Sci Paris 305 II: 467–472

Gidon M (1975) Sur l'allochtonie du "dôme de Remollon" (Alpes françaises du Sud) et ses conséquences. C R Acad Sci Paris 280: 2829–2832

Gidon M, Aprahamian J (1980–1981) Le rôle de la paléotectonique jurassique dans la structure des montagnes de Beaumont (zone dauphinoise du SE de Grenoble). Bull Bur Rech Geol Min Orléans 1: 23–33

Gignoux M, Moret L (1934) Les grandes subdivisions géologiques des Alpes françaises. Ann Géogr 244: 337–363

Gillcrist R, Coward M, Mugnier JL (1987) Structural inversion and its controls: examples from the Alpine foreland and the French-Alps. Geodin Acta Paris 1 (1): 5–34

Goffe B, Chopin C (1986) High-pressure metamorphism in the Western Alps: zoneography of metapélites, chronology and consequences. Schweiz Mineral Petrogr Mitt 66: 41–52

Gondolo A, Toutin-Morin N (1989) Le volcanisme de la Provence orientale et de la Corse: synthèse des données-actuelles. In: Toutin Morin N, Vinchon C (eds) Les Bassins permiens du Sud-Est. Synthèse géologique des Bassins Permiens français. Mém Bur Rech Geol Min Fr 128: 242–257

Graciansky PC de, Lemoine M (1988) Early Cretaceous extensional tectonics in the southwestern French Alps: a consequence of North-Atlantic rifting during Tethyan spreading. Bull Soc Géol Fr Paris (8) IV, No 4: 733–738

Graciansky PC de, Bourbon M, Charpal O de, Chenet PY, Lemoine M (1979) Genèse et évolution comparées de deux marges continentales passives: marge ibérique de l'océan Atlantique et marge européenne de la Téthys dans les Alpes occidentales. Bull Soc Géol Fr Paris 7, XXI, 5: 665–674

Graciansky PC de, Dardeau G, Lemoine M, Tricart P (1989) The inverted margins of the French Alps and foreland basin inversion. In: Cooper MA, Williams GD (eds) Inversion Tectonics. Geol Soc Lond Spec Publ 44: 87–104

Graham RH (1978) Quantitative deformation studies in the Permian rocks of Alpes Maritimes. Mem Bur Rech Geol Min Orléans 91: 219–238

Grand T (1988) Mesozoic extensional inherited structures on the European margin of the Ligurian Tethys. The example of the Bourg d'Oisans half-graben. Bull Soc Géol Fr 8 (4): 613–622

Greber C (1965) Flore et stratigraphie du Carbonifère des Alpes Françaises. Mém Bur Rech Geol Min Paris 21: 1–380

Guardia P, Ivaldi JP (1985) Les déformations schistogènes du tégument de l'Argentera (Alpes Maritimes): description, genèse et chronologie relative dans le cadre géodynamique des Alpes occidentales. Bull Soc Géol Fr Paris 8, I, 3: 353–362

Guardia P, Ivaldi JP (1987) Contrôle tectonique de la sédimentation paléogène sur le bord méridional du massif de l'Argentera (Alpes Maritimes). Géol Alp Grenoble Mém HS 13: 313–317

Guerin G, Pages V, Reverdy F (1990) Structures extensives jurassiques et déformation alpine (bordure ouest du Pelvoux). Rapp Inst Fr Pét et Ecole Nationale Supérieure du Pétrole et des Moteurs Rueil Malmaison. 37800, pp 1–101

Guillot F, Ploquin A, Raoult JF, Peruccio-Parison MD (1986) Les séries antepermiennes de Vanoise septentrionale (zone briançonnaise, Alpes de Savoie): lithologie et géochimie dans le Massif de Bellecôte: arguments pour un âge antehouiller. C R Acad Sci Paris 303, II, 12: 1141–1145

Haccard D, Beaudoin B, Gigot P, Jorda M (1989) Notice explicative Carte géol France (1/50.000) feuille LA JAVIE. Bur Rech Geol Min Orléans 918: 1–152 Carte géologique par Haccard D, Beaudoin B, Gigot P, Jorda M et al. (1989)

Haudour J, Sarrot-Reynaud J, Vernet J (1958) Observations nouvelles sur le houiller et la molièrésite dans le Massif du Mercantour et comparaison avec le Houiller du dôme de la Mure (Isère). C R Acad Sci Paris 247: 716–719

Hayward AB, Graham RH (1989) Some geometrical characteristics of inversion. In: Cooper MA, Williams GD (eds) Inversion tectonics. Geol Soc Lond Spec Publ 44: 17–39

Homewood P, Allen PA, Williams GD (1986) Dynamics of the molasse basin of Western Switzerland. Spec Publ Assoc Sedimentol 8: 199–217

Horrenberger JC, Michard A (1978) Analyse des mesostructures dans le Trias et les Schistes Lustrés du Val Maira: la tectonique tardive de compression N–S. Sci Géol Strasbourg 31 (1): 3–13

Horrenberger JC, Michard A, Werner P (1978) Le couloir de décrochement de Bersezio en Haute Stura. Alpes externes, Italie. Structure de compression subméridienne. Sci Géol Strasbourg 31 (1): 15–20

Jaillard E (1988) Une image paléogéographique de la Vanoise briançonnaise (Alpes françaises). Eclogae Geol Helv 81 (3): 553–566

Latouche L, Bogdanoff S (1987) Evolution précoce du Massif de l'Argentera: apport des éclogites et des granulites. Géol Alp 63: 151–165

Laubscher HP (1971) The large scale kinematics of the Western Alps and the Northern Apennines and its palinspatic applications. Am J Sci 271: 193–226

Laubscher HP, Bernoulli D (1977) In: Nairn AEM, Kanes WH, Stehli FG (eds) The ocean Basins and Margins, 4A. The Eastern Mediterranean. Plenum Press, New York, pp 1–22

Laubscher HP, Bernoulli D (1982) History and deformation in the Alps. In: Hsu KJ (ed) Mountain Building Processes. Academic Press, London, pp 169–180

Le Guen M, Maurin JC (1987) Tectonique et karstification calloviennes: contrôle structural des minéralisations Pb-Zn du Bathonien des Malines (Gard, France). C R Acad Sci Paris 305, II: 413–417

Lemoine M, Tricart P (1986) Les Schistes Lustrés piémontais des Alpes occidentales. Approche stratigraphique, structurale et sédimentologique. Eclogae Geol Helv 79 (2): 271–294

Lemoine M, Trümpy R (1987) Pre-oceanic rifting in the Alps. Tectonophysics 133: 305–320

Lemoine M, Bas T, Arnaud-Vanneau A, Arnaud H, Dumont T, Gidon M, Bourbon M, Graciansky PC de, Rudkiewicz JL, Mégard-Galli J, Tricart P (1988) The continental margin of the Mesozoic Tethys in the Western Alps. Mar Pet Geol 3: 179–199

Lemoine M, Dardeau G, Delpech PY, Dumont T, Graciansky PC de, Graham RH, Jolivet L, Roberts D, Tricart P (1989) Extension syn-rift et failles transformations jurassiques dans les Alpes occidentales. C R Acad Sci Paris 309, II: 1711–1716

Livermore RA, Smith AG (1985) Some boundary conditions for the evolution of the Mediterranean region. In: Stanley DJ, Wezel FC (eds) Geological Evolution of the Mediterranean Basin. Springer, Berlin Heidelberg New York, pp 84–98

Malaroda R (1966) Mylonites et paramylonites dans le massif de l'Argentera (Alpes Maritimes). Rend Acc Naz Lincei 8 (41): 151–162

Malaroda R (1970) Carta geologica del Massicio dell'Argentera alla scala 1/50.000. Mem Soc Geol Ital 9: 557–663

Mattauer M, Tapponnier P (1978) Tectonique des plaques et tectonique intracontinentale dans les Alpes franco-italiennes. CR Acad Sci Paris 287 D: 899–902

Mégard-Galli J, Faure JL (1988) Tectonique distensive et sédimentation au Ladinien supérieur-Carnien dans la zone briançonnaise. Bull Soc Géol Fr Paris 8, 5: 705–716

Ménard G, Thouvenot F (1984) Coupes équilibrées crustales: méthodologie et applications aux Alpes occidentales. Geodin Acta 1: 35–45

Menot RP (1988) Magmatismes paléozoïques et structuration carbonifère du massif de Belledonne (Alpes françaises). Mémoires et documents du Centre Armoricain d'Etude structurale des socles. Rennes 21: 1–264

Mercier D, Beaudoin B (1987) Révision du Carbonifère Briançonnais: stratigraphie et évolution du bassin. Géol Alp Mém HS 13: 25–31

Merle O, Brun JP (1984) The curved translation path of the Parpaillon Nappe (French Alps). J Struct Geol 6: 711–719

Messiga B (1984–1986) Il metamorphismo alpino nelle Alpi Liguri: Alcuni aspetti petrologici. Mem Soc Geol Ital 28: 151–179

Michard A (1967) Etudes géologiques dans les zones internes des Alpes Cottiennes. CNRS, Paris

Mugnier JL, Guellec S, Menard G, Roure F, Tardy M, Vialon P (1990) Crustal balanced cross-sections through the external Alps deduced from the Ecors profile. In: Roure F, Heitzmann P, Polino R (eds) Deep structures of the Alps. Mém Soc Géol Fr Paris 156. Mém Soc Géol Suisse Zürich 1, vol Spec Soc Geol Ital Roma, pp 203–216

Nicolas A, Polino R, Hirn A, Nicolich R (1990) ECORS-CROP traverse and deep structure of the Western Alps: a synthesis. In: Roure F, Heitzman P, Polino R (eds) Deep structure of the Alps. Mém Soc Géol Fr Paris 156. Mém Soc Géol Suisse Zurich, vol 1 Spec Geol Ital Roma, pp 15–27

Peruccio-Parison MD (1984) Problèmes pétrochimiques et structuraux en Vanoise septentrionale. Thèse 3ème cycle, Univ Paris XI Orsay, 358 pp

Platt JP, Lister GS, Cunningham P, Weston P, Peel F, Baudin T, Dondey H (1989) Thrusting and back thrusting in the Briançonnais domain of the Western Alps. In: Coward MP, Dietrich D, Park RG (eds) Alpine Tectonics. Geol Soc Lond Spec Publ 45: 135–152

Ramsay J, Hubert MI (1983) The techniques of modern structural geology. Academic Press, London, 307 pp

Ravenne C, Vially R, Riche P, Trémolières P (1987) Sédimentation et tectonique dans ce bassin marin Eocène supérieur-Oligocène des Alpes du Sud. Rev Inst Fr Pét 42 (5): 529–553

Ricou LE (1981) Glissement senestre des nappes penniques le long de la bordure Nord de l'Argentera: son rôle dans le jeu de l'arc alpin. C R Acad Sci Paris 292: 1305–1308

Ricou LE, Siddans AWB (1986) Collision tectonics in the Western Alps. In: Coward MP, Ries AC (eds) Collision tectonics. Geol Soc Lond Spec Publ 19: 229–244

Sacco F (1911) Il gruppo delle'Argentera. Mem Acc Sci Torino 2, 61: 457–515

Schmid SM, Aebli HR, Heller F, Zingg A (1989) The role of the Periadriatic Line in the tectonic evolution of the Alps. In: Coward MP, Dietrich D, Park RG (eds) Alpine Tectonics. Geol Soc Lond Spec Publ 45: 153–171

Siddans AWB (1980) Compaction, metamorphisme et structurologie des argilites permiennes dans les Alpes Maritimes (France). Rev Géol Dyn Géogr Phys 22 (4–5): 279–292

Stanley D (1961) Etudes sédimentologiques des Grès d'Annot et de leurs équivalents latéraux. Thèse, Grenoble IFP réf 6821

Sturani C (1963) La couverture sédimentaire de l'Argentera-Mercantour dans le secteur compris entre les Barricate et Vinato (Haute vallée de la Stura di Demonte, Italie). Trav Lab Géol Fac Sci Grenoble 39: 83–124

Suess E (1885–1888) Das Antliz der Erde. Temsky, Vienna

Tardy M, Deville E, Fudral S, Gellec S, Ménard G, Thouvenot F, Vialon P (1990) Interprétation structurale des données du profil de sismique reflexion profonde ECORS-CROP Alpes entre le front Pennique et la ligne du Canavese (Alpes occidentales). In: Roure F, Heitzmann P, Polino R (eds) Deep structures of the Alps. Mém Soc Géol Fr Paris 156, Mém Soc Géol Suisse Zurich, Vol 1 Spec Soc Geol Ital Roma, pp 217–226

Toutin-Morin N (1987) Les bassins permiens provençaux, témoins de l'orogenèse hercynienne et de l'ouverture de la Tethys dans le Sud-Est de la France. Ann Soc Géol Nord 106: 183–187

Toutin-Morin N, Delfaud J (1987) Les facteurs régissant le détritisme Permien dans le bassin du Bas Argeno (Provence orientale). Géol Alp Grenoble Mém HS 13: 47–56

Toutin-Morin N, Vinchon C (1989) Les Bassins Permiens du Sud–Est. Synthèse géologique des Bassins Permiens français. Mém Bur Rech Géol Min Fr 128: 114–121

Tricart P (1980) Tectonique superposées dans les Alpes occidentales au Sud du Pelvoux. Evolution structurale d'une chaîne de collision. Thèse, Univ Louis Pasteur, Strasbourg, 407 pp

Tricart P (1984) From passive margin to continental collision: a tectonic scenario for the Western Alps. Am J Sci New Haven 284: 97–120

Tricart P, Lemoine M (1986) From faulted blocks to megamullions and megaboudins: Tethyan heritage in the structure of the Western Alps. Tectonics Washington DC 5 (1): 95–110

Trümpy R (1975) Penninic-Austroalpine boundary in the Swiss Alps: a presumed former continental margin and its problems. Am J Sci 275 A: 209–238

Trümpy R (1976) Du Pélerin aux Pyrénées. Eclogae Geol Helv 69 (2): 249–264

Vann IR, Graham RH, Hayward AB (1986) The structure of moutain fronts. Struct Geol 8 (3/4): 215–227

Vanossi M (1984–1986) Geologia delle Alpi Liguri. Mem Soc Geol Ital XXVIII: 1–598

Vernet J (1964) La zone Pelvoux-Argentera. Etudes sur la tectonique alpine du socle dans la zone des massifs cristallins externes du Sud des Alpes occidentales. Bull Serv Carte Géol Fr 60 (275): 1–294

Vialon P (1966) Etude géologique du massif cristallin Dora Maira. Alpes Cottiennes internes (Italie). Trav Lab Geol Fac Sci Grenoble 4: 1–282

Vialon P, Rochette P, Ménard G (1989) Indentation and rotation in the Western Alpine arc. In: Coward MP, Dietrich D, Park RG (eds) Alpine tectonics. Geol Soc Lond Spec Publ 45: 329–338

Ziegler PA (1988) Evolution of the Arctic-North Atlantic and the Western Tethys. Am Assoc Pet Geol Tulsa 43: 1–198

Palinspastic Reconstruction of the Pre-Triassic Basement Units in the Alps: The Central Alps

O. Adrian Pfiffner

Abstract

Inversion of the displacement associated with Mesozoic rifting and Mesozoic-Cenozoic collision was performed to obtain a palaeogeographic map showing the pre-Triassic configuration of basement blocks containing rocks of the Variscan cycle. Late Variscan granites are volumetrically less important in the Alps compared to the Black Forest-Vosges areas. They are palaeogeographically arranged in three EW-trending belts which are some 50–100 km wide and extend over distances of 200–300 km. Parallel to these granite belts Permo-carboniferous volcaniclastic sequences were originally deposited in narrow furrows which can in part be identified as post-Variscan graben structures.

1 Introduction

The Central Alps correspond to the EW-striking segment of the Alpine chain of Switzerland and the adjoining areas. Here, the stack of Alpine nappes is exposed because of pronounced axial plunges. This nappe pile encompasses the deformed margins of the European and Adriatic plates and fragments of the Tethys ocean between. The nappe pile is generally subdivided into four nappe systems based on the facies of the Mesozoic sedimentary cover rocks: the Helvetic, Penninic and Austro-Alpine/South-Alpine nappes (see Fig. 1).

Figure 2 is a cross-section through the Central Alps along the transect of the European Geotraverse (EGT), which crosses eastern Switzerland. In addition to the general nappe structure of the Alps, it contains information on the deep structure derived from EGT seismic refraction experiments and from near-vertical reflection seismology collected within the Swiss National Research Program, NFP 20. As is evident from Fig. 2 the overall structure of the Alps is that of a bivergent thrust belt. Detached flakes of upper crustal crystalline basement rocks are thrust on top of each other and now form an imbricate stack of thrust sheets. The Mesozoic-Cenozoic sedimentary cover was largely detached from its crystalline basement to form nappes of its own.

The northernmost, structurally lowest **Helvetic zone** represents parts of the former European margin. The Helvetic nappes consist of Mesozoic and Cenozoic sediments which were detached from their crystalline basement. They overly the Infrahelvetic complex, in which crystalline basement (including rocks of the Variscan cycle) was also involved in the Alpine deformation. This basement is now exposed in the external basement uplifts of the Aar, Tavetsch, Gotthard, Mont Blanc and Aiguilles Rouges massifs.

The **Penninic zone** represents the thinned, southernmost margin of the European plate and fragments of the Tethyan ocean. The Penninic nappes overlie the Helvetic nappes and are truncated to the south by the Insubric Line (see Fig. 1). Together with the Helvetic nappes, they now form a north-vergent pile of intricately folded thrust sheets. The deepest exhumed portion of the Penninic zone experienced amphibolite grade Alpine overprint. This overprint renders the interpretation of the various lithologic units within the crystalline basement difficult, although geochronologic dating of orthogneisses suggests the presence of Variscan, Caledonian and older granitoids (see later).

Within the Central Alps, the **Austro-Alpine nappes** are the uppermost units overlying the Penninic zone. Due to extensive erosion by exhumation they are preserved as klippen only over much of the area considered here (Fig. 1). The Austro-Alpine nappes were part of the northern and NW Adriatic margin thinned in the Mesozoic.

The **Southern Alps** were also part of the thinned NW margin of the Adriatic plate. Their Alpine deformation resulted in a south-vergent stack of thrust sheets (Fig. 2). They thus make the Central Alps a bivergent collision orogen.

The aim of this work is to present a palinspastic map showing the distribution of the various rock suites of the Variscan cycle dispersed throughout the

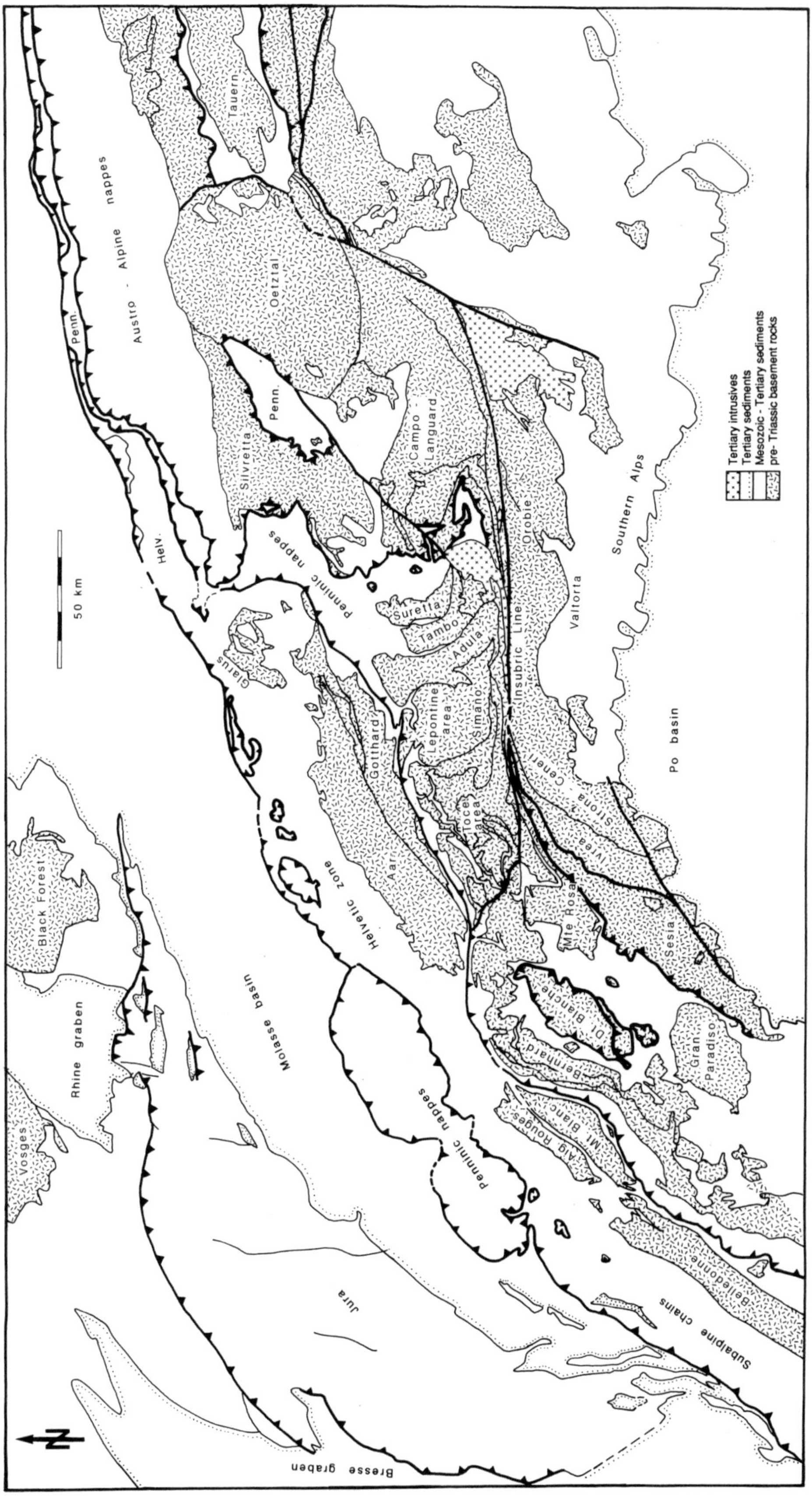

Fig. 1. Simplified tectonic map of the Central Alps and ajoining areas

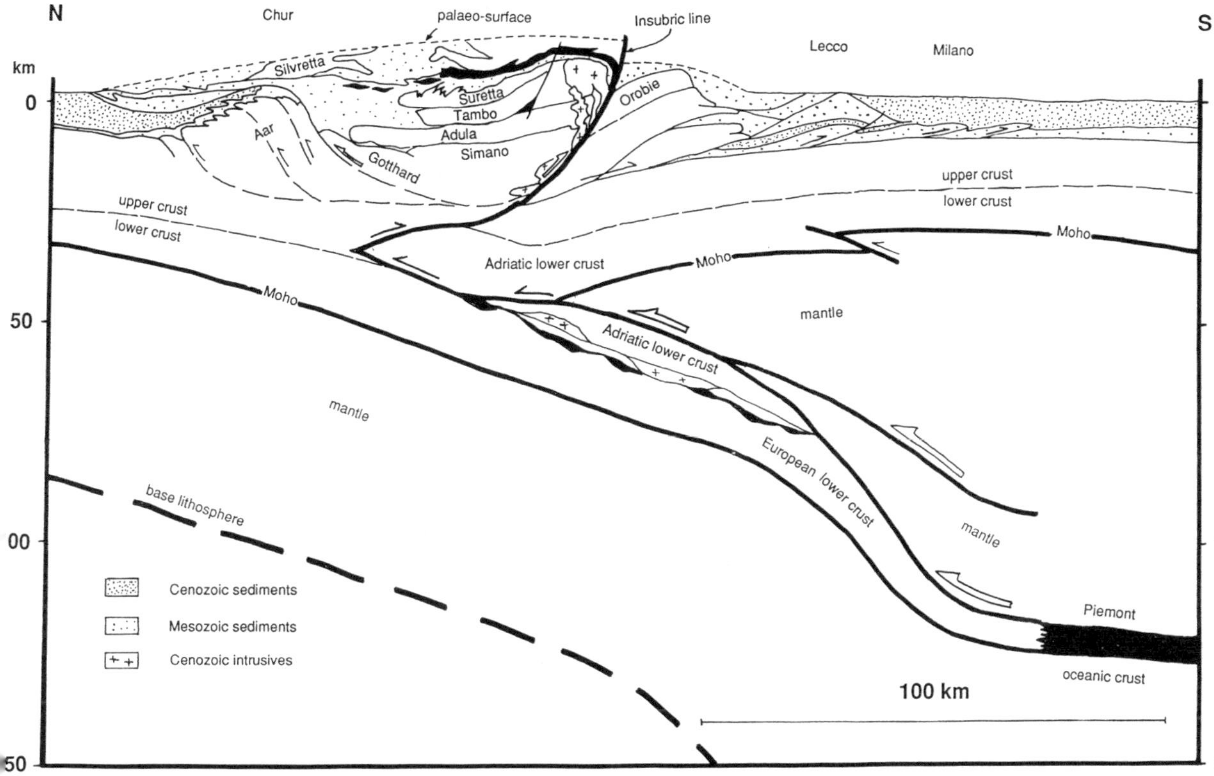

Fig. 2. Geologic profile through the Central Alps along the transect of the NFP 20 seismic lines. The profile shows the crustal structure which is characterized by subduction of lower crust and stacking of upper crust detached from its lower crust

Alpine nappes. To achieve this, the nature and occurrence of these rock suites within the Alpine nappes are first summarized. In the second section, the discussion focuses on how the palinspastic reconstruction was actually performed.

2 Variscan Units in the Central Alps

Variscan units occur within the basement of the Northalpine foreland, the basement nappes of the Helvetic, Penninic, and Austro-Alpine zones, as well as in the Southern Alps. They will be briefly discussed in that order.

2.1 Northalpine Foreland and Helvetic Zone

Pre-Triassic basement in the *Northalpine foreland* outcrops in two basement uplifts **(Vosges** and **Black Forest)** bordering the Rhine graben (Fig. 1). This graben formed in the Tertiary and is the result of a phase of E-W oriented extension affecting northern and central Europe (Illies and Greiner 1978; Ziegler 1987). The two basement uplifts of the Vosges and the Black Forest are also related to the Northalpine foreland basin which formed during the final collisional stage of the Alpine orogeny (e. g. Mugnier and Ménard 1986; Pfiffner 1986): the basement uplifts may represent bulges forming in response to flexural loading of the European lithosphere by collision with the Adriatic plate.

Upper Devonian to Lower Carboniferous sediments include the condensed Moldanubian sedimentary sequence of the orogenic belt of central Europe (Eisbacher et al. 1989). They were intruded by Variscan granites (330–315 Ma) and overlain by Permo-Carboniferous volcaniclastic sequences (Eisbacher et al. 1989). South of the Black Forest and Vosges an EW-trending graben filled with a Permo-Carboniferous fluvio-lacustrine sequence topped by red beds (Matter 1987) is known to exist in the subsurface beneath the Mesozoic-Cenozoic cover sediments of the Northalpine foreland. The graben was affected by, probably, synsedimentary transpression in Permian times (Diebold 1988).

Variscan rocks in the *Helvetic zone* of the Central Alps outcrop in the external basement uplifts of the

Infrahelvetic complex **(Mt. Blanc, Aiguilles Rouges, Aar** and **Gotthard massifs),** as well as the overlying Helvetic nappes (Fig. 1). The basement within the Infrahelvetic complex is overlain by a Mesozoic-Cenozoic cover sequence, which has been severely deformed during the Alpine orogeny. The present-day structure is characterized by a combination of folding and thrusting affecting both cover and crystalline basement (see Escher et al. 1987; Pfiffner et al. 1990 a, b). Alpine shortening includes a vertical component of uplift exceeding 10 km in places.

Rocks of the Variscan cycle include Late Variscan granites (313–255 Ma) and Permo-Carboniferous volcaniclastic sequences. The larger plutons form the backbone of the different basement blocks. The intrusive contacts were partly activated as faults in the Alpine cycle (Pfiffner et al. 1990 a). Generally speaking, the granites represent calc-alkaline suites (Oberhänsli et al. 1988).

Within the volcaniclastic sequence, an older sequence of sediments reaches in age up to dated Westphalian and Stephanian. This sequence is discordantly overlain by Triassic sediments and now forms relatively tight synclines within the crystalline basement. Examples are at St. Gervais (France) and Dorénaz (Switzerland) in the Aiguilles Rouges Massif and the Tödi area in the Aar Massif (for a review of the outcrops of Carboniferous sediments in the Central Alps, see Jongmans and Ritter 1960). The younger granite plutons (e.g. Central Aar granite, whose emplacement is dated at 298–296 Ma after Schaltegger and von Quadt 1990) intrude the volcaniclastic sequences (Oberhänsli et al. 1988). The older volcanic sequences are of calc-alkaline composition and are overlain discordantly by Mesozoic rocks, while younger mildly alkali-rhyolitic volcanics are more or less concordantly overlain by Mesozoic strata.

The *Helvetic nappes* consist essentially of detached Mesozoic cover sediments which were thrust northward onto the basement uplifts described above. Late Palaeozoic rocks are involved in nappe tectonics in two instances. A volcaniclastic, possibly Permian sequence, the "Verrucano", forms the base of the detached sediments of the Helvetic nappes **(Glarus nappe)** in eastern Switzerland (Fig. 1). It consists of conglomerates, red beds, layers of freshwater limestone, fanglomerates, as well as intercalations of rhyolitic and spilitic volcanic rocks. The "Verrucano" is generally conformably overlain by Mesozoic strata but the contact is locally an unconformity. A similar sequence is found along the northern contact of the Gotthard Massif in the Urseren-Garvera zone. Here, detachment of the Helvetic nappes followed a stratigraphically higher horizon, leaving the "Verrucano" attached to the crystalline substratum. The structure of the Glarus nappe s. l. in eastern Switzerland suggests that the detached "Verrucano" represents an inverted N-S oriented graben, while the Urseren-Garvera zone indicates a more WSW-ENE oriented basin.

2.2 *Penninic Nappes*

The Penninic nappes are composed of three different types:

1. Imbricate nappes made up principally of crystalline basement, stacked up to a manifold pile, with thin bands of remnants of Mesozoic carbonates and metapelites separating individual thrust sheets; they are in part intensely folded.
2. Nappes consisting of Mesozoic-Cenozoic sediments stripped off their basement (which now forms type 1 nappes).
3. Ophiolitic nappes containing fragments of Mesozoic oceanic crust and oceanic sediments.

The tpye 1 nappes, which contain the Variscan rocks, form a stack of thrust sheets in and east of the Lepontine area (see Pfiffner et al. 1990 a for a review) and in the Toce culmination (see Steck 1987; Klaper 1990), whereas west of the Simplon fault the overall structure of these nappes is dominated by the intensely and isoclinally folded Bernhard nappe (the equivalent of the external Briançonnais of the Western Alps; Escher 1988) and the Monte Rosa nappe.

Late Variscan granitoids are scattered throughout these nappes, but were severely overprinted by metamorphism in the lower units. U-Pb ages from granitic gneisses and migmatites reveal that sediments older than the Caledonian cycle (450 Ma) and from the time span between the Caledonian and Variscan orogeny are present in these high grade rocks (Köppel et al. 1981). S-type granites of Variscan age seem to dominate in the Lepontine area, while possibly I-type granites are found in the area of the Toce culmination. Late Variscan granitoids are clearly recognizable in the higher units. They include the Rofna porphyry (350 Ma) in the **Suretta** nappe, the Truzzo granite gneiss (303 Ma) in the **Tambo,** the Zerfreila granite gneiss (347–320 Ma) in the **Adula,** the Randa augen gneiss (Permian) in the **Siviez-Mischabel** digitation of the **Bernhard nappe** and the Monte Rosa granite (310 Ma) with its gneissic variety (260 Ma) in the **Monte Rosa nappe** (Hanson et al. 1969; Frey et al. 1976; Thélin 1987).

Permo-Carboniferous volcaniclastic sequences are also widespread, but their identification is rendered more difficult by Alpine overprint. The most prominent example is found in the **Bernhard nappe** (see Escher 1988 for a review): in the **Mont Fort** digitation of this nappe volcanic rocks (metagabbros, pillow lavas) are abundant in the lower part, while sedimentary rocks (meta-arkose, coal-bearing beds, conglomerate, schist) predominate in the upper part. The top of the sequence is characterized by Permo-Triassic detrital rocks resembling the "Verrucano". In the case of the **Siviez-Mischabel** digitation, volcaniclastics outcrop as a discontinuous band all along the nappe's outer edge. All these Late Palaeozoic formations are characterized by enormous thickness variations, which could indicate that they were deposited in narrow, steeply bound basins. For the Permo-Carboniferous of the **"Zone Houillère"**, a 250 km long NE-SW oriented graben structure must be assumed, which influenced the shape of the Alpine nappe structure significantly. Much smaller outcrops of rocks of partly doubtful Permo-Carboniferous age are scattered throughout the other Penninic nappes (see Jongmans and Ritter 1960). With the exception of the conglomerates within the **Lebendun nappe** and the "Verrucano" of the **Tasna nappe,** the thickness of these units never exceeds a few tens of metres.

2.3 *Austro-Alpine Nappes*

The Austro-Alpine nappes comprise huge thrust sheets involving large masses of crystalline basement rocks (Fig. 1). Thrusting in the Cretaceous was west-directed (Schmid and Haas 1989) and the whole nappe pile was transported piggyback to the north during Tertiary NS-oriented compression in the underlying Penninic nappes.

Variscan granitoids are particularly abundant in the Lower Austro-Alpine nappes of Graubünden. They include the Julier and the Bernina granites of the **Err-Bernina nappe.** The age of these granitoids is Carboniferous (305 Ma; Rageth 1984) and their chemistry is varied (granites, granodiorites, syenites and alkaline granites). A suite of Upper Palaeozoic volcanic rocks of the calc-alkaline series composed mainly of rhyolitic pyroclastics possibly predates the Julier granite (Mercolli 1989). The Arolla series of the Lower Austro-Alpine **Dent Blanche nappe** contains gabbros (250 Ma; Dal Piaz et al. 1977), granites of the calc-alkaline series (Arolla gneiss), as well as volcanic and clastic rocks of possible Late Palaeozoic age (Thélin and Ayrton 1983). In the **Sesia nappe** pre-Alpine, possibly Late Variscan granitoids are abundant, particularly in the "Gneiss minuti" complex (Compagnoni et al. 1977).

Late Variscan granitoids are seemingly less abundent in the Upper Austro-Alpine nappes and include orthogneiss bodies in the **Languard-Campo-Sesvenna nappe** (267–277 Ma; Del Moro and Notarpietro 1987). Volcaniclastic sequences of Permian age occur in the Upper Austro-Alpine nappes (Dössegger 1974; Eichenberger 1986) in the NW portion of the **Silvretta nappe** and the **Languard-Campo-Sesvenna nappe.** The Permian Ruina-Formation contains ignimbrites, tuffs and fanglomerates and was deposited in a graben structure. It is overlain by fluviatile sand- and silstones, as well as conglomerates and fanglomerates (with volcanic components) of the Permo-Triassic Chazforà-Formation. These rocks grade into bedded sandstones (channel fills) that contain evidence for marine influence (dolomitic cement, ripple marks).

2.4 *Southern Alps*

The Southern Alps, south of Switzerland are a south-vergent fold-and-thrust belt (Laubscher 1985; Roeder 1989), which also affected crystalline basement rocks (Fig. 2). WSW-directed thrusting has also been documented in the eastern Southern Alps (Doglioni 1987). Variscan rocks are abundant.

In the **Strona-Ceneri** zone of the western Southern Alps, Variscan granitoids include the Early Permian Baveno (276 Ma; Hunziker and Zingg 1980) and Montorfano (283 Ma; Boriani et al. 1987) granite suites which are considered post-metamorphic with regard to the Variscan orogeny (Zingg et al. 1990). Bordering these intrusive rocks to the east, the volcanic complex of the **Lugano** porphyry is also Permian: it overlies Upper Carboniferous sediments and predates Permo-Triassic clastics of the Servino formation (Stille and Buletti 1987). The volcanic rocks are of the calc-alkaline series and include basalts, andesites, dacites, rhyolites and subvolcanic granophyres. Another and larger volcanic complex, the **Bozen** (Bolzano) quartz-porphyry, is situated in the eastern Southern Alps. It is Early Permian and coeval with several granites associated with it (e.g. Brixen granite in the north, Cima d'Asta in the south).

The Collio formation of the western Southern Alps is also Early Permian and represents a volcaniclastic sequence. Late Permian fluviatile-continental sediments of the Lombardian Verrucano rest unconformably above it. In the eastern Southern Alps, the Late Permian is represented by the littoral-fluviatile

Gröden sandstone. The latest Permian is characterized by the evaporitic Bellerophon formation, which transgresses from east to west and is overlain by Triassic sediments of the Alpine cycle.

3 Palinspastic Map of the Central Alps

The relative movements between individual basement blocks are discussed in this section. The following points must be carefully considered:

1. The basement blocks were first broken up and displaced during Tethyan rifting in mainly Jurassic time. The general character of these movements is sinistral strike-slip between the Adriatic-African and European continents. According to plate reconstructions by Dewey et al. (1989), Africa moved ESE during that period with a rotation pole at 43.9° N/–9.08° E. Lemoine et al. (1989), based on Savostin et al. (1986) described more NE-SW oriented transform movements for Middle to Late Jurassic time with the rotation pole at 51.8° N/2.5° E. For the segment of the future Central Alps these movements generated a transtensional, transform fault dominated basin (Weissert and Bernoulli 1985). The amount of displacement between two points on the European and Adriatic plates associated with this strike-slip component is difficult to assess. It includes a component of stretching of continental crust as well as a component of rifting and generation of new oceanic crust. Estimates of total movements are bound to be crude and seem to be of the order of 100–500 km (Weissert and Bernoulli 1985; Gealey 1988; Schmid et al. 1990).
2. During the Alpine orogeny, the Mesozoic cover, which serves as basis for the reconstruction of the configuration of the various Tethys facies realms, was in many instances sheared off its crystalline basement. In some cases, the correlation between particular basement blocks and Mesozoic cover units is doubtful. Reconstruction of Variscan units for pre-Triassic time is therefore even more difficult than a reconstruction of the Mesozoic Tethys.
3. The overall plate motion between Africa and Europe since the middle Cretaceous deduced from Atlantic fracture zones and magnetic anomalies (Dewey et al. 1989) seems to be dominated by NE-SW convergence in the Upper Cretaceous, N-S to NNE-SSW convergence in Eocene to Middle Miocene time and NW-SE convergence since Miocene time. These plate motions, however, cannot be directly related to transport directions and relative movements between individual Alpine nappes. Those movement directions show a complex pattern, which includes the following elements: (1) west-directed thrusting in the Austro-Alpine nappes in the Cretaceous (Ring et al. 1988; Schmid and Haas 1989); (2) Cenozoic north to northwestward directed thrusting in the Helvetic and Penninic nappes (Pfiffner 1981; Ring et al. 1988; Pfiffner et al. 1990a); (3) Cenozoic southward directed backflow in the Tertiary (Merle et al. 1989); (4) Cenozoic strike-slip (Schmid et al. 1989); and (5) east-west extension by normal faulting (Mancktelow 1985; Selverstone 1988). In some instances it can be shown that the conspicuous nappe-internal strains are oblique to the direction of movement along the basal thrust faults (e.g. Pfiffner 1985 in the Helvetic nappes; Schmid et al. 1990 in the Penninic nappes).

Considering these difficulties, the result of the palaeogeographic reconstruction can only be a first approximation. In a general way, the reconstruction puts emphasis on those motions with large displacements, i.e. the movements between individual basement blocks. The nappe's internal, Alpine deformation was only retrodeformed in those special cases where it influences the general picture significantly. Moreover, by keeping the present-day shapes of the basement blocks, individual nappes are easy to recognize by the reader, rendering the reconstruction (and its shortcomings) more transparent.

The reconstruction shown in Fig. 3 on page 35 was done using all available structural data and will be discussed proceeding from north to south.

3.1 Northalpine Foreland and Helvetic Zone

The basement uplift of the **Vosges** and the **Black Forest** are separated by the Rhine graben. Neogene extension can be estimated to be of the order of 20 km, based on the reduced crustal thickness (25 km vs 30 km; cf. Werner et al. 1982). The Black Forest has been used as the fix point for the reconstruction.

NS-oriented shortening between the Black Forest and the **Aar Massif** can be estimated at 10 km in eastern Switzerland based on seismic data (Pfiffner et al. 1990b) and is expected to be higher (15 km) in central Switzerland, as indicated by the increased uplift of this basement complex (Pfiffner et al. 1990a). Transport directions along the major basal thrust fault of the Helvetic nappes is from south to north in eastern

Fig. 3. Palaeogeographic map showing the pre-Triassic basement configuration of the Central Alps

Switzerland (Pfiffner 1981) and in Central Switzerland as well according to field observations. The amount of shortening between **Gotthard** and Aar Massif can be estimated at about 20 km (Pfiffner 1986). No movements of any significance were assumed between Aar and **Aiguilles Rouges Massifs,** but the latter is taken to have been displaced right-laterally relative to the **Mont Blanc Massif** (Burkhard 1988) by 5–10 km along the Simplon-Rhone-Chamonix line.

3.2 *Penninic Nappes*

The basement nappes of **Simano** (including Leventina), **Lucomagno** and the Gotthard massif in the Lepontine area are separated by structures which can be interpreted as folds or faults. For the purpose of this reconstruction the Simano basement block was cut into two pieces and separated in a NS direction to compensate for some 30 km of shortening. This crude estimate is likely to be a minimum estimate.

The **Adula, Tambo** and **Suretta** basement blocks were rearranged in a NNW-SSE direction based on structural data by Löw (1987) and Mayerat (1991) which indicate NNW directed movements in a first phase of nappe emplacement. It has to be noted here that, in the Adula nappe, this first phase is accompanied by high-pressure conditions of hitherto poorly constrained age (Hunziker et al. 1989). The space between Adula and Tambo is meant to provide some space for the **Gruf** basement block, which outcrops next to the Tertiary Bergell intrusion and whose structural position and history remains somewhat enigmatic.

In the Toce area the lowermost unit (equivalent to the Simano discussed already) is the **Verampio** nappe which outcrops in a small inlier in Val Antigorio (Italy). Its link to the Gotthard Massif is speculative but has been evoked by most recent workers (Steck 1984; Leu 1986; Escher et al. 1987; Klaper 1990). For the purpose of this reconstruction, 10 km NS-shortening was assumed between these two units, which represents a minimum estimate.

The estimate of NS shortening between the Verampio and **Antigorio** basement blocks stems from similar arguments and using a NS-oriented trajectory (Merle et al. 1989). But within the Antigorio nappe, 20 km of horizontal, WSW-directed right-lateral shear (Merle et al. 1989) which might be related to Tertiary movements along the Simplon Line (Mancktelow 1985), have to be accounted for. This is done in Fig. 3 by simply cutting the Antigorio nappe into two pieces and shifting the pieces appropriately. The **Lebendun** nappe, representing a complexly shaped, thin sliver of conglomerates and gneisses must be positioned somewhere south of the Antigorio nappe (see also Steck 1984). A minimum estimate of 30 km shortening was used. Similarly (see also Steck 1984), the **Monte Leone** nappe was positioned 25 km south of the Lebendun. The shortening between the **Siviez-Mischabel** digitation of the **Bernhard** nappe and the Aar/Gotthard Massifs is more difficult to assess. A space of 80 km (subsequently stretched in the Jurassic and providing the space for the deposition of the shelf sediments of the Helvetic nappes and the Valais trough) corresponds to a reasonable estimate lying maybe on the short side. But the reconstruction must also take into account some 20 km of EW movements along the Simplon Line (Mancktelow 1985). The amount of shortening between the Siviez-Mischabel and the **Mont Fort** digitations of the Bernhard nappe is estimated at 25 km.

The Bernhard and **Monte Rosa** nappes are separated by a steep structure, the two nappes being more side by side in their present configuration (Escher et al. 1987; Escher 1988). Alpine age relative movements between the two basement blocks are ill-constrained and the spacing between the Bernhard and Monte Rosa basement blocks shown in Fig. 3 is likely to be a minimum. The **Moncucco** basement block is situated in the core of the south-vergent antiform defined by the Monte Rosa nappe. Its equivalent beneath the Simplon line is the **Berisal** nappe.

The basement complex in the core of the **Tauern** window underlying the Austro-Alpine nappes of the Eastern Alps is topped by Mesozoic cover rocks which were attributed either to the Penninic or the Helvetic Zone (see Lammerer 1988). In Fig. 3 the Tauern window is positioned such that its present day E-W distance from the Penninic basement nappes of Switzerland is kept constant and that it is aligned with the Simano nappe along strike. In this way the Tauern basement lies within the transition zone between the Helvetic and Penninic facies realms and its N-S distance from the Silvretta basement turns out to be equal to the one proposed by Ratschbacher and Frisch (this Vol.).

3.3 *Austro-Alpine Nappes*

The former position of the Austro-Alpine (and South-Alpine) nappes are more difficult to assess, because they were displaced over greater distances by the relative motion of the African and European plates in the Mesozoic. As pointed out earlier, these movements included sinistral strike-slip of the order of some 100 km related to rifting and were followed by dextral movements due to collision in Cretaceous and Cenozoic times. Collision-related movements exceeded rift-related movements and resulted in thrusting the leading edge of the African plate onto the European plate. The excess of the collision-related movements can be estimated from the overlap of the African (Apulian) plate onto the European plate. Most of the Austro-Alpine nappes thrust onto the European margin were subsequently eroded. The most northwestern remnants are found in the Iberg klippen of Central Switzerland (east of Lake Lucerne) and the **Dent Blanche** nappe (see Fig. 1). They were transported in a northerly and westerly direction in the Cretaceous and Cenozoic (Steck 1987; Ring et al. 1988). Thus the Iberg klippen have to be brought back to the SE behind the Penninic, i.e. European Suretta and Tasna basement blocks (e.g. on top of the **Margna** nappe) and the Dent Blanche basement together with the **Sesia zone** to a position SE of the Monte Rosa nappe. Minimum reasonable estimates are 150 km for the Iberg klippen and 100 km for the Dent Blanche-Sesia block. A similar displacement must be assumed for the **Silvretta** basement block.

The Lower and Upper Austro-Alpine nappes in SE Switzerland experienced northwesterly and then westerly directed transport (Schmid and Haas 1989). This is particularly the case for the displacement of the **Oetztal** basement block relative to the **Campo** and **Scarl** nappes (Fig. 3) along the Schlinig fault (Schmid and Haas 1989). The same transport direction was assumed for the **Err-Bernina** and the **Margna** basement blocks, taking 40 and 30 km respectively as a first-order approximation.

3.4 Southern Alps

The Southern Alps south of Switzerland were displaced dextrally along the Insubric line in Tertiary times (see Schmid et al. 1989 for a review). A minimum estimate of 40–50 km can be determined from the shape of the sheared Bergell intrusion. Within the Southern Alps the **Valtorta** and **Orobie** basement blocks were restored using south-directed thrusting (Laubscher 1985; Roeder 1989) and considering stacking of basement blocks (Laubscher 1985; Pfiffner et al. 1990b).

In the eastern part the reconstruction followed strictly the restoration performed by Doglioni and Bosellini (1987), who distinguished both, NS- and EW-oriented compression.

4 Conclusions

The pre-Triassic basement configuration in the Central Alps is constructed by inversion of the displacement which occurred during the Mesozoic rifting and the subsequent Alpine collision. The resulting palaeogeographic map shown in Fig. 3 clearly shows that our knowledge of the basement rocks is only fragmentary; much of the original volume of basement is in subsurface or eroded. Moreover, it must be realized that this reconstruction is a first-order approximation. Its strengths lie more in the relative than in the absolute positioning of basement units.

Nevertheless, the following trends emerge: Late Variscan granitoids, abundant in the Black Forest and Vosges basement complexes, are volumetrically less important in the Central Alps. They seem to be aligned in EW trending belts: the northern granite belt comprises the external massifs in the north (Aar, Gotthard, Mt. Blanc, Aiguilles Rouges), the central granite belt some Penninic and (lower) Austro-Alpine nappes (Dent Blanche, Monte Rosa, Monte Leone, Adula, Tambo, Suretta, Tauern, Err, Bernina), and the Southern-granite belt lies within the Southern Alps (Strona-Ceneri Zone and Bolzano area). In a similar way Permo-Carboniferous volcaniclastic sequences are concentrated in relatively narrow belts. The most prominent ones straddle the southern margins of the Vosges-Black Forest and the Aar Massif, and a third one extends throughout much of the Sothern Alps (including the Bolzano area). A narrow band, finally, is situated to the NW of the Bernhard nappe and extends into the Lebendun nappe and the Western Alps (see de Graciansky this Vol.).

Although the spacing between individual basement blocks as given in the palaeogeographic reconstruction in Fig. 3 are open to discussion, the relative order and configuration is much better constrained. In a general way, the reconstruction reflects estimates of displacements which are on the short side. Major unknowns are the former spacing between the Austro-Alpine/South-Alpine block and the Penninic units, and the one between Aar-Mt. Blanc Massifs and the Bernhard nappe.

Acknowledgements
This work benefited from fruitful discussions with I. Mercolli, M. Handy, E. Klaper and comments by P.C. de Gracianski, R. Hatcher, A. Escher and L. Ratschbacher. I thank I. Blaser, S. Sahli and W. Schaad for technical assistance.

References

Boriani A, Del Moro A, Giobbi Origoni E, Pinarellli L (1987) Rb/Sr systematics of the Hercynian plutonites of Massicio dei Lagi. Rend Soc Ital Mineral Petrol 42: 328–329

Burkhard M (1988) L'Helvétique de la bordure occidentale du massif de l'Aar (évolution tectonique et métamorphique). Eclogae Geol Helv 81(1): 63–114

Compagnoni R, Dal Piaz GV, Hunziker JC, Gosso G, Lombardo B, Williams PF (1977) The Sesia-Lanzo zone, a slice of continental crust with Alpine high pressure-low temperature assemblages in the Western Italian Alps. Rend Soc Ital Mineral Petrol 33: 281–334

Dal Piaz GV, De Vecchi G, Hunziker JC (1977) The Austroalpine layered gabbros of the Matterhorn and Mt. Collon-Dents de Bertol. Schweiz Mineral Petrogr Mitt 57: 59–88

Del Moro A, Notarpietro A (1987) Rb-Sr-Geochemistry of some Hercynian granitoids overprinted by eo-alpine metamorphism in the Upper Valtellina, Central Alps. Schweiz Mineral Petrogr Mitt 67: 295–306

Dewey JF, Helman ML, Turco E, Hutton DHW, Knott SD (1989) Kinematics of western Mediterranean. In: Coward MP, Dietrich D, Park RG (eds) Alpine Tectonics. Geol Soc Lond Spec Publ 45: 265–283

Diebold P (1988) Der Nordschweizer Permokarbon-Trog und die Steinkohlenfrage der Nordschweiz. Vierteljahresschr Naturforsch Ges Zür 133(1): 143–174

Doglioni C (1987) Tectonics of the Dolomites (Southern Alps, northern Italy). J Struct Geol 9(2): 181–193

Doglioni C, Bosellini A (1987) Eoalpine and mesoalpine tectonics in the Southern Alps. Geol Rundsch 76(3): 735–754

Dössegger R (1974) Verrucano and "Buntsandstein" in den Unterengadiner Dolomiten. Mitt Geol Inst ETH und Univ Zürich NF 235

Eichenberger U (1986) Die Mitteltrias der Silvretta-Decke (Ducankette und Landwassertal, Ostalpin). Mitt Geol Inst und ETH Univ Zürich NF 252

Eisbacher GH, Lüschen E, Wickert F (1989) Crustal-scale thrusting and extension in the Hercynian Schwarzwald and Vosges, Central Europe. Tectonics 8(1): 1–21

Escher A (1988) Structure de la nappe du Grand Saint-Bernard entre le val des Bagnes et les Mischabel. Rapp Géol Serv Hydrol Géol Natl 7: 1–26

Escher A, Masson H, Steck A (1987) Coupes géologiques des Alpes occidentales suisses. Rapp Géol Serv Hydrol Géol Natl 2: 1–11

Frey M, Hunziker JC, O'Neil JR, Schwander HW (1976) Equilibrium-disequilibrium relations in the Monte Rosa Granite, Western Alps: petrological, Rb-Sr and stable isotope data. Contrib Mineral Petrol 55: 147–179

Gealey WK (1988) Plate tectonic evolution of the Mediterranean-Middle East region. Tectonophysics 155: 285–306

Hanson GN, Grünenfelder M, Soptrayanova G (1969) The geochronology of a recrystallized tectonite in Switzerland – the Roffna gneiss. Earth Planet Sci Lett 5: 413–422

Hunziker JC, Zingg A (1980) Lower Palaeozoic amphibolite to granulite facies metamorphism in the Ivrea Zone (Southern Alps, northern Italy). Schweiz Mineral Petrogr Mitt 60: 181–213

Hunziker JC, Desmons J, Martinotti G (1989) Alpine thermal evolution in the central and the western Alps. In: Coward MP, Dietrich D, Park RG (eds) Alpine tectonics. Geol Soc Lond Spec Publ 45: 353–367

Illies H, Greiner G (1978) Rhinegraben and the Alpine system. Bull Geol Soc Am 89: 770–782

Jongmans WJ, Ritter E (1960) Die Karbonflora der Schweiz. Beitr Geol Karte Schweiz NF 108: 1–97

Klaper EM (1990) A discussion of contour maps in the Toce subdome of the Penninic Realm (Switzerland, Italy). Schweiz Mineral Petrogr Mitt 70: 349–360

Köppel V, Günthert A, Grünenfelder M (1981) Patterns of U-Pb zircon and monazite ages in polymetamorphic units of the Swiss Central Alps. Schweiz Mineral Petrogr Mitt 61: 97–119

Lammerer B (1988) Thrust-regime and transpression-regime tectonics in the Tauern window (Eastern Alps). Geol Rundsch 77(1): 143–156

Laubscher HP (1985) Large-scale, thin-skinned thrusting in the southern Alps: kinematic models. Bull Geol Soc Am 96: 710–718

Lemoine M, Dardeau G, Deplech PY, Dumont Th, de Graciansky PC, Graham R, Jolivet L, Roberts D, Tricart P (1989) Extension syn-rift et failles transformantes jurassiques dans les Alpes occidentales. CR Acad Sci 309 II: 1711–1716

Leu W (1986) Lithostratigraphie und Tektonik der nordpenninischen Sedimente in der Region Bedretto-Baceno-Visp. Eclogae Geol Helv 79(3): 769–824

Löw S (1987) Die tektono-metamorphe Entwicklung der nördlichen Adula-Decke, (Zentralalpen, Schweiz). Beitr Geol Karte Schweiz NF 161

Mancktelow N (1985) The Simplon Line: a major displacement zone in the western Lepontine Alps. Eclogae Geol Helv 78(1): 73–96

Matter A (1987) Faciesanalyse und Ablagerungsmilieus des Permokarbons im Nordschweizer Trog. Eclogae Geol Helv 80(2): 345–367

Mayerat AM (1991) Analyse structurale et tectonique du socle et de la couverture des nappes pénniques du Rheinwald (Grisons). Beitr Geol Karte Schweiz (in press)

Mercolli I (1989) The volcanic suite of the Julier area (Grisons), Part 1: Volcanic and tectonic evolution. Schweiz Mineral Petrogr Mitt 69: 423–433

Merle O, Cobbold PR, Schmid S (1989) Tertiary kinematics in the Lepontine dome. In: Coward MP, Dietrich D, Park RG (eds) Alpine tectonics. Geol Soc Lond Spec Publ 45: 113–134

Mugnier JL, Ménard G (1986) Le développement du bassin molassique suisse et l'évolution des Alpes externes: un modèle cinématique. Bull Cent Rech Explor Prod Elf-Aquitaine 10(1): 167–180

Oberhänsli R, Schenker F, Mercolli I (1988) Indications of Variscan nappe tectonics in the Aar Massif. Schweiz Mineral Petrogr Mitt 68: 509–520

Pfiffner OA (1981) Fold-and-thrust tectonics in the Helvetic Nappes (E Switzerland). In: McClay KR, Price NJ (eds) Thrust and nappe tectonics. Geol Soc Lond Spec Publ 9: 319–327

Pfiffner OA (1985) Displacements along thrust faults. Eclogae Geol Helv 78(2): 313–333

Pfiffner OA (1986) Evolution of the north Alpine foreland basin in the Central Alps. Spec Publ Int Assol Sediment 8: 219–228

Pfiffner OA, Klaper EM, Mayerat AM, Heitzmann P (1990a) Structure of the basement-cover contact in the Swiss Alps. In: Roure F, Heitzmann P, Polino R (eds) Deep structure of the Alps. Mém Soc Géol Fr 156: 247–262

Pfiffner OA, Frei W, Valasek P, Stäuble M, Levato L, DuBois L, Schmid SM, Smithson B (1990b) Crustal shortening in the Alpine orogen: results from deep seismic reflection profiling in the eastern Swiss Alps, line NFP 20-EAST. Tectonics 9(6): 1327–1355

Rageth R (1984) Intrusiva und Extrusiva der Bernina-Decke zwischen Morteratsch und Berninapass (Graubünden). Schweiz Mineral Petrogr Mitt 64: 83–109

Ring U, Ratschbacher L, Frisch W (1988) Plate-boundary kinematics in the Alps: motion in the Arosa suture zone. Geology 16: 696–698

Roeder D (1989) South-Alpine thrusting and trans-Alpine convergence. In: Coward MP, Dietrich D, Park RG (eds) Alpine tectonics. Geol Soc Lond Spec Publ 45: 211–227

Savostin LA, Sibuet J-C, Zonenshain LP, Le Pichon X, Roulet M-J (1986) Kinematic evolution of the Tethys belt from the Atlantic Ocean to the Pamirs since the Triassic. Tectonophysics 123: 1–35

Schaltegger U, von Quadt, A (1990) U-Pb zircon dating of the Central Aar Granite (Aar Massif, Central Alps). Schweiz Mineral Petrogr Mitt 70: 361–371

Schmid SM, Haas R (1989) Transition from near-surface thrusting to intrabasement decollement, Schlinig thrust, Eastern Alps. Tectonics 8(4): 697–718

Schmid SM, Aebli HR, Heller F, Zingg A (1989) The role of the Periadriatic line in the tectonic evolution of the Alps. In: Coward MP, Dietrich D, Park RG (eds) Alpine tectonics. Geol Soc Lond Spec Publ 45: 153–171

Schmid SM, Rück PH, Schreurs G (1990) The significance of the Schams nappes for the reconstruction of the paleotectonic and orogenic evolution of the Penninic zone along the NFP20 East traverse (Grisons, eastern Switzerland), In: Roure F, Heitzmann P, Polino R (eds) Deep structure of the Alps. Mém Soc Géol Fr 156: 263–288

Selverstone J (1988) Evidence for east-west crustal extension in the Eastern Alps: implications for the unroofing history of the Tauern window. Tectonics 7(1): 87–105

Steck A (1984) Structures de déformations tertiaires dans les Alpes centrales. Eclogae Geol Helv 77: 55–100

Steck A (1987) Le massif du Simplon – Réflexions sur la cinématique des nappes de gneiss. Schweiz Mineral Petrogr Mitt 67: 27–45

Stille P, Buletti M (1987) Nd-Sr isotopic characteristics of the Lugano volcanic rocks and constraints on the continental crust formation in the South Alpine domain (N-Italy – Switzerland). Contrib Mineral Petrol 96: 140–150

Thélin Ph, Ayrton St (1983) Cadre évolutif des événements magmatico-métamorphiques du socle anté-triasique dans le domaine pennique (Valais). Schweiz Mineral Petrogr Mitt 63: 393–420

Thélin Ph (1987) Nature originelle des gneiss oeillés de Randa (Nappe de Siviez-Mischabel, Valais). Mém Soc Vaudoise Sci Nat 104: 1–75

Weissert HJ, Bernoulli D (1985) A transform margin in the Mesozoic Tethys: evidence from the Swiss Alps. Geol Rundsch 74(3): 665–679

Werner D, Kahle HG, Ansorge J, Müller St (1982) Mass displacements and geothermics, within the upper mantle of the Rhinegraben rift system: a model to interpret data from gravimetry and seismology, In: Palmason G (ed) Continental and oceanic rifts. Am Geophys Union and Bull Geol Soc Am, Geodyn Ser 8: 283–292

Ziegler PA (1987) Late Cretaceous and Cenozoic intra-plate compressional deformations in the Alpine foreland – a geodynamic model. Tectonophysics 137: 389–420

Zingg A, Handy MR, Hunziker JC, Schmid SM (1990) Tectonometamorphic history of the Ivrea Zone and its relationship to the crustal evolution of the Southern Alps. Tectonophysics 182: 169–192

Palinspastic Reconstruction of the Pre-Triassic Basement Units in the Alps: The Eastern Alps

L. RATSCHBACHER AND W. FRISCH

Abstract

The pre-orogenic palaeogeographic zonation in the Eastern Alps is reconstructed by inversion of intra-orogenic displacements. We use palaeomagnetic data to constrain rigid block rotations around vertical axes and strain trajectories to elicit the orientations of horizontal displacements of Alpine nappes. The restoration uses a factorization into stages of orogenic wedge evolution: (1) a stacking-imbrication stage, which formed the Alpine wedge; (2) an unroofing stage, which caused wedge extension due to gravitational instability; and (3) a lateral extrusion stage, in which material moved from the compressed wedge toward a weakly constrained lateral margin.

Restoration of lateral extrusion deformation compensates for most of the 30° of post-Palaeogene rotations, rearranges the Southern Alps until the Periadriatic line becomes a straight line, and displaces the Austro-Alpine units E of the Tauern window westward until the presently exposed rocks in the Tauern window are covered by two thirds. All Miocene rifts in eastern Austria are closed. A minimum of 40% (ca. 150 km) E-W extension (parallel to the orogenic strike), 50% (ca. 100 km) N-S shortening at the position of the Tauern window, and 30% (ca. 70 km) N-S shortening along the eastern end of the Alps were recorded. Reconstruction of unroofing tectonics compensates for E-W extension portrayed by gravitational spreading-gliding nappes, core complexes, zones of sinistral wrenching, and the formation of the Gosau basins. Speculative minimum total E-W extension during this stage was 65% (ca. 80 km) in eastern Austria. Stacking-imbrication stage deformation shortened the Austro-Alpine basement units of the Eastern Alps by a minimum of 50% (ca. 385 km) in an E-W and 40% (ca. 185 km) in a N-S direction during transpressive convergence. The cumulative amount of contraction within the Austro-Alpine basement units is compatible with the amount of shortening determined by section balancing in the Northern Calcareous Alps (Austro-Alpine cover).

1 Introduction

Three approaches for reconstruction of pre-orogenic palaeogeographic zonations have been utilized in the past: (1) fitting of sedimentary facies and biogeographic zonations; (2) successive approximations based on palaeomagnetic rotations and magnetic anomalies in oceanic crust; and (3) inversion of intra-orogenic displacements. In this contribution, we use palaeomagnetic data to constrain rigid block rotations around vertical axes, and strain trajectories to elicit the orientations of horizontal displacements.

For a reconstruction based on a kinematic structural analysis of intraplate deformation, Choukroune et al. (1986) showed that it is necessary to split the total displacement pattern into a time sequence. We use three stages of orogenic wedge evolution (Platt 1986), which are all represented in the Eastern Alps (from old to young): (1) stacking-imbrication leading to the formation of the wedge; (2) unroofing during gravitational instability of the orogenic wedge; and (3) lateral extrusion [a combination of tectonic escape (Burke and Şengör 1986) and extensional collapse (Dewey 1988), which describes lateral motion of material compressed along an orogenic belt toward a weakly constrained lateral margin]. These stages have been described recently (Ratschbacher et al. 1989; Neubauer and Genser 1990; Ratschbacher et al. 1991 a, b). We retrodeform the Eastern Alps from the youngest (lateral extrusion) to the oldest stage (stacking-imbrication) to achieve the pre-Triassic configuration.

2 Tectonic Overview

2.1 *Tectonic Setting*

The classic subdivision of the Alps is used, i.e. from the external (northwestern) to internal (southeastern) units (Fig. 1): the Bohemian massif (Alpine fore-

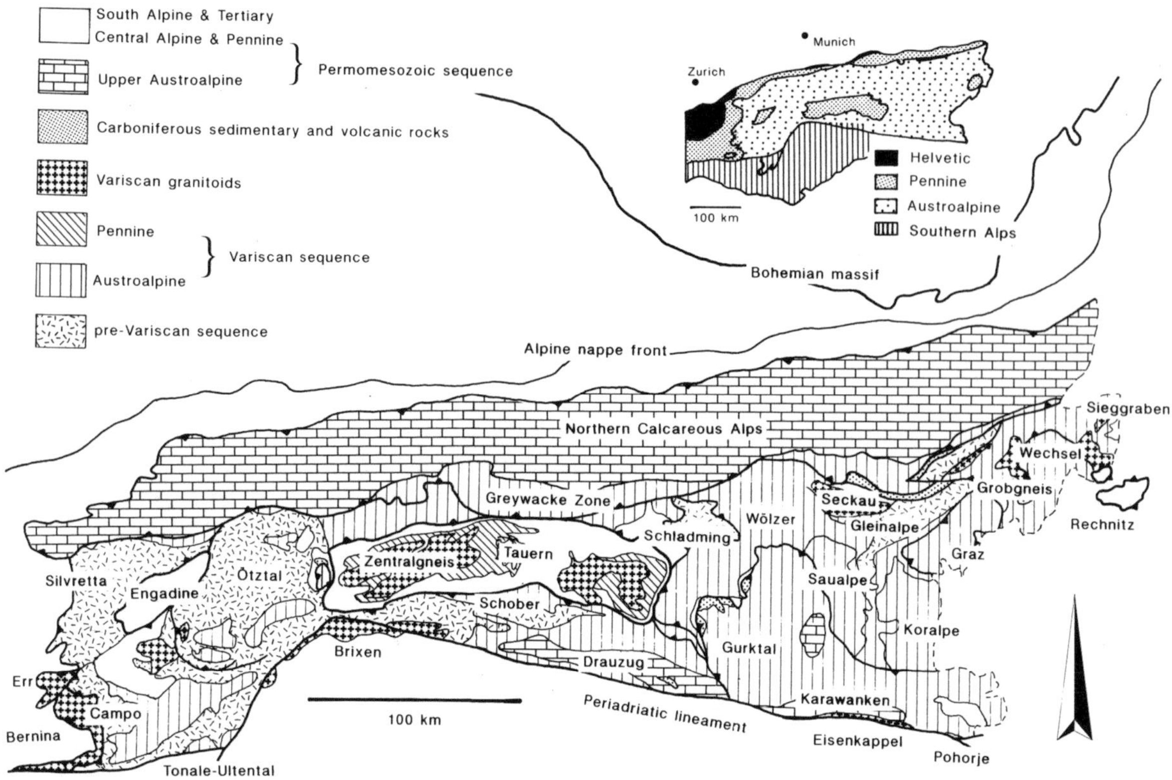

Fig. 1. Starting point for the palinspastic reconstruction of the pre-Triassic basement units in the Eastern Alps: Alpine tectonic configuration, distribution of basement units in the Eastern Alpine orogenic belt, and lithotectonic units referred to in text. The pre-Alpine lithotectonic subdivision is strongly simplified from Frisch et al. (1984)

land), and the Molasse, Helvetic, Penninic, and Austro-Alpine zones. The Austro-Alpine unit, exposed predominantly in the Eastern Alps, occupies the highest tectonic position and is subdivided into thrust systems each of which comprises pre-Alpine basement with Late Palaeozoic and Mesozoic sedimentary cover. The Northern Calcareous Alps occupy the highest position and form a wide belt of Mesozoic rocks in the N. The underlying series with prevailing crystalline rocks crop out in the central zone of the Eastern Alps. Penninic units, containing the ophiolitic suture between the European (Alpine foreland and Helvetic zone) and Apulian (Austro-Alpine and South Alpine units) plates, occur along the northern margin of the Northern Calcareous Alps (Rhenodanubian flysch belt) and in tectonic windows along the central zone (e.g. Tauern window). At the eastern end of the Alps, the Austro-Alpine unit is buried beneath Neogene sediments. The Periadriatic lineament bound the Eastern Alps in the S.

A crustal-extrapolated cross section through the Eastern and Southern Alps (Fig. 2) shows a bivergent thrust belt and indicates large-scale crustal decoupling. Thick-skinned thrusting and lateral flow of material toward the Carpathians compensate for imbrication in the deeper crust and lithospheric mantle (Roeder 1989; Ratschbacher et al. 1991 b).

2.2 *Characterization of Pre-Alpine Basement*

Variscan units occur in the basement of the Alpine foreland, the Penninic, and the Austro-Alpine zones. We briefly discuss them in this order. For details, we refer to sections in this volume and recent summaries of the basement evolution (e.g. Neubauer 1988 a, b; Franke 1989; Frisch et al. 1990).

N of the Eastern Alps, pre-Triassic basement crops out in the *Alpine foreland* (Bohemian Massif). In Austria, the Bohemian Massif essentially consists of medium- to high-grade metamorphic rocks of Precambrian to Palaeozoic age and extensive granitoid plutons of Variscan age (e.g. Scharbert and Schönlaub 1980; Matte et al. 1985). In the Moldanubian zone, which occupies the western part of the Bohemian Massif, deeper monotonous paragneisses contrast with higher variegated series containing graphitic

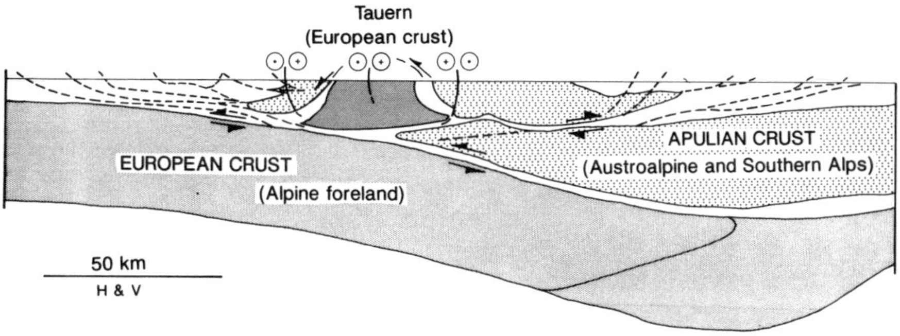

Fig. 2. Crustal-extrapolated N *(left)* – S *(right)* cross-section through the Eastern and Southern Alps. Crustal structure is characterized by decoupling of upper and lower crust and mantle, bivergent thrusting, and lateral motion (out of section) in the upper crust

marbles, ultrabasic and basic rocks, para- and orthogneisses, and granulites. Regional metamorphism caused metablastesis, anatexis, and the emplacement of the Variscan Moldanubian plutons. In the eastern part of the Bohemian Massif, the Moravian zone mainly consists of the pre-Devonian Thaya pluton and a series of low to medium grade metamorphic ortho- and paragneisses (e.g. Fuchs and Matura 1980). Locally, relics of a Late Carboniferous (?) and Permian overstep sequence are preserved ("Permian of Zöbing").

Probably *Helvetic basement* occurs as boulders in olistostrome deposits along the northern margin of the Penninic (Rhenodanubian) flysch. Rb/Sr whole rock data yielded ≥360 Ma (Thöni 1991). As a source area, Frasl and Finger (1988) postulated a basement uplift, the "Cetic massif", adjoining the Moldanubicum to the S.

The *Penninic basement,* restricted to the Venediger nappe of the Tauern window, comprises amphibolites, ultramafic rocks, plagioclase gneisses, and micaschists to phyllites intruded by large volumes of Variscan granitoids. Migmatites occur in broad halos around some of the granitoids. The pre-granitoid sequence (isotopic ages: 660–490 Ma, e.g. von Quadt 1989; Vavra and Hansen 1991) comprises metavolcanic rocks of an island-arc setting intercalcated with metasedimentary rocks (e.g. Vavra and Frisch 1989). Back-arc oceanic crust occurs as ophiolite slices (isotopic formation ages: 660–640 Ma, von Quadt 1989). The granitoids (isotopic ages around 320 Ma, e.g. Vavra 1989) form quartzdioritic to leucogranitic batholiths.

Austro-Alpine basement comprises fossiliferous, weakly metamorphosed sedimentary and volcanic rocks, low grade phyllitic rocks (quartzphyllite), and crystalline basement rocks, all with protolith ages in the range from Late Ordovician to Early Carboniferous; the crystalline basement includes older sequences with Precambrian relics. Differences in age, environment of formation, and degree of pre-Alpine metamorphism led to the interpretation of the basement as tectonostratigraphic terranes (Frisch and Neubauer 1989). The pre-Alpine units were stacked during Alpine orogeny, commonly with Permo-Mesozoic rocks intercalcated between the thrust units (central Alpine Permo-Mesozoic units). The Alpine metamorphic overprint locally reached medium grade (e.g. southeastern Ötztal unit, Thöni 1981; Koralpe, Frank 1987). A late Archean crustal component occurs in the basement of the Late Ordovician to Early Carboniferous Noric nappe of the Greywacke zone. Widespread gneisses and amphibolites (e.g. Ötztal unit, Muriden) contain Archean, Late Proterozoic, and Early Palaeozoic tectonometamorphic cycles.

Stratigraphic piles of different metamorphic grade, but equivalent biostratigraphically dated or inferred Late Ordovician to Early Carboniferous age, constitute the following Alpine nappes. The Noric nappe of the Greywacke zone, the Stolzalpen nappe of the Gurktal thrust system, and the Rannach-Hochlantsch-Laufnitzdorf nappe of the Palaeozoic-of-Graz thrust system comprise fossiliferous, very low to low grade metamorphic sequences. Monotonous, low grade, mainly siliciclastic rocks are contained in the Murau nappe (Gurktal thrust system), Schöckel nappe (Palaeozoic-of-Graz thrust system), and Innsbruck-Radstadt and corresponding quartzphyllite units. Medium grade equivalents are the widespread micaschist-marble complexes of the lower tectonic level in the Alpine edifice. Clastic sedimentary rocks dominated the Late Ordovician and Silurian sequence in all these units. Intercalated volcanic rocks evolved from basic subalkalic over acidic calc-alkalic to basic alkalic during this time. The Late Silurian to Middle Devonian sequence contains facies differentiation into a carbonate platform, pelagic limestone, and clastic sequences containing intraplate basalts (Heinisch 1988). The carbonate platform subsided into pelagic environments during the Late Devonian. Carboniferous orogeny is reflected by flysch sedimentation and orogenic deformation and metamorphism. Terrestrial conglomerates of Westphalian (B/C) and Stephanian age overstep the Variscan edifice. Permian detrital rocks with acid volcanic intercalcations and subordinate hypersaline rocks form the base of the Alpine sequence of the Northern Calcareous Alps and their equivalents.

The crystalline basement experienced medium to (rarely) high grade Carboniferous metamorphism (isotopic ages: 360–280 Ma, e.g. Neubauer 1988 b) accompanied by a few granitoid intrusions. Some areas, such as the basement of the Veitsch nappe of the Greywacke zone or the Bundschuh zone (western footwall of the Gurktal thrust system), recorded Devonian metamorphism and deformation. The Carboniferous sedimentary rocks of the Veitsch nappe form the sedimentary response to these pre-Carboniferous orogenic events (molasse-type deposits; Ratschbacher 1984, 1987 a). Lithostratigraphic units, whose age and tectonic setting are still under discussion are the eclogite-bearing units of the Koralpe and the Plankogel unit (central-eastern Austria). The Plankogel unit is interpreted as a metamorphosed ocean-floor and seamount complex containing ophiolitic slices in a tectonic mélange (Neubauer et al. 1989).

3 Pre-Triassic Reconstruction

Figure 1 shows strongly simplified the present-day tectonic configuration and the distribution of basement units in the Eastern Alpine orogenic belt. It serves as the starting point for the reconstruction. The pre-Alpine lithotectonic subdivision is simplified herein from Frisch et al. (1984). The units discriminated are: (1) pre-Variscan sequences (gneiss-amphibolite association, pre-Variscan granitoids, serpentinite-amphibolite association); (2) Variscan sequences (Ordovician to Early Carboniferous sedimentary and volcanic rocks in different tectonic settings and metamorphic grades); (3) Variscan granitoids; (4) Carboniferous sediments and vulcanic rocks unaffected by Variscan deformation and metamorphism. We used the outline of the Bohemian massif as the presently exposed Alpine foreland, and the Alpine nappe front (sole thrust of the folded Molasse) as a reference frame for the reconstructions.

Step 1: Lateral Extrusion Stage (Figs. 3, 4)

Figure 3 summarizes the major lateral extrusion structures which are related to late Oligocene-Miocene indentation by the Southern Alps. A structural analysis is presented in Ratschbacher et al. (1991 b). Indentation structures occur in the western Eastern Alps and comprise folds (brachyanticlines of the Tauern and Engadine windows, Laubscher 1983), thrusts (foreland-directed thrusting of the Ötztal unit onto the Northern Calcareous Alps and backthrusting within the Northern Calcareous Alps and along the Giudicarie belt), and strike-slip faults. Sets of sinistral, NE-trending fault zones contrast with dextral displacements along SE-trending fault zones, mainly the Periadriatic and related lineaments. The sets form conjugate pairs causing lateral extrusion of crustal blocks in front of the South Alpine indenter and into the "open" space provided by the evolving Carpathian loop. Gravitational spreading, with rifts and transtensional strike-slip faults in the eastern Eastern Alps, attenuated crustal thickness differences. Crustal decoupling partitioned deformation into brittle deformation in the upper plate, the Austro-Alpine unit, and ductile deformation in the lower plate, the Penninic unit. Major structures in the Penninic units are a subhorizontal mylonitic foliation with an E-W stretching lineation and ductile shear zones with a vertical mylonitic foliation and again an E-W stretching lineation (Ratschbacher et al. 1989, 1990).

Ratschbacher et al. (1991 b) performed a simple 2-D computer retro-deformation to obtain an estimate of the amount of deformation taken up by lateral extrusion in the Eastern Alps. They recorded a minimum of 40 % (ca. 150 km) E-W extension (parallel to the orogenic strike), 50 % (ca. 100 km) N-S shortening in front of the indenter tip (Tauern window), and 30 % (ca. 70 km) N-S shortening along the eastern edge of the Alps.

The present reconstruction, which is restricted to the area N of the Periadriatic lineament, includes the following steps: (1) The eastern Southern Alps are retracted until the Periadriatic line becomes a straight line. (2) The Austro-Alpine basement E of the Tauern window is displaced to the W until it overlaps the present outline of the Tauern window by about two-thirds. We use the following constraints: (a) compensation of 60–100 km sinistral strike-slip along the Salzachtal-Ennstal fault (Salzachtal-fault: offset of the Innsbruck and Radstadt-Ennstal quartzphyllite units; Ennstal fault: offset of the Greywacke zone, Fig. 3); (b) compensation of >100 km offset of the Schneeberg and Radenthein Austro-Alpine basement complexes along the decoupling zone between the Austro-Alpine and Penninic units in combination with dextral strike-slip displacements along the Periadriatic lineament and related faults N of it (e.g. Mölltal line, Fig. 3; Schimana 1986; Ratschbacher et

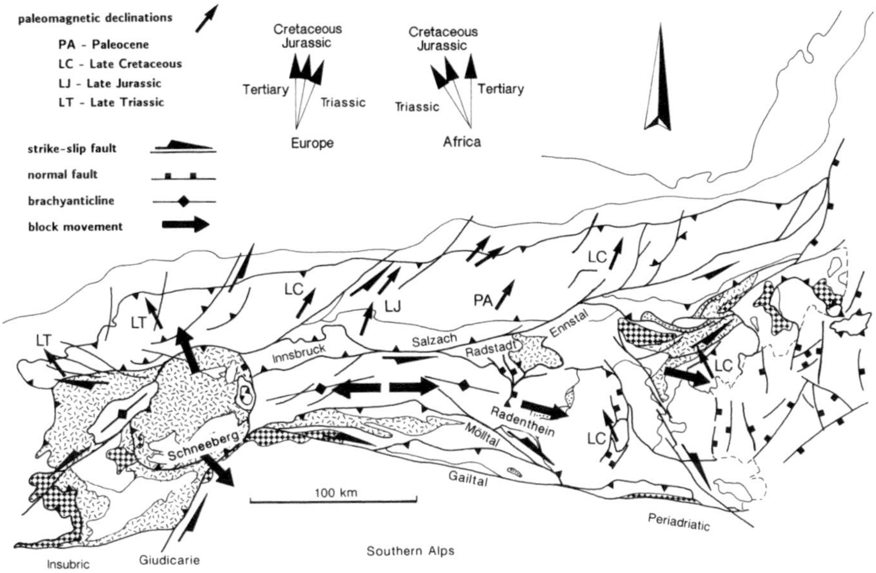

Fig. 3. Major lateral extrusion structures related to late Oligocene-Miocene northward indentation by the Southern Alps; structures simplified from Ratschbacher et al. (1991 b). Palaeomagnetic data from Mauritsch and Becke (1987). For clarity, signatures in this and the following maps are shown only for the pre-Variscan sequence and the Variscan granitoids

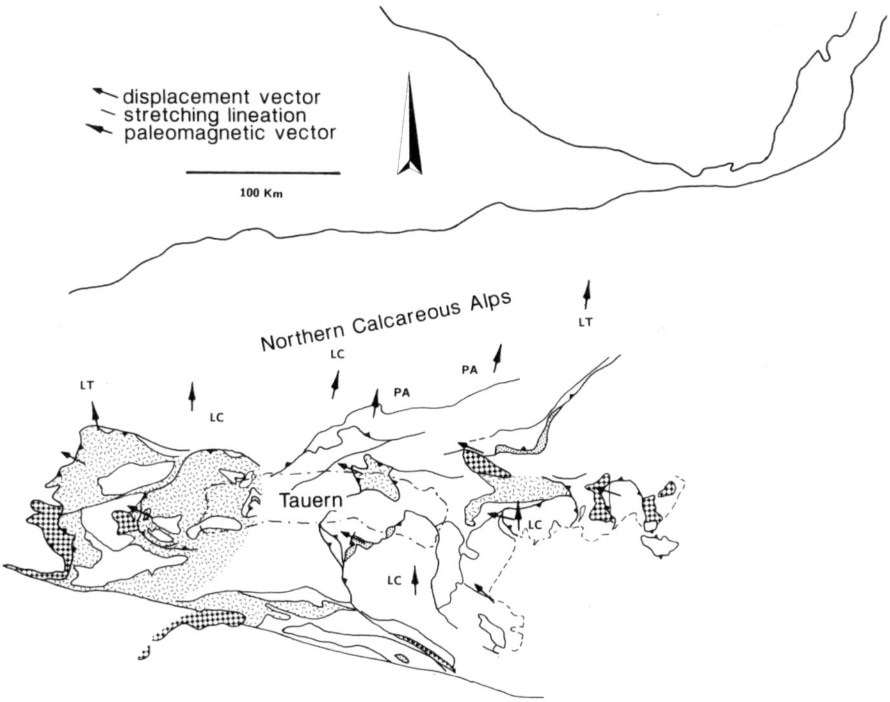

Fig. 4. Reconstruction of stage 3 (lateral extrusion) of Alpine orogenic wedge evolution. Compare displacement vectors of Cretaceous-Early Tertiary stacking-imbrication stage with the present-day configuration in Fig. 5; much of the present-day scatter has been eliminated. Reconstruction step compensates for most of the post-Palaeocene palaeomagnetic rotations in the Eastern Alps and also compensates to a large extent for unroofing of rocks presently exposed in the Tauern window

al. 1991 b). (3) The Ötztal unit is displaced to the SE to compensate for the thrusting onto and within the Northern Calcareous Alps. The solution was considered satisfactory when the Northern Calcareous Alps N and NE of the Ötztal unit were rotated counterclockwise by up to 30° to compensate for post-Palaeogene clockwise rotations as indicated by palaeomagnetic data from the Gosau deposits (Mauritsch and Becke 1987). The post-Triassic anticlockwise rotations NW of the Ötztal unit were compensated by up to 30° clockwise rotation. As a result, the dextral and sinistral strike-slip faults in the Northern Calcareous Alps NE respectively NW of the Ötztal unit enclose an acute, N-facing angle of 70°–80°, which approaches the angle typical for conjugate faults in upper crustal rocks for which a Mohr-Coulomb rheology is assumed [note that a line separating opposite rotations passes through the northern tips of the Southern Alps and the Ötztal unit (Ratschbacher et al. 1991 b)]. This stretches the Northern Calcareous Alps considerably in a N-S direction. (4) The two blocks framed by the conjugate sets of strike-slip faults E of the Tauern window (Fig. 3) are rotated clockwise until the post-Palaeocene palaeomagnetic directions (Mauritsch and Becke 1987) in the (central Alpine) Gosau deposits approach the directions of stable Europe. Then they are displaced back to the W, and overlaps are adjusted.

Figure 4 shows the pre-Late Oligocene reconstruction. Major uncertainities are: (1) the extent to which the rocks presently exposed in the Tauern window were exhumed by tectonic unroofing. Our reconstruction requires an unknown amout of erosional exhumation; (2) the reconstruction of the basement units S of the Tauern window. We left them in their present-day configuration, because little is known about the mechanics and age of movement along the block-bounding faults. Large-scale displacements probably occurred along strike-slip zones like the Defereggen-Antholz-Vals line (Kleinschrodt 1987; Kázmer and Kovács 1985), the Kalkstein line, the Debant-Isel fault zone, etc., from the Eocene to the end of the Miocene. Reconstruction of these displacements would probably close the hole between the Ötztal-Campo-Tonale basement, the Southern Alps, and basement units E of the Tauern window in our reconstruction.

Step 2: Unroofing Extension

This latest Cretaceous and Palaeocene stage is least well known from its structural reconstruction (Ratschbacher et al. 1989). In the Tauern window, it is related to uplift after cessation of thrusting and mainly affected rocks above the buoyant granitoid gneisses (Zentralgneis). Extension directions are grossly E-W and subhorizontal (e.g. Behrmann and Ratschbacher 1989). In the Austro-Alpine, unroofing structures are restricted to the area E of the Tauern window. Thrust geometries and mylonitic fabrics formed during initial stacking-imbrication (stage 1) were overprinted by distinct sets of extensional crenulation-cleavage which are widespread in the basement of the central and southeastern Eastern Alps. Klippen of Mesozoic cover rocks and low to very low-grade basement units (e.g. Gurktal and Palaeozoic-of-Graz thrust systems) contain E-directed normal faults, and thrusts were reactivated as normal faults or shear zones with a normal fault geometry (Ratschbacher and Neubauer 1989). The formation of the Gosau basins indicates extension at highest structural levels. Normal faulting is connected with strike-slip faulting (e.g. Neubauer and Genser 1990). Strike-slip structures studied so far are ductile and brittle-ductile shear zones in the southern Gailtal basement (Unzog 1991), in the Koralm basement (Krohe 1987), along the northern rim of the Palaeozoic-of-Graz thrust system, and within the Gleinalm basement (Neubauer 1988 a). Displacements along single shear zones reach 40 km (Neubauer 1988 a). The zones of normal faulting and strike-slip shear loosely outline crustal wedges expelled eastward by probably both gravitational forces (e.g. Palaeozoic-of-Graz spreading-gliding nappe, Ratschbacher et al. 1991 c) and tectonic push (e.g. escaping crustal wedges defined by the sinistral wedge corridor formed by the shear zones of the Gailtal-Koralpe-Gleinalpe region and probably a Late Cretaceous/Palaeogene forerunner of the Periadriatic lineament).

The reconstruction shifts the low-grade Gurktal and Palaeozoic-of-Graz basement thrust systems onto the eastern edge of the present-day Tauern window and the Koriden-Muriden (Koralpe-Saualpe) basement region (both in their present configuration), respectively. In the easternmost Eastern Alps, it unstrains the deformation portrayed by the metamorphic core complexes exposing Lower Austro-Alpine nappes of deep tectonic position (Wechsel and Fischbach windows) and creates a nearly coherent block of Variscan granitoids (the Grobgneis unit).

We do not present a separate figure for the unroofing stage. The reconstruction of the positions of the Gurktal and Palaeozoic-of-Graz thrust systems and the unstraining of deformation portrayed by the core-complexes in eastern Austria are incorporated in the reconstruction of step 3 (stacking-imbrication, see

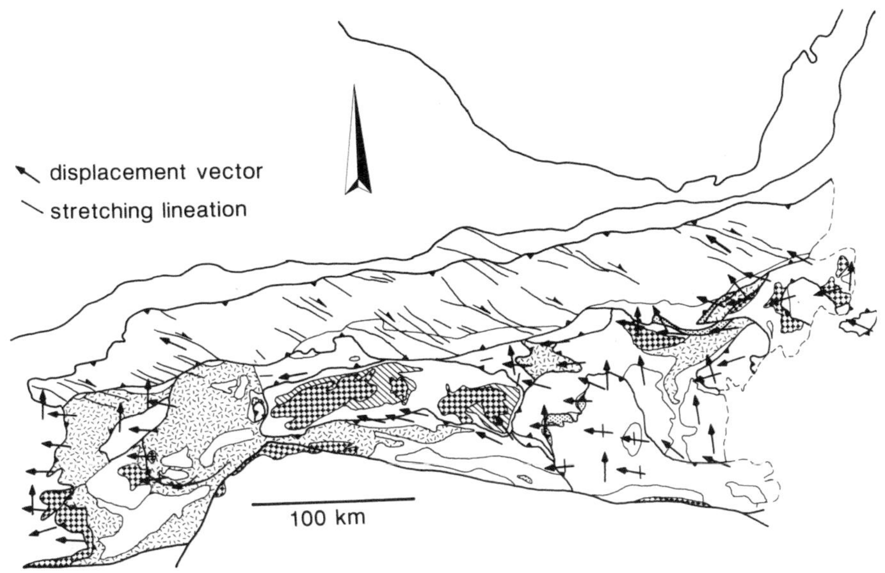

Fig. 5. Displacement vectors for major Alpine thrust systems in the central Eastern Alps (Ratschbacher et al. 1989 and references therein). Their orientations after compensation of rotations during stage 3 (lateral extrusion stage, Fig. 4) are used to translate Alpine nappes in their initial positions. In the Northern Calcareous Alps, strike-slip faults (note similar strikes of those faults and the displacement vectors in the central Eastern Alps) give the thrust-displacement directions

below). We calculated a minimum total E-W extension during this stage of 65% (80 km) for the area E of the Tauern window.

Step 3: Stacking-Imbrication Stage (Figs. 5, 6)

Stacking commenced in the upper and easternmost Austro-Alpine units coeval with subduction and accretion of the Penninic units (±125 Ma, e.g. Ratschbacher et al. 1989). Compression structures are thrust imbricates and ductile mylonites with a prominent, about E-trending stretching lineation parallel to the direction of thrusting. The sense of motion is indicated by shear-sense criteria (e.g. Schmid and Haas 1989). Displacement vectors in mylonites are plotted in Fig. 5 for major Alpine thrust units in the central Eastern Alps. Their orientations after compensation of rotations during stage 3 (lateral extrusion, see Fig. 4) are used to translate Alpine thrust systems in their initial positions. In the Northern Calcareous Alps, about WNW-trending maximum compression axes derived from palaeostress analysis and again WNW-striking strike-slip faults, and generally NE-trending fold axes indicate the displacement directions during thrusting. Note the difference in orientation between the displacement vectors plotted on the present-day map (Fig. 5) and those in the reconstruction compensating for stage 3 (lateral extrusion, Fig. 4). The better fit in the latter corroborates the validity of the reconstruction so far.

Stacking during stage 1 (stacking-imbrication) reflects the direction of convergence between the Austro-Alpine and Penninic units which was oblique to the trend of great parts of the continental margins. It was therefore described as dextral transpressive (e.g. Ratschbacher 1986; Schmid and Haas 1989). Dextral transpressive motions also explain the clockwise rotations recorded palaeomagnetically in the Northern Calcareous Alps since the Late Triassic (Mauritsch and Becke 1987). Three substeps of reconstruction take into account the structural expressions of transpression: (1) main phase imbrication in the Austro-Alpine units and accretion-underplating in the Penninic units were top-to-W (e.g. Ratschbacher et al. 1989); (2) Ring et al. (1988, 1989) documented a change in overthrust motion along the suture between the Austro-Alpine and Penninic units (Arosa zone, Engadine window) from top-to-W to top-to-N in the early Tertiary. Due to the presence of Eocene strata in the Engadine window, a minium of 80 km of top-to-N thrusting is required for the post-Eocene motion of the Austro-Alpine units; (3) large-scale folding, fold-imbrication, and formation of fold nappes, representing a Late Cretaceous N-S compression has been recognized in all Austro-Alpine units E of the Tauern window (e.g. Ratschbacher 1986, 1987b; Frank 1987; Neubauer 1987). All fold structures refold early stacking-related thrusts and shear zones and succeeded thrusting in a progressive sequence (Ratschbacher and Oertel 1987). Frank (1987) estimated shortening in the Koralpe fold

Fig. 6. Pre-Triassic palinspastic reconstruction. Some characteristic basement units are named. *Arrows* give mean Cretaceous-Early Tertiary displacement directions used for reconstruction (see Fig. 5). Note that the reconstruction does not include effects of Triassic/Jurassic transtension and Jurassic/Early Cretaceous formation of the Penninic ocean between Europe and Apulia. Incorporation of these effects would displace the Austro-Alpine pre-Triassic basement units several 100 km toward the SE

nappe to be up to 100 km. (1)–(3) have to restore the pre-contraction palaeogeography with Lower Austro-Alpine units N or W of the Upper Austro-Alpine units. Figure 6 shows the final reconstruction.

A check of the cumulative amount of stacking-imbrication related contraction within the Austro-Alpine basement units comes from the amount of shortening determined by section balancing in the Northern Calcareous Alps. We assume that the depositional area of the Northern Calcareous Alps was on the Austro-Alpine basement units and that remnants of Mesozoic strata on the basement are metamorphosed and tectonically reduced counterparts of the Northern Calcareous Alps. A minimum shortening of 65% has been estimated from balanced sections (Eisbacher et al. 1990; Linzer et al. 1990). This requires a deposition width in a NW-SE direction of more than 170 km on the basement. Compare this with the restored basement width in Fig. 6. Stacking-imbrication stage deformation shortened the Austro-Alpine basement units of the Eastern Alps by a minimum of 50% (ca. 385 km) in an E-W and 40% (ca. 185 km) in a N-S direction during transpressive convergence. Total N-S shortening of the Austro-Alpine units of the Eastern Alps is about 55% (ca. 230 km).

The final reconstruction (Fig. 6) overlaps and is spatially compatible with the reconstruction of the Central Alps (Pfiffner, this Vol.). We stress that this reconstruction is for pre-Triassic times and does not include effects of Triassic/Jurassic oblique rifting and Jurassic/Early Cretaceous formation of the Penninic ocean. Incorporation of these effects would displace the pre-Triassic basement units several hundred kilometres toward the SE.

4 Discussion

Many problematic tectonic events have not been considered in this reconstruction and this assigns semiquantitative character to it: (1) Jurassic transform faulting (e.g. Weissert and Bernoulli 1985; Channell et al. 1990). Transtensive motions during continental rifting and formation of the Penninic ocean are documented but not quantified so far. Successive approximation studies (see e.g. Dercourt et al. 1986; Neugebauer 1989) show differences in lateral, continental margin parallel displacements in order of several 100 km. Our restoration shows an irregular, NE-stepping Austro-Alpine continental margin, similar to the shape envisaged by Weissert and Bernoulli (1985). (2) Regional strain analysis established that several metamorphic nappes contain strong internal deformation. For this reconstruction, we refrained from restoring the original shape of the various nappes. Two facts support this decision: (a) strain data are very unevenly distributed regionally, and (b) characteristic shapes of pre-Triassic basement units would be lost or would be hard to identify on restored maps. For example, restoration of Tertiary ductile deformation of the Penninic basement in the Tauern window would result in an almost circular shape on the restored map. (3) Internal shortening in basement thrust-systems [e.g. Gurktal thrust system with two main Alpine nappes (Stolzalpe and Murau thrust sheets) which have been traced over most of the system]; this would again obscure the characteristic shapes of the basement units. A reconstruction of internal thrusting in the Gurktal thrust system, for example, would considerably stretch (>60%) its present-day outline (Fig. 6) in a NW-SE direction.

Acknowledgements
Field and experimental work on the structural evolution of the Eastern Alps on which this oversimplified reconstruction is based was funded by DFG (Fr610/7-2, Ra442/2-2). We appreciate many discussions on the Northern Calcareous Alps with Hans-Gert Linzer, Gerhard Eisbacher, and Blanka Sperner, and reviews by R. D. Hatcher, H.-P. Laubscher, F. Neubauer, and J. von Raumer.

References

Behrmann JH, Ratschbacher L (1989) Archimedes revisited. A structural test of eclogite emplacement models in the Austrian Alps. Terra Nova 1: 242–252

Burke K, Sengör AMC (1986) Tectonic escape in the evolution of the continental crust. In: Barazangi M, Brown L (eds) Reflection seismology: the continental crust. Am Geophys Union Geodyn Ser 14: 41–53

Channell JET, Brandner R, Spieler A, Smathers NP (1990) Mesozoic paleogeography of the Northern Calcareous Alps – evidence from paleomagnetism and facies analysis. Geology 18: 828–831

Choukroune P, Ballèvre M, Cobbold PR, Gautier Y, Merle O, Vuichard JP (1986) Deformation and motion in the Western Alpine arc. Tectonics 5: 215–226

Dercourt J, Zonenshain LP et al. (1986) Geological evolution of the Tethys belt from the Atlantic to the Pamirs since the Lias. Tectonophysics 123: 241–315

Dewey JF (1988) Extensional collapse of ororgens. Tectonics 7: 1123–1139

Eisbacher GH, Linzer GH, Meier L, Polinski R (1990) A depth-extrapolated structural transect across the Northern Calcareous Alps of western Tirol. Eclogae Geol Helv 83: 711–725

Frank W (1987) Evolution of the Austroalpine elements in the Cretaceous. In: Flügel HW, Faupl P (eds) Geodynamics of the Eastern Alps. Deuticke, Wien, pp 379–406

Franke W (1989) Tectonostratigraphic units in the Variscian belt of Central Europe. Geol Soc Am Spec Pap 230: 67–90
Frasl G, Finger F (1988) The "Cetic Massif" below the Eastern Alps – characterized by its granitoids. Schweiz Mineral Petrogr Mitt 68: 433–439
Frisch W, Neubauer F (1989) Pre-Alpine terranes and tectonic zoning in the Eastern Alps. Geol Soc Am Spec Pap 230: 91–99
Frisch W, Neubauer F, Satir M (1984) Concepts of the evolution of the Austroalpine basement complex (Eastern Alps) during the Caledonian-Variscan cycle. Geol Rundsch 73: 47–68
Frisch W, Ménot RP, Neubauer F, von Raumer JF (1990) Correlation and evolution of the Alpine basement. Schweiz Mineral Petrogr Mitt 70: 265–285
Fuchs G, Matura A (1980) Die Böhmische Masse in Österreich. In: Oberhauser R (Red) Der geologische Aufbau Österreichs. Springer, Berlin Heidelberg New York Wien, pp 121–143
Heinisch H (1988) Hinweise auf die Existenz eines passiven Kontinentalrandes im Altpaläozoikum der nördlichen Grauwackenzone – Ostalpen. Schweiz Mineral Petrogr Mitt 68: 407–418
Kázmér M, Kovács S (1985) Permian-Paleogene paleogeography along the eastern part of the Periadriatic lineament system: evidence for continental escape of the Bakony-Drauzug unit. Acta Geol Acad Sci Hung 28: 71–84
Kleinschrodt R (1987) Quarzgefügeanalyse im Altkristallin südlich des westlichen Tauernfensters (Südtirol/Italien). Erl Geol Abh 114: 1–82
Krohe A (1987) Kinematics of Cretaceous nappe tectonics in the Austroalpine basement of the Koralpe region (eastern Austria). Tectonophysics 136: 171–196
Laubscher HP (1983) The late Alpine (Periadriatic) intrusions and the Insubric line. Mem Soc Geol Ital 26: 21–30
Linzer HG, Frisch W, Ratschbacher L (1990) Deformation pathway of the Northern Calcareous Alps: subsidence and destruction. Terra Abstr 2: 16
Matte P, Maluski H, Echtler H (1985) Cisaillements ductiles varisques vers l'Est-Sud-Est dans les nappes du Waldviertel (Sud-Est du Massif de Bohème, Autriche). Données microtectoniques et radiométriques $^{39}Ar/^{40}Ar$. C R Acad Sci Paris t 301, II: 721–726
Mauritsch HJ, Becke M (1987) Paleomagnetic investigations in the Eastern Alps and the southern border zone. In: Flügel HW, Faupl P (eds) Geodynamics of the Eastern Alps. Deuticke, Wien, pp 282–308
Neubauer F (1987) The Gurktal thrust system within the Austroalpine region. Some structural aspects. In: Flügel HW, Faupl P (eds) Geodynamics of the Eastern Alps. Deuticke, Wien, pp 223–236
Neubauer F (1988a) Bau und Entwicklungsgeschichte des Rennfeld-Mugel- und des Gleinalm-Kristallins (Ostalpen). Abh Geol Bundesanst 42: 1–137
Neubauer F (1988b) The Variscan orogeny in the Austroalpine and Southalpine domains of the Eastern Alps. Schweiz Mineral Petrogr Mitt 68: 339–349
Neubauer F, Genser J (1990) Architektur und Kinematik der östlichen Zentralalpen – eine Übersicht. Mitt Naturwiss Ver Steiermark 120: 203–219
Neubauer F, Frisch W, Schmerold R, Schlöser H (1989) Metamorphosed and dismembered ophiolite suites in the basement units of the Eastern Alps. Tectonophysics 164: 49–62
Neugebauer J (1989) The Iapetus model: a plate tectonic concept for the Variscan belt of Europe. Tectonophysics 169: 229–256
Platt JP (1986) Dynamics of orogenic wedges and the uplift of high-pressure metamorphic rocks. Geol Soc Am Bull 97: 1037–1053
von Quadt A (1989) Sm-Nd and U-Pb systematics of zircon fractions, garnet and whole rock amphibolitic rocks of the Tauern window (Eastern Alps). Terra Abstr 1: 349
Ratschbacher L (1984) Beitrag zur Gliederung der Veitscher Decke (Grauwackenzone) in ihrem Westabschnitt (Obersteiermark, Österreich). Jahrb Geol Bundesanst 127: 423–453
Ratschbacher L (1986) Kinematics of Austro-Alpine cover nappes: changing translation path due to transpression. Tectonophysics 125: 335–356
Ratschbacher L (1987a) Stratigraphy, tectonics, and paleogeography of the Veitsch nappe (Graywacke zone, Eastern Alps, Austria): a rearrangement. In: Flügel HW, Sassi FP, Grecula P (eds) Pre-Variscan and Variscan events. Mineralia slovaca – Monography, 1987, Alfa, Bratislava, pp 407–414
Ratschbacher L (1987b) Strain, rotation, and translation of Austroalpine cover nappes. In: Flügel HW, Faupl P (eds) Geodynamics of the Eastern Alps. Deuticke, Wien, pp 237–243
Ratschbacher L, Neubauer F (1989) West-directed décollement of Austro-Alpine cover nappes in the Eastern Alps: Geometrical and rheological considerations. In: Coward MP, Dietrich D, Park RG (eds) Alpine tectonics. Geol Soc Lond Spec Publ 45: 243–262
Ratschbacher L, Oertel G (1987) Superposed deformation in the Eastern Alps: strain analysis and microfabrics. J Struct Geol 9: 263–276
Ratschbacher L, Frisch W, Neubauer F, Schmid SM, Neugebauer J (1989) Extension in compressional orogenic belts: the eastern Alps. Geology 17: 404–407
Ratschbacher L, Behrmann JH, Pahr A (1990) Penninic windows at the eastern end of the Alps and their relation to the intra-Carpathian basins. Tectonophysics 172: 91–105
Ratschbacher L, Merle O, Davy P, Cobbold P (1991a) Lateral extrusion in the eastern Alps, part I: boundary conditions and experiments scaled for gravity. Tectonics 10: 245–256
Ratschbacher L, Frisch W, Linzer HG, Merle O (1991b) Lateral extrusion in the eastern Alps, part II: structural analysis. Tectonics 10: 257–271
Ratschbacher L, Wenk HR, Sintubin M (1991c) Calcite textures: examples from nappes with strain-path partitioning. J Struct Geol 13: 369–384
Ring U, Ratschbacher L, Frisch W (1988) Plate-boundary kinematics in the Alps: motion in the Arosa suture zone. Geology 16: 696–698
Ring U, Ratschbacher L, Frisch W, Biehler D, Kralik M (1989) Kinematics of the Alpine plate-margin: structural styles, strain and motion along the Penninic-Austroalpine boundary in the Swiss-Austrian Alps. J Geol Soc Lond 146: 835–849
Roeder D (1989) South-Alpine thrusting and trans-Alpine convergence. In: Coward MP, Dietrich D, Park RG (eds) Alpine tectonics. Geol Soc Lond Spec Publ 45: 211–227
Scharbert S, Schönlaub HP (1980) Das Prävariszikum und Variszikum. In: Oberhauser R (Red) Der geologische Aufbau Österreichs. Springer, Berlin Heidelberg New York Wien, pp 3–20
Schimana R (1986) Geologische Entwicklung des Kristallins in der Umgebung von Radenthein (Nockgebiet, Kärnten). PhD Diss, Univ Wien
Schmid SM, Haas R (1989) The transition from near surface thrusting to intra-basement decollement, Eastern Alps. Tectonics 8: 697–718
Thöni M (1981) Degree and evolution of the Alpine metamorphism in the Austroalpine unit W of the Hohe Tauern in the light of K/Ar and Rb/Sr age determinations on micas. Jahrb Geol Bundesanst 124: 111–174

Thöni M (1991) Neue Rb/Sr-Daten an Mineralien und Gesteinen des Leopold von Buch-Denkmales und am Tonalit von Schaitten. Mitt Ges Geol Bergbaustud Österr 37: 157–162

Unzog W (1991) Alpidic left-lateral strike-slip tectonics in the Gailtal crystalline, Austria. Terra Abstr 3: 251

Vavra G (1989) Die Entwicklung des penninischen Grundgebirges im östlichen und zentralen Tauernfenster der Ostalpen. Geochemie, Zirkonmorphologie, U/Pb-Radiometrie. PhD Diss, Univ Tübingen

Vavra G, Frisch W (1989) Pre-Variscan back-arc and island-arc magnetism in the Tauern window (Eastern Alps). Tectonophysics 169: 271–280

Vavra G, Hansen BT (1991) Cathodoluminescence and U/Pb dating of zircons in pre-Mesozoic gneisses of the Tauern window: implications for the Penninic basement evolution. Geol Rundsch 80: 703–715

Weissert HJ, Bernoulli D (1985) A transform margin in the Mesozoic Tethys: evidence from the Swiss Alps. Geol Rundsch 74: 665–679

Part II

Pre-Mesozoic Basement of the Alps

Regional comparative studies

History of Geological Investigations in the Pre-Triassic Basement of the Alps

J. F. VON RAUMER and F. NEUBAUER

Abstracts

The authors want to underline that present-day research cannot be detached from a long tradition. Two centuries of observations made by field geologists, and new methods helped to attain our present level of knowledge. In this sense, the different sections recall the evolution of ideas accompanied by a growing sophistication of methods.

1 Introduction

An attempt to compare the pre-Permian units of the Alps with Caledonian and Variscan domains outside the Alpine realm is but a mere projection of current hypotheses on a more or less well-known geological edifice, pursuing a long tradition of geological observation and investigation into the Variscan structures of Europe and the Alps. The comparison permits the recognition of the main structures older than Alpine events and leads to the reconstruction of the main palaeotectonic trends. These have subsequently been modified to varying degrees according to their position relative to the zones of Alpine metamorphism. The map of Alpine metamorphic features (Frey at al. 1974; Niggli 1978) is used as a detailed guide to the distribution of Alpine metamorphism in the Alps, while the palinspastic reconstructions (Clar 1970; Gwinner 1971; Flügel 1977; Schönlaub 1979; Trümpy 1980; Debelmas et al. 1983; Wildi 1985; Neubauer 1988; Caron et al. 1989) form the basis for the positioning of pre-Triassic relics in their original geological framework.

The delineation of *pre-Alpine* structures has required additional effort, because it does not depend solely on palinspastic reconstruction, but has also necessitated a lot of comparative analysis in the Alps and in the foreland. The analysis draws upon two centuries of observation and interpretation in the Alps, which have influenced, and also been influenced by, the evolution of geological ideas in Europe since the 18th century (Bailey 1968; Pilger 1978; Hölder 1989).

2 Classic Precursors

At a point in time when Goethe (1749–1832) jotted down in papers and letters his impressions on trips through Switzerland, there were two opposed geological schools in Europe. The Neptunists, led by the famous G. A. Werner (1749–1817), considered that all rocks, including gneiss and granite, formed in the ocean. The Plutonists, on the other hand, with J. Hutton (1726–1797) at their forefront, explained the formation of volcanic rocks through fusion of the Earth's crust, followed by the ascent of the melts to the surface. For a century, these two trends were to influence observations in the Alps.

H. B. De Saussure (1740–1799), famous after 1787 for the second ascent of the Mont Blanc, was the first to explore the Alps as a mountaineer and geologist. His *Voyages dans les Alpes* (1796–1803) was a pioneering scientific account of the whole Alpine environment. Under the influence of the Neptunist school, he distinguished between the sedimentary terrains of the Jura and those of mountains with central massifs in fan structure on which the sediments lie. His maps reflect the romantic view of threatening mountains, where Man could barely survive.

In the *Western* and *Central Alps,* early publications dealt mainly with the description of the rocks and "fossils" (minerals and fossils), including the recognition of the Mont Blanc granite under the name "Protogine" by Jurine (1806), which was to have far-reaching consequences. According to Ebel (1808), the Alps have a longitudinal central zone of primitive rocks such as granite, gneiss and schist followed to the N and S by a zone of limestones, and beyond them sandstones, conglomerates and marls. As a typical example, he cited the Mont Blanc Massif in contact with the limestones of the Chamonix zone to the north and with those of the Ferret zone in the south. The author did, however, acknowledge that in different parts of the Alps, these zones may be repeated several times. The primitive rocks, representative of the pre-Triassic basement of the Alps, were carefully described by Necker (1828), who had already come to the conclu-

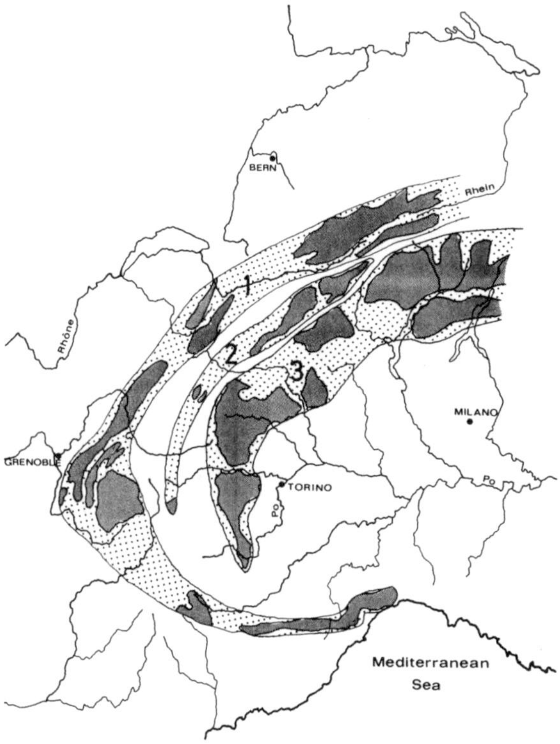

Fig. 1. Illustration of the Pre-Mesozoic basement in the Western and Central Alps, an early recognition of the Alpine division in three longitudinal zones "dauphinoise", "valaisane" and "piémontaise". (After Desor 1865 a, b)

sion that the Vallorcine granite was an intrusive body. Studer (1834) was the first to use the term "Centralmassen" for the "central masses", characterized by the presence of granites, gneisses and crystalline rocks, enveloped by sedimentary sequences. The "granite of the Alps", within these central masses, had successive envelopes of gneiss, schists and beds with "Petrefacten", and was fundamentally responsible for this zonation through the progressive transformation of sediments into schists and gneisses. The granite of the Alps was also considered to have uplifted the geological units, like trachytic domes; and thus brought to the surface the most transformed units within the bilateral structure. This interpretation was derived from the general observation that the granites were often in the central part of the central masses, the magma itself being the motor for the uplift, according to a view due to L. von Buch (1824 a), a disciple of Werner. In the course of his long journeys, he became an ardent Plutonist, stressing that the uplift of the Alps was due to magmatic activity. The geological map of the Alps by Studer (1851) shows virtually all the crystalline units, later to be considered as pre-Permian. Favre (1867) greatly amplified these geological and petrographic observations, and was the first to conclude that the crystalline basement of the central masses was unconformably overlain by sedimentary units (Aiguille de Belvedère, Aiguilles Rouges). This crystalline basement was considered to be composed of gneiss, the "Protogine" being representative of the oldest units, but many other rocks were described in detail. Desor (1865 a, b) made an important addition by showing on his map a well-differentiated distribution of the crystalline massifs into three longitudinal zones "dauphinoise", "valaisane" and "piémontaise", an early recognition of the Alpine division of pre-Triassic units (Fig. 1). Moreover, he made the interesting comment that the gneisses and schists of the crystalline massifs could represent the metamorphic equivalents of Devonian and Silurian rocks of the Graz area, and that they could even be older. There followed the great regional monographs by Theobald (1866) on the Grisons and by Gerlach (1871) on the Valais where he very rapidly mapped a considerable portion of the area, contributing abundant data on the distribution of the main rock types. Taramelli (1880) described the geological structures of the Tessin in a volume richly illustrated with etchings, which became famous amongst geologists of that time. The Central Massifs were investigated by von Fritsch (1873) for the Gotthard area and by Baltzer (1888) and von Fellenberg (1893) for the Aar Massif.

In the *Eastern Alps,* the geological investigations started with early descriptions of crystalline rocks in the late decades of the 18th century (Hacquet 1784) and the publication of geological maps began in the early 19th century (Unger 1843, von Morlot 1847). It was recognized early that the central zone of the Eastern Alps is composed of cupolas of crystalline rocks, mainly gneiss, which are surrounded by an inner rim of crystalline rocks like micaschist and marble, and an outer, higher rim mainly of quartz-phyllites ("transitional slate" and "transitional limestone") associated with fossiliferous Palaeozoic rocks. The Palaeozoic fossils are found in a high position within this framework: the Carboniferous was recognized by von Buch (1824 b), the Devonian by Unger (1843), the Silurian by von Hauer (1847), and the Ordovician by Stache (1874). Consequently, the rocks of the crystalline cupolas were interpreted as the oldest, having been covered by the "transitional" rocks during a later period. The recognition of fossiliferous Palaeozoic rocks in the Eastern Alps led to a different meaning for the term "Altkristallin", compared with the usage of French and Swiss geologists. While they used the term for all pre-Permian crystalline rocks in their parts of the Alps, in the Eastern Alps it came to mean supposedly pre-Palaeozoic rocks. Suess (1894) clarified the matter by stating that "petrographic dissimi-

a

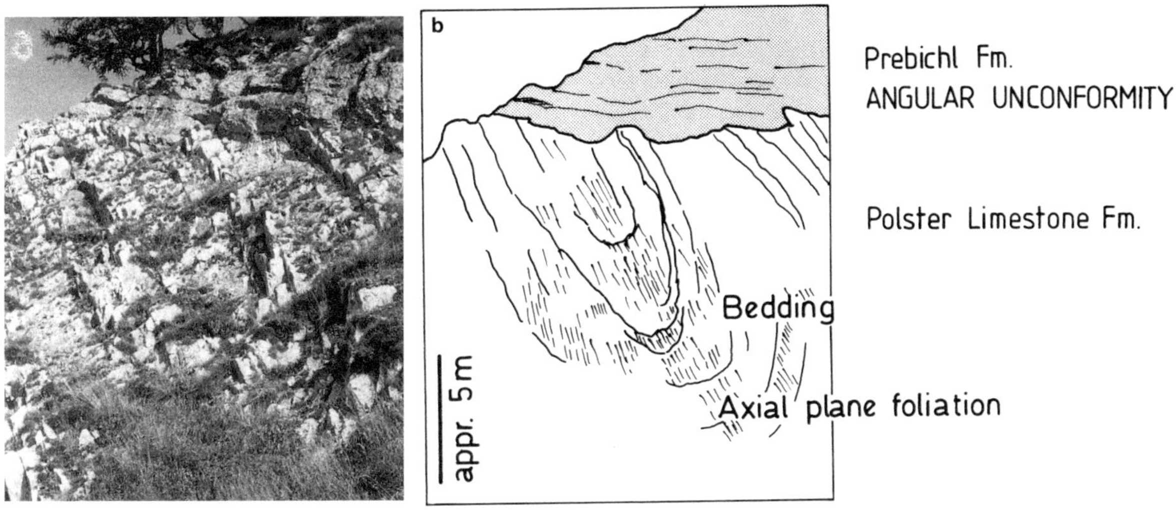

Fig. 2 a, b. Angular unconformity between Devonain limestones and Permian breccias due to the Late Carboniferous folding by the area of Polster, eastern Greywacke zone, Austria (Neubauer 1989)

larities are no guarantee for different age . . . A careful investigation might explain that such rocks are coeval formations in different metamorphic facies . . .". The discovery of an angular unconformity between rocks of the Greywacke zone in the footwall and Permo-Mesozoic rocks of the Northern Calcareous Alps in the hanging wall (Oestreich 1900; Fig. 2) clearly proves the difference between an Alpine evolution and a pre-Permian one.

At the end of the 19th century, research became more thorough, because Zirkel's adaptation of the microscope (1863) allowed precise, detailed and systematic petrographic description. Michel-Levy (1890) dispensed with any preconceived ideas linked to the term "Protogine", and definitely established the intrusive nature of the Mont Blanc granite. Mrazec (1892) published his doctoral thesis on the same granite, followed by Duparc and Mrazec's synthesis of their extensive work in the Mont Blanc Massif (1898). In these classic works, the reader is already confronted with the controversy on the interpretation of the "granulitic quartz" (i. e. microgranular quartz). For the latter authors, this form of quartz represents a "second magmatic introduction", while Grubenmann (1892) considers it to be due to "dynamometamorphism", i. e. recrystallization now attributed to Alpine metamorphism.

3 The 20th Century

After this long initial period, the 20th century brought new ideas on the formation of the Alps. Suess (1875) opposed the theory of uplift of the Alps, and instead explained their evolution through the contraction of the Earth leading to tangential movements. At about the same time, Heim (1871, 1878) came to similar conclusions in the Tödi area. Bertrand (1884) was the first to talk of a "masse de recouvrement" and in 1898 coined the term "nappe de recouvrement". Schardt (1893) initiated the use of the term "klippen" in relation to nappe tectonics, while Lugeon (1902, 1904) proposed the Helvetic domain as an example of the latter. Termier (1903) postulated the existence of large-scale nappes in the Eastern Alps and related the basement to two different tectonic units, the deeper Penninic unit and the higher Austro-Alpine unit. He also defined the three principal zones of the Alps as the "zone interne" (south of the Schistes Lustrés), the "zone axiale des Schistes Lustrés" and the "zone externe" (north of the Schistes Lustrés), where the "Variscan chain has left relics to the south and north", composed of "gneisses and micaschists of pre-Coal measures age".

This led Schardt (1908) to establish distinctions within the "massifs cristallins", representing pre-Triassic terrains in the central part of the Alps. He defined the "massifs à structure en éventail" (massifs with a fan structure, i. e. today's external massifs), and the "massifs en dôme ou nappes", corresponding to the "zone de la Dent Blanche", "zone gneissique du Mont-Rose", "zone du Grand Saint-Bernard", "nappe des gneiss du Monte Leone", "gneiss de Lebendun", "gneiss d'Antigorio", "gneiss de Crodo" and the "massifs gneissiques du Tessin", all these units being separated from one another by metamorphosed Mesozoic sediments. Detailed and precise mapping led to the brilliant publications in the "Matériaux de la carte géologique suisse" with accompanying maps (Dufour map and the Swiss Geological Atlas). The scientific journals of the five Alpine countries now included so many accounts of abundant observations made in the Alps that it is impossible to give an exhaustive list of them here.

The evolution of views on the Alps may be assessed by considering several great syntheses, one of the early ones being due to Suess (1909). At that time, the concept of nappe tectonics had made a deep impression, which is strongly reflected in Suess' writings. The pre-Alpine units are considered to be a basement, visible in four zones. "The massifs which constitute the Mont-Blanc zone are uplifted parts of the basement of the Alps, which have been folded with the latter". These massifs form the "Delphino-Savoie crystalline zone and the Mercantour massif". In a more internal position there follows the "axial subzone houillère", the "oriental or Vanoise subzone" and the "gneissic massifs and nappes of the Piémont zone". To the south lies the "Ivrea amphibolic zone". Two important syntheses were published almost at the same time. On the one hand, Kober (1923) considered the main granitic and gneissic masses of the Helvetic, Penninic and Lower Austro-Alpine domains to be the relics of Palaeozoic orogenic events with attendant metamorphism and deformation. On the other hand, Heim (1922) interpreted all the known elements at that time, adding many detailed observations, which remain today of great value. For him, the Crystalline basement ("Altkristallin") of the autochthonous, Penninic, Austro-Alpine and South-Alpine zones is composed of crystalline schists, whose intense metamorphism is related to a Caledonian or pre-Carboniferous orogenic event. He distinguishes a metamorphism with deformation of Lower Palaeozoic age, involving Archaean and Lower Palaeozoic rocks and a younger "Hercynian" episode of metamorphism and deformation. The conclusions of these monographs were combined in the Geological Guide of Switzerland by the Swiss Geological Society (1934) on the occasion of its 50th anniversary.

4 The Birth of Present-Day Methods

The need for more exact descriptions and for interpretation of mineralogical and structural observations on different scales has led to more intensive structural analysis. Sander (1911) began to systematically investigate the relationship between partial deformation and resulting textures. The "method of petrographic-tectonic analysis" (Sander 1923, 1925, 1926) was expanded into his famous "structural analysis of rocks" (Sander 1930, 1948). This became an essential tool of the Alpine geologist, and one of the supporting elements of Argand's geometrical approach (Argand 1911) through the statistical representation of observed tectonic features on different scales. Additional physico-chemical data and considerations led to a new generation of publications dealing with increasingly sophisticated structural petrology. While the foremost aim of most of these papers was a better definition of Alpine events, they indirectly afforded detailed information on the pre-Triassic basement. Oulianoff (1953) described examples of Alpine fold structures superimposed obliquely across the folded pre-Alpine basement (Ramsay 1967). Trümpy (1958) acknowledges the importance of pre-Permian events, as "there is no reason to exclude the region of the future Alps from the Hercynian orogeny which affected central and southern Europe".

During the time before World War I as well during the years between World War I and II, new ideas came from palaeontology to resolve the major structures, mainly by the Austrian school of F. Heritsch, Graz. Whereas the structures of large areas were resolved by use of macrofossils, basically valid up to recent times, the dating by fossils in quartz-phyllites and marbles, indicating Palaeozoic, Cambrian to Devonian formations, was rejected after World War II. Since that time, much progress has been made using conodonts, palynomorpha and sporomorpha. Since the early paper to Flügel and Ziegler (1957), more than 200 papers have led to a refinement of stratigraphy by the conodont method applied to carbonates (school of H. W. Flügel, Graz, mainly by Ebner, Flajs, Schönlaub (for a review, see Schönlaub 1979), and the school of G. B. Vai, Bologna). The conodont biostratigraphy provided evidence for continuous late Ordovician to Namurian A sedimentary sequences in classic Palaeozoic carbonate areas as well as in quartz-phyllite regions of the Eastern Alps and displaced the frontier of research to metamorphic rocks. Correspondingly, the work with macrofossils diminished dramatically, although the importance of macrofossils for large-scale, biogeographic relationships is recognized. Additionally, palynomorpha and sporomorpha stratigraphy was used to date siliciclastic metamorphic rocks of greenschist facies grade. This method provided evidence of Late Proterozoic rocks in the Penninic basement of the Eastern Alps, and the presence of early Ordovician in the South-Alpine and Austro-Alpine Palaeozoic sequences.

Geochronological methods have been used since the pioneer work of Jäger and Faul (1959) and Krummenacher and Evernden (1960). They published the first K/Ar and Rb/Sr data on pre-Triassic events in the Variscan basement of the Alps, followed by U/Pb ages from the Gotthard Massif (Grünenfelder 1962) and by the first Precambrian ages (Grünenfelder et al. 1964; Pasteels 1964). The geochronological evidence of Late Ordovician to Silurian magmatism and tectonothermal events in crystalline rocks (Grauert 1966) brought a new component into the discussion of the age of the Austro-Alpine basement and about the nature of the so-called Caledonian event, a discussion continuing up to recent years. Ménot (1987) was the first to determine the Cambrian age of the Chamrousse ophiolite and the Devonian age of the metavolcanic series of the Belledonne Massif, and Guillot et al. (1991) give the first Cambrian age on the monometamorphic rock series from the Vanoise region. A Pan-African age for the "old basement" was definitely confirmed in the Oetztal by Söllner and Hansen (1987). At present, analysis by the SRIMP method of single zircon grains reveals ages up to 3.5 b. y. (Gebauer et al. 1988) and allows distinction between magmatic and metamorphic periods.

Much work has been done on "geodynamics by chemical means" since the early work of Lodemann (1970). Although the discussion of the reliability is continuing, the broad use of geochemical properties shows remarkable differences in the composition of volcanic sequences as well as in metamorphic rocks. Combined with fieldwork and petrographic observations very different geodynamic settings for different basement sequences have been proposed over the last decade. As an early example, Stille (1980), in this manner, was able to interpret the amphibolites of the Berisal unit and of the St. Bernard nappe *s. l.* as having been derived from a Lower Palaeozoic island arc.

All research on the basement had been prepared by the better definition of Alpine overprint. Niggli (1960) published the first map of Alpine metamorphic zones, adding that pre-Alpine events could be of Palaeozoic or even Precambrian age, while Bearth (1963) came to similar conclusions in his study of the polymetamorphic Monte Rosa nappe. Simultaneously, Wenk (1962) discusses Alpine reactivation of the pre-Triassic basement of the Central Alps, where "the Variscan chain has been transformed through endogenic processes into Alpine crystalline basement". This gradual transformation due to increasing Alpine metamorphism is expressed by progressive recrystallization at the microscopic scale (Voll 1976) with attendant adaptation of crystalline phases and of the fluid components (Frey et al. 1980). The spectacular behaviour of the pre-Triassic basement during the Alpine orogenesis at the scale of the massifs has been convincingly demonstrated by Steck (1984) and illustrated in a transversal cross-section of the Alps by Escher et al. (1988) and corresponding geological map of the Penninic domain Escher (1988). Close collaboration between very different fields thus led to a modern interpretation of the Alpine edifice with its pre-Triassic relics. There remained the problem of matching these relics with the neighbouring Variscan basement.

5 An Element in the European Variscan Orogen

Detailed observation in the Alps over 200 years did not prevent Alpine geologists from casting their eyes on domains outside the Alps, acknowledging, consciously or not, that the relics of the Variscan basement of the Alps could not be detached from the Variscan European frame.

Ebel, as early as 1808, used the same colour for the Alpine crystalline massifs and outcrops of the neighbouring Variscan basement. We have referred at the beginning of this overview to Desor's work and suggestion that Pelaeozoic units in the Graz area and certain strongly metamorphosed rocks in the Western Alps might be equivalent (Desor 1865 a, b). This observation illustrates the main difference between Western and Eastern Alps. It is therefore not surprising that the Graz Palaeozoic units were to incite geologists to pursue the continuation of the Variscan edifice in the Eastern Alps.

Following the ideas on Variscan zonation of Suess (1909) and Kossmat (1927), Kober (1927, 1928) extended the Variscan palaeotectonic zones into the "Antivariszikum" of the Eastern Alps. Clar (1970) thought that one might find the continuation of the Sudetic-Moravian structures in the Eastern Alps. Tollmann (1982) links the Bohemian Massif with the Eastern Alps through Variscan nappes of great magnitude, a proposal to which Frisch et al. (1984) have applied a plate tectonic interpretation. For part of this region, Neubauer (1988) defined an orogenic zonal structure from an external flysch zone in the south to metamorphic internal zones in the north, on the basis of a Lower Carboniferous orogenic cycle with SE facing. Flügel et al. (1987) brought together the data concerning the Alpine and Mediterra near domains. Biogeographic relationships of the Eastern Alps to Urals and Tienshan confirm "travelled terranes" which docked the Variscides during the Carboniferous (Frisch and Neubauer 1989; Flügel 1990).

In contrast the western part of the Alps presented major difficulties for an attempt at correlation. Schwinner (1933) used geophysical data to define N-S structural trends common to southern Germany and the Alpine domain. On the other hand, Stille (1951) suggested a continuation of the NE/SW zones drawn by Kossmat (1927) and proposed the term "Antirheniden" for the south-facing Eastern Alps in reference to the Rhenoherzynikum to the north of the supposedly bilateral Variscan orogen. He was convinced that parts of the pre-Triassic basement had at least been affected by strong deformation of Variscan age. Faul (1962) considered the Variscan units of the Alps to be small entities with an evolution of isotopic ages common to the Variscan basement of Great Britain, Central Europe, including the Alps, and Asia. Krummenacher et al. (1965) are the first to discuss the resemblance between certain "series" of the external massifs and of the Massif Central. Similarly, von Raumer and Menot (1989) correlated the "granulitic and eclogitic gneisses" of the Massif Central, Vosges, Black Forest and external massifs, considering this lithological association to be a link between the Variscan foreland and the Alpine massifs, probably through Variscan nappes of considerable magnitude. An evolution in several phases for the Western Alps, starting in the Precambrian, is discussed by Laurent and Chessex (1969). Le Fort and Ehrström (1969) compare the evolution of the volcaniclastic series of the Chaillol (Belledonne) with that of series in the Vosges, and Carme (1971, 1975) correlates the structural evolution of the Belledonne Massif and the Cevennes, and the geochemical evolution of magmatic rocks from Belledonne and the Monts du Lyonnais. Jäger (1971, 1979) sees no difference between the isotopic evolution of Variscan relics in the Alps and in the European Variscides at large: this clearly indicates to what degree this identity is accepted.

All these general discussions led also to detailed comparisons between the Variscan relics in different Alpine zones. Dal Piaz et al. (1972), Boriani et al. (1974) and Dal Piaz et al. (1975) defined the main trends, while in the Helvetic domain a first attempt at comparison in the external massifs was undertaken by von Raumer (1976). Mercolli and Oberhänsli (1988), considering the domain between the Austro-Alpine, the Southern Alps and the external massifs, suggested correlations on a larger scale and a coherent magmatic evolution corresponding to different stages in the plate tectonic situation since the Lower Carboniferous. Frisch et al. (1987) and Frisch and Neubauer (1989) combined the geochemical characteristics of the basic rocks with "lithological associations". This lithostratigraphy, in which each lithological association corresponds to a specific evolution in time and in palaeotectonics, led to a reconstruction of the main trends of the palaeotectonic history of the Eastern Alps.

The establishment of the close link between the evolution of thermal gradients and regional metamorphism (Oxburgh and Turcotte 1974), and the integration of this evolution in a P-T path (Thompson 1981), further developed our understanding of the geological evolution. The present research is thus focussed on the age and nature of protoliths of different lithotectonic units and their P-T-time-deformation paths.

Recent years thus have seen the emergence of a sharp picture of Caledonian-Variscan structures in the Alps (von Raumer 1988). Frisch et al. (1990) prepared a first global view of basement evolution for the whole Alps, which should be refined by the works in this volume. Thus, two centuries of observation have given us an idea of the evolution of the basement in the Alps, but we must still acknowledge that our endeavours represent but small steps in this quest to understand.

Acknowledgement

We express our warmest thanks to St. Ayrton (Lausanne), who translated an early French version of this paper, and to R. Mason (London), who spent much time giving a final polish to the English version.

References

Argand E (1911) Les nappes de recouvrement des Alpes Pennines et leurs prolongements structuraux. Beitr Geol Karte Schweiz NF 31: 1–26

Argand E (1916) Sur l'arc des Alpes Occidentales. Eclogae Geol Helv 14: 145–191

Bailey EB (1968) Tectonic essays. University Press, Oxford

Baltzer A (1888) Der mechanische Contact von Gneiss und Kalk im Berner Oberland. Beitr Geol Karte Schweiz 20: 1–255

Bearth P (1963) Contribution à la subdivision tectonique et stratigraphique du cristallin de la nappe du Grand-St-Bernard dans le Valais (Suisse). Livre Mém Prof Paul Fallot Soc Géol Fr Mém h s (1960–1963), tome II: 407–418

Bertrand M (1884) Rapports de structure des Alpes de Glaris et du bassin Houiller du Nord. Bull Soc Géol Fr (3) 12: 318–330

Boriani A, Dal Piaz GV, Hunziker JC, von Raumer JF, Sassi FP (1974) Caratteri, distribuzione ed etá del metamorfismo prealpino nelle Alpi. Mem Soc Geol Ital XIII A: 165–225

von Buch L (1824a) Geognostische Gemälde von Südtirol. Leonhardts Mineral Taschenb 18

von Buch L (1824a) Über die karnischen Alpen. Leonhards Mineral Taschenb 18: 396–437

Carme F (1971) Donnés microstructurales pour une corrélation possible entre les Massifs Cristallins Externes alpins et les

Cevennes Septentrionales. C R Acad Sci Paris 273D: 2038–2041

Carme F (1975) Caracteres chimiques de deux ensembles de roches magmatiques préorogéniques du socle anté-houiller des Massifs Cristallins Externes alpins et des Monts du Lyonnais. C R Acad Sci Paris 281 D: 89–92

Caron Ch, Homewood P, Wildi W (1989) The original Swiss flysch: a reapprisal of the type deposits in the Swiss realm. Earth Sci Rev 26: 1–45

Clar E (1970) Bemerkungen für eine Rekonstruktion des variszischen Gebirges in den Ostalpen. Z Dtsch Geol Ges 122: 161–167

Dal Piaz GV, Hunziker JC, Martinotti G (1972) La zona Sesia-Lanzo e l'evoluzione tettonico-metamorfica delle alpi nordoccidentali interne. Mem Soc Geol Ital XI: 433–466

Dal Piaz GV, von Raumer JF; Zanettin E, Zanferrari A (1975) Geological outline of the Italian Alps. In: Squyres C (ed) Geology of Italy. Earth Sci Soc of the Lybian Arab Republic, Tripoli, pp 299–375

Debelmas J, Escher A, Trümpy R (1983) Profiles through the Western Alps. Am Geophys Union, Geodyn Ser 10: 83–96

Desor E (1865a) Der Gebirgsbau der Alpen. CW Kreidel, Wiesbaden, pp 1–151

Desor E (1865b) Sur la disposition des massifs cristallins des Alpes ou zones d'affleurement. Bull Soc Géol Fr (2) 22: 364–360

Duparc L, Mrazec L (1898) Recherches géologiques et pétrographiques sur le Massif du Montblanc. Mém Soc Phys Hist Nat Geneve XXXIII: 1–227

Ebel JG (1808) Über den Bau der Erde in dem Alpen-Gebirge. Orell Fuessli, Zürich

Elter G (1960) La zona pennidica dell'alta e media Valle d'Aosta e le unita limitrofe. Mem Ist Geol Mineral Univ Padova XXII: 1–114

Escher A (1988) Structure de la nappe du Grand Saint-Bernard entre le Val de Bagnes et le Mischabel. Rapp Geol Serv Hydrol Geol Nat 7: 1–26

Escher A, Masson H, Steck A (1988) Coupes géologiques des Alpes Occidentales Suisses. Rapp Geol Serv Hydrol Geol Natl 2: 1–11

Faul H (1962) Age and extent of the Hercynian complex. Geol Rundsch 52: 767–781

Favre A (1867) Recherches géologiques dans les parties de la Savoie du Piemont et de la Suisse voisins du Mont Blanc. Masson, Geneve

von Fellenberg E (1893) Geologische Beschreibung des westlichen Theils des Aarmassivs, enthalten auf dem nördlich der Rhone gelegenen Theile des Blattes XVIII der Dufourkarte. Beitr Geol Karte Schweiz 21: 1–367

Flügel HW (1977) Paläogeographie und Tektonik des alpinen Variszikums. Neues Jahrb Geol Palaeontol Monatsh 11: 659–674

Flügel HW (1990) Das voralpine Basement im Alpin-Mediterranen Belt - Überblick und Problematik. Jahrb Geol Bundesanst 133: 181–221

Flügel HW, Ziegler W (1957) Die Gliederung des Oberdevons und Unterkarbons am Steinberg westlich von Graz mit Conodonten. Mitt Naturwiss Ver Steiermark 87: 25–60

Flügel HW, Sassi FP, Grecula P (1987) Pre-Variscan and Variscan events in the Alpine-Mediterranean mountain belts. Alfa, Bratislava, pp 1–487

Frey M, Hunziker JC, Bocquet J, Dal Piaz GV, Jäger E, Niggli E (1974) Alpine Metamorphism of the Alps. Schweiz Mineral Petrogr Mitt 54: 247–290

Frey M, Bucher K, Mullis J (1980) Alpine metamorphism along the geotraverse Basel-Chiasso: A review. Eclogae Geol Helv 73: 527–546

Frisch W, Neubauer F (1989) Pre-Alpine terranes and tectonic zoning in the Eastern Alps. Geol Soc Am Spec Pap 230: 91–99

Frisch W, Neubauer F, Satir M (1984) Concepts of the evolution of the Austroalpine basement complex (Eastern Alps) during the Caledonian-Variscan cycle. Geol Rundsch 73: 47–68

Frisch W, Neubauer F, Bröcker M, Brückemann W, Haiss N (1987) Interpretation of geochemical data from the Caledonian basement within the Austroalpine basement complex. In Flügel HW, Sassi FP, Grecula P (eds) Pre-Variscan and Variscan events in the Alpine-Mediterranean mountain belts. Alfa, Bratislava, pp 209–226

Frisch W, Ménot RP, Neubauer F, von Raumer JF (1990) Correlation and evolution of the Alpine basement. Schweiz Mineral Petrogr Mitt 70: 265–285

von Fritsch K (1873) Das Gotthardgebiet. Beitr Geol Karte Schweiz 15: 1–154

Gebauer D, Quadt A, Compston W, Williams I, Grünenfelder M (1988) Archaische Komponenten in retrograden Eklogiten des Gotthardmassivs. Schweiz Mineral Petrogr Mitt 68: 485–490

Gerlach H (1871) Das südwestliche Wallis. Beitr Geol Karte Schweiz 9: 1–175

Grauert B (1966) RB-Sr age determinations on orthogneisses of the Silvretta (Switzerland). Earth Planet Sci Lett I: 139–142

Grubenmann U (1892) Über Gesteine des granitischen Kerns im östlichen Teil des Gotthardmassivs. Mitt Thurgau Naturforsch Ges 1892, 10: 125–144

Grünenfelder M (1962) Mineralalter von Gesteinen aus dem Gotthardmassiv. Schweiz Mineral Petrogr Mitt 42: 6–7

Grünenfelder M, Hofmänner F, Grögler N (1964) Heterogenität akzessorischer Zirkone und die petrographische Beurteilung ihrer Uran/Blei-Zerfallsalter II Präkambrische Zirkonbildung im Gotthardmassiv. Schweiz Mineral Petrogr Mitt 44: 543–558

Guillot F, Liégois JP, Fabre J (1991) Des granophyres du Cambrien terminal dans le Mont Pourri (Vanoise, zone briançonnais): première datation U-Pb sur zircon d'un socle des zones internes des Alpes françaises. C R Acad Sci Paris 313 II: 239–244

Gwinner M (1971) Geologie der Alpen. E Schweizerbarth, Stuttgart, pp 1–477

Hacquet B (1784) Mineralogisch-botanische Lustreise, von dem Berg Terglou in Krain, zu dem Berg Glokner in Tyrol, im Jahr 1779 und 81, 2nd edn. Kraus, Vienna, 149 pp

von Hauer F (1847) Versteinerungen von Dienten in Salzburg. Heidingers Ber 1: 10

Heim A (1871) Geologische Übersicht der Windgällen-Tödigruppe. Jahrb Schweiz Alpenclub 7: 385–414

Heim A (1878) Untersuchungen über den Mechanismus der Gebirgsbildung im Anschluß an die geologische Monographie der Tödi-Windgällen-Gruppe. Benno Schwab, Basel 1878, Vol 1, 345 pp

Heim A (1922) Geologie der Schweiz, Vol 2.2 Tauchnitz, Leipzig

Hölder H (1989) Kurze Geschichte der Geologie und Paläontologie. Springer, Berlin Heidelberg New York, 244 pp

Jäger E (1971) Die Geschichte des alpinen Raumes erarbeitet mit radiometrischen Altersbestimmungen. Verh Geol Bundesanst 87: 250–254

Jäger E (1979) Evolution of the European continent. In: Jäger E, Hunziker JC (eds) Isotopic geology. Springer, Berlin Heidelberg New York, pp 222–224

Jäger E, Faul H (1959) Age measurements of some granites and gneisses from the Alps. Bull Geol Soc Am 70: 1553–1558
Jurine (1806) Lettre à Monsieur Gillet-Laumont. J Mines 19 (113): 367–378
Kober L (1923) Bau und Entstehung der Alpen. Borntraeger, Berlin, pp 1–280
Kober L (1927) Das alpin-varistische Orogen. Centralbl Mineral Geol Palaeontol 1927 B: 153–159
Kober L (1928) Versuch einer Auflösung des varistisch-alpinen Baues. Centralbl Mineral Geol Palaeontol 1928 B: 76–82
Kossmat F (1927) Gliederung des varistischen Gebirgsbaues. Abh Saechs Geol Landesamt 1: 1–39
Krummenacher D, Evernden JF (1960) Déterminations d'age isotopique sur quelques roches des Alpes par la methode Potassium-Argon. Schweiz Mineral Petrogr Mitt 40: 267–277
Krummenacher D, Bordet P, Le Fort P (1965) Les massifs externes et leur series métamorphes, problemes de raccords stratigraphiques et géochronométriques. Schweiz Mineral Petrogr Mitt 45: 856–874
Laurent R, Chessex R (1969) Considerations sur le paléozoique dans les Alpes Occidentales. Eclogae Geol Helv 6: 11–18
Le Fort P, Ehrström CE (1969) Caracteres géochimiques et origine des formations métamorphiques des massifs de Chaillol et de Belledonne. Sci Terre 14: 215–238
Lodemann CKW (1970) Geochemie der Metamorphose im Saualpen-Kristallin (Ostkärnten). Neues Jahrb Geol Paläontol Abh 112: 188–215
Lugeon M (1902) Les grandes nappes de recouvrement des Alpes du Chablais et de la Suisse. Bull Soc Géol Fr (4) 1: 723–825
Lugeon M (1904) Les grandes nappes de recouvrement des Alpes Suisses. C R IX. Int. Geol Congr Wien 1: 477–492
Ménot RP (1987) Magmatismes paléozoiques et structuration carbonifere du Massif de Belledonne (Alpes Françaises). Mém Doc Centre Armoricain Etudes Structurales des Socles Rennes 1988 21: 1–465
Mercolli I, Oberhänsli R (1988) Variscan tectonic evolution in the Central Alps. Schweiz Mineral Petrogr Mitt 68: 491–500
Michel-Levy A (1890) Etude des roches cristallines et eruptives des environs du Mont Blanc. Bull Carte Géol Fr 1 (9): 1–26
von Morlot A (1847) Erläuterungen zur geologischen Übersichtskarte der nordöstlichen Alpen. Braumüller, Vienna, 213 pp
Mrazec L (1892) La protogine du Mont-Blanc et les roches eruptives qui l'accompagnent. F Taponnier, Geneve, pp 1–91
Necker LA (1828) Memoire sur la vallée de Valorsine. Mém Soc Phys Hist Nat Geneve IV: 1–37
Neubauer F (1988) Bau und Entwicklungsgeschichte des Rennfeld-Mugel und des Gleinalm-Kristallins (Ostalpen). Abh Geol Bundesanst 42: 1–137
Neubauer F (1989) Late Variscan structures of the Eastern Greywacke Zone (Eastern Alps). Neues Jahrb Geol Palaeontol Monatsh 1989: 425–432
Niggli E (1960) Mineralzonen der alpinen Metamorphose in den Alpen. Int Geol Congr Rep, 21st Sess Kopenhagen 13: 132–138
Niggli E (1978) Metamorphic map of the Alps. In: UNESCO (ed) Metamorphic belts of the world. UNESCO, Paris, pp 181–242
Oestreich K (1900) Ein alpines Längstal zur Tertiärzeit. Jahrb Geol Reichsanst 49: 165–212
Oulianoff N (1953) Superposition successive des chaînes de montagnes. Scientia 47: 1–5
Oxburgh ER, Turcotte DL (1974) Thermal gradients and regional metamorphism in overthrust terrains with special reference to the Eastern Alps. Schweiz Mineral Petrogr Mitt 54: 641–662
Pasteels P (1964) Mesures d'âges sur les zircons de quelques roches des Alpes. Schweiz Mineral Petrogr Mitt 44: 519–541
Pilger A (1978) Die tektonische Erforschung der Alpen zwischen 1787 und 1915. Clausth Geol Abh 32: 1–81
Ramsay JG (1967) Folding and fracturing of rocks. McGraw-Hill, New York, 568 pp
von Raumer JF (1976) Variszikum in den Zentral- und Westalpen. Nova Acta Leopold NF 224: 119–144
von Raumer JF (1988) Caledonian-Variscan structures in the Alps: an introduction. Schweiz Mineral Petrogr Mitt 68: 291–299
von Raumer JF, Ménot RP (1989) Evolution paléozoique du socle ouest-alpin et place des Massifs Cristallins Externes dans l'orogene Varisque. C R Acad Sci Paris 309 II: 397–402
Sander B (1911) Geologische Studien am Westende der Hohen Tauern I. Ber Akad Wiss Wien Denkschr 82: 257–319
Sander B (1923) Zur petrologisch-tektonischen Analyse I. Jahrb. Geol Reichsanst 75: 183–257
Sander B (1925) Zur petrologisch-tektonischen Analyse II. Jahrb Geol Reichsanst 75: 181–236
Sander B (1926) Zur petrologisch-tektonischen Analyse III. Jahrb Geol Reichsanst 76: 323–404
Sander B (1930) Gefügekunde der Gesteine. Springer, Berlin Heidelberg New York, 352 pp
Sander B (1948) Einführung in die Gefügekunde geologischer Körper. Springer, Wien
Saussure de HB (1779) Les Voyages dans les Alpes. S Fauche, Neuchâtel
Schardt H (1893) Sur l'origine des Préalpes Romandes. Eclogae Geol Helv 4: 129–142
Schardt H (1908) Géologie de la Suisse. Publications du dictionnaire geographique suisse, Neuchâtel, pp 1–58
Schönlaub HP (1979) Das Paläozoikum in Österreich. Abh Geol Bundesanst 33: 1–124
Schwinner R (1933) Variszisches und alpines Gebirgssystem. Geol Rundsch 24: 144–159
Söllner F, Hansen BT (1987) "Pan-afrikanisches" und "kaledonisches" Ereignis im Ötztalkristallin der Ostalpen. Jahrb Geol Bundesanst 130(4): 529–569
Stache G (1874) Über die Silurbildungen der Ostalpen, nebst Bemerkungen über die Devon-Karbon- und Permschichten dieses Gebietes. Z Dtsch Geol Ges 36: 277–378
Steck A (1984) Structures de deformations tertiaires dans les Alpes Centrales. Eclogae Geol Helv 77: 55–100
Stille H (1951) Das mitteleuropäische Variszische Grundgebirge im Bilde des gesamteuropäischen. Geol Jahrb Beih 2: 1–138
Stille P (1980) On the genesis of the amphibolites and the hornblendefelses in the Berisal complex. Mem Sc Geol XXXIV: 205–246
Studer B (1834) Geologie der westlichen Schweizer Alpen. K Gross, Heidelberg
Studer B (1851) Geologie der Schweiz. Stämpfli, Bern
Suess E (1875) Die Entstehung der Alpen. Braumüller, Wien
Suess E (1909) Das Antlitz der Erde. Braumüller, Wien
Suess FE (1894) Das Gebiet der Triasfalten im Nordosten der Brennerlinie. Jahrb Geol Reichsanst 44: 589–670
Taramelli T (1880) Il Canton Ticino meridionale ed i paesi finitimi. Beitr. Geol Karte Schweiz 17: 1–231
Termier P (1903) Les nappes des Alpes orientales et la synthese des Alpes. Bull Soc Géol Fr 4: 711–765

Theobald G (1866) Geologische Beschreibung der in Blatt XX des eidgen Atlasses enthaltenen Gebirge. Beitr. Geol Karte Schweiz 3: 1–359

Thompson AB (1981) The pressure-temperature (P-T) plane viewed by geophysicists and petrologists. Terra Cognita 1: 10–20

Tollmann A (1982) Großräumiger variszischer Deckenbau im Moldanubikum und neue Gedanken zum Variszikum Europas. Geot Forsch 64: 1–91

Trümpy R (1958) Remarks on the pre-orogenic history of the Alps. Geol Mijnbouw 20: 340–352

Trümpy R (1980) An outline of the Geology of Switzerland, part A. Wepf, Basel, pp 7–51

Unger (1843) Geognostische Skizze der Umgebungen von Grätz. In: Schreiner G (ed) Grätz. Ferstl, Graz, pp 69–82

Voll G (1976) The crystallization of quartz, biotite and feldspar from Erstfeld to the Bernhard Nappe, Swiss Alps, and its geological significance. Schweiz Mineral Petrogr Mitt 56: 641–647

Wenk R (1962) Das reaktivierte Grundgebirge der Zentralalpen. Geol Rundsch 52: 754–766

Wildi W (1985) Heavy mineral distribution and dispersal pattern in penninic and ligurian flysch basins (Alps, Northern Apennines). G Geol 3, 47: 77–99

Zirkel F (1863) Mikroskopische Gesteinsstudien. Sitzungsber Akad Wiss Wien Math.-naturwiss Kl 47(1): 226–270

Stratigraphy, Biogeography and Climatic Relationships of the Alpine Palaeozoic

H. P. SCHÖNLAUB

Abstract

The combined evidence from palaeomagnetism, climate-sensitive sediments and biogeography suggests that the plates on Earth have been in constant movement relative to the poles and relative to each other during the Phanerozoic and probably during older periods as well. Since the palaeogeographic position of the Alps during the Palaeozoic is not well-constrained by palaeomagnetic data, the best evidence to infer the palaeolatitudinal framework and to reconstruct the complex geological history for this area is provided by lithic palaeoclimatic indicators, e. g. the distribution of carbonate rocks and carbonate buildups, oolites, evaporites, coal beds, red beds, and certain fossil data.

During the Ordovician, the former supposed relationship with northern Africa decreased. Instead, the affinity to the warm water north European realm evidently became closer, suggesting that the Alps were positioned in a moderately warm climate and within the carbonate belt of not more than 45° S as opposed to the conclusions reached by Scotese and McKerrow (1990) and others.

In the Silurian, the relationship of faunas from the Eastern Alps to coeval faunas from southern, central and southwestern European regions continued. As for the Ordovician, the affinity with northern Africa was remarkably loose during the Silurian. Corals, ooids and other distinct faunal and lithic occurrences in the Eastern and Southern Alps indicate (1) a setting of about 35°S for Silurian rocks in the Alps and (2) the option of two separate terranes for this interval of time.

Similarily, during the Devonian, faunas and floras showed strong affinities wilt Hercynian-Bohemian and Uralo-Tienshan occurrences and were less closely related to northern Africa. Moreover, similarities did also exist with the Eifel Mountains, Belgium, France and Great Britain. Available faunal, floral and lithic evidences are best explained by supposing (1) a palaeolatitudinal setting within the tropical belt of some 30° or less, (2) the operation of equatorial gyres which aided the dispersal of many groups of organisms, and (3) the possible existence of at least two separate terranes during the Devonian.

In the sedimentary sequences, orogenic paroxysm was reached in the Middle Carboniferous, i. e. between the late Namurian *Gastrioceras* Zone and the Upper Miatchkovo. While Lower Carboniferous faunas of the Southern Alps are only of limited biogeographic significance, the famous "Carboniferous of Nötsch" located north of the Periadriatic line appears to be of great interest (Fig. 1). Its Early Namurian fossil assemblage is closely related to the Western Euro-

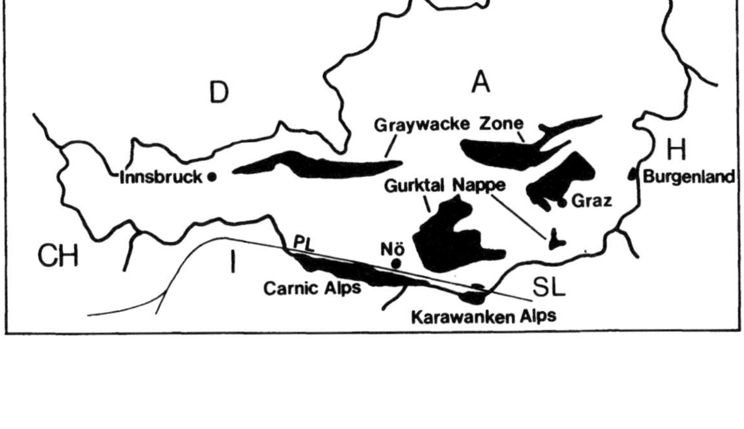

Fig. 1. Main regions with fossiliferous Palaeozoic strata in the Eastern Alps north and south of the Periadriatic line *(PL)*. *Nö* Carboniferous of Nötsch

pean Kohlenkalk facies and comparable with coeval occurrences in Spain and North America. Furthermore, exotic limestone clasts indicate the existence of a vast shallow platform environment in this or adjacent areas of the Alps of which, however, only small remnants have been preserved. In addition to others, the Veitsch Nappe might have belonged to this shelf facies which characterized the Central Alps during the Carboniferous in contrast to the Southern Alps. Consequently, both domains may have represented two different microplates during the Carboniferous Period, and thus seem to confirm the suggested fragmentation of the predecessors of the Alps during the Lower Palaeozoic. After the Carboniferous amalgamation, the biogeographic pattern of the Southern Alps closely matched those from the former settings in the Central Alps and migration paths developed between the Southern Alps and Western Europe. The major part of the post-Variscan fauna, however, exhibits a cosmopolitan aspect with similarities to coeval occurrences along the margin of the Tethys Sea. As for the Devonian, dispersal of faunas and floras was much aided by the oceanic circulation system.

From palaeoclimatic sensitive data, it is inferred that the Central Alps were mainly confined to a humid equatorial belt during the Carboniferous, with the Southern Alps located further to the south. Progressive northward movement during the Late Carboniferous and the Permian resulted in semi-arid and arid conditions that began north of the Periadriatic line apparently in the Lower Permian and in the Southern Alps in the Middle Permian. The Alps may thus have crossed the equator at different times between the Carboniferous and the Lower Permian depending on the plate position on either side of the Periadriatic line.

1 Introduction

In the Palaeozoic Era the relative position of continents and oceans can best be constrained by the combination of palaeomagnetism, climate-related lithic data and the biogeographical distribution of fossils.

In the last two decades, it was realized that the pre-Alpine central and southern Europe represents a continental collage, formed by accretion of separate, far-travelled, suspect blocks at different times during the Phanerozoic (e.g. Burrett 1972; Riding 1974; Badham and Halls 1975; Alvarez 1976; Lorenz 1976; Rau and Tongiordi 1981; Badham 1982; Gebauer 1986; Matte 1986; Ziegler 1986, 1989; Lefort 1989; Flügel 1990).

The present comprehensive study examines the Palaeozoic sediments and the enclosed fauna and flora of the Alps and its relationship with coeval communities from adjacent regions. It is compiled from various parts of the Eastern Alps of Austria and covers a complete stratigraphic record from the Ordovician to the end of the Permian. The bulk of the material has been collected from the Carnic Alps, the remainder from Middle Carinthia, the surroundings of Graz and the Greywacke Zone of Styria and Tyrol. In other parts of the Alps, available biostratigraphic control is sparse due to lack of fossils or metamorphism and thus they are excluded from this study (Fig. 1).

2 Palaeogeographical Constraints on the Alpine Palaeozoic

Recently, it has been suggested that the pre-Alpine basement was formed by accretion of several terranes (Frisch and Neubauer 1989). In fact, the existence of many, although volumetrically insignificant occurrences of mafic and ultramafic rocks, in different tectonic units of the Eastern Alps, might correspond to remnants of small oceans which may have separated these blocks, adjacent to the north African margin sometime during the Palaeozoic (Quadt 1985, 1987; Frank et al. 1987; Frisch and Raab 1987; Frisch et al. 1987; Giese 1988; Kraiger 1989; Poli 1989; Vavra 1989). This model, however, seems poorly founded: firstly, there is clearly a lack of true ocean floor basalts as opposed to within-plate basalts. Secondly, most, if not all, of the respective sequences represent high to low grade metamorphic unfossiliferous rocks with only few radiometric data yet available (see summaries in e.g. Frisch et al. 1984, 1990; Quadt 1985, 1987; Becker et al. 1987). Consequently, their age and primary geological setting has been a matter of ongoing discussion and inferred geodynamic models have been regarded as being preliminary since their conclusions have been partly controversial (Fig. 2).

2.1 *Late Proterozoic*

According to Gebauer et al. (1989), the metasedimentary precursors of the European Hercynides were deposited after the Pan-African Orogeny, i.e. between approx. 1000 and 600 Ma. Based on detrital zircons, crust-forming events, have been dated back as far as 3.84 billion years however, and are thus much older than previously assumed from Rb-Sr systemat-

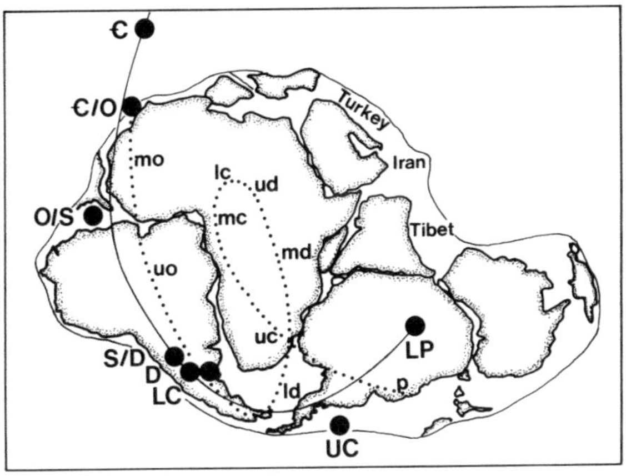

Fig. 2. Comparison between apparent polar wander path (APWP) for Gondwana during the Palaeozoic after Bachtadse and Briden 1990 *(dotted line)* and palaeoclimatically determined pole positions after Scotese and Barrett (1990) *(solid line)*. Abbreviations: *C* Cambrian; *C/O* Cambrian/Ordovician; *O/S* Ordovician/Silurian; *S/D* Silurian/Devonian; *D* Devonian; *LC* Lower Carboniferous; *UC* Upper Carboniferous; *LP* Lower Permian; *l* Lower; *m* Middle; *u* Upper. (After Scotese and Barrett 1990)

ics (Grauert and Arnold 1968; Grauert et al. 1973; Jäger 1977; Vidal 1977; Vidal et al. 1981; Söllner and Hansen 1987). The chronology of events recorded in individual zircon grains reveals a complex geological history that agrees well with the crustal evolution of the African craton and, hence, strongly supports the idea that the detrital sediments were supplied from Gondwana and not from Laurussia.

The oldest fossiliferous rocks of the Alps appear to occur in the Habach Group of the Penninic domain of the Hohe Tauern region of Salzburg (Reitz and Höll 1988). Another, probably coeval, microfossil assemblage was recently found in phyllitic rocks close to the northern margin of the Tauern window and may correspond to parts of the Habach group (Reitz et al. 1989). The latter, up to 1500 m thick, rock sequence comprises thick piles of metasediments and large volumes of metabasalts, meta-andesites and other calc-alkalic volcanics. These, according to different authors, either represent: (1) an older oceanic back-arc sheeted-dyke complex followed by subduction-related island arc volcanics in a continental setting, with argillites of the Habach phyllites on top of it (Kraiger 1989), or (2) represent an ensimatic island arc formed in an intra-oceanic environment (Frisch and Raab 1987).

The microfossil assemblage recovered from the black Habach phyllites and its equivalents are organic-walled, spheroidal acritarchs which appear to be diagnostic for the Late Riphean and Lower Vendian suggesting an age of approx. 670 Ma for the deposition of these sediments. Their restriction to this time interval, however, may be challenged due to new discoveries of large complex and ornamented acritarchs in 600 to 650 Ma old cherts and mudstones of the Pertatataka Formation in central Australia (Zang and Walter 1989). These shed new light on the evolution and biostratigraphic zonation of Late Proterozoic acritarchs (see comment by Knoll and Butterfield 1989). Moreover, new age constraints place the Riphean/Vendian boundary presently at approx. 620 million years as opposed to Haq and van Eysinga (1987).

Whether or not these findings are of climatic significance, or even display provincialism, is as yet, poorly understood in the Proterozoic. In the past, such ornamented acritarchs were reported from many localities on different plates favouring wide geographic and temporal distribution on continental plates and in a plate margin environment instead of colonizing volcanic islands in an oceanic setting as suggested by Frisch and Neubauer (1989) for the "Habach terrane".

2.2 Cambrian to Ordovician

Fossiliferous rocks of Cambrian age have yet not been recognized in the Alps. All previous reports on such occurrences were misleading since they were not based on true fossils (see Schönlaub 1979, pp. 11, 39).

Remarkably well-preserved acritarchs do, however, occur in phyllitic slates near the base of the Greywacke Zone in the vicinity of Kitzbühel, Tyrol (Reitz and Höll 1989, 1991) and in the Innsbruck quartzphyllite (Reitz and Höll 1990). They suggest an Early Ordovician age equivalent to the Tremadocian and Arenigian Series of the British succession.

The oldest megafossil assemblage of the Alps is of Upper Llandeilian age corresponding to the lower Berounian Series of Bohemia (Havlicek et al. 1987). It is derived from the locality Bruchnig on the Magdalensberg, north of Klagenfurt, Carinthia. The fossils comprise mostly brachiopods which occur in tuffaceous strata on top of basic metavolcanic and pyroclastic rocks. They represent mildly alkaline within-plate basalts which have been altered to spilites (Loeschke 1989a,b).

The second important fossil assemblage was recorded from arenaceous shales in the Carnic Alps and appears to be slightly younger, i. e. Caradocian in age. The highly diversified fauna comprises brachiopods, bryozoans, trilobites, cystoids and very rare hyolithes (Schönlaub 1971, 1988; Vai 1971;

Marek 1976; Vai and Spalletta 1980; Havlicek et al. 1987).

Interestingly, these two fossil sites, located to the north and the south of the Periadriatic line, differ significantly from coeval "typical" Mediterranean cold water associations, i. e. those from Bohemia and Morocco, although these regions and the Alps have some elements in common, for example, *Svobodaina ellipsoides, Gelidorthis meloui, Saukrodictya porosa, Aegiromena aquila aquila* and *Paterorthis paterina.* Instead, in the presence of warm water elements such as representatives of *Dolerorthis, Iberomena, Longvillia, Porambonites, Eoanastrophia* and others, they exhibit a closer affinity to Sardinia, the British Isles and North Europe. This indicates an invasion of North European warm water brachiopods as far south as the Alps, Sardinia, Montagne Noire and Spain (Havlicek 1976; Havlicek et al. 1987).

During the Hirnantian Stage, the supposed relationship with Baltoscandia can still be seen in the ostracod and echinoid fauna described by Schallreuter (1990) from the Carnic Alps. This time, corresponding roughly to the glacial maximum, is also characterized by a cold water influx from Gondwana (Jaeger et al. 1975). This is documented by the appearance of the Hirnantia Fauna which is generally concentrated in the higher latitudes of the Southern Hemisphere. However, exceptions do occur in a tropical belt and in northern low latitudes suggesting that this unique fauna was adapted to a glacially induced cold climate and consequently cooler waters at the close of the Ordovician.

The Upper Ordovician conodont fauna of the Alps is well known from detailed studies by Walliser (1964), Serpagli (1967) and Flajs and Schönlaub (1976) from the Uggwa Limestone of the Carnic Alps, and different limestone units of the Greywacke Zone of Styria. They were less well described using a few weakly metamorphosed occurrences in between (Buchroithner 1979; Neubauer 1979; Neubauer and Pistotnik 1984). It seems that this conodont association represents the *Hamarodus europaeus-Dapsilodus mutatus-Scabbardella altipes*-(HDS)-Biofacies of Sweet and Bergström (1984). Although their precise age within the uppermost Caradocian or early Ashgillian Series remains open, the conodont-bearing limestones clearly can be assigned to the *Amorphognathus ordovicicus*-Zone. According to Sweet and Bergström (1984) who tentatively revised the published conodont elements from the Carnic Alps in terms of the modern multi-element taxonomy, the Late Ordovician Uggwa Limestone is dominated by *Scabbardella altipes* (43 %), *Hamarodus europaeus* (17 %), *Amorphognathus* cf. *ordovicicus* (8 %) and *Dapsilodus mutatus* (2.4 %). Less abundant are *Plectodina alpina, Belodella pseudorobusta, "Prioniodus" ethingtoni* and *Strachanognathus parvus.* The occurrence of these species and the abundance of the others, in particular *Hamarodus europaeus,* varies from coeval faunas of Thuringia, Spain and France. It seems unclear as yet which factors are involved in these differences (Dzik 1989).

A comparison between this fauna from the Carnic Alps and the two others from the Greywacke Zone is difficult to assess due to probably minor differences in age and state of preservation (Flajs and Schönlaub 1976). In particular, this is true of the large collection derived from the limestone lenses underlying the thick acid volcanics of the Blasseneck Porphyroid in the surroundings of Eisenerz, Styria. Apparently, the revised conodont association represents the same general type as the one from the Carnic Alps in being equally dominated by *Amorphognathus* cf. *ordovicicus, Scabbardella altipes, Hamarodus europaeus, Dapsilodus mutatus* and perhaps *Plectodina alpina;* less abundant are *Belodella pseudorobusta, Panderodus* ssp. and certain elements which tentatively have been assigned to *Birkfeldia circumplicata.* Other differences between these two faunas were thoroughly reviewed by Flajs and Schönlaub (1976).

According to Bergström (1990), the "coefficient of similarity" (CS) between conodonts from Baltoscandia and the Mediterranean area has a value of 0.30 indicating moderate similarity between the two regions. For example, they share the occurrences of specimens of *Amorphognathus, Scabbardella* and *Dapsilodus* while others appear to be restricted to continental Europe or North Africa. Obviously, the distribution of Late Ordovician conodonts follows a similar pattern as inferred from megafossil assemblages and facies data. This led Sweet and Bergström (1984) to conclude that the Mediterranean Province was a cold water realm in a polar or subpolar latitudinal setting.

In the Alps, occurrences of carbonate sediments provide broad latitudinal constraints for the Upper Ordovician. Potentially useful, though only of limited climatic significance, is the distribution of limestones in the Carnic Alps, the Greywacke Zone and the Gurktal Nappe in between. According to Dullo (pers. comm. Nov. 1990) the up to 20 m thick carbonate units, in the local stratigraphical schemes named Wolayer and Uggwa Limestone respectively (Schönlaub 1985a), represent greyish and whitish grainstones to rudstones and occasionally also bafflestones with abundant debris of cystoids and bryozoans and less frequently trilobites and nautiloids. Cathodoluminescence studies have revealed the rare occurrence of coated grains. Moreover, of special significance are dogtooth-cements suggesting a vadose diagenetic environment for the Wolayer Limestone.

In a general, climatically based, latitudinal framework these carbonate units suggest a position within the confines of the larger "carbonate belt", i. e. between latitudes of about 45° N and S where it was moderately warm and where there was adequate light penetration rather than high water temperature (Ziegler et al. 1984). The nature of the corresponding sediments may thus have been the direct response to climatic changes during the Ordovician. For the Ashgillian Webby (1984) suggested a global climatic amelioration as the main cause for the increasing carbonate production. Alternatively, a progressive northward shift of the sedimentary basins into lower latitudes may also explain their temporal and spatial distribution (Young 1990). In the Ordovician of the Mediterranean Province, contemporary carbonates are widely distributed and have been reported from Sardinia (Vai and Cocozza 1986; Ferretti and Serpagli 1991), Montagne Noire, the Massifs of Mouthoumet and Agly of southern France (Engel et al. 1981), the Armorican Massif (Paris et al. 1981; Melou 1990; Robardet et al. 1990), the Pyrenees (Hartefelt 1970; Duran et al. 1984), Catalonia and other areas in Spain (Hammann 1976; Hafenrichter 1980; Duran et al. 1984), Portugal (Young 1985, 1988, 1990) and from the Anti-Atlas of Morocco (Destombes et al. 1985). Consequently, the Alpine occurrences of Upper Ordovician rocks suggest a position at considerably lower and more temperate latitudes than has been shown in the revised world maps of Scotese and McKerrow (1990). More precisely, available faunal and lithic data from the Upper Ordovician of the Alps rather indicate a position between approximately 40° and 50° S instead of being placed around 60° S.

Conclusions

1. During the Cambrian and Lower Ordovician, thick clastic sequences are the dominating sediments in northern Africa and in the adjacent southern and central European depocentres. Though these rocks are of no, or only limited, climatic significance, their inherited zircon population indicates Africa as source area (Fig. 3).
2. Carbonates first occur in the Lower Cambrian of Southern and Central Europe suggesting a low latitudinal position and close faunal relationships between the individual occurrences within the Mediterranean faunal realm (Sdzuy 1962; Freyer 1987). Yet, in the Alps the corresponding rocks have not been found. The oldest limestones are of Upper Ordovician age and occur in various parts of the Eastern Alps. Their fossil content and microfacies indicate a moderate climate in a temperate latitudinal setting.
3. Upper Ordovician fossils, in particular most brachiopods, cystoids, ostracods and conodonts, are more closely related to coeval warm water faunas

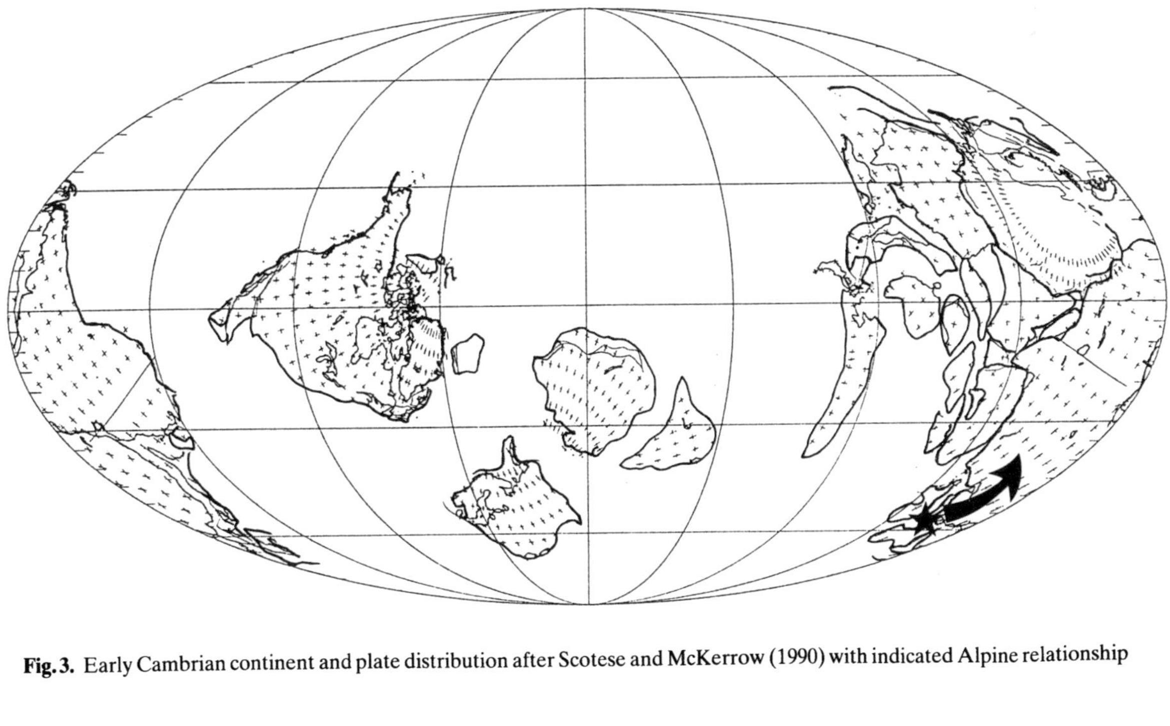

Fig. 3. Early Cambrian continent and plate distribution after Scotese and McKerrow (1990) with indicated Alpine relationship

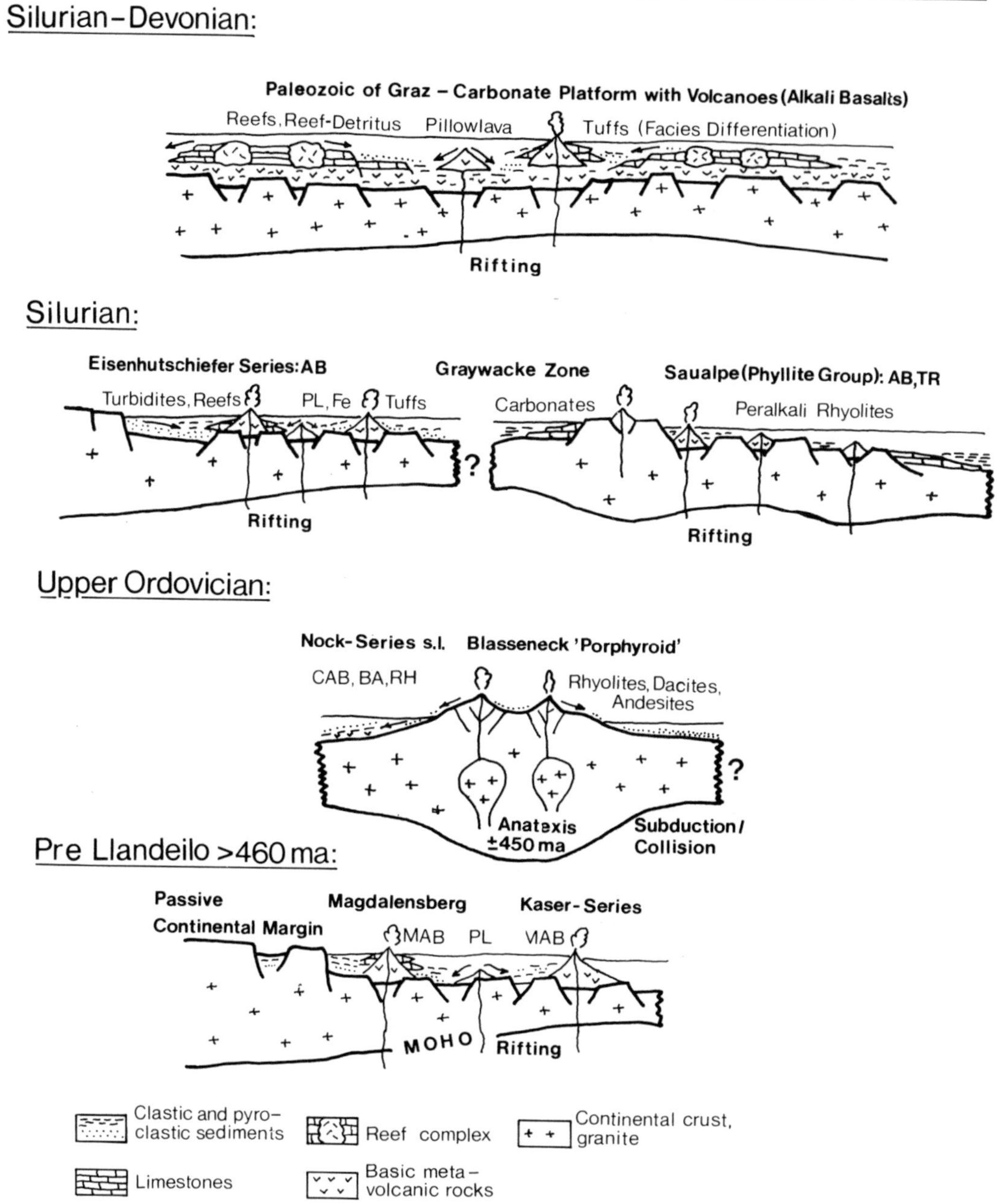

Fig. 4. Ordovician to Devonian rifting and collision/subduction-related volcanism in the Eastern Alps. Abbreviations: *AB* alkaline basalts; *BA* basaltic andesites; *CAB* calc-alkaline basalts; *FE* submarine hydrothermal iron ore deposits; *MAB* mildly alkaline basalts; *PL* pillow lava; *RH* rhyolites, *TR* trachytes. (Modified from Loeschke 1989a)

of northern Europe, Great Britain and Sardinia than to northern Africa. Exceptions are, however, the occurrences of the African brachiopod species *Paterorthis paterina* in the Caradocian, the Ashgillian Hirnantia fauna and the brachiopod *Clarkeia* sp. which indicate a temporary minor cold water influence from southern high latitudes.

4. Probably during the Llandeilian, a rifting-related basic volcanism occurred first recognized in Middle Carinthia but supposedly also occurring at other places in the Alps. Interestingly, this event seems to coincide with calc-alkaline igneous activity in the Ardennes, Wales and SE Ireland (Kokelaar et al. 1984) when Avalonia started to rift off from Gondwana (Cocks and Fortey 1982; McKerrow and Cocks 1986; Pickering 1989; Paris and Robardet 1990; Scotese and McKerrow 1990 with opposing statements) (Fig. 4). An analogous plate

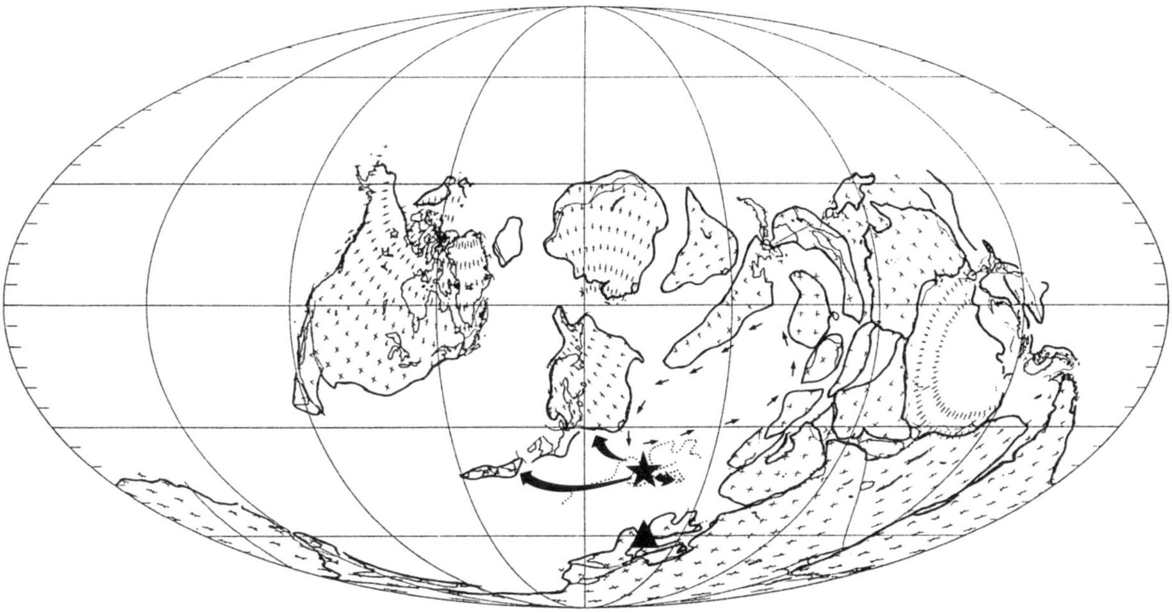

Fig. 5. Latest Ordovician (Ashgillian) palaeogeography. *Triangle* Position of Southern Alps according to Scotese and McKerrow (1990). The author's latitudinal setting is indicated by the *star.* Faunal relationships are shown by *thick arrows* and oceanic circulation system in the southern mid-European ocean by *short arrows.* (After Scotese and McKerrow 1990)

disruption and subsequent separation might well be assumed for certain parts of the Variscan Alps.

5. A second major magmatic event occurred in the Early Ashgillian and has been regarded as a collision-subduction-related process (Loeschke 1989a,b). In accordance with palaeomagnetic data from Gondwana (Bachtadse and Briden 1990) it seems reasonable to suggest that this event reflects the rapid northward movement of Africa and its final collision with an unknown microcontinent or terrane located to the north.
6. Our best estimate for the palaeolatitudinal position of the Late Ordovician of the Alps and its relationship with adjacent areas is illustrated on the ammended Scotese and McKerrow (1990) map for this time (Fig. 5). This plate configuration is based on the data from the Alps presented in the foregoing chapters and seems well constrained by sedimentary and faunal evidence from the West and Central European platform (Melou 1990; Paris and Robardet 1990; Robardet et al. 1990; Young 1990).

2.3 Silurian

Generally, three types of lithofacies, each with a distinct faunal assemblage, can be recognized:

1. Fossiliferous carbonate facies: the dominating lithologies are limestones and less frequently dolomites with a thickness of, at most, some 60 m (Fig. 6). Although the equivalents of the Lower Llandovery are missing, for the remaining of the Silurian the fossil record from many sections of the Carnic Alps and the Greywacke Zone has indicated a complete but slightly condensed sequence. Fossil assemblages consist of varying abundances of nautiloids, trilobites, bivalves, brachiopods and scarce graptolites as well as of conodonts, foraminifera, acritarchs, chitinozoans, scolecodonts and ostracods. During the last decades most, but certainly not all, groups have been revised or are being studied now. This facies is best represented by the Plöcken and Wolayer Groups of the Silurian of the Carnic Alps.
2. Graptolitic facies: it is characterized by black siliceous shales, cherts (lydites) and alum shales which prevail over quartzitic sandstones and greenish mudstones at the base and in the basal Pridolian, respectively. The thickness approximates that in the pure limestone facies. A continuous record of graptolites starting with the name bearer of the *Akidograptus acuminatus* Zone of the most basal Llandovery and ending in the Upper Lochkovian has indicated continuous sedimentation during the Silurian and across its boundary with the Devonian. In particular this is true for the Carnic Alps; in other areas, e. g., the Greywacke Zone such a continuous record has, as yet, not been demonstrated and it seems difficult to assess due to poor preservation of

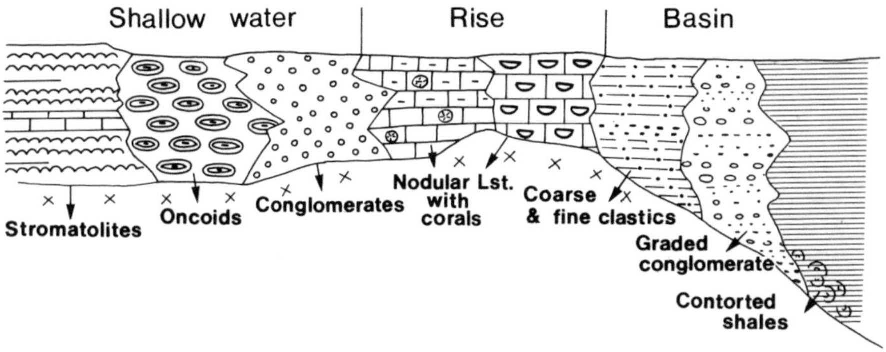

Fig. 6. Middle and Late Llandoverian facies variation in the Greywacke Zone of Tyrol and Salzburg. Note occurrences of climate sensitive stromatolites, algal onkoids, superficial ooids and solitary corals. (After Mostler 1970)

all collections. In the last 25 years all graptolite faunas were revised by Jaeger and numerous new ones have been collected; he named the corresponding strata Bischofalm Formation.

3. Transitional facies: it comprises a mixture of the above mentioned two main rock types, i. e. an alternation of black shales and marls with black or dark grey limestone beds. Its faunal content is very poor and consists of grapolites and few conodonts and nautiloids. In the local stratigraphic scheme of the Carnic Alps this facies was named Findenig Facies (see Fig. 1 of Schönlaub and Heinisch, p. 396 this volume).

All three main facies may intergrade to varying degrees depending on their setting in a distinct palaeogeographical and palaeotectonic environment with different amount of limestone production and fossil support. In particular, this is so of the region occupied by the Gurktal Nappe in Carinthia and parts of Styria with its dolomite and marble-rich facies, equivalent to several hundred metres thick clastic development in the same area (Neubauer 1979; Neubauer and Pistotnik 1984). Moreover, in some other fossiliferous sequences basic volcanics and tuffs are intercalated which indicate rifting associated with intracontinental volcanism at various times during the Silurian (for summary remarks, see Höll 1970; Ebner 1975; Buchroithner 1979; Neubauer 1979; Schönlaub 1979; Neubauer and Pistotnik 1984; Fritz and Neubauer 1988; Giese 1988; Leoschke 1989a,b).

Silurian faunas following the terminal-Ordovician mass extinction event are generally regarded as cosmopolitan and hence provide only little evidence to determine the latitudinal position of individual plates (Jablonski 1986). This evaluation, however, may change if a varied, highly diversified, fossil association is considered together with lithic data from the host rock.

According to Bergström (1990) most, if not all, post-Ordovician conodont localities lay between 40°S and 40°N palaeolatitude. In the Wenlock, continental Europe was located at the margin of this belt consistent with our conclusions drawn for the Late Ordovician(Fig. 7). For this time conodont evidence from the Carnic Alps and other areas of the Eastern Alps suggests close affinity to coeval faunas from central, southern and southwestern Europe. Comparable faunas from Britain and Gotland which occupied a more equatorial position seem, however, more diversified. These differences diminished towards the end of the Silurian. Bergström (1990) concluded that the Pridoli was a time of minimal conodont provincialism during which coeval faunas from the Alps, Bohemia and Nevada showed striking similarities at the generic level.

In the Silurian, the distribution of phyto- and zooplankton, i. e. acritarchs and chitinozoans displays a broad latitude-parallel-zonation. However, plotted on the new world maps of Scotese and McKerrow (1990) the old phytoplankton data of Cramer (1971) seem to reflect local environmental conditions rather than a biogeographic pattern (Colbath 1990).

Yet, from the Silurian of the Alps only few appropriate data are available (Bachmann and Schmid 1964; Priewalder 1976, 1987; Martin 1978). Accordingly, acritarchs from the Cellon section of the Carnic Alps suggest an intermediate position between the high latitude *N. carminae* and the tropical *Domasia-Deunffia* biofacies (Richardson et al. 1981; Priewalder 1987). This palaeolatitudinal setting is well-constrained by other data presented here. As regards chitinozoans too little is presently known from the Alps. According to Paris (1981, 1990) it appears that their Silurian distribution is essentially cosmopolitan.

Silurian trilobites from the Carnic Alps are closely related to Bohemia and other central European regions (Haas 1969; Alberti 1970). Affinities to Morocco may exist but are, as yet, not studied in detail. Interestingly, in the succeeding Devonian the apparent distinction with North Africa continued; instead,

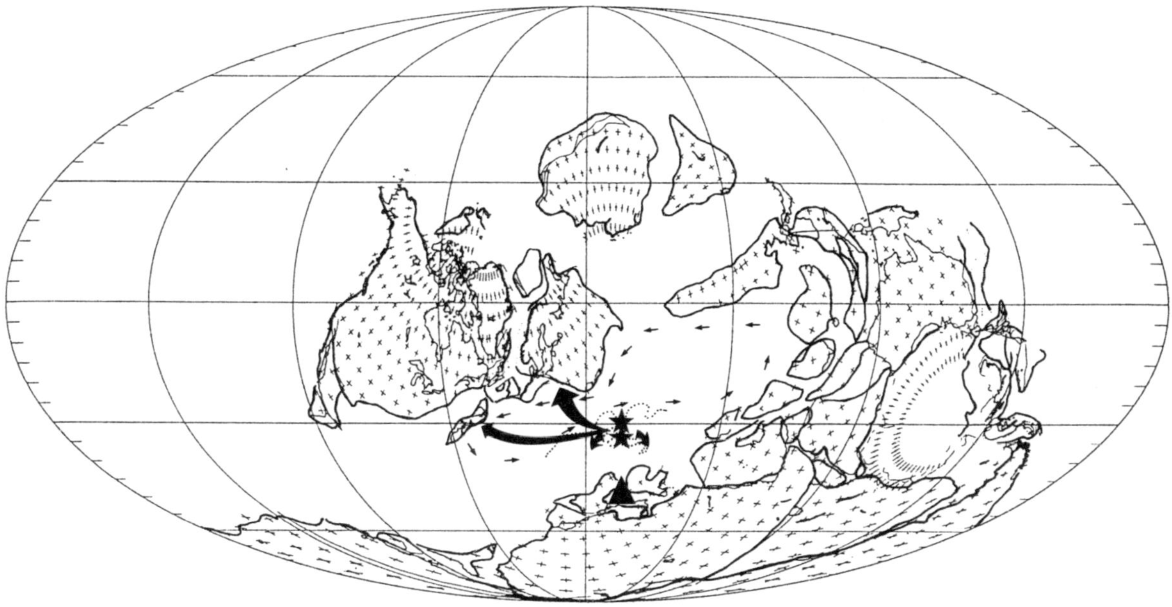

Fig. 7. Middle Silurian (Wenlockian) palaeogeography. *Triangle:* position of the Alpine Silurian as suggested by Scotese and McKerrow (1990). The author's latitudinal setting is shown by *two stars* for the Silurian of the Southern and Central Alps, respectively. Faunal relationships are shown by *heavy arrows,* the oceanic current system in the mid-European ocean by *short arrows.* (After Scotese and McKerrow 1990)

there appears a closer relationship with the Urals and Tienshan (Alberti 1969).

According to Koren (1979) most, if not all, Silurian to Lower Devonian graptolites occur within palaeolatitudes of some 30–40°N and 30°S. As noted by Berry (1979) Silurian graptolites show only very little endemism suggesting that interplate dispersal was possible and apparently occurred during the Silurian and Lower Devonian. Presumably, its distribution was mainly controlled by the character of the surface water plus ocean currents that overlaid the site at which graptolites are found. The distribution of graptolites may, thus, have very much depended on the size, shape and position of certain plates.

During the Ludlow and Pridoli an essentially uniform graptolite fauna developed in Europe. As pointed out by Jaeger (1976) and Barca and Jaeger (1990) the changing environment of this time is portrayed in strikingly similar and closely contemporaneous shifting lithofacies between northern Africa and Baltica which exhibit, with minor local modifications, a characteristic and continuous range from black graptolite shales to limestones, e.g. the well-known "Okkerkalk" of Thuringia. According to Jaeger (1975) and Alberti (1980) this change in facies was controlled by simultaneous sea-level rise and fall that affected a hypothesized single block along its passive margins.

In the Alps, the oldest nautiloids not described yet occur in the Ashgillian Uggwa Limestone of the Carnic Alps. From the Late Llandovery onwards, nautiloids became the dominating organisms in the carbonate facies of the Southern Alps with rich abundances of orthocerids in the Late Wenlock and Early Ludlow (Ristedt 1968, 1969). The diversified fauna seems closely related to Bohemia and Sardinia. With decreased numbers nautiloids continue into the Pridoli and Lower Devonian.

In contrast to the Carnic Alps, in the Greywacke Zone and in the Gurktal Nappe of Carinthia, very few nautiloids have been found despite the occurrence of very similar lithologies. This remarkable feature is, as yet, difficult to explain.

The distribution of other molluscs, in particular bivalves, corresponds largely to that of orthoconic nautiloids. According to Kriz (1979) the Silurian cardiolids from the Southern Alps and the Greywacke Zone inhabited the warm equatorial zone or were dispersed through south equatorial currents. However, plotted on the modified base maps proposed by Scotese and McKerrow (1990) the central and southern European faunas must either have occurred in slightly higher southern latitudes or the position of the respective plates is too high. In our opinion a position around 35°S would be the best estimate for the Silurian occurrences in the Alps.

This view is strongly supported by other evidence: in the Silurian corals were prominent constituents of the shallow water environment in the tropical belt.

After a crisis at the end of the Ordovician, several order among the Tabulata and the Rugosa diversified in the Early Silurian. During this period only weak provincialism is apparent at the generic level (Hill 1959; Kaljo and Klaamann 1973; Pickett 1975; McLean 1977; Oliver 1977).

Remarkably, rugose and tabulate corals occur abundantly in the Late Llandovery of Middle Carinthia (Strehl 1962; Buchroithner 1979), in the Upper Silurian (Ludlow) of Graz (Mensink 1953; Flügel and Schönlaub 1972; Ebner 1976) and are very rare in shallow water and locally superficial ooid-bearing limestones in the Late Llandovery of the Greywacke Zone of Tyrol (Mostler 1966, 1970; Al-Hasani and Mostler 1969). They are missing apparently in coeval strata of the Southern Alps. The inferred Silurian age of Heritsch (1929) is actually Lower Devonian (Lochkovian).

We find it hard to believe that these coral-bearing bioclastic limestones represent the fossil counterpart of modern cold-water coral reefs such as the common *Lophelia* reefs found presently as far north as beyond the Polar Circle, in the Barents Sea and on the shelf off Mid-Norway at depths between 250 and 300 m and at water temperatures of 6°C (Stromgren 1971; Mikkelsen et al. 1982). Rather, they display excellent environmental indicators controlled by such physical factors like light, temperature, suspended sediment, salinity, water agitation and other agents (Kinsman 1964; Scoffin et al. 1989 and others). Modern and ancient buildups cannot exist beyond the "Darwin Point" of about 35°-latitude (Grigg 1982), which is also the northernmost limit of corals in the present-day Pacific (Tribble and Randall 1986) and, more generally, carbonate production is light-limited to within 35° N and S (Ziegler et al. 1984). There is no objective reason against the idea that the Silurian of the Central Alps was positioned within these limits.

Conclusions

During the Silurian Period the Alpine occurrences of Silurian strata continued to shift into lower latitudes. Based on the evidence presented here the best position is estimated at approximate 30–35° S. Faunal relationships existed with southern Europe but apparently were closer to northern Europe. The affinities to southwestern Europe and northern Africa, however, decreased.

Palaeomagnetic data from Gondwana seem to support the assumption of a rapid northward movement (Bachtadse and Briden 1990; Kent and Van Der Voo 1990). It is associated with rifting-related volcanism through much of the Silurian. Interestingly, the Southern and Central Alps differ in two main aspects: the Silurian of the Central Alps, i. e. the development north of the Gailtal Fault, is characterized by warm water occurrences of rugose and tabulate corals in an environment which locally also contains superficial ooids, but these sequences yield only a few cephalopods as opposed to the Southern Alps with the opposite relationship suggesting, most plausibly, a farther south and slightly cooler environment. These differences may indicate two separate terranes or microcontinents prior to the Variscan deformation in the Alps.

2.4 Devonian

In the Alps, the Devonian Period is characterized by abundant shelly fossils, variably thick carbonate sequences, reef development and interfingering facies ranging from coastal sediments to carbonate buildups, slope deposits, condensed pelagic cephalopod limestones to deep ocean off-shore shales (Schönlaub 1971, 1980, 1985; Flügel and Schönlaub 1972; Burchette 1981; Kreutzer 1990 and others). In the Carnic Alps, for example, the ratio of thickness between shallow water limestones and contemporary cephalopod limestones is in the order of 1200 to 100 m and thus indicates differentially subsiding mobile basins affected by extensional tectonics (see Fig. 1 of Schönlaub and Heinisch, p. 396 this volume). Rifting-related volcanism in the Lower and Middle Devonian of the Greywacke Zone and the area near Graz, respectively, is well consistent with such a basin development (Fig. 8).

According to Boucot (1975, 1985, 1988) and his employment of a pangaeic palaeogeography, the biogeographic pattern of the Devonian changed from a highly provincial level in the latest Early Devonian attributed to a global regression to cosmopolitanism in the Late Devonian. At the beginning of the Devonian, the warm North Silurian realm was replaced by the Old World realm and the Eastern American realm. Both realms contrasted with the cool climate Malvinokaffric realm which was characterized, during the Lower and Middle Devonian, by the same characteristics as during earlier periods.

In the Alps, only a few groups from essentially two areas, i. e. the Carnic Alps and the surroundings of Graz, have been investigated in terms of biogeographic relationships.

During the Lower Devonian (Lochkovian to Emsian Stages) faunas and floras showed strong affinities

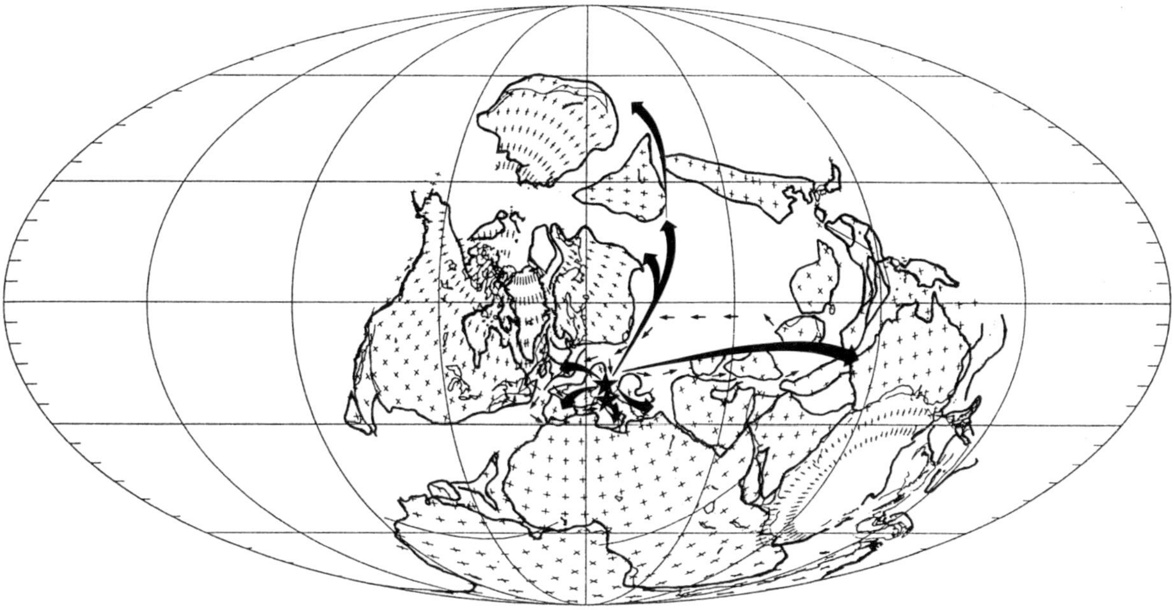

Fig. 8. Middle Devonian palaeogeography. *Two stars* Position of the Alpine Devonian of the Southern and Central Alps, respectively, as suggested by the author in this volume. Faunal and floral relationships are shown by *heavy arrows,* the equatorial gyre system in the southern mid-European ocean by *short arrows.* (After Scotese and McKerrow 1990; Oczlon 1990)

to both the Hercynian-Bohemian and to the Uralo-Tienshan Faunal Provinces and remarkably poor similarities with northern Africa (Alberti 1969, 1982). Such a relationship is less clearly displayed by planctic cosmopolitan groups like graptolites, conodonts and dacryoconarids the migration of which primarily was affected by ocean currents (Jaeger 1968, 1975, 1978, 1988; Flügel et al. 1977; Heckel and Witzke 1979; Klapper and Johnson 1980; Schönlaub 1980, 1985a; Alberti 1985) but, more distinctly, by the benthos. For example, brachiopods from the *Karpinskia* Beds are known from the Barrandian region, the Urals (Boucot 1975; Boucot et al. 1967; Vai 1973, 1975), Altai and Kazakhstan (for more details see Latz 1989); similarly, corals show a strong affinity to Extra-Malvinokaffric realms like the Urals and the Altai mountains but show no similarity to the Eastern North American realm (Ferrari 1968; Kodsi 1971; Kreutzer 1990; Pedder and Oliver 1990); the distinct *Hercynella*-gastropod community is typically of the Old World realm and occurs in the Upper Koneprusy Limestone of Bohemia and also in the Urals, although the relations with both regions appear less close than had been suggested previously (Jhaveri 1969; Blodgett et al. 1990); finally, trilobites although to some extend endemic, represent a mixed Hercynian-Uralo-Tienshan association wich exhibit only a weak affinity to northern Africa except for the Rabat area (Erben 1966, 1969; Alberti 1967, 1969, 1981; Bandel 1969; Alberti 1978; Ellermann 1989), and codiacean algae of the *Lancicula* Beds of Palla [(1966, 1967) (= Hercynella-Seewarte Limestone of Jhaveri 1969)] are found in Chios (Greece), Kusnetsk basin (southern Siberia) and New South Wales in Australia. According to Poncet (1990) they were dispersed by subtropical ocean currents.

From the Middle and Upper Devonian only few biogeographical relevant data are yet available. They confirm the general opinion of decline of endemism and provincialism and the appearance of cosmopolites at the beginning of the Middle Devonian. This has been inferred from the distribution of ammonoids in general (House 1971, 1973) and from the Carnic Alps (House and Price 1980; Korn 1992), trilobites (Feist 1992), brachiopods (Ferrari and Vai 1973), corals (Ferrari 1968; Küster 1987) and algae, e.g. the genus *Girvanella* and the problematical algae *Renalcis* (Kreutzer 1989; Rantitsch 1990).

The Palaeozoic outcrops in the surroundings of Graz (Fig. 9) have long been famous for the rich occurrences of fossils in particular brachiopods and rugose and tabulate corals; less abundant are trilobites, bivalves, gastropods and stromatoporoids (Flügel 1975a). Among many endemic species, mainly of corals, some are of biogeographic significance. In accord with sedimentological evidence from the underlying Dolomitsandstein-Formation (Fenninger and Holzer 1978) the Eifelian brachiopods from the Bar-

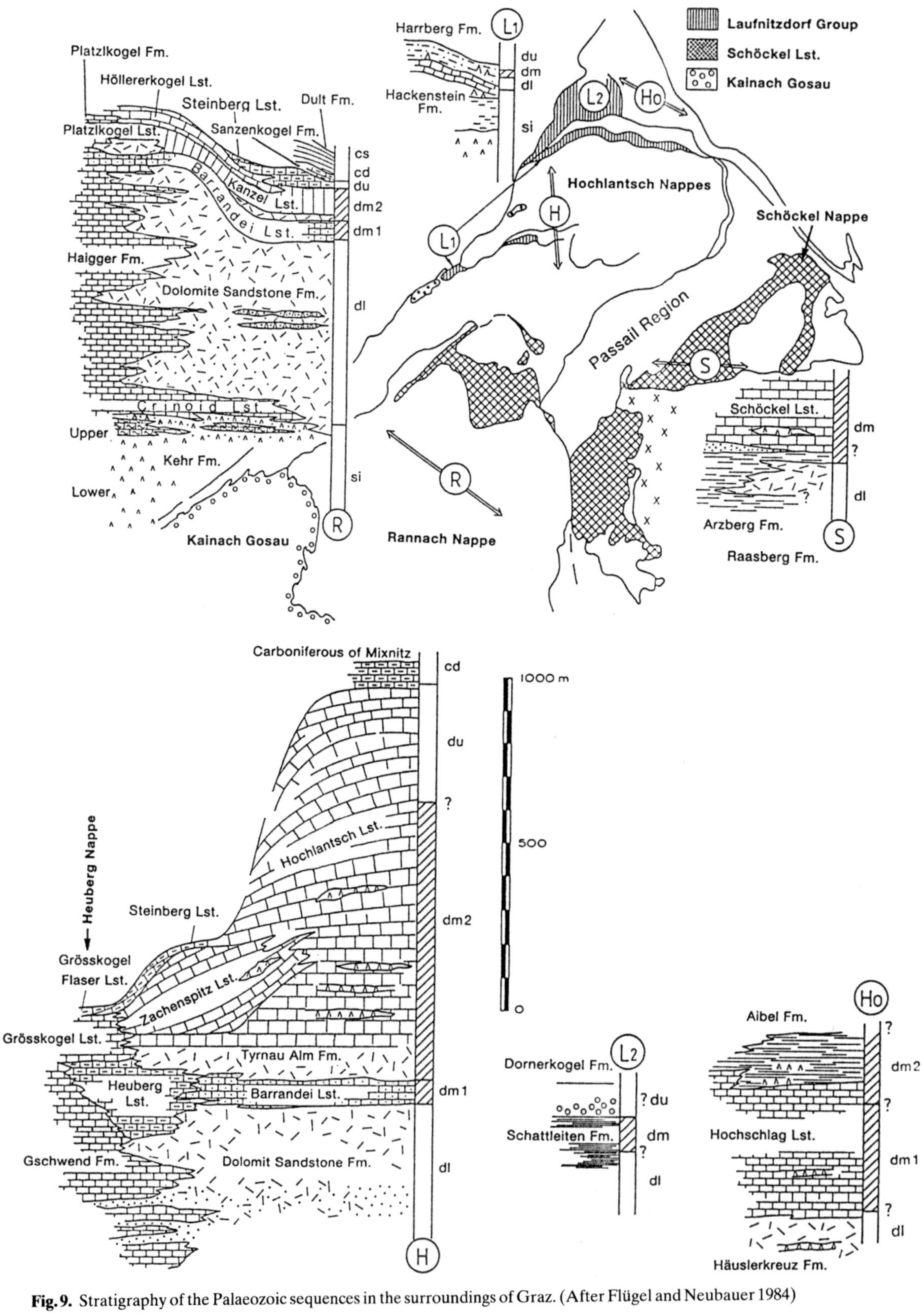

Fig. 9. Stratigraphy of the Palaeozoic sequences in the surroundings of Graz. (After Flügel and Neubauer 1984)

randei Formation indicate a strong Rhenish-Hercynian (e.g. neritic-pelagic) interchange, even in spite of taxonomic uncertainties, and thus immigration of Rhenish elements (Flügel 1971), while some others, e.g. the cosmopolitan genus *Zdimir,* suggests a relationship to the Ardennes, the Eifelian Hills, France, Bohemia, Urals, Siberia (Kusnetsk Basin) and Ferghana (Boucot and Siehl 1962). Whether or not there was a migration channel across Moravia as suggested by Chlupac (1971) and challenged by Schönenberg (1973) and Flügel (1975a) is not certain yet. Furthermore, corals from the succeeding Givetian *Calceola* Formation seem equally be related to the Eifelian Hills, but in addition they exhibit also some affinities to the Kusnetsk Basin of southern Siberia and Nepal (Flügel 1971). Rather of limited biogeographic significance are, however, the rarely occurring representatives of the genus *Calceola* which is a typical Old World realm coral (Oliver 1964; Hill and Jell 1969; Boucot 1988). Stromatoporoids, too, represent, to a high degree, cosmopolites during the Devonian (Stock 1990). As for corals and algae, however, they can be used as major indicators of warm temperatures of the Devonian seas (Heckel and Witzke 1979). In addition to these reports, the Eifelian strata of Graz contains one of the very few occurrences of an algal microflora in southern Europe. It consists of *Zeapora gracilis, Litanaia graecensis* and *Pseudopaleoporella lummatonensis* (Flügel 1959; Hubmann 1990). This flora indicates (1) a palaeolatitudinal setting within the tropical belt of some 30°, and (2) similarities to the east, e.g. the Urals, Siberia, Australia, and to the west, e.g. the Eifelian Hills, Belgium, France and Great Britain.

Conclusions

In order to test the proposed palaeogeographic models and to confirm the supposed palaeolatitude as the major control of the spatial faunal distribution the complete data base for the Devonian of the Alps can be employed.

There is general consent that the Devonian calcareous sediments were deposited in the tropical belt within palaeolatitudes of 30°S or less. Such an assumption is based on thick occurrences of carbonate buildups and associated limestones composed of typical microfacies types, fabrics and components (e.g. oncoids, coated grains, ooids), and all kinds of "guilds" (Fagerstrom 1987, 1988), e.g. abundant rugose and tabulate corals, massive and dendroid stromatoporoids, algae and stromatolites, suggesting warm agitated sea waters for most of the Devonian. Studies of the diagenetic environment and the cement stratigraphy closely support this inferred position (Galli 1985; Kreutzer 1989; Rantitsch 1990). Even stronger arguments in favour of a low latitudinal setting have been expressed by Fenninger and Holzer (1978 p. 139) who recorded tepee structures and vadose pisolites from the Dolomitsandstein Formation in the surroundings of Graz and concluded a hypersaline environment in an aride climate.

At first sight it appears difficult, if not impossible that these groups should have been capable of travelling over such long distances and/or crossing the supposedly wide and deep Mid-European ocean that separated, for example, Siberia and the Urals from the Alps during the Devonian. More easily, however, migration could be achieved between other central European regions, e.g. from the Alps to Bohemia, the Eifelian Hills, the Harz Mountains, the Ardennes and England. The hazards of fauna travelling, however, have long been recognized (Cloud 1961) and have been discussed at length for the widely distributed Lower Palaeozoic trilobites (e.g. Dean 1967; Palmer 1972; Whittington and Cisne 1973; Hughes 1972, 1973; Whittington 1973; Ross 1975; Cocks and Fortey 1982, 1990 and others).

Amongst the first, Ross (1975) emphasized the operation of coriolis forces in the Lower Palaeozoic. Comparable to the present-day ocean circulation systems in surface waters, they caused two equatorial gyres in which organisms (trilobites) with long-lasting larval stages were caught and transported along the margins of the seas to split eventually and to continue in other directions. However, this improved circulation pattern of favourable ocean surface currents operated not only during the Cambrian and Ordovician but aided the dispersal of trilobites and of many other groups during the Devonian as well, for example conodonts (Klapper and Johnson 1980), corals (Pedder and Oliver 1990), stromatoporoids (Stock 1990), algae (Poncet 1990) and probably many others too (Boucot 1988, 1990). Strong currents may even affect the bottom sediment and produce erosion and non-deposition or reworking and accumulations of clastic sediments; such deposits have been described in modern environments as contourites (Hollister and Heezen 1972; Lovell and Stow 1981). Independently from our study of faunal relationships between the Alps and adjacent or more distant areas for the Middle Devonian Oczlon (1990) convincingly postulated such a South Equatorial Current along the southern margin of Laurasia being deflected to the North Gondwana Current at the contact between Gondwana and Laurentia (see Scotese and McKerrow 1990). We believe that these two currents can be held responsible for the distinct exchange of faunas between Siberia, Kazakhstan, the Urals and the Alps on one side and Bohemia, Greece and Australia on the other. The surroundings of Graz may equally have been influenced by this current when it flowed through Central Europe and past the Eifel Hills and parts of Belgium. Likewise, such a current direction might also explain the low affinity between the Alps and northern Africa during the Devonian.

For the Devonian it may be speculated whether or not the surroundings of Graz were located in lower latitudes than the Southern Alps. Some evidence in favour of such an idea is indicated in climatic sensitive sediments and also in the close relationship to the Rhenish-Hercynian Province. In contrast to this development the Southern Alps represent a highly mobile subsiding basin affected by tensional tectonics throughout the Devonian. Possibly, thinning of the crust during this time is the result of the temporary southward movement of Africa relative to the south pole (Fig. 2) which described a rapid excursion ("loop") into central Africa.

2.5 Carboniferous

The Carboniferous Period of the Alps is generally subdivided into the final Variscan Series representing a Lower Carboniferous pelagic development in the Tournaisian and succeeding flysch deposits of Visean and Namurian age, and the post-Variscan transgressive cover sediments of Late Carboniferous and Permian age. Both groups of rocks are separated by the Variscan unconformity. Based on new and revised data on conodonts and fusulinids in the Southern Alps, the pre-Variscan strata were deformed between the late Namurian *Gastrioceras*-Zone and the Upper Miatchkovian of the late Middle Carboniferous in the Russian terminology (Kahler 1983; Schönlaub, unpubl.), the latter corresponding to the West European Westfalian D Substage.

From the older cycle only few biogeographically relevant data are yet available which mostly comprise cosmopolitan groups like goniatites and pelagic trilobites. According to Korn (in Schönlaub et al. 1988) and Korn (1991) across the Devonian/Carboniferous boundary a complete succession of ammonoids occur which indicate continuous pelagic sedimentation in an open marine pelagic environment comparable to many other places in the world, e. g. Rhenish Massif, Sauerland, Moravia, southern France or South China. Similarly, trilobites are related to those in Cornwall and north Devon as well as to the Rhenish Massif, Frankenwald, Montagne Noire, the Sudetan mountains, Poland, the Urals, Kazakhstan and southeast China (Feist 1992). Some of these faunas are characterized by blind or reduced eyes indicating benthonic forms of moderately deep waters; some, however, represent fully blind trilobites not known yet from elsewhere in the Variscan basin (Hahn and Kratz 1992). Nevertheless, loose relations do exist to those in Sauerland, Thuringia, Poland and England.

In contrast to these reports and, hence, of special interest is the "Carboniferous of Nötsch", north of the Gail valley and west of Villach in Carinthia. With regard to its lithology and the rich and diversified fossil content the Carboniferous of Nötsch has long been regarded as being unique and distinct within the entire Alps. The latest Visean or, more probably, Early Namurian fossil assemblage (Schönlaub 1985b) comprises brachiopods, trilobites, gastropods, bivalves, crinoids, corals, bryozoans, foraminifera, ostracods, plants and algae; yet, only a small part has been studied.

According to Hahn and Hahn (1973) the trilobite fauna is characterized by its special Tethyan aspect with some similarity to coeval occurrences in the Veitsch Nappe of the Greywacke Zone of Styria. Subsequently this view was rejected by Hahn and Hahn (1987) when they recovered additional trilobites showing a strong relationship with the Kohlenkalk of Belgium. They then concluded that there was a mixing of Asiatic-Australian, i. e. Tethyan and West-European trilobites. Based on additional rich material, however, Schraut (1990) finally emphasized a strong affinity to the Western European Kohlenkalk facies and even to North America, and that the similarities to Asia and Australia were less close. Even ostracods follow these suggested pathways and show affinities to the Cantabrian Mountains, the western Pyrenees, England and North America.

According to Flügel and Schönlaub (1990) in the Carboniferous of Nötsch as well as in the Hochwipfel Formation of the Southern Alps (Carnic Alps) there occur exotic limestone clasts of varying microfacies-types which indicate a shallow carbonate water setting of an open marine and restricted shelf environment during the Visean. Presumably, this platform development existed north of the Gailtal Line and adjacent to a supposed land area. Yet, no relics of this platform have been preserved. The only records are some limestone clasts and palaeoenvironmentally significant fossils such as the heterocoral *Hexaphyllia mirabilis* (Duncan), the algae *Pseudodonezella tenuissima* (Berchenko), the foraminifera *Howchinia bradyana* (Howchin) and abundant conodont faunas corresponding to the *Eumorphoceras*-Stage E2 of the basal Namurian. Recently in other parts of Carinthia apparently coeval limestone clasts of boulder size were found (Schlöser et al. 1990).

Litho- and biofacies of these exotic limestone clasts exhibit strong affinities to the Kohlenkalk Facies of various parts of Europe (Belgium, France, England, Poland), but also to Hungary, the eastern and southern Carpathians, the Pyrenees, southern Spain, northern Africa, the Donets Basin and the Urals (Poty 1981; Herbig 1986; Flügel and Herbig 1988; Ebner 1990; Henningsen and Herbig 1990; Schlöser et al. 1990). Moreover, the supposed setting on an active continental margin and its formation through successive erosion of an accretionary wedge during a collision of two different plates reflect a remarkable coincidence between the Eastern Alps and the western part of the Mediterranean.

Besides the lowermost Carboniferous, during which the end-Devonian climate prevailed, the available palaeoclimatic data from the Southern Alps, the Carboniferous of Nötsch and the Veitsch Nappe of the Greywacke Zone suggest an increase of temperature and humidity during the Visean. Of particular significance is a widespread emersion that occurred in the lengthy *Scaliognathus anchoralis*-conodont Zone, i.e. at the Tournaisian/Visean boundary prior to the deposition of transgressive cherts and the succeeding flysch deposits. It resulted in a variety of buried palaeokarst features like an extensive relief and small-scale disconformities, mixed faunas, coated fissures, collapse breccias, caves with internal fillings and mineralizations which recently have been recognized in the Carnic Alps and most probably also occurred in the Greywacke Zone and the surroundings of Graz.

In the Southern Alps, Late Palaeozoic shallow water sediments unconformably overlie the Variscan flysch and other basement rocks of varying age, i.e. different Silurian and Devonian strata. According to Kahler (1983), the oldest transgressive sediments are Middle Carboniferous in age and, more precisely, correspond to the *Fusulinella bocki*-Zone of the Upper Miatchkovo of the Moscow Basin. This late Palaeozoic cover comprises clastic and calcareous shallow marine sediments of the Auernig Formation in the Upper Carboniferous (Kasimovian and Ghzelian Stages) followed by various Lower Permian shelf and shelf edge deposits.

Upper Permian sediments rest disconformably upon the Lower Permian and its equivalents in the Dolomites, or, farther west on phyllites of the Variscan basement. They indicate a transgressive regime starting with red beds of the Gröden Formation and followed by the Bellerophon Formation of the Late Permian. This formation represents a carbonate ramp which gently dips to the southeast, but is located far east from the Permian shoreline exposed in the Dolomites of northern Italy in the west.

The marine post-Variscan sequences of the Southern Alps have long been famous for their abundant and highly diverse fossil groups. During the last few years the major part of the fauna and flora has been reinvestigated and new material was collected. Based upon these studies the following conclusions can be drawn: Fusulinids are of typical "Paleotethyan" and thus, apparently of cosmopolitan aspect showing similarities with coeval faunas in many other parts of the world, e.g. the Dinarides (Serbia, Velebit, Montenegro, Albania), the Bükk Mountains of Hungary, northern Africa (Tunis), Turkey (Anatolia), Iran (Elburs), Afghanistan, Indochina, South China, Japan as well as to the Moscow and Donets Basins, the Urals, Ferghana, Mongolia, Pamir, Greenland, northern California and Texas (Kahler 1939, 1955, 1974, 1983; F. and G. Kahler 1982); trilobites are closely related to the Karawanken Alps and the Cantabrian Mountains of northern Spain and less closely to the Urals, the Moscow and Donets Basins (Hahn and Hahn 1977, 1987, Hahn et al. 1989); brachiopods are equally related to these regions as they have many species in common as opposed to the weak links with North America (Gauri 1965; Ramovs 1972; Winkler Prins 1971, 1983, 1984); the ostracod fauna too suggests a close similarity with the Cantabrian Mountains of Asturia and reflects a shallow marine and low energy environment (Ruggieri 1966; Becker 1978; Fohrer 1990); sphinctozoans appear well comparable to those from New Mexico, Texas and the Cantabrian Mountains (Kügel 1987); the rich coral faunas have yet not been revised but it appears that it is closely related to Russia, East Asia and China (Heritsch 1936, 1943); in addition, Lower Permian faunas are of low diversity (Homann 1971); calcareous algae often occur as massive algal wackestones attributed to lense-shaped algal mud-mounds which consist of low diversity phylloid algae *(Epimastopora, Archaeolithophyllum, Eugonophyllum)* and the dasycladacean *Anthracoporella* and others (Buttersack and Boeckelmann 1984; Boeckelmann 1985) which appear of no biogeographic significance.

During the last twenty years in the Eastern Alps more than 60 localities with Upper Carboniferous and Early Permian plants were studied, together with revisions of old collections (for summary see Tenchov 1980; Boersma and Fritz 1990; Fritz and Boersma 1990). Besides implications for the palaeoclimate and for the local facies development no distinct palaeofloristic-biogeographic relationships can be inferred.

Yet, its main importance is the potential for correlating West-European continental with Tethyan marine sequences which has been demonstrated for a good deal from floras of the Carnic Alps.

Conclusions

As a response to the Variscan Orogeny dramatic changes affected the Alps during the Carboniferous Period. In the Southern Alps the climax of deformation occurred between the Late Namurian and the Late Westphalian Stages, or, in the Russian terminology, between the Early Bashkirian and the Middle or Late Moscovian Stages.

In the Central Alps, however, deformation and metamorphism evidently occurred earlier. This conclusion seems well founded from radiometric ages and from the transgressive molasse-type sediments within the Gurktal Nappe, the Carboniferous of Nötsch and the Veitsch Nappe of the Greywacke Zone. Moreover, we presented evidence that these scattered occurrences might represent the last remains of an originally vast shelf characterized by various platform sediments as opposed to the Southern Alps with contemporary flysch deposits.

During the Carboniferous this northern development was biogeographically more closely related to Western Europe and even to North America than to Eastern Europe or Asia. In particular, there appears a striking similarity with the Cantabrian Mountains, the western Mediterranean and to the "Kohlenkalk" regions of England, Belgium and Poland.

Consequently, we suspect that the Southern and Central Alps represented two different microplates during the Lower Carboniferous. This assumption confirms the suggested fragmentation of the predecessors of the Alps which has been concluded in the previous chapters from the analysis of older rocks and faunas. If at all and how much they were separated is presently difficult to decide. Yet, it is worth mentioning that reworked amphibolite clasts in the Badstub Breccie of the Carboniferous of Nötsch are metamorphosed tholeiitic ocean floor basalts (Teich 1982; Krainer and Mogessie 1991) suggesting sometime during the Palaeozoic an enigmatic oceanic crust in this area of the Alps.

Soon after collision and amalgamation of the two plates the biogeographic patterns of the Southern Alps began to match those from the former settings in the Central Alps indicating migration of faunas and floras into the newly established Southern Alps domain where they found remarkably favourable environmental conditions.

Most, if not all, suggested faunal and floral migration paths of fusulinids and other groups along the northern shelf margin of the Tethys Sea, the Ural Sea and the Arctic region to North America as well as to

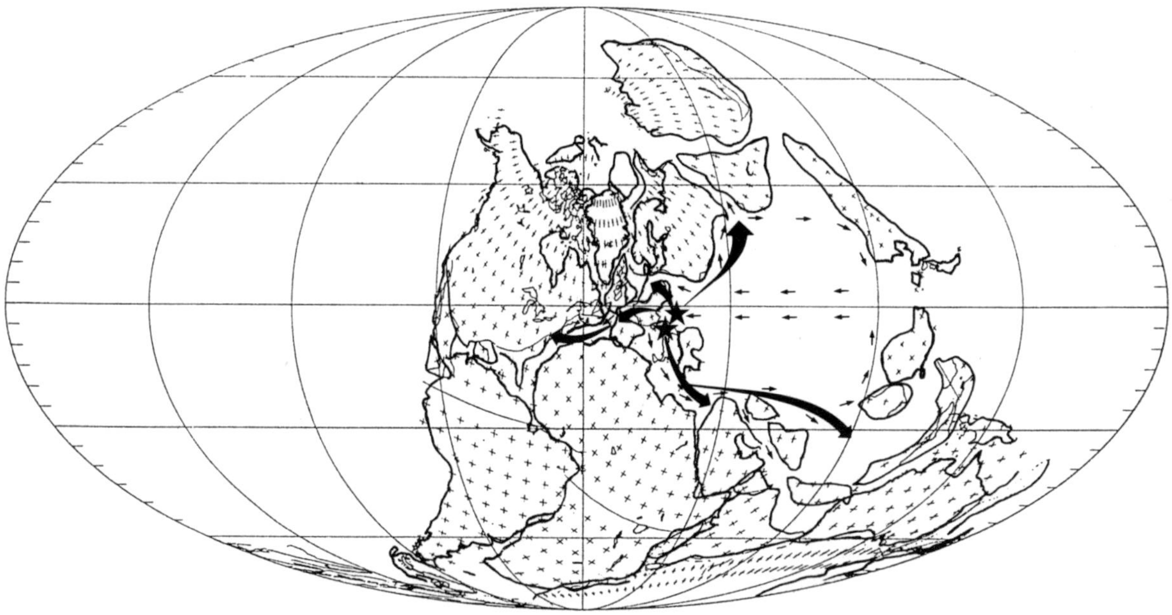

Fig. 10. Late Carboniferous (Westphalian) palaeogeography. *Two stars* Position of the Alpine Carboniferous of the Southern and Central Alps, respectively, as suggested by the author in this volume. Faunal and floral relationships are shown by *heavy arrows*, the equatorial gyre system by *short arrows*. (After Scotese and McKerrow 1990)

anlogous occurrences on the southern shelf appear well constrained by the revised World Maps of Scotese and McKerrow (1990) for the Late Carboniferous (Fig. 10). Possibly, dispersal of planctic groups was aided by warm subequatorial gyres which were blocked and deflected at the contact between Laurasia and Gondwana (Ziegler et al. 1981; Ross and Ross 1985; Kelley et al. 1990).

Potentially useful climate-sensitive sediments of Carboniferous age comprise in the Veitsch Nappe of the Greywacke Zone several tens of metres of graphite and related rocks as well as limestones and dolomites which supposedly formed in a temporary hypersaline environment (Ratschbacher 1984). Furthermore, at many localities plants occur in rich abundances and diversity; up to a few metres thick coal seams, however, are mainly restricted to the Carnic Alps and the Gurktal Nappe. In the former, they are interbedded with locally rich occurrences of corals, fusulinids, algal mud-mounds and oncoid limestones consistent with the inferred low latitudinal position close to the equator and humid climatic conditions for the Middle and Late Carboniferous of the Alps. Nonetheless, it should always kept in mind that the O_2-concentration of the Carboniferous atmosphere is still unclear and may have varied between 13 and 35% of the present 21% (Holland 1990).

According to Visser (1990) and Veevers and McPowell (1987), Caputo (1985), Caputo and Crowell (1985), Hambrey and Harland (1981) and Bouroz et al. (1978), the continental glaciation in the Southern Hemisphere started diachronously in the Namurian A or near the Namurian A/B boundary, i. e. in the Serpukhovian, and caused a high-latitude cooling and a contemporary equatorial warming episode. Kelly et al. 1990 assumed that this climatic alteration resulted in changes of latitudinal diversity patterns coupled with migration of different organisms, for example brachiopods. The well-known "Auernig cyclicity" in the Late Carboniferous of the Carnic Alps may certainly be explained as a glacial rebound although alternative proposals have also been made (e. g. Friedman 1989); evidently, it was of no consequence to the biogeographic distribution of fauna and flora of that region.

At the beginning of the Carboniferous the apparent polar wander path (APWP) shows a change in the drift direction from a Devonian southward movement to a continuous and rapid northward drift of Gondwana with minimum drift rates of 10 cm a^{-1} (van der Voo 1988; Bachtadse and Briden 1990; Kent and van der Voo 1990). This rapid movement of Africa over the south pole is held responsible for the final disappearance of the Mid-European or Rheic Ocean besides several other oceans and the collision between Gondwana and Laurussia in the Namurian (e. g. McKerrow and Ziegler 1972; Neugebauer 1988; Scotese and McKerrow 1990). As mentioned, the collision of the Southern Alps with the central part of the Eastern Alps can also be related to this motion; it occurred, however, slightly later at the end of the Namurian or at the beginning of the Westphalian Stage, i. e. in the Bashkirian or early Moscovian.

2.6 *Permian*

During the Permian, the facies pattern of the Carboniferous Period continued without any recognizable breaks. As for the latter, the marine facies was restricted to the Southern Alps but progressive shifting of the shoreline towards the west indicates an expanding and deepening sea as far west as the Italian Dolomites (Fig. 11).

In the Southern Alps, the marine Permian sediments are up to 1150 m thick. In the Lower Permian they comprise fossiliferous limestones and clastics which are developed in successive cycles. Generally, they contain abundant fossils, e. g. foraminifera, fusulinids, calcisponges, corals, bryozoans, brachiopods, gastropods, bivalves, ostracods, echinoderms, calcareous algae, plants and trace fossils suggesting a warm climate and predominantly full marine conditions on a shallow subsiding platform setting of the inner and outer shelf (Flügel 1974, 1980, 1981; Buggisch et al. 1976; Buggisch 1980).

The Permian fauna and flora show close affinities to adjacent areas in southeastern Europe (Slovenia, Velebit in Croatia, Montenegro, Chios), Turkey, Iran (Elburz), Afghanistan, Oman, East Asia (Vietnam, Thailand, Japan, China), the USSR (Moscow Basin, the Urals, Transcaucasia, Ferghana, Bashkiria) and also to Spain, Alaska, British Columbia, California, New Mexico, Texas, Kansas and other areas of the mid-West (Kahler 1939, 1955, 1962, 1973, 1974, 1980, 1983, 1989; Flügel 1966; Homann 1972; Flügel and Flügel-Kahler 1980).

The Permian climate of the Southern Alps can be inferred from several lines of evidence: the oldest Permian evaporites (gypsum, salt, bituminous sediments) and pedogenetic calcretes occur in the Middle Permian Gröden Formation suggesting semi-arid conditions for this time (Assereto et al. 1973; Buggisch et al. 1976; Buggisch 1978; Ori and Venturini 1981; Farabegoli and Viel 1982; Broglio Loriga et al. 1986, 1988; Conti et al. 1986; Massari et al. 1988). At the Bletterbach-Butterloch section of Radegno,

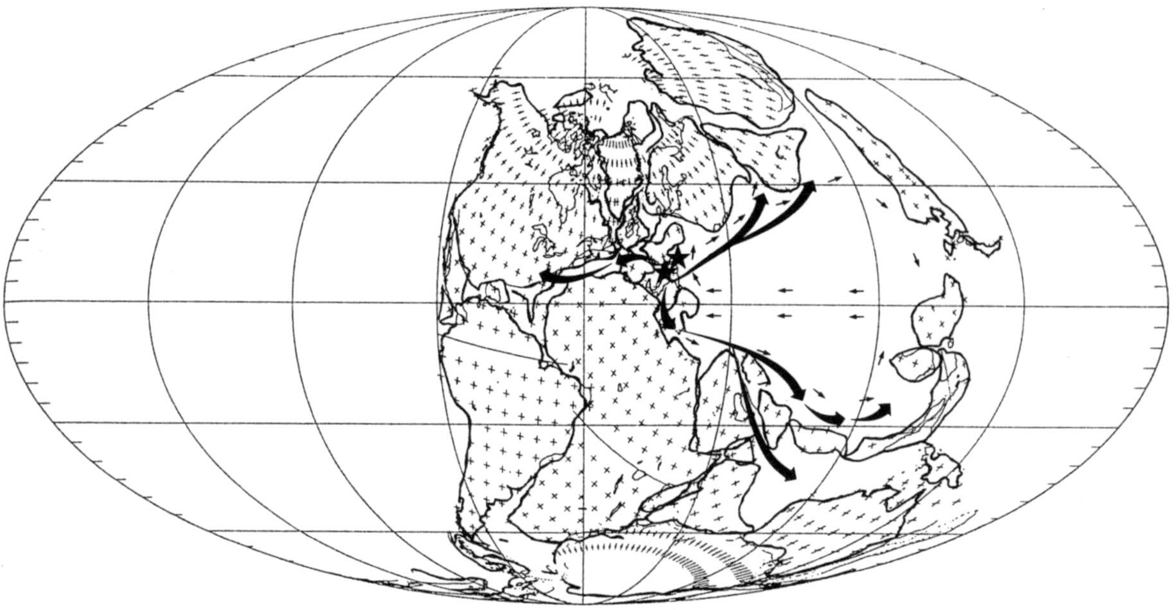

Fig. 11. Late Permian (Kazanian) palaeogeography. *Two stars:* position of the Alpine Permian of the Southern and Central Alps, respectively, as suggested by the author in this volume. Faunal and floral relationships are shown by *heavy arrows,* the equatorial gyre system of the Tethys by *short arrows.* (After Scotese and McKerrow 1990)

South Tyrol, they are associated with ooids, algal crusts and foraminifera. These evaporites grossly portray the "coal belt" which is characterized by centimetre to decimetre thick allochthonous coal layers interbedded with marine faunas and evaporitic layers suggesting a hot semi-arid but not arid climate (Leonardi 1948). Furthermore, Assereto et al. (1973) claimed that aridity is indicated by unaltered feldspars in sandstones, the stream network, rich occurrences of plants and tetrapod footprints as well as analysis of quartz in conglomerates (Dal Cin 1965). A seasonally arid climate has also been concluded from early diagenetic Pb-, Zn-, Cu-, and U-mineralizations and the formation of dolomite, magnesite and barite indicating agressive groundwaters (Wopfner 1984).

Towards the end of the Permian, the climatic pattern was even more pronounced. Due to the shallow Bellerophon Sea and a low-gradient topography, an extended evaporitic basin with high salinity developed in the Dolomites. Here anhydrite, gypsum, salt and aphanitic dolomite alternating with shale beds were deposited in sabhka-type regressive cycles (Assereto et al. 1973; Bosellini and Hardie 1973).

In contrast to this southern domain, north of the prominent Alpine geosuture of the Periadriatic Line continental deposits were widely deposited during the Permian, forming the base of the transgressive sequences of the Middle and Upper East Alpine tectonic units. Lack of fossils in these strata does not permit any precise age assignments nor any conclusive statements concerning paleobiogeography. Based on faunal or floral evidences only a few data are yet available from Lower Permian or uppermost Stephanian floras of Carinthia, i. e. from the localities Stangnock, Kötschach, Ulrichsberg, Christofberg, Wunderstätten and Pum, from the Kristberg Formation of the Montafon region of Vorarlberg, and from the "Perm von Zöbing" in the southeastern Bohemian Massif (Thiedig and Klussmann 1974; Vasicek 1977; Tenchov 1980; van Amerom et al. 1982; Boersma and Fritz 1990). Also, spores may be useful to date weakly metamorphosed rocks of Late Carboniferous or Lower Permian age or help to distinguish between Upper Permian and Lower Triassic sediments (Klaus 1974, 1980; Planderova and Pahr 1983).

The depositional framework of the Permian sequence north of the Periadriatic Line comprises various unfossiliferous clastic sediments of molasse-type ranging from conglomerates to fanglomerates, coarse-grained sandstones, shales, mudstones and magnesite. According to Niedermayr (1975, 1985), Erkan (1977), Niedermayr and Scheriau-Niedermayr (1982), Krainer (1984, 1985), Sylvester (1989) and others, they represent ephemeral braided and meandering stream deposits of alluvial plains, in the lower part, which are progressively replaced by coastal sabhkas and intertidal clastics above. This lithology suggests an overall arid to semi-arid climate with minor evaporation. The red colour results from primary clay minerals cemented by secondary dispersed hematite which originated from destroyed Fe-minerals (Krainer 1984; Sylvester 1989).

In the Northern Alps the evaporitic series include anhydrite and gypsum interbedded with shales and sandstones (Ramsau and Radmer type, respectively, according to Erkan 1977). Based upon spores an Upper Permian age has been proposed for these rocks (Klaus 1974). They overlie a thick unfossiliferous clastic complex which apparently was deposited in a predominantly continental and temporarily aquatic environment with minor marine incursions.

Riehl-Herwirsch (1972) concluded that a humid Late Carboniferous climate with karstification was followed by semi-humid conditions in the Rotliegend. Based on trace-element chemistry Kurat et al. (1974) even proposed a relative dry climate and lateritic weathering during the Permian. Rich floral occurrences in interbedded red and grey-greenish sediments of the basal Werchzirm Formation of Carinthia indicate that red bed sedimentation started in the lowermost Permian (Krainer 1990). These basal clastics are followed by predominantly red beds suggesting that the change from a semi-arid to an arid climate approximates the boundary between the Carboniferous and Permian Periods.

Conclusions

With regard to lithofacies and climate sensitive data the Permian sequences of the Eastern and Southern Alps reflect profound differences.

The distribution of climate sensitive data from the Southern Alps indicate a humid climate for the Carboniferous and Early Permian Periods. Evidently, at the beginning of the Middle Permian, this pattern changed when the dominating greyish-blackish (greenish) sediments abruptly changed to red beds, with temporarily appearing evaporites, suggesting semi-arid or even arid conditions that lasted at least to the end of the Permian (for details, see earlier).

In the Eastern Alps, north of the PeriadriaticLine, this change in climate may have occurred earlier in the Permian. Plant remains from the Laas and the Werchzirm Formations indicate the onset of arid conditions in the basalmost Rotliegend corresponding to the early Asselian in the marine sequence. An overall humid climate, however, prevailed throughout the Middle and Late Carboniferous.

Whether or not this different climate pattern reflects different latitudinal settings of the Southern and Central Alps, remains open. However, from the continuous northward shift of Gondwana through the Permian, it seems reasonable to assume and not merely to speculate, that a northern realm crossed the equatorial wet belt and approached an arid climatic belt earlier during the Permian than the Southern Alps located an unknown distance farther south.

References

Alberti GKB (1967) Neue obersilurische sowie unter- und mitteldevonische Trilobiten aus Marokko, Deutschland und einigen anderen europäischen Gebieten vol, 2. Senckenb Lethaea 48: 481–509

Alberti GKB (1969) Trilobiten des jüngeren Siluriums sowie des Unter- und Mitteldevons vol, I. Abh Senckenb Naturforsch Ges 520: 1–692

Alberti GKB (1970) Trilobiten des jüngeren Siluriums sowie des Unter- und Mitteldevons vol, II. Abh Senckenb Naturforsch Ges 525: 1–233

Alberti GKB (1980) Zur lithofaziellen Entwicklung, Fauna (Trilobiten, Tentaculiten) und Biostratigraphie des Paläozoikums (Silur-Devon) von Beni Afeur (Kleine Kabylei, Tell-Atlas, Algerien). Neues Jahrb Geol Paläontol Monatsh 1980: 277–286

Alberti GKB (1981) Beziehungen zwischen „hercynischen " Trilobiten-Faunen aus NW-Marokko und Deutschland (Unter- und Mittel-Devon). Nat Mus 111: 362–369

Alberti GKB (1982) Der Hamar Laghdad (Tafilalt, SE-Marokko), eine bedeutende Fundstätte devonischer Trilobiten. Nat Mus 112: 172–182

Alberti GKB (1985) Zur Tentakulitenführung im Unter- und Mittel-Devon der Zentralen Karnischen Alpen (Österreich). Cour Forschungsinst Senckenb 75: 375–388

Alberti H (1978) Ein Vertreter des Genus *Wolayella* Erben, 1966 (Trilobitae) aus dem Unter-Devon des Harzes. Aufschluß SB 28: 69–74

Al-Hasani N, Mostler H (1969) Zur Geologie der Spießnägel südlich Kirchberg (Nördliche Grauwackenzone, Tirol). Veröff Univ Innsbr 9, Alpenkdl Stud 5: 5–26

Alvarez W (1976) A former continuation of the Alps. Geol Soc Am Bull 87: 891–896

Amerom van HWJ, Angerer H, Mostler H (1982) Über eine Autuno-Stephanische Flora aus den Kristbergschichten im Montafon, Vorarlberg (Österreich). Jahrb Geol Bundesanst 124: 283–323

Amerom van HWJ, Flajs G, Hunger G (1984) Die „Flora der Marinelli Hütte" (Mittleres Visé) aus dem Hochwipfelflysch der Karnischen Alpen (Italien). Meded Rijks Geol Dienst 37–3: 1–41

Assereto R, Bosellini A, Fantini Sestini N, Sweet W (1973) The Permian-Triassic boundary in the Southern Alps (Italy). In: Logan A, Hills LV (eds) The Permian and Triassic systems and their mutual boundary. Alberta Soc Pet Geol 2: 176–199

Bachmann A, Schmid ME (1964) Mikrofossilien aus dem österreichischen Silur. Verh Geol Bundesanst-1964: 53–64

Bachtadse V, Briden JC (1990) Palaeomagnetic constraints on the position of Gondwana during Ordovician to Devonian times. In: McKerrow WS, Scotese CR (eds) Palaeozoic palaeogeography and biogeography. Geol Soc Mem 12: 43–48

Badham JPN (1982) Strike-slip orogens – an explanation for the Hercynides. J Geol Soc Lond 139: 493–504

Badham JPN, Halls C (1975) Microplate tectonics, oblique collision, and the evolution of the Hercynian orogenic system. Geology 3: 373–376

Bandel K (1969) Feinstratigraphische und biofazielle Untersuchungen unterdevonischer Kalke am Fuß der Seewarte (Wolayer See, zentrale Karnische Alpen). Jahrb Geol Bundesanst 112: 197–234
Barca S, Jaeger H (1990) New geological and biostratigraphical data on the Silurian in SE-Sardinia. Close affinity with Thuringia. Boll Soc Geol Ital 108 (1989): 565–580
Becker G (1978) Flachwasser-Ostracoden aus dem hohen Westfal Asturiens (Kantabrisches Gebirge, N-Spanien). 1. Palaeocopida. Senckenb Lethaea 59: 37–69
Becker LP, Frank W, Höck V, Kleinschmidt G, Neubauer F, Sassi FP, Schramm JM (1987) Outlines of the pre-Alpine metamorphic events in the Austrian Alps. In: Flügel HW, Sassi FP, Grecula P (eds) Pre-Variscan and Variscan events in the Alpine-Mediterranean mountain belts. Alfa, Bratislava, pp 69–106
Bergström SM (1990) Relations between conodont provincialism and the changing palaeogeography during the Early Palaeozoic. In: McKerrow WS, Scotese CR (eds) Palaeozoic palaeogeography and biogeography. Geol Soc Mem 12: 105–121
Berry WBN (1979) Graptolite biogeography: A biogeography of some Lower Palaeozoic plancton. In: Gray J, Boucot AJ (eds) Historical biogeography, plate tectonics, and the changing environment. Proc 37th Ann Biol Colloq, Oregon State Univ Press, Corvallis, Oregon, pp 105–115
Blodgett RB, Rohr DM, Boucot AJ (1990) Early and Middle Devonian gastropod biogeography. In: McKerrow WS, Scotese CR (eds) Palaeozoic palaeogeography and biogeography. Geol Soc Mem 12: 277–284
Boeckelmann K (1985) Mikrofazies der Auernig-Schichten und Grenzland-Bänke westlich des Rudnig-Sattels (Karbon-Perm, Karnische Alpen). Facies 13: 155–174
Boersma M, Fritz A (1990) Die Paläofloren Kärntens: Ober-Karbon/Unter-Perm. Carinthia II 180/SH 49: 133–172
Bosellini A, Hardie AL (1973) Depositional theme of a marginal marine evaporite. Sedimentology 20: 5–27
Boucot AJ (1975) Evolution and extinction rate controls. Development in palaeontology and stratigraphy vol, 1. Elsevier, Amsterdam
Boucot AJ (1985) Late Silurian-Early Devonian biogeography, provincialism, evolution and extinction. Philos Trans R Soc Lond B309: 323–339
Boucot, AJ (1988) Devonian biogeography: an update. In: Mc Millan NJ, Embry AF, Glass DJ (eds) Devonian of the world. Proc 2nd Int Symp Devonian System Calgary. Can Soc Petrol Geol 3: 211–227
Boucot AJ (1990) Silurian biogeography. In: McKerrow WS, Scotese CR (eds) Palaeozoic palaeogeography and biogeography. Geol Soc Mem 12: 191–196
Boucot AJ, Siehl A (1962) *Zdimir* Barrande (Brachiopoda) redefined. Notizbl Hess Landesamtes Bodenforsch 80: 117–131
Boucot AJ, Johnson JG, Talent JA (1967) Lower and Middle Devonian Faunal Provinces based on Brachiopoda. In: Oswald DH (ed) Int Symp Devonian System Calgary 1967, II. Alberta Soc Petrol Geol, pp 1239–1254
Boucot AJ, Johnson JG, Talent JA (1969) Early Devonian brachiopod zoogeography. Geol Soc Am Spec Pap 199: 1–113
Bouroz A, Einor OL, Gordon M, Meyen SV, Wagner RH (1978) Proposals for an international chronostratigraphic classification of the Carboniferous. C R 8eme Congr Int Stratigr geol Carbonif, 1: 36–52
Broglio Loriga C, Conti MA, Fontana D, Mariotti N, Massari F, Neri C, Nicosia U, Pasini M, Perri MV, Pittau P, Posenato R, Venturini C, Viel G (1986) Upper Permian sequence and P/T boundary in the area between Carnia and Adige Valley. In: Field Conference on Permian and Permian-Triassic boundary in the South-Alpine segment of the Western Tethys, Field Guide-Book. Soc Geol Ital, Brescia, pp 23–28
Broglio Loriga C, Neri C, Pasini M, Posenato R (1988) Marine fossil assemblages from Upper Permian to Lowermost Triassic in the western Dolomites (Italy). Mem Soc Geol Ital 34 (1986): 5–44
Buchroithner MF (1979) Biostratigraphie und fazielle Untersuchungen im Paläozoikum von Mittelkärnten. Carinthia II 169: 71–95
Buggisch W (1978) Die Grödener Schichten (Perm, Südalpen). Sedimentologische und geochemische Untersuchungen zur Unterscheidung mariner und kontinentaler Sedimente. Geol Rundsch 67: 149–180
Buggisch W (1980) Die Geochemie der Kalke in den Trogkofel-Schichten der Karnischen Alpen. In: Flügel E (ed) Die Trogkofel-Stufe im Unterperm der Karnischen Alpen. Carinthia II SH 36: 101–111
Buggisch W, Flügel E (1980) Die Trogkofel-Schichten der Karnischen Alpen – Verbreitung, geologische Situation und Geländebefund. In: Flügel E (ed) Die Trogkofel-Stufe im Unterperm der Karnischen Alpen. Carinthia II SH 36: 13–50
Buggisch W, Flügel E, Leitz F, Tietz G–F (1976) Die fazielle und paläogeographische Entwicklung im Perm der Karnischen Alpen und in den Randgebieten. Geol Rundsch 65: 649–690
Burchette TP (1981) European Devonian reefs: a review of current concepts and models. In: Toomey DF (ed) European fossil reef models. Soc Econ Paleontol Mineral Spec Publ 30: 85–142
Burrett C (1972) Plate tectonics and the Hercynian orogeny. Nature 239: 155–156
Buttersack E, Boeckelmann K (1984) Palaeoenvironmental evolution during the Upper Carboniferous and the Permian in the Schulter-Trogkofel area (Carnic Alps, northern Italy). Jahrb Geol Bundesanst-126: 349–358
Caputo MV (1985) Late Devonian glaciation in South America. Palaeogeogr Palaeoclimatol Palaeoecol 51: 291–317
Caputo MV, Crowell JC (1985) Migration of glacial centers across Gondwana during Palaeozoic Era. Geol Soc Am Bull 96: 1020–1036
Chlupac I (1971) Einige Erkenntnisse aus Vergleichsstudien des Paläozoikums der Ostalpen. Cas Mineral Geol 16: 175–185
Cisne JL (1973) Life history of an Ordovician trilobite *Triarthus eatoni*. Ecology 54: 135–142
Cloud PE (1961) Paleobiogeography of the marine realm. In: Sears M (ed) Oceanography. Am Assoc Adv Sci Publ 67: 151–200
Cocks LRM, Fortey RA (1982) Faunal evidence for oceanic separations in the Palaeozoic of Britain. J Geol Soc Lond 139: 465–478
Cocks LRM, Fortey RA (1990) Biogeography of Ordovician and Silurian faunas. In: McKerrow WS, Scotese CR (eds) Palaeozoic palaeogeography and biogeography. Geol Soc Mem 12: 97–104
Colbath GK (1990) Palaeobiogeography of Middle Palaeozoic organic-walled phytoplancton. In: McKerrow WS, Scotese CR (eds) Palaeozoic palaeogeography and biogeography. Geol Soc Mem 12: 207–213
Conti MA, Fontana D, Mariotti N, Massari F, Neri C, Nicosia U, Pasini M, Pittau D (1986) The Bletterbach-Butterloch section (Val Gardena Sandstone and Bellerophon Formation). In: Field Conference on Permian and Permian-Triassic boundary in the South-Alpine segment of the Western Tethys. Field Guide-Book. Soc Geol Ital, Brescia, pp 99–119

Cramer FH (1971) Distribution of selected Silurian acritarchs. Rev Espan Micropaleontol Num Extraord, pp 1–203
Dal Cin R (1965) Ricerche morfoscopiche sulle Arenarie di Val Gardena della Regione Dolomitica. Ann Univ Ferrara Sez 9: 163–172
Dean WT (1967) The distribution of Ordovician shelly faunas in the Tethyan region. In: Adams CG, Ager DV (eds) Aspects of Tethyan biogeography. Syst Assoc Publ 7: 11–44
Destombes J; Hollard H, Willefert S (1985) Lower Palaeozoic rocks of Morocco. In: Holland CH (ed) Lower Palaeozoic rocks of the world vol, 4 Wiley, New York, pp 91–336
Duran H, Gil Ibarguchi JI, Julivert M, Ubach J (1984) Early Paleozoic acid volcanism in the Catalonian coastal ranges (northwestern Mediterranean). In: Sassi FP, Julivert M (eds) Correlation of Prevariscan and Variscan events of the Alpine-Mediterranean mountain belt. Newsletter IGCP Proj 5: 33–43
Dzik J (1989) Conodont evolution in high latitudes of the Ordovician. Cour Forschungsinst Senckenb 117: 1–28
Ebner F (1975) Ein Beitrag zum Altpaläozoikum des Remschnigg (Steiermark). Verh Geol Bundesanst 1974: 281–287
Ebner F (1976) Das Silur/Devon-Vorkommen von Eggenfeld – ein Beitrag zur Biostratigraphie des Grazer Paläozoikums. Mitt Abt Geol Paläontol Bergb Mus Joanneum 37: 1–33
Ebner F (1990) Circummediterranean Carboniferous preflysch sedimentation. In: Venturini C, Krainer K (eds) Field workshop on Carboniferous to Permian sequence of the Pramollo-Nassfeld Basin (Carnic Alps). Proc, pp 20–32
Ellermann I (1989) Trilobiten aus dem Unterdevon der Karnischen Alpen, Österreich. Inaugural Diss, Univ Bonn, pp 1–50 (unpubl)
Engel W, Feist R, Franke W (1981) Le Carbonifere Ante-Stephanien de la Montagne Noire: Rapports entre mise en place des nappes et sedimentation. Bull Bur Rech Geol Min Ser 2, Sec 1/4, 1980–1981, pp 341–389
Erben HK (1966) Über die Tropidocoryphinae (Tril.)-Liefg. I. Neues Jahrb Geol Paläontol Abh 125: 170–211
Erben HK (1969) Faunenprovinzielle Beziehungen zwischen unterdevonischen Trilobiten der Karnischen Alpen und Zentralasiens. Carinthia II SH 27: 19–20
Erkan E (1977) Uran- und gipsführendes Permoskyth der östlichen Ostalpen. Jahrb Geol Bundesanst 120: 343–400
Fagerstrom JA (1987) The evolution of reef communities. Willey, New York, 600 pp
Fagerstrom AJ (1988) A structural model for reef communities. Palaios 3: 228–232
Farabegoli C, Viel G (1982) Il Permo-Scitico delle Alpi Meridionale. Relazione finale Contr Crest Samim-Pertusola, pp 1–48
Feist R (1992) Trilobiten aus dem Devon/Karbon-Grenzprofil an der Grünen Schneid, zentrale Karnische Alpen, Österreich. Jahrb Geol Bundesanst 135: 21–47
Fenninger A, Holzer H–L (1978) Die Genese der Dolomitsandstein-Folge des Grazer Paläozoikums. Mitt Österr Geol Ges 69: 109–162
Ferrari A (1968) Tetracoralli delle Alpi Carniche – La fauna di Monte Zermula. G Geol 34 (1966): 531–594
Ferrari A, Vai GB (1973) Revision of the Famennian rhynchonellid genus *Plectorhynchella*. G Geol 39 (1971): 163–220
Ferretti A, Serpagli A (1991) First record of Ordovician conodonts from southwestern Sardinia. Riv Ital Paleontol Stratigr 97: 27–34
Flajs G, Schönlaub HP (1976) Die biostratigraphische Gliederung des Altpaläozoikums am Polster bei Eisenerz (Nördliche Grauwackenzone, Österreich). Verh Geol Bundesanst-1976: 257–303
Flügel E (1966) Algen aus dem Perm der Karnischen Alpen. Carinthia II SH 25: 3–76
Flügel E (1974) Fazies-Interpretation der unterpermischen Sedimente in den Karnischen Alpen. Carinthia II 164: 43–62
Flügel E (1980) Die Mikrofazies der Kalke in den Trogkofel-Schichten der Karnischen Alpen. In: Flügel E (ed) Die Trogkofel-Stufe im Unterperm der Karnischen Alpen. Carinthia II SH 36: 51–99
Flügel E (1981) Lower Permian *Tubiphytes/Archaeolithoporella* buildups in the Southern Alps (Austria and Italy). In: Toomey DF (ed) European fossil reef models. Soc Econ Paleontol Mineral Spec Publ 30: 143–160
Flügel E, Herbig H–G (1988) Mikrofazies karbonischer Kalkgerölle aus dem Paläozoikum des Rif (Marokko): ein Beitrag zur Paläogeographie der westmediterranen Paläotethys im Karbon. Facies 19: 271–300
Flügel E, Schönlaub HP (1990) Exotic limestone clasts in the Carboniferous of the Carnic Alps and Nötsch. In: Venturini C, Krainer K (eds) Field workshop on Carboniferous to Permian sequence of the Pramollo-Nassfeld Basin (Carnic Alps). Proc, pp 15–19
Flügel HW (1959) *Zeapora* Penecke 1894: Eine Dasycladaceen-Gattung aus dem Mittel-Devon von Graz. Neues Jahrb Geol Paläontol Monatsh 1959: 145–152
Flügel HW (1971) Einige biostratigraphisch wichtige Rugosa aus den Calceola-Schichten des Hochlantsch (Grazer Paläozoikum). Mitt Naturwiss Ver Steiermark 100: 72–83
Flügel HW (1975a) Erläuterungen zur Geologischen Wanderkarte des Grazer Berglandes 1:100000. Geol Bundesanst Vienna, pp 1–288
Flügel E, Flügel-Kahler E (1980) Algen aus den Kalken der Trogkofel-Schichten der Karnischen Alpen. In: Flügel E (ed) Die Trogkofel-Stufe im Unterperm der Karnischen Alpen Carinthia II SH 36: 113–182
Flügel HW (1975b) Einige Probleme des Variszikums von Neo-Europa. Geol Rundsch 64: 1–62
Flügel HW (1984) Das Grazer Paläozoikum. In: Flügel HW, Neubauer F (eds) Steiermark-Geologie der österreichischen Bundesländer in kurzgefassten Einzeldarstellungen. Geol Bundesanst, pp 44–54
Flügel HW (1990) Das voralpine Basement im Alpin-Mediterranen Belt – Überblick und Problematik. Jahrb Geol Bundesanst 133: 181–221
Flügel HW, Neubauer F (1984) Steiermark – Geologie der österreichischen Bundesländer in kurzgefassten Einzeldarstellungen. Geol Bundesanst Wien, pp 1–127
Flügel HW, Schönlaub HP (1972) Nachweis von tieferem Unterdevon und höherem Silur in der Rannach-Fazies des Grazer Paläozoikums. Mitt Österr Geol Ges 63: 142–148
Flügel HW, Jaeger H, Schönlaub HP, Vai GB (1977) Carnic Alps. In: Martinsson A (ed) The Silurian-Devonian boundary. IUGS Ser A (5): 126–142
Fohrer B (1990) Verkieselte Flachwasser-Ostracoden und ihre Begleitfauna und -flora aus dem Oberkarbon der Karnischen Alpen (Nassfeld-Region, Kärnten, Österreich). Diplomarb, Inst Paläontol Univ Erlangen-Nürnberg, pp 1–170
Frank W, Kralik M, Scharbert S, Thöni M (1987) Geochronological data from the Eastern Alps. In: Flügel HW, Faupl P (eds) Geodynamics of the Eastern Alps. Deuticke, Wien, pp 272–281
Freyer G (1987) Über Archaeocyathinen-Funde und den lithologischen Aufbau des Unterkambriums im Gebiet von Torgau. Z Geol Wiss Berlin 15: 665–680
Friedman GM (1989) Recent developments in Carboniferous geology: some comments. Proc Geol Assoc 100: 239–240

Frisch W, Neubauer F (1989) Pre-Alpine terranes and tectonic zoning in the Eastern Alps. Geol Soc Am Spec Pap 230: 91–99
Frisch W, Raab D (1987) Early Paleozoic back arc and island arc settings in greenstone sequences of the central Tauern window (Eastern Alps). Jahrb Geol Bundesanst 129: 545–566
Frisch W, Neubauer F, Satir M (1984) Concepts of the evolution of the Austroalpine basement complex (Eastern Alps) during the Caledonian-Variscan cycle. Geol Rundsch 73: 47–68
Frisch W. Neubauer F, Bröcker M, Brückmann W, Haiss N (1987) Interpretation of geochemical data from the Caledonian basement complex. In: Flügel HW, Sassi FP, Grecula P (eds) Pre-Variscan and Variscan events in the Alpine-Mediterranean mountain belts. Alfa, Bratislava, pp 209–226
Frisch W, Ménot R–P, Neubauer F, Raumer von JF (1990) Correlation and evolution of the Alpine basement. Schweiz Mineral Petrogr Mitt 70: 265–285
Fritz A, Boersma M (1990) Die steinkohlenzeitliche Pflanzenwelt Kärntens in ihrer fossilen Überlieferung. Carinthia II 180/SH 49: 15–108
Fritz H, Neubauer F (1988) Geodynamic aspects of the Silurian and Early Devonian sedimentation in the Paleozoic of Graz (Eastern Alps). Schweiz Mineral Petrogr Mitt 68: 359–367
Galli G (1985) Depositional environment in the Devonian limestone succession of the Cima Ombladet (Carnic Alps, Italy). Facies 12: 97–112
Gauri KL (1965) Uralian stratigraphy, trilobites and brachiopods of the Western Carnic Alps (Austria). Jahrb Geol Bundesanst SB 11, pp 1–94
Gebauer D (1986) The development of the continental crust of the European Hercynides since the Archean based on radiometric data. In: Proc Eur Geotrav Worksh. 3rd Eur Sci Found Strasbourg, pp 15–22
Gebauer D, Williams IS, Compston W, Grünenfelder M (1989) The development of the Central European continental crust since the Early Archaean based on conventional and ion-microprobe dating of up to 3.84 b.y. old detrital zircons. Tectonophysics 157: 81–96
Giese U (1988) Lower Paleozoic volcanic evolution at the northwestern border of the Gurktal nappe, Upper Austroalpine, Eastern Alps. Schweiz Mineral Petrogr Mitt 68: 381–396
Grauert B, Arnold A (1968) Deutung diskordanter Zirkonalter der Silvrettadecke und des Gotthardmassivs (Schweizer Alpen). Contrib Mineral Pet 20: 34–56
Grauert B, Seitz MG, Soptrajanova B (1973) Age and origin of detrital zircons from the pre-Permian basement of the Bohemian Massif and the Alps. Contrib Mineral Pet 40: 195–230
Grigg RW (1982) Darwin point: a threshold for atoll formation. Coral Reefs 1: 29–54
Haas W (1969) Trilobiten aus dem Silur der Karnischen Alpen. Carinthia II SH 27: 23
Hafenrichter M (1980) The lower and upper boundary of the Ordovician system of some selected regions (Celtiberia, Eastern Sierra Morena) in Spain. Part II. The Ordovician/Silurian boundary in Spain. Neues Jahrb Geol Paläontol Abh 160: 138–148
Hahn G, Hahn R (1973) Trilobiten aus dem Unter-Karbon (Dinantium) von Nötsch (Österreich). Geol Palaeontol 7: 135–146
Hahn G, Hahn R (1977) Trilobiten aus dem Unter-Karbon der Veitsch (Steiermark, Österreich). Neues Jahrb Geol Paläontol Monatsh 1977: 137–143
Hahn G, Hahn R (1987) Trilobiten aus dem Karbon von Nötsch und aus den Karnischen Alpen Österreichs. Jahrb Geol Bundesanst 129: 567–619
Hahn G, Kratz R (1992) Eine Trilobiten-Fauna des tiefen Wassers aus dem Unter-Karbon der Karnischen Alpen (Österreich) – vorläufige Mitteilung. Jahrb Geol Bundesanst 135: 217–224
Hahn G, Hahn R, Schneider G (1989) Neue Trilobitenfunde aus der Waidegg-Formation (hohes Oberkarbon) der Karnischen Alpen (Österreich). Jahrb Geol Bundesanst 132: 645–664
Hammann W (1976) The Ordovician of the Iberian Peninsula – A review. In: Bassett MG (ed) The Ordovician system. Proc Palaeontol Assoc Symp, Birmingham 1974. Univ Wales Press and Nat Mus Wales, pp 387–409
Hambrey MJ, Harland WB (eds) (1981) Earth's pre-Pleistocene glacial record. Cambridge Univ Press, Cambridge
Haq BU, Eysinga van FWB (1987) Geological time table, 4th edn. Elsevier, Amsterdam
Hartefelt JJA (1970) Geology of the Upper Segre and Valira valleys, central Pyrenees, Andorra/Spain. Leidse Geol Meded 45: 167–236
Havlicek V (1976) Evolution of Ordovician brachiopod communities in the Mediterranean province. In: Bassett MG (ed) The Ordovician system. Proc Palaeontol Assoc Symp, Birmingham 1974. Univ Wales Press and Nat Mus Wales, pp 349–358
Havlicek V, Kriz J, Serpagli E (1987) Upper Ordovician brachiopod assemblages of the Carnic Alps, Middle Carinthia and Sardinia. Boll Soc Paleontol Ital 25: 277–311
Heckel PH, Witzke BJ (1979) Devonian World palaeogeography determined from distribution of carbonates and related lithic palaeoclimatic indicators. In: House MR, Scrutton CT, Bassett MG (eds) The Devonian system. Spec Pap Paleontol 23: 99–123
Heinisch H (1988) Hinweise auf die Existenz eines passiven Kontinentalrandes im Altpaläozoikum der Nördlichen Grauwackenzone – Ostalpen. Schweiz Mineral Petrogr Mitt 68: 407–418
Henningsen D, Herbig H–G (1990) Die karbonische Grauwacken der Malagiden und Menorcas im Vergleich (Betische Kordillere und Balearen, Spanien). Z Dtsch Geol Ges 141: 13–29
Herbig H–G (1986) Rugosa und Heterocorallia aus Obervisé-Geröllen der Marbella-Formation (Betische Kordillere, Südspanien). Paläontol Z 60: 189–225
Heritsch F (1929) Faunen aus dem Silur der Ostalpen. Abh Geol Bundesanst 23/2: 1–183
Heritsch F (1936) Die Karnischen Alpen. Monographie einer Gebirgsgruppe der Ostalpen mit variszischen und alpidischem Bau. Geol Inst Univ Graz, Graz, pp 1–205
Heritsch F (1943) Das Paläozoikum. In: Die Stratigraphie der geologischen Formationen der Ostalpen vol, I. Borntraeger, Berlin, pp 1–681
Hill D (1959) Distribution and sequence of Silurian coral faunas. J Proc R Soc N S W 42 NS 70: 151–173
Hill D, Jell JS (1969) On the rugose coral genera *Rhizophyllum* Lindström, *Platyphyllum* Lindström and *Calceola* Lamarck. Neues Jahrb Geol Paläonzol Monatsh 9: 534–551
Höll R (1970) Die Zinnober-Vorkommen im Gebiet der Turracher Höhe (Nock-Gebiet/Österreich) und das Alter der Eisenhut-Schieferserie. Neues Jahrb Geol Paläontol Monatsh 1970: 201–224
Holland HD (1990) Origins of breathable air. Nature 347: 17
Hollister CD, Heezen BC (1972) Geological effects of ocean bottom currents, western North Atlantic. In: Gordon AL (ed) Studies in physical oceanography vol, 1. Gordon, New York, pp 37–66
Homann W (1971) Korallen aus dem Unter- und Mittelperm der Karnischen Alpen. Carinthia II SH 28: 97–143

Homann W (1972) Unter- und tief-mittelpermische Kalkalgen aus den Rattendorfer Schichten, dem Trogkofel-Kalk und dem Tressdorfer Kalk der Karnischen Alpen (Österreich). Senckenb Lethaea 53: 135–313

House MR (1971) Devonian faunal distribution. In: Middlemiss FA, Rawson PF, Newall G (eds) Faunal provinces in space and time. Palaeontol Assoc, Seel House Press, Liverpool, pp 77–94

House MR (1973) Devonian Goniatites. In: Hallam A (ed) Atlas of palaeobiogeography. Elsevier, Amsterdam, pp 97–104

House MR, Price JD (1980) Devonian ammonoid faunas of the Carnic Alps, a review. In: Schönlaub HP (ed) Second European Conodont Symposium ECOS II, Field Trip A. Abh Geol Bundesanst 35: 14–15

Hubmann B (1990) Udoteaceen (Grünalgen) aus dem Grazer Paläozoikum/Österreich (Barrandeikalke, Eifelium). Facies 22: 147–158

Jablonski D (1986) Background and mass extinctions: the alternation of macroevolutionary regimes. Science 270: 129–133

Jaeger H (1968) Vorbericht über graptolithenstratigraphische Untersuchungen in den Karnischen Alpen, insbesondere an der Bischofalm. Anz Österr Akad Wiss Wien Math-Naturwiss Kl 1968 (7): 155–159

Jaeger H (1975) Die Graptolithenführung im Silur/Devon des Cellon-Profils (Karnische Alpen). Carinthia II 165/85: 111–126

Jaeger H (1976) Das Silur und Unterdevon vom thüringischen Typ in Sardinien und seine regionalgeologische Bedeutung. In: Franz-Kossmat-Symposium. Nova Acta Leopold NF 224: 263–299

Jaeger H (1978) Late Graptoloid Faunas and the Problem of Graptoloid Extinction. Acta Palaeontol Pol 23: 497–521

Jaeger H (1988) Devonian Graptoloidea. In: McMillan NJ, Embry AF, Glass DJ (eds) Devonian of the World. Proc 2nd Int Symp Devonian System, Calgary, Canada 3: 431–438

Jaeger H, Havlicek V, Schönlaub HP (1975) Biostratigraphie der Ordovizium/Silur-Grenze in den Südalpen – ein Beitrag zur Diskussion um die Hirnantia-Fauna. Verh Geol Bundesanst 1975: 271–289

Jäger E (1977) The evolution of the central and western European continent. In: Vidal P (ed) Limitation isotopique à l'age de l'évolution de la Croute Continentale en Europe Moyenne et Occidentale. Colloq Int Cent Natl Rech Sci 234: 129–141

Jhaveri RB (1969) Unterdevonische Gastropoden aus den Karnischen Alpen. Palaeontographica A 133: 146–176

Kahler F (1939) Verbreitung und Lebensdauer der Fusuliniden-Gattung *Pseudoschwagerina* und *Paraschwagerina* und deren Bedeutung für die Grenze Karbon/Perm. Senckenb Lethaea 21: 169–215

Kahler F (1955) Entwicklungsräume und Wanderwege der Fusuliniden im euroasiatischen Kontinent. Geologie 4: 178–188

Kahler F (1962) Stratigraphische Vergleiche im Karbon und Perm mit Hilfe der Fusuliniden. Mitt Geol Ges Wien 54: 147–161

Kahler F (1973) Beiträge zur Kenntnis der Fusuliniden der Ostalpen: Die Gattung *Quasifusulina* in den Karnischen Alpen. Palaeontographica A 141: 154–173

Kahler F (1974) Fusuliniden aus T'ien-schan und Tibet. Mit Gedanken zur Geschichte der Fusuliniden-Meere im Perm. Repts Sci Exp NW Prov China under Leadersh Sv Hedin, Sven Hedin Found, Stockholm, Publ 52 vol. Inv Palaeont 4: 1–148

Kahler F (1980) Zur Definition der Saalischen Phase im marinen Bereich der Südalpen. In: Flügel E (ed) Die Trogkofel-Stufe im Unterperm der Karnischen Alpen. Carinthia II SH 36: 259–260

Kahler F (1983) Fusuliniden aus Karbon und Perm der Karnischen Alpen und der Karawanken. Carinthia II SH 41: 1–107

Kahler F (1985) Oberkarbon und Unterperm der Karnischen Alpen. Ihre Biostratigraphie mit Hilfe der Fusuliniden. Carinthia II SH 42: 1–93

Kahler F (1989) Die Fusuliniden. In: Zapfe H (ed) Catalogus Fossilium Austriae, H II/b/1: Foraminifera Palaeozoica. Österr Akad Wiss Wien, pp 87–271

Kahler F, Kahler G (1982) Beiträge zur Kenntnis der Fusuliniden der Ostalpen: Oberkarbonische Fusuliniden der Karnischen Alpen. Palaeontographica A 177: 89–128

Kaljo DL, Klaamann E (1973) Ordovician and Silurian corals. In: Hallam A (ed) Atlas of palaeobiogeography. Elsevier, Amsterdam, pp 37–45

Kelly PH, Raymond A, Lutken CB (1990) Carboniferous brachiopod migration and latitudinal diversity: a new palaeoclimatic method. In: McKerrow WS, Scotese CR (eds) Palaeozoic palaeogeography and biogeography. Geol Soc Mem 12: 325–332

Kennedy WQ (1964) The structural differentiation of Africa in the Pan-African (+ 500 m. y.) tectonic episode. Univ Leeds Res Inst Afr Geol Annu Rep 8: 48–49

Kent DV, Van Der Voo R (1990) Palaeozoic plaaeogeography from palaeomagnetism of the Atlantic-bordering continents. In: McKerrow WC, Scotese CR (eds) Palaeozoic palaeogeography and biogeography. Geol Soc Mem 12: 49–56

Kinsman DJJ (1964) Reef coral tolerance of high temperatures and salinities. Nature 202: 1280–1282

Klapper G, Johnson JG (1980) Endemism and dispersal of Devonian conodonts. J Paleontol 54: 400–455

Klaus W (1974) Neue Beiträge zur Datierung von Evaporiten der Ober-Perm. Carinthia II 164: 79–85

Klaus W (1980) Microfloristic events towards Permo-Triassic boundary. Proc IV Int Palyn Conf Lucknow (1976–77) 2: 228–232

Knoll AH, Butterfield NJ (1989) New window on Proterozoic life. Nature 337: 602–603

Kodsi MG (1971) Korallen aus dem Unterdevon der Karnischen Alpen. Verh Geol Bundesanst 1971: 576–607

Kokelaar BP, Howells MF, Bevins RE, Roach RA, Dunkey PN (1984) The Ordovician marginal basin of Wales. In: Kokelaar BP, Howells MF (eds) Marginal basin geology. Geol Soc Spec Publ 16: 245–269

Koren T (1979) Late monograptid faunas and the problem of graptolite extinction. Acta Palaeontol Pol 24: 79–106

Korn D (1992) Ammonoideen vom Devon/Karbon-Grenzprofil an der Grünen Schneid, Karnische Alpen, Österreich. Jahrb Geol Bundesanst 135: 7–19

Kraiger H (1989) Die Habachformation – ein Produkt ozeanischer und kontinentaler Kruste. Mitt Österr Geol Ges 81: 47–64

Krainer K (1984) Sedimentologische Untersuchungen an permischen und untertriadischen Sedimenten des Stangalm-Mesozoikums (Kärnten/Österreich). Jahrb Geol Bundesanst 127: 159–179

Krainer K (1985) Zur Sedimentologie des alpinen Buntsandsteins und der Werfener Schichten (Skyth) Kärntens. Geol Paläontol Mitt Innsbr 14: 21–81

Krainer K (1990) Zur faziellen Entwicklung Pflanzenfossil-führender Oberkarbon- und Unterpermsedimente Kärntens. Carinthia II 180/SH 49: 109–132

Krainer K, Mogessie A (1991) Composition and significance of resedimented amphibolite breccias and conglomerates

("Badstubbreccia Complex") in the Carboniferous of Nötsch (Eastern Alps, Austria). Jahrb Geol Bundesanst 134: 65–81
Kreutzer LH (1989) Reef-basin distance in the Devonian of the Carnic Alps. Ann Soc Geol Belg 112: 159–163
Kreutzer LH (1990) Mikrofazies, Stratigraphie und Paläogeographie des Zentralkarnischen Hauptkammes zwischen Seewarte und Cellon. Jahrb Geol Bundesanst 133: 275–343
Kriz J (1979) Silurian Cardiolidae (Bivalvia). Sb Geol VED Paleontol 22: 1–157
Kügel H-W (1987) Sphinctozoen aus den Auernigschichten des Naßfeldes (Oberkarbon, Karnische Alpen, Österreich). Facies 16: 143–156
Kurat G, Niedermayr G, Korkisch J, Seemann R (1974) Zur Geochemie der postvariszischen Basis-Serien im westlichen Drauzug, Kärnten-Osttirol. Carinthia II 164: 87–98
Küster P (1987) Nachweis von Frasnium durch *Scruttonia julli* (Pedder, 1986) in den Flachwasserkalken der zentralen Karnischen Alpen. Münster Forsch Geol Paläontol 66: 33–56
Latz S (1989) Unterdevonische Brachiopoden aus dem Hellen Crinoidenschutt- und Riffkalk der Seewarte (Zentrale Karnische Alpen, Kärnten, Österreich). Inaug Diss, Univ Bonn, pp 1–127 (unpubl)
Lefort J-P (1989) Basement correlation across the North Atlantic. Springer, Berlin Heidelberg New York
Leonardi P (1948) Contributi alla conoscenza della Flora delle Arenarie di Val Gardena (Permiano medio-inferiore) dell'Alto Adige. La nuova flora di Redagno e una felce di Egna. Mem Ist Geol Univ Padova 16: 3–15
Loeschke J (1989a) Lower Palaeozoic volcanism of the Eastern Alps and its geodynamic implications. Geol Rundsch 78: 599–616
Loeschke J (1989b) Die paläotektonische Stellung der Vulkanite der Magdalensberg-Serie (Ober-Ordovizium, Gurktaler Decke, Kärnten, Österreich). Carinthia II 179: 491–507
Lorenz V (1976) Formation of Hercynian subplates, possible causes and consequences. Nature 262: 374–377
Lottes AL, Rowley DB (1990) Reconstruction of the Laurasian and Gondwanan segments of Permian Pangaea. In: McKerrow WS, Scotese CR (eds) Palaeozoic palaeogeography and biogeography. Geol Soc Mem 12: 383–395
Lovell JBP, Stow DAV (1981) Identification of ancient sandy contourites. Geology 9: 347–349
Marek L (1976) The distribution of the Mediterranean Ordovician: In: Bassett MG (ed) The Ordovician system, Proc Palеontol Assa Symp, Birmingham 1974. Univ Wales Press, Nat Mus Wales, Cardiff, pp 491–499
Martin F (1978) Sur quelques Acritarches Llandoveriens de Cellon (Alpes Carniques Centrales, Autriche). Verh Geol Bundesanst 1978: 35–42
Massari F, Conti MA, Fontana D, Helmold K, Mariotti N, Neri C, Nicosia U, Ori GB, Pasini M, Pittau P (1988) The Val Gardena Sandstone and Bellerophon Formation in the Bletterbach Gorge (Alto Adige, Italy), biostratigraphy and sedimentology. Mem Sci Geol 40: 229–273
Matte P (1986) Tectonics and plate tectonics model for the Variscan belt of Europe. Tectonophysics 126: 329–374
McKerrow WS, Cocks LRM (1986) Oceans, island arcs and olistostromes: the use of fossils in distinguishing sutures, terranes and environments around the Iapetus ocean. J Geol Soc Lond 143: 185–191
McKerrow WS, Ziegler AM (1972) Palaeozoic oceans. Nature 240: 92–94
McLean RA (1977) Biostratigraphy and zoogeographic affinities of the Lower Silurian Corals of New South Wales, Australia. Mém Bur Rech Géol Min 89: 102–107
McLean RA (1985) New Early Silurian rugose corals from the Panuara area, central New South Wales. Alcheringa 9: 23–34
Melou M (1990) Brachiopodes articulés de la coupe de l'Ile de Rosan (Crozon, Finistère). Formation des Tufs et Calcaires de Rosan (Caradoc-Ashgill). Geobios 23: 539–579
Mensink H (1953) Eine tektonische Detailuntersuchung im Raum nördlich Gratkorn. Mitt Naturwiss Ver Steiermark 83: 123–129
Mikkelsen N, Erlenkeuser H, Killingley JS, Berger WH (1982) Norwegian corals: radiocarbon and stable isotopes in *Lophelia pertusa*. Boreas 11: 163–171
Mostler H (1966) Zur Einstufung der „Kieselschiefer" von der Lachtal-Grundalm (Fieberbrunn, Tirol). Verh Geol Bundesanst 1966: 157–170
Mostler H (1970) Struktureller Wandel und Ursachen der Faziesdifferenzierung an der Ordoviz/Silur-Grenze in der Nördlichen Grauwackenzone (Österreich). Festbd Geol Inst 300 J Univ Innsbr, pp 507–522
Neubauer F (1979) Die Gliederung des Altpaläozoikums südlich und westlich von Murau (Steiermark/Kärnten). Jahrb Geol Bundesanst 122: 455–511
Neubauer F (1984) Das Murauer und Gurktaler Paläozoikum. In: Flügel HW, Neubauer F (eds) Steiermark-Geologie der österreichischen Bundesländer in kurzgefassten Einzeldarstellungen. Geol Bundesanst, pp 55–58
Neubauer F (1988) The Variscan orogeny in the Austroalpine and Southalpine domains of the Eastern Alps. Schweiz Mineral Petrogr Mitt 68: 339–349
Neubauer F, Pistotnik J (1984) Das Altpaläozoikum und Unterkarbon des Gurktaler Deckensystems (Ostalpen) und ihre paläogeographischen Beziehungen. Geol Rundsch 73: 149–174
Neugebauer J (1988) The Variscan plate tectonic evolution: an improved "Iapetus model". Schweiz Mineral Petrogr Mitt 68: 313–333
Niedermayr G (1975) Gedanken zur lithofaziellen Gliederung der postvariszischen Transgressions-Serie der westlichen Gailtaler Alpen. Mitt Geol Ges Wien 66/67: 105–126
Niedermayr G (1985) Fluvial Braidplain Passing into an Intertidal Belt at the Margin of the Tethys Sea in the Alpine Buntsandstein of the Drauzug in Carinthia and Eastern Tyrol (Austria). In: Mader D (ed) Aspects of fluvial sedimentation in the Lower Triassic Buntsandstein of Europe. Lecture Notes in Earth Sciences 4. Springer, Berlin Heidelberg New York, pp 487–496
Niedermayr G, Scheriau-Niedermayr E (1982) Zur Nomenklatur, Seriengliederung und Lithofazies der permo-skythischen Basisschichten des westlichen Drauzuges. Verh Geol Bundesanst 1982: 33–51
Oczlon MS (1990) Ocean currents and unconformities: the North Gondwana Middle Devonian. Geology 18: 509–512
Oliver WA Jr (1964) New occurrences of the rugose coral *Rhizophyllum* in North America. US Geol Surv Prof Pap 475-D, Art 153: D149–D158
Oliver WA Jr (1977) Biogeography of Late Silurian and Devonian rugose corals. Palaeogeogr, Palaeoclimatol Palaeoecol 22: 85–135
Ori GG, Venturini C (1981) Facies continentali Permiane nelle Alpi Carniche. Rend Soc Geol Ital 3: 21–24
Palla P (1966) *Lancicula gortanii*, n. sp. di Alga Codiacea del Devoniano inferiore della Carnia (Alpi Orientali). Natura 57: 89–102
Palla P (1967) New Lower Devonian Codiacean Algae from Carnic Alps. Natura 57: 94–112
Palmer AR (1972) Problems of Cambrian biogeography. Proc 24th Int Geol Congr 7: 310–315

Paris F (1981) Les Chitinozoaires dans le Paléozoique du Sud-Ouest de l'Europe. Mém Soc Géol Minéral Bretagne 26: 1–412
Paris F (1990) The Ordovician chitinozoan biozones of the northern Gondwana Domain. Rev Palaeobot Palynol 66: 181–209
Paris F, Robardet M (1990) Early Palaeozoic palaeobiogeography of the Variscan regions. Tectonophysics 177: 193–213
Paris F, Pelhate A, Weyant M (1981) Conodontes ashgilliens dans la Formation de Rosan, coupe de Lostmarc'h (Finistère, Massif armoricain). Conséquences paléogéographiques. Bull Soc Géol Minéral Bretagne 13: 15–35
Pedder AEH, Oliver Jr WA (1990) Rugose coral distribution as a test of Devonian palaeogeographic models. In: McKerrow WS, Scotese CR (eds) Palaeozoic palaeogeography and biogeography. Geol Soc Mem 12: 267–275
Pickering KT (1989) The destruction of Iapetus and Tornquist's oceans. Geology Today Sept/Oct 1989: 160–166
Pickett JW (1975) Continental reconstructions and the distribution of coral faunas during the Silurian. J Proc R Soc N S W 108: 147–156
Planderova E, Pahr A (1983) Biostratigraphical evaluation of weakly metamorphosed sediments of Wechsel Series and their possible correlation with Harmónia Group in Malé Karpaty Mts. Mineral Slow 15: 385–436
Poli S (1989) Pre-Hercynian magmatism in the Eastern Alps: the origin of metabasites from the Austroalpine basement. Schweiz Mineral Petrogr Mitt 69: 407–421
Poncet J (1990) Biogeography of Devonian Algae. In: McKerrow WS, Scotese CR (eds) Palaeozoic palaeogeography and biogeography. Geol Soc Mem 12: 285–289
Poty E (1981) Recherches sur les Tétracoralliaires et les Hétérocoralliaires du Viséen de la Belgique. Meded Rijks Geol Dienst 35: 1–161
Priewalder H (1976) Petrographisch-tektonische Untersuchungen in den Ennstaler Phylliten (Niedere Tauern, Steiermark) und deren Einstufung in das Silur durch Chitinozoen. Verh Geol Bundesanst 1976: 95–113
Priewalder H (1987) Acritarchen aus dem Silur des Cellon-Profils, Karnische Alpen, Österreich. Abh Geol Bundesanst 40: 1–121
Quadt von A (1985) Geochemische und U-Pb-Untersuchungen der unteren Schieferhülle (Ostalpen). Fortschr Mineral 62: 191–193
Quadt von A (1987) U-Pb- und Sm-Nd-Datierungen der Habach-Formation und des Basisamphibolits im Bereich der mittleren Hohen Tauern. Uni Aktuell Salzburg 10: 28
Ramovs A (1972) Mittelpermische Klastite und deren marine Altersäquivalente in Slowenien, NW Jugoslawien. Verh Geol Bundesanst 1972: 35–45
Rantitsch G (1990) Fazies und Diagenese devonischer Riffkalke des Seeberger Aufbruches (Kärnten, Österreich). Diplomarb, Inst Geol Paläontol Univ Graz, pp 1–120
Ratschbacher R (1984) Beitrag zur Neugliederung der Veitscher Decke (Grauwackenzone) in ihrem Westabschnitt (Obersteiermark, Österreich). Jahrb Geol Bundesanst 127: 423–453
Rau A, Tongiordi M (1981) Some problems regarding the Palaeozoic paleogeography in Mediterranean western Europe. J Geol 89: 663–673
Reitz E, Höll R (1988) Jungproterozoische Mikrofossilien aus der Habachformation in den mittleren Hohen Tauern und dem nordostbayerischen Grundgebirge. Jahrb Geol Bundesanst 131: 329–340
Reitz E, Höll R (1989) Unterordovizische Acritarchen aus der Nördlichen Grauwackenzone (Ostalpen). Jahrb Geol Bundesanst 132: 761–774
Reitz E, Höll R (1990) Biostratigraphischer Nachweis von Unterordovizium in der Innsbrucker Quarzphyllitserie (Ostalpen). Jahrb Geol Bundesanst 133: 603–610
Reitz E, Daneck T, Miller H (1989) Ein Nachweis jungproterozoischen Alters von Schwarzphylliten am Tauern-Nordrand (Salzburg, Österreich) und seine Bedeutung für den Bau der Hohen Tauern. Jahrb Geol Bundesanst 132: 751–760
Richardson JB, Rasul SM, Al-Ameri T (1981) Acritarchs, miospores and correlation of the Ludlovian-Downtonian and Silurian-Devonian boundaries. Rev Palaeobot Palynol 34: 209–224
Riding R (1974) Model of Hercynian foldbelt. Earth Planet Sci Lett 24: 125–135
Riehl-Herwirsch G (1972) Vorstellung zur Paläogeographie – Verrucano. Verh Geol Bundesanst 1972: 97–106
Ristedt H (1968) Zur Revision der Orthoceratidae. Abh Akad Wiss Mainz Math Naturwiss Kl 1968 (4): 213–287
Ristedt H (1969) Orthoceren als Leitfossilien des Silurs. Carinthia II SH 27: 25–28
Robardet M, Paris F, Racheboeuf PR (1990) Palaeogeographic evolution of southwestern Europe during Early Paleozoic times. In: McKerrow WS, Scotese CR (eds) Palaeozoic palaeogeography and biogeography. Geol Soc Mem 12: 411–419
Ross CA, Ross JRP (1985) Carboniferous and Early Permian biogeography. Geology 13: 27–30
Ross RJ (1975) Early Paleozoic trilobites, sedimentary facies, lithospheric plates, and ocean current. Fossils Strata 4: 307–329
Ruggieri G (1966) Nuovo genere di ostracode del Carbonifero superiore delle Alpi Carniche. Riv Ital Paleontol 72: 1–8
Sacchi R (1989) Pan African Rejuvenation. Proc 28th Int Geol Congr Wahington (Abstr) 3: 2–3
Schallreuter R (1990) Ordovizische Ostrakoden und Seeigel der Karnischen Alpen und ihre Beziehungen zu Böhmen und Baltoskandien. Neues Jahrb Geol Paläontol Monatsh 1990: 120–128
Schlöser H, Kullmann J, Loeschke J (1990) Korallen-führendes Unterkarbon auf der Brunnachhöhe (Nockgebiet, Gurktaler Decke, Österreich). Carinthia II 180: 643–650
Schmidt PW, Powell CMcA, Li ZX, Thrupp GA (1990) Reliability of Palaeozoic palaeomagnetic poles and APWP of Gondwanaland. Tectonophysics 184: 87–100
Schönenberg R (1973) Zur Frage der Verbindung von Sudetikum und ostalpinem Variszikum. Zentralinst Phys Erde 14: 437–450
Schönlaub HP (1971) Paleo-environmental studies at the Ordovician/Silurian boundary in the Carnic Alps. In: Colloque Ordovicien-Silurien Brest 1971. Mem Bur Rech Géol Min 73: 367–377
Schönlaub HP (1979) Die Nördliche Grauwackenzone. In: Das Paläozoikum von Österreich. Abh Geol Bundesanst 33: 76–97
Schönlaub HP (1980) Carnic Alps. Field Trip A. In: Schönlaub HP (ed) 2nd European Conodont Symposium ECOS II, Guidebook-Abstracts. Abh Geol Bundesanst 35: 5–57
Schönlaub HP (1985a) Das Paläozoikum der Karnischen Alpen. In: Arbeitstagung der Geologischen Bundesanstalt 1985. Geol Bundesanst 1985: 34–52
Schönlaub HP (1985b) Das Karbon von Nötsch und sein Rahmen. Jahrb Geol Bundesanst 127: 673–692
Schönlaub HP (1986) Significant geological events in the Paleozoic record of the Southern Alps (Austrian part). In: Walliser OH (ed) Global bio-events. Lecture Notes in Earth Sciences 8, Springer, Berlin Heidelberg New York, pp 163–167
Schönlaub HP (1988) The Ordovician-Silurian boundary in the Carnic Alps of Austria. In: Cocks LRM, Rickards RB (eds) A

global analysis of the Ordovician-Silurian boundary. Bull Br Mus Nat Hist (Geol) 43: 107–115
Schönlaub HP (1990) The collapse of the Carboniferous flysch trough in the Carnic Alps. In: Venturini C, Krainer K (eds) Field workshop on Carboniferous to Permian sequence of the Pramollo-Nassfeld Basin (Carnic Alps). Proc, p 14
Schönlaub HP, Feist R, Korn D (1988) The Devonian-Carboniferous boundary at the section „Grüne Schneid" (Carnic Alps, Austria): a preliminary report. Cour Forschungsinst Senckenb 100: 149–167
Schraut G (1990) Neue Trilobiten und andere Fossilien aus dem Karbon von Nötsch. Diplomarb (Teil 1), Fachbereich Geowiss Univ Marburg, pp 1–62
Scoffin TP, Tudhope AW, Brown BE (1989) Corals as environmental indicators, with preliminary results from South Thailand. Terra Nova 2: 559–563
Scotese CR, Barrett SF (1990) Gondwana's movement over the South Pole during the Palaeozoic: evidence from lithological indicators of climate. In: McKerrow WS, Scotese CR (eds) Palaeozoic palaeogeography and biogeography. Geol Soc Mem 12: 75–85
Scotese CR, McKerrow WS (1990) Revised world maps and introduction. In: McKerrow WS, Scotese CR (eds) Palaeozoic palaeogeography and biogeography. Geol Soc Mem 12: 1–21
Sdzuy K (1962) Neue kambrische Fossilien aus Bohrungen im Gebiet von Doberlug. Geologie 11: 1087–1101
Serpagli E (1967) I conodonti dell'Ordoviciano superiore (Ashgilliano) delle Alpi Carniche. Boll Soc Paleontol Ital 6: 30–111
Söllner F, Hansen BT (1987) „Pan-afrikanisches" und „kaledonisches" Ereignis im Ötztalkristallin der Ostalpen: Rb-Sr- und U-Pb-Altersbestimmungen an Migmatiten und Metamorphiten. Jahrb Geol Bundesanst 130: 524–569
Stock CW (1990) Biogeography of the Devonian stromatoporoids. In: McKerrow WS, Scotese CR (eds) Palaeozoic palaeogeography and biogeography. Geol Soc Mem 12: 257–265
Strehl E (1962) Die geologische Neuaufnahme des Saualpen-Kristallins. IV. Das Paläozoikum und sein Deckgebirge zwischen Klein St. Paul und Brückl. Carinthia II 152: 46–74
Stromgren T (1971) Vertical and horizontal distribution of *Lophelia pertusa* (Linne) in Trondheimsfjorden on the West Coast of Norway. K Nor Vidensk Selsk Skr 6: 1–19
Sweet WC, Bergström SM (1984) Conodont provinces and biofacies of the Late Ordovician. In: Clark DL (ed) Conodont biofacies and provincialism. Geol Soc Am Spec Pap 196: 69–87
Sylvester H (1989) Vergleich von Perm/Skyth-Profilen des Ober-Mittel- und Unterostalpins. Jahrb Geol Bundesanst 132: 791–821
Teich T (1982) Zum Chemismus der Badstubbrekzie im Unterkarbon von Nötsch in Kärnten. Carinthia II 162: 91–96
Tenchov YG (1980) Die paläozoische Megaflora von Österreich. Eine Übersicht. Verh Geol Bundesanst 1980: 161–174
Thiedig F, Klussmann D (1974) Limnisches Oberkarbon an der Basis der postvariszischen Transgressionsserie in den St. Pauler Bergen Ostkärntens. Mitt Geol Paläontol Inst Univ Hamb 43: 79–84
Torsvik HJ, Smethurst MA, Briden JC, Sturt BA (1990) A review of Palaeozoic palaeomagnetic data from Europe and their palaeogeographical implications. In: McKerrow WS, Scotese CR (eds) Palaeozoic palaeogeography and biogeography. Geol Soc Mem 12: 25–41
Tribble GW, Randall RH (1986) A description of the high-latitude shallow water coral communities of Miyaka-jima, Japan. Coral Reefs 4: 151–159
Vai GB (1971) Ordovicien des Alpes Carniques. In: Coll Ordovicien-Silurien Brest 1971. Mém Bur Rech Geol Min 73: 437–450
Vai GB (1973) Distribution of *Karpinskia* in the Lower Devonian of Mount Coglians (Carnic Alps). Trudy III Mem Simp Silura/Devona 2: 75–79
Vai GB (1975) Hercynian basin evolution of the Southern Alps. In: Squyres C (ed) Geology of Italy. Earth Sci Soc Libyan Arab Rep 1975: 293–298
Vai GB, Cocozza T (1986) Tentative schematic zonation of the Hercynian chain in Italy. Bull Soc Géol Fr 2: 95–114 (ed)
Vai GB, Spalletta C (1980) Field Trip A. The Uggwa Section In: Schönlaub HP (ed) 2nd European Conodont Symp ECOS II, Guidebook Abstr. Abh Geol Bundesanst Wien, pp 48–50
van der Voo R (1988) Paleozoic paleogeography of North America, Gondwana, and intervening displaced terranes: comparisons of paleomagnetism with paleoclimatology and biogeographical patterns. Geol Soc Am Bull 100: 311–324
van der Voo R (1990) The reliability of paleomagnetic data. Tectonophysics 184: 1–9
Vasicek W (1977) Perm von Zöbing. In: Exk Führer Arbeitstag Geol Bundesanst 1977 Waldviertel. Geol Bundesanst, Wien, pp 16–18
Vavra G (1989) Die Entwicklung des penninischen Grundgebirges im östlichen und zentralen Tauernfenster der Ostalpen. Geochemie, Zirkonmorphologie, U/Pb-Radiometrie. Tübinger Geowiss Arb A 6: 1–150
Veevers JJ, McPowell C (1987) Late Palaeozoic glacial episodes in Gondwanaland reflected in transgressive-regressive depositional sequences in Euramerica. Geol Soc Am Bull 98: 475–487
Vidal Ph (1977) Limitation Isotopique à l'age de l'évolution de la Croute Continentale en Europe Moyenne et Occidentale. Colloq Int Cent Natl Rech Sci 234: 227–238
Vidal Ph; Auvray B, Charlot R, Cogné J (1981) Precadomian relics in the Armorican Massif: their age and role in the evolution of the Western and Central European Cadomian-Hercynian belt. Precambrian Res 14: 1–20
Visser JNJ (1990) The age of the late Palaeozoic glacigene deposits in southern Africa. S Afr J Geol 93: 366–375
Walliser OH (1964) Conodonten des Silurs. Abh Hess Landesamtes Bodenforsch 41: 1–106
Webby PD (1984) Ordovician reefs and climate: a review. In: Bruton DL (ed) Aspects of the Ordovician system. Palaeontological contributions from the Univ Oslo 295: 89–100
Whittington HB (1973) Ordovician trilobites. In: Hallam A (ed) Atlas of palaeobiogeography. Elsevier, Amsterdam, pp 13–18
Whittington HB, Hughes CP (1972) Ordovician geography and faunal provinces deduced from trilobite distribution. Philos Trans R Soc Ser B 263: 235–278
Whittington HB, Hughes CP (1973) Ordovician trilobite distribution and geography. Spec Pap Palaeontol 12: 235–240
Winkler Prins CF (1971) Connections of the Carboniferous brachiopod faunas of the Cantabrian Mountains (Spain). In: Wagner RH (ed) The Carboniferous of northwest Spain, Pr II. Trab Geol 4: 687–694
Winkler Prins CF (1983) A general review of the Carboniferous brachiopods from the Cantabrian Mountains (North Spain). In: Lemos de Sousa MJ (ed) Contributions to the Carboniferous geology and palaeontology of the Iberian Peninsula. Porto, pp 69–91
Winkler Prins CF (1984) Brachiopods and the main classification of the Carboniferous. In: Sutherland PK, Manger WL (eds) Compte Rendu 2-Biostratigraphy, 9th ICC. Southern Ill Univ Press, Carbondale-Edwardsville, pp 47–51

Wopfner H (1984) Permian deposits of the Southern Alps as product of initial alpidic taphrogenesis. Geol Rundsch 73: 259–277

Young TP (1985) The stratigraphy of the Upper Ordovician of central Portugal. PhD Theses, Univ Sheffield

Young TP (1988) The lithostratigraphy of the Upper Ordovician of central Portugal. J Geol Soc Lond 145: 377–392

Young TP (1990) Ordovician sedimentary facies and faunas of southwest Europe: palaeogeographic and tectonic implications. In: McKerrow WS, Scotese CR (eds) Palaeozoic palaeogeography and biogeography. Geol Soc Mem 12: 421–430

Zang WL, Walter MR (1989) Latest Proterozoic plankton from the Amadeus Basin in central Australia. Nature 337: 642–645

Ziegler AM, Bambach RK, Parrish JT, Barrett SF, Gierlowski EH, Parker WC, Raymond A, Sepkoski JJ Jr (1981) Palaeozoic biogeography and climatology. In: Niklas KJ (ed) Paleobotany, paleoecology and evolutions vol 2. Praeger, New York, pp 231–266

Ziegler AM, Hulver ML, Lottes AL, Schmachtenberg WF (1984) Uniformitarianism and palaeoclimates: inferences from the distribution of carbonate rocks. In: Brenchley PJ (ed) Fossils and climate. Wiley, Chichester, pp 3–25

Ziegler PA (1984) Caledonian and Hercynian crustal consolidation of western and central Europe – a working hypothesis. Geol Mijnbouw 63: 93–108

Ziegler PA (1986) Geodynamic model for the Palaeozoic crustal consolidation of western and central Europe. Tectonophysics 126: 303–328

Ziegler PA (1989) Evolution of Laurussia – a study in Late Palaeozoic plate tectonics. Kluwer, Dordrecht

The Pre-Alpine Evolution of the Continental Crust of the Central Alps – An Overview

D. GEBAUER

Abstract

As for the European Variscides, the continental crust of the Central Alps is part of Gondwana. Geochronologically, this is manifested by a characteristic sequence of geological events presently found in the Brasilian shield or in Africa. However, in contrast to the latter shield areas, the Central Alpine and Variscan crusts consist, in their present erosion level, almost exclusively of recycled Gondwana crust, being geochronologically distinct from continental crust of the other supercontinent, Laurasia. Major Precambrian crust-forming events, as detected by ion-probe dating of detrital zircons from a paragneiss of the Gotthard Massif and a recent river sand from the Po delta, date back to 3.43 Ga and are concentrated around 2.6, 2.1, 1.0 and 0.65 Ga. Precambrian igneous rocks are, as notorious for the European Variscides, very rare in the Central Alps and geochronologically difficult to date. Often they are mafic rocks extracted from a suboceanic mantle, as for example 870-Ma-old gabbros of the Gotthard Massif, metamorphosed to eclogites in the Ordovician (468 Ma). Similar rocks are known from the Berisal complex (Simplon area), the Siviez-Mischabel unit (Valais) and from the southern part of the Penninic nappes. However, their 'ages' of ca. 1 Ga still need to be substantiated by further geochronological work, which is also the case for intermediate and felsic orthogneisses of the Upper Austro-Alpine Silvretta nappe.

There are a number of mantle melting events detected in mafic and ultramafic rocks of the Gotthard Massif, the southern steep belt (Lepontine area) and the Ivrea zone, which are probably related to oceanization and continental rift processes. They range from 3.17, 2.67, 2.45, 1.72, 1.27 Ga to Pan-African ages (650 and 670 Ma).

The post-Pan-African and pre-Ordovician evolution probably starts with rifting and oceanization processes in the course of which many precursors of the widespread metasediments were deposited. For the Central and Southern Alps this is documented by ages of ca. 600 Ma for the youngest detrital zircons of paragneisses metamorphosed first during the Ordovician. Rb-Sr whole-rock systematics of further paragneisses corroborate this finding. However, opposite to other parts of the European Variscides, Cambrian or Cambro-Ordovician formation of oceanic mafic rocks has been so far rarely dated unambiguously.

Ordovician orogenic activities are widely documented in the Central Alps. They include subduction zone HP-metamorphism at ca. 470 Ma (e.g. Gotthard Massif), regional amphibolite facies tectono-metamorphism and widespread granitic magmatism in the range of ca. 440–460 Ma followed by rapid uplift (conglomerates!). Evidence for Ordovician back-arc opening with formation of N-type MORBs exists for the external Aiguilles Rouge Massif.

Similar to the post-Pan-African evolution, post-Ordovician-Silurian opening phases are difficult to detect and have only been described so far for the external Belledonne Massif. The Carboniferous evolution with rarely detected HP metamorphism is documented, however, by numerous ages. These are usually associated with regional amphibolite facies metamorphism and mainly granitic subduction and collision zone magmatism.

Post-Carboniferous rifting events are best documented in the southern Alpine Ivrea zone, probably because of the unique exposure of rocks from the lower continental crust. They are documented by magmatic/metamorphic activities which probably were caused by successive underplating of mainly mafic magmas continuing at least into the Middle Jurrassic.

1 Introduction

From the geographic position of the Alps within the geological framework of Europe it is evident that the Alpine chain is built on continental crust consolidated at the end of the Variscan (Hercynian) orogenic cycle. Due to post-Carboniferous wrench faulting, a variable cover of pre- and post-Variscan sediments and the opening of the Mediterranean Sea and the Bay of Biscay, the Variscan crystalline basement outcrops discontinuously. Its largest exposed areas can be found on the Iberian Peninsula, in the French Massif Central and in the Bohemian Massif of Germany, the CSFR, Poland and Austria. Spanning the outer limits of this Gondwana-derived crust within Europe, one might start in southern Italy (Calabria), move to the W as far as southern Portugal, continue N into the British Isles – possibly as far N as the Scottish Dalradian (Rogers et al. 1989) – and turn to the SE via the Sudetes in Poland as far as the Carpathians in Rumania. Although largely disrupted and covered up by younger sediments, it is still obvious that the width of the Variscan mountain chain by far exceeds that of the Alpine chain. This is probably due to a combination of variably interacting reasons including a more intensive and longer duration of convergence and docking of more and larger microcontinental terranes when compared to the formation of the relatively narrow Alpine belt.

Thus, the position of the Alpine belt within the Variscan realm of Europe is undisputed and the pre-Triassic geological development of its continental crust can be closely matched with that outside the Alpine mountain chain. However, it is important to keep in mind that a uniform geological history for the entire continental crust of the European Variscides is unlikely and that considerable variations have to be expected both along and perpendicular to the chain axis. This means that, although this overview is based on geochronological data obtained within the Central Alps, many conclusions are also valid for other parts of the Alpine and Variscan domain. The term Central Alps is used here mainly for the Alps within the borders of Switzerland. However, whenever appropriate, also data from the Austrian, Italian and French Alps, i.e. from the Eastern, Southern and Western Alps have been included in the discussion. The fact that the Central Alps were variably overprinted by Alpine orogenic events, especially in the Lepontine domain, is obviously a great handicap in deciphering the pre-Alpine evolution. Therefore, a better understanding of the long pre-Alpine evolution of the continental crust of the Alps necessitates the comparison with the geological scenario outside of it.*

* All ages cited in the text are based on the decay constants as compiled and recommended by Steiger und Jäger (1977).

2 Precambrian Development of the Continental Crust of the European Variscides Outside the Alps

Generally, the Precambrian elements of the Variscan crust (Gebauer et al. 1989) carry many age features found today in the Brasilian Shield or in Africa, for example in the West African craton (e.g. Cahen et al. 1984). However, there is a fundamental difference to the Gondwana basement rocks presently exposed at the surface: in the Variscan basement we find, with a few exceptions, mainly its ubiquitous, recycled components occurring in variably metamorphosed post-Pan-African sedimentary and igneous rocks. In fact, we now know – due to the determination of primary crystallization ages of detrital zircons – more about the age distribution of the rocks in the source regions (Gondwana) in Cambro-Ordovician times than about the age distribution of the presently exposed rocks within the Gondwana basement. Naturally, also these provenances themselves will contain variable amounts of recycled material increasing with decreasing age.

Using U-Pb data of single zircons and/or zircon cores for establishing the quantitative distribution of primary magmatic ages of detrital zircons from various (meta)sediments of the European Variscides, semi-quantitative estimates of integrated crustal growth rates are possible (Gebauer and Williams 1990). As zircons are not only formed during first-order extraction of melts from the mantle but also during formation of metamorphic and mainly magmatic

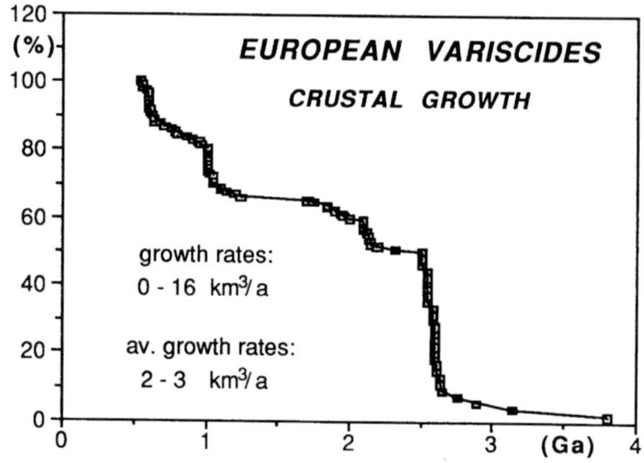

Fig. 1. Integrated crustal growth rates for the European Variscides based on primary ages of detrital zircons in Cambrian and Ordovician (meta)sediments (Gebauer et al. 1989). The average age of the continental crust, based on the data above, is 1.9 Ga. All ion microprobe data were obtained by SHRIMP at A.N.U. and are referenced to an age of 552 Ma for the SL3 standard (e.g. Compston et al. 1986)

rocks which were fully or partly recycled from pre-existing continental crust, the data have to be corrected for recycling. This can be approximated by deriving time-dependent recycling rates via the difference between Sm-Nd model ages on (meta)sediments and their stratigraphically known age of deposition (e.g. O'Nions et al. 1983; Allègre and Rousseau 1984; or Nägler et al. 1989). An additional correction is necessary because of the much higher abundance of zircon-poor mafic rocks in the Archaean crust when compared to the post-Archean crust. Applying these corrections to the ion microprobe data of Gebauer et al. (1989) results in the detection of four megacycles until the end of the Pan-African orogenic cycle (Fig. 1). The first, relatively short megacycle lasted from ca. 2.7 to 2.5 Ga. At the end of it ca. 50 % of the continental crust, existing after the Pan-African orogeny, was already present. Only a few percent seems to have been produced between about 3.8 and 2.7 Ga. A second megacycle took place between ca. 2.2 and 1.9 Ga. As is typical for the non-Australian part of Gondwana, no or only very minor growth of continental crust was observed in the time interval from 1.7 to 1.2 Ga. A third megacycle was detected for the time period between ca. 1.2 and 0.9 Ga. The last Precambrian megacycle, observed in many parts of Gondwana, especially Africa, is connected to the multistage Pan-African cycle (ca. 0.8 to 0.55 Ga). Each of the three post-Archean megacycles added about 15 % new crust to the existing one (Fig. 1).

Of course, there will be three-dimensional variations of integrated crustal growth rates in geographically different crustal segments of the European Variscides. However, from the ion probe data obtained so far on detrital zircon populations, no prominent differences were obtained in the different zircon populations between the southern part of the Massif Central and the Moldanubicum. Thus, it seems justified to extend the conclusions about crustal growth rates also to the Precambrian history of the Alps. As the Nd crustal residence ages of (meta)sediments of the European Variscides and the Alps, averaging ca. 1.7 Ga, are indistinguishable from the average age of the continental crust as derived from the ion probe zircon data, fast production of granitoid material after mantle melting has to be inferred.

A major point of dissent about crustal growth of the Variscan continental crust is whether all of the present crust consists of material which was either extracted from the mantle and/or eroded from source regions after about 700 Ma or whether there are still parts of the pre-700 Ma source region hidden at depth. At the present stage of knowledge the most probable scenario involves all three possibilities: consolidated Proterozoic and Archean continental crust, recycled material derived from this crust and newly formed continental crust since the onset of the Pan-African cycle. As estimates on the Pan-African cycle indicate crustal growth of ca. 16 % and as the amount of recycled detrital material should not exceed that of the source region, a minimum of 42 %, i.e. (100–16 %)/2, of pre-Pan-African continental crust should be expected at the end of the Pan-African and before the onset of the Caledonian-Variscan cycle. Clearly, such a relatively large quantity of continental crust is not exposed at the present level of erosion as the known areas with such old crust are geographically very limited, e.g. the Pentevrian, or Icartian (ca. 2 Ga) in northern Brittany and western Normandy (e.g. Leutwein et al. 1973; Vidal et al. 1981), the offshore northern margin of Spain in the Bay of Biscay (Postaire 1982; Guerrot et al. 1989) or some foliated granodiorites occurring as tectonic lenses in paragneisses of the Varied Group of the Moldanubian basement in Bohemia (Wendt et al. 1989). Recent ion probe (SHRIMP) data on zircons from the calc-alkaline, I-type Dobra gneiss (Moldanubian zone of the southern Bohemian Massif, north-eastern Austria) yielded an age of 1.37 ± 0.01 Ga for the formation of the granitic protolith (Gebauer and Friedl, in preparation). Thus, this Middle Proterozoic orthogneiss - possibly of volcanic origin (Frasl, 1970)- expands over similar areas as the Pentevrian crust within the northern Armorican Massif. In contrast to the Dobra gneiss, recent SHRIMP-data on zircons from two granodiorites of the Pentevrian crust (Guernsey) contain inherited cores that exceed the age of the comagmatic zircons (ca. 2.04 Ga) by about 100 Ma (Gebauer and D'Lemos, unpubl.). Therefore, these two areas represent the only volumetrically significant portions of pre- 1 Ga old, felsic igneous rocks within the European Variscides. Whereas a Gondwana derivation of the Pentevrian crust is very likely, the 1.37 Ga of the Dobra gneiss does not unambiguously argue for either a derivation from Gondwana or Laurentia. Thus, at the present stage of investigation the existence of a suture between these two supercontinents around the Moldanubian-Moravian boundary can not be excluded. The reason for the as yet limited amount of detected Precambrian crystalline basement rocks might be found in a stratigraphic or tectonic cover by post-Pan-African (meta)sedimentary and (meta)igneous rocks. The tectonic cover is probably due to Phanerozoic thrusting during closure of these sedimentary basins which were formed in the course of successive rifting and formation of Gondwana-derived microcontinents (e.g. Gebauer et al. 1989). Of course, Phanerozoic emplacement of melts, especially if produced by variable remelting and/or assimilation of Precambrian crust, would also produce a concentration of these rocks within the presently exposed middle and upper parts of the crust.

There are a number of pre-Pan-African age data on felsic and especially mafic rocks which are scattered all over the European Variscides including the Alps. They have no unique interpretation and therefore need either verification or falsification by independent dating techniques. The first data of this kind were conventional U-Pb zircon data published by Gebauer and Grünenfelder (1982) on various eclogitic mafic and ultramafic rocks from the Montagne Noire in southern France (1.9 Ga), the Gotthard Massif within the Central Alps of Switzerland (2.3 Ga) and the Moldanubicum of NE-Bavaria (>0.9 Ga). Peucat et al. (1982) also published U-Pb zircon data on similar rocks of the Armorican Massif in southern Brittany (1.3 Ga). Furthermore, N-Type MORB eclogites from two external massifs of the Alps, Argentera and Belledonne, revealed small amounts of probably inherited zircon with average ages of ca. 2.7 and 2.1 Ga respectively (Paquette et al. 1989). Various eclogites and metagabbros from different tectonic units of the Eastern Alps also carry Archaean to Late Proterozoic zircons (Neubauer et al. 1987; Gebauer and Grünenfelder unpubl. data; Paquette and Gebauer unpubl. data). Due to the fact that until about 1986 the possible existence of inherited zircon xenocrysts carrying inherited radiogenic lead was unknown for mafic rocks, the corresponding upper intercept ages were favoured to reflect the age of the usually non-continental protoliths. When compared to the frequent existence of inherited zircons in granitic melts, the presumed lack of zircon xenocrysts in mafic rocks was based on the assumption that zircons would either physically not survive in such melts or, if so, that their U-Pb systems would have been completely reset. Additionally, the possibility of the existence of inherited zircons was expected to be much lower or absent in oceanic protoliths which never passed through the relatively zircon-rich continental crust. Although all of these assumptions are still justified, it has recently been shown that inherited zircons may survive, both physically and isotopically, in mafic melts (Compston et al. 1986; Gebauer et al. 1988). This, however, should not automatically lead to the wrong conclusion that there are generally no Precambrian mafic protoliths within the European Variscides and the Alps. Instead, each case should be assessed independently (see also Sect. 3.3).

3 Precambrian Development of the Continental Crust and the Upper Mantle of the Central Alps

3.1 Provenance and Depositional Ages Derived from Metasediments

As primary magmatic ages obtained on single detrital zircon grains are only recently and to a still limited amount available for the Central Alps (Gebauer and Quadt 1991; Gebauer unpubl.), upper intercept ages on zircons from metasediments and meta(igneous) rocks can be used for the determination of approximate average ages of the rocks from which the zircons first crystallized. Similar to the results outside the Alps, such biased average ages often fall into the range 1.6 to 2.5 Ga (Grauert and Arnold 1968; Pidgeon et al. 1970; Köppel 1974; Nunes and Steiger 1974; Gulson and Rutishauser 1976; Köppel et al. 1980; Grünenfelder and Wenk pers. comm.). It is important to realize that the conventional upper intercept ages are multiply biased. This depends largely on the polymetamorphic histories of the individual mother and host rocks which usually causes multiple lead loss and/or new formation of zircon and therefore a lowering of the corresponding lead/lead ages. The degree of discordancy of the individual zircons after primary crystallization in the mother rock and before sedimentation into the present meta(sediment), or assimilation by the present meta(igneous) host rock, is thus of major importance. It might either increase or decrease the extrapolated upper intercept ages. This depends on the amount and timing of the lowering of the individual lead/lead ages as well as on the lower intercept age from which the upper intercept is projected. In addition, possible recent lead loss can change, i.e. usually lower the upper intercept ages. Thus, conventional upper intercept ages using the multigrain method cannot simply be regarded as real average provenance ages as various effects bias and often variably counteract each other. Only by chance will such an "age" represent the average provenance age and again only by chance might such an "age" coincide with a geological significant event. Of course, due to the large frequency of zircon-forming events in the possible provenances and the usually large upper intercept errors, the chances of a coincidence with a geological significant Proterozoic event are quite high.

The fact that Sm-Nd crustal residence ages as well as the upper intercept ages of detrital zircon populations are similar for Alpine and non-Alpine (meta)sediments (e.g. Voshage et al. 1987; Gebauer

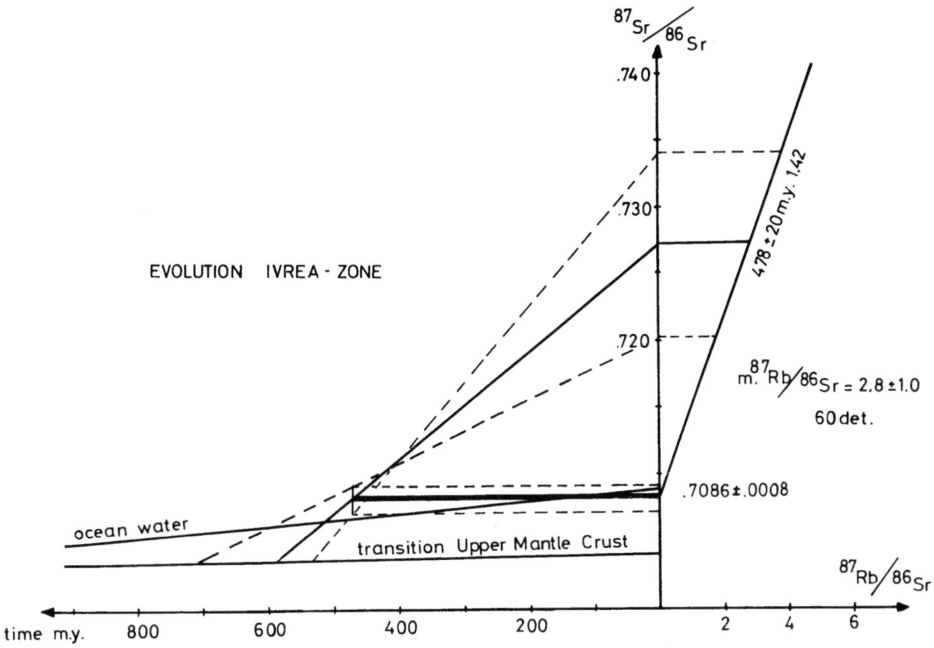

Fig. 2. Combined Compston-Jeffery/Nicolaysen diagram showing the evolution of the Ivrea zone metapelites. The mean $^{87}Rb/^{86}Sr$ ratio of 2.8 ± 1 was derived from 60 X-ray fluorescence determinations. (After Hunziker and Zingg 1980)

et al. 1989) suggests that the integrated growth curve for continental crust production, as derived for rocks outside the Alps (Fig. 1), is also valid for the Alps.

Depositional ages of meta(sediments) can be estimated via their Rb-Sr whole-rock systematics. Assuming that no significant change of Rb-Sr ratios in pelitic to psammitic sediments occurred during metamorphism(s), a maximum depositional age, based on the Sr-isotopic evolution of the mantle, can be estimated (Fig. 2). Doing this for existing data from all over the European Variscides, the obtained maximum depositional ages usually fall well below 650 Ma. As many rocks have been metamorphosed first in the Ordovician, most sedimentary precursors are probably of uppermost Proterozoic and Cambrian age. Ion microprobe data depicting the youngest detrital zircon grains of metasediments as well as the age of Caledonian metamorphism have confirmed such an interpretation of the Rb-Sr data for metasediments of the Moldanubicum and the French Massif Central (Gebauer et al. 1989). Aditionally, this could also be confirmed for a paragneiss of the Gotthard Massif containing detrital zircons ranging from 600 Ma to 3.43 Ga, the oldest age obtained within the Alps so far (Gebauer and Quadt 1991). For the Alpine realm under consideration, Rb-Sr-derived sedimentation ages were concluded for paragneisses of the Austro-Alpine Silvretta Complex (Grauert 1969) and for metapelites of the Ivrea zone in the Southern Alps (Fig. 2; Hunziker and Zingg 1980).

Direct proof for the existence of Proterozoic nonsedimentary continental crust of granitic composition is still lacking in the Central Alps, although evidence for it exists since 1964 (Grünenfelder et al. 1964). As suggested in Section 2, this might be due to extensive stratigraphic and tectonic coverage as well as Phanerozoic intracrustal differentiation of this Gondwana-derived Pan-African and older crust. Söllner and Hansen (1987) believed that they had found evidence for migmatization of granitic crust at ca. 670 Ma for the Oetztal crystalline complex of the Eastern Alps. However, the interpretation of their U-Pb zircon data pattern is not unique. The presence of detrital zircons, formed and/or variably reset in the course of Pan-African orogenic events within the provenance of the analyzed samples, can also explain the observed data pattern. Generally, such Pan-African detrital zircons seem to be rather common in meta(sediments) of the European Variscides (see also Sect. 2; Gebauer et al. 1989).

3.2 Precambrian Mantle Events (Lepontine Area, Ivrea Zone and Gotthard Massif)

Episodes of partial melting in the mantle can best be deciphered by U-Pb dating of zircons. This is due to the possibility of zircon to precipitate from melt segregates within peridotites (e.g. Gebauer 1991,

Gebauer et al., 1992b). A good example for such a process exists also in the Central Alps within the well-known garnet-peridotites of Alpe Arami. These peridotites occur together with minor Eo-Oligocene eclogites (Gebauer et al. 1992b) in sillimanite-bearing gneisses of the Cima Lunga/Adula nappe system within the southern steep belt of the Ticino subdome. Magmatically zoned zircons occur in small quantities within the peridotites. Ion-probe dating using SHRIMP (Gebauer et al. 1992b) yielded generally concordant U-Pb ages at 1.72 Ga which are interpreted as the age of zircon formation within trapped liquids crystallized within the peridotites. Interestingly, from the element distribution within these zircons as detected by cathodo-luminescence studies, it must be concluded that subordinate recrystallization of magmatic zircon in a trace element-rich milieu occurred already also at about 1.72 Ga. Pan-African lead loss at ca. 650 Ma could only be detected in relatively U-richer zircon domains reaching up to 1230 ppm of uranium. No lead loss or new formation of zircon could be found during Alpine HP metamorphism, by no means an uncommon observation for such zircons. Thus, Early to Middle Proterozoic mantle melting and coeval mantle metasomatism are manifested in this Alpine mantle fragment.

A Pan-African mantle-melting event leading to the magmatic formation of zircon was also found in a coarse-grained, 5-m-thick pyroxenite layer occurring within the Lower Layered Group of the Ivrea zone, not far from its contact with the Balmuccia peridotite. Cathodo-luminescence studies show that both magmatic as well as partially or completely recrystallized metamorphic domains exist within one crystal. As only two small grains could be analyzed by ion probe only four spots, two each, could be dated. The age data fall on a discordia trajectory from 670 to 265 Ma. In fact, only one data point containing both magmatic and metamorphic domains is about 77% discordant. Two data points obtained from metamorphic domains are concordant at 265 Ma and one data point from a magmatic domain is analytically concordant at 670 Ma. As neither the age relation of the pyroxenite to the enclosing gabbroic rocks nor the origin of the zircons, inherited or comagmatic, are clear at this stage of investigation, the obtained ages cannot be assigned safely to the history of the zircon's host or mother rock. In any case, the data infer partial mantle melting at the end of the Precambrian and Permian high-grade metamorphic overprinting of the same mantle or its cumulate partial melt products. The latter event has been inferred by a number of previous age data, while the 670 Ma age reflects the as yet oldest event in the history of the mafic and ultramafic Ivrea rocks, respectively their mantle sources. It conforms very well with the previously mentioned mantle-heating event as detected in the garnet-peridotite of Alpe Arami.

A series of further Precambrian mantle-melting events at 3.17, 2.67, 2.45, 1,27 Ga and 870 Ma were detected by SHRIMP dating in retrograded eclogites of the Gotthard Massif (e.g. Gebauer 1990). These are discussed in more detail in Section 3.3.2.

3.3 Precambrian Magmatic Rock Suites

Pre-Late Proterozoic magmatic rock suites are a rarely detected phenomenon within the European Variscides including the Alps. The Pentevrian (Icartien) of northern Brittany and lower Normandy, with ages around 2 Ga and 600 Ma, is still the best-studied Precambrian basement domain within the Variscan belt of Europe (e.g. Vidal et al. 1981). However, due to the very recent detection of 1.37 Ga old granitic crust in the Moldanubian part of Lower Austria (Gebauer and Friedl, in preparation) and the probability that similar rocks in this area as well as the adjacent Moravo-Silesian unit might also be of the same age, the Armorican Massif is not anymore the unique area for significant amounts of pre- 1 Ga old granitic basement rocks (see also Sect. 2). The Iberian Peninsula is likely to contain also larger amounts of exposed Precambrian basement units. However, due to the still few geochronological studies, only scattered occurrences of Pan-African igneous rocks are known so far (e.g. Schäfer et al. 1989). Interestingly, the oldest rocks of the Variscan belt of Europe were found in dredged samples from the continental margin of N Spain (e.g. Guerrot et al. 1989). Conventional U-Pb zircon data suggested Late Archaean protolith ages around 2.75 Ga for a mafic and acid granulite, and a granulite facies event at 1.86 Ga. Most other occurrences of exposed Precambrian rocks in the non-Alpine part of Europe (e.g. Massif Central or Bohemian Massif except the Moravo-Silesian unit) are either very small or still need geochronological verification. Thus, the detection of such unusually old rocks in the Alps – or at least good indications for them – are of major importance for the geological evolution of the continental and oceanic crust of this young belt. The next five sections will therefore deal with such unusually old rocks and possible candidates of them.

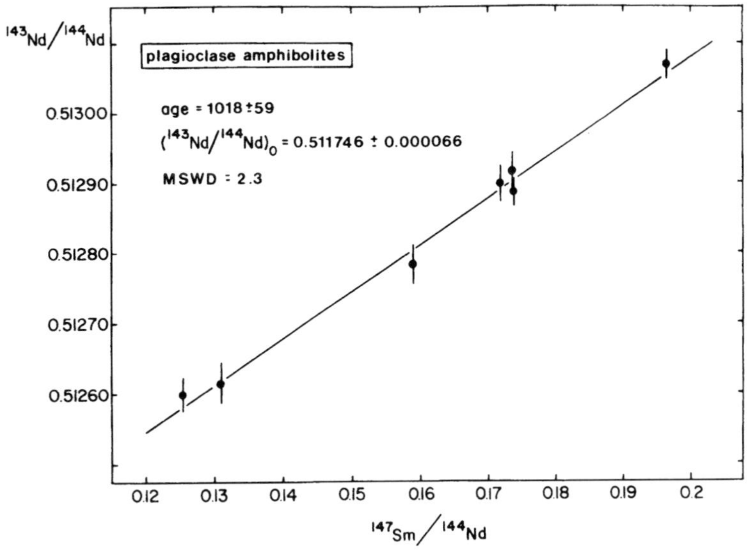

Fig. 3. Sm-Nd evolution diagram for seven whole-rock samples of plagioclase amphibolites from the Berisal complex (Simplon area). (After Stille and Tatsumoto 1985)

3.3.1 Berisal Complex

The Berisal complex (Simplon area of SW Switzerland and N Italy) is part of the Penninic nappe system metamorphosed under Alpine amphibolite facies conditions. It consists mainly of various metasedimentary gneisses and intercalated amphibolites. The latter have been analyzed for Rb-Sr whole rock by Stille (1980) and two parallel isochrons were obtained. One for "hornblendefelses" yielded 500 ± 80 Ma with an initial $^{87}Sr/^{86}Sr$ ratio of 0.7039 ± 0.0001 and another one for plagioclase amphibolites 530 ± 100 Ma with a significantly higher initial $^{87}Sr/^{86}Sr$ ratio of 0.7057 ± 0.0001. Both isochrons were interpreted to reflect the protolith ages of the various calc-alkaline rocks which, according to geochemical data, are cogenetic. The different initial Sr-isotopic ratios were interpreted to be a consequence of disequilibrium melting of a heterogeneous source. As Cambro-Ordovician protolith ages for amphibolitic and eclogitic rocks are a common feature for the European Variscides and the Alps (for a summary see Gebauer 1990a), the Rb-Sr data of Stille (1980) were very much in line with that known from U-Pb zircon dating of such rock types.

Sm-Nd whole-rock data from the same samples obtained by Stille and Tatsumoto (1985), however, yielded isochron ages which in part considerably exceeded the Rb-Sr dates. For the plagioclase amphibolites an isochron age of 1018 ± 59 Ma (Fig. 3) and for banded amphibolites, for which no Rb-Sr isochron information could be obtained, an isochron age of 1071 ± 43 Ma could be derived. The initial ε_{Nd} was in both cases slightly above +8, a rather high value for arc-related calc-alkaline tholeiitic to dacitic protoliths of that age. The "hornblendefelses" yielded an imprecise Sm-Nd isochron age of 475 ± 81 Ma, indistinguishable from the Rb-Sr isochron age of the same rocks. The ca. 1.05 Ga Sm-Nd isochron data caused Stille and Tatsumoto (1985) to re-interpret the Rb-Sr data of Stille (1980) which they now think to be due to either alteration processes or due to a "Caledonian orogenic and metamorphic event". Thus, the tholeiitic-dacitic rock suite is now interpreted to have formed ca. 1.05 Ga ago and the hornfelses are now interpreted as a 475 Ma old partial melt residue or cumulate formed from the 1.05 Ga old amphibolites. Without further age data, preferably using the U-Pb zircon method, it is difficult to argue on the basis of the existing data. However, there are a few observations which can be taken as support for yet other interpretations of the data base. Unfortunately, Sm-Nd whole-rock isochrons can give ages that are too old as has been shown for example by Göpel (1981) on Permian basalts of the Saar-Nahe basin (SW Germany), yielding an isochron age of 1580 Ma. According to Göpel (1981), the data array cannot be explained by a simple mixing model. Rb-Sr whole-rock data, however, matched the Permian emplacement age. Similarly, too old Sm-Nd isochrons were obtained, for example by DePaolo (1981) or Chauvel et al. (1985) and Stille and Buletti (1987) obtained an isochron age of 341 Ma for the Permian Lugano volcanics (Southern Alps) exceeding their true age by ca. 30% (see also Sect. 6.2). As to the inferred Sr-isotopic redistribution between the amphibolite whole-rock systems from the Berisal complex during a Caledonian metamorphic event, it is surprising that isochrons and even two parallel

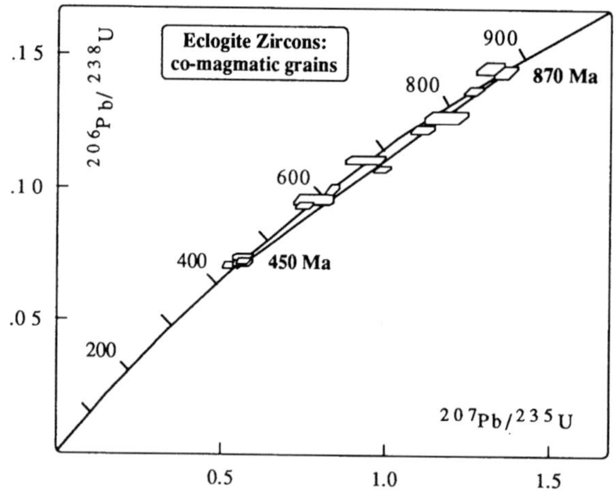

Fig. 4. Concordia diagram showing data points obtained by ion microprobe analyses on 20–30 μm big spots on the rare co-magmatic zircons of a meta-eclogite of the Gotthard Massif. The data align along the whole range of the discordia line for spots obtained within a single zircon grain as well as spots obtained amongst different zircon grains. The upper intercept age is interpreted to reflect the protolith age, the lower intercept the eclogite-forming event which is more accurately defined by the zircon xenocrysts (Fig. 5) and Sm-Nd data. (After Gebauer 1990 a)

isochrons have been obtained. For the plagioclase amphibolites, this isotopic equilibration of Sr is then required to have occurred between amphibolite samples separated by much more radiogenic gneisses. Contamination of the protoliths of the plagioclase amphibolites might better explain their higher Sr initial value when compared to the protoliths of the "hornblendefelses". Also, the rather high initial ε_{Nd}-value of +8 ca. 1 Ga ago fits less well to arc rocks than a value of 5.5–6 ca. 500 Ma ago. Thus, without further data using different decay schemes, it will be hard to derive a safe conclusion on the interpretation of especially the Sm-Nd data. If the latter do not reflect protolith ages, they could also be interpreted in terms of pseudo-isochrons based on ancient isotope fractionation and/or mixing within the source (mantle isochron).

Assuming the protoliths to have formed in the Cambro-Ordovician range, like many similar rock suites all over the European Variscides, then this could geodynamically be related to post-Pan-African opening of ocean basins and consequent convergence either in an intra-oceanic island arc system or in a continental margin setting. Unfortunately, the data of Stille (1980) and Stille and Tatsumoto (1985) do not give unequivocal evidence for the timing of the pre-Alpine metamorphic event(s). If the protoliths are indeed ca. 1.05 Ga old, then a poorly defined Caledonian metamorphic event might be deduced from the Rb-Sr data. A similar conclusion can be reached assuming the Berisal protoliths to be related to comparable rocks of the Gotthard Massif which were formed magmatically 870 Ma ago (Gebauer et al. 1988), as derived by ion-probe dating on zircon. In this case the Sm-Nd isochrons would be too old by ca. 150 and 200 Ma respectively. A high-grade, possibly also eclogite facies transformation could well have taken place in the Ordovician, as shown in the next section for the amphibolitized Gotthard eclogites by both Sm-Nd and U-Pb ion-probe zircon dating (Gebauer et al. 1988).

3.3.2 Gotthard Massif

Retrograded eclogites occur as lenses and layers up to ca. 100 m thick in pre-Alpine gneisses of the Gotthard

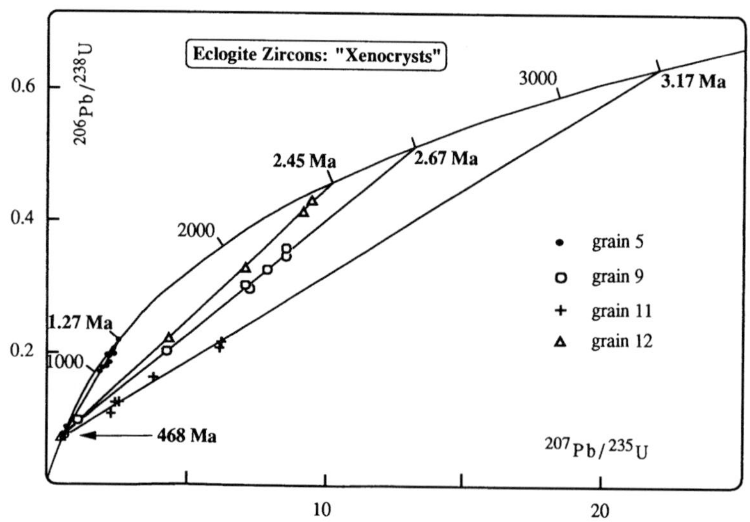

Fig. 5. Concordia diagram showing four distinct discordia lines obtained on individual zircon xenocrysts from a meta-eclogite of the Gotthard Massif. The closely matching lower intercepts reflect the eclogite-forming event of the host rock, in line with Sm-Nd data and the U-Pb data obtained on the co-magmatic grains (Fig. 4) of the same sample. The errors of the lower intercept ages vary between 3 and 38 Ma. (After Gebauer 1990 a)

Massif. The Alpine metamorphic overprint reached chloritoid grade and did not cause any significant disturbance of the U-Pb zircon systems. Based on major, trace and rare earth element data (Gebauer unpubl.), the protoliths of these former high-pressure rocks could have been calc-alkaline to tholeiitic melts of island arc or even ocean floor affinity.

U-Pb analyses (Gebauer et al. 1988; Gebauer 1990a) on over a dozen spots on the rare, clear, euhedral zircons (average U ca. 300 ppm), which are interpreted to have formed magmatically in the protolith, plot over the whole range of a discordia trajectory which intersects the concordia curve at ca. 870 Ma and around 450 Ma (Fig. 4). This wide range of discordance is found not only between grains but between different areas within single grains. Cathodoluminescence studies of the analyzed zircons show areas of euhedral, oscillatory zoning of probably magmatic origin (Fig. 6.4 in Gebauer 1990a) which give an age of ca. 870 Ma. Irregularly fluroscine areas, probably resulting from metamorphic rearrangements of trace elements, give ages at or close to the inferred age of eclogite facies metamorphism at ca. 468 Ma, as derived from the zircon xenocrysts as well as Sm-Nd data.

More than 95% of the zircons are irregularly shaped, often rounded and partly resorbed grains with a range in colour and clarity. Unlike the rare, euhedral 870 Ma old grains, their original magmatic zoning can be cut by the present grain boundaries (Fig. 6.6 in Gebauer 1990a), indicating post-crystallization resorption. Recrystallization, starting mainly from the grain surface, and/or new growth during the eclogite facies metamorphism are common. Most such zones and domains were almost or completely reset during the Ordovician HP metamorphism. Commonly, they are relatively U-rich and sometimes have lost variable amounts of radiogenic Pb during the Carboniferous overprinting. Four of the dozen grains of this type analyzed still contain up to almost 100% of their pre-eclogite radiogenic Pb (Fig. 5). Within-grain discordance lines defined by analyses of several areas on each of the four grains have lower concordia intersections of ca. 468 Ma (Fig. 5), the same age indicated also by the magmatic zircons. Errors in the intercept ages are between 3 and 38 Ma. The upper intersection ages are at ca. 3.17, 2.67, 2.45 and 1.27 Ga. None of the grains appear to have been affected by protolith formation at 870 Ma, extending the temperature range that zircons are known to survive from that of granitic melts to that of gabbroic melts. Similar conclusions have been reached previously by Compston et al. (1985/6) and Kinny (1986).

As none of the eclogite trace element analyses show any significant amount of crustal contamination, it is suggested that the inherited zircons were formed in the mantle. This would be consistent with their generally low U-contents typical for zircons in such rocks. They probably formed during partial melting events and incomplete melt extraction at 3.17, 2.67, 2.45 and 1.27 Ga and were incorporated into the mafic magma during the formation of the gabbroic protolith at 870 Ma. A second possibility might be that the irregular zircons, representing almost all of the zircon population, were brought into the mantle as a result of recycling of sediments along subduction zones. However, based on the geochemical data, such significant amounts of felsic material, unless very rich in U-poor zircon, could only have played a very minor part. Hence, the first hypothesis is preferred as also strong lead loss of old detrital zircon should be expected during subduction into the mantle. This, however, is not observed. In contrast, zircons crystallized in the mantle and stored there will continuously anneal their radiation damage and remain closed for U-Pb until they are brought into cooler crustal levels. For the Gotthard eclogites this was the case after protolith formation 870 Ma ago. Of course, zircons from gabbroic rocks or even plagiogranites of the oceanic crust might also have been brought back into the mantle via subduction zones. This possibility, however, does not change the implications of our preferred mantle contamination model. If the protoliths of the eclogites were tuffs and/or basaltic lava flows, incorporation of zircon xenocrysts of continental origin is possible, but again major and trace element data as well as the observed U-Pb zircon systematics do not support this.

The age of eclogite facies metamorphism at ca. 468 Ma is also supported by Sm-Nd analyses on whole rock and garnet, giving 461 ± 25 Ma (Gebauer et al. 1988; Gebauer 1990a). The initial ε_{Nd}-value at the time of emplacement of the protolith is about +1.3, a value within the lower range for rocks which originated in an arc environment and outside the typical range of MORB of that age. However, the inherited zircons suggest the presence of significant amounts of Archaean and Proterozoic trapped liquids in the source rock which would have enriched the originally depleted source at least at 3.17, 2.67, 2.45 and 1.27 Ga, progressively decreasing the Sm/Nd-ratio and lowering the rock's time-integrated ^{141}Nd growth. The average age of the zircons inherited from the mantle is given by conventional multi grain analyses as 2.23 Ga (Gebauer and Grünenfelder 1982).

Comparing the ages of the inherited zircons, as well as the formation age of the primary magmatic zircons, with ages of major crust-forming events, as discussed in Section 2, it seems that these possible mantle-melting events are older by ca. 100 to 300 Ma (cf. also with Fig. 1). If this tendency is indeed systematic, it suggests that the inherited and the co-magmatic zircons might have formed during mantle upwelling events in the course of oceanization processes, following and preceding major orogenic megacycles in the course of which large amounts of continental crust were newly formed. Again, this rather argues for a mantle origin of the analyzed zircons.

Based on ion-probe data on detrital zircons (Gebauer and Quadt 1991), deposition of the metasedimentary country rocks of the retrograded eclogites is post-Pan-African, implying tectonic juxtaposition of both rock types. Unfortunately, the available data do not indicate when and/or under which P-T conditions such a tectonic liaison has formed.

An important conclusion of the ion-probe study on the eclogitogenic Gotthard amphibolites is that the conventional analyses of zircon populations from the same sample may give geologically meaningless intercept ages. Post-eclogite facies lead loss, as shown by analyses on areas frequently high in U or close to internal or external grain surfaces, is clearly responsible for the conventionally obtained lower intercept age at 400±3 Ma. The Carboniferous amphibolite facies overprint is responsible for some analytically concordant but geologically meaningless ages less than 468 Ma, and in at least one case an Alpine or even more recent lead loss must have occurred. The conventionally measured upper intercept age at 2.23 ± 0.03 Ga is obviously the result of mixing inherited zircons of different ages with the few co-magmatic 870 Ma old zircons.

The SHRIMP data further demonstrate the importance of verifying conventionally measured zircon ages by other dating techniques, Sm-Nd or preferably ion microprobe analyses on zircon. The Caledonian eclogite facies metamorphism in the Gotthard Massif is worldwide the most reliably dated HP event, being defined by five independent zircon discordance lines in close agreement with Sm-Nd whole-rock and garnet data. In addition, it supports the microcontinent or terrane model (e.g. Gebauer and Grünenfelder 1982) which is based, at least in part, on the identification of differently old HP events in distinct domains of the European Variscides and the Alps.

3.3.3 Siviez-Mischabel

The Siviez-Mischabel unit in the Valais is part of the Saint Bernhard basement nappe. It also contains lenses and layers of amphibolites which possibly were, at least in part, retrograded from eclogites and which occur within non-pelitic paragneisses. In an attempt to date these rocks by the Sm-Nd whole-rock technique, Zingg (1989) reported an errorchron of 1026±46 Ma. This 'age' is based on four amphibolites and one metagabbro from the Siviez-Mischabel nappe as well as seven amphibolites from the Berisal complex (Stille and Tatsumoto 1985). Clearly, these data need confirmation (or falsification) by additional measurements, preferably applying also the U-Pb zircon technique.

3.3.4 Southern Part of the Penninic Nappes

Sm-Nd whole-rock analyses have also been carried out on amphibolites associated with ultramafics and paragneisses of numerous tectonic units of the southernmost parts of the Penninic nappe systems (Schenk-Wenger and Stille 1989). The MORB-type rocks are reported to define a ca. 1 Ga isochron with an initial Nd-isotopic composition typical for a 1 Ga old MORB mantle. These data were interpreted to reflect extensive formation of Proterozoic ophiolites, now partly exposed along the southern border of the Central Alps. However, as discussed for the Berisal amphibolites in Section 3.3.1, this interpretation is not unique and therefore the application of different dating techniques is required in order to give an unambiguous interpretation of these data.

3.3.5 Silvretta

First hints of possible Precambrian rocks including also intermediate and felsic intrusives were reported by Flisch (1987) for orthogneisses of the Upper Austro-Alpine Silvretta nappe (eastern Switzerland). Rb-Sr whole-rock errorchron results yielded 'ages' between 590 and 1000 Ma with most data falling between 800 and 950 Ma. The low initial Sr-isotopic composition at ca. 0.703 conforms with the I-type geochemistry of these rocks (Flisch 1987). Again, also these data need verification or falsification by independent techniques. This is all the more important because these rocks intrude high-grade, regionally deformed metasediments, implying that both sedimentation and tectonometamorphic overprint of the country rocks must have occurred before emplacement of the protoliths of these relatively oldest orthogneisses of the Silvretta nappe. Estimates of the depositional ages of the metasediments intruded by these possibly pre-Pan-African orthogneiss protoliths do not support the geological significance of the errorchron results of Flisch (1987). According to the data of Grauert (1969), deposition of the protoliths of the present metasedimentary country rocks is likely to be post-Pan-African, as is generally most probable for such rocks in the rest of the European Variscides.

4 The Caledonian Evolution Within the Central Alps

The term 'Caledonian' is used here strictly for a certain time intervall (ca. 400–500 Ma), which means that it should not be associated with the Caledonian fold belt of northern Ireland, Scotland and Scandinavia. In contrast to the Variscan fold belt, the Caledonides of northern Europe are founded on continental crust with strong affinities to the supercontinent Laurentia-Baltica. Thus, although strong convergence occurred in both orogenic belts in an overlapping time range, they are clearly separated in space and built upon continental crust with different palaeogeographic and geological histories. Only in the Appalachian orogen of North America do both belts seem to have merged into a single mountain range.

Naturally, the Caledonian age data are much better preserved outside the influence of Alpine metamorphism, and within the Alps they are obviously better preserved in areas of lower-grade Alpine overprint. The earliest reliable data of this kind were obtained by Grauert (1966) and Grauert and Arnold (1968) on the Flüela orthogneisses and surrounding paragneisses of the Upper Austro-Alpine Silvretta nappe, as well as on paragneisses of the Gotthard Massif in the Val Nalps region (Grisons). The data closely scatter at ca. 440 Ma for all the mentioned rock types. The interpretation on the paragneiss data (Rb-Sr whole-rock and U-Pb zircon suite method) is straightforward in terms of a first high-grade metamorphic overprint of previously unmetamorphic pelitic-psammitic sediments. However, no unambiguous interpretation can be given for the Rb-Sr orthogneiss data on the basis of the present knowledge of the behaviour of Rb-Sr whole-rock systems during high-grade regional metamorphism. Although the interpretation of Grauert (1966) that the 440 Ma isochron reflects the age of the granitic protoliths of the Flüela orthogneisses – a common interpretation for isochrons obtained on orthogneisses at that time – might well be realistic, an interpretation in terms of a metamorphic overprint cannot be excluded (e.g. Gebauer 1975; Paquette et al. 1984). However, even if the 440 Ma age on the Flüela orthogneisses reflects their common high-grade metamorphic transformation with the neighbouring (meta)sediments, it is still probable that the granitic source rocks were emplaced within the Caledonian cycle. This is based on their initial Sr-isotopic composition (ca. 0.712) 440 Ma ago and their high average Rb-Sr ratios (around 10) leading to an increase of ca. 0.01 for the $^{87}Sr/^{86}Sr$ initial ratio over a time period of only about 25 Ma. Thus, the protolith age of these granitoids very probably does not exceed 460 Ma. Additionally, the U-Pb systematics of zircons of these granitic orthogneisses also favour a Phanerozoic protolith formation (Grauert and Arnold 1968). Similar problems exist with the granitic Streifengneisses in the Gotthard Massif. Arnold (1970) obtained a Rb-Sr isochron age of 436±17 Ma which he also interpreted as the time of emplacement of the granitic protoliths of these very strongly deformed and often even mylonitic orthogneisses depicting unusually high initial Sr-isotopic compositions (ca. 0.714) 436 Ma ago. Recent ion microprobe data on zircons from the Streifengneiss confirm the rapid succession of post-eclogitic protolith formation and gneissification of these granitoids (Gebauer et al. unpubl.). For the possibly Precambrian protolith ages of structurally older orthogneisses within the Silvretta complex, see Section 3.3.5.

Metasedimentary rocks of the southern paragneiss zone of the Gotthard Massif between the Gotthard Pass and Airolo have also been reported by Nunes and Steiger (1974). Although their zircon data scatter due to a strong Variscan overprint, a Caledonian metamorphic event could still be distinguished. Rb-Sr whole-rock systems of these rocks scatter around reference lines at about 260 Ma which was interpreted to be a consequence of Alpine overprinting.

For a quick overview of conventional zircon data, obtained from the Central and Southern Alps outside the Lepontine domain, Fig. 6, taken from a summary paper of Köppel et al. (1980), is useful. Detrital zircons from paragneisses clearly show the highest U-Pb apparent ages up to about 1.9 Ga. Neglecting the Ivrea zone data, their data field is defined by upper intercept ages between ca. 1.6 and 2.4 Ga and lower intercept ages around 450 Ma. Concordant monazites from the Ceneri zone confirm this Caledonian high-grade metamorphic event. For the data of the Ivrea zone metasediments a rotation of the data arrays due to post-Caledonian, i.e. Carboniferous (Variscan), Permo-Triassic and/or possibly even Alpine open system behaviour is suggested here, modifying the interpretation of Köppel (1974). More details on this topic are given in Section 6.2.

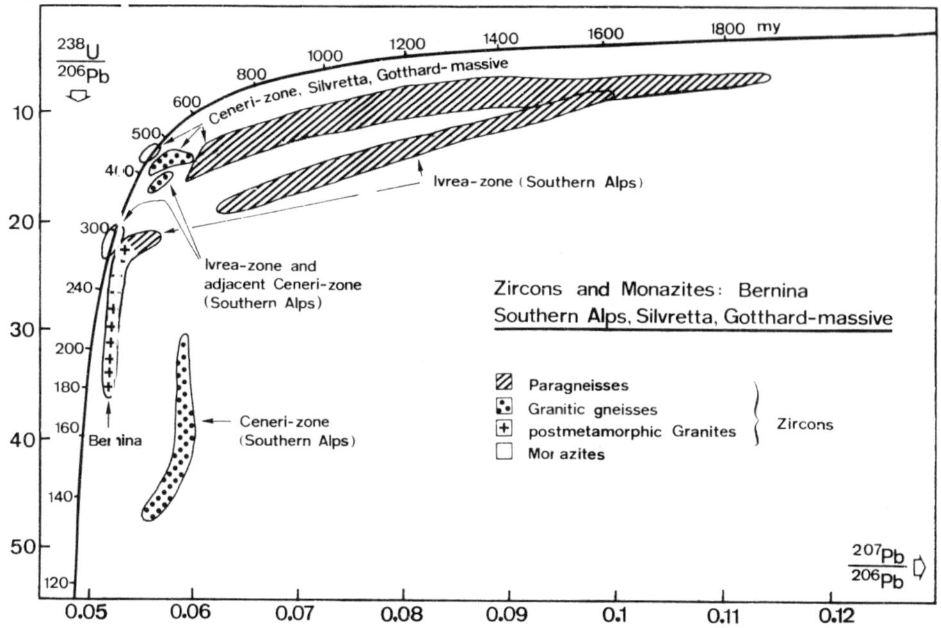

Fig. 6. Summary of conventional zircon and monazite data from geological units not or little affected by Alpine metamorphism. The data obtained from the Ivrea zone are interpreted here to be additionally affected by variable post-Carboniferous overprinting. (Köppel et al. 1980)

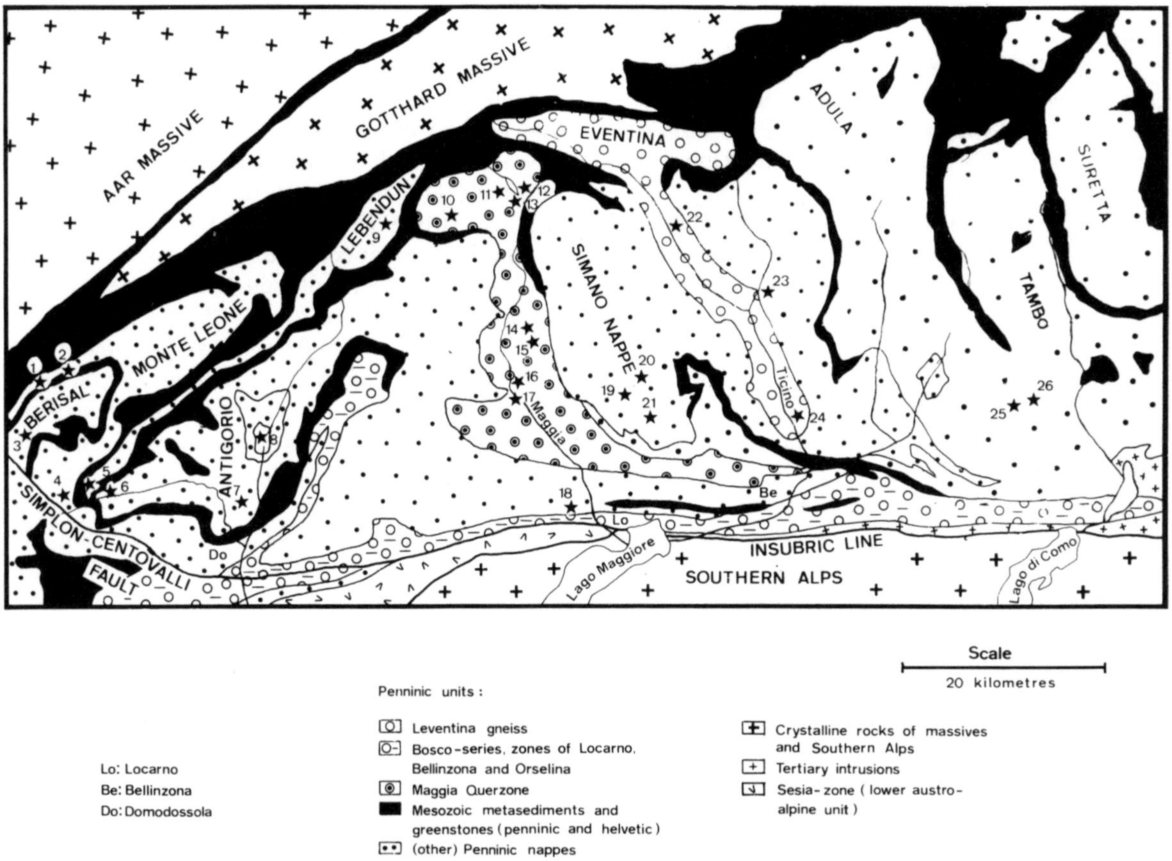

Fig. 7. Schematic geological map of the Central Alps. (After Köppel et al. 1980)

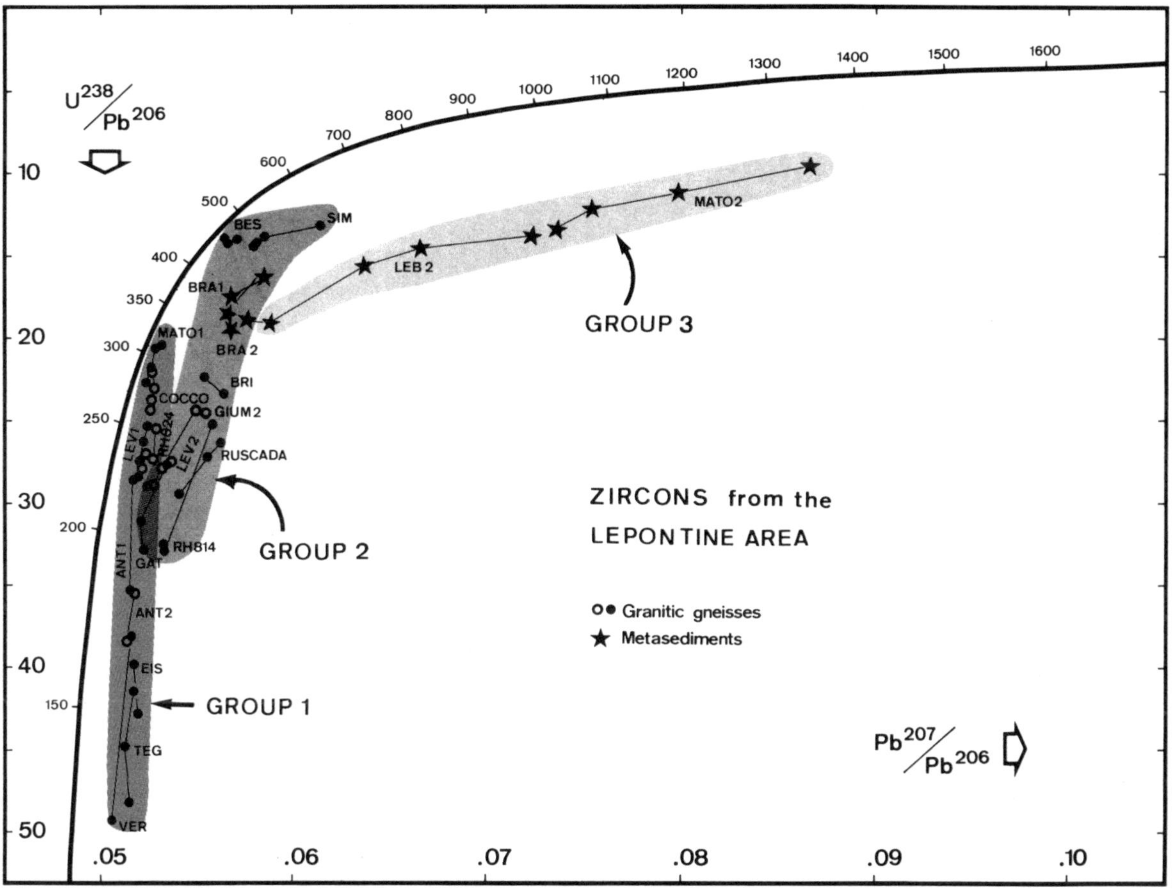

Fig. 8. Conventional zircon data points from the Lepontine. Sample locations are shown in Fig. 7. Data points connected with tie lines denote different size and/or magnetic fractions of a zircon population separated from one rock sample. (After Köppel et al. 1980)

Unfortunately, zircon data from granitic orthogneisses (dots) do not allow an independent datation of their protolith ages. This is due to variably interacting factors, like amount of inherited lead components together with post-crystallization lead loss(es). Nevertheless, from the position of the individual data fields and together with Rb-Sr whole-rock data, a Caledonian formation of these rocks is highly probable.

Proceeding to the Lepontine realm, the influence of the Alpine and possibly also Variscan, tectonometamorphic phases clearly re-opened most U-Pb systems in zircons and monazites, as illustrated by Köppel et al. (1980). As a consequence, the data points obtained on different size and magnetic fractions from a paragneiss from the Maggia nappe (Fig. 7, No. 12) scatter (Fig. 8, MATO 2) but still indicate that the average age of the detrital components is at least around 2 Ga as usually found in such rocks in the European Variscides. The U-Pb data from zircons extracted from augengneiss pebbles within presumed metaconglomerates of the Lebendun nappe (Fig. 7, No. 9) and the Maggia nappe (Fig. 7, No. 10) were interpreted by Köppel et al. (1980) to reflect the protolith ages of Caledonian granitic intrusives (Fig. 8, LEB 2 and BRA 2) containing variable amounts of inherited radiogenic lead. BRA 1 in Fig. 8 shows the position of zircon data from the matrix of augengneiss pebble BRA 2. Although the sedimentary origin of the "metaconglomerates" of the Maggia nappe is controversial – a tectonic origin has been proposed by M. Huber (pers. comm.) – the sedimentary origin of the Lebendun metaconglomerates seems to be undisputed. Thus, the probability that the zircons from the granitic augengneiss pebbles of the Lebendun metaconglomerate crystallized magmatically in a Caledonian protolith between about 400 and 500 Ma (Fig. 8) favours a post-Caledonian deposition of the conglomerate. On the other hand, the deposition must be older than the Carboniferous granites which intrude these metasediments. In addition, the weaker degree of metamorphism in the neighbouring Mesozoic Bündnerschiefer argues for a pre-Alpine high-grade metamorphism, leaving only the Variscan as the time of maximum transformation of the conglomerates. Thus, speculations that the basement rocks remained at depth from the end of the Caledonian orogenic events until the end of the Carboniferous (Variscan) cycle (e.g. Hunziker and Zingg 1980, for the Ivrea zone) are strongly contradicted. This is also true for the non-Alpine parts of the European Variscides.

Caledonian orthogneisses have been reported by Köppel et al. (1980) for granitic gneisses of the Berisal nappe and augengneisses of the Simano nappe (Fig. 7, No. 3 and 19). Interestingly, in both cases zircons seem to have remained closed systems during both Variscan and Alpine high-grade overprinting (Fig. 8, BES and SIM). They indicate protolith formation around 450 Ma ago, a common age for similar rocks in many parts of the European Variscides.

In summary, the Caledonian, and therein the Ordovician, subcycle in the Alps is not only of anorogenic nature, as suggested by various authors also for the non-Alpine part of the European Variscides (e.g. Schmidt and Söllner 1982; Matte 1986; Pin 1989). There is clear proof for compressional events starting with the formation of eclogites, probably in a subduction zone environment, at ca. 470 Ma. Regional metamorphic and tectonic activity, including granitic magmatism, is documented at around 440–460 Ma. Such a scenario, however, does not exclude the possibility of concurrent rifting, e.g. in a back-arc regime. Quick uplift must have occurred after collision, since Caledonian orthogneiss pebbles can be found in metaconglomerates, the matrix of which suffered a high-grade Carboniferous recrystallization. Thus, the Caledonian subcycle might be the result of collision of two terranes rifted off from Gondwana after the Pan-African orogeny. Most pre-Mesosoic metasediments within the Central Alps probably formed from detritus eroded from these terranes. Similarly, the mafic protoliths of many eclogites and amphibolites – a large part of the amphibolites might well represent retrograded eclogites – could have formed at various stages of oceanization after the Pan-African cycle. Such rocks are known from various external massifs of the Western Alps (e.g. Paquette et al. 1989; Ménot 1988) or from the Leventina nappe within the Lepontine area near Biasca (Gebauer, unpubl.).

5 The Variscan Development Within the Central Alps

5.1 *Rifting Events*

The post-Caledonian opening phases preceding the final Carboniferous collision events are even more difficult to decipher than the post-Pan-African opening phases before the Caledonian orogenic events. For the Alpine realm, Devonian to Lower Carboniferous rifting has been suggested by Ménot (1988) and Ménot et al. (1988) for parts of the Belledonne Massif, an external crystalline massif of the Western Alps east of Grenoble. This interpretation is based on zircon data obtained from two trondjemites (352 ± 56 Ma and 365 ± 17 Ma respectively) intruding a mafic-felsic bimodal rock suite. Other evidence for post-Caledonian and pre-Variscan, i.e. pre-Carboniferous events are strong uplift rates inferred from the presence of Caledonian orthogneiss pebbles in the metaconglomerates of the Lebendun nappe and possibly also the Maggia nappe (Köppel et al. 1980). The existence of post-Caledonian, i.e. post-Silurian opening phases, does not contradict the microcontinent model, which is based on the hypothesis of successive formation of Gondwana-derived microcontinents after the Pan-African orogeny and successive collision of these microcontinents until final closure of all basins in the Carboniferous (e.g. Gebauer et al. 1989). Naturally, rifting in back-arc settings in the course of Ordovician, Silurian and Devonian subduction processes has to be expected and is another factor complicating the understanding of the Variscan mosaic geology.

5.2 *Metamorphic Events*

First signs of Variscan convergence are reported for rocks of the Silvretta nappe yielding ages for a first Variscan metamorphism at 350–360 Ma (Flisch 1987). Similar results have been obtained in certain cases also outside the Alpine realm on granulite and eclogite facies rocks of the Bohemian Massif (e.g. van Breemen et al. 1982; Wendt et al. 1989; Gebauer 1991). Earlier data by Grauert (1969, 1981) on Silvretta granite gneisses might also be interpreted in terms of an Early Variscan event at ca. 360 Ma. According to Grauert (1969) and Flisch (1987), the most frequently observed Variscan metamorphic overprint took place around 300 Ma similar to most findings outside the Alps. No attempts to interpret the meaning of these rarely detected two Variscan metamorphic phases have been made so far. In any case, both rifting and first compressional events seem to have taken place in an overlapping time range around the Devonian-Carboniferous boundary in different geotectonic settings, arguing for a possible back-arc scenario. At the corresponding subduction site, paired metamorphic stages might have developed and been telescoped during either final collision or, more probably, extensional uplifting along normal faults at around 300 Ma, thereby retrograding a possible earlier higher pressure mineral paragenesis. The latter process might also be responsible for the emplacement of many S-type granites at that time.

The major Carboniferous metamorphic overprint is detected in para- and orthogneisses wherever the Alpine overprinting did not partially or fully reset the respective decay systems. Apart from the mentioned data from the Silvretta nappe, a number of Variscan Rb-Sr, K-Ar and U-Pb mineral ages have been obtained over the last three decades all over the Central Alps (e. g. Jäger 1962; Grauert 1969; Hunziker 1974; Köppel 1974; Boriani et al. 1981; Schaltegger 1986). Although these age data cannot be used directly to derive tectonic processes, it is still probable that many major deformational processes also took place in the time range between ca. 360 and 300 Ma. Again, the Lebendun metaconglomerates are crucial for an answer to this question since, apart from the Alpine overprint, they have only suffered an amphibolite facies tectono-metamorphic transformation in this time range. At this stage of knowledge it is difficult to state how pervasive a regional metamorphic deformation was in the different geological units of the Central Alps. Arguments from outside the Alps do not necessarily favour penetrative deformation in all pre-existing rocks (e. g. Gebauer et al. 1989). Rather, deformation seems to be restricted to, and strongly increasing into, major thrust zones. Subsequent normal faulting along ductile shear zones (e. g. Mattauer et al. 1988) is probably another as yet little studied structural factor during decompression and low-pressure metamorphism around 300 Ma ago.

5.3 Magmatic Events

As typically observed in the European Variscides, Variscan igneous rocks are mainly of granitic composition. They represent the most frequent rock type of the exposed crystalline basement. In the following sections they will be briefly reviewed according to their geographical occurrence within different tectonic units.

5.3.1 Aar Massif

Apart from the Lepontine area, the Aar Massif contains most of the granitic rocks of Variscan ages within the Central Alps. The calc-alkaline Aar batholith comprises various dioritic to granitic intrusions emplaced post-tectonically into high crustal levels, including also calc-alkaline Carboniferous volcanics. The Alpine metamorphic overprint only reached stilpnomelane grade. Conventional zircon data obtained on various fractions from a series of different rock types (Fig. 9; Schaltegger 1989) yielded upper intercept ages with little scatter at ca. 300 Ma. Interestingly, no inherited lead components were found in the studied zircon fraction, pleading for a predominantly

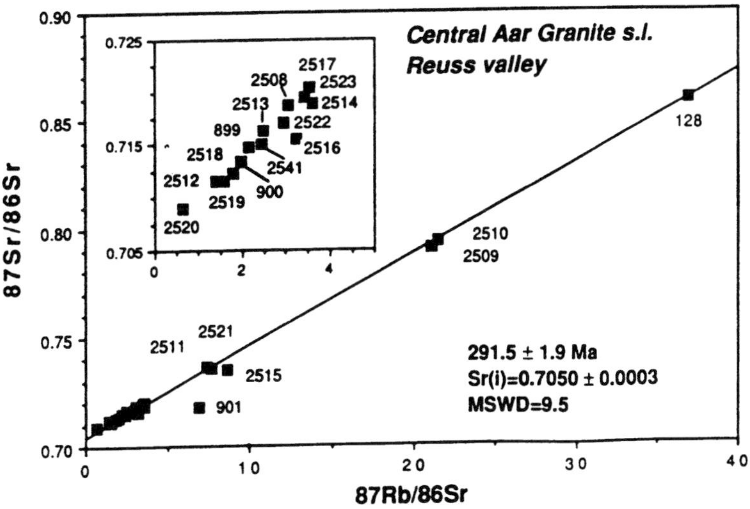

Fig. 9. Sr-evolution diagram of granites and granodiorites of the Aar batholith. The scatter of data of samples with $^{87}Rb/^{86}Sr < 9$ might mainly be due to post-crystallization open system behaviour. The statistical error of the isochron (± 1.9 Ma) probably underestimates the range of ages of the different intrusions which is also supported by zircon ages of 299 ± 2 Ma for different granites and granodiorites. (After Schaltegger 1989)

mantle origin for the multiply intrusive melts. This is in line with their relatively low initial Sr-isotopic values of ca. 0.7050.

A dioritic xenolith within the Central Aar granite yielded a zircon upper intercept age at 316 ± 8 Ma for the possibly oldest member of this magmatic rock suite. Rb-Sr whole-rock systems of various granite varieties, which were not reopened during a hydrothermal overprint at 230 Ma, also favour a magma evolution by crystal fractionation within a narrow time range at ca. 300 Ma (Schaltegger 1989).

The calc-alkaline nature of both plutonic and volcanic rocks within the Aar Massif has led Mercolli and Oberhänsli (1988) to conclude that these rocks formed above a subduction zone at an Andean-type continental margin.

5.3.2 Gotthard Massif

There are a number of smaller granitic complexes within the Gotthard Massif of which two have been dated. The Rotondo granite S of the Furka pass, previously thought to be of possibly Alpine age, has been analyzed for Rb-Sr whole rock and minerals and yielded an age of 269±11 Ma with an initial Sr-isotopic composition at 0.7105±0.0012 (Jäger 1979). Zircons of the smaller calc-alkaline Cacciola granite, ca. 5 km NE of the Rotondo granite, yielded a 292 ± 9 Ma upper intercept age (Oberli et al. 1981).

The same age, 292 + 5/–4 Ma, was obtained on comagmatic zircon grains of the Saedelhorn quartz diorite ca. 9 km to the SW of the Rotondo granite (Bossart et al. 1986). Interestingly, this rock contains rare zircon xenocryst of similar morphological and age criteria to those from the Caledonian Streifengneiss, the immediate country rock intruded by the quartz diorite dyke. Together with the low initial ε_{Nd}-value of –2.7, this argues for both partial or total magma extraction from Precambrian crustal reservoirs or for assimilation of wall rocks.

5.3.3 Lepontine Gneiss Complex

Similar to the Aar and Gotthard Massifs, Carboniferous granitoids play a major role within the various, relatively thin Alpine basement nappes. Due to the higher degree of Alpine deformation and metamorphism, the distinction between older, Caledonian orthogneiss protoliths and younger Carboniferous protoliths is not always unequivocal.

The Truzzo granite gneiss (Tambo nappe) has been interpreted as a high-level granite (Weber 1966) and yielded, together with other samples from the Tambo nappe, a Rb-Sr whole-rock age of 303 ± 14 Ma (Gulson 1973). Interestingly, the initial Sr-isotopic ratio of 0.704 ± 0.003 indicates a predominant mantle origin of the granitoid melts. Similarly, the Gruf migmatites, immediately W of the Alpine Bergell Massif, yielded a Rb-Sr whole-rock errorchron of 317 ± 30 Ma with an initial Sr-isotopic composition of 0.705 ± 0.002 (Gulson 1973).

Within the Bodengo Area, W of the Oligocene Bergell Massif, granitoid bodies (Soè and Garzelli) yielded Carboniferous errorchron ages with probably low initial Sr-isotopic ratios around 0.706 and 0.705, respectively (Hänny et al. 1975). Given the rather large age and initial Sr-isotopic uncertainties, it might still be justified to correlate these rocks geochemically with similar rocks of the Tambo nappe and the Aar Massif (see Sect. 5.3.1) which also show relatively low Sr-isotopic initial ratios at ca. 0.705 (Gulson 1973; Schaltegger 1989) and which have been interpreted to have formed in a subduction zone scenario. A predominant mantle origin of the Bodengo granitoid-gneiss protoliths is also supported by the very small or missing contents of inherited ragiogenic lead in the analyzed multigrain zircon fractions (Hänny et al. 1975).

Evidence for emplacement of numerous, small Variscan granitic bodies to the E of the Bergell Massif (Upper Valtellina) has also been reported by Del Moro and Notarpietro (1987). Although their Rb-Sr whole-rock and muscovite data are not aligned on a single isochron, parallel trends with slopes corresponding to Upper Carboniferous ages can be inferred. Thus, there is evidence that the numerous granitic occurrences immediately to the N of the Tonale line are not all of Alpine age, as previously thought, but also of Late Variscan age. Similar conclusions were made by Vocke and Hänny (1984) for mafic to acid composite intrusions (Sondalo and M. Serottini) of the same area within the Upper Valtellina.

Based on conventional zircon dating, Köppel et al. (1980) compiled data on Variscan granite gneisses from various tectonic units of the Penninic domain. A number of granitoids did not reveal inherited lead components in the analyzed zircon fractions. Such samples include rocks from the Leventina, Monte Leone, Lebendun and Antigorio nappe as well as from the "root zone" immediately N of the Insubric line separating Central from Southern Alps (Group 1 in Fig. 8 or left part of dotted field in Fig. 10). Although no initial Sr or initial Nd values have been published, one might still speculate that these rocks might also be of mainly mantle origin. The same could

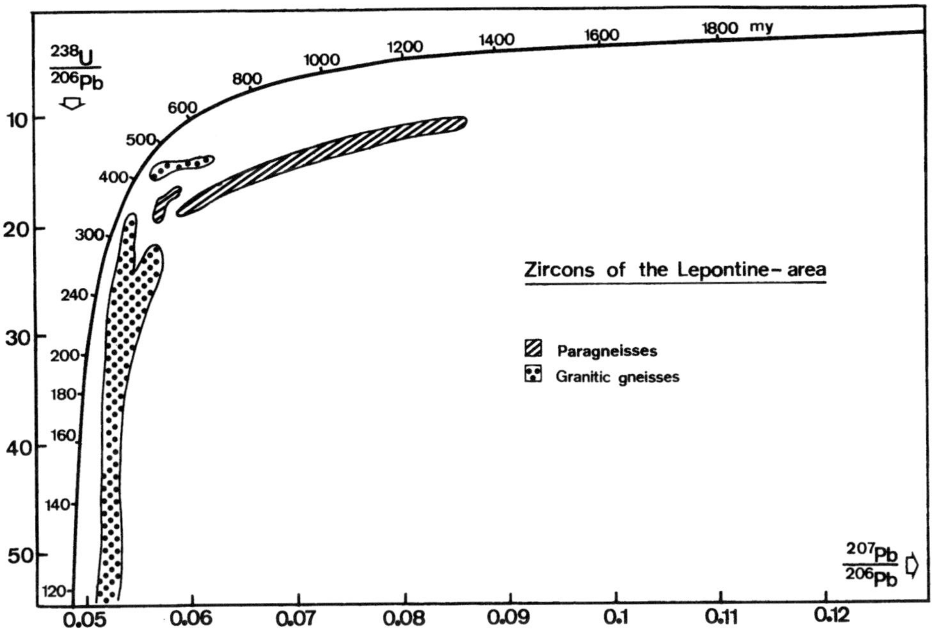

Fig. 10. Summary of zircon data of granitic gneisses and paragneisses from the Lepontine gneiss dome. All samples suffered Alpine amphibolite facies metamorphism resulting in variable loss of radiogenic lead from the analyzed zircon fractions. Thus, the data field of the paragneiss zircons is rotated obliterating the Caledonian regional metamorphic event. The *upper dotted field* depicts data from granitic orthogneisses of the Simano and Berisal nappe which seem to have suffered no or only very little post-Caledonian lead loss. The *elongated lower dotted field* represents data from zircons of Variscan granites containing no or little inherited radiogenic lead. In some cases this data field might include also Caledonian granitic rocks with very little or no inherited zircon lead. (After Köppel et al. 1981)

be true for further rocks of the Maggia and Leventina nappes for which only small amounts of an inherited lead component were found (Group 1 and parts of Group 2 in Fig. 8 or right part of dotted field in Fig. 10).

In a couple of cases inherited lead is of importance for granitoids from the Simano and Maggia nappe. For these, it was not possible to decide whether they are Variscan granitoids with relatively more, or Caledonian granitoids with relatively less, inherited radiogenic zircon lead (Fig. 8, Group 2). Single U-Pb zircon data by Iizumi et al. (1986) indicate that both Variscan and Caledonian granitic rocks, now orthogneisses, might be present in the Verzasca gneisses (Simano nappe). For an aplitic gneiss NW of Giumaglio (Maggia nappe), an upper intercept age of 270 to 280 Ma was obtained on monazite fractions which was interpreted as the time of granitic intrusion (Köppel et al. 1980). Monazites from the immediately surrounding paragneisses yielded a significantly higher upper intercept age between 285 and 305 Ma which the authors interpreted as possibly reflecting a period of regional Variscan metamorphism.

Figure 10 sums up the zircon data from the Lepontine (see explanation in legend).

6 The Southern Alps

The Ivrea zone – situated between Insubric line and Strona Ceneri zone in the westernmost part of the Southern Alps – represents a tilted cross-section through the lower crust. It consists mainly of mafic-ultramafic rock sequences as well as metasediments of amphibolite to granulite facies metamorphism. Numerous geochronological studies, carried out over more than 20 years, suggest an exceptionally complex geological history. Due to the many confusing 'ages' no generally accepted model is available at present. There are several reasons for this situation: the unusually complicated geological history, difficulties in understanding the geological, petrological and geochemical data bases which are a precondition for adequate sampling, difficulties in understanding the systematics of the applied dating techniques, etc. Thus, the Ivrea zone probably remains the most enigmatic crustal section within the European Variscides. The reader is asked to keep this unfortunate fact in mind when trying to comprehend the following summary. Due to the vast amount of isotopic data and the overview character of this work, not all data can be cited.

In contrast to the Ivrea zone, the adjacent Strona Ceneri- and Val Colla zones show the typical geological features found throughout the European Variscides, i. e. a series of high-grade metasedimentary gneisses enclosing amphibolites, orthogneisses and post-metamorphic Variscan granites.

6.1 Protolith Ages of Mafic/Ultramafic Rock Suites of the Ivrea Zone

6.6.1 Lower Layered Group (LLG)

The Lower Layered Group (LLG), forming the deepest unit of the Ivrea zone and reaching a thickness of up to 600 m, consists mainly of strongly deformed layers of pyroxenites and gabbros with several intercalated granulite-facies metasedimentary lenses (e. g. Rivalenti et al. 1975, 1980, 1984). The contact between LLG and the Balmuccia peridotite is generally tectonic. However, also magmatic contact relationships can be observed (e. g. Pin and Sills 1986). Sm-Nd whole-rock data by Pin and Sills (1986) yielded an isochron age of 596 ± 35 Ma for pyroxenites, gabbros and one peridotite with an initial ε_{Nd}-value of 1.7 ± 1.1. Although the authors were very cautious about the age significance of their ca. 600 Ma isochron, they favour the interpretation that this age represents the formation age of LLG which must thus have formed from a time-integrated, slightly depleted mantle source. The emplacement of the LLG into its present position was thought to have occurred at a later stage together with the residual Balmuccia peridotite.

Ion microprobe data by Gebauer et al. (1992 a), however, revealed the emplacements of Lower Oligocene as well as Upper Oligocene gabbroic rocks into the Balmuccia peridotite and into the LLG under lower crustal conditions. This casts serious doubts on the existing interpretation of the Sm-Nd data as well as on the previous geodynamic interpretations of at least the LLG of the Ivrea zone.

6.1.2 Upper Layered Group (ULG), Main Gabbro (MG) – Diorite Body and Metasediments

Upper Layered Group, Main Gabbro and Diorite Body (e. g. Rivalenti et al. 1980) follow E of the LLG in the direction of higher levels of the lower crust. Together with the rocks of the LLG they were usually interpreted as part of a single cogenetic complex. However, Ferrario et al. (1982) suggested that the Main Gabbro and Diorite Body might represent younger intrusive bodies and Pin and Sills (1986) concluded, based on geochemical and isotope data, that Main Gabbro and Diorite Body cannot have formed from the same liquids as the rocks of the LLG. Lead isotope data by Cumming et al. (1987), however, do not differentiate between these two magma reservoirs of the LLG and Main Gabbro-Diorite Body. Sm-Nd whole-rock data by Pin and Sills (1986) on rocks of the ULG-MG-Diorite Body scatter widely and do not allow any age conclusions. Voshage et al. (1987), however, believe they have found evidence for a ca. 600 Ma age for the ULG and the Main Group. These authors also present Sm-Nd whole-rock data and draw an isochron through the data points representing the peridotites of Balmuccia and Baldissero, a gabbro of the Main Gabbro Body, and an anorthosite and a gabbro of the ULG (Fig. 11). The isochron age yielded 607 ± 19 Ma with an initial ε_{Nd}-value of + 0.4 ± 0.3. Two gabbros from the MG and a monzodiorite plot below the 600 Ma isochron. Omitting the two data points of the two peridotites, which largely control the isochron age, would result in a smiliar scatter as obtained already by Pin and Sills (1986) on rocks of the same units. Thus, at the present stage of investigation the two parallel 600 Ma Sm-Nd isochrons of the LLG and the ULG-MG need confirmation, or partial or full correction by independent dating techniques. They leave a number of questions unanswered and they even create new ones. Assuming that the two isochrons are indeed geologically significant, the following queries arise: Is the mantle source for both LLG and ULG-MG the same? Are the differences in initial Nd isotopic compositions of the two units caused by an inhomogeneous mantle source, crustal contamination and/or partial remelting? Does the Balmuccia peridotite, falling by about one epsilon unit below the LLG isochron, still represent a melt residue for the LLG with which it is in both tectonic and magmatic contact or is it a melt residue for the ULG-MG, assuming that LLG and ULG-MG are not cogenetic? If degranitization of the paragneisses was indeed caused by the lower crustal intrusion of the mafic rocks of the Ivrea zone (e. g. Schmid and Wood 1976), it should have happened ca. 600 Ma ago. This is in contradiction to the 478 Ma Rb-Sr whole-rock isochron of Hunziker and Zingg (1980) which they interpreted as the age of granulite metamorphism of these metasediments probably caused by intrusion of mafic/ultramafic magmas into the pelitic series.

The use of the conventional zircon method on a diorite from the upper part of the MG (Pin 1986) resulted in an age of 285 Ma+7/–5 Ma (Fig. 12). Al-

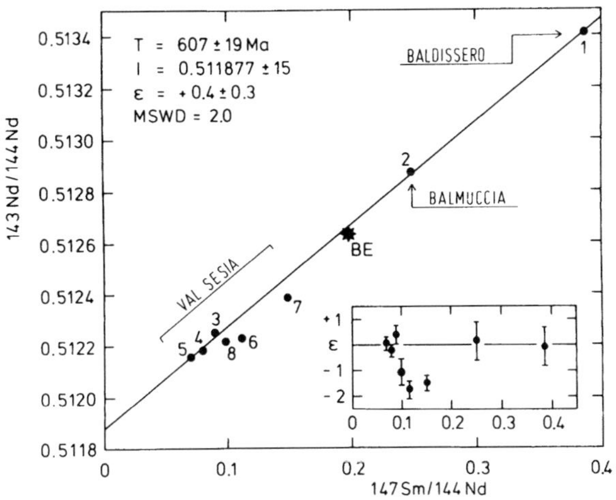

Fig. 11. Sm-Nd evolution diagram showing data points obtained from mafic and ultramafic rocks from Val Sesia, Balmuccia and Baldissero. Samples *6–8* were excluded for the isochron calculation. The isochron age of 607 Ma was interpreted by Voshage et al. to reflect a partial melting event in a homogeneous and almost chondritic mantle reservoir producing a suite of gabbros and ultramafic melt residues. (After Voshage et al. 1987)

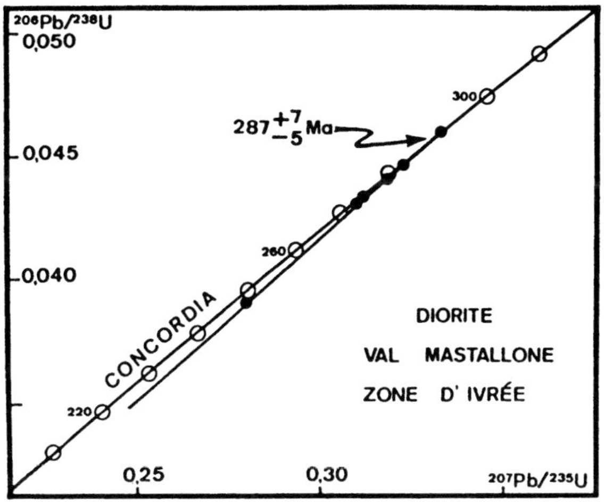

Fig. 12. Concordia diagram with the data *(solid dots)* of different zircon fractions from a diorite of the upper part of the MG. The data are interpreted to reflect the age of formation of the MG-Diorite Body. (After Pin 1986)

though this age is analytically much better substantiated than the two Sm-Nd isochrons – or errorchrons if all data points are considered – it still leaves a number of open questions.

Again, this uppermost Carboniferous age also does not conform – at least for the ULG-MG – with the interpretation of Hunziker and Zingg (1980) that their 478 Ma metasediment isochron is probably related to degranitization under granulite-facies conditions due to influx of mafic/ultramafic melts. In an attempt to solve this problem Pin (1989) suggested that the Caledonian Rb-Sr isochron at 478 Ma might be a result of Sr-isotopic equilibration during sedimentation or diagenetic dewatering. He supported this unusual interpretation with the fact that the initial isotopic composition of the amphibolite and granulite facies metapelites falls into the range of Sr-isotopic compositions of Lower Ordovician seawater. Obviously, this interpretation runs into general problems with the interpretation of Rb-Sr systematics of high-grade metasediments. It has been repeatedly demonstrated that such isochrons generally reflect the highest degree of transformation of the sediments (e.g. Gebauer and Grünenfelder 1974) which is also supported by lower intercept ages of detrital zircons from the same high-grade metamorphic samples using both conventional U-Pb and ion microprobe techniques (e.g. Gebauer et al. 1989). Apart from this, conventional zircon data, concordant monazite data (Köppel and Grünenfelder 1971) and Rb-Sr whole-rock data (Hunziker and Zingg 1980) on paragneisses of the neighbouring Strona Ceneri zone indicate a strong Caledonian metamorphic overprint. Additionally, there are no data available which demonstrate strong lead loss in zircons or new formation of monazite during sedimentation and/or diagenetic dewatering of detrital sediments. Thus, a Caledonian metamorphic transformation of the metapelites of the Ivrea zone is still very probable. However, the inferred superposed influx of mafic melts causing metamorphic peak conditions (Hunziker and Zingg 1980) is not supported by any of the geochronological data. Unfortunately, the conventional zircon data on two paragneiss samples from the Ivrea zone do not solve the problem of distinctly defining one of the possible metamorphic episodes (Ordovician, Carboniferous and post-Carboniferous). The data from a granulite facies paragneiss and a granulite facies pyriclasite (Köppel 1974) might best be explained with a first and variable Caledonian lead loss and possible new formation of zircon, a second Carboniferous and later post-Carboniferous overprint(s). The latter event(s) can be deduced from a number of Permian Sm-Nd mineral data (Polvé 1980; Voshage et al. 1987) as well as from zircon data (Wright and Shervais 1980; Gebauer and Grünenfelder unpubl.) which cluster around 240–250 Ma. These Permian events probably play a much more important role than previously thought. Additionally, the recently discovered Oligocene gabbroic rocks which were emplaced magmatically under lower crustal conditions into the Balmuccia peridotite and the adjacent LLG (Gebauer et al. 1992) argue for the possibility of an Alpine thermal influence in the Ivrea zone, although its limited regional extent is as yet unknown. Based on the two zircon suite data as well as monazite data from the same samples, Köppel (1974) concluded that the regional granulite to upper amphibolite facies metamorphism occurred before 295±5 Ma. In contrast, Teufel and Schärer (1989) believe that they have dated the amphibolite to granulite facies event at 280 Ma. This is based on U-Pb monazite ages on single grains and small size fractions.

The 285+7/–5 Ma diorite age of Pin (1986) is at present the best estimate for the magmatic formation of the Main Gabbro-Diorite Body, assuming that the diorites evolved from the gabbros of the MG. Unfortunately, not even such a consanguinity can be proved throughout in the field for all diorites. Thus, the possibility that the diorite sample dated by Pin (1986) might belong to a younger generation of diorites cannot be ruled out completely. If, however, the dated diorite is cogenetic with the MG-Diorite Body, which seems to be the most probable case, the latter must also be of Permo-Carboniferous age. This implies that the inferred degranitization of the neighbouring metasediments must have taken place in the Permo-Carboniferous. The 478 Ma isochron of the metasediments of the Ivrea zone would then have survived the magmatic emplacement of the gabbro-diorite complex, probably because the rocks had been dehydrated enough already during the Caledonian upper amphibolite facies metamorphism. Again, this argues against the hypothesis of Pin (1989) that the metasedimentary isochron is a product of Sr-Isotopic homogenization during sedimentation or diagenesis as such still water-rich sediments are much less likely to survive a first high-grade metamorphism in the Permo-Carboniferous than previously dehydrated paragneisses.

In summary, at this stage of investigation it seems that there is more than one granulite-facies event in the Ivrea zone. However, amongst the four age possibilities – Ordovician, Permo-Carboniferous, Permian and Oligocene – none is unequivocally proven. In addition, not even their regional extent within the Ivrea zone itself can be clearly defined, although it seems that especially the Permian and Oligocene possibilities seem to be locally more restricted than the two older, i.e. Ordovician and Permo-Carboniferous, candidates. Clearly, more work is needed to get closer to reality.

6.1.3 Amphibolites Within Metasediments

Amphibolites occur within the high-grade metasediments of the Ivrea zone as layers and lenses up to the 100 m range in thickness. They can be divided into two groups with N-type and E-type MORB geochemistry (e.g. Sills and Tarney 1984). Their MORB-type character is also confirmed by Pb-isotope data (Cumming et al. 1987). Sm-Nd whole-rock data on both types did not result in a single or two separate isochron ages. The data points show too much scatter for the limited spread of Sm-Nd ratios (Pin 1989). A tentative 'age' of 511 ± 28 Ma with an initial Nd-Isotopic composition of 3.4 ± 0.2 was derived by Pin (1989). However, before independent age information becomes available, these figures should be treated with great caution.

6.1.4 The Finero and Balmuccia Peridotites

Rb-Sr and Sm-Nd data on phlogopite-bearing peridotites from the Finero and Balmuccia ultramafic bodies as well as from phlogopite-free peridotites from Finero have been published by Voshage et al. (1987). A seven point Rb-Sr whole-rock isochron of phlogopite peridotites from Finero and Balmuccia yielded an age of 293 ± 13 Ma and an initial Sr-isotopic composition of 0.7063 ± 0.0006. The isochron was interpreted to be a result of influx of crustal-derived fluids causing the observed metasomatic mineralogical, geochemical and isotopic changes. The authors point to the fact that metasedimentary country rocks also occur within the mafic formations and thus could well have served as a source for the crustal fluids. In contrast to the Rb-Sr whole-rock systems, Sm-Nd whole-rock systems of the same phlogopite-bearing samples were probably not isotopically equilibrated 293 Ma ago. Hence, only three samples align on a regression line, yielding 266 ± 41 Ma with an initial ε_{Nd}-value at–5.3 ± 0.9. Three samples of phlogopite-free rocks from Finero, a gabbro, an amphibolite and an amphibole-bearing peridotite align along a similar regression line, yielding an 'age' of 270 ± 57 Ma, but with a much higher initial ε_{Nd}-value of+7.5 ± 1.8. Similar ages were suggested by Voshage et al. (1988) for ultramafic and mafic dykes within the Balmuccia peridotite based on Sm-Nd whole-rock errorchron results.

Viewing all the data points of the different rock types analyzed for Sm-Nd whole rock, it is impossible to state whether the Finero peridotite is cogenetic with those of Balmuccia and/or Baldissero. Instead, it seems that the data could well be the result of a combined effect of inhomogeneous mantle sources containing variably depleted and enriched domains, possible effects of crustal contamination, remelting effects and possible metamorphic/metasomatic open system behaviour. In any case, the 600 Ma 'event' derived by Voshage et al. (1987) for mantle differentiation and simultaneous formation of gabbroic melts remains highly speculative. Clearly, more work is needed to resolve the existing problems.

6.1.5 The ca. 250 Ma Episode Within the Ivrea Zone

A ca. 250 Ma magmatic-metamorphic episode, as detected in the Ivrea zone, seems to be absent in the non-Alpine part of the European Variscides. Similarly, the possible intrusion of large masses of mafic and intermediate melts as late as 285 Ma is an unusual feature for other parts of the European Variscides. It might be related to the unique exposure of lower crustal rocks within the Ivrea zone, i. e. such relatively young rocks might also exist at depth elsewhere in the non-Alpine part of the European Hercynides.

The first evidence for this Late Permian event was reported by Wright and Shervais (1980) on a 252 ± 1 Ma old zircon megacryst extracted from an albite granite pod occurring within a pegmatoidal pyroxenite at the contact between the Balmuccia peridotite and adjacent basic and acid granulites. Pin (1989) argues that this age probably relates to local partial melting events at the time of tectonic emplacement of the peridotite into the lower crust. Similar Rb-Sr ages have also been obtained on muscovites from pegmatites of the Ivrea zone (252 ± 10 Ma; Hunziker 1974) and the Lake Massif (259 ± 8 Ma; Boriani et al. 1981).

Additionally, there are Sm-Nd mineral data by Polvé (1983) and Voshage et al. (1987) yielding ages between ca. 230 and 270 Ma for granulite-facies rocks of the Ivrea zone. These ages can best be taken as an argument for at least amphibolite facies conditions at that time and the regional extent of these conditions seems to include at least basic and acid granulites between Val Sesia and Val Strona. An age of high-grade metamorphism at 265 ± 7 Ma was found within a metamorphically recrystallized zircon from a 5 m thick coarse-grained pyroxenite layer within the LLG close to the Balmuccia peridotite (Gebauer et al. 1992) supporting the previous conclusions of Polvé (1983).

Finally, there are a number of K-Ar mica ages in the Ivrea zone between about 230 and 180 Ma (e. g. McDowell and Schmid 1968; Hunziker 1974), which might be related to cooling during uplift, thermal peaks and/or hydrothermal activities. The latter might well be related to increased heat flow and thereby induced hydrothermal circulations during Triassic-Liassic rifting phases. However, due to the presence of lower crustal Oligocene gabbroic rocks in at least the Balmuccia peridotite and the LLG, partial Alpine resetting might be an additional factor causing the spread of mineral ages.

6.1.6 Permo-Triassic Rifting Events Within the Ivrea Zone

Conventional zircon data by Gebauer and Grünenfelder (unpubl.) revealed concordant ages at 238±1 Ma. The zircon fractions were extracted from a boudinaged gabbro grading into amphibolite at its contact to the paragneisses of the NE part of the Ivrea zone close to Ronco. Different fractions from both a deformed rim of the boudin near its contact to the paragneiss, as well as a relatively undeformed gabbroic part closer to the center of the boudin, yielded a total of five concordant ages at 238±1 Ma. This data set is interpreted to reflect the crystallization of euhedral zircons within the gabbroic protolith revealing slightly enriched REE patterns (Gebauer and Grünenfelder unpubl.). The latter are different to both the N- and E-type MORB amphibolites occurring within metasediments of the Ivrea zone (e. g. Sills and Tarney 1984). Thus, it seems that mafic rocks were emplaced as dykes during Permo-Triassic rifting events in the Ivrea zone and that deformation under amphibolite facies conditions post-dates these Permo-Triassic dykes or is at least contemporaneous.

Slightly younger zircon lead/lead ages of 225 ± 13 Ma (Kober method) were obtained by Stähle et al. (1991) on a syenite pegmatite in the Finero peridotite. Its mantle origin is documented by initial Sr and Nd data and it is also thought to be of rift-related origin.

6.1.7 Recent Ion Microprobe (SHRIMP I/II) data on rocks from the Sesia and Mastallone valley

Recent ion microprobe data on the comagmatic zircon population of a gabbro from the Main Gabbro unit (Mastallone valley; Gebauer and Schmid, unpublished) are analytically indistinguishable from the conventional zircon age of 285 Ma +7/-5 Ma for a diorite from the upper part of the Main Gabbro (Pin, 1986). Thus, the consanguinity of the two rocks is now well established.

The SHRIMP I/II data confirm furthermore that the emplacement of this large intrusive complex into the lower crust is coeval with the timing of a major granulite facies pulse in the metasedimentary country rocks. This can be derived from ages obtained in newly formed and/or metamorphically recrystallised zircon rims or domains that yield analytically indistinguishable ages to those of the zircons from a gabbro of the Main Gabbro unit (ca. 280 Ma). However, it is important to note that there existed also a series of post-280 Ma old granulite facies pulses that lasted at least until the Middle Jurassic (177 Ma). These additional, younger lower crustal overprints were recently detected in various rock types (Gebauer and Schmid, unpublished), e.g. in zircons from a metasedimentary rock to the W of the Balmuccia peridotite (ca. 230 Ma), in zircons from the MG itself (ca. 230 Ma) or in inherited zircons within an Alpine meta-gabbro-diorite from the LLG (Gebauer et al., 1992). The 177 Ma age was found in comagmatic zircons from a subsolidus equilibrated meta-gabbro in the Sesia valley. Thus, this meta-gabbro is as yet the youngest pre-Alpine intrusive rock within the Ivrea zone (Gebauer and Schmid, unpublished).

These newest SHRIMP data clearly favour the existence of multi-episodic rifting events and concurrent emplacements of mafic magmas into the lower crust until at least the Middle Jurassic. As a consequence of this, a series of metamorphic overprints caused subsolidus re-equilibration in the country-rocks of the newly added melts. Probably, all this happened on a variable scale in different parts of the Ivrea zone depending mainly on the volume and the location of the emplaced mafic melts. Considering the Sm-Nd whole-rock data from the Ivrea zone (and elsewhere), there is now even more evidence that they can only be treated with sufficient confidence if supported independently by other dating techniques.

6.2 Paragneisses, Orthogneisses and Post-Metamorphic Granitic Rocks of the Southern Alps Outside the Ivrea Zone

From the degranitization hypothesis applied to the metapelites of the Ivrea zone it has been argued that the granitic rocks of the Ceneri zone, including the Caledonian granitic orthogneisses, might have formed from these metasediments. However, Rb-Sr dating on granites to quartz diorites from the Baveno granite suite as well as rhyolites and rhyolitic tuffs from the Permian quartz porphyry province of the Ceneri zone (Hunziker and Zingg 1980) exclude this hypothesis. The initial Sr-isotopic ratios (0.7071 to 0.7087) are far too low to derive these rocks, yielding ages between 275 and 295 Ma (Köppel 1974; Köppel and Grünenfelder 1978; Hunziker and Zingg 1980; Boriani et al. 1981), from the metapelites of the Ivrea zone. However, the range of initial Sr-isotopic values is similar to the corresponding range of the MG-Diorite Body (Pin and Sills 1986; Voshage et al. 1987). Thus, a derivation of the post-metamorphic granitic rocks of the Ceneri zone from the main gabbros-diorites of the Ivrea zone through crystal fractionation is possible. The youngest rocks which might be related to these Permo-Carboniferous intrusives and extrusives have been described by Stille and Buletti (1987), who found a Rb-Sr ago of 262 ± 1 Ma for primary minerals, whole rock and glass of a vitrophyric dacite. These rocks belong to the Lugano volcanites. They are of calc-alkaline composition and range from rocks of basaltic through intermediate to rhyolitic composition. They are interpreted to have formed above an Andean-type subduction zone during the final closure between Laurasia and Gondwana (Buletti 1985). According to Sm-Nd and Rb-Sr whole-rock data, yielding false isochron ages ca. 80 Ma in excess of their stratigraphic age (Stille and Buletti 1987), crustal contamination of mantle melts seem to be a major factor in the genesis of these rocks.

Similar to the post-metamorphic granitic rocks of the Ceneri zone, the Caledonian granitic orthogneisses could not have formed by degranitization of metasediments, since their initial Sr-isotopic composition at the time of magmatic emplacement 466 ± 5 Ma ago was 0.7087 ± 0.0002 (Boriani et al. 1981). This is too low when compared to the paragneiss data of Hunziker and Zingg (1980), yielding 0.7107 ± 0.0018 478 ± 20 Ma ago for the Ceneri metasediments. Ion microprobe data on detrital zircons of these paragneisses indicated that they contain Pan-African detrital zircons (Grünenfelder et al. 1984) proving a post-Pan-African deposition of their pre-metamorphic precursors.

Acknowledgements

I would like to thank Profs. A. G. Milnes and V. Trommsdorff as well as an anonymous reviewer for their helpful comments on the first draft of the manuscript. Stimulating discussions with Dr. R. Schmid on the Ivrea zone were also of great help. Support of the Swiss National Science Foundation to obtain the ion-probe data is gratefully acknowledged.

References

Allègre CJ, Rousseau D (1984) The growth of the continents through geological time studied by Nd isotope analysis of shales. Earth Planet Sci Lett 67: 19–34

Arnold A (1970) On the history of the Gotthard Massif (Central Alps, Switzerland). Abstr Eclogol Geol Helv 63: 2930

Boriani A, Origoni Giobbi E, Del Moro A (1981) Composition, level of intrusion and age of the "Serie dei Laghi" orthogneisses (northern Italy – Ticino, Switzerland). Rend Soc Ital Mineral Petrol 38: 191–205

Bossart PJ, Meier M, Oberli F, Steiger RH (1986) Morphology versus U-Pb systematics in zircon: a high-resolution isotopic study of a zircon population from a Variscan dike in the Central Alps. Earth Planet Sci Lett 78: 339–354

van Breemen O, Aftalion M, Bowes RD, Dudek A, Misar Z, Povondra P, Vrana S (1982) Geochronological studies of the Bohemian Massif, Czechoslovakia, and their significance in the evolution of Central Europe. Trans R Soc Edin 73: 89–108

Buletti M (1985) Petrographisch-geochemische Untersuchungen im Luganer Porphyrgebiet. Ph D Thesis, University Bern, 157 pp

Cahen L, Snelling NJ, Delhal J, Vail JR (1984) The geochronology and evolution of Africa. Clarendon Press, Oxford

Chauvel C, Dupre B, Jenner GA (1985) The Sm-Nd age of Kambalda volcanics is 500 Ma too old! Earth Planet Sci Lett 74: 315–324

Compston W, Williams IS, Campbell IH, Gresham JJ (1985/86) Zircons xenocrysts from the Kambalda volcanics: age constraints and direct evidence for older continental crust below the Kambalda-Norseman greenstones. Earth Planet Sci Lett 76: 299–311

Compston W, Kinny PD, Williams IS, Foster JJ (1986) The age and Pb loss behaviour of zircons from the Isua supracrustal belt as determined by ion microprobe. Earth Planet Sci Lett 80: 71–81

Cumming GL, Köppel V, Ferrario A (1987) A lead isotope study of the northeastern Ivrea Zone and the adjoining Ceneri Zone (N-Italy): evidence for a contaminated subcontinental mantle. Contrib Mineral Petrol 97: 19–30

Del Moro A, Notarpietro A (1987) Rb-Sr geochemistry of some Hercynian granitoids overprinted by eo-Alpine metamorphism in the Upper Valtellina, Central Alps. Schweiz Mineral Petrogr Mitt 67: 295–306

DePaolo DJ (1981) A neodymium and strontium isotopic study of the Mesozoic calc-alkaline granitic batholiths of the Sierra Nevada and Peninsula Ranges, California. J Geophys Res 86: 10470–10488

Ferrario A, Garuti G, Sighinolfi GP (1982) Platinum and palladium in the Ivrea-Verbano basic complex, Western Alps, Italy. Econ Geol 77: 1548–1555

Flisch M (1987) Geologische, petrographische und isotopengeologische Untersuchungen an Gesteinen des Silvrettakristallins, Teil 1. Die Hebungsgeschichte der oberostalpinen Silvretta-Decke seit der mittleren Kreide, Teil 2. Diss, Universität Bern

Frasl G (1970) Zur Metamorphose und Abgrenzung der Moravischen Zone im niederösterreichischen Waldviertel. Nachr. Deutsch. Geol. Ges. 2, 55-61

Gebauer D (1975) Rb-Sr Gesamtgesteins- und Mineralsysteme sowie U-Pb Systeme in Zirkonen während der progressiven Gesteinsmetamorphose. Diss, ETH Zürich, 137 pp

Gebauer D (1990a) Isotopic systems – geochronology of eclogites. In: Carswell DA (ed) Eclogite facies rocks. Blackie, New York, pp 141–159

Gebauer D (1990b) Precambrian development of the continental crust of the European Hercynides. In: Freeman R, Mueller St (eds) Proc 6th EGT Worksh Eur Sci Found, pp 23–28

Gebauer D (1991) Two Palaeozoic high-pressure events in a garnet-peridotite of northern Bohemia (CSFR). Abstr for 2nd Int Eclogite field symposium, Spain 1991. Terra Abstr Suppl 6 to Terra Nova 3: 5

Gebauer D, Grünenfelder M (1974) Rb-Sr whole-rock dating of late diagenetic to anchimetamorphic, Palaeozoic sediments in southern France (Montagne Noire). Contrib Mineral Petrol 47: 113–130

Gebauer D, Grünenfelder M (1982) Geological development of the Hercynian Belt of Europe based on age and origin of high-grade and high-pressure mafic and ultramafic rocks. Extended Abstract. 5th Int Conf on Geochronology and isotope geology. Nikko, Japan, pp 111–112

Gebauer D, Quadt A (1991) Geochronological evidence for tectonic contacts between Caledonian meta-eclogite of the Gotthard Massif and its country-rocks (Central Alps, Switzerland). Abstr for 2nd Int Eclogite field symposium, Spain 1991. Terra Abstr Suppl 6 to Terra Nova 3: 6

Gebauer D, Williams IS (1990) Crust and mantle evolution of the European Hercynides. Abstr ICOG 7 Canberra. Geol Soc Aust 27: 38

Gebauer D, Quadt A, Compston W, Williams IS, Grünenfelder M (1988) Archean zircons in a retrograded, Caledonian eclogite of the Gotthard Massif (Central Alps, Switzerland). Schweiz Mineral Petrogr Mitt 68: 485–490

Gebauer D, Williams IS, Compston W, Grünenfelder M (1989) The development of the Central European continental crust since the Early Archean based on conventional and ion-micro-probe dating of up to 3.84 b.y. old detrital zircons. Tectonophysics 157: 81–96

Gebauer D, Schmid R, Quadt A, Ulmer P (1992a) Oligocene, pyroxene-rich metadiorites of different ages from the Ivrea zone and their geodynamic significance. Schweiz Mineral Petrogr Mitt 71,1

Gebauer D, Grünenfelder M, Tilton GR, Trommsdorff V, Schmid S (1992b) The geodynamic evolution of garnet-peridotites, garnet-pyroxenites and eclogites of Alpe Arami and Cima di Gagnone (Central Alps) from Early Proterozoic to Oligocene. Schweiz Mineral Petrogr Mitt 71,1

Göpel C (1981) Trace element abundances and 87Sr/86Sr-, 143Nd/144Nd ratios of Permian flood basalts, SW-Germany. Terra Cognita, Spec Issue Spring 1981: 80

Grauert B (1966) Rb-Sr age determinations on orthogneisses of the Silvretta (Switzerland). Earth Planet Sci Lett 1: 139–147

Grauert B (1969) Die Entwicklungsgeschichte des Silvretta-Kristallins auf Grund radiometrischer Altersbestimmungen. Diss, Univ Bern, 168 pp

Grauert B (1981) Das Alter der Schlingentektonik im Silvretta-Oetztalkristallin aufgrund radiometrischer Altersbestimmungen. Fortschr Mineral 59, Beiheft 1: 54–56

Grauert B, Arnold A (1968) Deutung diskordanter Zirkonalter der Silvrettadecke und des Gotthardmassivs (Schweizer Alpen). Contrib Mineral Petrol 20: 34–56

Grünenfelder M, Hofmänner F, Grögler N (1964) Heterogenität akzessorischer Zirkone und die petrographische Deutung ihrer Uran/Blei-Zerfallsalter. II. Präkambrische Zirkonbildung im Gotthardmassiv. Schweizer Mineral Petrogr Mitt 44 (2): 543–558

Grünenfelder M, Williams IS, Compston W (1984) Use of the ion-microprobe in deciphering complex U-Th-Pb systems in zircons from the pre-Alpine basement, Switzerland. Terra Cognita, Spec Issue ECOG VIII: 26–27

Guerrot C, Peucat JJ, Capdevila R, Dosso L (1989) Archean protoliths within Early Proterozoic granulitic crust of the West European Hercynian Belt: possible relics of the West African craton. Geology 17: 241–244

Gulson BL (1973) Age relations in the Bergell region of the south-east Swiss Alps: with some geochronological comparisons. Eclogal Geol Helv 66 (2): 293–313

Gulson BL, Rutishauser H (1976) Granitization and U-Pb studies of zircons in the Lauterbrunnen Crystalline Complex. Geochem J 10: 13–23

Hänny R, Grauert B, Soptrajanova G (1975) Palaeozoic migmatites affected by high-grade Tertiary metamorphism in the Central Alps (Valle Bodengo, Italy). Contrib Mineral Petrol 51: 173–196

Hunziker JC (1974) Rb-Sr and K-Ar age determination and the alpine tectonic history of the Western Alps. Mem Ist Geol Mineral Univ Padova 31: 1–54

Hunziker JC, Zingg A (1980) Lower Palaeozoic amphibolite to granulite metamorphism in the Ivrea Zone (Southern Alps, N-Italy). Schweiz Mineral Petrogr Mitt 60, 2/3: 181–214

Iizumi S, Meier M, Oberli F, Steiger RH (1986) Petrogenetics by morphology/U-Pb single zircon study. Terra Cognita 6: 152

Jäger E (1962) Rb-Sr age determination on micas and total rocks from the Alps. J Geophys Res 67 (13): 5293–5306

Jäger E (1977) The evolution of the Central and Western European continent. In: Vidal Ph (ed) Limitation isotopique à l'age de l'évolution de la croute continentale en Europe moyenne et occidentale. In: Cogné J (ed) La chaine varisque d'Europe moyenne et occidentale. Colloq Int Cent Natl Rech Sci, Rennes, 243, pp 227–239

Jäger E (1979) The Rb-Sr method. In: Jäger E, Hunziker JC (eds) Lectures in isotope geology. Springer, Berlin Heidelberg New York, pp 13–26

Kinny PD (1986) 3820 Ma zircons from tonalitic Amitsoq gneiss in the Godthaab district of southern West Greenland. Earth Planet Sci Lett 79: 337–347

Köppel V (1974) Isotopic U-Pb ages of monazites and zircons from the crustmantle transition and adjacent units of the Ivrea and Ceneri Zones (Southern Alps, Italy). Contrib Mineral Petrol 43: 55–70

Köppel V, Grünenfelder M (1971) A study of inherited and newly formed zircons from paragneisses and granitised sediments of the Strona-Ceneri-Zone (Southern Alps). Schweiz Mineral Petrogr Mitt 51, 2/3: 385–409

Köppel V, Grünenfelder W (1978) Monazite and zircon U-Pb ages from the Ivrea and Ceneri zones. Mem Ist Sci Geol Univ Padova 33: 257

Köppel V, Günthert A, Grünenfelder M (1980) Patterns of U-Pb zircon and monazite ages in polymetamorphic units of the Swiss Central Alps. Schweiz Mineral Petrogr Mitt 61: 97–119

Kröner A, Wendt I, Liew TC, Compston W, Todt W, Fiala J, Vankova V, Vanek J (1988) U-Pb zircon and Sm-Nd model ages of high-grade Moldanubian metasediments, Bohemian Massif, Czechoslovakia. Contrib Mineral Petrol 99: 257–266

Leutwein F, Power G, Roach P, Sonet J (1973) Quelques resultats géochronologiques obtenus sur des roches d'age précambrien du Cotentin. C R Acad Sci Paris 276: 2121–2124

Mattauer M, Brunel M, Matte Ph (1988) Failles normales ductile et grands chevauchements. Une nouvelle analogie entre l'Himalaya et la chaîne hercynienne du Massif Central français. C R Acad Sci Paris 306, II: 671–676

Matte Ph (1986) La chaîne varisque parmi les chaînes paléozoiques péri atlantiques, modèle d'evolution et position des grands blocs continentaux au Permo-Carbonifère. Bull Soc Geol Fr 8 (1): 9–24

McDowell FW, Schmid R (1968) Potassium-argon ages from the Valle d'Ossola section of the Ivrea-Verbano zone. Schweiz Mineral Petrogr Mitt 48: 205–210

Ménot RP (1988) The geology of the Belledonne massif: an overview (external crystalline massifs of the Western Alps). Schweiz Mineral Petrogr Mitt 68 (3): 531–542

Ménot RP, Peucat JJ, Paquette JL (1988) Les associations magmatiques acide-basique palèozoiques et les complexes leptyno-amphiboliques: les corrélations hasardeuses. Exemples du massif de Belledonne (Alpes occidentales) Bull Soc Geol Fr 8 (6): 917–926

Mercolli I, Oberhänsli R (1988) Variscan tectonic evolution in the Central Alps: a working hypothesis. Schweiz Mineral Petrogr Mitt 68: 491–500

Nägler ThF, Gebauer D, Schäfer H-J (1989) Nr-, Sr- and Pb-isotope geochemistry of two pre-Permian sedimentary profiles from the Spanish Meseta: evidence of both crustal recycling and crustal growth. Terra Abstr 1: 336

Neubauer F, Frisch W, Hansen BT (1987) New data on the evolution of the austroalpine basement: a U/Pb zircon study. Terra Cognita 7: 96

Oberli F, Sommerauer J, Steiger RH (1981) U-(Th)-Pb systematics and mineralogy of single crystals and concentrates of accessory minerals from the Cacciola granite, central Gotthard massif, Switzerland. Schweiz Mineral Petrogr Mitt 61: 323–348

O'Nions RK, Hamilton PJ, Hooker PJ (1983) A Nd isotopes investigation of sediments related to crustal development in the British Isles. Earth Planet Sci Lett 63: 229–240

Nunes PD, Steiger RH (1974) A U-Pb zircon, Rb-Sr and U-Th-Pb whole-rock study of a polymetamorphic terrane in the Central Alps, Switzerland. Contrib Mineral Pet 47: 255–280

Paquette JL, Marchand J, Peucat JJ (1984) Absence de tectonique cadomienne dans le complexe de Champtoceaux (Bretagne méridionale)? Comparaison des systèmes Rb-Sr et U-Pb d'un métagranite. Bull Soc Géol Fr 7, XXVI, 5: 907–912

Paquette JL, Menot RP, Peucat JJ (1989) REE, Sm-Nd and U-Pb zircon study of eclogites from the Alpine External Massifs (Western Alps): evidence for crustal contamination. Earth Planet Sci Lett 96, 1/2: 181–198

Peucat J-J, Vidal Ph, Godard G, Postaire B (1982) Precambrian U-Pb zircon ages in eclogites and garnet-pyroxenites from South Brittany (France) an old oceanic crust in the West European Hercynian belt? Earth Planet Sci Lett 60: 70–78

Pidgeon RT, Köppel V, Grünenfelder M (1970) U-Pb isotopic relationships in zircon suites from a para- and orthogneiss from the Ceneri Zone, southern Switzerland. Contrib Mineral Petrol 26: 1–11

Pin C (1986) Datation U-Pb sur zircons à 285 M. a. du complexe gabbro-dioritique du Val Sesia – Val Mastallone et âge tardihercynien du metamorphisme granulitique de la zone Ivrea-Verbano (Italie). C R Acad Sci Paris 303,II,9: 827–830

Pin C (1989) Essai sur la chronologie et l'évolution géodynamique de la chaine hercynienne d'Europe. Thèse, Université Blaise Pascal (Clermont-Ferrand II), 470 pp

Pin C, Sills JD (1986) Petrogenesis of layered gabbros and ultramafic rocks from Val Sesia, the Ivrea Zone, NW Italy: trace element and isotope geochemistry. In: Dawson JB, Hall J, Wedepohl KH (eds) The nature of the lower continental crust. Geol Soc Spec Publ 24: 231–249

Polvé M (1983) Les isotopes du Nd et du Sr dans les Lherzolite orogéniques: contribution à la détermination de la structure et de la dynamique du manteau supérieur. Thèse, Université de Paris

Postaire B (1982) Systématique Pb commun et U-Pb sur zircon. Application aux roches de haut grade métamorphique im-

pliquées dans la chaîne hercynienne (Europe de l'ouest) et aux granulites de Laponie (Finlande). Thèse, Université de Rennes, 1–71
Rivalenti G, Garuti G, Rossi A (1975) The origin of the Ivrea-Verbano basic formation (western Italian Alps). Whole-rock geochemistry. Boll Soc Geol Ital 94: 1149–1186
Rivalenti G, Garuti G, Rossi A, Siena F, Sinigoi S (1980) Existence of different peridotite types and of layered igneous complex in the Ivrea Zone of the Western Alps. J Pet 22: 127–153
Rivalenti G, Garuti G, Rossi A, Siena F, Finigoi S (1981) Chromian spinel in the Ivrea-Verbano layered igneous complex, Western Alps, Italy. Tschermaks Mineral Petrogr Mitt 29: 33–53
Rivalenti G, Rossi A, Siena F, Sinigoi S (1984) The layered series of the Ivrea-Verbano igneous complex, Western Alps, Italy. Tschermaks Mineral Petrogr Mitt 33: 77–99
Rogers G, Dempster TJ, Bluck BJ, Tanner PWG (1989) The implications of a high-precision U-Pb age from the Ben Vuirich granite for the evolution of the Scottish Dalradian. Terra Abstr 1: 11
Schäfer HJ, Gebauer D, Nägler ThF (1989) Pan-African and Caledonian ages in the Ossa Morena Zone (SW Spain): A U-Pb zircon- and Sm-Nd study. Terra Abstr: 350–351
Schaltegger U (1986) Voralpine und alpine Mineralbildung in der Gneiszone von Erstfeld (Sustenpass, Aarmassiv); der Mechanismus der K-Ar und Rb-Sr Verjüngung alpin umgewandelter Biotite. Schweiz Mineral Petrogr Mitt 66: 395–412
Schaltegger U (1989) Geochemische und isotopengeochemische Untersuchungen am zentralen Aaregranit und seinen assoziierten Gesteinen zwischen Aare und Reuss (Aarmassiv, Schweiz). Thesis, University Bern
Schenk-Wenger K, Stille P (1989) Geochemical and isotope evidence for an extensive Proterozoic ophiolite-suite in the Central Alps (Switzerland). Abstr Beih Eur J Mineral 1: 160
Schmid R, Wood BJ (1976) Phase relationships in granulitic metapelites from the Ivrea-Verbano zone (Northern Italy). Contrib Mineral Petrol 54: 255–279
Schmidt K, Söllner F (1982) Proposals for the Geodynamic Interpretation of the "Caledonian Event" in central and southern Europe with special reference to the Eastern Alps. In: Sassi F und Varga A. IGCP 5, Newsletter 4: 82–93
Sills JD, Tarney J (1984) Petrogenesis and tectonic significance of amphibolites interlayered with metasedimentary gneisses in the Ivrea Zone, Southern Alps, northwest Italy. Tectonophysics 107: 187–206
Söllner F, Hansen BT (1987) "Pan-afrikanisches" und "kaledonisches" Ereignis im Oetztal-Kristallin der Ostalpen: Rb-Sr- und U-Pb-Altersbestimmungen an Migmatiten und Metamorphiten. Jahb Geol Bundesanst 130 (4): 529–569
Stähle V, Frenzel G, Kober B, Michard A, Puchelt H, Schneider W (1990) Zircon syenite pegmatite in the Finero peridotite (Ivrea zone): evidence for a syenite from a mantle source. Earth Planet Sci Lett 101: 196–205
Steiger RH, Jäger E (1977) Subcommission on Geochronology: Convention on the use of decay constants in geo- and cosmochronology. Earth Planet Sci Lett 36: 359–362
Stille P (1980) On the genesis of the amphibolites and in the Berisal Complex (Simplon; Italy-Switzerland). Mem Sci Geol, Padua 34: 205–246
Stille P, Buletti M (1987) Nd-Sr isotopic characteristics of the Lugano volcanic rocks and constraints on the continental crust formation of the South Alpine domain (N-Italy-Switzerland) Contrib Mineral Petrol 96: 140–150
Stille P, Tatsumoto M (1985) Precambrian tholeiitic-dacitic rock-suites and Cambrian ultramafic rocks in the Penninic nappe system of the Alps: evidence from Sm-Nd isotopes and rare earth elements. Contrib Mineral Petrol 89: 184–192
Teufel S, Schärer U (1989) Unravelling the age of high-grade metamorphism of the Ivrea Zone: a monazite single-grain and small fraction study. Terra Abstr 1: 350
Vidal Ph (1977) Limitation isotopique à l'age de l'évolution de la croute continentale en Europe moyenne et occidentale. In: Cogné J (ed) La chaine varisque d'Europe moyenne et occidentale. Colloq Int Cent Natl Rech Sci, Rennes, 234: 129–141
Vidal Ph, Auvray B, Charlot R, Cogné J (1981) Precadomian relics in the Armorican Massif: their age and role in the evolution of the Western and Central European Cadomian – Hercynian belt. Precambrian Res 14: 1–20
Vocke RD, Hänny R (1984) Petrogenesis of the Sondalo – M. Serottini intrusions, Upper Valtellina, Northern Italy. Terra cognita 4: 24
Voshage H, Hunziker JC, Hofmann AW, Zingg A (1987) A Nd and Sr isotopic study of the Ivrea zone, Southern Alps, N-Italy. Contrib Mineral Petrol 97: 31–42
Voshage H, Sinigoi S, Mazzucchelli M, Demarchi G, Rivalenti G, Hofmann AW (1988) Isotopic constraints on the origin of ultramafic and mafic dikes in the Balmuccia peridotite (Ivrea Zone) Contrib Mineral Petrol 100: 261–267
Weber W (1966) Zur Geologie zwischen Chiavenna und Mesocco. PhD Thesis, ETH Zürich
Wendt JI, Kröner A, Todt W, Fiala J, Rajlich P (1989) 2 Ga zircon ages for the moldanubian basement and the age of the granulite facies metamorphism in southern Bohemia, CSSR. Terra Abstr 1: 4
Wright JE, Shervais JW (1980) Emplacement age of the Balmuccia Lherzolite Massif, NW Italy. 26th Int Geol Congr, Paris, Résumés 2: 804
Zingg MA (1989) Die Siviez-Mischabel Decke: Entstehung und Entwicklung eines Altkristallins und seiner Vererzungen (Wallis, Schweiz). Thesis, ETH Zürich

Ultramafic Rocks in the Pre-Mesozoic Basement of the Central and External Western Alps

H. R. Pfeifer, G. Biino, R. P. Ménot and P. Stille

Abstract

Ultramafic rocks, usually accompanied by metagabbroic and or metabasaltic mafic rocks typically occur as small lenses of several to some tens of metres in thickness in many pre-Mesozoic basement units of the Central and external Western Alps. Some of these ultramafic-mafic rock suites can be followed along strike for several tens of kilometres. Mineral assemblages are mainly metamorphic with rare structural or mineralogical relics from the mantle or lower crustal stage. Depending on the grade of the Alpine metamorphism, these rocks are either serpentinites or olivine-schists with various hydrous or anhydrous Mg-Ca-Al-silicates, such as diopside, tremolite, chlorite, talc and opaque spinel. The marginal parts of these lenses are often hydrothermally altered to talc-amphibole- and carbonate-rich rocks. When the ultramafic rocks are accompanied by former gabbros and basalts, an ophiolite-type oceanic origin can be inferred. Some ultramafics are related to larger masses of metagabbros, metadiorites or amphibole-rich gneisses which are interpreted as former calc-alkaline intrusions in the lower crust. In the case of the larger masses of pure ultramafic rocks, a mantle-wedge tectonically emplaced in the crust along large-scale shear zones seems to be the most likely explanation. From the regional point of view the external massifs and the Penninic Adula nappe contain by far the most occurrences of mafic-ultramafic rocks. As yet, only little is known about the age of most of these series, but many of them have ages in excess of 450 Ma.

1 Introduction

1.1 General Remarks

The pre-Mesozoic basement of the Central and the external Western Alps is relatively rich in small masses of metaperidotites (Figs. 1, 2; Table I and Appendix). Most of them are accompanied by amphibolites or greenschists which represent former gabbros and basalts. In many cases accompanying metasedimentary rocks (gneissic metapelites and metapsammites, sometimes marbles) can help to unravel the origin of a given series. In the field, these mafic – ultramafic – metasediment associations can often be followed over tens of kilometres (cf. Figs. 1 and 2). Due to the small size of the ultramafic bodies, which usually varies from a few metres to a few hundreds of metres, most maps that cover the area in question do not distinguish between ultramafic rocks and mafic rocks (e. g. Carte géologique de la France 1:250000; Bigi et al. 1990; Geologische Generalkarte der Schweiz 1:200000). However, there are a few overview publications (Evans and Trommsdorff 1970; Trommsdorff and Evans 1974; Nicolas 1984; Pfeifer and Serneels 1986; Mannoni et al. 1987; Pfeifer 1989), but some of them lack the exact location of the outcrops or are published in local journals that are difficult to find. This review intends to fill this gap and summarizes existing references and presents, in part unpublished field and laboratory data. For each occurrence the exact location, its tectonic situation, its country rocks, the actual ultramafic rock types and their likely protoliths as well as ages will be discussed. In order to avoid an excessively long reference list, publications older than 1980 will normally not be mentioned, but instead the reader will be referred to the review of Alpine ophiolites and the literature compilation of Dietrich (1980).

1.2 Rock Types

As is typical for orogenic type ultramafic complexes, all these rocks are polymetamorphic with rarely preserved relic structures or relic minerals from former mantle or lower crust stages. Depending on the grade of the different Palaeozoic and especially Alpine metamorphisms, the ultramafic rocks of the pre-Mesozoic basement of the Alps occur either as ser-

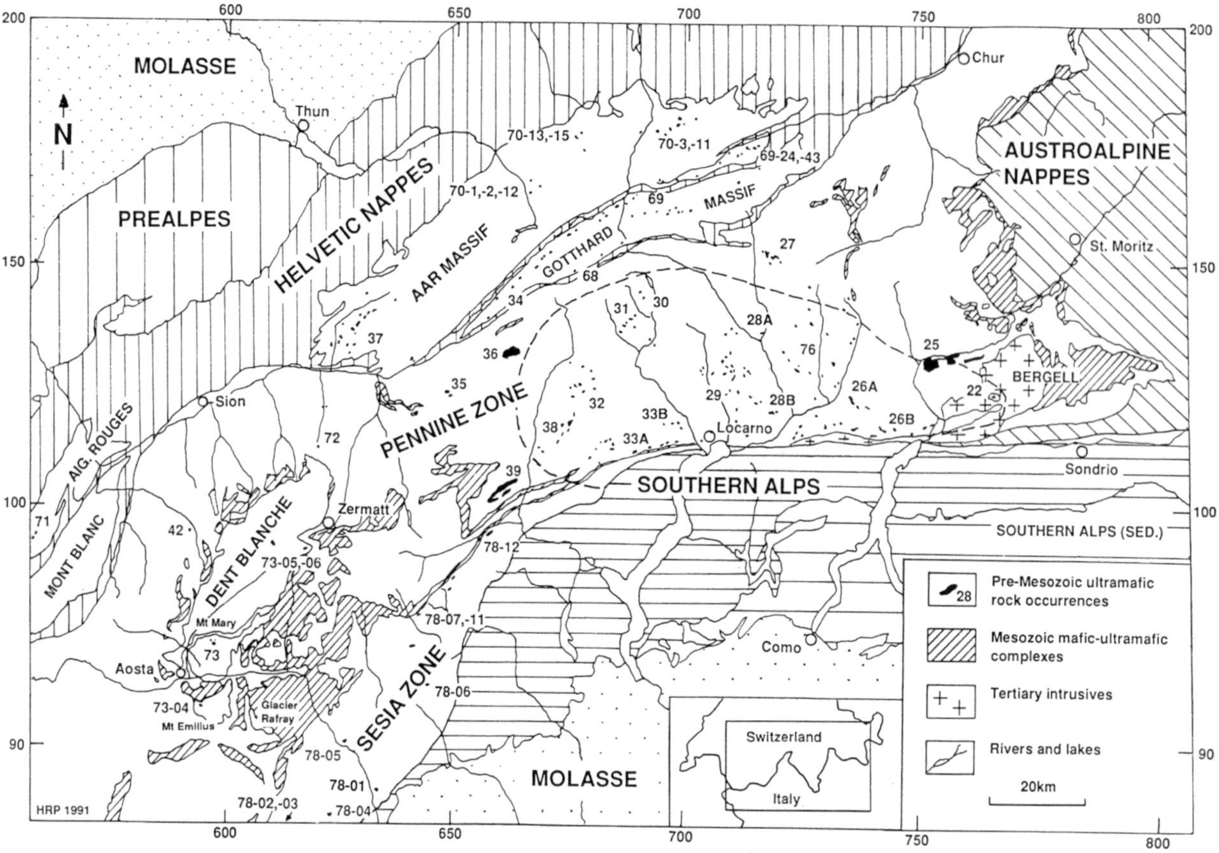

▲ **Fig. 1.** Distribution of pre-Mesozoic ultramafic rocks in the Central Alps. The area outlined by a *thick dashed line* marks the high-grade area of the Tertiary metamorphism, which is characterized by olivine schists. Outside of this central area, i. e. in the external massifs, the Western and Eastern Alps, serpentinites dominate

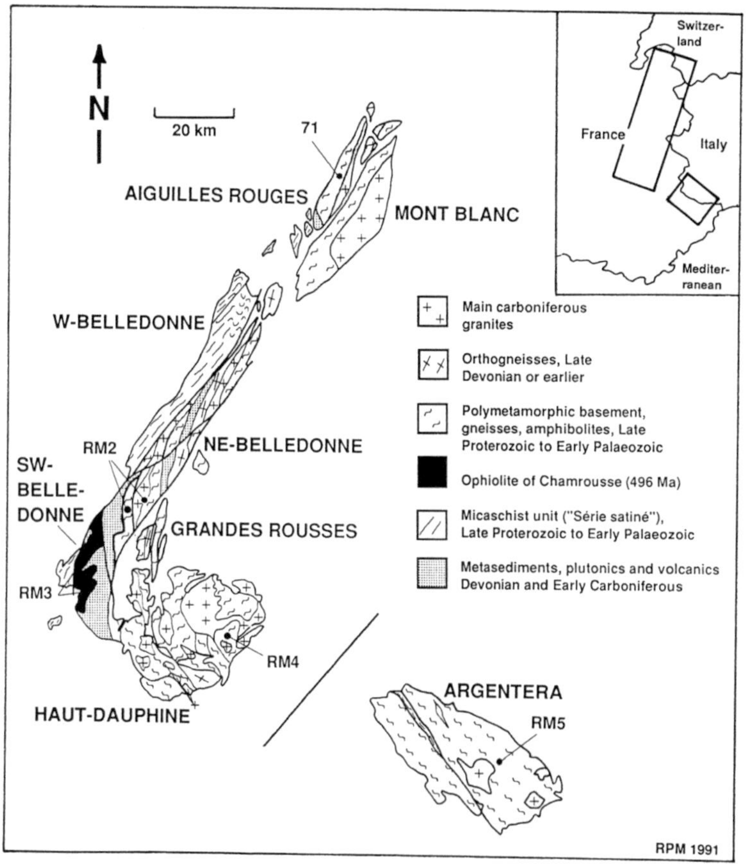

◀ **Fig. 2.** Distribution of pre-Mesozoic ultramafic rocks in the external Western Alps

Table 1. Characteristics of the major occurrences of pre-Mesozoic ultramafic rocks in the Central and external Western Alps

Tectonic unit	Occurrences	Typical size	Rock	Relic minerals[a]	Country rocks	Age	Origin
Aar	15	5×10 m	Lherzol. harzburg. serp.	Ol-Tc-MgAm	Metasediments gneisses	pre-Ordovician	Ophiolites Metam. differentation
Gotthard	30	10×20 m	Lherzol. harzburg. serp.	Ol-Tc-MgAm	Metabasalts,-gabbros, gneisses	pre-Ordovician	Ophiolites Lower crust intrusion
Aiguilles Rouges	1	5×10 m	Lherzol. harzburg. serp.	Ol-Px	Metabasalts, gneisses	?	Ophiolites
Belledonne	3	60 m×5 km	Pyroxenitic-dunitic	Ol-Cpx-Pl	Gabbros	496 Ma	Ophiolites
H. Dauphiné	1	1×1 m	Metasomatic peridotites	–	Migmatitic gneisses, pegmatites	?	Metam. differentiation
Argentera	1	1×1 m	Metasomatic peridotites	–	Migmatitic gneisses, pegmatites	?	Metam. differentiation
Monte Leone	1	1.5×4.5 km	Lherzolitic--pyrox. serp.	Cpx, Sp	Metabasalts, metasediments, gneisses	?	Ophiolite
Berisal	2	10×100 m	Hornblendites	Cpx	Metabasalts, gneisses	475 Ma	Lower crust intrusion
Bosco-Bellinzona	40	20×30 m	Harzburgitic serpentinites olivine schists	Sp	Metabasalts,- gabbros, gneisses, marbles	1000 Ma	Ophiolites
Cima Lunga-Adula	50	20×30 m	Harzburgitic olivine schists	Grt, Cpx, Sp	Metabasalts, eclogites gneisses, marbles	1700 Ma, 600 Ma	Ophiolites (no gabbros)
Chiavenna-Bondo	3	2.5×12 km	Lherzolitic olivine schists	Cpx, Sp	Metabasalts,- gabbros, pegmatites	?	Ophiolite
Bergell	16	30×50 m	Harzburgitic olivine schists	–	Amphibolites, migmatic gneisses	?	Mantle wedge, Contact metamorphism
Leventina-Simano	15	5×10 m	Harzburgitic olivine schists	–	Metabasalts, gneisses	?	Ophiolites
Antigorio	4	10×20 m	Harzb. olivine-schists; hornbl.	Grt, Sp, Pl	Metagabbros, -diorites gneisses	?	Ophiolites; lower crust intrusion (calc-alkaline)
Moncucco	2	1×7 km	Lherzolitic olivine schists	Opx	Gneisses, pegmatite	>200 Ma	Mantle wedge
Bernhard	4	20×30 m	Harzburgitic talc-carbonate schists	Srp	Gneisses	?	Mantle wedge
Sesia-Lanzo, Sec. Diorite-Kinzigite Dent Blanche	8	5×10 m	Various peri-dotites; pyrox.	Px	Gabbros, eclogites, gneisses	?	Ophiolites; lower crust intrusions

[a] Mineral abbreviations: Cpx, clinopyroxene; Grt, garnet; MgAm, magnesium-amphibole; Ol, olivine; Opx, orthopyroxene; Pl, plagioclase; Px, ortho- and clinopyroxene; Srp, serpentine; Sp, spinel; Tc, talc.

pentinites or as olivine-dominated schists. The sequence of index parageneses observable in the Western and Central Alps roughly corresponds to what would be expected from the idealized phase diagram (Evans 1977; Trommsdorff 1983; Pfeifer 1987). In the external massifs and the peripheral Penninic zones (Wallis, Grisons) lizardite-chrysotile (-diopside-chlorite-magnetite)-serpentinites grade into higher metamorphic antigorite- (olivine-diopside or tremolite-chlorite-magnetite-) serpentinites. In the upper amphibolite facies areas of the Lepontine Alps the dominant rock type is an olivine-talc- (chlorite-opaque spinel-) schist with variable amounts of stable tremolite, magnesium-amphiboles (anthophyllite, Mg-Cummingtonite, pyriboles), magnesite and enstatite (often relics; Trommsdorff and Evans 1974; Pfeifer 1981). Using the common nomenclature for mantle rocks, which uses the aluminium-rich phase to name the rocks, the latter type is sometimes called chlorite-peridotite (Rost 1971; Medaris and Carswell 1989). Due to the small size of most of these rock masses, hydrothermal metamorphism often led to metasomatic compositional and mineralogical changes, involving the influx of H_2O, CO_2, SiO_2, CaO and Al_2O_3 and the growth of hydrous silica- and aluminium-rich minerals as well as carbonates, especially in the marginal parts (Pfeifer 1978, 1981, 1987).

Mineral relics or typical outcrop structures can be used to deduce the upper mantle or lower crust precursors of these rocks. Typical mineral relics are pseudomorphs of chlorite and opaque spinel after ordinary Mg-Al-rich spinel. Large pigmented olivine crystals or larger recrystallized olivine domains in serpentinites most likely correspond to former mantle olivine. Large pigmented diopside crystals with typical diallage cleavage are considered to represent relics of mantle clinopyroxene. A great abundance of calcium- and aluminium-bearing phases (Ca-amphibole, diopside, chlorite, rarely garnet) indicates lherzolitic or pyroxenitic protoliths. Pure serpentinites and olivine-talc rocks are equivalents of harzburgites and dunites.

Bulk rock analyses of metamorphic ultramafic rocks are as yet a little used method of determining ultramafic protoliths. Pfeifer (1993) proposed a series of discrimination diagrams mainly based on CaO, Al_2O_3, TiO_2, Co, Sc and V. These diagrams allow us to distinguish between former pyroxenites, lherzolites, harzburgites and dunites, but only a few pre-Mesozoic occurrences have been chemically analyzed (see Sect. 3).

At outcrop scale, original layering is often preserved as layers rich in diopside/tremolite, chlorite and/or opaque minerals (former ortho- and clinopyroxenites) or simply as layers rich in opaque spinel (magnetite or ferrite-chromite). Due to the usually intense deformation which parallelized original cross-cut features, former dykes cannot be distinguished from former layers in most cases. The same is true for former basaltic or gabbroic dykes or sills. They usually occur as boudin-type inclusions. Some notable exceptions with preserved original cross-cutting features occur in the Chamrousse ophiolites, Western Alps and in the Gotthard Massif, see later. These mafic inclusions often occur as Ca-Al-silicate rich rocks (rodingites) which are typical of serpentinized rocks (Coleman 1977; Evans et al. 1979; Trommsdorff and Evans 1980 a; Rösli et al. 1991).

2 External Massifs

A comprehensive discussion of the palaeogeographic situation of these massifs is given by von Raumer et al. (this Vol.). Most of them contain several, often structurally aligned mafic and ultramafic-mafic zones, ranging up to 100 m in thickness. The Chamrousse mafic-ultramafic complex in the SW region of the Belledonne Massif, with ultramafic rocks outcropping on an area of several square kilometres is an exception. The numbers given in parentheses after a locality name refer to the maps of Fig. 1, 2 and Table I.

2.1 Aar, Gotthard and Tavetsch Massifs

In all three massifs the mafic-ultramafic zones occur in the oldest pre-Variscan part. Metamorphic recrystallizations and penetrative deformations have usually obliterated the original relationship between mafic and ultramafic rocks and their host country rock. Nevertheless the original lithostratigraphic environment (e.g. metabasalts or metagabbros, detrital metasediments) seems to have been preserved in many cases. In a few cases well preserved magmatic textures and internal foliation discordant to the Variscan directions in the surrounding gneisses have been observed (Abrecht et al. 1991; Biino 1992). There is a textural feature common to many ultramafic rocks described below: they exhibit centimetre-sized acicular aggregates composed of amphiboles (tremolite or Mg-amphibole) or are pseudomorphically replaced by talc, magnesite and magnetite.

In both massifs new geochemical and isotopic investigations were done by Biino and coworkers (Biino et al. 1993; Biino and Meisel 1993 a, b).

2.1.1 Aar Massif (No. 37, 70)

The western part of the massif, the Lötschen Valley (No. 37 A, Fig. 1) is especially rich in metre-size structurally aligned ultramafic lenses (Hügi et al. 1988) which are in part accompanied by mafic rocks of MORB-composition (today normal and garnet-amphibolites). Some of them contain antigorite-serpentinites with minor chrysotile, lizardite, talc, tremolite and carbonates, but most of these lenses are almost completely metasomatically altered to talc-carbonate-chlorite-tremolite rocks (soapstone).

In the Trift area east of Guttannen in the Hasli Valley/Bern (No. 70-1, -2, -12), Schenker and Abrecht (1987) describe four localities with small angular inclusions of mafic-ultramafic rocks embedded in a trondhjemitic pegmatoid matrix within a migmatitic gneiss series, which they refer to as "Schollenamphibolite" (cf. Abrecht 1980). The ultramafic inclusions contain in one case forsterite, talc and serpentine and in the other cases serpentine and amphiboles. This special mafic-ultramafic rock-type association is very common in the whole Aar Massif. It has also been described by Labhart (1965) in the Massa Valley N of Brig/Wallis (No. 37–51), containing serpentinite lenses of up to 10 m and by Pflugshaupt (1927) in the Maderaner valley E of Amsteg/Uri (No. 70-5 to -11).

In the northeastern part of the massif there are fewer ultramafic rocks. In the Sustenpass area, metre-sized chrysotile-lizardite-serpentinite lenses with marginal reaction zones containing anthophyllite, tremolite, fibrous serpentine and talc occur (No. 70-13, -14, -15). However, serpentine and anthophyllite do not seem to be in equilibrium (Schaltegger 1986). Garnet amphibolites are also present. The adjacent gneisses are interpreted as a metamorphic flysch series that has been cross-cut by granitic rocks of Ordivician or older age (Albrecht et al. 1991). In the Reuss Valley near Gurtnellen (No. 70-4) rocks containing anthophyllite, tremolite, chromite and phlogopite have been discovered (Biino 1992).

For the 10-km-long mafic-ultramafic rock zone in the west of the massif (Lötschen Valley) and the few lenses at Sustenpass unpublished isotopic and trace element geochemistry point to an ophiolite-type oceanic origin. The "Schollenamphibolites" is a phenomenon known from many basement complexes and Abrecht (1980) has interpreted it as a metamorphic differentiation phenomena related to the mobilization of quartzo-feldspatic gneisses under intense deformation.

2.1.2 Gotthard and Tavetsch Massifs (No. 34, 68, 69)

In these massifs the ultramafic rocks are more widespread than in the Aar Massif. Over 90% of the occurrences are situated along a narrow 2-km-wide zone within the northern pre-Variscan polymetamorphic metasedimentary gneisses which extend along the Rhone-Rhine line from Ernen-Mühlebach/Wallis to Trun/Grisons (Fig. 1, Appendix; Oberholzer 1955; Labhart 1977; Oberhänsli et al. 1988). Some occur along late Ordovician metagranitic gneisses ("Streifengneis") in the central part of the Gotthard Massif (Abrecht et al. 1991) and in the composite zones of the southern part of the massif (Hafner 1958; Labhart 1977).

Two different occurrences of ultramafic rocks can be recognized. The first type shows a close spatial relationship with metagabbroic mafic rocks which have well-preserved igneous textures. The second type occurs as dismembered ultramafic lenses usually hosted in metasedimentary gneisses and are accompanied by metagabbroic and metabasaltic mafic rocks. In both cases mafic rocks typically are garnet-amphibolites with occasional relics of a pre-Variscan high-pressure event (Abrecht et al. 1991; Biino 1992).

An example of the first type can be found at the Chastelhorn SE of Hospenthal, Urseren Valley (No. 69–11), where a metagabbro body of 1.5 km by 400 m, rich in magmatic relics contains small lenses of antigorite serpentinite of a few metres. The marginal zone of the gabbro is mylonitic and the serpentinites have peculiar ovoidal structures of several centimetres. The ultramafics probably represent a cumulative layer within an eclogitized gabbro which intruded an accretionary wedge (Abrecht et al. 1991).

The second type is more widespread and often preserves mantle structures and/or amphibolite facies mineralogy. In the field it is accompanied by Mg – and Fe-Ti-rich metabasalts or – gabbros. The series of antigorite-serpentinites lenses between Hospental/Uri and Val Nalps/Grisons (Chämleten/Hospental; southern Unteralp Valley; Fuorcla Paradis; Val Nalps; No. 69-7 to 69-20) belong to this type. A representative lithostratigraphic section is shown on Fig. 3. Many of these lenses contain layers of metapyroxenite or layers which are rich in chlorite and/or magnetite (former chromite layers). The rocks often contain two generations of Ca-pyroxene: an older one represented by relatively Na-, Al-, Cr- and Ti-rich diopsides and a younger more pure Ca-pyroxene.

In the large outcrop of Fuorcla Paradis (69-17), which comprises three lenses of a total length of almost 1 km and a width of 200 m, pyroxenitic dykes

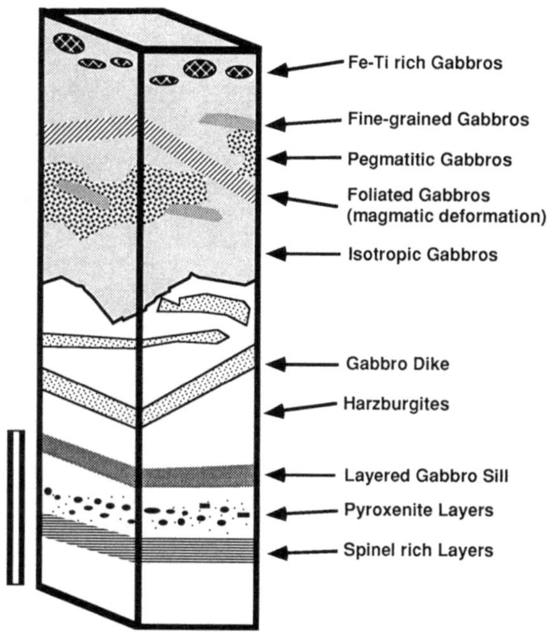

Fig. 3. Schematic lithostratigraphic section typical for mafic-ultramafic series of the eastern Gotthard Massif

(chlorite-diopside) and gabbroic dykes (coarse grained amphibolite) cross-cut the lenses. Zircons of the biotitic black wall zone yield U-Pb ages of 448–472 Ma (Ordovician; Grauert and Arnold 1968), whereas the Rb-Sr cooling age from the biotites indicates a Lower Permian age. Relatively high temperature minerals like olivine with forsterite 70–85, Cr-bearing spinels and tremolite have been found together with antigorite, diopside, Cr-chlorite, talc and magnesite at Gigestafel SE of Hospental/Urseren Valley and in Gotthard railway tunnel just below. Based on the lithostratigraphic environment, the inferred ultramafic protoliths and the geochemistry of the accompanying mafic rocks, Abrecht et al. (1991) concluded that this second type is derived from ophiolite type oceanic crust emplaced in a accretionary wedge during the Lower Palaeozoic.

In the northeastern Gotthard Massif and the Tavetsch Massif between Oberalp Pass and Trun/Grisons, many small serpentinite and talc-carbonate rock lenses occur in strongly migmatized and deformed metasedimentary gneisses (exact location: Appendix). Some petrographic relationships are described by Heim (1918), Parker (1921) and Niggli (1944). They have not yet been studied in detail with modern methods, but according to their lithostratigraphic environment, mineralogical and geochemical character, these lenses most likely belong to the second type.

2.2 *Aiguilles Rouges and Mont Blanc Massifs* (No. 78)

In the Aiguilles Rouges Massif NE of Lac Cornu/France, a zone of several kilometres in length contains around forty well-aligned lenses of serpentinized lherzolitic metaperidotites and metapyroxenites, ranging in size from a few metres to hundreds of metres (Pfeifer and von Raumer 1993). The parageneses include chrysotile/lizardite-chlorite-tremolite with frequent pigmented relics of olivine, clino- and orthopyroxene. Some lenses are hydrothermally altered to talc-tremolite-carbonate rocks. The adjacent rocks comprise mainly anatectic granitoid gneisses and eclogitic mafic rocks similar to the ones described from Lac Cornu (Liègois and Duchesne 1981).

Only a few small occurrences of ultramafic rocks have been observed in the Mont Blanc Massif, which resemble the mafic-ultramafic inclusions of the "Schollenamphibolite"-type (von Raumer pers. comm. 1991).

2.3 *Belledonne and Grandes Rousses Massifs* (No. RM2 and RM3)

According to the structural subdivision of the Belledonne Massif (Ménot 1988; von Raumer et al. this Vol.), the pre-Mesozoic ultramafic rocks occur in the SW and NE domains (Fig. 2). Mainly in the southern region of the NE domain (No. RM2), ultramafic rocks occur in the polymetamorphic basement together with amphibole- or biotite-gneisses in narrow bands of several kilometres in length. The size of the ultramafic lenses varies from several metres to several tens of metres (Kalsbeek 1961; Bordet 1961, 1969, 1977). Serpentinites, talc-schists and actinolite-chlorite schists are commonly accompanied by calcschists (Bordet 1961) and by metapyroxenites (Kalsbeek 1961). No detailed petrological and chemical data are available and therefore the origin of these rocks is currently unknown.

In the SW domain of the Belledonne Massif (No. RM3), the ultramafic rocks belong to the ophiolitic sequence of Chamrousse (see Ménot and Paquette this Vol. for a more complete description). This Early Palaeozoic ophiolite (Carme and Pin 1987; Ménot et al. 1988a) is overturned, therefore the metaperidotites and -pyroxenites are lying in the uppermost part of the layered complex. The ultramafic pile does not exceed 60 m in thickness. It displays millimetre to metre thick layers in which the following original rock-types occur: dunites, chromite-rich dunites, wherlites and diallagites (Choubert 1936; Den Tex 1950; Carme 1965, 1972, 1975; Ménot 1979, 1987; Bodinier et al. 1981; Carme and Pin 1987; Ménot et al. 1988a). This igneous layering indicates an early olivine-spinel-clinopyroxene sequence of crystallization before the occurence of the Ca-rich plagioclase which marks the transition into the layered gabbros (type B of Church and Riccio 1977). This layering is cross-cut by pyroxenitic dykes, several centimetres to several tens of centimetres thick.

The primary igneous minerals are represented by chromite and clinopyroxene (diallage and diopside-augite; Ménot 1987). Olivine is inferred from the bipyramidal pseudomorphs of antigorite (Den Tex 1950). The chemistry of the Cr-spinels has been investigated by Den Tex (1950), Legendre (1982) and Fischer et al. (1985). From the Cr, Fe^{3+}, Al and Mg contents and from the distribution of the platinoid elements (normalized to chondrites), the chromites of the Chamrousse complex are similar to that of the layered series from other ophiolites and the ocean floor.

Successive recrystallizations are related to intra-oceanic and Variscan metamorphic events, with the alpine imprint being very weak (zeolite facies, "anchizone"). The intra-oceanic stage leads to low pressure-high temperature parageneses (clinopyroxene and pargasite) and further to hornblende and actinolite-chlorite-serpentine, bastite assemblages (mineral chemistry in Ménot 1987). Epidote-amphibolite grade metamorphism occurred during the Variscan.

The layered ultramafic (and mafic) rocks are derived from E-MORB type magmas through fractional crystallization processes (Bodinier et al. 1981; Carme and Pin 1987; Ménot 1987; Ménot et al. 1988a; see Ménot and Paquette this Vol. for detailed discussion and references). The associated meta-pyroxenitic dykes show N-MORB type affinities and represent a later magmatic stage related to distinct mantle sources.

The ophiolite of Chamrousse is one of the most spectacular remnants of the Lower Palaeozoic oceanic crust. In spite of a strikingly different tectono-metamorphic evolution (Ménot and Paquette this Vol.), it can be correlated (1) with the mafic and ultramafic rocks of the polymetamorphic basement of the external crystalline massifs (the mafic igneous events have similar ages and geochemical characters) and (2) with the leptynitic and amphibolitic groups of the Western European Variscan belt (Bodinier et al. 1981; Pin and Carme 1987; Carme and Pin 1987; Ménot et al. 1988b).

No ultramafic rocks have been recognized in the Grandes Rousses Massif.

2.4 *Haut Dauphiné Massif* (No. RM4)

The ultramafic rocks seem to be restricted to the deep crustal, migmatitic, formations of the "Noyau" (Le Fort and Pécher 1971; Pécher and Barfety 1984). They are always closely associated with amphibolites. Three kinds of rocks have been described:

1. Rounded xenoliths showing an internal zonation;
2. Lenses of serpentinite;
3. Layers of pyroxenite.

The first type forms elongated zones along the boundaries between the amphibolites and the migmatitic acid gneisses (Pécher 1970; Pécher and Barfety 1984; Le Fort 1973, 1976). These dark-green xenoliths are thought to be relics of small, boudinaged ultrabasic rocks and their concentric structure is probably the result of chemical transfers during migmatitic processes. Up to five successive zones can be observed, from the inner to the outermost (Pécher 1970; Pécher and Barféty 1984):

1. An equant core of Fe-Mg and Ca-rich minerals (anthophyllite, tremolite, chlorite, phlogopite, carbonates, talc, serpentine, green hornblende, augite, sphene, apatite and Fe-oxides);
2. A shell of Fe-oxides (1 cm);
3. A first shell of green amphiboles oriented parallel to the rims (<0.5 cm);
4. A second shell of green amphiboles but with a radial orientation (1 cm);
5. A very narrow rim of biotite.

The second type is often associated with the xenolith alignments. The serpentinites contain only antigorite-chrysotile assemblages including some amphiboles as relics. Their primary nature is unknown. The significance of these different ultramafic rock types has not been investigated, but an origin similar to the "Schollenamphibolites" (see Sect. 2.2) seems likely.

The third type (RM4) only outcrops in the SE area of the Ht. Dauphiné Massif (Claphouse valley, Pécher and Barféty 1984) as a 30-m-thick layer of metapyroxenite which forms the basal part of the amphibolite unit. The pyroxenite is weakly layered and foliated. It has a high modal content of augite associated with minor intergranular sericitized plagioclase, oxides, sphene and chloritized biotite (Pécher and Barféty 1984).

2.5 *Argentera Massif* (No. RM5)

Due to their scarcity, very little is known about the ultramafic rocks from the Argentera Massif. Some occurrences have been reported by Faure Muret (1955) and Malaroda et al. (1970). They are represented by amphibole-talc-chlorite bearing serpentinites and by "pyroxenitic amphibolites" (diopside-actinolite-calcite). They outcrop as lenses within the migmatitic gneisses of the Malinvern-Argentera complex (Malaroda et al. 1970, p. 567). Moreover, Faure Muret (1955) described a swarm of mafic and ultramafic inclusions within the Valmasque granite. This swarm includes actinolitic inclusions surrounded by a biotitic rim, hornblenditic enclaves, and, partly amphibolitized, augitic inclusions. These different rock types could be remnants of a pre-migmatitic layered mafic-ultramafic complex.

3 Penninic Nappes

Here, only the Middle and Lower Penninic units contain ultramafic rocks in notable amounts (Fig. 1, Tab. I). The main units are the Monte Leone nappe, the Berisal complex, the Bosco-Isorno-Orselina-Bellinzona-Dascio-zone, the Adula and Cima Lunga nappe, the Chiavenna-Bondo zone, the Bergell complex, the Simano nappe, the Antigorio nappe and the Moncucco zone.

3.1 *Monte Leone Nappe* (No. 36)

The Monte Leone nappe contains one of the larger pre-Mesozoic ultramafic masses of the Central Alps, the mafic-ultramafic complex of Geisspfad pass in the eastern Binna Valley/Wallis. Embedded in metasedimentary gneisses it has a thickness of 300 m and occupies an area of 4.5 by 1.5 km. It has been extensively studied by Keusen (1972). The original layering of the ultramafic part which orginally consisted of layered spinel-lherzolites with minor layers of pyroxenites and dunites, is well-preserved. Today the lherzolitic rocks consist of antigorite-olivine-tremolite-diopside, the dunitic rocks of chrysotile-olivine-antigorite and the pyroxenitic rocks of chlorite-tremolite-diopside. At the border a several metre thick, in part rodingitic, mafic zone separates the ultramafic rocks from the sourrounding gneisses. Similar to certain Mesozoic serpentinite occurrences of the Alps, this complex contains, still not very well-understood, cross-cutting forsterite veins of centimetre size.

Whether or not the ultramafic complex from Geisspfad pass is really part of the Monte Leone nappe is still a matter of discussion. Milnes (1974) and Keusen (1972) observed strong petrographical, chemical and structural similarities with other ultramafic complexes and surrounding paragneisses from the Bosco-Isorno-Orselina-zone (e. g. Cima d'Agaro, Isorno Valley/Italy).

3.2 *Berisal Crystalline Complex* (No. 34)

Meta-ultramafic rocks occur predominantly in the eastern part of the Berisal crystalline complex. Ultramafic and mafic rocks (amphibolites) are typically 20 m by a few 100 m long layers hosted by gneisses, mainly representing pelitic and psammitic metasediments (Stille 1980). These sediments were deposited prior to the period of the Caledonian metamorphism, i. e. prior to about 450 Ma (Köppel et al. 1981).

The metamorphic history of the mafic and the ultramafic rocks has been studied by Stille and Oberhänsli (1987). The meta-ultramafic rocks are exclusively hornblende rocks (more than 90 vol% amphibole). They occur as small, lenticular, intensively squeezed and stretched intercalations in schistose gneisses. In addition, different generations of amphiboles and relic diopside were found (Stille and Oberhänsli 1987).

The parent equivalent of the hornblende rocks is still a matter of discussion. Stille (1980) suggested an origin from a hornblenditic-pyroxenitic magma. However, their strong chemical affinity to pyroxene-controlled cumulates recently caused a redefinition of their parent equivalents (Stille and Tatsumoto 1985). Major and trace element data of the amphibolites reflect both the characteristics of island arc and/or mid-ocean ridge basalts (MORB; Stille 1980, 1987; Stille and Tatsumoto 1985; Jaquet and Munger 1984). The Sm-Nd isotopic data, however, clearly points to a mid-oceanic ridge origin.

Although ultramafic and mafic rocks are closely associated, Sm-Nd isotopic data indicates that the hornblende-bearing rocks must have been formed during two different magmatic events, yielding isochron ages of 475 ± 80 Ma for the hornblende rocks and 1018 ± 57 Ma for the amphibolites (Stille and Tatsumoto 1985). They originated from mantle segments differently depleted in light rare earth elements. The initial Nd isotopic composition derived from the isochron of the hornblende felses corresponds to an epsilon Nd value of +6 which is intermediate between MORB-type and chondritic mantle values. The initial

Nd isotopic composition of the amphibolites, however, corresponds to an epsilon Nd value of + 8.2 indicating that their protoliths originated from a MORB-type mantle source. It is interesting to note that the same ages and isotopic signatures can be found again in mafic and ultramafic rocks of the Bosco-Isorno-Orselina-Bellinzona zone (see later; Stille 1987; Schenk and Stille 1989, 1991). Therefore, it has been suggested that the mafic and ultramafic rocks from the Berisal complex are closely related to those of the Bosco-Isorno-Orselina-Bellinzona zone (Stille 1987). This in turn would imply that the Berisal complex belongs to the Bosco-Isorno-Orselina-Bellinzona zone.

In the same tectonic unit, a small lens of a talc-chlorite-magnesite-rock of unknown origin occurs E of Gantner bridge of the old road to the Simplon pass (No. 37-14; Mannoni et al. 1987).

3.3 Monte Rosa Nappe, Gran Paradiso Massif

These two Upper Penninic basements units contain minor amounts of mafic rocks in their pregranitic basement (Bearth 1952; Prato 1971; Laduron 1976), but no ultramafic rocks seem to be associated. The few serpentinite lenses described from the Furgg zone from Val Loranco-Antrona/Italy and Stockknubel-Zermatt/Wallis by Wetzel (1972) are very likely tectonic wedges from the adjacent Mesozoic Antrona and Zermatt Saas zones (Lobrinus and Béglé 1985).

3.4 Bosco-Isorno-Orselina-Bellinzona-Dascio Zone (No. 32, 33A, 26B, 38)

One of the most prominent pre-Mesozoic mafic-ultramafic belt extends over 150 km from Bosco Gurin in the NW Maggia Valley/Ticino, through the Isorno Valley NE of Domodossola and through Valle Vigezzo/Italy and the Centovalli area to Bellinzona/Ticino in the east and ends in the Dascio-Novate area at Lake Como/Italy (reference group No. 32 and 33; Gansser and Dal Vesco 1962; Heitzmann 1975; Dietrich 1980; Trommsdorff and Evans 1980 a, b; Colombi 1983, 1989; Valenti 1983; Merz 1985; Colombi and Pfeifer 1986; Stille 1987; Schmidt 1989).

The mafic and ultramafic rocks are embedded in a rather inhomogeneous melange-type zone of 1 to 2 km thickness which includes light granitic, dark metapelitic rocks and occasionally marbles. The mafic rocks comprise metamorphic magnesium- and ferrogabbros as well as tholeiitic metabasalts of MORB type (Colombi and Pfeifer 1986; Colombi 1989; Schmidt 1989; Neuffer 1990; Colombi and Pfeifer 1992). The ultramafic rocks are mainly harzburgitic (tremolite often absent). All these features suggest an oceanic origin of this zone.

In the Bosco/Gurin area (Ticino) the rocks exhibit the transition from antigorite-serpentinites to talc- and magnesium-amphibole-bearing olivine schists, whereas the southeastern continuation of the zone is dominated by olivine schists with variable amounts of talc, Mg-amphibole, enstatite and tremolite. The latter mineral seems to be a product of alpine metasomatism. It occurs locally, with modal abundances increasing from the margins towards the centre of lenses. However, the relatively high Al-content combined with a low Ca-content of the common rock type which contains olivine-talc-chlorite-ferrite-chromite +/– Mg-amphibole or enstatite, suggests that the harzburgitic nature of these rocks is due to an almost complete early serpentinisation during which most of the original mantle calcium was lost (cf. Evans and Trommsdorff 1978; Neuffer 1990; Pfeifer 1993).

Sm-Nd isotopic data of tholeiitic metabasaltic amphibolites collected over the whole mafic-ultramafic belt points to a formation age of 985±21 Ma. The latest 5 m-Nd measurements in the corresponding ultramafics give an isochron of 900±80 Ma (Schenk 1993). The initial Nd isotopic composition deduced from the Sm-Nd isochron suggests an origin from a MORB-type depleted mantle source (Schenk and Stille 1989, 1991; Schenk 1993). The age and initial ratio deduced from the isochron are within error limits in line with those of the Berisal amphibolites. This indicates that the metabasalts from the Berisal complex formed at the same time in a similar MOR environment as the metabasalts from the Bosco-Isorno-Orselina-Bellinzona-Dascio zone.

3.5 Cima Lunga Zone (No. 28, 29)

In this Middle Penninic basement zone a large number of metaperidotite lenses occur together with amphibolites and eclogites. Many detailed studies have been undertaken during the last 20 years and for details on the more than 20 different localities the reader is referred to the extensive literature (ref. No. 29, Evans and Trommsdorff 1974, 1978, 1983; Ernst 1978, 1981; Pfeifer 1978, 1979, 1989; Stäuble 1978; Evans et al. 1979, 1981; Rost et al. 1979; Schläpfer 1979; Zingg 1979; Dietrich 1980; Frey et al. 1980; Heinrich 1983; Ernst and Ottonello 1984; Medaris and Carswell 1989). The ultramafic rocks are embedded in a series composed of metapelitic and metapsammitic gneisses, amphibolites and rare marbles. They often display well developed hydrothermal zonations (Pfeifer 1979). The rare garnet-diopside-enstatite-olivine rocks that can be found in this series ("garnet-lherzolites") contain the mineral fluorine titanian clinohumite attributed to the Early Alpine Upper Cretaceous high-pressure metamorphism (Evans and Trommsdorff 1978; Medaris and Carswell 1989). The metabasaltic rocks enclosed in or closely related to the ultramafic rocks have MORB characteristics. Those outside the ultramafic rocks seem to be metasedimentary in origin (Schläpfer 1979). Rodingitic Ca-enrichment is common and metagabbros are extremely rare. Although no direct connection to Mesozoic rocks exists, some authors believe that the whole complex is Mesozoic (Trommsdorff pers. comm. 1990). Most likely, Palaeozoic or older oceanic material and Mesozoic rocks are present in this very complex nappe unit. Gebauer et al. (1992) report on U/Pb data from zircons in ultramafic and mafic rocks of Alpe Arami (No. 29-12). Garnet-clinopyroxene-orthopyroxene-olivine-schists contain discordant zircons that have an upper intersection with the concordia curve at 1700 Ma and a lower intersection at 600 Ma. Zircons from garnet-rich metapyroxenites and eclogites are partly discordant partly concordant and yield intercepts at 2450, 1700 and 33 Ma (Gebauer pers. comm. 1991).

3.6 *Adula Nappe* (No. 26A, 27, 76)

This large composite nappe has the same tectonic position and a comparable lithology as the Cima Lunga zone. There are about 30 different 20 m-sized ultramafic lenses known from this unit. Most authors concentrated their research on the often well preserved eclogites (ref. No. 24, 25, 26, 27, Evans and Trommsdorff 1978; Deutsch 1979; Dietrich 1980; Codoni 1981; Heinrich 1982, 1983, 1986; Hoffmann and Bühl 1982; Pfeifer 1989; Santini 1992). The following valleys and localities show abundant ultramafic lenses (Pfeifer 1989): Val Carrassina/Olivone, Val Calanca, lower Valle Mesolcina, Monte Duria, Val Cama (cf. Appendix). Depending on their location with respect to the isograds of the Tertiary metamorphism all possible ultramafic rock types described in the introduction can be found, including serpentinites and garnet-clinopyroxene-schists. Hydrothermal alteration is very common.

Preliminary U-Pb isotopic data from one group of zircons included in mafic eclogites from the Adula form a line that intersects the concordia curve at 560 Ma and at around 145 Ma, other groups are concordant around 530 Ma (Santini 1992).

3.7 *Chiavenna-Bondo Complex* (No. 25)

This mafic-ultramafic complex is situated between the migmatitic Gruf unit and the upper Penninic Tambo nappe, NW of the Tertiary Bergell intrusion in northern Italy. It has a thickness of about 600 m and an outcrop area of 12 by 2.5 km (Schmutz 1976). It is deformed by several east-west trending large scale folds. Its exact tectonic history is not known. Some authors relate it to the Mesozoic Misox zone (Heinrich 1983), but it is cross-cut by presumably Palaeozoic pegmatites, several tens of metres-thick. The dominant rock type is an relatively Al-rich metalherzolite with clinopyroxene and spinel relics which exhibits a regional metamorphic zonation from north to south passing from antigorite-tremolite-serpentinites, through olivine-amphibole-talc schists to olivine-amphibole-enstatite schists. Along large-scale shear zones the ultramafic rocks are locally transformed into talc-carbonate rocks which have been extensively quarried for soapstone between Roman times and the 19th century (Mannoni et al. 1987). The accompanying mafic rocks point to an overturned piece of oceanic crust with the ultramafic rocks forming the present top and coarse grained metagabbroic amphibolites below, which are in turn underlain by metabasaltic banded amphibolites. Several thin calcsilicates layers (calcite-tremolite-quartz-diopside) are found inside the mafic part (Schmutz 1976), but no geochemical data exist for these mafic rocks.

3.8 *Leventina and Simano Nappes* (No. 28, 29)

From these two basement nappes which are separated by a 10-m-thick quartzite layer, only a few ultramafic rock bodies are known. The largest ones are located at Loderio, in the lower Val Blenio at the limit of the two nappes (reference group No. 28 of Dietrich 1980; No. 28-18 in the Appendix; Evans and Trommsdorff 1970; Pfeifer and Serneels 1986). Accompanied by amphibolites and metasedimentary gneisses, several small lenses ranging up to a few metres in size can be found. They are all more or less hydrothermally altered and exhibit the typical mineralogy of talc-magnesite-magnesioamphibole-chlorite. Another series of small metaperidotite lenses included in a 200-m-thick amphibolite-rich paragneiss layer of the Simano nappe is known from the Cramosina Valley E of Giornico (No. 28-2 to -4). In the Simano nappe outcropping in the Verzasca valley occasionally ultramafic lenses occur (No. 29-06, -15). They are in most cases almost completely transformed into talc-magnesite-Mg-amphibole rocks (E. Wenk pers. comm. 1986).

3.9 Antigorio Nappe (No. 30, 31, 33B, 38)

This very extensive basement unit to the west of the Ticino culmination mainly contains granodioritic gneisses and locally metasedimentary plagioclase-biotite gneisses or garnet-mica-schists. It is only at its southeastern and northeastern border in the Maggia Valley that this nappe contains some ultramafic-mafic lenses which range in size from a few metres to several tens of metres (reference group No. 30, 31 and 33 of Dietrich 1980; Colombi and Pfeifer 1986; Pfeifer and Serneels 1986). The southern ones are mainly olivine-enstatite-chlorite schists (Capoli/Avegno; Val Noccia/Avegno; Ri del Vo/Onsernone N and NE of Intragna; No. 33-12, -15; Kobe 1956; Rost et al. 1979; Pfeifer et al. 1990). They often contain chlorite pseudomorphs after garnet and the amphibolites contain some eclogitic relics, indicating a high-pressure event (Pfeifer et al. 1991). The northern ones (W of Peccia, NW of Fusio) contain ultramafic rocks which are strongly altered by metasomatic processes. Two types can be distinguished. The first type corresponds to the typical soapstone-type hydrothermally altered olivine-schists of the Lepontine Alps and consists of talc-magnesite-chlorite rocks (Corte Bena and Pizzo di Röd W of Fusio, No. 31-12, -13; Pfeifer and Serneels 1986). The second comprises coarse grained amphibole- and chlorite +/− biotite-dominated former olivine-plagioclase-hornblendites (Val Lavizzara, Valle di Peccia; No. 30, 31; Serneels 1988) which are embedded in amphibolites and amphibole gneisses of calc-alkaline character (in part metadiorites). The observable relationships between the hornblendites and the accompanying rocks suggest an origin of a first differentiation product of a calc-alkaline intrusion which possibly intruded into a lower crustal environment. Unfortunately, no age determination is available for these interesting rocks.

3.10 Moncucco zone (No. 39)

Lying at the western end of the Southern Steep belt of the Central Alps in a poorly understood tectonic position, the Moncucco ultramafic body is one of the few cases were related mafic rocks are completely absent. The main body (Moncucco-Pianai) is about 7 km long and 1 km thick. To the east a series of smaller lenses, each measuring about 100 × 500 m (main locality: Mondei) is possibly related to the main body by a fold hinge (Bearth 1956; Greller 1976; Müller 1976). Both parts are embedded in various metagranitic and metasedimentary rocks of the Camughera-Moncucco unit (Laduron 1976; Merlyn 1977). The dominating rock is a lherzolitic metaperidotite with the assemblage olivine-talc-tremolite-chlorite +/− magnesioamphibole and enstatite and in local shear zones antigorite-tremolite can be found (Greller 1976). Due to the minimum age of 200 Ma obtained from the white mica of the cross-cutting pegmatite of Mondei (Ferarra et al. 1962), this peculiar ultramafic body is probably a mantle wedge that was emplaced in the crust during the Palaeozoic.

3.11 Bernhard Super-Nappe (No. 37, 42, 72)

Very few occurrences of ultramafic rocks are known in this part of the Alps (Escher 1988). Zingg (1989) describes two lenses of a few metres in the metasedimentary gneisses and schists of the upper Turtmann Valley (No. 72-1, -2) consisting mainly of tremolite and chlorite. Other minor occurrences of ultramafic rocks of unknown mineralogy occur in the middle Turtmann Valley (No. 37-17, -18; Mannoni et al. 1987). In the gneisses attributed to the Mont Fort nappe above Evolène/Val d'Hérens, two talc-rich lenses of a few metres occur (No. 42-31, -33). The only other known ultramafic lens of non-Mesozoic origin in this part of the Penninic region is situated at Bonatchesse in the Bagnes Valley SE of Martigny (No. 42-35). Embedded in metasedimentary micaschists and gneisses, the main lens has dimensions of 30 × 20 × 20 m and consists of a rather homogeneous talc-magnesite-chlorite-rock that has been mined as high-quality soapstone (Denkel 1954; Mannoni et al. 1987). In a small boudin north of the main lens, Y. Gouffon (pers. comm. 1990) has found some serpentinite. No mafic rocks from these outcrops are known. Again, a mantle wedge origin seems to be the most likely hypothesis.

3.12 *Bergell Complex* (No. 22)

The Tertiary Bergell intrusion and its considerably older tectonic frame contains around 15, in part several tens of metres thick metaperidotite lenses (Wenk and Cornelius 1977). One of the bigger ones outcrops in the area of Bagni di Masino, in a window to the underlying Penninic units (No. 22-01). A 80-m-wide, 400-m-long metaperidotite body is surrounded by amphibolites and gneissic migmatites. It consists mainly of olivine schists with enstatite, tremolite and chlorite (Hansmann 1981). Another series of ultramafic rock bodies is located in the upper Codera Valley (No. 22-10 to -15) and in the Bondasca Valley (No. 22-02 to -09). Many of them form massive lenses of several tens of metres, but several are connected to spectacular breccias with zoned metaperidotite fragments of centimetre size, interpreted as a result of contact metamorphism (Artus 1959; Wenk 1982). Another occurrence of metre sized zoned ultramafic nodules has been reported from the San Fedelino granite in the Novate area in the western part of the Bergell complex (No. 22-16; Callegari and Monese 1961). The ultimate origin of all these ultramafic rocks remains speculative, they possibly derive from a Palaeozoic mantle wedge.

4 Austro-Alpine Nappes

4.1 *Sesia-Lanzo and Second Diorite-Kinzigite Zone* (No. 78)

The composite eclogitic micaschist unit of the lower Aosta Valley/Italy contains a few ultramafic rock occurrences. Pognante et al. (1980) describe a ultramafic-mafic rock association from Ivozio (No. 78-01). Surrounded by mafic eclogites, a several metres thick dismembered layer of serpentinized ultramafics can be followed in the field over several hundred metres. The predominating rock is a metapyroxenite containing two generations of diopside-rich pyroxene (cf. Sect. 1), Mg-Fe-chlorite, magnetite and sphene. Metalherzolitic rocks with antigorite-chlorite-diopside-magnetite with minor talc and carbonates occur as metre thick layers within the metapyroxenites. Conspicuous elongated pseudomorphs of fine-grained antigorite are interpreted as former olivines.

The second type of ultramafic rocks in the Sesia zone represented by minor amounts of pyroxenites or serpentinized olivine-rich layers is related to the gabbros included within the eclogitic micaschist unit. Going from southwest to northeast in the Sesia zone, these are Corio and Monastero near Lanzo, Cima Bossola, Cima Bonze, Croix Courma, Cima Pianone and Val Segnara (Bianchi and Dal Piaz 1963; Isler and Zingg 1974; Compagnoni et al. 1977; Stünitz 1989; Chabloz 1990; Venturini et al. 1991; 1993; No. 78-2, -07).

To the east of the Aosta Valley, frequent small slices of ultramafic rocks can be found within the Sesia gneisses just south of the contact with the Mesozoic ophiolites of the Piemonte zone. In the Vigezzo Valley S of Malesco/Italy such rocks occur one kilometre south of this contact (Reinhardt 1966). Careful mapping in all these areas indicates that the Sesia-ophiolite contact is strongly folded at a scale of several hundred meters and that basement slices kilometres in thickness are not uncommon within the Mesozoic ophiolites (Chabloz 1990; Halter 1992; Simic 1992; No. 78-08, -11). Therefore it is very likely that all these ultramafic occurrences are tectonic fragments of the adjacent Mesozoic ophiolite zones.

From the overlying Second Diorite-Kinzigite zone (IIDK zone) only one occurrence of metaperidotite is known (Dal Piaz et al. 1971; Beccaluva et al. 1978). In the upper Artogna valley there is a harzburgitic lens with a diameter of about 150 m embedded in marbles, amphibolites and metasedimentary gneisses (No. 78-06). In the core it contains olivine and enstatite but its marginal parts have been serpentinized.

The large ultramafic body of Lanzo-Balangero at the southwestern end of the Sesia zone is situated between the basement of the Sesia zone and the Mesozoic ophiolites of the Piemonte zone. It has been extensively studied by many authors (see references in Bodinier 1988). Originally considered as an integrated part of the pre-Mesozoic basement of the Sesia zone, more recent authors (e. g. Pognante et al. 1985; Rösli et al. 1991) consider the Lanzo ultramafic body as a piece of subcontinental mantle that has been emplaced in its current position during the opening of the Liguro-Piemonte Ocean (during the Jurassic).

4.2 *Dent Blanche Nappe System* (No. 73)

An increasing number of occurrences of ultramafic rocks are known from this complex nappe system which, using the terminology of Martinotti and Hunziker (1984), also includes the Mont Mary and Mt. Emilius klippes etc. Two types can be distinguished: (1) plagioclase-peridotites or pyroxenites as layers in the gabbros tectonically enclosed in the Valpelline series (occasionally serpentinized; Nicot

1977; Dal Piaz et al. 1977; No. 73-05, 73-06) and (2) amphibole-bearing olivine-enstatite – peridotites as small lenses within kinzigitic metapelitic gneisses (Benciolini 1989; Cesare et al. 1989; Canepa et al. 1990; Benciolini and Lardeaux 1993; No. 73-01, -04). Most of latter occurrences are interpreted as mantle slices emplaced in the lower continental crust during the Palaeozoic.

5 Conclusions

In this final section, common features of the presented ultramafic occurrences are discussed, Table I will serve as the base.

Rock Types. In more than 90 % of all cases the predominant ultramafic rocks are metamorphic peridotites and pyroxenites. In zeolite, greenschist and lower amphibolite facies terrains, serpentinites predominate. In the Central Alps they contain mainly antigorite which is the typical serpentine mineral of regional metamorphism. In amphibolite facies terranes (during Alpine time limited to the Lepontine Alps) olivine-schists with typical metamorphic forsterite contents of around 90 % dominate. In a few rare cases "garnet-peridotites" (garnet-pyroxene-olivine-schists) indicate high-temperature eclogite facies conditions, which are Cretaceous in most cases. Retrogression is usually marked by the formation of talc and chrysotile as pseudomorphs or in microfissures. In a few cases, the original rock presumably was a hornblendite (Bérisal, Val Lavizzara-Peccia). Almost invariably there are hydrothermal alteration zones containing various hydrous mafic silicates and carbonates.

Relic Features and Mantle Precursors. In terms of mineral relics, the most common case are pseudomorphs of opaque spinel (often magnetite) and chlorite. In serpentinites, pigmented large forsterite and diopside crystals represent the most common relics of the mantle or lower crust stage. In olivine schists, pseudomorphs of talc after almost any mineral are very common. At the coutcrop scale the most typical remnants of former layering are the chlorite- and/or tremolite-zones, interpreted as former pyroxenite layers or dykes. In larger lenses, zones rich in olivine or serpentine are usually considered to be former harzburgite- or dunite-layers. Magnetite-rich zones, corresponding to former chromite layers, are rare. Applying the chemical method of Pfeifer (1993) to some of the olivine-enstatite rocks of the Penninic Adula nappe and the Bellinzona zone, it turns out that their Al_2O_3/MgO ratio indicates a former lherzolitic parent rock and that their low CaO/MgO ratio is due to a Ca-depletion (most likely a serpentinisation process) that occurred before the last metamorphic event. The same is probably true for many other Ca-poor ultramafic rocks of the Central Alps.

Distribution and Size. Ultramafic rocks are most frequently found in the Gotthard Massif and in the Penninic nappes (Adula-Cima Lunga, Isorno-Orselina-Bellinzona) and they usually occur as small, highly deformed boudin-type lenses. In four cases kilometre-sized fragments have been preserved (Chamrousse, Geisspfad, Chiavenna, Moncucco).

Country Rocks. In about 80 % the ultramafic rocks are accompanied by metagabbros and metabasalts, accompanying carbonate rocks are rare. However there is almost always a direct contact between metapelitic or metapsammitic gneisses and more rarely to metagranitic gneisses.

Origin. For the most cases of ultramafic lenses embedded in narrow several kilometres-long mafic rock series (including former gabbros and basaltes) an oceanic ophiolite-type origin can be inferred (cf. Nicolas 1989). Therefore, these elongated mafic-ultramafic rocks series most likely correspond to Precambrian or Palaeozoic suture zones or accretionary wedges. All the mineralogical changes which occurred after the emplacement of the rocks in the crust, make an unambiguous determination of the mantle protoliths quite difficult. Consequently, palaeogeographic interpretations based on peridotite and its depletion in Al_2O_3 as applied for the actual oceans (Bonatti and Michael 1989) remain speculative.

The peculiar association of angular mafic and ultramafic inclusions in a pegmatoid matrix ("Schollenamphibolite") is typical for a number of external massifs but its exact origin is still controversial. For several occurrences in the Penninic and Austro-Alpine zone an origin as mafic-ultramafic layers of a lower crust calc-alkaline intrusion, seems likely (hornblendites of the Bérisal/Wallis and Peccia/Ticino; pyroxenitic rocks of within gabbroic rocks of Sesia and Dent Blanche). However, for the few cases in which mafic rocks are missing and thick packages of ultramafic rocks lie directly in gneissic country rocks (e.g. Moncucco/Val Antrona; Bonatchesse/Wallis; inclusions in kinzigites of the Dent Blanche and Sesia zone), a Pyrenean-type origin seems more likely, that is during extreme crustal shortening mantle material has been transported upwards along deep fault zones and emplaced in the lower or middle crust as ultramafic wedges (see e.g. Vielzeuf and Kornprobst 1984).

Ages. Due to the scarcity of suitable minerals in orogenic ultramafic rocks that can be dated (e.g. zircons, phlogopite, garnet) and low Sm and Nd contents, only a few dates exist for the pre-Mesozoic mafic-ultramafic rock complexes of the Alps. Any kind of mantle material has necessarily a complex and long history (e.g. metamorphic recrystallization in the mantle, melting and emplacement in the lower crust, orogenic metamorphism) and the interpretation of such isotopic dates remains speculative. Some of the Sm-Nd isotopic data point to an important, possibly high-pressure event some 450 to 500 Ma ago. The few U/Pb ages available point to a mantle melting event more than 1700 Ma ago and to a metamorphic recrystallization at around 600 Ma.

Acknowledgements
Constructive reviews by J. Abrecht (Bern), D. Marshall (Lausanne) and W. Schubert (Würzburg) are gratefully acknowledged. G. Biino was supported by the Swiss National Science Foundation (project no. 21-29901.90).

References

Abrecht J (1980) Zur Bildung migmatitischer Schollenamphibolite aus dem Altkristallin des Aarmassivs. Geol Rundsch 69: 695–725

Abrecht J, Biino G, Mercolli I, Stille P (1991) Mafic-ultramafic rock-association in the Aar, Gotthard and Tavetsch Massifs (Helvetic domain, Central Switzerland): marker of a ophiolitic pre-Variscan suture, reworked by polymetamorphic events? Schweiz Mineral Petrogr Mitt 71: 295–300

Artus F (1959) Über metamorphe Breccien basischer Gesteine im Granit des Bergeller Massivs. Hamb. Beitr. Miner. Kristallphys. Petrogen. 2: 1–39

Bearth P (1952) Geologie und Petrographie des Monte Rosa. Beitr Geol Karte Schweiz NF 96, 94 pp

Bearth P (1956) Zur Geologie der Wurzelzone östlich des Ossolatales. Eclogae Geol Helv 49: 267–278

Beccaluva L, Dal Piaz GV, Macciotta G, Zeda O (1978) The Austroalpine harzburgite body of the Artogna Valley (Italian Western Alps). Mem Ist Geol Mineral Univ Padova 33: 173–181

Benciolini L (1989) Discovery of an ultramafic slice in the Austroalpine Mt. Emilius Klippe (Italian Western Alps): metamorphic evolution of the upper mantle involved in the alpine orogenesis. Terra Abstr 1: 308

Benciolini L, Lardeaux JM (1993) Mineralogical evolution of a mantle slice involved in the Alpine orogenesis: an example from the Mt. Emilius nappe (Western Alps). Eur J Mineral (in press)

Bianchi A, Dal Piaz GB (1963) Gli inclusi dei "Micaschisti Eclogitic" della zona Sesia-Lanzo nella formazione porfirica permiana della zona del Canavese fra Biella e Oropa. G Geol 2(31): 39–76

Bigi G, Consentino D, Parotto M, Sartori R, Scandone P (1990) Structural model of Italy, 1:500000, sheet 1. Consiglio Nazionale delle Richerche, Florence

Biino G (1992) Prealpine evolution of the Aar, Tavetsch and Gotthard massifs (Swiss Alps). Ph D Thesis, Univ Berne

Biino G, Stille P, Abrecht J, Mercolli I (1993) Trace element and Sm-Nd isotopic studies of Proterozoic MOR and island are basalts in the Central Alps Contrib Mineral Petrol (in press).

Biino G, Meisel T (1993 a) Geochemistry of polymetamorphic ultramafics (major, trace, noble and rare earth elements): an example from the Helvetic basement (Central Alps. Switzerland). Mineral Petrol (in press).

Biino G, Meisel T (1993 b) Major, trace, noble and rare earth element distribution in polymetamorphic ultramafics (Aarmassif. Central Alps, Switzerland). Schweiz. Mineral Petrogr Mitt 73 (in press).

Bodinier JL (1988) Geochemistry and petrogenesis of the Lanzo peridotite body. Tectonophysics 149: 67–88

Bodinier JL, Dupuy C, Dostal J, Carme F (1981) Geochemistry of ophiolites from the Chamrousse complex (Belledonne massif. Alps). Contrib Mineral Petrol 78: 379–388

Bonatti E, Michael PJ (1989) Mantle peridotites from continental rifts to ocean basins to subduction zones. Earth Planet Sci Lett 91: 297–311

Bordet P (1961) Sur la géologie du massif de Belledonne (s. s.) C R Acad Sci Paris 252(3): 434–436

Bordet P (1969) Notice de la carte géologique de la France a 1/50000, feuille de Domène (XXXIII-33). Bur Rech Geol Min Orléans, 13 pp

Bordet P (1972) Notice de la carte géologique de France a 1/50000, feuille de Vizille (XXXIII-35). Bur Rech Geol Min Orléans, 35 pp

Bordet P (1977) Notice de la carte géologique de France a 1/50000, feuille de St Jean de Maurienne (XXXIV-34). Bur Rech Geol Min Orléans, 43 pp

Callegari E, Monese A (1961) Studio petrogenetico dei bordi di reazione metamorfico-metasomatici attorno a lenti ultrafemiche incluse nel granito di San Fedelino (Sondrio). Rend Soc Mineral Ital XVII: 147–192

Canepa M, Castellotto M, Cesare B, Martin S, Zaggia L (1990) The Austroalpine Mont Mary nappe (Italian Western Alps). Mem Ist Geol Mineral Univ Padova 42: 1–17

Carme F (1965) Sur deux formations, d'origine volcanique, des schistes cristallins anté-houillers de la chaine de Belledonne (Alpes françaises). C R Acad Sci Paris 260(9): 6401–6404

Carme F (1972) Notice de la carte géologique de France 1/50000, feuille de Vizille (XXXIII-35). Bur Rech Geol Min Orléans, 35 pp

Carme F (1975) Le massif basique de Chamrousse et du Tabor (chaine de Belledonne, Alpes occidentales): une variante originale de complexe alpinotype et un jalon majeur d'une ceinture basique briovérienne. C R Acad Sci, Paris D 281: 223–226

Carme F, Pin C (1987) Vue d'ensemble sur le magmatisme préorogenic et l'évolution métamorphique et tectonique varisque dans le sud de la chaîne de Belledonne (Massifs cristallins externes, Alpes françaises). C R Acad Sci Paris 304, II, 19: 1177–1180

Carte géologique de la France 1:250000. Feuilles 25 (Thonon-les-Bains; 29 (Lyon); 30 (Annecy); 34 (Valence); 35 (Gap). Bur Rech Geol Min Orléans

Cesare B, Martin S, Zaggia L (1989) Mantle peridotites from the Austroalpine Mt. Mary nappe (Western Alps). Schweiz Mineral Petrogr Mitt 69: 91–97
Chabloz O (1990) Etude géologique et pétrographique de la région de Carcoforo, Haute Val Sesia, Italie du nord. Diploma thesis, Univ Lausanne, 179 pp
Choubert G (1936) La serpentine du Tabor (extrémité sud du massif de Belledonne, Dauphiné) et les roches qui l'accompagnent. Trav Lab Géol Grenoble 18: 51–120
Church WR, Riccio L (1977) Fractionation trends in the Bay of Island ophiolite of Newfoundland: polycyclic cumulates sequences in ophiolites and their classification. Can J Earth Sci 14: 1156–1165
Codoni A (1981) Geologia e petrografia della regione del Pizzo Claro. Ph D Thesis, Univ Zürich 177 pp
Coleman R (1977) Ophiolites, ancient oceanique lithosphere. Springer, Berlin Heidelberg New York
Colombi A (1983) Contribution à l'étude géologique de la région de Bosco Gurin (Ti). Diploma Thesis, Univ Lausanne
Colombi A (1989) Métamorphisme et géochimie des roches mafiques des Alpes Ouest-centrales (géoprofil Viège-Domodossola-Locarno). Ph D Thesis, Mém Géol Lausanne 4, 195 pp
Colombi A, Pfeifer H-R (1986) Ferrogabbroic and basaltic meta-eclogites from the Antrona mafic-ultramafic complex and the Centovalli-Locarno region (Italy and southern Switzerland). First results. Schweiz Mineral Petrogr Mitt 66: 99–110
Colombi A, Pfeifer H-R (1994) Premesozoic ophiolites from the Orselina zone, Southern Steep Belt of the Central Alps, W of Locarno, Schweiz Mineral Petrogr Mitt (in prep)
Compagnoni R, Dal Piaz GV, Hunziker J, Gosso G, Lombardo B, Williams PF (1977) The Sesia-Lanzo zone, a slice of continental crust with alpine high pressure-low temperature metamorphism in the Western Italian Alps. Rend Soc Ital Mineral Pet 33: 281–334
Dal Piaz GV, Gosso G, Martinotti G (1971) La II zona diorito-kinzigitica tra la Valsesia e la Valle d'Ayas (Alpi occidentali). Mem Soc Geol Ital 10: 257–276
Dal Piaz GV, Vecchi GP, Hunziker JC (1977) The Austroalpine layerd gabbro of the Matterhorn and Mt. Collon-Dent de Bertol. Schweiz Mineral Petrogr Mitt 57: 59–88
Denkel U (1954) Etude géologique et pétrographique de la pierre ollaire de Bonatchesse. Diploma thesis, Univ Genève
Den Tex E (1950) Les roches basiques et ultrabasiques des Lacs Roberts et le Trias de Chamrousse (Massif de Belledonne). Leidse Geol Med XV: 1–204
Deutsch A (1979) Serpentinite und Rodingite der Cima Sgiu (NW Aduladecke, Ticino). Schweiz Mineral Petrogr Mitt 59: 319–347
Dietrich VJ (1980) The distribution of ophiolites in the alps. Ofioliti Spec Issue 1: 7–51
Ernst WG (1978) Petrochemical study of lherzolitic rocks from the Alps. J Petrol 19: 341–392
Ernst WG (1981) Petrogenesis of eclogites and peridotites from the Western and Ligurian Alps. Am Mineral 66: 443–472
Ernst WG, Ottonello G (1984) Synthesis of trace element geochemistry for five western alpine lherzolites. Ofioliti 9: 425–442
Escher A (1988) Structure de la nappe du Grand St. Bernard entre le Val de Bagnes et les Mischabels. Rapp Géol Serv Hydro géol Nat Suisse 7, 26 pp
Evans B (1977) Metamorphism of alpine peridotite and serpentinite. Annu Rev Earth Planet Sci 5: 397–447
Evans B, Trommsdorff V (1970) Regional metamorphism of ultramafic rocks in the Central Alps: parageneses in the system CaO- MgO- SiO_2- H_2O. Schweiz Mineral Petrogr Mitt 50: 481–492
Evans B, Trommsdorff V (1974) Stability of enstatite+talc, and CO_2-metasomatism of metaperidodite, Val d'Efra, Lepontine Alps. Am J Sci 274: 274–296
Evans BW, Trommsdorff V (1978) Petrogenesis of garnet lherzolite, Cima di Gagnone, Lepontine Alps. Earth Planet Sci Lett 40: 333–348
Evans B, Trommsdorff V (1983) Fluorine hydroxyl titanian clinohumite in alpine recrystallized garnet peridotite: compositional controls and petrologic significance. Am J Sci 283A: 355–369
Evans BW, Trommsdorff V, Richter W (1979) Petrology of an eclogite-metarodingite suite at Cima di Gagnone, Ticino, Switzerland. Am Mineral 64: 15–31
Evans B, Trommsdorff V, Goles G (1981) Geochemistry of high grade eclogites and metarodingites from the Central Alps. Contrib Mineral Pet 76: 301–311
Faure Muret A (1955) Etudes géologiques sur le massif de l'Argentera-Mercantour et ses enveloppes sédimentaires. Mém Carte Géol Fr, 336 pp
Ferarra G, Hirt B, Jäger E, Niggli E (1962) Rb-Sr and U-Pb age determinations on the pegmatite of I Mondei (Val Antrona, Ossola, Italian Alps) and some gneisses from the neighbourhood. Eclogae Geol Helv 55: 443–450
Fischer W, Amosse J, Piboule M (1985) Caractérisation géochimique du complexe mafique et ultramafique de Chamrousse par détermination du spectre des planitoïdes. Abstr IIIrd EUG, Strasbourg. Terra Cognita 5: 277
Frey M, Trommsdorff V, Wenk E (1980) Excursion No. VI, Alpine metamorphism of the Central Alps. In: Trümpy R (ed) Geology of Switzerland. Part B. Geological excursions. Wepf, Basel, pp 310–316
Gansser A, Dal Vesco E (1962) Beitrag zur Kenntnis der alpinen Wurzelzone. Schweiz Mineral Petrogr Mitt 42: 153 168
Gebauer D, Grünenfelder M, Tilton GR, Trommsdorff V, Schmid S (1992) The geodynamic evolution of the garnet-peridotite/eclogite association of Alpe Arami and Cima di Gagnone (Central Alps) from early Proterozoic to oligocene Schweiz Mineral Petrogr Mitt 72: 107–111
Geologische Generalkarte der Schweiz 1:200000. Sheets 5 (Genève-Lausanne); 6 (Sion); 8 (Tessin). Kümmerly & Frey, Bern
Grauert B, Arnold A (1968) Deutung diskordanter Zirkonalter der Silvrettadecke und des Gotthardmassivs (Schweizer Alpen). Contrib Mineral Pet 20: 34–56
Greller M (1976) Zur Geologie des Moncucco-Camughera W Domodossola. Diploma thesis, ETH Zürich, 133 pp
Hafner S (1958) Petrographie des südlichen Gotthardmassivs. Schweiz Mineral Petrogr Mitt 38: 288–291
Halter W (1992) Etude géologique et pétrographique à l'est d'Alagna, Haute Val Sesia, Italie du nord. Diploma Thesis, Univ Lausanne
Hansmann W (1981) Geologisch-petrographische Untersuchungen im südlichen Bergell. Diploma Thesis, ETH-Zürich, 150 pp
Heim A (1918) Der Talkbergbau von Disentis in Graubünden. Z prakt Geol 26: 2–11
Heinrich C (1982) Kyanite-eclogite to amphibolite facies evolution of hydrous mafic and pelitic rocks, Adula nappe, Central Alps. Contrib Mineral Petrol 81: 30–38
Heinrich C (1983) Die regionale Hochdruck-Metamorphose der Adula-Decke, Zentralalpen, Schweiz. Ph D Thesis 7282, ETH Zurich
Heinrich C (1986) Eclogite facies regional metamorphism of

hydrous mafic rocks in the Central Alpine Adula nappe. J Petrol 27: 123–154
Heitzmann P (1975) Metamorphose und Tektonik im südöstlichen Teil der lepontinischen Alpen. Schweiz Mineral Petrogr Mitt 55: 467–522
Hoffmann F, Bühl H (1982) Überprüfung zweier schweizerischer Ultramafitit-Vorkommen zur Gewinnung von Olivinsand als Gießereiformstoff. Schweiz Mineral Petrogr Mitt 62: 135–144
Hügi T, Ledermann H, Schläppi E (1988) Erläuterungen Geologischer Atals der Schweiz, 1:25000, Blatt 1268 Lötschental. Landeshydrologie u. -geologie. Kümmerly & Frey, Bern
Isler A, Zingg A (1974) Geologie der Sesia-Zone zwischen Rimella und der Valle Anzasca (Norditalien). Schweiz Mineral Petrogr Mitt 51: 81–96
Jaquet O, Munger F (1984) Etude pétrographique d'une région comprise entre le Nanztal et le Saastal (VS). Etude géochimique des roches vertes de la zone supérieure de Stalden. Diploma Thesis, Univ Lausanne
Kalsbeek F (1961) Petrology and structural geology of the Berlanche-Valloire area (Belledonne massif, France). Pub Dep Petrogr Miner Cyst, Univ Leyden 2, 5, 136 pp
Keusen HR (1972) Mineralogie und Petrographie des metamorphen Ultramafitit-Komplexes vom Geisspfad (Penninische Alpen). Schweiz Mineral Petrogr Mitt 52: 385–478
Kobe H (1956) Geologisch-petrographische Untersuchungen in der Tessiner Wurzelzone zwischen Vergeletto-Onsernone und Valle Maggia. Schweiz Mineral Petrogr Mitt 36: 83–236
Köppel V, Günthert A, Grünenfelder M (1981) Patterns of U-Pb zircon and monazite ages in polymetamorphic units of the Swiss Central Alps. Schweiz Mineral Petrogr Mitt 61: 97–119
Labhart TP (1965) Petrographische Untersuchungen am Südrand des Aarmassivs. Beitr Geol Karte Schweiz N F 124, 80 pp
Labhart TP (1977) Aarmassiv und Gotthardmassiv. Sammlung geologischer Führer. Borntraeger, Stuttgart, 173 pp
Laduron D (1976) L'antiforme de Vanzone. Ph D Thesis, Univ Cathol Louvain, Belgium, 214 pp
Le Fort P (1973) Géologie du Haut Dauphiné cristallin (Alpes françaises). Etude pétrologique et structurale de la partie occidentale. Thèse Nancy et Sci Terre 25, 373 pp
Le Fort P (1976) Notice explicative de la carte géologique de la France a 1/50000, feuille de La Grave (XXXIV-35). Bur Rech Geol Min Orléans, 44 pp
Le Fort P, Pecher A (1971) Présentation d'un schéma structural du Haut Dauphiné cristallin. C R Acad Sci Paris 273: 3–5
Légendre O (1982) Minéralogie et géochimie des planitoïdes dans les chromitites ophiolitiques. Comparaison avec d'autres types de concentration en planitoïdes. Thèse, Univ Paris VI
Liègois JP, Duchesne JC (1981) The lac Cornu retrograded eclogites (Aiguilles Rouges Massif, Western Alps, France): evidence of crustal origin and metasomatic alteration. Lithos 14: 35–48
Lobrinus S, Béglé P (1985) Géologie et pétrographie du Val Loranco (Zone d'Antrona, Novara, Italie). Diploma Thesis, Univ Lausanne, 61 pp
Malaroda R, Carraro F, Dal Piaz GV, Franceschetti B, Sturani C, Zanella E (1970) Carta geologica del Massiccio dell'Argentera alla scala 1/50000. Note illustrative. Mem Soc Geol Ital vol IX, pp 557–663
Mannoni T, Pfeifer H-R, Serneels V (1987) Giacimenti et cave della peitra ollare nelli Alpi. Atti del convegno "La pietra ollare dalla preistoria all' età moderna", Oct 1982, Museo Civico Archeologico "Giovo". Edizioni New Press, Como, pp 7–46
Martinotti G, Hunziker J (1984) The Austroalpine system in the Western Alps: a review. Mem Soc Geol Ital 29: 233–250
Medaris LG, Carswell DA (1989) The petrogenesis of Mg-Cr garnet peridotites in European metamorphic belts. In: Carswell DA (ed) Eclogite facies rocks. Blackie, Glasgow, pp 260–290
Ménot RP (1979) Lithostratigraphie et pétrographie du groupe amphibolitique de Séchilienne et du massif ultrabasique et basique de Chamrousse (Massif de Belledonne, Alpes occidentales). Géol Alp 55: 93–110
Ménot RP (1987) Magmatismes paléozoïques et structuration carbonifère du massif de Belledonne (Alpes françaises). Contraintes nouvelles pour les schémas d'évolution de la chaîne varisque Ouest-Européenne. Thèse Université Lyon 1, 364 pp. Mem Doc Cent Armor. Etudes Struct Socles, Rennes, 1988, 21, 465 pp
Ménot RP (1988) The geology of the Belledonne massif: an overview (external crystalline massifs of the western Alps). Schweiz Mineral Petrogr Mitt 68: 531–542
Ménot RP, Peucat JJ, Scarenzi D, Piboule M (1988 a) 496 Ma age of plagiogranites in the Chamrousse ophiolite complex (External Cristalline Massifs in the French Alps): evidence of a Lower Paleozoic oceanization. Earth Planet Sci Lett 88 (1–2): 82–92
Ménot RP, Peucat JJ, Paquette JL (1988 b) Les assoications magmatiques acides-basiques paleozoiques et les complexes leptyno-amphiboliques: les corrélations hasardeuses. Exemples du Massif de Belledonne (Alpes occidentales). Bull Soc Géol Fr 8/IV: 917–926
Merlyn M (1977) Structure et métamorphism du complexe Camughera-Moncucco entre la Valle Anzasca et le Val Brevettola (Prov. de Novara, Italie). Ph D Thesis, Univ Cathol Louvain, Belgium
Merz C (1985) Minéralogie et pétrographie de la région de Valle di Campo/Maggia. Diploma Thesis Univ Lausanne
Milnes AG (1974) Post nappe folding in the Western Lepontine Alps. Eclogae Geol Helv 67: 333–348
Müller R (1976) Zur Geologie der Antrona-Mulde östlich von Antronapiana. Diploma Thesis, Univ Zürich, 150 pp
Neuffer P (1990) Etude géologique et pétrographique de la Valle Arbedo (Tessin). Diploma Thesis, Univ Lausanne
Nicolas A (1984) Lherzolites of the western Alps: a structural review. In: Kornprobst J (ed) Kimberlites, vol II. The mantle and crust-mantle relationships. Dev Petrol 11B. Elsevier, Amsterdam, pp 333–345
Nicolas A (1989) Structures of ophiolites and dynamics of oceanic lithosphere. Kluwer, Dordrecht, 367 pp
Nicot E (1977) Les roches méso- et catazonales de la Valpelline (nappe de la Dent Blanche, Alpes Italiennes). 3th Cycle Thesis, Univ Paris VI
Niggli E (1944) Das westliche Tavetscher Zwischenmassiv und der angrenzende Nordrand des Gotthardmassivs. Schweiz Mineral Petrogr Mitt 24: 58–301
Oberhänsli R, Abrecht J, Mercolli I, Schenker F (1988) Geological map of the Aar Massif. Univ Bern (unpubl)
Oberholzer W (1955) Geologie und Petrographie des westlichen Gotthardmassivs. Schweiz Mineral Petrogr Mitt 35: 320–409
Parker RL (1921) Serpentin und Talk von Disentis-Surrhein. Schweiz Mineral Petrogr Mitt 1: 7–43
Pécher A (1970) Etude pétrographique de la partie orientale du massif des Ecrins-Pelvoux. Le socle ancien. Thèse 3ème Cycle, Univ Grenoble, 122 pp
Pécher A, Barfety JC (1984) Notice de la carte géologique de la France a 1/50000, feuille de Saint Christophe en Oisans. Bur Rech Geol Min, Orléans. 64 pp

Pfeifer HR (1978) Hydrothermal alpine metamorphism in metaperidotite rocks of the Cima di Gagnone zone, Valle Verzasca, Switzerland. Schweiz Mineral Petrogr Mitt 58: 400–405

Pfeifer HR (1979) Fluid-Gesteins-Interaktion in metamorphen Ultramafititen der Zentralalpen. Ph D Thesis, ETH Zürich 6379, 200 pp

Pfeifer HR (1981) A model for fluids in metamorphosed ultramafic rocks. III. Mass transfer under amphibolite facies conditions in olivine-enstatite rocks of the Central Alps, Switzerland. Bull Minéral 104: 834–847

Pfeifer HR (1987) A model for fluids in metamorphosed ultramafic rocks. IV. Metasomatic veins in metaharzburgites of Cima di Gagnone, Valle Verzasca, Switzerland. In: Helgeson HC (ed) Chemical transport in metasomatic processes. Nato Sl Ser, Reidel, Amsterdam, pp 591–632

Pfeifer HR (1989) Wenig bekannte Beispiele von ehemaliger Lavez-Ausbeutung in den südlichen Alpentälern. Minaria Helv 9: 8–54

Pfeifer HR (1993) Major and trace element discrimination diagrams to determine possible mantle protoliths of orogenic ultramafic rocks. Schweiz Mineral Petrogr Mitt (in press)

Pfeifer HR, Serneels V (1986) Exploitation de la pierre ollaire au Tessin et régions voisines: aspects minéralogiques et miniers. Quad informazione. Dipartemento del ambiente, Cantone del Ticino 11: 147–228

Pfeifer HR, Colombi A, Kobe H, Schmid S (1990) Preliminary version of the geological map sheet 1312, Locarno, Landeshydrologie u. -geologie, Bern (unpubl)

Pfeifer HR, Colombi A, Ganguin J, Hunziker J, Oberhänsli R, Santini L (1991) Relics of high-pressure metamorphism in different lithologies of the central Alps, an updated inventory. Schweiz Mineral Petrogr Mitt 71: 441–451

Pfeifer HR, von Raumer J (1993) Lherzolitic and pyroxenitic ultramafics from the Lac Cornu area (Aiguilles Rouges Massif, External Zone of the Alps, France). Abstr. Terra Nova Vol. 5, 1:187-188

Pflugshaupt P (1927) Beiträge zur Petrographie des östlichen Aarmassivs. Schweiz Mineral Petrogr Mitt 7: 321–378

Pin C, Carme F (1987) A Sm-Nd isotopic study of 500 Ma old oceanic crust in the Variscan belt of western Europe: the Chamrousse ophiolite complex, Western Alps (France). Contrib Mineral Petrol 96: 406–413

Pognante U, Compagnoni R, Gosso G (1980) Micro-mesostructural relationship in the continental eclogitic rocks of the Sesia-Lanzo zone (Italian Western Alps): A record of a subduction cycle. Rend Soc It Min Petr 36: 169–186

Pognante U, Rösli U, Toscani L (1985) Petrology of ultramafic and mafic rocks from the Lanzo peridotite body (Western Alps). Lithos 18: 201–214

Prato R (1971) Il settore centro-occidentale del massiccio del Gran Paradiso. Acc Sci Torino Atti 105: 453–467

Reinhardt B (1966) Geologie und Petrographie der Monte Rosa-Zone, der Sesia-Zone und des Canavese im Gebiet zwischen Vald'Ossola und Valle Loana (Prov. di Novara, Italien). Schweiz Mineral Petrogr Mitt 42: 553–678

Rösli U, Hoernes S, Köppel V (1991) Isotope data of metarodingites and associated rocks from the Lanzo and the Bracco ophiolitic massifs: indications on the evolution of the alpinotype ultramafic-mafic complexes. Schweiz Mineral Petrogr Mitt 71: 125–141

Rost F (1971) Probleme der Ultramafitite. Fortschr Mineral 48: 54–68

Rost F, Wannemacher J, Anton J (1979) Garnet ultramafitites in the Ivrea zone and other crystalline zones along the periadriatic lineament. Mem Ist Geol Mineral Univ Padova 33: 183–191

Santini L (1992) Geochemistry and geochronology of the basic rocks of the Penninic Nappes of the east central Alps (Switzerland). Ph D Thesis, Univ Lausanne 150 pp

Schaltegger U (1986) Voralpine und alpine Mineralbildung in der Gneiszone von Erstfeld (Sustenpass, Aarmassiv) der Mechanismus der K-Ar- und Rb-Sr-Verjüngung alpin umgewandelter Biotite. Schweiz Mineral Petrogr Mitt 66: 395–412

Schenk K, Stille P (1989) Chemical and isotopical evidences for a 1 Ga old ophiolite sequence in the Penninic realm of the Alps. Ber. Deutsche Min. Ges. 1: 160

Schenk K, Stille P (1991) Geochemistry of mafic-ultramafic rock associations in the south-penninic nappes of the Central Swiss Alps: evidence for Proterozoic MORB. Terra 3: 476

Schenk K (1993) Geochemistry of mafic-ultramafic rock associations in the south-Penninic nappes of the Alps. Ph. D. thesis, ETH Zürich (in prep).

Schenker F, Abrecht J (1987) Prä-aargranitische Anatexis, variszische Kontaktmetamorphose und alpidische Regionalmetamorphose im Oberhasli (zentrales Aarmassiv, Schweiz). Schweiz Mineral Petrogr Mitt 67: 13–26

Schläpfer E (1979) Metamorphose und Strukturen der Cima Lunga-Serie. IV. Lago d'Efra-Val Gagnone. Diploma Thesis, ETH Zürich

Schmidt M (1989) Petrography and structural evolution of ophiolitic remnants in the Bellinzona Zone, southern steep belt, Central Alps (CH/I). Schweiz Mineral Petrogr Mitt 69: 393–406

Schmutz HU (1976) Der Mafitit-Ultramafitit-Komplex zwischen Chiavenna und Val Bondasca (Provinz Sondrio, Italien; Kt. Graubünden, Schweiz). Beitr Geol Karte Schweiz, NF 149, 73 pp

Serneels V (1988) Les carrières de pierre ollaire de l'Alpe Sovenat (Peccia, Valle Maggia, Ticino). Minaria Helv 8b: 3–16

Simic D (1992) Etude géologique et pétrographique du haut Val Mastallone, Italie du nord. Diploma Thesis, Univ Lausanne

Stäuble J (1978) Metamorphose und Strukturen der Cima Lunga-Serie: II Val d'Agro. Diploma thesis, ETH Zürich

Stille P (1980) On the genesis of the amphibolites and hornblendefelses in the Berisal complex. Mem Ist Geol Mineral Univ Padova 34: 205–246

Stille P (1987) Geochemische Aspekte der Krustenevolution im zentral- und südalpinen Raum. Habilitationsschrift, ETH Zürich, 127 pp

Stille P, Oberhänsli R (1987) Petrology and Alpine metamorphic evolution of metabasic and ultrabasic rocks in the Berisal crystalline complex. Lithos 20: 169–180

Stille P, Tatsumoto M (1985) Precambrian tholeiitic-dacitic rock-suites and Cambrian ultramafic rocks in the Penninic nappe system of the Alps: evidence from Sm-Nd isotopes and rare earth elements. Contrib Mineral Pet 89: 184–192

Stünitz H (1989) Partitioning of metamorphism and deformation in the boundary region of the "Seconda Zona Diorítico-Kinzigitica", Sesia zone, Western Alps. Ph D Thesis 8817, ETH Zürich, 244 pp

Trommsdorff V (1983) Metamorphose magnesiumreicher Gesteine: Kritischer Vergleich von Natur, Experiment und thermodynamischer Datenbasis. Fortschr Mineral 61: 283–308

Trommsdorff V, Evans B (1974) Alpine metamorphism of peridotitic rocks. Schweiz Mineral Petrogr Mitt 54: 333–354

Trommsdorff V, Evans B (1980a) High grade rodingites from the Central Alps: metamorphism and geochemistry. Arch Sci 33: 181–184

Trommsdorff V, Evans B (1980b) Excursion to Alpe Albion (east of Bellinzona). Excurs. guide, Sympos. In: Bertrand J

(ed) Tectonic inclusions and associated rocks in serpentinites. Geneva, Sept 1979 Dept Mineral, Univ Geneve, 15 pp
Valenti G (1983) Geologia e petrografia della Zona di Bosco/Gurin (Ti). Diploma Thesis, ETH Zürich, 75 pp
Venturini G, Martinotti G, Hunziker J (1991) The protoliths of the "eclogitic micaschists in the Aosta Valley (Sesia Lanzo; western Alps). Mem Padova XLIII: 347–359
Venturini G, Armando G, Barbero M, Martinotti G, Hunziker J (1993) New field observations and revised interpretations of the lithostratigraphic subdivisions of the Sesia Zone. Schweiz Mineral Petogr Mitt 73 (in press)
Vielzeuf D, Kornprobst J (1984) Crustalsplitting and the emplacement of Pyrenean lherzolites and granulites. Earth Planet Sci Lett 67: 87–96
Wenk HR, Cornelius SB (1977) Geologischer Atlas der Schweiz I: 25000, Blatt No. 1296 Sciora. Schweiz Geol Komm Bern
Wenk HR (1982) A geological history of Bergell granite and related rocks. In: Transformists petrology. Theophrastus publications, Athens, pp 113–148
Wetzel R (1972) Zur Petrographie und Mineralogie der Furgg-Zone (Monte Rosa-Decke). Schweiz Mineral Petrogr Mitt 52: 161–236
Zingg A (1979) Metamorphose und Strukturen der Cima Lunga-Serie, III. Val d'Efra, Diploma Thesis, ETH Zürich
Zingg MA (1989) Die Siviez-Mischabel-Decke: Entstehung und Entwicklung eines Altkristallins und seinen Vererzungen. Ph D Thesis 8997, ETH Zürich, 170 pp

Appendix

The following list contains the coordinates of known occurrences of pre-Mesozoic ultramafic rocks in the Central and the external Western Alps. It is based on the compilation of soapstone quarries of Pfeifer and Serneels (1986) and Mannoni et al. (1987). The numbers given are locality numbers, not sample numbers, based on the literature compilation of Dietrich (1980).
Comments on the different columns:

Tectonic units: same as in the text. The numbers in parentheses are those marked on Fig. 1 and 2.

Numbering: based on a regional scheme, extended from Dietrich (1980), original Roman numbers are converted to Arabic numbers. For a certain tectonic unit there are often different groups of locality numbers.

22	Bergell (So)
24	Adula north-Tambo
25	Chiavenna (So)-Val Bondasca (GR)
26	Valle della Mera (So), Val Darengo (So), Monte Duria (So), Gana Rossa (GR), Val Cama/ Alpe di Mea (GR)
27	Cima Ciu, Val Carassino (TI)
28	Loderio (TI)
29	Alpe Arami (TI), Alpe Alai (TI), Val Verzasca, Cima di Gagnone (TI)
30	Valle Maggia north (TI)
31	Val Peccia (TI)
32	Bosco/Gurin (TI), Monte Larone (No), Baceno (No), Agaro (No), Val dell'Isorno (No)
33	Valle Vigezzo (No), Centovalli (TI) Valle Maggia South (TI)
34	Turbhorn (VS), Hohsandhorn (VS), Banhorn (VS), Goms (VS)
36	Geisspfad (VS)
37	Rhone and side valleys (VS)
38	Antrona-Valle Loana, Furgg Zone
39	Moncucco (No)
42	Zinal-Mauvoisin
52	Lanzo (Pi)
68	IX[a] Gotthard Massif (GR, TI, VS, UR)
69	X[a] Goms-Hospental-Disentis (VS, UR, GR)
70	XI[a] Aar Massif (BE, UR)
71	XII[a] Aiguilles Rouges Massif (VS)
72	XIII[a] Mt Vélan-Fionnay-Turtmanntal (VS)
73	XIV[a] Mt Collon (VS), Valpelline (Dent Blanche, Pi, Ao)
76	Val Calanca, Valle Mesolcina (GR)
77	Mont Blanc Massif
78	Sesia Zone (Pi)
RM2	NE Belledonne
RM3	SW Belledonne
RM4	Haut Dauphiné
RM5	Argentera

Region: Switzerland (cantons): TI Ticino/Tessin, GR Graubünden/Grisons, VS Valais/Wallis, UR Uri, BE Bern. Italy (provinces): Ao Aosta, Co Como, No Novara, Pi Piemonte, So Sondrio, Ve Vercelli. France ("départements"): 04 Alpes, 05 Hautes Alpes, 06 Alpes Maritimes, 38 Isère, 73 Savoie, 74 Haute Savoie.

[a] Original Roman numbers of Dietrich (1980).

Coordinates: in most cases given in Swiss kilometric coordinates (the first 3 numbers correspond to kilometres, the following three numbers to metres (Bern X 600000, Y 200000). For some occurrences in Italy international longitude-latitude coordinate are given in degrees (°) and minutes ('). For the French Alps Lambert-type coordinates are used.

Valley and locality names: They normally correspond to those used in the 1 : 25 000 map.

No.	Region	Valley	Locality	Coord. X	Coord. Y
Bergell complex (No. 22)					
22-1	So	Val Masino	Bagni/Corte Vecchia	766100	124300
22-02	GR	Val Bregaglia	Cantun 1/Val Bondasca	765600	131850
22-03	GR	Val Bregaglia	Cantun 2/Val Bondasca	765440	131720
22-04	GR	Val Bregaglia	Cantun 3/Val Bondasca	765600	131740
22-05	GR	Val Bregaglia	Camin 1/Val Bondasca	766400	132050
22-06	GR	Val Bregaglia	Camin 2/Val Bondasca	766500	131900
22-07	GR	Val Bregaglia	Sasc Fura l/V. Bondasca	765100	131600
22-08	GR	Val Bregaglia	Sasc Fura 2/V. Bondasca	765020	131500
22-09	GR	Val Bregaglia	Trubinasca/V. Bondasca	764340	131040
22-10	So	Val Codera	Bivacco Vaninetti	763900	129400
22-11	So	Val Codera	Sivigia	762600	128700
22-12	So	Val Codera	Valle Canina	761250	128950
22-13	So	Val Codera	Val del Conco 1	761000	128800
22-14	So	Val Codera	Val del Conco 2	760150	128600
22-15	So	Val Bregaglia	Val Casnaggina	760300	129650
22-16	So	Val Mera	Novate Cava del Pescarino	750000	121500
Chiavenna-Bondo complex (No. 25)					
25-01	GR	Val Bregaglia	Alla Predacia/Bondo	763930	132400
25-02	So	Val Bregaglia	Chiavenna-1	751975	132425
25-03	So	Val Bregaglia	Chiavenna-2	752030	132245
25-04	So	Val Bregaglia	Chiavenna-3	750530	129100
25-05	So	Val Bregaglia	Piuro-1	754810	132030
25-06	So	Val Bregaglia	Piuro-2	754100	131820
25-07	So	Val Bregaglia	Caneto/V. Chiavenna	758200	132500
25-08	So	Val Bregaglia	Foppate/Val Casnaggina	760300	131500
25-09	GR	Val Bregaglia	Lizöl/Val Casnaggina	761100	131700
25-10	GR	Val Bregaglia	Sot al Canta/V. Casnaggina	766200	131550
25-11	GR	Val Bregaglia	Denc dal Luf/V. Bondasca	762600	131450
Bellinzona zone (No. 26B)					
26-01	TI	Val Marobbia	Pianezzo	723150	114450
26-02	TI	Val Marobbia	Monte della Zotta	724500	115800
26-04	TI	Val Marobbia	Monte della Zotta	724600	116000
26-05	TI	Ticino-Bellinzona	Artore	723800	116900
26-06	TI	Val Arbedo	Tagliada	725450	118050
26-07	TI	Val Marobbia	Alpe d'Urno	727150	115500
26-08	TI	Val Arbedo	Alpe Gesero	729900	115800
26-09	TI	Val Arbedo	Motto della Costa	728000	117050
26-10	TI	Val Arbedo	Valle della Pira	728500	116750
26-11	TI	Val Arbedo	Valle della Pira	728750	116500
26-12	TI	Val Arbedo	Valle del Bucco	729450	116800
26-13	TI	Val Arbedo	Cadolcia	729800	117650
26-14	TI	Val Arbedo	Corno di Gesero	730800	116150
26-15	TI	Val Traversagna	Alpe Albion	731200	116300
26-16	TI	Val Traversagna	Val March	730600	118800
26-17	TI	Val Traversagna	Val d'Albionasca	732600	118400
26-18	TI	Val Traversagna	Val di Roggiasca	733500	118250
26-16	TI	Val Traversagna	Gana Rossa	735400	116800
26-24	Co	Valle di Gera/Mera	Burano	748800	115900
26-25	Co	Valle Mera	Peledo-Dascio	752000	117000
26-26	Co	Valle Mera	Dascio	753200	117450
Adula nappe (No. 26A, 27, 76)					
26-01	GR	Val Cama	E Alpe di Mea	737000	123300
26-02	GR	Val Cama	Bocchetta del Notaro	739650	122600
26-03	Co	Val Bodengo	Avert del Notaro	740130	122300

No.	Region	Valley	Locality	Coord. X	Coord. Y
26-17	Co	Valle Liro	Alpe Duria	740800	118050
26-18	Co	Valle Liro	Alpe Lanzero	741450	119280
26-19	Co	Valle Liro	Dosso Bello Duria	740200	117720
26-20	Co	Valle Liro	A. Duria	740630	118200
26-21	Co	Valle Liro	A. Duria	741280	117660
26-22	Co	Val Darengo	Borgo	743600	119750
26-23	Co	Val Darengo	Alpe Alterno	746100	120050
27-01	TI	Val Carrassino	Cima Sgiu	717100	154100
27-02	TI	Val Carrassino	Cima Bresciana	718950	151600
27-03	TI	Val Carrassino	Fornee	720250	153150
27-04	TI	Val Carrassino	NE Alpe Bresciana	719650	152600
27-05	TI	Val Soi	Coronella	721200	149600
76-01	GR	Val Calanca	Marscia d'Aion	732350	131650
76-02	GR	Val Calanca	Rossa	731100	137400
76-03	GR	Val Calanca	Alpe di Remia	727900	139650
76-04	GR	Val Calanca	Fil di Revi	727520	142480
76-05	GR	Val Calanca	Alpe di Stabi	727500	147000
76-08	GR	Valle Mesolcina	Trona, SW Soazza	736210	135730
76-10	GR	Valle Mesolcina	Piano di Verdabbio	733550	127550
76-11	GR	Valle Mesolcina	Piano di Verdabbio	733150	126950
76-13	GR	Val Calanca	Braggio, Mont di Fora	730100	129000
76-14	GR	Val Calanca	Bogan, Arvigo	727500	129400
76-15	TI	Val Pontirone	Pizzo di Borgeno	724500	139100
76-16	GR	Valle Mesolcina	San Vittore	729100	122500
76-17	GR	Valle Mesolcina	Bosch del Faed, Giova	728500	124200
76-18	GR	Valle Mesolcina	Piano di Verdabbio	732980	126850
Leventina and Simano nappes (No. 28A, 29)					
28-01	TI	Valle Leventina	Campi/Bodio	711750	137380
28-02	TI	Valle Leventina	Cagnago/V. Cramosina	709150	137200
28-03	TI	Valle Leventina	Cagnago/V. Cramosina	709300	137100
28-04	TI	Valle Leventina	Cagnago/V. Cramosina	709200	137200
28-05	TI	Valle Leventina	Or Grande/V. Cramosina	710600	138300
28-13	TI	Valle Leventina	A. di Grav/V. Cresciano	725100	130650
28-14	TI	Valle Leventina	A. di Simidi/V. Cresciano	724500	130080
28-15	TI	Valle Leventina	A. di Simidi/V. Crescinao	725000	129650
28-16	TI	Valle d'Osogna	Alpe di Piviciou	721900	133600
28-17	TI	Valle d'Osogna	Cima di Piancabella	723000	132000
28-18A	TI	Val Blenio	Loderio/Mt d. Bella	717300	137450
28-18B	TI	Val Blenio	Loderio/Mt d. Bella	717200	137450
28-18C	TI	Val Blenio	Loderio/Mt d. Bella	717080	137450
28-18D	TI	Val Blenio	Loderio/Mt d. Bella	716950	137320
29-06	TI	Valle Verzasca	Val Alnasca	705100	128800
29-07	TI	Valle Verzasca	Val Alnasca	705250	128850
29-08	TI	Valle Verzasca	E Pizzo Marcio	703950	128700
29-09	TI	Valle Verzasca	E Pizzo Marcio	704050	129100
29-10	TI	Valle Verzasca	E Pizzo Marcio	703950	129700
29-11	TI	Valle Verzasca	E Pizzo Marcio	703650	129700
29-13	TI	Valle Verzasca	Alpe Fornaro/V. Osola	696500	130750
29-14	TI	Valle Verzasca	Alpe Barddughe	710600	121500
29-15	TI	Valle Verzasca	Monti Eos	710600	124800
Cima Lunga zone (No. 28B, 29)					
28-06	TI	Valle Leventina	P. Rosso/V. Iragna	713350	128850
28-07	TI	Valle Leventina	P. Rosso/V. Iragna	713600	128500
28-08	TI	Valle Leventina	A. Piatto/V. Drosina	714800	124400
28-09	TI	Valle Leventina	Alpe di Lai/V. Moleno	716900	123350
28-10	TI	Valle Leventina	Prosecco/V. Moleno	716080	123120
28-11	TI	Valle Leventina	Piotella/V. Gnosca	720500	121600
28-12	TI	Valle Leventina	A. Arami/V. Gnosca	718900	121100
28-19	TI	Valle Leventina	S A. Sesaldora/Gnosca	719500	121650
29-01	TI	Val Agro/Verzasca	Alpe Lignasc	710300	129700
29-02	TI	Val Agro/Verzasca	S Cima di Rierna	709600	130400

No.	Region	Valley	Locality	Coord. X	Coord. Y
29-03	TI	Val d'Efra/Verzasca	Passo di Gagnone	708580	131700
29-04	TI	Val d'Efra/Verzasca	Guglia/Corte di Cima	708900	132350
29-05	TI	Val Gagnone/Verzasca	Lavesc	709500	132550
29-16	TI	Val Motto/Verzasca	Pizzo Scaiee	707750	131770
29-17	TI	Val Motto/Verzasca	W Cima di Gagnone	707980	131420
29-18	TI	Val Motto/Verzasca	W Cima di Gagnone	707970	131460
29-19	TI	Val Motto/Verzasca	S Cima di Gagnone	708400	131150
29-20	TI	Val Motto/Verzasca	Alpe di Motto	707500	131350
29-21	TI	Val d'Efra/Verzasca	NW Cima di Gagnone	708100	131850
29-22	TI	Val d'Efra/Verzasca	NNW Cima di Gagnone	708300	131930
29-23	TI	Val Agro/Verzasca	Alpe Corte Nuovo	708900	130080
29-24	TI	Val Agro/Verzasca	Alpe Corte Nuovo	709000	130150
29-25	TI	Val Agro/Verzasca	Alpe Corte Nuovo	709000	130400
29-26	TI	Val Agro/Verzasca	Corte Nuovo	708550	129200
29-27	TI	Val Agro/Verzasca	Corte Nuovo	708850	129150
29-28	TI	Val Agro/Verzasca	Pianca	710300	128400
Antigorio nappe (No. 30, 31, 33B)					
30-01	TI	Val Lavizzara	Piano, E Menzonio	692800	134600
30-02	TI	Val Lavizzara	Piano, E Menzonio	692780	134580
30-03	TI	Val Lavizzara	Pianello, Menzonio	692760	134450
30-04	TI	Val Lavizzara	Nelle France, Menzonio	692800	134760
30-05	TI	Val Lavizzara	Mognee, NE Menzonio	693200	135880
30-06	TI	Val Lavizzara	Isola, E Menzonio	693630	135400
30-07	TI	Val Lavizzara	E Menzonio	693500	135250
30-08	TI	Val Lavizzara	E Menzonio	693650	135150
30-09	TI	Val Lavizzara	Pianello/V. Cocco	693130	134250
30-10	TI	Val Lavizzara	P. Somazzi, Brontallo	690780	134100
30-11	TI	Val Lavizzara	P. Somazzi, Brontallo	690750	134000
30-12	TI	Val Lavizzara	Alpe Bena, Fusio	691830	144500
30-13	TI	Val Lavizzara	Pizzo di Rod, Fusio	690650	145200
31-01	TI	Valle di Peccia	Predera	688900	140100
31-02	TI	Valle di Peccia	Croso	688750	140170
31-03	TI	Valle di Peccia	Corte Grande	689380	139860
31-04	TI	Valle di Peccia	Stallareccio	687950	139450
31-05	TI	Val Bavona	Pizzo d'Oglie	687300	138600
31-06	TI	Valle di Peccia	Poida dei Muna	688280	139080
31-07	TI	Valle di Peccia	Vena Nuova	689940	138390
31-08	TI	Valle di Peccia	Canicc	690550	141500
31-09	TI	Val Bavona	Rosed	685100	136900
33-01	TI	Val Nocca/Maggia	Dunzio-Avegno	699700	117830
33-12	TI	Valle Maggia inf.	Capoli/Avegno	699200	118900
33-13	TI	Valle Onsernone	Ri del Vo/Auressio	696050	118690
33-14	TI	Valle Maggia inf.	Corte d. Cima/Aurigeno	698600	118900
33-15	TI	Valle Maggia inf.	N Salmone, Auressio	697750	118760
Bosco zone (No. 32)					
32-01	TI	V. di Bosco/Maggia	Bosco/Gurin	681250	130060
32-02	TI	V. di Bosco/Maggia	Zum Schwarza Brunna	680400	129480
32-03	TI	V. di Bosco/Maggia	Grossalp	678650	130300
32-04A	TI	Valle di Campo	Alpe Magnello	677700	126600
32-04B	TI	Valle di Campo	Alpe di Magnello	677370	126130
32-04C	TI	Valle di Campo	Alpe di Magnello	677240	126250
32-04D	TI	Valle di Campo	Alpe di Magnello	677100	126300
32-04E	TI	Valle di Campo	Alpe di Magnello	676800	126660
32-04F	TI	Valle di Campo	Alpe di Magnello	678380	126550
32-05	No	Valle di Campo	Alpe Venanzio	675800	126050
32-06	No	Valle di Campo	Alpe Colobiasca	676450	124000
32-07	TI	Valle di Campo	Fontanella, Cimalmot.	679440	126000
32-08	TI	Valle di Campo	Corte Nuovo/Campo	680850	127730
32-09	TI	Valle di Campo	Balm da Rossa/Cimalm.	678700	125300
32-17	No	Val Antigorio	Alpe Foglia	675850	131130

No.	Region	Valley	Locality	Coord. X	Coord. Y
Isorno zone (No. 32, 38)					
32-18	No	Val Isorno	Cima d'Agaro	674400	118320
38-02	No	Val d'Isorno	Altoggio	669750	112850
38-03	No	Val d'Ossola	Roldo	668500	111500
38-29	No	Val d'Isorno	Nava	669100	113800
38-31	No	Val d'Isorno	Alpe Solada	672600	113400
38-32	No	Val d'Isorno	Alpe al Lago	672400	114300
38-33	No	Val d'Isorno	Alpe al Lago	672850	115150
38-34	No	Val d'Isorno	Monte Navone	672100	114950
38-35	No	Val d'Isorno	Alpe Agarina	673200	119200
38-36	No	Val d'Isorno	Alpe Cortone	674850	119300
38-37	No	Val d'Isorno	Alpe Ratagina	674200	121500
38-38	No	Val d'Isorno	Passo della Forcola	675500	123000
Orselina zone (No. 33A)					
33-02	TI	Centovalli	Corcapolo	695660	114130
33-03	TI	Centovalli	Case Cavalli/Calezzo	695940	114300
33-04	TI	Centovalli	Verguno/Verdasio	693400	113720
33-05	TI	Centovalli	Verdasio	692900	113600
33-06	TI	Centovalli	Borgnone	690850	112800
33-07	TI	Centovalli	Sertoo/Costa-Camedo	689750	112900
33-08	No	Valle Vigezzo	Cima di Caneto	684900	113600
33-09	No	Valle Vigezzo	Btta di S. Antonio	684400	113400
33-10	No	Valle Vigezzo	Piano/Cima Trubbio	681100	113450
33-11	No	Valle Vigezzo	Eglio/Craveggia	682200	110600
33-21	TI	Val Maggia	Monteggia	703400	117300
33-22	TI	Val Maggia	Cardada	704750	116900
33-23	TI	Val Verzasca	Mte di Ditto	711700	116020
Monte Leone nappe (No. 36)					
36-01	VS	Binntal	Geisspfad Pass	661800	131800
Berisal Complex (No. 35, 37)					
35-04	VS	Steinental	Bortelgeltscher	652400	127050
35-05	No	Val Cairasca	Alpe Veglia	652890	126120
37-14	VS	Gantertal	Ganterbruecke	648600	127240
Bernhard nappe (No. 37, 42, 72)					
37-17	VS	Turtmanntal	Eggimatte	620880	121610
37-18	VS	Turtmanntal	Amosi	621000	121770
72-1	VS	Turtmanntal	Turtmanngletscher	619463	112413
72-2	VS	Turtmanntal	Turtmanngletscher	620038	111463
42-31	VS	Val d'Hérens	Evolène-1	605100	106440
42-32	VS	Val d'Hérens	Evolène-2	605080	106500
42-33	VS	Val d'Hérens	Evolène-3	605050	106210
42-35	VS	Val de Bagnes	Bonatchesse	591000	095920
Aar Massif (No. 37, 70)					
37-19	VS	Loetschental	Gletschertossen	629450	138930
37-20	VS	Loetschental	Wilerra-2	627850	137360
37-21	VS	Loetschental	Feesch	625100	136100
37-22	VS	Loetschental	Goppenstein	624650	135000
37-23	VS	Rhone	Steg	624000	129000
37-24	VS	Rhone	Hohtenn	624480	129750
37-28	VS	Loetschental	Underi Meiggu	623850	134620
37-29	VS	Loetschental	Oberi Meiggu	623200	134760
37-30	VS	Loetschental	Alti Pochi	624740	135000
37-31	VS	Loetschental	Marwig-1	625060	141280
37-32	VS	Loetschental	Marwig-2	625090	141220
37-33	VS	Loetschental	Marwig-3	625060	141140
37-34	VS	Loetschental	Willerra-1	627740	137520
37-35	VS	Loetschental	Wilerhorn	628480	136780
37-36	VS	Loetschental	Blattjigrat	626700	134200

No.	Region	Valley	Locality	Coord. X	Coord. Y
37-37	VS	Loetschental	Joligletscher-1	628120	134370
37-38	VS	Loetschental	Joligletscher-2	628150	134480
37-39	VS	Loetschental	Howitzen	628940	138680
37-40	VS	Loetschental	Mannlich	629780	139250
37-41	VS	Loetschental	Bellwadwald	629510	140530
37-42	VS	Loetschental	Resti	630380	139860
37-43	VS	Loetschental	Juidas	630530	140200
37-44	VS	Loetschental	Oigstchummuweng	631350	140970
37-45	VS	Loetschental	Oigstchumme	632290	141290
37-46	VS	Loetschental	Chrindelspitza	631000	145000
37-47	VS	Loetschental	Beichgrat	637400	143620
37-51	VS	Rhone	Naters Hegdorn	642900	131900
37-52	VS	Rhone	Naters Massaschlucht	642700	133100
70-01	BE	Haslital	Chammegg	667300	168150
70-02	BE	Haslital	Schaulauiberg/Gutta.	667200	165600
70-03	UR	Goeschenertal	Sustenjoch	679000	174450
70-04	UR	Reusstal	Richligen/Gurtnellen	691100	178100
70-05	UR	Maderanertal	Geschel/Rueteli	699500	180900
70-06	UR	Maderanertal	Fruttlauital	703050	180800
70-07	UR	Maderanertal	Griesserental	700400	179800
70-08	UR	Maderanertal	Lungental	699200	179500
70-09	UR	Maderanertal	Sellentobel	697700	178700
70-10	UR	Maderanertal	Blaetzen/Etzliboden	696800	177700
70-11	UR	Maderanertal	Bristlaui	694850	178400
70-12	BE	Haslital	E Ofenhorn	669300	166800
70-13	UR	Meiental	Sustenloch/Sustenpass	677475	176825
70-14	UR	Meiental	Silberberg/Sustenpass	677095	177180
70-15	UR	Meiental	Klein Sustli/Sustenp.	678250	177960
Moncucco zone (No. 39)					
39-01	No	Val Brevettola	Mondei	660950	103650
39-02	No	Val Brevettola	Aulamia	660000	102900
Gotthard Massif, Tavetscher Massif (No. 34, 68, 69)					
34-01	VS	Goms	Aegenenbach-1	668500	150280
34-02	VS	Goms	Aegenenbach-2	669000	150650
34-03	VS	Goms	Gintrige	669700	152100
34-04	VS	Goms	Gere-1	671500	153460
34-05	VS	Goms	Gere-2	671620	153580
34-06	VS	Goms	Oberbach	670300	152670
34-07	VS	Goms	Lichere	672250	155950
34-08	VS	Goms	Waljiwald	665150	148350
34-09	VS	Goms	Abiloub	663660	146500
34-10	VS	Goms	Spiswald-1	660050	144450
34-11	VS	Goms	Spiswald-2	660050	144350
34-12	VS	Goms	Spiswald-3	660100	144400
34-13	VS	Goms	Figulti	658300	140050
34-14	VS	Goms	Stock	665450	145100
34-15	VS	Goms	Spisswald	669560	143440
68-01	TI	Val Bedretto	Alpe di Monio	676900	149200
68-02	TI	Val Bedretto	A. Cavanna, Cas. d. Viei	682500	152900
68-03	TI	Val Bedretto	A. Sorescia, Sc. Orella	687300	155600
69-01	VS	Muttbachtal/Furka	Blauberg	675500	157000
69-02	UR	Urseren/Furka	Stotzige Firsten	677100	158100
69-03	UR	Urseren/Realp	Giltnasen/Isenma. alp	682300	160050
69-04	UR	Urseren/Realp	Uf den Stuedelen	683100	161500
69-05	UR	Urseren/Realp	Hoch Stock	683700	161000
69-06	UR	Urseren/Hospenthal	Stafel	685630	162750
69-07	UR	Urseren/Hospenthal	Chaemmleten	686400	163000
69-08	UR	Urseren/Hospenthal	Wannelen	687430	163250
69-09	UR	Urseren/Hospenthal	Gigenstafel	688000	163500
69-10	UR	Urseren/Andermatt	Mettlen/Gurschenbach	689600	164850
69-11	UR	Guspistal/Gotthard	Chastelhorn	689200	161000

No.	Region	Valley	Locality	Coord. X	Coord. Y
69-12	UR	Oberalp	Sattelegghubel	690350	165500
69-13	UR	Unteralp	Gafallengrat	692200	161600
69-14	UR	Unteralp	Portgerenstock	694700	161900
69-15	UR	Val Maighels/Oberalp	Piz Alpetta-Grat	697000	161500
69-16	GR	Val Curnera/Oberalp	Val Aulta/P. Alpetta	697400	162180
69-17	GR	Val Curnera/Oberalp	Fuorcla Paradis	698800	162700
69-18	GR	Val Nalps/Tavetsch	Fuorcla Serengia	699350	161900
69-19	GR	Val Nalps/Tavetsch	Piz Fuorcla	699900	162000
69-20	GR	Val Nalps/Tavetsch	Piz Vatgira	701750	162150
69-21	GR	Oberalp	Calmut/Rivas	695500	167900
69-22	UR	Urserern/Hospenthal	Gademli	685850	163200
69-23	GR	Oberalp	Prau Mulins/Calmot	696200	167650
69-24	GR	Vorderrheintal	Nual/Selva-Surrein	698900	168400
69-25	GR	Vorderrheintal	Mompe-Tujetsch/Dise.	705900	171350
69-26	GR	Vorderrheintal	Buretsch/Disentis	707250	172150
69-27	GR	Vorderrheintal	Mompe Medel/Disentis	708100	172300
69-28	GR	Vorderrheintal	Caischavedra/Disent.	711900	173500
69-29	GR	Vorderrheintal	Uldauna/Disentis	712600	173900
69-30	GR	Vorderrheintal	Plaun Mistral/V. Sumv.	713650	175000
69-31	GR	Vorderrheintal	Aclas da Reits/V. Sumv.	714050	175100
69-32	GR	Vorderrheintal	Crest Vaulas/Surrein	715350	174900
69-33	GR	Vorderrheintal	Cua/Surrein	715500	175350
69-34	GR	Vorderrheintal	Radunda/Surrein	716400	175400
69-35	GR	Vorderrheintal	Puzzastg/Surrein	714700	175050
69-36	GR	Vorderrheintal	Loch/Surrein	715800	175100
69-37	GR	Vorderrheintal	Las Plattas/Surrein	715900	174750
69-38	GR	Vorderrheintal	Crap/Surrein	716700	175550
69-39	GR	Vorderrheintal	Sulagval/Surrein	716200	175550
69-40	GR	Vorderrheintal	Parvials/Trun	718600	177300
69-41	GR	Vorderrheintal	Rosas Dadens/V. Sumv.	716400	171000
69-42	GR	Vorderrheintal	Runcahez/V. Sumvitg	716900	170850
69-43	GR	Val Medel	P. Caschleglia/V. Plat.	712000	168000
Aiguilles Rouges Massif (No. 71)					
71-1	74	Chamonix	Desert d'Arlève	554400	090300
Sesia zone and second Diorite-Kinzigite zone (No. 78)					
78-01	Pi	Valle d'Aosta	Ivozio	07°51′	45°33′
78-02	Pi	Valle T. Tesso	Monastero di Lanzo	07°26′	45°19′
78-03	Pi	Valle Malone	Corio	07°30′	45°19′
78-04	Pi	Val Savenca	Cima Bossola	07°44′	45°29′
78-05	Pi	Valle d'Aosta	Croix Courma	07°46′	45°40′
78-06	No	Val Artogna	Lago di Fondo	639000	069000
78-07	No	Val Carcoforo/Sesia	Cima Lampone	646330	082550
78-08	No	Val Carcoforo/Sesia	Le Piane	648650	085000
78-09	No	Val Carcoforo/Sesia	Malosso	648010	084000
78-10	No	Val Carcoforo/Sesia	Cima Piandone	650000	084800
78-11	No	Val Sesia/Alagna	Alpe Campo	640600	078600
78-12	No	Val Segnara	Ai Patalli	658300	095150
Dent Blanche Nappe System (No. 73)					
73-01	Ao	Valle T. Castello	Pointe de Genève	598300	071600
73-02	Ao	Valpelline	Foillou	595200	072700
73-03	Ao	Valpelline	Col de Foillou	596200	073150
73-04	Ao	Valle d'Arpisson	Arpisson Damon	07°21′	45°41′
73-05	Ao	Valtournanche	Matterhorn	616500	090500
73-06	VS	Val d'Arolla	Dent de Bertol	6075000	093500

French external massifs (without Aiguilles Rouges):

Massif	No.	Map sheet 1:50000	Coordinates x	(Lambert system) y
Belledonne NE	RM2	St. Jean-de Maurienne	898.5–903.0	334.0–345.0
Belledonne SW	RM3a	Domène	883.0	326.0–327.0
Belledonne SW	RM3b	Vizille	878.0–883.0	316.0–325.0
Belledonne SW	RM3c	Vizille and La Mure	877.0–878.0	305.0–307.0
Haut Dauphiné	RM4a	St. Christophe en Oisans	922.5	1993.0
Haut Dauphiné	RM4b	St. Christophe en Oisans	916.0	1989.0
Argentera	RM5	St. Martin de Vésubie	Exact location unknown	

Pre-Alpidic Ore Deposits in the Central, Eastern and Southern Alps

V. Köppel, F. Neubauer, and E. Schroll

Abstract

The pre-Mesozoic metallogenetic evolution of the Alps reflects a progressing cratonization of the crust. The first stage lasted from the upper Proterozoic to the Lower Ordovician and is characterized by the formation of oceanic crust and island arcs which are now preserved mainly in Penninic and Middle Austro-Alpine tectonic units. Besshi-type deposits (Fe-Cu-Zn), some with minor Pb, As, Sb, the important scheelite deposit at Mittersill and stratiform Au-occurrences are related to this early stage. Most of these mineralizations are associated with greenschists or amphibolites. The lead isotope signature of many of these deposits reveals the presence of a distinct mantle component. Significant amounts of detrital material were deposited in back-arc basins in proximity to a southern continent. Stratiform Pb-Zn (-Cu-Ag) deposits occur in sequences dominated by metasediments and -volcanics; they are spatially separated from the Fe-Cu deposits. The widespread, approximately 450 Ma old granitoids which are virtually devoid of any mineralizations terminated this first crust forming period of the Alps. The Ordovician to Lower Carboniferous sequences of the Upper Austro-Alpine units and the Innsbruck quartzphyllite are characterized by facies differentiation, waning magmatism mainly of an intraplate nature and the formation of platform carbonates. Cu is concentrated in the western Greywacke zone and in the Kreuzeck Mountains, whereas to the East and South Pb-Zn, magnesite, siderite, Sb, and Hg deposits occur. This distribution suggests a more advanced stage of cratonization during the Palaeozoic in the southern and eastern segments of the Alps than in the northwestern.

Polymetallic vein deposits, Li-pegmatites and probably also disseminated Au-mineralizations are associated with some of the Variscan magmatic complexes. The end of the Variscan cycle is marked by U, U-Cu and Pb-Zn mineralizations in volcanic rocks, sandstones, shales and carbonates which postdate the Variscan deformation.

The Pb-isotope signatures of Pb and Sb dominated deposits in Austro-Alpine units and in the Southern Alps are of typical crustal nature and are comparable to the signature observed in most metasediments of these units. They indicate a prolonged evolution of the lead in a crustal environment with high U/Pb and Th/Pb ratios.

The lead of Cu-dominated mineralizations in Austro-Alpine units and of all types of ores in the Penninic zone and external massifs is distinct and indicates a comparatively short crustal residence time. Metasedimentary lead indicates the presence of old as well as of relatively young components. The lead isotope patterns and the metal distribution support the idea of a relatively early cratonization of the Southern Alps and some Austro-Alpine domains whereas the Penninic and some of the more westerly domains of the Austro-Alpine units cratonized later during the Variscan orogenic cycle.

1 Introduction

Attempts to correlate pre-Alpidic ore deposits with the crustal evolution were presented by Höll and Maucher (1976) and by Pohl (1984). They, as well as the one presented here, rely to a large extent on speculations; many aspects of the Palaeozoic and earlier evolution of the Alpine crust are still open to debate; furthermore, the ages of many of the ore deposits as well as of their host rocks are poorly constrained. However, based on microfossils, Palaeozoic and even Precambrian depositional ages of greenschist to lower amphibolite facies metasediments have been reported. Geochemical characterization of magmatic rocks indicative of their geotectonic setting is progressing. Furthermore, our knowledge of the relations between ore forming processes and tectonic regime has improved considerably. Pb-, Sr-, S-, O- and C-isotope investigations provided information on the sources of metals and fluids and thereby placed also constraints on genetic models.

Table 1. The pre-Mesozoic mineralization of the Alps

	Central Alps Penninic domain external massifs	Eastern Alps Tauern window (Penninic)	Austro-Alpine units	Southern Alps Eastern	Western
Proterozoic – Lower Ordovician	Island arc systems		Back-arc systems Detrital input mainly from Early Proterozoic to Archean sources		
		W-(Mo, Bi, Cu, Pb, B, Be)	Cr		Fe-Cu-(Zn)
		Fe-Cu-Zn-Pb-Au-W	Fe-Cu-Zn- (As-Sb-Bi-Pb-Au-Sn)	Fe-Cu	
			Pb-Zn-Ag, Cu-Ag-Sb-As, Ni-Co-As-Bi	Fe-Cu-Zn-Pb-As-Sb	
Ordovician – – Lower Carboniferous	Rifting, intraplate magmatism, development of platforms formation of nappes, magmatism/metamorphism, deposition of flysch-type sediments				
	As-Au-W Pb-Zn-Ba-F Cu-Bi, Ni-Co-As, Au-As-Sb		Fe-Cu-Pb-Zn Sb-(Fe, As, W) Au-As, W, Pb-Zn-Ba Fe-As-W, Hg-As, Hg, magnesite, siderite	Cu-Zn-Pb-Sb-Hg-Ba-F	
Upper Carboniferous- – Permian	Magmatism related to subduction/collision and subsequent extensional tectonics deposition of molasse-type sediments and red beds				
	Mo-W, Pb-Zn-Ba-F		Li U-Cu	F-Ba-Pb-Ag-Zn-Cu Fe-As-Pb-Zn-Cu-Sb- Ag-Au	Fe-Ni-Cu, Cr Fe-As-Zn- Pb-Cu-Sb- Ag-Au-Ba
	U-Pb-Zn-Bi			U-Zn-Sb-As-Cu- Mo-Pb-B	U-Cu
	U-Cu			Pb-Zn, U-Cu	

In spite of the numerous contributions dealing with the pre-Triassic development of the Alps many questions regarding their metallogenic evolution still remain unanswered. We attempt to present an overview of the major types of ore deposits of the Central, Eastern and Southern Alps and to emphasize the geotectonic setting in which they were probably formed (Table 1). Well over 2000 occurrences of ores are known, the majority of which are hosted in pre-Mesozoic rocks of the Eastern and Southern Alps. Today, only Pb-Zn, W, siderite and magnesite are exploited; Au, Li and U were, and partly still are, explored for. The overview is by no means a complete one and it necessarily focuses on deposits the genesis of which is thought to be reasonably well understood. We furthermore summarize lead-isotope data which reflect the evolution of the crust and which delineate isotopic provinces. In order not to overburden the reference list we have cited, wherever possible, only the most recent publication which provides access to older publications.

For convenience we have subdivided the evolution of the Alps into three major periods (Tables 1, 2; Fig. 1). This suggests a synchroneity of the major events in all tectonic units which appears rather unlikely. For example, the tectonothermal events during the Late Cambrian and the Ordovician are not synchronous in different tectonic units, and therefore deposition of postorogenic sediments in these units started at different times.

2 Late Precambrian to Mid-Ordovician Cycle (Fig. 1 a)

Within the **Penninic** area of the **Central Alps** it is difficult to distinguish between pre-Ordovician and Ordovician to Lower Carboniferous terranes because of the Variscan and Alpidic metamorphism. Nevertheless, the presence of approximately 1000 Ma old amphibolites has been reported which constitute the oldest oceanic crust of the Alps (Stille and Tatsumoto 1985; Schenk and Stille 1991). Basic and acidic Cambrian to Ordovician magmatism is also known from this area as well as from the external massifs to the north.

Table 2. Selected pre-Mesozoic ore deposits and occurrences of the Eastern, Central and Southern Alps. The numbers refer to Fig. 1 a–c

A) Late Precambrian to Mid-Ordovician cycle (Fig. 1 a)			
a) Penninic domain			
1	W	Felbertal (Mittersill)	Höll and Schenk (1988) Thalhammer et al. (1989)
2	Au-W-As	Schellgaden	Göd (1981)
b) Austro-Alpine domain			
3	Fe-Cu-Zn	Oetztal	Vavtar (1988)
4	Pb-Zn	Oetztal	Vavtar (1988)
5	Pb-Zn-Ag-Cu	Monteneve (Schneeberg)	Niggli (1987)
6	Fe-Cu-Pb-Zn-As-Au	Kreuzeck Mts.	Friedrich (1963)
7	Pb-Zn-Ag	Duisitz (Schladminger Tauern)	Friedrich (1975)
8	Cu-Ag-Sb-As	Krombach (Schladminger Tauern)	Friedrich (1975)
9	Ni-Co-As-Bi	Zinkwand-Vöttern	Stumpfl and Tarkian (1979)
10	Cr	Kraubath	El Ageed et al. (1980)
11	Fe-Cu	Trattenbach	Tufar (1963)
12	Fe-Cu-As	Waldbach	Tufar (1979)
c) Southern Alps			
13	Fe-Cu-(Zn)	Val d'Ossola	Omenetto and Brigo (1974)
14	Fe-Cu(-Zn-Pb-As-Sb)	Agordo	Omenetto and Brigo (1974)
B) Mid-Ordovician to Lower Carboniferous cycle (Fig. 1 b)			
a) Penninic domain and external massifs			
15	Au	Disentis	Wenger and Steiger (1990)
16	Au	Amsteg	Wenger and Steiger (1990)
17	Pb-Zn-F-Ba	Goppenstein	Morel (1978)
18	Au-As-W	Salanfe	Woodtli et al. (1987)
b′) Austro-Alpine domain, weakly metamorphosed, fossiliferous sequences and quartzphyllites			
19	Fe-Cu(-Zn-Pb)	Tessenberg-Panzendorf	Tornquist (1935)
20	Fe-Cu(-Zn-Pb)	Rettenbach (Mittersill)	Unger (1972)
21	Fe-Cu(-Zn-Pb)	Walchen	Schlüter et al. (1984)
22	Fe-Cu(-Zn-Pb)	Kalwang	Schäffer and Tarkian (1984)
23	Fe-Cu(-Zn-Pb)	Naintsch	Weber (1990)
24	Sb(-Cu-As-Zn)	Arztal/Tulfes/Navistal	Haditsch and Mostler (1983)
25	Au	Zell	Schulz and Wenger (1980)
26	W	Tux	Raith (1989)
27	W	Kleinarltal	Höll and Maucher (1976)
28	Sb(-W-As)	Rabant	Weber (1989)
29	Hg	Stockenboi	Höll and Maucher (1976)
30	Hg	Christofberg	Höll and Maucher (1976)
31	Hg	Dallakkogel	Ebner and Weber (1982)
32	Cu-Sb-As-Hg	Röherbühel	Unger (1972)
33	Cu-Sb-As-Hg	Schwaz	Gstrein (1979)
34	Fe-Cu	Mitterberg (Mühlbach)	Weber et al. (1973)
35	Pb-Zn-Ag-Ba	Palaeozoic of Graz	Weber (1990)
36	Pb-Zn-Ag	Meiselding	Friedrich (1968)
37	Magnesite	Tux	Morteani (1989)
38	Magnesite	Hochfilzen	Morteani (1989)
39	Magnesite	Breitenau	Gollner (1983)
40	Siderite	Erzberg (Eisenerz)	Thalmann (1979)
41	Magnetite	Plankogel	Weber (1990)
b″) Austro-Alpine domain, micaschist-marble units			
42	Pb-Zn-Ag(-Ba)	Oberzeiring	Haditsch (1967)
43	Pb-Zn-Ag(-Ba)	Moosburg	Tufar (1979)
44	Pb-Zn-Ag(-Ba)	Ramingstein	Tufar (1981)
45	Magnesite	Radenthein	Angel and Awerzger (1948)
46	Siderite	Hüttenberg	Schulz et al. (1986)
47	Hematite	Waldenstein	Friedrich (1929)
c) Southern Alps			
48	Ba-F-Cu-Zn-Pb-Sb-Hg	Carnic Alps	Omenetto and Brigo (1974)
49	Ba-F-Cu-Zn-Pb-Sb-Hg	Southern Karawanken	Drovenik et al. (1980)

Table 2. Continued

C) Variscan syn- to post-orogenic cycle (Fig. 1 c)		
a) Penninic domain, external massifs and cover sequences		
Penninic domain		
50 U-Cu	Isérables	Eikenberg et al. (1989)
51 Pb-Zn-Cu	V. d'Anniviers	Woodtli et al. (1987)
52 Cu-Bi	Baicolliou	Woodtli et al. (1987)
53 Ni-Co-As	Turtmanntal	Woodtli et al. (1987)
54 Pb-Zn-Ag	Praz Jean	Woodtli et al. (1987)
Massifs		
55 U-Cu-Pb-Zn-Ni-Co-(Au)	Trun	Kramers (1973)
56 Mo-(W)	Baltschiedertal	Steck and Hügi (1970)
57 U-Cu-Pb-Zn-As-Bi	La Creusa (Marécotte)	Gilliéron (1988)
58 Ba-F-Pb-Zn	Les Trappistes	Wutzler (1983)
59 Mo	Alpeiner Scharte	Mostler (1982)
Cover sequences		
60 U-Cu	Mürtschenalp	Gilliéron (1988)
61 U-Cu	Affeier-Waltensburg	Gilliéron (1988)
b) Austro-Alpine domain		
62 Li-pegmatite	Weinebene	Göd (1989)
63 Li-pegmatite	St. Radegrund	Koller et al. (1983)
64 Magnesite	Veitsch	Pohl (1990)
65 Magnesite	Oberdorf	Pohl (1990)
66 Magnesite	Triebenstein	Pohl (1990)
67 U	Forstau	Erkan (1977)
68 Pb-Zn-Ba	Prinzenkogel	Tufar (1963)
69 Siderite-hematite-Ba-Cu	Hirschwang	Haditsch and Mostler (1982)
c) Southern Alps		
70 Fe-Ni	Val Sesia – Val Strona	Ferrario and Garuti (1990)
71 Fe-Ni	Palagnedra	Wenger and Steiger (1990)
72 Fe-As-Pb-Zn-Au, Pb-Zn-Cu, Fe-As-Pb-Zn-Cu-Sb-Ag-Au, Zn-Ba, Sb	Malcantone	Wenger and Steiger (1990)
73 Pb-Zn-Cu-Ba-F, F-Ba-Zn-Pb, Fe-Cu-Zn-Pb-As-Sb-Ag-Au	Area of Bolzano	Dessau and Perna (1966) Omenetto and Brigo (1974) Castaldo and Stampanoni (1975)
74 Pb-Zn-Cu-Ag-Ba-F	Lugano-Varese	Omenetto and Brigo (1974)
75 U-Zn-Pb-As-Cu-(Mo)	Novazza	Origoni Giobbi et al. (1982)
76 U	Val Vedello	Origoni Giobbi et al. (1982)
77 U	Zirovsky Vrh	Drovenik et al. (1980)
78 Cu	Skofje	Drovenik et al. (1980)
79 Pb-Zn	Korosi	Drovenik et al. (1980)
80 Pb-Zn	Puharje	Drovenik et al. (1980)
81 Zn	Tregiovo	Haditsch and Mostler (1982)
82 Cu-U	Oboiniggraben (Eisenkappel)	Petrascheck et al. (1976)

The Penninic basement of the **Eastern Alps** is exposed in the Tauern window. It includes ophiolite-like assemblages of the Stubach and Habach complex with sedimentary-volcanic formations indicative of magmatic arcs. Metasediments of the Habach group contain a major suite of calc-alkaline mafic and acid rocks which host the important scheelite deposit of Mittersill (Felbertal).

The **Austro-Alpine** amphibolite-gneiss association forms large thrust sheets in eastern Switzerland and western Austria and consists of amphibolites, orthogneisses and biotite-plagioclase paragneisses which contain detrital zircons of lower Proterozoic to Archean age. This association is interpreted as having been formed in a rift (Magetti and Flisch, this Vol.) or subduction environment (Frisch et al. 1990). In the

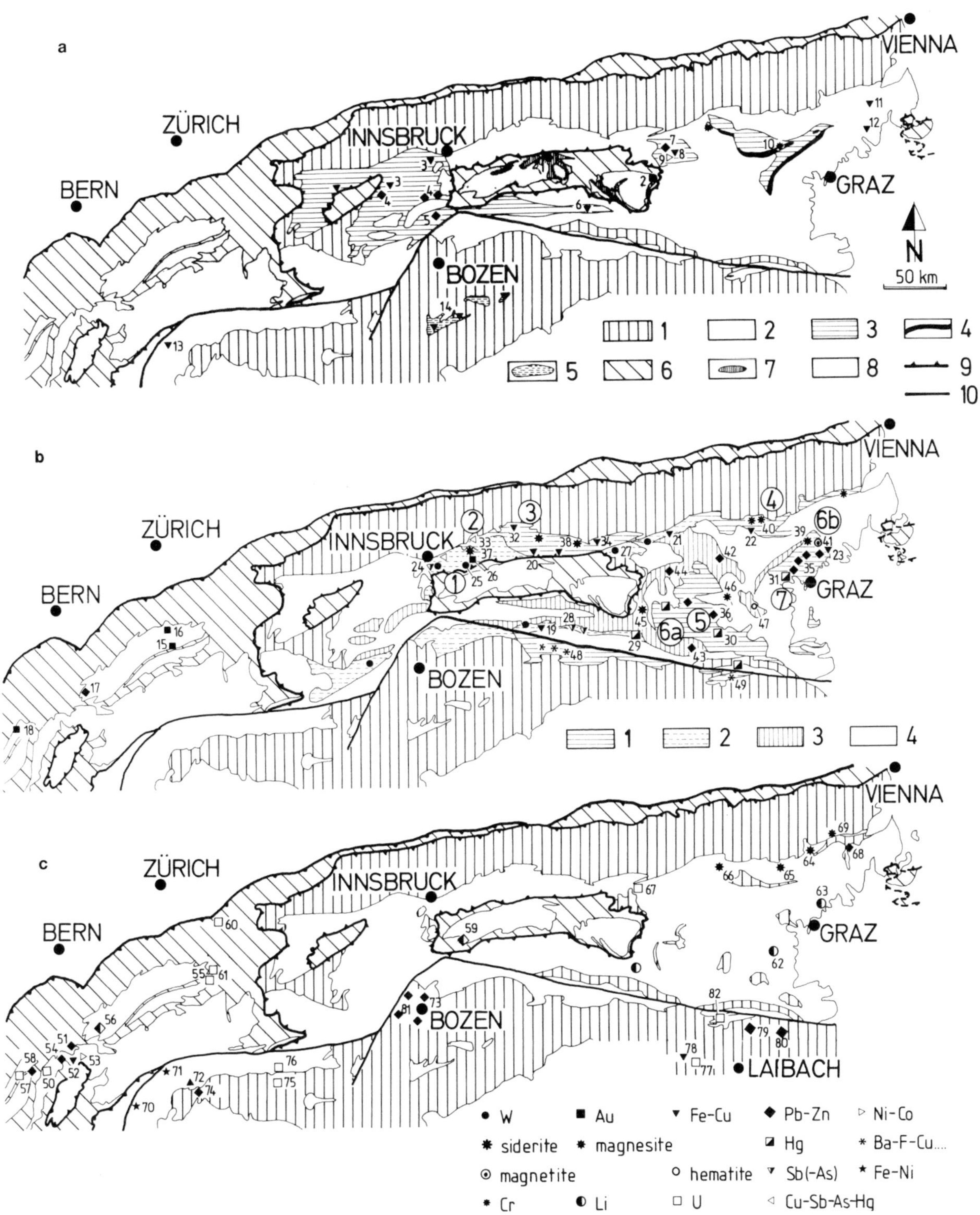

Fig. 1 a–c. Sketch map of the Central and Eastern Alps displaying the distribution of the ore deposits which are listed in Table 1. *Numbers* refer to Table 2. **a** Pre-Upper Ordovician cycle: *1* Austro-Alpine and Southalpine post-Lower Carboniferous cover sequences; *2–5* Austro-Alpine and Southalpine basement with *3* gneiss-amphibolite association; *4* Speik ophiolite complex and *5* Cambrian and Lower Ordovician quartzphyllites of Agordo-Cereda; *6* post-Upper Carboniferous Helvetic and Penninic cover sequences; *7* Habach complex (Upper Riphaean to Lower Vendium); *8* other Penninic and Helvetic basement units (pre-Upper Carboniferous); *9* major Alpidic thrust boundaries; *10* Periadriatic lineament. **b** Ordovician to Lower Carboniferous cycle: *1* fossiliferous sequences; *2* quartzphyllites; *3* micaschist-marble complexes (probably of the same age as 1); *4* other Austro-Alpine, Penninic and Helvetic basement complexes. Compare also explanation to **a.** ***Circled numbers*** mark locations of sections shown in Fig. 2. **c** Upper Carboniferous to Permian cycle. For explanation of symbols, see **a**

Silvretta nappe it comprises pre-450-Ma-old MORB-type amphibolites, volcanic arc metagranites, tonalitic gneisses and gabbros which may be according to Sm-Nd and Rb-Sr data as old as 900 Ma (Flisch 1987). The association suffered tectonothermal events at different times. The earliest metamorphism recorded in the Oetztal nappe was of high pressure and occurred not later than at the Cambrian-Ordovician boundary. It was followed by anatexis and a regional low pressure metamorphism during the Middle to Upper Ordovician which is the "Caledonian event". The tectonothermal cycle was completed by the Late Ordovician intrusion of collision-type granites which now constitute a second group of orthogneisses.

In the **Southern Alps** metasediments dominate over amphibolites and orthogneisses. The clastic metasediments contain mainly Early Proterozoic to Archean detritus as indicated by the crystallization ages of detrital zircons (Köppel and Grünenfelder 1971) and by the high μ ($={}^{238}U/{}^{204}Pb$) and W ($={}^{232}Th/{}^{204}Pb$) values of whole rock and feldspar leads. Phyllites from the eastern part of the Southern Alps yielded palynologic data which indicate a late Lower Cambrian to Early Ordovician (Tremadoc) age, i. e. approx. 560–490 Ma (Sassi et al. 1984).

It appears reasonable to assume that these future South and Austro-Alpine domains constituted foreland basins which were located close to the African continent. Towards the N, in the future Penninic realm, the input from the African plate diminuished and younger sources such as developing island arcs, were important contributors. 700 to 500 Ma ago island arc systems had developed such as in the Habach formation and the Gleinalm area. High concentration of some significant elements, e. g. Sc, mirror the distribution of Proterozoic to Early Ordovician andesitic to basaltic rock such as in the Habach formation, the Speik complex or in the Stubalpen-Gleinalm area (Schroll 1990). The existence of island arcs implies that subduction occurred. Evidence for a high-pressure metamorphism has been reported not only from Austro-Alpine units but also from the Southern Alps, from external massifs and Penninic nappes. Its age predates the widespread, 450-Ma-old acid magmatism.

Figure 1 a shows the distribution of ores pertaining to this early stage of crustal evolution of the Eastern Alps and Southern Alps. Major ore deposits are (the numbers refer to those in Figs. 1 a–c and in Table 2):

A. W-(Mo, Bi, Cu, Pb, B, Be) in the Penninic Habach complex at Mittersill (1),
B. Fe-Cu-Zn-Pb with Au and W at the eastern margin of the Tauern window, type Schellgaden (2),
C. Fe-Cu-Zn-(As, Sb, Bi, Pb, Au, Sn) in the Austro-Alpine Oetztal nappe (3), in the Southern Alps (Trentino, 14), Kreuzeck Mountains (6), and in the Wechsel area (11, 12)
D. Zn-Pb in the Oetztal nappe (4),
E. Fe-Cu-Zn in the Ivrea zone of the Southern Alps (13),
F. Cr in the East Alpine Speik ophiolite complex (Kraubath, 10),
G. Pb-Zn-Ag, Cu-Ag-Sb-As and Ni-Co in the Schladming Mountains (8, 9). The ages of these deposits are, however, not well established.

The important scheelite deposit near Mittersill (1) is associated with a suite of basic and acid magmatic rocks contained in phyllites of the **Penninic** Habach formation. The phyllites were biostratigraphically dated as Upper Riphaean to Lower Vendian (600–700 Ma) by Reitz and Höll (1988) and Reitz et al. (1989). Geochemical and isotope data suggest that the deposit was formed in an island arc environment (von Quadt 1992). Höll and Schenk (1988) drew attention to some striking similarities with features of the active hydrothermal field of the Taupo zone on the North Island of New Zealand. The stratiform Au-mineralizations, type Schellgaden, with Cu, Pb and W occur in a rock suite of probably similar age at the eastern end of the Tauern window (2).

The Speik complex in the eastern **Austro-Alpine** domain (10) represents a dismembered pre-Silurian ophiolite complex which contains chromite occurrences (El Ageed et al. 1980). It is conformably overlain by probably Silurian-Devonian micaschists and marbles. In the Oetztal nappe Fe-Cu with As-Zn and minor Sb-Bi-Ni-Co-Au-W-Sn occur in the north where amphibolites constitute major components of the amphibolite-gneiss association (3). In the southern part where amphibolites are rare, stratiform Pb-Zn deposits with minor Cu dominate (4, 5) (Vavtar 1988). In middle Austro-Alpine units (Tollmann 1977) of the Kreuzeck Mountains stratiform Fe-Cu-Zn-Pb-(Au) deposits are associated with amphibolites, acid metatuffites and metarhyolites of calc-alkaline character which suggests a back-arc environment (6) (Reimann and Stumpfl 1981).

Within the **Southern Alps** stratiform Fe-Cu deposits with minor Zn occur in the Ivrea zone where they are associated with amphibolites and metasediments (13). In the South Tyrol Fe-Cu or Fe-Cu-Zn-Pb-As-Sb (Omenetto and Brigo 1974) occur in graphite-bearing schists, quartzites and chlorite schists as principal host rocks the age of which has only been established for the deposit of Agordo (Valle Imperina) (14) which is located in

Cambrian to Tremadocian phyllites (Sassi et al. 1984); similar deposits occur elsewhere in lithologically comparable sequences (Omenetto and Brigo 1974).

The stratiform sulphide deposits with prevailing Fe, Cu and Zn of the Oetztal nappe and the Southern Alps resemble Besshi-type mineralizations. Possible modern analogues have been reported from the Escanaba trough of the Gorda Ridge in the NE Pacific (Zierenberg et al. 1989). If the comparison is valid then these deposits would have formed at spreading centres where comparatively high sedimentation rates prevailed. Seawater-derived fluids not only reacted with cooling basalts but also with varying proportions of sediments which would account for the presence of variable proportions of elements other than Fe, Cu, Zn and Au. One may, however, also envisage, for those deposits with noticeable amounts of elements other than Fe, Zn, Cu and Au, a back-arc environment with continental crustal material present.

The ore deposits of this early cycle of the evolution of the Alps also lead to the conclusion that island arcs were situated in the future Penninic as well as in parts of the future Austro-Alpine areas during the late Precambrian to early Ordovician. To the South, in higher units of the Austro-Alpine nappes, which now are dominated by the gneiss-amphibolite association, and the Southern Alps (e.g. Agordo, 14) back-arc rifting may have developed in environments with high sedimentation rates.

Late Ordovician collision-type granites intruded not only into the gneiss-amphibolite association of Austro-Alpine units but also in the external massifs, in the Penninic area and in the western part of the Southern Alps. This period of magmatism is apparently devoid of mineralization at the present erosional levels.

3 Mid-Ordovician to Lower Carboniferous Cycle (Fig. 1 b)

3.1 Southern Alps

Within the metamorphic basement of the Southern Alps the distribution of upper Ordovician to lower Carboniferous rocks is not well established. Gansser and Pantic (1988) reported microfloras indicating Ordovician to Silurian ages for the Edolo schists. Furthermore, lithological similarities of quartzphyllites with those of Austro-Alpine units exist. Based on palynologic evidence, Sassi et al. (1984) concluded that in the Southern Alps Cambrian quartzphyllites occur whereas the Austro-Alpine quartzphyllites are of Upper Ordovician to Silurian age.

The Carnic Alps contain a complete unmetamorphosed sequence of Upper Ordovician (Caradocian) to Carboniferous sediments with some intercalated volcanics (Schönlaub 1979). They rest, as an allochthonous cover, on Brixener quartz-phyllites which were metamorphosed during the Variscan. The sequence starts with conglomerates, psammites, pelites and some acid porphyroids. Carbonate deposition started during the Upper Ashgill and persisted during the Silurian and Devonian. An emersion period during the Lower Carboniferous produced an erosional surface which reaches into Devonian limestones. An extended mineralization with quartz, barite, fluorite, Cu, Zn, Pb, Sb, and Hg (48) is linked to this karst surface. Carboniferous sandstones constitute the hanging walls (Brigo et al. 1988). A comparable mineralization occurs within a similar sequence of the Karawanken (49) (Drovenik et al. 1980).

3.2 Eastern Alps

Within the Austro-Alpine units micaschist-marble complexes prevail in the East and South. They are thought to represent the metamorphic equivalents of fossil-bearing sequences which comprise the Greywacke zone in the north and various quartzphyllite units in the north (Innsbrucker quartzphyllite) and east of the Tauern window where they tectonically overly micaschist-marble complexes (e.g. Palaeozoic of Graz).

Normally no basement to these Palaeozoic units can be recognized. However, locally preserved slices of basement which were metamorphosed and deformed 540 to 500 Ma ago have been recognized in the eastern part of the Greywacke zone (Kaintaleck) (Neubauer 1991). According to rare fossil finds sedimentation probably started during the Lower Ordovician (Reitz and Höll 1988). Mafic calc-alkaline and felsic volcanic sequences indicate a back-arc environment (Loeschke and Heinisch this Vol.). Continuous stratigraphic sections (Fig. 2) normally start with Upper Ordovician to Silurian strata. The columns show that during the Palaeozoic facies differentiation is widespread and probably is related to a new pulse of rifting during the Silurian. Intraplate magmatism at passive margins developed in the western part of the Greywacke zone and during the Devonian also in the Palaeozoic of Graz. Alkaline mafic and mafic-keratophyric products were deposited in

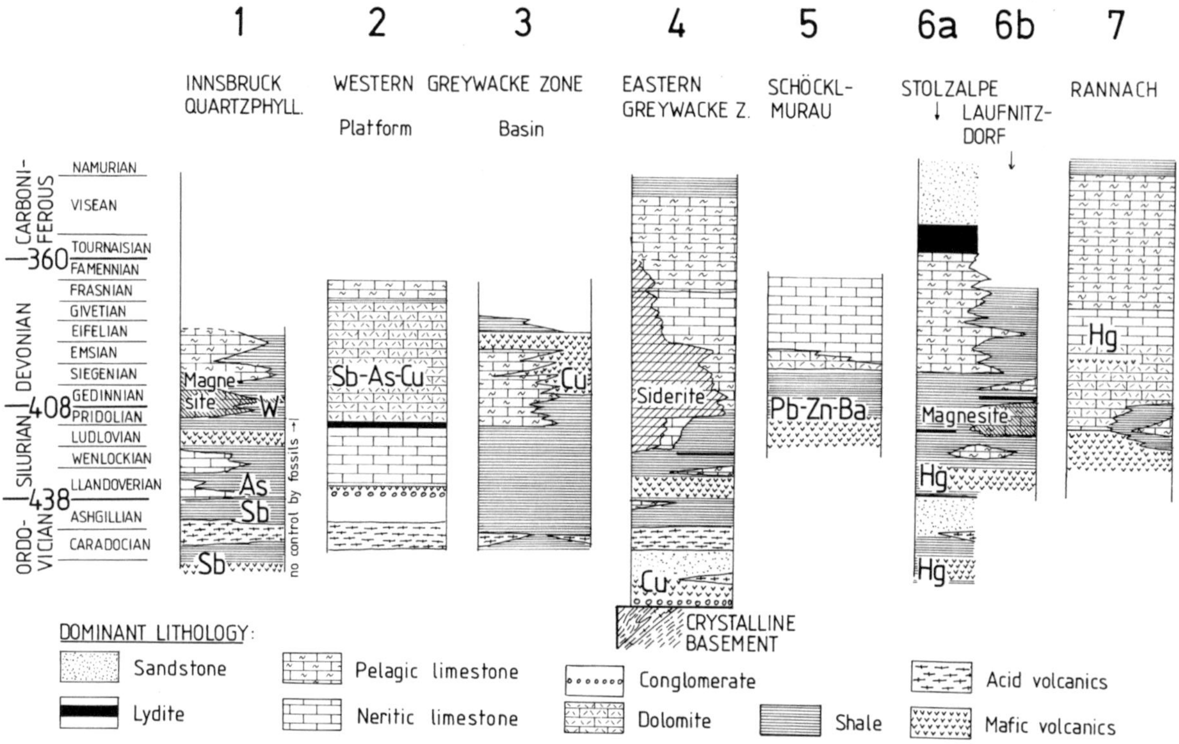

Fig. 2. Lithostratigraphic columns (in places biostratigraphically controlled) of Austro-Alpine nappes showing the position of the major types of mineralizations. The *numbers* refer to the approximate locations shown in Fig. 1 b

basins, whereas clastic wedges formed along the basin edges and intrabasinal highs (Pohl 1984; Fritz and Neubauer 1988; Heinisch 1988; Schlaegel 1988). A passive continental margin developed during the Lower and Middle Devonian with carbonate platforms and Pacific-type reefs, lagoonal deposits and pelagic limestones. Fauna and floras suggest widespread exchange along the margin of a major ocean basin (Schönlaub and Heinisch this Vol.). Pelagic limestone sedimentation persisted until the Lower Visean. During the Early Upper Carboniferous carbonate platforms subsided due to deposition of flysch-type sediments, partly of volcanic origin, and due to loading by thrust sheets.

There are, however, common tendencies noticeable in the development of the Palaeozoic sequences. The basal portion of the sections are composed of thick phyllites and rare conglomerates and of mafic and acidic volcanics. A conspicuous porphyroid horizon occurs in the Greywacke zone and in other upper Austro-Alpine and South-Alpine units, including the Carnic Alps. Clastic sediments usually prevail until the Lower Devonian. Carbonate sedimentation dominates from the Emsian to the Visean. Visean clastic sediments mark a period of renewed tectonic unrest. Quantitatively, the development of these lithologies varies considerably in different tectonic units, and carbonates are generally more abundant in the eastern areas.

The Upper Ordovician to Devonian was, thus, a period of hot spot magmatism coupled with incipient rifting during the Silurian which in certain domains of the middle to higher Austro-Alpine units proceeded during the Devonian to a rather advanced stage. On the whole, however, crustal consolidation appears to have prevailed especially in the lower tectonic units.

Ore deposits are widespread in the Palaeozoic of the Austro-Alpine nappes. Important stratiform types include the following associations: Fe-Cu-Zn-Pb (20–23), Sb-(Fe-As-W) (24, 28), Au-As (25), W (26–27) and Pb-Zn (35, 36). Strata-bound ores include important magnesite (39–39) and siderite (40) deposits and furthermore Cu-As-Sb-Hg (32, 33) and Hg-(Cu-Pb-Ag-Au) (29–31) associations.

Figure 1 b reveals a NE-SW zoning of the metals: Fe-Cu is widespread in the Greywacke zone, whereas Pb-Zn-Ba occur in the southeastern region where they are often associated with alkaline mafic volcanics of most probably Late Silurian to Early Devonian age. Stratiform Fe-Cu deposits with Zn and Pb (20–22) occur in the northern Greywacke zone, the highest Austro-Alpine nappe, in basins where they are associated with tholeiitic rocks (Fig. 2) which probably are related to back-arc rifting. The ages of these rock and the ores are poorly constrained; Lower Devonian can be considered to be their minimum age. In platform areas to the west, fahlore mineralizations with Cu-Sb-As-Hg are hosted in Devonian dolomites (33) where they form a stratiform mineralization below which they occur as vein type ores (Gstrein 1979).

In Upper Austro-Alpine units of the Kreuzeck Mountains, strata-bound stibnite-ores with pyrite, arsenopyrite and scheelite and elevated concentrations of Hg, Bi, Be and Au occur in black schists and metavolcanics within garnet-micaschists (28). According to Weber (1989) most of the occurrences are clearly of epigenetic origin and follow an Alpine shear zone. However, they could represent mobilizations of originally sedimentary ores. Further to the east the same horizon contains a stratiform and strata-bound cinnabar-arsenopyrite mineralization. Still further to the east, in the Goldeck Mountains, a strata-bound cinnabar mineralization is hosted in sericite quartzite and is considered to be genetically related to basic metavolcanics (29) (Lahusen 1972).

The mineralizations of the Lower Austro-Alpine Innsbrucker quartzphyllites show striking changes from stratigraphically lower to higher units (Fig. 2) (Haditsch and Mostler 1983). Stratiform stibnite mineralizations with minor Fe, Zn and Cu dominate in the lower part of the section where quartzphyllites, basic and acid metavolcanics prevail (24). Arsenic is typical of the central part of the section where carbonates appear besides phyllites and minor volcanics. Arsenopyrite, pyrite, pyrrhotite are the principal ore minerals of significant, stratiform gold mineralization; Cu, Zn, Pb, Ni and Sb form accessory sulfides (25) (Schulz 1984). Devonian dolomites and black schists constitute the upper part. The schists contain stratiform scheelite together with pyrite, graphite, apatite and tourmaline, whereas in the overlying dolomite magnesite (37) and scheelite occurs in clefts (26). The appearance of scheelite at the onset of carbonate deposition has also been reported from other quartzphyllite units (27) (Höll 1978).

Important magnesite (37–39) and siderite (40) deposits are contained within Devonian carbonates, with magnesite dominating in the west. They also occur in the micaschist-marble units (45, 46), which most likely represent metamorphic equivalents of the less metamorphosed Palaeozoic sequences such as the Greywacke zone. Major magnesite deposits also occur in Carboniferous carbonates (64–66). The discussion concerning the origin of these deposits is still going on (Pohl 1990). Sedimentary to diagenetic textures as well as geochemical data, including Sr-, O- and C-isotopes (Schroll et al. 1986; Frimmel 1988; Kralik et al. 1990) favour a concentration of Mg in a restricted marine environment. Widespread epigenetic features are the result of diagenetic as well as later metamorphic processes which were responsible for remobilisations and for a lowering of the $d^{18}O$ and $d^{13}C$ values and an increase in the $^{87}Sr/^{86}Sr$ ratios of successive carbonate-generations. Many Devonian siderite and magnesite [e.g. Erzberg (40), Hüttenberg (46), Palaeozoic of Graz (39)] are characterized by $\delta^{18}O$ and $\delta^{13}C$ values of about +19‰, respectively of −4‰ to −8‰. Sulphates which accompany some of the siderite and magnesite deposits exhibit $\delta^{34}S$ values of +25‰ which correspond to Devonian sea-water sulphate and demonstrates the resistence of sulphates to S-exchange under low grade metamorphic conditions (Schroll 1990). In contrast to magnesite, the high $^{87}Sr/^{86}Sr$ ratios of even the earliest siderite and the progressively higher ratios of successive generations require the presence of solutions containing crustal Sr. The question whether such solutions derived from dewatering processes due to compaction or other low-temperature processes is still unclear (Frimmel 1988).

Stratiform Pb-Zn-Ba deposits in Devonian sediments are tied to the development of local basins with shales and black schists (35, 36). A relationship with mafic volcanics probably exists (Weber 1990). Palaeozoic Pb-Zn deposits contain significant Ag concentrations and contrast thereby with the generally low Ag values of the counterparts hosted in Triassic carbonates. Finally, Devonian platform carbonates with minor mafic volcanics host Hg mineralizations (31).

3.3 *Penninic Zone and External Massifs*

In view of the relatively high-grade Tertiary metamorphism in all Penninic domains identification of Upper Ordovician to Lower Carboniferous units is difficult. The recently discovered low grade Au deposit in the Tavetsch Massif is hosted by sheared, quartz-rich rocks within sericite-muscovite-schists which probably represent hydrothermally altered tuffites (15) (Jaffe 1989; Knopf et al. 1989). A similar deposit was discovered in the northern part of the Aar Massif near Amsteg (16) (Wenger and Steiger 1990). Both are contained in presumably pre-Upper Carboniferous metavolcanics to metasediments.

Within the upper parts of the basement rocks of the Siviez-Mischabel nappe a number of stratiform to strata-bound Pb-Zn (54), Pb-Zn-Cu (51) and Cu-Bi (52) as well as vein type Ni-Co-As (53) deposits occur. Their ages, as well as the ages of the host rocks, are only poorly constrained. They may be related to an Early Carboniferous magmatism. Alpidic remobilization under greenschist to lower amphibolite facies conditions was, based on structural or isotopic evidence, postulated to have occurred in some cases (Schmutz 1984; Woodtli et al. 1987; Della Valle 1988; Zingg 1989).

Within the external Aar Massif a stratiform Pb-Zn-Ba-F deposit (17) occurs in crystalline basement rocks of pre-Upper Carboniferous age (Huttenlocher 1931). In the Aiguilles Rouge a skarn-type pre-Upper Carboniferous strata-bound Au-As-W deposit (18) is hosted by thin and discontinuous marble horizons in basement rocks (Woodtli et al. 1987).

4 Variscan Syn- to Post-Orogenic Cycle

(Fig. 1 c)

Calc-alkaline magmatism started at the beginning of the Upper Carboniferous and lasted until the Lower Permian. It is widespread in the Southern Alps, in the Penninic area of the Tauern window and the Swiss Alps, in the external massifs and also in some parts of Austro-Alpine nappes, e.g. Bernina, Err-Julier. Their geochemical signatures suggest that they have evolved in a Cordilleran situation above a subduction zone. Independent evidence for Variscan subduction is, however, lacking and therefore this magmatism may instead be related to collisions (Frisch and Neubauer 1989). Alkaline suites are probably somewhat younger and point to a subsequent phase of rifting. Volcanic rocks comprise basalts, andesites, dacites and rhyolites (Mercolli and Oberhänsli 1988).

4.1 External Massifs and Penninic Zone

In the Aar Massif a notable molybdenite deposit with minor scheelite occurs, along with similar but smaller mineralizations, in cupolas of the central Aar granite (56) (Steck and Hügi 1970). Geochemical and isotope data led Schaltegger (1990) to conclude that the host rock is a late orogenic granite not directly related to an active continental margin. The magma is of mantle origin and contains only a minor crustal component. Thus, this deposit does not belong to the subduction-related porphyry types but is rather related to extensional tectonics. Another molybdenite occurrence (Alpeiner Scharte) is associated with Variscan metagranitoids of the Tauern window (Mostler 1982).

4.2 Eastern Alps

In the Middle Austro-Alpine Koralpe spodumene bearing pegmatites occur (62) (Göd 1989). They constitute the biggest Li deposit in Europe. The mineralization is accompanied by elevated Sn and Be concentrations. The pegmatites were probably formed during or after a retrograde phase of the regional Variscan amphibolite facies metamorphism. Comparable Li-pegmatites are widespread in the Austro-Alpine region. Other Variscan metagranitoids are devoid of mineralizations.

4.3 Southern Alps

The ultramafic complex of the Ivrea zone hosts small Fe-Ni (Garuti et al. 1990) and layered chromite occurrences (70, 71) (Ferrario and Garuti 1990). The age of the ultramafics is not yet well established (compare Cumming et al. 1987). Zircon U-Pb data indicate that the complex is the result of multiple intrusions which may have started as early as 670 Ma ago. Subsequent intrusions occurred 290–260, 210–220 and possibly 20–30 Ma ago (Staehle et al. 1990; Gebauer et al. 1992).

Vein deposits are widespread in the Southern Alps and centre around Upper Carboniferous to Permian calc-alkaline volcano-magmatic complexes such as in the area of Bolzano and Trento (73) (Castaldo and Stampanoni 1975; Brigo 1976) and of Lugano-Varese (74) (Wenger and Steiger 1990). Petrochemical and isotope data indicate for the latter a mixed mantle and crustal origin probably in an Andean-type subduction environment (Stille and Buletti 1987). The mineralization occurs in basement rocks, Variscan granites and volcanics. They comprise a variety of vein types the principal ones being fluorite-barite veins with minor Pb, Zn and Cu, argentiferous galena and sphalerite with minor fluorite and barite and polymetallic veins with varying proportions of galena, sphalerite, pyrite, arsenopyrite, chalcopyrite, pyrrhotite, stibnite, Sb-sulfosalts, fluorite, barite, quartz and carbonates. Gold has been reported from the Lugano-Varese area where it occurs either in association with pyrite, arsenopyrite, galena and sphalerite or else in a complex Sb-bearing paragenesis with Ag-rich tetraedrite and Ag-Sb-sulfosalts. Brigo (1971) reported the occurrence of gold from the Bolzano Trento area in an extensive polymetallic (Cu-Ag-Pb-Zn) vein system which is associated to the diorite of Chiusa and which shows some features such as K-alteration which are typical of porphyry copper systems.

At Novazza (76) and Val Vedello (77) Upper Carboniferous to Permian rhyolitic ignimbrites with intercalated fluviatile sandstones and conglomerates contain U mineralizations accompanied by Zn, Sb, As, Mo, Pb, Cu and B in hydrothermally altered volcanics (Origoni Giobbi et al. 1982).

The **Penninic, Austro-Alpine** and **Southalpine** basement units were covered by post-Variscan molasse-like sequences which in the Southern Alps include volcanics. Deposition of this sequence did not commence everywhere synchroneously. In the eastern Greywacke zone deposition started during the Visean with coarse, detrital sediments. In other areas it started during the Upper Carboniferous or during the Permian. These sequences contain deposits which are typical of a continental environment. Most of the magnesite mineralizations hosted in Permian sediments are related to evaporites (Niedermayr et al. 1989). The Tregiovo shales of the Southern Alps host stratiform Zn-Pb-F deposits (81) (Klau and Mostler 1983), and the overlying sandstones contain U mineralization accompanied by Cu, Pb, Zn. These, as well as Pb-Zn mineralizations occur preferentially in association with coalified plant remains (Haditsch and Mostler 1982; Wopfner et al. 1983). During a marine transgression the Upper Permian Bellerophon limestone was deposited which hosts Pb-Zn-Ba mineralizations in lagoonal environments. U-Cu mineralizations in Permian sediments are also common in Austro-Alpine, Penninic and Helvetic nappes (67, 76, 77) (Gilliéron 1988).

5 Pb-isotopes

When summarizing the Pb-isotope data of the Alps one has to bear in mind that the available results do not cover all major ore- and rock types and that the data are regionally biased. Nevertheless, some important points emerge.

Mantle dominated leads are observed in the pre-Upper Ordovician terranes which host the scheelite deposit at Mittersill (1) (von Quadt 1985) and the gold deposits of the type Schellgaden (2) at the eastern end of the Tauern window (Köppel and Schroll 1983 a). The isotope signatures point to a mixture of lead derived from a depleted mantle and from continental crustal sources (Fig. 3). Such mixtures are characteristic of evolving island arcs or at divergent oceanic plate-boundaries in the presence of sediments. The Besshi-type Fe-Cu(-Zn) deposits and their amphibolite host rocks of the Ivrea zone (13) also contain lead with a mantle component (Cumming et al. 1987).

In contrast, lead of the pre-Upper Ordovician Pb-Zn deposits in Austro-Alpine units bear a continental signature (Fig. 4). Thus, contrasting environments existed already during the early stages of the formation of Alpine crust. Similarily, the lead isotope com-

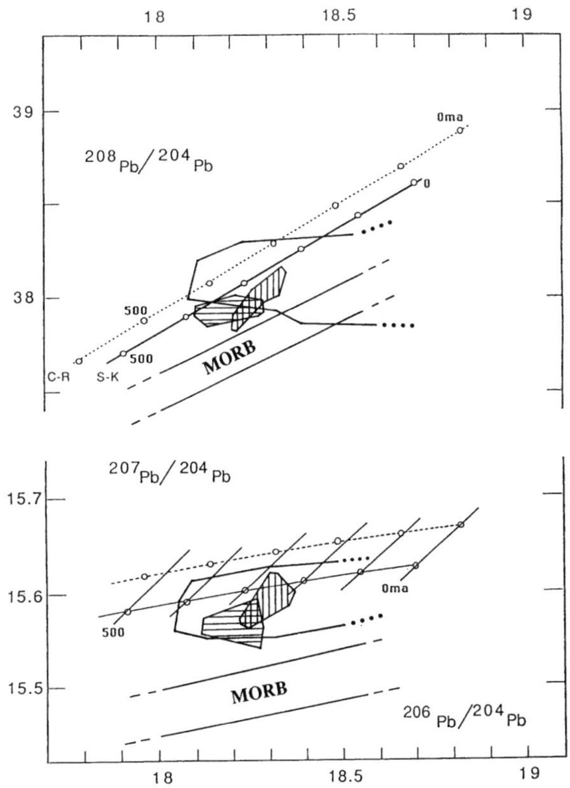

Fig. 3. Pb-isotope data fields of pre-Upper Ordovician ores which contain a mixed mantle-crust lead. *Horizontally stippled* data field: galena from Au Mineralizations of type Schellgaden (2) (Köppel and Schroll 1983 b). *Vertically stippled* data field: trace lead of pyrrhotite from stratiform Fe-Cu(-Zn) deposits of the Ivrea Zone (13) (Southern Alps) (Cumming et al. 1987). *Unmarked* data field: galena, Pb sulphosalts and scheelite from the W-deposit Mittersill (1) (von Quadt 1985). Later generations of ore minerals of the scheelite deposit at Mittersill contain an additional, metamorphically derived uranogenic Pb component from minerals with high U/Th ratios. *MORB:* the data field of mid-ocean ridge basalts from the Atlantic and Pacific is shown here for reference. *S-K* and *C-R* growth curves of average continental lead according to Stacey and Kramers (1975) and Cumming and Richards (1975) respectively

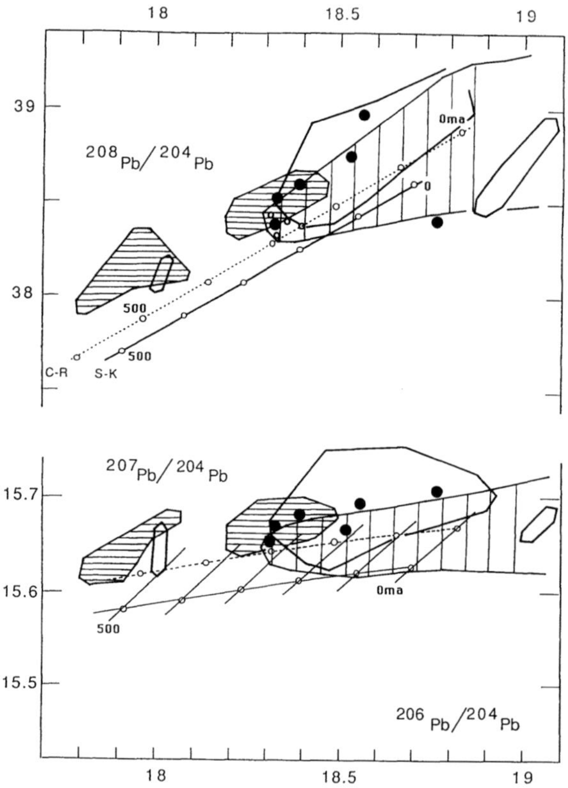

Fig. 4. Comparison of present-day isotope ratios of feldspar- and whole-rock leads from **Austro-Alpine** metasediments *(white field)* with PbS-data from stratabound and stratiform, pre-Ordovician to Triassic Pb-Zn and Cu-Zn-Pb deposits *(horizontally stippled area)* (Niggli 1987; Köppel and Schroll 1983 b, 1988). The *vertically stippled field:* trace lead of chalcopyrite and tetraedrite-tennanite of stratabound and vein deposits of Upper Ordovician to Devonian deposits (e. g. 32–34). *Filled circles* denote trace lead from stibnite (e. g. 28)

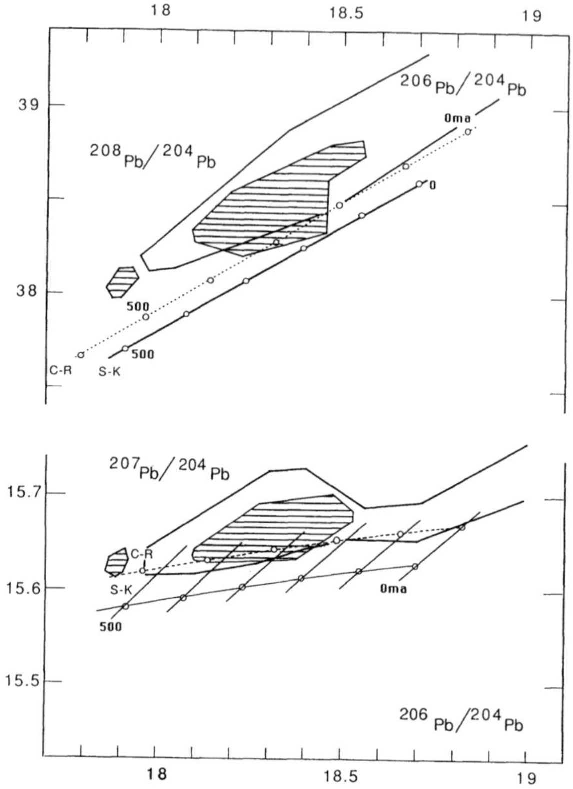

Fig. 5. Comparison of present-day feldspar- and whole-rock leads *(white field)* with PbS data from the **Southern Alps** ***(horizontally stippled fields)*** (Köppel and Schroll 1985, Cumming et al. 1987; and unpubl. data). The ore-lead data field with 206Pb/204Pb between 18.1 and 18.5 comprises vein mineralizations associated with the Upper Carboniferous to Permian magmatism of the Bolzano (73) and Lugano-Varese (74) areas, Permian (e. g., 81) to Triassic Pb-Zn deposits and the stratabound, polymetallic Ba-F mineralization of the Carnic Alps (48)

positions of Upper Ordovician or younger stratiform to strata-bound Fe-Cu-Zn-Pb, Pb-Zn and Sb deposits of the Eastern Alps point to a crustal origin of the lead.

The two galena data-fields of Fig. 4 suggest the existence of a pre- and a post-Ordovician age group. The one with model ages of about 600 Ma comprises the Fe-Cu-Pb-Zn-As-Au ores of the Kreuzeck Mountains (6) which occur in the gneiss-amphibolite association. It should, however, be noted that the lead of the biostratigraphically dated deposits of the Devonian of the Palaeozoic of Graz (35) also plot in the data field of the older group, whereas the lead of the Monte Neve deposit in pre-Upper Ordovician metasediments of the Oetztal nappe (5) yields model ages of about 400 Ma. Although the ages of many of these deposits are poorly constrained it seems probable that each of the data fields contains pre- as well as post-Upper Ordovician deposits. This would indicate that two isotopically distinct sources existed during the Late Proterozoic and most of the Palaeozoic. No correlation appears to exist between the tectonic positions of these deposits and the source type.

The lead isotope pattern of galena-bearing ores from the Southern Alps (Fig. 5) is similar to that of the metasediments and resembles also that of galena and rock-leads of the Eastern Alps (Fig. 4). The metasediments of both tectonic units were, therefore, the principal source beds to all types of Palaeozoic to Mesozoic ore deposits which contain significant proportions of galena. With respect to average continental leads the isotope ratios are characterized by relatively high $^{207}Pb/^{204}Pb$ and $^{208}Pb/^{204}Pb$ ratios and thus reflect a prolonged crustal residence time since at least the early Proterozoic in agreement with the U-Pb age patterns of detrital zircons indicating early Proterozoic to Archean crystallization ages. The lead isotope signature of the metasediments is thus inher-

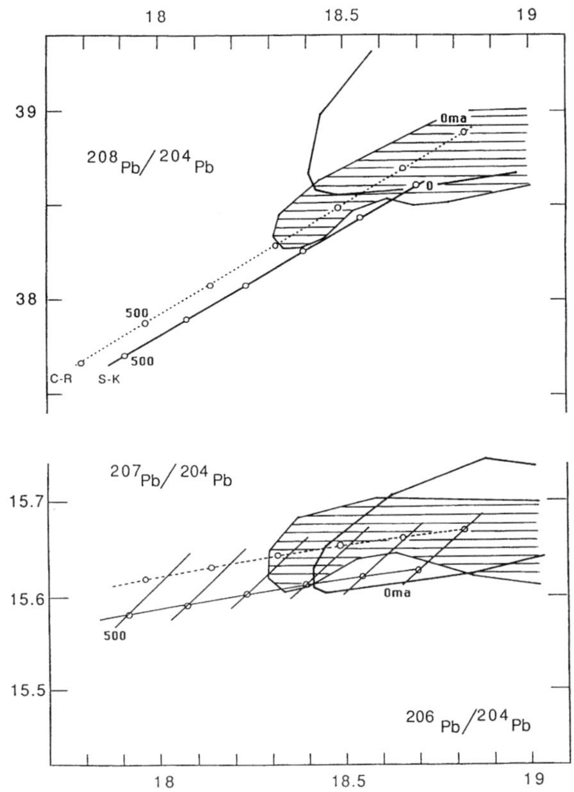

Fig. 6. Comparison of present-day feldspar and whole-rock leads *(white data field)* with PbS-lead from mineralizations in **Penninic** units including the Tertiary, vein-type Au-mineralizations of the Monte Rosa area and of the Hohe Tauern (Nunes and Steiger 1974; Köppel and Schroll 1983a; Curti 1987; Zingg 1989 and unpubl. data)

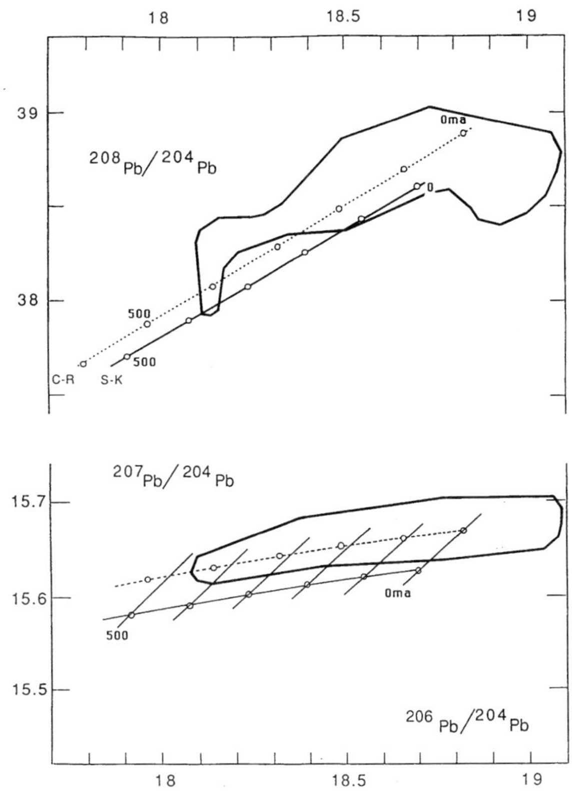

Fig. 7. Present-day feldspar-lead data from Caledonian, Variscan and Triassic acid magmatic rocks from the Southern Alps and Penninic units (Köppel and Schroll 1985; Cumming et al. 1987; Curti 1987; Zingg 1989)

ited from detrital material which derived from old continental terrains on which now is contained in U-poor minerals such as feldspars. The lack of highly radiogenic ore-lead, which is commonly observed in strata-bound Pb-Zn deposits hosted in Phanerozoic sediments covering old shield areas, argues for the absence of significant volumes of old, i.e. Early Proterozoic or older basement rocks within the Alps.

Cu deposits of the northern Greywacke zone contain an isotopically distinct lead with somewhat lower $^{207}Pb/^{204}Pb$ and $^{208}Pb/^{204}Pb$ ratios and thus with a less pronounced signature of old crustal material than the Pb-Zn dominated and the Sb-deposits (Fig. 4). This could reflect the presence of younger, i.e. Late Proterozoic source rocks or it could indicate the presence of a mantle component which may have been derived from leaching of associated mafic volcanics. The comparatively radiogenic lead of some of the chalcopyrite, tetraedrite and stibnite samples are typical for trace-lead of metamorphically recrystallized or remobilized sulphides.

The zoning of the dominant metals – Fe-Cu in the NW, Pb-Zn in the E and SE – is matched by a corresponding zoning of the μ (Fig. 8) and the W values of the lead isotope ratios. The zoning probably reflects an increasing dominance of metasedimentary metal source-beds containing old crustal material in the E and SE, whereas in the NW terrains containing younger crustal components and/or mantle-derived rocks constituted significant source rocks to the mineralizing fluids. The zoning therefore suggests that in general the Alpine domains in the E and SE had cratonized earlier than those which now occupy the NW part of the Eastern Alps.

Within the Penninic zone and the external massifs the ore-lead of galena-bearing deposits is characterized by relatively low $^{207}Pb/^{204}Pb$ and $^{208}Pb/^{204}Pb$ ratios (Fig. 6) similar to the Cu deposits of the northern

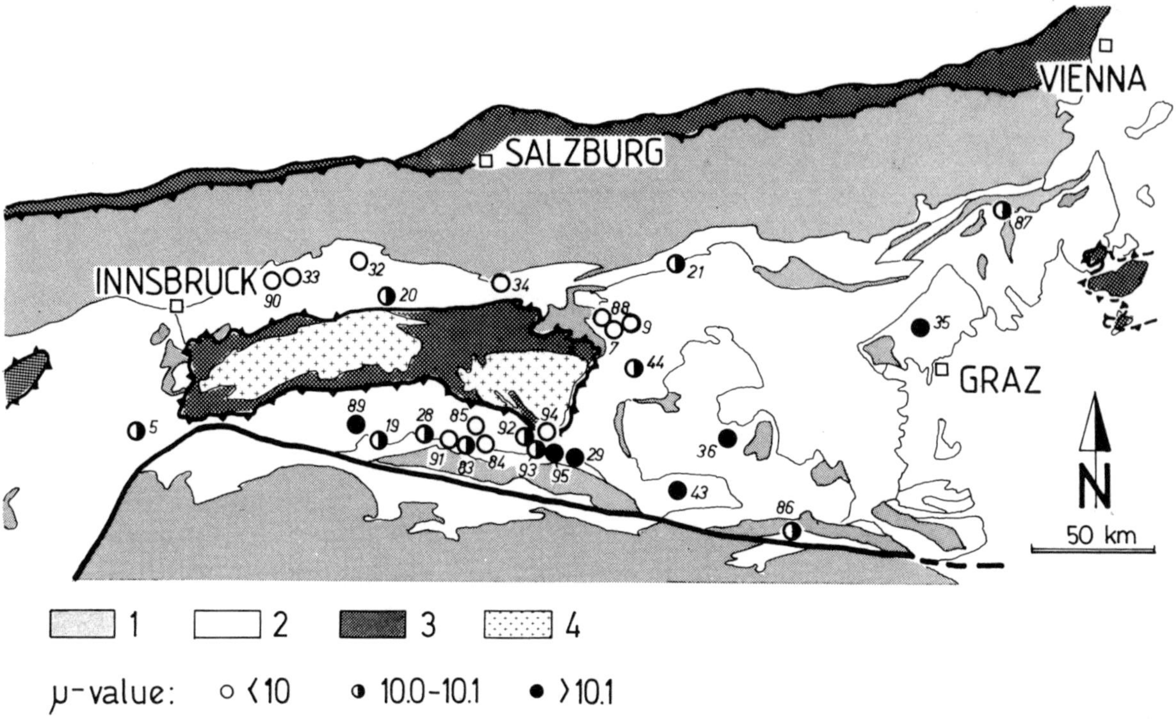

Fig. 8. Sketch map of the Eastern Alps showing the distribution of the second-stage μ-values (Stacey and Kramers 1975) of pre-Mesozoic ores. *1* Austro-Alpine and Southalpine cover sequences; *2* Austro-Alpine and Southalpine basement; *3* Penninic and Helvetic cover sequences; *4* Penninic basement. The *numbers up to 82* refer to ore deposits listed in Table 2. **Galena:** ***19*** Tessenberg-Panzendorf; *20* Rettenbach (Mittersill); *29* Stockenboi (Buchholzgraben); *35* Palaeozoic of Graz; *36* Meiselding; *43* Moosburg; *44* Ramingstein; *83* Striedner Plaike; *84* Kaser Wieserl (Gnoppitz); *85* Knappenstube; *86* Koprein/Leppengraber (Eisenkappel); *87* Arzberg/Fröschnitz; *88* Kreiteralm/Oberalm (Schladming); *89* Villgraten. **Tetraedrite-tennanite:** ***7*** Krombach (Schladming); *9* Zinkwand-Vöttern; *32* Röhrerbühel; *33* Schwaz; *34* Mitterberg (Mühlbach); *90* Brixlegg. **Stibnite:** ***28*** Abfaltersbach; *91* Gurserkammer; *92* Radlberg; *93* Lessnig; *94* Siflitz; *95* Guginock

Greywacke zone. Feldspar-lead of pre-Upper Carboniferous metasediments of the Penninic zone shows a similar tendency. However, the $^{207}Pb/^{204}Pb$ and $^{208}Pb/^{204}Pb$ ratios exhibit considerable scatter which indicates the input of old as well as of young detritus into Penninic basins.

The lead of magmatic rocks of the Southern Alps and of the Penninic domain (Fig. 7) show less variations in the $^{207}Pb/^{204}Pb$ and $^{208}Pb/^{204}Pb$ ratios and on the average lower values than the metasediments of the Southern Alps. In terms of the plumbotectonic model (Zartman and Doe 1981) their lead approximately follows the orogen growth curve which represents average continental crustal lead. The Penninic ore-lead, therefore, matches both, the metasedimentary as well as the magmatic leads.

The isotope data of trace-lead from mafic and ultramafic intrusions in the Ivrea zone indicate an origin from a contaminated mantle source. Cumming et al. (1987) suggested a contamination by fluids released from subducted sediments.

6 Conclusions

Table 1 summarizes the metallogenic evolution of the central and eastern segments of the Alps during the Late Proterozoic and Palaeozoic. The metallogeny reflects the increasing cratonization of the crust. The oldest mineralizations are Besshi-type (Fe-Cu-Zn) sulphide deposits some of which are accompanied by minor amounts of other metals, stratiform Au mineralizations and the important scheelite deposit at Mittersill. They are considered to be indicative of oceanic and back-arc spreading centres and island arc settings, respectively. Lead isotope ratios point to the presence of distinct mantle components in these mineralizations. However, pre-Ordovician Pb-Zn deposits with a continental lead isotope signature indicate the existence of sharply contrasting environments already during the early stages of the evolution of the Alpine crust.

The development of the Upper Austro-Alpine, Southalpine and quartzphyllite units of the Eastern Alps from the Upper Ordovician to the Devonian was characterized by conspicuous facies differentiation with recurrent, but gradually diminishing magmatism of mainly intraplate nature and a corresponding diversity of mineralizations. Cu, Sb, Au, W and partly Hg prevail in the western and Pb, Zn and Hg in the eastern part of the Eastern Alps. The lead isotope signature of galena-bearing and of stibnite deposits indicates an early Proterozoic to Archean crustal metal source, whereas the Cu-mineralizations originate from younger sources, in part possibly from mantle-derived rocks. These differences suggest that during the Palaeozoic cratonization was more advanced in the eastern and southeastern than in the western and northwestern segments. The accumulation of clastic Ordovician to Silurian sediments was succeeded during the Devonian by carbonate deposition which indicates a period of relative quiescence. In the west they host important magnesite deposits and siderite occurrences, and in the East substantial siderite and some major magnesite deposits. Their mode of formation is still being debated. Sr-Isotope data especially of siderite indicate that fluids other than evaporitic seawater participated in the formation of these deposits.

On the whole, the Variscan magmatism was, especially in the Penninic and Austro-Alpine domains, not very fertile. The questions why massive sulphide and porphyry copper deposits are virtually absent remains open to speculation. In the Southern Alps, however, polymetallic vein deposits, some of which contain Au and exhibit features typical of porphyry copper mineralizations, as well as U-mineralizations are associated with this magmatism. In external massifs of the central Alps Au-deposits were recently discovered in metavolcanics of presumably Lower Carboniferous age. The Mo-deposit in the Aar Massif is the only porphyry-type deposit in these parts of the Alps. It is apparently related to an extensional tectonic phase.

The late Variscan stage is characterized by mineralizations which typically pertain to cratonized environments such as the widespread U-Cu association in Permo-Carboniferous sandstones in all compartments of the Alps and Pb-Zn in shales and sandstones of the Southern Alps. The latter apparently herald the important diagenetic Pb-Zn deposits which are hosted in Triassic carbonates of the Southern and Eastern Alps which are probably related to an early stage of rifting at the onset of the Alpidic cycle.

When summarizing the most conspicuous qualitative changes of the mineralizations it appears that the first evolutionary cycle (Proterozoic to Lower Ordovician) was characterized by the dominance of W and Cu-Zn, in places also of Zn-Pb. During the second cycle (Ordovician to Lower Carboniferous) Au and Sb become more prominent, Hg appears, and magnesite and siderite constitute important deposits. Compared to the first cycle W has lost its dominance. During the third cycle (Upper Carboniferous to Permian), U, Mo and Li appear in notable concentrations for the first time.

The metallogenetic evolution of Alps continued during the Alpidic cycle. Deposition of carbonate-hosted, strata-bound Pb-Zn deposits attained a climax during Lower to Middle Triassic times. A further generation of Besshi-type Fe-Cu-Zn deposits was formed during the opening of the Tethys in Mesozoic times. Triassic and Jurassic carbonates host deposits of Fe-Oxydes and Mn (Huttenlocher 1934; Kündig and de Quervain 1953; Pfeifer et al. 1988). During the final stages of the regional Tertiary metamorphism quartz-gold veins formed in structurally favourable areas of lower amphibolite- to greenschist-facies metamorphism of the Hohe Tauern and in the Monte Rosa area (Curti 1987; Lattanzi et al. 1989; Diamond 1990).

Comparative ore- and rock-lead isotope studies demonstrate that metasediments were the principal metal sources to galena deposits of the Southern Alps and of the Austro-Alpine nappes, including those hosted in Triassic carbonates. The data defines a province which is isotopically characterized by the dominance of Early Proterozoic to Archean source material. In contrast, in the external massives, the Penninic realm and in some Austro-Alpine units such as the Greywacke zone and the Kreuzeck Mountains, the ore Pb-isotope pattern reflects a diminished input from old crustal sources in favour of Pb from younger sources. Whole-rock and feldspar-lead from the Penninic metasediments also indicate the presence of significant amounts of younger detrital material in comparison to the metasediments from the Southern Alps and the Austro-Alpine realm.

Acknowledgements
The authors wish to thank L. Brigo (Milano), M. Grünenfelder (Zürich), R. Höll (Munich) and L. Weber (Vienna) for critically reading the manuscript and for making many very helpful suggestions which improved not only its contents but also its readability.

References

Angel F, Awerzger A (1948) Die Magnesitlagerstätte auf der Millstätter Alpe bei Radentheim (Kärnten). Radex-Rundsch 1948: 91–95

Beck-Mannagetta P, Grill R, Holzer H, Prey S (1966) Erläuterungen zur geologischen und Lagerstätten-Karte 1:1 000 000 von Oesterreich, 94 pp

Brigo L (1971) Mineralizzazioni e metallogenesi nell'area della fillade quarzifera di Bressanone nelle Alpi Sarentine. St Tri Sci Nat A 48: 80–125
Brigo L (1976) Quadro metallogenico dell'Alto Adige. Boll Soc Geol Ital 95: 1137–1159
Brigo L, Dulski P, Möller P, Schneider H-J, Wolter R (1988) Strata-bound mineralizations in the Carnic Alps/Italy. In: Boissonas J, Omenetto P (eds) Mineral deposits within the European Community. Springer, Berlin Heidelberg New York, pp 485–498
Castaldo G, Stampanoni G (1975) Memoria illustrativa della carta mineraria d'Italia. Servizio Geologico d'Italia, Pinto, Roma, 213 pp
Cumming GL, Richards JR (1975) Ore lead isotope ratios in a continuously changing Earth. Earth Planet Sci Lett 28: 155–171
Cumming GL, Köppel V, Ferrario A (1987) A lead isotope study of the northeastern Ivrea zone and the adjoining Ceneri zone (N-Italy): evidence for a contaminated subcontinental mantle. Contrib Mineral Petrol 97: 19–30
Curti E (1987) Lead and oxygen isotope evidence for the origin of the Monte Rosa gold lode deposits (Western Alps, Italy): a comparison with Archean lode deposits. Econ Geol 82: 2115–2140
Della Valle G (1988) Contribution a l'interpretation metallogenique des fahlbandes et des veines polymetalliques du Val d'Anniviers et du Val de Tourtemagne (Valais). Thèse, Universitè de Lausanne, 168 pp
Dessau G, Perna G (1966) Le mineralizzationi a galena e blenda del Trentino-Alto Adige e lor contenuto in elementi accessori. Symp Int sui Giacimenti Minerari delle Alpi Trento-Mendola 3, pp 578–688, Saturnia, Trento
Diamond L (1990) Fluid inclusion evidence for the P-V-T-X evolution of hydrothermal solutions in late-Alpine gold-quartz veins at Brusson, Val d'Ayas, northwest Italian Alps. Am J Sci 290: 912–958
Drovenik M, Plenicar M, Drovenik F (1980) The origin of Slovenian ore deposits. Geologija 23: 1–157
Ebner F, Weber L (1982) Geochemische Prospektion auf Quecksilbervererzungen im Mitteldevon der Rannach-Fazies des Grazer Paläozoikums. Arch Lagerstättenforsch Geol Bundesanst 2: 47–62
Eikenberg J, Köppel V, Labhart T, Signer P (1989) U-Pb, U-Xe and U-Kr systematics of a greenschist facies metamorphic uranium mineralization of the Siviez-Mischabel nappe (Valais, Switzerland). Schweiz Mineral Petrogr Mitt 69: 331–344
El Ageed A, Saager R, Stumpfl EF (1980) Pre-Alpine ultramafic rocks in the eastern central Alps, Styria, Austria. Proc Int Ophiolite Symp Nicosia Cyprus. Ministry of Agricultural and Natural Resources, Geol Surv Dep, Nicosia, Cyprus, pp 601–606
Erkan E (1977) Uran- und gipsführendes Perm in den östlichen Ostalpen. Berg Hüttenmänn Monatsh 122: 9–17
Ferrario A, Garuti G (1990) Platinum-Group mineral inclusions in chromites of the Finero mafic-ultramafic complex (Ivrea-zone, Italy). Mineral Petr 41: 125–143
Flisch M (1987) Teil 1: Geologische, petrographische und isotopengeologische Untersuchungen an Gesteinen des Silvrettakristallins. Ph D Thesis, Universität Bern
Friedrich O (1929) Die Siderit-Eisenglimmer-Lagerstätte von Waldenstein in Ostkärnten. Berg Hüttenmänn Monatsh 77: 131–145
Friedrich O (1963) Die Erzlagerstättenkarte der Ostalpen. Radex Rundsch Radenthein 7/8: 371–407
Friedrich OM (1968) Die Vererzungen der Ostalpen, gesehen als Glied des Gebirgsbaues. Arch Lagerstättenforsch Ostalpen 8: 1–136
Friedrich OM (1975) Kurzbericht über die Vererzung der Schladminger Tauern. Arch Lagerstättenforsch Ostalpen 15: 117–127
Frimmel H (1988) Strontium isotopic evidence for the origin of siderite, ankerite and magnesite mineralizations in the Eastern Alps. Mineral Deposita 23: 268–275
Frisch W, Neubauer F (1989) Pre-Alpine terranes and tectonostratigraphic zoning in the Eastern Alps. Geol Soc Am Spec Pap 230: 99–100
Frisch W, Ménot RP, Neubauer F, von Raumer J (1990) Correlation and evolution of the alpine basement. Schweiz Mineral Petrogr Mitt 70 (2): 265–286
Fritz H, Neubauer F (1988) Geodynamic aspects of the Silurian and Early Devonian sedimentation in the Paleozoic of Graz (Eastern Alps). Schweiz Mineral Petrogr Mitt 68: 359–368
Gansser A, Pantic N (1988) Prealpine events along the eastern Insubric line (Tonale line, northern Italy). Eclogae Geol Helv 81: 567–577
Garuti G, Naldrett AJ, Ferrario A (1990) Platinum-group elements in magmatic sulfides from the Ivrea zone: their control by sulfide assimilation and silicate fractionation. Econ Geol 85: 328–336
Gebauer D, Schmid R, Quadt A, Ulmer P (1992) Oligocene, Permian and Panafrican zircon ages from rocks of the Balmuccia peridotite and of the Lower Layered Group in the Ivrea Zone. Schweiz. Mineral Petrogr Mitt (in press)
Gilliéron F (1988) Zur Geologie der Uranmineralisation in den Schweizer Alpen. Beitr Geol Schweiz, Geotechn Ser 77: 54
Göd R (1981) Ein Beitrag zur Petrographie und Geologie des Bergbaureviers Schellgaden. Mitt Ges Geol Bergbaustud Österr 27: 189–200
Göd R (1989) The spodumen deposit at "Weinebene", Koralpe, Austria. Mineral Deposita 24: 270–278
Gollner H (1983) Die stratigraphische Stellung der Breitenauer Magnesitvorkommen. Berg Hüttenmänn Monatsh 101: 1–12
Gstrein P (1979) Neuerkenntnisse über die Genese der Fahlerzlagerstätte Schwaz (Tirol). Mineral Deposita 14: 185–194
Haditsch JG (1967) Die Zeiringer Lagerstätten. Arch Lagerstättenforsch Ostalpen 6: 1–196
Haditsch JG, Mostler H (1982) Late Variscan and early Alpine mineralization in the Eastern Alps. In: Amstutz GC, El Goresy A, Frenzel G et al. (eds) Ore genesis. The state of the art. Springer, Berlin Heidelberg New York, pp 582–589
Haditsch JG, Mostler H (1983) The succession of ore mineralization of the lower Austroalpine Innsbruck quartzphyllite. In: Schneider H-J (ed) Mineral deposits of the Alps and of the Alpine Epoch in Europe. Proc IV ISMIDA, Berchtesgaden, 1981. Springer, Berlin Heidelberg New York, pp 51–59
Heinisch H (1988) Hinweise auf die Existenz eines passiven Kontinentalrandes im Altpaläozoikum der nördlichen Grauwackenzone – Ostalpen. Schweiz Mineral Petrogr Mitt 68: 407–418
Höll R (1978) Time and stratabound early Paleozoic scheelite, stibnite and cinnabar deposits in the Eastern Alps. Verh Geol Bundesanst Jg 1978 Heft 3: 369–387
Höll R, Maucher A (1976) The strata-bound ore deposits in the Eastern Alps. In: Wolf KH (ed) Handbook of strata-bound and stratiform ore deposits, vol 5. Elsevier, Amsterdam, pp 1–36
Höll R, Schenk P (1988) Metamorphosed hydrothermal eruption breccia (conglomerate) in the Felbertal scheelite deposit, Eastern Alps, Austria. Mar Geol 84: 273–282

Huttenlocher HF (1931) Die Blei-Zinklagerstätte von Goppenstein (Wallis). Beitr Geol Schweiz, Geotechn Ser Lief 16: 1–47

Huttenlocher HF (1934) Die Erzlagerstättenzonen der Westalpen. Schweiz Mineral Petrogr Mitt 14: 22–148

Jaffé FC (1989) Gold in Switzerland. Econ Geol 84: 1444–1451

Klau W, Mostler H (1983) Zn-Pb-F mineralization in Middle Permian of Tregiovo (Province of Bozen/Trient, Italy). In: Schneider H-J (ed) Mineral deposits of the Alps and of the Alpine Epoch in Europe. Proc IV ISMIDA, Berchtesgaden, 1981. Springer, Berlin Heidelberg New York, pp 70–80

Knopf DJ, Naert KA, Bell DR (1989) New type mineralization in the Swiss Alps: The Disentis gold occurrence. Min Mag Oct: 290–296

Koller F, Götzinger MA, Neumayer R, Niedermayer G (1983) Beiträge zur Mineralogie und Geochemie der Pegmatite des St. Radegunder Kristallins und der Gleinalpe. Arch Lagerstättenforsch Geol Bundesanst 3: 47–65

Köppel V, Grünenfelder M (1971) A study of inherited and newly formed zircons from paragneisses and granitised sediments of the Strona-Ceneri-zone (Southern Alps). Schweiz Mineral Petrogr Mitt 51: 385–409

Köppel V, Schroll E (1983 a) Bleiisotopen und Remobilisation von Erzlagerstätten. In: Petrascheck WE (ed) Ore mobilization in the Alps and in SE Europe. Oesterr Akad Wiss Schriftenr Erdwisse Komm 6: 39–51

Köppel V, Schroll E (1983 b) Lead isotopes of Palaeozoic stratabound to stratiform galena bearing sulfide deposits of the Eastern Alps (Austria); implications for their geotectonic setting. Schweiz Mineral Petrogr Mitt 63: 347–360

Köppel V, Schroll E (1985) Herkunft des Pb der triassischen Pb-Zn-Vererzungen in den Ost- und Südalpen, Resultate bleiisotopengeochemischer Untersuchungen. Arch Lagerstättenforsch Geol Bundesanst 6: 215–222

Köppel V, Schroll E (1988) Pb-isotope evidence for the origin of lead in strata-bound Pb-Zn deposits in Triassic carbonates of the Eastern and Southern Alps. Mineral Deposita 23: 96–103

Kralik M, Aharon P, Schroll E, Zachmann D (1990) Carbon and oxygen isotope systematics of magnesites: a review. In: Möller P (ed) Magnesites. Monogr Ser Mineral Deposits 28. Bornträger, Berlin, pp 197–223

Kramers JD (1973) Zur Mineralogie, Entstehung und alpinen Metamorphose der Uranvorkommen bei Trun, Graubünden. Beitr Geol Karte Schweiz, Geotech Ser Lief 52, 75 pp

Kündig E, de Quervain F (1953) Fundstellen mineralischer Rohstoffe in der Schweiz, Kümmerly & Frey, Bern, p 214

Lahusen L (1972) Schicht- und zeitgebundene Antimonit-Scheelit-Vorkommen und Zinnober-Vererzungen in Kärnten und Osttirol/Oesterreich. Mineral Deposita 7: 31–60

Lattanzi P, Curti E, Bastogi M (1989) Fluid inclusion studies of the gold deposits of the upper Anzasca valley, northwestern Alps, Italy. Econ Geol 84: 1382–1397

Maggetti M, Flisch M, Bollin R (1990) Bericht über die Exkursion der Schweizerische Mineralogische Petrographische Gesellschaft ins Silvretta-Kristallin und in den Westrand des Unterengadiner Fensters. Schweiz Mineral Petrogr Mitt 70: 121–157

Mercolli I, Oberhänsli R (1988) Variscan tectonic evolution in the Central Alps: a working hypothesis. Schweiz Mineral Petrogr Mitt 68: 491–500

Morel F (1978) L'Extension du filon Pb-Zn de Goppenstein (Valais, Suisse). Contribution à son extension géophysique et géochimique. Matér Géol Suisse, Sér Géotech 57, 92 pp

Morteani G (1989) Mg-metasomatic type sparry magnesites of Entachen Alm, Hochfilzen/Bürglkopf and Spiessnägel (Austria). In: Möller P (ed) Magnesites. Monogr Ser Mineral Deposits 28. Bornträger, Berlin, pp 105–114

Mostler H (1982) Untersuchungen von Erzlagerstätten im Innsbrucker Quarzphyllit und auf der Alpeiner Scharte. Arch Lagerstättenforsch Geol Bundesanst 1: 77–83

Neubauer F (1991) Late Proterozoic and early Paleozoic tectonothermal evolution of the Eastern Alps. In: Dallmeyer RD, Lecorche IP (eds) Paleozoic West-African orogens and their correlatives. Springer, Berlin Heidelberg New York, pp 290–297

Niedermayr G, Beran A, Brandstätter F (1989) Diagenetic type magnesites in Permo-Skythian rocks of the Eastern Alps. In: Möller P (ed) Magnesites. Monogr Ser Mineral Deposits 28. Bornträger, Berlin, pp 35–60

Niggli M (1987) Geochemische Untersuchungen metamorpher Pb-Zn-Lagerstätten und ihrer Nebengesteine. Diss, ETH Zürich

Nunes PD, Steiger RH (1974) A U-Pb zircon, and Rb-Sr and U-Th-Pb whole-rock study of a polymetamorphic terrane in the Central Alps, Switzerland. Contrib Mineral Petrol 47: 255–280

Omenetto P, Brigo L (1974) Metallogenesi nel quadro dell'orogene ercinico delle Alpi (con particolare riguardo al versante Italiano). Mem Soc Geol Ital 13: 339–362

Origoni Giobbi E, Bernasconi A, Ravagnani D (1982) Petrologic and metallogenetic investigations on the Collio Formation of the Novazza Uranium mine. Rend Soc Ital Mineral Petrol 38: 293–305

Petrascheck WE, Erkan E, Siegl W (1976) Type of uranium deposits in Austrian Alps. In: Jones MJ (ed) Geology, mining and extraction processes of uranium. Inst Min Met, London, pp 71–75

Pfeifer HR, Oberhänsli H, Epprecht W (1988) Geochemical evidence for a synsedimentary hydrothermal origin of Jurassic iron-manganese deposits at Gonzen (Sargans, Helvetic Alps, Switzerland). Marine Geol 84: 257–272

Pohl W (1984) Metallogenetic evolution of the East Alpine Paleozoic basement. Geol Rundsch 73: 131–147

Pohl W (1990) Comparative geology of magnesite deposits and occurrences. In: Möller P (ed) Magnesites. Monogr Ser Mineral Deposits 28. Bornträger, Berlin, pp 1–14

Raith JG (1989) Tourmaline rocks associated with stratabound scheelite mineralization in the Austroalpine crystalline complex, Austria. Mineral Petrol 39: 265–288

Reimann C, Stumpfl EF (1981) Geochemical setting of stratabound stibnite mineralizations in the Kreuzeck mountains, Austria. Trans Instn Min Metall 90 B: 126–132

Reitz E, Höll R (1988) Jungproterozoische Mikrofossilien aus der Habachformation in den mittleren Hohen Tauern und dem nordostbayerischen Grundgebirge. Jahrb Geol Bundesanst 131: 329–340

Reitz E, Daneck Th, Miller H (1989) Ein Nachweis jungproterozoischen Alters von Schwarzphylliten am Tauern-Nordrand (Salzburg, Österreich) und seine Bedeutung für den Bau der Hohen Tauern. Jahrb Geol Bundesanst 132/4: 751–760

Sassi FP, Kalvacheva R, Zanferrari A (1984) New data on the age of deposition of the South-Alpine phyllitic basement in the Eastern Alps. Neues Jahrb Geol Paläontol Monatsh 2: 741–751

Schäffer U, Tarkian M (1984) Die Genese der stratiformen Sulfidlagerstätte Kalwang der Grüngesteinsserie und einer assoziierten Banded Iron Formation. Tschermaks Mineral Petrogr Mitt 33: 169–186

Schaltegger U (1990) The Central Aar granite: highly differentiated calc-alkaline magmatism in the Aar Massif (Central Alps, Switzerland). Eur J Mineral 2: 245–259

Schenk K, Stille P (1991) Geochemistry of mafic-ultramafic rock associations in the South Penninic nappes of the Central

Swiss Alps: evidence for Proterozoic MORB. Terra Abstr 3/1: 476
Schlaegel P (1988) Geotektonische Neuinterpretation des basischen Magmatismus der nördlichen Grauwackenzone, Ostalpen – ein Überblick. Schweiz Mineral Petrogr Mitt 68: 419–432
Schlüter J, Tarkian M, Stumpfl EF (1984) Die stratiforme Sulfidlagerstätte Walchen, Steiermark, Österreich: Geochemie und Genese. Tschermaks Mineral Petrogr Mitt 33: 287–296
Schmutz L (1984) Mineralbildende Prozesse an einer Erzlagerstätte im Val d'Anniviers. Diss, Universität Basel
Schönlaub HP (1979) Das Paläozoikum in Österreich. Abh Geol Bundesanst 33: 1–124
Schroll E (1990) Die Metallprovinzen der Ostalpen im Lichte der Geochemie. Geol Rundsch 79: 479–493
Schroll E, Papesch W, Dolezel P (1986) Beitrag der C- und O-Isotopenanalyse zur Genese ostalpiner Sideritvorkommen. Mitt Österr Geol Ges 78: 181–191
Schulz O (1984) Zell am Ziller, a syngenetic old Palaeozoic gold deposit in the Innsbruck quartzphyllite belt. In: Wauschkuhn A, Kluth C, Zimmermann RA (eds) Syngenesis and epigenesis in the formation of mineral deposits. Springer, Berlin Heidelberg New York, pp 170–176
Schulz O, Wenger H (1980) Die Goldlagerstätte Zell am Ziller, Tirol. Jahrb Geol Bundesanst 123: 113–141
Schulz O, Schroll E, Dieber K, Fuchs H (1986) Zur Frage der Genese der Lagerstätten um Hüttenberg. Carinthia II 196 (96): 479–512
Stacey JS, Kramers JD (1975) Approximation of terrestrial lead isotope evolution by a two-stage model. Earth Planet Sci Lett 26: 207–221
Staehle V, Frenzel G, Kober B, Michard A, Puchelt H, Schneider W (1990) Zircon syenite pegmatite in the Finero peridotite (Ivrea Zone): evidence for a syenite from a mantle source. Earth Planet Sci Lett 101: 196–205
Steck A, Hügi Th (1970) Das Auftreten des Molybdänglanzes im westlichen Aarmassiv und Molybdängehalte von Gesteinen der gleichen Region. Schweiz Mineral Petrogr Mitt 50 (2): 257–276
Stille P, Buletti M (1987) Nd-Sr isotopic characteristics of the Lugano volcanic rocks and constraints on the continental crustal formation in the South Alpine domain (N-Italy – Switzerland). Contrib Mineral Petrol 96: 140–150
Stille P, Tatsumoto M (1985) Precambrian tholeiitic-dacitic rock suites and Cambrian ultramafic rocks in the Penninic nappe system of the Alps: Evidence from Sm-Nd isotopes and rare earth elements. Contrib Mineral Petrol 89: 184–192
Stumpfl EF, Tarkian M (1979) Schichtgebundene Sulfidvererzung in den Schladminger Tauern. Anz Österr Akad Wiss Math-Naturwiss Kl 1979 (5): 1–5
Thalhammer DAR, Stumpfl EF, Jahoda R (1989) The Mittersill scheelite deposit, Austria. Econ Geol 84: 1153–1171
Thalmann F (1979) Zur Eisenspatvererzung in der nördlichen Grauwackenzone am Beispiel des Erzberges bei Eisenerz und Radmer/Bucheck. Verh Geol Bundesanst 1978: 479–489
Tollmann A (1977) Geologie von Österreich, Bd 1. Die Zentralalpen. Deuticke, Wien
Tornquist A (1935) Die hochmetamorphe Kieslagerstätte von Tessenberg-Panzendorf in Osttirol. Sitzungsber Akad Wiss Math-Naturwiss Kl I. 144: 19–32
Tufar W (1963) Die Erzlagerstätten des Wechselgebietes. Johanneum Mineral Mitteilungsbl 1963 (1): 1–60
Tufar W (1979) Mikroskopisch-lagerstättenkundliche Charakteristik ausgewählter Erzparagenesen aus dem Altkristallin, Paläozoikum und Mesozoikum der Ostalpen. Verh Geol Bundesanst 78: 499–528
Tufar W (1981) Die Vererzungen der Ostalpen und Vergleiche mit Typlokalitäten anderer Orogengebiete. Mitt Österr Geol Ges 74/75: 265–306
Unger HJ (1972) Die Kupfer- und Schwefelkieslagerstätten (Alpine Kieslager) der nördlichen Grauwackenzone. Arch Lagerstättenforsch Ostalpen 13: 59–74
Vavtar F (1988) Die Erzanreicherungen im Nordtiroler Stubai-, Oetztal und Silvrettakristallin. Arch Lagerstättenforsch Geol Bundesanst 9: 103–154
von Quadt A (1985) Geochronologische, geochemische und isotopengeochemische Untersuchungen an Gesteinen der Habach-Formation, der Scheelitlagerstätte und des angrenzenden Altkristallins im Felbertal (Land Salzburg). Diss, ETH Zürich
von Quadt A (1992) U-Pb zircon and Sm-Nd geochronology of mafic and ultramafic rocks from the central part of the Tauern Window. Contrib Mineral Petrol 110: 57–67
Weber L (1989) Zur Geologie der Antimonvererzungen des Osttiroler Anteils der Kreuzeckgruppe. Arch Lagerstättenforsch Geol Bundesanst 10: 65–74
Weber W (1990) Die Blei-Zinklagerstätten des Grazer Paläozoikums und ihr geologischer Rahmen. Arch Lagerstättenforsch Geol Bundesanst 12: 1–289
Weber L, Pausweg F, Medwenitsch W (1973) Zur Mitterberger Kupfervererzung (Mühlbach/Hochkönig, Salzburg). Mitt Geol Ges Wien 65 (1972): 137–158
Wenger C, Steiger R (1990) Karte der Vorkommen mineralischer Rohstoffe der Schweiz, Blatt 1, Tessin-Uri (mit Erläuterungen). Schweiz Geotechn Komm, Kümmerly & Frey, Bern
Woodtli R, Jaffe F, von Raumer J (1987) Prospection minerale en Valais: Le projet Uromine. Beitr Geol Schweiz, Geotechn Ser 71: 179 pp
Wopfner H, Griesecke S, Koch J, Fels H (1983) New aspects on metal deposits of the Groeden sandstone (South Tyrol, Italy). In: Schneider H-J (ed) Mineral deposits of the Alps and of the Alpine Epoch in Europe. Proc IV ISMIDA, Berchtesgaden, 1981. Springer, Berlin Heidelberg New York, pp 60–69
Wutzler B (1983) Geologisch-lagerstättenkundliche Untersuchungen am Mont Chemin (nordöstliches Mont Blanc Massiv). Clausth Geol Abh 42, 104 pp
Zartmann RE, Doe BR (1981) Plumbotectonics – the model. Tectonophysics 75: 135–162
Zierenberg RA, Shanks WC, Koski RA, Morton JL, Normark WR, Kvenvolden KA (1989) Massive sulfide deposition and sediment alteration in the Escabana Trough Gorda Ridge, NE Pacific ocean. 28th Int Geol Cong, Washington, Abstr 3–448
Zingg M (1989) Die Siviez-Mischabel Decke: Entstehung und Entwicklung eines Altkristallins und seiner Vererzung. PhD Thesis, ETH Zürich

Palaeomagnetic Data from the Palaeozoic Basement of the Alps

H. J. Mauritsch

Abstract

Palaeomagnetic data from the Palaeozoic basement of the Alps (Silurian, Devonian, Carboniferous, Permian) show two distinct patterns of magnetization directions, the "European" and the "African". For this definition APWPs of Europe and Africa were used as the tectonic frame. All data which were found to be of primary origin and within the Alpine orogene show "African" affinity, outside "European". This pattern fits the Mesozoic and Cenozoic one and is therefore understood as being young in origin.

1 Introduction

When dealing with Palaeozoic palaeomagnetic data one has to be aware of the basic criteria of reliability of published palaeomagnetic results. Rocks serving as sources of the palaeomagnetic information show all kind of influences like metamorphism, chemical changes and strong deformation (Marton et al. 1987). Statistical criteria (Fisher 1953; Irving 1964; McElhinny 1973; McFadden and Jones 1981; McFadden and Lowes 1981; McFadden and McElhinny 1990), stability criteria (Tarling and Symons 1967; Wilson et al. 1968; Briden 1972; Symons and Stupavsky 1974) and the knowledge of the influence of rock deformation, i.e. the anisotropy of the magnetic susceptibility on the palaeomagnetic remanence vectors (Heller 1973; Henry 1973; Cogne and Perroud 1985; Lowrie et al. 1986) can help in understanding the history of magnetization processes. Radiometric dating of secondary processes like metamorphism, deformation and chemical alterations can help to reduce the uncertainties of the interpretation. Finally, the accuracy of conversion of palaeontologically determined ages to the absolute time scale can be of critical importance.

For tectonic interpretation, palaeomagnetic directions from autochthonous areas are used for comparison. In the case of the Alpine-Mediterranean realm, the reference directions are deduced from the apparent polar wander paths (APWP) for Africa and stable Europe (Fig. 1). However, although for stable Europe a reliable APWP based on actual palaeomagnetic results exists, at least for the Mesozoic (Westphal et al. 1986), data from Africa are sparse.

Bullard et al. (1965) tried to fit Atlantic bordering continents using the geometric fitting of the continental edges. An important step further was the reconstruction of the circum-Atlantic continents for Pangea by Van der Voo and French (1974). Various stages of the opening of the Atlantic Ocean, combined with available palaeomagnetic results of the different continents, were used to establish an optimized APWP from Upper Carboniferous to the present (Fig. 1). Further attempts have been undertaken since, using the more recent data sets, but so far the reference APWPs for both Europe and Africa are too imprecise for many time periods to allow detailed tectonic interpretations. For this purpose, a sector of the full reference path is really desirable (Lowrie et al. 1980; Marton 1987; Mauritsch and Becke 1987), but not available, so far, for the Palaeozoic and Early Mesozoic.

In the present review, average directions, valid for larger areas were calculated and compared with the APWP paths for Africa and Europe. This comparison permits assessments of African and European affinities as well as clockwise (CW) and counterclockwise (CCW) rotations (Table 1). Furthermore, it should be noted that the palaeoinclination bears a simple and unique relationship to the palaeolatitude at which the rocks were formed. The geocentric dipole model is a very good approach to the configuration of the Earth's magnetic field, at least for the Mesozoic and the Permian, but very little is so far known for the Palaeozoic. Shallow inclinations close to zero are a good control for possibly Palaeozoic palaeomagnetic directions, but also a source of uncertainty in the direction of rotation. Several comprehensive reviews (Lowrie et al. 1980; Van den Berg and Zijderveld 1982; Westphal et al. 1986; Marton and Mauritsch 1990) primarily dealt with the Mesozoic, the present review restricts the results to the Palaeozoic basement of the Alps.

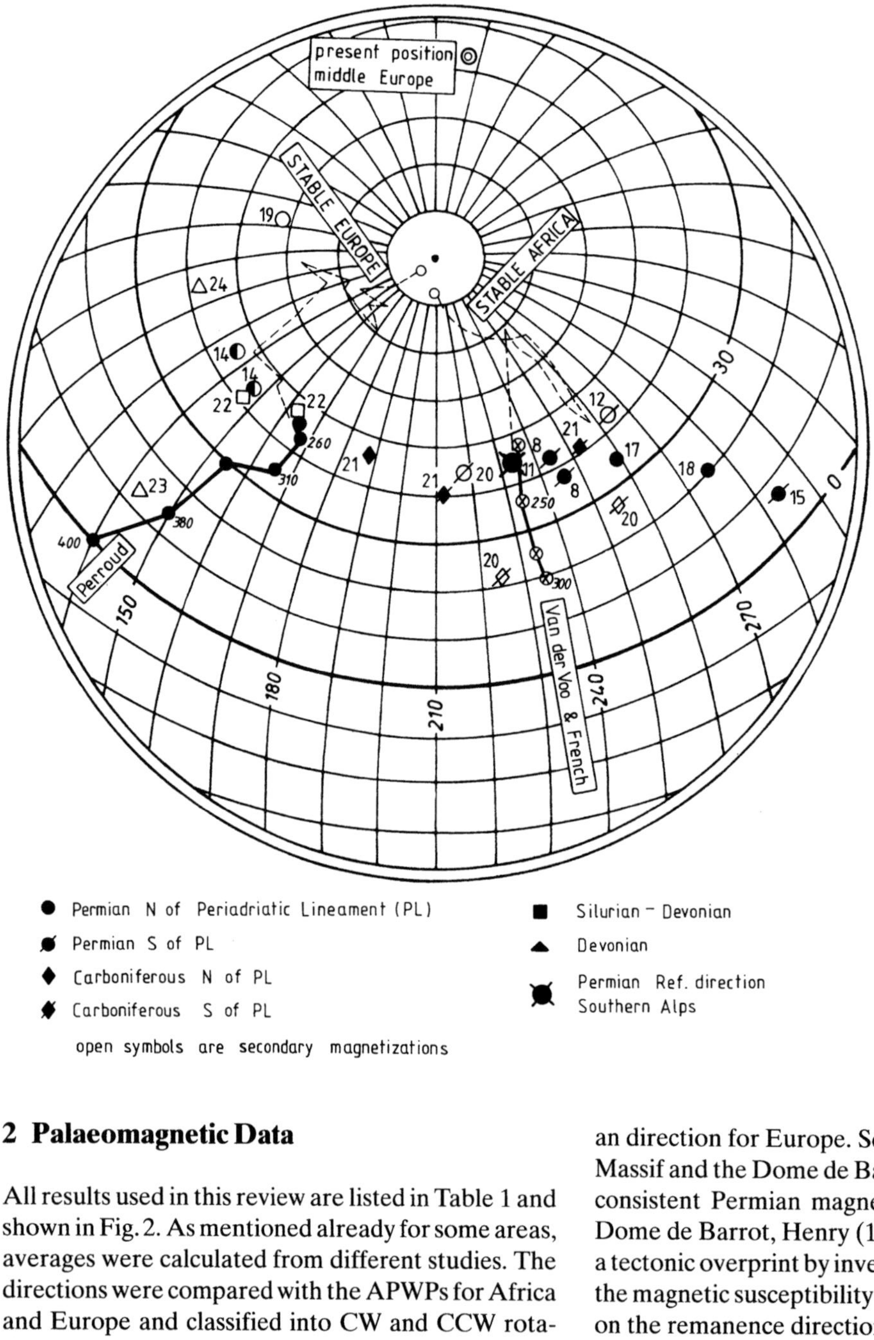

Fig. 1. Apparent polar wander path (APWP) for "stable Europe" and Africa and palaeomagnetic pole positions of the Palaeozoic basement of the Alps. For Europe reconstructions after Westphal et al. (1986), Krs (1982) and Perroud (1986), for Africa after Westphal et al. (1986), Van der Voo and French (1974). The *dashed* part of APWP belongs to the Mesozoic and Cenozoic. *Numbers* along the APWPs are absolute ages in Ma. Different poles with the same number indicate two or more distinct groups of directions which were not averaged for geological reasons

2 Palaeomagnetic Data

All results used in this review are listed in Table 1 and shown in Fig. 2. As mentioned already for some areas, averages were calculated from different studies. The directions were compared with the APWPs for Africa and Europe and classified into CW and CCW rotations. Special attention was paid to the age of the magnetization, classified as primary (p) or secondary (s) (Fig. 2). The magnetic age is given to the published accuracy, when available.

Starting in the southwest, in the Maritime Alps, there is a uniform magnetization direction for the Esterel (1), Argentera Massif (2) and the Dome de Barrot (3). In the Esterel Massif, the good agreement of results from sediments and volcanites (basalt, rhyolite) is a strong argument for a primary origin of the magnetization. The direction fits the expected Permian direction for Europe. Sediments in the Argentera Massif and the Dome de Barrot are the source for the consistent Permian magnetization direction. In the Dome de Barrot, Henry (1973) found indications for a tectonic overprint by investigating the anisotropy of the magnetic susceptibility and expected an influence on the remanence directions. The perfect agreement with the other Permian results of the Maritime Alps proves that these tectonic overprints are of minor significance.

Further to the north of the external zone of the Western Alps (Dauphinois), expected Permian directions were found in the western edge of the Belledonne Massif (6) (Petit Collet, Col du Barioz). From the internal zone (Brianconais and Piemontais), which is much more disturbed and displaced, Permian directions are published for the Vallon de Ponsonniere (5) and the Vallée de Guil (4). In Vallon de Pon-

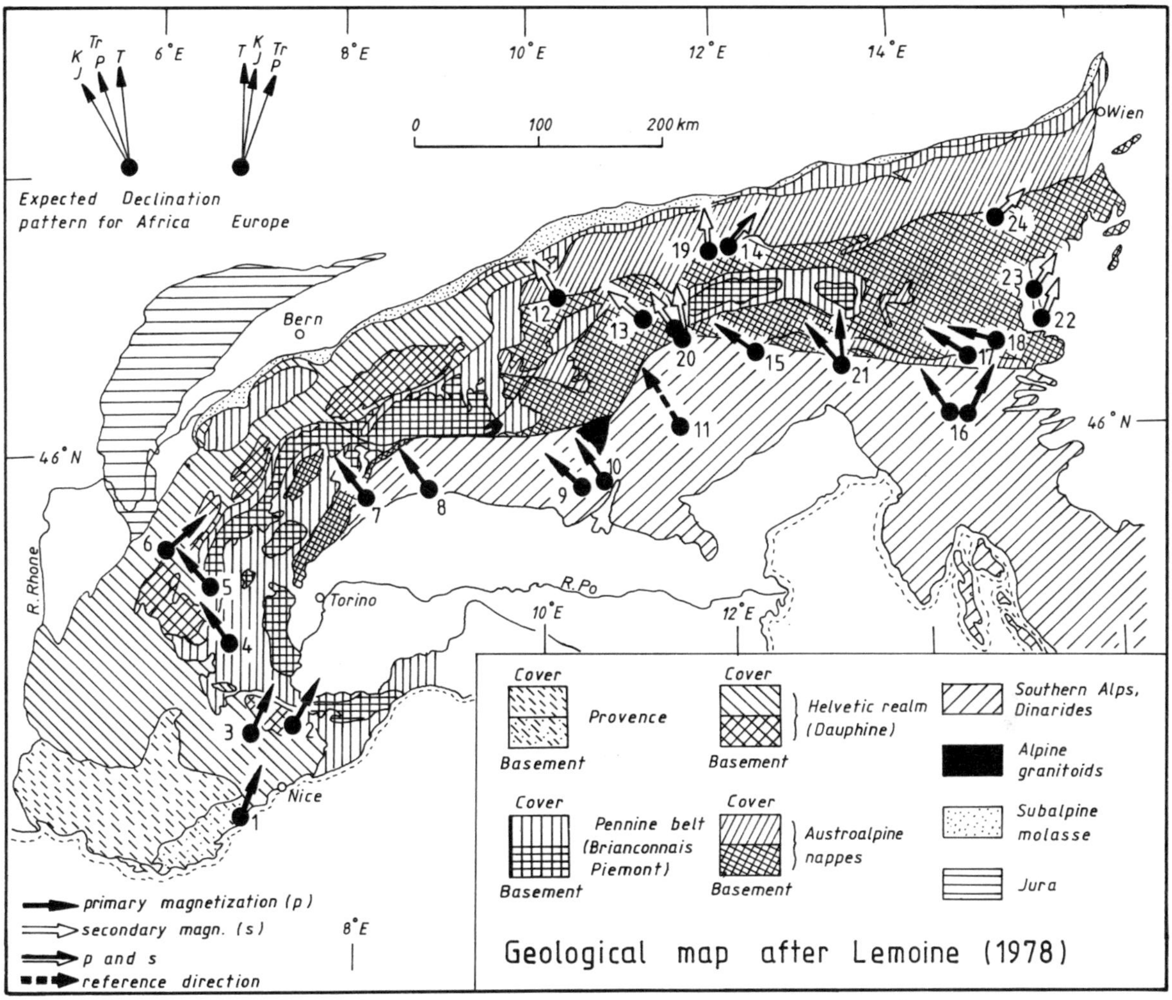

Fig. 2. Palaeodeclinations from the Palaeozoic basement of the Alps. *Site numbers* refer to the text and represent in case of site numbers 8, 9, 11, 15 and 20, 21 mean direction from a larger area and from different studies. In the text, Fig. 2–8 means site 8 in Fig. 2

sonniere, two basalt flows were investigated. The upper flow shows a meaningless steep direction, while the lower flow has an African Permian direction. The results of the Vallée de Guil, where sediments are the source of the remanence, are in good agreement with the former ones.

In the Ivrea-Verbano zone (7) metabasites were investigated which have suffered amphibolite facies metamorphism. The age of the metamorphism (150–250 Ma), i.e. the age of the magnetization, requires the Ivrea zone to have already been in its steep position when undergoing this metamorphism. In the southernmost part the Vallée Sesera, a stable Permo-Carboniferous direction was found.

For the Southern Alps (8, 9, 10, 11, 15, 16, 21), where volcanites and sediments were investigated, very consistent directions were found. The good agreement of both rock types, the proof of a stable and dominating depositional remanence in the sediments, and positive conglomerate tests clearly indicate a primary Permian direction. The average direction of the large complex of the Bozen quarzporphyry is taken as the most reliable Permian reference direction for the Southern Alps. In NW Slovenia (16) Soffel et al. (1983) found two rotation directions which seem to be related to the western and the eastern part of the investigated area. Since there is no evidence for a tectonic zonation in the required sense, the explanation remains incomplete.

North and south of the Periadriatic Lineament (PL) (21) metamorphosed rocks as well as the Periadriatic intrusives were investigated (Förster et al. 1975). Diabases from the Ötztal Massif (13), with absolute ages in the range of 488–94 Ma, show a surprisingly uniform direction after magnetic cleaning. From a palaeomagnetic point of view, and according

Table 1.

Local. No.	Sampling area	Age and lithology		Palaeodirections Decl.°	Incl.°	κ	α_{95}	Reference
1	Esterel Massif 43,4 6,8	Permian	Volcanites and sediments	206.5	-6.0	–	–	Zijderveld (1975)
2	Argentera Massif 44,05 7,12	Permian	Sediments	203	-21	31	5	Bogdanoff and Schott (1977)
3	Dome de Barrot 44,0 7,0	Permian	Sediments	208	-10	–	–	Henry (1973)
4	Vallee du Guil Brianconais 44,6 6,6	Permian	Rhyolites	156	-19	30	4	Roche and Westphal (1969) Westphal (1973)
5	Vallon de Ponsonniere Brianconais 45,0 6,5	Permian	Rhyolites, ignimbrites	138	-28	18	30	Westphal (1973)
6	Belledonne Massif 45,4 6,2	Permian	Ignimbrites	209	-15	66	8	Westphal (1973)
7	Ivrea-Verbano zone (Val Sessera) 45,7 8,2	Permo-carboniferous	Metabasite	129	-1	46	14	Heller and Schmid (1974)
8	Auccia, Ganna 45,7 8,6	Permian	Porphyrites	141.2	-17.8	95	8.6	Heiniger (1979) Becke and Mauritsch (unpubl.)
9	Bergamask Alps 45,8 10,4	Permian	Porphyrites, sand-siltstones	139.7	-14.1	88	8.2	Heiniger (1979) Becke and Mauritsch (unpubl.)
10	Eastern Lombardic Alps 45,8 10,5	Permian	Porphyrites, sand-siltstones	143.0	-13.5	100	7.2	Zijderveld and De Jong (1969) Becke and Mauritsch (unpubl.)
11	Dolomites 46,3 11,3	Permian	Porphyry, sand-siltstones	155.2	-17	25	4	Manzoni (1970) Becke and Mauritsch (1984)
12	Northern Calcareous Alps 47,1 9,9	U.-Permian	Diabase	130.4	-36.7	8	9.5	Soffel (1975)
13	Central Alps 46,8 11,1	Pre-Permian	Diabase	312.0	8.3	20	13.9	Förster et al. (1975)
14	Kaisergebirge 47,5 12,3	Permian-Triassic	Sand-siltstones	35.8	30.6	21	6.2	Soffel et al. (pers. comm.)
15	Carnian Alps 46,5 12,9	Permian	Sand-siltstones	295	10	–	–	Mauritsch and Becke (1983)
16	NW-Slovenia 46,0 14,4	M. Permian M. Triassic	Sand-siltstones	136.3	-8.2	25	14.0	Soffel et al. (1983)
	NE-Slovenia			225.9	-15.0	7	18.8	
17	Carinthia 46,7 14,4	Permian	Sand-siltstones	131.3	-19.2	19	15.6	Mauritsch et al. (1990)
18	Carinthia (St. Paul) 46,7 14,8	Permian	Sand-siltstones	109.7	-17.3	34	7.3	Mauritsch et al. (1990)
19	Brixental (Wörgl) 47,4 12,1	Permian	Sandstones	44.3	59.2	103	6.9	Soffel (1979)
20	N and S of Periadriatic Lineament	Permo-Triassic	Amphibolites	346	40	3	11.5	Förster et al. (1975)
	46,7 11,6	Permian	Diorites	150	30	8	20	
21	Naßfeld-Nötsch 46,5 13,4	Carboniferous	–	3.4	9.2	60	3.0	Heinz and Mauritsch (1980) Manzoni et al. (1989)
22	Graz-Palaeozoic (Sausal) 46,8 15,4	Silurian Devonian	Diabase, limestones	34.2	17.8	16	23.5	Flügel et al. (1980) Mauritsch (1981)
23	Graz-Palaeozoic (Steinb.) 47,1 15,3	Devonian	Limestones	52.8	-26.1	9	2.8	Flügel et al. (1980) Mauritsch (1981)
24	Präbichl 47,5 14,9	Devonian	Limestones	53.4	42.6	158	20	Mauritsch (1981) Neubauer (1989)

to the original paper, the dominating overprint could have occurred during the Carboniferous, requiring no stronger tilting in Late Hercynian and Alpine times. In the Periadriatic intrusives, sampled between Meran and Brixen (20), a Late Carboniferous to Early Permian remanence was found, which is in good agreement with those of the Southern Alps. The northern marginal zone of the intrusives seems to be effected by the Late Alpine Rensen Tonalite intrusion since the observed ages are at about 32 Ma. The

Schneeberger Zug (20) remanence directions were found in amphibolites, whose age are still under discussion. The age of the magnetization is most probably Upper Cretaceous.

Two investigations were carried out in Carboniferous limestones in the Carnian Alps and N of the Periadriatic Lineament near Nötsch (22). Manzoni et al. (1989) isolated a Carboniferous direction in Stephanian limestones, Heinz and Mauritsch (1980) found similar ones for the Hochwipfelschichten and the Visean limestones from Nötsch. While no rotation difference can be seen between N and S of the Periadriatic Lineament, there is a distinct deviation between the two investigations. The CW deviation of the N compared with the data S of Naßfeld could be due to tectonic deformation in the vicinity of the PL. Permian red sand- and siltstones are the source of a reliable magnetization direction in the Krappfeld and Lavanttal (17, 18) both belonging to the Gurktaler nappe.

From the Northern Calcareous Alps (NCA) some results exist from Permian intrusives and sediments. Near St. Anton (12) Soffel (1975) investigated diabases and established a direction which is thought to be of Permian age. In the area between Wörgl (19) and Scheffau (14), Permian red sandstones show directions which are not rotated in any direction. In both localities, a large number of samples show steep inclinations which give rise to suspicions of a Cretaceous overprint. In the case of unaltered, haematite-bearing sandstones, Permian signals were isolated. The observed directions could, in some sites, be Permian with varying, remaining Cretaceous overprints.

The Palaeozoic of Graz (23, 24) was investigated in several localities and includes limestones, dolomites, mafic tuffs, and diabases. In the southern part, a uniform direction was found for the Sausal and Remschnig. The palaeoinclination, which occurs just in one polarity, points to a later metamorphic overprint. In the northern part (Steinberg) a similar declination but a shallower inclination was found. For this locality absolute ages, established in micas, show a Jurassic metamorphism. Comparing the results of both areas (23, 24) with the APWP for Europe (Fig. 1), a Late Carboniferous to Early Permian overprint should be expected. Another alternative for understanding these directions is to assume an outcome of a Variscan and Alpine metamorphic direction as indicated by strongly varying radiometric results (98–176 Ma).

In folded Devonian limestones of the Präbichl area (25), a magnetization direction was found which is evidently due to an Alpine metamorphism. As in the Palaeozoic of Graz, it could again be the result of a Variscan and Alpine metamorphism since the first is proved by radiometric dating (Neubauer 1989) and the latter indicated by steep inclinations.

3 Conclusions

When looking at the final result in Fig. 2, it should be remembered that most of the arrows, marked in the figure, represent averages. This means that most of the local tectonic disturbances are smoothed out. A serious problem arises from insufficient information about the age of the magnetization in some results. The necessarily post-magnetization age of the rotations which cause deflection of the palaeomagnetic declination implies that only a time sequence of palaeomagnetic declinations may constrain the timing of rotations. Moreover, declinations found for each age interval are composite, i. e. they may reflect the net rotation of an area after the acquisition of remanence in a rock.

All these factors have to be taken into account when looking at the results of Fig. 2. With the exception of the directions in the Southern Alps and some of the Eastern Alps, average directions of one epoch, sometimes of one period, are shown. Therefore, it is more correct to talk about affinities to one or other of the patterns, than to talk about local tectonic features. In the present review we have the CCW "Africa" pattern and the CW "European" pattern as a frame of definition (Fig. 1). The discussed sites from Esterel-Argentera-Dome de Barrot and Belledonne Massif, belonging to the external zone of the Western Alps, show a uniform Permian CW deviation. These directions are in excellent agreement with the Permian direction of Europe (Krs 1982).

The internal zone of the Western Alps (Brianconnais and Piemontais), the Ivrea zone and the whole Southern Alps have a unique CCW rotation direction, according to the African pattern. A very interesting result is the continuity of this CCW rotation across the Periadriatic Lineament, N into the Ötztal Masse, the base of the NCA and the Gurktal nappe. Particularly the latter are a good example for the coincidence of rotations seen in the Palaeozoic basement as well in the Mesozoic overburden. For sites 12, 13, 20 it has to be particularly remembered that these directions refer to metamorphic events, since the geometric conditions of the overprinted unit are often uncontrolled. The same could be valid for the Permian results of Wörgl (19) and to a lesser extent for the ones of the Kaisergebirge (14); because two of the three published sections show mixed polarities, with shallow inclinations, which are thought to be primary.

In the Palaeozoic of Graz (GP) and the Präbichl (24) area a CW direction can be seen which is difficult to understand. It is thought that these directions are the result of secondary processes. Furthermore, the declinations of the GP are of minor significance since the palaeoinclinations are always close to zero.

Comparing the distribution of the primary Palaeozoic directions (Marton et al. 1987; Fig. 2) with the ones for the Mesozoic (Marton 1987; Mauritsch and Becke 1987) it is easily seen that the basement has rotated in full agreement with the Mesozoic overburden. The deformation at which we are looking is therefore believed to be young.

Acknowledgement
The author would like to thank W. Frisch, E. Marton and D. H. Tarling for critical reading of the manuscript and their comments for improving the article.

References

Becke M, Mauritsch HJ (1984) Rock and palaeomagnetic investigations on the quartzporphyry complex near Bolzano (Bozen) Italy. In: Sassi FP, Julivert M (eds) Int Geol Corr Prog 5, Padova. Newsletter 6: 11–19

Bogdanoff S, Schott JJ (1977) Etude paléomagnétique et analyse tectonique dans les schistes rouges permiens du Sud de l'Argentera. Bull Soc Géol Fr 19: 909–916

Briden JC (1972) A stability index of remanent magnetism. J Geophys Res 77: 1718–1720

Bullard EC, Everett JE, Smith AG (1965) The fit of the continents around the Atlantic. Proc R Soc Lond 258 A: 41–51

Cogne JP, Perroud H (1985) Strain removal applied to palaeomagnetic direction in an orogenic belt: the Permian red slates of the Alpes Maritimes, France. Earth Planet Sci Lett 72: 125–140

Fisher RA (1953) Dispersion on a sphere. Proc R Soc Lond A 217: 295–305

Flügel H, Mauritsch HJ, Heinz H, Frank W (1980) Paläomagnetische und radiometrische Daten aus dem Grazer Paläozoikum. Mitt Österr Geol Ges 71/72: 201–211

Förster H, Soffel H, Zinsser H (1975) Palaeomagnetism of rocks from the Eastern Alps from north and south of the Insubrian Line. Neues Jahrb Geol Paläontol Abh 149: 112–127

Fritz H (1988) Kinematics and geochronology of Early Cretaceous thrusting in the northwestern Palaeozoic of Graz (Eastern Alps). Geodin Acta 2(2): 53–62

Heiniger C (1979) Palaeomagnetic and rock magnetic properties of the Permian volcanics in the western Southern Alps. J Geophys 46: 397–411

Heinz H, Mauritsch HJ (1980) Paläomagnetische Untersuchungen an der "Periadriatischen Naht". Mitt Österr Geol Ges 71/72: 269–274

Heller F (1973) Magnetic anisotropy of granitic rocks of the Bergell Massif (Switzerland). Earth Planet Sci Lett 20: 180–188

Heller F, Schmid R (1974) Paläomagnetische Untersuchungen in der Zone Ivrea-Verbano (Prov. Novara, Norditalien): vorläufige Ergebnisse. Schweiz Mineral Petrogr Mitt 54: 229–42

Heller F, Lowrie W, Hirt AM (1989) A review of palaeomagnetic and magnetic anisotropy results from the Alps. Alpine tectonics. Geol Soc Spec Publ 4: 399–420

Henry B (1973) Studies of microtectonics, anisotropy of magnetic susceptibility, and palaeomagnetism of the Permian Dome de Barrot (France). Palaeomagnetic and palaeosedimentological implications. Tectonophysics 17: 61–72

Irving E (1964) Palaeomagnetism and its application to geological and geophysical problems. Wiley, New York, 399 pp

Krs M (1982) Implication of statistical evaluation of Phanerozoic palaeomagnetic data (Eurasia, Africa). J Cesk Akad Ved Rozp 92 Sesit 3: 55

Lemoine M (1978) Geological atlas of Alpine Europe and adjoining Alpine areas. Elsevier, Amsterdam, 584 pp

Lowrie W, Channel JET, Alvarez W (1980) A review of magnetic stratigraphy investigations in Cretaceous pelagic carbonate rocks. J Geophys Res 85 B7: 3597–3605

Lowrie W, Hirt AM, Kligfield R (1986) Effects of tectonic deformation on the remanent magnetization of rocks. Tectonics 5: 713–722

Manzoni M (1970) Palaeomagnetic data of Middle and Upper Triassic age from the Dolomites (Eastern Alps, Italy). Tectonophysics 10: 411–424

Manzoni M, Venturini C, Vigliotti L (1989) Palaeomagnetism of Upper Carboniferous limestones from the Carnic Alps. Tectonophysics 165: 73–80

Marton E (1987) Palaeomagnetism and tectonics in the Mediterranean region. J Geodyn 7: 33–57

Marton E, Mauritsch HJ (1990) Structural applications and discussion of a palaeomagnetic post-Palaeozoic data base for the central Mediterranean. Phys Earth Planet Interiors 62: 46–59

Marton E, Mauritsch HJ, Tarling DH (1987) Pre-Alpine palaeomagnetic results of the Alpine-Mediterranean belt: compilation, discussion, interpretation. In: Flügel HW, Sassi FP, Grecula P (eds) Pre-Variscan and Variscan events in the Alpine-Mediterranean mountain belts. Mineralia slovaca, Monography. Alfa, Bratislava, pp 351–360

Mauritsch HJ (1981) Palaeomagnetic results from some palaeozoic areas in the Eastern Alps for correlation purposes within IGCP Project No. 5. In: Karamata S, Sassi FP (eds) Int Geol Corr Prog 5, Padova. Newsletter 3: 117–118

Mauritsch HJ, Becke M (1983) A magnetostratigraphic profile in the Permian (Gröden beds, Val Gardena formation) of the Southern Alps near Paularo (Carnic Alps, Friuli, Italy). In: Sassi FP, Szederkényi T (eds) Int Geol Corr Prog 5, Padova. Newsletter 5: 80–86

Mauritsch HJ, Becke M (1987) Palaeomagnetic Investigations in the Eastern Alps and the Southern Border Zone. In: Flügel HW, Faupl P (eds) Geodynamics of the Eastern Alps. Deuticke, Vienna, pp 282–308

Mauritsch HJ, Reisinger J (1991) Paläomagnetische Untersuchungen in der Böhmischen Masse. Projekt S47/07-GEO. In: Höck V, Steinhauser P (eds) Österr. Beiträge zu Meteorologie und Geophysik. Mitt Zentralanst Wien, pp 25–42

Mauritsch HJ, Reisinger J, Agnoli F (1989) Paläomagnetische Ergebnisse aus dem Perm der Gurktaler Decke. Carinthia II, 179/99: 383–390

McElhinny MW (1973) Palaeomagnetism and plate tectonics. University Press, Cambridge, 357 pp

McFadden PL, Jones DL (1981) The fold test in palaeomagnetism. Geophys J R Astron Soc 67: 53–58

McFadden PL, Lowes FJ (1981) The discrimination of mean directions drawn from Fisher distributions. Geophys J R Astron Soc 67: 19–33

McFadden PL, McElhinny MW (1990) Classification of the reversal test in palaeomagnetism. Geophys J Int 103: 725–729

Neubauer F (1989) Late Variscan structures of the Eastern Greywacke Zone (Eastern Alps). Neues Jahrb Geol Palaeontol Monatsh 7: 425–432

Perroud H (1986) Palaeomagnetic evidence for tectonic rotations in the Variscan mountain belt. Tectonics 5: 205–214

Roche A, Westphal M (1969) Sur la direction d'aimantation des roches volcaniques permiennes de la vallée du Guil. C R Somm Séances Soc Géol Fr 7: 239

Soffel H (1975) Palaeomagnetism of the Permian effusives near St. Anton, Vorarlberg (Austria) and the anticlockwise rotation of the Northern Calcareous Alps through 60°. Neues Jahrb Geol Palaeontol Monatsh 6: 375–384

Soffel H (1979) Palaeomagnetism of Permo-Triassic red sandstones from the Northern Calcareous Alps. J Geophys 45: 447–450

Soffel H, Pohl W, Buser St (1983) Palaeomagnetism of Permo-Triassic rocks from northern Slovenia, Yugoslavia, and the eastern margin of the Adriatic Plate. Tectonophysics 91: 301–320

Symons DTA, Stupavsky M (1974) A rational palaeomagnetic stability index. J Geophys Res 79: 1718–1720

Tarling DH, Symons DTA (1967) A stability index of remanence in palaeomagnetism. Geophys J 12: 443–448

Van den Berg J, Zijderveld H (1982) Palaeomagnetism in the Mediterranean area. Geodyn Ser 7: 83–112

Van den Ende C (1977) Palaeomagnetism of Permian red beds of the Dome de Barrot (S France). PhD Thesis, State University of Utrecht, 171 pp

Van Hilten D, Zijderveld JDA (1966) The magnetism of the Permian porphyrites near Lugano (northern Italy/Switzerland). Tectonophysics 3: 429–466

Van der Voo R, French RB (1974) Apparent polar wandering for the Atlantic-bordering continents: Late Carboniferous to Eocene. Earth Sci Rev 10: 99–119

Westphal M (1973) Etudes palaeomagnetiques de quelques formations Permiennes et Triassiques dans les alpes occidentales (France). Tectonophysics 17: 323–335

Westphal M, Bazhenov M, Lauer J, Pechersky D, Sibuet J (1986) Palaeomagnetic implications on the evolution of the Tethys belt from the Atlantic Ocean to the Pamirs since the Triassic. Tectonophysics 123: 37–82

Wilson RL, Haggerty SE, Watkins ND (1968) Variation of palaeomagnetic stability and other parameters in a vertical traverse of a single Icelandic lava. Geophys J 16: 79–96

Zijderveld JDA (1975) Palaeomagnetism of the Esterel. PhD Thesis, State University of Utrecht

Zijderveld JDA, De Jong KA (1969) Palaeomagnetism of some Late Palaeozoic and Triassic rocks from the Eastern Lombardic Alps, Italy. Geol Mijnbouw 48: 559–564

Zijderveld JDA, Hazeu GJA, Nardin M, Van Der Voo R (1970) Shear in the Tethys and the Permian palaeomagnetism in the southern Alps, including new results. Tectonophysics 10: 639–661

Late Variscan Magmatic Evolution of the Alpine Basement

B. Bonin (Coordinator), P. Brändlein, F. Bussy, J. Desmons, U. Eggenberger, F. Finger, K. Graf, Ch. Marro, I. Mercolli, R. Oberhänsli, A. Ploquin, A. von Quadt, J. von Raumer, U. Schaltegger, H. P. Steyrer, D. Visonà and G. Vivier

Abstract

After having experienced the Variscan orogenic episodes, the pre-Mesozoic Alpine basement was subjected to large-scale shearing effects accompanying lithosphere distensional thinning, "Basin and Range"-like tectonics and high geothermal regimes. As a result of intrusion of mantle-derived magmas and induced crustal anatexis, almost all pieces of basement within the Alpine chain display contrasting magma associations.

After the Late Devonian major collision stage, peraluminous associations of crustal melts were associated and/or mixed with high-K basic magmas. Low-pressure regional metamorphic regimes ultimately developed "wet" and/or H_2O-deficient anatexis, which may or may not be induced by the intrusion of mantle-derived melts.

Lower to Middle Carboniferous high-K calc-alkaline suites indicate a post-collisional stage characterized by uplift and erosion in a short-lived transpressional and/or transtensional regime. They are followed by Late Carboniferous near-alkaline suites, yielding both alkaline and calc-alkaline characteristics and emplaced as volcanic-plutonic massifs in a major distensional regime. The Alpine basement was, therefore, emplaced increasingly in an intracontinental position between the Gondwana and Laurasia continental blocks.

Though subduction-related calc-alkaline suites were scarce, Late Carboniferous to Early Permian granitoids, emplaced in some specific areas, closely resemble the circum-Pacific Cordilleran batholiths. They may indicate a renewed ocean-continent plate margin convergence at the southern flank of the Variscan belt.

Mid-Permian post-orogenic and Late Permian to Triassic early anorogenic alkaline complexes constitute a part of the Western Mediterranean alkaline province. Associated thermal imprints were recorded in the basement mineral isotopic clocks. The anorogenic magmatic activity can be related both to continental consolidation of the European plate and to precursory stages of the formation of the Meso-Tethys oceanic basin.

1 Introduction

Mobile belts are crustal segments that have experienced strong rigid and/or plastic deformations. Together with the occurrence of regionally metamorphosed rocks, the most outstanding feature of mobile belts is the abundance of numerous magmatic rocks with differing compositional, textural and geological characteristics.

The Variscan orogenic belt of Europe evolved in connection with the Palaeozoic convergence of Gondwana in the south and Laurasia in the north and was consolidated in the Late Palaeozoic. In the Alpine segment of the Variscan belt, primary features were obscured by the superimposition at the same places of orogenic episodes of differing ages, namely Cadomian (Upper Proterozoic), Caledonian-Variscan (Palaeozoic *s. l.*) and Alpine (Cretaceous-Tertiary).

The aim of this chapter is to focus on the late Variscan magmatic evolution. '*Variscan*' is given a wide sense to encompass all events between the end of the Caledonian orogenic episode and the beginning of the Alpine orogeny, i. e. from Early Devonian to Permian-Triassic. The working hypothesis is based on the assumption that distinct tectonic regimes are likely to produce magmatic suites differing either by their sources and modes of differentiation, or by their styles of emplacement, or both (for reviews, see Pitcher 1983, 1987; Barbarin 1990). The objective is, from the consideration of petrological evidences, an attempt to unravel the geodynamic environments of the successive magmatic episodes. As far as the Variscan orogenesis is concerned, the major unsolved questions centre around the plate framework of the Variscan realm and, specifically, where oceans and subduction zones could have been located in the Late

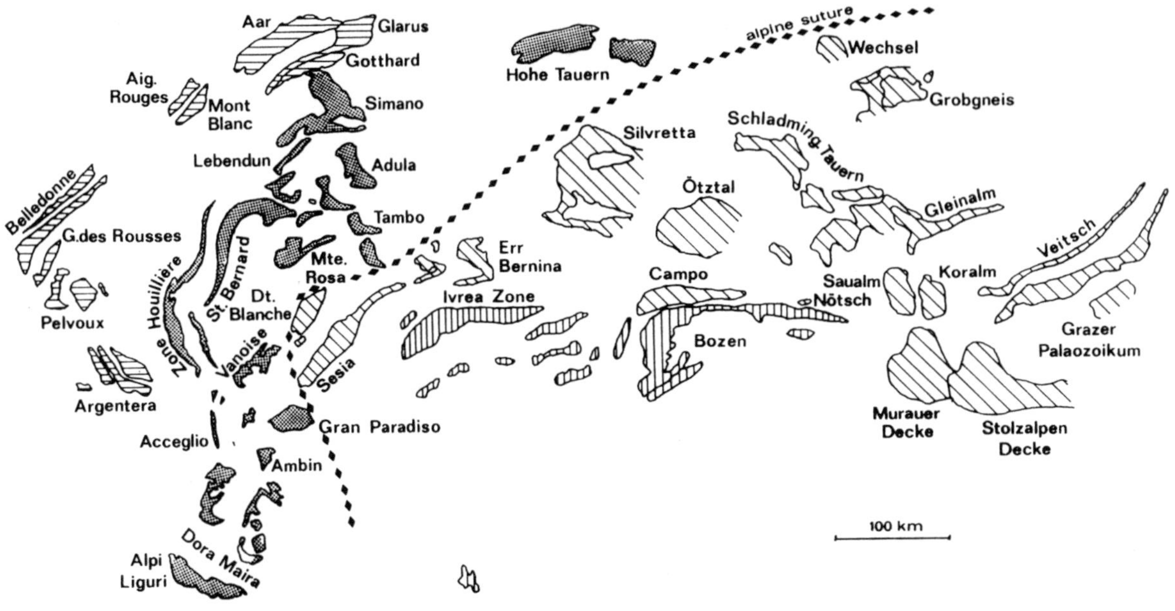

Fig. 1. Sketch map of the pre-Alpine basement units with location of the areas described. The pre-Mesozoic reconstruction has been taken from von Raumer and Neubauer (this Vol.), and is based on the palinspastic reconstructions of de Graciansky, Frisch and Neubauer and Pfiffner (all this Vol.) Hached areas: Horizontal – Helvetic Realm; crosshatched – Penninic Realm; Diagonal – Austro-Alpine Realm; Vertical – Southern Alps.

Palaeozoic (for reviews and contrasting hypotheses, see Matte 1986; Ziegler 1986; Neugebauer 1988).

The major difficulty raised in this study is that present Alpine structural units have nothing or little to do with previous Variscan structural units. Thus, Late Variscan magmatic bodies belonging to the same geodynamic suite can be encountered in several distinct Alpine units, while, in the same Alpine unit, magmatic formations can be related to distinct geodynamic environments.

The aim of this chapter is to provide an overview of the pattern and evolution of late Variscan granitic massifs and associated volcanites in the pre-Mesozoic basement of the Alps. First, various regional contributions were sent to the first author (B. B.), acting as the coordinator. Initials correspond to the co-authors of this chapter:

- French Crystalline Massifs (G. V.);
- Aiguilles Rouges and Mont Blanc External Massifs (F. B., P. B., C. M., J. V. R.);
- Aar Massif, Lower and Eastern Alps of Switzerland, and Lugano area of the Southern Alps (U. E., K. G., I. M., R. O., U. S.);
- Western Alps (J. D., A. P.);
- Tauern Window (F. F., H. P. S., A. V. Q.);
- Southern Alps (D. V.).

Then, after a first version had been reviewed and, as asked by both reviewers and editors, this new version has been entirely rewritten and reorganized, taking into account the original contributions in order to provide a homogeneous view.

In the first part, brief overviews of the different magmatic provinces will be presented from the more external Alpine units to the more internal ones, i. e. from 'external crystalline massifs' through Penninic and Austro-Alpine domains to the Southern Alps, and will be arranged in a clockwise manner, i. e. from southwest to northeast (Fig. 1).

In the second part, a comparative summary and speculations on a possible geodynamic framework are be provided for the interpretation of the late Variscan magmatic activity.

2 Distribution of the Magmatic Formations

Areas subjected to Variscan folding, metamorphism and magmatism occur within the entire Alpine chain, as well as outside the Alpine front. Most of the intra-Alpine Variscan outcrops constitute the basement of Alpine thrust sheets, where they underwent Cretaceous-Tertiary deformation and metamorphism at various scales and levels. Fragments of Permian-Carboniferous magmatic massifs are exposed in both external and internal domains, together with relics of a pre-existing continental crust of poorly known age. Al-

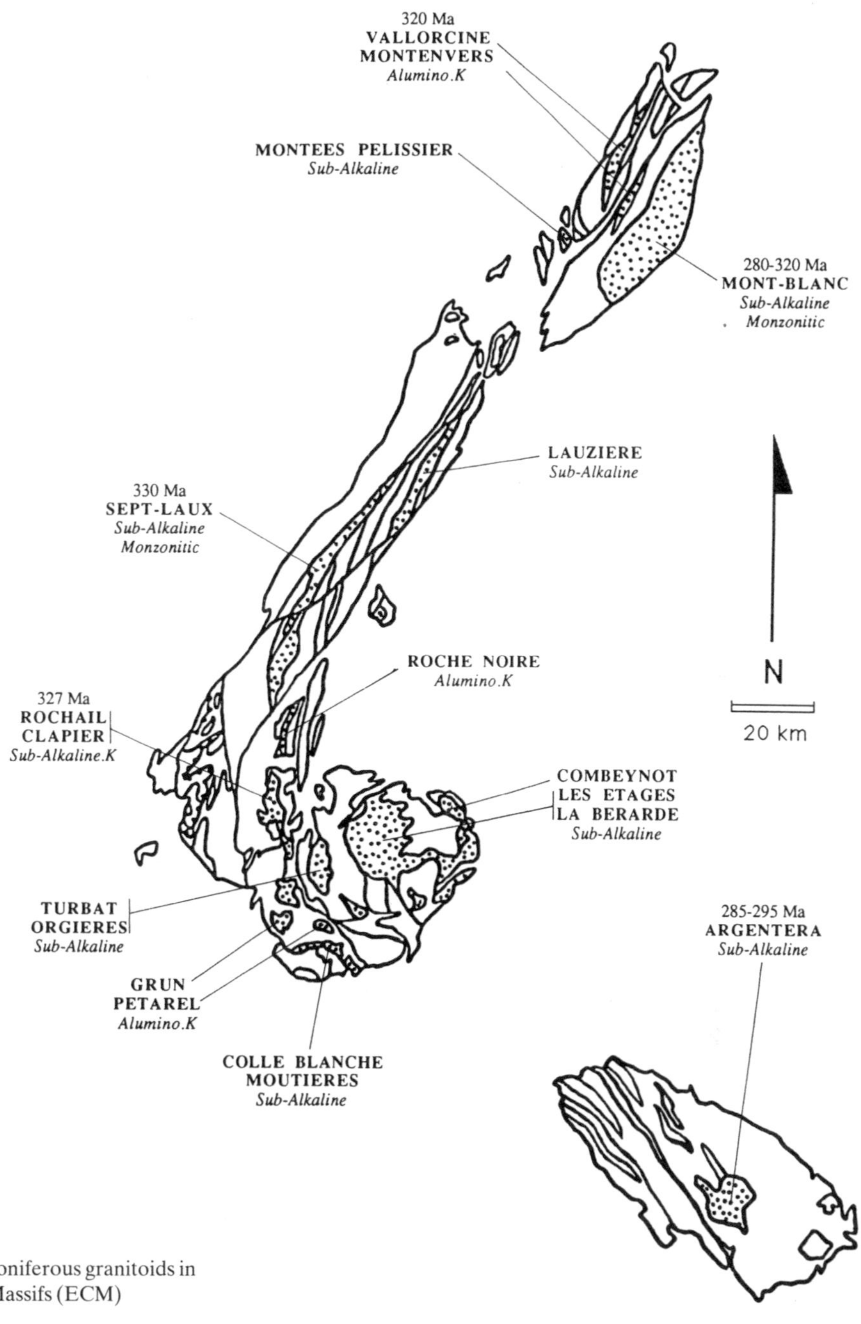

Fig. 2. Location of the main Carboniferous granitoids in the French External Crystalline Massifs (ECM)

though the present location of the Palaeozoic massifs is largely disturbed by Alpine tectonics, they seem to have formed parts of the southern flank of the Variscan fold belt (e. g. see reviews in Dallmeyer 1989).

2.1 *The External Crystalline Massifs*

The "external crystalline massifs" constitute the uplifted basement of thick Mesozoic sedimentary covers. Mostly Variscan in age, they are made up of Precambrian to Lower Palaeozoic metamorphosed formations and of late Variscan magmatic intrusions. In this section, we briefly examine successively from south to north the magmatic bodies in France, then from west to east in Switzerland (Fig. 2).

In France, the "external crystalline massifs" are all located in the "Dauphinois" realm of the western part of the Alpine arc and are exposed in its southeastern part (Argentera) and in the Dauphiné and Savoie historical provinces in the northwest. These massifs constitute discontinuous outcrops uncovered by erosion of their Mesozoic-Tertiary sedimentary cover as all as by uplift along recent normal faults.

After the Lower Palaeozoic distensional regimes that lead to the development of bimodal 'leptyno-amphibolite' associations and basic metamagmatites, the major Devonian-Carboniferous collision event was accompanied by the emplacement of calc-alkaline magmatic belts. Early synkinematic intrusives were converted into orthogneisses. Later Carboniferous granites intruded, at relatively shallow levels, a basement composed of paragneisses, migmatites and orthogneisses. These calc-alkaline granites yield generally weak magmatic fabrics and large compositional variations (granodiorite to monzogranite).

2.1.1 Argentera

Two late intrusive granite bodies occur within the Argentera area: the central granite and the Valmasque granite massifs intrude formations having undergone three successive migmatitic events (Faure-Muret 1955; Bogdanoff 1980). A sillimanite-andalusite contact metamorphism was superimposed on earlier regional metamorphic parageneses (Compagnoni et al. 1976) and was ultimately followed by a retromorphic albite-biotite-chlorite-sericite assemblage related to Alpine cataclasis.

The Valmasque granite constitutes an anatectic body, enriched in country rocks xenoliths (Faure-Muret 1955).

The central granite intrusive facies, a medium-grained garnet-bearing two-mica leucocratic granite, porphyritic in the core, is affected by zones of cataclasis and evolves towards a finer grain size in the borders and at the top. It contains as xenoliths Malinvern amphibolites and migmatites and develops a dyke swarm of microgranite and aplite. The monzogranite yields subalkaline affinities (Bogdanoff 1980) and has been dated at 285–293 Ma (Ferrara and Malaroda 1969). Intruding the same formations, the small Becco Rosso Massif, located in the northern part of Argentera, constitutes another piece of the same magmatic episode.

2.1.2 Haut-Dauphiné

Carboniferous granitoids constitute several plutons of variable sizes and occupy approximately half of the masif. They intrude gneisses and migmatites but associated contact metamorphism is weak. Three types of associations have been recognized.

1. In the western border, granitoids are characterized by the association of amphibole + biotite-rich mafic rocks and of leucocratic biotite-bearing acid rocks (Le Fort 1973; Aumaître et al. 1985). The same type of association is encountered in the Roche Noire granite of the Grandes Rousses Massif. The magmatic suite displays strong similarities with subalkaline potassic suites with shoshonitic affinities in other Variscan domains (Vittoz et al. 1987). Scarce volcanic equivalents can be found in the Grandes Rousses Massif and consist of K-rich dacitic to rhyolitic pyroclastic formations. A Stephanian age is inferred from fossil plants in sedimentary interlayers.
2. In the core of the massif, east of a major suture zone marked by Liassic sedimentary formations, the leucocratic, sometimes porphyritic and enclave-poor granites of Turbat-Lauranoure and La Bérarde-Promontoire constitute two major 20 km-long belts and are made up of composite intrusions. Because the overall geometry of the massif has been largely disturbed since their emplacement, caution is urged in interpreting the repartition of the magmatic suites in terms of zonation. However, from west to east, geochemical data indicate that magma sources are likely to be deeper and deeper.

The sources of the Carboniferous granitic magmatism are located in the lower crust, with a possible mantle contribution. According to Vittoz et al. (1987), uplift of the basement caused its melting under relatively dry conditions. The western granitoids could have been produced by the melting of greywacke under amphibolite conditions, while the central granites could originate from the same materials under granulite conditions (Vittoz et al. 1987). In the western belt of Haut Dauphiné, the large heat flux required for H_2O-deficient crustal anatexis was supplied by the emplacement of mantle-derived basic formations with lamproitic affinities.

3. In the eastern most part, within the 'Ultra-Dauphinois' structural unit, the Combeynot biotite subvolvus granite massif corresponds to a subvolcanic ring-complex, consisting of two granite units arranged concentrically, associated with volcanic and volcano-clastic formations and a dense dyke swarm. Emplaced at shallow levels, the Combeynot granite has a probable mantle origin, displays within-plate alkaline affinities and indicates a distensional geodynamic context.

2.1.3 Belledonne

Devonian-Carboniferous plutonic activity came after Barrovian metamorphic episodes. Early intrusives, converted into Al-K-rich orthogneisses, are related to crustal anatexis and were emplaced in a thickened crust. They characterize the end of a P-T-t path from a HP regime through increasing temperature and decreasing pressure (von Raumer 1987). Later Carboniferous granitoids were emplaced within a tectonically active domain and their margins have undergone synmetamorphic mylonitic effects at around 323 Ma (amphibole K-Ar data, Demeulemeester 1982). Two major granite provinces have been recognized.

1. In the northeastern part of Belledonne belt, the Sept-Laux granite constitutes a 100-km-long continuous band, displaced along faults of "synclinal médian" in the northeast and of Belle-Étoile in the southwest. The massif was diapirically emplaced during a phase of oblique collision (Vivier et al. 1987). Intrusive contacts within country gneisses are associated with dense peri-batholithic dyke swarms. Early facies yield syntectonic features (Carme 1970; Gasquet 1979).

The leucocratic medium-grained, sometimes porphyritic, monzogranite belongs to a slightly Al-enriched subalkaline suite. Fractionated REE patterns (La/Yb > 35) and lack of Eu anomaly are similar to porphyritic orthogneisses. Thus, as its composition resembles the older orthogneisses which it contains as xenoliths and because its early facies is also syntectonic, the Sept-Laux granite has been considered as an evolved product originating from continuing and/or renewed lower crustal anatexis. The radiometric age of 322 ± 43 Ma (Demeulemeester 1982) is, by far, too imprecise to constitute a good chronological marker.

2. The Lauzière magmatic suite which constitues a 30-km-long and 2-km-wide belt axis was emplaced during a late dextral transtensional episode. It is made up of early NS-trending monzosyenites and late 20°–30° N-trending leucocratic monzogranites. Characterized by compositional gaps, the magmatic suite yields potassic subalkaline affinities ($Na_2O + K_2O > 8$ wt%, $K_2O > 5$ wt%, SiO_2 depletion, Ca and Mg-enrichment relative to Variscan calc-alkaline suites).

It has been interpreted as resulting from varying stages of contamination of a mostly mantle-derived source for basic magmas and from mixing of basic subalkaline and granitic melts (Negga 1984; Aumaître et al. 1985). Transtensional tectonic regimes (Poncerry 1981) favoured the ascent of a magma composed of a mixture of crustal and mantle-derived melts (acid-basic associations of Aumaître et al. 1985).

Located in the south-western part of the massif and only north of the Fond-de-France fault zone, distinct leucogranites clearly do not belong to the Lauzière magmatic suite and may have a different source (Negga 1984).

2.1.4 Aiguilles Rouges

This external crystalline massif is located in the Helvetic realm of the western Alpine arc across the borders of France and Switzerland. Most of the data presented here are derived from different papers published during the last 20 years (von Raumer 1984a, b, 1987, 1988) and from a recent Ph. D. thesis (Brändlein 1991).

The Aiguilles-Rouges Massif is an Alpine tectonic dome consisting of pre-Triassic formations appearing as tectonic windows under the Helvetic nappe pile. This area was affected by Alpine deformation of variable intensity, up to mylonitization, as well as by Alpine metamorphism which increased southeasterly from anchizone in the Aiguilles Rouges to greenschist facies in the nearby Mont Blanc area. Despite this overprint, pre-Triassic formations are well preserved and reflect a complex history (von Raumer et al. this Vol.).

Located in the northern part of the massif, the Vallorcine granite is an about 15-km-long and 1-km-thick sheet-like intrusion, which extends from Vallorcine (France) in the SW to Miéville (Valais, Switzerland) in the NE. Pebbles of the granite can be found in nearby Upper Carboniferous sedimentary formations.

The original intrusive contact is still preserved along its NW margin and its northern end, where it splits into numerous microgranite dykelets. Along the tectonic SE contact, the isotropic core facies is converted progressively into mylonite along a major ductile shear zone, the ultimate stage being the well-known but yet undated (Permian or Alpine?) "Miéville ultramylonite", a dark, extremely fine-grained, slate-like rock (Kerrich et al. 1980). Other parts escape such a late deformation and display a vertical, SW–NE trending magmatic flow texture, recording the regional stress field during the emplacement of the magma. Shape, orientation and location of the Vallorcine intrusion seem therefore to have been tightly controlled by a major dextral strike-slip shear zone system, which was active or rather long periods as indicated by deformation of pre-, syn- and post-intrusion rocks.

The main granite facies consists of a large, dominantly porphyritic, coarsegrained, syeno- to alkali-feldspar-granite with several centimetre long K-feldspar megacrysts and locally abundant enclaves (i. e. mafic enclaves, restitic micaceous lenses and gneissic xenoliths). The common mineralogy includes biotite, pink magmatic andalusite, colourless topaz, brown tourmaline, as well as apatite, zircon, titanite, rutile, allanite, sillimanite, epidote, pyrite, hæmatite. Topaz and tourmaline crystallized during late pneumatolitic to hydrothermal stages within fracture and shear zones. Hydrothermal activity is also responsible for intense feldspar muscovitization, biotite chloritization, grain boundary recrystallization, etc.

Subsidiary facies include: (a) an early, medium-grained, pinite-cordieritebearing leucogranite, in the lowest oberserved level of the intrusion, (b) porphyritic or even-grained fine-grained microgranite, restricted to border zones, (c) aplitic to pegmatitic, tourmaline-bearing dykes, and (d) one observed lamprophyric dyke, (c) and (d) intruding wall-rocks SW of the main body. (a) and (b) have an alkali-feldspar granite modal composition (Fig. 8).

Zircon typology (Pupin 1980) of all granitic facies is typical of peraluminous crustal granites (Fig. 9). Some samples of the main facies yield a bimodal distribution with peak concentrations around IA = 200/IT = 200 and IA = 350/IT = 450. Numerous relic cores are further evidences of a crustal anatectin origin.

Chemically, the Vallorcine granite is a peraluminous K-rich granite with A/CNK > 1.1, indicating an S-type composition with normative Cor reaching values up to 5% and high K_2O/Na_2O ratios ranging from 1.25 to 1.65 (Fig. 10). The upper part of the body is distinctly more evolved than the lower part, which is interpreted as the result of in situ differentiation through plagioclase ± biotite fractionation. REE patterns (Fig. 16) display a regular fractionation (La/Yb_N = 9–22) with moderate Eu anomalies (Eu/Eu^* = 0.43–0.56) and ΣREE ranging between 75 and 200 ppm. The early facies is REE-depleted (ΣREE = 21 ppm) and displays a flatter pattern.

Most of Vallorcine compositions plot in the field of syn-collisional and/or late-orogenic granites (Pearce et al. 1984; Batchelor and Bowden 1985) (Figs. 12 and 14). Most characteristics suggest that the Vallorcine granite originated in the melting of metapelites, but some features (i.e. mafic enclaves, rapakivi K-feldspar, bimodal zircon populations, geochemical scatter) indicate involvement of more mafic materials in the source rocks, either amphibolites and metagreywackes, or mantle-derived melts.

A deep crustal origin (15–20 km, 850° C) is assumed for the granitic melt and anatexis may have been induced by a combination of thermal recovery of deepseated rocks and pressure release in a transtensional uplifting context along major dextral strike-slip zones. Rapid and shallow emplacement of the magma prevented major wall-rock assimilation.

Grey, porphyritic dacite to rhyodacite are spatially closely related to the Vallorcine granite and sometimes incorporate it as enclaves. Other rhyolites, distinctively red in colour, can be observed in the SE part of the Aiguiles Rouges Massif and, independent from the granitic body, are presumably younger.

2.1.5 Mont Blanc

This external massif is located across the borders between France, Italy and Switzerland. Most of the data presented here are derived from recent Ph. D. theses (Marro 1987, 1988; Bussy 1990).

The Mont Blanc Massif is an elongated 55×14 km SSW–NNE trending parallelogram, which extends from the Mont Blanc summit (4807 m) to Martigny (Valais, Switzerland), where it ends as a very thin strip. It has been dated at 316 ± 19 Ma (Bussy et al. 1989) by whole-rock Rb-Sr analysis of mafic enclaves. U-Pb dating on zircon has yielded a more precise age of 304 ± 3Ma (Bussy and Cliff, in prep.)

Intrusive contacts are preserved along the SW margin, while other roughly parallel contacts are tectonic with either basement gneisses (NW), or Mesozoic sedimentary formations (SE). Along the eastern margin, the granite displays intrusive contacts into gneisses and is partly limited by a presumably Permian erosion surface. As a result of successive deformation episodes, granitic rocks frequently display gneissic structures.

The chronological sequence of NS to N30° E-trending deformations, recorded by the whole massif, includes:

1. A pre-intrusion mylonitic deformation in the basement (Bellière 1988),
2. A magmatic flow texture within the granite,
3. Synplutonic lamprophyre dyke emplacement,
4. Leucocratic microgranite and aplite dyke emplacement,
5. A second mylonitic deformation, which have affected large volumes of the pluton (Bellière 1988).

A third mylonitic deformation with a distinct N45° E trend may be related to either a Late Variscan, or the Alpine orogeny.

The massif includes: (a) a major, coarse-grained, leucocratic, porphyritic central facies, rich in several cm-long K-feldspar, (b) a coarse even-grained border facies, (c) several m-thick, leucocratic, locally porphyritic, microgranitic dykes, and (d) aplitic dykelets, associated with (c). Types (a) to (c) have the same monzo-to syeno-granite composition (Fig. 8).

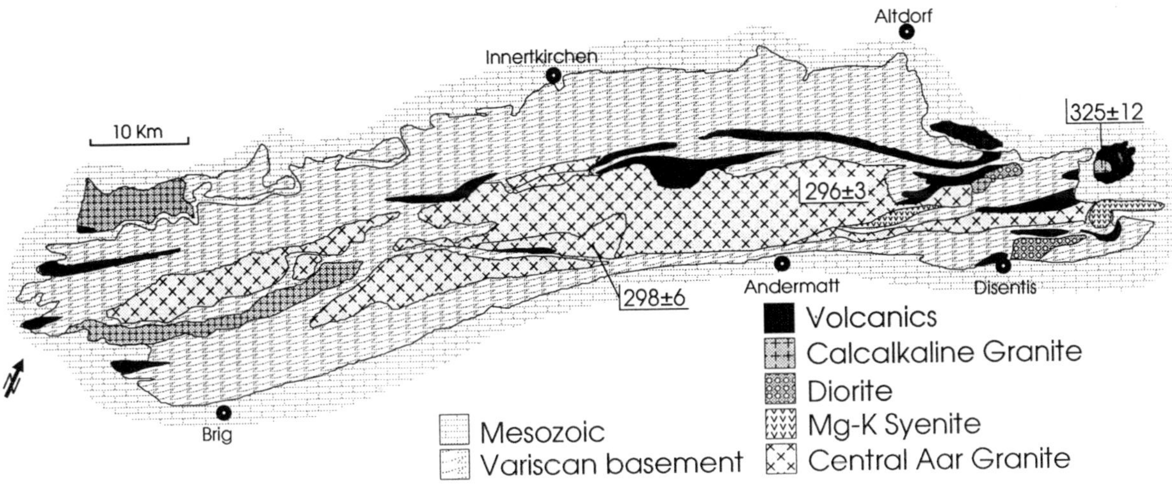

Fig. 3. Sketch map of the Aar Massif. Data on volcanites from Schenker (1986). Data on plutonites from Gnos (1988), Riesen (in prep.), Schaltegger (1984)

Within the pluton, Fe, Mg, Ti, Ca, Ba, Sr, V, and Zr steadily decrease from core to rim of the massif, in contrast to Si, K and Rb. This organization is attributed to flow differentiation during the emplacement of the magma.

All granitic facies are typical of Fe-K alkali-calcic granites, as evidenced by zircon morphology using the classification of Pupin (1988) (Fig. 9) and by Fe-rich biotite compositions. The Mont Blanc granite is an evolved metaluminous to slightly peraluminous (A/CNK = 0.95–1.50), K-rich (K_2O/Na_2O = 1.1–1.5, $Na_2O + K_2O$ = 7.6–8.9) granite yielding both calc-alkaline and alkaline affinities (Fig. 10). Both REE (Fig. 16) and other trace element concentrations are typical of late-orogenic granites.

Leucocratic dykes are highly differentiated acidic rocks, strongly depleted in compatible elements (e.g. Sr, Ba, Zn, V, Sc) (Fig. 15). Their distinctive V-shape REE patterns (Fig. 16) are similar to other late leucocratic facies of late Variscan batholiths (see review in Cocherie 1984).

Numerous magmatic mafic enclaves occur mainly within (a), together with monzodioritic, partly synplutonic lamprophyre. Two types (Bussy 1987) have been distinguished from mineralogical and chemical criteria (Figs. 10 and 12): hornblende + biotite-bearing Mg-enclaves, and biotite-bearing Fe-enclaves. Both record mixing and mingling processes between coeval granitic and basic magmas. Mafic enclaves, which form clusters along several hectometre-long swarms, are the products of the dismembering of basic to intermediate synplutonic dykes intruding the partially crystallized granitic melt.

As shown in R1–R2 diagrams (Fig. 12), the alkali-calcic Mont Blanc granite and its calc-alkaline Mg-enclaves and lamprophyres cannot belong to the same magma series, which suggests coeval, but not cogenetic felsic-mafic magma associations. Mantle-derived rocks, either incorporated in the lower crust through infracrustal underplating, or located within an upper mantle previously metasomatized by earlier subduction products, are potential sources for the mafic melts and, possibly, for at least a part of the felsic melt. Crustal source rocks for the other part of the granitic magma could be deep-seated dry granulites.

Melting of this composite source should have been induced by late-orogenic pressure release of a thickened thermally re-equilibrated lithosphere, in the context of transtensional strike-slip shear zones. Both felsic and mafic magmas were emplaced simultaneously in the upper crust (10–15 km). Due to strong uplift and erosion processes, the Mont Blanc pluton had already reached a 3–5-km-deep level when leucocratic dykes were emplaced.

A large, still undated, rhyolitic mass directly overlies the eastern margin of the Mont Blanc massif, which was at least partly unroofed at the time of volcanic activity. Locally, rhyolite yields a well-defined flow texture. Zircon typology indicates calc-alkaline affinities, even though the chemistry is similar to the granite, with the exception of higher Sr and Ba and lower Rb contents. No spatial geochemical coherence has been observed.

2.1.6 The Aar Massif

Entirely located in Switzerland, in the central part of the Helvetic realm, the Aar Massif has a wealth of Late Palaeozoic volcanic, plutonic and sedimentary

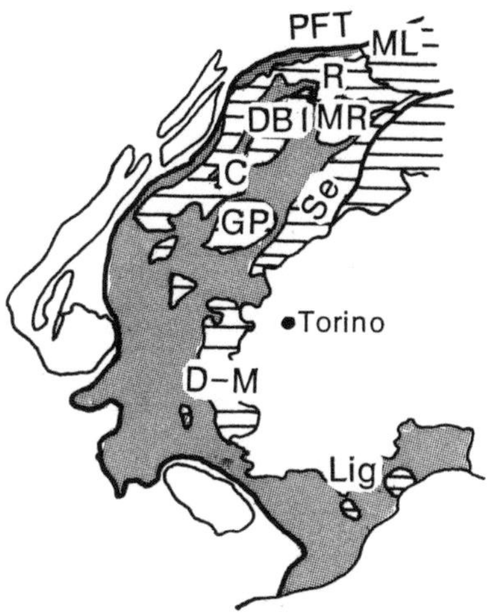

Fig. 4. Sketch map of the internal Western Alps. *Horizontal stripes:* Penninic, Austro-Alpine and South-Alpine basement units; *shaded zones:* Briançon cover, Valais, Zermatt and Combin zones; *unornamented:* external zones. Abbreviations: *C* Cogne "diorite"; *DBl* Dent Blanche nappe; *DM* Dora-Maira massif; *GP* Gran Paradiso massif; *Lig* western Ligurian Briançonnais zone; *ML* Monte Leone nappe; *MR* Monte Rosa massif; *PFT* Penninic frontal thrust; *R* Randa body; *Se* Sesia-Lanzo zone

formations (Fig. 3). Post-Variscan erosion stripped plutons from their cap-rocks but also spared volcanic equivalents in palaeotopographic lows. The following description, based on recent papers, theses and diplomas, is based on geochemical data (Fig. 10, 12, 14) and, if available, zircon typology according to Pupin (1976, 1980; Fig. 9).

Volcanites of calc-alkaline affinities were deposited on a pre-Carboniferous metamorphic basement complex and were deformed by late Variscan tectonic episodes, suggesting Variscan nappe tectonics (Mercolli and Oberhänsli 1988) or extensive regimes. They were later intruded by the central Aar granite, as shown by the intrusive contact between the granite and the volcano-clastic rocks of the Diechtergletscher Formation in the Trift area (Schenker 1986).

The central Aar granite seems to be the product of a single magmatic pulse during the post-Variscan and was only affected by Alpine tectonics. It yields an alkaline tendency, as shown by geochemical and zircon typological investigations by Eggenberger (1990). It contains enclaves of Mg-K-rich monzodioritic rocks (the Schöllenen "syenite" in the Reuss Valley).

Isotopic age determinations of Aar plutonic intrusions have confirmed two distinct suites:

1. An older calc-alkaline group of Lower Carboniferous rocks is exposed as a series of small intrusive bodies. Age determinations for this group are unfortunately very rare. A white mica of a pegmatite in the Tödi granite has yielded an age of 325 ± 12 Ma (Wuethrich 1965, recalculated with the new decay constants). The Schöllenen "syenite" has yielded a zircon U-Pb upper intercept age of 316 ± 27 Ma (Schaltegger and von Quadt 1990). New U-Pb data (Schaltegger and Corfu 1992) suggest that Mg-K-rich rocks are slightly older than calc-alkaline intrusions.
2. A younger group, the central Aar granite, yields alkaline tendencies and age determinations cluster around 290–300 Ma. Schaltegger and von Quadt (1990) found zircon U-Pb upper intercept ages of 298 ± 6 Ma for the Grimsel granodiorite and 296 ± 3 Ma for the Reuss Valley rocks. A similar age of 299 ± 2 Ma was given by two concordant zircon fractions of a microgranite intruding the Windgällen pyroclastic rocks (Schaltegger et al. 1991). Rb-Sr whole-rock ages are consistently younger than U-Pb data. Wuethrich (1965) has obtained a 4-whole-rock isochron of 287 ± 30 Ma (recalculated with new decay constants) for the central Aar granite.

3 The Penninic Domain of the Western Alps

Granitoids data, or considered, as late Variscan are less widespread in the Pennine realm than in the external units (Fig. 4). These granitoids have generally been converted into orthogneisses during Alpine tectonic episodes, which can reach extremely high grade facies, such as the famous coesite-bearing rocks in Dora-Maira (see Sandrone et al. this Vol.).

Radiometric measurements yielding late Variscan ages are available only for the Monte Rosa and Sesia granitoids. Other acid rocks of the internal Western Alps either are regarded as older than late Variscan or, if Upper Carboniferous to Permian in age, they are volcanic rocks. The Sapey gneisses are now recognized as belonging to a basement unit and are not Permian, but older than Late Variscan (Détraz 1984), and so are the Acceglio meta-porphyries (Monié 1990). In western Ligury, despite their apparent Variscan ages, all granitoids of the basement are also considered to have crystallized during pre-Variscan times (Del Moro and Visonà 1982).

The interpretation of the magmatic trends proposed here is based on a whole set of diagrams which test the effects on whole-rock compositions of metamorphic mineral changes.

3.1 Randa and Other Augengneisses of the Bernhard-Briançon Zone

In the Siviez-Mischabel basement subunit of the Bernhard zone, the Randa augengneiss forms a 45 km^2 main body and 20-km-long thin apophyses (Thélin 1983, 1987). A Rb-Sr isochron (Hunziker in ECOG 1969) yielded a Triassic apparent age (235 ± 16 Ma), which represents the lower limit for an age of emplacement assumed by Thélin (1987) to be Permian.

The main body has a laccolith shape which was converted into an asymmetric, probably reversed, synform in Alpine times. If there ever was a contact aureole, Alpine events have wiped it out entirely. The main rock facies is a mylonitic augengneiss without mafic enclaves, accompanied by mylonite, porphyritic granite gneiss and, as a rim facies, aplitic gneiss. Relic magmatic minerals indicate a two-mica granite with predominating muscovite. K-feldspar is perthite and plagioclase An_{3-17}. The mono- to poly-crystalline megaclasts ("eyes") average 2.5 cm in length and can reach up to 8 cm.

Chemically, the main body (c in Fig. 12) is a granite, yielding subalkaline to alkaline affinities, along with (a and b in Fig. 12) quartz enrichment and aluminous character which might reflect either mica concentrations, or feldspar breakdown and sericitization. The upper apophysis is an Al-rich monzogranite (d in Fig. 12): it is uncertain whether the apparent calc-alkaline affinity is primary or induced by chloritization effects.

Granitoids of similar characteristics can be found in the nearby Monte Leone nappe of the Central Alps and as minor, mylonitized, exposures in the same Siviez-Mischabel subunit. In Monte Leone, augengneisses yielding either sub-alkaline to K-alkaline, or apparently calc-alkaline chemistries, recall respectively the Randa main body and apophyses. In Siviez-Mischabel, a small, presumably Permian, retromorphosed monzogabbroic body is located at Stausee, in upper Turtmanntal (Wallis, Switzerland) (Sartori 1988).

In the Zone Houillère, mylonitic augengneisses in Tion (Switzerland) have been interpreted as either highly immature metasediments, or quartz-keratophyres (Thélin 1983; Desmons and Ploquin 1989). In the Aosta Valley, albite-rich leucogratic gneisses associated with acid and basic volcanic rocks have been described as belonging either to a basement slice (Zulauf 1963), or to a late Variscan laccolith (Caby 1974); a volcanic to subvolcanic origin cannot be precluded. Unfortunately, no chemical data are available.

3.2 Monte Rosa and Gran Paradiso

The Monte Rosa and Gran Paradiso metagranitoids cover very large areas of the order of, respectively, 250 and 450 km^2. The most common type is augengneiss derived from a porphyritic granitoid. Aplitic dykes cut through the whole bodies. In Monte Rosa, whole-rock Rb-Sr data (new decay constants) define isochrons of 320 Ma (Middle Carboniferous) and 260–240 Ma (Permian-Triassic boundary) with initial $^{87}Sr/^{86}Sr$ values of, respectively, 0.712 and 0.713 (Hunziker 1970; Hurford and Hunziker 1989). Monazite has also yielded an U-Pb age of 260 ± 5 Ma (Köppel and Grünenfelder 1975).

All granite are biotite monzogranites. The few available data plot in, or near to, the anatectic convergence field, with a possible subalkaline trend along with quartz enrichment (Fig. 12), in Gran Paradiso samples.

3.3 Cogne

The Cogne "diorite" is a small (25–30 km^2), shallow intrusion within rocks of the Zona Interna (a subzone of the Bernhard-Briançon zone). According to recent zircon U-Pb datings of a granophyre body in the adjacent Mont-Pourri massif (Guillot et al. 1991), the formations would be Late Cambrian-Early Ordovician in age. No radiometric evidence is so far available to support the commonly assumed Permian age of the Cogne body.

Magma mineral relics are hornblende, biotite, often zoned feldspars, quartz and accessory minerals. Mafic enclaves are frequent. Very few chemical data are available, as many theses completed in Turin have remained unpublished. They indicate tonalite and subordinate quartz-diorite compositions, following a calc-alkaline trend (Fig. 12; Desmons and Ploquin 1989).

3.4 Dora-Maira

In Val Pellice, small (max. 100-m-long) lenticular bodies of diorite, quartz-diorite and tonalite gneisses intrude the basement and the pre-Triassic sedimentary cover and are cut by rare basic dykes. They contain mafic enclaves (Sandrone et al. 1988). A Permian age is supported only by the assumed Permo-Carboniferous age of the intruded sedimentary cover. No magmatic mineral relics have been found. The composition is typically calc-alkaline (Fig. 12), undisturbed by metamorphism, and similar to Cogne "diorite", albeit yielding quartz-depleted chemistry.

Other undated augengneisses of the Dora-Massif (banded augengneisses of Sandrone et al. 1988) are possibly derived from late Variscan granitoids. Sandrone et al. (this Vol.) have distinguished five types: (a) metadiorite (Malanaggio "diorite"), (b) metagranodiorite, heavily transformed under eclogite conditions (Brossasco type), (c) primary muscovite-bearing porphyritic metagranite (Borgone type), (d) allanite-rich augen metagranite with locally preserved magmatic texture and assemblage (Freidour type), and (e) porphyritic metagranite (Sangone type).

Chemically, the basic to intermediate (a) suite follows the calc-alkaline trend defined for the other basic and intermediate rocks of the Bernhard-Briançon zone (Fig. 12). All acid (b) to (e) facies are enriched in silica and plot in the anatectic convergence field (Desmons and Ploquin 1989). However, from zircon morphology and REE patterns, two contrasting suites can be defined: (b) and (c) display a crustal anatectic origin, while (d) and (e) yield subalkaline to alkaline affinities (Sandrone et al. this Vol.).

3.5 Sesia Zone

In the internal Sesia unit, granitoids form small, up to ~3 km^2, scattered exposures (Reinsch 1977; Oberhänsli et al. 1985). The metamorphic facies is dominantly either eclogite, or amphibolite, metamorphic and deformational imprints ranging from weak to strong. Aplite and acid enclaves can be found.

A whole-rock Rb-Sr isochron yielded a 129 ± 15 Ma Early Cretaceous age with a Sr initial value of 0.7134. Zircon U-Pb datings (Paquette et al. 1989) assign to the plutonic activity a minimum age of 286 ± 2 Ma (Carboniferous-Permian boundary) if due account is taken of Alpine disturbances. Both U-Pb upper intercepts and Nd model ages indicate a 1.3 Ga-old Proterozoic crustal source, suggesting that Permian magmatism could be derived from the rejuvenation of Pan-African magmatism and/or metamorphism (Desmons in prep.).

Mineral relics include biotite, K-feldspar, quartz and accessory minerals. The composition (Fig. 12) is granodioritic, locally close to quartz-monzodioritic, and typically calc-alkaline. A slight shift towards quartz enrichment probably does not reflect alkali loss due to plagioclase breakdown, as there is no correlation between quartz contents and the degree of metamorphic reworking.

The acid orthogneisses of the external Sesia zone and of the Arolla nappe in the Dent Blanche unit, the "gneiss minuti", may be Late Variscan in age, but there is no supporting radiometric evidence. It is not clear to what extent their protoliths were volcanic or plutonic.

4 The Tauern Window

Late Variscan granitoids form a dome along the main axis of this Penninic window of the Eastern Alps. As they were affected at different degrees by Alpine deformation and metamorphism, they are commonly referred to as "Zentralgneise" (= central gneisses) in geological maps. Several discrete massifs, often named "cores" or "nappes", have been mapped throughout the Tauern window and display variable plutonic lithologies (Fig. 5). For age datings, see review in Finger et al., this Vol.

The most abundant felsic facies (Fig. 8), biotite granodiorite (about 60% of the Zentralgneise) and biotite granite (about 30%), are seldom fine-grained and more commonly medium- to coarse-grained, with K-feldspar megacrysts. Less felsic, sometimes amphibole-bearing, intermediate granitoids are mostly medium-grained and display tonalite (about 10–20% of the Zentralgneise), rarely quartz-syenite, quartz-monzonite or quartz-diorite compositions. Small gabbro and diorite bodies have locally been found.

Chemically, most Zentralgneise are relatively Na-rich and display either low-silica metaluminous, or silica-rich slightly peraluminous compositions (Fig. 11). They can, therefore, be ascribed to I-type granites, according to the Australian classification (Chappell and White 1974).

Two compositionally extended I-type suites with significantly different chemical parameters have been recognized in the eastern and western Tauern window (Finger and Steyrer 1988; Finger et al. 1990).
1. The western suite makes up the Zillertal-Venediger massif and consists dominantly of amphibole tonalite grading into K-feldspar-poor granodiorite, associated with minor bodies of cognate gabbro and diorite, and with granite and aplite. Data (Fig. 8) follow the low-K calc-alkaline trondhjemitic trend of Lameyre and Bowden (1982). Well-defined trends in Harker diagrams, characterized by Ca enrichment and K, Rb, Ba und Sr depletion, resemble those of mantle-derived plutonic suites, such as of island arcs (e. g. Chappell and Stephens 1988).
2. The eastern suite is a granodiorite-dominated association occurring in the Hochalm Massif. Fol-

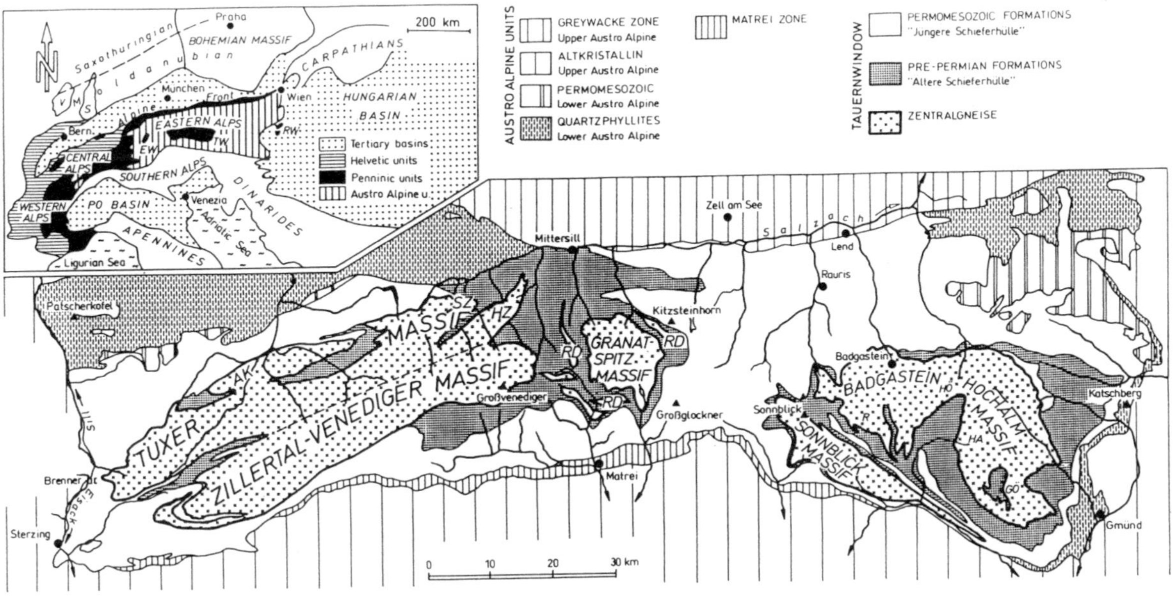

Fig. 5. Sketch map of the Tauern window. Abbreviations: *AK* Ahornkern; *Gö* Gößgrabenkern; *HA* Hochalm-Ankogelkern: *Hö* Hölltor-Rotgüldenkern; *HZ* Habachzunge; *R* Romatedecke; *RD* Riffldecke; *SZ* Sulzbachzungen; *inset: EF* Engadine window; *RF* Rechnitz window; *TF* Tauern window

lowing the medium-K calc-alkaline granodioritic trend (Fig. 8), it displays a trend marked by Ca depletion, slight Na, K, Rb enrichment, and large Ba and Sr enrichment. The chemical signature of this suite can be compared to many plutonic suites emplaced in active continental margins (e.g. Bateman and Chappell 1979; Lopez-Escobar et al. 1979).

The extremely K-rich granitoids of the Romate nappe (Bad Gastein massif) (Figs. 8, 11) are metaluminous and should, therefore, be ascribed to I-type and not S-type suites (high K_2O I-type, in the sense of Finger et al. 1990). As a rule, K-rich rocks, yielding monzonitic or syenitic tendencies (Romate nappe, Hochalm massif and Riffl nappe), were emplaced relatively early in the evolution of the Zentralgneise throughout the Tauern window.

K-rich and highly peraluminous S-type granites are represented in the Tauern window by cordierite granite of the Granatspitz Massif. However, it cannot be precluded that this granite is derived from a fractionated I-type melt, which has been largely contaminated by wall rocks. This is true also for other granites, yielding S-type affinities, in the Badgastein and Hochalm Massifs (Finger and Steyrer 1988).

A-type peralkaline granitoids, if not absent, are scarce. However, a distinct group of late stage felsic granites resembles A-type granites, in that these rocks have high contents of Y and HREE.

Seen as a whole, the Tauern window is certainly an I-type dominated granite terrain. Sr and Nd isotope data (Fig. 17), as well as zircon U-Pb datings (von Quadt and Finger in prep.), suggest that the Late Palaeozoic plutonism can be explained in terms of mixtures of mantle-derived Variscan melts and of old (Mid-Proterozoic crustal residence age) continenal sources. As most Zentralgneise display typical trace element signatures of volcanic arc granites (Fig. 14; see Finger and Steyrer 1990; Finger et al., this Vol.), an active continental margin environment should be envisaged in Late Palaeozoic times.

5 The Austro-Alpine Domain

In the Eastern Alpine nappe stack, abundant both plutonic and volcanic massifs of Variscan age have undergone metamorphic transformations at various scales (Fig. 6). Most of them were converted into orthogneiss lenses, sometimes with preserved primary magmatic features in the core.

The Err nappe contains calc-alkaline volcanic formations ranging in composition from andesite to rhyolite, which were deformed together with the basement during pre-Alpine episodes and were overlain unconformably by a Mesozoic sedimentary cover.

In the Julier-Bernina nappe, granodiorites associated with diorites were followed by the emplacement

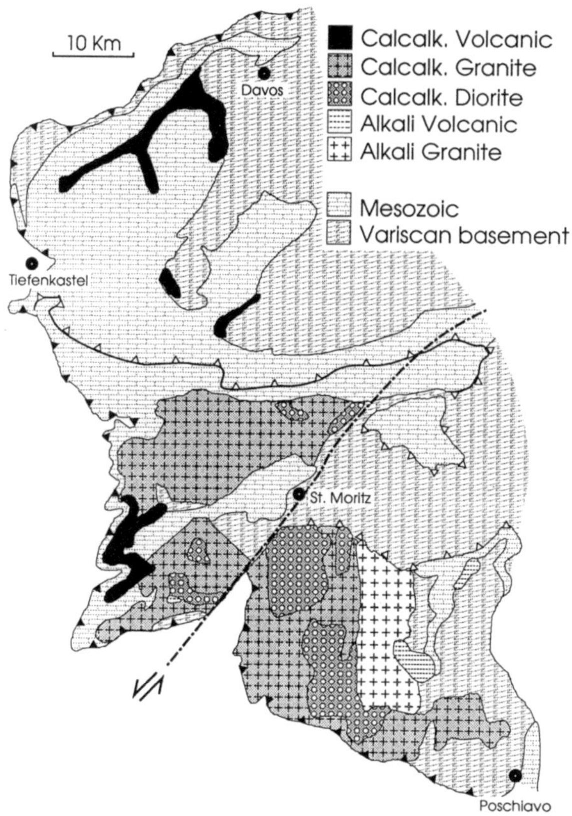

Fig. 6. Sketch map of the Eastern Alps of Switzerland. Data on volcanites from Graf (1987) and Mercolli (1985). Data on plutonites from Eggenberger (1990), Rageth (1984), Bühler (1983), Büchi (1987)

of a granite-syenite alkali series. Volcanic formations of the Diavolezza are highly silicic, potassic, peralkaline rhyolites, sometimes still retaining nice rheomorphic features.

Located at the southwestern front of the Silvretta thrust sheet, calc-alkaline und alkaline pyroclastites of the Upper Carboniferous-Lower Permian Ruina formations and of the Permian Chazforà formation are associated with terrestrial clastic sedimentation and synsedimentary faulting within northeast-southwesttrending grabens.

6 The Southern Alps

In the Southern Alps, Variscan magmatism is well represented by both intrusive and effusive formations, without Alpine metamorphic overprints (Fig. 7). Alpine tectonics, marked by faults almost parallel to the Insubric line, flank strips of metamorphic basement containing plutonic massifs emplaced at depths ranging from 15 to 4 km (Bellieni et al. 1979; Boriani et al. 1988 a) up to shallower Hercynian crust in which plutonic rocks intrude volcanites (Visonà et al. 1987; Bakos et al. 1990).

Plutonic rocks are more abundant in the western (Massiccio dei Laghi) and eastern (Bressanone, Cima d'Asta) sectors, while volcanites prevail in the central (Lombardo) sector (Fig. 7). Magmatic activity is limited to the Lower Permian, with plutonites in the 285–275 Ma range (Boriani et al. 1985; Sassi et al. 1985; Pinarelli et al. 1988; Bakos et al. 1990) and volcanites in the 272–261 Ma range (Stille and Buletti 1987; D'Amico and Del Moro 1988).

Clear, vertical contacts with country rocks indicate a permissive ascent (Bargossi et al. 1979; Boriani et al. 1988), with the exception of the initial phase at Mottarone-Baveno (Massicio dei Laghi) (Giobbi Origoni 1987). Relations with volcanites illustrate magma resurgence after cauldron subsidence (Boriani et al. 1988 a; Bakos et al. 1990) and the related distensional regime caused long, now EW-trending, basins to form at the surface, in which Permian sedimentary and volcanic formations were deposited (Wopfner 1984; Cadel 1984).

Plutons and smaller bodies are composite, the lower zones being intruded by acidic bodies, a sequence also generally respected in Permian volcanites. In QAP, R1–R2, AFM diagrams (Figs. 8, 12, 13), data plot in both calc-alkaline (low- and medium-K subseries) and anatectic series. Granite and granodiorite are the most frequent lithotypes, with significant amounts of tonalite, quartz diorite and quartz gabbro.

Geological (Boriani and Peyronel Pagliani 1968; Visonà et al. 1987) and scanty geochronological (Cumming et al. 1987) data substantiate two major intrusive episodes.

1. The first episode is characterized by abundant intermediate rocks (quartz diorite and quartz gabbro) with minor tonalite, granodiorite and granite. The association can be compared to compositionally expanded calc-alkaline orogenic suites (Pitcher 1979). Small composite bodies are associated with dyke swarms, which are also composite and display sometimes breccia structures. In the Massiccio dei Laghi and Bressanone areas, dyke swarms form belts separated by the main granite-granodiorite plutons.

In the Massiccio dei Laghi, intermediate rocks ("appinites", Boriani et al. 1974) were emplaced at depths and induced high-grade metamorphism in country schists (Burlini and Caironi 1988). Appinites yield a hydrous (hornblende + biotite) mineralogy, even in the most mafic lithotypes, and quartz and K-feldspar only in the most evolved types. Cummingtonite and biotite aggregates are interpreted as due to

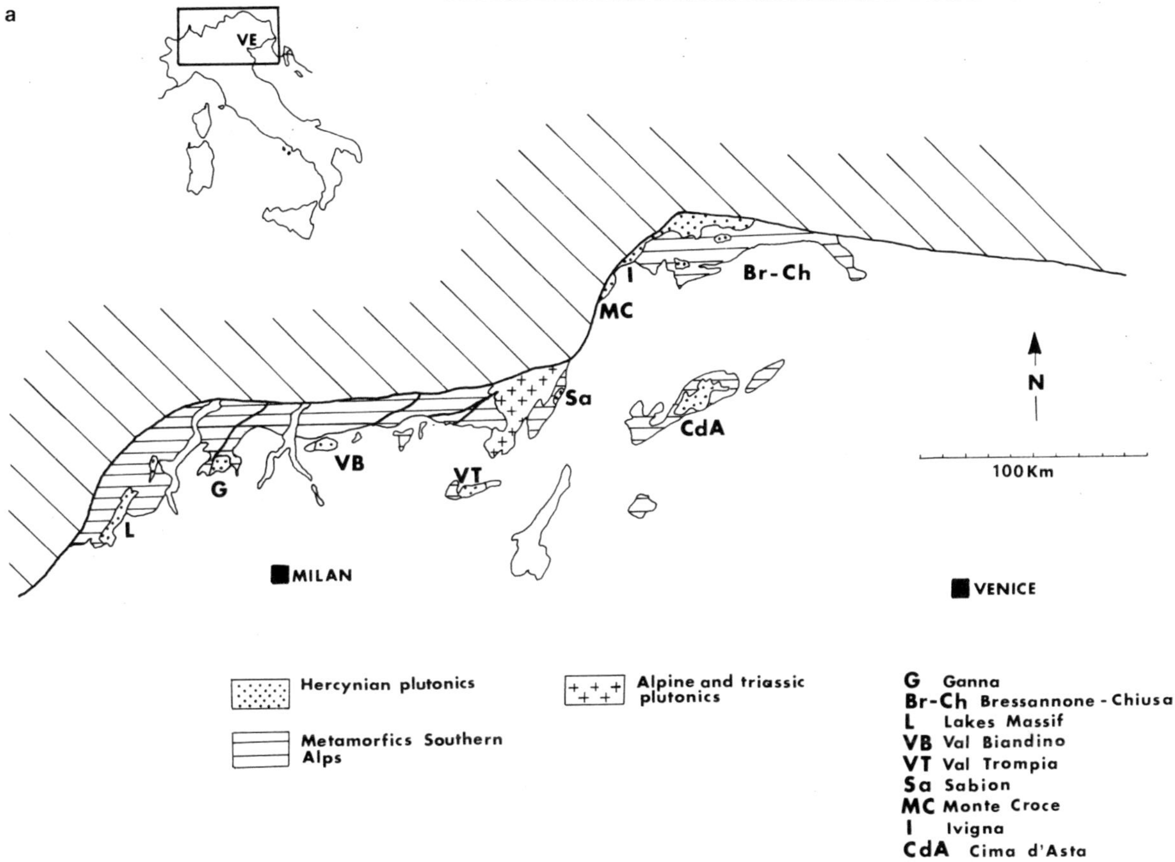

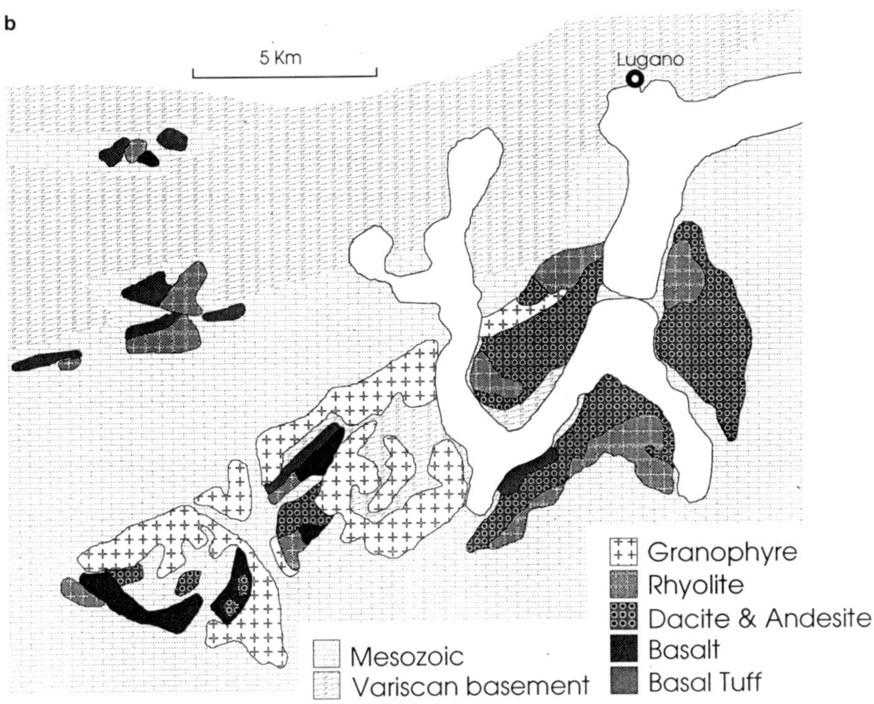

Fig. 7. a Sketch of Southern Alps and distribution of Variscan plutonic massifs. **b** Sketch map of the Lugano volcanic field. Data from Buletti (1985), Bakos et al. (1990)

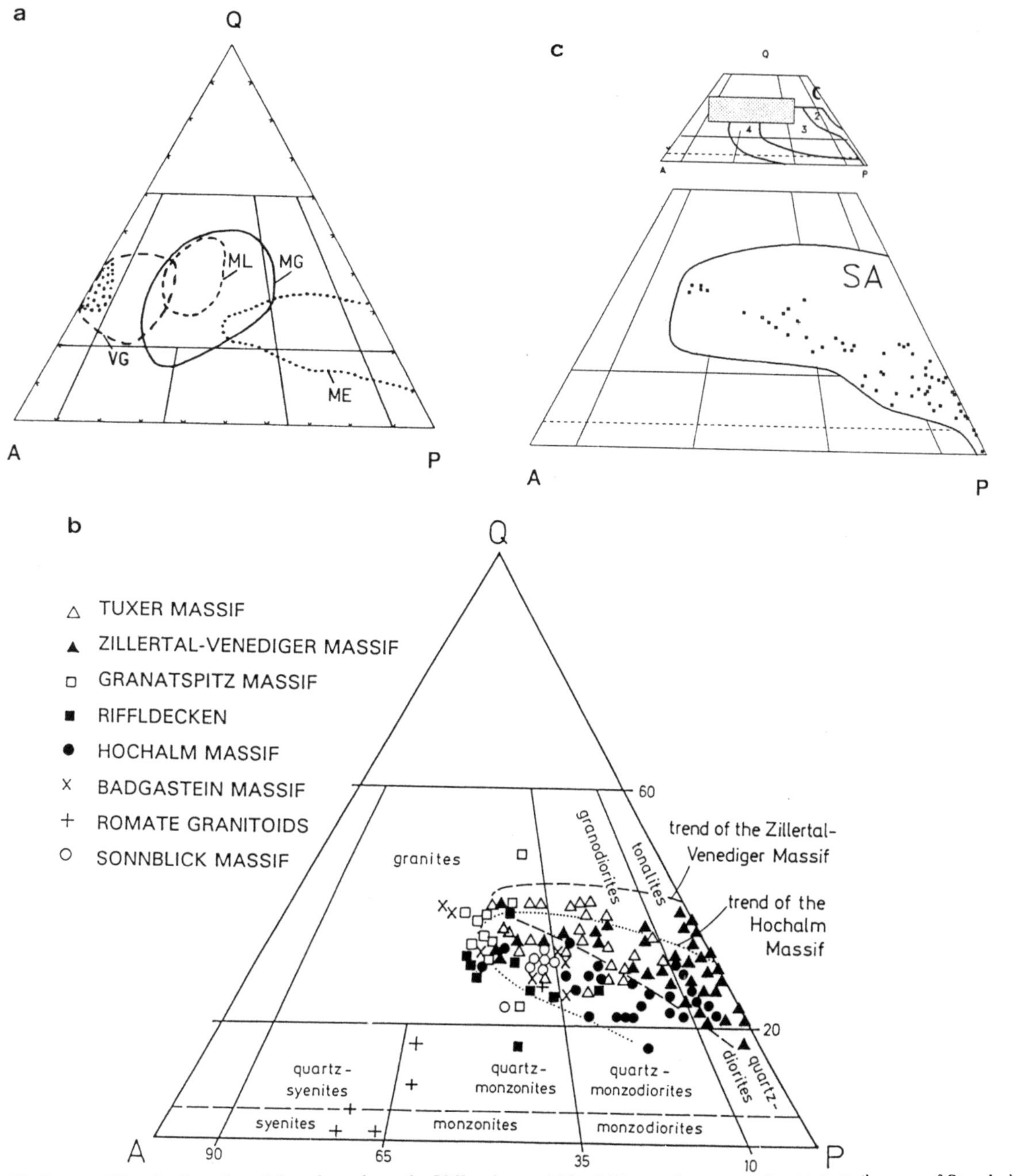

Fig. 8 a–c. a Distribution of modal analyses from the Vallorcine and Mont Blanc intrusions in the Q-A-P diagram of Streckeisen (1974). *MG* Mont Blanc granite, central+border facies together; *ML* Mont Blanc leucogranitic dykes; *ME* Mont Blanc Mg-rich and Fe-rich microgranular enclaves; *VG* Vallorcine granitic rocks, all facies together, 90% of the analyses fall in the *stippled area*. **b** Zentralgneise of the Tauern Window. Data were calculated from chemical analyses (Finger and Steyrer 1988 and unpubl.) converted into "mesonorm" (Mielke and Winkler 1979). **c** South-Alpine plutonic rocks (after Visonà, in Bonin et al. 1991). *SA* South-Alpine plutonic field. *Dots* first-stage rocks of the eastern area (Visonà et al. 1987). *Inset: dotted area,* anatectic granite field; *2* low-K calc-alkaline; *3* medium-K calc-alkaline; *4* high-K calc-alkaline fields (Lameyre and Bowden 1982)

pseudomorphosis of, respectively, orthopyroxene and garnet (Origoni Giobbi et al. 1975).

Intermediate and basic rocks, which have been reworked by, and included in, second-stage acidic plutonites, have been recognized in the Quarna pluton (western sector) by Burlini and Caironi (1988).

In the Bressanone area, the first-stage rocks are either deep-seated, with layered gabbro and minor diorite, or subvolcanic, with prevailing diorite and frequent granodiorite and granite (Visonà et al. 1987). Lithotypes with anhydrous mineralogy dominate, even in the granite differentiates, in which cordierite and plagioclase + orthopyroxene clusters are characteristic. In QAP diagram (Fig. 8), the suite follow the typical trend of Andean magmatism (Lameyre and Bowden 1982).

2. The second episode includes the most widespread and well-known plutonic massifs, composed mainly of various types of granite and granodiorite. Abundant metaluminous biotite- and hornblende-bearing facies are sometimes accompanied by small two-mica, cordierite- or garnet-bearing peraluminous masses (Visonà 1980; Bellieni et al. 1974; Boriani et al. 1988). In some cases (Cima d'Asta, Val Biandino), minor tonalite, diorite and gabbro, result in a compositionally restricted calc-alkaline suite (Pitcher 1979).

Magmatic activity ended with alkaline biotite subsolvus granites, such as the Baveno pink granite, the granite porphyry of the Lugano area (Fig. 7), minor fayalite-bearing leucogranitic dykes, and associated K-rich rhyolites (Lugano area, possibly Bolzano, Val Trompia, Valle Brembana, and Canavese areas), which have been interpreted as highly differentiated products of contaminated subcrustal melts (Del Moro 1987; Bakos et al. 1990).

Geochemical and Rb-Sr systematics (Del Moro 1987; Boriani et al. 1988b; Burlini and Caironi 1988; Pinarelli et al. 1988; Bakos et al. 1990) substantiate: (a) the presence of crustal anatectic magmas (two-mica cordierite-bearing leucogranites), (b) more abundant mantle component in the first-stage suites than in the second-stage group (initial Sr values = 0.7051–0.7131 vs 0.7086–0.7120), (c) the role played by AFC processes involving mantle and crustal melts, and (d) a probable volcanic-arc setting (Fig. 14).

Mantle-crust interactions, wall-rock assimilation and fractional crystallization constitute the major processes for magma generation and evolution, in plutonic massifs as well as in volcanic formations (Stille and Buletti 1987).

Even though palaeontological and palaeomagnetic evidences do not support the concept of a large Permian oceanic basin open within the European Variscan belt (e.g. Neugebauer 1988), chemical data suggest an environment with a continental active margin of Cordilleran-type (e.g. Pitcher 1988). Widespread porphyry-copper mineralizations constitute further arguments for such a geodynamic setting.

7 Towards a Geodynamic Interpretation?

Late Variscan magmatism within the Alpine belt appears to be continuous since Devonian up to Permian. Abundance and compositional characteristics of the magmatic suites are strongly differing in space and time from one area to another. However, they cannot be easily compared with the orogenic suites of Western Europe. As a matter of fact, a reasonable number of high-quality age determinations is unfortunately lacking for most of the massifs and many formations were considered as Late Variscan only on the basis of ill-defined criteria, such as the "monocyclic" metamorphic evolution of a massif unconformably overlain by a Mesozoic sedimentary cover, or the unproven assumption that red K-rich rhyolites should be Permian in age.

Unravelling the geodynamic implications of Late Variscan magmatic episodes would require therefore: (a) a good knowledge of the framework of Pangaea supercontinent at the end of Palaeozoic times (how many continental fragments?), and (b) unequivocal discriminations of the magmatic suites in terms of magma associations and of times of emplacement. Up to now, these tasks are far from being completed for the entire Alpine chain.

7.1 Petrological Constraints

If present-day settings are relatively easy to define, great difficulties have been encountered in older orogenic belts. During the last decades, considerable attention has been paid to past geodynamic environments in which granites were emplaced.

Actually, a granitoid is almost always associated with other rock types, such as basic-intermediate rocks and volcanic formations. The association constitutes a series, if coeval rocks come from cogenetic melts, and a suite, if they do not. Thus, a granitic massif can be used as a geodynamic indicator, if all pieces constituting the geological puzzle are assembled, i.e. rock types making up the massif as well as their wall rocks. Geological, structural, petrological, mineralogical and geochemical data should be collected together in order to provide a coherent framework.

7.1.1 Mineralogical and Geochemical Tools

Rock-forming minerals and whole-rock chemistries have been shown to provide useful indications of thermo-dynamic conditions prevailing during melt generation, evolution, crystallization, and rock con-

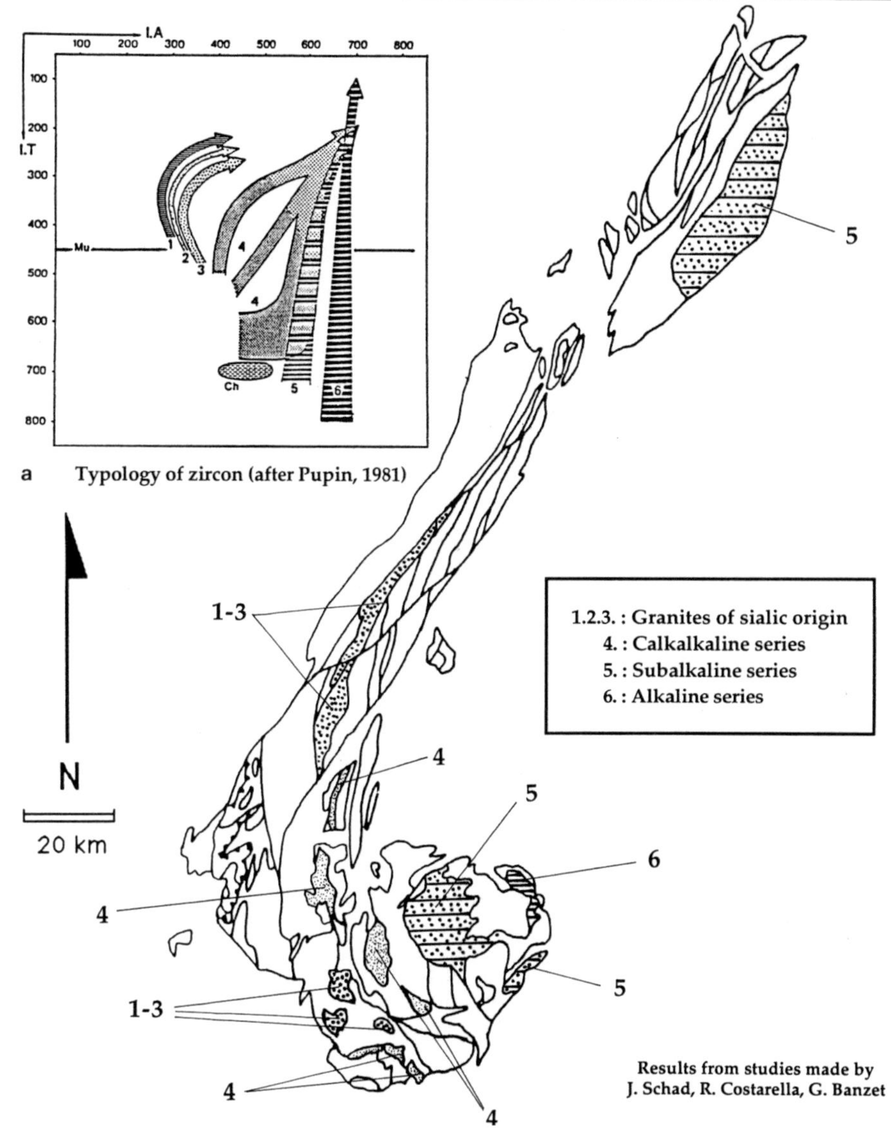

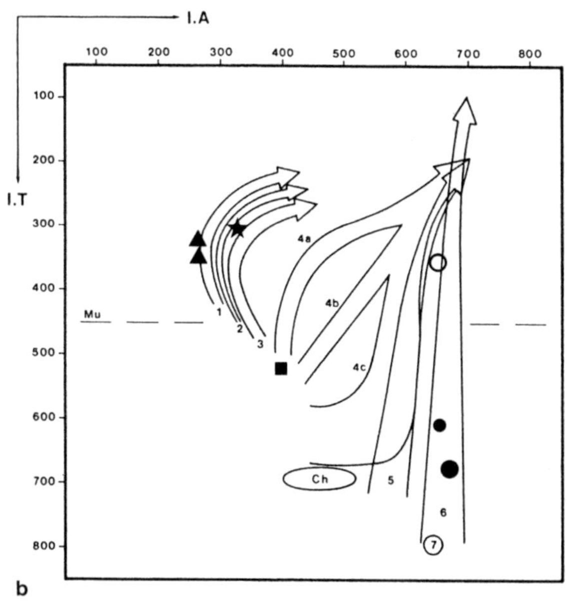

Fig. 9 a, b. Zircon typology, according to the method of Pupin (1976, 1980). **a** Characterization of the carboniferous granitoids of the French ECM by the typology of zircons (method from Pupin 1981). **b** Zircon typology of Vallorcine and Mont Blanc magmatic rocks in the classification of Pupin (1980). *Triangles* Vallorcine granite, main porphyritic facies (samples 4+114); *star* Vallorcine leucogranitic facies (1C2); *large filled circle* central facies of the Mont Blanc granite (unpubl. data from J. P. Pupin) (PK 4554); *small filled circle* border facies of the Mont Blanc granite (FB 893); *open circle* leucogranitic dyke of the Mont Blanc intrusion (FB 744); *square* Mont Blanc rhyolite (FB 727).

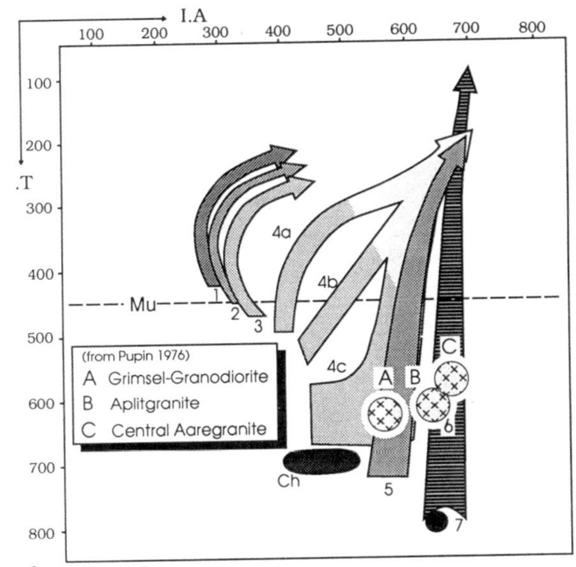

c

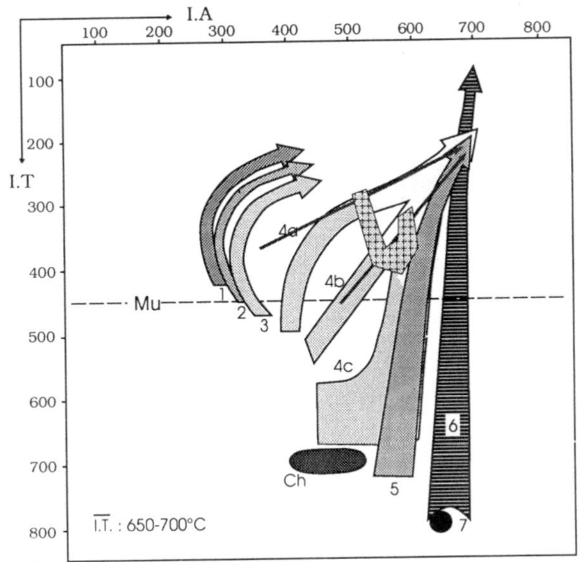

d

Aluminous anatectic granitoids
1: aluminous anatectic granites
2: autochthonous monzogranites and granodiorites
3: intrusive aluminous monzogranites and granodiorites

Hybrid granitoids of crustal and mantle origin
4: calcalkaline series granitoids
light dotted area: monzogranites & alkali-granites
5: sub-alkaline series granites

Granitoids of mantle or mainly mantle origin
6: alkaline series granites
7: tholeiitic series granites

Mu: limit of muscovite granites
Ch: area of magmatic charnokites

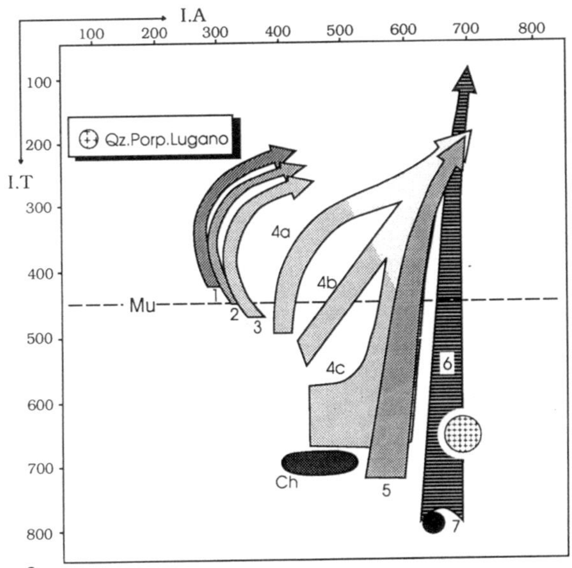

e

Fig. 9 c–e. Aar massif. **d** Eastern Alps of Switzerland. **e** Lugano volcanic field, Southern Alps (After Buletti 1985)

solidation and cooling. The most sensitive parameters used in this chapter are:

1. Modal data, reduced to QAP plot (Fig. 8).
2. Zircon morphology (Fig. 9), according to the method of Pupin (1976, 1980). The typological method defines zircon crystals on the basis of the relative development of prism faces (T index) and pyramid faces (A index).
3. Mineral chemistries. Classical geothermometers and barometers have been devised for specific mineral equilibria, involving the major minerals, such as feldspars, biotite, amphibole, anhydrous mafic silicates, and Fe-Ti oxides.
4. Whole-rock chemical parameters. Major as well as trace elements can be combined into simple or complex parameters. The classical TAS (total alkalies-silica (Fig. 10), A/CNK (molecular Al_2O_3/ $CaO + Na_2O + K_2O$) (Fig. 11), R1–R2 plots (Fig. 12), and AFM (alkali-iron-magnesium) (Fig. 13) have been used extensively to distinguish magmatic suites. As far as trace elements are concerned, Rb vs Y + Nb diagram (Pearce et al. 1984; Fig. 14), Rb-Ba-Sr triangle (Fig. 15), and REE chondrite-normalized patterns (Fig. 16) were devised in order to provide indications of sources and geodynamics settings.

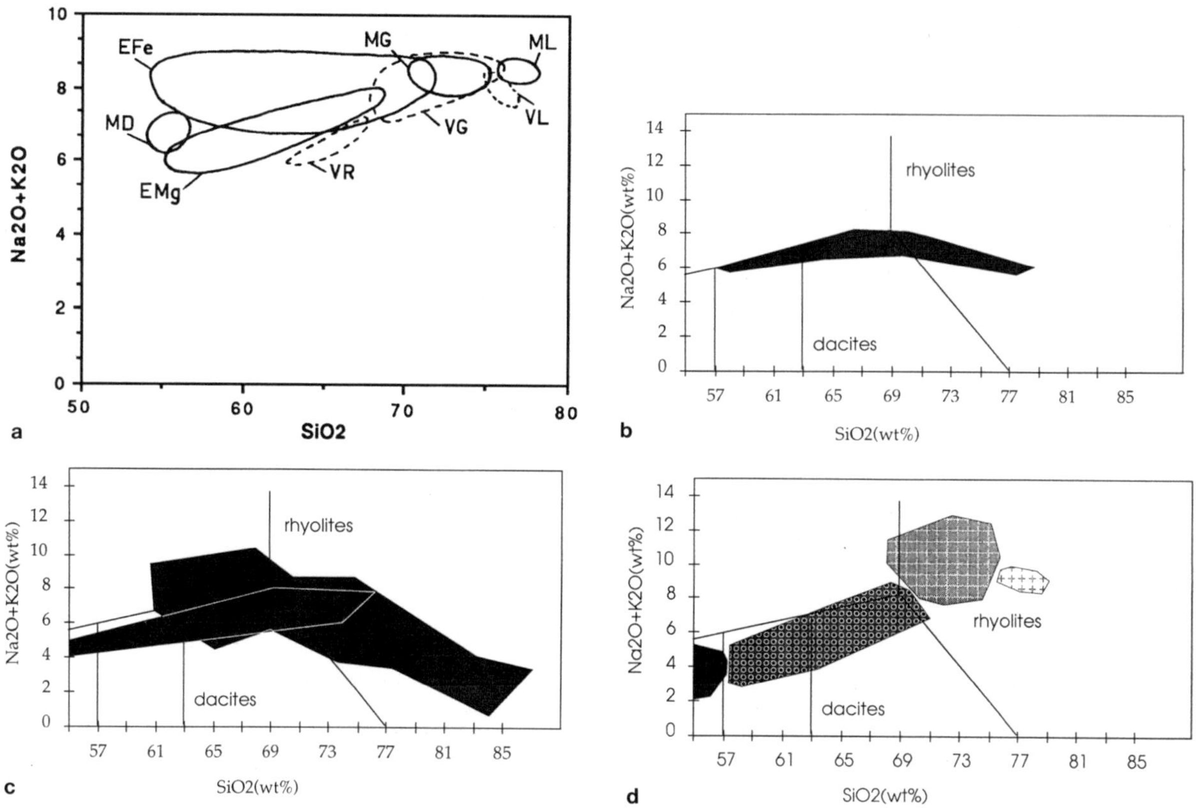

Fig. 10 a–d. Total alkalies-silica (TAS) diagram (after Le Maitre et al. 1984). **a** Chemical data from the Vallorcine and Mont Blanc magmatic rocks reported in the SiO_2-total alkalies diagram. *VG* main porphyritic facies of the Vallorcine granite; *VL* leucocratic facies; *VR* Vallorcine grey rhyolites; *MG* central and border facies of the Mont Blanc granite; *ML* Mont Blanc leucogranitic dykes; *MD* Mont Blanc lamprophyric dykes; *EMg* Mont Blanc Mg-rich microgranular enclaves; *EFe* Mont Blanc Fe-rich microgranular enclaves. **b** Aar Massif. **c** Eastern Alps of Switzerland. **d** Lugano volcanic field, South Alps (After Buletti 1985) Symbols as in Fig. 7b

7.1.2 Comparison of Available Data

All regional contributions have followed the same scheme of presentation: (a) distribution of granitic massifs, rock types; (b) field characteristics: shape, structure, fabrics; (c) petrographical and mineralogical data; (d) geochemical data; (e) discussion on petrogenesis and geodynamic environment.

All granitic massifs yield a data base, which has been compiled from published as well as from unpublished results. As the whole set of data exceeds the length of this volume, a brief summary has been given for each province.

An example of comparison of data, given in Table 1, concerns two nearby and yet contrasting granites: Vallorcine (Aiguilles-Rouges Massif) and Mont-Blanc. The different criteria used to define geodynamic settings are in agreement to uphold the concept that source rocks of the melts, modes of differentiation and crystallization of the magmas, levels of emplacement, etc. are completely different, even though these granites seem at a first glance rather similar (e.g. parallelogram shape, K-feldspar megacrysts, relation to large-scale shear zones, etc.).

7.2 The Basement of the Alpine Belt and of Its European Foreland

The geodynamic problem is obscured by the fact that the precise location of the different Alpine structural units, before Permian-Triassic dismembering of Pangaea and subsequent readjustment during Cretaceous-Tertiary tectonic episodes, remains an area of speculation.

In the basement of the Alpine chain, small fragments of oceanic and/or back-arc basin crust are preserved as sparse ophiolitic masses and/or eclogitic lenses, suggesting: (a) evolution of at least one oceanic basin and (b) subsequent subduction processes during Lower Palaeozoic times. Then, the late Variscan

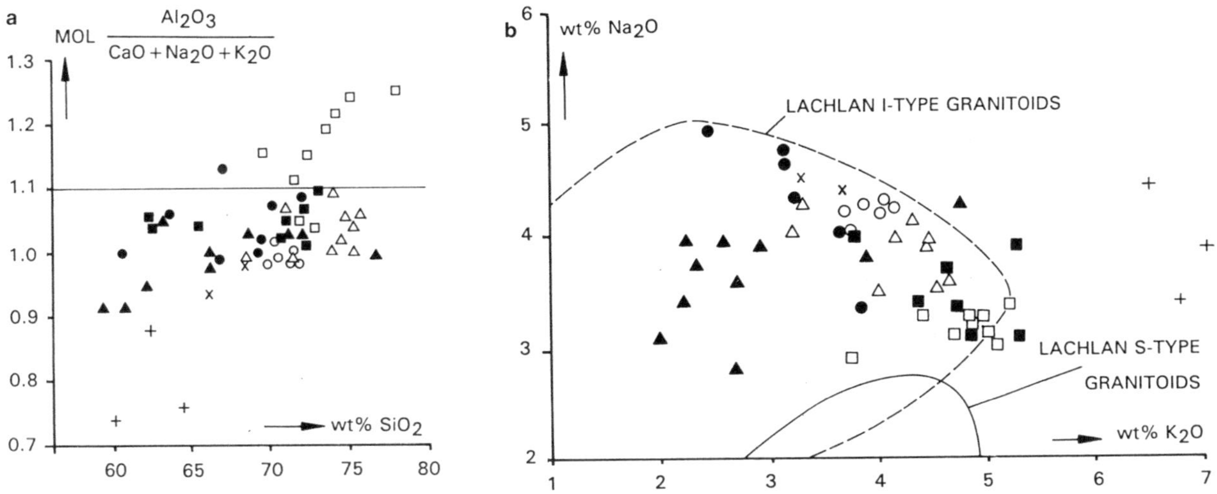

Fig. 11. **a** Molecular $Al_2O_3/CaO+Na_2O+K_2O$ versus SiO_2 and Na_2O versus K_2O plots of Zentralgneise, Tauern window. *Symbols* as in Fig. 8b. Shown for comparison are fields of S- and I-type granites of the Lachlan Fold belt (**b**). (After Chappell and White 1974; White and Chappell 1983)

period was characterized by the end of oceanic crust subduction, the closure of the oceanic basin(s), the collision of Gondwana with Laurussia and smaller terranes, and the amalgamation onto the Pangaea super-continent.

The presence of discrete terranes has been suspected from different points of view. As far as magmatic episodes are concerned, two major features should be pointed out.

7.2.1 Discrete Terranes? The "Caledonian Controversy"

In the Southern Alps as well as in southern France, orthogneissic bodies have been found yielding Ordovician to Silurian ages of emplacement and defining magmatic suites ranging from older orogenic calc-alkaline to younger post-orogenic alkaline ones. Though the term "Caledonian" is not a good one and was not always accepted, this suggests a short orogenic episode, resembling the Caledonian event in the British Isles.

Orogenic calc-alkaline massifs of the same age have been described in other parts of the belt, they record a long-lived subduction regime but do not uphold a complete orogenic cycle as post-orogenic alkaline suites are lacking. Thus, the "Caledonian" event is not recorded so far in the rest of the Alps and their European foreland, substantiating the view that the Southern Alps may represent a discrete terrane with its specific geodynamic evolution before the assembly of continental fragments onto Pangaea.

7.2.2 The Timing of the Last Orogenic Magmatic Events

In both external and internal zones (Helvetic and Penninic realms), magmatism is characterized by calc-alkaline and anatectic suites predating near-alkaline suites, with age distribution ranging from Late Devonian to Late Carboniferous (Stephanian), i.e. ~350 – 290 Ma. Permian magmatism, though present, was poorly developed within graben and troughs filled with molassic sediments.

In contrast, in Austro-Alpine nappes and Southern Alps, as well as in southern France, Corsica-Sardinia, and Morocco, additional and abundant (Permian (i.e. ~300 – 250 Ma) magmatism is particularly well developed and displays a clear succession of (high-K) calc-alkaline and alkaline suites. This provides a further indication supporting the concept of contrasting late Variscan geodynamic contexts within the Alpine chain.

7.3 Space and Time Relationships of Magma Associations

The following magma associations have been observed: (a) calc-alkaline suites with their low-K and high-K variants, frequently referred to as I-type plutons (e.g. Finger and Steyrer 1990) (b) crustal peraluminous granites, often referred to as S-type plutons, and (c) alkaline and subalkaline A-type suites (Fig. 18).

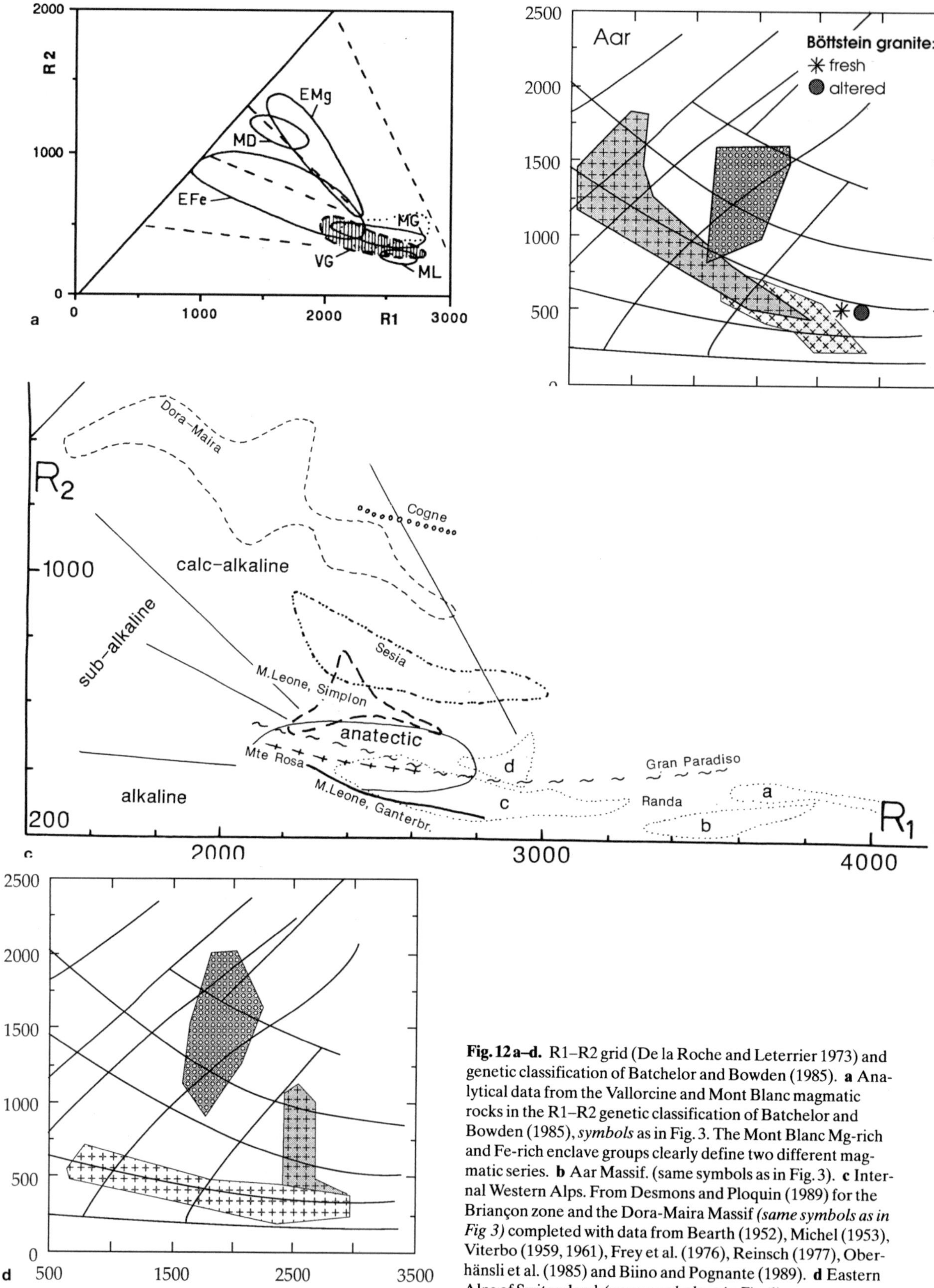

Fig. 12a–d. R1–R2 grid (De la Roche and Leterrier 1973) and genetic classification of Batchelor and Bowden (1985). **a** Analytical data from the Vallorcine and Mont Blanc magmatic rocks in the R1–R2 genetic classification of Batchelor and Bowden (1985), *symbols* as in Fig. 3. The Mont Blanc Mg-rich and Fe-rich enclave groups clearly define two different magmatic series. **b** Aar Massif. (same symbols as in Fig. 3). **c** Internal Western Alps. From Desmons and Ploquin (1989) for the Briançon zone and the Dora-Maira Massif *(same symbols as in Fig 3)* completed with data from Bearth (1952), Michel (1953), Viterbo (1959, 1961), Frey et al. (1976), Reinsch (1977), Oberhänsli et al. (1985) and Biino and Pognante (1989). **d** Eastern Alps of Switzerland *(same symbols as in Fig. 6)*.

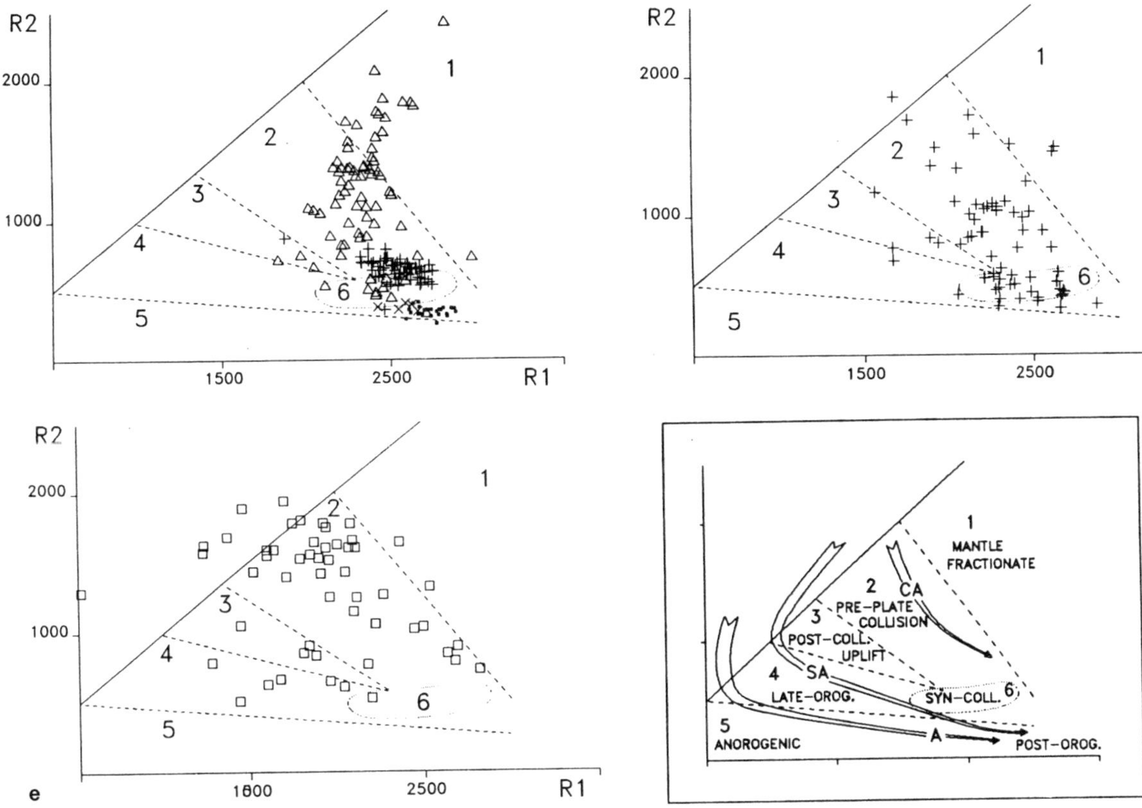

Fig. 12 e. Southern Alps. *1* Bressanone area. *Triangle* first-stage rocks; +second-stage granite and granodiorite; *x* fayalite granite (Visonà 1977, 1980, 1983, 1986, and unpubl.). *2* Central area (De Capitani and Liborio 1988; De Capitani et al. 1988). *3* Western area (Boriani et al. 1988; Burlini and Caironi 1988 a, b; Giobbi Origoni et al. 1988; Pinarelli et al. 1988). *Inset* Fields of Batchelor and Bowden (1985). *Ca* Calc-alkaline trend; *SA* subalkaline trend; *A* alkaline trend; *1* mantle fractionates; *2* pre-plate collision; *3* post-collision uplift; *4* late-orogenic; *5* anorogenic

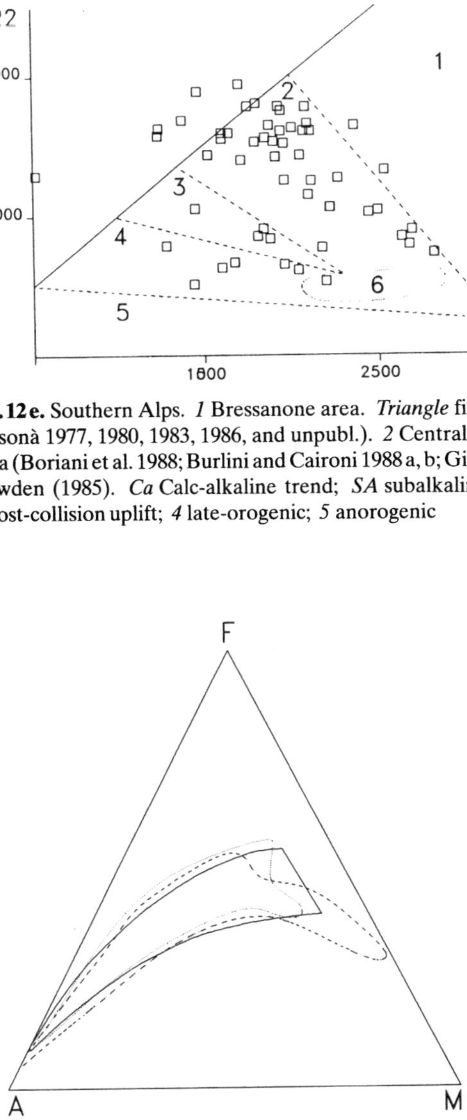

Fig. 13. AFM diagram for the Southalpine plutonic rocks. *Complete line* field of calc-alkaline rocks (after Green and Ringwood 1968). All western area intrusives fall within this field; *dotted line* central area intrusives; *dashed line* eastern area intrusives. (After Visonà, in Bonin et al. 1991)

Low-K calc-alkaline suites, with a diorite-tonalite-trondhjemite association, are rare. Their ages of emplacement range between 360 and 350 Ma at the Devonian-Dinantian boundary, i.e. very close to the major collision event and just after the culmination of Barrowian metamorphic episodes. Thus, they are not late Variscan *s. s.*

Normal-K and high-K calc-alkaline suites, with expanded compositional variations from (monzo)diorites through granodiorites to monzogranites, are widespread. Their ages of emplacement cover the whole Carboniferous with a maximum at ca. 330–320 Ma in the Mid-Carboniferous, with the development of large batholiths and eruption of associated volcanites. They fit closely enough the original definition of I-type granitoids by Chappell and White (1974) and belong to the H_{LO} group of "calc-alkaline rocks with mixed (crust+mantle) origin" of Barbarin (1990).

The interpretation of calc-alkaline I-type suites is not yet fully clear. Normal-K calc-alkaline associations can represent the "Cordilleran type" of Pitcher

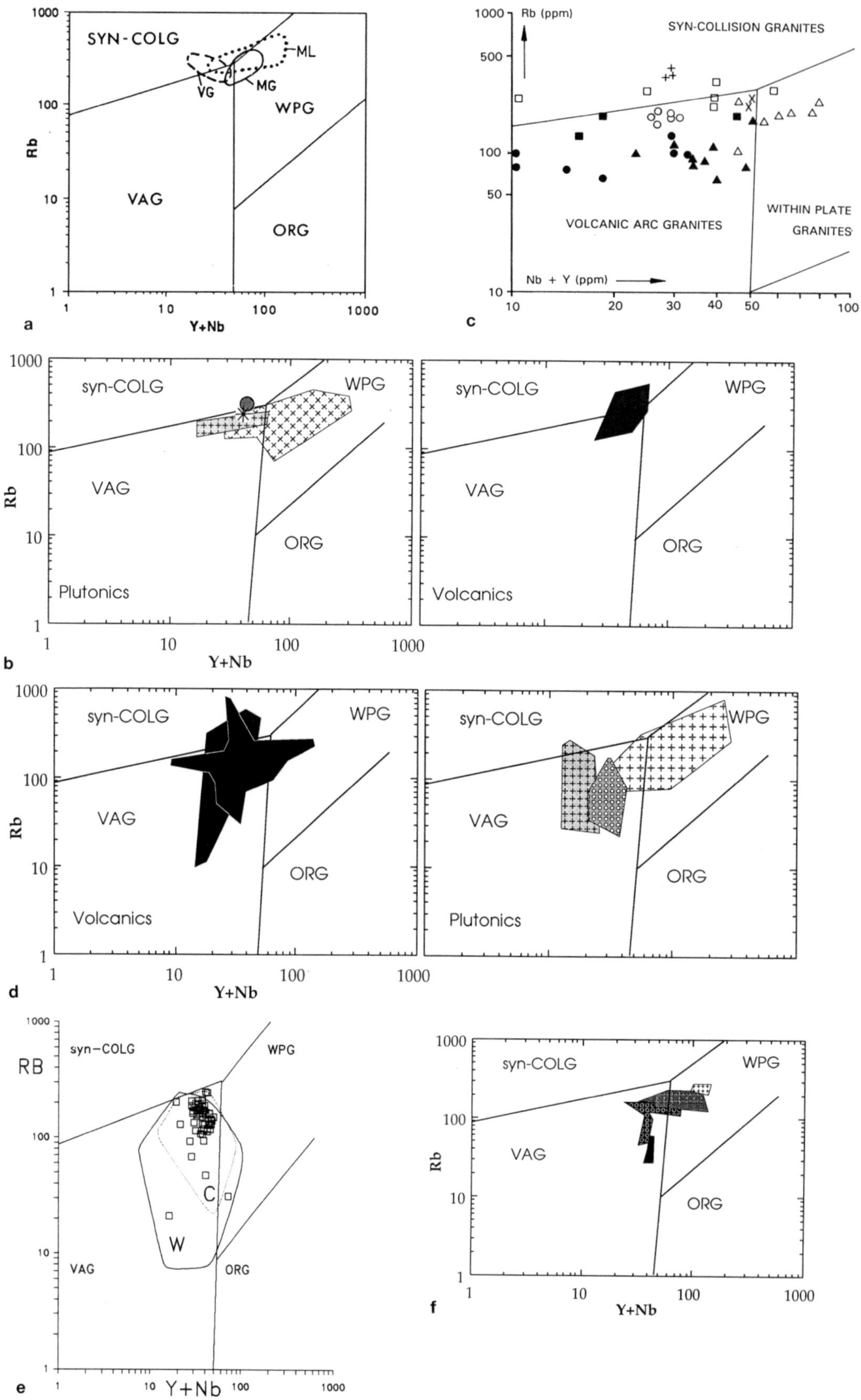

SYN-COLG
ML
VG
MG
WPG
VAG
ORG
Rb
Y+Nb
a
Rb (ppm)
SYN-COLLISION GRANITES
VOLCANIC ARC GRANITES
WITHIN PLATE GRANITES
Nb + Y (ppm)
c
syn-COLG
WPG
VAG
ORG
Plutonics
Volcanics
b
d
RB
W
C
e
f

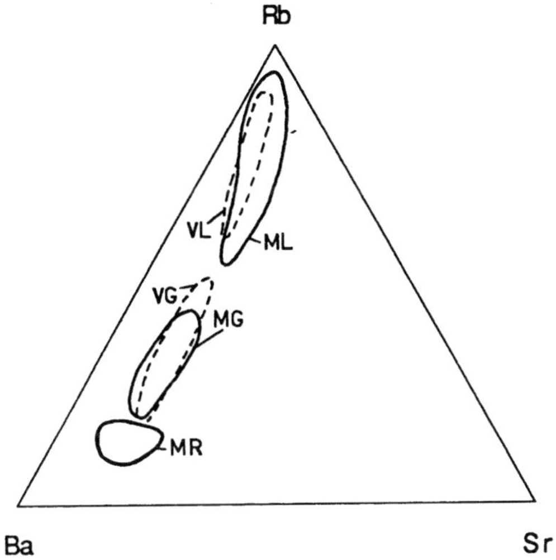

Fig. 15. Analytical data from the Vallorcine and Mont Blanc magmatic rocks in the *Sr-Ba-Rb triangle* (after El Bouseily et al. 1975). *MR* Rhyolite from the Mont Blanc massif, other *symbols* as in Fig. 3

(1983, 1987), but the high-K massifs can instead indicate a post-collisional regime, defined as "Caledonian type" by Pitcher. However, Finger and Steyrer (1990) have pointed out that almost all massifs are metaluminous in character and fall in the "volcanic are granites" (VAG) defined by Pearce et al. (1984). In the southern edge of the Alps, a continuing calc-alkaline activity is recorded during the Lower Permian.

Carboniferous anatectic suites comprise parautochthonous and allochthonous peraluminous granitoid bodies. As a rule, the peraluminous granodiorite-monzo-granite suite, which corresponds to S-type granitoids (Chappell and White 1974), predates the two-mica leucogranites, which are quite different (Barbarin 1990). They belong to the C_{ST}, C_{CA} and C_{CI} groups of "peraluminous rocks with crustal origin" of Barbarin (1990), and fall acordingly in the "syn-collisional granites" (SYNCOLG) of Pearce et al. (1984). As they are often associated with K-rich lamprophyric basic rocks, it was suggested that the heat input provided by basic magmas was sufficient to cause local crustal anatexis in their vicinity. The peraluminous granitoid association characterizes the "Hercynotype" of Pitcher (1983, 1987).

Late- to post-orogenic in character, the near-alkaline A-type suites can be subdivided into two contrasting groups:

1. The 300–290-Ma old subalkaline association of Aar, Mont Blanc and probably Saint Gothard is composed of an expanded suite which yield both markedly alkaline affinities (zircon morphology, Fe-enriched mineral chemistry, HFSE-enrichment characteristic of "within-plate granites") and calc-alkaline features (compositional range from granodiorite through monzo-granite to syenogranite, abundance of mafic enclaves). This suite closely resembles the Ploumanac'h association, for which the name of "Fe-rich subalkaline series" has been proposed (Barrière 1977).
2. The 270–240 Ma-old alkaline association is well developed in the Southern Alps and related massifs have also been described in Austro-Alpine, Penninic and external crystalline massif domains (Bonin et al. 1987). This magmatic A-type suite, with scarce basic rocks and a wealth of acidic (rhyolite-granite) rocks, constitutes the "anorogenic Nigeria-type" of Pitcher (1983, 1987) and the A group of "alkaline granitoids with a mantle origin" of Barbarin (1990). Like the 300–290 Ma-old association, they fall in the "within-plate granites" (WPG) field of Pearce et al. (1984) near the boundary with the VAG field. According to Bonin (1990), the 270-Ma-old granitoids constitute a discrete group of "post-orogenic" alkaline granites", while the 250–230 Ma-old granitoids define the "early anorogenic alkaline granites".

◄——

Fig. 14a–f. Trace-element genetic classification of Pearce et al. (1984). **a** Analytical data from the Vallorcine and Mont Blanc magmatic rocks in the genetic classification of Pearce et al. (1984). *VG* Vallorcine granitic rocks, all facies together; *MG* central and border facies of the Mont Blanc granite; *ML* leucogranitic dykes from the Mont Blanc intrusion. **b** Aar Massif *Symbols as in Fig. 3.* **c** Tauern window. *Symbols* as in Fig. 8b. **d** Eastern Alps of Switzerland *Symbols as in Fig. 6.* **e** Southern Alps. *Open square* eastern area (D'Amico et al. 1987; Messina et al. 1987; Visonà unpubl.); *field C* central area (same sources as in Fig. 12e); *field W* western area (same sources as in Fig. 12e; Bakos et al. 1990). **f** Lugano volcanic field, Southern Alps (Buletti 1985) *Symbols as in Fig. 7b*

8 Summary and Conclusion

Geological, petrological and geochemical data are consistent with the current models of magma evolution within a major orogenesis. As a possible geodynamic model, we propose this scheme, which can be summarized as follows:

1. The Late Devonian major collisional stage, recorded throughout the Variscan belt of Europe, was followed by the emplacement within a thick crust experiencing high geothermal gradients of peral-

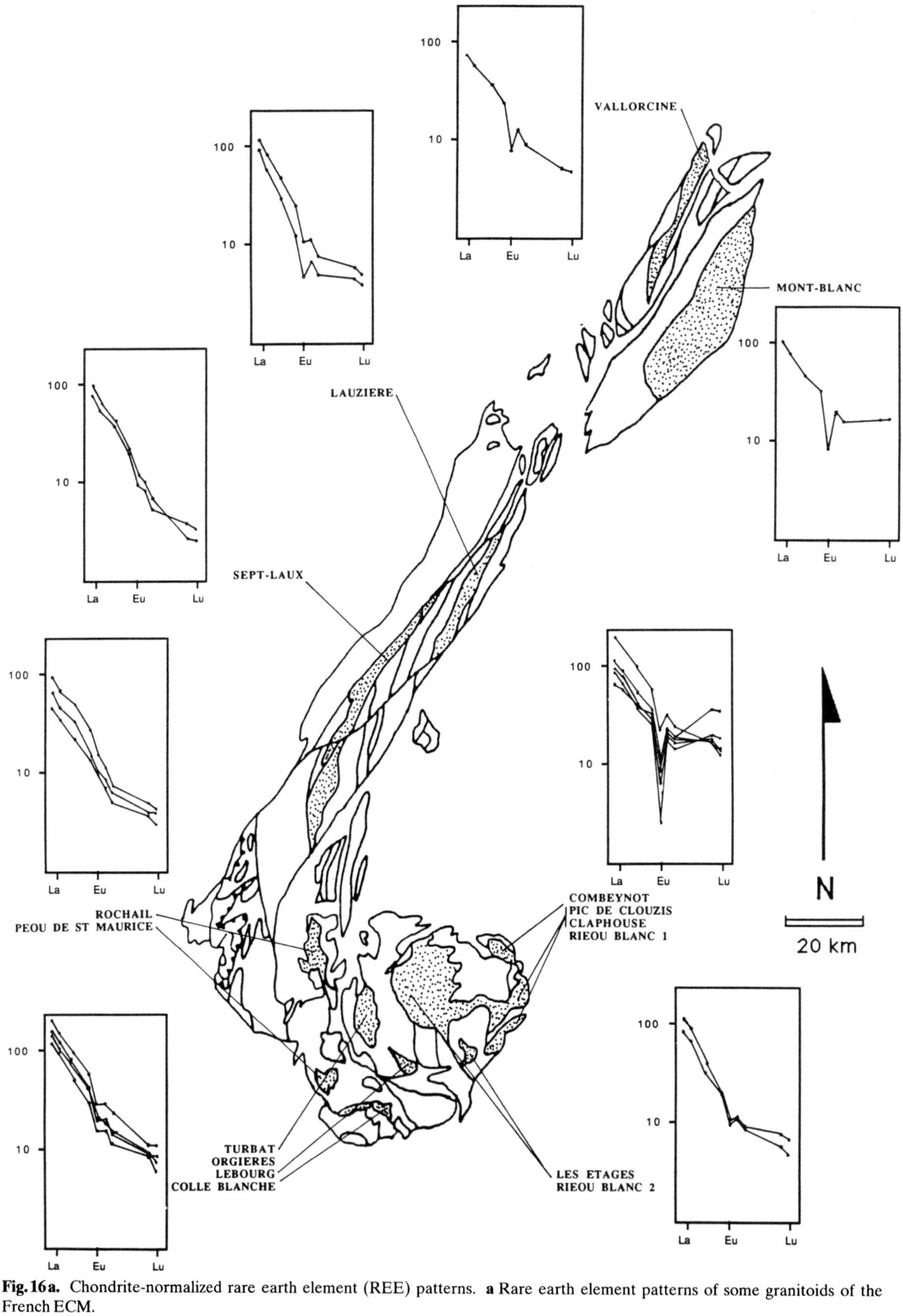

Fig. 16 a. Chondrite-normalized rare earth element (REE) patterns. **a** Rare earth element patterns of some granitoids of the French ECM.

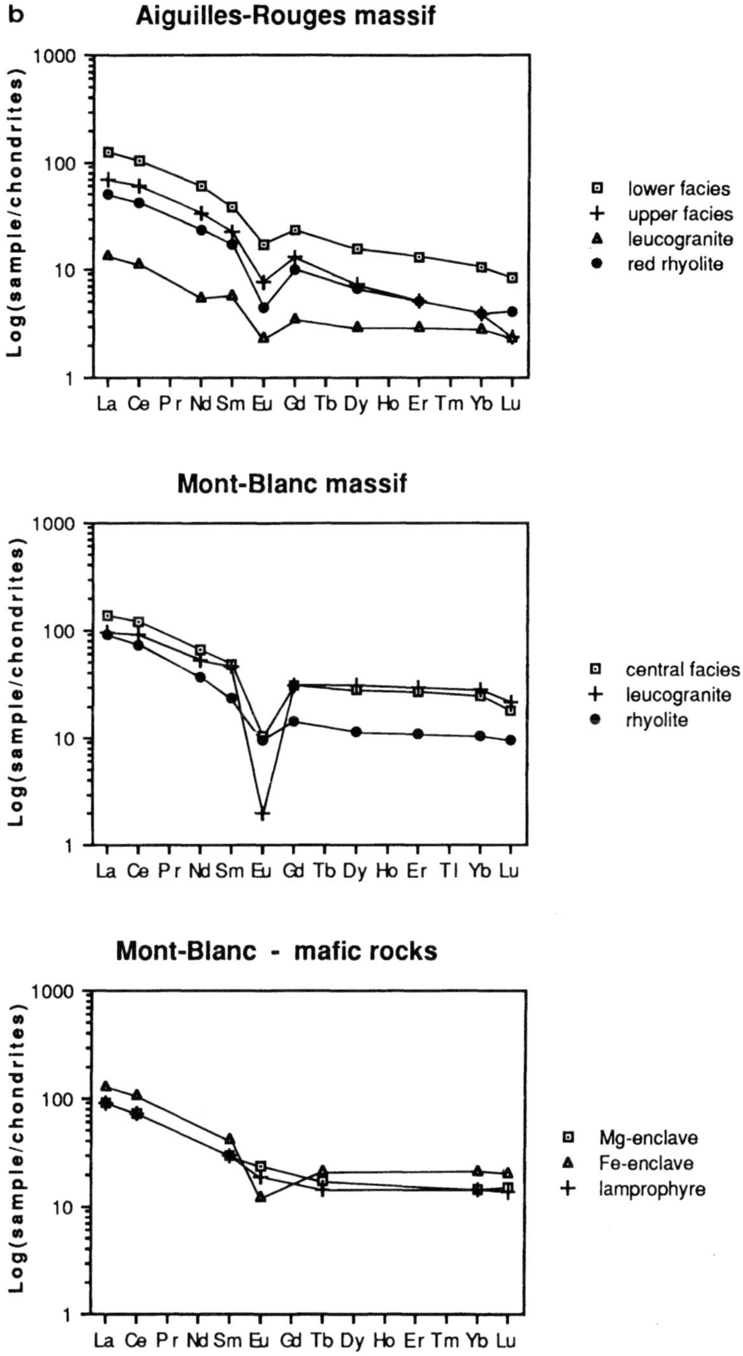

Fig. 16 b. Chondrite-normalized rare earth element (REE) diagrams for Vallorcine and Mont Blanc magmatic rocks. *Above* Leucogranitic facies and lower+upper parts of the main porphyritic facies of the Vallorcine granite, late red rhyolite from the SE part of the Aiguilles-Rouges massif; *middle, below* samples from the Mont Blanc intrusion. Data are from Table 1

muninous associations of crustal melts associated and/or mixed with high-K basic magmas. They characterize the "Hercynotype" of Pitcher, which corresponds to the development of crustal shortening and thickening and associated low-P regional metamorphism, with subsequent "wet" and/or H_2O-deficient crustal anatexis, which may or may not be induced by the intrusion of mantle-derived magmas.

2. Lower to Middle Carboniferous high-K calc-alkaline suites, emplaced as crystal mush (porphyritic facies with K-feldspar megacrysts) at shallow levels, represent the "Caledonian-type" of Pitcher and indicate a post-collisional stage characterized by rapid uplift and erosion, followed by molassic deposition in intracontinental basins created in short-lived (trans)pressional and/or tensional regime. They are followed rapidly by the Late Car-

Table 1. Comparison of geological and petrological data from the Vallorcine and the Mont Blanc granites (After F. Bussy 1990; P. Brändlein 1991; C. Marro 1987)

Criteria	Vallorcine Massif	Mont Blanc Massif
Rock types	Porphyritic coarse-grained syenogranite Zonation from core to rim Magmatic mafic enclaves Restitic supermicaceous lenses Gneissic xenoliths	Porphyritic coarse-grained syenogranite Zonation from core to rim Mafic enclaves (2 types)
Age of emplacement	Pre-Upper Carboniferous	304 ± 3 Ma (U-Pb zr)
Geological features	Sheet-like body 15-km-long, 1-km-thick Intrusive contacts (NW) Tectonic contacts (SE) Magmatic flow texture Dextral strike-slip shear zone	Elongated body 55-km-long, 14-km-wide Intrusive contacts (SW) Tectonic contacts (NW, SE) Magmatic flow texture Strike-slip shear zone
Mineralogy	Zoned feldspars	Zoned feldspars
Biotite	Calc-alkaline to alumino-potassic field Al-silicates Topaz, tourmaline	Fe-rich subalkaline field
Zircon	Crustal anatectic trend	Alkaline to subalkaline trend
Chemistry	S-type A/CNK > 1.1 K_2O/Na_2O = 1.25–1.65	I-type, with minor A-type 0.95 < A/CNK < 1.5 K_2O/Na_2O = 1.1–1.5
R1–R2	Anatectic field Enclaves: calc-alkaline field	Late-orogenic field Enclaves: calc-alkaline field
Rb–Y+Nb	Syn-collisional field	Within-plate field
REE patterns	Moderate Eu anomalies	Large Eu anomalies
Petrogenesis	Extreme acid/basic melt ratio	Moderate acid/basic melt ratio
Probable crustal sources	Metapelites	Dry granulites
Mantle sources	Contaminated mantle Weak contribution	Contaminated mantle Large contribution
Geodynamic setting	Post-collisional Transtensional uplift	Post-collisional Transtensional uplift

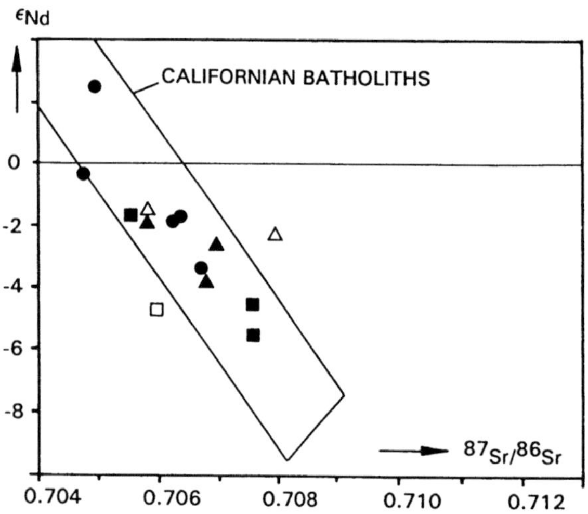

Fig. 17. Initial $^{87}Sr/^{86}Sr$ versus model εNd plots of Zentralgneise (von Quadt and Finger in prep.). Data fall within field of continental margin I-type batholiths. Shown for comparison is the field of California plutons (after De Paolo 1981). *Symbols* as in Fig. 8b

boniferous near-alkaline associations, yielding both alkaline and calc-alkaline characteristics and emplaced as volcanic-plutonic massifs in a major "Basin and Range"-like distensional regime.

Thus, the Carboniferous overall magmatic evolution from "Hercynotype" to "Caledonian type" associations in the Alpine basement clearly substantiates the view that the Variscan belt was emplaced increasingly in an intracontinental position between the Gondwana and Laurasia continental blocks.

3. Subduction-related calc-alkaline associations, defining the "Cordilleran I type", are accordingly very scarce during the Carboniferous. However, in some areas, specifically Eastern and Southern Alps, Late Carboniferous to Early Permian granitoids, displaying normal-K to high-K chemistries, are strongly reminiscent of the circum-Pacific Cordilleran batholiths. Finger and Steyrer (1990) have therefore speculated, that these Late Palaeozoic granitoids indicate a renewed ocean-continent

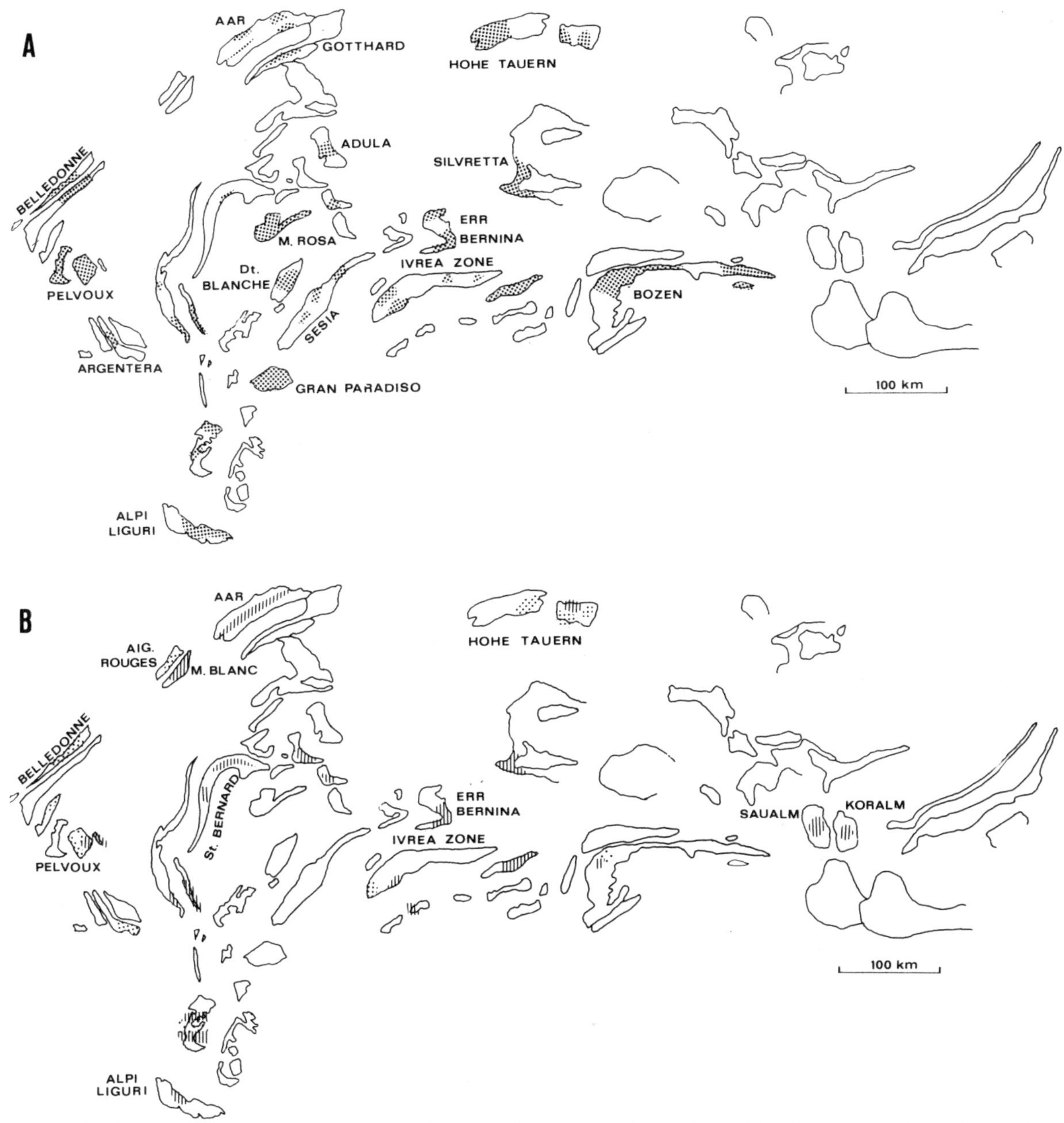

Fig. 18. Sketch maps of pre-Alpine basement units, with emphasis on space and time relationships of magma associations (see text). Location of the areas, see Fig. 1 (after von Raumer and Neubauer, this Vol.). **a** High-K calc-alkaline post-collisional suites and medium-K calc-alkaline Cordilleran suites. **b** ***dots*** Crustal anatexis syncollisional associations; *hatched* near-alkaline to alkaline post-orogenic suites

plate margin convergence at the southern flank of the Variscan belt.

4. Mid-Permian post-orogenic and Late Permian to Triassic early anorogenic A-type plutonic-volcanic complexes constitute a pat of the Western Mediterranean alkaline province. Associated thermal imprints were recorded in mineral ages (Ferrara and Innocenti 1974). The large alkaline magmatic activity of the Western Mediterranean province (Bonin et al. 1987) was related both to post-orogenic continental consolidation of the European plate, containing fragments of Gondwana aggregated to Laurasia, and to precursory stages of the formation of Meso-Tethys oceanic basin created from the Palaeo-Tethys ocean and propagating westwards within the Gondwana continental land.

References

Alberti A, Cernigoi R, Longo-Salvador G, Stolfa D (1983) I graniti di Cima d'Asta (Trentino) nel settore della vetta. Mineral Petrogr Acta 27: 34–49

Alberti A, Comin-Chiaramonti P, Crosilla F, Longo-Salvador G, Sinigoi S, Stolfa D (1980) Il massiccio granitico di Cimon Rava nel complesso plutonico di Cima d'Asta (Trentino), Rend Soc Ital Mineral Petrol 36: 107–122

Aumaitre R, Giraud P, Poncerry E, Vivier G (1985) Les granitoïdes sub-alcalins potassiques de la Lauzière massif cristallin externe de Belledonne: Geol Alpine 61: 17–32

Bakos F, Del Moro A, Visonà D (1990) The Hercynian volcano-plutonic association of Ganna (Lake Lugano, Central Southern Alps, Italy). Eur J Mineral 2: 373–383

Barbarin B (1990) Granitoids: main petrogenetic classifications in relation to the origin and tectonic setting. Geol J 25: 227–238

Bargossi GM, D'Amico C, Visonà D (1979) Hercynian plutonism in the southern Alps. IGCP 5, Newsletter 1, pp 9–32

Barrière M (1977) Le complexe de Ploumanac'h, Massif Armoricain (Essai sur la mise en place et l'évolution pétrologique d'une association plutonique subalcaline tardiorogénique). D Sc Thesis, UBO, Brest, 290 pp

Batchelor RA, Bowden P (1985) Petrogenetic interpretation of granitoid rock series using multicationic parameters. Chem Geol 48: 43–55

Bateman PC, Chappell BW (1979) Crystallization, fractionation, and solidification of the Tuolumne intrusive series, Yosemite National Park, California. Geol Soc Am Bull 1/90: 465–482

Bearth P (1952) Geologie und Petrographie des Monte Rosa. Beitr Geol Karte Schweiz NF 96, 94 pp

Bellieni G, Molin G, Visonà D (1979) The petrogenetic significance of the garnets in the intrusive massif of Bressanone and Vedrette di Ries (Eastern Alps, Italy). Neues Jahrb Mineral Abh 136: 238–253

Bellière J (1988) On the age of mylonites within the Mont-Blanc massif. Geodin Acta 2: 13–16

Biino G, Pognante U (1989) Palaeozoic continental-type gabbros in the Gran Paradiso nappe (Western Alps, Italy): Early-Alpine eclogitization and geochemistry. Lithos 24: 3–19

Boehm C (1986) Petrographische Untersuchungen im Gebiet Val Gliems – Val Russein, Graubünden. Diploma, Univ Bern

Bogdanoff F (1980) Analyse structurale dans la partie occidentale de l'Argentera-Mercantour (Alpes Maritimes). Thèse, Orsay, 316 pp

Bonin B (1990) From orogenic to anorogenic settings: evolution of granitoid suites after a major orogenesis. Geol J 25: 261–270

Bonin B, Platevoet B, Vialette Y (1987) The geodynamic significance of alkaline magmatism in the western Mediterranean compared with West Africa. In: Bowden, P, Kinnaird JA (eds) African geology reviews. Geol J 22: 361–387

Bonin B, Visonà D, Kekelia MA (1991) Hercynian plutonism in the Alpine-Mediterranean belts. In: Fluegel HW, Sassi FP (eds) Pre-Variscan and Variscan events in the Alpine Mediterranean belts. IGCP n 5 Final Volume (in press)

Boriani A, Peyronel Pagliani G (1986) Rapporti tra le plutoniti erciniche e le metamorfiti del "Massiccio dei Laghi" nella zona del Monte Cerano (bassa val d'Ossola). Rend Soc Ital Mineral Petrol 24: 111–142

Boriani A, Colombo A, Giobbi Origoni E, Peyronel Pagliani G (1974) The "Appinite Suite" of Massiccio dei Laghi (Northern Italy) and its relationship to the regional metamorphism. Rend Soc Ital Mineral Petrol 30: 893–917

Boriani A, Colombo A, Macera P (1985) Radiometric geochronology of Central Alps. Rend Soc Ital Mineral Petrol 40: 139–186

Boriani A, Burlini L, Caironi V, Giobbi Origoni E, Sassi A, Sesana E (1988a) Geological and petrological studies on the hercynian plutonism of Serie dei Laghi – geological map of its occurrence between Valsesia and Lago Maggiore (N Italy). Rend Soc Ital Mineral Pet 43: 367–384

Boriani A, Caironi V, Oddone M, Vannucci R (1988b) Some petrological and geochemical constraints on the genesis of the Baveno-Mottarone and Montorfano plutonic bodies. Rend Soc Ital Mineral Petrol 43: 385–394

Brändlein P (1991) Petrographische und geochemische Charakteristika des Vallorcine-Granits, Aiguilles-Rouges-Massiv (Westalpen, Schweiz). thesis University of Erlangen 95 pp. (unpubl.)

Buechi HJ (1987) Geologie und petrographie der Bernina IX: Das Gebiet zwischen Pontresina und Morteratschgletscher. Diploma, ETH Zürich

Buehler C (1983) Petrographische und geochemische Untersuchungen im Gebiet La Tescheppa-Lagrev (Julier-Bernina Decke). Schweiz Mineral Petrogr Mitt 63: 457–477

Buletti M (1985) Petrographisch-geochemische Untersuchungen im Luganer Porphyrgebiet. PhD Thesis, Univ Bern

Burlini L, Caironi V (1988) Geological and petrographical data on the Quarna pluton (Serie dei Laghi). Rend Soc Ital Mineral Petrol 43: 429–444

Bussy F (1987) Interactions physico-chimiques entre le granite du Mont-Blanc et ses enclaves. Schweiz Mineral Petrogr Mitt 67/3: 380–384

Bussy F (1990) Pétrogenèses des enclaves microgrenues associées aux granitoïdes calcoalcalins: exemple des massifs varisque du Mont-Blanc (Alpes occidentales) et miocène du Monte Capanne (Ile d'Elbe, Italie). Mém Géol Univ Lausanne 7: 309 pp

Bussy F, Schaltegger U, Marro C (1989) The age of the Mont-Blanc granite (Western Alps): a heterogeneous system dated by Rb-Sr whole rock determinations on its microgranular enclaves. Schweiz Mineral Petrogr Mitt 69: 3–13

Caby R (1974) Gneiss permocarbonifères d'origine granitique et volcanique dans la Zone Houillère et la Zone du Grand-Saint-Bernard en Val d'Aoste (Italie). Géol Alp 50: 39–44

Cadel G (1984) Geology and uranium mineralization of the Collio basin (Central Southern Alps, Italy). Uranium: 47–65

Callegari E, Fiora L, Compagnoni R, Crisci GM, Sandrone R (1982) Caratterizzazione geochimica degli ortogneiss granitici dei massicci del Gran Paradiso e del Dora Maira (Alpi Occidentali). Rend Soc Ital Mineral Petrol 38: 498–499

Carme F (1970) Successions granitiques et caracteres spécifiques de la granitisation hercynienne dans la chaîne de Belledonne. CR Acad Sci Fr (D) 271: 696–699

Castellarin A, Lucchini F, Rossi PL, Selli L, Simboli G (1988) The Middle Triassic magmatic-tectonic arc development in the Southern Alps. Tectonophysics 146: 79–89

Chappell BW, Stephens WE (1988) Origin of infra-crustal (I-type) granite magmas. Trans R Soc Edinb Earth Sci 79: 71–86

Chappell BW, White AJR (1974) Two contrasting granite types. Pac Geol 8: 172–174

Cocherie A (1984) Interaction manteau-croûte, son rôle dans la genèse d'associations plutoniques calco-alcalines, contraintes géochimiques (éléments en traces et isotopes du Strontium et de l'Oxyène). Thesis, Univ. Rennes I, 246 pp

Compagnoni R, Elter G, Lombardo B (1976) Eterogeneita stratigrafica del complesso degli "Gneiss minuti" nel Massiccio cristallino del Gran Paradiso. Mem Soc Geol Ital 13/1 (1974): 227–239

Cumming GL, Köppel V, Ferrario A (1987) A lead isotope study of the northeastern Ivrea Zone and the adjoining Ceneri Zone (N Italy): evidence for a contaminated subcontinental mantle. Contrib Mineral Petrol 97: 19–30

D'Amico C, Del Moro A (1988) Permian and Triassic Rb-Sr dating in the Permian rhyodacitic ignimbrites of Trentino (Southern Alps). Rend Soc Ital Mineral Petrol 43: 171–180

D'Amico C, Siena F (1977) Conoscenze petrochimiche e geochimiche del complesso plutonico di Cima d'Asta (Trentino). Mem Sci Geol 32: 3–21

D'Amico C, Franceschini C, Nannetti MC (1987) Geochimica di tonaliti, granodioriti, graniti del bordo SW di Cima d'Asta (Trentino). Mineral Petrogr Acta 30: 227–245

Dallmeyer RD (ed) (1989) Terranes in the Circum-Atlantic Paleozoic orogens. Geol Soc Am Spec Pap 230

De Capitani L (1981) Contributo alla conoscenza dei plutoni sudalpini: le masse intrusive della val Biandino (Como). Rend Soc Ital Mineral Petrol 38: 109–118

De Capitani L, Liborio G (1988) Trace element abundance in the val Biandino pluton (Southern Alps, Italy). Mem Sci Geol 40: 99–110

De Capitani L, Delitala MC, Liborio G, Mottana A, Nicoletti M, Petrucciani C (1988) K-Ar dating of the val Biandino plutonic complex (Orobic Alps, Italy). Mem Sci Geol 40: 285–294

De La Roche H, Leterrier J (1973) Transposition du tetraèdre minéralogique de Yoder et Tilley dans un diagramme chimique de classification des roches basaltiques. CR Acad Sci D 276: 3115–3118

De La Roche H, Leterrier J, Grandclaude P, Marchal M (1980) A classification of volcanic and plutonic rocks using R1–R2 diagram and major element analyses. Its relationships with current nomenclature. Chem Geol 29: 183–210

Del Moro A (1987) Sistematica Rb-Sr di alcune magmatiti tardoerciniche dell'area italiana. In: Boriani A, Boriani N (eds) Lezioni e seminari di Geocronologia e Geochimica isotopica. Ric Sci ed Educaz Permanente

Del Moro A, Visonà D (1982) The epiplutonic Hercynian complex of Bressanone (Brixen, Eastern Alps – Italy). Petrographic and radiometric data. Neues Jahrb Mineral Abh 145: 66–85

Del Moro A, Pardini G, Messiga B, Boggio M (1982) Dati petrologici e radiometrici preliminari sui massicci cristallini della Liguria occidentale. Rend Soc Ital Mineral Petrol 38: 73–87

Demeulemeester P (1982) Contribution à l'étude radiométrique à l'argon et au strontium des massifs cristallins externes (Alpes francaises). Thèse, Grenoble, 227 pp

DePaolo DJ (1981) A Nd and Sr isotopic study of the Mesozoic calc-alkaline batholith of the Sierra Nevada and Peninsular Ranges, California. J Geophys Res 86: 10470–10488

Desmons J, Ploquin A (1989) Chimisme du Briançonnais – Grand Saint-Bernard antémésozoïque (Alpes occidentales): une vue synoptique. Géol Alp 65: 1–31

Détraz G (1984) Etude géologique du bord interne de la Zone Houillère briançonnaise entre la vallée de l'Arc et le massif de Péclet-Polset (Alpes de Savoie). Thèse 3e cycle, Univ Grenoble, 163 pp

Didier J (1973) Granites and their enclaves. Elsevier, Amsterdam, 393 pp

Doesseger R (1974) Verrucano und "Bundsandstein" in der Unterengadiner Dolomiten. PhD Thesis, ETH Zürich

Dollinger H (1989) Petrographische und geochemische Untersuchungen des Altkristallins zwischen Nägelisgrätli und Oberaarjoch, Part 1. Diploma, Univ Bern

Duchesne JC (1990) Origin and evolution of monzodiorites related to anorthosites. Schweiz Mineral Petrogr Mitt 70: 189–198

ECOG (1969) Colloquium on the geochronology of Phanerozoic orogenic belts, Switzerland, Field-Trip volume

Eggenberger U (1990) Petrographische und geochemische Untersuchungen im Gebiet zwischen St Moritz und dem Val Bever (Unterostalpin/Err Decke); Geochemischer Vergleich mit anderen Varizischen Intrusiva (Aar/Baveno/ Böttstein/ Bernina/Err/Julier). Diploma, Univ Bern (unpubl)

El Bouseily AM, El Sokkary AA (1975) The relation between Rb, Ba and Sr in granitic rocks. Chem Geol 16: 207–219

Faure-Muret A (1955) Etudes géologiques sur le massif del' Argentera-Mercantour et ses enveloppes sédimentaires. Mém Carte Géol Fr 47: 1–336

Ferrara G, Innocenti F (1974) Radiometric age evidences of a Triassic thermal event in the southern Alps. Geol Rundsch 63: 572-581

Ferrara G, Malaroda R (1969) Radiometric age of granitic rocks from the Argentera Massif (Maritime Alps). Bull Soc Geol Ital 88: 311–320

Finger F, Steyrer HP (1988) Granite types in the Hohe Tauern (Eastern Alps, Austria) – some aspects on their correlation to Variscan plate tectonics processes. Geodin Acta (Paris) 2: 75–87

Finger F, Steyrer HP (1990) I-type granitoids as indicators of a Late Paleozoic convergent ocean-continent margin along the southern flank of the central European Variscan orogen. Geology 18: 1207–1210

Finger F, Friedl G, Matl H (1990) Three contrasting subtypes of I-type granitoid associations with "high Na_2O", "high CaO" and "high K_2O" characteristics. Mitt Österr Geol Ges 135: 233–239

Flisch M (1987) Geologische, petrographische und isotopengeologische Untersuchungen an Gesteinen des Silvretta Kristallins. PhD Thesis, Univ Bern (unpubl)

Franks G (1968) A study of upper Paleozoic sediments and volcanics in the northern part of the eastern Aar Massif. Eclogae Geol Helv 61: 49-140

Frey M, Hunziker JC, O'Neil JR, Schwander HW (1976) Equilibrium-disequilibrium relations in the Monte Rosa granite, Western Alps: petrological, Rb-Sr and stable isotope data. Contrib Mineral Petrol 55: 147–179

Gasquet D (1979) Etude pétrologique, géochimique et structurale des terrains cristallins de Belledonne et du Grand Chatelard traversés par les galeries E. D. F. Arc-Isère, Alpes Françaises. Thèse 3ème cycle, Univ Grenoble, 230 pp

Giobbi Origoni E (1987) Hercynian plutonism in the western of the southern Alps. In: Fluegel HW, Sassi FP, Grecula P (eds) Pre-variscan and variscan events in the Alpine-Mediterranean mountain belts – Mineralia slovaca Monography. Alfa, Bratislava, pp 237–250

Giobbi Origoni E, Bocchio R, Boriani A, Carmine M, De Capitani L (1988) Late-Hercynian mafic and intermediate intrusives of Serie dei Laghi (N Italy). Rend Soc Ital Mineral Petrol 43: 395–410

Gnos E (1988) Geologie und Petrographie des westlichen Brunnitals. Diploma, Univ Bern (unpubl)

Graf K (1987) Geologische Untersuchungen der Sedimente und Vulkanite der Landwasserkette zwischen Valbellahorn und Wiesen (Graubünden). Diploma, Univ Bern

Green TH, Ringwood AE (1968) Genesis of the calc-alkaline igneous rock suite. Contr Mineral Petrol 18: 105–162

Guillot F, Liégeois JP, Fabic J (1991) Des granophyres du Cambrien terminal dans le Mont Pourri (Vanoise, zone briançonnaise): première datation U Pb sur zircon d'un socle des

zones internes des Alpes françaises. CR Acad Sci Paris 313 (II): 239–244
Hibbard MJ (1981) The magma mixing origin of mantled feldspars. Contrib Mineral Petrol 76: 158–170
Hunziker JC (1970) Polymetamorphism in the Monte Rosa, Western Alps. Eclogae Geol Helv 63: 151–161
Hurford AJ, Hunziker JC (1989) A revised thermal history for the Gran Paradiso massif. Schweiz Mineral Petrogr Mitt 69: 319–329
Kerrich R, Allison I, Barnett R, Moss S, Starkey J (1980) Microstructural and chemical transformations accompanying deformation of granite in a shear zone at Miéville, Switzerland: with implications for stress corrosion cracking and superplastic flow. Contrib Mineral Petrol 73: 221–242
Köppel V, Grünenfelder M (1975) Concordant U-Pb ages of monazite and xenotime from the Central Alps and the timing of the high temperature Alpine metamorphism, a preliminar report. Schweiz Mineral Petrogr Mitt 55: 129–132
Lameyre J, Bowden P (1982) Plutonic rock type series: discrimination of various granitoid series and related rocks. J Volcanol Geother Res 14: 169–186
Le Fort P (1973) Géologie du Haut Dauphiné cristallin (Alpes françaises). Sci Terre Mém 25: 1–373
Le Maitre RW (1984) A proposal by the IUGS Subcommision on the systematics of igneous rocks for a chemical classification of volcanic rocks based on the Total Alkali Silica (TAS) diagram. Aust J Earth Sci 31: 243–255
Lopez-Escobar L, Frey FA, Oyarzun J (1979) Geochemical characteristics of central Chile granitoids. Contrib Mineral Petrol 70: 439–450
Marro C (1987) Histoire des granitoïdes du Mont-Blanc en Suisse. Bull Soc Frib Sci Nat 76: 73–128
Marro C (1988) Organisation géochimique et intrusion du granite du Mont-Blanc et de deux leucogranites. Schweiz Mineral Petrogr Mitt 68: 521–529
Matte P (1986) Tectonics and plate tectonic model for the Variscan belt of Europe. Tectonophysics 126: 329–374
Mercolli I (1985) The volcanic suite from the Julier area: a case history of Hercynian magmatism in the Alps. Habilitationsschrift, Univ Bern
Mercolli I, Oberhänsli R (1988) Variscan tectonic evolution in the Central Alps: a working hypothesis. Schweiz Mineral Petrogr Mitt 68 (3): 491–500
Messina A, Russo S, Fichera S, Minzoni N (1987) Stock intrusivo di Cima di Mezzogiorno (Val Vanoi, Cima d'Asta – Trentino): caratteri strutturali e petrochimici delle rocce. Rend Soc Ital Mineral Petrol 42: 103–122
Michel R (1953) Les schistes cristallins des massifs du Grand Paradis et de Sesia-Lanzo (Alpes franco-italiennes). Sci Terre Nancy 1, 287 pp
Mielke P, Winkler HGF (1979) Eine bessere Berechnung der Mesonorm für granitische Gesteine. Neues Jahrb Mineral Monatsh 10: 471–480
Monié P (1990) Preservation of Hercynian $^{40}Ar/^{39}Ar$ ages through high-pressure low-temperature Alpine metamorphism in the Western Alps. Eur J Mineral 2: 343–361
Nachit H, Razafimahefa N, Stussi JM, Caron JP (1985) Composition chimique des biotites et typologie magmatique des granitoïdes. CR Acad Sci Paris 301, II (11): 813–818
Negga HS (1984) Compartement de l'uranium lors des métamorphisms tardi-hercynien et alpin dans les massifs des Aiguilles Rouges et de Belledonne (Vallorcine et Lauzière), Alpes occidentales. Thèse de 3ème cycle, CREGU, Nancy, 400 pp
Neugebauer J (1988) The Variscan plate tectonic evolution: an improved "Iapetus model". Schweiz Mineral Petrogr Mitt 68: 313–334
Oberhänsli R, Hunziker JC, Martinotti G, Stern WB (1985) Geochemistry, geochronology and petrology of Monte Mucrone: an example of Eo-Alpine eclogitization of Permian granitoids in the Sesia-Lanzo Zone, Western Alps, Italy. Chem Geol 52: 165–184
Oberhänsli R, Schenker F, Mercolli I (1988) Indications of Variscan nappe tectonics in the Aar Massif. Schweiz Mineral Petrogr Mitt 68 (3): 491–500
Ohenstetter M, Rossi Ph (1987) Signification des complexes basiques-ultrabasiques, associés aux granitoïdes dans l'évolution de la chaîne varisque du Sud de l'Europe. Bull Soc Géol Fr 8, III/5: 801–810
Paquette JL, Chopin C, Peucat JJ (1989) U-Pb zircon, Rb-Sr and Sm-Nd geochronology of high- to very-high-pressure meta-acidic rocks from the western Alps. Contrib Mineral Petrol 101: 280–289
Peacock MA (1931) Classification of igneous rock series. J Geol 39: 54–67
Pearce JA, Harris N, Tindle AG (1984) Trace element discrimination diagrams for the tectonic interpretation of granitic rocks. J Petrol 25: 956–983
Peters TJ, Matter A, Bläsi HR, Gautschi A (1986) Sondierbohrung Böttstein, Geologie, Textband. Tech Ber 85–02
Pinarelli L, Boriani A, Del Moro A (1988) Rb-Sr geochronology of the Lower Permian plutonism in the Massiccio dei Laghi. Southern Alps (NW Italy). Rend Soc Ital Mineral Petrol 43: 411–428
Pitcher WS (1979) The nature, ascent and emplacement of granitic magmas. J Geol Soc Lond 136: 627–662
Pitcher WS (1988) Andean batholiths and marginal basins. Rend Soc Ital Mineral Petrol 43: 275–280
Pitcher WS (1983) Granite type and tectonic environment. In: Hsü K (ed) Mountain building processes. Academic Press, London, pp 19–40
Pitcher WS (1987) Granites and yet more granites forty years on. Geol Rundsch 76: 51–79
Pupin JP (1976) Signification des caractères morphologiques du zircon commun des roches en pétrologie. Base de la Méthode Typologique. Applications. PhD Thesis, Univ Nice, 394 pp
Pupin JP (1980) Zircon and granite petrology. Contrib Mineral Petrol 73: 207–220
Pupin JP (1985) Magmatic zoning of Hercynian granitoids in France based on zircon typology. Schweiz Mineral Petrogr Mitt 65: 29–56
Pupin JP (1988) Granites as indicators in paleogeodynamics. Rend Soc Ital Mineral Petrol 43 (2): 237–262
Rageth R (1984) Intrusiva und extrusiva der Bernina Decke zwischen Morteratsch und Berninapass (Graubünden). Schweiz Mineral Petrogr Mitt 64: 83–109
Reinsch D (1977) High pressure rocks from Val Chiusella (Sesia-Lanzo Zone, Italian Alps). Neues Jahrb Mineral Abh 130: 89–102
Riesen D Geologie und Petrographie des östlichen Brunnitals. Diploma, Univ Bern (in prep)
Sandrone R, Cordola M, Fontan D (1986) Pre-Variscan granitoids in the Doira-Maira Massif (Cottian Alps). Atti Accad Sci Torino 120, 11 pp
Sandrone R, Sacchi R, Cordola M, Fontan D, Villa IM (1988) Metadiorites in the Dora-Maira polymetamorphic basement (Cottian Alps). Rend Soc Ital Mineral Petrol 43: 593–618

Sartori M (1988) L'unité du Barrhorn (Zone Pennique, Valais, Suisse). Un lien entre les Préalpes médianes rigides et leur socle paléozoïque. Thèse, Lausanne, 150 pp

Sassi FP, Cavazzini G, Visonà D (1985) Radiometric geochronology in the Eastern Alps: results and problems. Rend Soc Ital Mineral Petrol 40: 187–224

Schaltegger U (1984) Geologie und Petrographie der Gneiszone von Erstfeld in der Umgebung des Sustenpasses, Aarmassiv. Diploma, Univ Bern

Schaltegger U, Quadt von A (1990) U-Pb dating of the Central Aar granite (Aar Massif, Central Alps). Schweiz Mienral Petrogr Mitt 70: 361–371

Schaltegger U, Corfu F, Krogh TE (1991) Precise U-Pb ages and initial Hf isotopic composition of late Hercynian mafic to acidic intrusives in the Aar massif, Central Alps. Terra Abstr 3: 37

Schaltegger U, Corfu F (1992) The age and source of late Hercynian magmatism in the Central Alps: evidence from precise U-Pb ages and initial Hf isotopes. Contrib Mineral Petrol, 111: 329–344.

Schenker F (1986) Spätpaläozoischer saurer Magmatismus und Beckenbildung im Aarmassiv unter kompressiv Tektonik. PhD Thesis, Univ Bern

Schenker F (1987) Hinweise für kompressive Tektonik während der Ablagerung von oberpaläozoischen Sedimenten und Vulkaniten im Aarmassiv. Bull Ver Schweiz Pet Geol Ing 54/124: 45–57

Stille P, Buletti M (1987) Nd-Sr isotopic characteristics of the Lugano volcanic rocks and constraints on the continental crust formation in the South Alpine domain (NItaly-Switzerland). Contrib Mineral Petrol 96: 140–150

Streckeisen A (1974) Classification and nomenclature of plutonic rocks. Geol Rundsch 63: 773–786

Thélin P (1983) Les gneiss œillés de la nappe du Grand Saint-Bernard: essai d'évaluation des critères susceptibles d'en préciser l'hérédité pré-métamorphique (Alpes valaisannes, Suisse). Thesis, Univ Lausanne, 485 pp

Thélin P (1987) Nature originelle des gneiss œillés de Randa. Bull Geol Lausanne 290: 75 pp

Vernon RH (1983) Restite, xenoliths and microgranitoid enclaves in granites. J Proc R Soc N S W 116: 77–103

Visonà D (1977) Contributo alla conoscenza del massiccio di Bressanone (Alpi Orientali). Rend Soc Ital Mineral Petrol 33/2: 647–665

Visonà D (1980) Lo stock granitico a cordierite del M Sella nel massiccio di Bressanone (Alpi Orientali). Rend Soc Ital Mineral Petrol 36: 91–106

Visonà D, Alberti F, Stefani C, Stenico L (1987) Le plutoniti di Chiusa, Dosso Lives e Luson: una serie calcalcalina ercinica nelle Alpi Orientali. Mem Sci Geol 39: 85–99

Viterbo C (1959) La composizione chimico-petrografica di alcune rocce tipiche del Gran Paradiso. Rend Soc Mineral Ital 15: 255–281

Viterbo C (1961) Caratteri chimico-petrografici di alcune intercalazioni famiche negli scisti del cristallino antico della zona Sesia-Lanzo (Alpi Piemontesi). Rend Soc Mineral Ital 17: 591–598

Vittoz P, Costarella R, Vivier G, Oliver RA (1987) Typologie des granitoïdes hercyniens du Haut Dauphiné par l'étude des terres rares et de la morphologie des cristaux de zircon: mise en évidence d'une zonation magmatique dans ce massif. Géol Alpine 63: 119–163

Vivier G, Menot RP, Giraud P (1987) Magmatismes et structuration orogénique paléozoiques de la chaîne de Belledonne (massifs cristallins externes). Le domaine sud-occidental. Geol Alpine 63: 25–53

von Raumer JF (1984 a) The external massifs, relics of Variscan basement in the Alps. Geol Rundsch 73: 1–31

von Raumer JF (1984 b) The Hercynian basement in the Helvetic realm and central Alps. Mem Soc Geol Ital 29: 57–69

von Raumer JF (1987) Les massifs du Mont-Blanc et des Aiguilles Rouges: témoins de la formation de croûte varisque dans les Alpes occidentales. Géol Alp 63: 7–24

von Raumer JF (1988) Caledonian-Variscan structures in the Alps: an introduction. Schweiz Mineral Petrogr Mitt 68: 291–299

White AJR, Chappell BW (1983) Granitoid types and their distribution in the Lachlan fold belt, southeastern Australia. In: Roddick JA (ed) Circum Pacific plutonic terranes. Geol Soc Am Mem 159: 21–34

Wopfner H (1984) Permian deposits of the Southern Alps as product of initial alpidic taphrogenesis. Geol Rundsch 73: 259–277

Wuethrich H (1965) Rb-Sr Altersbestimmung am alpin überprägten Aarmassiv. Schweiz Mineral Petrogr Mitt 45: 875–971

Zaugg F (1987) Geologische Untersuchungen der Sedimente und Vulkanite der Landwasserkette zwischen Sandhubel und Wiesen (Graubünden). Diploma, Uiv Bern (unpubl)

Ziegler P (1986) Geodynamic model for the Paleozoic crustal consolidation of Western and Central Europe. Tectonophysics 126: 303–328

Zulauf R (1963) Zur Geologie der tiefpenninischen Zonen nördlich der Dora Baltea im oberen Val d'Aosta (Italien). Truninger, Zürich, 150 pp

Late Palaeozoic – Early Mesozoic Plate Reorganization: Evolution and Demise of the Variscan Fold Belt

P. A. ZIEGLER

Abstract

Closure of the Proto-Tethys Proto-Atlantic Ocean was diachronous along the Variscan-Appalachian fold belt. Collision of Gondwana and Laurussia commenced in Famennian-Visean time in the western Mediterranean domain and propagated during the Late Carboniferous westwards and eastwards. By Westphalian time the Proto-Atlantic was closed. In the European Variscan fold belt, orogenic activity terminated at the end of the Westphalian. Development of Stephanian-Autunian wrench-fault systems, dissecting the Variscan fold belt, can be related to a change in the convergence direction between Gondwana and Laurussia during their terminal Alleghenian suturing phase. This was followed by a fundamental plate boundary reorganization and the development of a new tensional regime in the interior of Pangea. Late Permian and Triassic southward propagation of the Arctic-North-Atlantic rift system and westward propagation of the Tethys rift system involved tensional reactivation of Permo-Carboniferous crustal discontinuities. During the Mid-Jurassic a new divergent-transform plate boundary was established between Laurasia and Gondwana.

1 Introduction

This chapter attempts to provide the reader with a summary of plate motions which governed the Carboniferous suturing of Pangea and its Early Mesozoic break-up, and thus had repercussions on the geological framework of the basement in the Alpine domain.

Reconstructions presented here are based on the assumption that the Western and Central Mediterranean area was occupied during the Late Carboniferous and Early Permian by a coherent Variscan fold belt and thus by continental crust, whereas in the Eastern Mediterranean-Black Sea area the Variscan fold belt faced the Proto-Tethys Ocean, separating it from the northern, passive margin of the Gondwana subcontinent. The dimensions of the area occupied by continental crust in the Western and Central Mediterranean domain were defined on the basis of the Permo-Triassic trans-Atlantic fit of North America, Africa and Europe. For the Mesozoic break-up pattern, the Alpine-Mediterranean area was subdivided into such generally recognized major tectono-stratigraphic units as, for instance, the Italo-Dinarid promontory; dimensions of these units were kept constant during the Mesozoic opening of Tethys and were only reduced upon their involvement in the Cretaceous early Alpine collision (Ziegler 1988).

The data base, on which the Palaeozoic reconstructions given in this chapter were built, includes only limited information from the Alpine area. However, it should be kept in mind that major structural elements of the Variscan fold belt, as defined in the Alpine foreland, project into the autochthonous basement and the external massifs of the Alps. For these areas the evolution of the extra-Alpine parts of the Variscan fold belt may provide an analogue. On the other hand, Variscan basement complexes outcropping in the Penninic and Austro-Alpine nappes and in the South Alpine and Carnic domain are derived from locations that probably do not find equivalents in the Variscides of the European foreland; their geological record may therefore not conform to the evolutionary schemes developed by, for instance, Matte (1986) and Ziegler (1989, 1990). This concerns particularly the timing of detachment of potential intra-Alpine terranes from Gondwana, their accretion to the southern margin of Laurussia and ultimately the timing of their involvement in the Variscan orogeny. Available data indicate also that these areas were affected by the latest Carboniferous-earliest Permian phase of wrench-faulting that is so much in evidence in the Alpine foreland. The onset of crustal extension during Late Permian and Triassic times, culminating in Mid-Jurassic crustal separation between Africa and Europe is diachronous in the various parts of the Alpine and Mediterranean domain.

By providing a plate tectonic reference base, it is hoped that this chapter contributes towards a better understanding of the Late Palaeozoic and Early Mesozoic evolution of the Alpine domain. For detailed discussions of the subject treated here, and a more comprehensive reference list, the reader is referred to Ziegler (1988, 1989, 1990).

2 Variscan Suturing of Gondwana and Laurussia

Late Devonian to Early Permian convergence and collision of Gondwana and Laurussia mega-continents resulted in the consolidation of the Variscan fold belt of Europe, the Marathon-Ouachita-Appalachian fold belt of North America and the Mauretanides fold belt of northwest Africa. Considering an Early Permian continent assembly, these fold belts formed a coherent mountain system along which Gondwana and Laurussia were welded together, thus forming the core of the Permo-Triassic super-continent Pangea. Crustal shortening in the European Variscan fold belt ceased during the Late Westphalian, whereas the Appalachian-Mauretanides and Ouachita-Marathon fold belts became tectonically inactive at the end of the Early Permian.

2.1 *Variscan Orogeny*

Referring to Ziegler (1988, 1989, 1990), palaeomagnetic data and palaeoclimatological criteria indicate that Laurussia had remained more or less stationary during the Devonian whereby the equator crossed the Hudson Bay and southern Greenland. On the other hand, the Devonian clockwise rotational movement of Gondwana persisted even after its Famennian collision with Laurussia. Upon their collision, increased coupling between these two mega-continents is evident by their joint clockwise rotation and northward drift during the Early Carboniferous. Although northward movements ceased at the transition from the Early to the Late Carboniferous, clockwise rotation of Laurussia and Gondwana, amounting to a total of about 54 °, continued during the Late Carboniferous to Early Permian consolidation of the Hercynian megasuture (Figs. 1–3; Ziegler 1989; see also McKerrow and Scotese 1990).

Early Carboniferous lateral propagation of the collision front between Gondwana and Laurussia was combined with the gradual build-up of back-arc compressional stresses in the Variscan geosynclinal domain, causing folding and partial destruction of the Central Armorican-Saxothuringian Basin and gradual closure of the Cornwall and Rhenish back-arc oceanic basins of the Rhenohercynian zone (Fig. 1). At the transition to the Namurian, the Variscan orogenic front collided with the northern shelves of the Rhenohercynian Basin; this coincided with the termination of lateral movements along the Arctic-North Atlantic megashear and the beginning of crustal extension in the Norwegian-Greenland Sea area (cf. Figs. 1 and 2; Ziegler 1989, 1990).

As a consequence of the full-scale collision of Gondwana and Laurussia, the Variscan orogenic system of Europe entered during the Late Visean a Himalayan-type continent-to-continent collisional stage. Continued Late Carboniferous dextral-oblique convergence of Gondwana and Laurussia and progressive closure of the Proto-Atlantic and western Proto-Tethys oceans was accompanied by the rapid propagation of their collision front into the Appalachian and the Alpine-Central Mediterranean domains (Fig. 2). In the latter, timing of onset of syn-orogenic flysch deposition illustrates that multiple collisions of continental fragments and commensurate closure of intervening oceanic and/or para-oceanic basins occurred and that these were not synchronous throughout the area (Vai 1980). Correspondingly, the onset of the Variscan deformation phase varies in time and space in the different Alpine and Mediterranean parts of the Variscan fold belt. Similarly, the compressional overpowering of the individual back-arc extensional basins of the Variscan geosynclinal system is not fully synchronous.

During the Late Carboniferous, major crustal shortening was achieved in the European Variscan fold belt, as indicated by the occurrence of major nappe structures, in part involving basement, for instance in the Moldanubian area, the southern Massif Central, in Iberia and also in the Variscan externides (Julivert et al. 1977; Thiele 1977; Burg and Matte 1978; Meissner et al. 1981; Behr et al. 1982; Matte 1983, 1986; Burg et al. 1984; Cazes et al. 1986; Behr and Heinrichs 1987). Hovewer, the amount of crustal shortening that was achieved during the Variscan orogeny is difficult to quantify. Palaeomagnetic data suggest that post-Devonian crustal shortening between the Armorican Massif and the northern Variscan foreland does not exceed the palaeomagnetic margin of error and thus cannot be larger than some 500 km (Jones et al. 1979). Moreover, the Carboniferous apparent polar wander paths of Africa and Europe are still not sufficiently constrained to permit quantification of the total amount of crustal shortening in the Variscan orogen (Daly and Irving 1983; Aïfa et al. 1990). Yet, widespread occurrence of low-

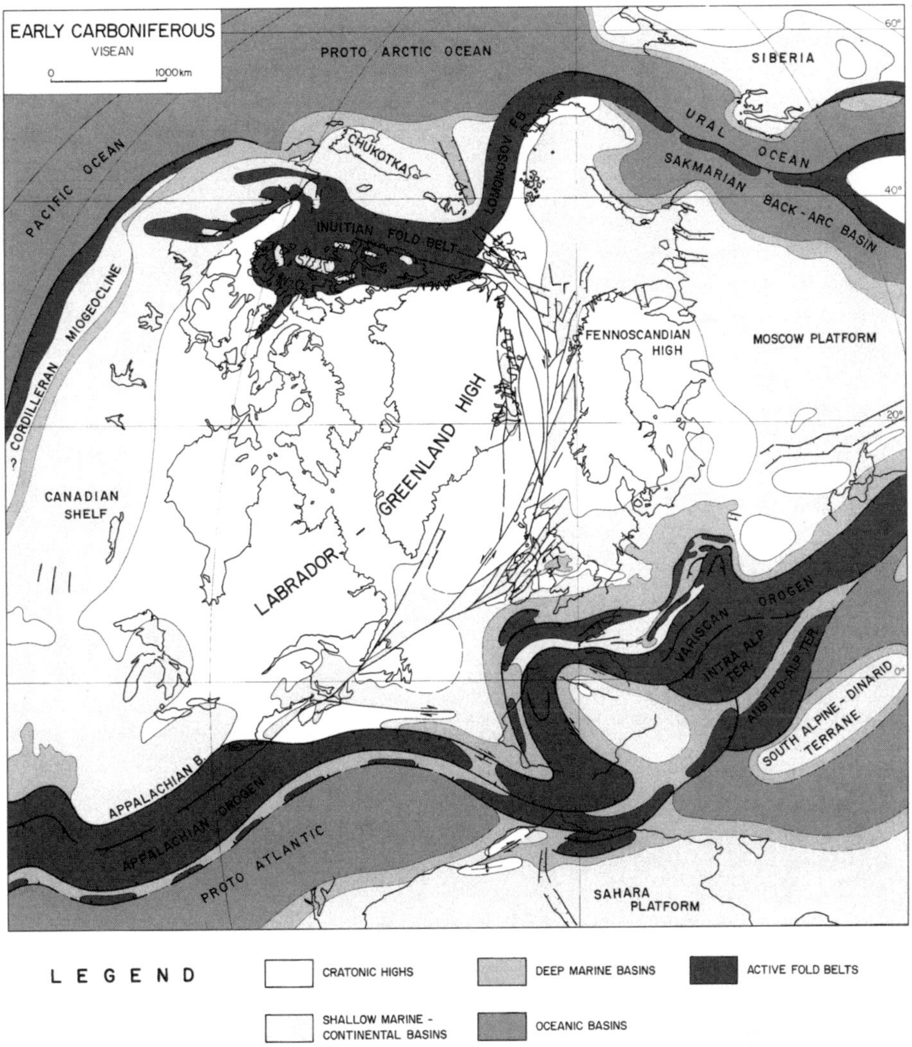

Fig. 1. Early Carboniferous framework of Arctic-North Atlantic domain. *B* Basin; *FB* fold belt; *TER* terrane

pressure metamorphic rocks and late- to post-orogenic calc-alkaline intrusives in the Variscan fold belt can be taken as indirect evidence that significant amounts of crustal shortening, involving crustal delamination, subduction and anatectic remobilization of lower crustal material and partial melting of upper mantle material, took place during the Variscan orogeny (Zwart and Dornsiepen 1978; Matte 1983, 1986; Ziegler 1984, 1986).

The occurrence of granitic intrusions in Cornwall and in the Harz Mountains, some 75 km south of the Variscan deformation front, suggests that the Benioff zone(s), along which lower crustal and upper mantle material was subducted, dipped northward from the Proto-Tethys suture(s) at a relatively gentle angle under the Variscan internides and possibly even under the Variscan externides (Shackleton et al. 1982); in this respect it should be kept in mind that the south-dipping subduction zone, which paralleled the northern margin of the Normannian-Mid-German High and which was active during the closure of the Rhenohercynian basin, formed a conjugate to the north-dipping Proto-Tethys subduction zone(s). In the western parts of the Variscides, the Carboniferous Proto-Tethys suture has to be sought to the south of the Aquitaine-Cantabrian block. In the central and eastern Variscides, the Proto-Tethys suture is difficult to locate owing to severe Alpine overprinting, limited outcrops of the Variscan basement complex, and uncertainties about the definition and dimensions of potential allochthonous terranes (Ziegler 1986, 1988; see also Pin 1990).

The complex, arcuate shape of the Variscan fold belt, involving north and south verging nappes, is

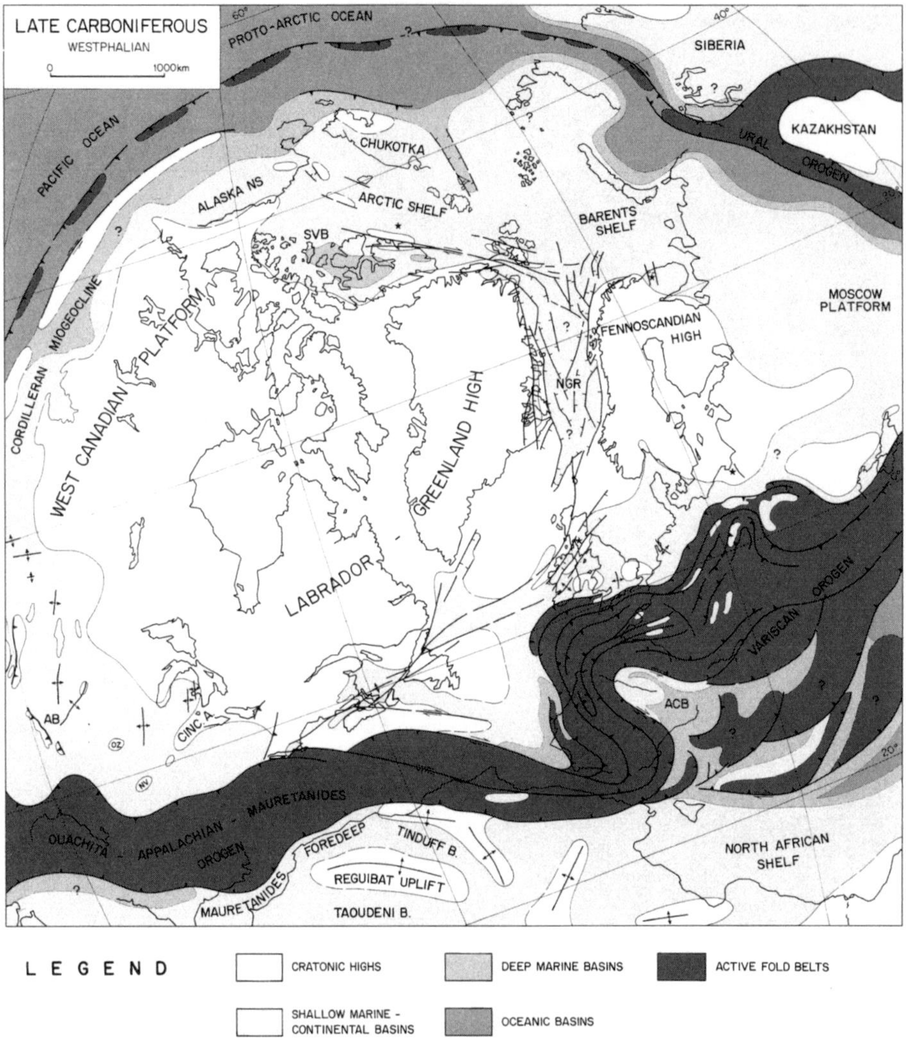

Fig. 2. Westphalian framework of Arctic-North Atlantic domain. *AB* Anadarko Basin; *ACB* Aquitaine-Cantabrian Basin; *B* basin; *CINC* Cincinnati Arch; *NGR* Norwegian-Greenland Sea rift; *NS* Alaska North Slope; *NV* Nashville Dome; *OZ* Ozark Uplift

partly the result of it being draped around a system of relatively rigid, internal micro-cratons (e.g. Aquitaine-Cantabrian and East Silesian block) and partly also of the irregular geometry of its northern and southern forelands (Fig. 2); these were shaped respectively by the pattern of the Devonian-Early Carboniferous back-arc rift systems and by the Ordovician to Carboniferous Gondwana rifts. During the Variscan orogeny, increasing space constraints between the African and European forelands caused transpressional northwestwards "escape" of the Aquitaine-Cantabrian block and accentuation of the Ibero-Armorican arc, and northeastwards "escape" of the Bohemian-Saxothuringian block and accentuation of the Sudetic arc. Development of these arcs involved major wrench-faulting, rotation of individual blocks and considerable oroclinal bending of thrust elements. This concept is supported by palaeomagnetic data (Perroud and Bonhommet 1981; Bachtadse et al. 1983; Perroud and van der Voo 1983; Edel et al. 1984; Lorenz and Nicholls 1984; Bachtadse and van der Voo 1986; Bonhommet and Perroud 1986; Edel 1987).

During the Late Carboniferous collision of Gondwana and Laurussia, progressive lithospheric thickening was accompanied by thrust propagation into the North African and European forelands in which the build-up of intra-plate stresses resulted in Westphalian deformation of the Sahara Platform and inversion of Carboniferous rifts in the British Isles (Fig. 2; see Ziegler 1988, 1990).

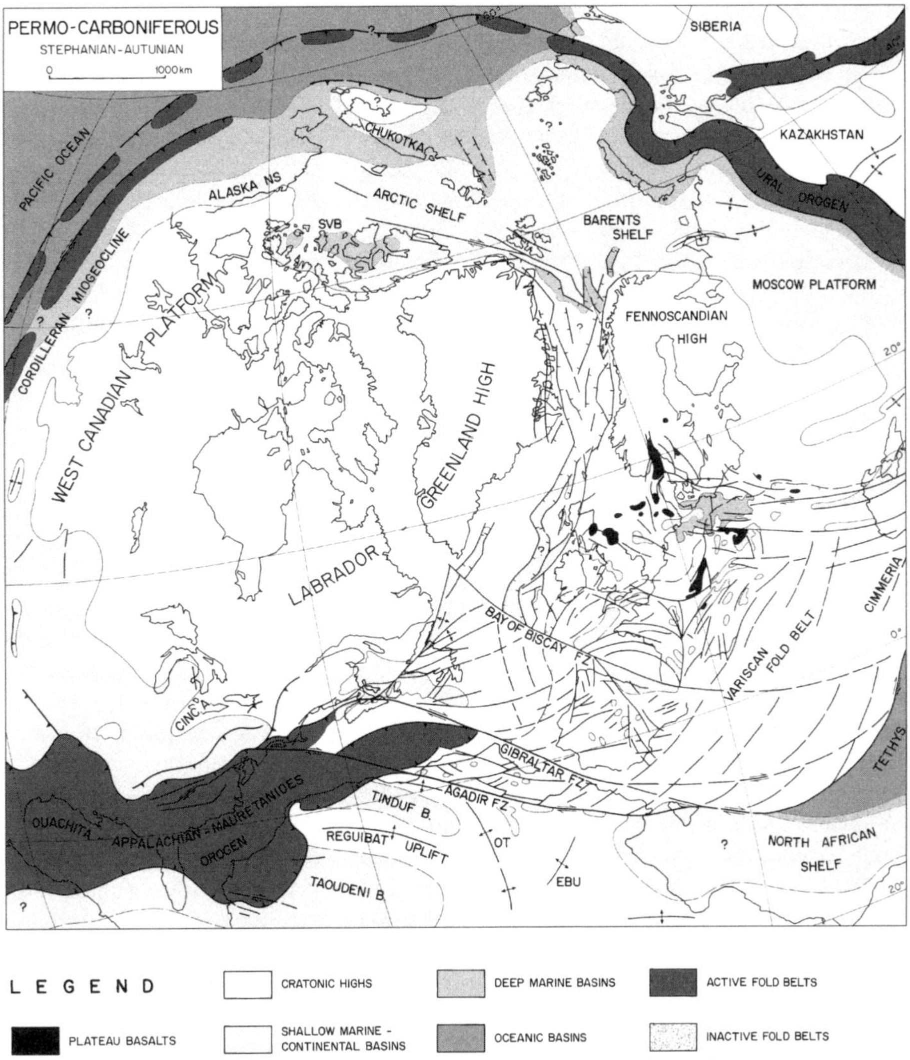

Fig. 3. Stephanian-Autunian framework of Arctic-North Atlantic domain. *B* Basin; *EBU* El Biot Uplift; *FZ* fracture zone; *NS* Alaska North Slope; *OT* Ougarta Trough

2.2 Alleghenian Orogeny and Dextral Translation Between Europe and Africa

During the Stephanian and Autunian terminal phases of the Variscan orogenic cycle, Gondwanda continued to converge with Laurussia. A further increase in coupling of these mega-continents is evident from their joint northward drift by some 8°. By Westphalian time the Proto-Atlantic Ocean had closed. This marked the onset of the Himalayan-type Alleghenian orogeny of the Appalachian-Mauretanides and Marathon-Ouachita segment of the evolving Hercynian megasuture. On the other hand, crustal shortening in the Variscan fold belt had terminated during the Late Westphalian. During the Stephanian-Autunian, the Variscan fold belt became dissected by a system of post-orogenic, conjugate fractures which developed in response to dextral translations between North Africa and Europe during the Alleghenian orogeny (Fig. 3; Ziegler 1982, 1989, 1990; Rast 1984). Development of this fracture system, which caused partial destruction of the Variscan fold belt, is thought to reflect a change in the convergence direction between Gondwana and Laurussia that may well have been a consequence of collisional lithospheric thickening having reached in the Variscan fold belt the point at which no further orthogonal crustal shortening could be achieved.

Main elements of this fault system are the dextral Tornquist-Teisseyre, Bay-of-Biscay, Gibraltar-Minas and Agadir fracture zones. Areas located between these four principal fracture systems became dissected by a complex set of conjugate shear faults. At the northwestern termination of the Tornquist-Teisseyre line, the Oslo Graben developed as a major pull-apart feature. Tensional stresses developing at the termination of the Bay-of-Biscay fracture zone, and of subsidiary fault systems crossing the British Isles, may have facilitated the southward propagation of the Norwegian-Greenland Sea rift system. The Gibraltar and Agadir fracture zones mark the northern termination of the Alleghenian orogen (Kent and Keppie 1988). Activity along these fault systems probably induced during the Late Autunian partial inversion of the Permo-Carboniferous basins of the Canadian maritime provinces (Maritime disturbance: Howie and Barss 1975; Gibling et al. 1987; McCutcheon and Robinson 1987).

The post-Variscan shear-fault system transsecting the Variscan fold belt is interpreted as representing an intra-continental diffuse plate boundary between the Africa-Western Proto-Tethys segment of the Gondwana plate and the European part of the Laurussian plate. This dextral fracture system linked the southern end of the Uralian orogen with the northern end of the Alleghenian orogen; it remained active until the late Early Permian consolidation of the Appalachian-Mauretanides-Ouachita-Marathon fold belt (Arthaud and Matte 1977; Ziegler 1988, 1990). Dimensions of the dextral translation between Europe and Africa are difficult to quantify, but should be equal to the amount of crustal shortening that was achieved in the Appalachian-Mauretanides fold belt during the late phases of the Alleghenian diastrophism. For the Appalachians, COCORP deep reflection seismic data indicate that Alleghenian crustal shortening amounted to at least 175 km (Cook et al. 1979, 1981; Cook 1983; Secor et al. 1986). For the entire Appalachian-Mauretanides megasuture, Late Carboniferous-Early Permian crustal shortening of the order of 300 to 400 km could be visualized; however, it appears improbable that Alleghenian crustal shortening could be an order of magnitude greater. In view of this, the Early Permian Pangea A-2 assembly (van der Voo and Peinardo 1984) is preferred to the Pangea-B assembly (Morel and Irving 1978; Daly and Irving 1983). The latter would require a counter-clockwise rotation of Gondwana relative to Laurussia, involving a 3500 km dextral transform motion between North Africa and southern Europe within the Stephanian-Autunian time span of some 35 Ma. This would correspond to plate movements at a rate of about 10 cm/year, which appear excessive, particularly in a continent-to-continent collisional setting. Yet, there is evidence for some dextral movements along the axis of the Appalachians during the terminal phases of the Alleghenian orogeny (Secor et al. 1986).

The Stephanian-Autunian disintegration of the Variscan fold belt is probably the combined effect of gravitational collapse of its overthickened crust, major dextral shear movements and possibly back-arc extension related to post-orogenic steepening and decay of the north-dipping Proto-Tethys subduction zone (Jowett and Jarvis 1984; Lorenz and Nicholls 1984; Ménard and Molnar 1988). However, as the Stephanian-Autunian fault systems affected not only the Variscan fold belt but also large parts of its northern foreland, it is likely that dextral shear between North Africa and Europe was the principal mechanism governing their development.

In Western and Central Europe, Stephanian-Autunian fault sytems and basin axes are multidirectional and irregularly spaced. The area affected by the Late Variscan fracture systems, including the Alpine and Mediterranean domains, extends some 5000 km in an east-west and about 3000 km in a north-south direction (Fig. 3). Wrench faulting and subsidence of associated pull-apart, transtensional and tectonically silled basins persisted over a time span of about 35 Ma. In time, a gradual, but systematic rotation of the principal horizontal compressional stress-axis from an initial north-south orientation to one directed east-west can be deduced from microtectonic analyses (Blès et al. 1989). In many basins this stress rotation caused syn- and post-depositional deformations that are indicative of convergent wrenching. Although the amount of upper crustal extension associated with the subsidence of these basins appears to be small on a regional scale, widespread magmatic activity is evident whereby in time the contribution of mantle-derived magmas played an increasingly important role. The Stephanian, and particularly the Autunian thermal pulses were apparently associated with a massive thinning of the subcrustal lithosphere as evident in the area of the future northern and southern Permian basins of northwest Europe; this can be deduced from their subsequent subsidence pattern (Ziegler 1990). Regional thermal uplift of areas affected by wrench faulting and volcanic activity caused deep truncation of Palaeozoic sediments in the Variscan foreland and erosional unroofing of the Variscan fold belt. Profound thermal destabilization of the lithosphere was accompanied by important lower crustal attenuation, particularly in areas of high volcanic activity, presumably involving upward displacement of the Moho by physico-chemical processes in areas where mantle-derived melts ponded at the crust-mantle boundary (e. g. Oslo Graben, northern Germany).

As a result of its Stephanian-Autunian fragmentation, the Variscan fold belt ceased to play the role of a cohesive tectonic and morphological unit shortly after ist consolidation. Crustal fractures developing during the Stephanian-Autunian phase of wrench faulting were reactivated time and again during the Mesozoic and Cenozoic evolution of Europe; as such, they are conspicuously evident in its present structural configuration. In Europe, the Late Variscan tectonism can be considered as a precursor of the Late Permian-Early Triassic plate boundary reorganization that underlies the break-up of Pangea (Ziegler 1990).

During the Late Carboniferous and Early Permian, Europe drifted northward out of an equatorial setting. Resulting palaeoclimatological changes are reflected by the termination of coal-measure deposition in northern areas at the transition to the Westphalian D and in southern areas at the end of the Stephanian (Paproth 1987). In Europe, the Stephanian-Autunian tectonism was associated with a regional regression. Intermittent marine transgressions reached the Variscan foreland basin during the Stephanian, presumably from the Moscow Platform. During the Stephanian and Autunian repeated transgressions advanced westward from the Tethys embayment along the axis of the Bay-of-Biscay fracture zone into Cantabria; this seaway closed at the end of the Autunian.

3 Late Permian-Early Mesozoic Development of Pangea

With the Alleghenian consolidation of the Marathon-Ouachita-Appalachian-Mauretanides fold belt, the Permo-Carboniferous suturing process of Gondwana and Laurussia came to an end. Together these mega-continents formed the core of the super-continent Pangea. However, orogenic activity continued in the Uralian system during the Late Permian-Early Triassic; by this time Siberia and Kazakhstan had become accreted to the eastern margin of the Laurussian subcontinent (Ziegler 1989). Moreover, Gondwana-derived continental terranes, now forming parts of China, continued to converge and collide during the Late Permian and Triassic with Kazakhstan and Siberia (A. M. Ziegler 1981, Sengör 1985; Parrish et al. 1986).

The Uralian subduction system probably was linked via an Arctic arc-trench system, skirting the northern margin of the Laurussian subcontinent, with the subduction system of the North American Cordillera (Zonenshain et al. 1987). In the latter, orogenic activity accelerated during the Late Permian and culminated at the transition to the Triassic in the Sonoma orogeny (Burchfiel and Davis 1972, 1975; Frazier and Schwimmer 1987; Walker 1988). Also, in the Andean system there is evidence for a major Late Permian and earliest Triassic orogenic event (Mesner and Woodridge 1964; Carlier et al. 1982). Contemporaneous crustal shortening in the Cape fold belt of South Africa accompanied the development of its foreland basin (Hälbich 1983). Furthermore, orogenic activity continued along the Pacific margins of Antarctica and Australia (St. John 1986). Similarly, the southern margin of the Eurasian subcontinent, facing the Proto-Tethys, remained active during Permian and Triassic times. On the other hand, the north-eastern margin of the Gondwanan part of Pangea was affected by rifting activity, culminating in the detachment of a number of continental fragments; these make up the composite Cimmerian terrane, which became accreted to Eurasia during the Late Triassic-Early Jurassic (Sengör 1984).

This indicates that locking of the Variscan megasuture between Gondwana and Laurussia, combined with continued sea-floor spreading in Panthalassa, was accompanied by increased activity along subduction zones fringing the "Pacific" margins of Pangea; this reflects as a first-order subduction system progradation from the interior of Pangea to its American, Antarctic and Australian peripheries. This plate-boundary reorganization was accompanied by a counterclockwise rotation of Pangea, amounting during the late Early and Late Permian to about 20° and during the Triassic to a further 17°. The rotation pole for this motion was located in the Gulf of Mexico (Ziegler 1990).

3.1 Permo-Triassic Plate Reorganization

Following its Early Permian consolidation, Pangea showed signs of inherent instability already during the Late Permian and even more so during the Triassic. Initially this was manifested in reactivation of the pre-existing Arctic-North Atlantic rift system and the inception of the Gondwana rifts, but later, during the Triassic, by the development of new interior rift system in the area of the Gulf of Mexico, the Central and North Atlantic and in the Mediterranean domain. Although crustal separation was not achieved along these rifts during the Late Permian and Triassic, they paved the way for the Jurassic, Cretaceous and Early Cenozoic break-up of Pangea. In contrast, the evolution of rift systems along the Afro-Arabian margin of

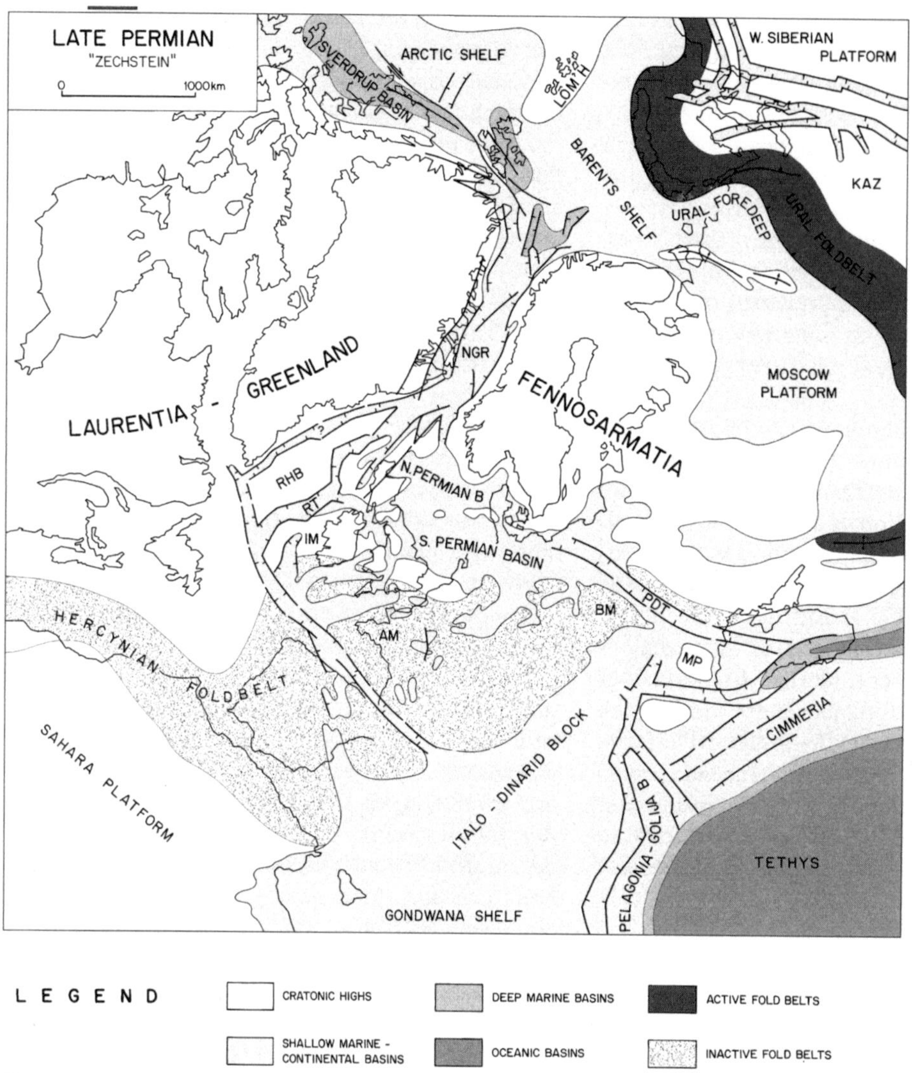

Fig. 4. Late Permian framework of Arctic-North Atlantic domain. *AM* Armorican Massif; *BM* Bohemian Massif; *KAZ* Kazhakstan; *MP* Moesian Platform; *PDT* Polish Drobrugea Trough; *RHB* Rockall-Hatton Bank; *RT* Rockall Trough

the Gondwana subcontinent culminated in Late Permian to Early Triassic crustal separation between the Arabian Peninsula and the continental Central and East Iranian blocks, and with this in the development of the Mesozoic Tethys sea-floor spreading axis. Although the southeastern margin of the Eurasian subcontinent remained active during the Permian and Triassic, back-arc extension in the Black Sea area, probably related to the decay of the Variscan subduction system, culminated in Late Permian-Early Triassic sea-floor spreading and the separation of the Variscan-deformed Cimmerian terrane (Cimmeria *sensu stricto*) from the southern margin of Fennosarmatia (Fig. 4; Adamia et al. 1981; Khain 1984; Belov 1986; Kazmin and Sborshchikov 1989). Similar processes, in combination with the slow development of the Tethys sea-floor spreading axis, may underlie the evolution of the Late Permian and Triassic graben systems of the central and western Mediterranean area, which came into evidence shortly after the Stephanian-Autunian wrench faults became inactive.

The late Permian and Triassic evolution of Pangea reflects a fundamental re-organization of plate boundaries which heralds its Mesozoic break-up. Particularly in the domain of the Variscan and Appalachian fold belts, a change is evident in the regional stress pattern from Late Carboniferous-Early Permian compression to Triassic extension. The underlying causes are seen in the insulating effect that the wide expanse of the coherent continental lithosphere of Pangea had on downwelling branches of the pre-existing asthenospheric convection systems active during the Permo-Carboniferous suturing phases of Pangea. This insulating effect probably caused the decay of these long-standing downwelling cells and the development of new upwelling systems beneath the core of Pangea (Pavoni 1985, 1988). Such a reorganization of the asthenospheric convection cells probably entailed a change in the polarity of drag forces exerted by the convecting asthenosphere on the overlying lithosphere (Ziegler, 1992). In addition, deviatoric tensional forces developed in the lithosphere above the new upwelling mantle cells (Dewey 1988). A corresponding change in the stress regime of the lithosphere is reflected by the development of the Triassic and Early Jurassic rift systems in the interior of Pangea (Fig. 4).

In the Arctic-North Atlantic, Central Atlantic and Mediterranean domains, the major Mesozoic and Cenozoic break-up axes of Pangea coincide to a large extent with the trace of Palaeozoic fold belts. Exceptions are the Labrador Sea-Baffin Bay and the area between Greenland and the Rockall-Hatton Bank where rifts, preceding opening of these oceanic basins, developed entirely within Precambrian basement and cross-cut its structural grain. On the other hand, the main break-up axes of Gondwana are all located within Precambrian basement. In Western and Central Europe, and also in the North Atlantic domain, Late Carboniferous and Early Permian crustal discontinuities were reactivated tensionally during the Triassic and played an important role in the localization of rift systems; at the same time new fault patterns also came into evidence (e. g. North Sea rift). In the Appalachians, wrench and thrust faults, active during the final consolidation of this fold belt, were tensionally reactivated during the Triassic rifting phase. This illustrates that at least some of the major break-up axes of Pangea were preconditioned by crustal discontinuities which had developed during its suturing phases (Ziegler 1990).

The Late Permian evolution of Western and Central Europe was dominated by the relaxation of lithosperic thermal anomalies which were introduced during the Stephanian-Autunian phase of wrench faulting. This is evident, for instance, from the thermal subsidence of the northern and southern Permian basins in northwest and Europe (Fig. 4; Ziegler 1990). On the other hand, in the Mediterranean domain, Late Permian times correspond to a period of development of major rift systems as seen in the Dinarides (Pamic 1984). On a regional scale, rifting activity increased sharply at the transition from the Permian to the Triassic. During the Triassic, the Arctic-North Atlantic rift system propagated southward into the North and Central Atlantic domain and the Tethys rift system propagated westward into the North Atlantic area where it interfered with the Arctic-North Atlantic rift system (Fig. 5). At the same time Western and Central Europe became transsected by a multidirectional graben system that is characterized by a very low level of volcanic activity. Despite the inception of this new stress regime, lithospheric cooling processes continued apparently, causing the foundering of broad areas. By Late Triassic time, areas affected by regional crustal stretching extended from the Grand Banks to the Tethys margin over a distance of some 3000 km and from North Africa to Poland over about 2500 km.

As a consequence of the aforementioned Permo-Triassic counterclockwise rotation of Pangea, northern Siberia moved into a circumpolar position during the Late Permian, and by the Late Triassic it straddled the North Pole. At the same time, Antarctica moved out of a location centred on the South Pole. By the Late Permian, continental ice sheets were present in both northern Siberia and in Antarctica; during the Late Triassic, only northern Siberia was occupied by grounded ice sheets (Ustritsky 1973; Smith and Briden 1977; Scotese et al. 1980; Hambrey and Harland 1981; Caputo and Crowell 1985; Veevers and Powell 1987; Scotese and McKerrow 1990).

The marked cyclicity of the Late Permian series is suggestive of continued glacio-eustatic sea-level fluctuation (Martin 1981). The global low stand in sea level at the Permian-Triassic boundary, and its cyclical rise during the Triassic (Schopf 1974; Forney 1975; Vail et al. 1977; Haq et al. 1987; Holser and Magaritz 1987; Ross and Ross 1987) are though to be of a tectono-eustatic nature. These long-term sea-level changes are probably related to the decay of Permian sea-floor spreading axes and the development of new ones during the Triassic in conjunction with the Permo-Triassic reorganization of asthenospheric convection cells and lithosperic plate boundaries (Schwan 1980; Bally 1982; Heller and Angevine 1985). The short-term cyclicity of Triassic sedimentation may, at least in part, be of a glacio-eustatic nature, whereby flooding and evaporation of large rifted basins in the Central Atlantic domain possibly also played a role during the Late Triassic (Biddle 1984; Ziegler 1988).

The combined effect of globally rising sea-levels, progressive decay of Early Permian lithospheric thermal anomalies and subsidence of the Early Mesozoic rift systems brought about regional transgressive conditions which persisted throughout Triassic and Early Jurassic times. With this the area of sedimentation increased gradually in Europe whereby a systematic overstepping of basin margins is evident (Ziegler 1990).

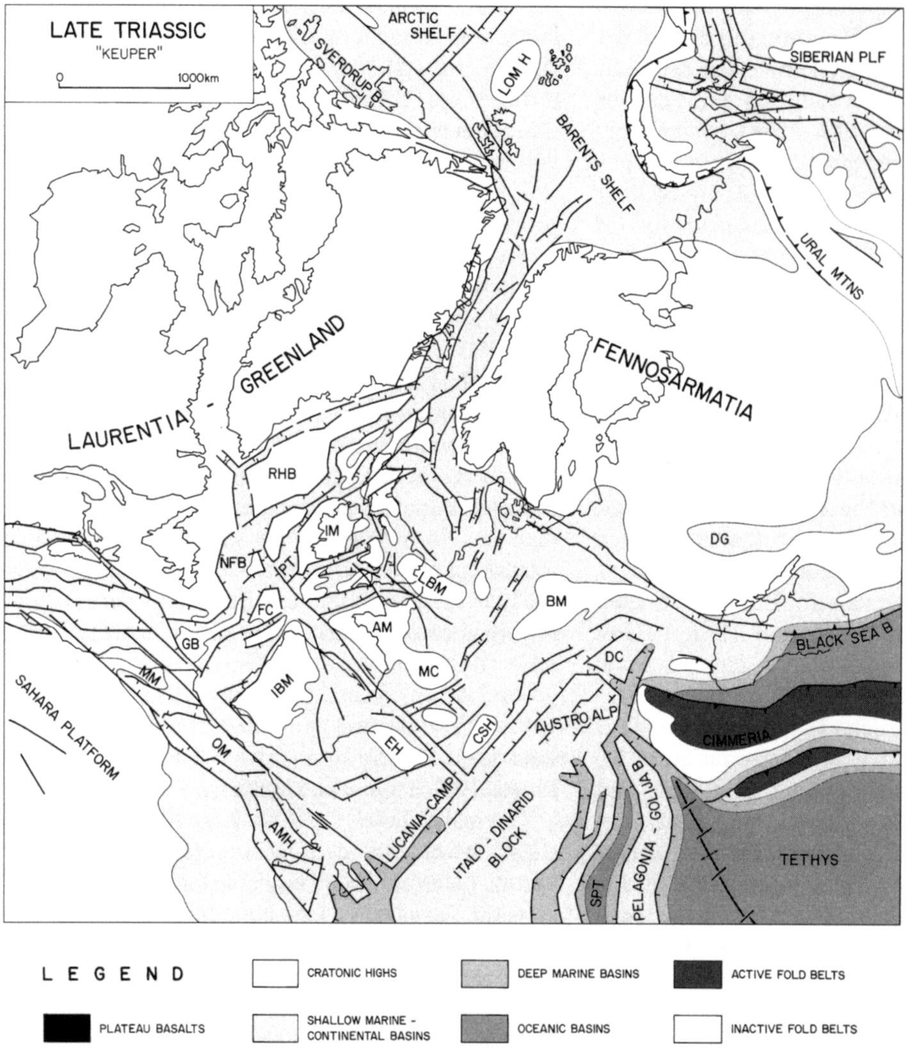

Fig. 5. Late Triassic framework of Artic-North Atlantic domain. *AM* Armorican Massif; *AMH* Ain-M'lila High; *BM* Bohemian Massif; *CSH* Corsica-Sardinia High; *DC* Dacides Block; *DG* Donets Graben; *EH* Ebro High; *FC* Flemish Cap; *GB* Grand Banks; *IM* Irish Massif; *IBM* Iberian Meseta; *LMB* London-Brabant Massif; *MC* Massif Central High; *MM* Morocco Meseta; *NFB* East Newfoundland Basin; *OM* Oran Meseta; *PLF* Platform; *PT* Porcupine Trough; *RHB* Rockall-Hatton Bank; *SPT* Sub-Pelagonia Trough

3.2 Jurassic Crustal Separation Between Gondwana and Laurasia

In the Late Triassic and Early Jurassic, tectonic activity intensified along the Tethys-Central Atlantic-Caribbean rift/wrench system. At the same time, the Tethys sea-floor spreading axis, which had come into evidence already by the Early Triassic, gained in importance and propagated westward. By Late Triassic time the Pelagonia-Golija block became partly separated from the Italo-Dinarid platform and, during the Early Jurassic, the Hellenic-Dinarid-Pienid-Transylvanian oceanic basin began to open (future Vardar Suture zone, Fig. 5; Pamic 1984; Spray et al. 1984; Debelmas and Sandulescu 1987; Ziegler 1988). At the same time, the Cimmerian terrane *(sensu strictu)* was deformed compressionally and the Black Sea back-arc oceanic basin closed (Sengör 1984); with this, compressional stresses were then applied to the Sarmatian platform where the Donets Graben was partly inverted (Konashow 1980).

During the Early Jurassic, the Central and Western Mediterranean area continued to be affected by rift and wrench tectonics. In the Alpine domain, rifting activity became increasingly important in the South Alpine and Austro-Alpine realms, as well as in the Penninic, Helvetic and Dauphinois areas. This led to the gradual individualization of the North and

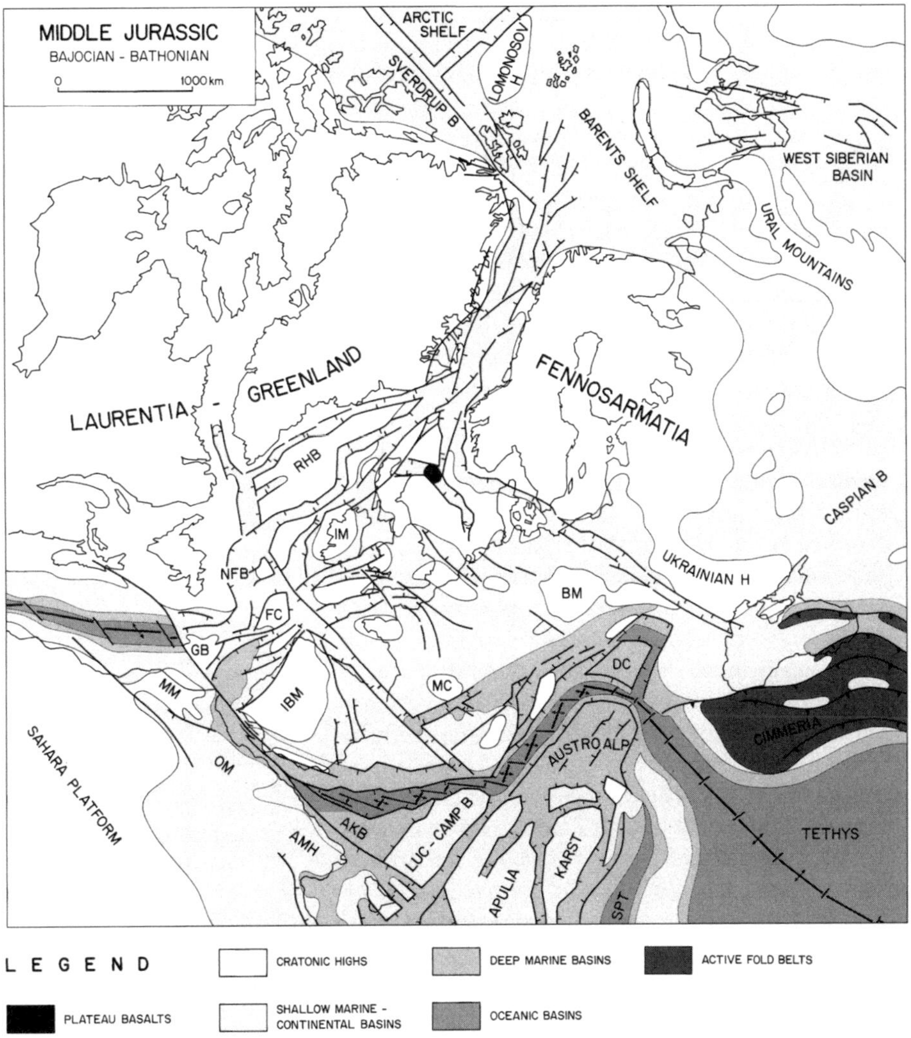

Fig. 6. Middle Jurassic framework of Arctic-North Atlantic domain. *AM* Armorican Massif; *AMH* Ain M'lila High; *BM* Bohemian Massif; *CSH* Corsica-Sardinia High; *DC* Dacides Block; *DG* Donets Graben; *EH* Ebro High; *FC* Flemish Cap; *GB* Grand Banks; *IM* Irish Massif; *IBM* Iberian Meseta; *LMB* London-Brabant Massif; *MC* Massif Central High; *MM* Morocco Meseta; *MFB* East Newfoundland Basin; *OM* Oran Meseta; *RBH* Rockall-Haton Bank; *SPT* Sub-Pelagonia Trough

South Penninic throughs (Fig. 4; Trümpy 1980; Lemoine et al. 1986; Lemoine and Trümpy 1987). Continued crustal extension and wrench-faulting along the western and northern margins of the Italo-Dinarid platform went hand-in-hand with its further fragmentation. Similarly, the Algerian and Moroccan system of wrench-induced pull-apart basins, which had come into evidence during the Triassic, continued to subside. During Late Triassic and Early Jurassic times, the subsidence of these basins was accompanied by volcanism that is indicative of deep crustal fracturing.

In the Central Atlantic, Early Jurassic crustal extension led to the rapid subsidence of its axial rift zone whereas its flanks were gradually uplifted in response to progressive lithospheric thinning and decompression of the asthenosphere; this was accompanied by extensive intrusive and extrusive magmatic activity affecting wide areas around the future divergent plate boundary (Bertrand and Westphal 1977; Emery and Uchupi 1984; Hutchinson et al. 1986; Manspeizer 1988; McHone 1988). In the Gulf of Mexico, evidence of Late Triassic and Early Jurassic rifting is provided by the occurrence of continental red beds, associated with dykes and sills. However, in this area the pattern of rifts and pull-apart basins is difficult to define due to a great overburden of post-rift sediments (Scott et al. 1961; Pindell 1985; Frazier and Schwimmer 1987).

By the Early Bajocian, crustal failure occurred in the Central Atlantic and a first oceanic basin began to open between the Equatorial and the Azores fracture zones (Pitman and Talwani 1972; Vogt 1973; Emery and Uchupi 1984). This was followed by the onset of sea-floor spreading in the Western Tethys domain where the Alboran-Piedmont-South Penninic-Mid-Pannonian oceanic basin came into evidence (Fig. 6; Laubscher and Bernoulli 1977; Bernoulli et al. 1979; Dercourt et al. 1985; Ziegler 1988). At the same time, the Pienid-Transsylvanian sea-floor spreading system was abandoned, but the Hellenic-Dinarid basin continued to widen. The sea-floor spreading axes of the nascent Central Atlantic and Piedmont-Penninic oceans were linked via the sinistral Azores-Gibraltar-Maghrebian fracture zone. During the Mesozoic opening of the Central Atlantic, much of the sinistral displacement between Africa and Europe was taken up along this fracture zone whereby only a relatively minor amount was compensated by oblique sea-floor spreading in the Piedmont-Penninic basin. In the Caribbean, crustal separation between the North and the South American cratons was not achieved until the Kimmeridgian (Pindell 1985; Ross and Scotese 1988). Following the establishment of this new, divergent, transform plate boundary, which transsected Pangea in an east-west direction, the kinematic regime in the Tethys and Arctic-North Atlantic domains changed, and the gradual break-up of Laurasia commenced (Ziegler 1988, 1990).

Acknowledgements
This chapter is an excerpt from the author's "Geological Atlas of Western and Central Europe, 2nd Edition" which was published by Shell International Petroleum Mij. B. V., Den Haag, The Netherlands in December 1990. The author is indebted to Shell International for giving permission to reprint the text figures contained in this work.

References

Adamia Sh A, Chkhotua T, Kekelia M, Lordkipanidze M, Shavishvili I (1981) Tectonics of the Caucasus and adjoining regions: implications for the evolution of the Tethys ocean. J Struct Geol 3(4): 437–447
Aïfa T, Feinberg H, Pozzi J-P (1990) Devonian-Carboniferous palaeopoles for Africa: consequences for Hercynian geodynamics. Tectonophysics 179: 287–304
Arthaud F, Matte P (1977) Late Palaeozoic strike-slip faulting in southern Europe and North Africa; results of a right-lateral shear zone between the Appalachians and the Urals. Geol Soc Am Bull 88: 1305–1320
Bachtadse V, van der Voo R (1986) Paleomagnetic evidence for crustal and thin-skinned rotations in the European Hercynides. Geophys Res Lett 13 (1): 161–164
Bachtadse V, Heller F, Kröner A (1983) Paleomagnetic investigations in te Hercynian Mountain Belt of Central Europe. Tectonophysics 91: 285–299
Bally AW (1982) Musing over sedimentary basin evolution. In: Sir Peter Kent, Bott MHP, McKenzie DP, Williams CA (eds) The evolution of sedimentary basins. Philos Trans R Soc Lond 305A: 325–338
Behr HJ, Heinrichs T (1987) Geological interpretation of DEKORP 2–5: a deep seismic reflection profile across the Saxothuringian and possible implications for the late Variscan structural evolution of Central Europe. Tectonophysics 142: 173–202
Behr HJ, Engel W, Franke W (1982) Variscan wildflysch and nappe tectonics in the Saxothuringian zone (northeast Bavaria, West Germany). Am J Sci 282: 1438–1470
Belov AA (1986) The Paleozoic of the Caucasus and the problem of Paleo-Tethys. Geotectonics 20 (3): 230–237
Bernoulli D, Kälin O, Patacca E (1979) A sunken continental margin of the Mesozoic Tethys: the northern and central Apennines. Symp Sédimentation Jurassic W. européen. ASF Publ Spec 1: 197–210
Bertrand H, Westphal M (1977) Comparaisons géologiques et palaeomagnétiques des tholeiites du Maroc et de la côte orientale de l'Amérique du Nord; implications pour l'ouverture de l'Atlantique. Bull Soc Géol Fr (7) XIX (3): 513–520
Biddle KT (1984) Triassic sea level changes and the Ladinian-Carnian boundary. Nature Lond 308: 631–633
Blès JL, Bonijoly D, Castaign C, Gris Y (1989) Successive post-Variscan stress fields in the European plate (French Massif Central and its borders): comparison with geodynamic data. Tectonophysics 113: 105–121
Bonhommet N, Perroud (1986) Apport du paléomagnétism à la comprehension de l'orogénèse hercynienne en Europe occidentale. Bull Soc Géol Fr (8) II (1): 35–42
Burchfiel BC, Davis GA (1972) Structural framework and evolution of the southern part of the Cordilleran orogen, western United States. Am J Sci 272: 97–118
Burchfiel BC, Davis GA (1975) Nature and controls of Cordilleran orogenesis, western United States: extension of an earlier synthesis. Am J Sci 275: 363–396
Burg J-P, Matte PJ (1978) A cross-section through the French Massif Central and the scope of variscian geodynamic evolution. Z Dtsch Geol Ges 129: 429–460
Burg J-P, Leyreloup A, Marchand I, Matte Ph (1984) Inverted metamorphic zonation and large-scale thrusting in the Variscan Belt: an example in the French Massif Central. In: Hutton DHW, Sanderson DJ (eds) Variscan tectonics of the North Atlantic region. Geol Soc Lond Spec Publ 14: 47–61
Caputo M, Crowell JC (1985) Migration of glacial centers across Gondwana during Paleozoic era. Geol Soc Am Bull 86: 1020–1036
Carlier G, Gardin G, Laubacker G, Marocco R, Mégard F (1982) Present knowledge of magmatic evolution of the eastern Cordillera of Peru. Earth Sci Rev 18: 253–283
Cazes M, Mascle A, Torreilles G, Bois C, Damotte B, Matte P, Raoult J-F, Pham VN, Hirn A, Galdeano X (1986) Large Variscan overthrusts beneath the Paris Basin. Nature (Lond) 323: 144–147
Cook FA (1983) Some consequences of palinspastic reconstruction in the southern Appalachians. Geology 11: 89–98
Cook FA, Albaugh DS, Brown LD, Kaufman S, Oliver JE, Hatcher RD Jr (1979) Thin-skinned tectonics in the crystalline southern Appalachians: COCORP seismic-reflection

profiling of the Blue Ridge and Piedmont. Geology 7: 563–567

Cook FA, Brown LD, Kaufman S, Oliver JE, Petersen TA (1981) COCORP seismic profiling of the Appalachian orogen beneath the coastal plain of Georgia, Part I. Geol Soc Am Bull 92: 738–748

Daly L, Irving E (1983) Paléomagnetism des roches Carbonifères du Sahara Central; analyse et aimanations juxtapos.es: configurations de la Pangée. Ann Geophys 1 (3) : 207–216

Debelmas J, Sandulescu M (1987) Transformante nordpenninique et problèmes de correlation palinspastique entre les Alpes et les Carpates. Bull Soc Géol Fr (8) III (2): 403–408

Dercourt J, Zonenshain LP, Ricou L, Kazmin V, le Pichon X, Knipper AL, Grandjacquet C, Sborshchikov IM, Boulin J, Sorokhtin O, Geyssant J, Lepvrier C, Biju-Duval B, Sibuet J-C, Savostin LA, Westphal M, Lauer J (1985) Présentation de 9 cartes paléogeographiques au 1/20000000 s'étendant de l'Atlantique au Pamir pour la période du Lias à l'Actuel. Bull Soc Geol Fr (8) I (5): 637–652

Dewey JF (1988) Extensional collapse of orogens. Tectonics 7 (6): 1123–1139

Edel JB (1987) Paleomagnetic evolution of the Central Massif (France) during the Carboniferous. Earth Planet Sci Lett 82: 180–192

Edel JB, Coulon M, Hernot MP (1984) Mise en évidence par le paléomagnetisme d'une importante rotation antihoraire des Vosges méridionales entre le Viséen terminal et le Westphalien supérieur. Tectonophysics 106: 239–257

Emery KO, Uchupi E (1984) The geology of the Atlantic Ocean. Springer, Berlin Heidelberg New York, 1050 pp

Forney GG (1975) Permo-Triassic sea-level change. J Geol 83: 773–779

Frazier WJ, Schwimmer DR (1987) Regional stratigraphy of North America. Plenum Press, New York, 719 pp

Gibling MR, Boehner RC, Rust BR (1987) The Sidney Basin of Atlantic Canada: an Upper Paleozoic strike-slip basin in a collisional setting. In: Beaumont C, Tankard AJ (eds) Sedimentary basins and basin-forming mechanisms. Can Soc Petrol Geol Mem 12: 269–285

Hälbich IW (1983) Tectogenesis of the Cape Fold Belt (CFB). In: Söhnge APG, Hälbich IW (eds) Geodynamics of the Cape fold belt. Geol Soc S Afr Spec Publ 12: 165–175

Hambrey MJ, Harland WB (eds) (1981) Earth's pre-Pleistocene glacial record. University Press, Cambridge, 1044 pp

Haq BU, Hardenbol J, Vail PR (1987) Chronology of fluctuating sea levels since the Triassic (250 million years ago to present). Science 235 (4793): 1156–1167

Heller PL, Angevine CL (1985) Sea-level cycles during the growth of Atlantic-type oceans. Earth Planet Sci Lett 75: 417–426

Holser WT, Magaritz M (1987) Events near the Permian-Triassic boundary. Mod Geol 11 (2): 155–180

Howie RD, Barss MS (1975) Paleogeography and sedimentation in the Upper Paleozoic, eastern Canada. Can Soc Petrol Geol Mem 4: 47–57

Hutchison DR, Klitgord KD, Detrick RS (1986) Rift basins of the Long Island Platform. Geol Soc Am Bull 97: 688–702

Jones M, van der Voo R, Bonhommet N (1979) Late Devonian to Early Carboniferous palaeomagnetic poles from the Armorican Massif, France. Geophys J R Astro Soc 58: 287–308

Jowett EC, Jarvis GT (1984) Formation of foreland rifts. Sediment Geol 41: 51–72

Julivert M, Marcos A, Perez-Estaun A (1977) La structure de la chaîne Hercynienne dans le secteur libérique et l'arc Ibéro-Armoricain. In: La chaîne varisque d'Europe moyenne et occidentale. Colloq Int Cent Natl Rech Sci Rennes 243: 429–440

Kazmin VG, Sborshchikov IM (1989) The history of Tethys in the Late Paleozoic-Early Mesozoic and the reconstruction of Pangea. Int Geol Rev 31: 649–661

Kent DV, Keppie JD (1988) Silurian-Permian paleocontinental reconstruction and tectonics. In: Harris AL, Fettes DJ (eds) The Caledonian-Appalachian orogen. Geol Soc Lond Spec Publ 38: 469–480

Khain VE (1984) The Alpine-Mediterranean fold belt of the USSR. Episodes 7(3): 20–29

Konashow VC (1980) Expression of early Kimmerian folding phase in the Donets basin. Geotectonics 14 (4): 268–272

Laubscher HP, Bernoulli D (1977) Mediterranean and Tethys. In: Nairn AEM, Kanes WH, Stehli FG (eds) The ocean basins and margins, vol 4A. The eastern Mediterranean. Plenum Press, New York, pp 1–28

Lemoine M, Trümpy R (1987) Pre-oceanic rifting in the Alps. Tectonophysics 133: 305–320

Lemoine M, Bas T, Arnaud-Vanneau A, Arnoud H, Gidon M, Bourbon M, de Graciansky P-C, Rudkiewicz J-L, Megard-Galli J, Tricart P (1986) The continental margin of the Mesozoic Tethys in the Western Alps. Mar Petrol Geol 3 (3): 179–199

Lorenz V, Nicholls IA (1984) Plate and intraplate processes of Hercynian Europe during the Late Palaeozoic. Tectonophysics 107: 25–56

Manspeizer W (1988) Triassic-Jurassic rifting and opening of the Atlantic: an overview. In: Manspeizer W (ed) Triassic-Jurassic rifting, continental break-up and the origin of the Atlantic Ocean and passive margins. Developments in Geotectonics 22, Part A. Elsevier, Amsterdam pp 41–79

Martin H (1981) The Late Palaeozoic Gondwana glaciation. Geol Rundsch 70 (2): 480–496

Matte Ph (1983) Two geotraverses across the Ibero-Armorican Variscan arc of Western Europe. In: Rast N, Debelmas FM (eds) Profiles of orogenic belts. Am Geophys Union, Geodyn Ser 10: 53–81

Matte Ph (1986) La chaîne varisque parmi les chaînes paléozoiques péri-atlantiques, modèle d'évolution et position des grands blocs continentaux au Permo-Carbonifère. Bull Soc Géol Fr (8) II (1): 9–24

McCutcheon SR, Robinson PT (1987) Geological constraints on the genesis of the Maritime Basin, Atlantic Canada. In: Beaumont C, Tankard AJ (eds) Sedimentary basins and basin-forming mechanisms. Can Soc Petrol Geol Mem 12: 287–297

McHone JG (1988) Tectonic and paleostress patterns of Mesozoic intrusions in eastern North America. In: Manspeizer W (ed) Triassic-Jurassic rifting, continental breakup and the origin of the Atlantic Ocean and passive margins. Development in Geotectonics 22, Part A. Elsevier, Amsterdam, pp 607–620

McKerrow WS, Scotese CR (1990) Palaeozoic, palaeogeography and biogeography. In: McKerrow WS, Scotese CR (eds) Palaeozoic, palaeogeography and biogeography. Geol Soc Lond Mem 12: 435 pp

Meissner R, Bartelsen H, Murawski H (1981) Thin-skinned tectonics in the northern Rhenisch Massif, Germany. Nature (Lond) 290: 399–401

Ménard G, Molnar PN (1988) Collapse of a Hercynian Tibetan Plateau into a Late Palaeozoic European Basin and Range Province. Nature (Lond) 334: 235–237

Mesner JL, Woodridge LCP (1964) Maranhao Paleozoic basin and Cretaceous coastal basins. North Brazil. Bull Am Assoc Petrol Geol 48: 1475–1512

Morel P, Irving E (1978) Tentative paleocontinental maps for the Early Phanerozoic. J Geol 86: 535–561

Pamic JJ (1984) Triassic magmatism of the Dinarides in Yugoslavia. Tectonophysics 109: 273–307

Paproth E (1987) L'Europe, de la fin du Westphalien à l'Autunien. Ann Soc Géol Nord CV1: 77–85
Parrish JM, Totman-Parrish J, Ziegler AM (1986) Permian-Triassic paleogeography and paleoclimatology and implications for tetrapsid distribution. In: Hotton N III, MacLean PD, Roth JJ, Roth EC (eds) The ecology and biology of mammal-like reptiles. Smithonian Institution Press, Washington, pp 109–131
Pavoni N (1985) Die pazifisch-antipazifische Bipolarität im Strukturbild der Erde und ihre geodynamische Deutung. Geol Rundsch 74: 251–266
Pavoni N (1988) Das bipolare geotektonische Modell. Ein Versuch zur Beschreibung der großräumlichen Bewegungsvorgänge im Erdmantel. Geogr Rundsch 40 (10): 58–64
Perroud H, Bonhommet N (1981) Palaeomagnetism of the Ibero-Armorican Arc and the Hercynian orogeny in Western Europe. Geogr Rundsch 40(10): 58–64
Perroud H, van der Voo R (1983) Palaeozoic evolution of the Armorican plate on the basis of palaeomagnetic data. Tectonophysics 91: 271–283
Pin Ch (1990) Variscan oceans: ages, origins and geodynamic implications inferred from geochemical and radiometric data. Tectonophysics 177: 215–227
Pindell JL (1985) Alleghennian reconstruction and subsequent evolution of the Gulf of Mexico, Bahamas and Proto-Caribbean. Tectonics 4 (1): 1–39
Pitman III WC, Talwani M (1972) Sea floor spreading in the North Atlantic. Geol Soc Am Bull 83: 619–646
Rast N (1984) The Alleghenian orogeny in eastern North America. In: Hutton DHW, Sanderson DJ (eds) Variscan tectonics of the North Atlantic region. Geol Soc Lond Spec Publ 14: 197–218
Ross CA, Ross RP (1987) Late Paleozoic sea levels and depositional sequences. Cushman Found Foraminiferal Res Spec Publ 24: 137–149
Ross MI, Scotese CR (1988) A hierarchical tectonic model of the Gulf of Mexico and Caribbean region. Tectonophysics 155: 139–168
Schopf TJM (1974) Permo-Triassic extinctions: relation to sea floor spreading. J Geol 82: 129–143
Schwan W (1980) Geodynamic peaks in Alpinotype orogenies and changes in ocean-floor spreading during Late Jurassic-Late Tertiary times. Am Assoc Petrol Geol Bull 64: 359–373
Scotese CR, McKerrow WS (1990) Revised world maps and introduction. In: McKerrow WS, Scotese CR (eds) Palaeozoic Palaeogeography and biogeography. Geol Soc Lond Mem 12: 1–21
Scotese CR, Snelson S, Ross WC, Dodge LA (1980) A computer animation of continental drift. J Geomag Geoelectr 32 (Suppl III): SIII61–SIII70
Scott KR, Hayes WE, Fietz RP (1961) Geology of the Eagle Mills formation. Gulf Coast Assoc Geol Soc Trans 11: 1–14
Secor DT Jr, Snoke AW, Dallmeyer RD (1986) Character of the Alleghanian orogeny in the southern Appalachians. Part III. Regional tectonic relations. Geol Soc Am Bull 97: 1345–1353
Sengör AMC (1984) The Cimmeride orogenic system and the tectonics of Eurasia. Geol Soc Am Spec Pap 195, 82 pp
Sengör AMC (1985) The story of the Tethys: how many wives did Okeanos have? Episodes 8 (1): 3–12
Shackleton RM, Ries AC, Coward MP (1982) An interpretation of the Variscan structure in SW England. J Geol Soc Lond 139: 533–541
Smith AG, Briden JC (1977) Mesozoic and Cenozoic paleocontinental maps. Cambridge Earth Science Series, Cambridge University Press, Cambridge, 63 pp
Spray JG, Bébien J, Rex DC, Roddick JC (1984) Age constraints on the igneous and metamorphic evolution of the Hellenic-Dinaric ophiolites. In: Dixon JE, Robertson AHF (eds) The geological evolution of the eastern Mediterranean. Geol Soc Lond Spec Publ 17: 619–628
St. John B (1986) Antarctica – Geology and Hydrocarbon potential. In: Halbouty MT (ed) Future pertroleum provinces of the world. Am Assoc Petrol Geol Mem 40: 55–100
Thiele O (1977) Zur Tektonik des Waldviertels in Niederösterreich (südliche Böhmische Masse). Nova Acta Leopold NF 244: 67–82
Trümpy R (1980) Geology of Switzerland a guide book. Part A. An outline of the Geology of Switzerland. Wepf, Basel, p 104
Ustritsky VI (1973) Permian climate. In: Logan A, Hills L (eds) The Permian-Triassic systems and their mutual boundary. Can Soc Petrol Geol Mem 2: 733–744
Vai GB (1980) Pré-Paléozoique et Paléozoique pré-Varisque. In: Fragnani G, Zuffardi P (eds) Introduction à la géologie génerale d'Italie. Publ 26th Congr Géol Int Paris G-13, pp 53–60
Vail PR, Mitchum RM Jr, Todd RG, Widmier JM, Thompson S, Sangree JB, Bubb JN, Hatledid WG (1977) Seismic stratigraphy and global changes of sea level. In: Payton CE (ed) Seismic stratigraphy application to hydrocarbon exploration. Am Assoc Petr Geol Mem 26: 49–212
van der Voo R, Peinardo J (1984) A paleomagnetic reevaluation of Pangea reconstructions. In: van der Voo R, Scotese CR, Bonhommet N (eds) Reconstructions from Paleozoic paleomagnetism. Am Geophys Union, Geol Soc Am Tulsa, pp 11–26
Veevers JJ, Powell C McA (1987) Late Paleozoic glacial episodes in Gondwanaland reflected in transgressive-regressive depositional sequences in Euramerica. Geol Soc Am Bull 98: 475–487
Vogt PR (1973) Early events in the opening of the North Atlantic. In: Tarling DH, Runcorn SK (eds) Implications of continental drift to Earth Sciences, vol 2. Academic Press, London, pp 693–712
Walker JD (1988) Permian and Triassic rocks of the Mojave desert and their implication for timing and mechanisms of continental truncation. Tectonics 7: 685–709
Ziegler AM (1981) Paleozoic paleogeography. In: McElhinny MW, Valencio DA (eds) Paleoreconstruction of the continents. Geodyn Ser, Am Geophys Union, Geol Soc Am 2: 31–37
Ziegler PA (1982) Geological atlas of Western and Central Europe. Elsevier, Amsterdam, 130 pp, 40 plates
Ziegler PA (1984) Caledonian and Hercynian crustal consolidation of Western and Central Europe – a working hypothesis. Geol Mijnbouw 63 (1): 93–108
Ziegler PA (1986) Geodynamic model for the Paleozoic crustal consolidation of Western and Central Europe. Tectonophysics 126: 303–328
Ziegler PA (1988) Evolution of the Arctic-North Atlantic and the Western Tethys. Am Assoc Petrol Geol Mem 43: 198 pp, 30 plates
Ziegler PA (1989) Evolution of Laurussia – a study in Late Palaeozoic plate tectonics. Kluwer Academic, Dordrecht, 102 pp
Ziegler PA (1990) Geological atlas of Western and Central Europe, 2nd edn. Shell Internationale Petroleum Mij BV, Geol Soc London, 260 pp, 54 encl
Ziegler, P A. (1992) Plate tectonics plate moving mechanisms and cifting. Tectonophysics. 215: 9–34
Zonenshain LP, Kuzmin MI, Konov MV (1987) Absolute reconstruction of continent configuration during the Palaeozoic and early Mesozoic. Geotektonika 1987 (2): 16–27 (in Russian)
Zwart HJ, Dornsiepen UF (1978) The tectonic framework of Central and Western Europe. Geol Mijnbouw 57: 627–654

Part III

Specific Descriptions of Pre-Alpine Basement

Chapter 1

The Helvetic Realm

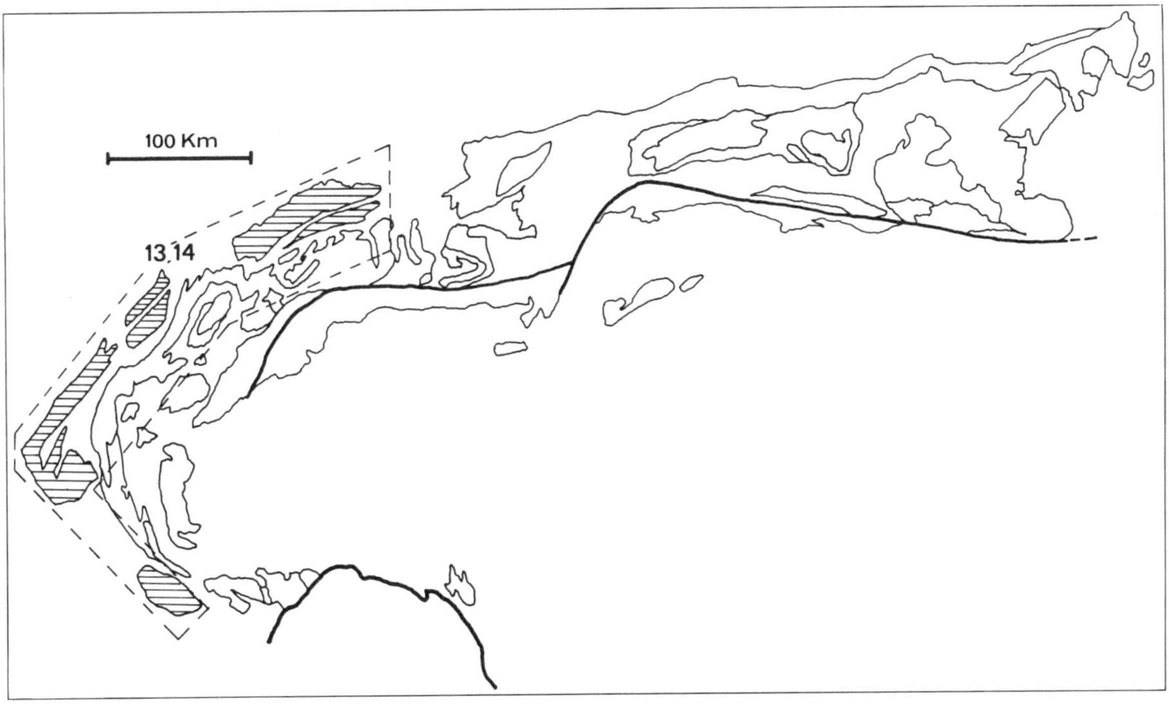

The Pre-Alpine Evolution of the External Massifs

J. F. von Raumer, R. P. Ménot, J. Abrecht and G. Biino

Abstract

Relics of the pre-Mesozoic geology are preserved in the five crystalline basement complexes of the External Massifs. Precambrian metamorphic units and Precambrian to Palaeozoic sediments, interlayered with volcanics and ultramafic rocks, suffered in most areas a polymetamorphic history. After formation of eclogitic mineral assemblages (possible Ordovician), the geotectonic evolution followed a collisional type of metamorphism (Ky-Sill-And) accompanied by crustal thickening. Rising geothermal gradients since the Devonian may be the expression of updoming and erosion and/or post-orogenic collapse. Strike-slip since the Early Carboniferous was accompanied by anatexis, guided the formation of transtensional sedimentary troughs, the nappe stacking of metamorphic units that had not been subjected to the main metamorphic event, and the intrusion of Late-Variscan granite bodies.

The External Massifs, during the Variscan evolution, were part of the internal zone of the Variscides corresponding to the Moldanubian zone of the extra-Alpine Variscan domain.

1 Introduction

Pre-Permian continental crust is, at present, exposed only in a number of tectonic windows that constitute the External Massifs along the N and NW rim of the Alps. Together with their Mesozoic and Tertiary cover they form the Helvetic or Dauphinois realm (Fig. 1). Exposed basement areas include from SW to NE: the Argentera Massif, the Haut Dauphiné – Grandes Rousses areas, the Belledonne Massif, the Mont Blanc – Aiguilles Rouges Massifs, and the Aar, Tavetsch and Gotthard Massifs. Minor parts of the Mesozoic-Tertiary sequence form the more or less autochthonous cover of the basement, the major part, however, forms today the substratum of the classical Helvetic and Dauphinois nappe piles.

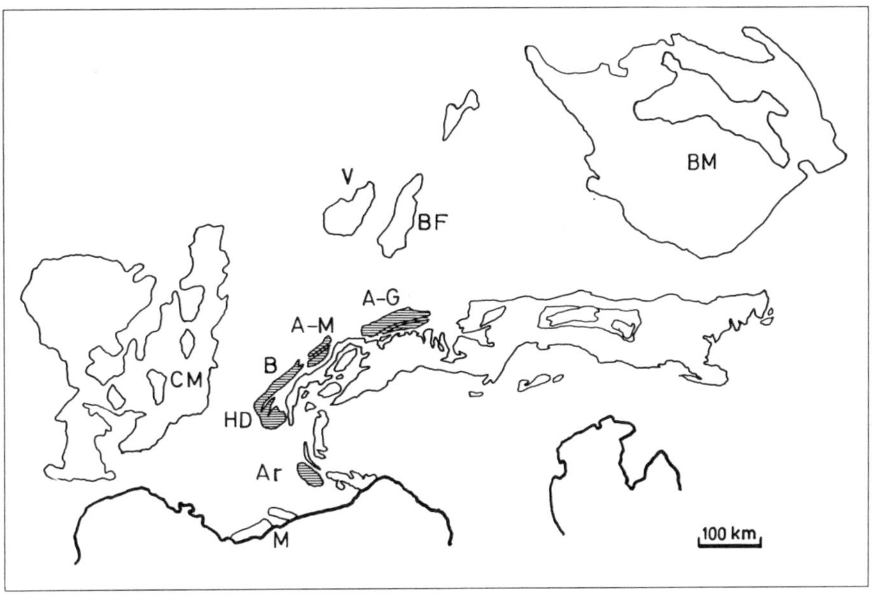

Fig. 1. The external massifs *(shaded)* and their location relative to the framework of the European Variscan basement. Helvetic domain: *Ar* Argentera Massif; *HD* Haut Dauphiné Massif; *B* Belledonne Massif; *A–M* Aiguilles Rouges and Mont Blanc Massifs; *A–G* Aar- and Gotthard Massifs (details, see Fig. 2). Variscan framework: *BF* Black Forest; *BM* Bohemian Massif; *CM* French Central Massif; *V* Vosges; *M* Maures and Esterel Massifs

The Variscan and pre-Variscan evolution of the external massifs is highly complex. However, although respective P-T-t-d paths can differ from massif to massif, parallels in their evolution and lithologies have been noted (Krummenacher et al. 1965; von Raumer 1976, 1987). Many important tectono-thermal features characteristic of the European Variscan belt are also observed in the external massifs (von Raumer 1988; von Raumer and Ménot 1989) despite differences due to both pre-Alpidic and, more importantly, Alpidic orogenic events. A short consideration of these Alpidic overprints will preceed our more detailed treatment of the basement units.

2 Alpine Overprints

From recent cross-sections through the Alps (Pfiffner 1986; Escher et al. 1987) and the Metamorphic Map of the Alps (Frey et al. 1980) it is evident that large portions of the pre-Mesozoic crust were not only affected by Alpine metamorphism but were also involved in Alpine thrust tectonics. In almost every part of the external massifs, pre-Alpine basement outcrops are separated by late Variscan and more prominent Alpine tectonic lineaments respectively marked by the occurrence of Late Variscan and Mesozoic sediments (e.g. Argentera: "zone d'écrasement La Valetta – Mollières"; Belledonne: "synclinal médian"; Mont Blanc – Aiguilles Rouges: "synclinal" de Chamonix; Aar-Tavetsch-Gotthard: "Urseren-Garvera zone" and "Tavetsch-Disentis zone"; all zones indicated in Fig. 2 A–F by the occurrence of late Variscan to Mesozoic sediments separating pre-Mesozoic crystalline basement). Metamorphism of the basement and its Mesozoic cover ranges from very-low grade to greenschist facies conditions (Argentera, Haut Dauphiné and Belledonne: Gratier et al. 1973; Apprahamian 1988; Aiguilles Rouges: von Raumer 1974; Pilloud 1991; Mont-Blanc: von Raumer 1984b; Aar: Steck and Burri 1971; Gotthard: Niggli and Niggli 1965; Steck 1976; Merz 1989). Complications resulting from this Alpine overprint make it much more difficult to understand the pre-Alpine features hidden in the Alpine domain.

In the more external massifs (e.g. Aiguilles Rouges), the basement appears as a dome structure already visible in the cross-sections of Heim (1922) and Badoux (1972), whereas nappe-like structures of Alpine age are evident in the Haut Dauphiné, Mont Blanc, Aar, Tavetsch and especially the Gotthard Massifs. The large-scale updoming and the formation of basement folds are reflected in small-scale structures that vary with lithology and crustal level. During Tertiary compressional tectonics, the Aar-Massif was overthrust by some 10–12 km (Pfiffner 1986) of sediments, (the estimated thickness of the Helvetic cover), consistent with the P-T conditions recorded by Aar basement rocks for the Tertiary orogenic phase. In contrast, the most internal side of the Gotthard Massif records temperatures of around 530–560 °C and pressures up to 5–7 kbar (Klaper and Bucher-Nurminen 1987; Le Goff 1989). Thus, the Gotthard Massif underwent metamorphic conditions different from the Aar Massif.

The structural evolution of the basement is the result of small-scale conjugate strike-slip, flattening and extension, where the maximum elongation is often vertical (Steck 1984, 1987). In the Aiguilles Rouges area the longest axis of the deformation ellipsoid dips SE (von Raumer 1987). In the Fibbia granite of the central Gotthard Massif, which is metamorphosed to the upper greenschist facies, progressive flattening with a subvertical extension has produced an average shortening in the order of 65 % in a NW direction (Marquer 1990). In the Aiguilles Rouges Massif, maximum shortening is estimated to be about 30 % in NW-SE direction. Vertical stretching of 300–1000 % has been observed in KF-augengneisses and Permian rhyolites of the Mont Blanc area (von Raumer 1984b).

The regional vertical stretching in the Mont Blanc area as well as in the central external massifs has produced countless horizontal tension gashes that yield the beautiful crystals the Alpine domain is famous for (Stalder et al. 1973). Fluid inclusion studies on quartz from such fissures (Poty et al. 1974; Baumer et al. 1982) indicate crystallization temperatures that rise from 340 °C in the Haut Dauphiné to 400 °C in the Mont Blanc area and 400–505 °C in the Gotthard area. Pressures in the Mont Blanc area and the Aar Massif, during this late stage of evolution, lie between 2.5 and 3 kbar. Age determinations carried out on such fissure minerals yielded ages between 8.5 and 18 Ma (Rb/Sr, K/Ar, Leutwein et al. 1970; Arnold 1972; Purdy and Stalder 1973), and are in general agreement with the thermal evolution model proposed by Jäger et al. (1967), Monié and Maluski (1983) and Demeulemeester et al. (1986).

3 Basement Units

Despite their similarity, the western (i.e. Argentera Massif, Haut-Dauphiné – Grandes-Rousses areas, Belledonne Massif, and Mont Blanc – Aiguilles

Fig. 2 A–C. Geological outlines and main geological units of the external massifs. Large areas are underlain by pre-Variscan basement *(pVb)*, which is undifferentiated except for the orthogneisses and, in the Argentera, metabasic units. Variscan migmatites (cordierite) are observed in all external massifs and are followed by the Variscan intrusives and the late Variscan rhyolites. **A** Argentera Massif; **B** Haut Dauphiné; **C** Belledonne Massif (*inset,* see Fig. 2 D);

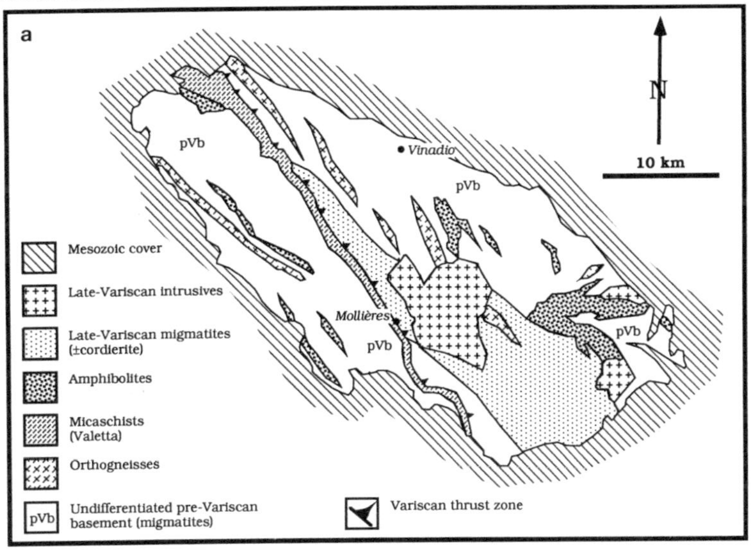

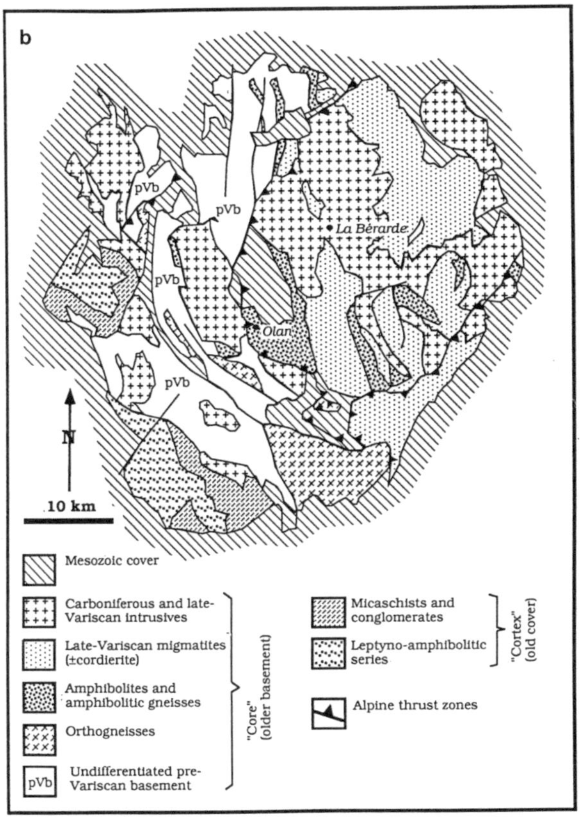

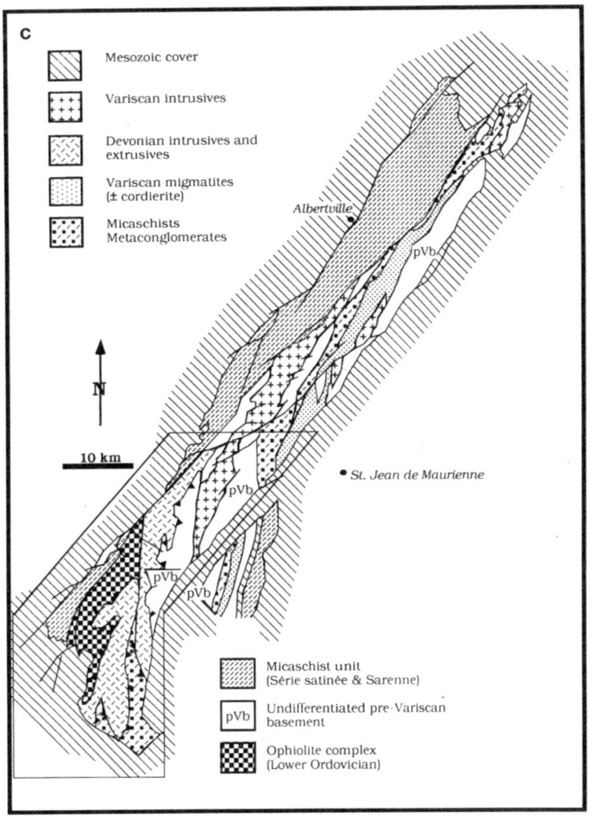

Rouges Massifs) and central massifs (i.e. Aar, Tavetsch and Gotthard Massifs) represent distinct geographical sectors with specific evolutionary histories that correspond to different crustal levels and different segments in the orogen. This subdivision is therefore used to emphasize the differences in the lithologies and pre-Alpine evolution of the massifs.

Two main types of basement area can be differentiated:

1. Those areas that show a polymetamorphic evolution.
2. A few areas that show a monometamorphic evolution and, locally, include fossiliferous epizonal metasediments.

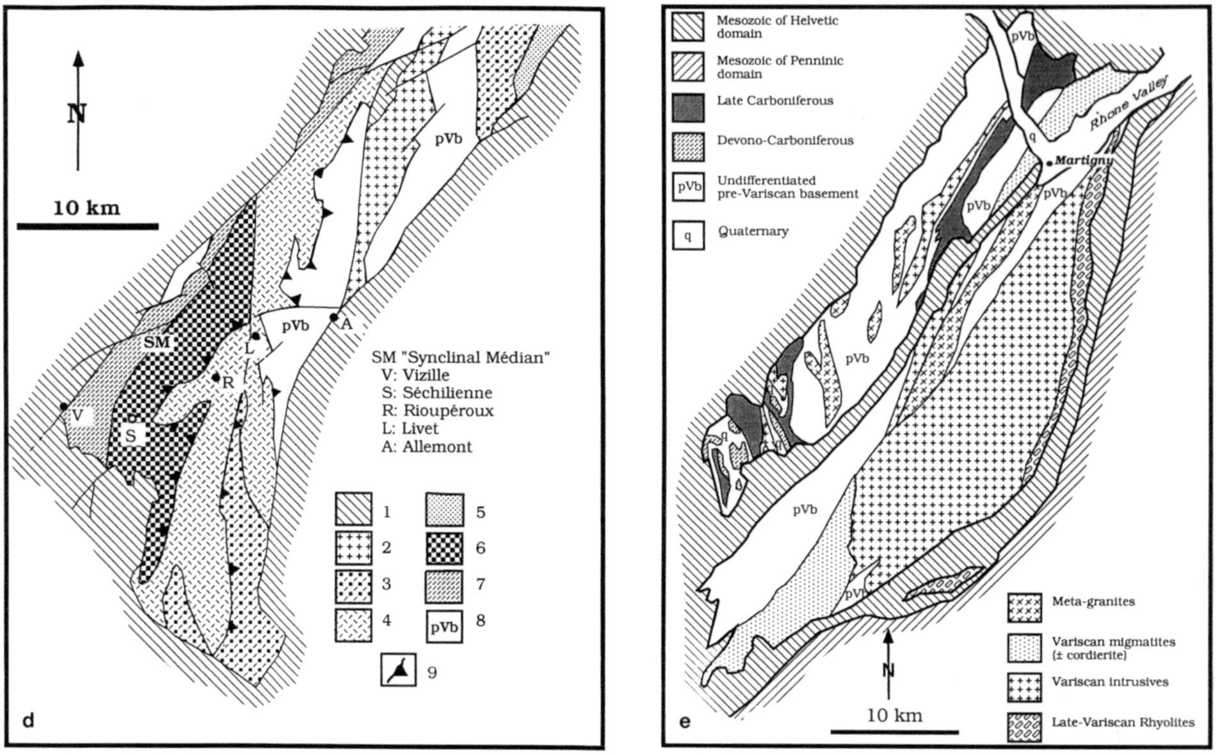

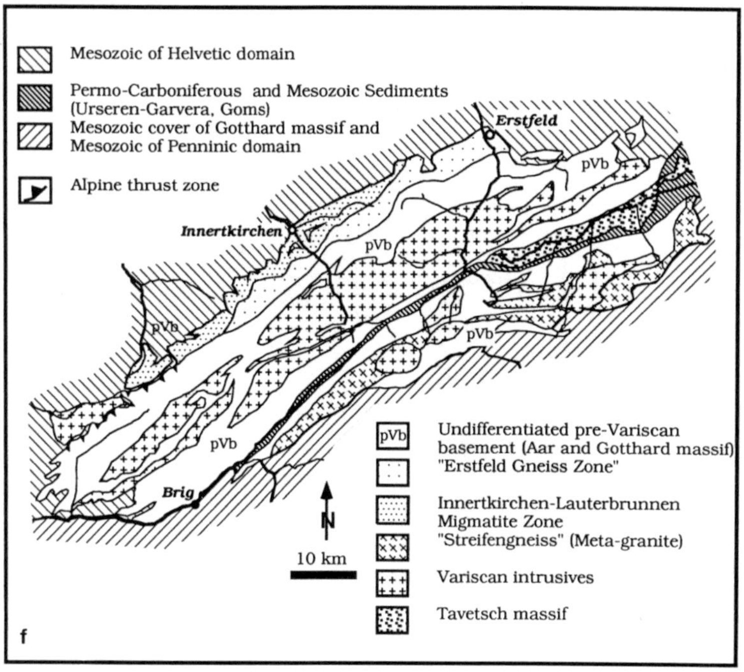

Fig. 2 D–F detailed map of the lower Romanche valley area (Belledonne, Fig. 2 C). *1* Post-Dinantian sedimentary cover; *2* Variscan intrusives; *3* Taillefer series (Dinantian); *4* Rioupéroux-Livet igneous complex (Devonian); *5* Variscan migmatites (±cordierite); *6* Chamrousse ophiolite (Early Ordovician); *7* "Serie satinée" (age?); *8* undifferentiated pre-Variscan basement; *9* main Variscan thrusts. **E** Mont Blanc – Aiguilles Rouges Massifs. The central zone of Mesozoic of the Helvetic domain separates the Aiguilles Rouges Massif in the NW from the Mont Blanc Massif in the SE. **F** Aar, Tavetsch and Gotthard Massifs. The Permo-Carboniferous and Mesozoic sediments of the Goms-Urseren-Garvera zone separate the Aar and the Tavetsch Massif in the N from the Gotthard Massif in the S

3.1 Polymetamorphic Basement

Most areas of the external massifs consist of metasediments of Late Proterozoic to Early Palaeozoic age which record a polymetamorphic evolution. Since their protoliths are interpreted as platform sediments (von Raumer 1987), the presence of older structures serving as a basement should be admitted. Such older crustal elements of probably Cadomian – Pan-African age could be hidden among granitoid gneisses and migmatites containing amphibolites, quartzites and lenses of calc-silicate rocks. First evidence for older rocks comes from the Gotthard area (Arnold 1970b), and Gebauer et al. (1988) discuss the Late Riphean age of eclogite protoliths and Abrecht et al. (1991) a Precambrian age of HP-metamorphism from the same region. The presence of an Archean and Lower Proterozoic hinterland is reflected by upper intercept ages of detrital zircons (U/Pb zircon data; Gebauer et al. 1988; Paquette et al. 1989; Schaltegger 1992; see also Gebauer this Vol.).

Despite their high-grade tectonothermal overprint, among the metasediments different sequences can be recognized. A typical succession comprises metagreywackes interlayered with quartzites and metapelites and rare carbonate rocks, the latter usually appearing as calc-silicate lenses or nodules. A different lithologic series is composed of mica schists with a few amphibolite layers and diopside marbles and accompanied by banded or massive metagreywackes and acidic gneisses. Our comparative studies suggest that both sequences could represent an Upper Proterozoic age. These different lithologies have been observed in all five massifs (Argentera: Faure Muret 1955; Malaroda 1968; Bogdanoff 1986; Bogdanoff et al. 1991; Haut Dauphiné: Le Fort 1973; Belledonne-Grandes Rousses: Tobi 1959; Kalsbeek 1961; Carme 1970a, 1971; Giorgi et al. 1979; Vivier et al. 1987; Mont Blanc/Aiguilles Rouges: Bellière 1958; von Raumer 1976; Aar-Gotthard: Niggli 1944; Oberholzer 1955; Hafner 1958; Steiger 1962; Stalder 1964; Labhart 1965, 1977; Steck 1966; Arnold 1970a; Abrecht 1980; Schaltegger 1986a; Schenker 1987). They have been regionally compared by Krummenacher et al (1965) and von Raumer (1981) and a preliminary palaeogeographic scheme has been developed by von Raumer (1984a, 1987). Part of the metasediments is now interpreted as former platform sediments deposited in a rifting environment since the Late Proterozoic to the Ordovician. Amphibolites and eclogites normally occur as small bodies, and their chemical characteristics suggest that the associated magmatites in the western massifs crystallized in an extensional continental environment (Liégois and Duchesne 1981; Paquette 1987; Paquette et al. 1989; von Raumer et al. 1990; Ménot and Paquette this Vol.) or in an oceanic environment in the Central Alps massifs (Biino et al. 1993). Granitoid banded gneisses with K-feldspar and sillimanite nodules are locally abundant (e.g. Argentera Massif) and could represent acid volcanic rocks of unknown age. New observations in the Aiguilles Rouges (Chiaradia 1992) plead for an older basement including earliest Cambrian covered unconformably by K-feldspar augengneisses and fine-grained granitoid gneisses, interpreted as representing a metavolcanic suite of island arc character. K-feldspar augengneisses, from the Mont Blanc area, on the other hand, yielded Ordovician ages (460 Ma, U/Pb zircon, Bussy and von Raumer 1993) and are interpreted as Palaeozoic granite bodies (comparable to the Streifengneiss in the Gotthard area and the "gneiss de Crupillouse" from the Haut Dauphiné area, both identified as orthogneisses in Fig. 2). The metasedimentary series appear to represent a rather reduced cover sequence but have experienced a complex series of metamorphic events leading to their present high amphibolite facies grade.

Metapelitic formations, comparable to flysch-type sediments, occur in the St. Gervais area (S. Aig. Rouges), the "Série satinée" (W Belledonne), the "Schistes verts" (NE Belledonne), the Huez formation (Grandes Rousses) and probably the Valetta series (Argentera). The "Schistes verts" also include bimodal volcanics (Negga et al. 1986). In contrast to most of the lithologies mentioned here, their polymetamorphic evolution is related to supracrustal conditions. Their depositional age is poorly constrained but is thought to be either Late Proterozoic to Early Palaeozoic (Carme 1970a; Simeon 1979; Bogdanoff 1986) or Visean ("Schistes verts": Bordet and Bordet 1963). Cambrian to Ordovician microfossils have been discovered in the black and green schists of the Huez Formation (Giorgi et al. 1979).

3.2 Monometamorphic Areas

Exceptions to this long lasting polymetamorphic evolution are found in the monometamorphic areas that mainly lie in the SW domain of the Belledonne Massif (Ménot 1986, 1987; Fig. 2D). In this area (Romanche valley) four different lithostratigraphic units with distinct poly- and monometamorphic histories are involved in a Late Visean nappe stacking (Carme 1973a, b; Ménot 1986, 1987).

1. The **Allemont-Rochetaillée Formation** of Late Proterozoic to Early Palaeozoic age (Carme and Pin 1987) mainly comprises gneisses and amphibolites. Although their lithology and geochemistry is comparable to that of the polymetamorphic basement, they do not show evidence of high pressure metamorphism. Instead, medium-pressure metamorphism is locally followed by the formation of anatectic melts (Ménot 1987). General retrogression to greenschist facies grade is related to thrusting and strike-slip faulting (Carme 1970b; Ménot 1987). Alternatively, the anatectic stage may be (1) earlier than (Carme 1974) or (2) unrelated to (Carme and Pin 1987) the

Barrovian metamorphism. In the latter case, the anatectic gneisses would belong to a tectonic slice inserted within the Allemont-Rochetaillée Formation.

2. The most important unit, the **ultramafite and gabbro complex of Chamrousse-Séchilienne,** has yielded crystallization ages of 496 ± 6 Ma (U/Pb zircon from plagiogranites; Ménot et al. 1984, 1988a) and 497 ± 24 Ma (Sm/Nd; Pin and Carme 1987). Today, the structurally inverted ophiolite body (Carme 1965a, 1975; Ménot 1979, 1987; Ménot et al. 1987) shows a continuous downward gradation from an ultrabasic and gabbroic sequence (Chamrousse Formation) into the volcanic and volcano-sedimentary sequences (Séchilienne Formation) (Ménot 1987, 1988). The complex geochemical evolution (Bodinier et al. 1981) suggests formation of the ophiolite in either a transitional environment between attenuated continental crust and oceanic crust (Ménot 1987) or an oceanic back-arc environment (Pin and Carme 1987). The ophiolite does not show evidence of the polymetamorphic evolution recorded in the basement. Rather, metamorphic recrystallization is related to early intra-oceanic extensional tectonics and to Late Devonian to Early Carboniferous tectonic events (Ménot 1987; Ménot et al. 1987; Carme and Pin 1987; Guillot et all. 1992).

3. The **Rioupéroux-Livet plutonic and volcanic complex** comprises leptynitic and amphibolitic units and plutonic rocks of trondhjemitic composition, associated with mainly acidic volcanics, volcaniclastic rocks and greywackes (Carme 1965a; Ménot 1986, 1987). The trondhjemites have been dated at 352 ± 56 and 365 ± 17 Ma (U/Pb zircon, Ménot 1987; Ménot et al. 1988b). The complex now forms five main tectonic units (Ménot 1986). Deformation is younger than the trondhjemite intrusions and older than 324 ± 12 Ma (cooling age, K/Ar; Ménot et al. 1987), and was associated with the development of low- to medium-grade mineral assemblages. The igneous complex suggests either an active margin environment (Carme 1975b; Carme and Pin 1987) or one of continental thinning and rifting (Ménot 1987). A southward extension of this Devonian to Dinantian complex may also outcrop in the westernmost area of the Haut Dauphiné massif ("cortex"; Le Fort and Pecher 1971; Fig. 2B).

4. The **Taillefer series** consists of metapelites, -arenites and -conglomerates and metaspilites and -keratophyres (Carme 1965b). The epizonal monometamorphic imprint is weak and only well-marked in shear zones which are related to mid-Carboniferous nappe tectonics (Ménot 1987). The original setting is still ambiguous (Carme and Pin 1987) but the series has been considered to represent an intra-continental pull-apart basin (Ménot 1987). A possible Visean age (Gibergy 1968) would make the series a time-equivalent of the metapelitic formation from the SW edge of the Aiguilles Rouges (Bellière and Streel 1980; Devonian-Carboniferous of Fig. 2E).

3.3 Pre-Alpine Metamorphic and Structural Evolution

3.3.1 Western Alps

Most parts of the Western Alps are underlain by high grade polymetamorphic rocks and only locally do their low grade counterparts appear (see earlier).

For most of the *polymetamorphic basement* of the Western Alps a multistage metamorphic evolution has been described, especially in the Aiguilles Rouges – Mont Blanc (Bellière 1958, 1980; von Raumer 1983), Belledonne (Tobi 1959; Carme 1973a; Vivier et al. 1987; Ménot 1987; Carme and Pin 1987) and Argentera Massifs (Bortolami and Sacchi 1968; Bogdanoff 1986). The earliest relics of this polyphase metamorphic and structural evolution document an HP and HT event. Meta-eclogites have been observed in the Argentera (Latouche and Bogdanoff 1987; 710–760 °C/12–14 kb; 710–760 °C/9–11 kb), Belledonne, Mont Blanc and Aiguilles Rouges (Liégois and Duchesne 1981; 780 °C/11 kb) areas. Radiometric data on eclogitization give Early Silurian ages in the Argentera Massif and Early Devonian ages in the Belledonne and Aiguilles Rouges Massifs (U/Pb, zircons; Paquette 1987; Paquette et al. 1989). The occurrence of eclogite layers accompanying ultramafics over several kilometres has led von Raumer (1988) to suggest that large parts of the basement were metamorphosed under lower crustal conditions. Granulitic mineral assemblages have been reported from the Argentera Massif (Roccati 1903) and from the Gotthard area (Arnold 1970a), and coronitic textures in the Argentera- (Latouche and Bogdanoff 1987) and the Aiguilles Rouges-areas indicate metamorphism to granulitic assemblages.

Most rocks of the polymetamorphic basement are of amphibolite facies grade and reveal a medium pressure evolution following a Barrovian type of P-T-t path such as that observed in the Aiguilles Rouges (von Raumer and Schwander 1985; Joye 1989; Schulz and von Raumer 1993; Fig. 3). Early kyanite-bearing mineral assemblages (500–550 °C/10 kb) are overprinted by sillimanite-bearing assemblages which today appear in most outcrops. Depending on the erosion level, mineral assemblages with cordierite-silli-

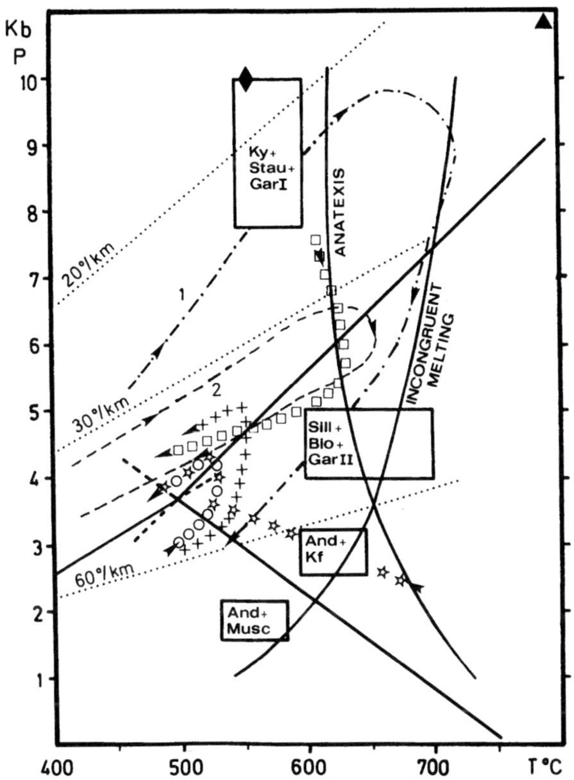

Fig. 3. Pressure – temperature (P-T) evolution of some representative units of the external massifs through time. Al-silicate triple point after Holdaway (1971); limits of anatexis and dry melts after Thompson (1982). P-T-data from the Aiguilles Rouges: *triangle:* minimum PT-conditions of formation of Lac-Cornu eclogites (Liégois and Duchesne 1981); *filled diamond:* P-T-determination (garnet-plagioclase) in metapelites (von Raumer 1983); *rectangles:* mineral paragenesis: data from metapelitic rocks (von Raumer and Schwander 1985; von Raumer 1987) *1* P-T-t-path from metapelites (Joye 1989); *2* P-T-t-path from amphibolites (Joye 1989). P-T-paths from the Belledonne area (Ménot 1987): *open squares* gneiss formations from Allemont; *crosses* Chamrousse ophiolite, Devonian-Dinantian orogenic evolution; *circles* metasediments of Livet formation; *stars* trondjhemites of the Livet formation (the polymetamorphic basement corresponds in its evolution to the Aiguilles Rouges basement)

manite-K-feldspar, anatectic first melts, and incongruent cordierite-bearing granitoid melts appear at the thermal peak (Aiguilles Rouges: 650–700 °C/4–6 kb; von Raumer and Schwander 1985; von Raumer 1987; Argentera: Bortolami and Sacchi 1968; Bogdanoff 1986). A late overgrowth of andalusite is locally recorded (Aiguilles Rouges; von Raumer 1984a). P-T conditions for the entire evolution have been substantiated using geothermometers (Joye 1989) in amphibolites and garnet bearing gneisses (Fig. 3). The resulting P-T path reflects a general change towards higher geothermal gradients accompanied by decompression. This amphibolite facies metamorphism is Devonian in age (NE Belledonne: 373 Ma; Demeulemeester 1982; Argentera: 375–350 Ma; Monié and Maluski 1983; Aiguilles Rouges: 370 Ma; Thöni 1989).

This metamorphic evolution is accompanied by deformation observable in the Argentera (Bogdanoff 1986) and the Aiguilles Rouges areas (Joye 1989; Schulz and von Raumer 1993). Two main periods of deformation can be distinguished. During the first, probably of Devonian age, early intrafolial folds were overprinted by the main schistosity which is accompanied by a strong stretching lineation and a great number of non-cylindrical folds. This deformation occurred in response to an north-directed tangential shear. During the second deformation, probably of Carboniferous age, the earlier folds were overprinted by large-scale folds which, depending on the area, display vertical to horizontal fold axes and result from dextral shear. This deformational sequence should be viewed as a continuous and dynamic process, in which earlier fold generations formed during the evolution of Variscan nappes were progressively rotated into the direction of shear. The resulting mylonites and mylonitic gneisses show HT mineral parageneses and have been observed by the authors in the Argentera, Haut Dauphiné and Belledonne areas.

During metamorphic culmination anatectic mobilization along the shear zones (317 ± 2 Ma, U/Pb monazite, Bussy and von Raumer 1993) promoted the intrusion of Carboniferous granites (e.g. Vallorcine granite; shear zone dated 310 Ma, Thöni 1989). The emplacement of the Mont Blanc granite (305 ± 2 Ma, Bussy 1992) has also been interpreted to have been facilitated by the formation of a pull-apart structure between shear zones that were active by at least the Early Carboniferous. Contact metamorphism from the Mont Blanc granite produced biotite whereas an aureole of different quartz-recrystallization has been observed in the neighbourhood of the Vallorcine granite, being probably the product of shear continuing during intrusion.

Other formations of the polymetamorphic basement, such as the "Série satinée" and the Allemont-Rochetaillée formations (Belledonne), underwent a more supracrustal history. Within the "Série satinée", the mineral assemblage suggests a decrease from medium- (?) to low-grade metamorphic conditions (Carme 1970b; Simeon 1979). The youngest recrystallization is related to Early to Mid-Carboniferous strike-slip tectonics (Karlsbeek 1961; Carme 1970b; Gasquet et al. 1981). The Allemont-Rochetaillée unit underwent Barrovian metamorphism followed by a local anatexis (Ménot 1987), (see 3.2 for discussion).

Table 1. Summary of the lithostratigraphic units and lithologies along three profiles in the Aar Massif

Main geologic and tectonic units	Western Aar Massif: Lauterbrunnen Valley, Gastern Valley, Lötschen Valley – Rhone Valley	Central Aar Massif: Sustenpass, Grimselpass	Eastern Aar Massif: Erstfeld, Maderaner Valley, Val Russein, Punteglias, Tavetsch
Mesozoic cover	Triassic to Tertiary sediments	Permian to Tertiary sediments	Triassic to Tertiary sediments
Innertkirchen-Lauterbrunnen Crystalline Zone (ILC)	Migmatites (granodioritic melts; inclusions of layered biotite-gneisses, amphibolites, marbles, calcsilicate rocks). No dikes!		
Variscan intrusives	Gastern Granite (GaGr)		
Erstfeld Gneiss Zone (EGZ)		Carboniferous sediments (conglomerates)	
		Bio-plag gneisses (± calcsilicate rocks, marbles, metaconglomerates, ultrabasic lenses), Granitoid gneisses, (orthogneisses), partly migmatitic, mafic dykes;	Bio-plag gneisses, granitoid gneisses (orthogneisses), partly migmatic (± calc-silicate rocks, marbles).
		Mesozoic sediments *("Färnigen Zone")*	
Guttannen Unit (GU)	Muscovite-gneisses, ser-chl-schists, banded amphibolites ± ultrabasic rocks (serpentinites), ± calcsilicate rocks.		
	Bio-gneisses, amphibolites ("Schollenamphibolites"), migmatitic bio-plag-gneisses, ± calcsilicate rocks. Mafic and felsic dikes, pegmatites.	Chl-ser-gneisses, bio-plag-gneisses (± calc-silicate rocks, ± amphibolites, ± ultrabasic rocks). Migmatites (metatexitic). Mafic and felsic dikes.	Bio-plag-gneisses, ser-chl-gneisses, amphibolites ("Schollenamphibolites"), ± ultrabasic lenses), gabbroic rocks. Dioritic intrusions, mafic dykes.
Maderaner Group	Volcanic rocks (Rhyolites)	Volcano-sedimentary rocks (*"Trift-Formation":* rhyolites, pyroclastic rocks, shallow water sediments, metaconglomerates)	Carboniferous volcanics (*"Windgällen- Formation":* Rhyolites, pyroclastic rocks, microgranitic stocks). *"Tscharren-Formation", "Bifertengrätli-Formation"*
Variscan intrusives	Central Aar Granite (=Bietschhorn Granite) Baltschieder Granite	Mittagflue Granite Central Aar Granite	Central Aar Granite, Brunni Granite Düssi Diorite
Cavardiras Group	Volcanic rocks (Pyroclastic rocks)	Volcano-sedimentary rocks (*"Diechtergletscher-Formation":* rhyolites, pyroclastic rocks, shallow water sediments, metaconglomerates)	Volcano-sedimentary rocks (*"Bifertenfirn-Formation", "Val Gliems-Formation":* rhyolites, pyroclastic rocks, shallow water sediments, metaconglomerates)
Ofenhorn-Stampfhorn Unit (OSU)	Bio-plag-gneisses, granitoid gneisses ("Augengneisses", orthogneisses), amphibolites ("Schollenamphibolites"), migmatites, ± ultrabasic rocks, gabbros, Meta-granitoids (bio-tonalites, granodiorites, granites).	Bio-plag-gneisses, bio-gneisses, amphibolites ("Schollenamphibolites", ± ultrabasic lenses). Migmatites (metatectic, tonalitic to quartzdioritic melts). Granitic stocks, mafic dykes.	
Variscan intrusives		Grimsel Granodiorite Southern Aar Granite	Giuv Syenite Southern Aar Granite, Russein Diorite
Southern Gneiss Zone (SGZ)	Ser-gneisses, Ser-"Augengneisses", Ep-Bio-Ser-gneisses; Migmatites ("Schollenamphibolites"), Ultrabasic lenses.	Granitic intrusions, orthogneisses ("augengneisses"), Mafic to ultramafic rocks, syenite "restites". (agmatites). *"Gletsch Migmatite Zone"*	Bio-plag-gneisses, ser-chl-gneisses, mafic intrusive rocks (gabbroic to dioritic)
Mesozoic cover	Permo-Carboniferous to Upper Jurassic	Carboniferous to Upper Jurassic sediments ("Disentis Zone")	Permo-Carboniferous to Upper Jurassic

Some units of the Romanche valley area (Fig. 2D) show a *monometamorphic evolution* of Early to Mid-Carboniferous age (Ménot 1987; Ménot et al. 1987). The present classification, based on the mono- and polymetamorphic evolution of the rocks, deals with the metamorphic overprints related to the orogenic (subduction-collision) processes. In this regard, the Chamrousse ophiolite should be ranged within the monometamorphic formations, despite of the described succession of two main recrystallization events, as the earliest stage developed in connection with an intraoceanic tectonics, and only the latest (Dinantian) event is associated to the orogenic incorporation of the ophiolite. The Early Ordovician ophiolite of Chamrousse suffered neither the eo-Devonian HP event nor the Devonian MP and anatectic stages. The medium-grade regional metamorphism (550 °C, 4–5 kb) is particularly well developed in the uppermost part of the ophiolite stratigraphy (volcanic and sedimentary series of Séchilienne). It overprints earlier metamorphic parageneses related to intra-oceanic tectonies (i.e. of Early Ordovician age) (Ménot 1987).

In the Rioupéroux-Livet volcanic and plutonic complex, the main metamorphic recrystallization took place after intrusion of the trondhjemites (at ca. 360 Ma) and before late metamorphic cooling (324 ± 12 Ma; K/Ar hornblende, Ménot et al. 1987). The mineral assemblages point to medium- and low-grade conditions, respectively, for the Livet and Rioupéroux tectonic units (Carme 1973a; Ménot 1986).

The Taillefer series is weakly recrystallized (Carme 1973b) with epizonal parageneses preferentially developed within shear zones related to Mid-Carboniferous nappe tectonics.

In summary overthrusting of mesozonal formations (Chamrousse and Allemont-Rochetaillée) upon epizonal ones (Taillefer, Rioupéroux) followed the metamorphic climax but occurred prior to the 324 Ma cooling age of amphiboles. The Carboniferous age of the crust stacking and metamorphism implies that most of the Romanche valley formations were involved in orogenesis only in Late Devonian-Carboniferous times. Finally, in the Belledonne massif, polymetamorphic and monometamorphic pre-Alpine basements are juxtaposed along major strike-slip faults of late Variscan age.

3.3.2 Central Alps

3.3.2.1 Tectonostratigraphic Units

The tectonostratigraphic subdivision of the Aar, Tavetsch and Gotthard basement generally follows NE-SW-trending units (Fig. 2F), although the absence of certain units gives rise to differences appearing between eastern, central and western cross sections (Table I). For geographic and tectonic names the reader should consult the Tectonic Map of Switzerland 1:500000.

Following Abrecht and Schaltegger (1988) and Table 1, we recognize that lithostratigraphic units with different P-T-t-histories are juxtaposed along mylonite zones, locally followed by younger sediments. From north to south the main units in de **Aar Massif** comprise the migmatitic Innertkirchen-Lauterbrunnen zone (ILC), the Erstfeld Gneiss zone (EGZ), the Guttannen unit (GU) and the southernmost Ofenhorn-Stampfhorn unit (OSU) (Fig. 2F, Table I). The ILC comprises a migmatized series of metasediments (mainly sandstones to arkoses with intercalations of marls and limestones). The remaining units comprise mainly biotite-plagioclase gneisses, biotite-schists with some calcsilicate lenses, migmatitic granitoid gneisses, and mafic and ultramafic lenses. In the western part of the massif a gradual transition between the ILC and the GU has been described (Rutishauser 1975). A thin zone of paragneisses and volcano-sedimentary rocks (Val Gliems formation, Oberhänsli et al. 1988), including important subaerial volcanic units, is exposed along the southern border of the main mass of the Aar granite (Variscan intrusives, Fig. 2F) and extends into the eastern part of the massif.

The **Tavetsch Massif,** a thin basement zone along the SE border of the Aar Massif, is clearly separated from the latter only at its eastern end by the Mesozoic "Disentis Zone". This zone is interpreted as the relic sedimentary cover of the Aar Massif. The Tavetsch Massif mainly consists of acid gneisses, amphibolites, rare metagabbros and ultramafic lenses, and pegmatites. Due to the intense Alpine deformation original structures are rare. The massif is not exposed in the Urseren Valley, but probably reappears as a smaller basement slice to the W of the Furka ("Goms Massif").

The **Gotthard Massif** is separated from the Aar Massif to the N and Tavetsch Massif to the ENE by Permo-Carboniferous and Mesozoic units that define an important tectonic line of Alpine age between Brig (Valais)-Furka-Urseren-Oberalp Pass ("Urseren-Garvera Zone"). Along its southern margin the massif is either overlain by a Mesozoic sedimentary series

Table 2. Summary of lithostratigraphic units and lithologies along two profiles in the Gotthard Massif

Main geologic and tectonic units	Gotthard Massif W Gotthard road (Wallis, Bedretto Valley, Distelgrat)	Gotthard Massif E Gotthard Road (Gotthard Pass, Urseren, Unteralp Valley, Nalps Valley, Sumvitg
Sedimentary cover	Permo-Carboniferous to Mesozoic rocks *("Furka-Urseren Zone")*	Permo-Carboniferous to Mesozoic rocks *("Urseren-Garvera Zone")*
Gurschen Gneiss Zone (GGZ)	Metasedimentary gneisses (2-mica-gneisses, Bt-pl-gneisses, calc-silicate lenses, mafic to ultramafic layers and inclusions (amphibolites, gabbroic rocks, ultrabasic rocks, eclogites). Quartzdioritic to dioritic intrusions. "Paradis Gneiss": inhomogeneous granitic gneiss (abundant inclusions)	
"Streifengneiss" Orthogneisses (Ordovician)	Orthogneisses (granitoid); Intrusive contacts against paragneisses, sometimes with migmatitic character *("Mischgneisse")*	
Variscan intrusives	Granitic to dioritic intrusions: Sädelhorn Diorite, Rotondo Granite	Fibbia Granite, Gamsboden Granite, Medels Granite, Cristallina Granodiorite)
Metasedimentary Zones S of Variscan intrusions:		
"Guspis Zone" – "Motta Naira Zone" – "Distelgrat Zone"	Bt-ms-gneisses, amphibolites, gabbros, ± ultrabasic lenses	
"Borel Zone"		Amphibolites, Bt-rich gneisses, quartzites
"Tenelin – Val Nalps Zone"		Bt-ms-gneisses, Grt-bearing gneisses, amphibolites, quartzites.
"Giubine Series"		Bt-rich gneisses and -schists, 2-mica-gneisses, Grt-bearing gneisses, Hbl-schists, calc-silicate rocks
	Homogeneous 2-mica-gneisses.	
"Hüenerstock Gneiss" *"Sorescia Gneiss"* *"Cavanna Series"* *"Prato Series"* (incl. *"Corandoni Zone"*)	Bt-rich gneisses, 2-mica-gneisses (± alk fsp) Ms-gneisses, Bt-schists, Hbl-schists, amphibolites, ultrabasic rocks	
"Tremola Series"	Hbl-bearing rocks *("Garbenschiefer")*, mica schists and -gneisses	Mica-schists and -gneisses, Hbl-schists ("Garbenschiefer"), Chl-bearing gneisses and -schists
Sedimentary cover	Mesozoic sediments ("Ultrahelvetic" cover of Gotthard massif)	

usually referred to as the "Gotthardmassivische Bündnerschiefer" of the Helvetic realm, or is in contact with sediments of the Penninic domain. The main geologic units are given in Table 2.

The oldest units in the Gotthard Massif are probably represented by large masses of monotonous biotite-plagioclase gneisses that form a continuous zone of variable thickness along its northwestern boundary ("Gurschengneiss Zone"; pVb in Fig. 2F). Within the gneisses are lenses and layers of amphibolites, eclogites, gabbros, ultrabasic rocks, marbles and calc-silicate rocks. This whole sequence was interpreted as an accretionary prism (Abrecht et al. 1991). Minor granitic to granodioritic gneisses of more igneous character are also quite frequent. The mafic-ultramafic lenses are aligned along two conspicuous lineaments approximately parallel to the northern contact of the massif. Along the southern margin of the central and eastern parts of the massif, several elongate zones containing distinct, generally metasedimentary lithologies occur. Unlike the Aar Massif, the Gotthard Massif is characterized by an abundance of Ordovician granitoid rocks (termed "Streifengneiss" after its strong linear, pre-Variscan texture).

Contacts between "Streifengneiss" and migmatitic gneisses have mistakenly been interpreted as gradational ("Mischgneise"), implying emplacement at deep crustal levels. However, discordant contacts with migmatites and rocks of the "Gurschengneiss Zone" demonstrate its intrusive character.

3.3.2.2 Metamorphism and Tectonics

Determination of the metamorphic conditions pertinent to the evolution of the basement rocks is greatly hampered by the lack of adequate lithologies with critical mineral assemblages and, less importantly, by the Tertiary retrogressive overprint. These Tertiary effects generally increase from NW to SE, with rocks least affected by Alpine retrogression occurring in the northernmost part. However, even in the Gotthard area, pre-Alpine mineral assemblages are astonishingly well preserved.

In most areas amphibolite facies mineral assemblages predominate and indicate low pressure, high temperature conditions with local anatexis. Based on petrographic, textural, and chemical evidence, the Innertkirchen-Lauterbrunnen zone formed by in-situ anatexis of a sandy to arkosic metasediment series containing intercalations of marls and limestones (Rutishauser 1973, 1975). Melts formed from the detrital sediments are granodioritic in composition with corderite (pinite) as the typical Mg-Fe phase. In the Susten area (Central Aar Massif) large blocks of EGZ-type gneisses are preserved within the ILC and were interpreted by Schaltegger and Labhart (1983) as relic precursors of the ILC migmatites. Typical mineral assemblages indicating amphibolite facies conditions are found in the Sustenpass area where semi-pelitic gneisses contain prismatic and fibrolitic sillimanite as an index mineral. In places, partial melting accompanied the formation of cordierite (Schaltegger 1986b, 1990b). Although Alpine retrograde assemblages predominate in the units farther south, relic amphibolite-grade assemblages are also observed. Kyanite with muscovite pseudomorphed after fibrolitic sillimanite or muscovite + chlorite after cordierite in the EGZ SE of Erstfeld. Similar textures also occur in the central part of the Aar Massif near Guttannen where cordierite is typically related to partial melting of alkali-feldspar bearing biotite-plagioclase gneisses. Rocks having undergone partial melting are also widespread in the SW and NE portions of the massif. The fibrolite/sillimanite belongs to an earlier, (i.e. pre-migmatitic), high-T metamorphic event. In the eastern part of the Aar Massif sillimanite and cordierite are reported from the pelitic rocks in the Etzli Valley, where they were interpreted by Weber (1904) to be the product of contact-metamorphism caused by the Variscan Giuf Syenite intrusion. Olivine (± Cr spinel) of high-T metamorphic origin is occasionally preserved in ultrabasic rocks of the GU in the central Aar Massif and in serpentinites within mafic/ultramafic associations in the Maderaner Valley (eastern massif; Staub 1911). The only report of an earlier medium pressure event in the Aar Massif is that of Hügi et al. (1988) who describe staurolite and kyanite from mica schists in the northern Lötschental (western Aar massif).

In contrast to the Aar Massif, relict mineral assemblages indicating older high-grade metamorphic events are more frequent in the Gotthard area. Arnold (1970a) reported granulite facies assemblages in an ultramafic lens and the quartzdiorite surrounding it, both of which were enclosed by rocks retrogressed to the amphibolite-facies. Polymetamorphic garnet-amphibolites and garnet-bearing ultramafics in the eastern and central Gotthard Massif still contain relic high-pressure assemblages (omphacite, garnet, quartz, kyanite, amphibole) overprinted at granulite- (orthopyroxene, diopside, plagioclase, garnet) and amphibolite-facies (amphibole, plagioclase) conditions (Abrecht et al. 1991). An important feature of pre-Alpine deformation is the linear texture of the "Streifengneiss" and some of the surrounding gneisses which predates the 315 Ma intrusion of the Medels Granite, and the presence of a pre-Streifengneiss deformation and local migmatization in the metasedimentary gneisses. Variscan large-scale folds with steeply plunging fold-axes ("Schlingentektonik") are inferred from the geologic map of the Val Maighels and are published by Huber (1943).

3.3.2.3 Age of Events

Most geochronologic data are available from the Gotthard area (Gebauer this Vol.). U/Pb on zircons (440 Ma Bossart et al. 1986, Sergeev and Steiger 1993) and a Rb/Sr whole rock isochron (420 ± 17 Ma, Arnold 1970) define the Late Ordovician age of the protolith of the "Streifengneiss". Arnold (1970a) demonstrated that the linear texture is not confined to the "Streifengneiss" but is also found in the "Augengneisses", and that the migmatitic gneisses had been imprinted before the intrusion of the Variscan plutons (Medels granite: 315 Ma; Grünenfelder and Hafner 1968). This indicates a pre-Variscan age for at least some of the migmatites, possibly related to the granulite-facies metamorphism. U/Pb zircon analyses from biotite-plagioclase gneisses of the "Gurschengneiss Zone" indicate a high grade metamorphic event at around 440 Ma (Grauert and Arnold 1968). An ultramafic lens analyzed by the same authors yielded nearly concordant zircon ages between 448 and 472 Ma. The country rock, a biotite-plagioclase-gneiss with sillimanite and kyanite, gave discordant zircon ages with a lower intercept of around 450 Ma and a minimum crystallization age of 1200–1500 Ma (Arnold 1970b). Additional data are discussed in more detail by Gebauer (this Vol.).

Amphibolite facies conditions that reached partial melting occur throughout the Aar Massif, but a pre-Alpine metamorphic zonation is far from established.

So far only the Val Gliems pelites, and the Borel zone of the Gotthard massif seem to be monometamorphic. Structural evidence indicates pre-migmatitic metamorphism and folding of mafic rocks in the SW Aar Massif (Labhart 1965, 1967). South to the Grimselpass area medium scale structures of possible "Schlingenbau" type are crosscut by granodioritic intrusions which predate the Aar granite (Schaltegger and von Quadt 1990). "Schlingenbau" -structures have also been described from the EGZ in the Susten area and attributed to pre-Alpine deformation phases (Labhart and Schaltegger 1983). Recent age determinations (U/Pb zircon data, Schaltegger 1991) from the northern Aar Massif differentiate between Archaean detrital zircons which underwent Pan-African metamorphism around 620 Ma before they suffered Caledonian metamorphism at 455 Ma and formation of anatectic first melts at 445 Ma.

Intrusion of late Variscan granitoids has been demonstrated in both the Aar (Wüthrich 1965; Schaltegger 1986b, 1990b) and Gotthard (Grünenfelder and Hafner 1968; Jäger and Niggli 1964) massifs. Their detailed evolution is discussed by Bonin et al. (this Vol.).

3.4 *Metamorphic Evolution of Basement*

In the central and western parts of the external massifs different lithologic sequences are metamorphosed to the high amphibolite facies grade. The lithologies comprise series of Proterozoic to Early Palaeozoic age, and distinct lithological sequences can be differentiated. Although their metamorphic evolution is recorded in the preservation of relic structures, their evolution through time is still debated.

We envisage the possible existence of a Precambrian continental crust (Fig. 4) covered with thin sediments ("Moldanubian type" after the nomenclature of Hirschmann 1991) – metagreywackes and metapelites with some quartzites and metacarbonates interlayered with metabasics and acid metavolcanics.

The zircons (Schaltegger 1992) indicate formation of Proterozoic sediments in the domain of the Aar Massif. Rifting processes then led to the formation of oceanic crust of unknown width. Oldest metamorphic relics are eclogites dated as Ordovician to Devonian (Gebauer et al. 1988; Paquette 1987; Paquette et al. 1989). The diachronous evolution (i.e. subduction/collision and eclogite facies/granulite facies metamorphism) occurred at different times in the different massifs. Hence, Gebauer (this Vol.), refers to stepwise accretion and eclogite formation since the Ordovician. Whether the zircons of the meta-eclogites really retained the memory of their eclogite metamorphism or only a later thermal event could be questionable. Their metamorphism to granulites may also be recorded in their ages (Abrecht et al. 1991). Caledonian metamorphism, 455 Ma, and anatectic first melts, 455 Ma (U/Pb zircon data; Schaltegger 1992) have been proved recently for the Aar massif, and K-feldspar augengneisses from the Mont Blanc massif yielded Ordovician ages (460 Ma, U/Pb zircon, Bussy and von Raumer 1993). Ordovician granitoids are widespread in the area.

After an early period of rifting and ocean crust formation, the external domain was consequently a convergent margin setting during the Early Palaeozoic and was situated in a large zone of collision by the Devonian (Fig. 4).

The following metamorphic evolution is recorded in a Barrovian type of P-T-t-path that ended with a thermal peak and the formation of anatectic and incongruent cordierite-bearing melts, and was accompanied by the formation of high temperature mylonites and shear zones (see 3.3.1). Schulz and von Raumer (1992) discuss a strike-slip-uplift model to explain the metamorphic evolution through Variscan time. This metamorphism overprints pre-existing metamorphic rocks and is recognized in many of the monometamorphic series. Where the monometamorphic rocks are mostly Palaeozoic metasediments, this metamorphism is the product of collision. Nappe stacking during the Carboniferous brought together the polymetamorphic and monometamorphic domains which formerly represented independent crustal blocks or different structural levels.

The final period of metamorphism accompanied the formation of intracontinental strike-slip-zones, locally high temperature mylonites, transtensional sedimentary basins, and pull-apart structures that control the emplacement of late Variscan granites.

4 Magmatic Events

Basic rocks appear in all five massifs as amphibolites and meta-eclogites. Chemical data from the amphibolites (von Raumer et al. 1990) and eclogites (Liégois and Duchesne 1981; Paquette et al. 1989) of the Aiguilles Rouges reveal the tholeiitic character of these metabasic rocks, which appear to have formed in a continental environment. Given its age (450–475 Ma; U/Pb zircons; Paquette et al 1989) and geodynamic significance, this basic magmatism cannot be compared to the meta-ophiolitic association (496 Ma) of

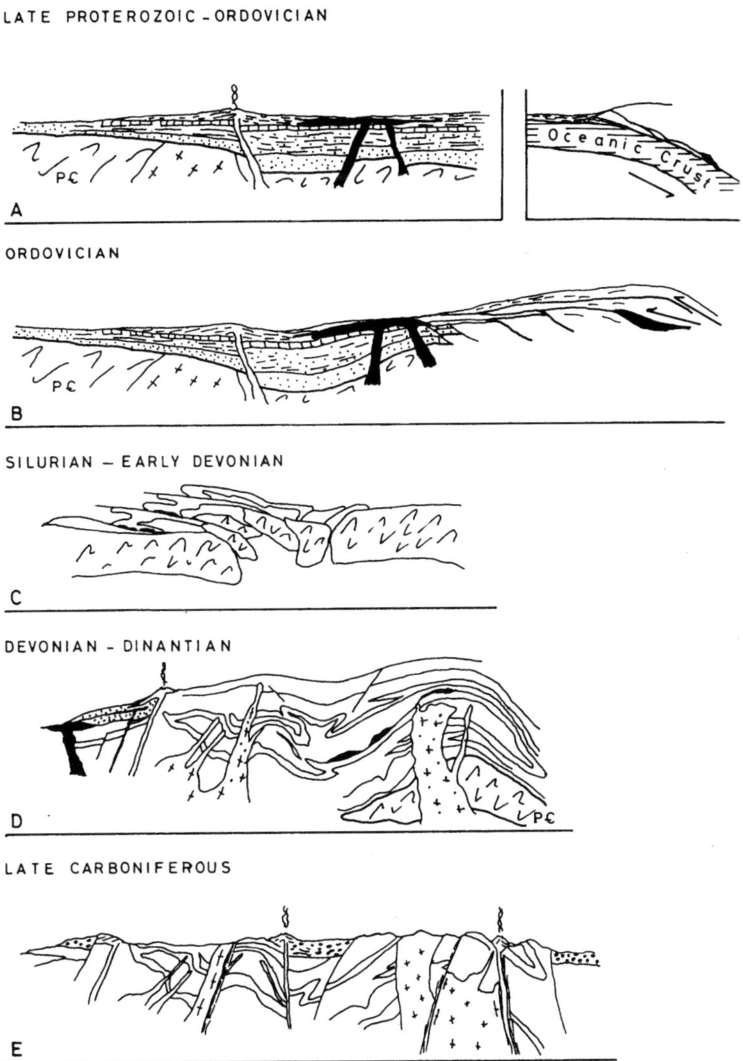

Fig. 4 A–E. Schematic evolution of the pre-Mesozoic basement of the Helvetic domain (after von Raumer 1984a, 1987). **A** Late Proterozoic-Ordovician – sediments with acidic and basic Precambrian and Palaeozoic volcanics underlain by a supposed Precambrian basement. *On the right:* supposed lateral existence of oceanic crust and subduction zones at the same time period. **B** Ordivician (and younger): subduction and obduction of oceanic crust. Perhaps early nappes corresponding to the Taconian phase? **C** Silurian – Early Devonian – first stages of collision and formation of large-scale nappes with transport of ultrabasic rocks and eclogites to higher crustal levels, Kyanite-bearing assemblages in metapelites and mylonites. Early fold generations. **D** Devonian – Dinantian – rising geothermal gradients and formation of mineral assemblages with sillimanite ± cordierite. Anatectic first melts and cordierite-bearing incongruent melts. Higher surface erosion rates with detrital sediments being collected in transtensional basins. **E** Late Carboniferous strike-slip movements, pull-apart and/or transtensional basins with detrital sediments and volcanics, Late Variscan granites controlled by the strike-slip structures

Chamrousse (SW Belledonne; see further details and discussion, Ménot and Paquette (this Vol.). In the Gotthard massif the mafic-ultramafic association can be split into two groups (Biino et al. 1993). The older group, probably of Upper Proterozoic age, shows a chemical and isotope signature characteristic of an oceanic sequence. The younger group, probably pre-Caledonian in age, is charaterized by an island arc affinity. A special feature, typical for the Aar (Abrecht 1980), Mont Blanc and Argentera areas, is the appearance of "Schollenamphibolite" or brecciated amphibolites (agmatites) in a granitoid to trondhjemitic matrix, which might have originated during the strike slip of Carboniferous age mentioned above.

K-feldspar-bearing amphibolites represent a more monzonitic trend of basic rocks (similar to Durbachites) which are observed in the Argentera, the Haut Dauphiné (gneiss d'Olan), the Mont Blanc/Aiguilles Rouges and the Aar Massifs (parts of the syenite of Giuf). Since they appear as xenoliths in the cordierite-bearing granitoids mentioned later, they may in part represent time equivalents of the Rioupéroux and Livet complexes (SW Belledonne) of Devonian – Early Carboniferous age (Ménot et al. 1988b). The latter complex comprises a mafic-silicic association of continental affinity (Ménot 1987; Ménot et al 1992). More detailed discussions of these basic rocks accompany Ménot and Paquette (this Vol.).

Ultrabasic rocks, mostly altered to serpentinites, occur in all massifs (see details in Pfeifer et al. this Vol.). Best known are those from the Chamrousse area (Belledonne) which represent part of an Early Ordovician ophiolitic association (Carme 1965a, 1975; Bodinier et al. 1981; Ménot 1979; Ménot et al. 1988a). Generally, however, their metamorphic overprint prevents most occurrences from being satisfactorily assigned to as ophiolitic setting. Those observed in the Aiguilles Rouges area (former lherzolites; Pfeifer and von Raumer 1993) appear to have a close relationship to the eclogites and could have entered their host rock before the regional Devonian anatexis. Ultramafic rocks from the Aar and Gotthard Massif preserve high-T metamorphic assemblages (forsterite, clinopyroxene, chromite), have an oceanic origin and are related to gabbroic and N-P-MORB eclogites (Abrecht et al. 1991; Biino et al. 1993).

The **granitoid rocks** can be generally subdivided into five main groups.

1. *Orthogneisses* of grantitoid or dacitic composition from the Aiguilles Rouges area (Favre pers. comm.) appear as K-feldspar augengneisses and fine-grained acidic gneisses. They either represent effusive and subvolcanic series of probable Cambrian to Ordovician age (island-arc volcanics; Chiaradia 1992), or they are strongly deformed granites of Caledonian age (e.g. protolith of "Streifengneiss" from the Gotthard area, K-feldspar augengneisses from the Mont Blanc massif). Such rocks are found in larger areas in the Argentera, Mont Blanc/Aiguilles Rouges and Gotthard Massifs. They also include the "gneiss oeillés de Crupillouse" of the Haut Dauphiné area.

2. The *cordierite-bearing granitoids* form different groups according to their age of intrusion and occur in a zone from the Mont Blanc area to the northern part of the Aar Massif (Innertkirchen-Gastern, Rutishauser 1973). They are also found in the Belledonne area and occupy large areas of the Argentera Massif. The granitoids are closely related to Variscan events, following the early nappe stacking and forming at the thermal peak of regional metamorphism (erosional unloading and/or, more probably, post-orogenic collapse and crustal thinning; probably since the Devonian). The earliest granitoids are anatectic first melts and cordierite- and/or garnet-bearing incongruent melts. These first melts are sometimes preserved as orthogneisses as a result of strong mylonitic deformation of Early to Mid-Carboniferous age (Belledonne: Bonnet 1982; Vivier et al. 1987). They may be followed by discordant, cordierite-bearing granitoids of Late Carboniferous age (e.g. Vallorcine-granite, Brändlein 1991; Gastern granite).

3. *Meta-trondhjemites* of SW-Belledonne reflect extensional magmatism represented by the Devonian Rioupéroux-Livet complex. These crustal and mantle-derived granitoids are closely associated with mafic rocks (Carme 1965a; Ménot 1986, 1987) and, according to their chemical character, are considered to represent a continental rift environment (Ménot 1987; Ménot and Paquette this Vol.; Ménot et al. 1992).

4. *Late Variscan granites* are well known from the entire external domain (Aumaître et al. 1985; Schaltegger 1986b, 1990b; Marro 1987, 1988; Vittoz et al. 1987; Vivier et al. 1987; Mercolli and Oberhänsli 1988; Bussy 1990; Bonin et al. this Vol.). Schenker (1987) discussed the close succession of Variscan nappe-like thrusting and emplacement of granite plutons. The close relationship between late Variscan strike-slip systems and the intrusion of granite bodies has been discussed by Joye (1989). The overall evolution of these late Variscan granites in general is given by Bonin et al. (this Vol.).

5. *Late magmatic events* are represented by rhyolite dykes of probable Permian age, which are related to the late granites. These crosscut older structures and follow the main tectonic lines of the late Variscan crustal blocks.

5 Ore Deposits

The tectono-thermal evolution of the external massifs since Cambrian times has led to the formation of a variety of ore deposits. Although their primary origin is often uncertain, skarn-ores, pegmatitic-pneumatolytic ores and hydrothermal concentrations can be recognized.

5.1 Skarns

Diopside-scheelite skarns occur in the Aiguilles Rouges and also in the Aar Massif (Frey and von Raumer 1977a; Della Valle et al. 1987; von Raumer 1987; Thierrin 1987). W-concentrations, in the Aiguilles Rouges, owe their origin to volcano-sedimentary lithologies (von Raumer and Chiaradia 1992). Skarns are confined to carbonate horizons of probably Late Proterozoic to Early Cambrian age in the neighbourhood of Late Variscan granite bodies (Chiaradia 1992, 1993).

Magnetite skarns are also restricted to Palaeozoic carbonates and formed adjacent to the Late Variscan Mont Blanc granite in the Mont Chemin area (Hugi et al. 1948).

5.2 Pegmatitic-Pneumatolytic Ores

Molybdenite and wolframite/scheelite also occur in the neighbourhood of the late Variscan granites.

Traces of these minerals are found in the late leucocratic differentiates of the Mont Blanc granite and, in the Aar Massif, occur in aplitic-pegmatitic veins related to the Central Aar granite (Steck and Hügi 1970; Thierrin 1987). They are also known in the central and eastern areas of Haut Dauphiné and in Argentera Massifs.

5.3 *Hydrothermal Concentrations*

Most ore deposits are of hydrothermal origin with arsenopyrite-gold and galena-sphalerite-chalcopyrite-pyrite parageneses predominating (Huttenlocher 1934; Ladame 1935; Rickenbach and von Känel 1953), the latter often accompanied by baryte and fluorine. Many of these hydrothermal concentrations are confined to the Palaeozoic carbonates (e.g. Mont Blanc/Aiguilles Rouges areas; von Raumer 1987), but larger veins of baryte-fluorine-quartz with mainly galena and sphalerite occur in the Mont Chemin area/Mont Blanc Massif (Wutzler 1983) and at various sites in the Aar Massif (Huttenlocher 1930; Jenni 1973; Saheurs 1975; Labhart 1977; Morel 1978). Abrecht (1989) described a hydrothermal manganiferous sulphide assemblage in Carboniferous volcanic rocks of the central Aar Massif.

Of special interest are concentrations of pichblende found along the southern border zone of the Aar Massif (Labhart 1967) and in the neighbourhood of the Late Variscan granites (Labhart and Rybach 1974; Frey and von Raumer 1977b; Negga et al. 1986).

All of the ore sites show considerable alterations as a result of Late Variscan and Alpine overprints, giving rise to a multitude of secondary products.

6 Summary and Correlation with the Variscan Foreland

The external massifs share their evolution with units known from bore-holes in northern Switzerland (Peters et al. 1987a, b) as well as with the Variscan basement of the French Massif Central, Vosges and Black Forest (Martin and Huber-Aleffi 1984; Müller et al. 1984; Dufour et al. 1985; Hirschmann 1985; Flöttmann 1988; Wickert 1988; Stenger et al. 1989; Wimmenauer and Stenger 1989; Gardien et al. 1990; Mazurek 1991; Echtler and Chauvet 1991; Mazurek and Peters 1992). Comparable lithologies are observed in the "upper gneiss nappe" from the French Massif Central (Ledru et al. 1989). Von Raumer and Ménot (1989) suggest a close relationship existing between all the "gneiss units with granulites and eclogites" in the Vosges-Black forest area, the French Massif Central and the Alpine Massifs. The existence of highly mobilized K-feldspar augengneisses and paragneisses containing lenses of eclogite, granulite, and serpentinite (former Lherzolite) in the domain of the Aiguilles Rouges finds parallels in the Mont du Lyonnais (Gardien et al. 1990) and in the Gföhl- and Drosendorf-units (Matte et al. 1990) of the Bohemian Massif. We are thus tempted to see parallels, for a given time period, a situation near a major suture bound to the stacking of crustal slices of the region concerned. The Gotthard area, in contrast, could have had a situation in the neighbourhood of a major, pre-Variscan suture with formation of oceanic crust, old lithologies resembling astonishingly those of the Lake-district (Ceneri-gneisses) in the Southern Alps. The external massifs, in consequence, cover already a larger domain in the cross-section of the former Variscan fold belt, and the following evolution is proposed.

The deposition of mainly detrital sediments during the Upper Proterozoic accompanied rifting and was followed by the deposition of more varied sediments which, partly, may include Lower Palaeozoic units. Rifting, locally, led to the formation of oceanic crust.

Convergence and formation of eclogites and later of granulites is documented for the Gotthard-area since the Ordovician, and during the same time period the Aar Massif underwent metamorphism and anatectic mobilization, whereas in the Mont Blanc and Aiguilles Rouges areas magmatism is related to an island arc situation. During the Silurian, the more external parts of the external massifs underwent metamorphism under mid crustal conditions which probably was related to subduction and the beginning of collision. Since the Devonian, the same zones were the locus of Variscan nappe stacking indicating collision between Gondwana and Laurasia. Crustal thickening was compensated by updoming and erosion and post-orogenic collapse and crustal thinning during the Devonian and Early Carboniferous. This sequence of events fits well a palaeotectonic situation in the large zone of Variscan accretion and collision (Matte 1991) identified from Sardinia, the Maures Massif, the Southern Black Forest, and the Bohemian Massif.

During the Lower Carboniferous large-scale tangential tectonics accompanied by strike-slip movements resulted in a thermal peak and in the nappe stacking of metamorphic units as well as those that had not been subjected to the main metamorphic event. Contemporaneously, volcanics and detrital sediments were deposited in transtensional basins with pull-apart structures facilitating the intrusion of the Late Variscan granitoids. The external massifs, in consequence, have to be interpreted as a puzzle-stone situated in the large suture of accretion and collision between Gondwana and Laurasia.

Acknowledgements
We express our gratitude to the reviewers, as the manuscript profited from the remarks of V. Köppel (Zürich), and from the patient work of R. D. Nance (Athens) transforming the manuscript into its present English version. The critical remarks and questions of D. Gebauer (Zürich) and P. Matte (Montpellier) helped to polish up the final version of this paper. We acknowledge the financial support over many years and for different projects (actual project 2.0–5.603; 2.0-29901.90) by the Swiss National Science Foundation.

References

Abrecht J (1980) Zur Bildung migmatitischer Schollenamphibolite aus dem Altkristallin des Aarmassivs. Geol Rundsch 69: 695–725
Abrecht J (1989) A hydrothermal manganiferous sulfide assemblage in Carboniferous volcanic rocks of the central Aar massif (Switzerland). Schweiz Mineral Petrogr Mitt 69: 117–128
Abrecht J, Schaltegger U (1988) Aplitic intrusions in the Central Aar massif basement: geology, petrography and Rb/Sr data. Eclogae Geol Helv 81(1): 277–239
Abrecht J, Biino G, Mercolli I, Stille P (1991) Mafic-ultramafic rock-associations in the Aar, Gotthard and Tavetsch Massifs (Helvetic domain, central Switzerland): marker of a ophiolitic pre-Variscan suture, reworked by polymetamorphic events? Schweiz Mineral Petrogr Mitt 71(2): 295–300
Apprahamian J (1988) Cartographie du métamorphisme faible à très faible dans les Alpes françaises externes par l'utilisation de la cristallinité de l'illite. Geodyn Acta 2–1: 25–32
Arnold A (1970a) Die Gesteine der Region Nalps-Curnera im nordöstlichen Gotthardmassiv, ihre Metamorphose und ihre Kalksilikateinschlüsse. Beitr Geol Schweiz NF 138: 1–128
Arnold A (1970b) On the history of the Gotthard Massif (Central Alps, Switzerland). Eclogae Geol Helv 63: 29–30
Arnold A (1972) Rb-Sr-Untersuchungen an einigen alpinen Zerrklüften des Cristallina-Granodiorites im östlichen Gotthardmassiv. Schweiz Mineral Petrogr Mitt 52: 537–551
Aumaitre R, Giraud P, Poncerry E, Vivier G (1985) Les granitoïdes sub-alcalins potassiques de la Lauzière, massif cristallin externe de Belledonne. Géol Alp 61: 17–32
Badoux H (1972) Tectonique de la nappe de Morcles entre Rhône et Lizerne. Beitr Geol Karte Schweiz NF 143: 1–78
Baumer A, Klee W, Lapraz D, Mullis J, Oberhänsli R (1982) Etudes cristallochimiques de l'apatite d'une fente alpine, La Fibbia, Massif du St-Gothard. Schweiz Mineral Petrogr Mitt 62: 353–363
Bellière J (1958) Contribution à l'étude pétrogénétique des schistes cristallins du massif des Aiguilles Rouges (Haute Savoie) Ann Soc Géol Belg 81: 1–198
Bellière J (1980) Massif du Mont Blanc et des Aiguilles Rouges. Structure et pétrologie du socle. Géol Alp 56: 237–249
Bellière J, Streel M (1980) Roches d'âge viséen supérieur dans le massif des Aiguilles Rouges (Haute Savoie). C R Acad Sci Paris 290 D: 1341–1343
Biino G G, Stille P, Abrecht J, Mercolli I (1993) Trace element and Sm-Nd isotopic study of Proterozoic MOR and island arc basalts in the Central Alps. Contr Mineral Petrol (submitted)
Bodinier JL, Dupuy C, Dostal J, Carme F (1981) Geochemistry of ophiolites from the Chamrousse complex (Belledonne Massif, Alps). Contrib Mineral Pet 78: 379–388
Bogdanoff S (1986) Evolution de la partie occidentale du massif cristallin externe de l'Argentera. Place dans l'arc alpin. Bull Bur Rech Géol Min Ser Geol Fr 4: 433–453
Bogdanoff S, Ménot R P, Vivier G (1991) Les massifs cristallins externes des alpes occidentales françaises, un fragment de la zone interne varisque. Sci Géol Bull 44:237:285
Bonnet D (1982) Les mégacristaux de feldspaths alcalins dans les formations porphyroïdes des massifs cristallins externes de Belledonne et des Grandes Rousses (Alpes françaises) Thèse 3ème cycle, Université de Grenoble
Bordet C (1961) Recherches géologiques sur la partie septentrionale du Massif de Belledonne (Alpes françaises). Mém Serv Carte Géol Fr 1961: 1–218
Bordet P, Bordet C (1963) Belledonne – Grandes Rousses et Aiguilles Rouges – Mont Blanc: quelque données nouvelles sur leurs rapports structuraux. Livre Mém Prof Paul Fallot Soc Géol Fr 1961: 311–316
Bortolami G, Sacchi R (1968) Osservazioni geologico-petrologiche sui medi valloni de S. Anna e Rio Freddo (Massicio cristallino dell'Argentera). Mem Soc Geol Ital VII: 37–64
Bossart P J, Meier M, Oberli F, Steiger RH (1986) Morphology versus U-Pb systematics in zircon: a high-resolution isotopic study of zicron population from a Variscan dike in the Central Alps. Earth Planet Sci Letters 78:339-359
Brändlein P (1991) Petrographische und geochemische Charakteristika des Vallorcine-Granits, Aiguilles-Rouges-Massiv (Westalpen, Schweiz). Thesis Univ Erlangen, 95 pp (unpubl)
Bussy F (1990) Pétrogenèse des enclaves microgrenues associées aux granitoïdes calco-alcalins: exemple des massifs varisque du Mont Blanc (Alpes occidentales) et miocène du Monte Capanne (Ile d'Elbe, Italie) Mém Géol (Lausanne) 7: 1–309
Bussy F (1992) U-Pb zicron datting of the Mont-Blanc granite and its microgranular enclaves. Abstract, Jahrestagung 1993, Basel p. 40.
Bussy F, Schaltegger U, Marro Ch (1989) The age of the Mont Blanc granite (Western Alps): a heterogeneous isotopic system dated by Rb-Sr whole rock determination on its microgranular enclaves. Schweiz Mineral Petrogr Mitt 69: 3–13
Bussy F and von Raumer J (1993) U-Pb dating of Palaeozoic event in the Montr-Blanc crystalline massif, Western Alps. Terra Nova 1993 Vol. 5, suppl. I, 1993, 187–188
Carme F (1965a) Sur deux formations, d'origine volcanique, des schistes cristallins antéhouillers de la chaîne de Belledonne (Alpes françaises). C R Acad Sci Paris 260: 6401–6404
Carme F (1965b) Existence de deux formations détritiques remarquables (conglomérats et flyschs) dans les schistes cristallins anté-houillers du Taillefer (chaîne de Belledonne, Alpes françaises) C R Acad Sci Paris 260: 6656–6659
Carme F (1970a) Tectoniques superposées et granitisations successives dans les massifs cristallins externes des Alpes; arguments nouveaux en faveur du rôle majeur d'une orogenèse anté-hercynienne dans la formation des terrains cristallins de Belledonne et des Aiguilles Rouges. C R Acad Sci Paris 271: 1733–1736

Carme F (1970b) Caractères fondamentaux de l'architecture d'ensemble des chaînes de Belledonne et des Aiguilles Rouges. Rôle des décrochements. C R Acad Sci Paris 271: 2232–2235
Carme F (1971) Les phases successives de déformation continue dans l'ensemble Belledonne-Aiguilles Rouges (Massifs cristallins externes, Alpes françaises). C R Acad Sci Paris 273: 1771–1774
Carme F (1973a) Précisions sur le métamorphisme majeur des schistes cristallins de la chaîne de Belledonne; essai de zonéographie et distribution verticale des zones au niveau de la coupe de la Romanche. C R Acad Sci Paris 277 D: 2133–2136
Carme F (1973b) Précisions nouvelles sur l'ampleur de la tectonique tangentielle hercynienne dans la chaîne de Belledonne. C R Acad Sci Paris 277 D: 2309–2312
Carme F (1974) Sur une paragenèse tardive à disthène, sillimanite prismatique et zoisite dans les anatexites à pinite des environs d'Allemont (Chaîne de Belledonne); essai d'interprétation pétrogénétique. C R Acad Sci Paris 278 D: 565–567
Carme F (1975) Le masif basique de Chamrousse et du Tabor (chaîne de Belledonne, Alpes occidentales): une variante originale de complexe alpinotype et un jalon majeur d'une ceinture basique briovérienne. C R Acad Sci Paris 281 D: 351–354
Carme F, Pin C (1987) Vue d'ensemble sur le magmatisme préorogénique et l'évolution métamorphique et tectonique varisques dans le sud de la chaîne de Belledonne (Massifs cristallins externes, Alpes françaises) C R Acad Sci Paris 304 II 19: 1177–1181
Chiaradia M (1992) The Scheelite-skarn of Salanfe (Valais, Switzerland). Thesis Fribourg 221 pp (unpubl)
Chiaradia M (1993) The Scheelite-skarn of Salanfe (Valais, Switzerland). Schweiz Min Petr Mitt 73,1: 41–51
Della Valle G, Gex P, von Raumer JF (1987) Le gîte d'or. d'arsenic et de tungstène de Salanfe. Beitr Geol Schweiz Geotech Ser 72: 118–128
Demeulemeester P (1982) Contribution à l'étude radiométrique à l'argon et au strontium des Massifs cristallins externes (Alpes françaises). Thesis, Grenoble 1982, 227 pp
Demeulemeester P, Roques M, Giraud P, Vivier G, Bonhomme M (1986) Influence du métamorphisme alpin sur les âges isotopiques Rb/Sr et K/Ar des biotites des massifs cristallins externes (Alpes françaises). Géol Alp 62: 31–44
Dufour E, Lardeaux JM, Coffrant D (1985) Eclogites et granulites dans les Monts du Lyonnais: une évolution métamorphique plurifacielle éohercynienne. C R Acad Sci Paris 300 II: 141–144
Echtler H und Chauvet A (1991) Carboniferous convergence and subsequent crustal extension in the Southern Schwarzwald (SW Germany). Geodin Acta 5:37-49
Escher A, Masson H, Steck A (1987) Coupes géologiques des Alpes occidentales suisses. Rapp Géol Serv Hydrol Géol Natl 2: 1–11
Faure Muret A (1955) Etudes géologiques sur le massif de l'Argentera-Mercantour et ses enveloppes sédimentaires. Mém Carte Géol Fr 47: 1–336
Flöttmann Th (1988) Strukturentwicklung, P-T-Pfade und Deformationsprozesse im zentralschwarzwälder Gneiskomplex. Frankf Geowiss Arb A 6: 1–206
Frey A, von Raumer JF (1977a) Über einen Scheelit-Neufund. Schweiz Strahler 4–7: 296–298
Frey A, von Raumer JF (1977b) Uranmineralien aus dem Val du Trient. Urner Mineralienfreund 1977–1: 1–4
Frey M, Bucher K, Mullis J (1980) Alpine metamorphism along the geotraverse Basel-Chiasso – a review. Eclogae Geol Helv 73(2): 527–546
Gardien V, Tegvey M, Lardeaux M, Miseri M, Dufour E (1990) Crust-mantle relationships in the French Variscan chain: the example of the southern Monts du Lyonnais unit (eastern French Massif Central). J Metam Geol 8: 477–492
Gasquet D, Giraud P, Ploquin A, Vivier G (1981) Géochimie des mylonites et relations entre les Rameaux interne et externe du massif de Belledonne (Alpes françaises). C R Acad Sci Paris 292 II: 607–610
Gebauer D, Quadt A, Compston W, Williams I, Grünenfelder M (1988) Archaische Komponenten in retrograden Eklogiten des Gotthardmassivs. Schweiz Mineral Petrogr Mitt 68: 485–490
Gibergy P (1968) Découverte de "grès à trous" renfermant des débris d'organismes dans les schistes noirs du Valbonnais (série cristallophyllienne des massifs cristallins externes des Alpes françaises). C R Acad Sci Paris 267 D: 1251–1254
Giorgi L, Giraud P, Vachard D (1979) Sur la présence de micro-organismes d'âge cambrien dans les schistes cristallins du versant occidental du massif cristallin externe des Grandes Rousses (Alpes occidentales). C R Acad Sci Paris 288 D: 1079–1082
Gratier JP, Lejeune B, Vergne JL (1973) Etude des déformations de la couverture et des bordures sédimentaires des massifs cristallins externes de Belledonne, des Grandes Rousses et du Pelvoux. Thèse, Université de Grenoble, 289 pp
Grauert B, Arnold A (1968) Deutung diskordanter Zirkonalter der Silvrettadecke und des Gotthardmassivs. Contrib Mineral Petrol 20: 34–56
Grünenfelder M, Hafner S (1968) Über das Alter und die Entstehung des Rotondogranits. Schweiz Mineral Petrogr Mitt 42: 169–207
Guillot S, Ménot R P, Lardeaux J M (1992) Tectonique intra-océanique distensive dans l'ophiolite paléozoïque de Chamrousse (Alpes occidentales). Bull Soc Géol France 163,3:229-240
Hafner S (1958) Petrographie des südwestlichen Gothardmassivs (zwischen St. Gotthardpaß und Nufenenpaß). Schweiz Mineral Petrogr Mitt 38: 255–362
Heim A (1922) Geologie der Schweiz. Tauschnitz, Leipzig
Hirschmann G (1985) Lithostratigraphic units in the gneisses of the Central Schwarzwald. Excursion guide Schwarzwald. 2nd Int Symp on Observation of the continental crust through drilling, Seeheim 1985: pp 8–15
Hirschmann G (1991) Zu den Korrelationsmöglichkeiten des Erzgebirgsproterozoikums mit den Profilen des Moldanubikums. (in press)
Holdaway MJ (1971) Stability of andalusite and the aluminium silicate phase diagram. Am J Sci 271: 97–131
Huber HH (1943) Physiographie und Genesis der Gesteine im südöstlichen Gotthardmassiv. Schweiz Mineral Petrogr Mitt 23: 72–260
Hügi T, Ledermann H, Schläppi E (1988) Blatt Lötschental Nr 1268, 1:25000, Erläuterungen. Kümmerly u. Frey, Bern
Hugi E, Huttenlocher HF, Gassmann F, Fehlmann H, Ladame GC, Hügi Th, Wohlers J (1948) Die Eisen- und Manganerze der Schweiz. Die Magnetitlagerstätten. Beitr Geol Schweiz Geotech Ser 13(4): 1–116
Huttenlocher H (1930) Die Blei-Zink-Lagerstätten von Goppenstein (Wallis). Beitr Geol Karte Schweiz NF 16(2): 85–116
Huttenlocher H (1934) Die Erzlagerstätten der Westalpen. Schweiz Mineral Petrogr Mitt 14: 20–144
Jäger E, Niggli E (1964) Rubidium-Strontium-Isotopenanalysen an Mineralien und Gesteinen des Rotondo-Granits und ihre geologische Interpretation. Schweiz Mineral Petrogr Mitt 44: 61–81

Jäger E, Niggli E, Wenk E (1967) Rb-Sr-Alterbestimmungen an Glimmern der Zentralalpen. Beitr Geol Karte Schweiz NF 134: 61–81

Jenni J (1973) Die Vorkommen von Bleiglanz, Kupferkies und Zinkblende des Bristenstocks (Kt. Uri). Beitr Geol Karte Schweiz Geotech Ser 53: 1–44

Joye JB (1989) L'évolution pression-température-déformation dans le massif des Aiguilles Rouges, massif externe alpin. Thesis, Fribourg 1991, 134 pp (unpubl.)

Kalsbeek F (1961) Petrology and structural geology of the Berlanche-Valloire area (Belledonne massif, France). Pub Dep Petrogr Crist Univ Leyden 2(5): 1–136

Klaper EM, Bucher-Nurminen K (1987) Alpine metamorphism of pelitic schists in the Nufenen Pass area, Lepontine Alps. J Metam Geol 5: 175–194

Krummenacher D, Bordet P, Le Fort P (1965) Les massifs externes alpins et leurs séries métamorphiques. Schweiz Mineral Petrogr Mitt 45: 855–874

Labhart TP (1965) Petrographische Untersuchungen am Südrand des Aarmassivs. Beitr Geol Karte Schweiz NF 124

Labhart TP (1967) Die Uranvererzungen am Südrand des Aarmassivs bei Naters (Kt. Wallis, Schweiz). Beitr. Geol Schweiz Geotech Ber 43: 1–30

Labhart TP (1977) Aarmassiv und Gotthardmassiv. Bornträger, Berlin

Labhart TP, Rybach L (1974) Granite und Uranvererzungen in den Schweizer Alpen. Geol Rundsch 63: 135–147

Labhart T, Schaltegger U (1983) Das Aarmassiv und seine Bedeckung zwischen Wasen und Innertkirchen. Eclogae Geol Helv 76: 117–123

Ladame GC (1935) Le gisement de galène et de spatfluordes Trappistes. Beitr Geol Karte Schweiz Geotech Ser 19: 1–34

Latouche L, Bogdanoff S (1987) Evolution précoce du massif de l'Argentera: apport des eclogites et des granulites. Geol Alp 63: 151–164

Ledru P, Lardeaux JM, Santallier D, Autran A, Quenardel JM, Floc'h JP, Lerouge G, Maillet N, Marchand J, Ploquin A (1989) Ou sont les nappes dans le Massif Central Français? Bull Soc Géol Fr (8)5–3: 605–618

Le Fort P (1973) Géologie du Haut Dauphiné cristallin (Alpes françaises). Etude pétrologique et structurale de la partie occidentale. Sci Terre 25: 1–373

Le Fort P, Pecher A (1971) Présentation d'un schéma structural du Haut Dauphiné cristallin. C R Acad Sci Paris 273 D: 3–5

Le Goff E (1989) Conditions pression-température de la déformation dans les orthogneiss: Modèle thermodynamique et exemples naturels. PhD Thesis, University Rennes, Mém Doc Centre Armoricain Etudes Structurales des Socles, Rennes 29: 1–303

Le Goff E, Marquer D (1992) P-T constraints in oligoclase garnet-bearing orthogneiss: the example of the Fibbia granite (Gotthard massif, Central Swiss Alps). Schweiz Mineral Petrogr Mitt 72 (in press)

Leutwein F, Poty B, Sinet J, Zimmermann JL (1970) Age de cavités du granite du Mont Blanc. C R Acad Sci Paris 271: 156–158

Liégois JP, Duchesne JC (1981) The Lac Cornu retrograded eclogites (Aiguilles Rouges Massif, Western Alps, France): evidence of crustal origin metasomatic alternation. Lithos 14: 35–48

Malaroda R (1968) Atlante delle facies petrografiche del massicio cristallino dell'Argentera. Mem Soc Geol Ital VII: 29–36

Marquer D (1990) Structures et déformation alpines dans les granites hercyniens du massif du Gotthard (Alpes centrales suisses). Eclogae Geol Helv 83(1): 77–97

Marro Ch (1987) Histoire des granitoides du Mont Blanc en Suisse. Bull Soc Frib Sci Nat 76: 73–128

Marro Ch (1988) Organisation géochimique et intrusion de granite du Mont Blanc et de deux leucogranites. Schweiz Mineral Petrogr Mitt 68: 521–529

Martin D, Huber-Aleffi A (1984) Das Kristallin des Südschwarzwaldes. NAGRA Tech Ber 84–30: 1–226

Matte Ph (1991) Accretionary history and crustal evolution of the Variscan belt in Western Europe. Tectonophysics 180: 1–28

Matte Ph, Maluski H, Rajlich P, Franke W (1990) Terrane boundaries in the Bohemian massif: result of large-scale Variscan shearing. Tectonophysics 177: 151–170

Mazurek M (1992) Phase equilibria and oxygen isotopes in the evolution of metapelitic migmatites: a case study from the pre-Alpine basement of northern Switzerland, Contrib Mineral Petrol 109: 494–510

Mazurek M, Peters T (1992) Petrographie des kristallinen Grundgebirges der Nordschweiz und Systematik der herzynischen Granite. Schweiz Min Petr Mitt 72:11-35

Ménot RP (1979) Lithostratigraphie et pétrographie du groupe amphibolique de Séchilienne et du massif ultrabasique et basique de Chamrousse (Massif de Belledonne, Alpes occidentales). Géol Alp 55: 93–110

Ménot RP (1986) Les formations plutono-volcaniques dévoniennes de Rioupéroux-Livet (Massifs cristallins externes des alpes françaises): nouvelles définitions lithostratigraphiques et pétrographiques. Schweiz Mineral Petrogr Mitt 66: 227–256

Ménot RP (1987) Magmatismes paléozoiques et structuration carbonifere du Massif de Belledonne (Alpes Françaises). These, Université de Lyon, Mém Doc Centre Armoricain Etudes Structurales des Socles, Rennes, 1988 21: 1–465

Ménot RP (1988) The geology of the Belledonne massif: an overview (external crystalline massifs of the Western Alps). Schweiz Mineral Petrogr Mitt 68: 531–542

Ménot RP, Peucat JJ, Piboule M, Scarenzi D (1984) Cambro-ordovician age for the ophiolitic complex of Chamrousse-Tabor (Belledonne massif, French external alpine domain). Ofioliti 10(2–3): 527

Ménot RP, Bonhomme MG, Vivier G (1987) Structuration tectono-métamorphique carbonifère dans le massif de Belledonne (Alpes occidentales françaises). Apport de la géochronologie K/Ar des amphiboles. Schweiz Mineral Petrogr Mitt 67: 273–284

Ménot RP, Peucat JJ, Scarenzi D, Piboule M (1988 a) 496 My age of plagiogranites in the Chamrousse ophiolite complex (external crystalline massifs in the French Alps): evidence of a lower Paleozoic oceanization. Earth Planet Sci Lett 88: 82–92

Ménot RP, Peucat JJ, Paquette JL (1988 b) Les formations leptyno-amphiboliques paléozoïques et les Complexes Leptyno-Amphiboliques: les corrélations hasardeuses. Exemples des Massifs Cristallins Externes alpins. Bull Soc Géol Fr (8) IV, 6: 917–926

Ménot RP, Peucat JJ, Paquette JL (1992) Isotopic and trace element evidences for a Devonian continental rifting event in the pre-alpine basement of Belledonne (Western Alps). Chem Geol (in press)

Mercolli I, Oberhänsli R (1988) Variscan tectonic evolution in the Central Alps. Schweiz Mineral Petrogr Mitt 68: 491–500

Merz C (1989) L'intrusif Medel-Cristallina (massif du Gothard oriental). Partie 1: déformations alpines et relations socle-couverture. Schweiz Mineral Petrogr Mitt 69: 55–71

Monié P, Maluski H (1983) Données géochronologiques 39Ar-40Ar sur le socle antépermien du massif de l'Argentera-Mercantour (Alpes-Maritimes, France). Bull Soc Géol Fr (7) XXV 2: 247–257

Morel F (1978) L'extension du filon Pb-Zn de Goppenstein (Valais, Suisse). Beitr Geol Schweiz Geotech Ser 57: 1–92

Müller WH, Huber M, Isler A, Kleboth P (1984) Erläuterung zur geologischen Karte der zentralen Nordschweiz 1:100000. Ber NTB 84-02, NAGRA, Baden

Negga HS, Sheppard SMF, Rosenbaum JM, Cuney M (1986) Late Hercynian U-vein mineralization in the Alps: fluid inclusion and C, O, H isotopic evidence for mixing between two externally derived fluids. Contrib Mineral Petrol 93: 179–186

Niggli E (1944) Das westliche Tavetscher Ziwschenmassiv und der angrenzende Nordrand des Gotthardmassivs. Schweiz Mineral Petrogr Mitt 24: 58–301

Niggli E, Niggli CR (1965) Karten der Verbreitung einiger Mineralien der alpidischen Metamorphose in den Schweizer Alpen (Stilpnomelan, Alkali-Amphibol, Chloritoid, Staurolith, Disthen, Sillimanit). Eclogae Geol Helv 58: 335–368

Oberhänsli R, Schenker F, Mercolli I (1988) Indications of Variscan nappe tectonics in the Aar massif. Schweiz Mineral Petrogr Mitt 68: 509–520

Oberholzer W (1955) Geologie und Petrographie des westlichen Gotthardmassivs. Schweiz Mineral Petrogr Mitt 35: 320–409

Paquette JL (1987) Comportement des systèmes isotopiques U-Pb et Sm-Nd dans le métamorphisme éclogitique. Chaîne hercynienne et chaîne alpine. Mém Doc Centre Armoricain Etudes Structurales des Socles, Rennes 14: 1–190

Paquette L, Ménot RP, Peucat JJ (1989) REE, Sm-Nd and U-Pb zircon study of eclogites from the Alpine External massifs (Western Alps): Evidence for crustal contamination. Earth Planet Sci Lett 96: 181–198

Peters T, Matter A, Isenschmid C, Blaesi HR, Meyer J, Ziegler HJ (1987a) Sondierbohrung Leuggern – Geologie. NAGRA Tech Ber NTB 86-04, Baden

Peters T, Matter A, Isenschmid C, Blaesi HR, Meyer J, Ziegler HJ (1987b) Sondierbohrung Kaisten – Geologie NAGRA Techn Ber NTB 86-05, Baden

Pfeifer H R and von Raumer (1993) Lherzolitic and pyroxenitic ultramafics from the Lac Cornu area (Aiguilles Rouges Massif, External Massif, External zone of the Alps, France). Abstract Terra Nova 1993 Vol. 5, suppl. 1, 1993, 382–383

Pfiffner AO (1986) Evolution of the north Alpine foreland basin in the Central Alps. Spec Publ Int Assoc Sediment 8: 219–228

Pilloud Cl (1992) Structures de deformation alpines dans le synclinal de Permo-Carbonifère de Salvan-Dorenaz (massif des Aiguilles Rouges). Mem Géol (Lausanne) 9: 1–100

Pin C, Carme F (1987) A Sm-Nd isotopic study of 500 Ma old oceanic crust in the Variscan belt of Western Europe: the Chamrousse ophiolite complex, Western Alps (France). Contrib Mineral Petrol 96: 406–413

Poty BP, Stalder HA, Weisbrod AM (1974) Fluid inclusions studies in quartz from fissures of Western and Central Alps. Schweiz Mineral Petrogr Mitt 54: 717–752

Purdy JW, Stalder HA (1973) K-Ar age of fissure minerals from the Swiss Alps. Schweiz Mineral Petrogr Mitt 53: 79–98

von Raumer JF (1974) Zur Metamorphose amphibolitischer Gesteine im Altkristallin des Mont Blanc- und Aiguilles-Rouges-Massivs. Schweiz Mineral Petrogr Mitt 54: 471–488

von Raumer JF (1976) Variszikum in den Zentral- und Westalpen Nova Acta Leopold NF 224: 147–176

von Raumer JF (1981) Variscan events in the Alpine region Geol Mijnbouw 60: 67–80

von Raumer JF (1983) Die Metapelite von Emosson (Aiguilles-Rouges-Massiv) als Beispiel spätkaledonisch-frühvariszischer Metamorphose im Altkristallin des helvetischen Bereichs. Schweiz Mineral Petrogr Mitt 63: 421–455

von Raumer JF (1984a) The Hercynian basement in the Helvetic Realm. Western and Central Alps. Mem Soc Geol Ital 29: 57–69

von Raumer JF (1984b) The External Massifs, relics of Variscan Basement in the Alps. Geol Rundsch 73: 1–31

von Raumer JF (1987) Les massifs du Mont Blanc et des Aiguilles Rouges: témoins de la formation de croûte varisque dans les Alpes Occidentales. Géol Alp 63: 7–24

von Raumer JF (1988) Caledonian-Variscan structures in the Alps: an introduction. Schweiz Mineral Petrogr 68: 291–299

von Raumer JF, Ménot RP (1989) Evolution paléozoique du socle ouest-alpin et place des Massifs Cristallins Externes dans l'orogene Varisque. C R Acid Sci Paris 309 II: 397–402

von Raumer JF, Schwander HW (1985) Garnet evolution in pre-Variscan pelitic rocks from the Lake Emosson area, Aiguilles Rouges Massif, Western Alps. J Metam Geol 3: 467–479

von Raumer JF, Galetti G, Oberhänsli R, Pfeifer HR (1990) Amphibolites from Lac d'Emosson/Aiguilles Rouges (Switzerland): tholeiitic basalts at a transition zone between continental and oceanic crust. Schweiz Mineral Petrogr Mitt 70(3): 419–435

von Raumer J and Chiaradia M (1992) Bedeutung vor- und frühpaläozoischer Lithologien für Wolfram-Mineralisationen. Zbl Geol Paläonotol I, 1/2:87–91

Rickenbach E, von Känel F (1953) Die Arsen-Goldlagerstätte von Salanfe (Wallis). Beitr Geol Schweiz Geotech Ser NF 31: 1–52

Roccati A (1903) Ricerche petrografiche sulle valli del Gesso (Valle della Meris e Rocca Val Miana). Atti R Acc Sci Torino 38: 929–940

Rutishauser H (1973) Die quantitative Erfassung von Migmatiten im Aufschlußbereich (Erläutert am Beispiel des Lauterbrunner Kristallins). Schweiz Mineral Petrogr Mitt 53: 99–124

Rutishauser H (1975) Die Bedeutung des Jungfrau-Gipfelgesteins für die Zonierung des Aarmassivs. Eclogae Geol Helv 68(3): 511–515

Saheurs JP (1975) Geochemische Prospektion im Gebiet der Blei-Zink-Baryt-Lagerstätten bei Trachellauenen. Beitr Geol Schweiz Geotechn Ser 55: 1–91

Schaltegger U (1986a) Voralpine und alpine Mineralbindung in der Gneiszone von Erstfeld (Sustenpass, Aarmassiv) der Mechanismus der K-Ar- und Rb-Sr-Verjüngung alpin umgewandelter Biotite. Schweiz Mineral Petrogr Mitt 66: 395–412

Schaltegger U (1986b) Geochemie und RB-Sr-Systematik der Aarmassiv-Granite zwischen Grimsel und Reusstal. Schweiz Mineral Petrogr Mitt 66: 462–466

Schaltegger U (1990a) The Central Aar granite: highly differentiated calc-alkaline magmatism in the Aar massif (Central Alps, Switzerland) Eur J Mineral 2:245–259

Schaltegger U (1990b) Post-magmatic resetting of Rb-Sr whole rock ages – a study in the Central Aar granite (Central Alps, Switzerland). Geol Rdsch 79:709–724

Schaltegger U (1991) Die polymetamorphe Geschichte des Aarmassivs – neue U-Pb-Resultate. Schweiz Mineral Petrogr Ges Jahrestagung Chur 1991 Abstracts: p. 20

Schaltegger U, Labhart T (1983) Das Aarmassiv und seine Bedeckung zwischen Wassen und Innertkirchen. Eclogae geol Helv 76: 117–123

Schaltegger U, von Quadt A (1990) U-Pb zircon dating of the Central Aar Granite (Aar Massif, Central Alps). Schweiz Mineral Petrogr Mitt 70: 361–371

Schenker F (1987) Hinweise für kompressive Tektonik während der Ablagerung von oberpaläozoischen Sedimenten und Vulkaniten im Aarmassiv. Bull Ver Schweiz Pet Geol Ing 54/124: 45–57

Schulz B, von Raumer JF (1992) Syndeformational uplift of Variscan high pressure rocks (Val de Bérard, Aiguilles Rouges, Western Alps) Zeitschr Dtsch Geol Ges 144 (in print)

Sergeev S A and Steiger R H (1993) High-reolution U-Pb single zircon dating of Variscan and Caledonian magmatic cycles in the Gotthard Massif, Central Swiss Alps. Terra Cognita Abstract (in press)

Simeon Y (1979) Etude pétrologique, géochimique et structurale des terrains cristallins de Belledonne entre l'Arc et l'Isère (Alpes françaises). Thèse, Université de Grenoble, 224 pp

Stalder HA (1964) Petrographische und mineralogische Untersuchungen im Grimselgebiet. Schweiz Mineral Petrogr Mitt 44: 187–398

Stalder HA, de Quervain F, Niggli E, Graeser S (1973) Die Mineralfunde der Schweiz. Wepf, Basel

Staub R (1911) Geologische Beschreibung der Gebirge zwischen Schächental und Maderanertal. Beitr Geol Karte Schweiz NF 32: 1–84

Steck A (1966) Petrographische und tektonische Untersuchungen am Zentralen Aaregranit und seinen altkristallinen Hüllgesteinen im westlichen Aarmassiv im Gebiet Belalp-Grisighorn. Beitr Geol Karte Schweiz NF 130: 1–100

Steck A (1968) Die alpidischen Strukturen in den zentralen Aaregraniten des westlichen Aarmassivs. Eclogae Geol Helv 61: 19–48

Steck A (1976) Albit Oligoklas-Mineralgesellschaften der Peristeritlücke aus alpinmetamorphen Granitgneisen des Gotthardmassivs. Schweiz Mineral Petrogr Mitt 56: 269–292

Steck A (1984) Structures de déformations tertiaires dans les alpes centrales. Eclogae Geol Helv 77: 55–100

Steck A (1987) Le massif du Simplon – Réflexions sur la cinématique des nappes de gneiss. Schweiz Mineral Petrogr Mitt 67: 27–45

Steck A, Burri G (1971) Chemismus und Paragenesen von Granaten aus Granitgneisen der Grünschiefer- und Amphibolitfazies der Zentralalpen. Schweiz Mineral Petrogr Mitt 51: 534–538

Steck A, Hügi T (1970) Das Auftreten des Molybdänglanzes im westlichen Aarmassiv und Molybdängehalte von Gesteinen der gleichen Region. Schweiz Mineral Petrogr Mitt 50: 534–538

Steiger RH (1962) Petrographie und Geologie des südlichen Gotthardmassivs zwischen St. Gotthard- und Lukmanierpaß. Schweiz Mineral Petrogr Mitt 42: 381–577

Stenger R, Baatz K, Klein H, Wimmenauer W (1989) Metamorphic evolution of the pre-Hercynian basement of the Schwarzwald (Federal Republic of Germany). Tectonophysics 157: 117–121

Thierrin J (1987) Introduction à la géologie de la partie occidentale du massif de l'Aar. Beitr Geol Schweiz Geotech Ser 72: 25–27

Thöni M (1989) In: Joye JB (1989) Evolution tectonométamorphique varisque du massif des Aiguilles Rouges, massif cristallin externe alpin Thesis Fribourg 134 pp (unpubl)

Thompson AB (1982) Dehydration melting of pelitic rocks and the generation of H2O-undersaturated granitic liquids. Am J Sci 282: 1567–1595

Tobi AC (1959) Petrological and geological investigation in the Merdaret-Lac de Crop region (Belledonne massif). Thèse Leiden, Leidse Geol Meded 24: 182–282

Vittoz, P, Costarella R, Vivier G, Oliver RA (1987) Typologie des granitoides hercyniens du Haut Dauphiné par l'étude des terres rares et de la morphologie des cristaux de zircon: mise en évidence d'une zonation magmatique dans ce massif. Géol Alp 63: 119–163

Vivier G, Ménot RP, Giraud P (1987) Magmatismes et structuration orogénique paléozoiques de la chaîne de Belledonne (massifs cristallins externes). Le domaine nord-oriental. Géol Alp 63: 25–53

Weber F (1904) Über den Kali-Syenit des Piz Giuf. Beitr Geol Karte Schweiz NF 14 Lfg

Wickert F (1988) Paleozoic evolution of continental crust in the Beaujolais-Lyonnais area, north-eastern part of the Massif Central, France. Geol Rundsch 77: 467–482

Wimmenauer W, Stenger R (1989) Acid and intermediate HP metamorphic rocks in the Schwarzwald (Federal Republik of Germany) Tectonophysics 157: 109–116

Wüthrich H (1965) Rb-Sr-Alterbestimmungen am alpin metamorph übersprägten Aar-Massiv. Schweiz Mineral Petrogr Mitt 45: 875–971

Wutzler B (1983) Geologisch-lagerstättenkundliche Untersuchungen am Mont Chemin. Clausth Geol Abh 42: 1–104

Geodynamic Significance of Basic and Bimodal Magmatism in the External Domain

R. P. Ménot and J. L. Paquette

Abstract

This chapter deals with the geodynamic significance of the pre-Westphalian mafic magmatic events which are recorded in the basement of the Western and Central Alps. It is based on the available petrological, chemical and geochronological data concerning the meta-igneous rocks. This review leads to the following results: (1) The mafic and bimodal magmatisms show evidence of two successive periods of pre-orogenic lithospheric extension in the Cambrian-Ordovician and Devonian-Dinantian times. Their chemical characteristics testify to attenuated continental crust environment, sometimes reaching to narrow oceanic basins. The diversity of their post-magmatic orogenic evolution implies a Late Variscan tectonic "collage". (2) Most of the external crystalline massifs can be correlated with the westernmost segment of the European Variscan fold belt, when the Aar-Gotthard Massifs display striking affinities with the basement of the Eastern and Southern Alps.

1 Introduction

Previous studies on the crystalline basement of the Western Alps have pointed out the lithological continuity between the different massifs (Krummenacher et al. 1965; von Raumer 1981; Bogdanoff 1986). The lithological subdivisions proposed by Krummenacher and coworkers were mainly supported by the relative abundance of a mafic igneous component. On the one hand, the "série satinée" and "série brune" were considered as metasedimentary series but including a few interlayered mafic igneous rocks, and on the other, the magmatic origin of the amphibolitic and leptynitic "série verte" was obvious, as testified by premetamorphic textural and mineralogical relics. "Leptynite" is the common term for meta-igneous silicic rocks.

However, such a lithological subdivision did not take into account both the age of emplacement and the subsequent tectonic and metamorphic evolution of the rocks. With regard to these criteria, the previously defined "série verte" included amphibolitic and/or meta-igneous mafic and silicic formations of Early Carboniferous, Devonian, Early Palaeozoic and even older ages. Moreover, it involved mono- and polymetamorphic materials which have been metamorphosed under various P-T conditions at different times. Such diversity was pointed out for the Belledonne massif (Ménot 1987; Ménot et al. 1988b), thus, this massif could be compared with the strike-slip composite terranes as defined by Keppie (1985) and Hutton (1987). In the Belledonne, segments that are more or less internal with regard to the zonation of the Palaeozoic orogenic belt, have been fused together by a Mid- to Late Carboniferous tectonic event (see review in Ménot 1987, 1988).

So, in that context, the present work summarizes the available data concerning meta-igneous rocks from the different external crystalline massifs (ECM) of the Western Alps. It also tries to give a synthetic overview of the successive magmatic mafic and bimodal events, considered as geodynamic witnesses for the pre-Alpine orogenic history. Two main magmatic events (a) Precambrian to Early Palaeozoic, and (b) Devonian to Dinantian ages will be distinguished.

2 Late Precambrian to Early Palaeozoic Magmatic Events

Representative materials of these earliest events crop out as partially retrogressed high-grade, polymetamorphic rocks (eclogites, granulites and amphibolites) or as monometamorphic medium-grade amphibolites. The former are known within most of the ECM but, at present, the latter have been described only in the SW area of Belledonne.

2.1 *The Polymetamorphic Igneous Formations*

Generally, they occur as lenses and layers reaching thicknesses of more than 100 m within the metasedimentary gneissic basement, mainly metapelites and metagreywakes, and are commonly associated with serpentinites.

2.1.1 Aar – Gotthard

In the Aar Massif, metabasites are known as mafic and ultramafic associations in the central and eastern areas and everywhere as amphibolitic enclaves within pegmatitic trondhjemitic matrices ("Schollenamphibolites") (Abrecht 1980). Olivine and spinel bearing parageneses are thought to be indicative of a high temperature (HT) metamorphic event of Variscan or pre-Variscan age (Arnold 1970; Schenker and Abrecht 1987).

In the Gotthard Massif, amphibolites, metagabbros and serpentinites occur within metasedimentary gneisses which also include marbles and calc-silicates. According to Gebauer et al. (1988), this association is of tectonic origin. In the central and eastern areas, some high pressure (HP) parageneses have been recorded and they are partially overprinted by granulite and amphibolite conditions (Abrecht and Biino 1991). Ion-probe U-Pb analyses of euhedral zircons (Gebauer et al. 1988) reveal a Late Proterozoic magma crystallization age (870 Ma) and a Late Ordovician age (468 Ma) for the eclogite facies metamorphism. The latter result is supported by whole-rock – garnet Sm-Nd data (461 +/– 25 Ma: Gebauer et al. 1988 and this Vol.). In addition, inherited zircon xenocrysts give older ages at around 3.17, 2.67, 2.45 and 1.27 Ga. These data, together with that obtained on country rocks, will be more precisely discussed by Gebauer (this Vol.).

Few publications have been made on the chemistry and the inferred geological significance of meta-igneous rocks from Aar and Gotthard Massifs. In the latter, trace and rare earth element (REE) patterns and low initial εNd values (+1.3) suggest a calc-alkaline nature for the protoliths, indicative of an island arc environment (Gebauer et al. 1988).

2.1.2 Aiguilles Rouges

In the Aiguilles Rouges Massif, retrogressed eclogites have been described by Brière (1920); Corbin and Oulianoff (1928); Delhemmes (1954); Belliere (1958, 1980) and von Raumer (1976). The surrounding polymetamorphic gneisses are considered as the original country rocks of the mafic intrusions (von Raumer 1984b, 1988). Eclogites and garnet-bearing amphibolites have been studied respectively in the southern (Lac Cornu: Liegeois and Duchesne 1981; Paquette et al. 1989) and northern (Lac Emosson: von Raumer et al. 1990) regions.

Eclogitic P-T conditions (780 °C and 0.11 GPa: Liegeois and Duchesne 1981) were followed by medium to low pressure recrystallizations, as witnessed by the successive parageneisses within the surrounding gneisses (ky-st-grt; bt-sil-grt-crd; and-Kfs) (Belliere 1958; von Raumer 1984a; von Raumer and Schwander 1985; von Raumer and Ménot 1989).

Following Paquette et al. (1989), the HP conditions are probably related to an eo-Devonian, subduction-collision event that leads to recrystallization of mafic protoliths emplaced in Late Ordovician times (453 + 3/ –2 Ma, U-Pb zircon ages).

No magmatic relics have been described in the metabasites but according to major and trace elements (including REE), their protoliths display tholeiitic affinities (Liegeois and Duchesne 1981; Paquette et al. 1989; von Raumer et al. 1990). To explain a relatively wide range of compositions, Liegeois and Duchesne (1981) proposed a two-fold origin suggesting that some protoliths represent "magmatic residual liquids formed by low pressure fractional crystallization" whilst others are thought to be volcano-sedimentary materials. For von Raumer et al. (1990), the relatively wide range of composition of the volcanic and subvolcanic rocks could be related either to various magmatic processes which occurred during the magma emplacement (flow and gravitational differentiation; exchanges with pelitic country rocks) and either to post-magmatic alteration.

With regards to REE distributions, basaltic protoliths from the northern and southern areas differ and are respectively comparable to P- and N-types of MORB (Paquette et al. 1989; von Raumer et al. 1990). N-MORB-normalized abundances (Fig. 1) are characterized by an enrichment of the large ion lithophile elements (LILE) (especially Th) and a nearly flat pattern of the Nb to Yb section [high field strength (HFS) elements]. If alkalies are removed during post-magmatic alteration, Th is commonly considered as a less mobile element (Pearce 1983). A high Th content is indicative of a crustal component provided either by intracrustal contamination of the mafic magma or by metasomatism of a sub-continental lithosphere (Pearce 1983; Dupuy and Dostal 1984). The initial Nd isotope composition of the Lac

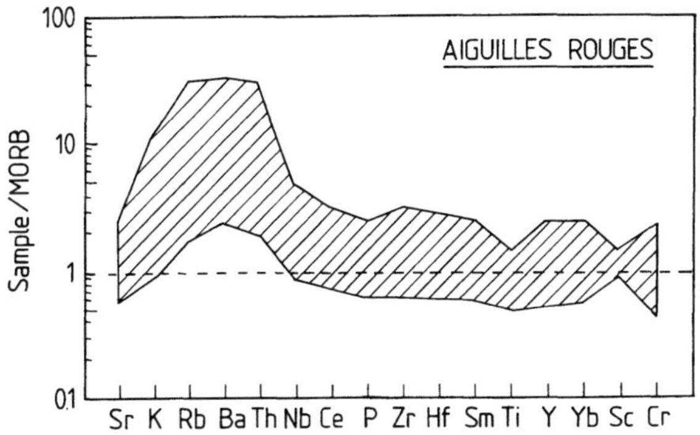

Fig. 1. MORB-normalized spidergram (Pearce 1983): Aiguilles Rouges retrogressed eclogites (data: Paquette et al. 1989; von Raumer et al. 1990)

Cornu basaltic protoliths ($E_{Nd(500)} = +6.5$) may also indicate crustal influences (Paquette et al. 1989).

According to both field evidence and geochemical data, the Ordovician mafic magmatism of the Aiguilles Rouges Massif may be inferred to a continental rift zone ending with the generation of oceanic crust (Liegeois and Duchesne 1981; Paquette et al. 1989; von Raumer et al. 1990).

2.1.3 Northeastern Domain of the Belledonne

Strong mesozonal to epizonal retrogression of high grade metabasites leads to the common occurrences of plagioclase amphibolites (see review in Vivier et al. 1987). Amphibolites are often associated with metafelsic rocks (leptynites). Eclogitic relics are scarce and mainly restricted to the Beaufortin area in the northernmost part of the region (Brière 1920), thus P-T calibrations of the eclogitic stage are lacking. However, a U-Pb zircon Eo-Devonian age (395 +/− 2 Ma) has been defined for this high-grade event (Paquette et al. 1989).

According to their chemistry (trace elements and Sm-Nd isotopes: Paquette et al. 1989), the mafic protoliths of the Beaufortin area should be compared with basaltic liquids and not with cumulates which are very scarce. They closely resemble those of the Aiguilles Rouges Massif: probably sub-contemporaneous emplacement ages (473 + 28/− 31 Ma: U-Pb zircon ages; Paquette et al. 1989), similar trace element and REE distribution (Fig. 2) and initial E_{Nd} ranging from + 8 to + 6. Likewise, continental influence is indicated by (a) the scattering of E_{Nd} values, inferred by a very low crustal contamination (under 4–5%) and by (b) the presence of old-inherited cores within some zircon grains (2.5 Ga). Such inherited zircons may have been incorporated during the emplacement of basic sills.

In the southernmost regions of the NE domain of Belledonne, very few detailed geochemical studies have been performed. But from unpublished data (Fig. 3: RPM), the amphibolites seem to be quite similar to that of the Beaufortin area and probably represent the southern extension of the latter. This is supported by field evidence (Vivier et al. 1987). Consequently, an attenuated continental crust environment or a narrow oceanic basin are favoured as original settings for such mafic and bimodal magmatism.

In the south western domain of Belledonne, the lithology of the Allemont-Rochetaillée tectonic unit may be compared to the polymetamorphic gneissic basement of the NE domain: following Carme and Pin (1987), the amphibolites display comparable emplacement ages (600–500 Ma) and similar chemical patterns, N- and T-type MORB affinities and scattered initial E_{Nd} (+5.5 to +9). However, a subduction-related extensional setting has been proposed by the previous authors. But as these amphibolites are devoid of any HP metamorphic relics, the Allemont-Rochetaillée unit could correspond to a shallower level of a comparable crust section (Ménot 1987; von Raumer et al. this Vol.).

A southwards extension of the Allemont-Rochetaillée unit could be present in the western part of the Ht. Dauphiné Massif, in the "Cortex" area.

2.1.4 Central and Eastern Domains of the Ht. Dauphiné (or Pelvoux): the "Noyau"

Following Le Fort and Pécher (1971), the Ht. Dauphiné Massif is commonly subdivided into two contrasting domains: the central and eastern domain which suffered a polycyclic, or polymetamorphic, evolution under deep crustal conditions and the western domain which only underwent less severe recrystallizations (Le Fort and Pécher 1971) or even a monocyclic evolution (Buffet 1980). The relationships of these units (basement-cover or nappe piling) are still ambiguous.

Within the "Noyau", amphibolites have been described within the migmatitic gneisses of the Lavey formation and within the acid granulitic gneisses of the Peyre Arguet lineament.

The Lavey amphibolites outcrop as intercalations within metasediments (greywackes, sandstones and marbles) and are associated with metafelsic rocks and some ultramafic occurences (Le Fort 1973; Pécher 1984). Two migmatization events are recorded in the Lavey formation, thus amphibolites appear as massive, layered or agmatic facies. If only the less remobi-

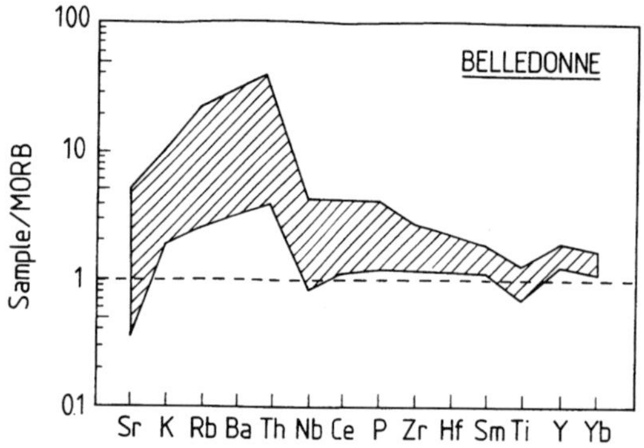

Fig. 2. MORB-normalized spidergram (Pearce 1983): NE Belledonne (Beaufortin) retrogressed eclogites (data: Paquette et al. 1989)

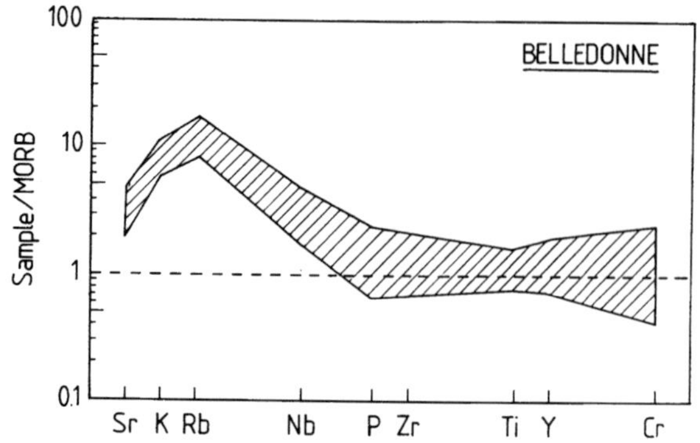

Fig. 3. MORB-normalized variation diagram (Pearce 1983): NE Belledonne amphibolites (unpublished, data: Ménot)

lized metabasites and metafelsites are taken into account, their chemistry is close to that of the spilite-keratophyre series, with a striking Na-rich character (Le Fort 1973). According to available trace elements data (RPM unpublished), N to T-types MORB affinities can be seen and N-MORB normalized variation diagrams (Fig. 4) enhance a slight Nb enrichment relative to Zr and Y contents.

The Peyre Arguet meta-eclogites form masses cutting accross the layering of the granulitic gneisses and leptynites. The Peyre Arguet formation could be considered as a lithological equivalent of the Lavey formation, but which has been partially preserved from migmatization (Pécher 1984). As yet, no geochemical data is available.

2.1.5 Argentera (Mercantour)

From the work of Faure-Muret (1955), Malaroda et al. (1970) and Bogdanoff (1980, 1986), three main migmatitic formations, separated by tectonic contacts of pre-Alpine age, have been distinguished in the Argentera Massif: from west to east, the plagioclase-rich migmatites of Anelle, the Kfs-Ab migmatites of Rabuons and the Kfs (+/– crd) migmatites of Malinvern-Chastillon. Kyanite, sillimanite and garnet parageneses testify to the severe metamorphic conditions. These migmatitic formations originate from a series of sediments (greywackes, sandstones and pelites) and enclose some orthogneissic siliceous intercalations (granodiorites and granites). Metabasites also occur as amphibolites, mafic granulites, eclogites and coronitic metagabbros. Two different groups of mafic rocks have been defined by Latouche and Bogdanoff (1987) with regard to their distinctive early metamorphic history: (a) eclogites which belong to an older generation that suffered an HP event (700 °C; 0.14 GPa) followed by retrogression under granulitic conditions (see later) and (b) granulites and coronitic metagabbros, arising from later intrusions, which only underwent the granulitic stage (700 °C, 0.9 GPa and 600 °C, 0.7–0.8 GPa).

Geochronological and geochemical investigations have been carried out on the eclogites from the eastern migmatitic formation (Malinvern-Chastillon) (Paquette et al. 1989). In the field, these metabasites differ from those of the Belledonne and Aiguilles Rouges Massifs: they appear as lenses of hectometric size or as metric dismembered masses which display an internal mineralogical banding of probably igneous origin. Thus most of the eclogite protoliths were originally parts of mafic layered intrusions.

U-Pb zircon data (Paquette et al. 1989) give the following results: (a) emplacement and crystallization of mafic protoliths during Early Ordovician times (471 +40/– 29 Ma); (b) eclogite facies metamorphism of Late Silurian age (424 +/– 4 Ma); and (c) presence of an ancient inherited component testified to 2.4 and 2.7 Ga upper intercepts. Some younger ages (351 ± 1 Ma and 327 ± 1 Ma) have been tentatively interpreted, as either reflecting another eclogite facies event (351 Ma) or as successive recrystallizations during the retrograde evolution. This second interpretation fits with the 39 Ar/40 Ar cooling ages, 375 – 350 Ma and 330 – 315 Ma on muscovite and biotite respectively (Monié and Maluski 1983). Mafic protoliths are heterogeneous and correspond to mainly plagioclase-rich cumulates and to basaltic liquids ranging from N-type to predominant E-Type MORB compositions (Fig. 5). On variation diagrams, the basalts are much more enriched in LILE than their equivalents from the Aiguilles Rouges and NE Belledonne Massifs, whilst a slight negative slope characterizes the Nb to Yb distribution. This relative

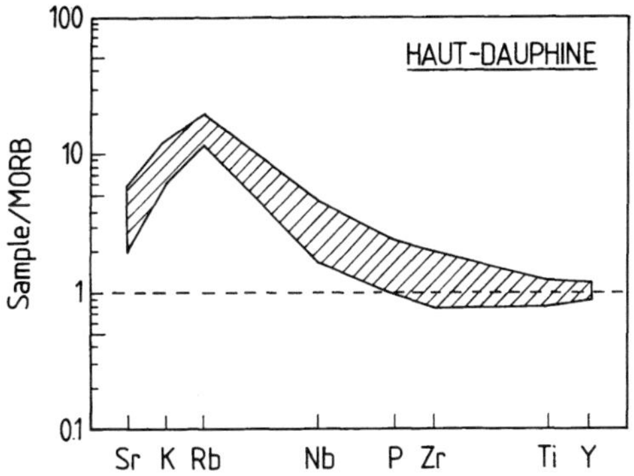

Fig. 4. MORB-normalized variation diagram (Pearce 1983): Haut Dauphiné "Noyau" amphibolites (data: major elements: Le Fort 1973 and unpublished trace element: Ménot)

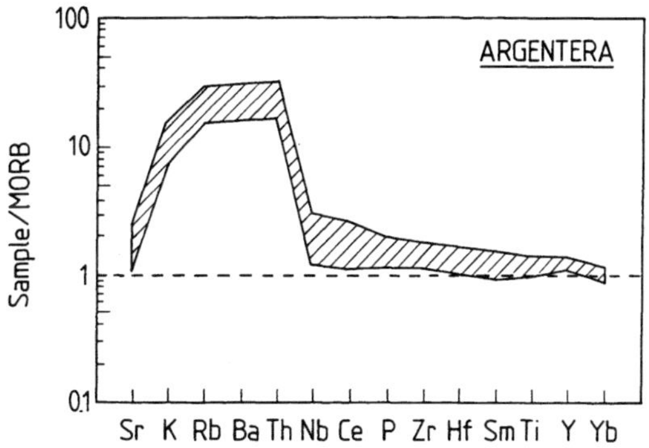

Fig. 5. MORB-normalized variation diagram (Pearce 1983): Argentera retrogressed eclogites (data: Paquette et al. 1989; and unpublished: Ménot)

enrichment of Nb relative to Zr and of Zr with respect to Y could be indicative of a subcontinental origin for the mantle source (Pearce 1983).

Initial E_{Nd} values are fairly similar in both N and E-types MORB protoliths and vary respectively from + 5.9 to + 8 and from + 6.1 to 8.2. These data indicate a depleted mantle source extraction but also infer interactions between this (these) source(s) and the continental crust (see later). Finally the influence of continental contamination, pointed out from trace element and isotope chemistry, is further corroborated by evidence of an old inherited component in zircon. It is probable that therefore, the protoliths of the eclogites from the eastern Argentera were emplaced into segments of an attenuated continental crust.

2.1.6 Concluding Remarks

Emplacement ages for the mafic and bimodal magmatism of the ECMs range from Late Proterozoic (Gotthard) to Ordovician (Aiguilles Rouges, NE Belledonne, Argentera). With regards to their location, the mafic protoliths have been involved at different times in subduction-collision processes, as suggested by the various ages for the HP metamorphic event (Late Ordovician: Gotthard; Late Silurian-Early Devonian: Aiguilles Rouges, NE Belledonne, Argentera).

According to their chemical characteristics, two distinct types of mafic rocks have be distinguished. They are indicative of island arc (Gotthard) and continental rifts to narrow oceanic basins [Aiguilles Rouges, NE Belledonne, Ht. Dauphiné (?), Argentera] environments.

When detailed geochemical data are available (Aiguilles Rouges, NE Belledonne, Argentera), the basaltic protoliths show obvious affinities with tholeiitic suites comparable to E, T and N-type MORBs of the present-day oceanic floor. The range of composition (E- to N-type) can be explained by various degrees of partial melting (Bertrand et al. 1982; Pearce 1983; Dupuy and Dostal 1984; Fodor and Vetter 1984). But basalts also display a well-marked LILE enrichment resulting from mantle source – continental crust interactions. The scattering of the initial ε_{Nd} values (discussed by Paquette et al. 1989) is possibly related to both the diversity of the mantle sources and to crustal contamination. But the latter process is a complex one and implies a selective contamination either through a sub-continental mantle source metasomatism and/or during the intracrustal emplacement. The presence of inherited zircons bearing further evidence for such a continental contribution.

Finally, in the three above-mentioned massifs, the Early Palaeozoic basaltic magmatism is consistent with a major distensive event leading to continental thinning and to narrow zones of oceanic rifting and spreading. Meta-igneous mafic and felsic rocks may be compared to the "Groupes leptyno-amphiboliques" (Santallier et al. 1988) of the west European Variscan belt (Bodinier et al. 1981; Ménot et al. 1988b; Bouchardon et al. 1989; von Raumer and Ménot 1989).

2.2 *Monometamorphic Igneous Rocks*

Monometamorphic igneous rocks of Early Palaeozoic age occur only within the Chamrousse ophiolite which outcrops in the southwestern domain of the Belledonne Massif (Lower Romanche Valley).

The Chamrousse-Séchilienne ophiolite complex appears as a tectonic slab (25 × 5 km) overlying the Devonian and Dinantian formations of Rioupéroux-Livet and Taillefer formations. The overthrust may be linked to a late Visean tectonic event, as deduced

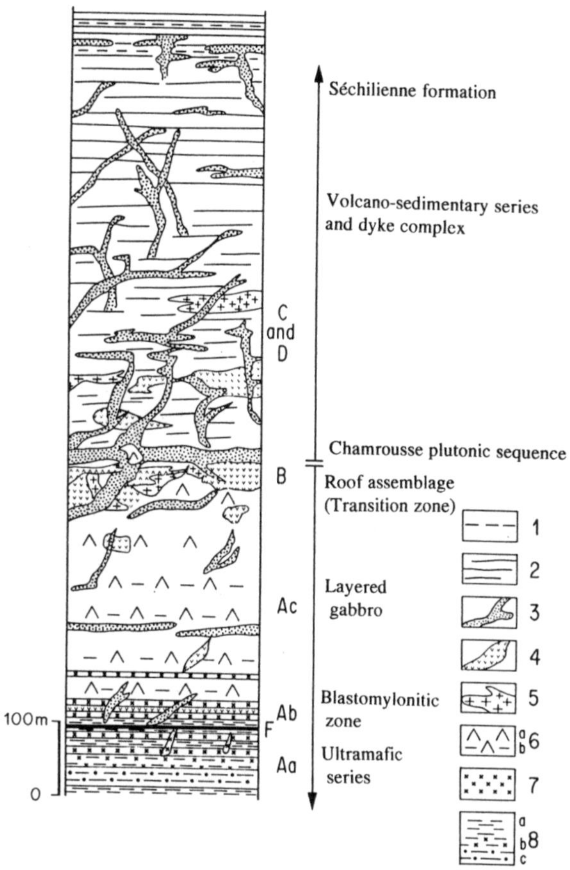

Fig. 6. Schematic reconstruction of the Chamrousse ophiolite with the primitive lithostratigraphic polarity (actually overturned). *1* Metapelites; *2* amphibolites and leptynites of igneous (Na-granites and keratophyres) and of sedimentary (arenites) origin; *3* metabasalts and metadolerites: *4* Isotropic metagabbros (fine-grained and pegmatoid); *5* metaplagiogranites and related rocks; *6* metacumulates (*a* massive; *b* layered); *7* metapyroxenites, dikelets and cumulates; *8* serpentinites *(a)* with Ca-pyroxenes *(b)* or with spinels *(c)*. *A* Plutonic sequence (*Aa* ultramafites; *Ab* blastomylonitic zone; *Ac* gabbroic cumulates); *B* transition zone or roof assemblage; *C* volcano-sedimentary series (Séchilienne Formation); *D* dyke complex (no scale for the dikes)

by K/Ar amphibole cooling ages; it post-dates the thermal peak of the Barrovian mesozonal metamorphism of Late Devonian to Dinantian age (Carme and Pin 1987; Ménot et al. 1987).

Thus, compared with the amphibolitic series described in Section 2.1, the tectonic and metamorphic evolution of the ophiolite is devoid of the eo-Devonian HP event and of the subsequent Devonian and Early Carboniferous anatexis and granitization (von Raumer et al. this Vol.).

The more striking characteristics of this well-preserved part of an ancient immature, slow-spreading ridge can be summarized as follows:

1. *A Late Cambrian to Early Ordovician age* for the ophiolitic complex: this is supported by crystallization ages of plagiogranites (496 ± 6 Ma; U-Pb zircon: Ménot et al. 1984a, 1988a) and of gabbros (497 ± 24 Ma; Sm-Nd whole rock: Pin and Carme 1987).
2. *The present-day upside-down position and relatively weak thickness* of the ophiolitic pile. The plutonic sequence (Chamrousse formation) grades downwards to a volcanic and sedimentary series (Séchilienne formation; Fig. 6). Moreover the inverted "lithostratigraphy" testifies to the downward internal succession of the cumulates, from ultramafites to gabbros and roof assemblage (Carme 1965, 1975b; Ménot 1979, 1987; Bodinier et al. 1981; Pin and Carme 1987; Ménot et al. 1988a). This being linked either to the mineral crystallization sequence and/or to the cryptic evolution of the mineral chemistry. The inverted "lithostratigraphy" is also demonstrated by the textural and petrological polarity of the dykes (Ménot 1979).

The approximative thickness of both the Chamrousse and Séchilienne formations (fms) is less than 2000 m. Even if the lowermost section of the ultramafic sequence is now lacking, this value is rather low when compared to that of "normal" oceanic lithosphere.

3. *Very strong syn-accretion tectonics* (Den Tex 1950; Carme 1965, 1975b; Ménot 1979, 1987; Pin and Carme 1987) leads to ubiquitous flaser gabbros, to mylonites and ultramylonites occurring as cross-cutting shear zones within the cumulates. The foliated rocks are intruded by undeformed masses and dikes of gabbroic, doleritic and basaltic nature. According to recent field investigations (Guillot et al. 1992), the syn-magmatic tectonic event involves two successive stages: both generations of ductile shear zones testify to lithospheric extensional regimes and may be respectively related to the stretching and the thermal subsidence of the oceanic crust during successive events of rifting and cooling. The intra-oceanic origin of such tectonics is implied (1) by the time relationships of the successive magma injections and of the phases of deformations (see later), (2) by the HT-LP conditions of the associated syn-kinematic recrystallization and (3) by the thermal decrease observed from the earlier to the second deformational event.
4. *The complex and polyphased nature of the ophiolitic accretion.* Both field evidence and geochemical data point to a complex process of ocean ridge accretion, induced by various, successive or contemporaneous, processes. These being:

a) Multistage accretion: from the time relationships of intra-oceanic deformation and successive magma

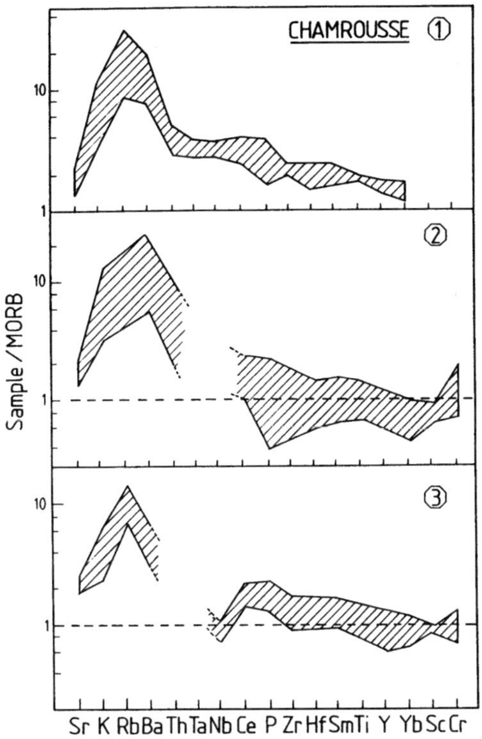

Fig. 7. MORB-normalized variation diagram (Pearce 1983): Chamrousse ophiolite, effusive rocks from the 3 successive stages (data: Bodinier et al. 1981; Ménot 1987; Pin and Carme 1987). *(1)*, *(2)* and *(3)* respectively for the first (amphibolites of Sechilienne series), the second (isotropic metagabbros) and the third (metabasalts and metadolerites), stages

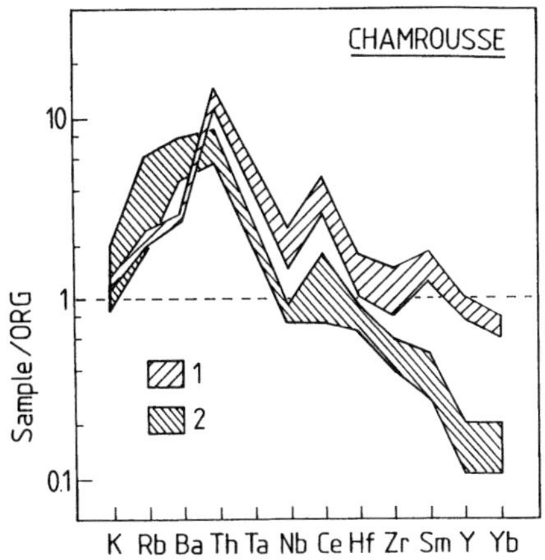

Fig. 8. ORG-normalized variation diagram (oceanic ridge granite: Pearce et al. 1984): Chamrousse ophiolite plagiogranites: *1* and *2* respectively enriched and depleted plagiogranites (data: Ménot 1987; Ménot et al. 1988a)

pulses, three main episodes of emplacement have been distinguished (Carme 1975b; Bodinier et al. 1981; Ménot 1979, 1987; Pin and Carme 1987; Ménot et al. 1988a). The first is mainly represented by ultramafic-mafic layered series, plagiogranites (Chamrousse fm.), basalts and keratophyres (Séchilienne fm.), the second by isotropic gabbros, plagiogranites, dolerites and basalts and the third by pyroxenitic and basaltic dikes respectively intruding the lowest and highest sections of the original ophiolitic sequence, Séchilienne fm. included.

The compositions of the successive batches range from E-, through T-, to N-type MORB and such a time-related evolution could be controled by an increasing degree of partial melting of the mantle (dynamic partial melting: 3–19%) (Bodinier et al. 1981). Further the E-type MORB affinities of the cumulates are also supported by the Ol-Cpx-Plg-Opx-Hn crystallization order.

Trace element variation diagrams (Fig. 7) highlight the time-dependent decrease in the Th and lithophile contents and the significant Nb depletion of the latest melts (mafic dykes). These chemical features could reflect a transitional environment from an attenuated continental lithospheric segment to a true oceanic crust, with a possible subduction zone influence on the latest stages of magma genesis (Ménot 1987; Ménot et al. 1988a).

b) Heterogeneous mantle sources and magma mixing. Using trace elements and Sm-Nd isotopes, Pin and Carme (1987) confirmed the time evolution from less depleted to very depleted ($E_{Nd(500)} = +5$ to $+9$) magma sources. They favoured a complex model involving: (i) the melting of a composite mantle, previously and heterogeneously enriched, that leads to three main types of melts (i.e. E-type MORB, N-type MORB and supra-subduction zone basalts); (ii) the subsequent mixing of these three end members.

Such a model substantiates an oceanic back arc basin origin for the Chamrousse ophiolite.

c) The existence of distinct and localized magma chambers: this can account for the diversification of the mafic rocks as it allows for contemporaneous but separated differentiation trends (continuous or periodical replenishment, open and closed systems) (Ménot 1987).

d) Variation of the physical parameters affecting the fractional crystallization processes. Although the plagiogranites differentiated from a single mafic source of E-MORB composition, they have to be ranged in two main types having distinct REE and trace element distributions, i.e. relatively enriched and depleted plagiogranites (Fig. 8). These two kinds of patterns can be explained by a change of the P-T parameters, and more specially of the P_{H_2O}, controlling the process of crystal fractionation. The enriched plagiogranites (earlier) are produced by Ol-Cpx-Opx-Plg-FeTioxides-Apat. and the depleted variety (later) by Plg-Hn-FeTi oxides-Apat-Zircon segregation respectively. Transition from "dry" to "wet" fractionated mineral assemblages is closely linked to the syn-accretion tectonics which supplies for the increase of P_{H2O} (Ménot 1987; Ménot et al. 1988 a).

To sum up, the Chamrousse ophiolite represents a time-equivalent to the eclogite protoliths of the polymetamorphic igneous formations of the ECMs (Aar-Gotthard excepted?). It indicates too a major extensional tectonic event in the Late Cambrian to Early Ordovician times. But its post magmatic orogenic history differs strongly: the ophiolite is devoid of any eo-Devonian high pressure imprint and of Devonian deep-seated recrystallization and remobilization. It was probably involved, only during Early Carboniferous times, in the collision belt, a long time ago after its obduction (Ménot et al. 1988 a, b).

2.3 Conclusions

In the Western Alps, the pre-Alpine basement displays numerous remnants of an important extension-related magmatic event of Early Palaeozoic age. Most of these remnants now occur as retrogressed eclogites, granulites and amphibolites, the polymetamorphic evolution of which extended from Late Silurian to Mid-Carboniferous. In contrast, the Chamrousse ophiolite recrystallized under less severe conditions and then only at the onset of Carboniferous times. The former are quite similar to the "leptynitic and amphibolitic groups" (Santallier et al. 1988) of the westernmost segment of the European Variscides; they bear a comparable geodynamic significance but, in the Alpine region, they clearly indicate an ensialic setting.

Although most parts of the ECM may be linked to the innermost section of the Variscan chain, two areas can be marked off: (1) the Chamrousse ophiolite which has a similar emplacement age but an obviously different tectonic and metamorphic evolution, is a very peculiar remnant of the Early Palaeozoic oceanization and seems to have no other equivalent within the Western European Palaeozoic fold belt. Its present day location, close to the polymetamorphic basement formations, implies a late orogenic major crustal shortening event; (2) the Aar-Gotthard Massif, the polymetamorphic igneous formations of which seem to display older emplacement and recrystallization ages than leptynitic and amphibolitic groups and to be representative of a quite different tectonic setting (arc magmatism). The origin of the Aar-Gotthard Massifs must be reappraised within the global framework of the pre-Alpine basement (von Raumer et al. this Vol.).

3 Devonian and Dinantian Magmatic Events

3.1 Devonian Igneous Formations

As yet, Devonian mafic and bimodal igneous suites are only known in the south-western edge of the Belledonne Massif. They possibly outcrop in the Haut Dauphiné "cortex" as a leptynitic and amphibolitic series (Valgaudemar) but, as they have not been yet dated, they could be time-equivalents to the Allemont-Rochetaillée formation (Late Proterozoic to Early Palaeozoic). In the Belledonne Massif, they have been defined, as the Rioupéroux and Livet formations, by Carme (1965) and later described (Carme 1972; Scarenzi 1984; Ménot 1986). The Rioupéroux-Livet fms. outcrop mainly on both sides of the lower Romanche Valley and they appear as a tectonic window below the overlying Late Proterozoic to Early Palaeozoic formations of Allemont-Rochetaillée and Chamrousse-Séchilienne (Carme 1972, 1973 a, b; Ménot 1986).

The Rioupéroux and Livet formations occur as a present-day tectonic pile consisting of five main units (Scarenzi 1984; Ménot 1986). In spite of subsequent dismembering, they can be considered as parts of a unique plutonic and volcanic complex as testified by petrological and geochronological data: meta-igneous rocks from the different units display similar compositions and comparable emplacement ages (Ménot et al. 1988 b). Within this framework, the Livet and Rioupéroux fms. have been respectively interpreted as the deeper and uppermost sections of the igneous complex. Within the former, plutonic sills and stocks are prominent and intrude a leptynitic and amphibolitic series. Whilst, the Rioupéroux units consist of mainly volcanic rocks.

Igneous protoliths are now represented by amphibolitic and leptynitic gneisses, meta-trondhjemites, -keratophyres and -basalts, according to more or less severe metamorphism. Emplacement ages of trondhjemites have been defined in the Livet fm. (352 +/- 55 Ma, with an upper intercept age of 2717 +/- 12 Ma) and in the Rioupéroux fm. (362 +/- 4 Ma) (U-Pb zircon: Ménot et al. 1984 b, 1988 b). The leptynitic and amphibolitic series of Livet have been intruded by trondhjemites and therefore is older. This was hypothetically related to the onset (Early Devonian?) of the magmatic event (Ménot 1987).

Prograde metamorphism developed during Early Carboniferous times, later than the trondhjemites intrusion (ca 360 Ma) but before the 324 Ma cooling age (Ménot et al. 1987). It leads to epizonal and mesozonal, locally polyphased, metamorphism in the Rioupéroux and Livet fms. respectively, but a number of magmatic relics can still be recognised (Carme 1972, 1973 a; Ménot 1986).

From field relationships, the composition of the igneous materials changes from a bimodal silicic-mafic pole (striped amphibolites and leptynites) towards a more siliceous one (meta-trondhjemites, -keratophyres and subordinate metabasalts). The contemporaneous emplacement of mafic and acidic magmas is underlined by the preserved field evidence of uncompleted magma mixing (Ménot 1986).

The geotectonic significance of the Devonian complex is still being debated. From chemical constraints,

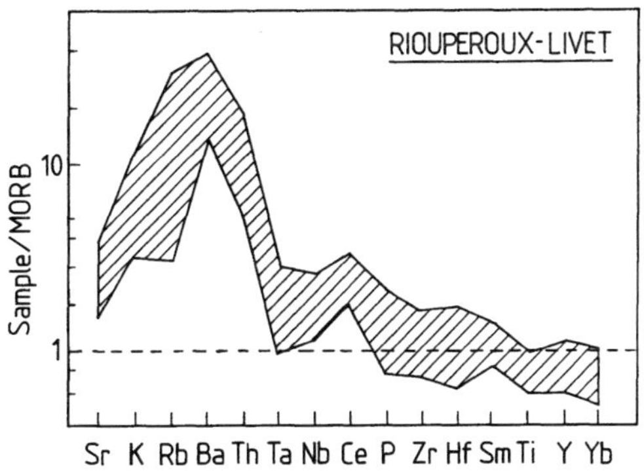

Fig. 9. MORB-normalized variation diagram (Pearce 1983): Rioupéroux and Livet volcanic and plutonic complex, metabasaltes and amphibolites (data: Ménot 1987)

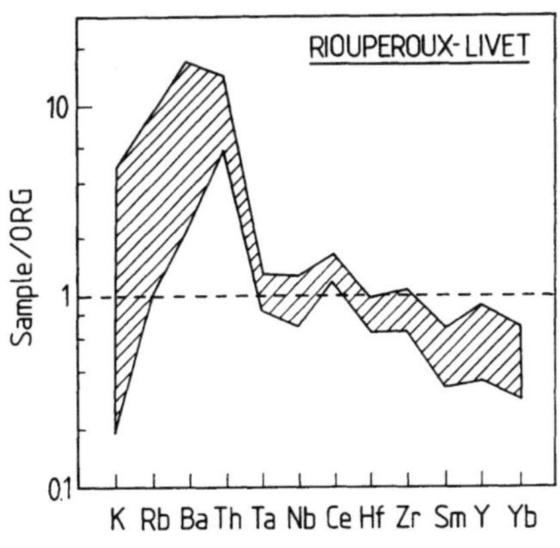

Fig. 10. ORG-normalized variation diagram (Pearce et al. 1984): Rioupéroux and Livet volcanic and pluonic complex, meta-trondhjemites, -keratophyres and leptynites (data: Ménot 1987)

Carme (1975 a) and Carme and Pin (1987) argued for an active margin environment ($E_{Nd(350)}$ = 2.7 to 6.1). A conclusion that was also supported by Scarenzi (1984) and Ménot et al. (1984 b) from preliminary results, but more recently, it has been proposed that both mafic and silicic magmas were emplaced within an attenuated continental crust (Ménot et al. 1988 b). Amphibolites and metabasalts are very similar in composition to enriched tholeiites with E-type MORB affinities and are characterized by high Th and lithophiles contents (Fig. 9). Moreover, both siliceous intrusives and volcanics and leptynites display a rather wide chemical range and arise either from differentiation of enriched mafic magmas and/or from magma mixing of crustal and mantle melts (Ménot 1987; Ménot et al. 1988 b). These acidic igneous rocks are also characterized by a well-marked lithophile enrichment and by the lack of a large negative Nb anomaly (Fig. 10). Finally selective continental contamination is enhanced by trace element and isotope data (unconsistent and uncorrelated Th contents, presence of old-inherited zircons and scattering of the initial E_{Nd} values) (Ménot et al. 1988 b).

3.2 Dinantian Igneous Formations

These are badly documented. Visean microfossils have been found in sediments only in the southern edge of the Aiguilles Rouges Massif (Belliere and Streel 1980) and in the Taillefer Massif, a boundary zone between the SW Belledonne and the Haut Dauphiné Massifs (Gibergy 1968). In the Taillefer Massif, bimodal volcanics are interlayered with sediments of mainly detrital composition (Carme 1965, 1972, 1975 a; Carme and Pin 1987; Ménot 1987). Otherwise, silicic and mafic meta-igneous rocks, together with metasediments, are also known in the NE domain of Belledonne, i.e. the "Schistes verts" or "Schistes verts supérieurs" series (Krummenacher et al. 1965; Negga et al. 1986; Vivier et al. 1987). They were commonly considered as equivalent in time to the Taillefer fm., but as no dating and no geochemical data are as yet available, they will not be included in the present discussion.

The Taillefer series lies in stratigraphic unconformity upon the Rioupéroux formation, but the contacts commonly display a complete tectonic concordance. It sometimes occurs as tectonic slices inserted between the Chamrousse-Séchilienne and the Rioupéroux formations (Carme 1972, 1973 a).

Metamorphic recrystallization is generally weak and biotite and actinolite only grew within localized shear zones, which are linked to the Upper Visean thrust tectonics (Carme 1973a; Carme and Pin 1987; Ménot et al. 1987).

The geochemical characteristics of the metavolcanics remain ambiguous, with both continental tholeiites and calc-alkaline basaltic affinities (Carme 1975a). Hence these volcanics testify to "neither a clearly extensional nor a typical active margin (orogenic) context" (Carme and Pin 1987). Our unpublished data emphasize the chemical similarities of the Dinantian (Taillefer) and Devonian (Rioupéroux-Livet) igneous rocks and a magmatic continuity (parentage) has been suggested (Ménot 1987). In this light, the Dinantian event results from continental rift tectonics and its complex and "atypical" chemical patterns could be explained by magma genesis and differentiation in a multiprocess evolution, involving distinct sources, magma mixing and contamination, as exemplified by the chemistry of the Early Permian dyke complex from NW Corsica (Fumey-Humbert 1989).

Within the Aar-Gotthard Massifs, Late Carboniferous (Westphalian D-Stephanian: Jongmanns 1960) plutonic and volcanic rocks, with calc-alkaline affinities, are considered as witnesses of a continental cordillera (Mercolli and Oberhänsli 1988; Oberhänsli et al. 1991). In this light, a subduction-related compressional regime has been postulated for these massifs throughout Carboniferous times (Schenker and Abrecht 1987; Mercolli and Oberhänsli 1988; Oberhänsli et al. 1988). Furthermore, Mercolli and Oberhänsli (1988) propose a working hypothesis which underlines tight similarities for the evolution of the pre-Mesozoïc basement in the Aar-Gotthard and in the Lower Austro-Alpine, Austro-Alpine and Southern Alps areas.

3.3 Concluding Remarks

In the ECMs, the Devonian and Dinantian bimodal magmatic events are only known in the southern area of the Belledonne Massif. Their geodynamic significance is still being debated (active margin or continental rifting), because of a probably complex magmatic history leading to "atypical" magmas. Thus, more detailed chemical studies are now being carried out to solve this uncertainty. At present, and according to our interpretation, the Rioupéroux-Livet and Taillefer fms. are thought to be the witnesses of a major extensional event during Devonian and Early Carboniferous times. Such tectonics (crustal thinning) differs drastically from the contemporaneous evolution of the NE domain of the Belledonne and of most parts of the ECMs, which involved an important crustal thickening and a subsequent uplift (medium to low pressure metamorphism, anatexis and granite emplacement) (von Raumer et al. this Vol.).

The Devonian to Dinantian magmatic extension-related event of the SW Belledonne area may be correlated to its time-equivalents of Armorica (Rolet et al. 1986) and of the NE Massif Central (Brévenne and Beaujolais series) (Beurrier et al. 1981; Pin et al. 1982; Sider et al. 1986; Sider and Ohnenstetter 1986; Ohnenstetter and Sider 1988). In the former region, the rifting event only developed in a continental environment but, in the latter, it evolved to a narrow oceanic basin (Brévenne ophiolite). The plutonic and volcanic complex of Rioupéroux-Livet could be representative of an intermediate setting in such a continental-oceanic lithosphere transition.

4 Conclusion

The available petrological, chemical and geochronological data concerning the meta-igneous rocks of the external crystalline massifs lead to the following conclusions:

1. According to their chemical affinities (N-type and E-type MORB with a slight crustal influence), the protoliths are representative of lithospheric extensional environments. Both mafic and bimodal igneous series testify to segments of attenuated continental crust, locally reaching to true oceanic basins, albeit narrow (Chamrousse ophiolite). Passive margin environments have been proposed, but a back-arc setting could also be invoked during the last stage of the accretion of the Chamrousse ophiolite.
2. According to their emplacement ages, two pre-orogenic extensional events have been pointed out: (a) the earlier, dated from Cambrian to Ordovician times, is now represented either by high grade polymetamorphic rocks (eclogites and granulites from Aiguilles Rouges, NE Belledonne, Argentera and probably Ht Dauphiné) or by the mesozonal, monometamorphic formation of Chamrousse-Séchilienne (SW Belledonne); (b) the second, Devonian to Dinantian in age, has only been recognized in the southern region of Belledonne as the Rioupéroux-Livet and Taillefer formations. But time-equivalent series of igneous rocks could probably be shown to be of the same age in other areas of the ECMs (southern part of the Aiguilles Rouges and Ht. Dauphiné "cortex").

These two successive episodes of rifting are well-known outside the Alpine domain. The Early Palaeozoic episode leads (a) to the Appalachian (Mattinson 1976; Dunning and Krogh 1985), Scottish (Dewey and Shackleton 1984), Norwegian (Dunning and Pedersen 1988) and NW Iberian (Peucat et al. 1990) ophiolites and (b) to the "leptynitic and amphibolitic groups" of the Western European Variscides. The precise geodynamic significance of the latter (continental rifts, back-arc basins, etc.) is still debated (see review in Santallier et al. 1988; Bouchardon et al. 1989). The Devonian-Dinantian distension-related magmatism has been mapped in central Brittany, Vosges, and NE Massif Central, etc. In these areas, igneous rocks are characterized by obvious intracontinental affinities. But ophiolites of Devonian age have also been described in the Variscan basement (Lizard complex, SW England: Davies 1984; Sudetic Poland: Pin et al. 1988).

3. According to their post-magmatic, i.e. orogenic evolution, two types of Early Palaeozoic metabasites can be considered. On the one hand, most of these show a long-lived, polyphased tectonic and metamorphic history. Thus, they have been involved in the subduction-collision processes during the Late Silurian up to Visean time span. Moreover, the ages of the HP metamorphic imprint, that testifies to the onset of the orogenic process, range from 424 Ma (Argentera) to 395 Ma (Aiguilles Rouges and NE Belledonne). Such a range could reflect the diachronism of the closure event linked to a complex pattern of multiple microplate convergences. On the other hand, the orogenic structuring of the Chamrousse ophiolite only began in Late Devonian times (Ménot 1988; von Raumer et al. this Vol.).

Such contrasted orogenic evolution implies that the Early Palaeozoic protoliths, precursors of the eclogites and granulites of the ECMs as well as the Chamrousse amphibolites, were emplaced within different areas with regard to the future zonation of the Palaeozoic chain, i.e. in the inner and outermost zones respectively. The strikingly different evolution of these Early Palaeozoic rocks reveals the time-related evolution of the orogenic processes towards the outer zones. This is emphasized from the results from the events in Devonian times.

4. The Devonian extension-related magmatism (pre-orogenic) of Rioupéroux-Livet (SW Belledonne) highlights the tectonic heterogeneity of the Belledonne Massif. This magmatic event obviously differs from the time equivalent anatectic event and earliest granite emplacement (now orthogneisses) that occurred in the NE domain of Belledonne (Vivier et al. 1987; Ménot 1988; Bonin et al. this Vol.). In the NE Belledonne, the late orogenic anatexites and granites arise from the partial melting of a thickened continental crust (Barrovian metamorphism and subsequent HT-LP context: von Raumer et al. this Vol.).

In this light, the Belledonne Massif appears as a tectonic "collage" which is related to late orogenic, Mid Carboniferous strike-slip faults (see Ménot 1987, 1988).

5. These results (1 to 4) do not fit with the few available data for the Aar-Gotthard metabasites: they are therefore considered to be remnants of an ancient volcanic arc and both the ages of emplacement and of the HP metamorphic event appear to be somewhat older (ca. 870 Ma and 470 Ma respectively). Some contemporaneous formations are to be found to the north, in the Black Forest and Vosges areas. So the Aar-Gotthard metabasites could be interpreted within a more general framework, which also includes the Eastern Alps. They could be evidence for an earlier orogenic zone to that invoked for the eclogites and granulites of the external crystalline massifs. Furthermore, the Late Variscan tectonic framework appear to be quite different as Mercolli and Oberhänsli (1988) argue for a subduction-related compressional regime during Carboniferous time when local and post-collision transtensional events are recorded in the ECMs.

To sum up, the petrological, chemical and geochronological characters of the Pre-Stephanian basaltic and bimodal magmatic events present a strong argument to correlate the Alpine ECMs (Aar-Gotthard excepted) and the westernmost segment of the European Variscan chain. When both the magmatic features and the subsequent tectonic and metamorphic history are taken into account, they become a powerful tool for the comprehension of the building of the Variscan fold belt.

Acknowledgements

Thanks to R. Oliver (I. L. L Grenoble) for reviewing and improving the English version of this paper and to R. Oberhänsli and an anonymous reviewer for helpful comments.

References

Abrecht J (1980) Zur Bildung migmatitischer Schollenamphibolite aus dem Altkristallin des Aarmassivs. Geol Rundsch 69: 695–725

Abrecht J, Biino GG (1991) Metamorphic evolution of the pre-Ordovician eclogite of the Gotthard massif, Helvetic domain, Central Switzerland. EUG VI, Strasbourg, Abstr in Terra Nova 3,(1): 95

Arnold A (1970) On the history of the Gotthard massif (Central Alps, Switzerland). Eclogae Geol Helv 63: 29–30
Bellière J (1958) Contribution à l'étude pétrogénétique des schistes cristallins du massif des Aiguilles Rouges (Hte Savoie). Mém Soc Géol Belg 81: 204
Bellière J (1980) Le massif du Mont Blanc et des Aiguilles Rouges. Structures et pétrologie du socle. Geol Alpine 56: 237–249
Bellière J, Streel M (1980) Roches d'âge viséen supérieur dans le massif des Aiguilles Rouges (Hte Savoie). C R Acad Sci Paris 290 D: 1341–1343
Bertrand H, Dostal J, Dupuy C (1982) Geochemistry of early Mesozoïc tholeiites from Marocco. Earth Planet Sci Lett 58: 225–239
Beurrier M, Piboule M, Durand C (1981) Gisement, pétrographie et géochimie des plagiogranites dévono-dinantiens des Monts de Tarare et du Beaujolais méridional (Massif central français). C R Acad Sci Paris 290 D: 147–150
Bodinier JL, Dupuy C, Dostal J, Carme F (1981) Geochemistry of ophiolites from the Chamrousse complex (Belledonne massif, Alps). Contrib Mineral Petrol 78: 379–388
Bogdanoff S (1980) Analyse structurale dans la partie occidentale de l'Argentera-Mercantour. Thèse, Université Paris Sud, Orsay, 316 pp
Bogdanoff S (1986) Evolution de la partie occidentale du massif cristallin externe de l'Argentella. Place dans l'arc alpin. Bull Bur Rech Geol Min Géol Fr 4: 433–453
Bouchardon JL, Santallier D, Briand B, Ménot RP, Piboule M (1989) Eclogites in the French Palaeozoic Orogen: geodynamic significance. Tectonophysics 169: 317–332
Brière Y (1920) Les éclogites françaises – leur composition minéralogique et chimique, leur origine. Thèse, Université de Paris, Giard M, Brière E (eds), 139 pp
Buffet G (1980) Notice de la carte géologique de la France à 1/50.000, feuille de Saint Bonnet. Bur Rech Geol Min, Orléans
Carme F (1965) Sur deux formations, d'origine volcanique, des schistes cristallins anté-houillers de la chaîne de Belledonne (Alpes françaises). C R Acad Sci Paris 9 260: 6401–6404
Carme F (1972) Notice de la carte géologique de France 1/50.000, feuille de Vizille (XXXIII–35). Bur Rech Geol Min, Orleáns
Carme F (1973 a) Précisions sur le métamorphisme majeur des schistes cristallins de la chaîne de Belledonne; essai de zonégraphie et distribution verticale des zones au niveau de la coupe de la Romanche. C R Acad Sci Paris D 277: 2133–2136
Carme F (1973 b) Précisions nouvelles sur l'ampleur de la tectonique tangentielle hercynienne dans la chaîne de Belledonne. C R Acad Sci Paris D 277: 2309–2312
Carme F (1975 a) Caractères chimiques de deux ensembles de roches magmatiques pré-orogéniques du socle anté-houiller des Massifs Cristallins Externes alpins et des Monts du Lyonnais. C R Acad Sci Paris D 281: 89–92
Carme F (1975 b) Le massif basique de Chamrousse et du Tabor (chaîne de Belledonne, Alpes occidentales): une variante originale de complexe alpinotype et un jalon majeur d'une ceinture basique briovérienne. C R Acad Sci Paris D 281: 223–226
Carme F, Pin C (1987) Vue d'ensemble sur le magmatisme pré-orogénique et l'évolution métamorphique et tectonique varisques dans le sud de la chaîne de Belledonne (Massifs cristallins externes, Alpes françaises). C R Acad Sci Paris 304 II 19: 1177–1180
Corbin P, Oulianoff N (1928) Les roches basiques de la région du lac Cornu (Aiguilles Rouges) et la question de leur origine. Bull Soc Géol Fr 4, 28: 43
Davies GR (1984) Isotopic evolution of the Lizard complex. J Geol Soc Lond 141: 3–14
Delhemmes R (1954) Etude géologique du massif cristallin des Aiguilles Rouges (Alpes de Savoie). Trav Lab Geol Fac Sci Grenoble 32: 67–94
Den Tex E (1950) Les roches basiques et ultrabasiques des Lacs Roberts et le Trias de Chamrousse (Massif de Belledonne). Leidse Geol Meded XV: 1–204
Dewey JF, Shackleton RM (1984) A model for the evolution of the Grampian tract in the early Caledonides and Appalachian. Nature 312: 115–121
Dunning GR, Krogh TE (1985) Geochronology of ophiolites of the Newfoundland Appalachians. Canad J Earth Sci 22, 11: 1659–1670
Dunning GR, Pedersen RB (1988) U-Pb ages of ophiolites and arc-related plutons of the Norwegian Caledonides: implications for the development of the Iapetus. Contrib Mineral Petrol 98: 13–23
Dupuy C, Dostal J (1984) Trace element geochemistry of some continental tholeiites. Earth Planet Sci Lett 67: 61–69
Faure-Muret A (1955) Etudes géologiques sur le massif de l'Argentera-Mercantour et ses enveloppes sédimentaires. Mém Carte Géol France, 336 pp
Fodor RV, Vetter SK (1984) Rift-zone magmatism: petrology of basaltic rocks transitional from CFB to Morb. Contrib Mineral Petrol 88: 307–321
Fumey-Humbert F (1989) Le magmatisme hypovolcanique tardi-hercynien; l'exemole du réseau filonien de Balagne (Nord-Ouest de la Corse). Thèse, Université de Saint Etienne, 237 pp
Gebauer D, Quadt A, Compston W, Williams IS, Grünenfelder M (1988) Archean zircons in a retrograded Caledonian eclogite of the Gotthard massif (Central Alps, Switzerland). Schweiz Mineral Petrogr Mitt 68(3): 485–490
Gibergy P (1968) Découverte de "grès à trous" renfermant des débris d'organismes sans les schistes noirs du Valbonnais (série cristallophylienne des massifs cristallins externes des Alpes françaises). C R Acad Sci Paris 267: 1251–1254
Guillot S, Ménot RP, Lardeaux JM (1992) Technique intra-océanique distensive dans l'ophiolite paléozoïque de Chamrousse (Alpes occidentales). Bull Soc Géol 163 (3): 229–240
Hutton DHW (1987) Strike slip terranes and a model for the evolution of the British and Irish Caledonides. Geol Mag 5: 151–164
Jongmanns W (1960) Die Karbonflora der Schweiz. Beitr Geol Karte Schweiz NF 108
Keppie JD (1985) The Appalachian collage. In: Gee DG, Sturt BA (eds) The Caledonide Orogen – Scandinavia and related areas. Wiley, New York, pp 1217–1226
Krummenacher D, Bordet P, Le Fort P (1965) Les massifs externes alpins et leurs séries métamorphiques. Problèmes de raccords statigraphiques et géochronométriques. Schweiz Mineral Petrogr Mitt 65: 855–874
Latouche L, Bogdanoff S (1987) Evolution précoce du massif de l'Argentera: apport des éclogites et des granulites. Géol Alp 63: 151–164
Le Fort P (1973) Géologie du Haut Dauphiné cristallin (Alpes françaises). Etude pétrologique et structurale de la partie occidentale. Sci Terre 25: 373
Le Fort P, Pécher A (1971) Présentation d'un schéma structural du Ht Dauphiné cristallin. C R Acad Sci Paris 273 D: 3–5
Liegeois JP, Duchesne JC (1981) The lac Cornu retrograded eclogites (Aiguilles Rouges Massif, Western Alps, France): evidence of crustal origin and metasomatic alteration. Lithos 14: 35–48

Malaroda R, Carraro F, Dal Piaz JV, Sturani C, Zanella E (1970) Carta geologica del Massicio dell'Argentera alla scala 1/50.000. Note illustrative. Mem Soc Geol Ital IX: 557–663

Mattinson JM (1976) Age of zircons from the Bay of Island ophiolite complex, western Newfoundland. Geology 393: 3

Ménot RP (1979) Lithostratigraphie et pétrographie du groupe amphibolitique de Séchilienne et du massif ultrabasique et basique de Chamrousse (Massif de Belledonne, Alpes occidentales). Géol Alp 55: 93–110

Ménot RP (1986) Les formations plutono-volcaniques dévoniennes de Rioupéroux-Livet (Massifs cristallins externes des Alpes françaises): nouvelles définitions lithostratigraphiques et pétrographiques. Schweiz Mineral Petrogr Mitt 66(2): 227–256

Ménot RP (1987) Magmatismes paléozoïques et structuration carbonifère du massif de Belledonne (Alpes françaises). Contraintes nouvelles pour les schémas d'évolution de la chaîne varisque Ouest-Européenne. Thèse, Université Lyon 1, 364 pp; Mém Doc Centre Armor Etudes Struct Socles, Rennes, 1988, 21, 465 pp

Ménot RP (1988) The geology of the Belledonne massif: an overview (External crystalline massifs of the western Alps). Schweiz Mineral Petrogr Mitt 68: 531–542

Ménot RP, Peucat JJ, Piboule M, Scarenzi D (1984a) Cambro-ordovician age for the ophiolitic complex of Chamrousse-Tabor (Belledonne massif, French external alpine domain). Meet Ophiolites through time, Nancy; Ofioliti 10, 2–3: 527 (Abstr)

Ménot RP, Peucat JJ, Scarenzi D, Piboule M (1984b) A continental (?) volcanic arc of Devono-Carboniferous age in the French alpine external massifs: the Rioupéroux and Livet formations (Romanche valley, Belledonne massif). Meet Ophiolites through time, Nancy; Ofioliti 10, 2–3: 528 (Abstr)

Ménot RP, Bonhomme MG, Vivier G (1987) Structuration tectonométamorphique carbonifère dans le massif de Belledonne (Alpes occidentales françaises). Apport de la géochronologie K/Ar des amphiboles. Schweiz Mineral Petrogr Mitt 67(3): 273–284

Ménot RP, Peucat JJ, Scarenzi D, Piboule M (1988a) 496 Ma age of plagiogranites in the Chamrousse ophiolite complex (external crystalline massifs in the French Alps): evidence of a Lower Paleozoic oceanization. Earth Planet Sci Lett 88 (1–2): 82–92

Ménot RP, Peucat JJ, Paquette JL (1988b) Les formations leptyno-amphiboliques paléozoïques et les Complexes Leptyno-amphiboliques: les corrélations hasardeuses. Exemple des Massifs cristallins externes alpins. Bull Soc Géol Fr (8) IV 6: 917–926

Mercolli I, Oberhänsli R (1988) Variscan tectonic evolution in the Central Alps: a working hypothesis. Schweiz Mineral Petrogr Mitt 68: 491–500

Monié P, Maluski H (1983) Données chronologiques $^{39}Ar/^{40}Ar$ sur le socle anté-perrmien du massif de l'Argentera-Mercantour (Alpes Maritïmes, France). Bull Soc Géol Fr 7, XXV: 247–257

Negga HS, Sheppard SMF, Rosenbaum JM, Cuney M (1986) Late Hercynian U-vein mineralization in the Alps: fluid inclusion C, O, H isotopic evidence for mixing between two externally derived fluids. Contrib Mineral Petrol 93: 179–186

Oberhänsli R, Schenker F, Mercolli I (1988) Indications of Variscan nappe tectonics in the Aarmassif. Schweiz Mineral Petrogr Mitt 68(3): 509–520

Oberhänsli R, Krähenbühl U, Stille P (1991) Contrasting REE characteristics in metalamprophyres from Variscan massifs of the Central Swiss Alps and REE patterns in lamprophyres from the Variscan terranes of Western Europe. Schweiz Mineral Petrogr Mitt 71: 53–62

Ohnenstetter M, Sider H (1988) Contraintes géochimiques apportées par le magmatisme sur le développement du bassin marginal ensialique du Beaujolais au Dévonien. Bull Soc Geol Fr 8 IV 3: 499–510

Paquette JL, Ménot RP, Peucat JJ (1989) REE, Sm-Nd and U-Pb zircon study of eclogites from the Alpine External massifs (Western Alps): evidence for crustal contamination. Earth Planet Sci Lett 96: 181–198

Pearce JA (1983) Role of sub-continental litosphere in magma at active continental margins. In: Hawkesworth CJ, Norry MJ (eds) Continental basalts and mantle xenoliths. Shiva, Nantwich, UK, pp 230–249

Pearce JA, Harris NBW, Tindle AG, Aldiss DT (1984) Trace element discrimination diagrams for the tectonic interpretation of granitic rocks. J Petrol 25(4): 956–83

Pécher A (1984) Notice de la carte géologique de la France à 1/50.000, feuille de Saint Christophe en Oisans. Bur Rech Geol Min, Orléans

Peucat JJ, Bernard-Griffiths J, Gil Ibarguchi JI, Dallmeyer RD, Ménot RP, Cornichet J, Iglesias Ponce de Leon M (1990) Geochemical and geochronological cross section of the deep Variscan crust: the Cabo Ortegal high-pressure nappa (north-western Spain). Tectonophysics 177: 263–292

Pin C, Dupuy C, Peterlongo JM (1982) Répartition des Terres Rares dans les roches volcaniques basiques dévono-dinantiennes du Nord-Est du Massif Central. Bull Soc Géol Fr 7, XXIV, 4: 669–674

Pin C, Carme F (1987) A Sm-Nd isotopic study of 500 Ma old oceanic crust in the variscan belt of western Europe: the Chamrousse ophiolite complex Western Alps (France). Contrib Mineral Petrol 96: 406–413

Pin C, Majerowicz A, Wojciechowska I (1988) Upper Paleozoic oceanic crust in the Polish Sudetes: Nd-Sr isotope and trace element evidence. Lithos 21: 195–209

Raumer von JF (1976) Variszicum in den Zentral- und Westalpen. Nova Acta Leopold 224, 45: 119–144

Raumer von JF (1981) Variscan events in the Alpine region. Geol Mijnbouw 60: 67–80

Raumer von JF (1984a) The External Massifs, relics of Variscan basement in the Alps. Geol Rundsch 73(1): 1–31

Raumer von JF (1984b) The Hercynian basement in the Helvetic Realm (Western and Central Alps). In: Dal Piaz GV (ed) Alpi Occidentali. Mem Soc Geol Ital 29: 57–69

Raumer von JF (1988) Caledonian – Variscan structures in the Alps – an introduction. Schweiz Mineral Petrogr Mitt 68: 291–299

Raumer von JF, Schwander HW (1985) Garnet evolution in pre-Variscan pelitic rocks from the Lake Emosson area, Aiguilles Rouges Massif Western Alps. J Metam Geol 3: 467–479

Raumer von JF, Ménot RP (1989) Evolution paléozoïque du socle ouest-alpin et place des Massifs Cristallins Externes dans l'orogène varisque. C R Acad Sci Paris 309 II: 397–402

Raumer von JF, Galetti G, Pfeifer HR, Oberhänsli R (1990) Amphibolites from Lake Emosson/Aiguilles Rouges, Switzerland: tholeiitic basalts of a Paleozoic continental rift zone. Schweiz Mineral Petrogr Mitt 70: 419–435

Rolet J, Legall B, Darboux JR, Thonon P, Gravelle M (1986) L'évolution géodynamique dévono-carbonifère de l'extrêmité occidentale de la chaîne hercynienne d'Europe sur le transect Armorique-Cornwall. Bull Soc Géol Fr 8 II 1: 43–54

Santallier D, Briand B, Ménot RP, Piboule M (1988) Les Complexes leptyno-amphiboliques: définitions et implications. Bull Soc Géol Fr 8 IV 1: 3–12

Scarenzi D (1984) Les granites sodiques et les roches associées de l'ophiolite de Chamrousse-Tabor et du groupe de Rioupéroux-Livet (Isère, france); pétrographie et géochimie. Thèse, Université Grenoble, 259 pp

Schenker F, Abrecht J (1987) Prä-aargranitische Anatexis, variszische Kontaktmetamorphose und alpidische Regionalmetamorphose im Oberhasli (zentrales Aarmassiv, Schweiz). Schweiz Mineral Petrogr Mitt 67: 13–26

Sider H, Ohnenstetter M (1986) Field and petrological evidences for the development of an ensialic marginal basin related to the Hercynian orogeny in the north eastern Massif Central. Geol Rundsch 75/2: 421–443

Sider H, Leistel JM, Gagny Cl (1986) Reflexions sur l'évolution géodynamique de la chaîne hercynienne au paléozoïque dans le Nord-Est du massif Central français. Bull Soc Géol Fr 8 II 4: 637–644

Vivier G, Ménot RP, Giraud P (1987) Magmatisme et structuration orogénique, paléozoïques de la chaîne de Belledonne (Massif cristallins externes alpins). Le domaine nord-oriental. Géol Alp 63: 25–53

Chapter 2
Penninic Realm

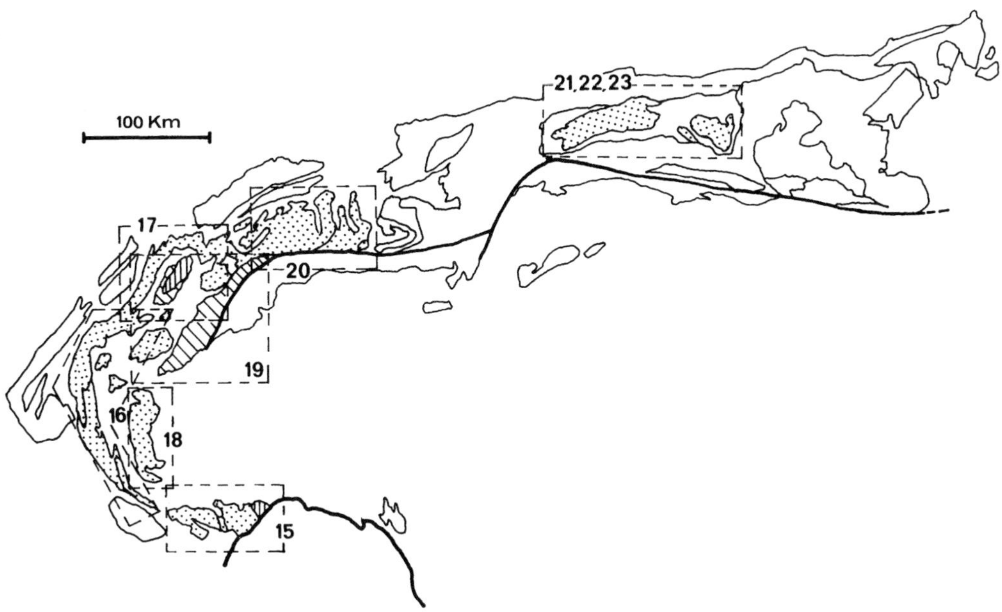

Elements of the Palaeozoic History of the Ligurian Alps

L. Cortesogno, G. Dallagiovanna, L. Gaggero and M. Vanossi

Abstract

The pre-Triassic formations of the Ligurian Alps build up: (1) the pre-Namurian basement; (2) the Permian-Carboniferous tegument; (3) the Upper Permian Verrucano.

1. The pre-Namurian basement is made up of two main rock sequences: I, which shows a pre-Alpine polymetamorphic evolution and II, which is affected by only one Variscan foliation. The basement of the lower Briançonnais units is only represented by orthogneisses II, which are stratigraphically covered by Permian-Carboniferous formations.

The upper Briançonnais rootless Calizzano-Savona nappe is mainly composed of older polymetamorphic, tightly associated rocks of the I sequence (paragneisses, orthogneisses I and amphibolites) and by the younger orthogneisses II.

Rarer rock types occur; they belong either in the older I sequence (more or less retrograded eclogites, granulites) or in the younger II sequence (anatexites, metagabbros). Paragneisses and amphibolites are believed to derive from pre-Ordovician, originally interfingered basalts and fine detrital sediments; most of the orthogneisses I might represent nearly contemporaneous acidic intrusions.

The first tectonometamorphic event (between Ordovician and Silurian times?) produced re-equilibration of nearly all pre-existing rocks under amphibolite facies conditions (with low to intermediate thermal gradient). Only a minor part of these rocks might have been driven into lower crust levels, where migmatites (from sediments), eclogites and granulites (from basalts) were generated.

During the Silurian – (?) Early Carboniferous interval, fine, probably very thick, sediments were deposited, while at deeper levels monzogranitic masses (actually orthogneisses II) were emplaced.

The second tectonometamorphic event took place before Namurian times; it overprinted both older metamorphites and later magmatites under amphibolite facies conditions (with intermediate to high thermal gradient); the post-Ordovician sedimentary cover was correspondingly transformed under upward decreasing metamorphic conditions.

2. Late Variscan progressive basement exhumation was accompanied by conspicuous erosion, which removed nearly all the Siluro-Devonian cover; Permian-Carboniferous tegument began to be laid down, under tectonic conditions that appear to have been – at least in the uppermost crust – mainly extensional.

Late Variscan formations in the Briançonnais belt consist of sediments and volcanics, which filled up continental graben and half-graben, bordered by two systems of active faults, presently trending N 30° and N 110°. The volcanic sequence, whose petrochemical characters are principally calc-alkaline although sometimes potassic sub-alkaline with shoshonitic affinity, was built up during three main, mostly explosive, events. Rhyolitic, then andesitic, finally again rhyolitic products were erupted, for a total volume of at least 1000 km^3.

3. The last, Late Permian faulting phase, devoid of volcanic activity, produced erosion of inner sectors and deposition of disconformable Verrucano sediments. The continental Verrucano arenaceous to conglomeratic deposits grade into the Scythian marine quartzites.

1 Introduction

This chapter synoptically presents up-to-date knowledge about pre-Triassic formations of the Ligurian Alps, assembling data and hypotheses as well as unpublished results of recent research. More complete information about regional geology is given in Vanossi et al. (1986) and references therein; for older papers see Boni et al. (1971).

In the Ligurian Alps, except for the Dauphinois zone, which is not dealt with in this chapter, pre-Triassic formations only occur in the Briançonnais (Fig. 1, No. 4, 5, 6) and Prepiedmont (Fig. 1, No. 11) zones, i.e. in the palaeo-European continental margin, which they widely occupy.

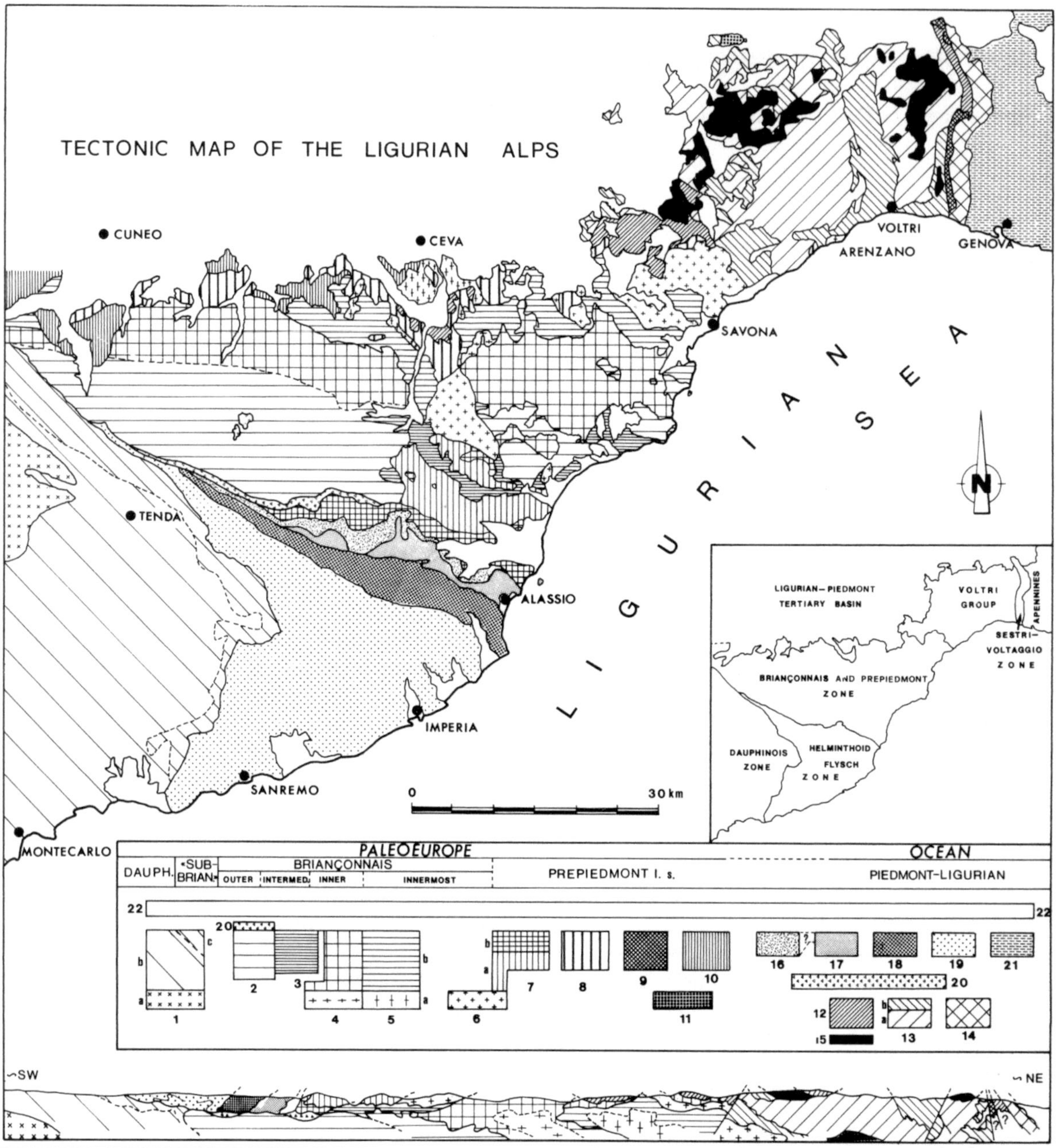

Dauphinois and «Sub-briançonnais» units

1a = Crystalline basement *(pre-Stephanian)* of the Argentera-Mercantour massif.
1b = Sedimentary cover *(late Carboniferous - Oligocene ?)*.
1c = Olistostromes and associated "sub-briançonnais" slices on the top of Dauphinois sequences.

Briançonnais units

2 = Outer and intermediate lower *(Ormea p.p., M.Carmo nappes)* and upper *(Caprauna-Armetta klippe)* units.
3 = Internal, upper rootless units *(C.Volte and Castelvecchio-Cerisola cover nappes)*.
4 = Internal, lower units *(Ormea p.p. and Mallare nappes)*:
  a = pre-Namurian basement *(Barbassiria)*;
  b = tegument and cover.
5 = Innermost lower units *(Pamparato-Murialdo nappe; Deviglia and Bc. Garofano klippen)*:
  a = pre-Namurian basement *(Nucetto, C.Dardella)*;
  b = tegument and cover.
6 = Innermost, upper rootless units *(Calizzano-Savona and Bagnaschino nappes: Arenzano "massif")*: pre-Namurian basement.

Prepiedmont units

7 = Prepiedmont units on the external Briançonnais margin *(C.Tuberto and Arnasco-Castelbianco nappes)*:
  a = pre-radiolarites sequences;
  b = radiolarites and post-radiolarites sequences.
8 = Triassic-Liassic units on the internal Briançonnais margin *(M.Sotta, Balzo, Villanova, Frabosa, Naso di Gatto, Cogoleto klippen and slices)*.
9 = Triassic-Liassic unit *(M.Gazzo-Isoverde)* in the Sestri-Voltaggio Zone.
10 = Internal Prepiedmont units with ophiolitic olistoliths *(Montaldo, Viglioni, Mombasiglio nappes)*.
11 = T. Visone (Valosio) crystalline basement.

Oceanic pre-flysch units

12 = Montenotte and Cravasco-Voltaggio ophiolitic units.
13 = Voltri Group ophiolitic units:
  a = metagabbros, eclogites, serpentinites *(Beigua, Ponzema, S.Luca-Colma, Varazze units)*;
  b = calcschistes and prasinites *(Voltri-Rossiglione, Alpicella, Ortiglieto, Palmaro-Caffarella units)*.
14 = M.Figogna ophiolitic unit.
15 = Lherzolites of Erro-Tobbio unit.

Flysch units

16,17,18,19, respectively = Colla Domenica-Leverone, Borghetto d'Arroscia-Alassio, Moglio-Testico and Sanremo-M.Saccarello Helmintoid flysch units.
20 = Helminthoid flyschs basal complex and olistostromes on the top of outer Briançonnais sequences.
21 = Northern Apennine Ligurian flysch units *(Gottero-Val Lavagna p.p. and Antola nappes)*.

Late and post-orogenic covers

22 = Marine *(Oligocene-Pliocene)* and continental *(Quaternary)* deposits.

A basic discrimination is traditionally made between Late Carboniferous-Permian *tegument,* lacking pre-Alpine metamorphism, and pre-Namurian *basement,* showing at least Variscan metamorphic parageneses and, commonly, a longer and more complex pre-Alpine metamorphic history (Cortesogno et al. 1982; Cortesogno 1986).

2 The Alpine Frame

Before drawing the pre-Triassic evolution, the Alpine tectonostratigraphic pattern is briefly outlined, as it has a significant bearing on the reconstruction of Palaeozoic history.

Alpine orogeny accreted huge volumes of rocks as tectonostratigraphic units by outward (presently southward or southwestward) thrusts, which were then backfolded, and finally deformed by the latest events. As a rule, both Alpine deformation and metamorphism decrease from older to younger phases and, within each phase, from inner to outer zones (Messiga et al. 1982; Vanossi et al. 1986). In the innermost zones Alpine metamorphic evolution is characterized by low temperature, intermediate pressure parageneses (Nam/Lws + Ab + (Chl) + Pmp) which are transitional between the Ab-Lws and low temperature blueschist facies ($250° \leq T \leq 300$ °C and $P \geq 7$ kb), and overprinted by low pressure re-equilibration (destabilization of Lws/Nam and phengites). In outermost sectors the climax never exceeded zeolite conditions. As a consequence, pre-Alpine fabric (bedding in the tegument, schistosities in the basement) may be either completely transposed, even twice or more by Alpine tectonics, or almost completely preserved.

That is why the same formation of the tegument can be designated either by metamorphic, e.g. "paragneiss" or sedimentary "arkosic (meta-)conglomerate", terminology. To avoid ambiguity, here terminology relative to pre-Alpine lithologic characters is used.

The present Alpine structure of the Briançonnais and Prepiedmont zones may be schematically depicted as follows.

◄

Fig. 1. Main Alpine units of the Ligurian Alps (after Vanossi et al. 1986, modified on the basis of unpublished data of G. Dallagiovanna, A. Di Giulio, B. Galbiati, M. Vanossi)

2.1 *Lower Briançonnais Units*

Some units (Nos. 2, 3, 4, 5 in Fig. 1) at the bottom of the nappe pile have been transported and more or less widely superposed, apparently without losing either their initial contiguity, or, at least in part, their Ceno-Mesozoic covers; their palaeogeographic reconstruction looks fairly simple: the lowest ones pertain to the external Briançonnais domain, whilst intermediate and internal Briançonnais nappes are found toward the top of the sequence.

2.2 *Upper Rootless Units*

Nappes bearing pre-Triassic formations form small bodies, so that the whole pile of units is nowhere visible. Correlations among isolated klippen rely on the analogies between: (a) relative geometric position; (b) stratigraphic series; (c) tectono-metamorphic Alpine evolution. Since palaeogeographic reconstruction for allochtonous basement-tegument slices is often inferred from the features of the present cover, mistakes may arise from possible cover substitutions, which are very difficult to detect. Surely no nappe rests on Piedmont-Ligurian (i.e. oceanic) units, thus suggesting their probable pertinence to the palaeo-European margin.

The following units are to be considered:

a. *Calizzano-Savona unit s. l.* (No. 6 in Fig. 1): it is represented by the basement sequences of Calizzano, Savona, Pallare, Loano, Bagnaschino, Arenzano isolated "massifs". Their lithologies and the Alpine metamorphic history allow the hypothesis, based on their present tectonic position, that they pertain to the innermost Briançonnais and to the contiguous Prepiedmont domains. On these bases, the key role of these terranes is evident, as they are the only clue to the Briançonnais-Prepiedmont basement.

b. *C. Tuberto unit* (No. 7 *pro parte* in Fig. 1). This is a Prepiedmont slab mainly consisting of post-Palaeozoic cover and of minor Permian tegument which allows correlations with the Briançonnais one. Furthermore, as it locally represents the stratigraphic cover of the Calizzano "massif" (Dallagiovanna 1989), it shows one of the few correlations known between a Mesozoic cover and a pre-Variscan polymetamorphic basement.

c. The *Torrente Visone unit* (N. 11 in Fig. 1), also known as *Valosio "massif"* (Forcella et al. 1973), is a slab of probable Prepiedmont pertinence tectonically underlying the "Voltri Group" metaophiolites, with which it underwent Alpine metamorphism in eclogite-blueschist conditions.

3 The Pre-Namurian Basement

The pre-Namurian basement is represented by two principal rock sequences: **I,** which shows a pre-Alpine polymetamorphic evolution, and **II,** which is affected by only one Variscan schistosity.
1. The basement of the lower Briançonnais units (Nucetto, Costa Dardella and Barbassiria "massifs") (Nos. 4, 5 in Fig. 1) is represented by orthogneisses II which constitute the original stratigraphic base of Late Carboniferous up to Eocene sequences. The upper rootless Calizzano-Savona unit (No. 6, Fig. 1) is constituted by paragneisses and micaschists, orthogneisses and migmatites (I), metabasites (I) and (II), orthogneisses and migmatites (II).
2. The Prepiedmont basement, outcropping in the Torrente Visone unit (ex Valosio "massif"), is formed by ortho- and paragneisses and by silicate marbles and metabasites.

The major difficulties in reconstructing the lower Palaeozoic events in Ligurian Alps derive from: (a) metamorphic-deformative Alpine events, often obliterating both lithological features and geometrical field relations; (b) poor chronological constraints and data.

3.1 The Briançonnais Basement

3.1.1 Lithologies

In the Briançonnais domain (Calizzano-Savona unit) the basement is composed of the following main lithologic groups.

1. Orthogneisses I and Migmatites I
They include rocks of acidic magmatic origin which have been associated with paragneisses since the beginning of their polymetamorphic history: their paragenetic evolutions (see paragneisses) are, in fact, well comparable (see also Messiga 1981).

Geometric relationships with the paragneisses and textural features allow the following lithotypes to be distinguished, probably having a somewhat different origin.
a) *Plutonic Bodies.* They represent the largest orthogneissic bodies, up to some hundreds of metres thick. The orthogneisses are coarse to medium-grained, sometimes augen-textured, generally biotite-rich and commonly show interbedded leucogranitic to aplitic layers.
b) *Migmatites I.* Interlayered decimetric to decametric bands of orthogneisses and paragneisses locally occur, with a total thickness of some hundreds of meters; they are characterized by flattened, centimetric, almost monomineralic biotite lenses and are considered to be metamorphosed migmatites.
c) *Dykes.* Orthogneissic dykes are commonly found lying parallel to the schistosity of paragneisses; they range from granodioritic to leucogranitic in composition; local relics of the primary magmatic textures survive penetrative schistosities.
d) *Volcanites.* Decametric lenses, showing augen textures with large phenoclasts of K-feldspars and lenticular quartz, probably represent volcanic or subvolcanic metarhyolites.

2. Paragneisses
They range in composition from metagreywacke to metapelite (micaschist) with minor quartzitic layers; carbonates are lacking.

In alumina-rich micaschists and paragneisses S_1 schistosity is characterized by the stability of syn- and post-kinematic coexistence of kyanite and sillimanite (in the Arenzano Massif sillimanite was not found) with staurolite, almandine, biotite, plagioclase (An_{25}), muscovite and rutile.

In orthogneisses granoblastic K-feldspar coexists with muscovite and quartz. Plagioclase + garnet + biotite + muscovite + cordierite(?) assemblages and very rare andalusite megablasts are found to grow near intrusive contacts with the orthogneisses II.

S_1 foliation is transposed on S_2 pre-Alpine schistosity, associated with kilometric folds. During the syn- post-S_2 re-equilibration, kyanite, sillimanite, cordierite and garnet from paragneisses are not stable and biotite + muscovite + oligoclase represent the equilibrium assemblage. Fibrolite grows locally rimming biotite and garnet.

3. Metabasites I
They include widespread amphibolites and rare eclogites, retrograded eclogites and granulites.

Amphibolites are interlayered as decimetric to pluridecametric bands in the paragneisses, often within metapelites or quartzites. Rare ultrabasic lenses occur in the amphibolites.

Eclogites, together with very rare granulites, generally showing alteration to amphibolitic parageneses, are found as small lenses commonly interlayered with anatectic migmatites II.

Trondhjemitic, sometimes garnet-bearing, mobilizates are found within eclogitic or amphibolitic bodies.

Both eclogites and amphibolites have basaltic composition with tholeiitic affinities (Fig. 2), but, probably owing to secondary mobilizations, chemical

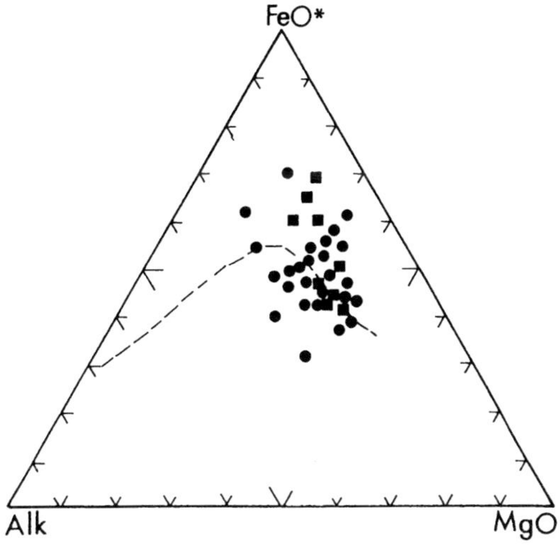

Fig. 2. AFM diagram for metabasites I. *Solid squares* eclogites; *Solid circles* amphibolites. Tholeiitic-calc-alkaline fields boundary (Irvine and Baragar 1971)

data do not seem to bear any reliable tectonic significance.

In the Arenzano unit, hornblende marbles are thought to derive from basic volcanogenic materials polluted by carbonates, possibly during spilitization processes and/or sedimentary reworking.

a) *Amphibolites*
Amphibolites are characterized by two main phases of metamorphic re-equilibration and penetrative schistogenous events in amphibolite facies.

Hornblende(1) + garnet + plagioclase + rutile and plagioclase (An>30) + diopside + clinozoisite + rutile assemblages are commonly found to develop in a compositional banding folded and transposed on the S_2 schistosity.

Re-equilibration syn- post-kinematic with the S_2 gives hornblende(2) + oligoclase (An≈25) ± clinozoisite ± biotite + titanite assemblages. Pseudomorphous aggregates of plagioclase, hornblende, biotite and cordierite(?) on garnets are characteristic features.

Greenschists assemblages (actinote, epidote $X_{Ps} \geq 25$ and perhaps chlorite) are found on spaced, brittle cleavage.

b) *Eclogites and retrograded eclogites*
Eclogite facies assemblages typically show coexistence of omphacite (Jd_{40-45}), garnet (Alm_{30-40}, Grs_{15-45}, Prp_{3-10}, Adr_{10-20}, Sps_{1-10}), zoisite and rutile.

Omphacite and garnet show polyphase pre- and post-kinematic growth (with respect to S_1) and strong compositional zoning. A phengitic white mica included in the core of zoned garnets seems to represent a relic of pre-eclogitic prograde metamorphism. Zoisite, Na-Ca amphibole and white mica are stable hydrated phases showing syn- and post-kinematic textures (S_1). Thermobarometric evaluations from pre-kinematic (Na-cpx inclusions and garnet core) and post-kinematic (Na-cpx at the contact and garnet rim) couples give temperatures from 750 to 770 °C for pressures in the range of 10–15 kb, pointing out increasing metamorphism. The later decompressional phases lack important pervasive schistosities and lead to the alteration of omphacite into symplectitic aggregates of diopside+albite. At increasing degrees of re-equilibration, zoisite, sometimes showing corrosion textures, is replaced by clinozoisite with increasing Fe^{3+} content, whereas oligoclase (An_{25-30}) grows replacing albite.

The garnet partially survives, possibly as a metastable phase, whereas rutile is partially replaced by ilmenite. Complete re-crystallization leads to diopside and/or hornblende + clinozoisite + plagioclase + ilmenite assemblages. As the re-equilibration proceeds, the titanium content progressively decreases from Na-Ca to Ca-amphiboles, as well as their Al^{IV} content, thus suggesting both temperature and pressure decrease.

Finally, the S_2 schistosity is followed by re-equilibration in green hornblende + plagioclase (An_{25}) ± Fe-clinozoisite ± biotite + titanite assemblages; cordierite is supposed to form in pseudomorphic aggregates with biotite, plagioclase and hornblende replacing garnet, but it never survives Alpine metamorphism.

c) *Granulites*
Eclogitic rocks are tightly associated with garnet, orthopyroxene, plagioclase and clinopyroxene-bearing granulitic layers. During post-S_2 alteration, orthopyroxene is largely replaced by cummingtonite. Granulite parageneses might have developed in anhydrous systems at the thermal peak during the decompressive phases.

4. Ortogneisses II

Orthogneisses II occur as large bodies in the Calizzano-Savona unit and form the whole basement of the Barbassiria, Nucetto, Lisio and Costa Dardella "massifs" (Cortesogno et al. 1986). According to radiometric data from Barbassiria "massif" (Del Moro et al. 1981), the intrusion age is older than 327 Ma. Variscan and Alpine events have obliterated original contacts; nevertheless, granitic bodies commonly still include, near their margins, metapelitic xenoliths, centimetres to tens of decimetres in size. Orthogneisses II commonly show K-feldspar decimetric megacrysts; the original magmatic texture, only seldom preserved, shows primary muscovite, in agreement with the peraluminous chemical composition (A/CNK always$\gg$1.1). Orthogneisses II are affected by only one pre-Alpine penetrative foliation, which corresponds to the S_2 schistosity of the paragneisses, orthogneisses I and metabasites I. The metamorphic conditions also compare well: orthogneisses II show biotite + muscovite + K-feldspar assemblages; oligoclase + quartz recrystallization in the pre-Alpine schistosity and very rare fibrolite growth on micas occur.

5. Migmatites II (Anatexites)
Rare anatexitic bodies occur as lenses and dykes, associated with eclogites (metabasites I) and with metagabbros (metabasites II). Anatexites were originated from a paragneissic protolith. Melanosomes of refractory composition become enriched in aluminiferous phases (polyphase biotite, sillimanite, garnet); biotite 2 and muscovite grow on S_2 schistosity, followed by post-kynematic andalusite megablasts. The leucosomes range from granitic to quartz-granitic in composition and become in part mobilized as granitic dykes. Like the orthogneisses II, the leucosomes show only one pre-Alpine foliation; this means that the anatectic event took place between the tectonic phases D_1 and D_2, thus meaning that it was almost contemporaneous with the emplacement of orthogneisses II and metabasites II.

6. Metabasites II
Lenses of metagabbros – sometimes displaying still-preserved magmatic cumulus textures – are associated with the anatectic migmatites II. The metagabbros, originally olivine gabbros with minor allotriomorphic pyroxenes, were re-equilibrated under amphibolite facies conditions and affected by pervasive foliation during the second pre-Alpine event: the primary phases are completely replaced by colourless magnesio- and tschermakitic-hornblende and oligoclase.

3.1.2 P-T-t Evolution

The tight association of paragneisses and amphibolites since the first-recognized schistogenous event and their parallel metamorphic evolution strongly suggest their derivation from penecontemporaneous, originally interfingered volcano-sedimentary sequences. These same sequences might have been contemporarily transformed, under higher grade metamorphic conditions, into eclogites and anatexites. In fact, similar basaltic chemical compositions characterize both eclogites and amphibolites, and no evidence exists that the metamorphic evolution of amphibolites passes through an eclogitic stage nor that eclogites evolve towards garnet-amphibolite facies. In the same way, anatexites derive from paragneissic protoliths, while paragneisses associated with amphibolites lack any migmatitic feature.

The two different metamorphic trends are shown in Fig. 3. The strong decompressive regime following the eclogitic event develops under essentially non-deformative conditions. The granulite-type assemblages developed during this decompressive event. The partial melting of metasediments (migmatites II) and possibly of metabasites (trondhjemitic mobilizates) is the consequence of increasing temperatures. A temperature rise might, in turn, be correlated with local intrusion of basaltic magmas at deeper levels and with emplacement of large granitic bodies at relatively shallow depths.

The second deformative phase developed, with lowered pressures and temperatures, affecting all metamorphic sequences and intrusive bodies which were folded together and re-equilibrated under low pressure-intermediate temperature amphibolite facies, with high thermal gradient conditions, characterized by andalusite and, seemingly, cordierite stability.

Late pre-Alpine greenschist-type assemblages, developed during late brittle phases, prelude and accompany the exhumation of the basement.

3.1.3 Magmatic Events

It cannot be excluded that a minor part of orthogneisses I, together with a part of paragneisses, belongs to a Precambrian basement. Probably, however, most of orthogneisses I (Fig. 4), co-genetic with migmatites I and with granodioritic and leucogranitic dykes, represent an early Palaeozoic acidic magmatism, developed both as deep intrusions and as minor syn-sedimentary volcanics. A more important basalto-tholeiitic effusive activity (metabasites I) alternated with the detrital sedimentation. The uplift phases affecting the metamorphosed sequences are marked by intrusions, probably at deeper levels, of basaltic magmas giving cumulitic gabbros (metabasites II) and, at shallower levels, of monzogranitic (Fig. 4) rocks (orthogneisses II). At the same time, anatectic melts originated from paragneisses (granites) and, possibly, from metabasites (trondhjemitic mobilizates).

3.2 The Torrente Visone (Ex-Valosio) Unit

The strong Alpine polyphase metamorphism makes it difficult to interpret the primary features as well as

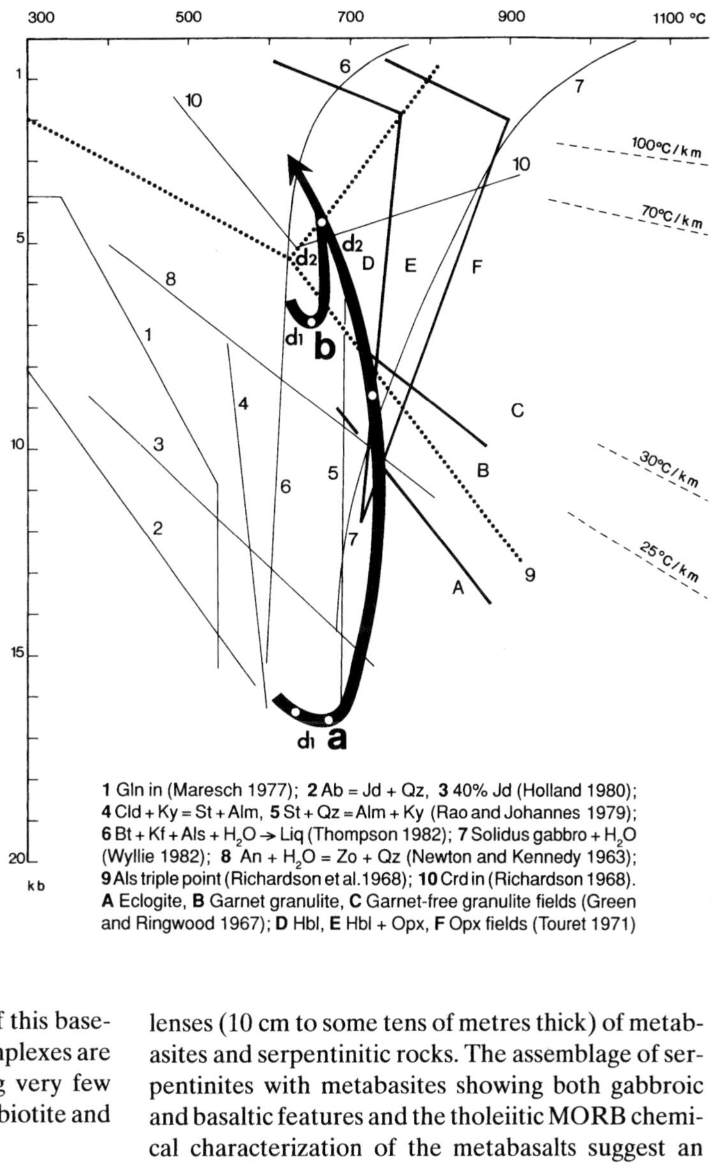

1 Gln in (Maresch 1977); 2 Ab = Jd + Qz, 3 40% Jd (Holland 1980); 4 Cld + Ky = St + Alm, 5 St + Qz = Alm + Ky (Rao and Johannes 1979); 6 Bt + Kf + Als + H_2O → Liq (Thompson 1982); 7 Solidus gabbro + H_2O (Wyllie 1982); 8 An + H_2O = Zo + Qz (Newton and Kennedy 1963); 9 Als triple point (Richardson et al. 1968); 10 Crd in (Richardson 1968). A Eclogite, B Garnet granulite, C Garnet-free granulite fields (Green and Ringwood 1967); D Hbl, E Hbl + Opx, F Opx fields (Touret 1971)

Fig. 3. Pressure, temperature, time, deformation diagram based on the paragenetic and tectonic evolution of *(path a)* eclogitic rocks and *(path b)* amphibolites and associated paragneisses. d_1, d_2: penetrative schistogenous main phases; *white dots* on a and b paths: main paragenetic re-equilibration phases. Stability of phases and reactions simplified from Gaggero (1990)

the pre-Alpine metamorphic characters of this basement unit. Two different metamorphic complexes are however easily recognized, both showing very few traces of a pre-Alpine foliation marked by biotite and muscovite.

a) The topographically underlying metamorphic complex, including augen gneisses, paragneisses and quartzschists with minor layers of metabasites, is interpreted (Cabella et al. 1989) as a sequence of metarhyolites characterized by large phenocrysts of K-feldspar, quartz and minor plagioclase, interbedded with meta-arenites and quartzites. Some arenitic beds are rich in, probably volcanogenic, K-feldspar and quartz and, locally, in tourmaline. Thin levels (few metres thick) of calcareous-dolomitic marbles are folded with the meta-arenites suggesting a marine shelf sedimentary environment. The layers and the boudins of metabasites seemingly derive from transposed basaltic dykes.

b) The topographically higher metamorphic complex consists of detritic carbonate sediments with interbedded thin metapelitic layers and small to large lenses (10 cm to some tens of metres thick) of metabasites and serpentinitic rocks. The assemblage of serpentinites with metabasites showing both gabbroic and basaltic features and the tholeiitic MORB chemical characterization of the metabasalts suggest an ophiolitic affinity for these rocks. The petrographic features of the metabasites and serpentinitic rocks furthermore suggest that they underwent earlier carbonatation processes within a sedimentary sequence.

Beyond this, the metabasalts of both complexes show strong chemical similarities suggesting a common genetic environment.

3.3 Regional Comparisons

On the whole, the pre-Namurian basement of the Ligurian Alps is fairly comparable with basements known in other Alpine domains, from the Helvetic zone (von Raumer 1986, 1987; von Raumer et al. this Vol.), to the Penninic zone (Desmons and Ploquin

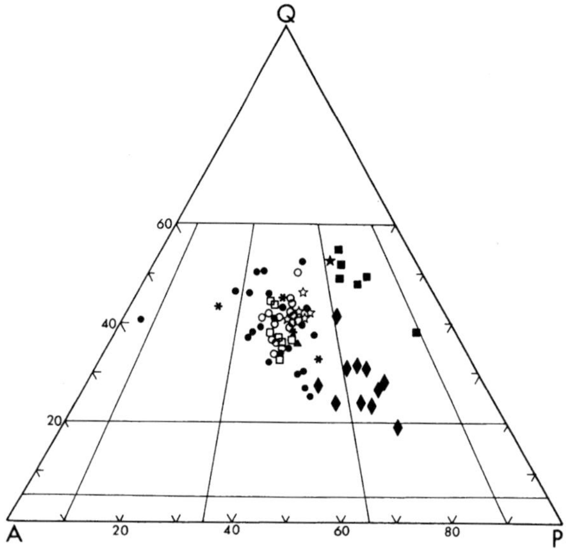

Fig. 4. AQP diagram. *Solid squares* granodioritic dykes (ogn I); *solid circles* Calizzano unit orthogneisses II (ogn II); *open circles* Barbassiria "massif" orthogneisses II (ogn II); *asterisks* Nucetto massif orthogneisses II (ogn II); *solid star* granitic dyke (ogn II) in migmatites II; *solid diamonds* Borda Granodiorites; *open stars* Rio Castorello granophyres; *open squares* Barbarissiria granopyhres

1989, Thélin 1989), to the South-Alpine zone (Boriani and Rivalenti 1984; Vai et al. 1984) and to the Austro-Alpine zone (Sassi and Zirpoli 1980; Sassi et al. 1987; Maggetti and Galetti 1988; Maggetti et al. 1990). Consistent comparisons may also be made with the Sardinian basement (Carmignani et al. 1982), as well as with extra-Alpine zones (Bouchardon et al. 1989). Similarities are found both in lithologic sequences and in magmatic and tectono-metamorphic events; thus, the chronology already proposed or settled for such regions may be used as a guide to reconstructions in our area.

Outcrops of a Precambrian basement are locally known from different Alpine sectors (Maggetti et al. 1990; von Raumer et al. this Vol.), but generally it can only be stated that the oldest metamorphites are pre-Variscan, or, in some instances, pre-Caradoc.

The oldest sedimentary sequences were detrital, arenaceous-pelitic sediments, nearly always, except for the western South-Alpine zone and for the Penninic Zone in the Tauern window, associated with carbonate deposits of presumed shelf environment. They are now represented by gneisses, micaschists and marbles and are dubitatively attributed to the Cambrian-Ordovician interval, or to pre-Caradoc times.

A basic, mainly effusive, essentially Ordovician magmatism is known everywhere. In most cases it produced intraplate, either tholeiitic or alkaline, basalts. However, possibly ophiolitic suites are locally quoted from the western edge of the Belledonne Massif (Ménot and Paquette this Vol.) and from the Austro-Alpine basement. An acidic magmatism is also recorded everywhere, with both effusive (porphyroids) and intrusive (orthogneisses) more or less coeval bodies; their radiometric ages, from different places, range between Late Ordovician and Early Silurian.

Basic and acidic magmatisms are, in turn, considered by most authors as approximately contemporaneous (bimodal activity accompanying a rifting phase); in some geodynamic models (e.g. von Raumer et al. this Vol.), however, basic magmatism produces earlier ocean floor sequences, while acidic activity is connected with subsequent formation of magmatic arcs above subduction zones.

Evidence of post-Caradoc and pre-Namurian deposits, often associated with mostly Visean, both basic and acidic volcanics, is known everywhere, except in western South-Alpine zone. Mostly Visean, migmatitic bodies, the probable equivalents of our migmatites II, are well known in the Western Alps (von Raumer 1987) and in Sardinia (344 ± 7 Ma: Del Moro et al. 1975). As to pre-Alpine metamorphic events, their relative chronology seems to be substantially similar in all regions (Boriani et al. 1974).

A Variscan polyphase metamorphism, with variable grade in different zones, is reported from everywhere. Its radiometric ages mainly correspond to Early Carboniferous times, with possible prolongations either below (Western Alps) or above (Sardinia, up to 310–290 Ma). Compared to this event, the migmatites can be either relatively precursory (Sardinia) or late (Western Alps).

In some sectors a pre-Variscan metamorphic event has been locally discovered: lenses of metabasites, with eclogitic and/or granulitic parageneses, included in the Visean migmatites (Sardinia: Ghezzo et al. 1977); small eclogitic, sometimes granulitic, masses (Western Alps: Liégeois and Duchesne 1981); pre-Variscan amphibolitized eclogites, perhaps produced by a single biphase event (Austro-Alpine zone). A pre-Variscan metamorphism of unknown grade is also reported in the western South-Alpine zone (Ebner et al. 1987; Sassi and Zirpoli 1980; Sassi et al. 1987) and in the Schwarzwald (Klein and Wimmenauer 1984).

This older metamorphism is thought to be Ordovician, but it might be pre-Caradoc; in the Silvretta nappe its age is 1000 Ma (Maggetti et al. 1990). Depending on regions and authors, it falls either before or after the acid magmatic event. Among extra-Alpine regions, earliest Silurian, eclogitic parageneses are known in the Central Massif and Armorican

Massif zones (Bouchardon et al. 1989), where protoliths, emplaced mostly during early Palaeozoic times, represent different geotectonic environments, from intraplate to normal and transitional MORBs, to even calc-alkaline rocks.

Finally, opinions differ in considering the two metamorphic events either as sharply separated in time ("Caledonian" and "Variscan" respectively), or as both belonging to a single megacycle; in a nearly continuous evolution, it would connect Ordovician magmatic and metamorphic processes with (Devono) Visean, more properly "Variscan", ones.

3.4 A Possible Pre-Namurian History

3.4.1 The Briançonnais Zone

There is not an adequate number of constraints to define a univocal pre-Namurian history for the Briançonnais zone of the Ligurian Alps; the following is just one possible solution.

A certainly Precambrian basement is not known, although it might be represented by part of the orthogneisses I and by the paragneisses associated with migmatites I.

The Palaeozoic history (Cambrian-Ordovician?) begins with the deposition of detrital sediments (presently paragneisses and micaschists) in an unknown environment; no carbonatic sedimentation occurs, just like in South-Alpine and in some Penninic regions. During more or less the same span of time, tholeiitic basalts (presently eclogites, granulites and amphibolites) effused, interstratified with fine-grained detrital sediments. Somewhat later or earlier, depending on places and on the geodynamic model, the acidic magmatic activity occurs, producing migmatites I, granitoid intrusions, granodioritic dykes and rhyolitic flows, all together forming most of the present orthogneisses I. Most probably, rhyolites were primarily interbedded with sediments.

The magmatic, bimodal activity might indicate an extensional regime, but little evidence exists for oceanization; it is more likely that just an intracontinental rifting was produced.

Later history is only documented by tectono-metamorphic and magmatic events. As in the western South-Alpine region, no sedimentary record is known for the interval between S_1 development and Late Carboniferous continental deposits. However, the previous existence of some thousand of metres of cover can be inferred (Sect. 4.1) from the lithologic composition of Westphalian coarse detrital formations.

The dating of the first tectono-metamorphic event is prevented by lack of radiometric or reliable relative data; in addition, it is not possible to demonstrate that all syn-S_1 mineral assemblages originated through only one tectono-metamorphic phase (Late Ordovician? Early Silurian?). If so, two main contemporaneous metamorphic trends, affecting originally similar rock sequences, should be admitted. In a plate tectonic model they might be identified either in two distinct, differently dipping subduction zones, reaching different structural levels, or with a unique subduction zone, marked by a curved T-P path.

A first trend, affecting various primary lithotypes, should bring to high grade ($T \approx 650\,°C, 6 \leq P \leq 7$ kb) amphibolitic parageneses for intermediate thermobaric gradients during tectonic phase 1. A second trend leads, at lower T-P ratios, to polyphase eclogitic parageneses showing deformative events (schistosity, metamorphic banding) and prograde re-equilibration up to a thermobaric maximum at least at $T \approx 750\,°C$ and $P \approx 17$ kb. The decompressive phase, shown by diopside-albite replacing omphacite, and first amphibolitization phases, probably together with slightly increased temperatures, allow local granulitic mineral assemblages (Opx + Grt + Plg) to develop.

The increasing thermobaric gradient occurring at the end of the tectonic climax (Devonian?) is related to magmatic activity: basic melts, which differentiate gabbroic bodies (metabasites II) and probably induce partial melting processes in the paragneisses (migmatites II) are emplaced; acidic melts (orthogneisses II) widely occur as intrusions and dykes in paragneisses. S_1 schistosity in both eclogites and amphibolites is cross-cut by leucocratic mobilizates (affected by the S_2 schistosity); they are garnet-bearing quartz-dioritic in composition when associated with eclogites and shift to a more acidic plagiogranitic or granitic composition when occurring in amphibolites.

As the decompressional phase indicates a load decrease, equal to a thickness of about 30–40 km, it may be admitted that partial exhumation and obduction, perhaps together with tectonic transport, occurred.

During the second tectonic phase (Visean?), anatectic gneisses and eclogites were transported toward higher levels and, together with gneisses, amphibolites and associated magmatic bodies, were folded and re-equilibrated under low-pressure amphibolite facies conditions. The latest stages of this event are marked by some basaltic to andesitic dykes crosscutting the S_2 foliation but still re-equilibrated under amphibolitic conditions. Finally, the development of greenschist mineral assemblages precede the exhumation and the Namurian erosion of the basement.

3.4.2 The Prepiedmont (Torrente Visone Unit) Zone

The Prepiedmont Torrente Visone (ex-Valosio) unit strongly differs from the Briançonnais zone both in its primary lithologic characteristics and its pre-Alpine metamorphic evolution. In particular, the presence of ophiolite-type suites and, moreover, of two carbonate sequences makes a striking contrast with lower Palaeozoic in the Briançonnais zone.

Should the Torrente Visone really pertain to the "Sardo-Corso" block, it would follow that the late Variscan break-up led to the incorporation of some Insubric fragments into the future European margin.

4 The Permo-Carboniferous Tegument

This chapter mainly deals with the Briançonnais domain; spatial-temporal data have been obtained from lower, rooted, units: therefore, the palaeogeographic restoration may be considered fairly reliable.

The tegument is composed of interfingered continental deposits and volcanic products; as a rule, sediments mainly occur in the lower parts of the stratigraphic column, while volcanites are conspicuous in the upper and younger sequences. Two unconformities bound the tegument: the lower one marks the top of the basement affected by a pre-Alpine tectono-metamorphic history; the upper one limits the Verrucano sediments.

Ages are poorly known: only late Westphalian-Stephanian floras have been locally found in the Ollano formation (Bloch 1966). Regional correlations (Cabella et al. 1988), mainly based on the volcanic episodes, suggest a chronostratigraphic range between Namurian and Late(?) Permian, depending on the age attributed to the Verrucano deposits (Fig. 5).

In the frame of a tripartite palaeogeographic distinction (outer, intermediate and inner sectors) in the Briançonnais domain during the late Variscan phases, the evidenced formations (Cortesogno et al. 1988c) are briefly introduced in the following paragraphs.

All rock units lack pre-Alpine metamorphism, but suffered Alpine metamorphic events to a variable extent, depending on their position in the nappe pile. The following presentation disregards Alpine history, focusing on the primary lithology of the Permo-Carboniferous formations. Map (Fig. 6) and columns (Fig. 7) may help to appreciate their extension and the stratigraphic relationships in the different sectors.

4.1 Sedimentary Formations

4.1.1 Lisio Formation

Locally representing the bottom of the tegument, it consists of fine – to coarse – detrital sediments, probably deriving from the erosion of the underlying orthogneisses II; arkoses with quartzitic layers may attain even 200 m in thickness. The age, unknown, is supposed to range between early Westphalian and Namurian.

4.1.2 Ollano Formation

Probably corresponding to the classic Briançonnais "Houiller productif", it is represented by conglomeratic-sandy-pelitic fluvial-lacustrine sequences

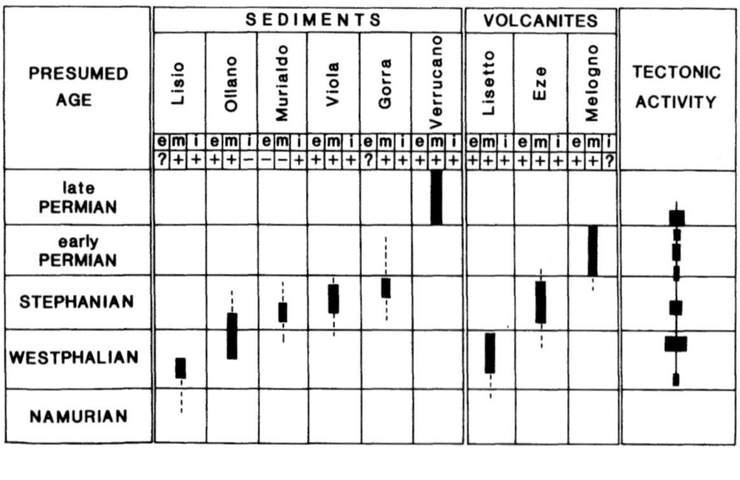

Fig. 5. The Ligurian Briançonnais: highly hypothetical ages of various Permian-Carboniferous formations and of faulting phases accompanying their emplacement in a continental environment, by sedimentation or volcanic activity. Probable primary presence (+) or absence (–) of each formation in external *(e)*, intermediate *(m)* and internal *(i)* sectors is also recorded (Cortesogno et al. 1988b)

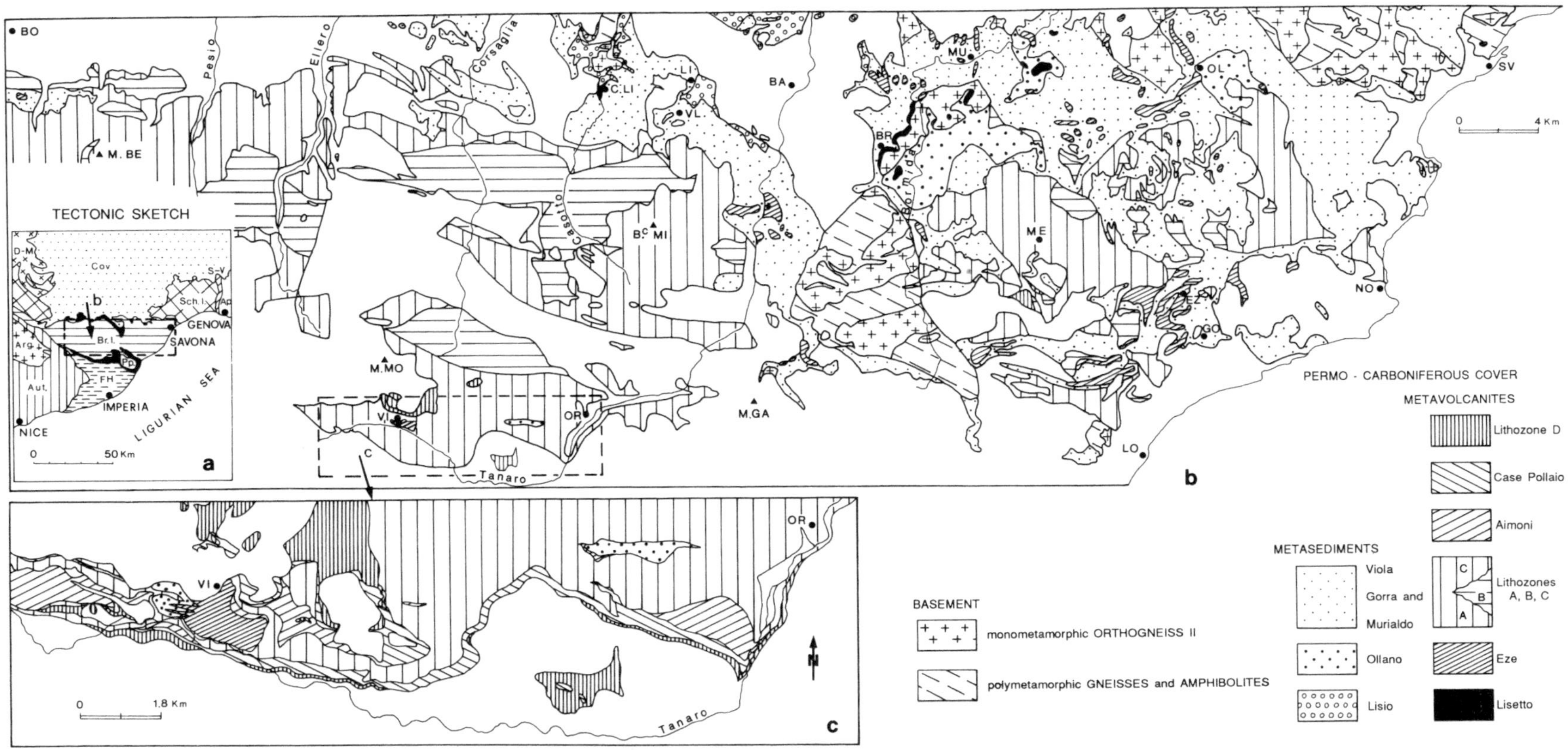

Fig. 6. General schematic *(top)* and southwestern detailed *(bottom)* map of pre-Namurian basement and of Permian-Carboniferous tegument outcrops in the Ligurian Briançonnais. Aimoni and Case Pollaio Members have been separated in the detailed map only. *BA* Bagnasco; *BO* Boves; *Bc. MI* Bric Mindino; *BR* Barbassiria; *C. LI* Case Lisetto; *EZ* Eze; *GO* Gorra; *LI* Lisio; *LO* Loano; *ME* Melogno; *M. BE* Monte Besimauda; *M. GA* Monte Galero; *M. MO* Monte Mongioie; *MU* Murialdo; *NO* Noli; *OL* Ollano; *OR* Ormea; *SV* Savona; *VI* Viozene; *VL* Viola. **Tectonic sketch.** *COV.* Oligocene to Quaternary cover; *AP.* Nothern Apennine units; *Sch. l.* "Schistes lustrés" (oceanic domain); *D.-M.* Dora-Maira "massif"; *FH* Helminthoid Flysch; *Pp* Prepiedmont *s. l.* units; *Br. l.* Ligurian Briançonnais; *Aut.* Autochthonus and paraautochthonus of external zones; *Arg.* Argentera-Mercantour

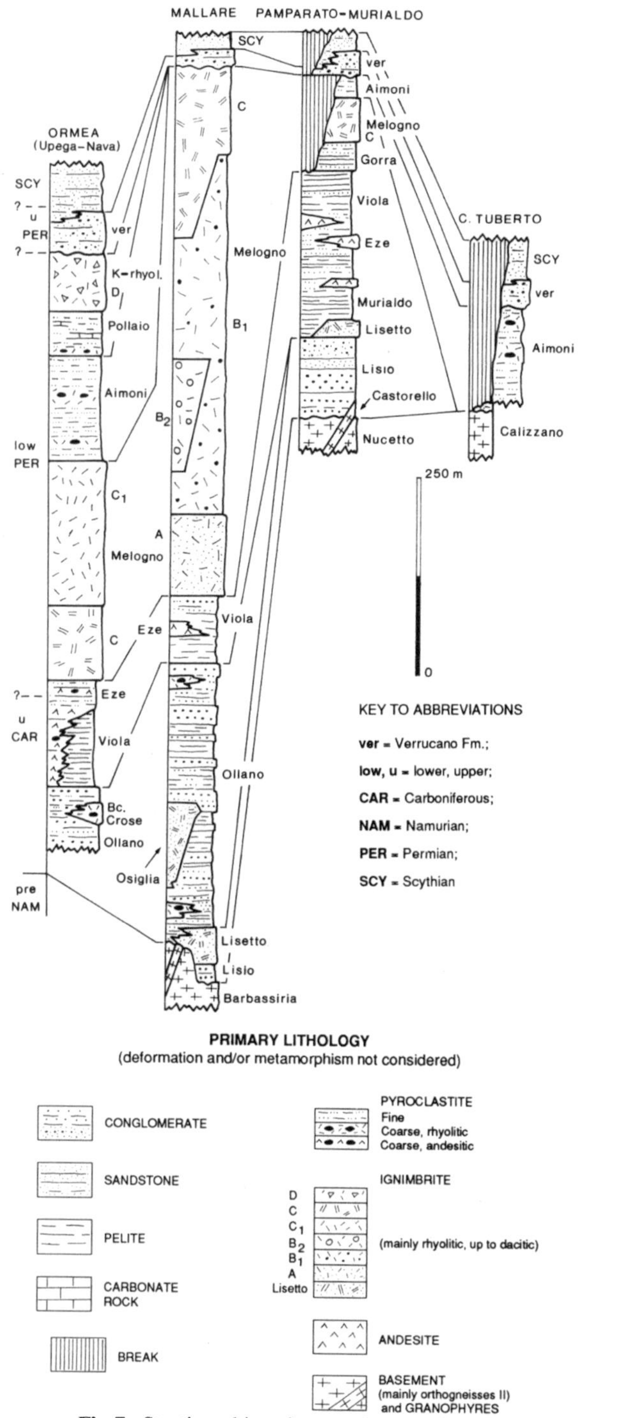

Fig. 7. Stratigraphic columns of some Ligurian Briançonnais and Prepiedmont (C. Tuberto) Permian-Carboniferous series, arranged from outer *(left)* to inner zones. The Meso-Cenozoic top of the sequences has been omitted

(Vanossi 1970), sometimes with graphitic lenses; thickness varies either gradually or abruptly, up to some hundreds of metres.

4.1.3 The Murialdo Formation; the Viola Schists; the Gorra Schists

They correspond to fine-grained metasediments, differentiated on the ground of different areal location, partially different lithologies and lack of good correlations. The Murialdo formation outcrops only in the inner Briançonnais zone and is typically represented by black graphitic phyllites. In the ubiquitous Viola schists, black phyllites are associated with light-coloured sandy metapelites, the composition of which (quartz, micas, chloritoid, tourmaline, rutile) may indicate highly weathered continental deposits. The Gorra schists, from the inner and intermediate sectors, may reach some hundreds of metres; interbedded with quartz-micaceous, fine, white sediments, comparable to the Viola schists, there are pyroclastites and reworked acidic volcanic products.

All these fine-grained sedimentary sequences contain andesitic volcanites pertaining to the Eze formation.

4.2 Volcanic Formations

Three main volcanic episodes, perhaps partly overlapping (Fig. 5), are known. As no radiometric data are available, their ages are inferred from their relative stratigraphic position: the earliest episode is the better dated, by means of pyroclastic levels interbedded in the Ollano formation. Limited volumes of volcanites represent the two earlier episodes, but the third and youngest one is very important (assessed to be about 1000 km^3) (Figs. 6, 7).

4.2.1 The Case Lisetto Metarhyolites

They rarely but evenly outcrop in the upper part of the Lisio formation or at different levels within the Ollano formation as ignimbrites ("Osiglia Porphyries") or pyroclastic agglomerates ("Bric Crose Tuffs") (Cortesogno et al. 1984b). Textures are characterized by large (1–3 cm) K-feldspar and quartz phenoclasts (porphyritic index: P.I. 30–40) with plagioclase (0–10%) and biotite (0.5–10%); rare xenoliths (quartzites, garnet micaschists, muscovitic-biotitic metapelites) are found. The composition is mainly rhyolitic with minor rhyodacites, near the minimum melt, and is possibly related to anatectic processes in the crust.

4.2.2 The Eze Formation

It is mainly represented by lavas and pyroclastites of intermediate composition interbedded as lenses of variable size (from a few cubic metres to some hundreds of cubic metres) in the continental sequences of the Murialdo formation, Viola and Gorra schists. In the Calizzano subunit both discordant dykes and lava flows directly covering the polymetamorphic basement are found (Cortesogno et al. 1983). Lavas show a porphyric texture (P.I. 5–10, max. 20 in the dykes); plagioclase and biotite or minor hornblende and clinopyroxene are the phenocryst phases; microphenocrysts of olivine and orthopyroxene are rarely observed. Ti-magnetite, ilmenite and apatite are always present and sometimes abundant. Rare basaltic flows occur in the Viola schists of the intermediate Briançonnais sector, whereas a more leucocratic trend appears in the outer sector (the Viozene area).

Pyroclastic levels are mainly tuffs and ashes, grading into adjacent sediments; even very coarse pyroclastic fall deposits, up to 250 m thick, have been observed in the Viozene area (Cabella et al. 1988).

Chemical composition varies from andesitic, rarely basaltic andesitic, to trachytic and dacitic, but trachyandesites are prevailing. Both the FeO_{tot}/MgO ratios (<2 in the internal Briançonnais zones, generally about 2 in the external areas) and Al_2O_3 content suggest a high K, subalkaline trend.

A more evident shoshonitic character is shown by trachyandesitic to trachytic and rhyolitic terms found as clasts in Liassic breccias and considered as probable representatives of the intermediate Late Carboniferous volcanism in the Prepiedmont zones.

4.2.3 The Melogno Porphyroids

Unknown in the inner zones, probably owing to both the original absence and late Permian erosion (Fig. 11), they largely outcrop in the intermediate and outer Briançonnais zone (Fig. 6), directly covered by the Verrucano deposits; their presumed age is Early Permian. The formation is mainly composed of ignimbrites, in which a certain number of lithozones and members have been distinguished (Lorenzoni 1958; Guillaume 1969; Cortesogno et al. 1984c; Cabella et al. 1988). A complete idealized sequence (see Fig. 7 for actual columns in different sectors) would show, from the bottom, the following terms.

A Lithozone: Thin and very discontinuous levels of light coloured, fine grained, frequently silicified rhyolites.

B Lithozone: Up to several hundreds of metres of mainly rhyolitic-rhyodacitic ignimbrites (B_1 facies), with coarse (up to 5 cm) and very abundant phenocrysts (P.I. up to 80) of quartz and K-feldspar, with minor plagioclase, biotite and rare hornblende.

Locally the ignimbrites are replaced by rhyodacitic breccias in a rhyolitic matrix, by granophyric massive bodies (extrusion domes?) or (B_2 facies) are interbedded with rhyodacitic/dacitic lavas and breccias.

C Lithozone: It represents the most common facies; mainly rhyolitic, fine to medium-grained, greenish ignimbrites, locally reworked, up to many hundreds of metres thick. Phenocrysts (P.I. = 20–35) of K-feldspar and quartz, with minor plagioclase and biotite occur.

C_1 Lithozone: Up to a few hundreds of metres of ignimbrites similar to C, with more conspicuous K-feldspar modal content.

Aimoni member: Up to some hundreds of metres of variegated, fine to exceedingly coarse pyroclastic fall deposits, with minor dacitic ignimbrites.

Case Pollaio member: Up to 100 m of black ash deposits, interbedded with rare, often silicified, carbonate beds, of probable evaporitic origin, and with danburite-bearing levels; they are also interbedded with ignimbrites of the D lithozone.

D Lithozone: Up to a few hundreds metres of K-rhyolitic fine to medium-grained ignimbrites, characterized by compact texture, dominant reddish, but locally green or black colour, and high K-feldspar ratio in the phenocrysts. The petrographic and chemical data evidence the bimodal dacitic and rhyodacitic-rhyolitic composition of this Permian volcanism; the general calc-alkaline affinity of the lithozones A, B, C is also pointed out by $FeO_{tot}/MgO > 2$ ratios and medium to low K_2O contents. The D lithozone strongly differs, showing an alkaline characterization and high K_2O content.

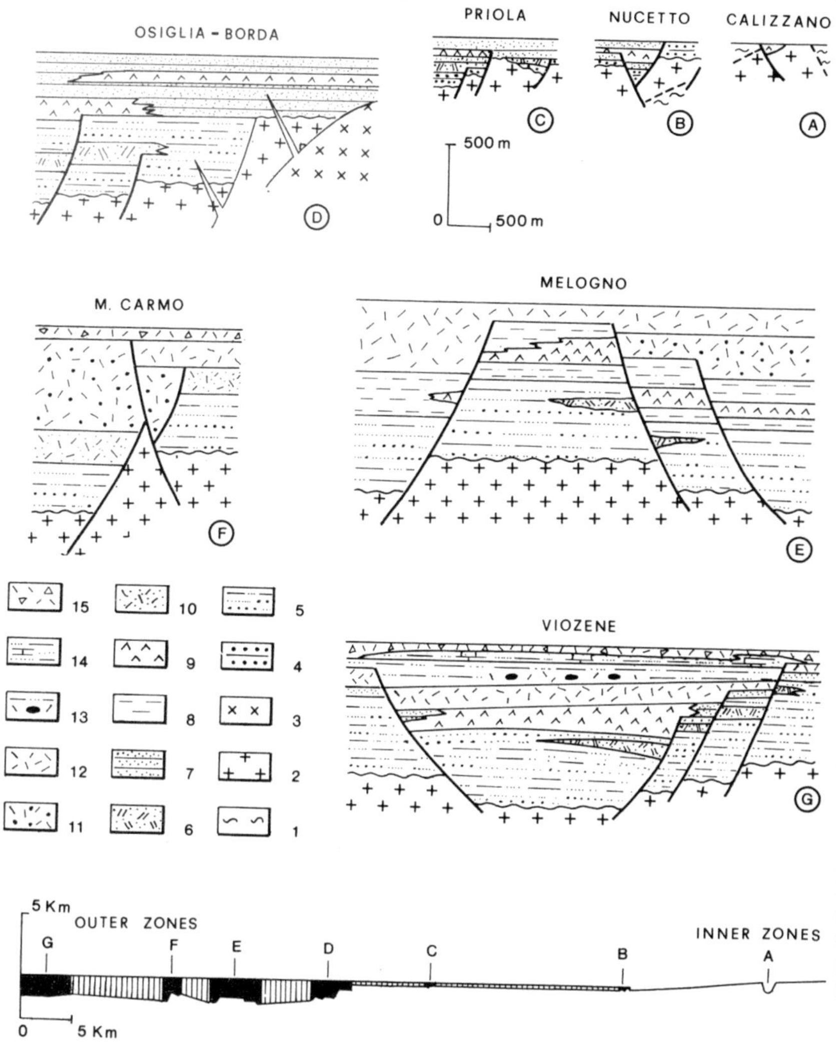

Fig. 8. Highly speculative palaeogeographic setting in the Ligurian Briançonnais domain at the end of Early Permian times. *1* Pre-Namurian polymetamorphic basement (gneisses and amphibolites); *2* Pre-Namurian monometamorphic granitoids (orthogneisses II); *3* Late-Variscan Borda Granodiorites; *4 to 6* Namurian-Stephanian formations (*4* Lisio Fm.; *5:* Ollano Fm.; *6* C. Lisetto metarhyolites); *7 to 9* Stephanian-Autunian formations (*7* Viola schists; *8* Gorra schists; *9* Eze Fm.); *10 to 15* Early Permian Melogno porphyroids (*10* A Lithozone; *11* B Lithozone; *12* C Lithozone; *13* Aimoni Member; *14* C. Pollaio Member; *15* D Lithozone). **Lower cross-section.** *A* to *G schemes* have been projected onto the same vertical plane; intervening distances might have been even much wider. *Black and vertical hatching:* presumed thickness of Permian-Carboniferous tegument, respectively in *A* to *G sectors* and in interposed areas. In regions more internal than *B* only erosion is presumed.

4.3 Intrusive and Subvolcanic Magmatism

Extensive late Variscan plutonic bodies do not appear in the Ligurian Alps; small and rare outcrops are known in the inner Briançonnais:

Borda Granodiorites: They are represented (Fig. 4) by small bodies (no more than 0.1 km^3 in the outcrop) and dykes intruded in the Viola schists. Mineral phases are plagioclase, hornblende, quartz and biotite. Allanite is a common occurrence. Chemical features show the calc-alkaline affinity.

Rio Castorello Granophyres: They occur as fine to medium-grained dykes, crosscutting both Lisio formation and orthogneisses II; they are characterized by leucocratic appearance and granophyric textures, and are considered to be the ultimate products of crystal fractionation from a parental granodioritic magma (Cortesogno et al. 1984 a).

4.4 The Late Variscan Evolution

In the outer Briançonnais domain, where Alpine events did not obliterate older geometric relationships, detailed mapping and structural analysis (Cabella et al. 1988) have pointed out that tegument formations display both sudden and gradual variations in thickness which cannot be due to the Alpine multiple folding: they are easily understood if two sets of faults, presently trending N 30° and N 110°, are introduced, defining graben and half-graben structures. The existence of pull-apart basins could not be proved.

The following steps have been tentatively suggested on the basis of available data (Cabella et al. 1988).

The faults offered an easy way to rising magmas; volcanic and pyroclastic products piled up in subsiding continental basins, in an almost flat region. Climate evolved from wet during Stephanian times, to semiarid, during Early Permian times. The close rela-

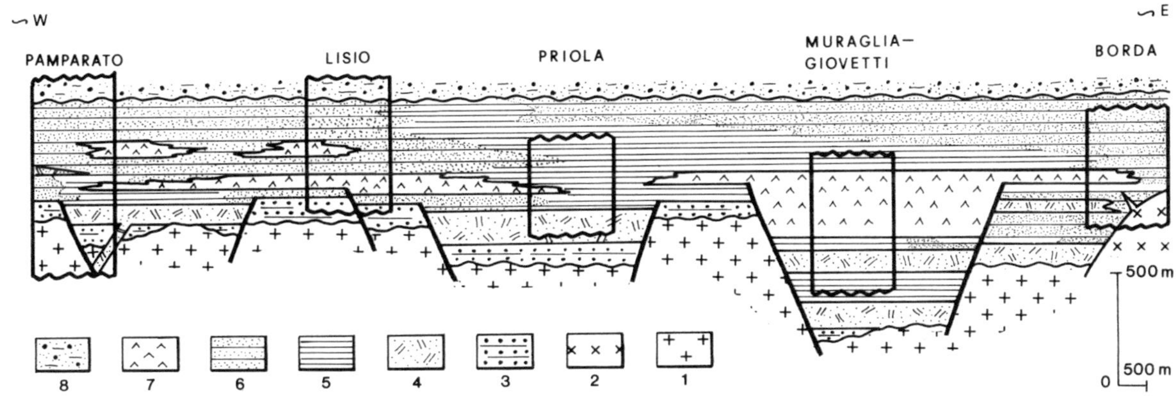

Fig. 9. Very hypothetical E–W palaeogeographic distribution of Permian-Carboniferous tegument formations inferred from five sections (their sequences are in turn deduced form present highly deformed rock bodies). Note N–S trending system of normal faults, introduced mostly to explain sudden disappearance and thickness variations of volcanic complexes. Compare with Fig. 8. *1* Pre-Namurian monometamorphic granitoids (Orthogneisses II); *2* Late-Variscan Borda Granodiorites; *3* Lisio Fm.; *4* C. Lisetto metarhyolites; *5* Murialdo Fm.; *6* Viola schists; *7* Eze Fm.; *8* Verrucano

tionship between tectonics and volcanism is also outlined by rejuvenation or birth of new faults at each of the three main volcanic phases: Westphalian-Stephanian (calc-alkaline rhyolites); middle (?) Stephanian (high K, sub-alkaline to "shoshonitic" andesites); Early Permian (calc-alkaline rhyolites and dacites, K-rhyolites). Only the last faulting phase (Late Permian), producing suitable gradients for erosion and transport of the Verrucano clastic components, was not accompanied by volcanism: in the Ligurian Alps the late Variscan magmatism seems thus to end by the end of Early Permian times.

Although great uncertainties, due to Alpine complications, arise, the graben model (Fig. 8 G, from Cabella et al. 1988) may also be applied to intermediate and inner zones (Figs. 8 A–E, and 9). From these schemes we can derive a discrimination between the innermost zones, behaving as source-areas, and intermediate and outer ones, trapping sediments in a multiply activated fault system. As inner zones constitute the limit between the Briançonnais and Prepiedmont Alpine domains, Late Variscan palaeogeography exerts a prominent control on Alpine depositional and structural features.

5 Lithoclasts in Carboniferous Conglomerates and Their Bearing on Older History

5.1 A Possible Reconstruction of a Pre-Namurian Cover That Has Disappeared

In the few places where the stratigraphic contact between tegument and underlying basement was preserved, the conglomerates of the Ollano formation rest on pre-Alpine orthogneisses II. The age (Variscan, probably Sudetic) and the grade of the metamorphism in the orthogneisses imply that, just after the Sudetic phase (end of Early Carboniferous?) and before the deposition of upper Westphalian conglomerates they were covered by a pile, at least 5 km thick.

As that pre-Namurian cover does not outcrop anywhere, it may follow that it did not disappear by tectonic transport, but that it was eroded. If so, it may be inferred that: (a) although lacking palaeontological evidence, a post-Variscan detrital tegument, older than late Westphalian (perhaps early Westphalian or Namurian, as it is in the classic Briançonnais area), should exist somewhere; (b) information on the lithological features of the pre-Namurian cover can be deduced from the Ollano conglomerates.

Recently (Cortesogno et al. 1988c), two main groups of pebbles have been distinguished: ubiquitous and rare.

1. The rare pebbles are confined in just a few levels that represent the product of in situ reworking of volcanics and volcanic breccias (Bric Crose Tuffs); they are mainly volcanic elements (most frequently

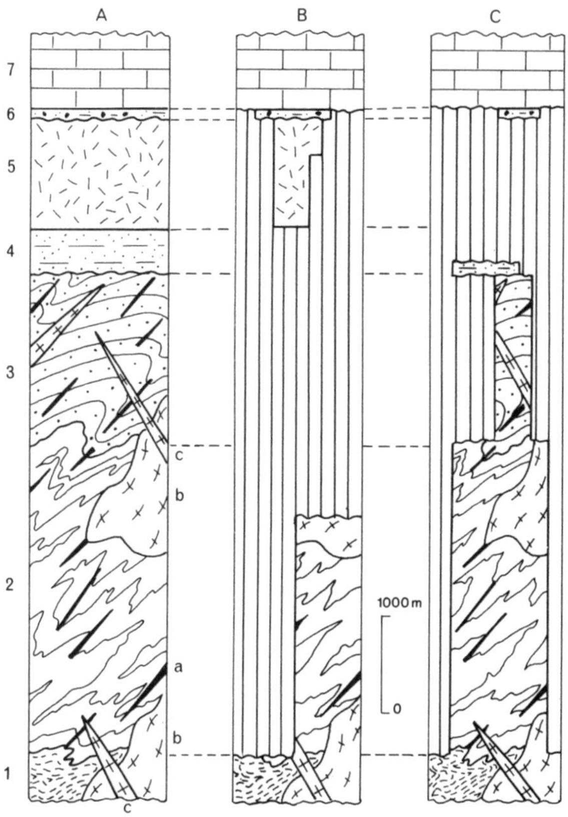

Fig. 10. Hypothetical stratigraphic columns of source areas of Permian-Carboniferous sediments. **A** Ideal complete column: *1* polymetamorphic basement, with Variscan metamorphism in amphibolite facies conditions; *2* polymetamorphic basement, with Variscan metamorphism in greenschist facies conditions; *3* monometamorphic basement, with Variscan metamorphism from greenschist to anchizone facies conditions; *2+3* "disappeared cover"; *a* quartz veins; *b* orthogneisses II; *c* post-Sudetic acidic dykes and plutons; *4* (Upper?) Westphalian-Stephanian clastic sediments (Houiller); *5* Lower Permian volcanites; *6* Upper Permian Verrucano; *7* Triassic sediments. **B** and **C:** Possible columns for source areas, respectively of Upper Carboniferous (X zones) and of Verrucano (Y zones) deposits

rhyolites and rhyodacites; sometimes dacites, very seldom quartz-latites and andesites).

Volcanics also contain xenoliths, which give information on the Ollano formation basement. They are represented by: (a) plutonites (aplites, leucogranites, granodiorites) lacking Variscan metamorphism and therefore late orogenic, though they were emplaced relatively early: they might be coeval and possibly cogenetic with the plutonic bodies described in Section 4.3; (b) metasediments (pelites, sometimes graphite-rich; siltites and fine sandstones; cherts; no carbonatic elements have been found) and metarhyolites (just one occurrence), all affected by one greenschist to subgreenschist schistogenous pre-Alpine metamorphic event. They might represent relics of the upper part (Silurian to early Visean?) of the sequence covering the older basement.

2. The abundant ubiquitous pebbles, giving information on the source areas, are mainly represented by polycrystalline quartz, with structures indicating that they originated from older veins; there are also subordinate pebbles of paraschists (quartz-micaschists, muscovitic micaschists sometimes with garnet or rare clinozoisite; biotite-muscovite-quartz-micaschists; paragneisses) with pre-Alpine polymetamorphic evolution, like that affecting the Calizzano-Savona unit (Sect. 3.1).

The great amount of information from pebbles may help to reconstruct the sequence corresponding to the "disappeared cover" originally interposed between the old polymetamorphic basement and the Late Carboniferous detrital deposits (Fig. 10A). Polymetamorphic rocks from the higher levels of the ancient basement, with pre-Variscan metamorphism, represent the bottom of the sequence (mainly detrital, fine-grained, easily erodible metasediments). An uncomformity separates a second sequence, with detrital and fine-grained elements lacking pre-Variscan metamorphism, deposited during early-middle Palaeozoic times and accompanied by rare rhyolites. The ancient basement and its post-S_1 cover were injected with quartz veins and intruded with granitic bodies (orthogneisses II) before the main Variscan regional metamorphic event.

5.2 *Late Variscan Source Areas*

By the end of Carboniferous times, the outcropping ancient basement should have represented source-areas. Later on, these zones ("X zones") might have been covered by Early Permian ignimbrites and/or by Verrucano and/or by Triassic sediments (Fig. 10B).

Again, by the end of Permian times, other different source areas ("Y zones") fed the Verrucano deposits (Sect. 6) with the eroded coarse sediments of the Ollano formation. The substratum of these "Y zones" could have been the ancient basement, either directly outcropping, or covered by Palaeozoic pre-Namurian sequences (Fig. 10C).

X and Y zones cannot have coincided, but the present location of late Variscan deposits and other evidence (Cortesogno et al. 1988a), not quoted here, indicate that they should both have been located in internal sectors.

In the Ligurian Alps, the core of the Calizzano "massif", with the old polymetamorphic basement directly covered by late Variscan andesitic lavas (Cortesogno et al. 1983), may correspond to the X zone features, whereas its southern border (Dallagiovanna 1989), as well as the Arenzano "Massif" (Cortesogno and Forcella 1978), both showing Triassic deposits on the ancient basement, depict the possible characters of Y zones.

Tentative evaluations of the area of X and Y zones and of the thickness of the eroded sequences, based upon some assumptions, have been made (Cortesogno et al. 1988c). For the Y zones, the calculated area and thickness are respectively 2×10^3 km^2 and 0.5 km, while for X zones the corresponding values are 3×10^3 km^2 and 5 km. Thus, assuming an average erosion rate of 0.5 mm/y, the complete removal of the "disappeared cover" would have taken about 10 Ma, less than the interval between the middle Westphalian and Autunian.

6 The Change from Late Variscan to Alpine History: The Verrucano Sedimentation

In the different sectors of the Briançonnais and Prepiedmont domains, the continental detrital deposits known as "Verrucano", always following the end of Permian volcanism, may lie on different tegument formations or, very locally, even on the oldest basement: on a regional scale, their base is therefore an erosional, disconformal surface. As they are gradually followed by marine Triassic deposits, their sedimentation indicates the beginning of Alpine history. Lack of fossils prevents accurate chronological determinations: a Late Permian age, by comparison with the well-known South-Alpine stratigraphy, has been tentatively proposed (Cassinis et al. 1986).

In an idealized stratigraphic column five levels can be recognized (Cortesogno et al. 1988c); however the poor lateral continuity only seldom allows to find all of them superposed:

5. Quartzitic sandstones with conglomeratic lenses (probable scour and fill structures), grading into Early Triassic quartzites; up to 20 m;
4. Heterometric white conglomerates, mainly formed by polycrystalline quartz pebbles, from reworked underlying Carboniferous conglomerates; up to 150 m;
3. Transitional level; up to 20 m;
2. Heterometric red conglomerates, with mainly K-rhyolitic, frequently silicified pebbles from underlying D lithozone (top level of Permian Melogno porphyroids); up to 50 m;
1. Basal arkosic sandstones, reworking directly underlying Permian ignimbrites; up to 20 m.

Some features of the lower levels (abundant detrital hematite in matrix, sometimes associated with phosphates: Cortesogno et al. 1987; scarce detritic feldspars; predominant silicified pebbles resisting to alteration) indicate that the source area, without high relief, was submitted to important weathering under warm, oxidizing and, at least periodically, wet conditions.

The pebble sizes of several centimetres, the very poor sorting and the low degree of roundness point to a moderate distance of transport. As the red conglomerates are found with homologous features in all the Briançonnais and Prepiedmont zones, it is possible to say that the D-lithozone ignimbrites, by now only locally preserved, were originally widespread.

As for the upper levels, the predominance of quartz pebbles inherited from Westphalian conglomerates, coupled with the almost complete lack of pebbles from the lower A, B, C lithozones of the Melogno Porphyroids, depict a peculiar distribution of tegument formations (Fig. 11), which follows the horst and graben pattern originated during previous Permo-Carboniferous times (Sect. 4).

The erosion of the tegument was not preceeded nor accompanied by ductile deformations, as Late Carboniferous-Early Permian folding was not recognized anywhere in the Ligurian Alps; it follows that the clastic material which supplied the Verrucano sedimentation derived from relief originated by reactivation of some late Variscan faults. The lack of pebbles from the metamorphic basement suggests movements without considerable vertical throw, possibly within a tensional (or transtensional) system.

7 Summary

A possible pre-Mesozoic history of the Ligurian Briançonnais domain is summarized in Fig. 12 (outer to inner zones from left to right).

a. Metagranitoid rocks (orthogneisses I), alternated with paragneisses and showing the same pre-Alpine polymetamorphic history, might in part derive from a Precambrian basement and, in part, represent the products of an acidic magmatic activity, which gave, during a Cambro-Ordovician rifting, both intrusions (No. 3: granites, migmatites I, granodioritic to leucogranitic dykes) and minor synsedimentary volcanics (No. 2).

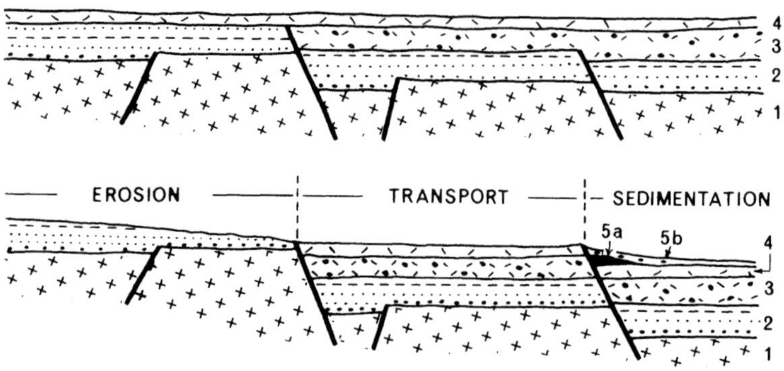

Fig. 11. The pebble lithology of the Verrucano leads to the following idealized scheme. After the filling up of the grabens existing in the basement *(No. 1)* with continental Upper Carboniferous deposits (Ollano Fm.; *No. 2*), the tectono-volcanic activity produced Lower Permian ignimbrites (Melogno Porphyroids, *No. 3*) of variable thickness. Only the topmost K-rhyolites *(No. 4)* were poured out everywhere.
Late Permian fault reactivation was accompanied by erosion in inner sectors, which delivered pebbles from K-rhyolites, forming the Verrucano basal red level *(No. 5a)* and reworked quartz pebbles from Carboniferous conglomerates, building the thickest Verrucano sequences *(No. 5b)*

More important and nearly contemporaneous basaltic eruptions (No. 4, now metabasites I, i.e. eclogites and amphibolites) alternated with arkoses and pelites (No. 1, presently represented by the already mentioned paragneisses and by micaschists, sometimes showing anatectic phenomena).

b. The first tectonometamorphic event might have started between Ordovician and Silurian times. As the various metamorphic assemblages (eclogites and anatectic migmatites; amphibolites and paragneisses) originated from similar protoliths, it might be inferred that the volcano-sedimentary sequences and the granitic intrusives were transformed (Nos. 1′, 2′, 3′, 4′) by metamorphic re-equilibration at different crustal levels. At the extreme metamorphic conditions sedimentary rocks underwent partial anatexis (migmatites II), while associated basites became eclogites and granulites (P = 17 kb; T = 700 °C, calculated on mineral pairs in eclogites; P = 10 kb; T = 750 °C, calculated on the phase stability fields of granulitic assemblages).

At the same time, the greatest (?) part of the sediments and of the associated basites underwent paragenetic metamorphic re-equilibration under amphibolite facies conditions (with low to intermediate thermal gradient) and was transformed into paragneisses (Ky + Sil + Grt) and amphibolites (Grt + Hbl + Di).

During the Silurian-Devonian (and Early Carboniferous ?) interval, fine, probably some thousands of metres thick, sediments (No. 5) were deposited, while at deeper levels large monzogranitic masses (No. 7) and dykes (No. 6) and minor gabbroic lenses and basaltic dykes were emplaced.

c. A later tectonometamorphic event (Variscan orogenesis) overprinted under amphibolite facies conditions (with intermediate to high thermal gradient: P = 6–7 kb; T = 670 °C) both older metamorphites – which were once more transformed (Nos. 1″, 2″, 3″, 4″) – and later magmatites: monzogranites and gabbro-basaltic bodies became respectively orthogneisses II (Nos. 6′ and 7′) and metabasites II. The post-Ordovician cover was correspondingly transformed (No. 5′) under upward decreasing metamorphic conditions.

The paragenetic evolution, which attains in the basement the andalusite stability field, suggests a general uplift, followed by a rarely recorded greenschist facies re-equilibration. The progressive exhumation of the basement was accompanied by a conspicuous erosion, which removed nearly all the Siluro-Devonian cover: in fact, this is presently recorded only by pebbles in Upper Carboniferous conglomerates; these clasts consist of pelites, sandstones and radiolarian cherts, which underwent only greenschist to sub-greenschist facies tectonometamorphic re-equilibration.

d. After the Early Carboniferous Variscan main phases, during the Late Carboniferous and Early Permian times detrital sequences and volcanic products were piled up in subsiding continental basins, in an almost flat region, under climate evolving from wet to semiarid conditions. The following steps have been identified:

1. Arkosic sequences (Lisio formation, No. 8) were sedimented in low-standing areas;
2. Block-faulting tectonism led to the birth of intracontinental graben and half-graben structures

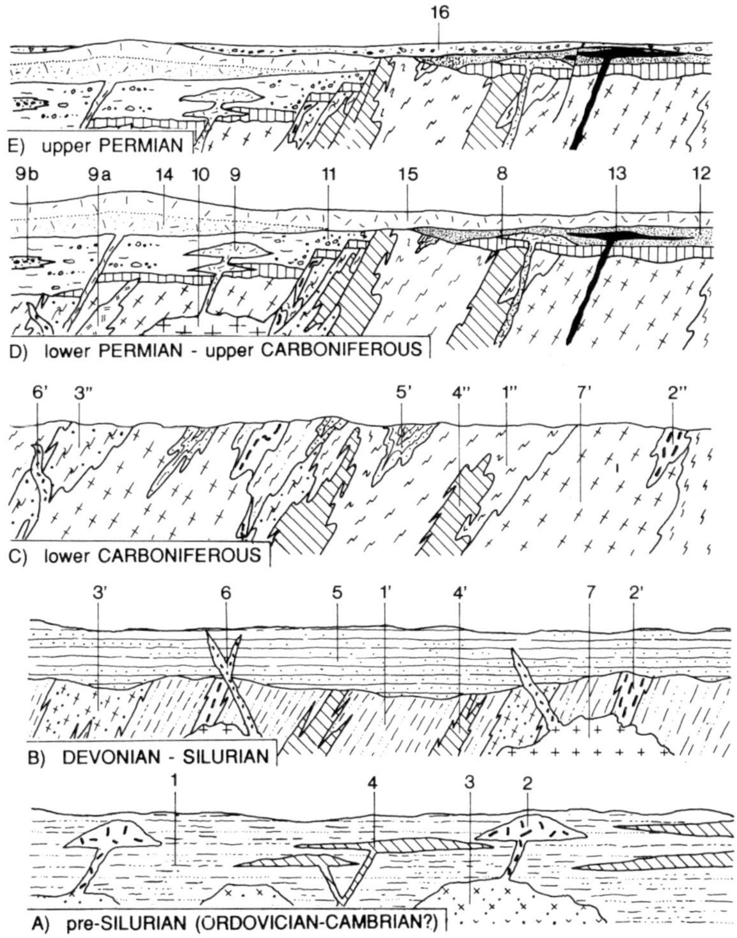

Fig. 12. Proposed pre-Triassic evolution of Ligurian Briançonnais. Length of sections: about 200–300 km. See text for explanation.

and of an early magmatic activity, giving acidic extrusives (C. Lisetto rhyolitic ignimbrites, No. 9) and subvolcanic granophyres (rio Castorello, No. 9 a);

3. Conglomerates, sandstones and plant debris bearing pelites (Ollano formation, No. 11) of late Westphalian-Stephanian age, locally interlayered with ignimbrites and pyroclastites (Osiglia rhyolites, Bric Crose Tuffs, No. 9 b) progressively filled up the subsiding grabens. During the Stephanian-(?)Autunian interval quartzitic arenites finely interbedded with more or less reworked rhyolitic pyroclastites (Gorra schists) were deposited in the internal and intermediate zones; black, lacustrine (?) pelites (Murialdo formation) are localized in the internal zone, while fine-grained quartz-argillitic continental sediments (Viola schists, No. 12), deriving from severe weathering, though dominating in internal and intermediate zones, appear ubiquitously. In the same time span, andesitic to shoshonitic, central, often explosive, volcanic activity widely occurred (Eze formation, No. 13) and sediments were intruded by large tonalitic dykes (No. 10);
4. During Lower Permian times extensive and thick rhyolitic ignimbrites and minor dacitic pyroclastites and lavas were outpoured (Melogno porphyroids, No. 14), progressively invading inner sectors, where only the latest K-rhyolitic products (No. 15) where probably emplaced.

e. The last, Late Permian faulting phase, devoid of volcanic activity, produced erosion of inner sectors and disconformable deposition, mainly in intermediate and outer zones, of the Verrucano continental arenaceous to conglomeratic sediments (No. 16). These grade upward into the Scythian marine quartzites.

In the *Prepiedmont* zone (Torrente Visone unit) two non-coeval sequences from different geodynamic environments represent the probable remnants of the Siluro-Devonian covers.

The presumed older sequence consists of rhyolites interbedded with quartzitic to arkosic sediments and with rare dolomitic levels; basalto-tholeiitic bands are interpreted as later dykes. The second sequence is represented by detrital carbonate sediments interbedded with metapelitic layers and with lenses (from 10 cm to some tens of metres thick) of tholeiitic basalts, gabbros and ultrabasic rocks. Basic and ultrabasic rocks, likely of ophiolitic affinity, are possibly reworked (olistoliths?) in this terrigenous sequence. Rare relics suggest a pre-Alpine metamorphic event in the biotite stability field.

References

Bloch JP (1966) Le Permien du domaine briançonnais ligure. Essai de chronologie des formations antétriasiques In: Atti del Symposium sul Verrucano. Soc Tosc Sci Nat, Spec Issue, Pisa, pp 99–115

Boni A, Cerro A, Gianotti R, Vanossi M (1971) Note illustrative della Carta Geologica d'Italia: Foglio 92–93 (Albenga-Savona). Tip L Salomone, Roma 144 pp

Bonin B (1988) From orogenic to anorogenic environments: evidence from associated magmatic episodes. Schweiz Mineral Petrogr Mitt 68: 301–311

Boriani A, Rivalenti G (1984) Crosta profonda e significato delle rocce basiche e ultrabasiche dell'Ivrea-Verbano in un secolo di studi. Mem Soc Geol Ital 24, Suppl E, Spec Issue, Cento anni di geologia italiana: 113–131

Boriani A, Dal Piaz GV, Hunziker JC, von Raumer J, Sassi FP (1974) Caratteri, distribuzione ed età del metamorfismo prealpino nelle Alpi. Mem Soc Geol Ital 13 (1): 165–225

Bouchardon JL, Santallier D, Briand B, Ménot RP, Piboule M (1989) Eclogites in the French Paleozoic Orogen: geodynamic significance. Tectonophysics 169: 317–332

Cabella R, Cortesogno L, Dallagiovanna G, Vannucci R, Vanossi M (1988) Vulcanismo, sedimentazione e tettonica nel Brianzonese ligure esterno durante il Permo-Carbonifero. Atti Tic Sci Terra 31: 269–326

Cabella R, Cortesogno L, Gaggero L (1989) Il basamento cristallino del Torrente Visone. Rend Soc Geol Ital (in press)

Carmignani L, Cocozza T, Ghezzo C, Pertusati PC, Ricci CA (1982) Lineamenti del basamento Sardo. In: Guida alla geologia del Paleozoico Sardo, Guide geologiche Regionali della Societá Geologica Italiana, Pisa, pp 11–23

Cassinis G, Gianotti R, Vanossi M (1986) Confronti nella successione tardo-ercinica delle Alpi liguri e delle Alpi meridionali. Mem Soc Geol Ital 28: 331–352

Cortesogno L (1986) Metamorfismo e magmatismo prealpini nel basamento e nel tegumento delle Alpi liguri. Mem Soc Geol Ital 28: 79–95

Cortesogno L, Forcella F (1978) Il massiccio cristallino di Arenzano, frammento di crosta continentale Brianzonese al margine meridionale del Gruppo di Voltri. Rend Soc Ital Mineral Petrol 34/2: 307–350

Cortesogno L, Gianotti R, Oxilia M, Vannucci R, Vanossi M (1982) Genesi ed evoluzione dello zoccolo pre-mesozoico in alcuni settori del Brianzonese ligure interno. Rend Soc Ital Mineral Petrol 38: 219–260

Cortesogno L, Oddone M, Oxilia M, Vannucci R, Vanossi M (1983) Le metavulcaniti a chimismo andesitico del Permo – Carbonifero Brianzonese (Alpi Marittime). Rend Soc Ital Mineral Petrol 38 (2): 581–606

Cortesogno L, Oddone M, Oliveri F, Vannucci R (1984a) Geochimica e petrogenese dei granofiri permo – carboniferi del Brianzonese ligure interno (Alpi Marittime). Mem Soc Geol Ital 28: 397–407

Cortesogno, Gianotti R, Oddone M, Vannucci R, Vanossi M (1984b) Contributo alla conoscenza delle meta-vulcaniti tardo – erciniche del Brianzonese Ligure (Alpi Marittime): 1. I porfidi di Osiglia ed i clasti di vulcaniti nella formazione di Ollano. Rend Soc Ital Mineral Petrol 39: 575–592

Cortesogno L, Gianotti R, Vanucci R, Vanossi M (1984c) Le volcanisme permo – carbonifère du Briançonnais ligure (Alpes Maritimes) dans le cadre des phases tardives de l'orogenése hercynienne. Sci Géol Bull 37: 37–50

Cortesogno L, Oddone M, Oliveri F, Vanucci R (1986) Considerazioni petrogenetiche sugli ortogneiss granitoidi di Nucetto e Barbassiria (Alpi Marittime). Rend Soc Ital Mineral Petrol 41: 75–81

Cortesogno L, Gaggero L, Lucchetti G (1987) Phosphate mineralization in a permo-triassic sequence (Giogo di Toirano, Italy). Neues Jahrb Mineral Monatsh 7: 305–313

Cortesogno L, Gaggero L, Vannucci R (1988a) Updating of knowledge on the Calizzano Massif granitoids (Western Liguria). Rend Soc Ital Mineral Petrol 43 (2): 623–626

Cortesogno L, Dallagiovanna G, Vanucci R, Vanossi M (1988b) Volcanisme, sédimentation et tectonique pendant le Permo-Carbonifère en Briançonnais ligure: une revue. Eclogeae Geol Helv 81 (2): 487–510

Cortesogno L, Dallagiovanna G, Vanossi M (1988c) Paléogeographie du Briançonnais ligure d'après les lithoclastes des formations permo-carbonifères. Géol Alp Mém h-ser 14: 9–28

Dallagiovanna G (1989) Testimonianze e significato di una trasgressione permo-triassica sul basamento cristallino del Brianzonese ligure interno. Boll Soc Geol Ital 107: 445–451

Del Moro A, Di Simplicio P, Ghezzo C, Guasparri G, Rita F, Sabatini G (1975) Radiometric data and intrusive sequence in the Sardinian batholith. Neues Jahrb Mineral Abh 126 (1): 28–44

Del Moro A, Pardini G, Messiga B, Poggio M (1981) Dati petrologici e radiometrici preliminari sui massicci cristallini della Liguria occidentale. Rend Soc Ital Mineral Petrol 38 (1): 73–87

Desmons J, Ploquin A (1989) Chimisme du Briançonnais-Grand Saint-Bernard ante – mésozoique (Alpes Occidentales): une vue synoptique. Géol Alp 65: 1–31

Ebner F, Neubauer F, Stattegger K (1987) The Caledonian event in the Eastern Alps: a review. In: Flügel HW, Sassi FP, Grecula P (eds) Pre-Variscan and Variscan events in the Alpine-Mediterranean mountain belts, Alpha, Bratislava, pp 169–182

Forcella F, Mottana A, Pasquaré G (1973) Il Massiccio cristallino interno di Valosio (Gruppo di Voltri, provincia di Alessandria). Mem Soc Geol Ital 12: 482–528

Gaggero L (1990) Eventi magmatici e metamorfici antealpini nel basamento Brianzonese ligure: dati petrochimici e minerochimici. Thesis, University of Genoa (unpubl)

Ghezzo C, Memmi I, Ricci CA (1977) Le granuliti e le eclogiti della Sardegna Nord-Orientale. In: Guida alla Geologia del Paleozoico Sardo, Guide Geologiche Regionali. della Societá Geologica Italiana, Pisa, pp 151–156

Guillaume A (1969) Contribution à l'étude géologique des Alpes Liguro-piémontaises. Doc Lab Géol Fac Sci Lyon 30 (2), 658 pp

Irvine TN, Baragar WRA (1971) A guide to the chemical classification of the common volcanic rocks. Can J Earth Sci 8: 523–548

Klein H, Wimmenauer W (1984) Eclogites and their retrograde transformation in the Schwarzwald (Federal Republic of Germany). Neues Jahrb Mineral Monatsh H 1: 25–38

Liégeois JP, Duchesne JC (1981) The Lac Cornu retrograded eclogite (Aïguilles Rouges massif, Western Alps, France): evidence of crustal origin and metasomatic alteration. Lithos 14: 35–48

Lorenzoni S (1958) Studio chimico-petrografico della serie permiana del Bric Colmè (Alpi Marittime) e delle mineralizzazioni uranifere connesse. Studi Ric Div Geomineral Ital 1/2: 435–498

Maggetti M, Galetti G (1988) Evolution of the Silvretta eclogites: metamorphic and magmatic events. Schweiz Mineral Petrogr Mitt 68: 467–484

Maggetti, Flisch M, Bollin R (1990) Bericht über die Excursion der Schweiz. Mineralogischen und Petrographischen Gesellschaft ins Silvretta-Kristallin und in den Westrand des Unterengadiner Fensters (11–14 September 1989). Schweiz Mineral Petrogr Mitt 70: 121–157

Messiga B (1981) Evidenze strutturali e paragenetiche dell'evoluzione polifasica pre-alpina del Massiccio cristallino di Savona. Rend Soc Ital Mineral Petrol 37 (2): 739–754

Messiga B, Oxilia M, Piccardo GB, Vanossi M (1982) Fasi metamorfiche e deformative alpine nel Brianzonese e nel pre-Piemontese-Piemontese esterno delle Alpi Liguri: un possibile modello evolutivo. Rend Soc Ital Mineral Petrol 38 (1): 261–280

von Raumer JF (1986) The Hercynian basement in the helvetic realm western and central Alps. Mem Soc Geol Ital 29: 57–69

von Raumer JF (1987) Les massifs du Mont Blanc et des Aiguilles Rouges: témoins de la formation de croûte varisque dans les Alpes Occidentales. Géol Alp 63: 7–24

Sassi FP, Zirpoli G (1980) Stato delle conoscenze sulla storia pre-ercinica delle Alpi Orientali. Mem Soc Geol Ital 20: 11–20

Sassi FP, Zanferrari G, Zirpoli G (1987) The Caledonian event in the Eastern Alps: a review. In: Flügel HW, Sassi FP, Gracula P (eds) Pre-Variscan and Variscan events in the Alpine-Mediterranean mountain belts, Alpha, Bratislava, pp 431–434

Thélin P (1989) Essai de chronologie magmatico-métamorphique dans le socle de la nappe du Grand Saint-Bernard: quelques points de repère. Schweiz Mineral Petrogr Mitt 69: 193–204

Vai GB, Boriani A, Rivalenti G, Sassi FP (1984) Catena Ercinica e Paleozoico nelle Alpi Meridionali. Mem Soc Geol Ital 24, Suppl E, Spec Issue, Cento anni di geologia italiana, pp 133–154

Vanossi M (1970) Studio sedimentologico della Formazione di Ollano (Carbonifero medio-superiore della Liguria Occidentale). Atti Ist Geol Univ Pavia 21: 3–36

Vanossi M, Cortesogno L, Galbiati B, Messiga B, Piccardo GB, Vanucci R (1986) Geologia delle Alpi Liguri: dati, problemi, ipotesi. Mem Soc Geol Ital 28: 5–75

Passing Through the Briançon Zone

J. Desmons and D. Mercier

En passant par la zone briançonnaise
Avec mes sabots . . .
Passing through the Briançonnais zone
With my clogs . . .
(Variation on an old song)

Abstract

In the Briançon-Bernhard zone the pre-Alpine units, from the Ruitor Massif to the Acceglio Zone, show a variety of lithological, metamorphic and mineralogical characteristics. Younger and ancient basements are distinguished. The *younger basement* sequences, which include the Mt. Pourri-Bellecôte Massif, the Zona interna, the Ambin formation and the Acceglio basement, are defined on the basis of their monocyclic metamorphic imprints: only Alpine phases are recorded. A recently acquired U-Pb zircon result indicates a Late Cambrian age for the protoliths. The *ancient basement* sequences are characterized by remnants of pre-Alpine amphibolite and eclogite assemblages, the minimum age of which is Upper Proterozoic-(Lower Cambrian?), i.e. Upper Pan-African. The ancient basement is further subdivided into a more external Sapey-Ruitor type and a more internal Chasseforêt-Clarea type where the Alpine imprint has been pervasive.

In the Zone Houillère, the Upper Carboniferous (Silesian) basin had a S-to-N drainage pattern. The series includes a lower, coal-bearing, part consisting of Namurian to lower Stephanian sandstones and pelites, unconformably overlain by an upper part containing various conglomerates and sandstones of an assumed middle Stephanian to Upper Permian age, which grades into quartzites assigned to Scythian.

A clastic series lithologically similar to the uppermost, Permian to Scythian, part of the Zone Houillère overlies some basement units, especially the younger type.

It is proposed to correlate the younger basement with the Métailler-Mt. Fort unit of the Swiss Bernhard zone, and both types of the ancient basement units with the Pontis and Siviez-Mischabel units. Emphasis is laid on the Briançon-Bernhard units pertaining to Gondwana until the beginning of the Alpine times.

1 Introduction

The Briançon-Bernhard zone forms the leading edge of the internal basement units of the Western Alps, behind the Penninic front. The zone underlies sheets of Combin calc-schists and ophiolites. These have been eroded in the northwestern and the western parts of the Briançon zone, leaving a few klippen, but they are more widely preserved between, approximately, the town of Briançon and the Ligurian Alps.

The Briançon zone is structurally subdivided into two parts: (1) westward, the Zone Houillère and the classical Briançonnais, with Upper Carboniferous to lower Cainozoic sequences, and (2) on the internal side, basement units and various Permian to lower Cainozoic cover sequences.

Stratigraphically the Briançon zone consists of: (1) a post-Variscan cover, Permo-Triassic to lower Cainozoic in age, of variable thickness (from a few tens of metres to more than 2000 m in the western part of the Chasseforêt Massif), which will not be discussed in this chapter, (2) Upper Carboniferous and Lower Permian beds making the Zone Houillère, and (3) older sequences, the basement sequences. In France and Italy the latter do not form parallel strips as in the Swiss part of the zone, but they are mostly distributed in individual basement massifs. This dissection is due to the superimposition of N-S and E-W structural directions in that N-S trending part of the Alpine belt. The massifs consist of nappes or stacked sheets. From N to S in the French-Italian Western Alps the massifs are (Fig. 1): the Ruitor Massif, the Northern Vanoise (or Pourri-Bellecôte) Massif and the Zona interna, the Southern Vanoise (or Chasseforêt) Massif, with the Sapey Zone along both N and S Vanoise Massifs, the Ambin Massif, a small part of the Acceglio Zone and the west-Ligurian Massifs: Calizzano, Savona, Arenzano, Pallare, Barbassiria and Nucetto-Lisio-Costa Dardella. The Ligurian part of the zone is dealt with by Cortesogno et al. (this Vol.), whereas the Swiss portion, i.e. the Bernhard zone, is the topic of the chapter by Thélin et al. (this Vol.).

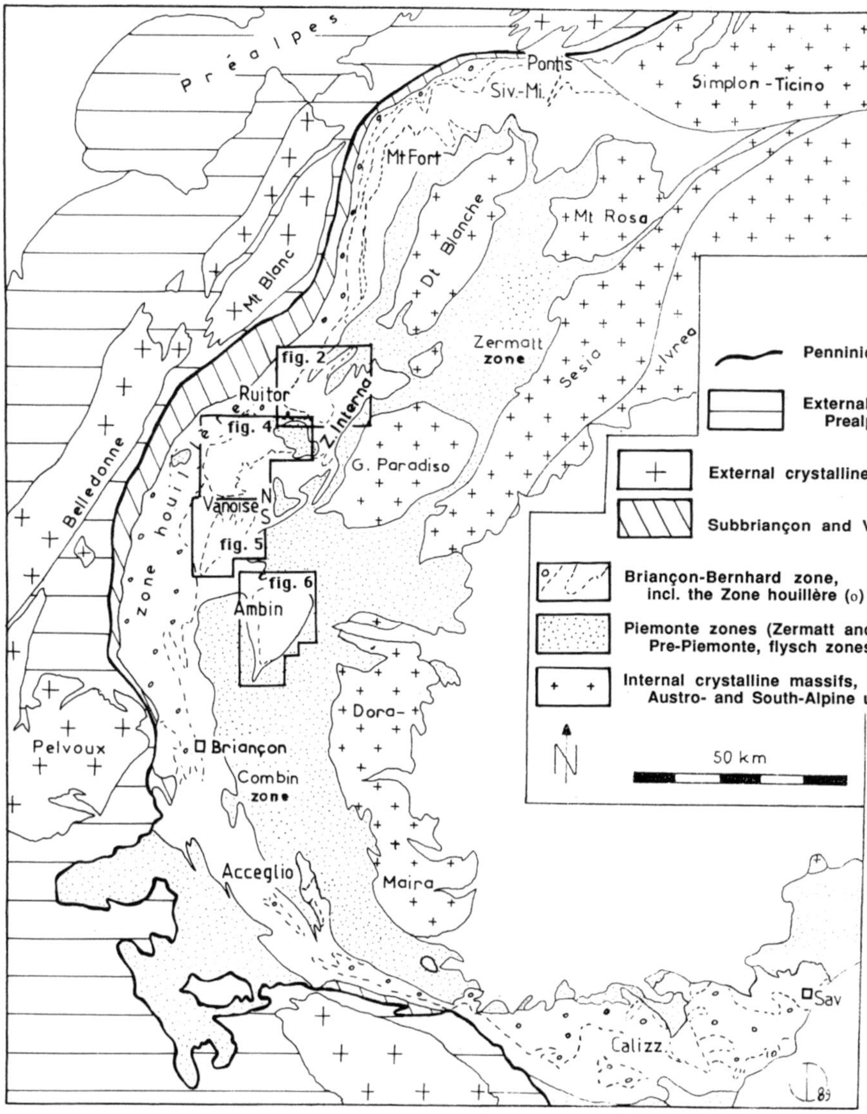

Fig. 1. Sketch map of the Western Alps with location of Figs. 2, 4–6. *Calizz.* Calizzano; *Sav.* Savona; *Siv.-Mi.* Siviez-Mischabel

In the conventional picture the Briançon zone, like the other Penninic sialic units, is considered as belonging to the Eurasian plate. However, according to Radelli and Desmons (1987, 1988) it derived from a sialic arc which, from Jurassic times, migrated away from the margin of Gondwana and was eventually accreted to Eurasia during the Cainozoic orogeny. In deep seismic profiles (ECORS-CROP Program 1987) it appears as a layered lithospheric wedge thrust on the Eurasian margin duplex. Low-angle reflectors internal to the Briançon zone possibly indicate intracrustal redoubling, but there are no bore-hole data and any interpretation of seismic reflectors in such a metamorphic basement would be premature.

Obtaining data on the Briançon basement massifs has been hampered by Alpine recrystallization and deformation. These imprints have been pervasive in many places, operating a strong grain-size reduction: many rocks are now tectonites, pre-Alpine mineral relics are scarce or their identification as pre-Alpine relics has been much debated. The progress of the knowledge is marked by the early cartographic, petrographical and chemical investigations of Termier (1891), the identification of the pre-Alpine age by Gignoux (1929), the first radio-isotopic measurements of Bocquet (Desmons) et al. (1974) and the studies of many other authors whose names will be mentioned later.

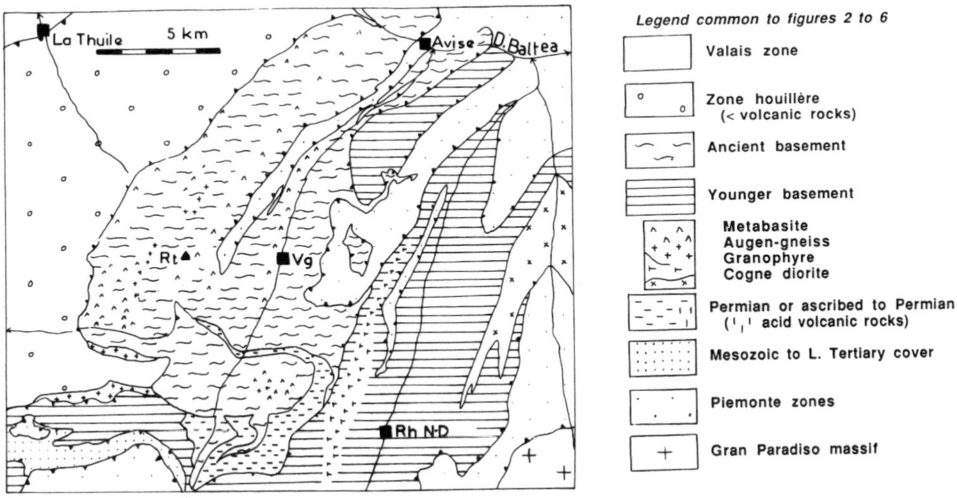

Fig. 2. Sketch-map of the Ruitor Massif (after Baudin, 1987; Debelmas et al. 1991). *Rh N-D* Rhêmes Notre-Dame; *Rt* Ruitor; *Vg* Valgrisanche

In the following, the structural position, lithology and metamorphic evolution of each massif will be reviewed briefly, as well as the lithology and sedimentary evolution of the Zone Houillère. A classification of the massifs and correlations with the Briançon-Bernhard zone in Switzerland and in western Ligury will be proposed. The chemical characteristics that will be mentioned are taken from Desmons and Ploquin (1989, with previous references) who compiled and re-interpreted the available chemical data. The corresponding chemical diagrams can be found in this latter paper and, as concerns the late Variscan granitoid rocks, in Bonin et al. (this Vol.).

2 Ruitor Massif

West- and southward, the Ruitor Massif (Fig. 2) lies below the backfolded Zone Houillère (Mercier and Baudin 1990; see also Guillot 1987, p. 209) and the northern Vanoise. Eastward, it is backfolded upon the Zona interna. This attitude post-dates an Alpine NW-vergent thrusting phase. Ophiolitic rocks and calcschists of the Piemonte zone are tectonically pinched within the massif, forming the Avise "syncline". The main part of the Combin nappe bounds the massif to the east and north-east. Together with a few Mesozoic carbonates, small outcrops of quartz-rich and/or albite ± ankerite rocks which have been interpreted as Lower Permian in age, represent the cover of the massif. However, rock types similar to these Permian layers do exist within undisputed basement sequences, e.g. in the Ambin and Southern Vanoise Massifs and a Permian age of all occurrences is still uncertain. A few mafic dykes intruding the Ruitor rocks are considered as late Variscan. The visible thickness greatly exceeds 2000 m.

The Becca di Toss klippe, east of Valgrisanche, has been successively ascribed to the Dent Blanche nappe on the basis of its structural position upon the Combin calcschists (Hermann 1938; Staub 1942), and to the Ruitor basement on the basis of lithological similarities (Dal Piaz 1965). The latter interpretation implies a backfolded structure which is not yet entirely elucidated (Baudin 1987; Platt et al. 1989).

The massif contains (Fig. 3) metasedimentary, metabasic and metagranitoidic rocks [Fabre 1961; Dal Piaz and Govi 1965; Caby 1968; Bocquet (Desmons) 1974 a, c; Desmons 1977; Eteradossi 1983; Baudin 1987; Desmons and Ploquin 1989]. The composition of the metasedimentary rocks is that of shales, sandstones, greywackes and occasionally marbles. The metabasic rocks have a tholeiitic composition and, often interlayered with acidic rocks of possible plagiogranitic affinity, form banded complexes. A few lenticular cumulate bodies have a composition similar to alkaline rocks and contain very rare cumulate meta-pyroxenite. Trace element contents (summarized by Desmons and Hunziker 1988) point to MORB and within-plate affinities. The metagranitoids are sub-alkaline (that is, close to alkaline; equivalent to the calc-alkaline of some authors) and contain zircons with anatectic morphological features (Baudin 1987).

Quartz-K-feldspar-muscovite-tourmaline pegmatites and quartz-keratophyres are subordinate. So far no migmatites have been identified, perhaps be-

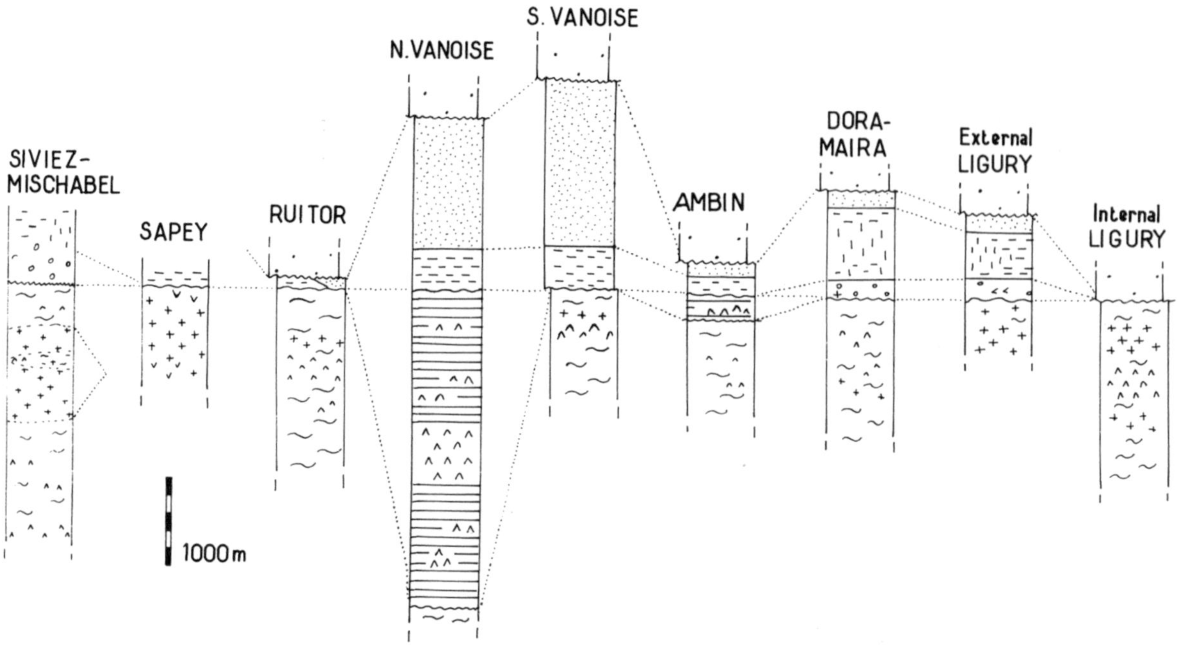

Fig. 3. Schematic lithostratigraphic columns of the basement massifs. *Patterns:* see legend to Fig. 2. Note that the name of Dora-Maira has to be hyphenated: it originates in the *Dora* Riparia and *Maira* (not Maïra!) rivers bounding the massif N and S, respectively

cause of the Alpine transformations. The contacts between the rock types are now, in many places, parallel to an Alpine foliation plane associated with greenschist facies, which transposed most of the pre-Alpine planar features. The Al-drift in the chemical composition of the augen-gneisses probably results also from Alpine mobilization (Desmons and Ploquin 1989).

Two or three pre-Alpine metamorphic phases have been distinguished. In addition to plagioclase of which there is no relict, the main, middle-pressure amphibolite facies phase produced, in metasedimentary rocks, coarse-grained muscovite ± biotite$_I$ ± garnet ± staurolite ± kyanite schists and gneisses, and, in metabasites, hornblende ± biotite$_I$ ± clinopyroxene amphibolites. These minerals, which are still present as relics, were accompanied by plagioclase and possibly sillimanite which is known from the Briançon-Bernhard zone near the Grand Saint Bernard pass (Bearth 1960–1963).

The amphibolite phase has been followed by a biotite$_{II}$-bearing phase where kyanite was no longer stable and which indicates either deep greenschist or low-pressure amphibolite facies, that is, the same evolution as deduced in western Ligury (Messiga 1981; Cortesogno 1984).

Eclogite relics (Dal Piaz and Govi 1965) are considered as either older than the amphibolite assemblages, in agreement with Swiss and Ligurian occurrences (Desmons in Debelmas et al. 1991), or Alpine in age (Caby and Kienast 1989; Caby in Debelmas et al. 1991). Apart from these very rare eclogite relics nothing can be said as to any pre-amphibolite metamorphic facies.

Pre-Alpine deformation is recognizable only in pre- or syn-amphibolite facies minerals such as helicitic garnet, tourmaline and muscovite, in relict fold hinges and as two foliation planes, the first containing amphibolite facies minerals and the second, oblique to the first, containing biotite$_{II}$.

Compared with other Briançonnais massifs, the Alpine metamorphic and deformational imprint in the Ruitor Massif is in many places weak: relics of pre-Alpine minerals are various and widespread in the western and central part of the massif, contrasting with the eastern parts. This fact has been interpreted as evidence of a tectonic partition of the massif into two (Caby in Debelmas et al. 1991), the boundary lying east of the Avise syncline. However, the question is not settled, as a transitional contact is supported by the field observation that both transformation and deformation gradually increase eastward and by the lithologic similarities between both E and W parts.

Microprobe data collected by Desmons et al. (1977), Laduron and Desmons (1981), Cigolini (1981) and Desmons (1992) show the following average compositions of the pre-Alpine minerals: muscovite $Si_{\sim 6.05-6.46}$, Fe-Mg-biotite, clinopyroxene (salite), pargasitic or tschermakitic to Mg-hornblende and garnet (in schists ~ alm_{60-69} prp_{4-11} sps_{4-14} grs_{7-23}, with a normal zonation; in metabasites ~ alm_{55-60} prp_{5-11} sps_{2-7} grs_{28-35}.

Radiometric measurements on minerals [Bocquet (Desmons) et al. 1974; Desmons and Hunziker in prep.] so far yielded Alpine ages or late Palaeozoic reset ages or meaningless ages. Permian mineral ages are interpreted as indicating resetting by a thermal event in Permian time. On the basis of lithological and metamorphic similarities with the Siviez-Mischabel unit (Zingg 1989) and the Berisal complex in the Central Alps (Stille 1980; Stille and Tatsumoto 1985), upper Proterozoic (to Cambrian?) ages can be regarded as possible times for the formation of the sedimentary and magmatic protoliths and of the eclogite to amphibolite facies metamorphisms, the biotite-greenschist (or L-P amphibolite) phase being also pre-Variscan (lowermost Palaeozoic?).

The massif has been affected by an Alpine glaucophane-epidote metamorphic phase, followed by a greenschist phase with a late blastesis of albite and green to brownish biotite. At least three Alpine deformational phases have been identified (Baudin 1987; Platt et al. 1989).

3 Sapey Zone

The Sapey Zone (Figs. 2, 4 and 5) forms no massif but a narrow, discontinuous, zone of outcrops situated between, on one side, the Zone Houillère and, on the other side, the N and S Vanoise. In the most extensive part of the zone the thickness ranges from 100 to 1000 m. Much petrological, chemical and structural work has still to be done on this zone.

The Sapey Zone (Fig. 3) has lithological similarities to the Ruitor Massif, except in the relative amounts of the lithotypes (Ellenberger 1958; Fabre 1961; Détraz 1984). Metagranitoids are predominant. They have been considered to be derived from a Lower Permian very shallow migmatization event (Ellenberger 1958) and later suggested to be granitoids intruded in the Permian [Bocquet (Desmons) 1974a]. They have now been recognized as pre-late Variscan granitoids (Détraz 1984). Their contact with the Zone Houillère is often marked by mylonite. The tectonic emplacement of the Sapey Zone has been considered by Détraz (1984) to have resulted from a mid-Permian (Saalian) thrusting event. Guillot (1987) suggested, instead of converging movements, a distensive regime during the Stephanian, inducing uplift of the old basement and thus explaining also the coarse sediment discharge in the adjacent Stephanian layers of the Zone Houillère. On the other hand, Alpine tectonics is likely to have played a prominent or secondary part in the thrusting of the Sapey gneisses upon the Zone Houillère.

The metagranitoids are of various lithofacies, mostly foliated augen-gneisses. They are peraluminous and probably anatectic, an origin also indicated by the zircon morphology (Schade, cited by Baudin 1987, p. 47). Na and Ca-loss, K and Si-increase are related to (Alpine?) feldspar breakdown and white mica formation. The associated rock types include various fine-grained rocks, the ill-named leptynites of the French literature. (Leptynite is used in some French literature either for – restricted meaning – a granulite-facies, white mica-poor, K-feldspar and quartz gneiss with a centimetre-spaced planar cleavage, or – general meaning – for a leucocratic, mica-poor gneiss). Some of these were derived from mylonitized augen-gneisses or from meta-arkoses. Others, possibly derived from pyroclastics or even sub-alkaline basic rocks, show a secondary aluminous character. The discovery of kyanite-garnet ± biotite gneisses (Détraz and Loubat 1984) has been the determining factor in the recognition of the Sapey pre-Alpine age. Thus, a pre-Alpine middle-pressure amphibolite facies is definitely present. The present mineral assemblages partly result from Alpine metamorphic phases, which formed albite-chlorite assemblages with stilpnomelane and rare Fe-rich Na-amphibole.

A few mineral chemical data (Desmons 1992) concern muscovite ($Si_{6.2-6.3}$), Fe-Mg-biotite and garnet (~ alm_{60-65} prp_{2-23} sps_{1-14} grs_{6-30}).

Radiometric determinations include, in addition to a few Alpine ages, a K-Ar Permian age [Bocquet (Desmons) et al. 1974] and a 320 Ma (Visean-Namurian boundary) $^{39}Ar-^{40}Ar$ age of muscovite from a granitoid (Debelmas et al. 1989a).

4 Northern Vanoise or Pourri-Bellecôte Massif

As the following chapter will show, both N and S Vanoise massifs differ strongly from each other as to their lithology and metamorphic evolution. The Pourri-Bellecôte Massif (Fig. 4) includes the Mts. Bellecôte, Pourri and Grand-Bec areas. It lies in a thrust position on the Zone Houillère. The connection with the Chasseforêt Massif is concealed by the thick Mesozoic-lower Tertiary cover rocks of the

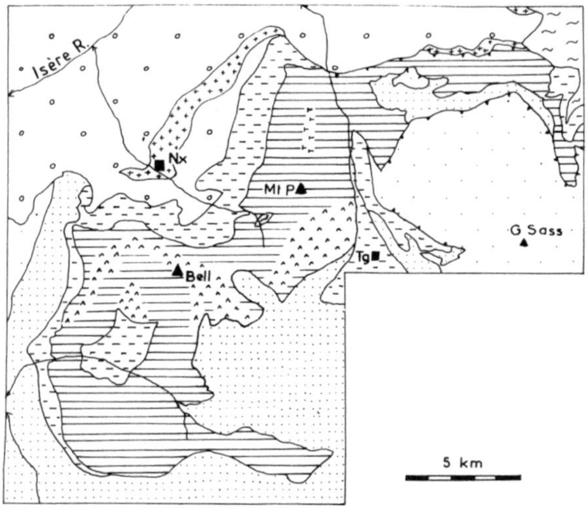

Fig. 4. Sketch map of the Pourri-Bellecôte or Northern Vanoise Massif (after Debelmas et al. 1989b; Guillot 1987). *Patterns:* see legend to Fig. 2. *Bell* Bellecôte; *G Sass* Grande Sassière; *Mt P* Mont Pourri; *Nx* Nancroix; *Tg* Tignes. Exposures of Permian-Triassic quartz-conglomerates and quartz-mica-schists north of the Mt. Archeboc, on the northern slope of the Northern Vanoise, have been mistaken for Sapey gneisses (Caby, in Debelmas 1980, Debelmas et al. 1991)

Pralognan-Col de la Vanoise area. Northeastwards, beyond the Sassière klippe of Combin calc-schists and ophiolites, the Pourri-Bellecôte Massif continues into the Zona interna.

The western and southern parts of the Pourri-Bellecôte Massif (Bellecôte, Mio, Grand Bec) have been recently mapped and both their structure and rock chemistry have been studied (Peruccio-Parison 1984; Guillot 1987, from which most data mentioned are taken). Only reconnaissance work has been done in the northeastern, Mt. Pourri, part [Ellenberger 1958; Bocquet (Desmons) 1974a; Marion 1984; Desmons and Fabre 1988; Guillot unpubl. 1988].

The Massif (Fig. 3) consists of folded tectonic slices, with a Permian to Lower Triassic cover, more than 400 m thick, in unconformable, depositional, contact (Guillot and Raoult 1984). The visible thickness of the pre-Permian basement is of the order of about 3600 m. The lower half of the Bellecôte sequence is formed by basic rocks with some interlayered quartz-albite rocks, overlain by a magmatic body comprising gabbro and minor albitite, intruded by veins and sills of other basic rocks. The upper half consists of sedimentary rocks, locally rich in organic matter, also including abundant metabasic veins and sills. In the magmatic body the metabasites are low-Ti tholeiites, the gabbro being a Mg-rich cumulate. The metabasites from the volcanic-sedimentary sequence are Ti-rich tholeiites with a spilitic trend. The metasedimentary rocks, pelitic in composition, were derived either from quartz-albite pyroclastics and were deposited not far from the eroded source, or from a mixture of quartz-pelites and Na-argillites. Toward the top metagreywackes are present, which were derived from both acidic magmatic rocks or migmatites and from medium-grade metamorphic rocks. A granophyre body [Ellenberger 1958; Bocquet (Desmons) 1974a; Marion 1984; Guillot et al. 1991] with micropegmatitic structure, occurs N of Mt. Pourri. Its composition is alkaline.

The mineral assemblages in these rocks are of greenschist facies, with relics of glaucophane-epidote facies. From regional considerations, all these assemblages can reliably be considered as Alpine in age. The organic matter apparently has not undergone temperatures higher than 250–300 °C (Guillot 1987). A hornblende-plagioclase (now altered to albite) assemblage from the metabasites, not being associated with a foliation, possibly is late magmatic. Thus, any pre-Alpine regional amphibolite facies is missing in these rocks. This type of basement has been called younger basement by Desmons and Ploquin (1989).

Rare polymetamorphic rocks (garnet-muscovite schist and hornblende-bearing schist) found S and E of Mt. Pourri testify to a relict amphibolite metamorphic facies and, thus, to pieces of an underlying older basement or to a separate ancient basement slice.

Chemical data are available on pre-Alpine minerals: garnet ($alm_{45}\,py_{<2}\,sp_{30}\,gross_{25}$), actinolitic Ca-amphibole, micas, etc. (Guillot 1987; Desmons and Fabre 1988; Desmons 1992).

A pre-Namurian, Upper Ordovician to Silurian, minimum age had been proposed for the magmatic and volcaniclastic sequences [Bocquet (Desmons) 1974b; Guillot 1987; Desmons and Fabre 1988]: that is, an age between a possible late Pan-African amphibolite-facies regional metamorphic event in the ancient basement and the Silesian of the Zone Houillère sedimentary rocks. U-Pb radiometric measurement of zircon from the Pourri granophyre body has recently yielded a Late Cambrian age (Guillot et al. 1991). Termier (1891, 1927), followed by Ellenberger (1958), had assumed that the Northern Vanoise sequences were the exact time equivalents of the Zone Houillère, but the lithology and chemical composition of both differ and this opinion is no longer considered.

5 Zona Interna

On its NE side the Northern Vanoise goes on to the Zona interna or Savaranche Massif (Fig. 2), its Italian equivalent in Val di Rhêmes and Val Savaranche (Dal Piaz 1928; Caby 1968; Cigolini 1981; Debelmas et al. 1991; Desmons pers. data). The Zona interna is separated from the Ruitor Massif by thin sporadic strips of Permian-Triassic cover rocks and it is partly backfolded on the Combin zone. The knowledge concerning the Zona interna is still poor. No recent map of the entire Zona interna is available (for the western part, see the Ste-Foy-Tarentaise sheet 1:50000). Its internal tectonic structure may be that of stacked slices piled up in Alpine times.

Metasedimentary rocks, locally coal-bearing, are interlayered with metabasites and with quartz-albite ± K-feldspar-rich rocks which also form a thick body, granophyric in microstructure (Dal Piaz, cited by Caby 1968) and alkaline in composition. Chemical data are too few to allow a reliable interpretation of the magmatic-sedimentary sequence. It is not yet clear whether remnants of polymetamorphic schists occur as in the Northern Vanoise. The Alpine metamorphic overprint is of a slightly higher grade than in the Northern Vanoise Massif, with greenish-brown biotite and garnet.

The "Cogne diorite" intruded the metasedimentary rocks which have been transformed locally into hornfelses. The Cogne diorite actually is of calc-alkaline, tonalitic-quartz-dioritic, composition, and is unrelated to the granophyre. Mafic enclaves are olivine basalt in composition. Two-mica granite is subordinate. The body contains hornblende, rare clinopyroxene, biotite, zoned feldspars and titanite, and does not show any distinct trace of pre-Alpine metamorphism. For this reason and because it intrudes rocks that were interpreted as Upper Carboniferous-Lower Permian, the Cogne diorite has often been considered to be Permian in age. There are no radiometric age data.

6 Southern Vanoise or Chasseforêt Massif

As said above, the contact between both Northern and Southern Vanoise (or Chasseforêt) basement massifs is concealed by Mesozoic to Tertiary sedimentary rocks. On its western side the Chasseforêt basement (Fig. 5) dips under Permian schists and quartzites which are identical to the Pourri-Bellecôte cover (Fig. 3), suggesting the prolongation of the Pourri-Bellecôte basement unit in depth, sutured in the Modane-Chavière fault zone. Various cover sequences have been thrust and backthrust upon the basement and their attitude is mostly inverted (Platt and Lister 1985; Platt et al. 1989).

The metasedimentary rocks (Fig. 3) are greywackes or shales in composition. Most metabasites are interlayered with leucocratic gneisses. Relics of ophitic structure are rare (the gabbro of Platt and Lister 1985). Alternation of the banded metabasites with metasedimentary rocks seems to result from tectonic deformation. The mebabasites are Ti-rich tholeiite in composition, with a strong spilitic trend. A decametric, slightly porphyric (a few mm in grain size), metagranitoid of a typical leucocratic composition, is associated with rare muscovite-quartz-K-feldspar veins and biotite lamprophyric rock [Bocquet (Desmons) 1974 a].

All of these rocks received a very strong Alpine tectono-metamorphic imprint and most of them are tectonites. In spite of Gignoux' (1929) opinion that they are pre-Upper Carboniferous in age, they have been considered to be Alpine metamorphic rocks derived from Permo-Carboniferous protoliths by Termier (1927), who first mapped the massif, and by Ellenberger (1958), who worked out the stratigraphy of the Mesozoic cover. Bocquet (Desmons) (1974 b, c; see also Desmons 1977) has shown conclusively that both protoliths and first metamorphic imprint are pre-Alpine in age and this opinion now is at last generally accepted (e.g. Debelmas et al. 1989 a).

The pre-Alpine relict minerals are garnet, muscovite and rare hornblende, allowing a pre-Alpine amphibolite facies to be inferred. Wet chemical [Bocquet (Desmons) 1974 a] and microprobe (Desmons 1992) mineral analyses show that the pre-Alpine white mica is Fe-muscovite and that the garnet is normally zoned and has the approximate composition of $alm_{70}\ prp_{11}\ sps_8\ grs_{11}$ (in metasedimentary rocks) and $alm_{52-55}\ prp_7\ sps_5\ grs_{28-32}$ (in metabasites).

Radiometric K-Ar measurements [Bocquet (Desmons) et al. 1974; Desmons and Hunziker in prep.) yielded Alpine and Permian ages, but so far no older ages have been obtained. The Permian results have been interpreted to date a thermal resetting.

Two tectonic sheets which differ lithologically are superposed (Platt and Lister 1985; Landès 1988). The lower one, devoid of any cover sequence, is rich in muscovite-garnet schists with abundant quartz veins parallel to the main foliation plane. The upper one is rich in metabasites and contains the metagranitoid body. Jadeitic Na-pyroxene, previously considered as Alpine in age but which could also be derived from pre-Alpine minerals, has so far been found only in the upper slice. Estimated thickness of the lower sheet, which is truncated at its base by a thrust plane, is

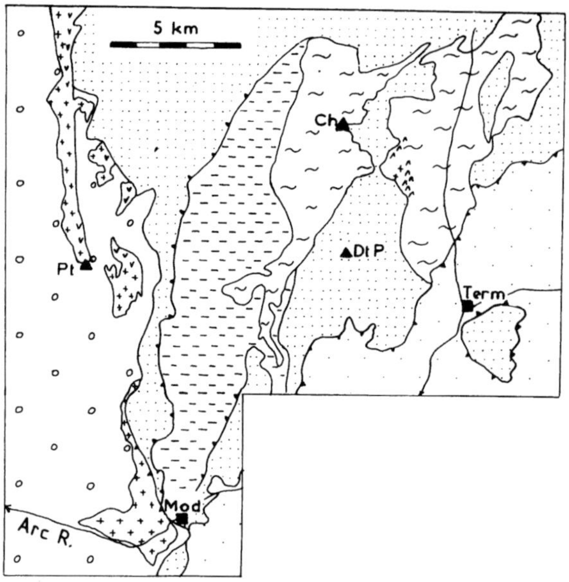

Fig. 5. Sketch map of the Chasseforêt, or Southern Vanoise, Massif (after Landès 1988; Debelmas et al. 1989a). *Patterns:* see legend to Fig. 2. *Ch* Chasseforêt; *Dt P* Dent Parrachée; *Mod* Modane; *Pt* Péclet; *Term* Termignon

~ 1000 m and that of the upper sheet 100–150 m (from Platt and Lister 1985). Both sheets contain pre-Alpine mineral relics and there is, thus, no possibility that the upper one represents a younger basement of the Pourri-Bellecôte type.

The pre-Alpine parageneses have been largely obliterated by the Alpine processes. Traces of pre-Alpine structures have been preserved only as helicitic structure in garnet and as muscovite microlithons. Most present rock types are garnet-glaucophane metabasites, prasinites-ovardites and albite-phengite-chlorite-green biotite schists.

7 Ambin Massif

The Ambin Massif (Fig. 6) is exposed in a window surrounded by Combin rocks and, to the northwest, by evaporites of the "gypsum nappe" which is believed to have been the glide sole of the Combin nappe. Argand (1911) and Hermann (1938) first connected it with the Dora-Maira internal crystalline massif with which it shares many lithological similarities (see the chap. by Sandrone et al. this Vol.). Following Staub (1942) it is considered to belong to the Briançon zone owing to the characteristics of the Mesozoic cover, to its position under, or backfolded upon, Combin, not Zermatt rocks and to its being the northern prolongation of the Pelvo d'Elva subunit in the Acceglio zone.

It is made up of (Fig. 3) (Lorenzoni 1965; Gay 1970): (1) an old basement unit, the Clarea formation; (2) a younger basement unit, the Ambin formation, first considered to be the Permian-Carboniferous cover of the Clarea formation, but also suggested to be older than Upper Carboniferous (Desmons and Fabre 1988); (3) an Upper Permian cover unit, the Etache formation; and (4) a Mesozoic to lower Cainozoic cover distribution in allochthonous slices and consisting of carbonate rocks of a type similar to that of the inner Chasseforêt Massif.

7.1 The Clarea Formation

About 1400 m in apparent thickness, the Clarea basement bears lithological resemblance to Chasseforêt rocks. It comprises metasedimentary rocks (Na-rich greywacke and pelite; rare amphibole-epidote marble) and some, mostly banded, metabasites which have the composition of low Ti-tholeiites, some of them with a strong spilitic trend. No granitoid intrusion has been found.

Hornblende, muscovite, garnet and rare brown biotite are relics of amphibolite facies pre-Alpine metamorphic phase(s). Probable relics of a pre-Alpine deformational phase are represented at the mesoscopic scale by interfolial hinges (Callegari et al. 1980). The rocks are now quartz-albite-mica schists, gneisses and prasinites-ovardites, having been affected by the Alpine events.

Chemical data concerning pre-Alpine minerals have been given by Gay (1972a, wet analyses) on muscovite ($Si_{3.12}$), Fe-Mg-biotite, Mg-hornblende to tschermakite and garnet (~ alm_{57} prp_{14} sps_4 grs_{17} $andr_7$).

Rb-Sr and K-Ar radiometric measurements of biotite and muscovite separates [Bocquet (Desmons) et al. 1974; partly re-done by Monié 1990, by the $^{39}Ar/^{40}Ar$ method and with similar results] point to a Variscan (Early Carboniferous) closure age, a Permian partial resetting and Ar-inheritance in the biotite. In view of the metamorphic evolution revealed by Ruitor and west-Ligurian rocks, the Variscan age, if it is not only an apparent age, is likely to correspond to a re-opening under higher heat flow conditions.

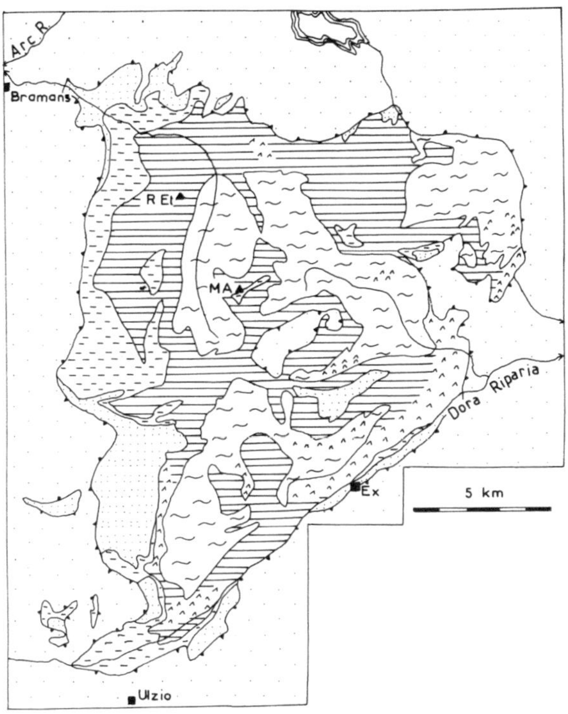

Fig. 6. Sketch map of the Ambin Massif (after Gay 1972b). *Patterns:* see legend to Fig. 2. *Ex* Exilles; *MA* Mont d'Ambin; *R Et* Rocher d'Etache

7.2 The Ambin Formation

Overlying the Clarea formation with a contact variously considered as mylonitized or depositional, the Ambin formation, 150 m in average thickness, contains: (1) clastic sediments which are richer in K than the Clarea sediments (arkose, conglomerate, shale, locally coal-bearing); (2) leucocratic quartz-albite gneisses (the ill-called "leptynites"), variously interpreted as either rhyolites, pyroclastics or granophyres (Callegari et al. 1980), which have an alkaline composition, modified by a loss of alkaline elements; (3) metabasites of basaltic to andesitic composition, which are associated with the leucocratic gneisses and represent metagabbro and metavolcanic or pyroclastic rocks (Pognante et al. 1983). The affinity of the metabasites is tholeiitic with a medium Ti-content, but a calc-alkaline trend is also possibly present.

No trace of any regional amphibolite facies metamorphism has been found. This is the reason why the protoliths in the Ambin formation have once been ascribed to Permo-Carboniferous time. More recently, Desmons and Fabre (1988) suggested an age similar to the Pourri-Bellecôte Massif. No radiometric measurements have yet been published on rocks of the Ambin formation.

A strong Alpine overprint generated jadeite + quartz, glaucophane, phengite, garnet and, later, chlorite, albite and greenish biotite.

8 Acceglio Zone

The Acceglio zone, defined by Debelmas and Lemoine (1957) on the basis of its lacunar Mesozoic sequence (a characteristic that is also found in the Chasseforêt Massif), is exposed on the inner side of the Briançon zone and in a half window within Combin calc-schists (Fig. 1). It consists of several narrow, highly tectonized, sheets of varying stratigraphic sequence. Two of these sheets (Pelvo d'Elva and Sagnères) contain, at the stratigraphic bottom, small bodies of biotite-garnet gneisses (Michard 1967; Lefèvre and Michard 1976; Lefèvre 1982) showing a relict pre-Alpine foliation. These gneisses are thus to be interpreted as remnants of an ancient, polymetamorphic basement, as suggested by Bocquet (Desmons) (1974a).

In the Pelvo d'Elva thrust sheet monometamorphic interlayered metabasites and leucocratic gneisses (lawsonite gneisses of Lefèvre and Michard 1976) are found between the pre-Alpine gneisses and acid volcanic rocks assumed to be (Upper?) Permian. These banded rocks had been interpreted as being Permian in age. Recently, Desmons and Fabre (1988) claimed that they may be older than Upper Palaeozoic, thus similar in age to the Northern Vanoise and Ambin formation. Radiometric measurements on white mica (Monié 1990) support such a pre-Carboniferous age.

The composition of the metabasites is tholeiitic with medium Ti-content, modified by feldspar breakdown and muscovite-chlorite formation. The leucocratic gneisses have a quartz-keratophyre composition, also modified by feldspar breakdown. The compositions compare well with those of the Ambin formation.

South of the Val Maira near Pradleves and Cavaliggi, tiny outcrops of crystalline and Triassic rocks may represent the connection with the west Ligurian exposures.

9 "Ecailles Intermédiaires" (Intermediate Slices) near Briançon

Near the town of Briançon the "intermediate slices" or "4th slice" (Fig. 1) are located either, like the Acceglio zone, on the inner side of the Briançon zone, or upon the Permian to Mesozoic rocks of the external Briançon zone. These slices have been interpreted either as Briançon basement slices or (Staub 1942) as Dent Blanche klippen.

The best outcrop, about 1.5 km^2, lies W of Briançon, at Serre-Chevalier-Eychauda. It shows, above an Acceglio-type lower unit consisting of Mesozoic (-Tertiary?) and Permo-Triassic Verrucano, an upper unit comprising various rock types similar to Ruitor rocks: fine-grained mica-quartz schist, coarse-grained (>1 cm) garnet-biotite-muscovite ($\sim Si_{6.6}$) gneiss (one analysis yielded a greywacke composition) and Mg-hornblende-plagioclase ± biotite ± garnet ($\sim alm_{60}\, prp_{10}\, sps_3\, grs_{27}$) amphibolite interlayered with plagioclase-rich leucocratic gneiss. One foliation plane is connected with these pre-Alpine minerals. The Alpine overprint has been comparatively weak and includes albite, chlorite and fine-grained white mica.

10 Zone Houillère (Upper Carboniferous-Permian) [section written by D. Mercier]

The Zone Houillère (Fig. 1), defined by Gignoux and Moret (1934), includes Upper Carboniferous and Permian continental deposits and magmatic rocks. It results from an Alpine polyphase piling-up of nappes, most series are incomplete and their geometric relationships appear strongly perturbed. However, on the basis of tectonic and sedimentary evidence, the formations are grouped in two series separated by a major unconformity (Asturian phase):
1. A very thick lower series which consists of sandstones and pelites, rich in coal and dated from Namurian to lower Stephanian times (Gréber 1965);
2. An upper series constituted by varicoloured conglomerates and sandstones, and characterized by a gradually more arid climate. The presumed age range from middle Stephanian to Lower Triassic.

10.1 *The Briançonnais Coal Measures* **(Namurian to Lower Stephanian)**

The Alpine thrusting phases uncoupled the Carboniferous series from its primary basement, which remains unknown. The Ruitor Massif (Fig. 2) can no longer be considered as this missing basement (as assumed by Fabre 1961; Fabre et al. 1987), since it actually lies upon the Zone Houillère and has a Mesozoic autochthonous cover devoid of Carboniferous beds (Mercier and Baudin 1990). Similar geometric relationships are found southwards, between the Zone Houillère, the Sapey Zone or the "4th slice" (Fig. 1).

From the Namurian up to the lower Stephanian inclusively (Feys 1963), a very active (>2500 m in thickness) clastic deposition occurred in the future Zone Houillère. The clasts (mainly polycrystalline quartz, also paraschists and -gneisses displaying a polyphase pre-Alpine metamorphic imprint) possibly derived from Briançon basement types. The distinctions made between the southern and the northern parts (Mercier and Beaudoin 1987) are illustrated in Fig. 7.

In the whole Zone Houillère, but especially near Briançon, the Upper Carboniferous series has been intruded by subvolcanic, porphyric, rocks, mainly sills but also dykes which show cross-cutting relationships. These magmatic rocks range between diorite and granodiorite, rarely granite and belong to the calc-alkaline lineage (Piantone 1980). The intrusion time possibly spreads from Westphalian to Early Permian.

10.2 *The Upper Conglomeratic Formations* **(Middle Stephanian to Lower Triassic?)**

The highly-coloured continental clastic or volcaniclastic sedimentary rocks of the upper series strongly contrast with the underlying monotonous Upper Carboniferous series. A major regional unconformity individualizes a lower complex and an upper complex which is known from the entire Zone Houillère (Fabre and Feys 1966; Fabre et al. 1987; Schade 1989).
1. The *lower complex* (middle Stephanian – Permian?) is exposed along the present-day borders of the Zone Houillère. Important variations are noted between both internal and external borders (Fig. 7):
a. On the internal border, the lower complex is represented by conglomeratic formations bearing crystalline basement clasts, which lie unconformably on the Upper Carboniferous coal measures. The lower part is dated from the middle Stephanian (Fabre 1961).
b. On the external border of the Zone Houillère, the lower complex is mainly volcaniclastic, containing conglomerate, tuffaceous sandstone and pelite, rhyolitic ignimbrite and andesite-dacite. The space and time relationships between these formations are poorly known: the use of chrono-stratigraphic names is, thus, not recommended.
2. The thin (≤50 m) upper complex (Permian–Lower Triassic) is represented: (1) near Briançon, by the Verrucano conglomerates and by variegated pelites, and

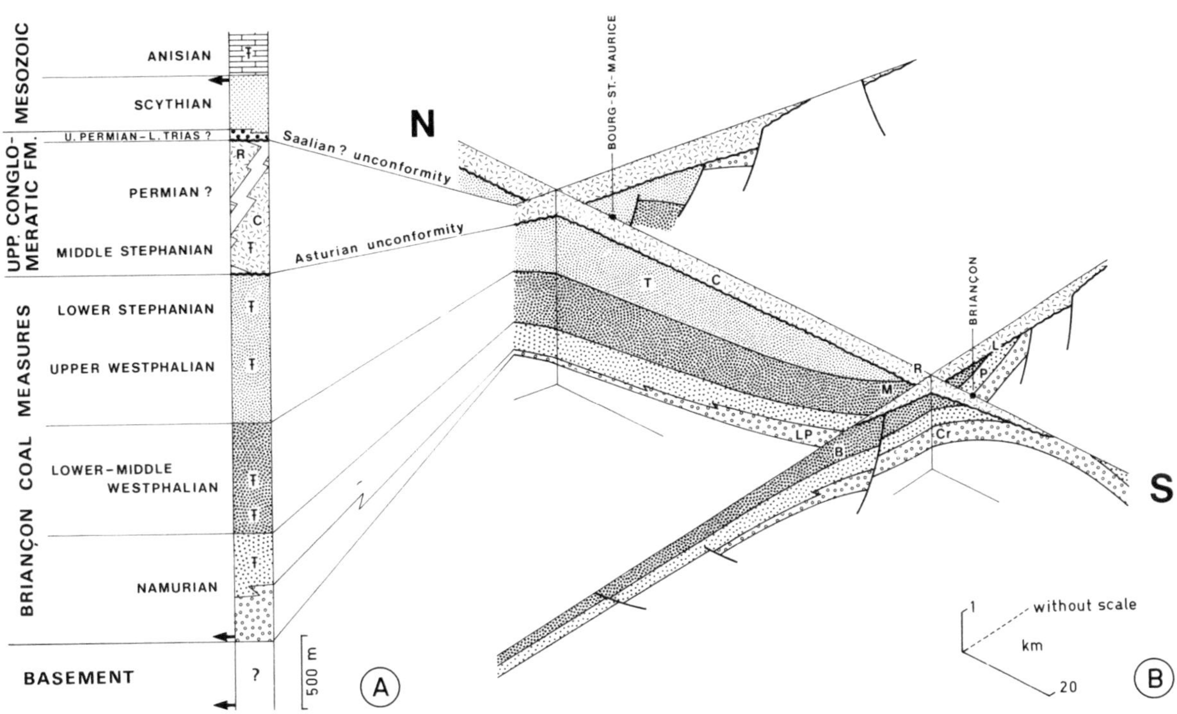

Fig. 7. The Zone Houillère. *A* schematic lithostratigraphic column (*arrows* indicate the major Alpine overthrusts); *B* hypothetical diagram showing the relationships between the formations. *B:* Benoîte formation: Sandstone, pelite and coal seams; *C* Courchevel formation: coarse conglomerate, arkose; *Cr* Cristol formation: conglomerate and coarse sandstone; *L* Rocher du Loup formation: conglomerate, arkose and pelite; *LP* La Praz formation: coarse sandstone; *M* Madeleine formation: sandstone, pelite and coal seams; *P* Pasquier formation: coal-bearing sandstone; *R* Rochachille formation: volcani-clastic rocks; *T* Tarentaise formation: coal-bearing pelite and sandstone

(2) in the Arc and Isère valleys, by ankeritic sandstones and pearly sericite schists. These facies in all places grade into the white, Lower Triassic (?), quartzites.

10.3 Palaeogeographic Outline and Evolution of the Basin

Between the Namurian and the lower Stephanian a south-to-north transition has been detected, from a braided-fluviatile to a flood-plain environment, with sporadic lacustrine episodes. This bears evidence of a N-S orientated, northward flowing, hydrographic network, which is also supported by the SSW → NNE current directions measured in various places. Correlatively, coal was more abundant northward (Mercier and Beaudoin 1984). In addition the coarse clastic deposits, which at first prevailed along the axis of the Upper Carboniferous basin, were overlain by gradually finer-grained facies, indicating a broadening of the basin and a southward migration of the alluvial-plain environment, perhaps generated by a tilting process initiated during the upper Westphalian.

From the middle Stephanian onwards and as a result of the Asturian tectonic events, coarse fluvial deposits again invaded almost the whole basin, unconformably overlaying the lower deposits. Varicoloured beds and a few acid volcanic rocks heralded the climate and morphology that were to prevail during the Permian period. At the time, the Zone Houillère was broken up in response to the Saalian tectonic processes. Mountainous areas were eroded under oxidizing conditions, while subsiding sectors were filled with calc-alkaline and, later, K-rich sub-alkaline, volcaniclastic material.

By the end of the Permian the Zone Houillère, together with the more internal basement units, formed a wide subsiding platform which was to be flooded during the Lower Triassic by the Tethys Sea.

11 Permian *s. l.* of the Internal Briançon Zone

The lack of any Upper Carboniferous cover is a common feature of all the above-described basement exposures. In contrast, layers of assumed Permian age are found on top of the younger basement massifs or their possible continuation at depth, and perhaps, locally, on top of ancient basement types. The contact of those Permian layers on the basement is, at least in places, described as depositional (in Northern Vanoise: Guillot and Raoult 1984; in the Ambin Massif: Gay 1970).

During the Alpine orogeny the Permian rocks have been transformed into quartz-mica schists, meta-conglomerates and -magmatic rocks. The metasedimentary rocks are subdivided into: (1) Upper Permian-Lower Triassic rocks (called the Permo-Triassic, i. e. Permo-Scythian), with abundant quartz-rich schists, mica schists and metaconglomerates of a Verrucano facies; and (2) Lower Permian-upper Stephanian (called the Permo-Carboniferous or Stephano-Permian) with quartz-rich and light- to varicoloured pelitic sedimentary and also volcaniclastic rocks, deposited subaerially in a continental environment. Tourmaline is a common accessory mineral which can also be found in veins.

These rocks are similar to the Stephano-Permian and Permian rocks described in the Zone Houillère and point to a palaeogeographic environment common to both the Zone Houillère and basement units, mainly of the younger type.

In addition to the mentioned volcaniclastic rocks, intrusives considered as Permian cross-cut various basement rocks. One can mention the so far undated Cogne diorite in the Zona interna and dykes of intermediate composition intruding the Ruitor basement.

12 Discussion and Conclusions

12.1 *Summarized Lithostratigraphy*

The lithologic and chemical characteristics of the *ancient basement* give a coherent picture. The granitoids are sub-alkaline and of a probably anatectic origin. Tholeiitic basic rocks have high to medium Ti-contents and MORB affinity, banded amphibolites of unclear origin are abundant and ultramafic cumulates, associated with alkaline metabasites, are extremely rare. The sedimentary rocks, mostly shales and greywackes, were probably deposited in a marine environment.

In the *younger basement,* metabasites form both gabbro-basalt bodies, which are low-Ti tholeiitic in composition and associated with some acid rocks, and high- to medium-Ti tholeiitic volcanics which are interlayered with sedimentary rocks. The composition of the latter, which have been deposited in a marine environment, is mostly pelitic. Some granophyric intrusions in the younger basement are alkaline in composition. There is no evidence of an ocean-type floor for the basin, as the metabasites do not form an ophiolite sequence. An attenuated sialic crust is thus suggested.

The *"Permian s. l.,"* assumed to be Stephano-Permian to Upper Permian-Lower Triassic in age, consists of volcaniclastic and clastic rocks which overlie the Upper Carboniferous of the Zone Houillère and especially the younger basement.

In the *Zone Houillère* post-Sudetic, Namurian to lower Stephanian, clastic rocks were deposited in an intracontinental basin, the drainage of which was directed S-N (in the present-day coordinates). Volcanic rocks of an as yet undetermined age, which are sub-alkaline and K-rich in the west-Ligurian Alps (Cortesogno et al. 1984), have a calc-alkaline trend in the French Western Alps.

12.2 *Classification of the Pre-Alpine Units*

The above-described basement units can be classified on the basis of their litho-biostratigraphic content and pre-Alpine metamorphic evolution. In addition to the Upper Carboniferous rocks of the Zone Houillère and to the Permian *(s. l.)* rocks overlying Carboniferous and pre-Carboniferous sequences, two groups are distinguished (cf. Desmons and Fabre 1988; Desmons and Ploquin 1989): (1) younger basement units, bearing no recognizable pre-Alpine metamorphic imprint; and (2) ancient basement units, showing relics of pre-Alpine amphibolite-facies metamorphism.

At least three successive pre-Alpine metamorphic phases have been identified (Fig. 8): an early eclogite, a prevalent middle-pressure amphibolite, a late low-pressure amphibolite and/or biotite-bearing greenschist facies. Granulite facies relics have not been found so far. The ancient basement may have constituted a cratonic deep crust (cratonic eclogite after Griffin and O'Reilly 1987), later uplifted and hydrated to amphibolite (an almost isothermal, or an adiabatic decompression?), and later to greenschist facies level (an adiabatic P and T decrease and a later T increase?). Desmons and Hunziker (1988) suggested that the lower crust underlying this ancient basement may have been similar to the Ivrea zone granulites. This, however, would apply to times younger than the eclogite and the kyanite-bearing amphibolite facies, i. e. for the biotite$_{II}$-bearing phase. It would also imply

Table 1. Classification of the Briançon basement units and proposed correlations with the Swiss classification

	Ancient basement		Younger basement	Upper Carboniferous
French-Italian Western Alps	Ruitor Sapey ("écailles intermédiaires"?) Acceglio p. p.	S Vanoise Clarea	N Vanoise Zona interna Ambin Formation Acceglio p. p.	Zone Houillère
Swiss Alps	Pontis	Siviez-Mischabel?	Mont Fort	Zone Houillère
Ligurian Alps	Nucetto	Calizzano-Savona?	Costa Dardella p. p.?	Ollano Formation

an evolution in the situation of this basement, from cratonic-cool (eclogitic) to tectonically active-hot (granulitic), a phenomenon that remains to be further documented. Owing to the strong Alpine deformation and transport by thrusting nothing can be said concerning the attitude of the pre-Alpine structures.

A further subdivision of the ancient basement massifs is based on the Alpine metamorphic features and structural position, likely to reflect the pre-Alpine configuration. These characters, together with the lithostratigraphic features already summarized, lead to the classification scheme and correlations with the Swiss and west Ligurian parts of the Briançon-Bernhard zone (Cortesogno 1984; Escher 1988; Thélin 1989) that are proposed in Table 1. The following remarks are added.

1. Owing to the very complex Alpine thrusting, folding and back-folding events, it is hazardous in the present state of knowledge to propose palaeopositions for the three groups of units. However, if the contacts between the younger and ancient basement are thrust contacts, it should be inferred from the structural superposition of units in the Ambin Massif and the Acceglio zone that the original palaeogeographical position of these younger basement units was more internal than the ancient basement. The Pourri-Bellecôte Massif is much thicker than all other younger basement units (Fig. 3) and thus palaeogeographically different. Even if it is found to be equivalent to the latter ones in the time of deposition and magmatism, perhaps it is not equivalent in its structural position. Its first Alpine structural relationships with the adjoining units are obscured by the later thrusting and back-folding phases.

2. In the Dora-Maira Massif it is not clear whether the upper sheet, that of ancient basement, belongs to the internal crystalline massifs or to the Briançon basement. In the last case the upper position would have been acquired through backthrusting on the lower, younger, sheets. A question remains concerning the age of the graphite-bearing schists or Pinerolo unit. It underlies the Permian and Triassic units which are similar to the Pourri-Bellecôte and Zona interna cover sequences and lies above a possible basement of its own. An Upper Carboniferous age is commonly assumed owing to the abundance of carbonaceous matter, but palaeontological evidence is lacking.

As to the tiny ancient basement remnants found at the bottom of both Pourri-Bellecôte Massif and Acceglio unit, they represent either their own basement or, as in the Ambin Massif, pieces of a basement sheet that they passed or came to rest upon.

3. Coming back to the present-day relationships of the basement units, it must be noted that, depending on the W-E profile considered, not all basement groups are present at the outcrop level. An ancient basement of Chasseforêt-Clarea-type is missing from the Ruitor-Gran Paradiso profile northward and on the Chasseforêt-Gran Paradiso profile the younger group is not represented or only by Permian-Triassic rocks along the Modane-Chavière fault (which according to Platt et al. 1989, is the trace of an oversteepened backthrust). Does this mean that the units did not have any lateral continuity?

12.3 Ages of the Protoliths and of the Pre-Alpine Metamorphic Phases

More radiometric data are needed for both magmatic protoliths and pre-Alpine metamorphic phases. In particular, the ages of most granitoids are unknown. The following can be said concerning the chronology (Fig. 8). The Zone Houillère shows that the Late Variscan (Neo-Variscan of Ledru et al. 1989, in the French Central Massif) was not accompanied by any noticeable low-grade metamorphism. The younger basement shows that also amphibolite-facies Variscan metamorphism (Meso-Variscan of Ledru et al. 1989) was missing from at least part of the future Briançon zone.

From zircon U-Pb measurement Guillot et al. (1991) have now given evidence of a Late Cambrian protolith age in the younger basement. If during Variscan times all rocks of the future Briançon zone

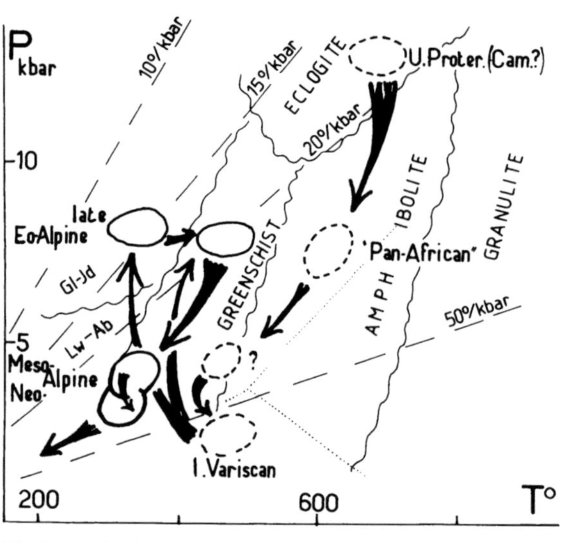

Fig. 8. Synthetic *P-T-t* diagram of the Briançon basement. Four *P-T* gradients are indicated as *dashed straight lines*: high-*P*, medium (two different gradients) and high-*T*. Pressure and temperature estimates, based on mineral assemblages, are approximate.

The pre-Alpine path of the ancient basement includes the following succession (see text): pre-Variscan eclogite, pre-Variscan kyanite-amphibolite, pre-Variscan or Variscan high-T greenschist facies, Permian high-T facies (according to isotopic resetting of minerals and the presence of andalusite in veins in the Swiss Bernhard zone). The Alpine facies include the succession: glaucophane-jadeite and/or glaucophane-epidote assemblages, various greenschist and/or lawsonite-albite-chlorite to pumpellyte-actinolite assemblages. The path of the younger basement type starts with Meso-Alpine glaucophane-epidote assemblages. The *arrows* simply connect the circled fields and do not represent paths between these fields

did belong to one domain, the *youngest possible* metamorphic age of the ancient basement thus seems to be Lower Cambrian, i.e. late Pan-African (or Sardic according to Tollmann 1986; the term Caledonian should not be applied, as the Caledonian orogenic domain did not include the future Alpine areas).

This is only a minimum age and no evidence so far opposes a longer evolution, starting with as yet undated lower Proterozoic steps. Proterozoic ages have actually been obtained for mantle separation of basic material in other Penninic units (Stille and Tatsumoto 1985; Schenk-Wenger and Stille 1989).

Such a chronology emphasizes the different evolution of both external crystalline massifs and Penninic crystalline massifs. The first were related to the European Variscides, accreted to Laurussia and submitted to intense metamorphism and deformation in Variscan times. The Penninic crystalline massifs were accreted to Eurasia one orogenesis later and show only local thermal Variscan effects.

12.4 Final Remarks

The Briançon basement compares well with Austro-Alpine and South-Alpine basement in lithology, chemistry and metamorphic evolution (Desmons 1986; Radelli and Desmons 1987). It does not compare with the Tauern crystalline complexes which, owing to their Variscan granitization and metamorphic imprint, resemble the external basement (Desmons and Radelli 1990), or rather the Ultrahelvetic/Subpenninic of the Central Alps: the Gotthard-Simano units (Milnes 1974; these Subpenninic units were called Infrapenninic in 1980 by Trümpy owing to euphony. Both terms are unfortunate, for they convey the false implication of these units being the lower part of the Penninic instead of lower than the Penninic).

Further detailed investigations remain to be done, the cover characteristics helping to distinguish the palaeogeographical origins. The few available palaeomagnetic data support this belonging of the Briançon and all Penninic units to the Gondwanan margin of the Tethys domain (e.g. Heller et al. 1989). Cortesogno et al. (1982) emphasized the similarities of granitoid rocks in the Ligurian Briançon, South- and Austro-Alpine zones, and Cabella et al. (1988) those of the Permo-Carboniferous volcanic rocks. Proterozoic (or lower Palaeozoic) eclogite followed by amphibolite and lower grade facies have been inferred for both Austro-Alpine and South-Alpine crystalline units (e.g. Frisch et al. 1984; Sassi et al. 1987; Borghi 1989). Rocks similar to the Briançon ancient basement are found especially in the Middle and Lower Austro-Alpine. The younger basement type partly recalls the Upper Austro-Alpine.

As concerns the Zone Houillère, which lies in a lower position with respect to the other Briançon units, no exact counterpart is found in the South-Alpine or Austro-Alpine: the Graz Palaeozoic and the Greywacke zone, which belong to the Upper Austro-Alpine nappes, include rocks of both upper and lower Palaeozoic. Only the Veitsch nappe in the Greywacke zone, which consists of Carboniferous rocks (Ratschbacher 1987; Frisch and Neubauer 1989), somewhat resembles the Zone Houillère. However, long-distance correlations may not be valid as concerns the Alpine dissection and overstacking of Gondwana.

It must be emphasized also that the ancient and younger basement units as well as the Zone Houillère bear evidence that the Gondwanan margin has been repeatedly submitted to stresses, leading to basin subsidence and repeated dissection.

Acknowledgements

J. F. von Raumer is thanked for having invited J. D. to participate at the "3e cycle" in Fribourg and inviting both of us to contribute to this book. H. Foster corrected the English in an earlier version of the manuscript. The careful reviews carried on by the two referees are acknowledged.

References

Argand E (1911) Les nappes de recouvrement des Alpes pennins et leurs prolongements structuraux. Beitr Geol Karte Schweiz N F 31: 1–25

Baudin T (1987) Etude géologique du massif du Ruitor (Alpes franco-italiennes): évolution structurale d'un socle briançonnais. Thesis, Grenoble University

Bearth P (1960–63) Contribution à la subdivision tectonique et stratigraphique du cristallin de la nappe du Grand-St-Bernard dans le Valais (Suisse). Mém h s Soc Géol Fr 2: 407–418

Bocquet (Desmons) J (1974a) Etudes minéralogiques et pétrologiques sur les métamorphismes d'âge alpin dans les Alpes françaises. Thesis, Grenoble University

Bocquet (Desmons) J (1974b) Le socle briançonnais de Vanoise (Savoie): arguments en faveur de son âge anté-alpin et de son polymétamorphisme. C R Acad Sci Paris 278 D: 2601–2604

Bocquet (Desmons) J (1974c) Il metamorfismo prealpino nella Vanoise (Savoia) e in altri settori dello zoccolo brianzonese. Mem Soc Geol Ital 13(1): 271–284

Bocquet (Desmons) J, Delaloye M, Hunziker JC, Krummenacher D (1974) K-Ar and Rb-Sr dating of blue amphiboles, micas and associated minerals from the Western Alps. Contrib Mineral Petrol 47: 7–26

Borghi A (1989) Evoluzione metamorfica del settore nord-est della Serie dei Laghi (Alpi meridionali – Canton Ticino). Rend Soc Geol Ital 11: 165–170

Cabella R, Cortesogno L, Dallagiovanna G, Vannucci R, Vanossi M (1988) Vulcanismo, sedimentazione e tettonica nel Brianzonese ligure esterno durante il Permo-Carbonifero. Atti Tic Sci Terra 31: 269–326

Caby R (1968) Contribution à l'étude structurale des Alpes occidentales: subdivisions stratigraphiques et structure de la zone du Grand-Saint-Bernard dans la partie sud du Val d'Aoste (Italie). Géol Alp 44: 95–111

Caby R, Kienast JR (1989) Meso-Alpine high-pressure assemblages and excavation of the Ruitor Briançonnais basement (Savoie, Val d'Aoste, Graie Alps). Terra Abstr 1: 266

Callegari E, Sacchi R, Bovo S, Torassa G (1980) Osservazioni strutturali sul versante italiano del Massiccio di Ambin (Alpi graie). Boll Soc Geol Ital 99: 395–404

Cigolini C (1981) Garnet chemistry and zonation in the Italian sector of the Grand Saint Bernard nappe. Atti Accad Sci Torino 115: 331–344

Cortesogno L (1984) Metamorfismo e magmatismo prealpini nel basamento e nel tegumento delle Alpi liguri. Mem Soc Geol Ital 28: 79–94

Cortesogno L, Gianotti R, Oxilia M, Vanossi M, Vannucci R (1982) Genesi ed evoluzione delle zoccolo premesozoico di alcuni settori del Brianzonese ligure interno. Rend Soc Ital Mineral Petrol 38: 219–260

Cortesogno L, Gianotti R, Vannucci R, Vanossi M (1984) Le volcanisme permo-carbonifère du Briançonnais ligure (Alpes maritimes) dans le cadre des phases tardives de l'orogenèse hercynienne. Sci Géol Bull 37: 37–50

Cortesogno L, Dallagiovanna G, Vannucci R, Vanossi M (1988) Volcanisme, sédimentation et tectonique pendant le Permo-Carbonifère en Briançonnais ligure: une revue. Eclogae Geol Helv 81: 487–510

Dal Piaz Gb (1928) Geologia della catena Herbetet-Grivola-Grand Nomenon. Mem Ist Geol Univ Padova 7: 1–83

Dal Piaz GV (1965) Il lembo di ricoprimento della Becca di Toss: struttura retroflessa della zona del Gran San Bernardo. Mem Accad Patavina 77: 107–136

Dal Piaz GV, Govi M (1965) Osservazioni geologiche sulla "Zona del Gran San Bernardo" nell'alta Valle d'Aosta. Boll Soc Geol Ital 84: 1–23

Debelmas J, Lemoine M (1957) Calcschistes piémontais et terrains à faciès briançonnais dans les hautes vallées de la Maira et de la Varaita. C R Somm Soc Géol Fr: 38–40

Debelmas J (1980) Carte geologique de France (1/250000), feuille Annecy. Bur Rech Géol Min, Orléans

Debelmas J, in collaboration with Desmons J, Ellenberger F, Goffé B, Fabre J, Jaillard E, Pachoud A (1989a) Notice explicative, Carte géologique de France (1/50000), feuille Modane. Bur Rech Géol Min, Orléans

Debelmas J in collaboration with Antoine P, Barféty JC, Dabrowsky H et al. (1989b) Carte géologique de France (1/50000), feuille Moûtiers et notice explicative. Bur Rech Géol Min, Orléans

Debelmas J, Caby R, Desmons J (1991) Notice explicative, Carte géologique de France (1/50000), feuille Sainte-Foy-Tarentaise. Bur Rech Géol Min, Orléans 43 pp

Desmons J (1977) Mineralogical and petrological investigations of Alpine metamorphisms in the internal French Western Alps. Am J Sci 277: 1045–1066

Desmons J (1986) The Alpine metamorphisms and their environments in the Western Alps: unsolved problems. Schweiz Mineral Petrogr Mitt 66: 29–40

Desmons J (1989) Different metamorphic evolutions in the Alpine-Apennine ophiolites (France-Italy-Switzerland-Austria). Chem Geol 77: 229–250

Desmons J (1992) The Briançon basement (Pennine Western Alps): mineral composition and polymetamorphic evolution. Schweiz mineral petrogr Mitt 72: 37–55

Desmons J, Fabre J (1988) Contribution à la connaissance pétrographique du Mont Pourri (Savoie, France); conséquences structurales. Géol Alp 64: 13–26

Desmons J, Hunziker JC (1988) Genesis of layered amphibolites: the case of the Penninic zones of the Western Alps. Bol Dep Geol Uni-son 5 (1–2): 53–61

Desmons J, Ploquin A (1989) Chimisme du Briançonnais – Grand Saint-Bernard anté-mésozoïque (Alpes occidentales): une vue synoptique. Géol Alp 65: 1–31

Desmons J, Radelli L (1990) Essai de corrélation tectonique des Alpes W aux Alpes E. Réun Sci Terre Grenoble, Soc Géol Fr: 41 (Abstr)

Desmons J, Laduron D, de Béthune P (1977) Grenats zonés de la nappe du Grand Saint-Bernard et de la nappe piémontaise (Alpes occidentales). Mém Inst Géol Univ Louv, Livre Jub Félix Kaisin Jr 29: 327–347

Desmons J, Hunziker JC, Delaloye M (1982) Unconvincing evidence against the blocking temperature concept. Comments on "^{40}Ar-^{39}Ar dating of high pressure metamorphic micas from the Gran Paradiso area (Western Alps): evidence against the blocking temperature concept" by C. Chopin and H. Maluski. Contrib Mineral Petrol 80: 386–390

Détraz G (1984) Etude géologique du bord interne de la Zone houillère briançonnaise entre la vallée de l'Arc et le massif de Péclet-Polset (Alpes de Savoie). Thèse 3e cycle, Grenoble University

Détraz G, Loubat H (1984) Faciès à disthène, staurotide et grenat dans un micaschiste appartenant à l'unité des "gneiss du Sapey" (Vanoise, Alpes françaises). Géol Alp 60: 5–12

ECORS-CROP Program (1987) Premiers résultats de la traversée des Alpes occidentales par sismique réflexion verticale. C R Acad Sci Paris II 305: 1461–1470

Ellenberger F (1958) Etude géologique du pays de Vanoise. Mém explic Carte géol Fr, 561 pp

Ellenberger F (1966) Le Permien du pays de Vanoise. Atti Symp Verrucano, Pisa, Soc Tosc Sci Nat, pp 170–211

Escher A (1988) Structure de la nappe du Grand Saint-Bernard entre le val de Bagnes et les Mischabel. Geol Ber 7, Geol Ber Landeshydrol-geol, Bern, 26 pp

Escher A, Masson H, Steck A (1988) Coupes géologiques des Alpes occidentales suisses. Mém Géol Lausanne 2, 12 pp

Eteradossi O (1983) Les roches vertes orthodérivées du massif du Ruitor (Alpes franco-italiennes): pétrographie et géochimie. Dipl Et Approf, Grenoble University

Fabre J (1961) Contribution à l'étude de la Zone Houillère en Maurienne et en Tarentaise (Alpes de Savoie). Mém Bur Rech Géol Min 2, 315 pp

Fabre J, Feys R (1966) Les séries bariolées du massif de Rochachille. Leurs rapports avec le "Verrucano" de Briançon et les "Permiens" de Maurienne et de Tarentaise. Atti Symp Verrucano, Pisa, Soc Tosc Sci Nat, pp 143–169

Fabre J, Schade J, Baudin T, Desmons J, Mercier D, Perruccio-Parison D (1987) Relics of pre-Mesozoic events in the Briançon zone (northern French Alps). In: Flügel HW, Sassi FP, Grecula P (eds) Pre-Variscan and Variscan events in the Alpine-Mediterranean mountain belts. Mineralia slovaca Monogr. Alfa, Bratislava, pp 183–208

Feys R (1963) Etude géologique du Carbonifère briançonnais (Hautes-Alpes). Mém Bur Rech Géol Min 6, 387 pp

Frisch W, Neubauer F (1989) Pre-Alpine terranes and tectonic zoning in the eastern Alps. Geol Soc Am Spec Pap 230: 91–100

Frisch W, Neubauer F, Satir M (1984) Concepts of the evolution of the Austroalpine basement complex (Eastern Alps) during the Caledonian-Variscan cycle. Geol Rundsch 73: 47–68

Gay M (1970) Le massif d'Ambin et son cadre de schistes lustrés (Alpes franco-italiennes). Evolution paléogéographique anté-alpine. Bull Bur Rech Géol Min (2) 3: 5–81

Gay M (1972 a) Le massif d'Ambin et son cadre de schistes lustrés (Alpes franco-italiennes). Evolution métamorphique. Arch Sci Genève 25: 5–100

Gay M (1972 b) Le massif d'Ambin et son cadre de schistes lustrés (Alpes franco-italiennes). Evolution structurale. Arch Sci Genève 25: 165–214

Gignoux M (1929) Les problèmes géologiques de la région Vanoise-Mont Pourri (Savoie). Trav Lab Géol Grenoble 15: 98–142

Gignoux M, Moret L (1934) Les grandes subdivisions géologiques des Alpes françaises. Ann Géogr 244: 337–363

Gréber C (1965) Flore et stratigraphie du Carbonifère des Alpes françaises. Mém Bur Rech Géol Min 21, 380 pp

Griffin WL, O'Reilly SY (1987) Is the continental Moho the crust-mantle boundary? Geology 15: 241–244

Guillot F (1987) Géologie de l'Antépermien de Vanoise septentrionale (Zone briançonnaise interne, Alpes occidentales, Savoie, France). Thesis, Lille University

Guillot F, Raoult JF (1984) Permien et base du Trias en Vanoise septentrionale: données nouvelles et hypothèses (zone briançonnaise interne, Alpes françaises). Ann Soc Géol Nord 104: 183–192

Guillot F, Liégeois JP, Fabre J (1991) Des granophyres du Cambrien terminal dans le Mont Pourri (Vanoise, zone briançonnaise): première datation U-Pb sur zircon d'un socle des zones internes des Alpes françaises. C R Acad Sci Paris 313, II: 239–244

Heller F, Lowrie W, Hirt AM (1989) A review of palaeomagnetic and magnetic anisotropy results from the Alps. In: Coward MP, Dietrich D, Park RG (eds) Alpine tectonics. Geol Soc Lond Spec Publ 95: 399–420

Hermann F (1938) Note illustrative per la Carta geologica delle Alpi nord-occidentali. Allegretti Di Campi, Milano

Laduron D, Desmons J (1981) Résorption et néocroissances dans les grenats d'un micaschiste du Ruitor (Alpes penniques franco-italiennes). Mém Ist Géol Univ Louv 31: 335–347

Landès B (1988) Les unités briançonnaises de la bordure orientale du massif de la Vanoise méridionale (Savoie). Thesis, Grenoble University

Ledru P, Lardeaux JM, Santallier D, Autran A, Quenardel JM, Floc'h JP, Lerouge G, Maillet N, Marchand J, Ploquin A (1989) Où sont les nappes dans le Massif Central français? Bull Soc Géol Fr (8) 5: 605–618

Lefèvre R (1982) Les nappes briançonnaises internes et ultra-briançonnaises dans les Alpes cottiennes méridionales. Thesis, Paris-Sud-Orsay University

Lefèvre R, Michard A (1976) Les nappes briançonnaises internes et ultra-briançonnaises de la bande d'Acceglio (Alpes franco-italiennes). Une étude structurale et pétrographique dans le faciès des schistes bleus à jadéite. Sci Géol Bull 29: 183–222

Lorenzoni S (1965) Studio geo-petrografico del versante italiano del Massicio d'Ambin. Mem Ist Geol Mineral Univ Padova 25, 88 pp

Marion R (1984) Contribution à l'étude géologique de la Vanoise, Alpes occidentales. Le massif de la Grande Sassière et la région de Tignes-Val d'Isère. Trav Dép Sci Terre Chambéry 2, 172 pp

Mercier D, Baudin T (1990) Chevauchement alpin précoce du cristallin du Ruitor sur la zone houillère briançonnaise (Savoie, Val d'Aoste). 13e Réun Sci Terre Grenoble, Soc Géol Fr, p 90 (Abstr)

Mercier D, Beaudoin B (1984) Le Briançonnais au Carbonifère: modalités de la subsidence d'une gouttière méridienne. C R Acad Sci Paris 298: 125–128

Mercier D, Beaudoin B (1987) Révision du Carbonifère briançonnais: stratigraphie et évolution du bassin. Géol Alp Mém hs 13: 25–31

Messiga B (1981) Evidenze strutturali e paragenetiche dell'evoluzione polifasica prealpina del Massiccio cristallino di Savona. Rend Soc Ital Mineral Petrol 37: 739–754

Michard A (1967) Etudes géologiques dans les zones internes des Alpes cottiennes. Cent Natl Rech Sci, Paris

Milnes AG (1974) Structure of the Penninic zone (Central Alps): a new working hypothesis. Geol Soc Am Bull 85: 1727–1732

Monié P (1990) Preservation of Hercynian $^{40}Ar/^{39}Ar$ ages through high-pressure low-temperature Alpine metamorphism in the Western Alps. Eur J Mineral 2: 343–361

Peruccio-Parison MD (1984) Problèmes pétrochimiques et structuraux en Vanoise septentrionale (Savoie). Thèse 3e cycle, Paris-Sud-Orsay University

Piantone P (1980) Magmatisme et métamorphisme des roches intrusives calco-alcalines du Carbonifère briançonnais entre Arc et Durance: minéralogie, pétrographie, géochimie. Thèse 3e cycle, Grenoble University

Platt JP, Lister GS (1985) Structural history of high-pressure metamorphic rocks in the southern Vanoise massif, French Alps, and their relation to Alpine tectonic events. J Struct Geol 7: 19–35

Platt JP, Lister GS, Cunningham P, Weston P, Peel F, Baudin T, Dondey H (1989) Thrusting and backthrusting in the Briançonnais domain of the western Alps. In: Coward MP, Dietrich D, Park RG (eds) Alpine tectonics. Geol Soc Lond Spec Publ 45: 135–152

Pognante U, Castelli D, Bogliotti C, Callegari E (1983) Caratteri petrografici e petrochimici di alcuni gabbri ed ortogneiss aplitici tardo-paleozoici del Massiccio d'Ambin, zona Brianzonese interna (Alpi occidentali). Rend Soc Ital Mineral Petrol 39: 275–280

Radelli L, Desmons J (1987) Pennique, Téthys et orogenèse crétacé moyen dans les Alpes. C R Acad Sci Paris II 305: 1375–1378

Radelli L, Desmons J (1988) Cinématique des Alpes. Rapp Dép Soc Auteurs

Ratschbacher L (1987) Stratigraphy, tectonics, and paleogeography of the Veitsch nappe (Graywacke zone, Eastern Alps, Austria): a rearrangement. In: Flügel HW, Sassi FP, Grecula P (eds) Pre-Variscan and Variscan events in the Alpine-Mediterranean mountain belts. Mineralia slovaca Monogr Alfa, Bratislava, pp 407–414

Sassi FP, Zanferrari G, Zirpoli G (1987) The Caledonian event in the Eastern Alps: a review: In: Flügel HW, Sassi FP, Grecula P (eds) Pre-Variscan and Variscan events in the Alpine-Mediterranean mountain belts. Mineralia slovaca Monogr Alfa, Bratislava, pp 431–434

Schade J (1989) Le Permien des Alpes françaises. In: Chateauneuf JJ, Farjanel G (coords) Synthèse géologique des bassins permiens français. Mém Bur Rech Géol Min 128: 122–138

Schenk-Wenger K, Stille P (1989) Geochemical and isotope evidence for an extensive Proterozoic ophiolite-suite in the Central Alps (Switzerland). Ber Dtsch Mineral Ges 1: 160 (Abstr)

Staub E (1942) Gedanken zum Bau der Westalpen zwischen Bernina und Mittelmeer. Mitt Geol Inst Eidg Techn Hochsch Univ Zürich B 7

Stille P (1980) On the genesis of the amphibolites and hornblendefelses in the Berisal complex (Simplon; Italy, Switzerland). Mem Sci Geol Padova 34: 205–246

Stille P, Tatsumoto M (1985) Precambrian tholeiitic-dacitic rock suites and Cambrian ultramafic rocks in the Pennine nappe system of the Alps: evidence from Sm-Nd isotopes and rare earth elements. Contrib Mineral Petrol 89: 184–192

Termier P (1891) Etude sur la constitution géologique du Massif de la Vanoise (Alpes de Savoie). Bull Serv Carte Géol Fr II 20: 367–513

Termier P (1927) Que la série cristallophyllienne de la Vanoise et du Mont-Pourri (Alpes de Savoie) est bien d'âge permien ou carbonifère. C R Acad Sci Paris 184: 641–645

Thélin P (1989) Essai de chronologie magmatico-métamorphique dans le socle de la nappe du Grand Saint-Bernard: quelques points de repère. Schweiz Mineral Petrogr Mitt 69: 193–204

Thélin P, Ayrton S (1983) Cadre évolutif des événements magmatico-métamorphiques du socle anté-triasique dans le domaine pennique (Valais). Schweiz Mineral Petrogr Mitt 63: 393–420

Tollmann A (1986) Geologie von Österreich, Bd 3. Deuticke, Wien

Trümpy R (1980) Geology of Switzerland, a guide-book, Part A. Schweiz Geol Komm, Wepf, Basel

Zingg MA (1989) Die Siviez-Mischabel-Decke: Entstehung und Entwicklung eines Altkristallins. Schweiz Mineral Petrogr Mitt 69: 130 (Abstr)

The Pre-Alpine Basement of the Briançonnais (Wallis, Switzerland)

P. Thélin, M. Sartori, M. Burri, Y. Gouffon, and R. Chessex

Abstract

The basement lithostratigraphy within the Middle Penninic domain in Wallis, Switzerland (formerly the Grand St. Bernard nappe), is described in order to try to produce reconstruction of the pre-Alpine history. The combined effect of pervasive Alpine deformation and metamorphism (mainly Tertiary greenschist facies overprint) as well as the lack of biostratigraphic and radiometric data make this attempt difficult.

Polymetamorphic basements, affected successively by pre-Alpine metamorphism(s) and by Alpine metamorphism, are distinguished from monometamorphic basements, that apparently were only subject to this latter disturbance. The Pontis and Siviez-Mischabel nappes comprise both poly- and mono-metamorphic units while the rocks of the "Zone Houillère" and the Mont Fort nappe do not seem to have suffered Pre-Alpine metamorphism.

Upper Proterozoic to lower Palaeozoic(?) siliciclastic meta-sediments of the polymetamorphic basement form a thick coarse-grained series overlaid by a dominantly pelitic series. Carbonate rocks (skarns, marbles) do not exist. A wide variety of magmatites (calc-alkaline granitoids, pyroxenites, felsic and mafic volcanites) are presumed to have been generated in different geodynamic environments from upper Proterozoic to lower Palaeozoic time.

Two metamorphic preserved windows allow a partial reconstitution of the pre-Alpine metamorphic history: (1) outcrops of eclogites and retroeclogites (Siviez-Mischabel nappe), usually associated with banded amphibolites, show HP mineral assemblages (pyrope-rich almandine and omphacite) which are overprinted by an amphibolite-facies event (syn-Variscan?); (2) the most internal part of the Ruitor zone (Pontis nappe) is essentially composed of garnet-staurolite micaschists and aluminum silicate-bearing schists (andalusite-sillimanite). Textures and mineral chemistry support a possible evolution in staurolite-almandine subfacies and P-T estimates lead to a polyphase metamorphic trajectory, retrograde in pressure (10→2.5 kb) and thermally prograde (470→580 °C). Correlation with Variscan climax is likely, even if radiometric data is still lacking.

Post-metamorphic late Carboniferous to Early Triassic fluviatil deposits form a thick sequence within the "Zone Houillère" and other parallel grabens(?). The most southern pull-apart(?) basin that is filled up with a thick volcano-sedimentary series (Carboniferous(?) Early Triassic) is largely incorporated into the Mont Fort nappe. Permian gabbros and subalkaline granitoids form sills and laccoliths intruding both poly- and mono-metamorphic basements.

The major uncertainties still concern: (1) the precise age of all rocks and series, except those belonging to the "Zone Houillère"; (2) the initial nature of the polymetamorphic magmatites and the geodynamic context during their ascent; (3) the age and the geodynamic meaning of the different pre-Alpine metamorphic events (HP, HT), especially the age of Variscan(?) paroxysm (amphibolite-facies).

1 Introduction

Abundant basement rocks occur within the Middle Penninic domain (Grand St. Bernard nappe in the widest sense), but have rarely been considered in global syntheses on the Variscan. This is obviously not an easy task, due to the combined effect of quasi-ubiquitous Alpine deformation and metamorphism. Mapping, petrographic and structural analysis have, however, revealed a number of important points concerning the pre-Alpine history of these basement units.

The area considered here stretches from the Simplon pass in the east to the southern part of the Aosta Valley, parallel to the Alpine chain (Fig. 1 A, p. 301). Accounts of geological investigation in this region are to be found in Müller (1982) and in Thélin (1989).

"Windows" in which the preserved pre-Alpine basement may be studied are rare. Biostratigraphic data only exist for the "Zone Houillère", and few radiometric analyses are available. On the other hand, Alpine metamorphism and deformation are sufficiently well known to be summarized. The main features of the Alpine history are, in short:

1. The pre-Alpine basement appears in four nappes formerly assembled under the denomination of the Grand St. Bernard nappe. These are from NW to SE: the "Zone Houillère", the Pontis nappe, the Siviez-Mischabel nappe and the Mont Fort nappe (Escher 1988; Fig. 2). The extension and geometry of the basement which forms the main part of these nappes is given in Fig. 1 A, B (p. 301 and p. 302).
2. Analysis of the Alpine covers, both autochthonous, and allochthonous, reveals that the four basement fragments are related and were all situated in shallow continental crust (less than 5–6 km depth) during the Early Triassic.
3. From the Early Triassic to the Jurassic the continental crust in which they were incorporated was thinned through rifting. The basement fragments under consideration were then part of the European passive margin (Briançonnais domain *s. l.*). Stampfli and Marthaler (1990) consider this zone to represent the upper plate due to the simple shearing and concomitant thinning of the crust.
4. The main Alpine deformation (i. e. the nappe formation) started in the Eocene. On the nappe scale, Alpine deformation was mainly guided by pre-Alpine structures: the "Zone Houillère" and the Mont Fort nappe represent inverted Permo-Carboniferous basins (Fig. 2).
5. Alpine blueschist to greenschist facies overprint, essentially of Tertiary age, affected all these units. Amphibolite facies conditions where reached only in the easternmost zones (Berisal unit- Fig. 1 A, p. 301). The most representative Alpine minerals are listed in Table 1.

In basement rocks where Alpine metamorphism did not exceed greenschist conditions, systematic presence or absence of relict minerals typical of amphibolite assemblage allows discrimination between **polymetamorphic** (Alpine and pre-Alpine) and **monometamorphic** (only Alpine) units. In the metamorphic Alpine context, these terms do not bear the same significance as in the extra-Alpine area (one or more pre-Triassic metamorphism). "Polymetamorphic basement" means here one or more pre-Triassic events of at least amphibolite grade, whereas "monometamorphic basement" refers to a non or weakly metamorphosed unit before Alpine events.

2 Tectonostratigraphic Units and Lithologies

2.1 The "Zone Houillère"

The "Zone Houillère" can be traced from France and Italy into Switzerland where it forms a band, one to two kilometers wide, with almost continuous outcrops right to the alluvial valley (Fig. 1 A, B, p. 301, p. 302). Upstream from Sion, Quaternary deposits cover most of it, until the Turtmanntal is reached.

2.1.1 From the Aosta Valley to the Rhone Valley

In this area, the Zone Houillère is divided almost in the middle by a thin strip of Triassic gypsum and cornieule (Gerlach 1871; Christ 1925, Oulianoff and Trümpy 1958), into external and internal parts.

The external part characteristically includes large bodies of Triassic quartzites and dolomitic limestones. The Carboniferous sequence exhibits a very chaotic structure, with discontinuous layers rendering their mining very problematic (Christ 1925, Wehrli 1925; Lombard 1942, 1949). They are comprised of sandstones with a pelitic matrix, grey quartzites, shales and siliceous schists, and a few conglomerates (Burri 1975; Burri and Jemelin 1983). No fossil has ever been found in them.

In the internal part, surface and underground observations reveal a much less intricate structure. Along the Rhone Valley, the following lithostratigraphy has been established (Christ 1925; Lombard 1942, 1949):

1. At the bottom, sandstone layers about 10 cm thick are separated by thin strips of shale.
2. In the middle part, a thick shale sequence is mainly composed of shales, graphitic schists and slates. Sandstones are rare.
3. At the top, sandstones are again dominant with a small shale fraction.

The coal-bearing layers of this internal part that have occasionally been worked are concentrated within the middle shale sequence, except near Sion, where they are associated with the lowermost sandstones. Conglomerates are uncommon, but locally their abundance increases suddenly, almost to the exclusion of all other rock-types, suggesting the presence of old deltas (Christ 1925). Pebbles of white mica-gneiss, quartz and pelitic schists have been observed. The sandstones contain quartz, albite, sericite and graphite. Abundant large white mica flakes and K-feldspar fragments are considered to be detrital (Vallet 1950).

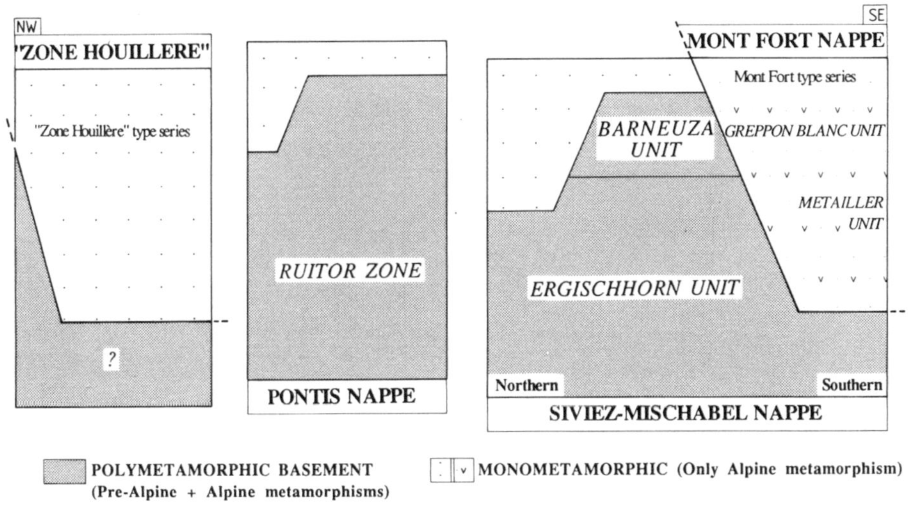

Fig. 2. Distribution of the poly- and mono-metamorphic basement units within the Middle Penninic Alpine nappes. Polymetamorphic units record HP (pre-Variscan?) and HT (Variscan?) pre-Alpine events. Monometamorphic rocks form terrestrial series of Zone Houillère type (Late Carboniferous to Early Triassic) and volcano-sedimentary series of Mont Fort type (Carboniferous? to Early Triassic)

Table 1 Main assemblages related to Palaeogene Alpine metamorphism

Lithology	Palaeogene Alpine Metamorphism		
	Prograde met.	Climax	Retrogression
Felsic	Grt-Gln-Qtz	Qtz-Ab-Bt/Chl-Phe-Ep	Chl-Qtz-Ab-Stp
Pelitic	Grt-Gln-Cld-Qtz	Phe-Bt/Chl-Grt-Qtz-Ab-Ep	Chl-Qtz-Ab-Stp
Mafic	Grt-Gln	Hbl/Act-Grt-Bt/Chl-Ab--Ep-Sph	Chl-Ab-Stp-Cal

a met = metamorphism

Jongmans (1960) found in this internal part of the Zone Houillère 24 plant species from 10 genera indicating a Stephanian A age tending towards Stephanian B.

In the high Entremont Valley, the sandstones grade upwards into white sericitic quartzites with conglomerate layers which could be of Permian age (Burri 1983a, b). At the entrance to the Anniviers Valley, Jäckli (1950) found conglomeratic quartzites in the same position.

2.1.2 East of the Turtmanntal

The Zone Houillère is still divided into two parts. The external part, with grey quartzites in metre-thick layers and black shales, decreases from a thickness of 200 m in the Turtmanntal to a trail of lenses in the Vispertal (Werenfels 1924; Staub 1927, 1934) which cease to be distinguishable from the shaly levels of the Valais domain (Burri 1979).

A thin Triassic band separates this external part from an internal one which thickens eastwards (the "lower Stalden zone", Escher 1988). The latter is comprised of thin levels of graphitic schists and grey quartzites, locally conglomeratic, but the major part of this series is composed of albitic gneisses and chlorito-sericitic schists, with a little carbonate, and light-coloured, locally conglomeratic quartzites. This series, in which prasinites are common, is considered to be Permo-Carboniferous (Bearth 1973, 1980). With increasing metamorphic grade past the Simplon fault, this series appears in the form of a greenish garnet-bearing gneiss with the prasinites converted to amphibolites (Jeanbourquin and Burri 1989).

2.1.3 The Magmatic Rocks

On Italian territory, Caby (1974) discovered and described the magmatic Costa Citrin complex with an outcrop of some 20 km^2. This is composed of metagranites with grey, globular quartz grains, big perthitic K-feldspars and biotite, sodic microgranites with large quartz grains, potassic porphyries and various acid volcanics whose chemical composition varies from potassic or sodic rhyolite to dacite, and finally various prasinites which represent former microdiorites emplaced as sills in the enclosing Carboniferous. This complex is contained in a sandy-shaly series of Westphalian or Stephanian age. It is overlain by feldspathic sandstones formed through the reworking of the rhyolites and microgranites, passing upwards into a banded albitic lithology and then into light-coloured prasinites of probable tuffaceous origin. It is possible that the tectonic unit in which these rocks are contained does not extend into Swiss territory.

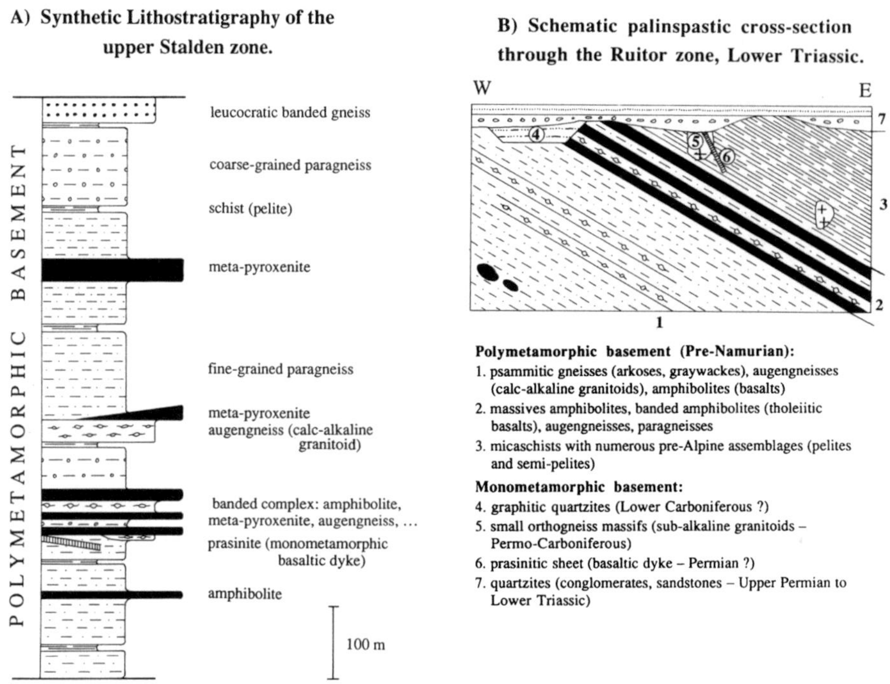

Fig. 3 A, B. Synthetic lithostratigraphy of the upper Stalden zone and palinspastic cross-section through the Ruitor zone

In the lower part of the Turtmanntal, a schistose lens rich in oligoclase-andesine has been considered to represent a former diorite, while another contains calcic oligoclase, hornblende and biotite (Argand 1911, Jäckli 1950). Finally, in the eastern part, lenses of aplite with quartz, albite and potassic feldspar are sometimes hardly distinguishable from the albitic gneisses (Bearth 1973, 1980).

2.1.4 Tectonics

There is no certain relic of Variscan tectonics in the Zone Houillère. On what these Carboniferous rocks were laid down is not known. Hercynian directions (Oulianoff 1957; Argand 1911) have apparently no more significance than the unconformities locally observed between the Carboniferous beds and the overlying Mesozoic.

2.2 *The Pontis Nappe*

The Pontis nappe includes a polymetamorphic basement, a monometamorphic Permo-Carboniferous basement and Triassic fragments of a Sub-Briançonnais cover (Fig. 1 A, B, p. 301 and p. 302). The polymetamorphic basement which forms the main part of the nappe is fragmented, due to Alpine tectonics, into three distinct units: from SW to NE, the Ruitor zone, the upper Stalden zone and the Berisal unit (see Burri 1983 a, b; Thélin 1983, 1989; Thélin and Ayrton 1983; Escher 1988; Escher et al. 1988).

2.2.1 The Polymetamorphic Basement (Fig. 3 A, B)

Metasediments are mainly characterized by very thick, monotonous, strictly siliceous, detrital rocks. Different varieties of paragneisses may however be distinguished, in the upper Stalden zone for instance, expressing contrasting conditions of detrital sedimentation, of unknown age, predominantly composed of sandstones, greywackes and conglomerates. The pre-Alpine assemblage includes muscovite-brown biotite-almandine with prograde zoning ± plagioclase (almandine amphibolite subfacies).

Assuming the Ruitor zone to be in normal position, it comprises, from bottom to top: (a) paragneisses of predominantly psammitic character (the most exter-

Fig. 1. **A** Tectonic map of the Middle Penninic units between the Ruitor and the Simplon area. Basements only have a signature. Te limits of the nappes partly cross-cut the Mesozoic covers (see the tectonic sketch map). Mainly after Escher (1988), Burri (1983 b) and Gouffon (1992). *a–d* Location of the cross-sections (Fig. 1 B).

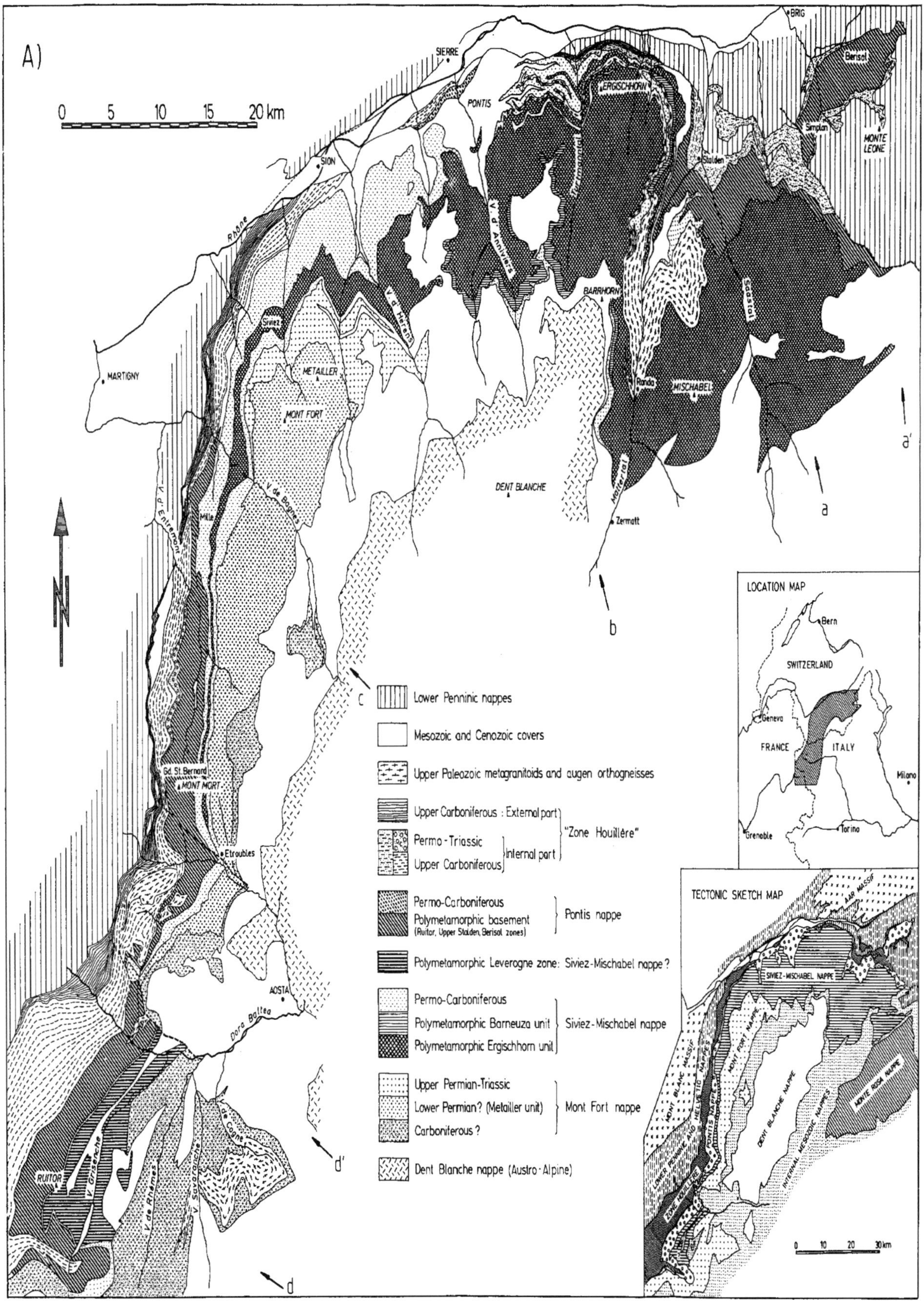
A)
0 5 10 15 20 km
LOCATION MAP
Bern
SWITZERLAND
Geneva
FRANCE
ITALY
Milano
Grenoble
Torino
TECTONIC SKETCH MAP
Lower Penninic nappes
Mesozoic and Cenozoic covers
Upper Paleozoic metagranitoids and augen orthogneisses
Upper Carboniferous : External part
Permo-Triassic
Upper Carboniferous
Internal part
"Zone Houillère"
Permo-Carboniferous
Polymetamorphic basement
(Ruitor, Upper Stalden, Berisal zones)
Pontis nappe
Polymetamorphic Leverogne zone: Siviez-Mischabel nappe ?
Permo-Carboniferous
Polymetamorphic Barneuza unit
Polymetamorphic Ergischhorn unit
Siviez-Mischabel nappe
Upper Permian-Triassic
Lower Permian? (Metailler unit)
Carboniferous ?
Mont Fort nappe
Dent Blanche nappe (Austro-Alpine)
SIERRE
SION
MARTIGNY
BRIG
PONTIS
ERGISCHHORN
Berisal
Simplon
MONTE LEONE
Stalden
BARRHORN
Randa
MISCHABEL
Zermatt
DENT BLANCHE
METAILLER
MONT FORT
Siviez
Gd. St. Bernard
MONT MORT
Etroubles
AOSTA
Dora Baltea
RUITOR
V. Grisenche
V. de Rhêmes
V. Savaranche
Rhône
V. d'Anniviers
V. d'Hérens
V. de Bagnes
V. d'Entremont
Saastal
Mattertal
a
a'
b
c
d
d'
N
SIVIEZ-MISCHABEL NAPPE
AAR MASSIF
MONT BLANC MASSIF
MONT FORT NAPPE
DENT BLANCHE NAPPE
MONTE ROSA NAPPE
PONTIS NAPPE
INTERNAL MESOZOIC NAPPES
0 10 20 30 km

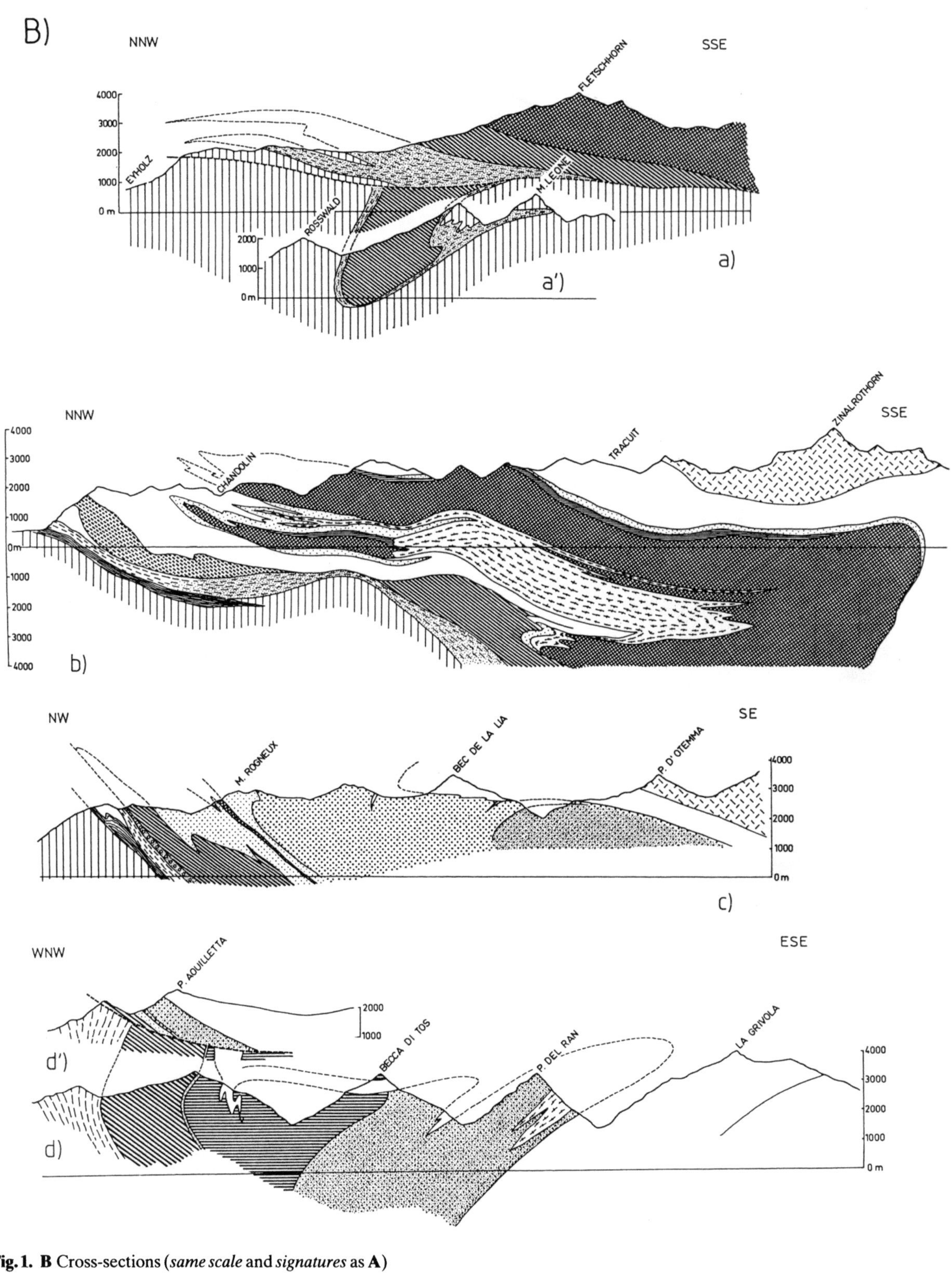

Fig. 1. B Cross-sections (*same scale* and *signatures* as **A**)

Table 2 Published radiometric data

Ages (Ma)	Methods	Rocks	Authors
Pontis Nappe (Berisal Unit)			
1070 ± 43	Sm/Nd whole rock	Banded amphibolite (tholeiite-dacite sequence?)	Stille and Tatsumoto (1985)
546 ± 21	Sm/Nd whole rock	Ultramafic rocks (komatiite?)	Stille and Tatsumoto (1985)
Siviez-Mischabel Nappe			
978 ± 59	Sm/Nd whole rock	Banded amphibolite	Zingg (1989)
475 ± 81	Sm/Nd whole rock	Ultramafic rocks (komatiite?)	Zingg (1989)
484 ± 24	U/Pb on zircons	Acid igneous rocks	Zingg (1989)
406 ± 49	U/Pb on zircons		
275 ± 12	U/Pb on pitchblendes	Uranium mineralization	Eikenberg et al. (1989)

nal part of this zone); (b) a banded complex, essentially alternating amphibolites, paragneisses and augengneisses; (c) a thick pelitic series forming the most internal part of this zone.

The latter offers an exceptional metamorphic window for pre-Alpine assemblages perfectly preserved in pelitic rocks. Oulianoff and Trümpy (1958), Bocquet (1974), Burri (1983a, b), Fabre (1987), Thélin (1989, 1992) present the available data. These metapelites are situated north of the Grand Saint Bernard pass. Staurolite-garnet micaschists form the major part of the outcrops which extend laterally for about 400 m. They are accompanied by metre-thick layers of dark schists bearing Al-silicates (kyanite, andalusite, sillimanite) (see Sect. 3.2).

Magmatic rocks are common, either as lenses or as concordant layers within the paragneisses.

Acid magmatic rocks: the polymetamorphic basement contains numerous parallel bands of augengneisses of remarkable thickness (ca. 30–300 m) and continuity, probably repeated tectonically. They can be traced equally well in the Berisal unit, the upper Stalden zone and the Ruitor zone. Thélin (1983, 1989) concludes that they represent porphyric granitoids of calc-alkaline granodioritic and monzogranitic composition. Zircon typology (Alkalinity index, I. A. = 483; Temperature index, I. T. = 362 according to Pupin's coordinates 1980) confirm the hybrid origin of the protolith. The megaclasts are single crystals (chessboard albite, low microline, quartz) or augen-shaped polycrystalline aggregates of the same minerals. The polymetamorphic history of these rocks is evidenced by a relic assemblage with plagioclase (An.max 20 wt.%)-zoned garnet-brown biotite-muscovite. The geodynamic context in which they were emplaced, on the basis of Nb, Ti and Zr contents, is compatible with that of an active continental margin or of an island arc.

The emplacement age of these granitoids is unknown. Comparison is warranted with the augen gneisses at Emosson (Aiguilles Rouges massif, Valais), considered by von Raumer (1987) possibly to represent discordant Ordovician/Silurian granites or acid volcanics of Cambrian/Ordovician age. The second view is founded on a comparison with the Blasseneck porphyries in Austria which are considered to be of late Ordovician age (Loeschke 1989).

Ultramafic and mafic magmatic rocks: the polymetamorphic basement exhibits a considerable abundance and a wide variety of these rocks, essentially massive amphibolites with large amphiboles, garnet- and/or plagioclase-bearing amphibolites, and banded amphibolites. They form lenticular boudins or continuous concordant layers with a sill-like aspect, occasionally of kilometric extent. The host rocks are generally metaclastics (sandstones, greywackes) and they are often closely associated with the augengneisses already described. Abundant data are to be found in Stille (1980), Stille and Tasumoto (1985), Stille and Oberhänsli (1987), Thélin (1983, 1989) and Sartori et al. (1989). From a number of Rb-Sr, Sm-Nd (Table 2), and REE analyses, Stille and co-workers suggest the following petrogenetic interpretation: (a) the mafic and ultramafic rocks are of magmatic origin and constitute a series with a calc-alkaline trend, (b) the banded and the plagioclase-bearing amphibolites are derived from a tholeiitic to dacitic sequence, emplaced in Proterozoic times (ca. 1020–1070 Ma); (c) the massive amphibolites with large amphiboles could represent pyroxenites, emplaced through mantle diapirism around 475 Ma (ca. middle Ordovician).

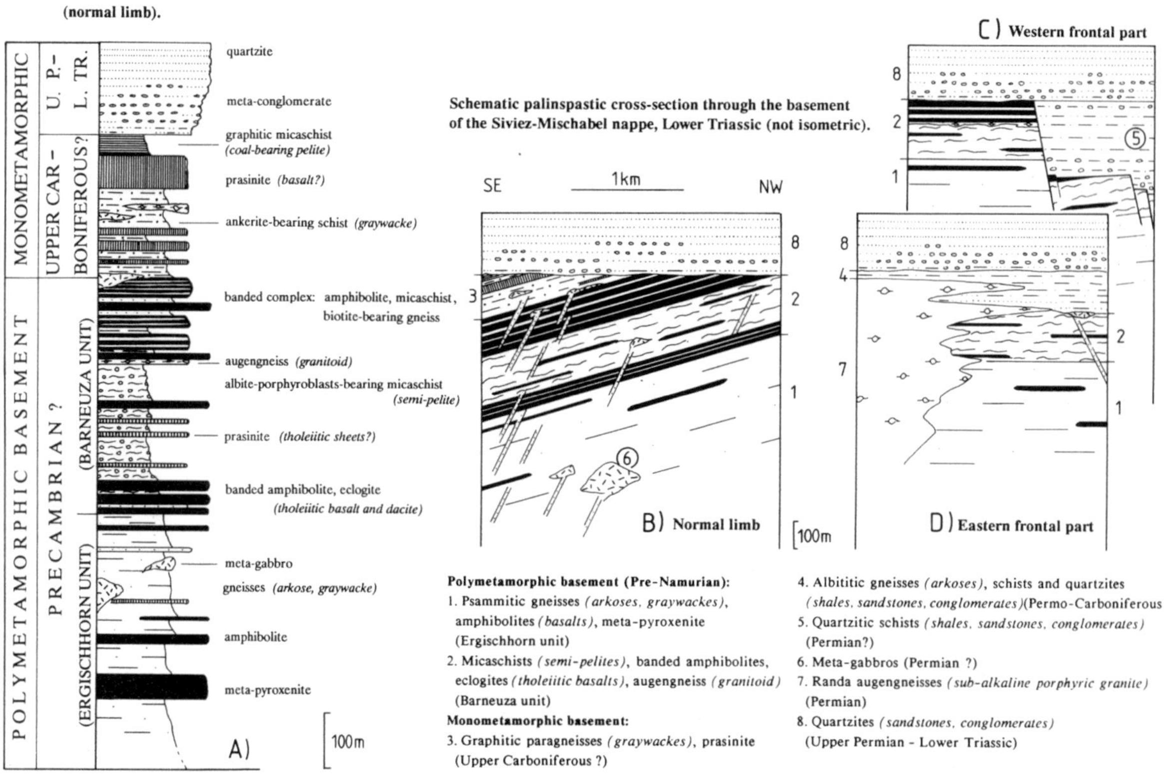

Fig. 4. A Synthetic lithostratigraphy of the Siviez-Mischabel basement. **B–D** Interpretative and schematic palinspastic cross-sections through the basement of the Siviez-Mischabel nappe, Early Triassic

2.2.2 The Monometamorphic Basement
(Figs. 1 A and 3 B)

Rocks lacking pre-Alpine metamorphic minerals form limited extent; they have not received much attention in recent years.

Metasediments form the tectonic slice of Niouc, at the entrance to the Anniviers Valley, which is comprised of metaconglomerates and psammitic gneisses, presumably Upper Carboniferous to Permian in age. Monometamorphic quartzites, locally graphitic (Carboniferous?), occur within the external part of the Ruitor zone and on its limits (Fig. 3 B). Conglomerates of "Verrucano" type overlie both the northern end of the Ruitor zone and the Niouc wedge; they are transitional to thick massive, white quartzites (Early Triassic).

Magmatic rocks are also present. Namely the small subalkaline granitic intrusions of Mont Flassin and Vedun (Ruitor zone, Aosta Valley) of Permo-Carboniferous age according to Caby (1974). They are intersected by rare prasinitic sheets of Permian or younger age.

2.3 The Siviez-Mischabel Nappe

The Siviez-Mischabel nappe constitutes the main part of the Grand St. Bernard nappe. It forms a fold-nappe of more than 40 km extent with a polymetamorphic core, a monometamorphic basement [Carboniferous(?)-Early Triassic] and a Mesozoic to Paleogene cover (Fig. 1 A, B). Investigations by Bearth (1963), Thélin (1983, 1989), Thélin et al. (1990), Marthaler (1984), Sartori (1988), Sartori and Thélin (1987), Burri (1983 a, b), Escher (1988), Escher et al. (1988) demonstrate the existence of a normal limb, a frontal zone and an inverted limb.

The internal half of what has been considered to be the "Ruitor Massif" may possibly constitute the southernmost extension of the basement of the Siviez-Mischabel nappe (Gouffon 1992). It will now be referred to as the Leverogne polymetamorphic zone.

2.3.1 The Polymetamorphic Basement (Figs. 1 A, 4)

The sum of the above-mentioned contributions confirms the lateral continuity (from the Aosta Valley to the Saastal) of the polymetamorphic basement. It could be divided into two distinct lithostratigraphic entities:

a. The Ergischhorn unit is the main component of the basement. It structurally forms the homogeneous core of the nappe. It is composed of siliceous clastic metasediments with various types of associated amphibolites. The distribution of these lithologies presents no mappable continuity.
b. The "stratiform" Barneuza unit overlies the latter and exhibits contrasting lithologies with marker horizons of very great regional extension, from bottom to top: (1) banded amphibolites; (2) micaschists; (3) banded complex.

The *metasediments* of the polymetamorphic basement are uniformly siliceous; no marble or calc-silicate gneiss has ever been found in them.

The Ergischhorn unit is predominantly composed of monotonous paragneisses whose composition and aspect vary gradually, without any clear-cut boundaries. Most typical are grey-greenish micaceous gneisses and schists with thin quartz bands or nodules, aphanitic, massive gneisses and well banded gneisses. The pre-Alpine more or less retrogressed assemblage includes zoned almandine, muscovite, brown biotite, plagioclase and K-feldspar. These very thick metasediments (ca. 1000 m) are derived from immature, often coarse clastic sediments, such as micaceous sandstones, arkoses and conglomerates.

The metasediments of the Barneuza unit are essentially pelitic. They are especially well developed in its middle part, forming augen-micaschists with albite porphyroblasts (Sartori and Thélin 1987). Complete Alpine recrystallization/neoformation of greenschist grade is pervasive. The marker horizon attaining a maximum thickness of 250 m and an east-west extension of at least 100 km could be derived from semipelites. In the banded uppermost complex, the metasediments are represented by a gneiss with Alpine green biotite or, locally, of micaschists with large relic garnets. These rocks might also be derived from semi-pelites, or even greywackes.

Rocks of magmatic origin are represented by a wide spectrum of greenstones (ultramafics, orthoamphibolites, eclogites and retroeclogites) and by various acid rocks (augengneisses, meta-aplites, metapegmatites).

Ultramafic and mafic rocks have been investigated, in particular in the Turtmanntal, by Zingg (1989), Sartori (1990), Rahn (1991) and Thélin et al. (1990). Ultramafic rocks, derived from pyroxenites and/or basalts with komatiitic affinities, form levels about a metre thick, either as continuous bands or lenses; they are only to be found in the Ergischhorn unit. Amphibolites exist in both Ergischhorn and Barneuza units. They are essentially garnet-bearing amphibolites and banded amphibolites which form lenses or boudins within the paragneisses and the micaschists. The banded amphibolites are composed of decimetric to metric layers of dark garnet-amphibolite and eclogite and light albite-quarz-garnet-bearing gneiss, both deriving from effusive material ("leptynitic and amphibolitic series"). They form the most continuous level at the base of the Barneuza unit. Eclogites have been found mainly within this level and also at the top of the Ergischhorn unit (Bearth 1980; Thélin et al. 1990). These eclogites form lenses enveloped in retroeclogites and in amphibolites (Fig. 4 A). The eclogitic assemblage with omphacite-garnet is earlier than an amphibolite facies assemblage (see Sect. 3.1). This fact, to which one may add other geological data, leads to the conclusion that this eclogitization must be attributed to a pre-Alpine HP metamorphism. Geochemical investigations on amphibolites demonstrate a tholeiitic composition and discriminating geochemical diagrams show affinities with T-MORB or WPB (Thélin et al. 1990). Prasinites occur in now concordant, up to 2-m-thick levels. They lack any relictual metamorphic assemblage and could derive from basaltic sheets. In some places, they are associated with monometamorphic gabbroic bodies (see Sect. 2.3.2).

Acid magmatic rocks are widely distributed within this polymetamorphic basement but have not been studied in detail. Bearth (1980) provides some petrographic data.

2.3.2 The Monometamorphic Basement (Fig. 4)

Monometamorphic rocks, mainly metasedimentary, envelop the polymetamorphic basement and have been described to some extent by Argand (1909), Trümpy (1966), Vallet (1950), Jäckli (1950), Bearth (1963, 1964), and Sartori (1990) amongst others. Presumably Carboniferous and Permian strata are distributed in distinct and discontinuous units, while the Permo-Triassic quartzites form a regular, continuous envelope. In all, this amounts to about 10 % of the total volume of the nappe.

Metasediments

Three distinct sedimentary series are defined as follows:

1. Pelites, sandstones and greywackes (Carboniferous?): these form a characteristically rust-coloured series within the eastern part of the normal limb of the nappe (Fig. 4 B); graphite, pyrite and ankerite-bearing schists, albitic schists and quartzites are the main components. There are abundant mafic rocks (prasinites, metagabbros) and acid rocks (metaquartz porphyries). Locally, the latter clearly present the features of dykes.
   The sedimentary component is essentially characterized by fine-grained volcano-clastics alternating with pelites rich in organic matter. The series is very similar to the lowermost unit of the Mont Fort nappe (see Sects. 2.4.1 and 4.3) of which it may represent the eastern extension.
2. Arkoses, conglomerates, greywackes (late Carboniferous-Permian?): this series is composed of micaschists, chloritic schists, albitic gneisses and metaconglomerates. It only occurs in the front of the nappe (Figs. 2, 4 C, D), from the Aosta Valley in the southwest to the Matter Valley in the east. Conglomerates are the main rock type near its base, with pebbles of quartz and metarhyolite. Greenstones are scarce and appear mainly at the base of the series. Thin layers of quartz porphyries are recognizable. The thickness of this monotonous series reaches several hundred metres but is the consequence of tectonic transposition at the front of the nappe.
3. Conglomerates, arkoses and siliceous sandstones (Late Permian to Early Triassic): the classic association of Briançonnais Verrucano-type conglomerates (Trümpy 1966) and white tabular quartzites forms a virtually continuous horizon around the nappe. The series sometimes begins with dark green quartz-rich schists containing occasional conglomerates and decimetric carbonate lenses. The conglomerates with pink quartz pebbles grade into the white quartzites. Thin levels of quartz porphyries appear within the conglomerates. Total thickness locally reaches 200 m in the normal limb. Variations in thickness observed elsewhere are most likely the consequence of Alpine tectonic deformation. This series lies unconformably on all the other components of the basement, except perhaps in the frontal part of the nappe, where a continuous transition with the underlying "Permian" schists is possible. Its top, on the other hand, shows a concordant transition to the carbonate sediments of the Middle Triassic (Sartori 1990).

Mafic Igneous Rocks

Small bodies of metagabbro, some tens to hundreds of metres across, intrude the two units of the polymetamorphic basement and the "Carboniferous" series of the normal limb (Sartori 1990). Their texture is well preserved but essentially Alpine minerals: albite, epidote and hornblende replace the magmatic plagioclase, pyroxene and amphibole. Chemical data, although still insufficient, indicate gabbroic to monzogabbroic compositions. Veins of albitite and hornblende-rich rocks are associated with these bodies, as well as, possibly, certain discordant prasinites in the polymetamorphic basement.

The youngest strata into which these gabbros intruded are the graphitic micaschists considered to be of Carboniferous age. They are to be linked with the gabbros and mafic volcanics of the Mont Fort nappe (Schaer 1959 a) as well as with the Randa metagranite, which all represent Early Permian bimodal magmatism (Thélin and Ayrton 1983).

In the Siviez-Mischabel nappe, the intrusive gabbroic bodies bind together the different components of the pre-Alpine basement.

Acid Igneous Rocks

The Randa augengneisses occur in the eastern part of the Siviez-Mischabel nappe (Figs. 1 A, 4 D), within its frontal part and its inverted limb, and reach thickness of up to 1000 m in the Matter Valley. They are the products of Alpine cataclasis and high greenschist metamorphism of predominantly porphyric granite with a Permian (ca. 270 Ma) age of intrusion (Thélin 1987). The main body constitutes a laccolith with sill-like extensions within Permo-Carboniferous metagreywackes. It presents all the megascopic characteristics of an intrusive granite: primary discordant contact, veins of aplite, micrographitic border facies. Mafic inclusions have never been observed. The large augen are essentially replacement feldspars (perthitic microcline and chessboard albite). They contain abundant magmatic relics (primary inclusions of plagioclase crystals, corroded dihexagonal quartz grains, primary biotite, pseudo-rapakiwi mantles). Ba-quantification and a thermometric study indicate multistage growth of the original alkali-feldspar, from the orthomagmatic to the subsolidus and deuteric stages.

Geochemical data confirm the magmatic origin of the Randa augengneisses and afford evidence on the original granitoid, aluminous subalkaline hybrid type (Thélin 1987; Desmons and Ploquin 1989).

Zircon typology (Pupin 1980) gives I. A. = 618 and I. T. = 353, which confirms a hybrid origin within the field of subalkaline granites. It must, however, be

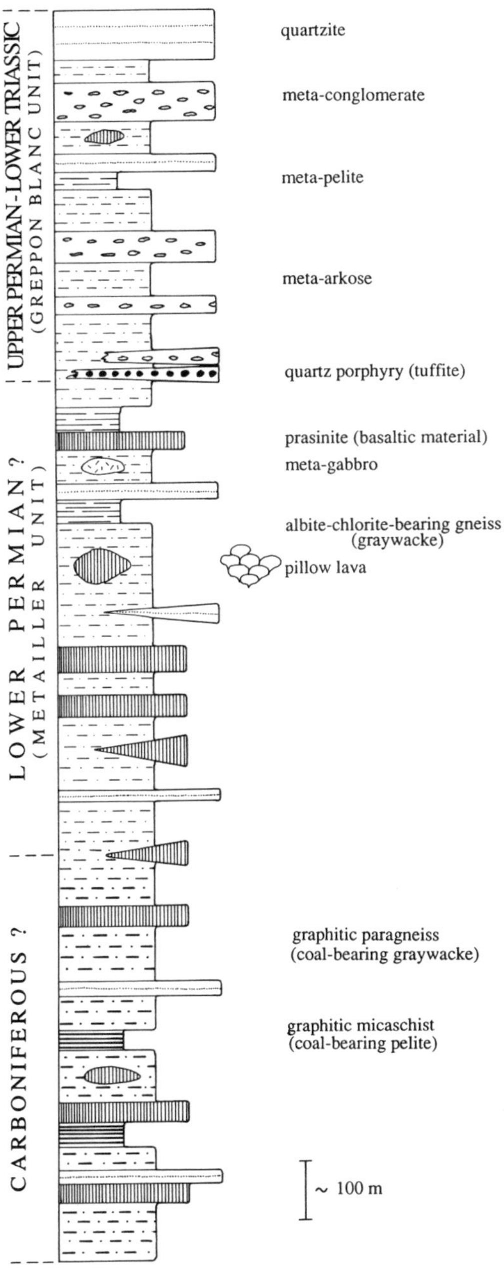

Fig. 5. Synthetic lithostratigraphy of the Mont Fort nappe basement

kept in mind, that the Randa granitoid is completely lacking in mafic enclaves.

In the normal limb of the nappe, a microcline-bearing augengneiss (about 5–50 m thick), separates the augen-micaschists with albite porphyroblasts from the banded complex (Barneuza unit). This is the controversial orthogneiss of Thyon (Wegmann 1923, Vallet 1948; Schaer 1959 a).

2.4 The Mont Fort Nappe

The Mont Fort nappe is the most internal unit of the Middle Penninic nappe pile in the Wallis (Figs. 1 A, B and 2). It includes a Palaeozoic basement and a reduced cover of Triassic to Cretaceous rocks (Allimann 1987, 1989; Escher 1988).

2.4.1 Lithostratigraphy

The basement only appears in the western part of the region described here, starting in the Hérens Valley (Schaer 1959 a; Bearth 1963; Caby 1968; Gouffon 1992). It is formed of three superposed lithological units, all monometamorphic. Passage from one lithology to another is generally progressive and continuous.

Metasediments (Fig. 5)

a. Grey pelites and greywackes (Carboniferous?): in its southwestern part, the core of the Mont Fort nappe is composed of rusty paragneisses and micaschists which systematically contain a small quantity of organic matter and often pyrite. Layers and lenses of prasinites are intercalated in these rocks whose thickness attains 1000 m.

These metapelites and metagreywackes represent the northern extremity of the "zona Interna" (Demons and Mercier this Vol.). Notwithstanding that certain authors consider the latter, as well as its southerly extension, the northern Vanoise zone, to be of pre-Namurian age (Fabre et al. 1987), maybe as old as late Ordovician-Silurian age (Desmons and Fabre 1988; Demons and Mercier this Vol.), we conclude that these sediments are more likely to be of Visean to late Carboniferous age (see Sect. 4.3).

b. Greywackes, sandstones, pelites (Early Permian?): the middle lithological unit (the Métailler unit) constitutes the main part of the Mont Fort nappe. Its stratigraphic base (inverted series) exhibits sometimes a transition to the above unit over a vertical distance of 50 to 100 m. Otherwise, this contact is underlined by a layer of metabasics and/or quartzite. The dominant lithology is a gneiss bearing albite, white mica, chlorite, quartz, epidote with occasional carbonate, glaucophane-crossite, chloritoid and garnet. Layers of graphitic schists, quartzites, meta-conglomerates and common mafic rocks constitute intercalations in this unit, which reaches a maximum apparent thickness of 4 to 5 km.

This clastic succession (greywackes, conglomerates, sandstones) ist taken to be of Early Permian age

because of its transitional contact with the next unit (see also Sect. 2.4.3).
c. Sandstones, arkoses, conglomerates and pelites (Late Permian to Early Triassic): the youngest of these units (the Greppon Blanc unit) forms the northern, frontal part of the nappe, essentially in an inverted limb position and attains a thickness of 1000 m. This is a continental detrital series with conglomerates, arkoses and sandstones occasionally with a considerable amount of shale; these are now schists compositionally situated between the four mineral poles, quartz, sericite, chlorite and albite, almost always ankerite-bearing. The stratigraphic base of this series is characterized by abundant acid volcanics alternating with conglomerates. A quartz porphyry level of some extent has moreover been retained to mark the limit with the middle unit (Schaer 1959b), but this contact may generally be considered to be transitional. Upwards, the series grades into the tabular white quartzites. This is the argument for placing this unit in the Late Permian to Early Triassic.

Acid Igneous Rocks (Fig. 5)
The transition beween the two last series is characterized by the presence of acid volcanic rocks of rhyolitic composition with an alkaline trend. Described by Schaer (1959b) as quartz porphyries, these rocks very likely represent former volcanic tuffs. They often alternate with argillaceous or conglomeratic sediments, outcropping in a few metres thick beds, reaching 50 m in the eastern buttresses of the summit of the Métailler.

Mafic Igneous Rocks (Fig. 5)
Greenstones are common in the middle unit (Métailler) and in the lower lithostratigraphic unit. In the latter, due to the increase in deformation in the southern part of the nappe, no protolithic structure has been detected. Mafic rocks in the Métailler unit represent basaltic flows and/or volcaniclastic deposits. Pillow structures and gabbroic textures have been observed (Schaer 1959a), in spite of their transformation into prasinite through Alpine metamorphism.

2.4.2 Geochemistry of the Metabasics in the Métailler Unit

A trace element geochemical study has been undertaken in an attempt to define the nature of the basic protolith and the palaeotectonic context in which the Métailler unit formed. These questions have been the subject of various speculations and they remain to some extent enigmatic. Present data, although still in limited number, may help to a better understanding of these problems. The analytical results, as well as a more detailed discussion, will be published in an article in preparation (Chessex in prep.).

This analysis has revealed that most of the investigated samples are derived from basaltic magma of transitional to subalkaline (tholeiitic) type. Those richest in Nb exhibit an alkaline tendency.

Diagrams relative to the petrotectonic context of the Métailler unit (Fig. 6) are not entirely conclusive. However, it seems more probable that the analyzed basalts (within-plate basalts, WPB) belong to a within-plate environment. In any case, they bear no resemblance to those of ophiolitic complexes, such as those of the Mediterranean domain for instance (Pearce 1980; Pfeifer et al. 1989).

Element distribution does not allow unequivocal discrimination, for WPB, between oceanic and continental environments. It is obviously essential to determine precisely the type of sedimentation related to the magmatic activity.

3 Pre-Alpine Metamorphism

Pre-Alpine almandine amphibolite facies can be uniformly recognized (Bearth 1963) within the polymetamorphic rocks (amphibolites, paragneisses, orthogneisses). Two spectaculary preserved metamorphic windows permit views of the pre-Alpine P-T events. They are the eclogites of the Siviez-Mischabel nappe and the Mont Mort metapelites, within the Ruitor zone (Pontis nappe).

3.1 The Eclogites of the Siviez-Mischabel Nappe

These eclogites provide the following data on the polymetamorphic history (see Sect. 2.3.1; Thélin et al. 1990; Rahn 1991). The eclogitic omphacite – almandine (pyrope-rich) – phengite predates an amphibolite facies paragenesis with pargasite-oligoclase-zoisite, itself followed by a greenschist facies retrogression (Tertiary Alpine event). The chemistry of the garnet, omphacite and phengite sets these rocks within the B eclogites (Coleman et al. 1965) and assigns the P-T conditions to a bracket of ca. 650 °C and 15–20 kb (Fig. 7 A). The geodynamic context in which this HP metamorphism took place is yet not known. Nevertheless, a pre-Alpine HP event may be documented in banded amphibolites ("leptynitic and amphibolitic series") of the Middle Penninic base-

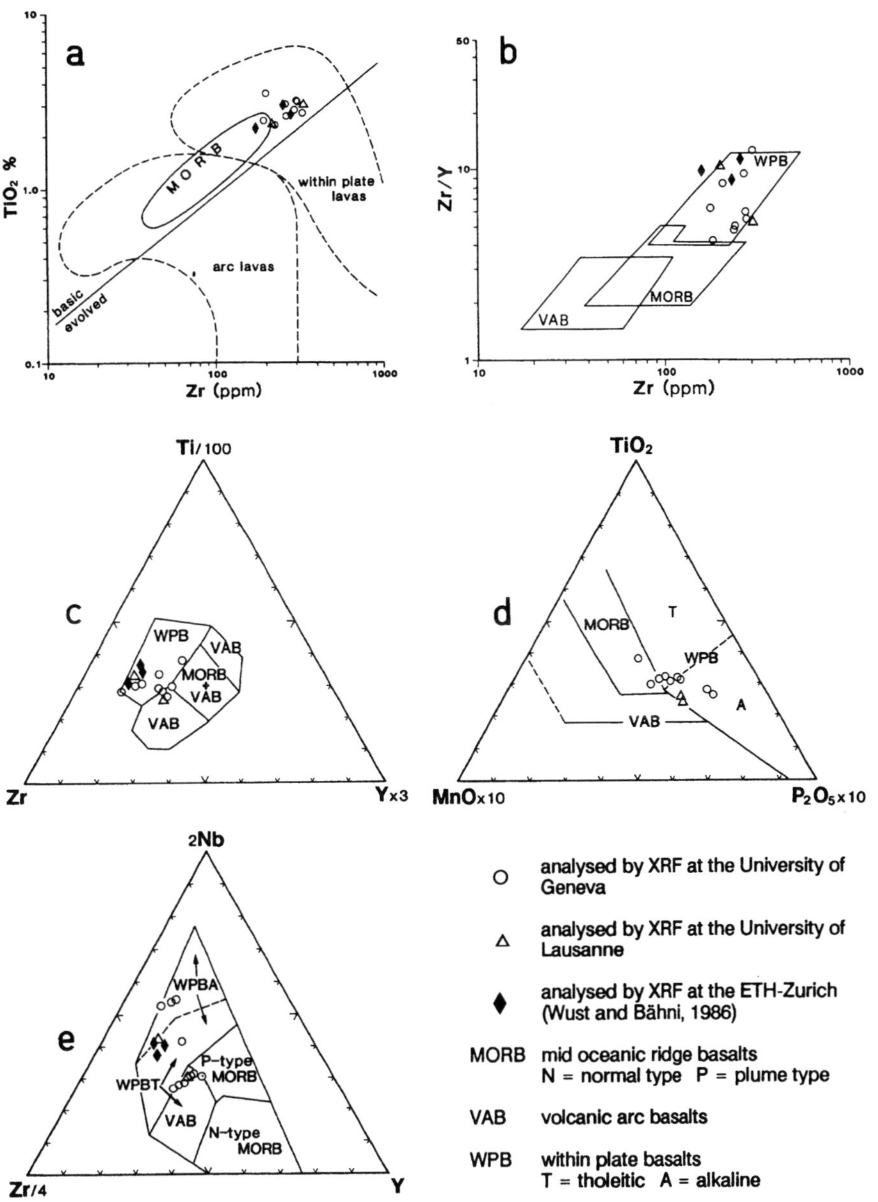

Fig. 6 a–e. Tectonic discrimant diagrams related to the metabasics of the Métailler unit: **a** TiO2-Zr (Pearce and Cann 1973); **b** Zr/Y-Zr (Pearce and Norry 1979); **c** Ti-Zr-Y (Pearce and Cann 1973); **d** TiO2-MnO-P2O5 (Mullen 1983); **e** Nb-Zr-Y (Meschede 1986)

ment, as better known in other pre-Alpine basements (Liégois and Duchesne 1981; Gebauer et al. 1988; Maggetti and Galetti 1988; Manby and Thiedig 1988; Paquette et al. 1989; Frisch et al. 1990).

3.2 The Mont Mort metapelites

These metapelites in the Grand St. Bernard pass area have been studied by Oulianoff and Trümpy (1958), Bocquet (1974), Burri (1983a, b), Thélin (1992). Three rock types with the same chemical composition can be distinguished (see Sect. 2.2.1):

1. Two-mica schists ± zoned almandine;
2. Two-mica schists ± zoned almandine ± staurolite;
3. Annite-andalusite schists ± sillimanite.

In brief, the following evolution of successive assemblages may be proposed:

a) Phases in: kyanite-muscovite 1-biotite 1-oligoclase;
b) Phases in: staurolite-zonedalmandine-mucovite 2-biotite 2 (this assemblage constitutes the main relict schistosity);
c) Phases out: staurolite-zonedalmandine-micas-oligoclase;
d) Phases in: annite-andalousite-sillimanite.

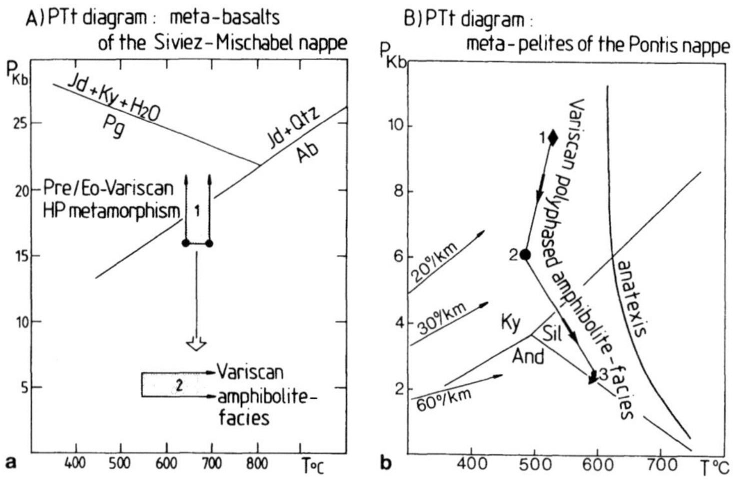

Fig. 7. A P-T-t diagram related to the eclogites of the Siviez-Mischabel nappe (simplified after Thélin et al. 1990; see also text). *1* Eclogitic facies; *2* amphibolite facies. **B** P-T-t diagram related to the Mont Mort metapelites (Pontis nappe) (simplified after von Raumer 1987; Thélin 1992); *1* kyanite-plagioclase In; *2* staurolite-muscovite In; *3* andalusite-annite In

Cordierite and K-feldspar are absent. Fig. 7B shows a possible P-T evolution. Resorting to various geothermo-barometers (notably by Holdaway et al. 1988), an equally retrodynamic and prothermic P-T path from 480 °C and 6 kb to 575 °C for 2.5 kb is likely. Referring to von Raumer (1983, 1987) and to Pin and Peucat (1986), this evolution could be compatible with the Variscan geodynamics in continental environments, the collision-obduction-anatexis. This latter advance could express two major stages: (a) Barrovian metamorphism (kyanite-staurolite-garnet) induced by crustal piling up (setting of Variscan nappes); (b) low pressure metamorphism (andalusite) with increase of the geothermal gradient induced by the rise of anatectic masses.

4 Synthesis and Interpretation

The basement components of the four nappes considered here are related and were initially part of the same palaeogeographic domain in the Early Triassic. This working hypothesis is supported by evidence from the basement rocks themselves, from the Alpine covers and from basement-cover relations.

4.1 Relative Ages

Unconformities and relationships lead to a relative chronology of sedimentary, magmatic and metamorphic events in these basement fragments (Bearth 1963; Trümpy and Dösseger 1972; Thélin and Ayrton 1983; Thélin 1989):

Polymetamorphic Basement (Pontis and Siviez-Mischabel nappes)

a) Sedimentation of a thick siliceous, clastic series; sedimentation of a pelitic series, both lithologies separated by banded amphibolites ("leptyninitic and amphibolitic series") and accompanied by mafic volcanics.
b) Emplacement of pyroxenites; intrusion of calc-alkaline granitoids.
c) HP eclogite facies metamorphism, followed by HT amphibolite facies conditions. The time interval between these two events and the characteristics of related deformations are not known.
d) Erosion(?).

Monometamorphic Basement

e) Sedimentation of fluviatil deposits in the external units (Zone Houillère, Pontis nappe and frontal part of the Siviez-Mischabel nappe); Stephanian A plant fragments in coal levels within the Zone Houillère.
f) Sedimentation of a volcano-sedimentary series in the internal units; coeval bimodal magmatism including intrusive and effusive acid (sub-alkaline granitoids, rhyolites and tuffs) as well as mafic (gabbros, basaltic lavas, tuffs) rocks (south-eastern normal limb of the Siviez-Mischabel nappe, Mont Fort nappe).
g) Widespread deposition of very mature conglomerates and siliceous sandstones with concomitant acid volcanism (rhyolites, tuffs) in all the nappes.

4.2 Absolute Age Dating

Sm/Nd and U/Pb radiometric data available from the literature are summarized in Table 2.

Table 3. Tentative chronology within the monometamorphic basements, sediments deposition and magmatic activities

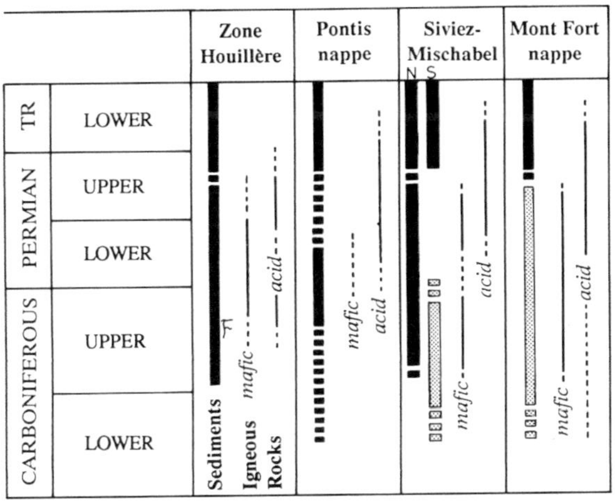

Dark columns: Upper Carboniferous to Lower Triassic sediments of Zone Houillère type; dottet columns: presumed Permo-Carboniferous volcano-sedimentary series, ages disputed (see text); abbreviations: F=dated flora, N, S: northern frontal part and southeastern normal limb of the Siviez-Mischabel nappe

4.3 Hypothetical Ages

The combination of relative ages and avaiblable absolute data leads to the following hypotheses:

Age of the polymetamorphic rocks: the deposition of a thick, monotonous detrital series and of a pelitic series is coeval with mafic volcanism (ca. 1000 Ma?) and predate acid, mafic and ultramafic intrusions (ca. 500 Ma?). The rocks of the polymetamorphic basement should be in this way of late Proterozoic to early Palaeozoic age. They suffered pre-Alpine HP and HT metamorphism episodes.

Age of the pre-Alpine metamorphic episodes: evidence from the Middle Penninic domain described here does not lead to any precise conclusions for the age of the HP and HT metamorphic episodes whose effects are visible in the polymetamorphic basement. They are later than the acid, mafic and ultramafic intrusives dated at ca. 475–500 Ma by Stille (1980) and Zingg (1989) and earlier than the proven Stephanian A of the Zone Houillère in Wallis. The metamorphic evolution recorded by the pre-Alpine preserved assemblages compares well with the metamorphic events ranging from Ordovician to Devonian-early Carboniferous time as documented in certain units of the external crystalline massifs and of the Austro-Alpine nappes (Frisch et al. 1990). This is also supported by preserved Variscan $^{40}Ar/^{39}Ar$ ages (340–360 Ma) reported by Monié (1990) on muscovites and biotites from the polymetamorphic crystalline basement of the Ambin Massif. This basement still belongs to the Briançonnais zone and is correlated with the Ruitor and Siviez-Mischabel polymetamorphic units (Gouffon 1992).

Age of the monometamorphic rocks: the reliable indications are the plant fragments of the Zone Houillère and the Early Triassic age of the top of the quartzitic series. We suggest the following ages for the series contained in each nappe (Table 3).

Coarse detrital sediments in the Pontis nappe and in the external part of the Siviez-Mischabel nappe may be compared with those of the Zone Houillère. In the best exposed French segment of the Zone Houillère, Namurian to Early Triassic series have been recognized (Desmons and Mercier this Vol.). The most internal monometamorphic series of the Mont Fort and Siviez-Mischabel nappes present, on the other hand, a quite different lithostratigraphical feature: fine-grained sediments, abundance of volcanic material. For this reason, their western extension in Italian and French Alps (zona Interna, Northern Vanoise) have been recently considered of early Palaeozoic age (Desmons and Mercier, this Vol.). Within the Wallis area however, the following arguments point rather to a Carboniferous to Permian age: (a) the monometamorphic series lie on a polymetamorphic basement recording HP and HT metamorphisms likely to represent pre-Variscan and Variscan events (Siviez-Mischabel nappe); (b) the series extends with apparently continuity from presumed Carboniferous and Permian schists to Early Triassic quartzites (Fig. 5).

4.4 Geodynamic and Palaeogeographic Contexts

Polymetamorphic basement: the sparse data derived from the studied basement units do not allow a reliable reconstitution of their pre-Variscan and Variscan history.

The polymetamorphic basement is very constant in the two Alpine nappes in which it occurs (Pontis and Siviez-Mischabel nappes). It contains the same kind of sedimentary series (compare Figs. 3 and 4) and virtually the same effusive and intrusive rocks. The basement of these two nappes probably formed a single pre-Alpine tectonic unit.

Although the pre-Alpine metamorphic history is legible through the screen of Alpine events, not tec-

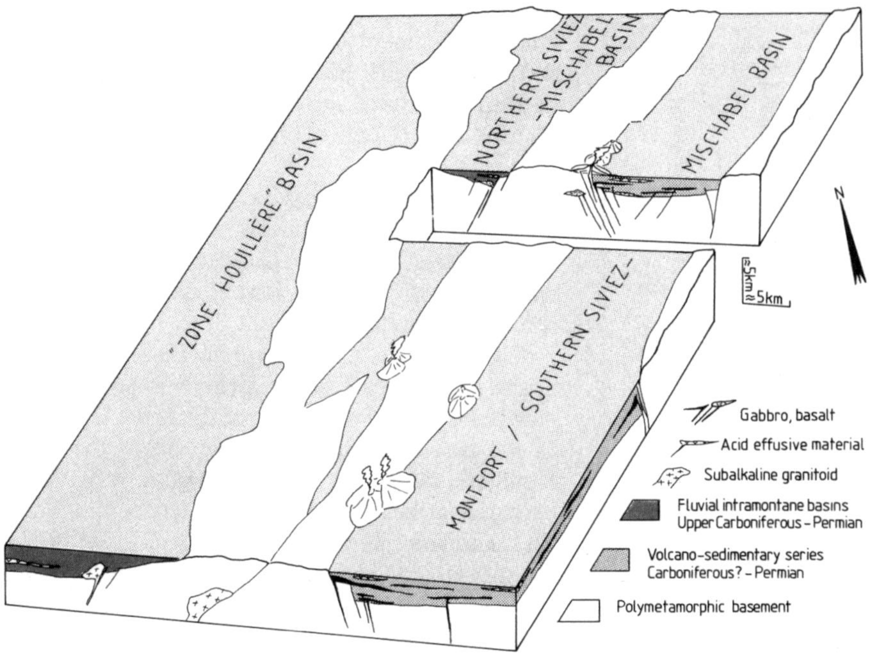

Fig. 8. Hypothetical palaeogeographic relations for the Middle Penninic units (Wallis) at Permian time

tonic structure of certain pre-Alpine age has been found. This situation complicates attempts at geodynamic reconstruction.

Monometamorphic basement: in the present state of knowledge, the bulk of the monometamorphic series probably derives from at least three distinct basins, from NW to SE (Fig. 8):

1. the Zone Houillère basin, whose basement ist not known;
2. the northern Siviez-Mischabel basin, still connected to the nappe;
3. the southern Siviez-Mischabel basin, largely incorporated into the Mont Fort nappe.

(1) and (2) are intramontane basins, probably grabens. Demons and Mercier (this Vol.) provide sedimentological analysis of the French Zone Houillère.

Geodynamic signification is much less clear for (3) which exhibits signs of intense magmatic activity. The sediments of the thick Métailler unit (Early Permian?) include an important terrigenous component which is evidence for an intracontinental environment. However, the presence of metapillows with variolitic texture could indicate that at least part of the subaquatic basaltic activity took place at a relatively important depth. The Métailler unit thus appears to be due to the local, rapid and relatively short-lived development of a basin of rift or pull-apart type with thinning of the continental crust stopping short before oceanization. Such a view is in good agreement with the general evolution of the Variscan orogen since Stephanian times (Arthaud and Matte 1977; Ziegler 1984), with the implication that a modification "of plate movements induced a phase of essentially anorogenic faulting (taphrogenesis). It caused the development of shear faults and related pull-apart structures". These conditions were sustained throughout the Early Permian, also characterized by intense bimodal magmatic activity.

Even if it seems reasonable to consider seriously this age and geodynamic evolution for the Métailler unit, definitive proof remains elusive.

4.5 Conclusion

The pre-Alpine history is only partially legible through the screen of Alpine deformation and metamorphism within the Middle Penninic domain in Wallis. Nevertheless, very well-exposed sections provide important observations on basement geometry and lithology.

Both the Pontis and Siviez-Mischabel nappes comprise a metasedimentary polymetamorphic basement which suffered HP and HT metamorphisms likely to correlate with pre-Variscan and Variscan events. It comprises small bodies of calc-alkaline granitoids and of pyroxenites as well as a wide variety of volcanites.

Sedimentary basins lying on the metamorphosed old basement are filled with two distinct series: a thick fluviatile and continental sequence of late Carboniferous to Early Triassic age mainly developed in the Zone Houillère and a volcano-sedimentary series of Carboniferous(?) to Early Triassic age mainly exposed in the Mont Fort nappe. Post-Variscan sub-alkaline granitoids and gabbros intrude these Middle Penninic basements.

Acknowledgements.
The authors are grateful to St. Ayrton and C. Bauchau (University of Lausanne) for translating the original manuscript into English and to R. Sartori for the final editing.

References

Allimann M (1987) La nappe du Mont Fort dans le Val d'Hérens. Bull Soc Vaudoise Sci Nat 78: 431–444

Allimann M (1989) Les brèches de la région d'Evolène (Nappe du Mont Fort, Valais, Suisse). Schweiz Mineral Petrogr Mitt 69: 237–250

Argand E (1909) L'exploration géologique des Alpes pennines centrales. Bull Soc Vaudoise Sci Nat 45: 217–276

Argand E (1911) Sur la tectonique de la grande zone permo-houillère des Alpes occidentales. Eclogae Geol Helv 11: 747–751

Arthaud F, Matte P (1977) Late Paleozoic strike-slip faulting in Southern Europe and Northern Africa: result of a right-lateral shear zone between the Appalachians and the Urals. Geol Soc Am Bull 88: 1305–1320

Bearth P (1963) Contribution à la subdivision tectonique et stratigraphique du Cristallin de la nappe du Grand-St-Bernard dans le Valais (Suisse). Livre Mém Prof Paul Fallot Soc Géol Fr Mém hs tome II: 407–418

Bearth P (1964) Geologischer Atlas der Schweiz 1:25'000, Blatt 43: Randa mit Erläuterungen. Schweiz Geol Komm

Bearth P (1973) Geologischer Atlas der Schweiz 1: 25'000, Blatt 61: Simplon mit Erläuterungen. Schweiz Geol Komm

Bearth P (1980) Geologischer Atlas der Schweiz 1:25'000, Blatt 71: St. Niklaus mit Erläuterungen. Schweiz Geol Komm

Bocquet J (1974) Il metamorfismo prealpino nella Vanoise (Savoia) e altri settori dello zoccolo brianzones. Mem Soc Geol Ital 3: 271–284

Burri M (1975) Sur les structures de la zone houillère dans les vallées des Drances (Vs). Eclogae Geol Helv 68: 285–299

Burri M (1979) Les formations valaisannes dans la région de Visp. Eclogae Geol Helv 72: 789–802

Burri M (1983a) Description géologique du front du Saint Bernard dans les vallées de Bagnes et d'Entremont (Valais). Bull Lab Géol Univ Lausanne 270: 1–88

Burri M (1983b) Le front du Grand St-Bernard du val d'Hérens au val d'Aoste. Eclogae Geol Helv 76: 469–490

Burri M, Jemelin L (1983) Notice explicative de la feuille Sembrancher (Feuille 77 de l'Atlas). Atlas géologique de la Suisse 1. 25'000. Comm Géol Suisse

Caby R (1968) Contribution à l'étude structurale des Alpes occidentales: subdivisions stratigraphiques et structures de la zone du Grand-Saint-Bernard dans la partie sud du Val d'Aoste (Italie). Géol Alp 44: 95–111

Caby R (1974) Gneiss permo-carbonifères d'origine granitique et volcanique dans la zone houillère et la zone du Grand-Saint-Bernard en Val d'Aoste (Italie). Géol Alp 50: 39–44

Chessex R Les métabasites de l'unité du Métailler (nappe du Mont Fort, Valais). Arch Sci Genève (in prep)

Christ P (1925) Das produktive Karbon der Schweizeralpen, Teil II: die walliser Anthrazitlagerstätten und der walliser Anthrazitbergbau während der Jahre 1917–1924. Beitr Geol Schweiz Geotech Ser XI, 162 pp

Coleman RG, Lee DE, Beatty LB, Brannock WW (1965) Eclogites and eclogites: Their Differences and Similarities. Geol Soc Am Bull 76: 483–508

Desmons J, Fabre J (1988) Contribution à la connaissance pétrographique du Mont Pourri (Savoie, France); conséquences structurales. Géol Alp 64: 13–26

Desmons J, Ploquin A (1989) Chimisme du Briançonnais-Grand Saint-Bernard Anté-Mésozoïque (Alpes Occidentales): une vue synoptique. Géol Alp 65: 1–31

Eikenberg J, Köppel V, Labhart T, Signer P (1989) U-Pb, U-Xe and U-Kr systematics of a greenschist facies metamorphic uranium mineralization of the Siviez-Mischabel nappe (Valais, Switzerland). Schweiz Mineral Petrogr Mitt 69: 331–344

Escher A (1988) Structure de la nappe du Grand Saint-Bernard entre le val de Bagnes et les Mischabel. Rapp Géol Serv Hydrol Geol Natl Berne 7: 1–26

Escher A, Masson H, Steck A (1988) Coupes géologiques des Alpes occidentales suisses. Rapp Géol Serv Hydrol Geol Natl Berne 2: 1–11

Fabre J (1987) Evénements tectoniques et sédimentaires pré-alpins dans la zone briançonnaise (Zone Houillère) des Alpes occidentales. In: Sassi FP, Bourrouilh R (eds) IGCP Proj 5 Newslett 7, Padova, Bordeaux, pp 143–147

Fabre J, Schade J, Baudin T, Desmons J, Mercier D, Perrucio-Parison MD (1987) Relics of pre-Mesozoic events in the Briançon zone (northern French Alps). In: Flügel HW, Sassi FP, Grecula P (eds) Pre-Variscan and Variscan events in the Alpine-Mediterranean mountain belts. Mineralia slovaca Monogr, Alfa, Bratislava, pp 183–208

Frisch W, Ménot RP, Neubauer F, von Raumer JF (1990) Correlation and evolution of the Alpine basement. Schweiz Mineral Petrogr Mitt 70: 265–285

Gebauer D, Quadt A, Compston W, Williams IS, Grünefelder M (1988) Archean zircons in a retrograded Caledonian eclogite of the Gotthard Massif (Central Alps, Switzerland). Schweiz Mineral Petrogr Mitt 68: 485–490

Gerlach H (1871) Das Südwestliche Wallis mit den angrenzenden Landesteilen von Savoien und Piemont. Beitr Geol Karte Schweiz IX

Gouffon Y (1992) Géologie de la "nappe" du Grand St-Bernard entre la Doire Baltée et la frontière suisse (Vallée d'Aoste, Italie). Mém Géol 12: 147 pp Lausanne

Holdaway MJ, Dutrow BL, Hinton RW (1988) Devonian and Carboniferous metamorphism in west-central Maine: The muscovite-almandine geobarometer and the staurolite problem revisited. Am Mineral 73: 20–47

Jäckli R (1950) Geologische Untersuchungen in der Stirnzone der Mischabeldecke zwischen Réchy, Val d'Anniviers und Visp (Wallis). Eclogae Geol Helv 43: 31–93

Jeanbourquin P, Burri M (1989) La zone de Sion-Courmayeur dans la région du Simplon. Rapp Géol Serv Hydrol Geol Natl Berne 11: 1–35

Jongmans WJ (1960) Die Karbonflora der Schweiz. Beitr Geol Karte Schweiz NF 108, 97 pp

Lombard A (1942) Excursion à la mine de Chandoline près de Sion. Eclogae Geol Helv 35: 158–159

Lombard A (1949) La formation et la mise en place des anthracites du Valais. Ann Mines Paris 138: 3–14

Liegois JP, Duchesne JC (1981) The Lac Cornu retrograded eclogites (Aiguilles Rouges massif, Western Alps, France): evidence of crustal origin and metasomatic alteration. Lithos 14: 35–48

Loeschke J (1989) Lower Paleozoic volcanism of the Eastern Alps and its geodynamic implications. Geol Rundsch 78: 599–616

Maggetti M, Galetti G (1988) Evolution of the Silvretta eclogites: metamorphic and magmatic events. Schweiz Mineral Petrogr Mitt 68: 467–484

Manby GB, Thiedig F (1988) Petrology of eclogites from Saualpe, Austria. Schweiz Mineral Petrogr Mitt 68: 441–466

Marthaler M (1984) Géologie des unités penniques entre le val d'Anniviers et le val de Tourtemagne (Valais, Suisse). Eclogae Geol Helv 77: 395–448

Meschede M (1986) A method of discriminating between different types of mid-ocean ridge basalts and continental tholeiites with the Nb-Zr-Y diagram. Chem Geol 56: 207–218

Monié P (1990) Preservation of Hercynian $^{40}Ar/^{39}Ar$ ages through high-pressure low-temperature alpine metamorphism in the Western Alps. Eur J Mineral 2: 343–361

Mullen ED (1983) $MnO/TiO_2/P_2O_5$: a minor element discriminant for basaltic rocks of oceanic environments and its implications for petrogenesis. Earth Planet Sci Lett 62: 53–62

Müller R (1982) Die tektonische Erforschung der Penninischen Alpen. Eclogae Geol Helv 75: 447–479

Oulianoff N (1957) La zone du Carbonifère entre Isérables et Bramois (vallée du Rhône). Eclogae Geol Helv 50: 482–491

Oulianoff N, Trümpy R (1958) Atlas géologique de la Suisse 1: 25'000, Feuille 33: Grand St. Bernard (avec notice explicative). Comm Géol Suisse

Paquette JL, Ménot RP, Peucat JJ (1989) REE, Sm-Nd and U-Pb zircon study of eclogites from the Alpine External Massifs (Western Alps): Evidence for crustal contamination. Earth Planet Sci Lett 96: 181–198

Pearce JA (1980) Geochemical evidence for the genesis and eruptive setting of lavas from Tethyan ophiolites. In: Panayiotou A (ed) Proc Int Ophiolite Symp Cyprus 1979, Rep of Cyprus Geol Surv Dep Nicosia, pp 261–272

Pearce JA, Cann JR (1973) Tectonic setting of basic volcanic rocks determined using trace elements analysis. Earth Planet Sci Lett 19: 290–300

Pearce JA; Norry MJ (1979) Petrogenetic implications of Ti, Zr, Y and Nb variations in volcanic rocks. Contrib Mineral Petrol 69: 33–47

Pfeifer HR, Colombi A, Ganguin J (1989) Zermatt-Saas and Antrona Zone: a petrographic and geochemical comparison of polyphase-metamorphic ophiolites of the West-Central Alps. Schweiz Mineral Petrogr Mitt 69: 33–47

Pin C, Peucat JJ (1986) Ages des épisodes de métamorphisme paléozoïques dans le Massif central et le Massif armoricain. Bull Soc Géol Fr II(3): 461–469

Pupin JP (1980) Zircon and granite petrology. Contrib Mineral Petrol 75: 207–220

Rahn M (1991) Eclogites from the Minugrat, Siviez-Mischabel nappe (Valais, Switzerland). Schweiz Mineral Petrogr Mitt 71: 415–426

von Raumer JF (1983) Die Metapelite von Emosson (Aiguilles Rouges Massiv) als Beispiel spätkaledonisch-frühvariszischer Metamorphose im Altkristallin des helvetischen Bereichs. Schweiz Mineral Petrogr Mitt 63: 421–455

von Raumer JF (1987) Les massifs du Mont Blanc et des Aiguilles Rouges: Témoins de la formation de croûte varisque dans les Alpes occidentales. Géol Alp 63: 7–24

Sartori M (1990) L'unité du Barrhorn (Zone Pennique, Valais, Suisse). Mém Géol Lausanne, 6: 156 pp

Sartori M, Thélin P (1987) Les schistes oeillés de Barneuza (Nappe de Siviez-Mischabel, Valais, Suisse). Schweiz Mineral Petrogr Mitt 67: 229–256

Sartori M, Bugnon PC, Frey M, Ganguin J, Masson H, Steck A, Thélin P (1989) Compte-rendu de l'excursion commune de la SSMP et de la SGS le long du profil Rawil-Zermatt 9/10/11 octobre 1988. Schweiz Mineral Petrogr Mitt 69: 261–281

Schaer JP (1959 a) Géologie de la partie septentrionale de l'Eventail de Bagnes. Arch Sci Genève 12: 473–620

Schaer JP (1959 b) Les porphyres quartzifères et les roches volcaniques prétriasiques de la nappe du Grand St-Bernard, Valais, Suisse. Geol Rundsch 46: 147–158

Stampfli GM, Marthaler M (1990) Divergent and convergent margins in the North-Western Alps: confrontation to actualistic models. Geodin Acta Paris 4: 159–184

Staub W (1927) Zur Tektonik des Gebirges zwischen Turtmanntal und Simplonpass. Eclogae Geol Helv 20: 247–252

Staub W (1934) Kleintektonik und Oberflächengestalt auf Westseite des Vispertales. Eclogae Geol Helv 27: 517–519

Stille P (1980) On the genesis of the amphibolites and hornblendefelses in the Berisal Complex (Simplon, Italy-Switzerland). Mem Inst Geol Univ Padova, tome XXXIV: 205–246

Stille P, Oberhänsli R (1987) Petrology and Alpine metamorphic evolution of the metabasic and ultrabasic rocks in the Berisal crystalline complex. Lithos 20: 169–180

Stille P, Tatsumoto M (1985) Precambrian tholeiitic-dacitic rock-suites and Cambrian ultramafic rocks in the Pennine nappe system of the Alps: Evidence from Sm-Nd isotopes and rare earth elements. Contrib Mineral Petrol 89: 184–192

Thélin P (1983) Les gneiss oeillés de la nappe du Grand Saint-Bernard: essai d'évaluation des critères susceptibles d'en préciser l'hérédité pré-métamorphique (Alpes valaisannes). Thèse, Univ Lausanne, 483 pp

Thélin P (1987) Nature originelle des gneiss oeillés de Randa. Bull Lab Géol Univ Lausanne 290: 1–75

Thélin P (1989) Essai de chronologie magmatico-métamorphique dans le socle de la nappe du Grand Saint-Bernard: quelques points de repère. Schweiz Mineral Petrogr Mitt 69: 193–204

Thélin P (1992) Les métapélites du Mont Mort: une fenêtre métamorphique (nappe des Pontis, Valais). Bull Soc Vaudoise Sci Nat 82: 97–116

Thélin P, Ayrton S (1983) Cadre évolutif des événements magmatico-métamorphiques du socle anté-triasique dans le domaine pennique (Valais). Schweiz Mineral Petrogr Mitt 63: 393–420

Thélin, P, Sartori M. Lengeler R, Schaerer JP (1990) Eclogites of Paleozoic or early Alpine age in the basement of the Penninic Siviez-Mischabel nappe, Wallis, Switzerland. Lithos 25: 71–88

Trümpy R (1966) Considérations générales sur le "Verrucano des Alpes suisses". In: Tongiorgi M, Rau A (eds) Atti del symposium sul Verrucano, Pisa 1965. Soc Tosc Sci Nat Pisa, pp 212–232

Trümpy R, Dösseger R (1972) Permian of Switzerland. In: Cuvillier J, Schürman P (eds) Rotliegend, essays of European Lower Permian. Int Sediment Petrogr Ser, Leiden, Brill, XV: 189–215

Vallet JM (1948) Sur la nature "des orthogneiss" de Thion (Val d'Hérens, Valais). Arch Sci Genève 1: 409–412

Vallet JM (1950) Etude géologique et pétrographique de la partie inférieure du Val d'Hérens et du Val d'Hérémence (Valais). Schweiz Mineral Petrogr Mitt 30: 322–476

Wegmann E (1923) Zur Geologie der St-Bernard Decke im Val d'Hérens. Bull Soc Neuchâtel Sci Nat 47: 3–63

Wehrli L (1925) Das produktive Karbon der Schweizeralpen, Teil I: Übersicht und Geschichte des Bergbaues von seinen Anfängen bis Mitte 1917 mit besonderer Berücksichtigung der Anthrazite des Wallis. Beitr Geol Schweiz Geotech Ser XI, 110 pp

Werenfels A (1924) Geologische und petrographische Untersuchungen des Vispertales. Beitr Geol Karte Schweiz 26/III, pp 81–125

Wust GH, Baehni LA (1986) The distinctive tectonometamorphic evolution of two basement complexes belonging to the Grand-Saint-Bernard Nappe (Val de Bagnes, Valais). Schweiz Mineral Petrogr Mitt 66: 53–71

Ziegler PA (1984) Caledonian and Hercynian crustal consolidation of Western and Central Europe – a working hypothesis. Geol Mijnbouw 63: 93–108

Zingg MA (1989) Die Siviez-Mischabel-Decke: Entstehung und Entwicklung eines Altkristallins. Schweiz Mineral Petrogr Mitt 69: 130

The Dora-Maira Massif

R. Sandrone, P. Cadoppi, R. Sacchi, P. Vialon

Abstract

Shaped as an "ellipsoid" of about 70 × 25 km, the Dora-Maira Massif is the crystalline basement of the Cottian Alps. It consists of an upper complex formed of pre-Carboniferous metasediments and metabasites, and a lower complex composed of (Permo)-Carboniferous metasediments. Both complexes contain meta-intrusives of granitic to dioritic composition, which are mainly regarded as Variscan. The sedimentary protoliths in the pre-Carboniferous basement are represented mostly by pelites with subordinate, more or less dolomitic limestones, whereas coarser clastic facies predominate in the (Permo)-Carboniferous cover. The Mesozoic sequence overlying the massif is similar to that of the other Penninic units.

The Alpine, tectono-metamorphic overprint is very strong throughout the massif: disregarding the "Polymetamorphic Unit" with its very high-pressure assemblages, the conditions of the early-Alpine metamorphic event in the massif are comparable with those of the other internal units of the Western Alps. The eclogite and blueschist facies stages were followed – and in some cases almost completely obliterated – by recrystallization into the greenschist facies. The Alpine metamorphism was accompanied by tectonic events including at least three phases of ductile deformation.

1 Introduction

Shaped as an "ellipsoid" of about 70 × 25 km, the Dora-Maira Massif is the crystalline basement of the Cottian Alps. The French authors class it as one of the *"massifs cristallins internes"* along with Monte Rosa and Gran Paradiso. Argand (1911) assigned its lower unit (comprising the *"Pinerolese"* Carboniferous sequences and some augengneisses) to the Penninic "Nappe No. 4" (Grand St. Bernard), and its upper unit (essentially basement rocks with augengneisses) to the Penninic "Nappe No. 5" (Monte Rosa). The topic is controversial, as there are two schools of thought. The first (Argand 1911) suggests that an upper, "Piedmontese" unit belonging to the Monte Rosa nappe was carried over a lower, more internal "Briançonnais" unit of the Grand St. Bernard nappe. This interpretation was adopted, inter alia, by Michard (1967) and by the 1:1000000 Structural Model of Italy (Sturani, in Ogniben et al. 1975). The alternative, autochthonous (or, to be more precise, unitary) view holds that there is a single crystalline basement covered by the late Palaeozoic terranes. This was the opinion of the surveyors of the 1:100000 Geological Map of Italy (sheets 55, 67 and 78–79) at the beginning of the century. It was then adopted by Vialon (1966) and has recently been further substantiated by additional evidence of some affinities in the magmatism of the two units (Sandrone et al. 1988).

Due to the age of some of the protoliths, Dora-Maira is generally described as a "Variscan" Massif, in spite of the mainly Alpine age of both its structure and metamorphism.

The memoirs of Vialon (1966, with 1:100000 geological map) and Michard (1967) are today the main source of information along with some papers produced in the 1980s, mainly centred on the Alpine-age tectonic and metamorphic evolution. Furthermore, Chopin (1984) made a remarkable discovery that aroused a great deal of interest in the massif, namely the identification, within pyrope-, phengite-, talc-, rutile- and kyanite-bearing quartzites cropping out in a small area near Martiniana Po (Fig. 1), of coesite within the pyrope porphyroblasts, pointing to the presence of rock volumes having attained extreme pressure conditions corresponding to depths of the order of 100 km during the Alpine-age evolution.

The Mesozoic sequence overlying the massif (Michard 1967) is similar to that of the other Penninic units. In Susa Valley where it is rather well exposed, a "siliceous" (?) Lower Trias grading downward into a sequence of probably Permian age (Vialon 1966) is followed (Tallone 1990) by dolostones (Middle Trias), grey-bluish marbles (? Malm) and Schistes Lustrés (Upper Cretaceous), with ophiolitic slices mainly preserved along thrust planes (Marthaler et

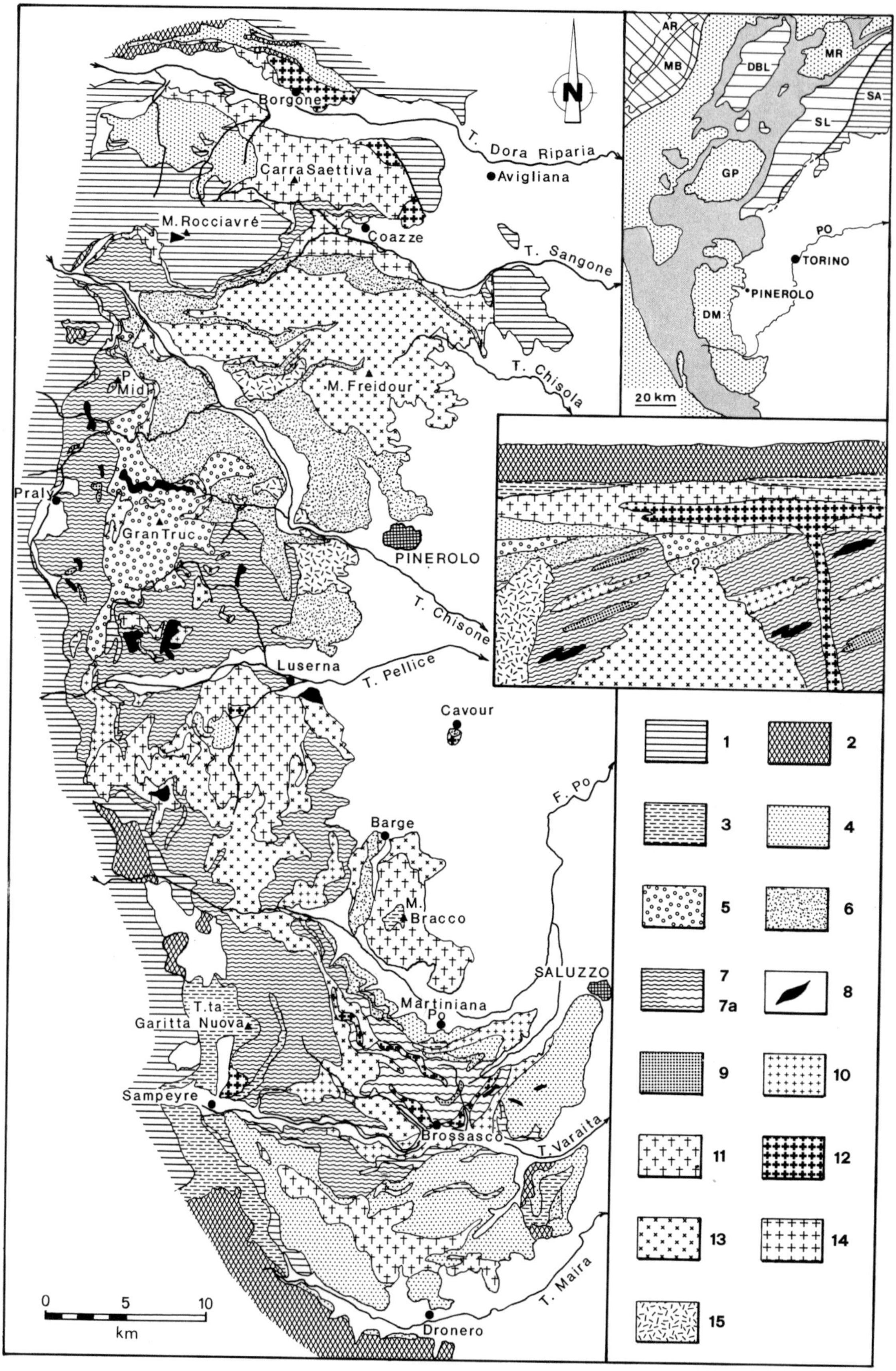

Borgone
T. Dora Riparia
Carra Saettiva
Avigliana
M. Rocciavré
Coazze
T. Sangone
M. Freidour
T. Chisola
P. Midi
Praly
Gran Truc
PINEROLO
T. Chisone
Luserna
T. Pellice
Cavour
F. Po
Barge
M. Bracco
SALUZZO
T.ta Garitta Nuova
Martiniana Po
Sampeyre
Brossasco
T. Varaita
T. Maira
Dronero
0 5 10
km
AR
MB
DBL
MR
SA
SL
GP
PO
TORINO
PINEROLO
DM
20 km
1
2
3
4
5
6
7
7a
8
9
10
11
12
13
14
15

al. 1986). Schistes Lustrés also occur within the basement, preserved along thrust planes, mainly in the central-southern portion of the massif (Fig. 1).

2 Alpine-Age Overprint

The Dora-Maira Massif belongs to the metamorphic, axial zone of the Alpine chain; its Alpine-age tectono-metamorphic overprint is therefore very strong, so as to make the Variscan palimpsest almost illegible.

As a consequence and a confirmation, the radiometric ages calculated using several methods are always <125 Ma (Monié and Chopin 1991 and references therein) with the exception of an age of 2000 ± 500 Ma (whole rock, Sm-Nd model age; Tilton et al. 1989).

The evolution of the Alpine metamorphism is more complex here than in the other internal units, although the usual trend from eclogitic (early-Alpine event of Cretaceous age) to blueschist- to greenschist-facies (meso-Alpine event of Eocene-Oligocene age) assemblages is apparent.

In the central-northern sector of the massif, the conditions of the early-Alpine event are comparable with those of the other internal units (Borghi and Sandrone 1990; Cadoppi 1990 with references). Pressures of about 14–15 kb and temperatures around 500 °C have been proposed for the eclogitic event. The subsequent formation of glaucophane probably occurred at P ≥ 9–10 kb.

In the southern portion of the massif, on the other hand, in a relatively small area (ca. 50 km^2) between the Po and Varaita valleys, Chopin et al. (1991 with references to earlier work) identified an unusual, coesite-bearing (as already mentioned) unit (Fig. 1) climaxing around 700 °C and 30 kb. The main rock-types are mica-rich "quartzites" ("whiteschists" in Kienast et al. 1991) with pyrope and coesite, sodic whiteschists, garnet-kyanite micaschists, marbles, kyanite-eclogites, and metagranites.

The eclogitic and glaucophanic stages were followed and in some cases almost completely obliterated by recrystallization in the greenschist facies. This recrystallization, too, comprised two decompression stages: the first followed the early-Alpine metamorphism perhaps without a time break, under isothermal conditions, or slightly rising temperature and at intermediate pressure; a subsequent stage, preceded by a cooling phase, attained P-T values around 550 °C and 3 kb (Borghi and Sandrone 1990).

The Alpine metamorphism was accompanied by a tectonic evolution consisting of at least three phases of ductile deformation. The axes of the early (obviously one of the first but probably not the earliest Alpine structure in the massif), ubiquitous "cross-folds" trend about E–W; their axial plane dip N in the northern sector and S in the central-southern sector. The second-phase folds have axes trending N–S and W-dipping axial planes, and are mainly concentrated along the western margin of the massif. The third phase displays a rather spaced-out crenulation cleavage connected with folds of variable size. This is es-

Fig. 1. Geological sketch map of the Dora-Maira Massif. **Upper inset** simplified tectonic map of the Western Alps. *Oblique ruling* Helvetic domain (*AR* Aiguilles Rouges; *MB* Monte Bianco). *Dotted area* Penninic basement (*MR* Monte Rosa, *GP* Gran Paradiso, *DM* Dora-Maira massifs). *Grey area* Piemonte ophiolite nappe. *Spaced horizontal ruling* Austro-alpine system (*DBL* Dent Blanche nappe; *SL* Sesia Lanzo zone). *Dense horizontal ruling* Southern Alps *(SA)*. **Lower inset** diagrammatic cross-section. *Symbols* as in the map. Legend to map: *1* Piemonte ophiolite nappe (undifferentiated) and minor bodies within the Dora-Maira Massif; *2* Mesozoic cover (marble and metadolomite); *3* micaceous quartzite (e.g. Monte Bracco); phengite-quartzite grading into quartzitic micaschist and phengite-schist (*"Ensemble de Sampeyre"*, Vialon 1966) (Permo-Trias?); *4* monometamorphic, (Permian?) chloritoid ± garnet micaschist grading into phengite-(± amphibole) quartzite (Susa Valley); micaschist and fine-grained gneiss with probable relics of acidic volcanics (Maira Valley); *5* fine-grained albite gneisses (*"gneiss minuti; "faciès devenant charbonneux"*, Vialon 1966) with thin lenses of quartzite (Permian?); *6* graphite-bearing micaschist, meta-arenite, meta-conglomerate, quartzite (*"Complesso Grafitico del Pinerolese"* auct.) (Carboniferous?); *7* polymetamorphic, garnet-chloritoid ± glaucophane micaschist and fine-grained gneiss with relics (muscovite, garnet) of a medium-grade, pre-Alpine assemblage; *7a* area of occurrence of very high-pressure assemblages (pyrope + coesite); *8* eclogite-facies metabasite (mostly re-equilibrated into greenschist facies) within *7;* fine-grained metabasite (greenschist facies) intercalated in *4* and *5*; *9* impure, micaceous marble, sometimes with relics of diopside. *10* Gneiss with large (2–6 cm), polycrystalline nodules (mainly made of albite and relics of K-feldspar) and relics of a relatively high-T assemblage (*"gneiss amygdalaires"*, Vialon 1966); *11* flaggy, fine-grained phengitic augengneiss (*"Pietra di Luserna"* auct.), tourmaline-rich leucocratic gneiss (Susa Valley), augengneiss with relics of magmatic texture (Sangone Valley) with thin intercalations of "silvery micaschists" (*"porphyroïdes arkosiques"*, Vialon 1966); *12* various type of metagranitoids with well preserved magmatic texture and parageneses: porphyritic, aluminous, muscovite bearing metagranite (*Borgone* type); peraluminous muscovite-garnet metagranite (Rocca di Cavour); porphyric metagranite to granodiorite (*Brossasco* type); *13* augengneiss with locally preserved, magmatic texture and assemblage, and with alkaline affinity (*Freidour* type); *14* porphyric metagranite to monzogranite with subalkaline affinity (*Sangone* type); *15* metaintrusive of dioritic to granodioritic composition with sometimes well-preserved magmatic texture (Chisone Valley) (*"Malanaggio diorite"* auct.)

sentially the picture given by Vialon in 1966 and not much modified by later workers, at least in the northern portion of the massif. The early "cross" folds and lineations are now viewed as transport directions connected with thrust tectonics (discussion and references in Borghi et al. 1984). The shear indicators generally suggest transport to the west (Laurent and Etchecopar 1976). Westerly dips are predominant so that lower and lower units are crossed from west to east. This setting developed in early-Alpine time according to Vialon (1966).

3 The Rocks

Polymetamorphic rocks belonging to an ancient basement were described by several workers in the massif. These are characterized by medium-grade, pre-high-pressure, relict assemblages (Vialon 1966, p. 158), lacking in the Mesozoic cover and therefore referable to a pre-Alpine event. Pre-Alpine relics are comparatively rare; the mineralogical assemblages are thus assumed to be of Alpine age in the following description unless otherwise specified.

The structurally higher complex (Argand's nappe No. 5) forms the outer shell being mainly exposed in the western part of the massif. Unlike the lower complex, it comprises rocks to be viewed as polymetamorphic (based on the reasoning outlined already). It is formed by:

1. Micaschists and fine-grained gneisses: phengite, garnet, chloritoid and rutile ±glaucophane assemblage with relics (muscovite, garnet, chloritoid pseudomorphs after staurolite) of a medium-grade, pre-Alpine assemblage;
2. Eclogite-facies metabasites, usually as lenses and boudins in (1), less commonly in (3). These also form two large bodies (1 to 2 km across) in Pellice Valley (Fig. 1);
3. Impure, generally micaceous marbles, sometimes with relics of diopside;
4. Gneisses characterized by large (2–6 cm), polycrystalline nodules mainly made of albite with relics of K-feldspar (*"gneiss amygdalaires"*, Vialon 1966). These rocks also bear relics of a relatively high-T assemblage (red biotite, muscovite, plagioclase, garnet) and were interpreted as metagranites by some workers (e.g. Sandrone et al. 1986);
5. A discontinuous talc level intercalated in (1), and associated with (3) and with (6) (Sandrone et al. 1990, with references);
6. The *"Porphyroïdes arkosiques"* (Vialon 1966): a problematic sequence made essentially of flaggy, fine-grained (*"Pietra di Luserna"*) augengneisses and of tourmaline-rich leucocratic gneisses (Susa Valley; Cadoppi 1990), with thin (up to a few metres) intercalations of *"micascisti argentei"* ("silvery micaschists"), i. e. schists almost entirely composed of phengite and quartz ±Mg-chlorite. Other features of this sequence are its structural location at the roof of the massif (outside the Po-Varaita divide) and its general shape, which approaches that of a flat body. Some of us (Vialon 1966) viewed it as a Permian volcanic-sedimentary sequence and some (Cadoppi 1990 with references) as mainly metagranite (in fact, both lithologies are probably present). This sequence is strongly reminiscent of the equally controversial Roffna Gneiss in the Suretta nappe (see e. g., on one side, Hanson et al. 1969 with references, and on the other side, Vocke et al. 1987 with references);
7. Pre-Alpine, Late to post-Variscan, supposedly Permian to Permo-Trias cover sequences including (7a) chloritoid- garnet micaschists grading into mica-(±amphibole)-rich quartzites (Susa Valley), (7b) fine-grained gneisses with probable relics of volcanics (Maira Valley; Bonioli et al. 1992) and (7c) quartzites (±conglomerates) grading into phengite-quartzites and quartz-rich micaschists well developed in the southern part of the Massif (Varaita and Maira Valleys; *"Esemble de Sampeyre"*, Vialon 1966).

A monometamorphic sequence attributed by Argand to the "Nappe No. 4" (Grand St. Bernard) is exposed at a lower structural level in the Pinerolo area and near the Po Valley outlet. This includes:

8. Graphite-bearing micaschists associated with meta-arenites and metaconglomerates (*"Série graphitique"*, Vialon 1966). A Carboniferous age is traditionally assigned to these rocks on the ground of lithological similarity with terranes of the Briançonnais zone of the Grand St. Bernard nappe;
9. Fine-grained albite gneisses (*"gneiss minuti"; "faciès devenant charbonneux"*, Vialon 1966) overlying (8) in clear stratigraphic continuity and generally regarded as Permian (though the polarity of the sequence cannot be determined). It should be pointed out that in the absence of fossils and radiometric data attribution to these rocks of a pre-Triassic age is based solely on the lithology and rather slender evidence of contact metamorphism induced by supposedly Palaeozoic granitoids.

Besides these rocks there are a great deal of intrusives generally referred to the Variscan event, metamorphosed in Alpine time. On textural, mineralogical and geochemical grounds we were able to distinguish several types, namely (see also the next section):

a) Widespread, allanite-rich, augen metagranite with locally preserved, magmatic texture and assemblage (*Freidour* type);
b) A porphyric metagranite intruding polymetamorphic rocks in lower Sangone Valley (*Sangone* type);
c) At the southernmost end of the massif, a jadeite-bearing metagranodiorite (*Brossasco* type);
d) A porphyric metagranite with primary muscovite in the Lower Susa Valley (*Borgone* type);
e) Bodies of diorite heavily transformed by the Alpine metamorphism, intruding the graphite bearing metasediments of the lower Chisone Valley ("*Malanaggio* diorite"); minor bodies preserving mineralogical evidence (glaucophane) of the early, high-pressure phase were found within polymetamorphic rocks in the Pellice Valley (Sandrone et al. 1988).

4 Geochemistry

The available geochemical data on the magmatic rocks of the massif are summarized hereafter. These rocks, especially the acidic and intermediate types, are generally thought of as Variscan in age. A few are ascribed to a pre-Variscan cycle, namely (1) the parent rock of the thin amphibolite intercalations within the polymetamorphic basement and (2) the parent rock of the *gneiss amygdalaires* seen by some of us as granitic in nature (which accounts for these rocks being dealt with in Sect. 4.2). On the same assumption (granitic protolith) we made use of the data concerning other problematic rocks such as the tourmaline-rich leucogneisses. The two quoted (Sect. 3) bodies of mafic rocks, along with a minor occurrence of gabbro described by Borghi et al. (1989) should also be mentioned. As to the age of these rocks, it is only known that they predate the early-Alpine metamorphism as indicated by relic assemblages.

Though the data are rather scarce and do not allow building of models, we think they are worth presenting in order to complete the state of the art account.

4.1 Intermediate and Basic Rocks

There are very few data for the basic rocks: six early analyses confined to the major elements (Grill 1925, 1929; Pagliani 1939; Grill et al. 1955), and some unpublished findings (again the major elements) by Bellion (1982) and Borghi (1983).

In the R_1R_2 diagram (De la Roche et al. 1980; Fig. 2) the points representing these rocks are spread along the projection of the silica saturation plane (line $R_1 = R_2$) and range from the field of mugearites to both that of the picrites, and that of the gabbros and diorites. Alkali mobility during the metamorphic processes may have been responsible for this dispersion. In the AFM diagram (Fig. 3), the points fall into the field defined by Desmons and Ploquin (1989) for the basic rocks of the Briançonnais-Grand St. Bernard basement.

The intermediate rocks in the polymetamorphic basement range from gabbro-diorites to diorites (Fig. 2), whereas those in the Pinerolo district are slightly more evolved and display a tonalitic to granodioritic chemistry (Sandrone et al. 1988). Their calcalkaline trend is evident in the AFM diagram (Fig. 3). According to Sandrone et al. (1988), the REE pattern (Fig. 4a) and the petrochemical features of the various types of metadiorite suggest

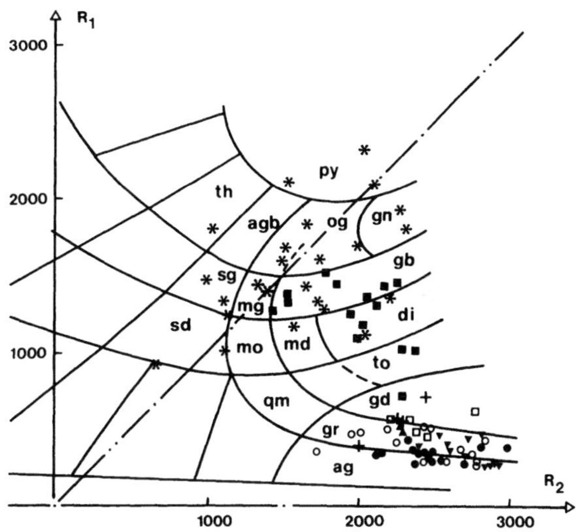

Fig. 2. Cumulative R_1R_2 diagram (De La Roche et al. 1980) for Dora-Maira rocks. Symbols and abbreviations: *asterisk* metabasics; *solid square* metadiorites; *cross* gneiss amygdalaires; *inverted solid triangle* augengneisses and metagranites (*Freidour* type); *solid triangle* porphyric metagranites (*Sangone* type); *open square* porphyric metagranites and granodiorites (*Brossasco* type); *open circle* porphyric aluminous metagranites (*Borgone* type); *solid circle* fined-grained augengneisses, leucocratic (± tourmaline-rich) gneisses and metagranitoids *(porphyroïdes arkosiques)*; *ag* alkaline-granite; *gr* granite; *qm* quartz-monzonite; *gd* granodiorite; *to* tonalite; *di* diorite; *gb* gabbro-diorite; *md* monzo-diorite; *mo* monzonite; *mg* monzo-gabbro; *sd* syeno-diorite; *sg* syeno-gabbro; *th* theralite; *agb* alkali-gabbro; *og* olivine-gabbro; *gn* gabbro-norite; *py* pyroxenite. $R_1 = 4Si - 11(Na + K) - 2(Fe + Ti)$; $R_2 = 6Ca + 2Mg + Al$

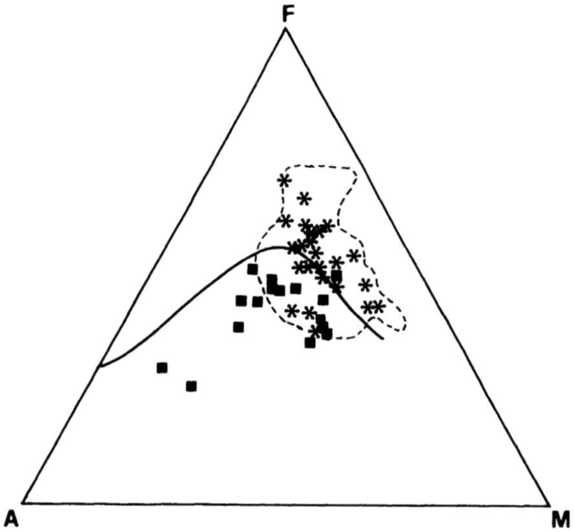

Fig. 3. Conventional AFM diagram for basic and intermediate rocks. *Symbols* as in Fig. 2. The *area* within the *dashed line* represents the field of basic rocks from the basement of Briançonnais – Gran San Bernardo zone. (After Desmons and Ploquin 1989).

that these rocks were formed from different degrees of partial melting of an amphibolitic protolith.

4.2 Acidic Rocks

These are distinctly granitic and alkali-granitic terms (Fig. 2), though some less differentiated terms are also present. The average composition of the *gneiss amygdalaires* is granodioritic, while their aluminous character could be seen as secondary due to loss of alkalis during metamorphism. By comparison with the other granitoids, the *gneiss amygdalaires* have a higher REE content, greater LREE as opposed to HREE fractionation, and a less pronounced, negative anomaly in Eu. An anatectic-crustal origin is suggested by their zircon typology (Cadoppi 1990). This evidence is not conclusive, however. The bulk of zircon characters could be the outcome of recrystallization during a high-T metamorphic phase.

Of the orthogneisses referable to the Variscan magmatic cycle, the least differentiated are the *Brossasco*-type metagranodiorites, and the Sangone Valley porphyric metagranites (*Sangone* facies). The REE patterns of these two granitoid types resemble that of the *gneiss amygdalaires:* marked LREE as opposed to HREE fractionation, slight negative Eu anomaly (Fig. 4).

Freidour-type meta-intrusives display a granite to alkaligranite composition (Fig. 2), with a subalkaline and weakly peraluminous character. The scanty REE (Fig. 4b) data show that the LREE-HREE fractionation of the *Freidour* facies in Sangone Valley (Cadoppi 1990) is slightly different from the *Freidour* augengneiss in Pellice Valley (Sandrone et al. 1986). These albeit slight differences could represent another type of magmatic evolution during the crystallization phases, or primary differences in the source.

The *Borgone* metagranite and some granitic facies from the *porphyroïdes arkosiques* display a major and trace element composition and a REE pattern similar to those of the *Freidour* granitoids, though their total REE content is lower (Fig. 4b).

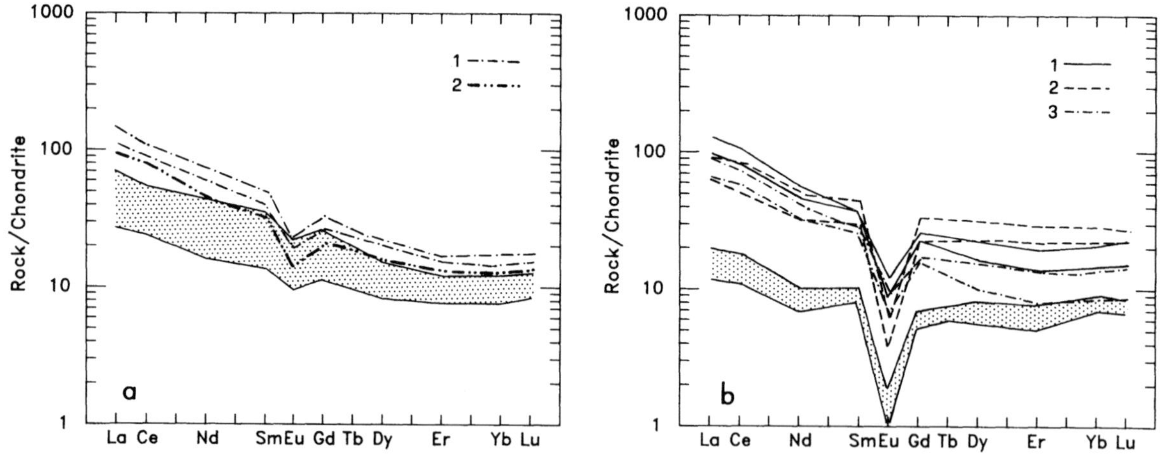

Fig. 4. Chondrite normalized REE patterns for **a** intermediate and **a** and **b** acidic rocks. **a** *Dotted area* metadiorites; *1* gneiss amygdalaires; *2* porphyric metagranites (*Sangone* type). **b** *Dotted area* tourmaline-rich leucogneisses; *1 Freidour* type augengneisses and metagranites (outcropping in Sangone Valley); *2 Freidour* type augengneisses and metagranites (outcropping in Pellice Valley); *3* augengneisses, fine-grained augengneisses *(porphyroïdes arkosiques)* and *Borgone* metagranites

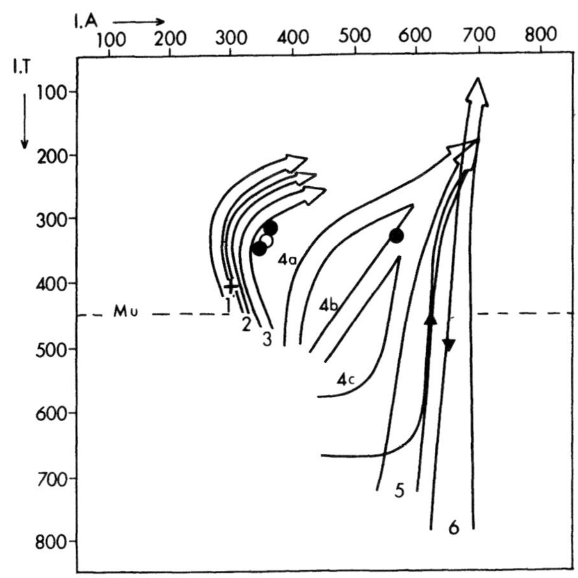

Fig. 5. A, T points for some of the granitic facies of the Dora-Maira Massif in the typologic evolution diagram of zircon populations, compared with typical distribution areas for different magmatic suites (after the petrogenetic classification of Pupin 1980). *Arrows* represent typologic differentiation trends. *1* Aluminous leucogranite; *2* (sub)autochthonous monzogranite and granodiorite; *3* intrusive aluminous monzogranite and granodiorite; *4 a, b, c* calc-alkaline series granite; *5* subalkaline series granite; *6* alkaline series granite. *Mu* Limit of the muscovite granite (I. T. < 450). *Symbols* as in Fig. 2

The flat pattern and strong Eu anomaly of the tourmaline-rich leucogneisses, on the other hand, coupled with their low total REE content, are compatible with their origin from greatly differentiated and variously boron-enriched magmas during the final crystallization phases (Alderton et al. 1980).

Zircon typology of samples from distinct granitoid groups (Cadoppi 1990) has provided partial confirmation of the characters indicated by the petrographic and chemical data. The *Borgone* metagranite zircon populations are mainly low-T types characteristic of aluminous granites (S_2–S_3) of anatectic-crustal origin (Fig. 5). The tourmaline-rich leucogneisses display similar characters, though with lower T, strongly metamict zircons, suggesting that these rocks developed from the crystallization of fluid-rich residual melts. The *Sangone* and *Freidour* metagranitoid zircons are mostly type P_2–P_4. It may thus be supposed that these rocks crystallized at higher temperature from a magma whose alkaline or sub-alkaline character is not revealed by their geochemistry.

The picture that thus emerges points to the presence of two distinct magmatic series: an alkaline-subalkaline *Freidour* series, and the *Borgone* facies with its more frankly crustal character. The data available, however, are too few to show whether this duality is a constant feature of the Dora-Maira Massif.

5 Discussion and Conclusions

A reconstruction of the pre-Alpine evolution of the Dora-Maira Massif, subject to the limits imposed by the heavy Alpine tectonic and metamorphic changes, can now be attempted.

The initial "situation" was that of an amphibolite-facies basement (the polymetamorphic Dora-Maira s. s.), primarily composed of metapelites (garnet micaschists), marbles, basic rocks, and granitoids (the *gneiss amygdalaires*). Neither the age of the protoliths (Lower Palaeozoic? Proterozoic? Brioverian?) nor that of metamorphism (Variscan? pre-Variscan?) are known.

During the Carboniferous, a detrital sequence rich in organic matter was deposited on this eroded, already metamorphic basement. Emplacement of granitic magmas – the soundest evidence of a Variscan event – also occurred (cf. the indications of contact metamorphism and a U-Pb zircon age of 303 ± 1 Ma obtained on the *Brossasco* metagranodiorite; J.-L. Paquette, pers. comm. 1989, in Chopin et al. 1991).

The nature of this event is the fundamental, unsolved problem. The main point is that rocks of Devonian-Dinantian age, i. e. belonging to the Variscan cycle, have not been detected in the massif. It is thus impossible to check the effects of the metamorphism linked to the Variscan cycle. Consideration can be given to a scenario in which the main metamorphism is Variscan; an uplift and erosion phase separates it from the deposition of a molasse (the *Pinerolese* Carboniferous) now recording the thermal metamorphism induced by the post-kinematic Variscan granites and diorites. This model is inspired and supported by the affinity of the *Série Graphitique* with the *Zone Houillère* of the Briançonnais.

In an alternative scenario, the main metamorphism is regarded as pre-Variscan. It was followed, after an erosion phase, by deposition of a detrital sequence, whose age could be Middle/Lower Carboniferous (the *Série Graphitique* once again) to Devonian (the associated fine-grained gneiss sequence shown as No. 9 in Sect. 3). At the crustal level now open to observation, the subsequent Variscan event is characterized by weak metamorphic effects (obliterated by and/or indistinguishable from Alpine effects), and by abundant emplacement of granite plutons over a time period long enough to allow the development of a still partly recognizable differentiation. This scenario, of course, would increase the age of the *Série Graphitique,* whose parallel may be sought in certain Taillefer (Belledonne) sequences.

Granitic magmatism is a late to post-Variscan event in both scenarios, in substance, the closing of the cycle. An event, none the less, with a number of uncertainties springing both from the obliterating effect of the Alpine tectonics (primary relations virtually never preserved), and the lack of radiometric data. There does exist one peculiar layered sequence containing metasediments, metavolcanics and metagranitoids (we are referring to units 6 and 7b; see Sect. 3). It may be supposed that this sequence formed in a rift domain, into which shallow, granitic bodies (e.g. *Borgone* granite, the granite facies of the *porphyroïdes arkosiques* sequence) intruded their volcanites. The rest of the granitoids, especially the widespread *Freidour* type with its mild alkaline trend, are apparently referable to a preceding phase. It needs hardly be stressed that this chronological succession from granites with an alkaline trend to calcalkaline granites (some with a peraluminous trend) raises difficulties when it is compared with the usual geodynamic patterns.

Bonin (1988) has put forward a "general" arrangement for the Variscan magmatism. Its key feature is a "critical period" at 270 Ma (Middle Permian) marking the transition to an alkaline trend connected with the establishment of a tensional regime. Comparison of the Dora-Maira granitoids with Bonin's arrangement is an interesting conjectural exercise. The granite suite resembling that of the "critical period" is the *Freidour* type. The *Borgone* granites find an equivalent in Bonin's "Lower Permian anatectic" suite, and their associated "volcanites" in the calcalkaline "Eo-Permian" magmatism near the Stefanian-Autunian boundary (Pitcher's Caledonian type 1982). A considerable upheaval, as can be seen, though here again the lack of radiometric datings must not be overlooked: granites regarded as Variscan may actually be older.

It is even more difficult to insert the supposedly Permian dioritic magmatism, which is thought to terminate the Variscan event, into any kind of arrangement. It does not resemble the Permian – Upper Triassic Monzoni and Mont Collon magmatism (Dal Piaz et al. 1977). According to Borghi et al. (1989) a Jurassic age could be envisaged within the compass of the model proposed by Lemoine et al. (1987) where a mantle magmatism in a thin crust is regarded as the product of the early opening phases of the Ligurian-Piedmontese Tethys. In the absence of datings, this can be no more than supposition: there is no certainty, indeed, that the diorites are not even older than the other granitoids.

References

Argand E (1911) Les nappes de recouvrement des Alpes pennines et leurs prolongements structuraux. Mat Carte Géol Suisse 31

Alderton DMH, Pearce JA, Potts PJ (1980) Rare earth mobility during granite alteration: evidence from southwest England. Earth Planet Sci Lett 49: 149–165

Bellion F (1982) Studio petrografico e geochimico delle metabasiti del Massiccio cristallino Dora-Maira. Thesis, Univ Torino

Bonin B (1988) From orogenic to anorogenic environments: evidence from associated magmatic episodes. Schweiz Mineral Petrogr Mitt 68: 301–311

Bonioli L, Cadoppi P, Sacchi R (1992) Occurrence of metavolcanics in the southernmost Dora-Maira Massif (Italian Western Alps). In: Carmignani L, Sassi FP (eds) Contributions to the geology of Italy with special regard to the Paleozoic basements. A volume dedicated to Tommaso Cocozza. IGCP Proj 276, Newsletter 5: 237–240

Borghi A (1983) Studio geologico della media Val Chisone (versante sinistro). Thesis, Univ Torino

Borghi A, Sandrone R (1990) Structural and metamorphic constraints to the evolution of a NW sector of Dora-Maira Massif (Western Alps). Mem Soc Geol Ital 45 (in press)

Borghi A, Cadoppi P, Porro A, Sacchi R, Sandrone R (1984) Osservazioni geologiche nella Val Germanasca e nella media Val Chisone (Alpi Cozie). Boll Mus Reg Sci Nat Torino 2: 503–530

Borghi A, Cadoppi P, Poli G, Sacchi R, Sandrone R (1989) Metaintrusives in the north-central part of the Dora-Maira Massif (Western Alps). Period Mineral 58: 171–172

Cadoppi P (1990) Geologia del basamento cristallino nel settore settentrionale del Massiccio Dora-Maira (Alpi Occidentali). PhD Thesis, Univ Torino

Chopin C (1984) Coesite and pure pyrope in high-grade blueschists of the Western Alps: a first record and some consequences. Contrib Mineral Petrol 86: 107–118

Chopin C, Henry C, Michard A (1991) Geology and petrology of the coesite bearing terrain, Dora Maira massif, Western Alps. Eur J Mineral 3: 263–291

Dal Piaz GV, De Vecchi GP, Hunziker JC (1977) The Austroalpine layered gabbros of the Matterhorn and Mt. Collon-Dents de Bertol. Schweiz Mineral Petrogr Mitt 57: 59–88

De la Roche H, Leterrier J, Grandclaude P, Marchal M (1980) A classification of volcanic and plutonic rocks using R_1R_2-diagram and major-element analyses – Its relationships with current nomenclature. Chem Geol 29: 183–210

Desmons J, Ploquin A (1989) Chimisme du Briançonnais – Grand Saint-Bernard anté-mésozoïque (Alpes Occidentales): une vue synoptique. Géol Alp 65: 1–31

Grill E (1925) Contributo alla conoscenza litologica della Valle Germanasca. Mem R Accad Naz Lincei Cl Sci Fis VI 1: 498–532

Grill E (1929) Rocce e minerali dei giacimenti di talco del Pinerolese. Atti Soc Toscana Sci Nat 39: 187–221

Grill E, Pagliani G, Sacchi L (1955) La genesi del talco della Valle della Germanasca (Alpi Cozie). Ist Lombardo Sci Lett Cl Sci 88: 442–490

Hanson GN, Grünenfelder M, Soptrayanova G (1969) The geochronology of a recrystallized tectonite in Switzerland – the Roffna Gneiss. Earth Planet Sci Lett 5: 413–422

Kienast JR, Lombardo B, Biino G, Pinardon JL (1991) Petrology of very high pressure eclogitic rocks from the Brossasco-Isasca complex, Dora Maira Massif, Italian Western Alps. J Metamorph Geol 9: 19–34

Laurent P, Etchecopar A (1976) Mise en évidence à l'aide de la fabrique du quartz d'un cisaillement simple à déversement ouest dans le Massif de Dora-Maïra (Alpes Occidentales). Bull Soc Géol Fr 18: 1387–1393

Lemoine M, Tricart P, Boillot G (1987) Ultramafic and gabbroic ocean floor on the Ligurian Tethys (Alps, Corsica, Apennines): In search of a genetic model. Geology 15: 622–625

Marthaler M, Fudral F, Deville E, Rampnoux JP (1986) Mise en évidence du Crétacé supérieur dans la couverture septentrionale de Dora Maïra, région de Suse, Italie (Alpes Occidentales). Conséquences paléogéographiques et structurales. CR Acad Sci Paris II 302: 91–96

Michard A (1967) Etudes géologiques dans les zones internes des Alpes Cottiennes. Cent Natl Rech Sci, Paris

Monié P, Chopin C (1991) $^{40}Ar/^{39}Ar$ dating in coesite-bearing and associated units of the Dora Maira massif, Western Alps. Eur J Mineral 3: 239–262

Ogniben L, Parotto M, Praturlon A (eds) (1975) Structural model of Italy. CNR, Roma (Quaderni de "La Ricerca Scientifica", vol 90)

Pagliani G (1939) Su alcune rocce e minerali della Miniera di talco delle Fontane (Valle della Germanasca). Atti Soc Ital Sci Nat 78: 25–39

Pitcher WS (1982) Granite type and tectonic environment. In: Hsü KJ (ed) Mountain building processes. Academic Press, London, pp 19–40

Pupin JP (1980) Zircon and granite petrology. Contrib Mineral Petrol 73: 207–220

Sandrone R, Cordola M, Fontan D, Sacchi R (1986) Pre-Variscan granitoids in the Dora-Maira Massif (Cottian Alps). Atti Accad Sci Torino 120: 179–189

Sandrone R, Sacchi R, Cordola M, Fontan D, Villa IM (1988) Metadiorites in the Dora-Maira polymetamorphic basement (Cottian Alps). Rend Soc Ital Mineral Petrol 43: 593–608

Sandrone R, Borghi A, Carossi G, Morsetti C, Tagliano C, Zucchetti S (1990) Geometry of the Fontane talc deposit and the structural evolution of the area (Dora-Maira Massif, Western Alps). Boll Assoc Min Subalpina 27: 45–62

Tallone S (1990) Il Dora-Maira settentrionale e le sue coperture mesozoiche: relazioni tra litostratigrafia, struttura ed evoluzione metamorfica. PhD Thesis, Univ Torino

Tilton GR, Schreyer W, Schertl HP (1989) Pb-Sr-Nd isotopic behaviour of deeply subducted crustal rocks from the Dora Maira Massif, Western Alps, Italy. Geochim Cosmochim Acta 53: 1391–1400

Vialon P (1966) Etude géologique du Massif cristallin Dora-Maïra (Alpes cottiennes internes - Italie). Trav Lab Géol Grenoble, Mém 4, 293 pp

Vocke RDJr, Hanson GN, Grünenfelder M (1987) Rare earth element mobility in the Roffna Gneiss, Switzerland. Contrib Mineral Petrol 95: 145–154

Evolution of Austro-Alpine and Upper Penninic Basement in the Northwestern Alps from Variscan Convergence to Post-Variscan Extension

G.V. Dal Piaz

Abstract

The oldest protoliths of the Austro-Alpine nappe system from the Western Alps consist of the intimate association of metapelites, tholeiitic metabasites and marbles which closely recalls the lithology and metamorphic evolution of the lower crustal section of the South-Alpine Ivrea-Verbano kinzigitic complex. This association may tentatively be interpreted as an Early Palaeozoic accretionary wedge, later deformed, thermally reworked and consolidated during the collisional Variscan orogeny; it was further characterized by Permian igneous underplating, low-P/high-T re-equilibration and production of anatectic melts. Similar high-T paragneisses are also the oldest protoliths in the Penninic Monte Rosa and Gran Paradiso nappes, although lesser amounts of mafic rocks and especially of marbles are recorded here; likewise the Penninic basement displays a late evolution characterized by cordierite-rich anatectic melts.

Lithospheric extension, uplift and thermal perturbance were demonstrably active during the Late Palaeozoic, when some basement sections operated as wall/roof rocks for huge intrusions of Late Carboniferous granites (Monte Rosa and Gran Paradiso) or Permian granitoids and ensialic cumulus gabbros (lower Austro-Alpine units), or when they were covered by Permian-Early Triassic clastic deposits. Late Palaeozoic bimodal magmatic activity and graben/pull-apart opening are also extensively recorded in the surrounding Grand St. Bernard, Canavese and South-Alpine domains and elsewhere in the whole Alpine chain. These Late Palaeozoic processes have traditionally been referred to the Late Variscan orogeny as evidence of an ensialic mobile belt or a trench-arc system. Both these models are rejected here and most of the "Late Hercynian orogenic events" are re-interpreted as post-Variscan metamorphic and magmatic signatures of Late Palaeozoic onset of lithosphere extension and thermal perturbance through simple shear mechanisms along deep low-angle detachment zones, evolving to asymmetric continental rifting. This may account for mantle melting by adiabatic decompression, underplating of gabbros, fast transfer of heat and fluids into the lower crust allowing production and removal of anatectic melts, and low-P/high-T re-equilibration of uplifting lower crustal segments.

1 Introduction

The collisional chain of the Western Alps consists of a thick pile of thin basement nappes with minor interbeddings of Mesozoic ophiolitic slices and post-Variscan cover sequences. Its upper section, which is exposed along the structural low of the Aosta Valley, northwestern Italy, exhibits: (1) the capping Austro-Alpine thrust system (Dent Blanche *s. l.* and Sesia-Lanzo nappes) and (2) the underlying Upper Penninic Monte Rosa/Arcesa-Brusson/Gran Paradiso basement nappes, separated by (3) the Mesozoic ophiolite- and calcschist-bearing Piedmont zone (Fig. 1). The latter is formed of: (1) the overlying greenschist facies and tectonically composite Combin zone, which includes some capping, locally Na-amphibole-bearing ophiolitic and metasedimentary sheets which are underlain by and locally interbedded with ophiolite-free thin décollement skins consisting of Permian-Cretaceous continental to marine cover sequences (Frilihorn and Bettaforca-Cime Bianche units; Marthaler 1984; Dal Piaz 1988); (2) the underlying eclogitic Zermatt-Saas ophiolitic nappe and southern homologues (e. g. Bearth 1967; Dal Piaz and Ernst 1978).

These continental and ophiolitic nappes show extensive Alpine signatures such as polyphase deformations, Eo-Alpine eclogitic metamorphism (130/90/65 Ma) and/or Meso-Alpine (Eocene-Early Oligocene) greenschist facies metamorphism (Dal Piaz et al. 1972; Frey et al. 1974; Hunziker 1974; Niggli 1978; Hunziker and Martinotti 1987; Polino et al. 1990; and references therein). Mineral assemblages indicating P-T-prograde to P-retrograde paths in the Eo-Alpine eclogitic units and the available P-T values

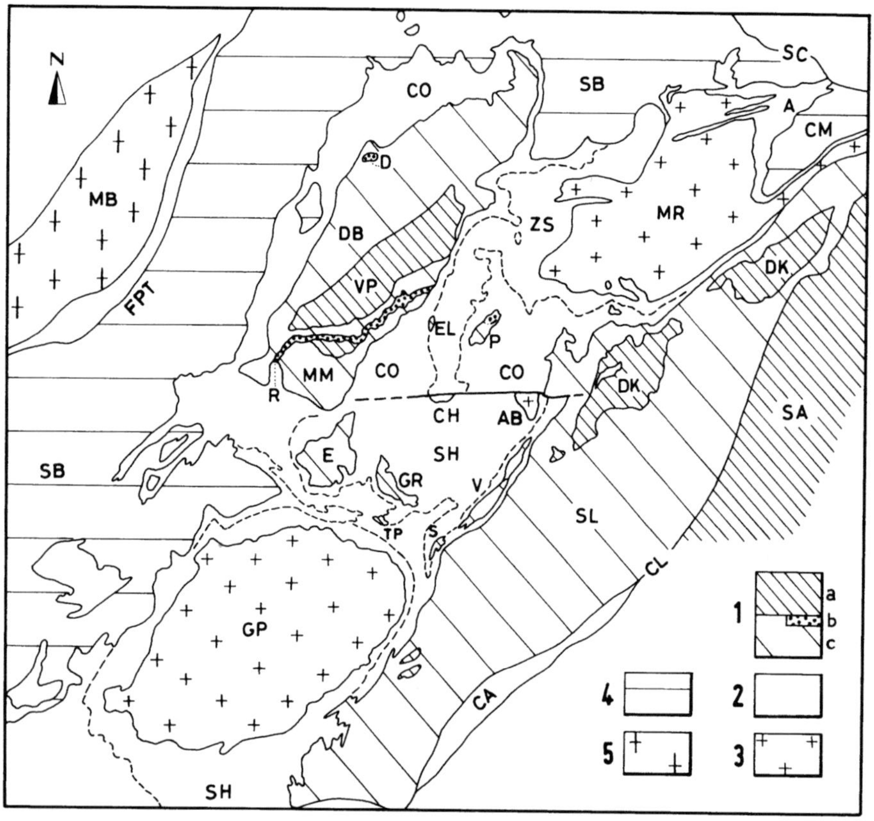

Fig. 1. Map of northwestern Alps indicating main tectonic units. *1a* Lower continental crust of Upper Austro-Alpine Valpelline *(VP)* and 2nd diorite-kinzigitic *(DK)* units and South-Alpine Ivrea zone *(SA)*; *1b* Lower Austro-Alpine Roisan *(R)* and Dolin *(D)* zones, including ophiolite-free Mesozoic cover and basement mylonites; *1c* Lower Austro-Alpine eclogitic (*SL* Sesia-Lanzo; *S* Santanel; *TP* Tour Ponton; *GR* Glacier-Rafray; *E* M. Emilius; *CH* Chatillon; *EL* Etirol-Levaz) and greenschist facies, locally Na-amphibole-bearing (*DB* Dent Blanche *s. s.; MM* M. Mary; *P* Pillonet; *V* Verres) basement nappes; *2* Piedmont ophiolitic nappe system (*A-ZS* Antrona and Zermatt-Saas eclogitic units; *SH* eclogitic southern homologues; *CO* greenschist, locally Na-amphibole-bearing Combin zone, and undifferentiated basal décollement sheets including ophiolite-free Permian-Cretaceous cover sequences; *3* Upper Penninic Monte Rosa *(MR)*, Arcesa-Brusson *(AB)* and Gran Paradiso *(GP)* nappes; *4* Middle and Lower Pennine basement and cover nappes, including Grand St. Bernard/Briançonnais nappe system *(SB)*; *5* Helvetic zone (*MB* M. Blanc massif); *FPT* frontal Penninic thrust; *SC* Simplon-Centovalli fault; *CL* Canavese fault system; *CA* Canavese zone

are consistent with the occurrence of a subduction-related steady-state thermal low operating at the Europe/Adria convergent plate margin since the Cretaceous (Dal Piaz et al. 1972; Ernst 1973; Hunziker 1974; Polino et al. 1990, and references therein).

Despite the heavy masking effect of the Alpine orogeny on the pre-Alpine memory of the basement rocks, textural and mineral records of some pre-Alpine protoliths are still preserved in the Austro-Alpine and Upper Penninic basement nappes as microscopic to large-scale relics, hence, the present attempt at reconstructing the pre-Alpine puzzle of the internal northwestern Alps. Nevertheless, this effort is hindered by numerous uncertainties. As discussed by Polino et al. (1990) restoration of the basement units back in space from their present position in the Alpine nappe stack to the Europe/Adria passive margins and Late Variscan setting cannot be univocally solved. Major uncertainties depend on the procedures of nappe pile de-imbrication, number and location of the oceanic gap(s), and the way of reconstructing the evolution of continental rifting. Further uncertainties regard the geodynamic significance of some petrogenetic processes occurring during the Permian which, traditionally considered as Late Variscan orogenic events, may be re-interpreted as first records of the Europe/Adria divergent plate margin.

Regional evidence and interpretations on the Italian northwestern basement may be helpfully integrated within the geological history of the whole Alpine area (e.g. von Raumer 1988, Frisch et al. 1990, and references therein).

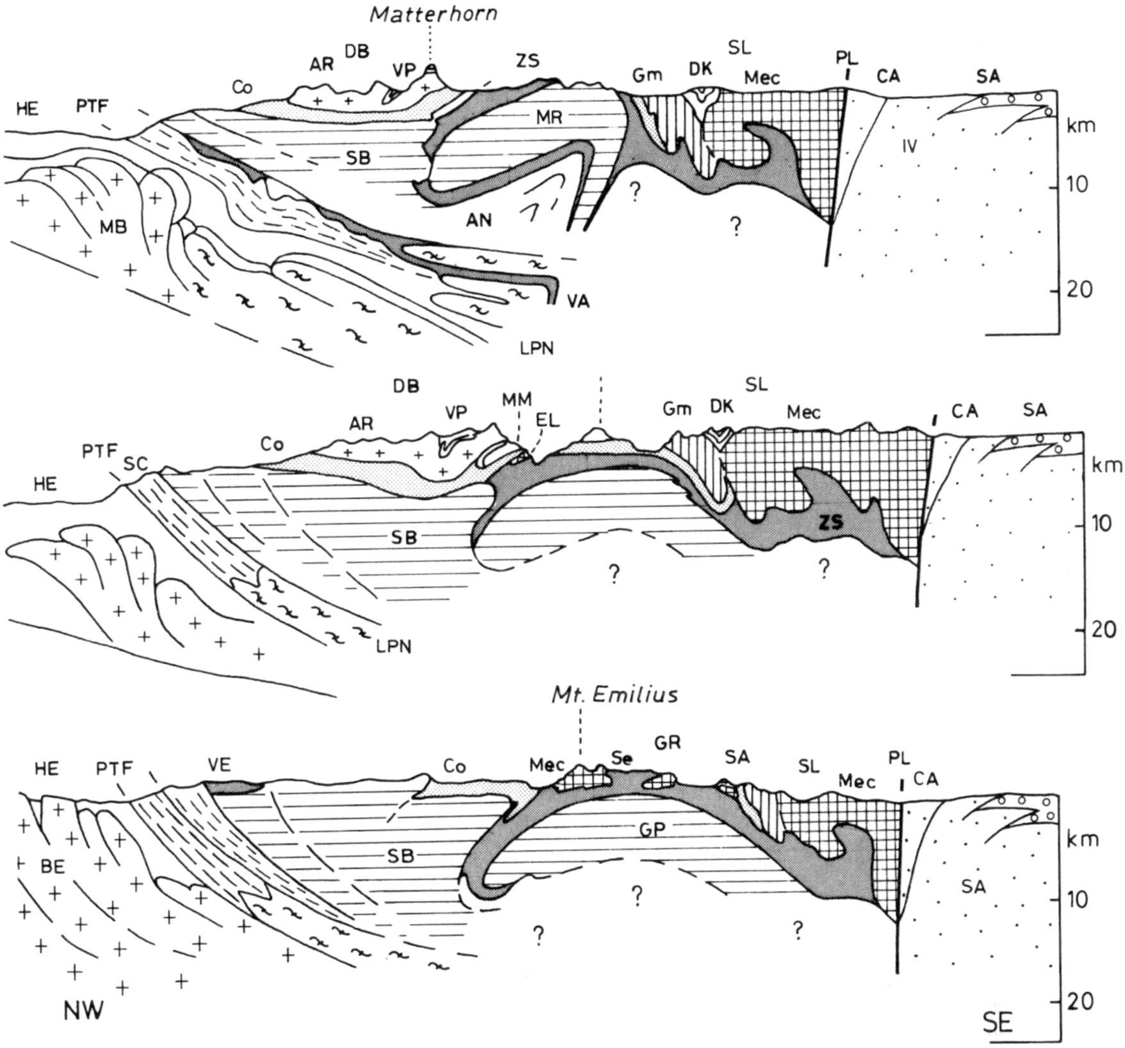

Fig. 2. Geological cross-sections along Italian northwestern Alps (sequence of nappes from *top* to *bottom*): Canavese *(CA)* and South-Alpine *(SA)* units, including Ivrea-Verbano zone *(IV)*; Austro-Alpine composite thrust system: Sesia-Lanzo internal nappe *(SL)*; Dent Blanche *(DB)*, M. Mary *(MM)*, Etirol-Levaz *(EL)*, Pillonet northern klippen and Mt. Emilius, Glacier-Rafray *(GR)* and Santanel *(SA)* southern klippen, including eclogitic micaschist *(Mec)* and Gneiss minuti *(Gm)* complexes, Arolla unit *(AR)* and capping 2nd diorite-kinzigitic *(DK)* and Valpelline *(VP)* units; Piedmont ophiolitic nappe system: greenschist facies Combin zone *(Co)*, eclogitic Zermatt-Saas nappe *(ZS)* and southern equivalents *(Se)*; Antrona unit *(AN)*; Upper (internal) Penninic Monte Rosa *(MR)* and Gran Paradiso *(GP)* basement nappes; Middle Penninic Grand St. Bernard *(SB)* composite nappe system; Lower (external) Penninic décollement cover nappes of Sion-Courmayeur zone *(SC)*, Valais/Versoyen ophiolites *(VA, VE)* and basement nappes *(LPN)*; *PTF:* Penninic thrust front; Ultrahelvetic-Helvetic system *(HE)*, including Mt. Blanc *(MB)* and Belledonne *(BE)* basement units

2 The Austro-Alpine Thrust System

The Austro-Alpine thrust system of the Western Alps may be first subdivided into the internal (SE) Sesia-Lanzo and the external (NW) Dent Blanche *s. l.* nappes, the latter including the northern (Dent Blanche *s. s.*, Mt. Mary, Etirol-Levaz, Pillonet) and southern (Mt. Emilius, Glacier-Rafray, Tour Ponton, Santanel, Chatillon-St. Vincent, Grun) klippen, separated by the Aosta-Ranzola fault and the Dora Baltea river (Figs. 1 and 2; Dal Piaz et al. 1972; Compagnoni et al. 1977; Ballèvre et al. 1986).

Further subdivisions may be put forward when the Alpine metamorphic signatures on the Austro-Alpine nappes and their ophiolitic tectonic sole are evaluated. This allows identification of two main groups of basement/ophiolite nappe couples which are characterized by contrasting Eo-Alpine evolutionary histories (Dal Piaz et al. 1972; Kienast 1983; Ballèvre et al. 1986; Dal Piaz and Lombardo 1986). The former group includes the northern klippen of the Dent Blanche system (except for the deeper Etirol-Levaz slice; Kienast 1983; Ballèvre et al. 1986) and the underlying ophiolitic slices, i. e. the composite Combin zone: these units are both eclogite-free and regionally exhibit an extensive Meso-Alpine greenschist facies imprint, although a few blueschist relics of debated Eo-Alpine age are reported (Dal Piaz 1976; Ayrton et al. 1982; Polino et al. 1990). The second group encompasses the Eo-Alpine eclogitic units as a whole: the Zermatt-Saas nappe and southern ophiolitic homologues are tectonically coupled with the overlying eclogitic southern klippen of the Austro-Alpine system and the Etirol-Levaz northern slice, and in turn are underlain by the Eo-Alpine high-P/low-T basement nappes of the Upper Penninic Monte Rosa/Arcesa-Brusson/Gran Paradiso system.

Lastly, the Sesia-Lanzo nappe and the Dent Blanche *s. s.* /Mt. Mary northern klippen are vertically composite (Stutz and Masson 1938; Diehl et al. 1952; Carraro et al. 1970), consisting of fragmentary upper units – known as the 2nd diorite-kinzigitic zone (Novarese 1929, 1931) and Valpelline series (Argand 1909, 1911, Stutz and Masson 1938) – and of widely exposed lower units (Fig. 2). They are characterized by a contrasted Alpine metamorphic imprint on partly analogous pre-Alpine protoliths.

The upper units consists of lower continental crust and are free of Permian-Mesozoic covers and Late Palaeozoic acidic plutons. The Alpine imprint, mostly recorded as greenschist facies alteration, concentrates along mylonites and shear zones, allowing preservation to a large extent of former lithology, high-grade fabrics and mineral assemblages.

The lower units display lower/mid to upper crustal protoliths, as high- to locally intermediate-grade metamorphic rocks, Late Palaeozoic intrusive granitoids and ensialic gabbros of Permian or undefined post-Variscan age. Minor slices of Triassic-Cretaceous covers are mainly recorded in the Dent Blanche (Mt. Dolin, including Permian beds?), Mt. Mary (Roisan zone) and Pillonet northern klippen (Diehl et al. 1952; Dal Piaz 1976; Ayrton et al. 1982; Canepa et al. 1990), these being often associated with very thick mylonitic shear zones.

The lower units were pervasively recrystallized and deformed during the Eo-Alpine eclogite to blueschist and/or Meso-Alpine greenschist facies metamorphic evolution (Dal Piaz et al. 1971, 1972; Hunziker 1974; Compagnoni et al. 1977; Ayrton et al. 1982; Kienast 1983; Ballèvre et al. 1986; Hunziker and Martinotti 1987, and references therein). Among the basement rocks of the Austro-Alpine lower units the main Alpine derivatives consist of: (1) eclogitic ortho- and paraschists, metabasites and marbles, traditionally known as the eclogitic micaschist complex (Stella 1894; Franchi 1902; Compagnoni et al. 1977); (2) greenschist facies massive metagranitoids, grading to fine-grained orthogneisses and mylonites (Arolla series, Dent Blanche nappe *s. s.;* Gneiss Minuti complex, external north-western Sesia-Lanzo), and albite-bearing retrogressed paraschists. The latter rocks mostly occur in the Mt. Mary and Pillonet northern klippen (Dal Piaz et al. 1971; Dal Piaz 1976; Canepa et al. 1990) and are again reported as Gneiss Minuti in the 1:100000 maps of the Italian Geological Survey; (3) eclogitic to greenschist facies ensialic metagabbros and related mylonites.

2.1 *Upper Units*

Disregarding here Alpine reworking, the Upper Austro-Alpine units consist fo silicic to mafic granulites, high-T pelitic paragneisses, tholeiitic amphibolites, marbles, and scarce slices of mantle peridotite/serpentinite (Masson 1938; Diehl et al. 1952; Bertolani 1971; Dal Piaz et al. 1971, 1972; Hunziker 1974; Compagnoni et al. 1977; Nicot 1977; Niggli 1978; Beccaluva et al. 1979; Vuichard 1987; Cesare et al. 1989; Ridley 1989; Stünitz 1989; Canepa et al. 1990). These metamorphic rock types occur as intimate interbanding on all scales and are traditionally known as the diorite-kinzigitic complex. The lithological similarity of these protoliths with the South-Alpine lower crust of the Ivrea-Verbano (or 1st diorite-kinzigitic) zone was first stressed by Novarese (1929, 1931) and later confirmed (Carraro et al. 1970).

The most common pre-Alpine assemblages are: Grt-Opx-Qtz; Bt-Grt-Qtz-Kf-Sil ± Pl; Qtz-Kf ± Pl ± Ms; Pl-Grt; Hbl-Pl ± Cpx ± Grt ± Bt. P/T values on the postulated equilibrium conditions of these metamorphic assemblages range from 8 kb at 750 °C in the Valpelline unit (Nicot 1977; Dal Piaz et al. 1983) to 6 kb at 700–750 °C in the 2nd diorite-kinzigitic unit (Vuichard 1987). These estimates are consistent with those reported from the Ivrea-Verbano zone (Zingg 1983 and references therein).

The age of protoliths (Cambrian-Early Ordovician?) and high-grade regional fabric (Early Palaeozoic and/or Variscan) may be tentatively inferred through comparison with the debated timing of the Ivrea-Verbano zone (Dal Piaz et al. 1972; Hunziker

1974; Compagnoni et al. 1977; Niggli 1978; Hunziker and Zingg 1980; Zingg 1983; Boriani et al. 1990; Zingg et al. 1990). The kinzigitic fabric predates the Permian (Pin 1986) deep-seated intrusion of gabbro/ultramafic bodies (i. e. "main mafic formation", Rivalenti et al. 1981, 1984) which partly re-equilibrated under granulite facies conditions and slowly cooled down (Voshage et al. 1990 and references therein). Alternatively it is believed that underplating of the Ivrea-Verbano continental crust by these gabbro bodies may be responsible for a granulite re-equilibration of Late Palaeozoic age (e. g. Pin and Vielzeuf 1983).

After the granulite imprint, the Upper Austro-Alpine basement was invaded by patches, concordant intercalations and discordant dykes of anatectic pegmatoids (Qtz-Kf-Pl-Bt ± Ms ± Grt), possibly coeval with cordierite growing in the surrounding kinzigites (Diehl et al. 1952). The age of these processes is also unknown; they may be interpreted as post-thickening Late Variscan collapse products at decreasing pressure (e. g. Menard and Molnar 1988) or as the thermal record of post-Variscan extensional events. The second hypothesis may be supported through comparison with Permian anatectic processes on kinzigitic migmatites occurring at the roof contact of the mafic igneous body of the Ivrea-Verbano zone (Bürgi 1987; Voshage et al. 1990; Zingg et al. 1990, and references therein). Only the latest uplifting steps of the Upper Austro-Alpine Valpelline and 2nd diorite-kinzigitic basement from the Variscan consolidation to the Permian-Jurassic crustal attenuation and rifting are documented by Rb-Sr and K-Ar cooling ages on metamorphic micas from paragneisses and anatectic pegmatoids yielding 240–180 Ma (Dal Piaz et al. 1972; Hunziker 1974). This perfectly recalls the overall cooling history of the Ivrea-Verbano zone (Hunziker 1974; Hunziker and Martinotti 1987; Zingg et al. 1990), pointing to a common late pre-Alpine thermal evolution.

2.2 Lower Units

Most of the pre-Alpine textural and mineral relics preserved in the Eo-Alpine eclogitic micaschist complex are recorded within poorly deformed discrete rock volumes of the Sesia-Lanzo, Mt. Emilius, Santanel paraschists and the Sesia-Lanzo metagranitoids. The first relict occurrences were reported from the classical Mt. Mucrone area, as intrusive granitoids (Bt-Qtz-Pl site, Hbl, Kf, Allanite) and high-grade paragneisses (Bt-Grt-Kf-Qtz-Sil) (Dal Piaz et al. 1972; Compagnoni et al. 1977; Niggli 1978). Systematic field surveys later revealed significant occurrences of pre-eclogitic granulites and/or high-T amphibolite facies metamorphic rocks with silicic and mafic composition. They are reported from the Sesia-Lanzo nappe (Isler and Zingg 1974; Lardeaux 1981; Lardeaux et al. 1982; Spalla et al. 1983; Lardeaux and Spalla 1991), the Mt. Emilius (Bearth et al. 1980; Dal Piaz et al. 1983; Benciolini 1989; Pennacchioni 1989) and Santanel (Battiston et al. 1987) eclogitized southern klippen and the Grun slice, east of St. Vincent (Bino and Compagnoni 1988). The blueschist Vasario-Sparone subunit from the southern Sesia-Lanzo nappe – a slice of deep continental crust tectonically interposed between the internal eclogitic micascists and the external Gneiss Minuti shear zone – preserves garnet-rich paragneisses with pegmatoid veins and metabasic bodies recording pre-Alpine high-T conditions (6 kb, 780 °C; Pognante et al. 1987). Disregarding the blueschist imprint, this subunit may be also interpreted as a southernmost fragment of the 2nd diorite-kinzigitic unit (Carraro et al. 1970).

Relict lithological associations from the Lower Austro-Alpine eclogitic units allow identification of at least three main pre-Alpine protolithic groups: (1) granulites, high-grade paragneisses and related mafic rocks and marbles recording a single regional fabric of pre-granitic or unknown age; (2) intrusive granitoids probably yielding, as a whole, the Permian Rb-Sr whole-rock isochron age of the Mt. Mucrone pluton, Sesia-Lanzo (Dal Piaz et al. 1978; Oberhänsli et al. 1985); (3) minor ensialic gabbro bodies of undefined post-Variscan (Permian?) age. Similar protolithic groups occur in the eclogite-free northern basement units, mostly represented by retrogressed paragneisses and related mafic interbeddings (Mt. Mary, Pillonet), Late Palaeozoic granitoids and Permian gabbro bodies (Arolla series, Dent Blanche nappe *s. s.*, Pillonet).

The first group of protoliths from the eclogitic basement complexes occur as pods of different dimensions, which mostly escaped the high-P deformation with intense foliation development recorded in the surrounding micaschists. This group consists of silicic to mafic granulites, high-grade amphibolite facies metapelites and igneous basic rocks, marbles and minor ultramafics; it also includes conformable anatectic pegmatoids and transecting leucocratic dykes, mostly converted to Eo-Alpine Grt-Ph-bearing quartzites. This group is lithologically very similar to the sliced lower continental crust (kinzigitic complexes) occurring in the Upper Austro-Alpine units and South-Alpine Ivrea-Verbano zone (Compagnoni et al. 1977; Dal Piaz et al. 1983): this is known since Novarese (1929) expressed the opinion that the 2nd diorite-kinzigitic unit and the Mt. Emilius klippe were equivalent lithologically, except for the eclogitic imprint.

The analogy is further supported by P-T estimates on silicic and mafic granulites from the eclogitic Mt. Emilius (6–8 kb, 700–750 °C: Dal Piaz et al. 1983) and Sesia-Lanzo (7–9 kb, 700–800 °C: Lardeaux and Spalla 1991) nappes, through application of independent thermobarometers. Moreover, the Sesia-Lanzo granulitic assemblages, locally including cordierite, display pre-eclogitic depressuring re-equilibration under an increasing thermal gradient (4–5 kb, 600 °C; Lardeaux and Spalla 1991); P-T estimates exceed the maximally relaxed geotherm for reasonable heat supply after thrusting (Thompson and England 1984). Decompressional signatures are also locally recorded in some mafic granulites and ultramafic rocks of the Mt. Emilius klippe (Dal Piaz et al. 1983; Benciolini 1989).

The Mt. Emilius ultramafic body was discovered by Benciolini (1989) in the Becca di Nona-Arpisson area. It consists of mantle-derived spinel-bearing websterite and minor peridotite with protogranular to porphyroclastic texture, later uplifted and deformed along a large mylonitic shear zone. The latter evolution developed under low-P/high-T conditions before the Eo-Alpine eclogitic imprint; its relative timing with respect to the Permian intrusions recorded elsewhere is unknown. A small lens-shaped body of norite and cumulus ultramafics displaying pre-Alpine granulitic signatures is reported from the eclogitic basement of the Etirol-Levaz unit (Kienast 1983). As suggested by Benciolini (1989), these bodies may tentatively be correlated to the Permian mafic/ultramafic intrusives from the Ivrea-Verbano zone which were emplaced at the mantle/lower crust transition zone (Voshage et al. 1990, and references therein).

Kinzigites, migmatitic paragneisses, minor mafic rocks and scarce marbles are again the main pre-granitic protoliths reported from the Lower Austro-Alpine nappes which escaped the Eo-Alpine eclogitic reworking, such as the Pillonet (Bt-Grt-Qtz-Pl-Kf; Hbl), Mt. Mary (Bt-Ms-Qz-Pl; Bi-Grt-Kf-Sil) and Dent Blanche *s. s.* northern klippen (Diehl et al. 1952; Dal Piaz and Sacchi 1969; Dal Piaz 1976; Canepa et al. 1990). They are poorly and randomly preserved within the surrounding greenschist facies, foliated to mylonitic Alpine derivatives. The pre-granitic relics from the Dent Blanche Arolla series are represented by high-grade paragneisses and fine-grained metabasites which occur as a kilometric roof pendant capping the Mt. Morion granite and as small xenoliths randomly recorded elsewhere. Further, the lower unit of the Mt. Mary nappe, as a peculiar protolith, exhibits relics of garnet-free two-mica-bearing paragneisses and micaschists which are not typically associated with marble and metabasite bodies. They locally preserve from single occurrences to large amounts of andalusite prophyroblasts, partly replaced by fine white mica aggregates, and a few pseudomorphs after cordierite, both postdating the Variscan (or older?) regional fabric. No inrusive bodies are exposed here and in the surrounding areas; therefore these minerals record regional evolution under low-P conditions.

The intrusive groups are first represented by the Permian Mt. Mucrone and related eclogitic metagranitoids from the internal Sesia-Lanzo zone (Dal Piaz et al. 1971, 1972; Callegari et al. 1976) and by the greenschist facies metagranitoids, augengneisses and mylonitic albite-rich, fine-grained, homogeneous to banded gneisses prevailing in the external side of the northwestern Sesia-Lanzo zone and in the Arolla unit of the Dent Blanche nappe *s. s.*, probably recording a former composite batholith intruded within the Variscan basement and later dismembered (Diehl et al. 1952; Dal Piaz et al. 1971, 1972; Callegari et al. 1976; Compagnoni et al. 1977; Ayrton et al. 1982). Minor occurrences of these igneous rocks are exposed in the Pillonet (Dal Piaz 1976) and Mt. Mary (Masson 1938, Canepa et al. 1990) klippen, mostly as extensively transposed flattened bodies parallel to the surrounding retrogressed paraschists. The former intrusive contact is locally preserved in both the eclogitic and greenschist facies lower units, indicating that, during the Permian, some slices of lower/intermediate continental crust had already been uplifted to relatively shallow crustal levels where the epiplutonic bodies were emplaced.

Huge bodies of largely unaltered gabbro and related cumulus ultramafics with magmatic brown mica yielding cooling ages at around 250 Ma (Dal Piaz et al. 1977) occur in the Matterhorn and Mt. Collon-Dents de Bertol area (Arolla unit, Dent Blanche nappe). These rocks are characterized by subsolidus reactions between olivine and plagioclase, recorded by coronitic Opx and spinel/amphibole symplectite. Further eclogitic, greenschist facies and mylonitic gabbro bodies of undefined (relative to the granite intrusion) post-Variscan age are reported from northern and southern Sesia-Lanzo nappe (Mt. Pinter, Alpe Chasten, Corio and Monastero) and Mt. Emilius and Pillonet klippen (Bianchi et al. 1965; Dal Piaz and Sacchi 1969; Dal Piaz et al. 1971; Dal Piaz 1976; Compagnoni et al. 1977; Spalla et al. 1983, 1990; Pennacchioni 1989; Stünitz 1989).

The Permian Matterhorn/Mt. Collon-Dents de Bertol gabbros are tectonically embodied within the Arolla augengneisses and metagranitoids which are the greenschist facies Alpine representative of intrusive protoliths of undefined Late Palaeozoic age. This is supported by wide boundary mylonitic zones which display extreme grain-size reduction of the dominantly quartz and altered calcic plagioclase framework of the juxtaposed silicic and mafic igneous rocks (Dal Piaz et al. 1977). However, a few dykes of possibly cognate very fine-grained metalamprophyre and brown amphibole-rich metagabbro cut the Arolla augengneisses, forming the megascopic pressure shadow of the Matterhorn gabbro body in the Leone glacier area. This suggests that the former magmatic emplacement of the Matterhorn gabbro and/or cognate apophyses occurred beneath and within (?) a thinning, fragmented, upper crustal lid rather than exclusively at the mantle/lower crust transition (Dal Piaz and Ernst 1978). In any case, these ensialic gabbro bodies, which cooled down to the closing temperature of the isotope system of magmatic micas at around 250 Ma, support the hypothesis that an extensional plate margin was already operative before the Permian/Triassic boundary (Dal Piaz et al. 1977; Dal Piaz and Ernst 1978).

3 Monte Rosa, Arcesa-Brusson and Gran Paradiso Nappe System

Two main groups of pre-Alpine gneissic and granitic protoliths have been recognized within the Alpine eclogitic and greenschist facies basement rocks of the Upper Penninic Monte Rosa and Gran Paradiso nappes (Bearth 1952; Dal Piaz 1966, 1971; Dal Piaz and Lombardo 1986, and references therein). They are:

1. Pre-granitic high-grade pelitic gneisses, veined by pods and streaks of anatectic pegmatites which in the Monte Rosa area display altered cordierite; this complex encompasses boudinaged intercalations of metabasites with basaltic composition which are randomly distributed (Gran Paradiso) or mostly concentrated along basement shear zones including minor pure to silicate-rich marbles (Monte Rosa, Furgg zone).
2. Late Variscan porphyritic granites (310 Ma: Rb-Sr whole-rock isochron, high initial Sr ratio; Hunziker 1970; Frey et al. 1976) and related dykes of aplite and pegmatite.

Former intrusive contacts of granite bodies sharply cutting through the folded high-grade schistosity of the surrounding paragneisses are locally well preserved on a megascopic scale (Monte Rosa: Dufour Spitze, Anzasca and upper Ayas valley; Bearth 1952; Dal Piaz 1966; Gran Paradiso: Piantonetto, Piantelessio and Soana valley; Callegari et al. 1969; Battiston et al. 1987). Therefore, the regional metamorphism which predates the Late Carboniferous granite intrusion is here of Variscan age or older. Instead, the related contact metamorphism seems poorly recorded; it is reported mostly from the central Gran Paradiso massif where rare relics of hornfelses containing red biotite, plagioclase, K-feldspar, sillimanite, corundum and green spinel are preserved (Callegari et al. 1969; Compagnoni and Prato 1969; Dal Piaz and Lombardo 1986).

In the middle Ayas valley, south of the Aosta-Ranzola fault, the basement rocks of the Arcesa-Brusson window are exposed beneath the Zermatt-Saas ophiolites (Fig. 1). The main pre-Alpine protoliths consist of porphyritic granites and pre-granitic high-grade paragneisses with numerous mafic boudins, i.e. the same pre-Alpine complexes occurring in the Monte Rosa and Gran Paradiso nappes which here represent a convincing evidence of their buried connection. The analogy also comprises the Alpine polyphase reworking which, in the Arcesa-Brusson window, is recorded by Eo-Alpine eclogitic relics from mafic rocks and by Meso-Alpine biotite-in greenschist-facies mineral assemblages. The latter imprint is recorded everywhere in the present lithologies, mainly consisting of coarse augengneisses to mylonitic granites, albite-bearing micaschists, albite-epidote-rich amphibolites and prasinites. The former intrusive contact is obliterated through large-scale transposition, as in the peripheric sectors and internal shear zones of the Monte Rosa and Gran Paradiso nappes.

Typical relict assemblages from high-grade metapelites of the Monte Rosa nappe are Bt-Sil-Grt-Qtz-Kf-Pl and Grt-Bt-Ms-Sil-Pl (Bearth 1952; Dal Piaz 1966, 1971). Cores of pre-eclogitic brown hornblende are locally reported from the metabasic rocks of the Anzasca valley (Laduron 1976). Some of these pre-Alpine relics occur in the pre-granitic complex of the Arcesa-Brusson window. Further relics are recorded in pre-granitic metapelites and associated metabasites from the northern sector of the Gran Paradiso nappe (Bt-Kf-Pl-Sil; Bt-Ms-Pl-Grt-Sil-And?; brown Hbl) (Callegari et al. 1969; Compagnoni and Prato 1969; Compagnoni et al 1976; Dal Piaz and Lombardo 1986). Temperature estimates on pre-Alpine mineral pairs from the Upper Penninic basement approximate the thermal conditions of the kinzigitic complexes. Physical conditions similar to those reported from the Upper Austro-Alpine kinzigites may be inferred also by analogy of mineral compatibilities. This likeness may be further supported by widespread occurrences of migmatites, cordierite-bearing pegmatites and possible andalusite which record here the late evolution of the pre-granitic regional metamorphism.

4 Attempts at Pre-Alpine Restoration

4.1 Traditional Restoration Models

Any tentative restoration of the Austro-Alpine and Upper Penninic basement nappes back in space and time through a passive margin setting up to a proper Late Variscan scenario is a highly speculative effort. This is due to the absence of any univocal solution about the de-imbrication procedures of the present nappe pile, Eo-Alpine thrust mechanisms and rifting evolution, and to the need to overcome additional difficulties such as the Alpine polyphase imprint, scattered distribution of relics and inadequacy of chronological data.

Classical restorations of the Alpine Tethys are mostly based on simple unfolding or pure de-imbrication of the western Alpine nappe stack and on symmetric models of the Europe/Adria divergent plate margin. This allowed the well known accommodation of Mesozoic sedimentary basins and related basement sole as a horizontal sequence of Europe-deriving Helvetic, North (Lower) Penninic, Middle-Penninic, South (Upper) Penninic, and Africa (Adria)-deriving Austro-Alpine, Canavese and South-Alpine domains, separated by the Piedmont ocean (Fig. 3 A). Alternative models suggest the occurrence of two or even more oceanic gaps, without changing the horizontal train of the continental blocks which is inferred from the vertical sequence of the present nappe pile (e.g. Dal Piaz et al. 1972; Trümpy 1980; Laubscher and Bernoulli 1982; Platt 1986; Mattauer et al. 1987). Whatever oceanic reconstruction may be, models that allocate the subduction-related high-P basement nappes in the Europe (Penninic) and Adria (Austro-Alpine) passive margins require syn-collisional mechanisms for Eo-Alpine high-P thrust formation (e.g. Dal Piaz et al. 1972; Mattauer et al. 1987), an unusual solution with respect to the modern plate tectonic scenario. Accordingly, restoration of the pre-Alpine basement and tentative location of the main magmatic and metamorphic relics from the Italian northwestern Alps are sketched in Fig. 3 B; this recalls former schemes suggested by Dal Piaz et al. (1975) and Boriani et al. (1976). Figure 3 B simply shows the lateral juxtaposition of the main Alpine units without evaluating the proper depth of each cover-free lower/intermediate crustal blocks at a specific time, as may be inferred from petrological estimates; this is prevented by inadequacy of radiometric data.

As sketched in Fig. 3 B, the Late Carboniferous granite plutons which sharply cut the high-grade Variscan fabric of the wall rocks were mostly emplaced within medium-upper levels of the Monte Rosa/Gran Paradiso and Helvetic crustal blocks, closely following the formation of cordierite-bearing assemblages and/or crustal anatectic melts (e.g. Boriani et al. 1976; von Raumer 1988; Frisch et al. 1990). Bimodal magmatic activity of documented or supposed Permian age is widely recorded in the Austro-Alpine (Compagnoni et al. 1977), Canavese (Biino et al. 1986) and South-Alpine edge (e.g. Vai et al. 1984; Boriani et al. 1988, 1990; Pinarelli et al. 1988; Zingg et al. 1990); instead, it occurs from the Late Carboniferous onwards in the Middle Penninic Grand St. Bernard/Briançonnais realm and Pinerolese zone (Thelin and Ayrton 1983; Fabre et al. 1987; Escher 1988; Sandrone et al. this Vol.) and in the Dauphinois Grand Rousses massif (Banzet et al. 1985). Stefanian-Lower Permian volcanic and subvolcanic products are reported from the Ligurian Alps, at the southernmost edge of the Briançonnais domain, recording three main magmatic episodes (Vanossi et al. 1986): they are represented by calc-alkaline rhyolites derived from anatectic crustal supplies and by mantle-derived sub-alkaline lati-andesites with shoshonitic affinity (Cabella et al. 1988; Cortesogno et al. 1988). This magmatism is associated with graben, half-graben or pull-apart basins with Late Carboniferous and/or Permian clastic infillings (e.g. Cadel 1986; Vanossi et al. 1986; Fabre et al. 1987; Cortesogno et al. 1988).

Figure 3 B does not indicate that the Permian gabbro bodies were actually emplaced at different structural levels, as the mantle/crust boundary and granulite environment of the Ivrea-Verbano zone (Rivalenti et al. 1981, 1984) and possibly the Mt. Emilius klippe (Pennacchioni 1989), the rising Sesia-Lanzo granulites and the shallower granitic crust of the Dent Blanche Arolla series, coevally to the onset of Late Palaeozoic volcano-sedimentary basins. The gabbro and granite plutons of the Canavese zone may be correlated to the Late Palaeozoic bimodal igneous activity occurring in the presently surrounding Austro-Alpine and South-Alpine blocks.

The pre-granitic basement displays the same relict signatures in most of the Austro-Alpine and Upper Penninic units. As regards the Austro-Alpine thrust system, the kinzigitic upper units and metasedimentary protoliths of the eclogitic/greenschist lower units

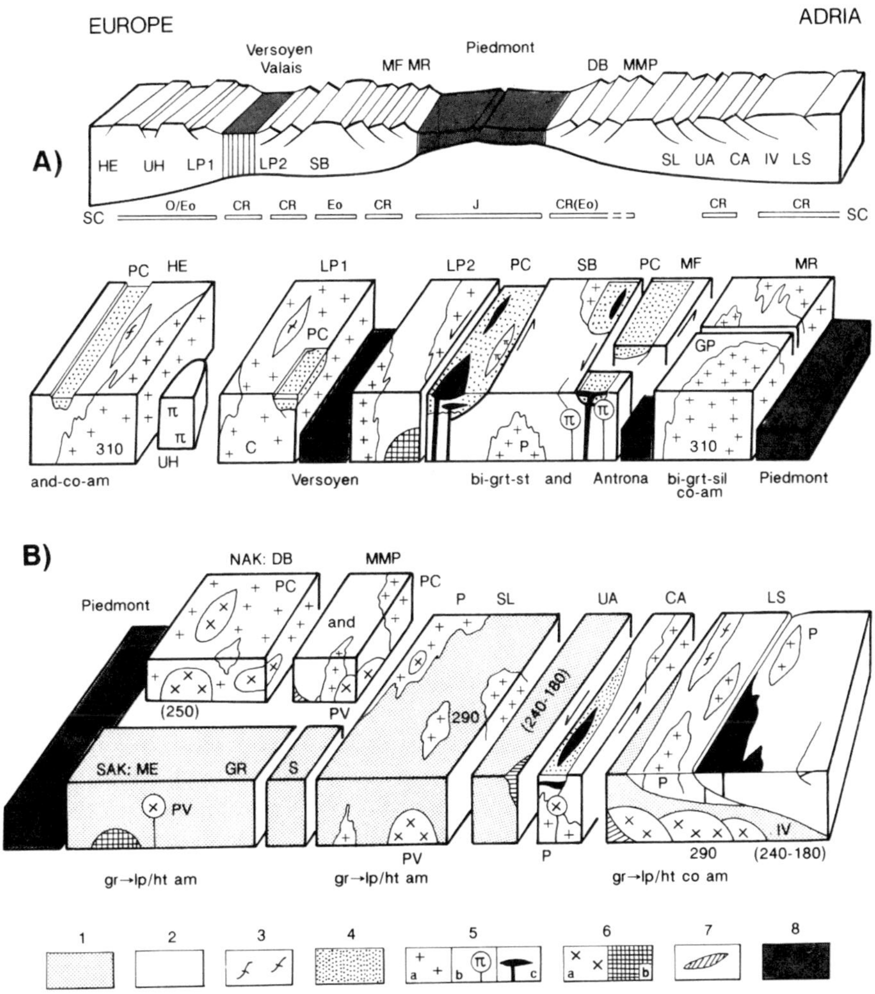

Fig. 3. A Classical restoration of Europe and Adria (Africa) passive margins. Age of top sedimentary cover *(SC): J* Late Jurassic (oceanic deposits, capped by Late Cretaceous flysch); *CR* Late Cretaceous; *Eo* Eocene; *O* Oligocene. **B** Inventory of pre-Alpine relics schematized as crustal blocks laterally juxtaposed according to present vertical sequence of Western Alpine nappes; not to scale. Details and references in text. Palaeogeographic domains/crustal blocks: Europe: Helvetic *(HE)*, Ultrahelvetic *(UH)*, Lower (external) Penninic *(LP1* Verampio, Antigorio; *LP2* M. Leone, Adula), Middle Penninic Grand St. Bernard *(SB)* including Mont Fort *(MF)*, Upper Penninic Monte Rosa/Gran Paradiso *(MR, GP)*. Adria: *NAK* northern; (*DB* Dent Blanche; *M. MP* M. Mary-Pillonet) and *SAK* southern; (*ME* Mt. Emilius; *GR* Glacier-Rafray; *S* Santanel) Austro-Alpine klippen, Sesia-Lanzo *(SL)*, Upper Austro-Alpine units *(UA)*, Canavese *(CA)*, South-Alpine Ivrea-Verbano *(IV)* and Laghi Series *(LS)*. Pre-Alpine lithology: *1* formerly lower crustal sections consisting of intimate association of metapelite/metabasite/marble possibly recording an Early Palaeozoic accretionary wedge, and characterized by granulite to depressuring high-T mineral assemblages; *2* Variscan high-grade metapelites, including minor metabasites and scarce marbles, mostly characterized by low-P/high-T evolution with cordierite-rich anatectic melts, and by Late Palaeozoic and/or Mesozoic covers; *3* Variscan orthogneisses; *4* Late Carboniferous and/or Permian-Eo-Triassic basins; *5* Late Palaeozoic intrusive granitoids *(a)*, subvolcanic *(b)* and volcanic *(c)* bodies; *6* gabbro *(a)* and ultramafic *(b)* bodies; *7* mantle slices (roughly in present position); *8* Mesozoic ophiolitic units. Pre-Alpine metamorphic signatures: *gr* granulite; *bi-grt-sil/st* biotite-garnet-sillimanite or staurolite; *lp/ht* low-P/hig-T late Variscan or post-Variscan re-equilibration; *am* anatectic melts; *co* cordierite; *and* andalusite; *PC* Permian-Carboniferous; *P* Permian; *PV* intrusions postdating the Variscan fabric; *310* Rb/Sr isochron and U/Pb ages (Ma); *(240–180) cooling ages*

exhibit the same association of high-grade metapelites, igneous metabasites and marbles. Granulite fabrics, locally sharply cut by Late Palaeozoic plutons, and Late- and/or post-Variscan low-P/high-T amphibolite facies re-equilibrations occur in both groups of units, strengthening their close analogy with the Ivrea-Verbano kinzigite/metabasite/marble sequence intruded by the Permian main mafic formation. Older fabrics seem to be lacking in this sector of the Western Alps.

The lithology and pre-granitic regional metamorphism of the Monte Rosa and Gran Paradiso basement roughly recall the foregoing scenario, although lesser amounts of mafic rocks and especially of marble interbeddings are recorded here.

As previously reported, the high-T depressuring paths of the Lower Austro-Alpine and Penninic basement are not yet defined by radiometric data. Dismembered crustal segments may have been diachronously uplifted to progressively shallower levels; this locally occurred at least from the Late Carboniferous (Monte Rosa/Gran Paradiso) or Permian (Lower Austro-Alpine units), because then these units represented the host-rocks of epi-granitic intrusions and shortly afterwards were overlain by clastic covers. Other Lower Austro-Alpine crustal sections, which are not chronologically constrainted by intrusions of Late Palaeozoic granites, should have undergone both granulite- and low-P amphibolite facies evolution even during post-Variscan crustal attenuation and rifting, as inferred petrologically by Lardeaux and Spalla (1991) from consistent P-T estimates.

4.2 *Alternative Models*

Alternative restoration models of the Alpine Tethys have recently been envisaged by assuming that the early formation of the high-P Penninic and Austro-Alpine basement nappes within the Cretaceous subduction zone may be referred to pre-collisional mechanisms of accretion (Platt 1986) and tectonic erosion (Polino et al. 1990), through comparison with modern orogenic wedges. As discussed by these authors, any possibility of allocating the high-P Penninic and Austro-Alpine nappes on the former passive margin of the European lower plate is excluded because these basement slices had already entered the subduction zone and acquired Eo-Alpine metamorphic signatures when the Tethyan ocean was still open. This is due to the need to preserve a proper deposition sole for the Upper Cretaceous supra-ophiolotic flysch prisms (discussion in Laubscher and Bernoulli 1982; Platt 1986; Hunziker et al. 1989; Polino et al. 1990). Consequently, the Eo-Alpine basement nappes may have been sliced only by intra-oceanic microcontinents of the lower plate (e.g. Platt 1986) and/or the active continental margin of the Adria upper plate before the Europe/Adria continental collision (Polino et al. 1990). Figure 4A, B shows sketches of these restoration options.

The multi-oceanic models envisage the occurrence of one or more microcontinents as formerly separated feeders, through collisional delamination, for the generation of the thin Eo-Alpine basement nappes, coupled with intervening ophiolitic slices. The available models again preserve the traditional sequence of the continental domains reported already and the lower plate extraction of the whole Penninic units (e.g. Ernst 1973; Trümpy 1980; Platt 1986; Mattauer et al. 1987). Therefore, basement restoration in Fig. 3B may be still suitable, although the huge crustal volumes in the colliding microcontinents which, being neither scraped off from the subducting lower plate nor accreted to the exhuming orogenic wedge, definitely disappeared in the mantle, are not evaluated here. Mass balancing of the assumed missing continental crust in the Alpine lithospheric root is discussed by Laubscher (1990).

The tectonic erosion model (Polino et al. 1990) supplies, on the contrary, innovating solutions for the basement restoration within the former continental margins. Accordingly, the whole Eo-Alpine high-P Austro-Alpine units, as well as the Penninic Monte Rosa, Gran Paradiso and other eclogitic basement nappes, may be allocated within the ocean-facing edge of the Adria (Africa) upper plate margin (Fig. 4B): these high-P basement nappes are envisaged to have been tectonically sampled from the toe of the active margin when indented by topographic elevations of the subducting oceanic plate, and then to have been incorporated within the Cretaceous orogenic wedge. Instead, the Arolla lower unit of the Dent Blanche nappe *s. s.*, the Mt. Mary klippe and the Austro-Alpine upper units which escaped the severe eclogitic imprint may be regarded, together with the Canavese zone and the cover-bearing units of the Grand St. Bernard system, as fragmentary records of the Eo-Alpine wedge roof and its internal buttress. This is the premise for discussing palaeogeographic inferences on the Adria active margin summarized in Fig. 4B. First, this model does not necessarily require a noticeable loss of continental crust deep down in the mantle, because the former crustal masses of the Penninic and Austro-Alpine zones may have mostly been preserved within the exhuming orogenic wedge. Further, it allows new options for allocating the high-P/low-T Penninic and Austro-Alpine basement units with respect to the eclogite-free ones. The restoration in Fig. 4B is attempted according to the following constraints and general postulates: (1) the Eo-Alpine high-P basement units are all allocated on the ocean-facing edge of the Adria active margin (or alternatively at its sole) where they may have been indented, tectonically eroded and dragged down along the subduction complex; (2) the former mutual relationships between the eclogitic Sesia-Lanzo/Mt. Emilius and Monte Rosa/Gran Paradiso nappes are tentatively in-

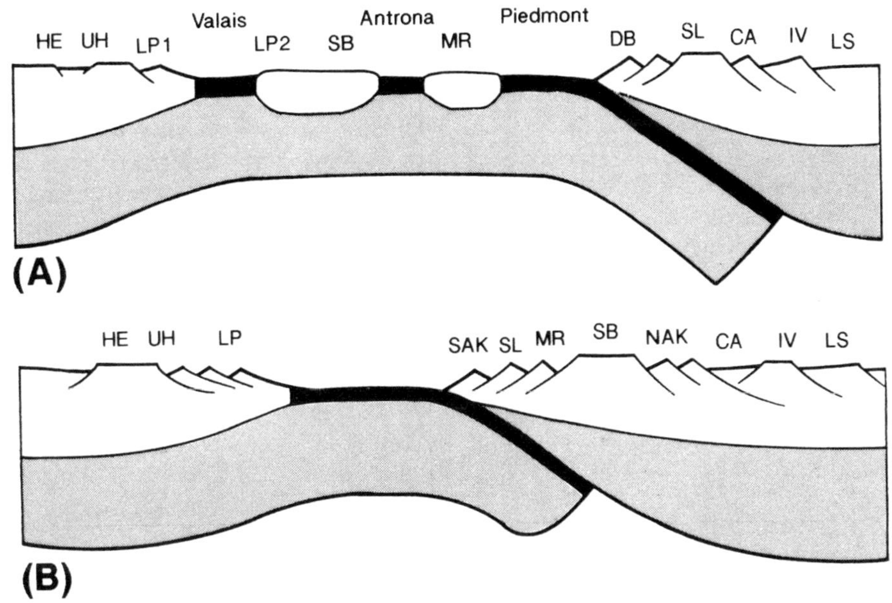

Fig. 4 A, B. Alternative restoration of Penninic basement nappes envisaged **A** as remnants of lower plate microcontinents or **B** as ocean-facing units of Adria active margin. Details and references in text; *symbols* and abbreviations as in Fig. 3

ferred from the timing of their metamorphic climax, yielding 130/100 and 100/90 Ma respectively; (3) the mutual position of the eclogite-free northern Dent Blanche klippen and the underlying Grand St. Bernard nappe system is depicted according to the present vertical position within the nappe pile and envisaging the tectonic insertion in between of the Combin zone as mid-Tertiary collisional event.

Instead, the scheme of Fig. 4B does not account for the pervasive pre-Alpine dismembering of the Adria active margin, which is hypothesized to have been the ancestral configuration that, when inverted, facilitated the nappe-forming processes of the pre-collisional Cretaceous orogeny (Polino et al. 1990). As discussed later, this setting may formerly have been aquired through Late Palaeozoic simple shear mechanisms of asymmetric lithosphere attenuation, deep-seated detachments and Mesozoic continental rifting.

5 Discussion and Conclusion

Despite the Alpine imprint and whatever restoration may be envisaged for the pre-Alpine relics, lower and mid-crustal sections of undefined age (Variscan or older?) may be seen as the former supplier for the Upper Austro-Alpine slices and for the southern Lower Austro-Alpine and Upper Penninic, high-P nappes from the northwestern Alps, mainly consisting of protolithic high-grade paragneisses and silicic to mafic granulites. Later, fragments of this lower continental crust were diachronously uplifted to progressively shallower structural levels. This is recorded by pre-granitic (Monte Rosa/Gran Paradiso) to Permian (Southern Alps, Austro-Alpine?) high-T/low-P metamorphic re-equilibration and anatectic processes, retrograde mineral alteration, Late Carboniferous (Upper Penninic) to Permian (Lower Austro-Alpine and South-Alpine) intrusion of epiplutonic granitoids, and emplacement of mafic igneous bodies of Permian (Dent Blanche and Ivrea-Verbano zone) or undefined post-Variscan age (Sesia-Lanzo and Pillonet). Large intrusive granitoid plutons are not recorded, and were probably never emplaced, in the kinzigitic basement of the Upper Austro-Alpine 2nd diorite-kinzigitic and Valpelline units, which probably represent independent rock volumes coming from different sections or structural levels of the Adria plate, at least at the time of the Late Palaeozoic magmatic activity.

The age of protoliths and pre-granitic high-grade regional fabric in the Upper Penninic and Lower Austro-Alpine units from the northwestern Alps and that, not necessarily coeval, of the granite-free Upper Austro-Alpine granulite basement, cannot at present be defined because the available geochronological data (240–180 Ma) on pre-Alpine metamorphic assemblages only represent cooling ages. The intimate association of paragneisses, marbles and metabasites defining the kinzigitic sequence of the the Upper and southern Lower Austro-Alpine units may have originated within the same Cambrian-Early Ordovician

(?) accretionary wedge which has been tentatively hypothesized for the kinzigitic complex of the Ivrea-Verbano zone (Sills and Tarney 1984; Boriani et al. 1990; Zingg et al. 1990). This model may be supported by the oceanic affinity of the metabasite interbeddings (Mazzucchelli and Siena 1986), occurence of Mn-rich quartzites (Ravinella, Strona valley; Bertolani 1967; Albrecht et al. 1978) and debated radiometric data from the Ivrea-Verbano zone (Hunziker and Zingg 1980; Zingg et al. 1990, and references therein).

The pre-Alpine evolution of the Alpine basement recorded by metamorphic signatures at decreasing temperature and by silicic to mafic magmatic activity has been traditionally referred to the Caledonian and/or Variscan orogenic cycles (e.g. Boriani et al. 1976; D'Amico and Mottana 1976; Hunziker and Zingg 1980; Vai et al. 1984; Sassi et al. 1985; Pinarelli et al. 1988; Boriani et al. 1990). Whatever the timing may be, the general absence of the regionally expected interference pattern of two Palaeozoic fabrics gives rise to basic doubts on the existence of two pre-Alpine orogenic cycles in the internal northwestern Alps.

A second doubtful point regards the exhaustion age of the Variscan convergence and the significance of the late pre-Alpine evolution history. According to Stille's original definition of the Variscan term, a sequence of deformation events occurring from Devonian to Permian times inclusive, the already discussed late pre-Alpine evolution of basement rocks from the internal northwestern Alps should still be referred to the Variscan orogenic cycle, as reported in most recent literature on the Late Palaeozoic Eastern Austro-Alpine and South-Alpine magmatic activity (e.g. Vai et al. 1984; Sassi et al. 1985; Frank et al. 1987; Boriani et al. 1990; Voshage et al. 1990). Nevertheless, in the Grand St. Bernard domain, the timing of processes forming the main amphibolite facies Palaeozoic fabric is constrained by sedimentary records definitely dating the Hercynian unconformity back to the Namurian or Westphalian (e.g. Gb. Dal Piaz 1939; Boriani et al. 1976; Vanossi et al. 1986; Fabre et al. 1987; Escher 1988, and references therein). This situation is repeated elsewhere from numerous basement complexes of the whole Alpine chain overlain by Late Palaeozoic sedimentary and volcanic covers (e.g. Oberhauser 1980; Frisch et al. 1990 and references therein). This suggests that caution should be used when assigning the "Late Hercynian" low-P thermal evolution, commonly reported as post-kinematic, and especially the Latest Palaeozoic magmatic activity to true Variscan orogenic events. Caution should also be used in handling the geochemical features of the Latest Palaeozoic magmatic rocks as unique geodynamic marker; this would have prevented the unrealistic hypothesis of a subducting oceanic (or continental) lithosphere generating a Late Palaeozoic magmatic arc, which has been suggested for the Central Alpine (Mercolli and Oberhänsli 1988) and South-Alpine Permian igneous bodies (e.g. Vai et al. 1984; Stille and Buletti 1987; Visonà et al. 1987; Boriani et al. 1988, 1990 pro parte; Baccos et al. 1990).

Further doubts also come from the global behaviour of the Alpine area during the Late Palaeozoic. When the once amalgamated Variscan belt in the future Alpine foreland was suffering collapse and the operation of Late Carboniferous and/or Permian strike-slip to wrench faulting was coupled with noticeable magmatic activity (e.g. Arthaud and Matte 1977; Rau and Tongiorgi 1981; Ziegler 1982; Bixel and Lucas 1983; Menard and Molnar 1988), the Alpine domain records tensive to transtensive deformations and crustal attenuation which may be seen as precursor processes of the opening of Tethys (e.g. Dal Piaz et al. 1977; Dal Piaz and Ernst 1978; Thelin and Ayrton 1983; Ziegler 1986; Zingg et al. 1990), rather than as presumed evidence of Late Variscan orogenic events. Tensional geodynamic regimes had been active in the western Alpine Penninic basement possibly since the Late Carboniferous, as inferred from graben opening, wide granitoid intrusions and bimodal magmatic activity. Lithosphere attenuation and transtensional crustal dismembering are thougth to have facilitated generation and emplacement of Permian-Carboniferous shoshonitic melts in the Dauphinois domain (Banzet et al. 1985). Tensional regional mechanisms were already operating in the Austro-Alpine during the Permian, as undoubtedly shown by huge intrusions of granitoids and ensialic emplacement of gabbro bodies (Fig. 5). Melting of hot asthenolithic bodies rising under tensional conditions beneath the thinning lithosphere or, more probably, of heated sections of the overlying lithospheric mantle may be envisaged as sources for the crustal igneous underplating of the Permian gabbro batholits (Dal Piaz et al. 1977; Dal Piaz and Ernst 1978). Associated thermal perturbance and fluid release may have facilitated melting of lower crustal sections, whilst fractionated products may have been supplied from the gabbro magma chambers.

Asymmetric rift models have recently been proposed for the restoration of Mesozoic sedimentary basins and Tethyan opening (De Graciansky et al. 1989; Stampfli and Marthaler 1990). A strongly dismembered continental crust is thought to have developed through listric faults and mylonitic zones since the early rifting stages. The size and shape of the en-

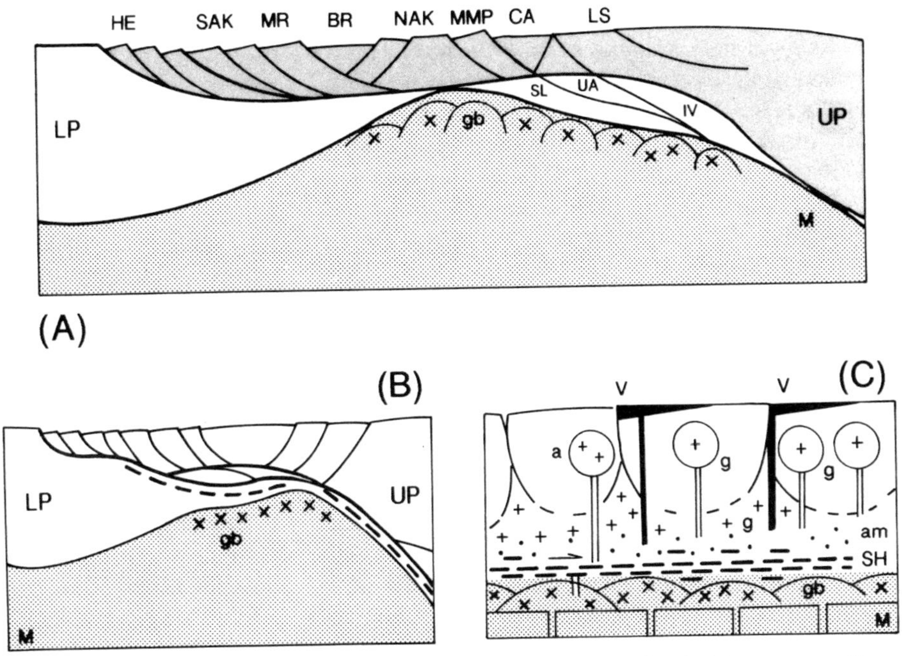

Fig. 5 A, B. Tentative Permian allocation of discussed crustal blocks as discrete allochthon segments and underlying metamorphic core complexes within an attenuating lithosphere through deep-seated low-angle shear zones; not to scale. For details, see in text. *UP, LP* Upper- and lower extensional plates; *HE* Helvetic; *SAK* southern Austro-Alpine klippen; *MR* Monte Rosa; *BR* Grand St. Bernard/Briançonnais; *NAK, MMP* Dent Blanche/Mt. Mary/Pillonet northern Austro-Alpine klippen; *CA* Canavese; *LS* Laghi Series; possibly incised and excised deep-seated lower crustal units: *SL* Sesia-Lanzo ± northern Austro-Alpine klippen (alternative location); *UA* Upper Austro-Alpine; *IV* Ivrea-Verbano; *gb* layered gabbro/ultramafic bodies; *M* upper mantle. **C** Summarizing sketch of Late Palaeozoic processes of melting and magma emplacement. Mantle-derived gabbro magma chambers *(gb)* located at mantle *(M)* /thinning crust transition may supply fractionated melts and fast transfer of heat, facilitating partial melting of overlying dismembered crust. *SH* Deep-seated shear zones; *am* crustal anatectic melting; *v* volcanics; *g* granites; *a* appinitic precursors

visaged upper plate "allochthons, riders, extensional duplex and detached bodies" (e.g. Wernicke and Burchfiel 1982; Gibbs 1984) may be consistent with the dimensions of the future Eo-Alpine basement nappes (Polino et al. 1990), especially when complex incisement/excisement geometries along low angle detachment zones may be expected (Lister and Davis 1989; Fig. 5). This setting may eliminate, a priori, some of the pending difficulties on delamination and nappe-forming mechanisms in the Alpine stacking models.

The deep crustal expression of these processes has recently been demonstrated on the basis of metamorphic and structural signatures from the South-Alpine Ivrea-Verbano zone (Hodges and Fountain 1984; Brodie and Rutter 1987; Handy 1987; Schmid et al. 1987; Rutter and Brodie 1988; Brodie et al. 1989; Zingg et al. 1990) and the Austro-Alpine Mt. Emilius and Sesia-Lanzo nappes (Benciolini 1989; Lardeaux and Spalla 1991).

According to Benciolini (1989), the pre-eclogitic low-P/high-T mylonites from the Mt. Emilius mantle ultramafic body may be envisaged as record of lithosphere attenuation through crust/mantle ductile detachments and may tentatively be correlated to the Permian-Mesozoic shear zones and low-angle normal faulting which assisted the post-Variscan extensional evolution of the South-Alpine basement. Similarly, the pre-Alpine granulite and post-granulite depressuring evolution of the Sesia-Lanzo basement has been interpreted, through a wide set of petrological data, as deep-crustal evidence of the Permian-Jurassic extensional regime which facilitated the opening of the Tethyan basin (Lardeaux and Spalla 1991). Nevertheless, due to the absence of radiometric dating, the granulite metamorphic imprint may alternatively be referred to a Late Variscan post-thickening collapse, as discussed by Lardeaux and Spalla (1991), and the relict high-grade fabric possibly to even older episodes, at least in the case of Lower Austro-Alpine slices in which the Permian granitoid intrusive bodies indicate that a relatively shallow crustal level was quickly attained by contact paraschists.

These tensional mechanisms and related thermal implications may be applied to the overall crustal infrastructure and underlying mantle of the Europe/Adria divergent plate margin, and is thought here to be appropriate also to Late Palaeozoic times, in light of an extensional re-interpretation of the "Late Hercynian orogenic events". Lithosphere attenuation and extension through simple shear mechanisms along deep low-angle detachment zones in modern and ancient settings have been proposed as models for the generation of metamorphic core complexes (Crittenden et al. 1980; Lister and Davis 1989), low-P/high-T metamorphic re-equilibration (Wickham and Oxburgh 1985; Sandiford and Powell 1986; Thompson and Ridley 1987; Harley 1989; De Saint Blanquat et al. 1990; Lardeaux and Spalla in press), dehydration reactions and granitic melts (e.g. Thompson 1982), and mantle melting by adiabatic depressuring supplying fast transfer of heat into the lithosphere (e.g. Wells 1979). Moreover, these extensional processes may solve some space problems as to how the huge volumes of silicic and mafic intrusions were accommodated in the Late Palaeozoic continental crust (e.g. Hutton et al. 1990) and further questions as to how the noticeable basement uplift inferred from the granulite to low-P metamorphic evolution was accomplished with a relatively scarce amount of erosional products. Accordingly, the Latest Palaeozoic history of the whole Alpine basement may be seen as early signature of the Europe/Adria divergent plate margin, i.e. the metamorphic and magmatic counterpart of the onset of lithosphere extension and associated thermal perturbance.

In the internal northwestern Alps, this model may account for low-P/high-T metamorphic re-equilibration under high heat flow and high fluid activity within the uplifting former deep crust, also facilitating the growth of andalusite- and/or cordierite-bearing mineral assemblages and the generation of granitic or bimodal melts from continental crust and underlying mantle (Fig. 5). Large quantities of fluids, which are provided by dehydration reactions as melts crystallize, may facilitate reactivation of some shear zones, assisting further steps of retrogressive alteration, dynamic recovery and recrystallization. The absence, in the Alpine nappe stack, of an adequate network of relict detachment faults, such as those recently reported from the Southern Alps, may be only fictitious contrasting proof, depending on complete recovery during the Alpine polyphase metamorphism. Moreover, some of the Alpine eclogitic and greenschist facies mylonites recorded in the Penninic and Austro-Alpine systems may actually be pre-Alpine extensional shear zones, inverted and recrystallized in contrasting P-T conditions during Alpine convergence.

In conclusion, the rising cover-free lower crustal sections of the Valpelline/southern Austro-Alpine/Sesia-Lanzo units and Ivrea-Verbano zone may even be envisaged as post-Variscan metamorphic core complexes capped by a dismembered and allochthon-rich upper plate crustal lid progressively thinning through pre-Alpine extensional unroofing (Fig. 5), i.e. the Canavese, northern Austro-Alpine, Monte Rosa and Grand St. Bernard basement and related Permian and/or Mesozoic sedimentary covers. Such a highly fragmented setting may in turn facilitate, during the Cretaceous orogeny, the decoupling of the future high-P lower crustal units from those which operated as the roof of the Eo-Alpine orogenic wedge and as crustal suprastructure, forming the sole of constructed fore-arc basins (Polino et al. 1990).

The post-Variscan extensional model is suitable elsewhere, as for instance:

1. The Penninic Briançonnais zone of the French Alps, where the Upper Carboniferous-Permian sedimentary and magmatic records may be integrated within an extensional setting, possibly compared to the basin-and-range province (Fabre et al. 1987).
2. The Late Palaeozoic evolution of the western South-Alpine basement, as envisaged by Zingg et al. 1990, where: the late growth of cordierite in the southwestern kinzigitic edge indicates depressuring conditions; the depleted nature of paragneisses with increasing temperature suggests large removal of anatectic melts; the emplacement of Permian I-type granitic plutons, rhyolitic/andesitic volcanics and calc-alkaline appinitic precursors may be related to tensional/transtensional mechanisms and to lower-crustal sources, like the deep-seated magma chambers of the underplating mafic igneous complex (Sighinolfi and Gorgoni 1978; Zingg 1983; Boriani et al. 1990, pro parte; Voshage et al. 1990; Zingg et al. 1990).
3. The calc-alkaline bimodal intrusive and volcanic activity of Permian age widely recorded from the Canavese zone (Biino et al. 1986) to the eastern South-Alpine and Austro-Alpine units and previously reported as Late Variscan events (e.g. D'Amico and Mottana 1976; Oberhauser 1980; Vai et al. 1984; Sassi et al. 1985; Visonà et al. 1987; Frisch et al. 1990, and references therein).
4. The post-kinematic "low-P and high-T metamorphic event" regionally recorded in the Oetztal/Silvretta and further units from the eastern Austro-Alpine basement by andalusite, minor cordierite and biotite-sillimanite aggregates which replace Variscan garnet; the growth of these minerals in fact postdates the last folding phase referred to the Variscan regional metamorphism and predates the first Alpine fold generation (e.g. Gregnanin and Piccirillo 1972);

5. The low-P/high-T metamorphic evolution of the central South-Alpine basement, where fibrolite replacing biotite and the growth of andalusite are coupled with extensional shear zones (Siletto et al. 1990) and with pegmatites yielding ages of 240–200 Ma (Mottana et al. 1985, and references therein).

Furthermore, this model may also convincingly solve the intricate question and mysterious significance of the "Permian thermal/metamorphic event", early hypothesized by Ellenberger (1958) as "metasomatic granitization" in the Briançonnais domain and then documented in some Penninic units of the Western and Eastern Alps through consistent or debated radiometric and isotope data (Hunziker 1970; Frey et al. 1976; Niggli 1978; Frank et al. 1987).

The hypothesized Late Palaeozoic evolution model of the internal northwestern Alpine basement is not yet supported by geochronologic data adequately defining, as in the Ivrea zone, the boundary between the exhaustion of the Variscan convergence and the onset of the lithosphere extension evolving to continental rifting. For instance, there is now a pressing need to estimate radiometrically the high-grade fabric, granulite assemblage and late low-P/high-T reequilibration, as well as the growing age of andalusite from the Mt. Mary nappe and of cordierite from the Valpelline unit, where the absence of Permian granitoids prevents any attempt at relative timing.

Acknowledgements
The author is very grateful to reviewers J. M. Lardeaux and St. Schmid, and to L. Benciolini, B. Cesare, G. Gosso, J. von Raumer and M. I. Spalla for helpful comments or stimulating discussions. Italian MURST funding is acknowledged.

References

Albrecht J, Peters TJ, Sommerauer J (1978) Manganiferous mineral assemblages of Ravinella di Sotto, Valle Strona. Mem Ist Geol Minerol Univ Padova 33: 215–222
Argand E (1909) L'exploration géologique des Alpes Pennines centrales. Bull Soc Vaudoise Sci Nat 45: 217–276
Argand E (1911) Les nappes de recouvrement des Alpes occidentales et les territoires environnants. Mat Carte Geol Suisse spec 64
Arthaud F, Matte P (1977) Late Palaeozoic strike-slip faulting in southern Europe and northern Africa: result of a right-lateral shear zone between Appalachians and the Urals. Bull Soc Geol Am 88: 1305–1320
Ayrton ST, Bugnon C, Haarpainter T, Weidmann M, Frank E (1982) Géologie du front de la nappe de la Dent Blanche dans la région des Monts-Dolins, Valais. Eclogae Geol Helv 75: 269–286
Baccos F, Del Moro A, Visonà D (1990) The Hercynian volcano-plutonic association of Ganna (Lake Lugano, Central Southern Alps, Italy). Eur J Mineral 2: 373–383
Ballèvre G, Kienast JR, Vuichard JP (1986) La nappe de la Dent Blanche (Alpes occidentales): deux unités austroalpines indépendantes. Eclogae Geol Helv 79: 57–74
Banzet G, Lapierre H, Lefort P, Pecher A (1985) Le volcanisme carbonifère-supérieur du Massif des Grandes Rousses (zone Dauphinoise-Alpes externes françaises): un magmatisme à affinité shoshonitique lié à la fracturation crustale tardivarisque. Geol Alp 61: 33–60
Battiston P, Benciolini L, Dal Piaz GV, De Vecchi P, Marchi G, Martin S, Polino R, Tartarotti P (1987) Geologia di una traversa dal Gran Paradiso alla Zona Sesia-Lanzo in alta Val Soana, Piemonte. Mem Soc Geol Ital 29 (1984): 209–232
Bearth P (1952) Geologie und Petrographie des Monte Rosa. Beitr Geol Karte Schweiz 96: 94 pp
Bearth P (1967) Die Ophiolite Zone von Zermatt-Saas Fee. Beitr Geol Karte Schweiz 132: 130 pp
Bearth P, Dal Piaz GV, Elter G, Gosso G, Martinotti G, Nervo R (1980) Il lembo di ricoprimento del Monte Emilius, Dent Blanche s. l. Mem Accad Sci Torino Cl Sci Phys 144: 227–241
Beccaluva L, Dal Piaz GV, Macciotta G, Zeda O (1979) The Austroalpine harzburgite body of the Artogna valley (Italian Western Alps). Mem Ist Geol Mineral Univ Padova 33: 173–181
Benciolini L (1989) Evoluzione tettonico-metamorfica del corpo ultramafico di Arpisson (Lembo Austroalpino del Mt. Emilius, Alpi occidentali). Tesi dott Sci Terra, Univ Padova
Bertolani M (1967) Rocce manganesifere tra le granuliti della Valle Strona (Novara). Period Mineral 36: 1011–1032
Bertolani M (1971) La petrografia della cosiddetta seconda zona kinzigitica nelle alte valli del Mastallone e del Sermenza (Valsesia). Rend Soc Ital Mineral Petrol 27: 367–391
Bianchi A, Dal Piaz GB, Viterbo C (1965) Le masse di anfiboliti gabbriche di Corio e Monastero e di altre localita della Zona Sesia-Lanzo (Alpi occidentali). Mem Accad Sci Torino Cl Sci Phys 4 (3): 1–36
Biino G, Compagnoni R (1988) La scaglia di Grun in Val d'Aosta: un lembo Austroalpino composito, incluso nelle meta-ofioliti della Zona Piemontese. Boll Soc Geol Ital 107: 101–107
Biino G, Compagnoni R (1989) The Canavese Zone between the Serra d'Ivrea and the Dora Baltea river (Western Alps). Eclogae Geol Helv 82: 413–427
Biino G, Castelli D, Rossetti P (1986) Plutonisme acide et basique dans le socle de la Zone du Canavèse. CR Acad Sci Paris 303: 1473–1476
Bixel F, Lucas C (1983) Magmatisme, tectonique et sédimentation dans les fosses stéphano-permien des Pyrénées occidentales. Rev Géol Dyn Géogr Phys 24: 329–342
Boriani A, Dal Piaz GV, Hunziker JC, von Raumer J, Sassi FP (1976) Caratteri, distribuzione ed età del metamorfismo prealpino nelle Alpi. Mem Soc Geol Ital 13/1 (1974): 165–225
Boriani L, Caironi V, Oddone M, Vannucci R (1988) Some petrological and geochemical constraints on the genesis of the Baveno-Mottarone and Montorfano plutonic bodies. Rend Soc Ital Mineral Petrol 43: 385–394
Boriani L, Burlini L, Sacchi R (1990) The Cossato-Mergozzo-Brissago line and the Pogallo line (Southern Alps, Northern Italy) and their relationships with late-Hercynian magmatic and metamorphic events. Tectonophysics 182: 91–102
Brodie KH, Rutter EH (1987) Deep crustal extensional faulting in the Ivrea zone of northern Italy. Tectonophysics 140: 193–212

Brodie KH, Rex D, Rutter EH (1989) On the age of deep crustal extensional faulting in the Ivrea zone, Northern Italy. Geol Soc Lond Spec Publ 45: 203–210

Bürgi A (1987) Mineralogische und petrographische Untersuchungen im Val Strona di Postua (Prov. Vercelli, Italien). Diplome, Min Petr Inst Univ Bern

Cabella R, Cortesogno L, Dallagiovanna G, Vannucci R, Vanossi M (1988) Vulcanismo, sedimentazione e tettonica nel Brianzonese ligure esterno durante il Permo-Carbonifero. Atti Tic Sci Terra 31: 269–236

Cadel G (1986) Geology and Uranium mineralization of the Collio basin (central Southern-Alps). Uranium 2: 215–240

Callegari E, Compagnoni R, Dal Piaz GV (1969) Relitti di strutture intrusive erciniche e scisti a sillimanite nel Massiccio del Gran Paradiso. Boll Soc Geol Ital 88: 59–69

Callegari E, Compagnoni R, Dal Piaz GV, Frisatto V, Gosso G, Lombardo B (1976) Nuovi affioramenti di metagranitoidi nella Zona Sesia-Lanzo (Alpi occidentali). Rend Soc Ital Mineral Petrol 32: 97–111

Canepa M, Castelletto M, Cesare B, Martin S, Zaggia L (1990) The Austroalpine Mont Mary nappe (Italian Western Alps). Mem Sci Geol Padova 42: 1–17

Carraro F, Dal Piaz GV, Sacchi R (1970) Serie di Valpelline e II Zona diorito-kinzigitica sono i relitti di un ricoprimento proveniente dalla Zona Ivrea-Verbano. Mem Soc Geol Ital 9: 197–224

Cesare B, Martin S, Zaggia L (1989) Mantle peridotites from the Austroalpine Mt. Mary nappe (Western Alps). Schweiz Mineral Petrogr Mitt 69: 91–97

Compagnoni R, Prato R (1969) Paramorfosi di cianite su sillimanite in scisti pregranitici del Massiccio del Gran Paradiso. Boll Soc Geol Ital 88: 537–549

Compagnoni R, Elter G, Lombardo B (1976) Eterogeneità stratigrafica del complesso degli "Gneiss minuti" nel Massiccio cristallino del Gran Paradiso. Mem Soc Geol Ital 13/1 (1974): 227–239

Compagnoni R, Dal Piaz GV, Hunziker JC, Gosso G, Lombardo B, Williams PF (1977) The Sesia-Lanzo zone, a slice of continental crust with Alpine high pressure-low temperature assemblages in the Western Italian Alps. Rend Soc Ital Mineral Petrol 33: 281–334

Cortesogno L, Dallagiovanna G, Vannucci R, Vanossi M (1988) Volcanisme, sédimentation et tectonique pendant le Permo-Carbonifère en Briançonnais ligure. Eclogae Geol Helv 81: 487–510

Crittenden MD, Coney PJ, Davis GH (eds) (1980) Tectonic significance of metamorphic core complexes of the North American Cordillera. Geol Soc Am Mem 153: 490 pp

Dal Piaz GB (1939) La discordanza ercinica nella zona pennidica e le sue conseguenze nei riguardi della storia geologica delle Alpi. Boll Soc Geol Ital 58: 105–152

Dal Piaz GV (1966) Gneiss ghiandoni, marmi ed anfiboliti antiche del ricoprimento Monte Rosa nell'alta valle d'Ayas. Boll Soc Geol Ital 85: 103–132

Dal Piaz GV (1971) Nuovi ritrovamenti di cianite alpina nel cristallino antico del Monte Rosa. Rend Soc Ital Mineral Petrol 27: 437–477

Dal Piaz GV (1976) Il Lembo di ricoprimento del Pillonet, Falda Dent Blanche nelle Alpi occidentali. Mem Ist Geol Minerol Univ Padova 31: 1–61

Dal Piaz GV (1988) Revised setting of the Piedmont zone in the northern Aosta Valley, Western Alps. Ofioliti 13: 157–162

Dal Piaz GV, Ernst WG (1978) Areal geology and petrology of eclogites and associated metabasites, of the Piemonte ophiolitic nappe, Breuil-St. Jacques area, Italian Western Alps. Tectonophysics 51: 99–126

Dal Piaz GV, Lombardo B (1986) Early Alpine eclogite metamorphism in the Penninic Monte Rosa-Gran Paradiso basement nappes on the northwestern Alps. Geol Soc Am Mem 164: 249–265

Dal Piaz GV, Sacchi R (1969) Osservazioni geologiche sul lembo di ricoprimento del Pillonet, Dent Blanche s. l. Mem Soc Geol Ital 8: 835–846

Dal Piaz GV, Gosso G, Martinotti G (1971) La II zona diorito-kinzigitica tra la Valsesia e la Valle d'Ayas (Alpi occidentali). Mem Soc Geol Ital 10: 257–276

Dal Piaz GV, Hunziker JC, Martinotti G (1972) La Zona Sesia-Lanzo e l'evoluzione tettonico-metamorfica delle Alpi nordoccidentali interne. Mem Soc Geol Ital 11: 433–466

Dal Piaz GV, von Raumer J, Sassi FP, Zanettin B, Zanferrari A (1975) Geological outline of the Italian Alps. In: Squires C (ed) Geology of Italy. Earth Sci Soc Lybian Rep, pp 299–375

Dal Piaz GV, De Vecchi G, Hunziker JC (1977) The Austroalpine layered gabbros of the Matterhorn and Mt. Collon-Dents de Bertol. Schweiz Mineral Petrogr Mitt 57: 59–88

Dal Piaz GV, Hunziker JC, Stern WB (1978) The Sesia-Lanzo Zone, a slice of subducted continental crust. Geol Surv Open-File Rep 78–701: 83–86

Dal Piaz GV, Lombardo B, Gosso G (1983) Metamorphic evolution of the Mt. Emilius klippe, Dent Blanche nappe, Western Alps. Am J Sci 283 A: 438–458

D'Amico C, Mottana A (1976) Hercynian plutonism in the Alps. Mem Soc Geol Ital 13/1 (1974): 49–118

De Graciansky PC, Dardeau G, Lemoine M, Tricart P (1989) The inverted margin of the French Alps and foreland basin inversion. Geol Soc Spec Publ Lond 44: 87–104

De Saint Blanquat M, Lardeaux JM, Brunel M (1990) Petrological arguments for high-temperature extensional deformation in the Pyrenean Variscan crust (Saint Barthélémy massif, Ariège, France). Tectonophysics 177: 245–262

Diehl E, Masson R, Stutz A (1952) Contributi alla conoscenza del ricoprimento Dent Blanche. Mem Ist Geol Univ Padova 17: 1–52

Ellenberger F (1958) Etude géologique du pays de Vanoise. Mém Carte Géol Fr 1958: 561 pp

Ernst GW (1973) Interpretative synthesis of metamorphism in the Alps. Geol Soc Am Bull 84: 2053–2078

Escher A (1988) Structure de la nappe du Grand Saint-Bernard entre le val de Bagnes et les Mischabel. Office fed protect environment Bern, Rapp Geol 7: 26 pp

Fabre J, Schade J, Baudin T, Desmon J, Mercier D, Perruccio-Parison MD (1987) Relics of pre-Mesozoic events in the Briançon zone (Northern French Alps). In: Flügel HW, Sassi FP, Grecula P (eds) Pre-Variscan and Variscan events in the Alpine-Mediterranean mountain belts. Mineralia Slovaca Monography, Alfa, Bratislava, pp 183–208

Franchi S (1902) Über Feldspath-Uralitisierung der Natron-Thonerde-Pyroxene aus des eklogitischen Glimmerschiefern der Gebirge von Biella. Neues Jahrb Mineral Geol Palaeontol 2: 112–126

Frank W, Kralik M, Sharbert S, Thoeni M (1987) Geochronological data from the Eastern Alps. In: Flügel HW, Faupl P (eds) Geodynamics of the Eastern Alps. Deuticke, Vienne, pp 272–281

Frey M, Hunziker JC, Frank W, Bocquet J, Dal Piaz GV, Jaeger E, Niggli E (1974) Alpine metamorphism of the Alps: a review. Schweiz Mineral Petrogr Mitt 54: 247–290

Frey M, Hunziker JC, O'Neil JR, Schwander HW (1976) Equilibrium-disequilibrium relations in the Monte Rosa granite, western Alps: petrological, Rb-Sr and stable isotope data. Contrib Mineral Petrol 55: 147–179

Frisch W, Menot RP, Neubauer F, von Raumer JF (1990) Correlation and evolution of the Alpine basement. Schweiz Mineral Petrogr Mitt 70: 265–285
Gibbs AD (1984) Structural evolution of extensional basin margins. J Geol Soc Lond 141: 609–620
Gregnanin A, Piccirillo EM (1972) Litostratigrafia, tettonica e petrologia negli scisti austridici di alta e bassa pressione nell'area Passiria-Venosta (Alto Adige). Mem Ist Geol Mineral Univ Padova 30: 1–53
Handy MR (1987) The structure, age and kinematics of the Pogallo Fault Zone, a deep crustal dislocation in the Southern Alps, northwestern Italy. Eclogae Geol Helv 80: 593–632
Harley SL (1989) The origins of granulites: a metamorphic perspective. Geol Mag 126: 215–247
Hodges AK, Fountain DM (1984) Pogallo line, South Alps, northern Italy: and intermediate crustal level, low angle normal fault? Geology 12: 151–155
Hunziker JC (1970) Polymetamorphism in the Monte Rosa, western Alps. Eclogae Geol Helv 63: 151–161
Hunziker JC (1974) Rb-Sr and K-Ar age determination and the Alpine tectonic history of the western Alps. Mem Ist Geol Mineral Univ Padova 31: 1–54
Hunziker JC, Martinotti G (1987) Geochronology and evolution of the western Alps: a review. Mem Soc Geol Ital 29 (1984): 45–56
Hunziker JC, Zingg A (1980) Lower Paleozoic amphibolite to granulite facies metamorphism in the Ivrea zone (Southern Alps, Northern Italy). Schweiz Mineral Petrogr Mitt 60: 181–213
Hunziker JC, Desmons J, Martinotti G (1989) Alpine thermal evolution in the central and western Alps. Geol Soc Spec Publ Lond 45: 353–367
Hutton DHW, Dempster TJ, Brown PE, Becker SD (1990) A new mechanism of granite emplacement: intrusion of active extensional shear zones. Nature 343: 452–455
Isler A, Zingg A (1974) Geologie der Sesia-Zone zwischen Rimella und der Valle Anzasca (Nord-Italien). Schweiz Mineral Petrogr Mitt 54: 81–96
Kienast JR (1983) Le métamorphisme de haute pression et basse température (eclogites et schistes bleus): données nouvelles sur la pétrologie de la crouûte océanique subductée et des sédiments associés. Thèse Etat, Univ Paris VI, 384 pp
Laduron D (1976) L'antiforme de Vanzone. Etude pétrologique et structurale dans la Valle Anzasca (Province de Novara, Italie). Mém Inst Géol Univ Louv 28: 1–121
Lardeaux JM (1981) Evolution tectono-métamorphique de la zone Nord du massif de Sesia-Lanzo (Alpes occidentales): un example d'eclogitization de croûte continentale. Thèse 3e cycle, Univ Paris VI
Lardeaux JM, Spalla MI (1991) From granulites to eclogites in the Sesia zone (Italian Western Alps): a record of opening and closure of the Piedmont ocean. J Metamorphic Geol 9: 35–59
Lardeaux JM, Spalla MI Tectonic significance of P-T-t paths in metamorphic rocks: examples from ancient and modern orogenic belts. 75 Congr Soc Geol It, Milan 1990, Mem Soc Geol Ital 45 (in press)
Lardeaux JM, Gosso G, Kienast JR, Lombardo B (1982) Relations entre le métamorphisme et la déformation dans la zone Sesia-Lanzo (Alpes occidentales) et le problème de l'eclogitization de la croûte continentale. Bull Soc Géol Fr 24: 793–800
Laubscher HP (1990) Deep seismic data from the Central Alps: mass distribution and their kinematics. Mem Soc Géol Fr 156: 335–343
Laubscher HP, Bernoulli D (1982) History and deformation of the Alps. In: Hsu KJ (ed) Mountain building processes. Academic Press, London, pp 170–180
Lister GS, Davis GA (1989) The origin of metamorphic core complexes an detachment faults during Tertiary continental extension in the Northern Colorado River region, U. S. A. J Struct Geol 11: 65–94
Marthaler M (1984) Géologie des unités penniques entre le Val d'Anniviers et le Val de Tourtemagne (Valais, Suisse). Eclogae Geol Helv 77: 395–448
Masson R (1938) Geologisch-petrographische Untersuchungen im unteren Valpelline, Provinz Aosta, Italien. Schweiz Mineral Petrogr Mitt 18: 54–212
Mattauer M, Malavielle J, Monié P (1987) Une coupe lithosphérique des Alpes occidentales dans l'hypothèse où Sezia n'est pas d'origine sud-alpine. CR Acad Sci Paris 304: 43–48
Mazzucchelli M, Siena F (1986) Geotectonic significance of the metabasites of the kinzigitic series, Ivrea-Verbano zone (Western Italian Alps). Tschermaks Mineral Petrogr Mitt 35: 99–116
Menard G, Molnar P (1988) Collapse of a Hercynian Tibetan plateau into a Late Paleozoic European Basin and Range province. Nature 334: 235–237
Mercolli I, Oberhänsli R (1988) Variscan tectonic evolution in the Central Alps. Schweiz Mineral Petrogr Mitt 68: 491–500
Mottana A, Nicoletti M, Petrucciani C, Liborio G, De Capitani L, Bocchio R (1985) Pre-alpine and alpine evolution of the South-Alpine basement of the Orobic Alps. Geol Rundsch 74: 353–366
Nicot (1977) Les roches méso- et catazonales de la Valpelline (Nappe de la Dent Blanche, Alpes italiennes). Thèse 3e cycle, Univ Paris VI
Niggli E (ed) (1978) Metamorphic map of the Alps 1:1000000 and explanatory text. Subcomm cartogr metam belts, UNESCO, Paris, 242 pp
Novarese V (1929) La zona del Canavese e le formazioni adiacenti. Mem Descr Carta Geol Ital 22: 65–212
Novarese V (1931) La formazione diorito-kinzigitica in Italia. Boll R Uff Geol Ital 56 (7): 1–62
Oberhänsli R, Hunziker JC, Martinotti G, Stern WB (1985) Geochemistry, geochronology and petrology of Monte Mucrone: an example of Eo-Alpine eclogitization of Permian granitoids in the Sesia-Lanzo zone, western Alps, Italy. Chem Geol 52: 165–184
Oberhauser R (ed) (1980) Der geologische Aufbau Österreichs. Geol Bundesanstalt, Wien, 699 pp
Pennacchioni G (1989) Struttura ed evoluzione metamorfica alpina del M. Emilius (Alpi occidentali). Tesi Dott Sci Terra, Univ Padova
Pin C (1986) Datation U-Pb sur zircons a 285 M. a. du complexe gabbro-dioritique du Val Sesia-Val Mastallone et age tardihercynien du métamorphisme granulitique de la zone Ivrea-Verbano (Italie). CR Acad Sci Paris 303: 827–829
Pin C, Vielzeuf D (1983) Granulites and related rocks in Variscan Median Europe: a dualistic interpretation. Tectonophysics 93: 47–74
Pinarelli L, Boriani A, Del Moro A (1988) Rb-Sr geochronology of the Lower Permian plutonism in the Massiccio dei Laghi, Southern Alps (NW Italy). Rend Soc Ital Mineral Petrol 43: 411–428
Platt JP (1986) Dynamics of orogenic wedges and the uplift of high-pressure metamorphic rocks. Geol Soc Am Bull 97: 1037–1053
Pognante U, Talarico F, Benna P (1987) Incomplete blueschist re-crystallization in high-grade metamorphics from the

Sesia-Lanzo unit (Vasario and Sparone subunit, Western Alps): a case history of metastability. Lithos 21: 129–142
Polino R, Dal Piaz GV, Gosso G (1990) Tectonic erosion at the Adria margin and accretionary processes for the Cretaceous orogeny of the Alps. Mem Soc Géol Fr 156: 345–367
Rau A, Tongiorgi M (1981) Some problems regarding the Palaeozoic paleogeography in Mediterranean Western Europe. J Geol 89: 663–673
von Raumer JF (1988) Caledonian-Variscan structures in the Alps – an introduction. Schweiz Mineral Petrogr Mitt 68: 291–299
Ridley J (1989) Structural and metamorphic history of a segment of the Sesia-Lanzo zone, and its bearing on the kinematics of Alpine deformation in the western Alps. Geol Soc Spec Publ Lond 45: 189–201
Rivalenti G, Garuti G, Rossi A, Siena F, Sinigoi S (1981) Existence of different peridotite types and of a layered igneous complex in the Ivrea zone of the Western Alps. J Petrol 22: 127–153
Rivalenti G, Rossi A, Siena F, Sinigoi S (1984) The layered series of the Ivrea-Verbano igneous complex, Western Alps, Italy. Tschermaks Mineral Petrogr Mitt 33: 77–99
Rutter EH, Brodie KH (1988) The role of tectonic grain size reduction in the rheological stratification of the lithosphere. Geol Rundsch 77: 295–308
Sandiford M, Powell R (1986) Deep custal metamorphism during continental extension: modern and ancient examples. Earth Planet Sci Lett 79: 151–158
Sassi F, Cavazzini G, Visonà D, Del Moro A (1985) Radiometric geochronolgy in the Eastern Alps: results and problems. Rend Soc Ital Mineral Petrol 40: 187–224
Schmid ST, Zingg SM, Handy M (1987) The kinematics of movements along the Insubric Line and the emplacement of the Ivrea Zone. Tectonophysics 135: 47–66
Sighinolfi GP, Gorgoni C (1978) Chemical evolution of high-grade metamorphic rocks – anatexis and remotion of material from granulite terrains. Chem Geol 22: 157–176
Siletto GB, Spalla MI, Tunesi A, Nardo M, Soldo L, Centemerlo M, Tognini P (1990) Cartografia strutturale di un settore di basamento sudalpino sulla sponda orientale del Lago di Como. Riassunti posters 75 Congr Soc Geol Ital Milano 1990, pp 135–136
Sills JD, Tarney J (1984) Petrogenesis and tectonic significance of amphibolites interlayered with metasedimentary gneisses in the Ivrea zone, Southern Alps, northwest Italy. Tectonophysics 107: 187–206
Spalla MI, De Maria L, Gosso G, Miletto M, Pognante U (1983) Deformazione e metamorfismo della Zona Sesia-Lanzo meridionale al contatto con la falda Piemontese e con il massiccio di Lanzo, Alpi occidentali. Mem Soc Geol Ital 26: 499–514
Spalla MI, Lardeaux JM, Dal Piaz GV, Gosso G (1991) Métamorphisme et tectonique à la marge externe de la croûte Austro-Alpine: ré-examen des unités de la zone Sesia (Alpes occidentales). Mem Sci Geol Padova 43 (in press)
Stampfli GM, Marthaler M (1990) Divergent and convergent margins in north-western Alps. Confrontation to actualistic models. Geodyn Acta 4: 159–184
Stella A (1894) Relazione sul rilevamento eseguito nell'anno 1893 nelle Alpi occidentali (Valli dell'Orco e della Soana). Boll R Comit Geol Ital 25: 343–371
Stille P, Buletti M (1987) Nb-Sr isotopic characteristics of the Lugano volcanic rocks and constraints on the continental crust formation in the South Alpine domain (N-Italy-Switzerland). Contrib Mineral Petrol 96: 140–150
Stünitz (1989) Partitioning of metamorphism and deformation in the boundary region of the "Seconda zona diorito-kinzigitica", Sesia-Zone, Western Alps. Diss 8817, ETH Zürich
Stutz AH, Masson R (1938) Zur Tektonik der Dent Blanche Decke. Schweiz Mineral Petrogr Mitt 18: 40–53
Thelin P, Ayrton S (1983) Cadre évolutif des évenements magmatico-métamorphiques du socle anté-triassique dans le domaine pennique (Valais). Schweiz Mineral Petrogr Mitt 63: 393–420
Thompson AB (1982) Dehydration and partial melting of pelitic rocks and the generation of H_2O-undersaturated granitic liquids. Am J Sci 282: 1567–1595
Thompson AB, England PC (1984) Pressure-temperature-time paths of regional metamorphism. II. Their inference and interpretation using mineral assemblages in metamorphic rocks. J Petrol 24: 929–955
Thompson AB, Ridley JR (1987) Pressure-temperature-time (P-T-t) histories of orogenic belts. Philos Trans R Soc Lond A321: 27–45
Trümpy R (1980) An outline of the geology of Switzerland. Schweiz Geol Komm, Wepf, Basel, pp 1–110
Vai GB, Boriani A, Rivalenti G, Sassi FP (1984) Catena ercinica e Paleozoico nelle Alpi Meridionali. In: Cento anni di geologia italiana. Soc Geol Ital, vol giubilare, Roma, pp 133–154
Vanossi M, Cortesogno L, Galbiati B, Messiga B, Piccarolo G, Vannucci R (1986) Geologia delle Alpi Lugiri: dati, problemi, ipotesi. Mem Soc Geol Ital 28: 5–75
Visonà D, Alberti F, Stefani C, Stenico L (1987) Le plutoniti di Chiusa, Dosso Lives e Luson. Una serie calcalcalina ercinica nelle Alpi orientali. Mem Sci Geol, Padova 39: 85–99
Voshage H, Hoffmann AW, Mazzucchelli M, Rivalenti G, Sinigoi S, Raczek I, Demarchi G (1990) Isotopic evidence from the Ivrea Zone for a hybrid lower crust formed by magmatic underplating. Nature 347/6295: 731–736
Vuichard JP (1987) Conditions P-T du métamorphisme antéalpin dans la "seconde zone diorito-kinzigitique" (Zone Sesia-Lanzo, Alpes occidentales). Schweiz Mineral Petrogr Mitt 67: 257–271
Wells PRA (1979) Thermal models for the magmatic accretion and subsequent metamorphism of continental crust. Earth Planet Sci Lett 46: 253–265
Wernicke B, Burchfiel BC (1982) Modes of extensional tectonics. J Struct Geol 4: 105–115
Wickham ER, Oxburgh ER (1985) Continental rifts as a setting for regional metamorphism. Nature 318: 330–333
Ziegler PA (1982) Geological atlas of Western and Central Europe. Elsevier, Amsterdam, 130 pp
Ziegler PA (1986) Geodynamic model for the Palaeozoic crustal consolidation of Western and Central Europe. Tectonophysics 126: 303–328
Zingg A (1983) The Ivrea and Strona-Ceneri zones (Southern Alps, Ticino and North Italy). A review. Schweiz Mineral Petrogr Mitt 63: 361–392
Zingg A, Handy MR, Hunziker JC, Schmid SM (1990) Tectonometamorphic history of the Ivrea Zone and its relationship to the crustal evolution of the Southern Alps. Tectonophysics 182: 169–192

The Lepontine Area, a Geochronological Summary

V. Köppel

Abstract

Within the subcontinental mantle of the Lepontine area evidence for a 1725-Ma-old magmatism is still preserved. The formation of oceanic crust and island arcs started about 1000 Ma ago. Orthogneisses constitute the dominating rock type of the area. The majority of them are of Variscan age, but some are of Caledonian origin. The depositional age of the pre-Permian metasediments is not well contrained. Zircon U-Pb ages of metapsammites and metapsephites indicate that some of the pre-Variscan metasediments were deposited after the emplacement of the Caledonian granites. It seems reasonable to assume that also pre-Orodovician metasediments are present.

1 Introduction

The Lepontine area forms part of the Penninic domain and exposes the deepest tectonic units of the Alps. It consists of north-vergent nappes and the Maggia transverse belt. East of the Leventina the nappes dip towards E and W of the Valle Antigorio westwards. The Teritary (20–30 Ma) high-temperature metamorphism reached upper amphibolite facies conditions and post-dates the formation of the nappes and the Maggia transverse belt. The Permian to Mesozoic sedimentary cover and the basement rocks exhibit the same deformational inprints (Wenk 1956). The basement rocks consists predominantly of orthogneisses, and paragneisses constitute the second most important rock type. Amphibolites play a minor role. Tectonic fragments of ultramafic rocks which are often accompanied by marbles and calc-silicate rocks occur as boundins especially in metasediments of the Adula nappe. High-pressure mineral assemblages indicate subduction related metamorphism (Heinrich 1986), which is probably related to an Alpine high pressure metamorphism (Gebauer et al. 1992).

Many of the orthogneisses and migmatites have in the past been considered to represent the products of a pre- to syn-kinematic Alpidic magmatism. Intrusive contacts to Mesozoic sediments have, however, never been observed, and therefore the opinion among geologists as to presence of Alpine orthogneisses was divided.

2 Pre-Alpine History

Deciphering the pre-Alpine history of the rocks constituting the Lepontine core of the Penninic domain is obviously a difficult task. Only by analyzing conservative radioactive parent-daughter systems such as U-Pb-zircon or Sm-Nd-whole-rock systems was it possible to gain some insight into the pre-Alpidic history of the Lepontine gneiss cores.

The oldest rocks of the Lepontine area are the garnet peridotites of Alpe Arami. Ionprobe-data indicate that the zircon crystallized 1724 ± 40 Ma ago (Gebauer et al. 1992) from trapped partial melts of the peridotite. The data furthermore indicate that shortly after the crystallization a first metamorphism occurred. A second Pan-African metamorphism caused some loss of lead in the zircons. Thus, the subcontinental mantle of the Alps has a long and complicated history.

Stille (1980) and Stille and Tatsumoto (1985) reported Rb-Sr, respectively Sm-Nd whole-rock ages of amphibolites from the Berisal complex (Fig. 1). The Rb-Sr results indicate an age of about 500 Ma for hornblendefelses and plagioclase amphibolites, whereas Sm-Nd whole rocks yielded isochron ages of 1018 ± 59 Ma for plagioclase amphibolites, 1071 ± 43 for banded amphibolites and, in agreement with the Rb-Sr results, 475 ± 81 Ma for hornblendefelses. The discrepancy between the Rb-Sr and Sm-Nd amphibolite ages remains to be explained. The authors argue that the Rb-Sr system is more mobile under metamorphic conditions than the Sm-Nd system. Compared to the Sm-Nd data the Rb-Sr isochrons are less

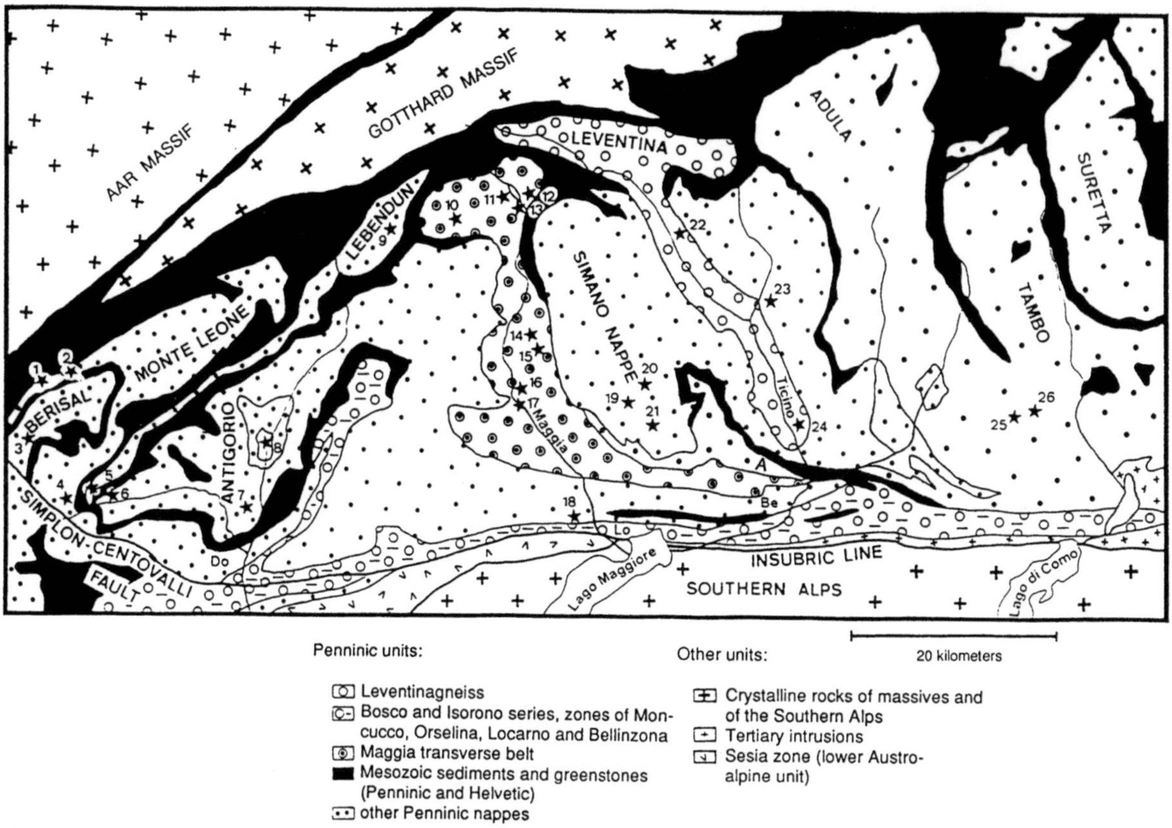

Fig. 1. Lepontine area showing the approximate locations of samples referred to in the text. *A* Alpe Arami; *Be* Bellinzona; *Do* Domodossola; *Lo* Locarno. The *numbers* refer to zircon and monazite samples reported in Köppel et al. (1981)

well defined and could possibly result from incomplete resetting 400 to 500 Ma ago. Gebauer (this Vol.) mentions further possible explanations for this discrepancy. The bimodal nature (dacitic and calc-alkaline tholeiitic) of the banded amphibolites is indicative of an island arc environment. Schenk-Wenger and Stille (1989) also reported Sm/Nd whole rock data indicating an age of about 1000 Ma for amphibolites of the Bosco and Isorno series W of. V. Antigorio and Moncucco, Orselina and Bellinzona zones between Bellinzona and V. Antigorio. On the basis of the initial 143Nd/144Nd and the REE-patterns which show only a weak enrichment of the LREE in comparison to N-MORB, the author concluded that the amphibolites are of MOR origin and represent a Proterozoic ophiolite series.

The 475 ± 81 Ma old komatiitic hornblendefelses of the Berisal complex were also derived from a depleted mantle source (Stille and Tatsumoto 1985). A Caledonian magmatism is evidenced by near-concordant zircon U-Pb ages at 450 Ma of the Berisal gneiss (3).[1] The zircons contain inherited Precambrian lead-components which are considered to indicate a contamination by continental crustal material such as sedimentary material containing detrital zircons (Allègre et al. 1974; Köppel et al. 1981).

Similar to the Berisalgneiss, zircons from the Simanogneiss (Val Blenio) (23) also yielded U-Pb ages which indicate a crystallization age of about 450 Ma. Again these zircons contain an inherited, Precambrian lead-component (Köppel et al. 1981). The Verzasca gneiss from Brione (close to 20) which forms in the Verzasca valley the upper part of the Simano nappe, also yielded single zircon U-Pb ages that are nearly concordant at 450 Ma, whereas at Corippo (S of 21) discordant ages indicate a Variscan crystallization age (Iizumi et al. 1986).

Zircon U-Pb age pattern from a metasediment of the Maggia nappe (12) is quite similar to those observed from areas which were not affected by high-grade Tertiary metamorphism. A discordia intercepts the concordia curve at 2000 Ma and at about 300–350 Ma. Of special interest are zircon U-Pb ages

[1] Number in parentheses refer to Fig. 1.

of metapsephites. Zircons from an augengneiss pebble from an undisputed metaconglomerate of the Lebendun nappe (9) indicate a Caledonian age for the formation of the augengneiss. Another augengneiss pebble, collected from a metaconglomerate of the Maggia nappe close to the Poncione di Braga (10), and the zircons from the matrix (10) yielded similar results (Köppel et al. 1981; Gebauer this Vol.). It should, however, be mentioned that the origin of this metaconglomerate is disputed. Huber (1981) suggested that the conglomeratic appearance could be the result of intensive shearing. Günthert (1954) and Wenk and Günthert (1960) described the polymict nature of this conglomerate which is a strong argument against a tectonic origin. The depositional age of these post-Caledonian sediments cannot be precisely constrained. In view of their higher metamorphic grade in comparison to the neighbouring Mesozoic metasediments and because Variscan granites intruded the metasediments a pre-Variscan sedimentation is strongly indicated. The results thus suggest the presence of post-Ordovician, pre-Permian sediments in the Lepontine area.

According to their zircon age pattern, granitic rocks of Variscan age include the following gneisses: Matorello (Maggia nappe) (11), Cocco (Maggia transverse zone) (15), Leventina, Antigorio (6, 7), Lebendun (5), Eisten (Monte Leone nappe) (1), Verampio (8) and Tegna (Mergoscia zone) (18) (Allègre et al. 1974; Köppel et al. 1981). An upper intercept age of 270 to 285 Ma of discordant monazites from the Cocco gneiss suggests a late Variscan origin of this granite (Köppel et al. 1981). Zircon suites from other gneisses yielded upper intercept ages of >340 Ma and thus could indicate Caledonian crystallization ages or the presence of old lead-components in Variscan zircon populations. The migmatites at Giumaglio (Valle Maggia) (17) are, according to their monazite ages, also of Variscan origin. The upper intercept age of 270–280 Ma of monazites from aplitic intrusions is indistinguishable from that of the Cocco gneiss whereas the upper intercept of monazites of the metasediments is distinctly higher at 285 to 305 Ma and probably represents a metamorphic age (Köppel et al. 1981).

Migmatites and granitic gneisses of the eastern Lepontine gneiss region (Bodengo area, Adula nappe) (25, 26) yielded a Variscan Rb-Sr whole rock errorchron and zircon U-Pb age pattern indicative of Variscan magmatism (Hänny et al. 1975). Gulson reported Rb-Sr whole rock data for the Tambo nappe and the Gruf migmatites which both indicate a Variscan origin of the magmatites. Interestingly, the initial $^{87}Sr/^{86}Sr$ are low and vary between 0.704 and 0.706. The Roffna granite porphyry of the upper Penninic Suretta nappe is located outside the Lepontine area. The zircon ages indicate a crystallization age of 300 Ma (Hanson et al. 1969).

The pre-Alpidic evolution of the Lepontine crust is in many ways similar to that of the surrounding terranes. The Caledonian and especially the Variscan magmatism are widespread within the Lepontine realm. Many of the zircon populations apparently contain no, or very small proportions of inherited zircons which might be indicative of the absence of inherited sedimentary material from Proterozoic to Archean continental sources. In this respect Pb-isotopes and initial Sr and Nd ratios are of interest. A general overwiev of rock – and ore-leads from the central, eastern and southern parts of the Alps is presented by Köppel et al. (this Vol.). Feldspar-lead from the Penninic domain is characterized by highly variable μ and W values which indicate a provenance from old as well as relatively young, i. e. Mid- to Late Proterozoic sources. In contrast, lead from Austro-Alpine units and the Southern Alps is normally of the high μ and W type indicating thereby the dominance of early Proterozoic to Archean source terranes. Initial $^{87}Sr/^{86}Sr$ and $^{143}Nd/^{144}Nd$ studies on orthogneisses from the Lepontine area also exhibit considerable variations and indicate the presence of mantle-derived Variscan granites in the Simano nappe at Corippo (Valle Verzasca). The ε-Nd values, however, normally indicate a crustal source age of 1400 Ma (Steiger and DePaolo 1986). The low initial Sr values reported from the Tambo gneisses and from the Bodengo and Gruf areas may either point to a mantle component or to a lower crustal origin of the magma.

3 Conclusion

In conclusion, one may state that the crustal evolution of the Lepontine terrane resembles that of the Penninic domain and probably also that of the external massives. The presence of 1000-Ma-old MOR-type metabasalts and island arc suites appears to be unique. However, this age is not unreasonably high in view of the 600- to 700-Ma-old Penninic. Habach-phyllites (Reitz and Höll 1988), or of the results (800–1000 Ma?) reported on the older intrusive rocks of the Austro-Alpine Silvretta nappe (Maggetti et al. 1990) and of the protolith age (870 Ma) of a Caledonian eclogite in the Gotthard massive (Gebauer et al. 1988). The Lepontine area is furthermore characterized by an abundance of acid magmatic rocks, most of which appear to relate to the Variscan magmatism. Some bear isotopic signatures which indicate that mantle material was involved in their generation. Metasediments, some of which are of Palaeozoic age, are less abundant. Their lead-isotope signature indicate that besides Early Proterozoic or Archean sources younger, i.e. Mid- to Late Proterozoic sources were of equal importance. It appears that the Lepontine domain cratonized mainly during the Variscan metamorphic/magmatic cycle similar to the entire Penninic realm.

References

Allègre CJ, Albarède F, Grünenfelder M, Köppel V (1974) $^{238}U/^{206}Pb$-$^{235}U/^{207}Pb$-^{232}-$Th/^{208}Pb$ zircon geochronology in Alpine and non-Alpine environment. Contrib Mineral Petrol 43: 163–194

Frey M, Hunziker JC, Frank W, Bocquet J, Dal Piaz GV, Jäger E, Niggli E (1974) Alpine metamorphism of the Alps, a review. Schweiz Mineral Petrogr Mitt 54: 247–290

Gebauer D, Quadt A, Compston W, Williams IS, Grünenfelder M (1988) Archean zircon in a retrograded, Caledonian eclogite of the Gotthard massif (Central Alps, Switzerland). Schweiz Mineral Petrogr Mitt 68; 485–490

Gebauer D, Grünenfelder M, Tilton G, Trommersdorff V, Schmid S (1992) The geodynamic evolution of garnet-peridotites, garnet-pyroxenites and eclogites of Alpe Arami and Cima di Gagnone (Central Alps) from the Early Proterozoic to Oligocene. Schweiz Mineral Petrogr Mitt 72 (in press)

Gulson BL (1973) Age relations in the Bergell region of the South-East Swiss Alps: with some geochemical comparison. Eclogae Geol Helv 66/2: 293–313

Günthert A (1954) Beiträge zur Petrographie und Geologie des Maggia-Lappens (NW Tessin). Schweiz Mineral Petrogr Mitt 56: 105–143

Hänny R, Grauert B, Soptrajanova G (1975) Paleozoic migmatites affected by high-grade Tertiary metamorphism in the Central Alps (Valle Bodengo, Italy). Contrib Mineral Petrol 51: 173–196

Hanson GN, Grünenfelder M, Soptrayanova G (1969) The geochronology of a recrystallized tectonite in Switzerland – the Roffna Gneiss. Earth Planet Sci Lett 5: 413–422

Heinrich CA (1986) Eclogite facies regional metamorphism of hydrous mafic rocks in the central Alpine Adula nappe. J Petrol 27: 123–154

Huber IM (1981) Geologisch-strukturelle Untersuchungen im Oberen Maggia-Gebiet (Tessin, Schweiz). Mitt Geol Inst ETH und Universität Zürich NF328: 220 pp

Iizumi S, Meier M, Oberli F, Steiger RH (1986) Petrogenetics by morphology/U-Pb single zircon study. Terra Cognita 6: 152

Köppel V, Günthert A, Grünenfelder M (1981) Patterns of U-Pb zircon and monazite ages in polymetamorphic units of the Swiss Central Alps. Schweiz Mineral Petrogr Mitt 61: 97–119

Maggetti M, Flisch M, Bollin R (1990) Bericht über die Exkursion der Schweizerischen Mineralogischen und Petrographischen Gesellschaft ins Silvretta Kristallin und in den Westrand des Unterengadiner Fensters. Schweiz Mineral Petrogr Mitt 70: 121–157

Reitz E, Höll R (1988) Jungproterozoische Mikrofossilien aus der Habachformation in den mittleren Hohen Tauern und dem nordostbayerischen Grundgebirge. Jahrb Geol Bundesanst 131: 329–340

Schenk-Wenger K, Stille P (1989) Geochemical and isotope evidence for an extensive Proterozoic ophiolite suite in the Central Alps (Switzerland). Berichte der DMG Beiheft zum EJM 1(1): 160

Steiger RH, DePaolo DJ (1986) Genealogy of the basement of the Central Alps. Terra Cognita 6: 127

Stille P (1980) On the genesis of the amphibolites and hornblendefelses in the Berisal complex (Simplon; Italy-Switzerland). Mem Sci Geol Univ Padua 34: 205–246

Stille P, Tatsumoto M (1985) Precambrian tholeiitic-dacitic rock suites and Cambrian ultramafic rocks in the Pennine nappe system of the Alps: evidence from Sm-Nd isotopes and rare earth elements. Contrib Mineral Petrol 89: 184–192

Wenk E (1956) Die lepontinische Gneissregion und die jungen Granite der Valle della Mera. Eclogae Geol Helv 49: 251–265

Wenk E, Günthert A (1960) Über metamorphe Psephite der Lebendun-Serie und der Bündnerschiefer im NW Tessin und Val d'Antigorio. Ein Diskussionsbeitrag. Eclogae Geol Helv 53(1): 179–188

Evolution of the Penninic Basement of the Eastern Alps

W. Frisch, G. Vavra, and M. Winkler

Abstract

Penninic basement is exposed in the Venediger nappe system of the Tauern window. Three lithostratigraphic units are distinguished: (1) the ophiolitic Stubach group which consists of splinters derived from back-arc oceanic crust, (2) the Habach-Storz group which is dominated by volcanic protoliths and represents remnants of an island arc probably formed on top of the back-arc oceanic crust, and (3) the Zentralgneis group which is composed of large volumes of Variscan granitoids. The island arc experienced maturation during its evolution in the Late Proterozoic and Early Palaeozoic. Carboniferous high-grade metamorphism and granitoid formation reflect a collisional event and collision- and subduction-related magma generation. S-type, originally cordierite-bearing, high-K granitoids, I-type high-K granitoids, and I-type medium-K granitoids form a sequence of decreasing age. Permian magmatism with a tendency to the A-type indicates an intraplate setting and signalizes the end of the Variscan orogenic cycle. The evolution of the basement in the Tauern window shows the progressive buildup of continental crust by long-lived subduction activity, starting in an intra-oceanic environment. The overall picture is reminiscent of granite-greenstone belts.

1 Introduction

In the Eastern Alps, Penninic basement is only exposed in the Tauern window where it is restricted to the lower mega-unit, the Venediger nappe system (Fig. 1). This basement is part of the Variscan orogenic belt. Alpine deformational and metamorphic overprint has obscured much of the pre-Alpine evolution and spatial relationships so that reconstructions are charged by uncertainties. Modern techniques, however, have rapidly increased knowledge about the nature and age of protoliths and about pre-Alpine tectono-metamorphic events during the last decade. Interpretation on the nature of protoliths is mainly based on studies of trace element chemistry, radiometric dating, and evolution paths of zircon crystals.

The basement of the Venediger nappe system is divided into the following lithologic associations (Fig. 1): (1) the ophiolitic association of the Stubach group; (2) the Late Proterozoic to Early Palaeozoic, mainly calc-alkaline magmatic and sedimentary association of the Habach-Storz group representing volcanic island arc rocks; and (3) the Variscan, predominantly calc-alkaline granitoid association of the Zentralgneis group which intruded the rocks of the former two associations. A short review of the basement in the Tauern window is given in Frisch et al. (1990).

2 Nature, Geotectonic Environment, and Age of Protoliths

2.1 The Ophiolitic Association (Stubach Group)

The pre-granitoid sequences predominantly consist of amphibolites and gneisses intercalated with metapelites and metacherts (Habach-Storz group). Associated are ophiolitic slices of the Stubach group, of which the Stubachtal ultramafic/mafic complex is the largest representative (Frisch and Raab 1987). The Ochsner ultramafic body in the Greiner synclinorium in the westernmost Tauern window (Lammerer 1986) and a number of serpentinite splinters and associated crustal rocks elsewhere are also considered to be parts of the ophiolitic assemblage.

The Stubachtal complex contains the metamorphic derivatives of ultramafic cumulates, gabbro, and minor amounts of basalt and acid magmatic rocks (Frisch and Raab 1987; Vavra 1989). According to their normative composition, the protoliths of the ultramafic rocks were mainly dunite, harzburgite, lher-

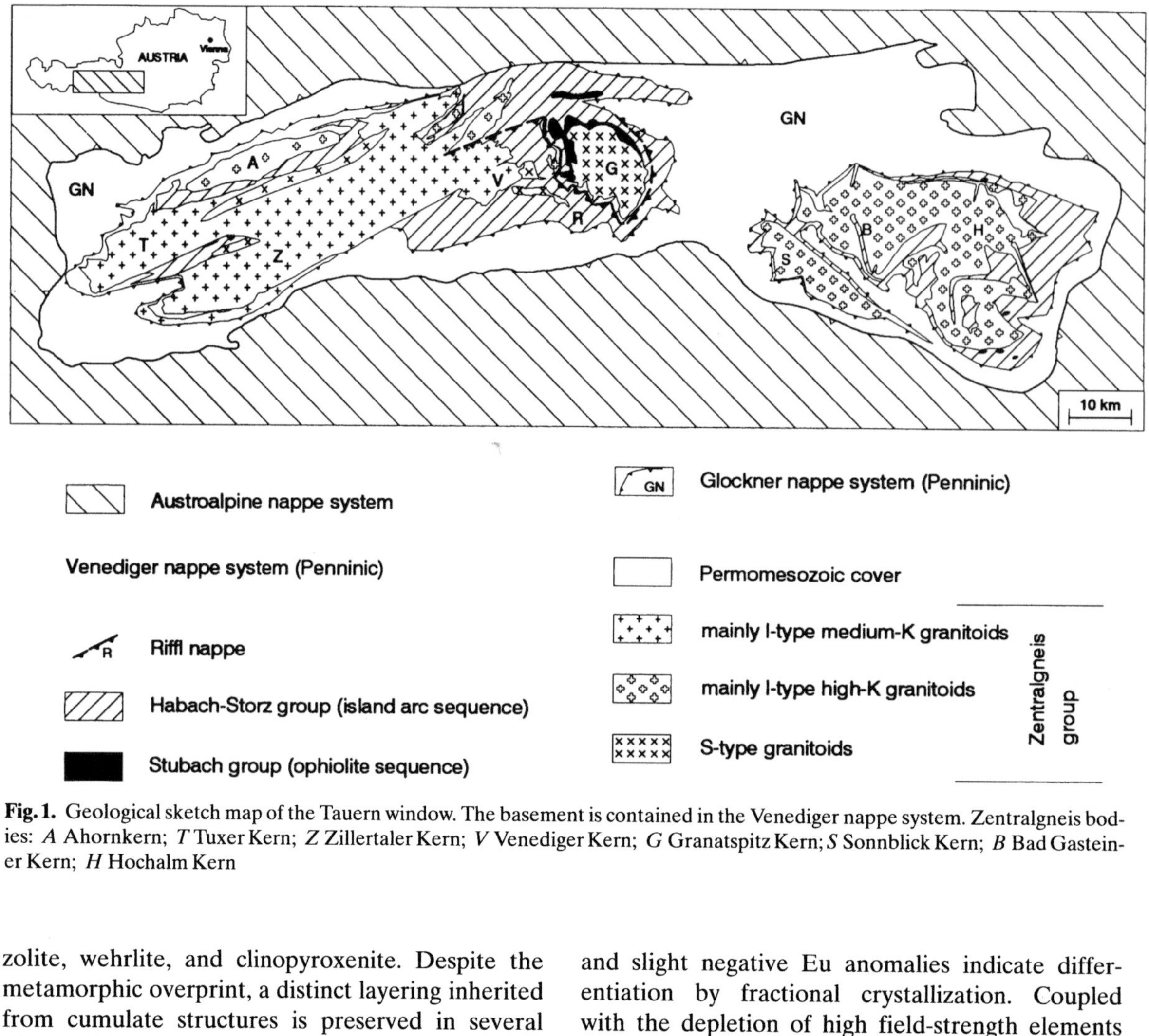

Fig. 1. Geological sketch map of the Tauern window. The basement is contained in the Venediger nappe system. Zentralgneis bodies: *A* Ahornkern; *T* Tuxer Kern; *Z* Zillertaler Kern; *V* Venediger Kern; *G* Granatspitz Kern; *S* Sonnblick Kern; *B* Bad Gasteiner Kern; *H* Hochalm Kern

zolite, wehrlite, and clinopyroxenite. Despite the metamorphic overprint, a distinct layering inherited from cumulate structures is preserved in several places. The ultramafic suite is interpreted by Neubauer et al. (1989) in terms of a cumulate complex formed above the petrologic Moho.

The basic magmatic rocks have been metamorphosed to amphibolite. They show geochemical characteristics transitional between mid-ocean ridge basalt and island arc tholeiite (Fig. 2a, e, g; Table 1a; Frisch and Raab 1987; Neubauer et al. 1989; Vavra and Frisch 1989). The concentrations of the high field-strength elements such as P, Zr, and Ti and the rare earth element patterns are close to those of normal-type mid-ocean ridge basalts, but show slight depletion. A slight depletion of light, relative to heavy, rare earth elements indicate a depleted mantle source. Large ion lithophile elements such as Sr, K, Rb, and Ba are enriched relative to mid-ocean ridge basalt. This is interpreted in terms of a primary subduction-related component and, in part, also in terms of secondary enrichment during intrusion of the Variscan granitoids. Low Cr contents and slight negative Eu anomalies indicate differentiation by fractional crystallization. Coupled with the depletion of high field-strength elements relative to mid-ocean ridge basalt, this speaks in favour of a supra-subduction zone setting (Pearce et al. 1984b).

Plagioclase gneisses related to ocean floor metabasalts, cumulate rocks, and serpentinites in the eastern Tauern window show the trace and rare earth element patterns of ocean ridge granites with the same chemical pecularities as the metabasaltic amphibolites: tholeiitic affinity (Fig. 2e, g; Table 1b), enrichment in large ion lithophile elements and slight depletion of high field-strength elements relative to ocean ridge granite (Fig. 2b; Vavra 1989). They are therefore interpreted as acidic differentiates of the ocean floor basalts. Fe-rich garnet micaschists are associated with the tholeiitic metavolcanic rocks and considered to represent the sedimentary cover of the ocean floor.

The lithological association of the Stubachtal complex is typical of oceanic crust including an ultramafic cumulate sequence as known from the lowermost

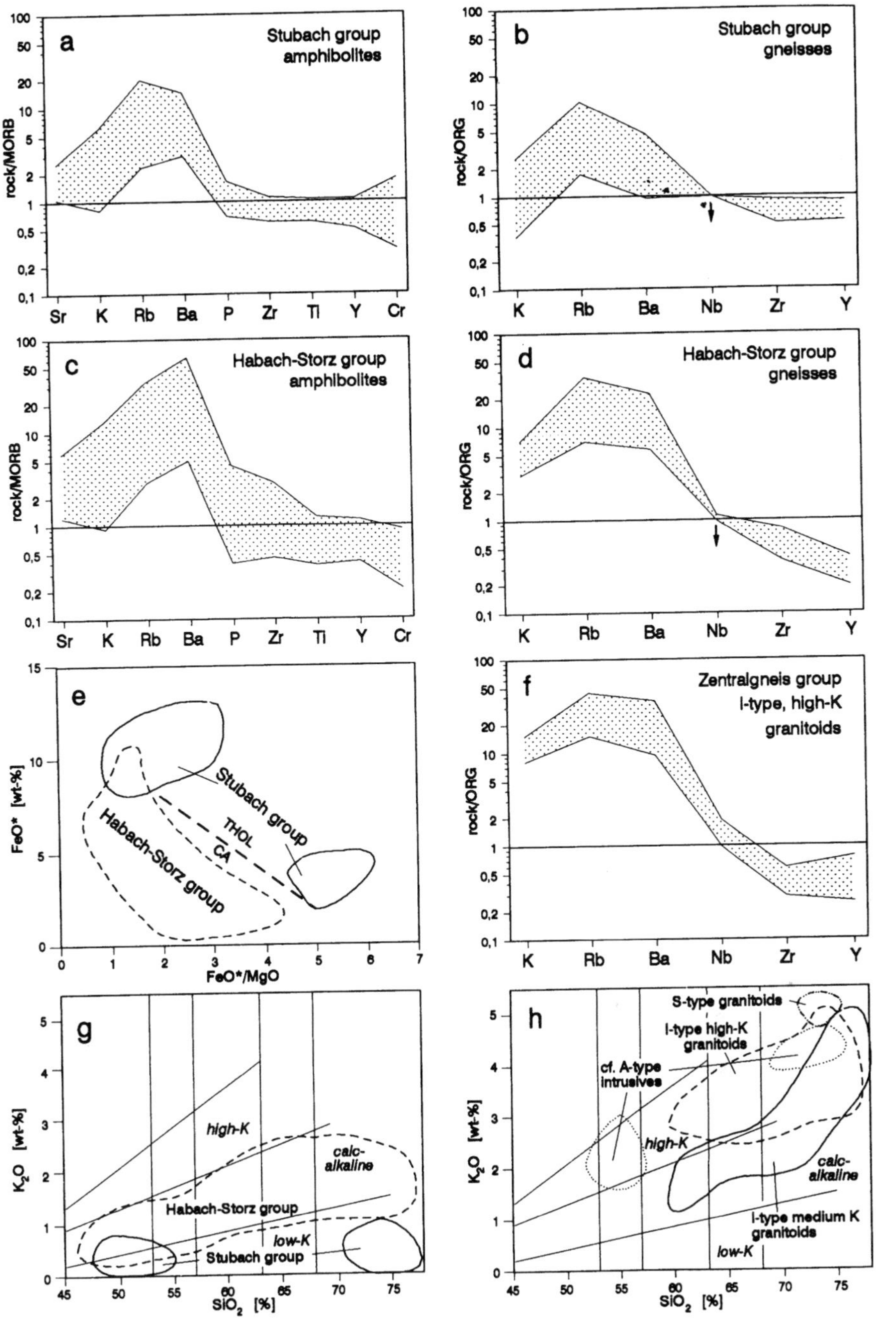

Fig. 2 a–h. Chemical characteristics of Tauern window basement rock assemblages. **a, c** Mid-ocean ridge basalt (MORB) normalized diagrams after Pearce et al. (1981), metabasaltic rocks. **b, d** Ocean ridge granite (ORG) normalized diagrams after Pearce et al. (1984 a), acidic metavolcanic rocks (the *arrows with Nb* indicate values below detection limit). **f** ditto, I-type high-K granitoids in the eastern Tauern window. **e** Diagram after Miyashiro and Shido (1975) for the discrimination of tholeiitic *(THOL)* and calc-alkaline *(CA)* rocks, metavolcanic rocks. **g, h** Diagram after Ewart (1982) for low-, medium- (calc-alkaline), and high-K magmatic series. Data from Frisch and Raab (1987), Finger and Steyrer (1988, only diagram **h**), Vavra (1989), Vavra and Frisch (1989), Winkler (1993). Single analyses may plot outside the fields

part of the crustal profile. Several reasons speak in favour of a back-arc/supra-subduction zone setting of the ophiolite (Neubauer et al. 1989):

1. The trace element pattern of its basic rocks suggests a subduction-related component of the ocean floor basalts (Frisch and Raab 1987; Vavra and Frisch 1989).
2. The intimate association with the volcanic arc sequence of the Habach-Storz group is evidence that the ophiolitic sequence formed the basement of the volcanic arc and therefore occupied a supra-subduction zone position. In the Felbertal scheelite mine area, amphibolites with the characteristics of the Stubach ophiolitic metabasalts form a continuous sequence with the overlying Habach-Storz group amphibolites (Frisch and Raab 1987).
3. The lack of pre-Alpine high pressure metamorphism (see later) is interpreted in terms of

Table 1. Selected chemical analysis of Tauern window basement rocks. FeO* is total iron. LOI, loss of ignition (Fe^{+3}-corrected, * indicates H_2O content. No figure, not determined. Figures are in weight percent and ppm, respectively

	a[a]	b	c[b]	d	e	f	g	h[c]	i	k	l	m	n	o
SiO_2	49.83	74.81	49.22	51.95	57.24	62.77	71.24	72.82	72.92	74.17	60.57	73.01	55.15	71.11
TiO_2	1.22	0.33	0.98	0.52	0.47	0.64	0.30	0.28	0.17	<0.04	0.77	0.23	0.94	0.49
Al_2O_3	14.56	12.02	18.04	16.97	16.68	16.71	14.32	15.31	14.82	14.95	17.13	13.98	18.44	14.16
FeO*	11.28	4.49	7.11	8.64	7.68	4.65	1.62	1.87	1.33	0.41	6.08	2.47	6.51	2.60
MnO	0.23	0.09	0.14	0.15	0.15	0.09	0.04	0.04	0.04	0.02	0.10	0.04	0.12	0.06
MgO	6.40	0.81	4.67	6.19	4.15	2.43	0.57	0.62	0.38	0.21	2.87	0.42	3.91	0.73
CaO	9.13	2.29	9.62	9.39	7.29	4.77	2.21	1.30	1.86	1.06	5.79	2.07	7.27	1.89
Na_2O	4.10	3.56	4.05	3.81	2.92	4.12	4.37	3.18	4.23	6.81	3.20	3.76	2.53	3.59
K_2O	0.18	0.30	1.01	0.40	0.49	1.90	2.54	5.34	3.21	0.99	2.06	2.58	2.97	4.37
P_2O_5	0.10	0.03	0.55	0.05	0.06	0.16	0.10	0.13	0.03	0.05	0.13	0.07	0.20	0.12
LOI/H_2O*	2.07	0.77	1.91	1.64	1.51	1.30	2.62	0.62*	0.61	1.11	0.71*	0.98*	1.86	0.73
Sum	99.10	99.50	97.30	99.71	98.64	99.54	99.93	101.51	99.60	99.82	99.41	99.61	99.90	99.85
Ba	114	<48	934	152		691	832	442	1162	212	399	2644	718	590
Cr	164	<70	<54	115	12	<37	<54	<37	<47	<47	42	<37	52	<37
Nb	<6	<10	9	<6		9	10	14	9	12	6	8	8	21
Ni	47	<37	<20	<20	8	<23	<20	<23	<37	<37	<23	<23	<23	<23
Rb	<4	<7	43	11	13	50	86	286	100	21	72	62	124	211
Sr	265	156	1138	221	272	568	227	108	447	172	319	247	578	165
V	308	23	186	257		85	32	20	<17	<17	114	<17	159	39
Y	23	51	27	<15	14	22	34	53	18	21	26	7	28	49
Zn	145	35	52	152		63	31	43	34	<11	56	30	91	53
Zr	84	269	210	39	67	172	170	125	123	45	182	174	158	201
La	<5	15	35			28		25	30		29			39
Ce	10	34	76			56		52	54		52			81
Nd	<10	16.5	37.4			23.1		25	20.9		25			30.8
Sm	3	8	8			5		5	4		5			8
Eu	1.3	1.3	1.9			1.1		<1	1		1.2			1
Dy	6	11	6			3		4	2		4			7
Yb	3	8	2.7			2.5		1.6	1.6		1.8			4.3
Lu	0.5	1.2	0.5			0.2		0.2	0.2		0.3			0.5
Th	<0.5	3.7	3.3			8		16	10.5		10			24
U	<1	3	<1			7		5	<1		2			11

[a] Stubach group. a: Tholeiitic metabasalt. Sample 1930, Frisch and Raab (1987) and Frisch (in Vavra 1989). Stubachtal. b: Tholeiitic metarhyolite. Sample LT437, Vavra (1989). Ankogel Mts.
[b] Habach-Storz group. c: Metabasalt. Sample 2245, Frisch (in Vavra 1989). Reisseck Mts. d: Metabasalt. Sample 2346, Frisch and Raab (1987). Felber Tauern. e: Metaandesite. Sample H1, Frisch and Raab (1987). Felber Tauern. f: Acid metaandesite. Sample C75, Vavra (1989). Hafner Mts. g: Metarhyodacite. Sample 2222, Frisch (in Vavra 1989). Hafner Mts.
[c] Zentralgneis group. h: S-type metagranite. Sample G2, Winkler (1993). Granatspitz Mts. i: I-type, high-K metagranite. Sample BK304, Vavra (1989). Hochalm Mts. k: Metaleucogranite. Sample LI269, Vavra (1989). Hochalm Mts. l: Metatonalite. Sample P4, Winkler (1993). Venediger Mts. m: I-type, medium-K metagranite. Sample R9, Winkler (1993). Venediger Mts. n: Metalamprophyre. Sample C39, Vavra (1989). Hochalm Mts. o: Metagranite. Sample C56, Vavra (1989). Hafner Mts.

obduction of hot oceanic crust, which is most common in supra-subduction zone environments.

4. The predominance of deep-crustal rocks is also in agreement with an obduction emplacement mode, since offscraping during subduction mainly involves high-level rocks such as sediment, basalt, and diapir serpentinite.

Radiometric dating indicates a Late Precambrian (late Sinian; for time scale, see Harland et al. 1989) age of formation. Gabbro amphibolite from the Stubachtal complex yielded U/Pb zircon and Sm/Nd mineral isochron ages in the range of 660–640 Ma, which are interpreted as formation ages (von Quadt 1989).

2.2 The Arc Association (Habach-Storz Group)

The island arc sequence (von Quadt 1984; Steyrer and Höck 1985; Frisch and Raab 1987; Vavra and Frisch 1989) comprises metavolcanic and metasedimentary rocks, referred to as the Habach group in the central part of the Tauern window (Frasl 1958), and as the Storz group in its eastern part (Exner 1971). The Habach and Storz groups are almost identical in their lithologies, and therefore the term Habach-Storz group is used here.

The Habach-Storz group forms a sequence made up of amphibolite, plagioclase gneiss, micaschist or phyllite, and quartzite. Calc-alkaline volcanic protoliths predominate (Fig. 2 e, g). The amphibolites derive from basalts and andesites (Table 1 c–e). Biotite plagioclase gneisses are the most prominent lithology and represent more acidic calc-alkaline volcanic (mainly pyroclastic) rocks (Table 1 f, g) and their detrital derivatives (Frisch and Raab 1987). Graphite-bearing micaschists are prominent in some areas. Graded bedding in original sandstone, arkose, or greywacke in a few places (e. g. the Greiner synclinorium in the westernmost part of the window; Lammerer 1986) reveal deposition of turbidites in local basins. Some of the quartzitic rocks are graphitic, still fine-grained metachert (Frasl 1958).

The widespread amphibolites derive from mainly medium-K calc-alkaline basalts and andesites (Fig. 2 c, g; Table 1 c–e; Steyrer and Höck 1985; Frisch and Raab 1987; Vavra and Frisch 1989). Low-K tholeiitic basalts and abyssal equivalents (Briegleb 1987) are mainly restricted to the basal part of the sequence above the Stubachtal ophiolite complex in the Felbertal scheelite deposit area. They are genetically related to the ophiolite and probably form part of the basement of the island arc. Higher up in the same profile are primarily intercalated low-K intermediate to acidic rocks within the island arc sequence (Frisch and Raab 1987).

The amphibolites of the calc-alkaline suite are intimately associated with plagioclase gneisses with which they form banded series. Their calc-alkaline nature is expressed in their chondrite-normalized rare earth element pattern, showing strong enrichment in the light rare earth elements and fractionation of the heavy rare earth elements, and by the depletion in iron compared with the tholeiitic amphibolites of the Stubach group (Table 1 a, c). Most of the amphibolites are characterized by enrichment in P and Zr relative to other high field-strength elements (Fig. 2 c). Frisch and Raab (1987) and Vavra and Frisch (1989) correlated the high P and Zr contents with strongly sheared rock and explained this nexus in terms of passive enrichment of these elements during deformation, P and Zr being retained in the stable mineral phases apatite and zircon, respectively. Vavra (1989) stresses that the high P and Zr contents are a primary feature, at least in those cases in which the rock did not experience high strains, and metabasaltic rock has higher P and Zr contents than intruding gabbroic rock. High contents in Zr result from an enriched mantle source indicative of ensialic (Pearce 1983) and some mature ensimatic island arc settings (Gill 1987).

The plagioclase gneisses that represent acidic volcanic rocks of the calc-alkaline suite are chemically characterized by strong depletion of Y and heavy rare earth elements (Fig. 2 d; Tab. 1 f, g). This indicates fractionation of these elements by amphibole, which commonly crystallizes at the high H_2O activity of calc-alkaline melts (Pearce and Norry 1979). The rare earth element pattern has the same shape as that of the associated amphibolites but shows higher concentrations of the light rare earth elements (Vavra 1989).

The metavolcanic amphibolites and gneisses of the Habach-Storz group span the whole range from basic to acidic in composition and are essentially of the medium-K type with a tendency to low-K type (Fig. 2 g). Intermediate to acidic high-K rocks of the Habach synclinorium are similar to late Variscan magmatism (see later). Vavra (1989) therefore concluded that they do not belong to the Habach-Storz group as has been proposed by Frasl (1958) and Steyrer (1983).

The more acidic (gneissic) rocks predominate over the more basic (amphibolitic up to about 60% SiO_2) rocks. Typical of island arc settings, some of the acidic rocks represent reworked volcanic material, as indicated by detrital zircon crystals. Originally intermediate compositions were changed to acidic ones by enrichment in quartz and the decomposition of mafic minerals during transport.

There is general agreement that the Habach-Storz group represents a subduction-related magmatic zone. Steyrer (1983) and Steyrer and Höck (1985) suggested an active continental margin setting, Kraiger (1989) a continental island arc. Frisch and Raab (1987) and Vavra and Frisch (1989) proposed an intra oceanic island arc setting on back-arc oceanic crust. Vavra (1989), more specifically, proposes an island arc setting in an essentially oceanic realm with long-lived subduction activity from the Late Proterozoic to the Early Carboniferous. During this period great volumes of crust were produced and the lithospheric mantle metasomatically enriched. This process led to the formation of primitive continental crust at a late stage.

The following characteristics speak in favour of an oceanic basement of the island arc:

1. The lack of evidence for a continental basement, which is revealed by the lack of pre-island arc ensialic rocks, Pb isotope data (Köppel 1983), and the lack of detrital sedimentary protoliths derived from a continental source.

2. The presence of oceanic basement with supra-subduction zone characteristics in the Stubach group and its intimate association with the island arc sequence.
3. The volumetric importance of basic-intermediate protoliths in some regions which possibly represent basal parts of the sequence (central Tauern window).

Features characteristic of a mature ensimatic island arc, an ensialic island arc, or an active continental margin are:

1. The overall predominance of acidic rocks.
2. The Zr enrichment relative to other high field-strength elements.
3. The original presence of important volumes of high-K acidic rocks which were transformed to granitoids during Variscan anatexis (see next Sect.)

In accordance with the available data, the Habach-Storz group is best interpreted as an island arc sequence which initiated on top of supra-subduction zone (back-arc) oceanic crust and evolved with time into a mature stage. At an early stage basic to intermediate volcanic rocks dominated, while at an advanced stage large volumes of acidic, dacitic to rhyolitic rocks contributed to the sequence, indicating maturation of the island arc. There is no clear evidence of slices of older crustal material, though its existence cannot be fully excluded. Isotopic data from Variscan S-type granitoids indicate the existence of a Proterozoic continental source (Schermaier et al. 1990) to which, however, the Tauern rock assemblage may be allochthonous.

Acritarchs and cyanobacteria from metapelites ("Habach phyllites") in the northern central Tauern window attributed to the Habach group (Frasl 1958) certify Late Proterozoic (Late Riphean to Early Vendian) sedimentation (Reitz and Höll 1988; Gilg et al. 1989). The volcanic arc was active since the end of the Proterozoic or the beginning of the Palaeozoic at least. A protolith age in the range between 640 and 500 Ma (late Sinian to Cambrian, depending on the amount of Alpine Pb loss) is documented by U/Pb zircon data from metadacite in the eastern Tauern window (Vavra and Hansen 1991). These ages conform with the late Sinian age of the ophiolitic basement (see earlier).

An Early Ordovician (492±2 Ma) U/Pb zircon upper intercept age from a cumulate metapyroxenite in the basal part of the Habach-Storz group in the central Tauern window is interpreted as an intrusion age (von Quadt 1989). However, nearby amphibolite from the late Sinian Stubachtal ophiolite complex yielded similar U/Pb zircon upper intercept and Sm/Nd whole rock, garnet and zircon isochron ages (von Quadt 1987), the latter certainly indicating a metamorphic event (von Quadt 1987). Since zircon is not expected to be a primary phase in these basic rocks, it is suggested that the U/Pb zircon ages represent a metamorphic (anatectic) event initiating the growth of zircon, in line with the Sm/Nd data (Vavra 1989).

The evolution of normal Pb in ores and feldspar with model ages of 400 Ma remains below the growth curve for normal crustal lead (von Quadt 1984). This implies continued supply of a mantle component during the Early Palaeozoic.

2.3 The Variscan Calc-Alkaline Granitoid Association (Zentralgneis Group)

Extensive granitoid magmatism of the batholith type took place in the Carboniferous. The predominant rock types belong to the calc-alkaline series and are of the I-type (for granitoid types, see Chappell and White 1974; White and Chappell 1983), the occurrence of S-type granitoids is limited (Finger and Steyrer 1988; Finger et al. this Vol.). Minor magmatic activity occurred in the Permian. Alpine deformation and metamorphism transformed the granitoids into gneisses ("Zentralgneise"). We assume that a large proportion of the early anatectic granitoids (see later) was also foliated by Variscan time. In the following only the protolithic magmatic names are used.

The batholiths are made up predominantly of tonalite, granodiorite, and granite but also include minor volumes of diorite, monzonite, and syenite (e.g. Frasl 1958; Karl 1959; Schmidegg 1961; Exner 1964; Lammerer 1986). They are exposed in dome-like or cylindrical antiformal structures called "cores" ("Kerne"; see Fig. 1) separated by narrow synclines. A detailed description and chemical and petrogenetic characterization of the manifold intrusive bodies is given by Finger et al. (this Vol.) These authors state that the intrusive I-type calc-alkaline suite which mainly encompasses tonalites and granodiorites was preceded by potassium-rich anatexites of mainly granitic but also syenitic and monzonitic composition. A late generation of leucocratic granites yield within-plate (A-type) characteristics.

Winkler (1993) discerned two distinct groups within the I-type granitoids. In total, four groups of granitoid magmatism can be distinguished in terms of their chemical and petrogenetic characteristics:

1. Early S-type high-K granitoids;
2. I-type high-K granitoids;

3. I-type, mainly medium-K granitoids;
4. Late-stage granitoids with a tendency towards A-type.

The early *S-type granitoids* (Fig. 2h; Table 1h) form the Granatspitz Kern and minor parts of the Zentralgneis complex of the western Tauern window (Fig. 1). These leucocratic high-K granitic rocks are peraluminous and contain pseudomorphs after cordierite (Frasl 1967; Morteani 1971; Lammerer 1986) underlining their crustal character. The trace element characteristics (Pearce et al. 1984a) are those of syn-collisional granites (Finger and Steyrer 1988). The zircon typology after Pupin (1985) classifies them as alkali-rich aluminous anatectic granites (Winkler 1993). Their crustal, anatectic character is emphasized by an inherited component in a coarse-grained zircon population (Cliff 1981) and by Nd isotope data (von Quadt 1987).

The *I-type high-K granitoids* (Fig. 2f, h; Table 1i) are exposed in vast areas and comprise most of the Zentralgneis complex in the eastern Tauern window (Finger and Steyrer 1988; Dölzlmüller et al. 1989; Holub and Marschallinger 1989; Vavra 1989; Marschallinger and Holub 1990) and large areas in the northern part of the Zentralgneis complex in the western Tauern window (Winkler 1993; Fig. 1). They are metaluminous granitoids, some of them with a tendency to medium-K character. The mainly granodioritic to granitic rocks are characterized by high contents of alkalis and Sr.

The granitoids of the eastern Tauern window have extensive migmatitic zones at the contacts with the gneisses and amphibolites of the Habach-Storz group (e.g. Cliff et al. 1971; Exner 1971; Holub and Marschallinger 1989). They formed in situ and did not migrate into higher crustal levels. Their trace and rare earth element patterns show a remarkable similarity to those of the dacitic rocks of the Habach-Storz group (Fig. 2d, f; Vavra 1989). Differences include higher Nb/La and Nb/Ce ratios and some other characteristics indicative of mature stages of magmatic arcs or a contribution from subcontinental lithosphere (Brown et al. 1984). The morphological evolution of zircon crystals as revealed by cathodo-luminescence images of sections through the crystal centre shows that the granitoids were formed by anatexis. The zircon crystals of the medium-K metadacites (plagioclase gneisses) of the Habach-Storz group display the same characteristic overgrowth of anatectic zones over magmatic cores as in the granitic to granodioritic Zentralgneis samples (Vavra 1989, 1990). The gneisses show an increasing amount of anatectic melt phase (leucosome) approaching the granitoids. In transition zones along margins of diatectic granodiorite, homogenization of leucosome with residual rock is observed (Vavra 1989). The metadacites experienced metatexis, whereas the granitoids represent diatectic rocks.

The more potassium-rich (mainly high-K) character of the granites and granodiorites compared with the Habach-Storz group rhyodacitic and dacitic gneisses (Fig. 2g, h, Table 2g, i) led Vavra (1989) to the following conclusion: original high-K dacites and rhyolites of the island arc sequence, according to their composition close to the granitic eutectic composition, were diatectically melted to form granitoids, whereas the medium-K gneisses only underwent partial melting. Geochemical composition and zircon morphology exclude the possibility that assimilation of metapelites contributed to the high-K character of the rocks. This implies that the acidic rocks of the island arc contained important volumes of high-K types which were transformed to granitoids during the Variscan orogeny. The I-type character of the diatectic granitoids is therefore inherited from the island arc protoliths. The importance of high-K acidic rocks in the island arc sequence of the Habach-Storz group shows that the island arc in fact reached a mature stage.

I-type, predominantly *medium-K granitoids* (Fig. 2h; Table 1l, m) form the main part of the Zentralgneis complex in the western Tauern window (Fig. 1). These predominantly tonalitic to granodioritic, metaluminous rocks have intrusive contacts. They have lower ($Na_2O + K_2O$) and Sr contents than the previous group, and a more tholeiitic trend in the AFM diagram (iron emphasized calc-alkaline suite; Winkler 1993). Sr_0 values are between 0.706 and 0.708. The zircon typology (Pupin 1985) is in accordance with a medium-K, calc-alkaline character. The zircon crystals generally do not show anatectic growth zones. In the migmatitic area east of the Venediger Kern (Riffl nappe; see later) apophyses of this granitoid group intrude granitoids of the high-K group and are therefore considered to be younger than these (Schermaier 1991). Both I-type groups plot as volcanic arc granites in the diagrams after Pearce et al. (1984a) (Finger and Steyrer 1988).

Late-stage high-level intrusive rocks with an *A-type* (Sylvester 1989) tendency encompass dioritic rocks (in part lamprophyre dykes; Exner 1971) and leucocratic and porphyric granite (Steyrer 1983; Vavra 1989; Finger et al. this. Vol.). They are of high-K character (Fig. 2h; Table 1n, o) and show high contents in Nb, U, Th, and heavy rare earth elements. They are attributed to a post-collisional, within-plate setting. Some leucogranites from the eastern Tauern window, according to their chemical composition and zircon morphology, represent anatectic minimum melts and are distinct from this more alkaline magmatism (Table 1 k; Vavra 1989).

U/Pb zircon and Rb/Sr whole-rock ages of Zentralgneis samples cluster around 330–310 Ma (for summaries, see Frank et al. 1987; Finger et al. this Vol.). They are interpreted as intrusion ages and the climax of anatexis. The clustering of the data speaks in favour of a collisional event around the Early/Late Carboniferous boundary. A number of Permian Rb/Sr ages are most probably slightly disequilibrated Carboniferous ages (Cliff 1981).

The Granatspitz S-type granite yielded Early Carboniferous U/Pb zircon ages of ca. 330 Ma (Cliff 1981), supported by a Rb/Sr isochron age of 325 ± 33 Ma (Cliff 1977; recalculated with lambda = 1.42). An age around 330 Ma is inferred for the anatectic event producing the S-type granitoids.

Tonalite and granodiorite from the southeast Tauern window (I-type high-K group) yielded U/Pb zircon ages of 314 ± 7 and 313 ± 10 Ma, respectively (Cliff 1981). A leucocratic granite in the region of the Felbertal scheelite mine yielded a Rb/Sr whole-rock age of 316 ± 10 Ma (Pestal 1983). The lower intercept from U/Pb zircon dating in an anatectically overprinted metadacite of the eastern Tauern window indicates that the peak of anatexis was reached around 320 Ma (Vavra 1989).

Winkler (1993) showed that 18 different samples from the I-type medium-K granitoids of the Zillertaler Kern (Besang et al. 1968; Jäger et al. 1969; Borsi et al. 1978; Satir and Morteani 1982) define a Rb/Sr errorchron of 310 ± 8 Ma with a Sr_0 value of 0.7065. Another group of 11 samples from the Northern Sulzbach Zunge and parts of the Tuxer Kern (Besang et al. 1968; Jäger et al. 1969) again define an errorchron of 310 ± 9 Ma with a Sr_0 value of 0.7076. Most of the granitoids forming the great batholiths are therefore inferred to be of early Late Carboniferous age.

A late-stage porphyric granite with intraplate affinity in the eastern Tauern window yielded an intrusion age of 256 +58/−32 Ma (U/Pb zircon age, upper intercept; Vavra 1989). This data proves late-orogenic mid- to Late Permian magmatism reflecting, according to its intraplate affinity, crustal consolidation.

3 Metamorphic and Structural Evolution

Information about pre-Alpine metamorphism in the Venediger nappe system of the Tauern window is poor due to Alpine overprinting. Alpine metamorphism was mainly of middle and upper greenschist facies grade and reached amphibolite facies grade in the central parts of the pre-Alpine complexes of the western and the eastern window (Selverstone and Spear 1985; Droop 1985).

Pre-Variscan metamorphism around 490 Ma (Early Ordovician) is indicated by U/Pb zircon and Sm/Nd mineral isochron ages in metabasic rocks of the central Tauern window (see earlier; von Quadt 1987, 1989). Variscan metamorphism reached anatectic grade in most of the Tauern window and had its climax around 330–310 Ma. This is indicated by K/Ar dating of amphibole (ages around 325–310 Ma interpreted as metamorphic cooling: Pestal 1983) and a zircon discordia lower intercept in anatectically overprinted gneisses near 320 Ma (Vavra 1989) and in amphibolite (von Quadt 1987). The vast migmatitic halos around the Late Carboniferous anatectic granitoids demonstrate the connection between metamorphism and melt formation. Satir and Friedrichsen (1986) calculated Variscan temperatures of up to approximately 620 °C in the western Tuxer Kern using oxygen isotope data on several mineral pairs. Pseudomorphs after andalusite described by Cornelius (1944) from near a granitoid contact demonstrate the low-P character of the metamorphism.

Droop (1983) and Zimmermann et al. (1990) suggested that certain high-pressure assemblages in the Tauern window are of pre-Alpine age. The localities, however, are situated close to the Alpine eclogite zone so that the pre-Alpine age of the high-pressure event is highly questionable.

The existence of a Variscan nappe structure was suggested by Frisch (1977, 1980) in the central and western Tauern window. It is characterized by thrusting of the migmatitic Riffl nappe (Cornelius 1941) including the Zillertal-Venediger Zentralgneis Kern over the pre-granitic envelope of the Granatspitz Kern. The Alpine or Variscan age of the Riffl nappe is the subject of controversy. The lack of slices of Mesozoic rocks along the thrust plane speaks in favour of a Variscan age of thrusting.

Intrusion of Variscan granitoids into the island-arc sequence caused broad dome structures to form separated by narrow synforms of the pre-existing supracrustal rocks. The lobate-cuspate morphology of the boundary between granitoids and their envelope is typical of lithological boundaries with high competence contrast (Ramsay 1967) and was formed in its present shape during Alpine deformation. However, the dome-shape of individual batholitic granite bodies was already a Variscan feature. This is demonstrated by localized deposition of terrestrial and shallow-marine Permian and Triassic sediments in the schist synclinoria, whereas most of the granitoid highs were loci of erosion up to the Jurassic (Frisch 1977).

In some localities in the migmatites, where undeformed or nearly undeformed Variscan granitoids cross-cut strongly deformed Variscan granite gneisses, a Variscan schistosity can be distinguished. In most other places, it is indistinguishable from Alpine structures.

4 Ore Deposits

The basement complex of the Tauern window is characterized by mineralizations, which are related to granitic members of the Variscan granitoids. The previously mined gold veins in the Zentralgneis area of the eastern part of the window and the scheelite mineralizations of the Felbertal mine and related occurrences are of economic importance.

The Felbertal tungsten deposit (central Tauern window) is controlled by a Carboniferous leucocratic granite (see earlier) and related rocks (Briegleb 1987). The mineralization is mainly concentrated in the basal parts of the Habach Group sequence, a fact which caused Höll and Maucher (1976) to interpret it in terms of a syngenetic, exhalative-sedimentary deposit controlled by the basic volcanism.

The gold-bearing quartz veins in the eastern part of the window, an emerald deposit in the Habach valley and other beryllium occurrences, as well as molybdenum and arsenic mineralizations are also controlled by the Variscan granitoids.

5 Geodynamic Evolution and Position within the Variscides

The Penninic basement of the Tauern window developed from supra-subduction zone (back-arc) ocean floor through stages from primitive to progressively maturing island arc to a collisional stage with extensive granitoid formation. The remnants of the ophiolitic basement contain a high percentage of rocks from the lower crust such as cumulate rocks and isotropic gabbros. The largest ophiolitic body, the Stubachtal complex, is not affected by the underlying Granatspitz granitoids and is therefore considered to have been obducted during Variscan movements. Lack of relics indicating high-pressure metamorphism are in line with such a mode of emplacement, as opposed to one involving subduction and accretion.

Although radiometric methods determined only late Sinian to Cambrian protolith ages for the island arc volcanic rocks, it is suggested that the island arc persisted until the Variscan orogeny in the Carboniferous. This suggestion is based on the lack of an alternative scenario for the post-Cambrian period and the recognition of the fact that large volumes of rock from the mature stage of the island arc, transformed to granitoids during the Variscan orogeny. Sr isotope systematics show that anatectic granites derived from material not much older than the anatectic event (Pestal 1983). The long-lasting (ca. 250–300 Ma) magma generation obove the subduction zone caused a high heat gradient in the thickening crust, which thus qualified for extensive anatectic melting.

The high-K chemistry of a large proportion of the island arc sequence does not necessarily imply the presence of continental basement. In the southwest Pacific intra-oceanic mature island arcs, many of the subduction-related dacites and most of the rhyolites are of the high-K type (Wilson 1989).

The Variscan magmatic cycle lasted over ca. 80 Ma from the Early Carboniferous to the Late Permian, with its climax in the Late Carboniferous. The granitoids show an evolution from crustal anatexites through intrusive subduction-related melts to late-stage within-plate magmatism with an alkaline affinity. The early S-type anatexites are collisional granites with an older crustal component alien to the rocks known from the Tauern window basement (Schermaier 1991). The younger anatexites, which comprise a large volume of the Tauern batholith, inherited their I-type character from the mature island arc rocks from which they originate. The intrusive, subduction-related melts are less potassium emphasized and may have been generated in an active continental margin setting on the western margin of the Palaeotethys ocean, as proposed by Finger and Steyrer (1990), as a result of the collision of the island arc. The Permian intraplate (cf. A-type) magmatic activity indicates consolidation of the crust and fading of the subduction-related component in the underlying mantle.

Finger and Steyrer (1990) discuss Variscan granitoids incorporated in the Alpine orogen in context with other central European Variscan granitoids. They state that the predominantly I-type granitoids of the Alps contrast with a contemporaneous belt of S-type granitoids in the Moldanubian zone of the

Variscides and compare this situation with circum-Pacific continental margin settings with an outer I-type and inner S-type paired belt. Finger and Steyrer (1988) attribute to most of the Tauern window granites a Cordilleran I-type character and propose a Cordilleran-type tectonic setting at the western margin of the Palaeotethys embayment. However, there is no indication for the establishment of an active continental margin between the consolidating Variscan belt and the Palaeotethys. On the contrary, the record in the supracrustal rocks clearly supports a passive continental margin setting.

Winkler (1993) demonstrated that the anatectic granitoids of the Tauern window, i. e. both the S-type and the I-type high-K granitoids, show chemical affinities to the Moldanubian granitoids. This K-, Mg-, and Sr-emphasized plutonism is a frequent phenomenon in a belt stretching from Corsica via the "external massifs" (Helvetic zone) of the Alps and the Vosges-Schwarzwald to the Bohemian Massif (Rossi et al. 1990). It is of crustal derivation, related to metamorphism and deformation, and characterized by Sr_0 values > 0.71 (Matte 1986). This belt also encompasses immediately post-collisional, calc-alkaline granitoids with distinctly lower contents of K, Mg, and Sr. These mainly granodioritic intrusives derive from the lower crust with a mantle component, intruded into higher crustal levels, show no direct correlation with regional metamorphism, and have Sr_0 values generally < 0.71 (Matte 1986). Granitoids with the same characteristics occur in a belt from the Pyrenees via Corsica-Sardinia to the Austro-Alpine–South-Alpine zone of the Alps. This I-type magmatism including the occurrences in the Tauern window has the chemical characteristics of *"Caledonian-type"* rather than "Cordilleran-type" as defined by Pitcher (1982). Moreover, the immediately post-orogenic, uplift regime of the I-type medium-K granitoids of the Tauern window corresponds well to the proposed geotectonic setting of Caledonian-type granitoids.

The Tauern window basement evolution shows the progressive buildup of continental crust by long-lived subduction activity, starting in an intra-oceanic environment. The rock associations, their modes of formation, and the evolutionary process resemble those of Archean granite-greenstone belts. Frisch and Raab (1987) and Frisch et al. (1990) proposed equivalent scenarios for both types of crustal assemblages. The intra-oceanic island arc was termed Habach terrane by Frisch and Neubauer (1989). In contrast to all other basement complexes of the Alps the Habach terrane has no early Precambrian memory. Its evolution was markedly different from that of the Austro-Alpine basement which frames the Tauern window.

Acknowledgements

Pauline O'Shea upgraded the English text which is gratefully acknowledged.

References

Besang C, Harre W, Karl F, Kreuzer H, Lenz H, Müller P, Wendt I (1968) Radiometrische Altersbestimmungen (Rb/Sr und K/Ar) an Gesteinen des Venediger-Gebietes (Hohe Tauern, Österreich). Geol Jahrb 86: 835–844

Borsi F, Del Moro A, Sassi FP, Zanferrari A, Zirpoli G (1978) New geopetrological and readiometric data on the Alpine history of the Austridic continental margin south of the Tauern window (Eastern Alps). Mem Ist Geol Mineral Univ Padova 32: 3–17

Briegleb B (1987) Geologische Verhältnisse im Bereich der Scheelitlagerstätte im Felbertal, Pinzgau (Land Salzburg). Uni-aktuell Salzburg 10/87: 10

Brown CG, Thorpe RS, Webb PC (1984) The geochemical characteristics of granitoids in contrasting arcs and comments on magma sources. J Geol Soc Lond 141: 413–426

Chappell BW, White AJR (1974) Two contrasting granite types. Pac Geol 8: 173–174

Cliff RA (1977) Rb-Sr isotopic measurements on granite-gneisses from the Granatspitzkern, Hohe Tauern, Austria. Verh Geol Bundesanst 1977: 101–104

Cliff RA (1981) Pre-Alpine history of the Penninic zone in the Tauern window, Austria: U-Pb and Rb-Sr geochronology. Contrib Mineral Petrol 77: 262–266

Cliff RA, Norris RJ, Oxburgh ER, Wright RC (1971) Structural, metamorphic and geochronological studies in the Reisseck and southern Ankogel Groups, the Eastern Alps. Jahrb Geol Bundesanst 114: 121–272

Cornelius HP (1941) Zur Geologie des oberen Felber und Matreier Tauerntals und zur Altersfrage der Tauernzentralgneise. Ber Reichsst Bodenforsch Wien 1941: 14–20

Cornelius HP (1944) Beobachtungen am Nordostende der Habachzunge (Venedigermassiv, Hohe Tauern). Ber Reichsamt Bodenforsch Wien 1944: 25–31

Dölzlmüller H, Finger F, Schermaier A (1989) Die Ausbildung der Zirkone im Granitgneis des Hohen Sonnblicks (Penninikum, Ostalpen) und ihre Aussage zur Gesteinsgenese. Mitt Österr Geol Ges 81: 33–45

Droop GTR (1983) Pre-alpine eclogites in the Penninic basement complex of the Eastern Alps. J Metamorphic Geol 1: 3–12

Droop GTR (1985) Alpine metamorphism in the south-east Tauern Window, Austria: 1. P-T variations in space and time. J Metamorphic Geol 3: 371–402

Ewart E (1982) The mineralogy and petrology of Tertiary-Recent orogenic volcanic rocks: with special reference to the andesitic-basaltic compositional range. In: Thorpe RS (ed) Andesites. Wiley, Chichester, pp 25–95

Exner C (1964) Erläuterungen zur Geologischen Karte der Sonnblickgruppe 1 : 50 000. Geol Bundesanst, Wien, 170 pp

Exner C (1971) Geologie der peripheren Hafnergruppe (Hohe Tauern). Jahrb Geol Bundesanst 114: 1–119

Finger F, Steyrer HP (1988) Granite types in the Hohe Tauern (Eastern Alps, Austria) – some aspects on their correlation to Variscan plate tectonic processes. Geodin Acta 2: 75–87

Finger F, Steyrer HP (1990) I-type granitoids as indicators of a late Paleozoic convergent ocean-continent margin along the

southern flank of the central European Variscan orogen. Geology 18: 1207–1210
Frank W, Kralik M, Scharbert S, Thöni M (1987) Geochronological data from the Eastern Alps. In: Flügel HW, Faupl P (eds) Geodynamics of the Eastern Alps. Deuticke, Wien, pp 272–281
Frasl G (1958) Zur Seriengliederung der Schieferhülle in den mittleren Hohen Tauern. Jahrb Geol Bundesanst 101: 323–472
Frasl G (1967) Glimmerpseudomorphosen nach Cordierit im Zentralgneis des Granatspitzkernes, Hohe Tauern. Mineral Mitteilungsbl 1/2: 11–17
Frisch W (1977) Der alpidische Internbau der Venedigerdecke im westlichen Tauernfenster. Neues Jahrb Geol Paläontol Monatsh 1977: 675–696
Frisch W (1980) Tectonics of the Tauern window. Mitt Österr Geol Ges 71/72: 65–71
Frisch W, Neubauer F (1989) Pre-Alpine terranes and tectonic zoning in the Eastern Alps. Geol Soc Am Spec Pap 230: 91–100
Frisch W, Raab D (1987) Early Paleozoic back-arc and island-arc setting in greenstone sequences of the central Tauern window (Eastern Alps). Jahrb Geol Bundesanst 129: 545–566
Frisch W, Ménot P, Neubauer F, Raumer J v (1990) Correlation and evolution of the Alpine basement. Schweiz Mineral Petrogr Mitt 70: 265–285
Gilg HA, Höll R, Kupferschmid MP, Reitz E, Stärk H, Weber-Diefenbach K (1989) Die Basisschieferfolge in der Habachformation im Felber- und Amertal (Tauernfenster, Salzburg). Mitt Österr Geol Ges 81: 65–91
Gill GB (1987) Early geochemical evolution of an oceanic island arc and backarc: Fiji and the South Fiji Basin. J Geol 95: 589–615
Harland WB, Armstrong RL, Cox AV, Lorraine EC, Smith AG, Smith DG (1989) A geologic time scale. Cambridge University Press, Cambridge, 263 pp
Höll R, Maucher A (1976) The stratabound ore deposits in the Eastern Alps. Handbook of stratabound and stratiform ore deposits, vol 5. Elsevier, Amsterdam, pp 1–30
Holub B, Marschallinger R (1989) Die Zentralgneise im Hochalm-Ankogel-Massiv (östliches Tauernfenster). Teil I: petrographische Gliederung und Intrusionsfolge. Mitt Österr Geol Ges 81: 5–31
Jäger E, Karl F, Schmidegg O (1969) Rubidium-Strontium Altersbestimmungen an Biotit-Muskowit Granitgneisen (Typus Augen-Flasergneise) aus dem nördlichen Großvenedigerbereich. Tschermaks Mineral Petrogr Mitt 13: 251–272
Karl F (1959) Vergleichende petrographische Studien an den Tonalitgraniten der Hohen Tauern und den Tonalitgraniten einiger periadriatischer Instrusivmassive. Jahrb Geol Bundesanst 102: 1–192
Köppel V (1983) Summary of lead isotope data from ore deposits of the Eastern and Southern Alps: some metallogenetic and geotectonic implications. In: Schneider HJ (ed) Mineral deposits of the Alps and of the Alpine epoch in Europe. Springer, Berlin Heidelberg New York, pp 162–168
Kraiger H (1989) Die Habachformation – ein Produkt ozeanischer und kontinentaler Kruste. Mitt Österr Geol Ges 81: 47–64
Lammerer B (1986) Das Autochthon im westlichen Tauernfenster. Jahrb Geol Bundesanst 129: 51–67
Marschallinger R, Holub B (1990) Die Zentralgneise im Hochalm-Ankogel-Massiv (östliches Tauernfenster, Österreich). Teil II: zirkontypologische und geochemische Charakteristik. Mitt Österr Geol Ges 82: 19–48
Matte P (1986) Tectonics and plate tectonics model for the Variscan belt of Europe. Tectonophysics 126: 329–374
Miyashiro A, Shido F (1975) Tholeiitic and calc-alkalic series in relation to the behaviour of titanium, vanadium, chromium, and nickel. Am J Sci 275: 265–277
Morteani G (1971) Gliederung und Metamorphose der Serie zwischen Stilluptal und Schlegeistal (Zillertaler Alpen, Nordtirol). Verh Geol Bundesanst 1971: 287–314
Neubauer F, Frisch W, Schmerold R, Schlöser H (1989) Metamorphosed and dismembered ophiolite suites in the basement units of the Eastern Alps. Tectonophysics 164: 49–62
Pearce JA (1983) Role of sub-continental lithosphere in magma genesis at active continental margins. In: Hawkesworth CJ, Norry MJ (eds) Continental basalts and mantle xenoliths. Shiva, Nantwich, pp 230–249
Pearce JA, Norry MJ (1979) Petrogenetic implications of Ti, Zr, Y, and Nb variations in volcanic rocks. Contrib Mineral Petrol 69: 33–47
Pearce JA, Alabaster T, Shelton AW, Searle MP (1981) The Oman ophiolite as a Cretaceous arc-basin complex: evidence and implications. Philos Trans R Soc Lond A 300: 299–317
Pearce JA, Harris NBW, Tindle AG (1984 a) Trace element discrimination diagrams for the tectonic interpretation of granitic rocks. J Petrol 25: 956–983
Pearce JA, Lippard J, Roberts S (1984 b) Characteristics and tectonic significance of supra-subduction zone ophiolites. Geol Soc Lond Spec Publ 16: 77–94
Pestal G (1983) Beitrag zur Kenntnis der Geologie in den mittleren Hohen Tauern im Bereich des Amer- und des Felbertales (Pinzgau/Salzburg). Thesis, Univ Wien, Wien, 116 pp
Pitcher W (1982) Granite type and tectonic environment. In: Hsü K (ed) Mountain building processes. Academic Press, London, pp 19–40
Pupin JP (1985) Magmatic zoning of Hercyanian granitoids in France based on zircon typology. Schweiz Mineral Petrogr Mitt 65: 29–56
Quadt A v (1984) Geochemische und U-Pb-Untersuchungen der unteren Schieferhülle (Ostalpen). Fortschr Mineral 62: 191–193
Quadt A v (1987) U-Pb- und Sm-Nd-Datierungen der Habach-Formation und des Basisamphibolits im Bereich der mittleren Hohen Tauern. Uni-aktuell Salzburg 10/87: 28
Quadt A v (1989) Sm-Nd and U-Pb systematics of zircon fractions, garnet and whole rock amphibolitic rocks of the Tauern window (Eastern Alps). Terra Abstr 1: 349
Ramsay J (1967) Folding and fracturing of rocks. McGraw Hill, New York, 568 pp
Reitz E, Höll R (1988) Jungproterozoische Mikrofossilien aus der Habachformation in den mittleren Hohen Tauern und dem nordostbayerischen Grundgebirge. Jahrb Geol Bundesanst 131: 329–340
Rossi P, Cocherie A, Johan V (1990) The geodynamic significance of Mg-K plutonism as constrained by mineralogical and geochemical data from Corsica (France) and the Bohemian Massif (Czechoslovakia). Abstr Vol Symp IGCP 233 Göttingen 1990
Satir M, Friedrichsen H (1986) Geochronological and stable isotope investigations on polymetamorphic rocks from the Eastern Alps (Western Tauern Window, Austria). Neues Jahrb Miner Abh 154: 313–334
Satir M, Morteani G (1982) Petrological study and radiometric age determination of the migmatites in the Penninic rocks of the Zillertaler Alpen (Tyrol/Austria). Tschermaks Mineral Petrogr Mitt 30: 59–75

Schermaier A (1991) Geologisch-petrographische Untersuchungen zur präalpidischen Entwicklung des Tauernfensters am Ostrand des Venedigermassivs (Hohe Tauern). Jahrb Geol Bundesanst 134: 345–367
Schmidegg O (1961) Geologische Übersicht der Venediger-Gruppe. Verh Geol Bundesanst 1961: 35–56
Selverstone J, Spear F (1985) Metamorphic P-T paths from pelitic schists and greenstones in the southwest Tauern Window, Eastern Alps. J Metamorphic Geol 3: 439–465
Steyrer HP (1983) Die Habachformation der Typlokalität zwischen äußerem Habachtal und Untersulzbachtal (Pinzgau/Salzburg). Mitt Österr Geol Ges 76: 69–100
Steyrer HP, Höck V (1985) Geochemistry of the metabasites in the Habach formation (Salzburg, Hohe Tauern, Austria). A preliminary report. Ofioliti 10: 441–456
Sylvester JP (1989) Post-collisional alkaline granites. J Geol 97: 261–280
Vavra G (1989) Die Entwicklung des penninischen Grundgebirges im östlichen und zentralen Tauernfenster der Ostalpen – Geochemie, Zirkonmorphologie, U/Pb-Radiometrie. Thesis, Univ Tübingen, Tübingen, 170 pp
Vavra G (1990) On the kinematics of zircon growth and its petrogenetic significance: a cathodoluminescence study. Contrib Mineral Petrol 106: 90–99
Vavra G, Frisch W (1989) Pre-Variscan back-arc and island-arc magmatism in the Tauern Window (Eastern Alps). Tectonophysics 169: 271–280
Vavra G, Hansen BT (1991) Cathodoluminescence studies and U/Pb dating of zircons in pre-Mesozoic gneisses of the Tauern Window: implications for the Penninic basement evolution. Geol Rundsch 80: 703–715
White AJR, Chappell BW (1983) Granitoid types and their distribution in the Lachlan fold belt, southeastern Australia. Geol Soc Am Mem 159: 21–34
Wilson M (1989) Igneous petrogenesis. A global tectonic approach. Unwin Hyman, London, 466 pp
Winkler M (1993) Genese und geodynamische Stellung der Zentralgneise im Tauernfenster. Thesis, Univ Tübingen, Tübingen
Zimmermann R, Franz G, Hammerschmidt K, Quadt A v (1990) Polymetamorphic evolution of the Penninic basement (lower Schieferhülle) in the Tauern window: constraints from structural and petrologic data. Terra Abstr 2: 35–36

The Habach-Formation and the Zentralgneis – A Key in Understanding the Palaeozoic Evolution of the Tauern Window (Eastern Alps)

V. Höck

Abstract

The pre-Permian Habach formation is a complicated metamorphosed sequence of magmatic and sedimentary rocks within the Tauern window. It can be separated into three subunits:

1. Ophiolites including the Basisamphibolit;
2. An island arc volcanic sequence;
3. The Eiser sequence (Biotitporphyroblastenschiefer).

The ophiolites consist of metamorphosed ultramafic and mafic plutonics as well as volcanics. Their chemistry resembles that of basalts from a marginal basin with some MORB characteristics. The island arc sequence consists of basaltic to rhyolitic rocks overlain by sediments. The volcanics display a geochemical pattern consistent with an island arc volcanic series formed on continental crust. The Eiser sequence comprises pelitic and psammitic metasediments interlayered by basic to acidic metavolcanics. The few analyses available indicate a possible oceanic island arc sequence.

Age data from all subunits of the Habach formation based on radiometric data and scarce fossil findings range from the Upper Proterozoic to the Carboniferous. The evolution of the Habach formation despite the uncertainty of the age dating fits a model of a long-lasting active continental margin on which different terranes were accreted during the Upper Proterozoic to the Palaeozoic.

1 Introduction

The Palaeozoic evolution of the Eastern Alps has an important bearing on the reconstruction of the Variscan orogen in Europe, but this area is often omitted from tectonic or palaeogeographic maps of the Variscan belt in Central Europe (e.g. Matte 1986). This is due to the incorporation of the Alpine Palaeozoic rocks into the Alpine orogenesis with the corresponding metamorphism and deformation. On the other hand, several papers have been published in the last few years devoted to the (pre)Variscan evolution in the Alps only (Frisch et al. 1984; von Raumer 1981, 1986, 1987). The present chapter should help in closing a gap in the knowledge of the Palaeozoic rocks within the Penninic area and contribute in developing a more coherent picture of the (pre)Variscan evolution in the Alps.

2 Geological Setting and Lithology

Within the Penninic windows of the Eastern Alps, the Tauern window is the only one containing large coherent areas of pre-Permo-Mesozoic rocks. Their distribution is shown in Fig. 1. According to Frasl (1958) they can be divided apart from the Zentralgneise into two formations: the Habach formation and the Altkristallin formation, the latter being largely affected by a strong pre-Alpine metamorphism, while the Alpine events overprinted both formations. A comprehensive review of the metamorphic evolution of the Habach formation in the central part of the Tauern window has been given recently by Grundmann (1989). The pre-Permo-Mesozoic formations were intruded at the change from the Lower Carboniferous to the Upper Carboniferous by a sequence of tonalites, granodiorites and granites (Cliff 1981) which are termed "Zentralgneise" according to their Alpine metamorphism and deformation (Finger et al., this Vol.). The close spatial relationship between the Zentralgneise and both the pre-Permian formations are shown in Fig. 1. Two coherent areas of Palaeozoic rocks can be depicted: a larger one in the western and central part of the Tauern window, a smaller but more segmented one in the eastern part.

The delimitation between the Altkristallin formation and the Habach formation was always a difficult and yet unsolved problem. Frasl (1958) included originally the amphibolites and some gneisses of the "Zwölferzug", both showing distinct signs of a pre-

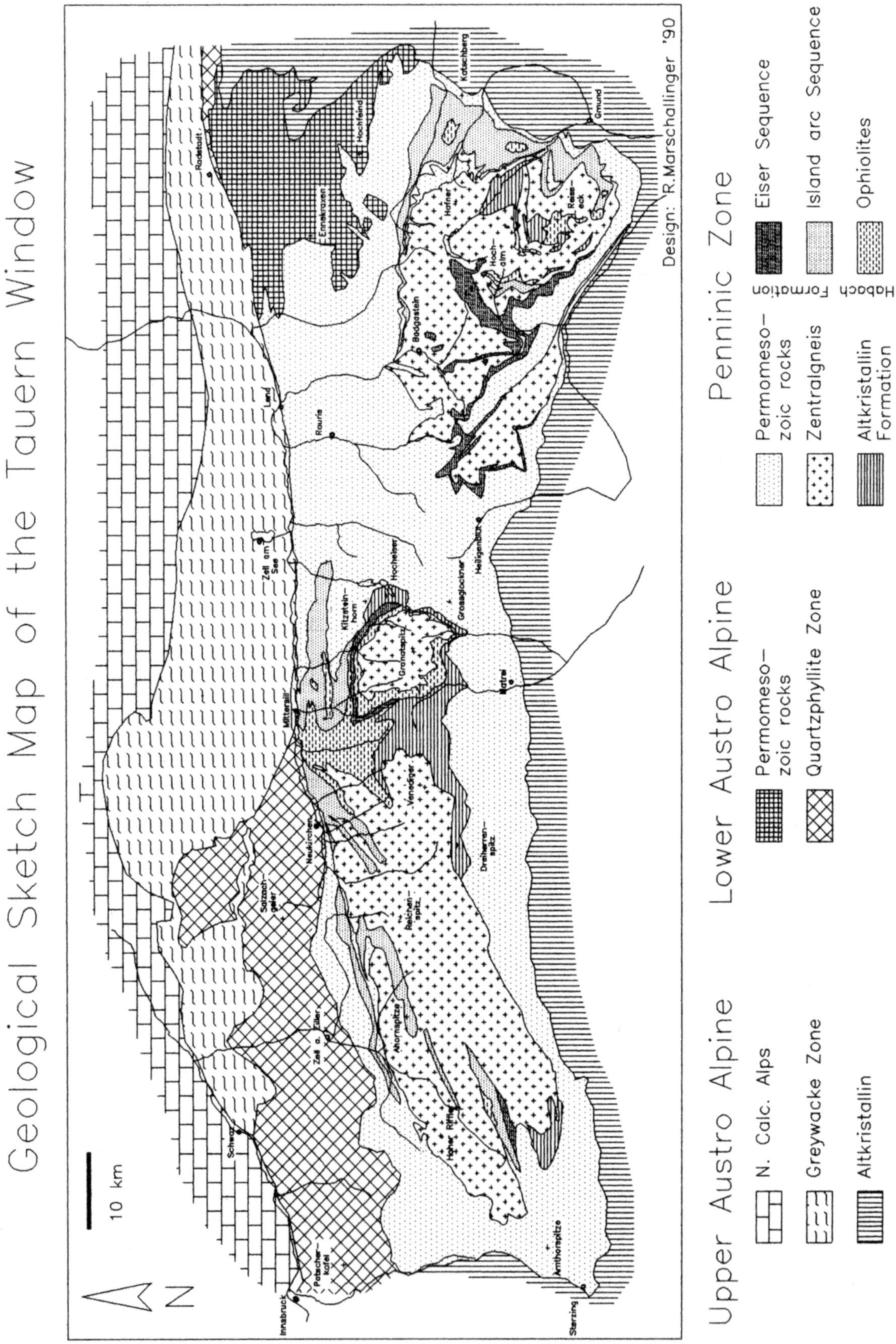

Fig. 1. Geological sketch map of the Tauern window showing the distribution of the different subunits of the Habach formation

Alpine amphibolite-grade metamorphic event into the Altkristallin. But in 1966 Frasl and Frank extended the range of the Altkristallin formation and delineated an "amphibolite sequence" and a "Serie der Alten Gneise". The criterion for separating these sequences was again the evidence of a high-grade, probably Variscan metamorphism restricted to the Altkristallin formation. Nevertheless they conceded some lithological accordance between the amphibolites of the Altkristallin formation and the Habach formation: "Regarding the origin of the amphibolite series, we can well imagine that during the Variscan orogeny the Lower Palaeozoic rocks underwent only an epizonal or contact metamorphism in the higher levels, which is then the Habachseries, while the same rocks metamorphosed farther in depth became the mesozonal amphibolite series." (Frasl and Frank 1966, p. 35). Accordingly, the rocks of the amphibolite sequence (Basisamphibolit/Zwölferzug) will be treated together with the Habach formation in this chapter, which deals in principle with the lithology and geochemistry. Consequently, here, only the gneisses (Serie der alten Gneise) including orthogneisss, paragneisses, migmatites, micaschists with some amphibolite lenses will be considered as Altkristallin formation (geological sketch map, Fig. 1).

The nomenclature of the lithostratigraphic units within the Tauern window is not completely internally consistent as documented by the problem of delimiting the Altkristallin formation from the Habach formation. Also, terms such as formation, group, complex, et cetera are often used in a different way. A general revision of all lithostratigraphic terms transforming the stratigraphic concepts in the Tauern window into a consistent system according to the rules of the International Subcommission on Stratigraphic Classification is far beyond the scope of this paper and will be published elsewhere (Höck in prep.). To avoid further confusion the old traditional terms will be used throughout this chapter.

Generally speaking, the Habach formation can be described as a metamorphosed (Alpine event) volcano-sedimentary sequence with lavas and tuffs ranging from basic to acidic chemistry. Ultrabasics (serpentinites and hornblendites) as well as mafic plutonics (metagabbros) occur occasionally. Metamorphosed shales prevail over sandstones and greywackes among the sediments. Frasl (1958) defined the Habach formation as "Lower Palaeozoic geosynclinal metamorphosed rock sequence, consisting of metasediments and various magmatites", a short and clear definition still valid with some restrictions.

The first systematic grouping of the Habach formation was developed by Höll (1975) in the vicinity of the newly discovered scheelite mine in the Felber Valley. Three groups have been subdivided: the Basisschiefer sequence, the eruptive rocks and the Habach phyllites. Later, this grouping was modified by Kraiger (1987, 1989) when studying the continuous outcrops of a 2-km-long adit of the scheelite mine. He divided, especially, the eruptive rocks into the lower magmatic sequence (LMS) and the upper magmatic sequence (UMS). The former was interpreted as dismembered ophiolite, the latter as part of a calc-alkaline island arc sequence. The Habach phyllites would be part of the sedimentary cover of the UMS. Despite its, originally, only local validity, a conception similar to that of Kraiger can be applied to the Habach formation as a whole: again three groups can be distinguished: (1) ophiolites, (2) an island arc sequence and (3) the Eiser sequence identical with the "Biotitprophyroblastenschiefer-Folge" of Pestal (1983).

The ophiolites comprise the LMS (Kraiger 1987, 1989) and the "Basisamphibolit" (Cornelius and Clar 1939; Frank 1969, Frisch and Raab 1987) the latter being termed Stubach group by Frisch and Neubauer (1989). It frames the Zentralgneis core of the Granatspitz (Fig. 1). Some layered ultrabasic rocks of the Stubach valley with wherlites, dunites, lherzolites and clinopyroxenites (Petrakakis 1977, 1978) are included here. The amphibolites themselves consist partly of layered gabbros and partly of metamorphic lavas. The position of the Zwölferzug, a sequence of garnet-bearing amphibolites and amphibolbearing gneisses strongly metamorphosed by a pre-Alpine metamorphism (Frasl 1958) is debatable. It has some similarities with the Basisamphibolit and is assigned here to the ophiolites for simplicity. A modern detailed lithological and petrological study of the whole group has not been completed.

The LMS is formed by small lenticular ultrabasic bodies (serpentinites and hornblendites) which are followed by relatively thick metagabbros (Grobkornamphibolite). Most fine-grained amphibolites are part of a possible sheeted dyke complex (Kraiger 1989), but dykes are only sporadically recognizable. Because of the intense metamorphism and deformation (upper greenschist to amphibolite facies) there is no clear distinction between dykes and basaltic lavas.

Whereas the Basisamphibolit is restricted to the immediate surroundings of the Zentralgneis of the Granatspitze (the Zwölferzug being a little further to the north), ophiolite-type rocks comparable to the LMS are found in a wide area in the middle part of the Hohe Tauern (Felber Valley, Hollersbach Valley and southern part of the Habach Valley). The Greiner ultramafics (Lammerer 1986) in the western part of the Hohe Tauern, and ultrabasic and gabbroic of the

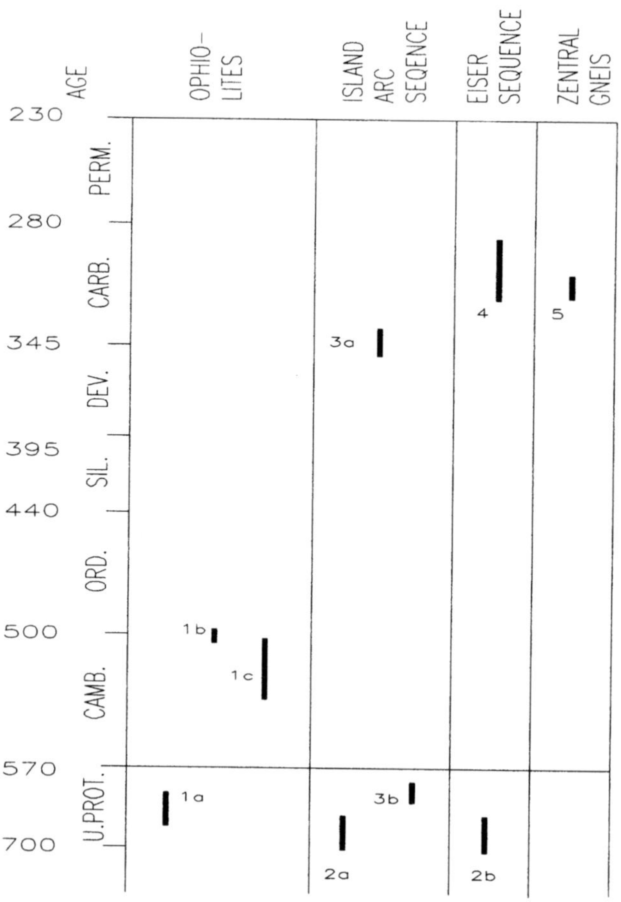

Fig. 2. Stratigraphic scheme of the Habach formation. Data sources: **1** von Quadt (1985): *1a* Basisamphibolit; *1b* Zwölferzug; *1c* ophiolites; **2** Reitz and Höll (1988), palaeontological dating with acritarchs: *2a* island arc sequence; *2b* Eiser sequence; **3** Vavra (1989): *3a* island arc sequence (Habach Valley); *3b* island arc sequence (Eastern Tauern window); **4** Franz et al. (1990), leaf fragments; **5** Cliff (1981)

Storz formation in the easternmost sections (Exner 1971; Stadlmann 1990) between Hochalmspitze and Katschberg are also probably part of the ophiolites.

The island arc sequence consists of a variety of metamorphosed basic, intermediate and acidic lavas and tuffs. Today they are found as amphibolites, prasinities, biotite-epidote gneisses, albite gneisses and meta-agglomerates as well as chlorite gneisses and chlorite micaschists containing some sedimentary material. Detailed lithological and petrographic descriptions of this sequence have been given recently by Steyrer (1983) for the Habach Valley and Kraiger (1989) for the Felber valley. Apart from their occurrence in the central part of the Tauern window between the Krimml Valley and the Kaprun Valley comparable rocks are widely distributed in the western part of the Tauern window (e.g. Schönach syncline, Greiner syncline) and in large parts of the Storz formation (Exner 1971) in the Eastern Tauern window. Geological and field evidence especially in the Habach Valley (Steyrer 1983) suggests a primary stratigraphic relationship between the metavolcanics and the overlaying sediments: black phyllites (=Habach phyllites) and associated micaschists, quartzites and greywackes.

The Eiser sequence (=Biotitporphyroblastenschiefer by Cornelius und Clar 1939; Pestal 1983; Höck and Pestal 1990) is largely identical with the Basisschiefer *sensu* Höll (1975) in the vicinity of the scheelite mine. In this area it is unconformably overlying the Basisamphibolit (Pestal 1983) and considered to be an independent unit consisting of biotite-rich schists, greywackes, graphitic quartzites, garnet-bearing micaschists and acidic, basic and some intermediate volcanics. Apart from their distribution on top of the Basisamphibolit, these rocks are abundant in the Eastern Tauern window south of Badgastein at the southern rims of the Zentralgneis. Equivalents at the western end of the Tauern window could be the Furtschagelschiefer and their related rocks.

Despite radiometric age determinations and recent findings of preserved fossils the age of the Habach formation and their associate units is still problematic. A compilation of age data related to the different subunits and the Zentralgneis is given in Fig. 2. The most reliable data are the U-Pb ages on zircons from the ophiolites (von Quadt 1984, 1985) ranging from 500 to 540 Ma. Data from the Basisamphibolit, Zwölferzug, hornblendites and gabbros are included here: according to palaeontological results (Reitz and Höll 1988) based on acritarchs and cyanobacteria, at least parts of Habach phyllites and, as a consequence, also parts of the island arc sequence, should be of Upper Proterozoic age (Vendian). On the other hand Vavra (1989) reported U-Pb data on zircons from acidic gneisses in the Eastern part 600 Ma and 330 Ma from the Habach valley. Both data come from rocks which are believed to be part of the Island arc sequence. Franz et al. (1990, 1991) described Permo-Carboniferous leaf fragments (cf. *Callipteridium* sp.) from graphitic schists. Such graphitic schists are believed to be part of the Habach formation. They occur in the island arc – and in the Eiser sequence. In Fig. 2 the leaf-bearing graphite schists are assigned to the Eiser sequence, but their actual position is not clear yet. However, the Habach formation must be older than the intrusion of the Zentralgneise (320 Ma and younger: e.g. Cliff 1977, 1981; Cliff and Cohen 1980; Pestal 1983), because intrusion relations between the Zentralgneise and the Habach formation are frequently observed.

The Zentralgneise form a complicated intrusion sequence ranging from tonalites and granodiorites to various types of granites, also quartz-monzonitic types occur. Particularly in the Eastern Tauern window multiple intrusions have been demonstrated by Holub and Marschallinger (1989) and Marschallinger and Holub (1991). A compilatory overview of available geochemical data of the Zentralgneise by Finger and Steyrer (1988) showed that I-type granites (in the Western and Eastern Tauern window) prevail over S-types (Granatspitz core in the central part). The protoliths of the overwhelming majority of the Zentralgneise present calc-alkaline geochemistry and are interpreted as volcanic arc granite *sensu* Pearce et al. 1984a (cf. Finger and Steyrer 1988; Marschallinger and Holub 1991). For a more detailed description of the Zentralgneise the reader is referred to the chapter by Finger et al. (this Vol.).

3 Geochemistry

In the following paragraphs only a short geochemical characteristic of the subunits of the Habach formation will be given supported by selected geochemical analyses (Table 1) and some diagrams (Figs. 3–7). A more detailed documentation of the geochemistry of the Habach formation will be published elsewhere (Höck in prep.).

The Basisamphibolit as well as the fine-grained amphibolites of the LMS show geochemical signatures with close similarities to MOR-basalts (Table 1, analyses 1–3). In the Zr/Ti/Y (Fig. 3a), in the Ti vs Zr (Fig. 5a) and in the Zr/Y vs Zr (Fig. 3b) diagrams according to Pearce and Cann (1973) and Pearce and Norry (1979) respectively, all analyses plot into or close to the fields compatible with MORB. Unfortunately these fields cannot be clearly separated from those occupied by volcanic arc basalts (VAB). This fact is also highlighted (Fig. 4a) in the Rock/MORB diagram (cf. Pearce 1980, 1983) as the elements Ta to Cr plot close to (often slightly below) unity, displaying a MORB-like signature. The LIL elements including Sr to Th show a distinct enrichment. This could be partly caused by some hydrothermal alterations which could be particularly true for the distinct Rb-enrichment of the LMS-amphibolites (Fig. 4a) in the area of the scheelite mine (Kraiger 1987). Some enrichment of the LIL elements outside the mine is also common. Since there is no convincing evidence for a significant shift in element concentration during the regional metamorphism as shown by Kraiger (1987), this enrichment is believed to reflect petrogenetic processes during magma genesis and is interpreted as a "subduction component" in the sense of Pearce (1982, 1983) and Pearce et al. (1984b): the LIL elements, highly mobile in aqueous fluids, released in a downgoing slab, were transported upwards in the overlying mantle wedge and incorporated into the partial melt or brought directly by some melt batches derived from the subduction zone (Pearce 1983). However, the chondrite normalized REEs (Fig. 4b) show a flat distribution pattern approximately ten times enriched, typical for MORBs. All these features suggest that a back-arc basin situated above a former subduction zone could be a suitable tectonic environment for generating these ophiolites.

In contrast to the MORB type signatures, the island arc volcanics (Table 1, analyses 4–9) show completely different patterns. In the Ti/Zr/Y diagram (Fig. 6a) they plot in the border area of the fields for volcanic arc basalts and within plate basalts. Furthermore, they occupy in the Zr/Y vs Zr diagram (Fig. 6b) the field of the WPB with a distinct variation in the Zr/Y ratio. The relative enrichment of Zr over Y expressed in both diagrams argues for a contribution of some within-plate component in addition to the volcanic arc generation of these magmas. The high Zr/Y ratios indicate source variations rather than fractionation of a phase retaining Y (e.g. garnet). In the last respect, the Ti vs Zr diagram (Fig. 5b) containing

Table 1. Selected geochemical analyses from the Habach formation (1–3: ophiolites, 4–9: basic to intermediate volcanics of the island arc sequence, 10: calc-alkaline dyke in the ophiolites). Data sources: Frisch and Raab (1987): F 1930; Kraiger (1987): U411, U191, U5, U334; Stadlmann (1990): TS114, TS186; Vavra (1989): AN245, V2245

	1 F 1930	2 U 411	3 AN 245	4 V 2245	5 U 191	6 STD 70	7 TS 114	8 U 5	9 TS 186	10 U 334
SiO_2	49,83	50.18	51.34	49.22	51.39	52.21	53.27	54.00	55.61	61.17
TiO_2	1.22	1.13	1.10	0.98	0.70	0.67	0.78	0.94	1.12	0.99
Al_2O_3	14.56	15.68	15.32	18.04	20.66	13.12	18.03	15.49	17.13	16.90
FeO^a	11.25	8.65	10.38	7.11	5.84	13.83	7.83	5.28	5.73	5.76
MnO	0.23	0.14	0.20	0.14	0.09	0.14	0.16	0.13	0.08	0.13
MgO	6.40	8.33	6.64	4.67	2.22	8.92	4.60	6.24	3.61	1.86
CaO	9.13	11.29	10.03	9.62	9.72	2.99	5.64	6.07	5.37	5.47
Na_2O	4.10	2.40	2.95	4.05	2.31	0.28	4.94	3.60	4.31	4.79
K_2O	0.18	0.14	0.30	1.01	2.76	3.66	1.04	1.54	2.96	1.32
P_2O_5	0.10	0.09	0.10	0.55	0.59	0.28	0.12	0.25	0.59	0.32
Loi	2.07	1.45	0.70	1.46	3.25	3.71	2.13	6.30	1.61	0.88
Total	99.07	99.48	99.06	96.85	99.53	99.81	98.87	99.84	98.45	99.59
Nb	–	2	–	9	9	9	6	11	13	18
Zr	84	73	56	210	159	181	73	142	158	273
Y	23	26	18	37	16	19	32	24	18	29
Sr	265	148	224	1138	2195	238	233	471	1120	490
Rb	–	28	–	43	105	134	30	48	80	167
Ni	47	86	45		22	81	22	63	42	13
Cr	164	395	160		41	284	72	186	42	122
Ba	114	69	105	934	1367		331	290	1222	781
Ta		0.09			0.52					0.94
Hf		2.04			5.29					6.50
La	<5	2.9	<5	35	73.9					52.6
Ce	10	11.4	7	76	147.8					121.6
Nd	<10		<10	37.4	60.0					52.9
Sm	3	4.7	3	8	12.6					5.3
Eu	1.3	1.0	0.7	0.7	2.6					2.2
Tb		0.34			0.38					1.0
Yb	3	2.7	2.4	2.2	1.9					2.8
Lu	0.5	0.57	0.3	0.5	0.26					0.32

[a] Total Fe as FeO.

analyses with a chemistry from basic to acid, is of some interest. There is almost no positive Ti/Zr correlation, the Ti concentration decreases with only a small variation of Zr in intermediate volcanics. This trend continues in acidic rocks with a tendency for decreasing Zr also. Such a distribution is consistent with a fractionation model in which minerals such as amphibole, biotite and magnetite play an important role as fractionating phases. A fractionation pattern like that is not only seen in the overall sequence but can be visualized also in individual sections of the island arc sequence as e.g. described by Seemann and Koller (1989). This type of fractionation is common in island arc environments.

The chondrite-normalized REE pattern (Fig. 7 b) shows a typical enrichment of the light REEs between 20 and 200 times chondrite and a corresponding relative depletion in the heavy REEs. Again the rock/MORB diagram (Fig. 7 a) of basic and intermediate volcanics highlights the geochemical signatures: The zig-zag pattern caused by the different enrichment of the elements is typical for island arc volcanics generated on a continental crust (Pearce 1983) as is particularly emphasized by the high Ce/Nb and Sm/Hf ratios with Ta, Nb, Zr and Hf concentrations significantly above unity (cf. Pearce 1983). The general trend is found everywhere in the island arc sequence, but is not so nicely expressed in some rocks from the Storz formation (Stadlmann 1990).

In areas clearly assigned to the Eiser sequence, only a few well documented geochemical analyses are available. Two sets of data from amphibolites published by Gilg et al. (1989 in Fig. 6) are compatible with the idea, that they have formed in an oceanic island arc. The Rock/MORB patterns have similar zig-zag variation as the continental island arc volcanics but the Ta, Nb, Zr and Hf concentrations are well below unity (Pearce 1982). The Rock/chondrite patterns (Fig. 7 b) are similar to those from the island arc sequence, but show a lower overall enrichment.

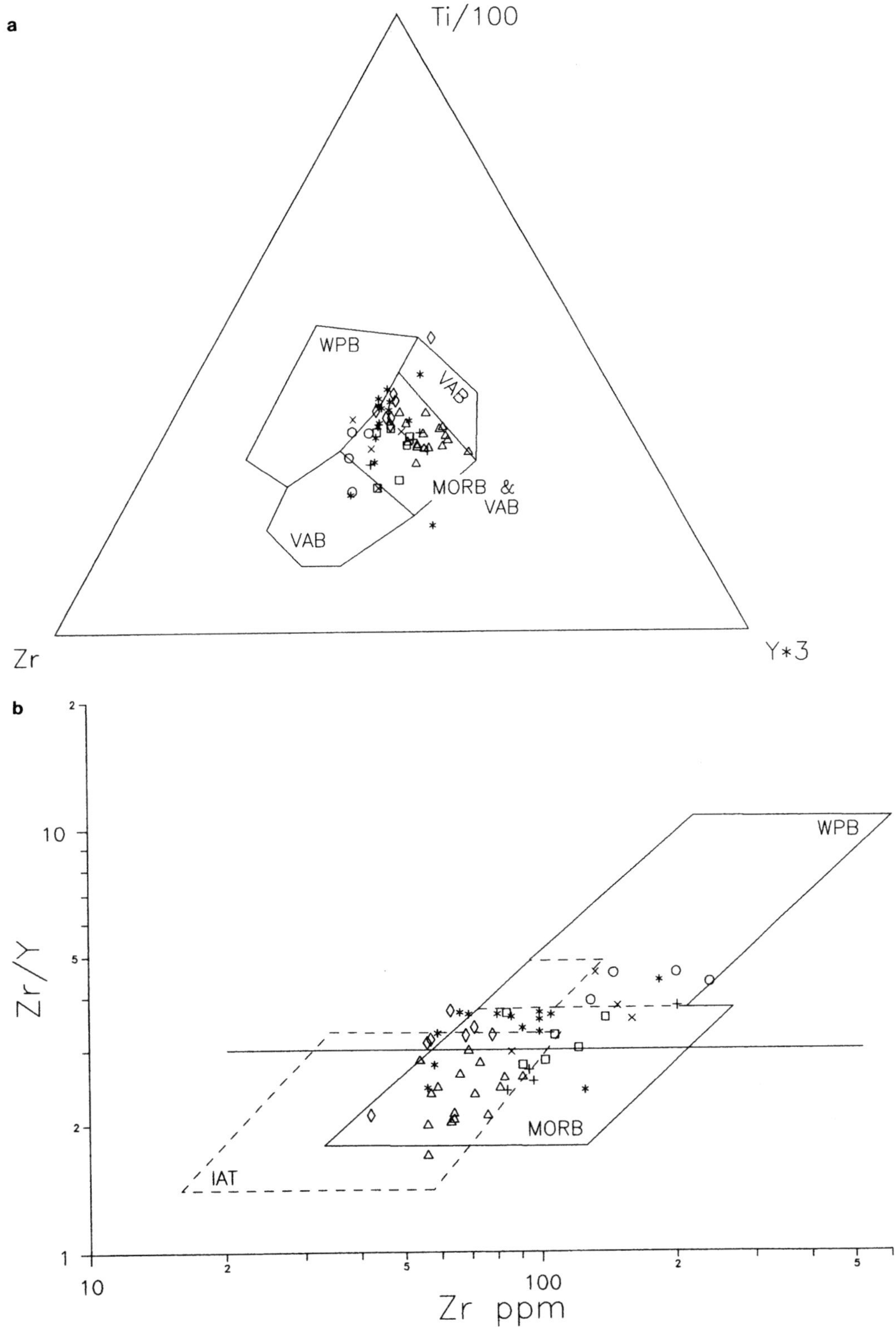

Fig. 3. a $Tix10^{-2}/Zr/Yx3$ diagram of the ophiolites. Most analyses fall in the field of MORB+VAB. Symbols: *open triangle* Felber Valley: *open square* Basisamphibolit; *diamond* Ankogel (Eastern Tauern window); *asterisk* Reißeck (Eastern Tauern window); *open circle* Felber Valley/Amer Valley; *cross* Habach Valley; *X* Hollersbach valley. Data sources: Pestal (1983); Steyrer and Höck (1985); Frisch and Raab (1987); Kraiger (1987); Vavra (1989); Seemann and Koller (1989); Stadlmann (1990) and unpublished analyses. **b** Zr/Y vs Zr diagram depicting the ophiolites. Most analyses plot into the field of MORB. Some analyses from the Amer Valley and the Hollersbach Valley show relative enrichment in Zr and relatively high Zr/Y ratios. *Symbols* and data sources as in **a**

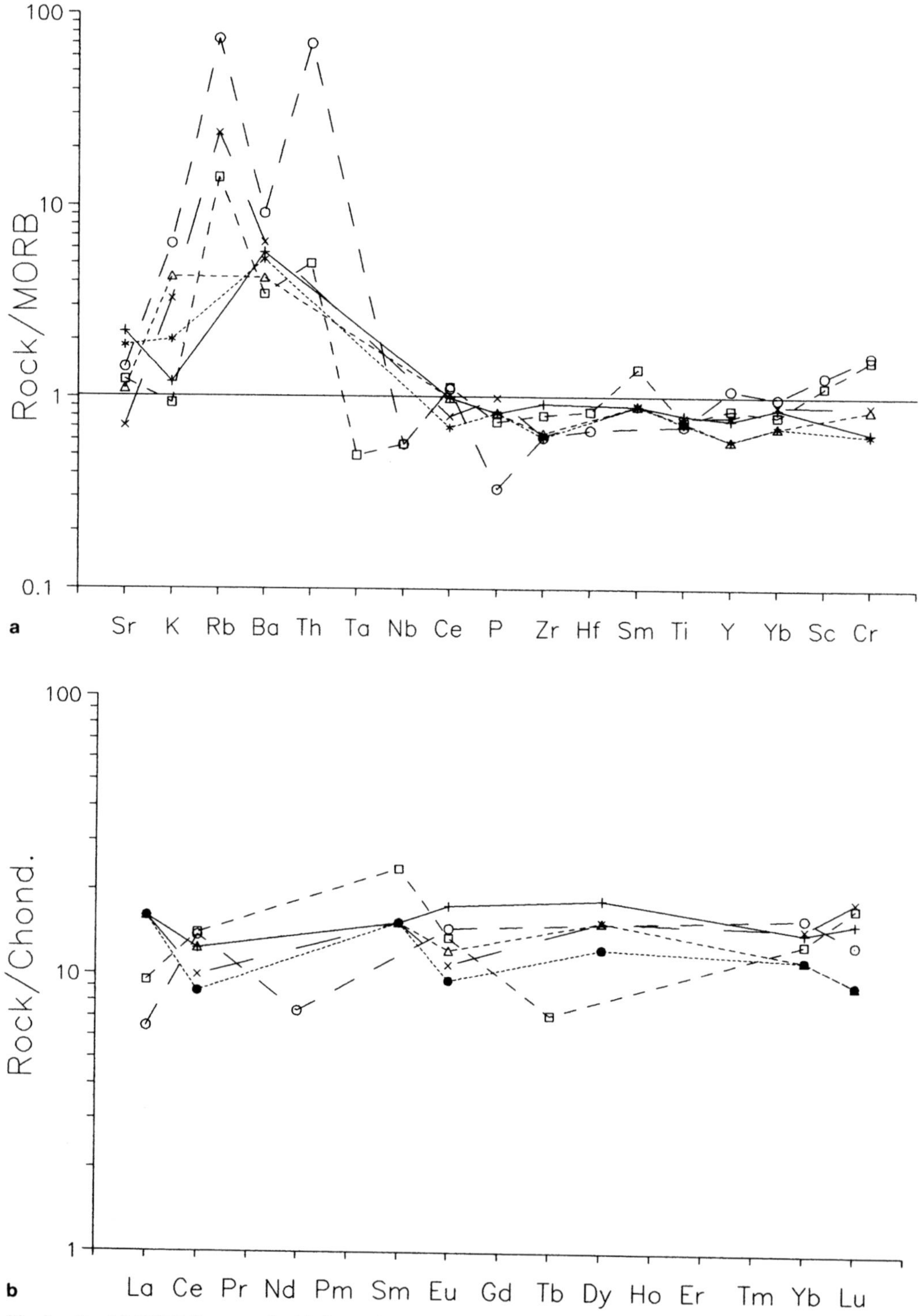

Fig. 4. a Rock/MORB diagram of ophiolites. Elements from Sr to Th are significantly enriched, especially Rb and Th (in the scheelite mine). Elements from Ta to Cr are close to, often slightly below unity. *Symbols: open circles, open squares, X* individual samples from the Scheelite mine, Felber Valley; *cross* Basisamphibolit; *open triangle* Reißeck; *asterisk* Ankogel. Data sources as in Fig. 3 a. **b** Rock/chondrite normalized pattern. They show a flat distribution pattern 6 to 15 times enriched, typical for ocean floor basalts. Symbols as in **a.** Data sources as in Fig. 3 a

Frisch and Raab (1987) presented geochemical evidence that more amphibolites in the Habach formation occur with oceanic island arc signature rather than continental island arc signatures. Whether these rocks can be incorporated into the classification presented here, or must be assigned to a separate group remains an open problem yet.

The geochemistry of the Zentralgneise will not be discussed here in detail (see also Finger et al. this Vol.; the compilation by Finger and Steyrer 1988 and the paper by Marschallinger and Holub 1991 for the Eastern Tauern window). It should be mentioned here that the majority of the Zentralgneise corresponds geochemically to volcanic arc granites (high-K I-types) intruding already thick continental crust. The actual distribution and significance of granites with S-type affinities remains to be clarified.

4 Discussion

The geochemical data of the ophiolite sequences (Basisamphibolit, LMS) indicate their formation in a back-arc or marginal oceanic basin producing basalts with some MORB characteristics. The enrichment of the LIL elements, not caused by hydrothermal alteration, can be explained with the position of such a basin above a subduction zone. This is in accordance with earlier findings by Steyrer and Höck (1985), Frisch and Raab (1987), Seemann and Koller (1989) and Vavra and Frisch (1989). According to the U/Pb age dating by von Quadt (1985), this basin might have formed 500 to 540 Ma ago. The island arc sequence on the other hand displays, as shown in the previous chapters, a marked continental signature. A thin continental crust or a transition between continental and oceanic crust could possibly be the tectonic environment of some parts of the island arc sequence, particularly in the Eastern Tauern window (Storz formation, according to Exner 1971), where the continental signature is not so clearly developed (Stadlmann 1990). It must be noted that a relatively close spatial relation exists between the ophiolites and the island arc sequence. The former are underlying the latter in the Central Tauern window and overlying in the Storz formation respectively (probably in an overturned section). In addition dykes frequently found in the ophiolites of the LMS resemble geochemically the island arc volcanics (Table 1, analysis 10). The most obvious suggestion would be to interprete them as feeder dykes of the island arc volcanics implying, at least in part, a younger age for the island arc rocks compared with the ophiolites. The palaeontological evidence, on the contrary, (Reitz and Höll 1988, see also Fig. 2) indicates an Upper Proterozoic age for the sediments and also for parts of the island arc rocks, compared with the probably Cambrian age of the ophiolites. A partition of the island arc sequence into an older and a younger group could be a possible solution for that contradiction in age relationships, but is not justfied by the field evidence indicating an undivided uniform geological unit (Steyrer 1983; Kraiger 1987; Stadlmann 1990). This problem will remain an open question until new reliable age data and new lithological, petrographical and geochemical evidence are available.

A similar problem arises with the interpretation of the Eiser unit. Apart from the hitherto rather weak indication of an oceanic island arc environment, the age dating is also problematic. Reitz and Höll (1988) describe acritarchs from the Upper Proterozoic ranging possibly into the Cambrian. A younger Palaeozoic flysch sequence is the preferred interpretation by Frank (pers. comm.) based on the lithology and the transgressive position on top of the Basisamphibolit in the Felber Valley.

The Zentralgneis magmas, however, intruded in an already well developed continental crust, as can be judged from the types of intrusion and the geochemistry (Finger et al. this Vol.; Finger and Steyrer 1988; Marschallinger and Holub 1991). A continous accretion of ophiolites, fragments of continental and oceanic island arcs with some sediments lasting from the Upper Proterozoic (?) to the Lower Carboniferous (?) could have slowly consolidated a former active continental margin forming a thick continental crust.

5 Geodynamic Evolution

At the present stage of knowledge, summarized in the preceding sections, it is not possible to construct a coherent model of the geodynamic evolution. This is mainly due to the problematic age dating of the Habach formation as a whole and the unclear age relationship between the several subunits of the Habach formation. Apart from the age dating problems it seems safe to say that the Habach formation formed along a probably long-lasting active continental margin adjacent to a large open ocean, on which, similar to the Pacific continental margin in North America (e.g. Dickinson 1981), several rock units as larger or smaller terranes have been accreted: ophiolites as remnants of one or more marginal basins, island arcs of continental and/or oceanic provenance together with their sedimentary cover.

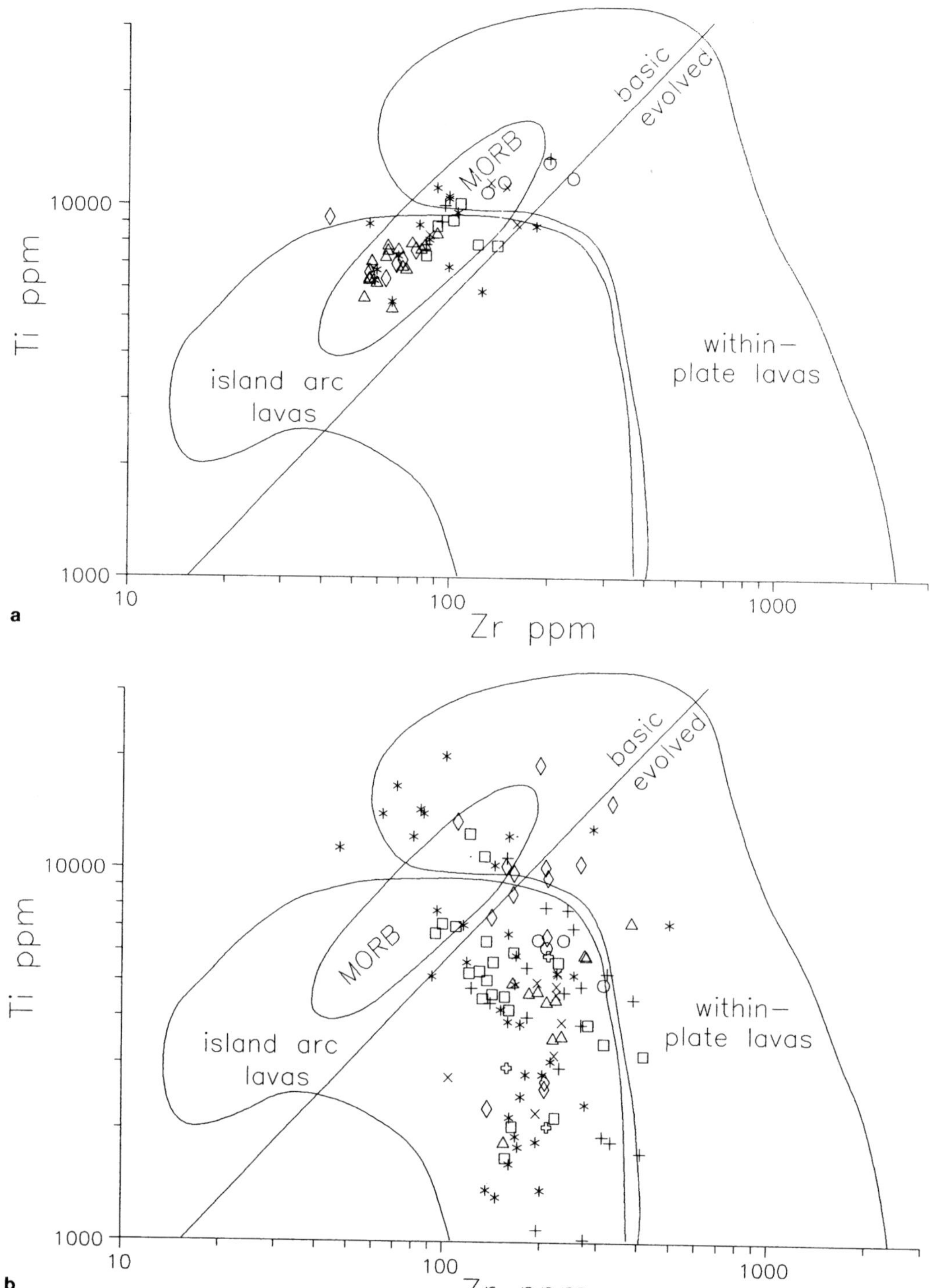

Fig. 5. **a** Ti/Zr diagram (Pearce 1980) of the ophiolites. The *straight line* separates basic from evolved basalts. Most analyses plot in the MORB field. Analyses from the Amer Valley and the Hollersbach Valley plot in the MORB section overlapping with within-plate basalts. *Symbols* and data sources as in Fig. 3 a. **b** Ti/Zr diagram of the island arc sequence. Rocks from basic to acidic chemistry are plotted in the diagram. No systematic positive correlation of Ti and Zr can be observed. Zr varies only slightly, while Ti decreases mostly from basic to acidic chemistry. Further explanations, see text. *Symbols: open square* Felber Valley (scheelite mine); *triangle* intermediate to acid dykes in the LMS (Felber valley); *cross* Habach Valley; *circle* Tauernkogel; *X* Weinbühel; *asterisk* Storz formation; *diamond* Ankogel; *full crosses* Reißeck; data sources as in Fig. 3 a

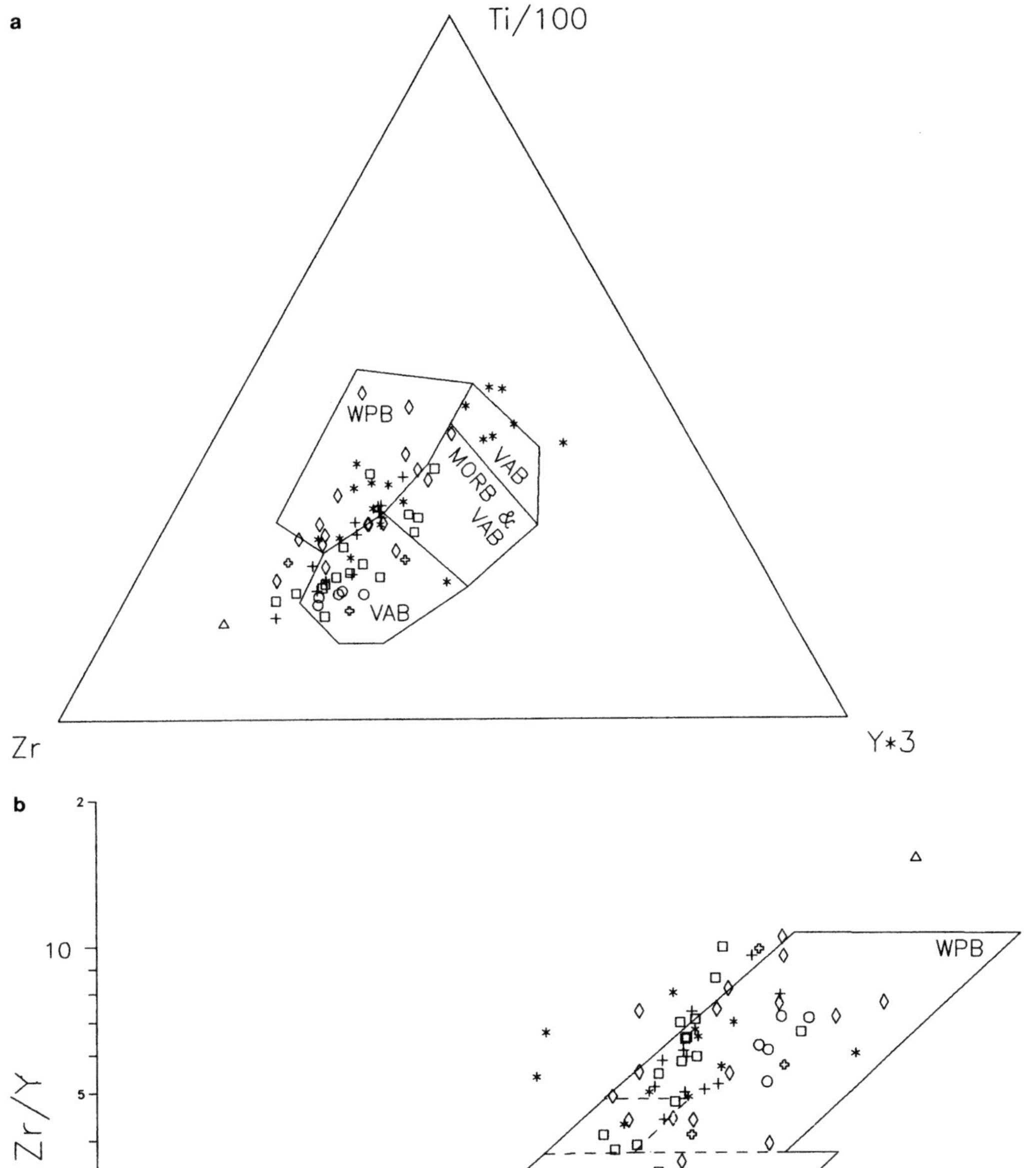

Fig. 6. a Tix10^{-2}/Zr/Yx3 diagram of the island arc volcanics (basic to intermediate). *Fields* as in Fig. 3 a. All data occupy an array, which ranges from volcanic arc basalts *(VAB)* to within-plate basalt *(WPB)*, thus showing signatures of both environments. *Symbols* as in Fig. 5 b, data sources as in Fig. 3 a. **b** Zr/Y vs Zr diagram of the island arc volcanics (basic to intermediate). *Fields* as in Fig. 3 b. The analyses show a wide variation in Zr/Y ratio, mostly above 4, thus occupying+he WPB-field. This also demonstrates some within-plate components in the island arc volcanics. *Symbols* as in Fig. 5 b, data sources as in Fig. 3 a

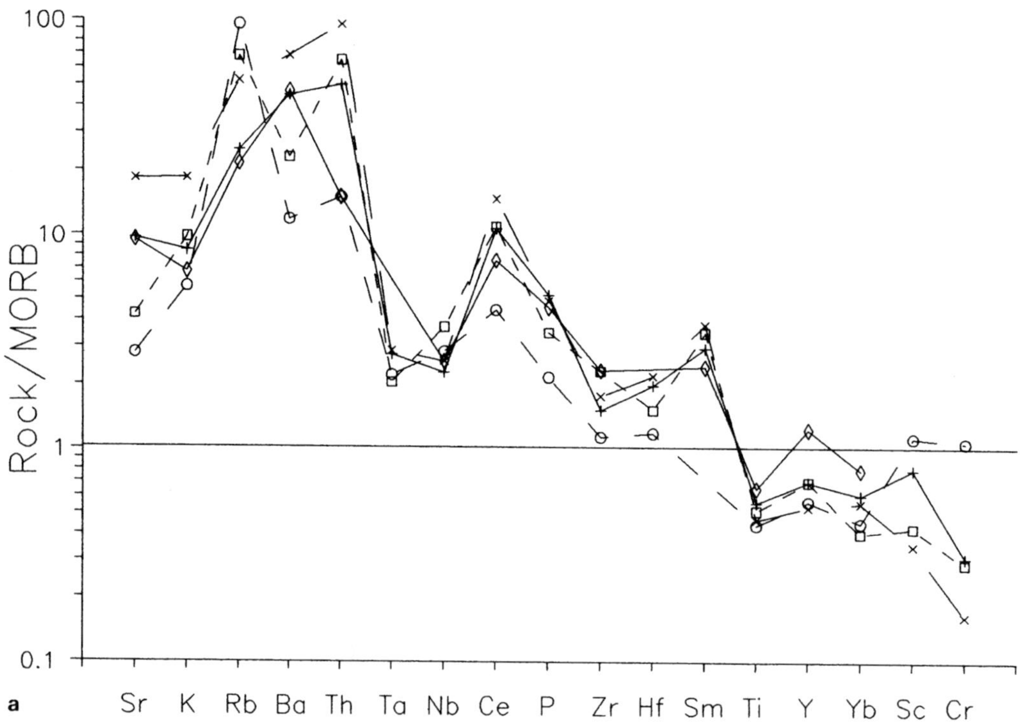

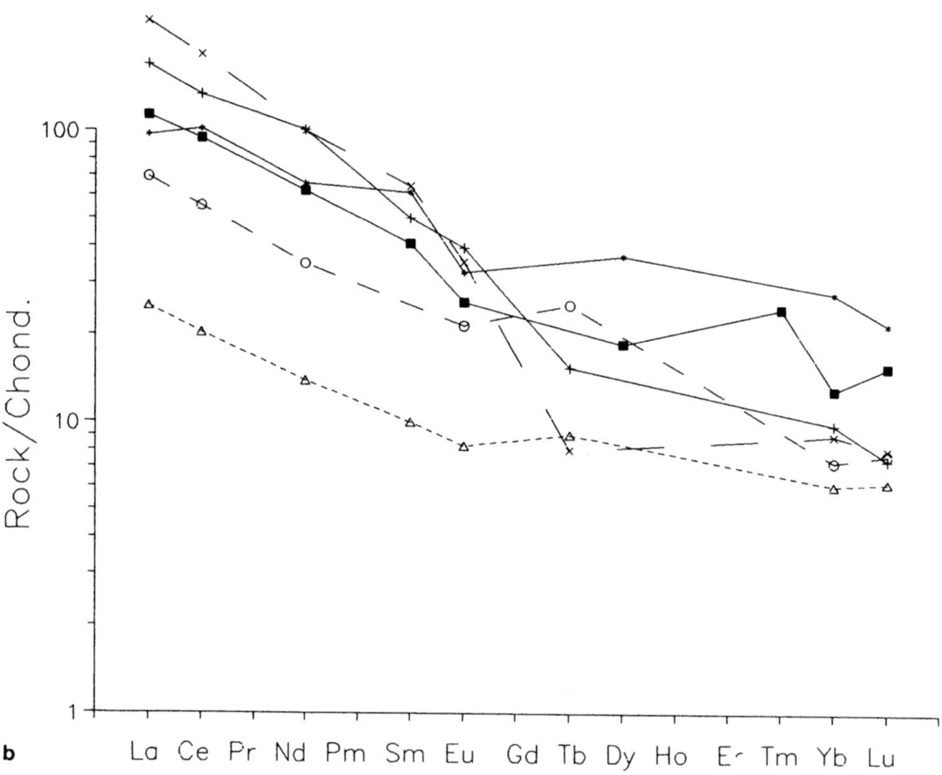

Fig. 7. **a** Rock/MORB diagram of the island arc volcanics (basic to intermediate chemistry). The zig-zag pattern combined with the high Ta, Nb and Zr, Hf normalized values argues for island arc volcanics generated on a continental crust. *Symbols: open circles, open squares, crosses, X* individual samples from Felber Valley (scheelite mine); *diamond* Reißeck; data sources as in Fig. 3a. **b** Rock/chondrite diagram of some island arc volcanics (basic to intermediate chemistry). The volcanics from the scheelite mine in the Felber Valley, from the Ankogel area and from the Reißeck show an enrichment of the LREE between 80 and 200 times chondrite and a relative depletion of the HREE. Two samples from the Eiser unit (=Basisschiefer) taken from Gilg et al. (1989) show a similar trend but a lower overall enrichment. *Symbols: cross, X* Felber Valley (scheelite mine); *full square* Reißeck; *asterisk* Ankogel; *open circle* and *open triangle* Eiser unit (Basisschiefer: Gilg et al. 1989), Data sources as in Fig. 3a

These accretion processes which may have lasted from the Upper Proterozoic to the Lower Carboniferous (?) led finally to the formation of a thick continental crust in which the magmas of the Zentralgneise generated along a Variscan subduction zone.

The direction of this subduction zone beneath the continental margin and the nature of the continent behind this margin is another problem which needs more investigation. For the Variscan time a northward directed subduction zone has been proposed by several authors (Mercolli and Oberhänsli 1988; Finger and Steyrer 1990; Finger et al. this Vol.). This will not be discussed here in general, but one point of the whole topic not mentioned in these models deserves some reasoning: one important problem associated with the question of the subduction zone is, whether the Habach formation can be related to similar formations in the Austro-Alpine unit or others outside the Alps. It is difficult to find comparable units within the Bohemian Massif but within the Alps, the most likely candidates for such a comparison can be found in the Lower Austro-Alpine frame at the eastern margin of the Alps (e.g. Pahr 1980). If such a relation could be verified, a close relation of the Habach formation to the northern rim of the Gondwana continent, the Austro-Alpine being a part of it, could be established. Such a palaeogeographic model would be more consistent with a long-lasting southwards-directed subduction zone beneath Gondwana. For this reasons a crucial point in unravelling the palaeogeographic position of the Habach formation and the Zentralgneise and their tectonic evolution must come from a careful comparison of appropriate units in the Austro-Alpine with the Habach formation.

Acknowledgements
I am grateful to Drs. M. Delaloye, F. Koller, F. Neubauer, G. Pestal and J. von Raumer for numerous discussions and critical reviews of the manuscript. R. Marschallinger helped kindly with the data handling and the figure drafting. Prof. Richter (Institute of Petrology, University of Vienna) provided the XRF equipment for the chemical analyses. This work has been supported by the Austrian Science Foundation (FWF grant S 4704) and the Austrian National Bank (Jubiläumsfonds grant 3620).

References

Cliff RA (1977) Rb-Sr isotopic measurements on granite-gneisses from the Granatspitzkern, Hohe Tauern, Austria. Verh Geol Bundesanst 1977: 101–104

Cliff RA (1981) Pre-Alpine history of the Pennine Zone in the Tauern Window, Austria: U-Pb and Rb-Sr geochronology. Contrib Mineral Petrol 77: 262–266

Cliff RA, Cohen A (1980) Uranium-lead isotope systematics in a regionally metamorphosed tonalite from the Eastern Alps. Earth Planet Sci Lett 50: 211–218

Cornelius HP, Clar E (1939) Geologie des Großglocknergebietes, Teil I. Abh Zweigst Wien Reichsst Bodenforsch 25: 350 S

Dickinson RW (1981) Plate tectonics and the continental margin of California. In: Ernst WG (ed) The geotectonic development of California, pp. 1–28

Exner C (1971) Geologie der peripheren Hafnergruppe (Hohe Tauern). Jahrb Geol Bundesanst 114: 1–119

Exner C (1980) Das Kristallin östlich der Katschbergzone. Mitt Österr Geol Ges 71–72, 1978/1979: 167–189

Finger F, Steyrer HP (1988) Granite types in the Hohe Tauern (Eastern Alps, Austria) – some aspects on their correlation to Variscan plate tectonic processes. Geodin Acta 2: 75–87

Finger F, Steyrer HP (1990) I-type granitoids as indicators of a late Palaeozoic convergent ocean-continent margin along the southern flank of the central European Variscan orogen. Geology 18: 1207–1210

Frank W (1969) Neue Forschungen im Umkreis der Glocknergruppe. Wiss Alpenver Hefte 21: 95–108

Franz G, Mosbrugger V, Menge R (1990) Pteridophyll leaf fragments from the Maurertal/Großvenediger. Mitt Österr Mineral Ges 135: 24–25

Franz G, Mosbrugger V, Menge R (1991) Carbo-Permian pteridophyll leaf fragments from an amphibolite facies basement, Tauern Window, Austria. Terra Nova 3: 137–141

Frasl G (1958) Zur Seriengliederung der Schieferhülle in den Mittleren Hohen Tauern. Jahrb Geol Bundesanst 101: 323–472

Frasl F, Frank W (1966) Zur Mineralogie und Geologie des Landes Salzburg und der Tauern. Aufschluß 15: 30–58

Frisch W, Neubauer F (1989) Pre-Alpine terranes and tectonic zoning in the Eastern Alps. Geol Soc Am Spec Publ 230: 91–100

Frisch W, Raab D (1987) Early Palaeozoic back-arc and island arc settings in greenstone sequences of the central Tauern window (Eastern Alps). Jahrb Geol Bundesanst 129: 545–566

Frisch W, Neubauer F, Satir M (1984) Concepts of the evolution of the Austroalpine basement complex (Eastern Alps) during the Caledonian-Variscan cycle. Geol Rundsch 73(1): 47–68

Gilg HA, Höll R, Kupferschmied MP, Reitz E, Stärk H, Weber-Diefenbach K (1989) Die Basisschieferfolge der Habachformation im Felber- und Amertal (Tauernfenster, Salzburg). Mitt Österr Geol Ges 81: 65–91

Grundmann G (1989) Metamorphic evolution of the Habach formation. A review. Mitt Österr Geol Ges 81: 133–149

Holub B, Marschallinger R (1989) Die Zentralgneise im Hochalm-Ankogel-Massiv (östliches Tauernfenster) Teil I: Petrographische Gliederung und Intrusionsfolge. Mitt Österr Geol Ges 81: 5–31

Höck V, Pestal G (1990) Exkursion E4: Habachformation, Altkristallin im Habachtal und Felbertal. Mitt Österr Mineral Ges 135: 129–137

Höll R (1975) Die Scheelitlagerstätte Felbertal und der Vergleich mit anderen Scheelitvorkommen in den Ostalpen. Abh Bayer Akad Wiss math-naturwiss Kl 157 A: 1–114

Kraiger H (1987) Geologie, Petrographie und Geochemie der Habachformation am Beispiel des Unterfahrungsstollens der Scheelitlagerstätte Felbertal, Salzburg. PhD Thesis, Univ Salzburg, 280 pp

Kraiger H (1989) Die Habachformation – ein Produkt ozeanischer und kontinentaler Kruste. Mitt Österr Geol Ges 81: 47–64

Lammerer B (1986) Das Autochthon im westlichen Tauernfenster. Jahrb Geol Bundesanst 129: 51–67

Marschallinger R, Holub B (1991) Die Zentralgneise im Hochalm-Ankogel-Massiv (östliches Tauernfenster) Teil II. Geochemische und zirkontypologische Charakteristik. Mitt Österr Geol Ges 82: 19–48
Matte P (1986) Tectonics and Plate Tectonics Model for the Variscan Belt of Europe. Tectonophysics 126: 329–374
Mercolli I, Oberhänsli R (1988) Variscan tectonic evolution in the Central Alps: a working hypothesis. Schweiz Mineral Petrogr Mitt 68: 491–500
Pahr A (1980) Das Semmering- und Wechselsystem: In: Oberhauser R (ed) Der geologische Aufbau Österreichs. Springer, Berlin Heidelberg New York Wien, pp 315–320
Pearce JA (1980) Geochemical evidence for the genesis and eruptive setting of lavas from Tethyan ophiolites. In: Panayiotou A (ed) Ophiolites. Cyprus 1979, Ministry of Agriculture and Natural Resources, Geol Surv Dep Nicosia, pp 261–272
Pearce JA (1982) Trace element characteristics of lavas from destructive plate boundaries. In: Thorpe RS (ed) Andesites. Wiley, Chichester, pp 525–547
Pearce JA (1983) Role in the sub-continental lithosphere in magma genesis at active continental margins. In: Hawkesworth CJ, Norry HJ (eds) Continental basalts and mantle xenoliths. Shiva, Nantwich, pp 230–249
Pearce JA, Cann JR (1973) Tectonic setting of basic volcanic rocks determined using trace element analyses. Earth Planet Sci Lett 19: 290–300
Pearce JA, Norry MJ (1979) Petrogenetic implications of Ti, Zr, Y and Nb variations in volcanic rocks. Contrib Mineral Petrol 69: 33–47
Pearce JA, Harris NBW, Tindle AG (1984a) Trace element discrimination diagrams for the tectonic interpretation of granitic rocks. J Petrol 25: 956–983
Pearce JA, Lippard SJ, Roberts S (1984b) Characteristics and tectonic significance of supra-subduction zones ophiolites. In: Kokelaar BP, Howells MF (eds) Marginal basin geology. Geol Soc Lond Spec Publ No 16, pp 77–94
Pestal G (1983) Beitrag zur Kenntnis der Geologie in den Mittleren Hohen Tauern im Bereich des Amer- und Felbertales (Pinzgau, Salzburg). PhD Thesis Univ Wien, 117 pp
Petrakakis K (19877) Zur Geologie des Stubach-Ultramafitit-Komplexes. Mitt Ges Geol Bergbaustud Österr 24: 47–57
Petrakakis K (1978) Der Stubachtal-Ultramafitit-Komplex (Salzburg, Österreich). Tschermaks Mineral Petrogr Mitt 25: 1–32
Quadt A von (1984) Geochemische und U-Pb-Untersuchungen der Unteren Schieferhülle (Ostalpen). Fortschr Mineral 62: 191–193
Quadt A von (1985) Geochronologische, geochemische und isotopenchemische Untersuchungen an Beispielen der Habachformation, der Scheelitlagerstätte und des umgebenden Altkristallins im Felbertal (Land Salzburg). PhD Thesis, ETH Zürich, 241 pp
Raumer JF von (1981) Variscan Events in the Alpine Region. Geol Mijnbouw 60: 67–80
Raumer JF von (1986) The Hercynian basement in the Helvetic realm, Western and Central Alps. Mem Soc Geol Ital 29: 57–69
Raumer JF von (1987) Mont-Blanc- und Aiguilles-Rouges-Massiv, Zeugen variszischer Krustenbildung in den Westalpen. Neues Jahrb Geol Paläontol Monatsh 1987: 493–512
Reitz E, Höll R (1988) Jungproterozoische Mikrofossilien aus der Habachformation in den mittleren Hohen Tauern und dem nordostbayerischen Grundgebirge. Jahrb Geol Bundesanst 131: 329–340
Seemann R, Koller F (1989) Petrographic and geochemical investigations in the Knappenwand Area, Habach formation (Tauern Window, Austria). Mitt Österr Geol Ges 81: 109–122
Stadlmann T (1990) Geologie, Petrographie und Geochemie der Storzformation im Typusgebiet zwischen Mur- und Pöllatal (östliche Hohe Tauern, Salzburg/Kärnten). Diploma Thesis, Univ Salzburg, 106 pp
Steyrer HP (1983) Die Habachformation der Typlokalität zwischen äußerem Habachtal und Untersulzbachtal (Pinzgau/Salzburg). Mitt Österr Geol Ges 76: 69–100
Steyrer HP, Höck V (1985) Geochemistry of the metabasites in the Habach formation (Salzburg, Hohe Tauern, Austria). A preliminary report. Ofioliti 10: 441–456
Vavra G (1989) Die Entwicklung des penninischen Grundgebirges im östlichen und zentralen Tauernfenster der Ostalpen – Geochemie, Zirkonmorphologie, U/Pb-Radiometrie. Tübinger Geowiss Arb Reihe A Paläontol Stratogr 6: 150 S
Vavra G, Frisch W (1989) Pre-Variscan back-arc and island-arc magmatism in the Tauern Window (Eastern Alps). Tectonophysics 169: 271–280

The Zentralgneise of the Tauern Window (Eastern Alps): Insight into an Intra-Alpine Variscan Batholith

F. Finger, G. Frasl, B. Haunschmid, H. Lettner, A. von Quadt, A. Schermaier, A.O. Schindlmayr, and H. P. Steyrer

Abstract

Geological, petrographic and geochemical data of the "Zentralgneise" (central gneisses) of the Tauern window (Eastern Alps) are presented. The rocks are interpreted as metamorphosed formation of late Palaeozoic granitoids, which once formed a batholitic terrain on the southern flank of the Variscan orogen. The primary structures of this batholitic terrain and its plutonic evolution are investigated. Field evidence suggests that the plutonic activity set in with widespread anatexis of pre-existing continental crust in the late Viseau to early Namurian, during which a series of K_2O-rich magmas with dominantly granitic, but also syenitic and monzonitic compositions were produced. Subsequently, large volumes of calc-alkaline I-type granitoids intruded (dominantly granodiorites and tonalites). Most of the Zentralgneise represent such calc-alkaline rocks. A late generation of leucocratic granites approaches within-plate (A-type) granites in composition. The change of magmatypes with time is accompanied by a rapid uplift of the batholitic terrain. Plate tectonic implications for the evolution of the Variscan orogen are discussed.

1 Introduction

This paper attempts to describe the present state of knowledge about a voluminous metamorphic assembly of late Palaeozoic granitoid rocks in the Penninic Tauern window, the "Zentralgneise". In focus is mainly the information, which is relevant for understanding the primary state and origin of these granitoid rocks. The metamorphic and tectonic evolution during the Alpine orogeny is handled only very briefly. For more detailed information on this latter subject see Frank et al. (1987 a) and further references therein.

The descriptions presented in the following sections are, on one hand, based on a careful review of published data. On the other hand, the authors were able to make use of several, new and as yet unpublished, data of their own, most of which were gathered during the past year in the frame of a research project of the Geological Institute of the Salzburg University, supervised by Prof. G. Frasl.

2 Geological Environment of the Zentralgneise

The Zentralgneise are part of the Hohe Tauern mountain range of Austria, where a deep structural level of the Alpine orogen crops out as a tectonic window below the voluminous Austro-Alpine nappe pile (Fig. 1). Approximately half of the Tauern window is made up of Permo-Mesozoic formations, the other half consists of pre-Permian rocks. The Zentralgneise form, in turn, approximately two thirds of these pre-Permian rocks.

Most of the Zentralgneise are exposed as large antiforms, which represent the very deepest structural units of the Tauern window, and dome up along its central axis. Apart from these large cupolas or cores, there are also some minor lens-shaped or nappe-like bodies of Zentralgneise intercalated between pre-existing rocks.

Generally, the Zentralgneise are surrounded by older rocks ("Ältere Schieferhülle"), some of which seem to have been, however, metamorphosed only during the Alpine orogeny (Habach formation; Frasl 1958). Others, e. g. the rocks of the Riffl nappe, represent an old metamorphic basement ("Altkristallin" formation; Frasl 1958). The boundaries between the Zentralgneise and the "Ältere Schieferhülle" are often tectonic and constitute a preferred location for Alpine shear zones. Nevertheless, there are also some areas, where the original relations between the magmatic protoliths of the Zentralgneise and their pre-existing roof rocks are preserved. In these cases, one can often observe a migmatitic aureole, but there are also several discordant intrusive contacts.

On the other hand, Permo-Triassic or Jurassic transgressive sediments are locally preserved on top

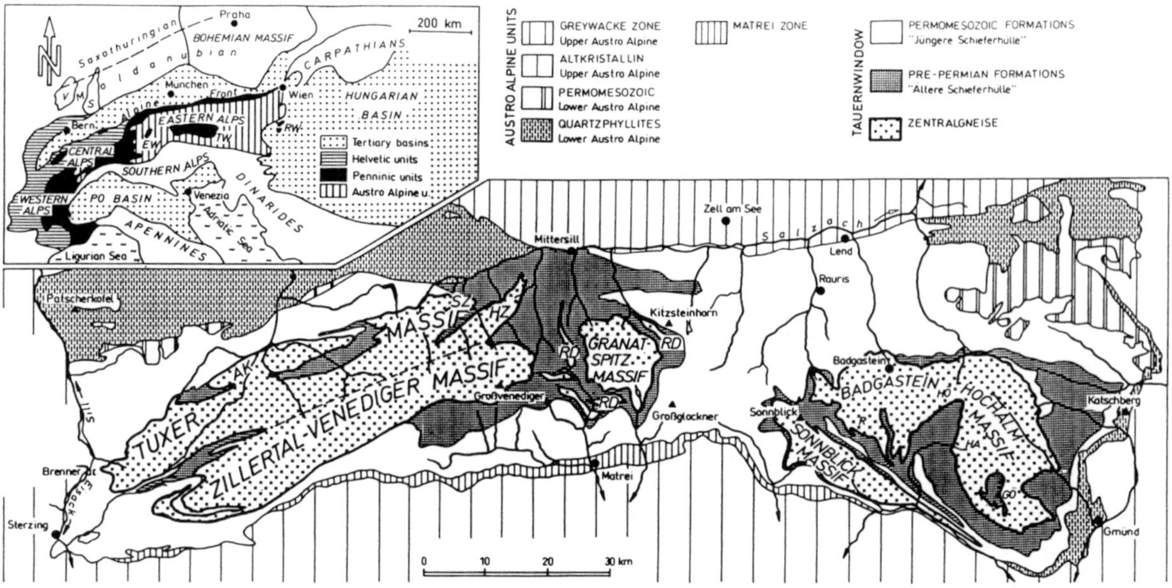

Fig. 1. Geological sketch map of the Tauern window featuring the different massifs, cores and nappes of Zentralgneise (*AK* Ahorn core; *GÖ* Gößgraben core; *HA* Hochalm-Ankogel core; *HÖ* Hölltor-Rotgülden core; *HZ* "Habachzunge"; *R* Romate nappe; *RD* Riffl nappe; *SZ* "Sulzbachzungen"; **inset.** ***EW*** Engadin window; *RW* Rechnitz window; *S* Black Forest; *TW* Tauern window; *V* Vosges)

of the Zentralgneise, which means that parts of the plutonic masses must have already been unroofed and eroded before the Alpine orogeny set in.

Together with the other rocks of the Tauern window, the Zentralgneise underwent greenschist- to amphibolite-facies metamorphism during Alpine times. This metamorphic overprint caused a penetrative recrystallization of the primary granitoid mineralogy and led to the growth of new metamorphic parageneses such as albite/oligoclase + epidote instead of earlier An-rich plagioclase ("filled feldspars"), or TiO_2-poor biotite + rutile ("Sagenit") instead of primary biotite (see e.g. Raith 1971). Amphibole relics are often replaced by metamorphic biotite. Chlorite and phengite are also very common newly formed minerals in the Zentralgneise, often clearly replacing earlier mafic phases.

Due to the influence of Alpine deformation, most Zentralgneise show a more or less pronounced gneissic texture. In some places no rest of a magmatic fabric is preserved.

Despite the fact that most of the Mesozoic rock formations of the Tauern window are regarded as "Penninic" with reference to their position within the Alpine trough system, it is widely agreed that the Zentralgneise were situated in the Mesozoic more to the north, in or closer to the Helvetic realm. Probably, they occupied a comparable palaeogeographic position in the Alpine belt as the external granitoid massifs of Switzerland.

The reconstruction of the palaeogeographic position of the Zentralgneise in the Variscan fold belt is still more difficult. An important fact is that the Zentralgneise occur in close primary relation with a preexisting low- or unmetamorphosed pelitic-volcanic sequence of presumably early Palaeozoic age (Habach formation *sensu* Frasl 1958). This indicates that the plutonic protoliths of the Zentralgneise formed in an area south of the highly metamorphic Moldanubian core region of the Variscan orogen. In this respect, the initial environment of the Zentralgneise is more comparable to the Pelvoux-Belledonne external massif of the Western Alps than to the Aar Massif e.g. which displays more lithological similarities to the Moldanubian region (von Raumer 1988).

3 The Zentralgneise of the Western Tauern Window

The western part of the Tauern window exposes three main cores of Zentralgneise, the *Ahorn Massif*, the *Tuxer Massif* and the *Zillertal-Venediger Massif* (Figs. 1 and 2).

The Ahorn Massif (Thiele 1974; see further references therein) dips below the pre-Mesozoic metasediments and -volcanics of the *Schönach synform*

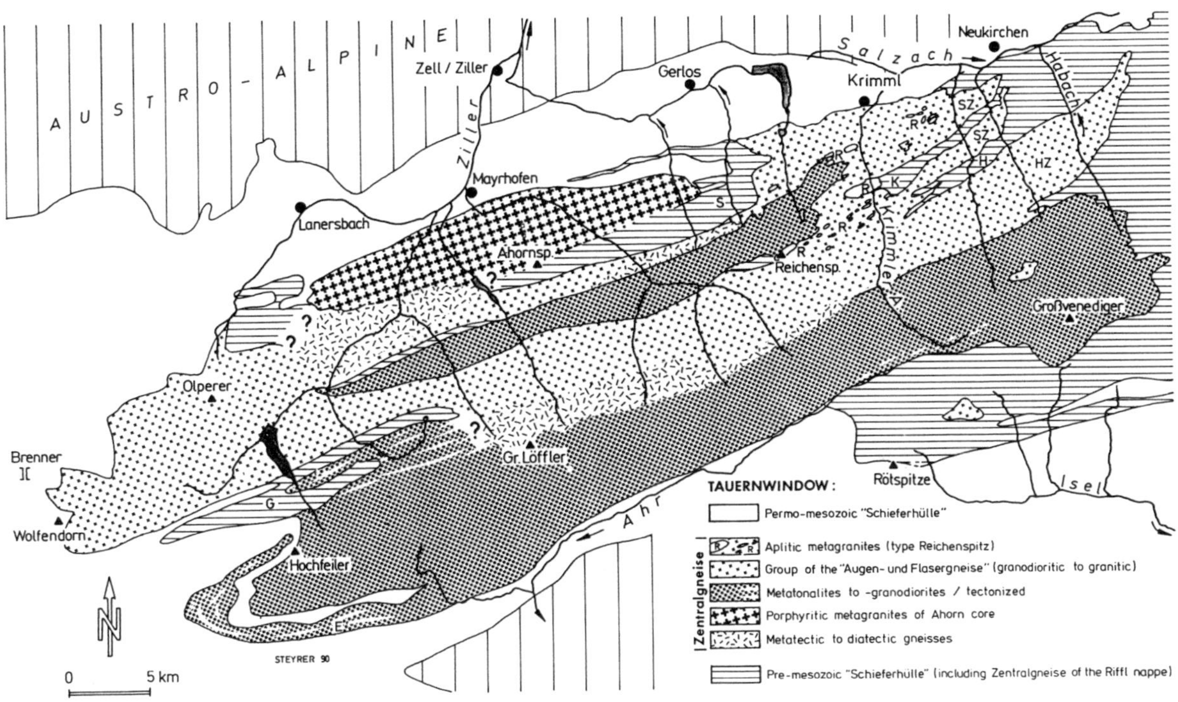

Fig. 2. Geological sketch map of the pre-Alpine basement in the western Tauern window (compiled by A. Schermaier, A. O. Schindlmayr and W. Arming according to mapping results of Frasl 1953; Kupka 1953; Schmidegg 1961; Morteani 1971; Raith et al. 1978; Karl and Schmidegg 1979; Frisch 1980; Lammerer 1986 and own field observations). *E* "Eisbrugglamelle"; *G* Greiner synform; *H* Habach synform; *HZ* "Habachzunge"; *K* Knappenwand synform; *R* stocks of Reichenspitz granite; *S* Schönach synform; *SZ* "Sulzbachzungen"

(Fig. 2), and consists mainly of a biotite-rich, porphyritic Augen gneiss with zoned K-feldspar idiocrysts up to 5 cm.

The Tuxer Massif exposes several types of coarse-grained light metagranites and -granodiorites (*"Augen- und Flasergneise" sensu* Karl 1959) and a thick layer of tonalitic gneiss (Morteani 1971). Furthermore, small stocks and dykes of *aplitic granites* (*type Reichenspitze;* Karl 1961) with sharp intrusive contacts to the country rocks were mapped.

The Zillertal-Venediger Massif, which is separated from the Tuxer Massif by a fault zone (Karl and Schmidegg 1979) is made up dominantly of a medium-grained *metatonalitic to metagranodioritic* lithology (Christa 1931; Karl 1959; Schmidegg 1961). Locally the tonalitic variants are associated with small bodies of diorites and gabbros and often contain dark cognate enclaves. There are usually continuous transitions from tonalite into granodiorites of similar grain-size.

In the Zillertal Alps (area of the Großer Löffler; see Fig. 2) the tonalites and granodiorites seem to be closely associated with meta- to diatectic gneisses and "Schollen" migmatites, continuous transitions have been observed (Morteani 1971; Raase 1972; Lammerer 1986). Towards the eastern end of the Zillertal-Venediger core, the contacts to the country rocks become increasingly sharp, probably due to a higher emplacement level. In the inner Krimmler Ache Valley, the tonalite intruded K-feldspar- and biotite-rich anatexites and Augen gneisses. These rocks obviously represent an early generation of Zentralgneise.

4 The Zentralgneise of the Central Tauern Window

The *Granatspitz core* forms the lowest tectonic unit (Figs. 1 and 3). It consists mainly of a leucocratic, medium- to coarse-grained, more or less porphyritic two-mica metagranite with pseudomorphs after cordierite. The metagranite usually displays tectonic contacts to its country rocks, which are mainly banded amphibolites and biotite-schists. Intrusive contacts, which are locally preserved (Cornelius 1930), indicate, however, that the cover series of the Granatspitz core represents the old (primary) country rocks of the metagranite.

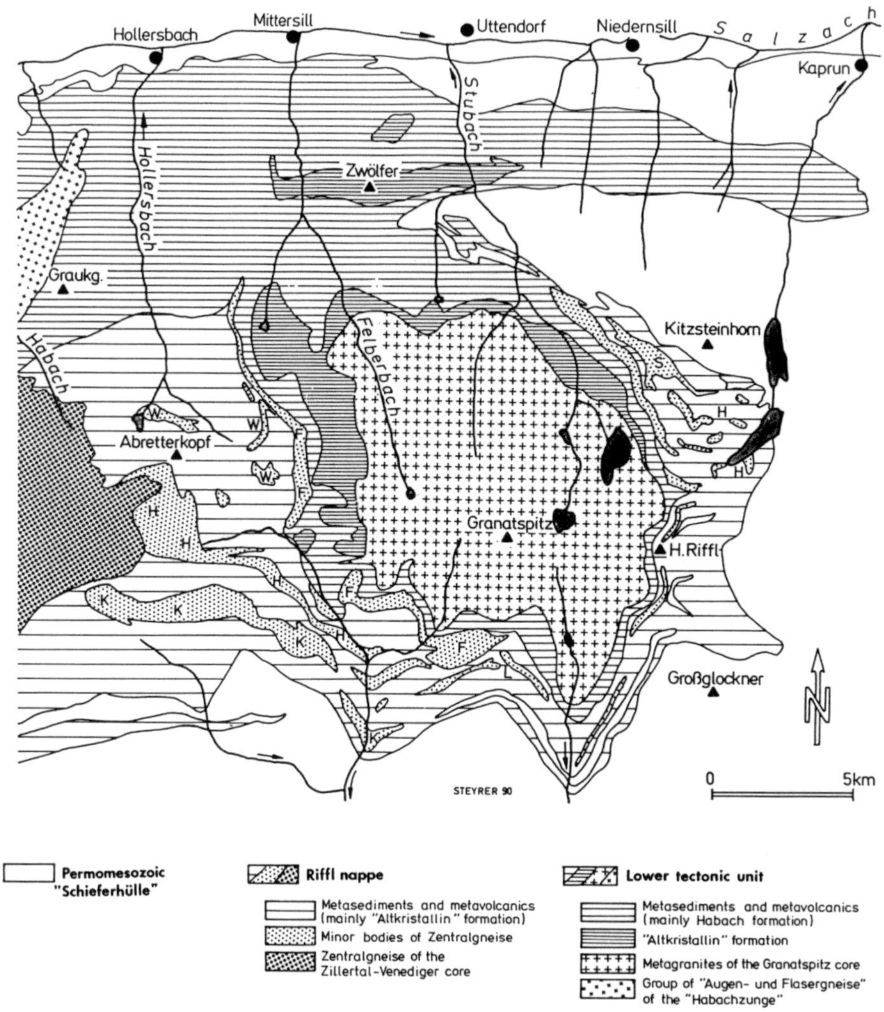

Fig. 3. Geological (and tectonic) sketch map of the pre-Alpine basement in the central Tauern window (compiled by A. Schermaier according to mapping results of Fuchs 1958; Frasl and Frank 1966; Frank et al. 1987 b and own field observations). Tectonic interpretation mainly according to Frisch (1980). Granitoids of the Riffl nappe: *F* Felbertauern Augen gneiss; *H* Hochweißenfeld gneiss; *K* Knorrkogel gneiss; *L* Luckenkogel gneiss; *S* Scharkogel gneiss; *W* Weißeneck-Dichtensee gneiss

The (par)autochthonous cover of the Granatspitz core is overlain by a series of anatectic-migmatitic biotite schists, gneisses and amphibolites (*Riffl nappe* according to Cornelius 1930; Cornelius and Clar 1939; *"Serie der Alten Gneise"* after Schmidegg 1961; Frasl and Frank 1966). This, up to several kilometre thick rock pile, is regarded as allochthonous, but the time period of the overthrusting event (Alpine or Variscan), as well as the internal tectonic structures have been the focus of discussion for several decades (Frisch 1977, 1980; see references therein).

Within the migmatitic series of the Riffl nappe several bodies of Zentralgneise have been mapped (see Fuchs 1958; Frank et al. 1987 b; note Fig. 3). These can be briefly described as follows:

The *Felbertauern Augen gneiss* (Cornelius 1942 a; Fuchs 1958) is a light, coarse-grained metagranite. It contains K-feldspar phenocrysts up to a size of several centimetres, which are usually deformed to thin eyes.

The *Hochweißenfeld gneiss* (Cornelius and Clar 1939; Fuchs 1958; Schermaier 1991) is a medium- to coarse-grained metagranite with a relatively high biotite content (up to 20 %) and 1–2 cm large, grey porphyritic K-feldspars. It is situated within an aureole of metatectic and diatectic gneisses and amphibolites. Because of the generally continous transitions into these migmatites, the Hochweißenfeld gneiss is regarded as deep-seated magma which formed "in situ" by crustal anatexis. At the eastern rim of the Venediger Massif, the anatectic structures of the Riffl nappe are cross-cut by several leucotonalitic and granodioritic dykes, which correspond in composition to, and are therefore thought to be derived from the plutonic masses of the Zillertal-Venediger core (Schermaier 1991). The leucocratic granodioritic gneiss of the *Weißeneck-Dichtensee* area (Fuchs 1958) is also interpreted as a (highly felsic) variant of the Zillertal-Venediger granitoid suite. It shows agmatic contacts to the migmatites of the Riffl nappe, indicating a high intrusive level. Consequently, the widespread deep-crustal anatexis of the Riffl nappe should be signifi-

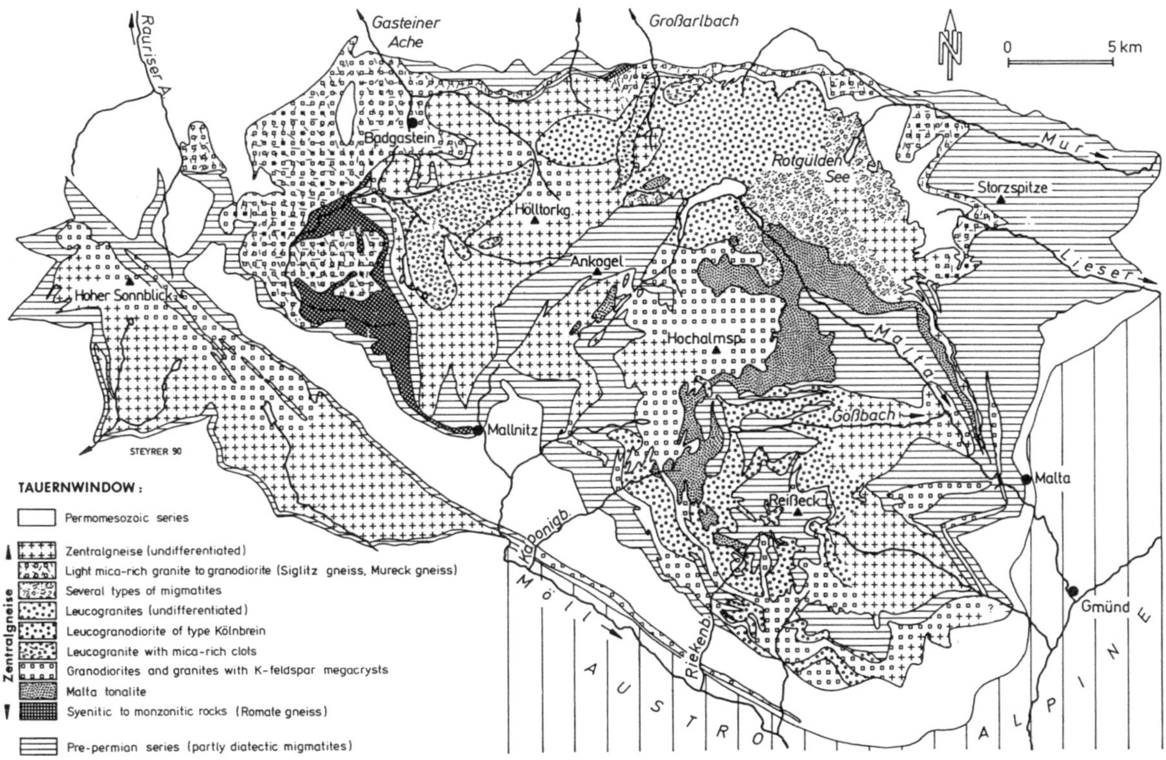

Fig. 4. Geological sketch map of the pre-Alpine basement in the eastern Tauern window (compiled by B. Haunschmid according to mapping results of Angel and Staber 1952; Exner 1957, 1964, 1983; Cliff et al. 1971; Alber 1976; Marschallinger and Holub 1990 a and own field observations). Note that a lithological subdivision of different types of Zentralgneise is not possible everywhere, because several areas are still insufficently mapped. Such areas of Zentralgneise are outlined as "undifferentiated"

cantly older than the tonalitic to granodioritic plutonic series of the Zillertal-Venediger Massif, which intruded the migmatites only after they were uplifted and cooled.

The *Knorrkogel gneiss* (Cornelius 1942 b) and its possible equivalent, the *Luckenkogel gneiss* (Benedict 1952) also show continuous transitions into their country rocks and might therefore be, at least partly, of a similar origin to the Hochweißenfeld gneiss.

The light *Scharkogel gneiss* (Cornelius and Clar 1939) contains phenocrysts of K-feldspar also. The gneiss was interpreted by Frank (1969) as a metavolcanic rock.

5 The Zentralgneise of the Eastern Tauern Window

The Zentralgneise of the eastern Tauern window crop out in the *Sonnblick Massif* and the *Badgastein-Hochalm Massif* (Figs. 1, 4). The two massifs are separated by a synform of mainly Mesozoic series (synform of Mallnitz; Exner 1964).

The Sonnblick Massif is built up mainly of a coarse porphyritic metagranite (Exner 1964).

The pre-Permian wall rocks of the Sonnblick Massif have been broadly correlated with the Habach formation of the central Tauern window by Alber (1976).

The Badgastein-Hochalm Massif forms a somewhat assymmetric dome with its deepest units outcropping in the southeast (Gößbach Valley). The massif contains about 80 % metagranitoids and 20 % older roof rocks. The latter occur in synforms, which subdivide the Badgastein-Hochalm Massif in several subcores: the Badgastein core in the west, the Hölltor-Rotgülden core with the Hochalm core in the northern and central part, and the Göß core in the southeast.

A classification of the metagranitoid rocks of the eastern Tauern window is probably more complex than in the western Tauern window. There is a pronounced compositional and structural diversity of the granitoid rocks, and the area visibly consists of a great number of rather small individual plutons. The relative age of different types of intrusions is only locally clarified (Frasl et al. 1987; Marschallinger and Holub

1990b). Roughly, the following five main plutonic lithologies can be distinguished:

1. A distinct type of metatonalite (*Malta tonalite,* Marschallinger and Holub 1990a) extends as a relatively large, narrow body from the Malta valley in the north to the western Reißeck mountains in the south. This medium-grained biotite-tonalite often displays a rather inhomogenous, more or less migmatitic internal texture ("Schlieren" texture). According to field evidence, the metatonalite belongs to one of the earliest intrusions in the area (Cliff et al. 1971). The rock intruded a series of diatectic migmatites along the Malta valley (Frasl et al. 1987).

2. A coarse syenitic to monzonitic metagranitoid (*Romate gneiss;* Angel and Staber 1952), with typically zoned perthitic K-feldspars, forms a body running N-S between Badgastein and Mallnitz. Amphibole is the dominant mafic mineral. Isolated enclaves in the Hochalm area suggest, that the Romate gneiss represents an intrusion which is older than the tonalitic body mentioned above (Marschallinger and Holub 1990b).

3. Porphyritic biotite-metagranodiorites and -metagranites with K-feldspar megacrysts are abundant over the entire region. A first subtype of mainly granodioritic composition (3a) builds a sizeable pluton around the Hochalmspitze, termed *"Hochalmporphyrgranit"* by Marschallinger and Holub (1990a), and a pluton south of the Reißeck (see also Angel and Staber 1952; Exner 1957; Cliff et al. 1971). Field evidence suggests that both bodies intruded the surrounding Malta tonalite. Zircon dating (Cliff 1981) indicates, however, that there was no long time interval between these younger intrusions and the Malta tonalite (see Sect. 6).

Other, smaller bodies of porphyritic biotite-metagranites (3b) occur just south of the Ankogel and are older than the metatonalite (see Cliff et al. 1971). These rocks have been summarized as *"Großelendflasergneis"* by Marschallinger and Holub (1990a).

Further, small bodies of a more fine-grained porphyritic metagranite (3c) occur near the southern margin of the Hochalm massif between the Kaponigbach Valley and Rieckenbach Valley. They appear to be younger than the porphyritic biotite-granodiorite pluton of the Hochalmspitze (Cliff et al. 1971).

Another porphyritic type of mainly granodioritic composition (3d) locally grading, however, into tonalitic and granitic variants, crops out around the eastern Gößbach Valley to the Malta Valley. These *Göß granitoids* (Marschallinger and Holub 1990a) are fine- to medium-grained and often rather poor in K-feldspar megacrysts (around 2 cm in size) when compared to the porphyritic granitoids mentioned already. They show intrusive contacts into the surrounding diatectic migmatites, which separate them from the granitoids of the Hochalm core.

An isolated lens of a porphyritic metagranite (3e) occurring within the old gneisses and migmatites of the tectonically higher Storz nappe just NNE of Malta (Exner 1983) has been interpreted by Vavra (1989) as a subvolcanic intrusion of Permian age (see Sect. 6).

4. A group of leucocratic metagranites and metagranodiorites is widespread in the Badgastein-Hochalm area. They appear to be generally younger than the porphyritic biotite-granitoids of group 3. A special type of leucocratic, fine- to medium-grained metagranite with typical mica clots (small xenoliths) has been mapped by Angel and Staber (1952) as *"Forellengneis"* in the Hölltor-Rotgülden-Hochalm core in the north and west of the Ankogel (4a). This type of "hybrid leucogranite" is somewhat older than another type (4b) of a mostly massive leucocratic metagranodiorite, which lacks such mica clots (Exner 1983; *Kölnbrein leucogranite* after Marschallinger and Holub 1990a). Both types of leucogranites are, however, likely to be closely related intrusions (Marschallinger and Holub 1990a). The Kölnbrein leucogranite, which occurs mainly N and NE of the Ankogel from the Hölltor area up to the Rotgülden Valley, can locally cross-cut older lithologies in such a penetrative way that "Schollen" migmatites are produced, e.g. along the Malta Valley and around the Rotgülden Lake (Exner 1983). In the Badgastein area, a similar-looking leucogranite (4c) shows gradual transitions into a coarse porphyritic metagranite with K-feldspars up to 10 cm (Riesenaugen gneiss, Exner 1957). Another type of fine-grained leucocratic metagranite (4d) forms small bodies immediately south of the Hochalmspitze and is thought to be one of the youngest intrusions among the Zentralgneise of this area (Cliff et al. 1971). In the central Gößbach Valley (approximately 8 km SE of the Hochalmspitze) a more coarse-grained equigranular leucogranite (4e) crops out. The extension of this body toward the east has not yet been mapped and its relationship to other bodies of Zentralgneise is not yet resolved.

5. Light mica-rich metagranites to metagranodiorites. The area of Badgastein is volumetrically dominated by such a lithology (*Siglitz gneiss;* Exner 1957). North of Badgastein, toward the east, this type forms a lamella near the margin of the Badgastein-Hochalm Massif. Toward the east, in about the same high structural level, another narrow lamella of a similar looking metagranitoid continues along the northern margin of the Hölltor-Rotgülden-Hochalm core (*Mureck*

Table 1. Compilation of Rb-Sr whole rock isochrons and U-Pb zircon ages of Zentralgneise (numbers in Ma)

	Rock Type	Location	Zircon	Rb-Sr whole rock	References
Badgastein – Hochalm Massif	Malta tonalite	4.8 km SSW Hochalmsp., upper Kaponig valley	314 ± 7		Cliff (1981)
	Porphyritic meta-granodiorite	4.2 km SSE Hochalmsp., upper Gößbach valley	313 ± 10		Cliff (1981)
	Porphyritic metagranite	2.5 km NNW Malta	256 + 58, –34		Vavra (1989)
	Metagranites and granodiorites	Area S of Hochalm and Reißeck mountains		273 ± 9	Hawkesworth (1976)
	Metaleucogranite	Between Gößbach and Kaponig valley		252 ± 6	Cliff et al. (1971)
	Metagranites with feldspar megacrysts	Between Badgastein, and Mallnitz		251 ± 11	Lambert (1964)
Granatspitz Massif	Various metagranitoids	Central section		325 ± 33	Cliff (1977)
	Various metagranitoids	Southern section		226 ± 26	Cliff (1977)
	Metagranites	Central section	333 ± 28		von Quadt (unpubl.)
	Metagranitoids	Southern and central section	320 – 330		Cliff (1981)
Riffl nappe	Hochweissenfeld metagranite	S Abretterkopf	308 ± 12		von Quadt (unpubl.)
Tuxer Massif	"Augen- und Flasergneise"	"Sulzbachzungen" and "Habachzunge"		257 ± 16	Jäger et al. (1969)
	"Augen- und Flasergneise"	Stollen beyond Olperer		245 ± 21	Satir (1974)
	Aplites, type Reichenspitz	"Sulzbachzungen" E Reichenspitz Krimmler Ache valley		292 ± 6	Besang et al. (1968)
Zillertal-Venediger Massif	Migmatites and gneisses	8 km W Gr. Löffler		283 ± 28	Satir and Morteani (1982)

Note that the Rb-Sr ages are all calculated with $\lambda = 1.42 \times 10^{-11} a^{-1}$

gneiss, Becke 1908). Both the Siglitz and Mureck gneiss, are very schistous and show a layered texture with numerous concordant, dark, fine-grained lens-shaped inclusions. Both can also contain porphyritic K-feldspars, whose number and size vary, however, considerably.

6 Radiometric Data

Several Rb-Sr whole-rock (WR) isochrons are published for various types of Zentralgneise (Table 1) and range from ca. 330 Ma to 220 Ma in age. At least for some of these isochrons (e.g. the "young" 226 Ma isochron for the southern Granatspitz Massif), a metamorphic disturbance of the Rb-Sr WR system is evident (Cliff 1981).

However, there are other Rb-Sr WR isochrons, which provide much more reliable formation ages, e.g. the 325 Ma isochron for the central Granatspitz Massif, which could be well confirmed by zircon dating (see Cliff 1981), and the 292 Ma isochron for aplitic rocks of the Venediger area. The precise significance of the single isochrons of Table 1 remains in most cases unclear.

Considering the comprehensive geochronological work of Cliff et al. (1971) and Cliff (1977, 1981) on various lithologies of Zentralgneise, it must be concluded that zircon dating is generally much more likely to provide "magmatic" ages than the Rb-Sr WR method. Five of the six zircon ages, which have been as yet reported for Zentralgneise (Table 1), fall between 335 and 300 Ma. This suggests that the bulk of Zentralgneise are Upper Carboniferous intrusions. However, the possibility that there are also some Permian plutons can, of course, not be ruled out with these few data. Moreover, there is the 256 Ma zircon age of Vavra (1989), which is, however, somewhat crucial because it obviously includes a very large uncertainty (see Table 1). Considering that Permian plutons are also present in other parts of the Alpine basement (see e.g. Gebauer, this Vol.), it would not be surprising to find them in the Tauern window as well.

7 Primary Modal Compositions of the Zentralgneise

As most of the Zentralgneise experienced significant mineralogical and structural alterations during the Alpine orogeny (see Sect. 2), and also because some of the rocks are rather coarse-grained, it is often a problem to determine the primary modal compositions by conventional microscopy of thin sections. Therefore, we prefered to calculate primary modal compositions on the basis of major element analyses (if possible of large low-strain samples), using the "Mesonorm" computer program of Mielke and Winkler (1979). Several tests showed that the method closely approaches those values which could be obtained by microscopy in case of "unproblematic" low deformed and low metamorphosed even grained variants of Zentralgneise.

Streckeisen nomenclature: in both eastern and western Tauern window, the bulk of Zentralgneise scatters within a typical calc-alkaline range between tonalitic and granitic compositions.

The Zentralgneise of the Zillertal-Venediger Massif feature (with only few exceptions of early stage K-feldspar-rich and quartz-poor granitoids) a calc-alkaline trondhjemitic trend in the sense of Lameyre and Bowden (1982) with tonalites as the dominant rock type, but also with granodiorites and even some few granitic end members (Fig. 5 a).

Conversely, the *Tuxer core* consists predominantly of granodiorites and K-feldspar-poor granites. Many of the granitoids of the Tuxer core fit the calc-alkaline trondhjemitic trend of the Zillertal-Venediger core but there is also a considerable number of granites, especially in the eastern part of the Tuxer core, which clearly plot below this trend.

Only a few analyses are available from the *Ahorn core.* These lie in the range of the old Augen gneisses of the Zillertal-Venediger core.

In the *eastern Tauern window* (Fig. 5 c), the Zentralgneise follow mainly a calc-alkaline granodioritic trend (Lameyre and Bowden 1982) with granodiorite as the dominant rock type, but also with tonalitic and granitic end members. However, a second overlapping field from syenitic to granitic is spanned by the early stage intrusions of the Badgastein-Hochalm Massif (Romate gneiss, "Großelendflasergneis"; see Sect. 5).

In the *central Tauern window,* the variability of the Zentralgneise is much less extended (Fig. 5 b). The late stage intrusions of the Weißeneck-Dichtensee area are quartz-rich granodiorites and they thus fit the trondhjemitic trend of the Zillertal-Venediger Massif. This accords with the concept of Schermaier (1991) who interpreted the Zentralgneise of the Weisseneck-Dichtensee type as intrusions derived from the Zillertal-Venediger core (see Sect. 4). Most other granitoids of the central Tauern window have granitic composition. The Hochweißenfeld gneiss, which formed early in the evolution history of the Zentralgneise displays a monzonitic tendency.

Mafic components: thin section observations as well as "Mesonorm" calculations reveal that biotite is by far the most common primary mafic constituent. It is present in all of the Zentralgneise. Many of the Zentralgneise (ca. 20–30 %) have, however, only rather low biotite contents and must therefore be described as leucocratic rocks in the sense of Streckeisen (1967).

On the other hand, there is hardly any evidence for the presence of large amounts of primary muscovite and highly peraluminous phases in general. An important exception is the Granatspitz Massif, where a primary content of muscovite+cordierite in the range of 3–5 % can be estimated from calculations and observations.

Approximately 10 % of the Zentralgneise provide evidence for the presence of primary amphibole. A primary amphibole content of ca. 5–10 % can be envisaged for the tonalitic end members of the Zillertal-Venediger Massif. In the case of the Romate gneisses, the primary amphibole content can be estimated at 10–20 %. Small amounts of primary amphibole might also have occurred in the Hochweißenfeld gneiss of the central Tauern window and in the gneisses of the Ahorn core. Strangely, there is hardly any evidence for the presence of amphibole in the tonalites of the eastern Tauern window.

8 Chemical Data

8.1 *Major and Trace Elements*

A lot of chemical work has been carried out during the past decades on the Zentralgneise (e.g. Angel 1954; Karl 1959; Gundlach et al. 1967; Morteani 1971; Prosser 1975; Vavra 1989; Marschallinger and Holub 1990 a). An overview on the major and trace element characteristics of the main types of Zentralgneise is given in Finger and Steyrer (1988).

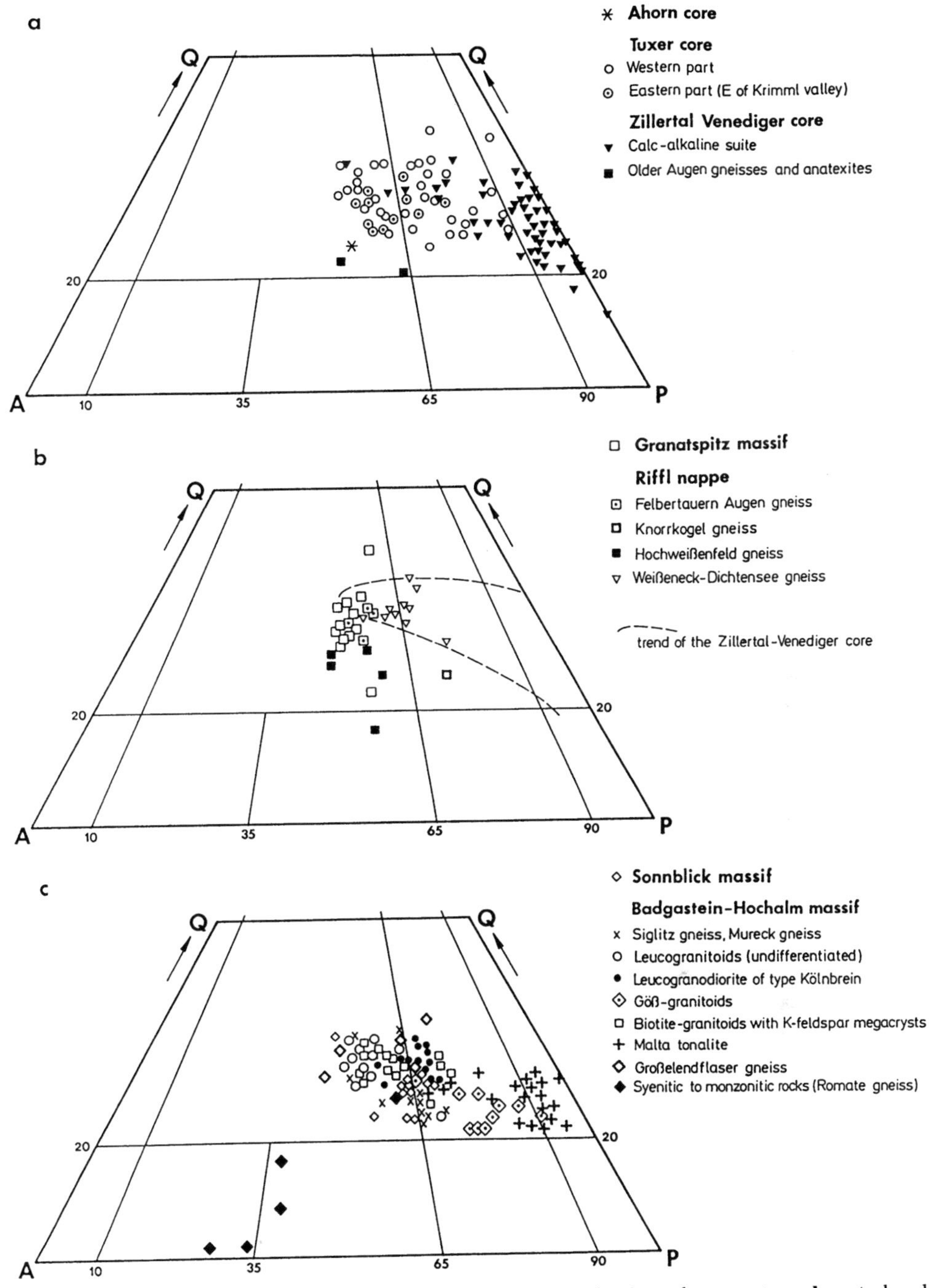

Fig. 5 a–c. QAP-diagrams (Streckeisen 1967) with plots of different types of Zentralgneise from the **a** western, **b** central and **c** eastern Tauern window. Most plots are calculated from chemical analyses (see text in Sect. 7; X^{Kfsp} was generally assumed to be 0.15; data source: Angel 1954; Exner 1957; Karl 1959; Gundlach et al. 1967; Jäger et al. 1969; Morteani 1971; Raith 1971; Raase 1972; Raab 1983; von Quadt 1985; Finger and Steyrer 1988 and unpubl.; Feitzinger 1989; Reicherter 1989; Vavra 1989; Marschallinger and Holub 1990 b; Schermaier 1991 and unpubl.; Haunschmid unpubl.)

The most important chemical signatures can be summarized as follows: most of the Zentralgneise are metaluminous or weakly peraluminous and relatively rich in sodium (see Table 2) and can, therefore, be broadly described as I-type granitoids in terms of the Australian classification of Chappell and White (1974). Only some variants are more aluminous and approach an S-type composition. One important example for a possible *S-type granite suite* is the Granatspitz Massif (Table 2), which might, consequently, be derived from melting of aluminous (meta)sedimentary sources. Peralkaline granites have not yet been found in the Tauern window.

In the trace element discrimination diagrams of Pearce et al. (1984) for the tectonic interpretation of granitic rocks, most of the Zentralgneise fall into the fields of volcanic arc granites.

The *I-type granitoids* among the Zentralgneise *can be subdivided into at least three different suites* (White and Chappell 1983). On one hand, there are the K-feldspar-rich, often syenitic or monzonitic types of Zentralgneise, which occur throughout the Tauern window. These high-potassic variants are also characterized by high Sr and Ba contents (e.g. 4, 7, 8, 12 in Table 2; see also Fig. 6). On the other hand, there is the rather "primitive" CaO-rich assembly of the Zillertal-Venediger core. The trace element patterns of this suite show remarkably low Sr, Rb and Ba concentrations (1–3 in Table 2). Many granitoids of the Tuxer Massif display corresponding patterns and appear to be part of the same plutonic suite (5 in Table 2).

The bulk of Zentralgneise of the eastern Tauern window displays, in contrast, a calc-alkaline trend, which is significantly higher in Na_2O, Sr, Ba and lower in CaO when compared to the trend of the primitive Zillertal-Venediger suite (11, 13, 16–19 in Table 2).

Furthermore, there is a *distinct group of leucocratic intrusions* in the Badgastein-Hochalm area (14, 15, 20 in Table 2), which seem to form an independent plutonic suite characterized by low CaO, Ba, Sr and high K_2O, Rb, Y, Th and U contents ("younger Hochalm leucogranites" in the sense of Finger and Steyrer 1988). Generally, these rocks display some affinities with S-type (syn-collision) granites (e.g. high K_2O and high Rb), but even more with within-plate granites, especially if the trace element patterns of Pearce et al. (1984) are used as reference (high Y, Yb, negative Ba anormaly; see Fig. 6). Granites with similar chemical features can also be observed in the central and western Tauern window (6, 9 in Table 2).

8.2 REE Data

REE analyses have been recently carried out for the main types of Zentralgneise (Table 2). Chondrite-normalized REE-patterns have the following signatures: the two analyzed samples of the Zillertal-Venediger and Tuxer core (1, 5 in Table 2; calc-alkaline suite), have almost identical REE patterns (Fig. 7a) with $La_{cn} \approx 100$ and $(La/Lu)_{cn} \approx 10$ and a slight negative Eu anomaly.

The LREE and HREE contents of the Zentralgneise of the central Tauern window (Fig. 7a) are not significantly different from those of the two samples of the western Tauern window. However, the high-K I-type granitoids, Hochweißenfeld gneiss and Knorrkogel gneiss (7, 8 in Table 2), have no Eu anomaly, whereas a very pronounced negative Eu anomaly can be observed for the Felbertauern augengneiss, which probably belongs to the group of younger leucocratic granites with A-type affinities (see Sect. 8.1; 9 in Table 2 and Fig. 7a).

In the eastern Tauern window, three main types of REE patterns can be distinguished (Fig. 7b): type one can be observed in the case of the calc-alkaline granitoids of the eastern Tauern window (13, 16–19 in Table 2). It is rather low in HREE (Lu_{cn} = 3–5) when compared to the calc-alkaline granitoids of the western Tauern window (Lu_{cn} = 7–11), has no Eu anomaly, and $La_{cn} \approx 100$ (only the Malta tonalite is more enriched in LREE).

The second type is found in the ultrapotassic Romate gneiss and has similar values of La_{cn} as the type one (≈ 100), but is significantly enriched in Nd and Sm, and low in HREE.

The third type of pattern could be recognized for the suite of younger leucocratic granites (14, 15, 20 in Table 2) and has a remarkably flat LREE pattern ($La_{cn} \approx 100$, $Sm_{cn} \approx 50$), a significant negative Eu anomaly ($Eu/Eu^* = 0.16$–0.29) and is strongly enriched in HREE (Lu_{cn} = 18–20).

8.3 Nd Isotope Data

Nd isotope analyses have been performed for some of the main types of Zentralgneise in order to gather further information on the magmatic genesis of the rocks. A comprehensive presentation and discussion of these results will be given in a forthcoming paper by von Quadt and Finger (in prep.).

Calculated $\varepsilon Nd_{T\text{-}320}$ model initial ratios are listed in Table 2. The values range between +1.32 and −5.58. This indicates that both magma sources with mantle

Table 2. Representative chemical analyses of main types of Zentralgneise

		Western Tauern window						Central Tauern window					Eastern Tauern window								
		Zillertal-Venediger Massif				Tuxer Massif		Riffl nappe			Granatsp. Massif	Sonnbl. Massif	Bad Gastein Massif				Hochalm Massif				
	No.:	1	2	3	4	5	6	7	8	9	10	11	12	13	14	15	16	17	18	19	20
	Bez:	HP158	F15	F25	AS85	F13	AQ114	F11	AS03	AS13	139	F1	F21	HA15	F20B	F14	F9	F6	HA52	2916C	F7
RFA	SiO2	60.73	66.23	72.15	60.85	73.93	75.45	62.37	65.60	72.39	69.76	71.44	62.39	68.50	72.54	74.54	60.60	63.54	70.20	72.91	73.23
RFA	TiO2	0.57	0.56	0.20	0.76	0.23	0.11	0.65	0.48	0.38	0.24	0.30	0.54	0.38	0.30	0.21	0.48	0.69	0.23	0.24	0.27
RFA	Al203	16.40	15.77	14.37	15.80	14.30	13.02	17.86	16.46	12.85	14.83	14.85	15.83	15.00	13.98	13.12	18.39	16.63	15.20	14.50	13.85
RFA	FeOtot	6.55	3.82	2.04	5.20	1.89	1.34	3.20	2.98	2.22	1.65	2.05	2.40	2.07	1.92	1.49	4.24	4.08	1.67	1.39	1.89
RFA	MnO	0.20	1.45	0.05	0.08	0.00	0.04	0.05	0.06	0.04	0.03	0.02	0.05	0.03	0.07	0.05	0.23	0.15	0.04	0.14	0.02
RFA	MgO	3.34	1.73	0.42	3.87	0.33	0.08	1.92	1.33	1.50	0.42	0.72	1.61	1.05	0.47	0.29	2.43	2.23	0.45	0.47	0.28
RFA	CaO	5.82	4.00	1.89	2.93	1.40	0.95	2.58	2.77	1.65	1.28	1.94	2.21	2.21	1.59	1.17	3.98	3.32	1.96	2.07	1.25
RFA	Na2O	3.08	3.59	3.81	3.35	4.73	3.57	3.88	4.02	2.81	3.01	4.13	3.87	4.42	3.43	3.23	4.64	3.35	4.32	4.12	3.69
RFA	K2O	1.98	2.73	3.91	5.61	3.48	4.30	5.31	3.81	5.02	5.09	3.87	7.05	3.69	4.53	4.68	3.16	3.94	3.30	3.21	4.73
RFA	P2O5	0.24	0.23	0.06	0.57	0.07	0.03	0.26	0.21	0.24	0.23	0.14	0.50	0.17	0.08	0.05	0.36	0.34	0.10	0.07	0.08
RFA	L.O.I.	1.08	0.78	1.35	0.70	1.02	0.68	1.86	1.96	0.65	0.84	0.85	1.84	1.60	0.79	0.82	1.50	1.64	1.60	1.03	0.87
RFA	tot	99.99	100.89	100.25	99.72	101.38	99.57	99.94	99.68	99.75	97.38	100.31	98.29	99.12	99.70	99.65	100.01	99.91	99.07	100.15	100.16
RFA	A/CNK	0.92	0.98	1.03	0.93	1.01	1.06	1.06	1.04	0.98	1.16	1.02	0.88	0.98	1.04	1.05	1.01	1.06	1.07	1.03	1.03
RFA	Nb	13	13	11	24	13	4	22	15	25	1	16	17	bdl	16	15	14	16	bdl	7	16
RFA	Zr	152	142	133	380	164	90	315	172	193	93	135	244	165	162	125	200	187	142	122	189
RFA	Y	27	18	12	35	18	35	24	13	34	11	16	12	bdl	49	58	19	12	11	7	47
RFA	Sr	287	264	192	347	128	49	450	528	188	81	461	1074	834	168	92	882	716	520	605	119
RFA	Ba	456	641	1014	1970	812	347	1775	1301	400	339	709	2736	1420	431	197	1492	1867	1127	1420	511
RFA	Rb	65	109	100	182	102	190	188	140	213	249	179	368	158	240	281	108	118	117	67	244
RFA	Cr	17	21	–	127	bdl	3	55	102	40	7	23	37	25	7	bdl	49	37	bdl	3	3
RFA	Ni	11	6	–	47	1	bdl	23	8	18	bdl	6	40	16	18	20	23	27	bdl	bdl	16
RFA	Th	6	11	–	33	11	–	22	12	50	16	30	55	28	28	31	13	18	19	12	20
NAA	Ba	382	–	–	–	774	–	1735	1101	390	–	–	2632	1381	431	273	1609	2012	1018	1321	601
NAA	Rb	60	–	–	–	105	–	192	129	198	–	–	356	147	241	257	98	136	100	73	244
NAA	Cr	23	–	–	–	2	–	29	84	35	–	–	35	14	6	4	10	38	3	3	3
NAA	Th	6	–	–	–	11	–	23	11	48	–	–	58	23	30	31	9	20	9	12	23
NAA	U	1.1	–	–	–	2.6	–	4.9	2.3	8.0	–	–	12.3	6.5	15.3	15.0	3.0	1.8	2.3	1.6	10.3
NAA	Hf	4.8	–	–	–	6.5	–	9.0	4.2	6.2	–	–	9.1	4.0	7.1	6.2	7.4	7.6	3.0	4.9	7.6
NAA	Sc	20.0	–	–	–	4.0	–	6.9	5.2	7.4	–	–	7.3	4.4	6.3	4.8	4.8	6.9	3.1	3.3	5.3
NAA	Co	17.0	–	–	–	2.2	–	9.4	5.1	5.7	–	–	7.7	4.3	3.7	2.0	7.9	11.9	1.8	1.8	2.3
NAA	Ta	0.85	–	–	–	0.85	–	1.64	1.04	2.52	–	–	1.12	0.52	2.50	2.30	0.31	0.71	1.01	0.29	2.30
NAA	La	27.6	–	–	–	34.2	–	42.2	37.1	46.0	24.1	–	36.7	27.8	32.7	25.6	46.0	83.4	31.5	36.9	40.5
NAA	Ce	51.5	–	–	–	62.9	–	76.4	69.4	92.4	50.8	–	77.1	56.6	62.6	55.5	75.1	146.4	57.5	65.9	81.4
NAA	Nd	22.8	–	–	–	24.2	–	34.4	27.5	39.6	22.3	–	46.3	25.8	31.7	28.5	23.0	50.0	21.6	24.3	34.4
NAA	Sm	6.70	–	–	–	5.15	–	6.94	5.29	10.20	5.05	–	11.72	4.92	9.23	9.15	4.38	9.38	3.77	4.74	10.76
NAA	Eu	1.41	–	–	–	0.80	–	1.92	1.22	0.95	0.58	–	1.95	1.14	0.75	0.46	1.31	1.90	1.02	1.28	0.94
NAA	Tb	0.73	–	–	–	0.55	–	0.74	0.41	0.98	0.67	–	0.48	0.25	1.08	1.14	0.34	0.67	0.36	0.37	1.16
NAA	Yb	2.35	–	–	–	1.80	–	2.17	1.07	2.22	1.48	–	0.60	0.63	4.31	4.95	1.09	0.92	0.90	0.92	4.22
NAA	Lu	0.37	–	–	–	0.28	–	0.32	0.17	0.36	0.21	–	0.12	0.08	0.65	0.68	0.18	0.17	0.15	0.15	0.06
	εNd(320)	−3.78	–	–	–	−1.66	–	−4.63	−1.93	−5.58	−4.83	−1.83	−4.39	–	–	–	−1.94	+2.70	–	−3.42	−0.60

Major elements in weight % oxides; trace elements in ppm. Abbreviations: bdl = below detection limits; L. O. I. = loss on ignition; A/CNK = mol Al_2O_3/CaO+Na_2O+K_2O; XRFA-data source: Finger and Steyrer (1988 and unpubl.), except for samples HA... = Haunschmid (unpubl.), AS... = Schermaier (unpubl.). INA analyses were carried out by Haunschmid and Lettner, Nd isotope analyses by mass spectrometry at the ETH Zürich (von Quadt unpubl.). Sample locations: 1: metatonalite, southern rim of Venediger core; 2: metagranodiorite, Ahrntal, near Kasern; 3: metagranite, path to Thüringer Hütte, near waterfall; 4: biotite-rich Augen gneiss, inner Krimml valley; 5: "Augen- und Flasergneis", mittlere Gfrorene Wand Spitze, near Olperer; 6: "Augen- und Flasergneis", quarry in the outermost Obersulzbach valley; 7: Hochweißenfeld gneiss, S of Abretterkopf; 8: Knorrkogel gneiss, Matreier Tauern valley, Raneburg; 9: Felbertauern Augen gneiss, south of Felbertauern, 2 km S St. Pöltener Hütte; 10: metagranite, W Tauernmoos lake; 11: metagranite, Fleißtal; 12: Romate gneiss, near Mallnitz; 13: Siglitz gneiss, 8 km south Badgastein, Sportgastein/Naßfeld; 14: Riesenaugen gneiss, south Badgastein, Radhausberg tunnel; 15: leucogranite, tunnel from Gastein valley to Mallnitz; 16: "Hochalmporphyrgranit", NNW Hochalmspitze, near Osnabrücker Hütte; 17: Göß granitoid, Malta valley, Koschach quarry; 18: Malta tonalite, Malta valley, bifurcation to Katowitzer Hütte; 19: Kölnbrein leucogranite, northern Hochalm massif, Rotgüldensee; 20: leucogranite, Gößbach valley, Untere Thomanbauernalm

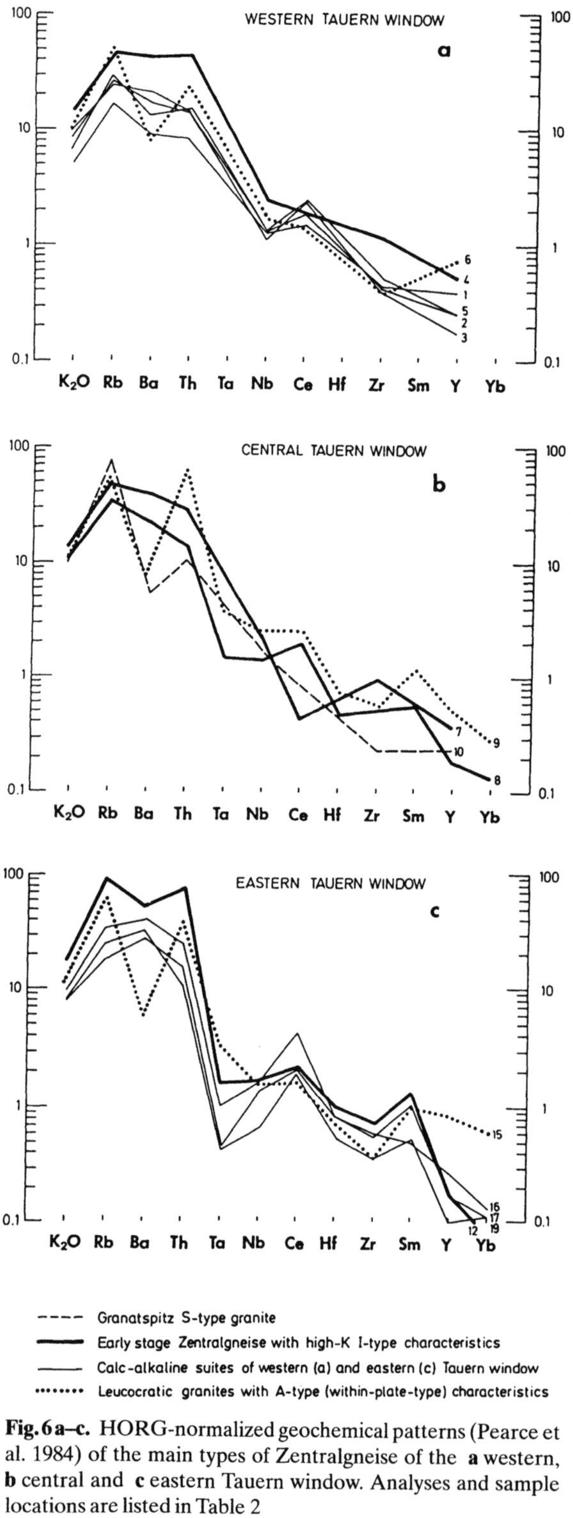

Fig. 6 a–c. HORG-normalized geochemical patterns (Pearce et al. 1984) of the main types of Zentralgneise of the **a** western, **b** central and **c** eastern Tauern window. Analyses and sample locations are listed in Table 2

isotopic characteristics (i. e. upper mantle and/or very fresh crust) and ancient crustal magma sources must have been involved. The lowest influence of old crust is generally found in the eastern Tauern window. Only the earliest K_2O-rich intrusions of this area (Romate gneiss) seem to include significantly high amounts of old (probably mid-Proterozoic) crust.

The Zentralgneise of the central Tauern window have, with only one exception (Knorrkogel gneiss, 8 in Table 2), low εNd values and are therefore likely to be issued from partial melting of crust pre-existing for a long time (with or without minor influxes of juvenile mantle melts). T_{DM} values of 1.4–1.7 Ga point to a mid-Proterozoic residence age of the crustal source. It should be noted here that the U/Pb zircon dating method has provided additional evidence for old crustal components in most Zentralgneise of the central Tauern window (Cliff 1981). In case of the Knorrkogel gneiss and the Hochweißenfeld gneiss, a mid-Proterozoic memory is preserved (von Quadt unpubl.).

Old crustal sources must have been also involved in the plutonic genesis of the Zentralgneise of the western Tauern window, although the influx of sources with mantle isotopic compositions appears to be larger in this area than in the central Tauern window.

9 Magmatic Evolution of the Zentralgneise

An important field observation is that the earliest members of the Zentralgneise have often, and throughout the Tauern window, diatectic textures and/or are closely associated with meta- to diatectic country rocks, which document a widespread regional metamorphic event up to anatexis at the beginning of the batholitic evolution. This field evidence, together with isotopic and chemical data, suggests that the earliest Zentralgneise were magmas, which formed fully or were at least strongly influenced by partial melting of the surrounding pre-existing continental crust. The dominance of high-K_2O I-type granites, combined with the remarkable scarcity of typical S-type granites, implies that the crustal source contained either only few highly evolved aluminous (meta)sediments, or that contributions of mantle melts helped to suppress the S-type component of the forming anatectic magmas.

In the eastern Tauern window, the exposed prebatholitic crust seems to consist predominantly of immature sedimentary and volcanic material of an Early Palaeozoic or late Proterozoic magmatic arc (see e. g.

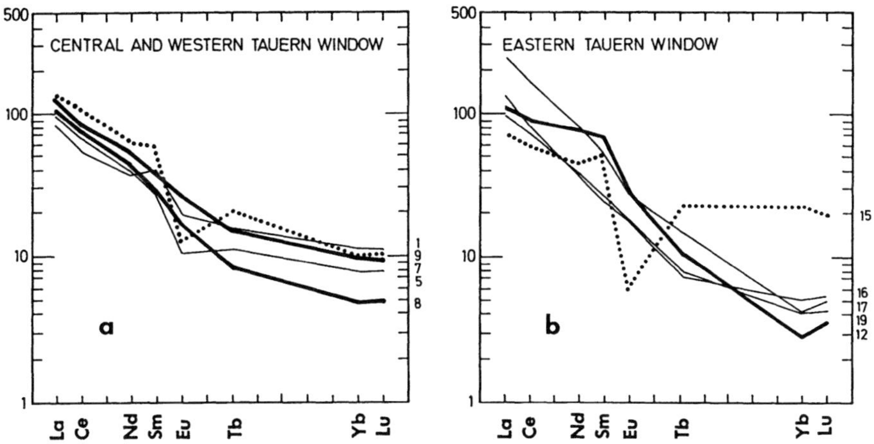

Fig. 7 a, b. Chondrite-normalized REE patterns of the main types of Zentralgneise of the **a** western and central Tauern window and **b** the eastern Tauern window. For normalizing values, see Wakita et al. (1971). *Symbols* as in Fig. 6. Analyses and sample locations are listed in Table 2

mapping results of Vavra 1989). However, the existence of a significantly older crustal basement can be indirectly infered on the basis of Nd isotope data (Sect. 8.3). The occurrence of early-stage Zentralgneise with a clear syenitic tendency might indicate some contributions of within-plate-type (?) mantle melts. It seems possible that such mantle melts actually triggered the regional migmatization.

The major volumes of Zentralgneise of the eastern Tauern window represent, however, clearly a calc-alkaline plutonic suite (see Sects. 5, 8), which intruded after the anatectic event and after the syenites. This could mean that a subduction-related mantle became an increasingly important magma source with time. The oldest members of this calc-alkaline assembly of the eastern Tauern window have been dated at ca. 315 Ma (see Sect. 6). The age of the syenites is still unknown.

A very similar situation is documented in the central and western Tauern window, where early migmatites and K_2O-rich granitoids of dominantly crustal derivance are post-dated by the voluminous calc-alkaline plutonic suite of the Zillertal-Venediger Massif (see Sect. 4). The early stage of crustal anatexis must be there of late Viscan to early Namurian age (see formation ages of the Hochweißenfeld gneiss and Granatspitz Massif in Table 2).

A last generation of leucocratic granites with affinities to within-plate (A-type) granites appears to complete the plutonic evolution history of the Zentralgneise.

Apart from the systematic change of magma-types with time, it could be observed throughout the Tauern window that the younger plutonic members of the Zentralgneise solidified generally as high-level plutons and often just within the older migmatitic series. This means that the migmatitic crust was uplifted and cooled in the meantime. Unfortunately, the ages of these younger subvolcanic intrusions are neither well-known in the eastern nor in the central and western Tauern window. A 292 Ma isochron of Besang et al. (1968) for late-stage aplitic intrusions in the Tuxer massif seems to provide one of the most reliable age values currently available for the "subvolcanic period" (see Sect. 6). This age, if correct, would require a very rapid uplift history of the earlier migmatitic crust, which in turn could be well explained when a concurrent detachment of upper crust is proposed. Large parts of the Tuxer and the Ahorn core must have been, however, upper crust during the entire Palaeozoic, because many Zentralgneise of these areas are high level plutons which solidified not within migmatitic basement but within non- or very low-grade metamorphic volcanic and sedimentary strata of Late Proterozoic or Early Palaeozoic age.

It is interesting that the magmatic evolution path of the Hohe Tauern batholitic terrane is, in many ways, similar to that of the batholite of Corsica, which is situated in a comparable structural position at the southern flank of the Variscan fold belt. There, an earlier (late Visean to early Namurian) suite of high-K_2O granitoids is also followed by a series of calc-alkaline intrusions, the latest of which are, moreover, similar to the young leucocratic intrusions of the Tauern window (compare data in Cochérie et al. 1984; Ferre 1989). The youngest intrusions of the Corsican batholite are alkaline rocks of mid-Permian to Triassic age (Bonin 1988). A comparable alkaline suite has not been recognized in the Tauern window.

10 Geodynamic Setting of the Zentralgneise in the Variscan Orogen – A Brief Discussion

It is widely accepted today that the Palaeozoic basement units of the Alps once formed parts of the southern flank of the Variscan fold belt. However, the exact locations of single units relative to each other and to the Variscides on the north prior to the Alpine tectonics are still largely unknown.

Like the southern Variscan units of Western Europe, the Tauern window comprises fragments of ophiolites (see e.g. Neubauer et al. 1989), which are probably remnants of a late Proterozoic or Early Palaeozoic ocean situated in the northern Gondwana foreland (a possible eastern continuation of the "Massif Central ocean" of Matte 1986). Furthermore, the Tauern window exposes large volumes of late Proterozoic/early Palaeozoic volcanic arc basalts and andesites, which are possibly related to northward subduction of the above-mentioned ocean (see e.g. Vavra and Frisch 1989).

As outlined in Section 8 of this work, the Zentralgneise also display, from the chemical point of view, features of "volcanic arc magmatism". This fact lead Finger and Steyrer (1988) to propose that the plutonic protoliths of the Zentralgneise were located along the same subduction system that also fed the volcanism before. It was speculated that both magmatic events might have occurred along a long-living northward dipping subduction system south of the Moldanubian unit, the plutons being generated during a final and mature stage of subduction. This concept was largely based on a model proposed by Ziegler (1986) for the evolution of the Variscan fold belt, which operates with a large Variscan ocean between Gondwana and Laurasia and a stepwise northward drift and attachment of Gondwana-derived terranes to a Laurasian Cordillera before the actual Variscan megacontinent-collision occurred.

However, recent years have brought new faunal, facies and palaeomagnetic evidence that the megacontinents Gondwana and Laurasia were linked and formed a nascent Pangea, when the late Variscan plutonism set in (see e.g. review in Neugebauer 1988). Indeed, it can hardly be disputed today that the Late Palaeozoic plutonism of the Tauern window occurred in a Variscan post-collision environment rather than in a pre-collision stage with reference to the continent collision event of Gondwana and Laurasia. One might, therefore, argue that the I-type plutonism of the Tauern window is also collision- rather than subduction-related, the I-type features being merely a consequence of a collision-induced anatectic remagmatization of infra-crustal I-type sources, e.g. the voluminous volcanic arc material of the older suture (see e.g. Vavra 1989).

An alternative model can, nevertheless, be constructed as follows: it appears possible that the plutonic protoliths of the Zentralgneise were situated above a late Variscan subduction zone, which descended during the Upper Carboniferous along the northern Palaeotethys margin below the nascent Pangea continent (Fig. 8). This geodynamic situation could have evolved as a consequence of the Late Palaeozoic westward drift of the Gondwana continent relative to Laurasia and the opening of the

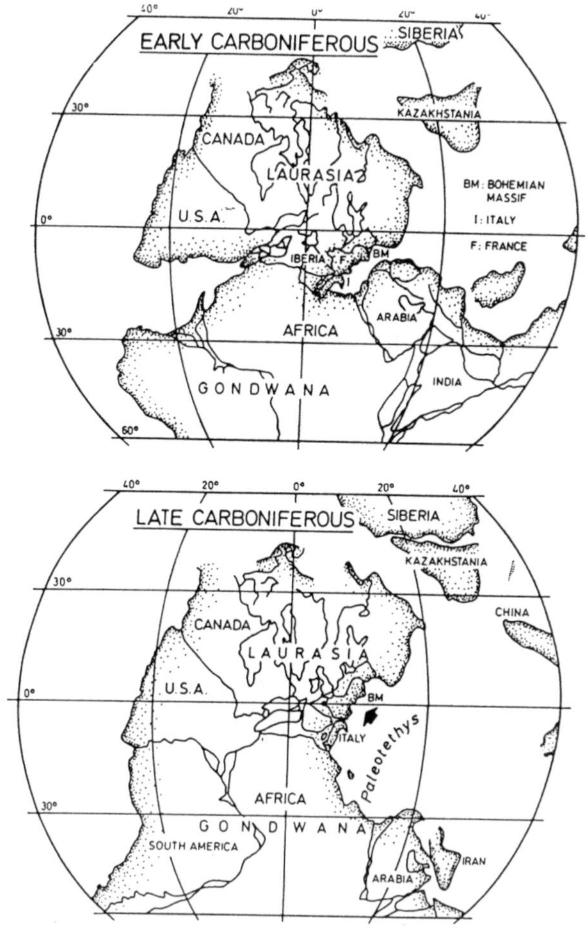

Fig. 8. Palaeogeographic base maps for the Early and Late Carboniferous according to Scotese (1984). It is hypothetized that the Palaeotethys ocean was subducted in the late Palaeozoic below the Variscan fold belt, which is in front of the northern Laurasian part of the nascent Pangea megacontinent *(black arrow)*. This Cordilleran-type plate configuration probably evolved as a consequence of the Carboniferous westward drift of Gondwana relative to central Europe. The map for the Early Carboniferous suggests a more or less intracontinental (=collisional) position of the Variscan fold belt between Gondwana and Laurasia

Pangea gulf (Arthaud and Matte 1977). One argument in favour of such a late Variscan Palaeotethys subduction system is that the Variscan orogen exhibits a significant active continent margin type zonation of late Palaeozoic granitoids (Finger and Steyrer 1990) with a southern belt of predominantly I-type plutons in the Alps (outer arc?) and a northern belt of predominantly S-type plutons in the Moldanubian unit (inner arc?).

A Upper Palaeozoic northwards subduction of the Palaeotethys plate has also been suggested for the Caucasian area (Adamia et al. 1990). Therefore, it seems quite possible that the Variscan fold belt was involved in a long-stretched magmatic arc system, active along the Eurasian margin of the Palaeotethys (Fig. 8).

Unlike the Moldanubian unit with its complex nappe structures, the Tauern window has, as yet, provided no clear evidence for Variscan continent-collision-type tectonics. This might be due to the fact that the Alpine orogeny has destroyed or at least blurred such collision-type structures. It might, however, also simply be due to the fact that the Palaeozoic units of the Tauern window were situated in a southeastern section of the Variscan fold belt, where the continents did not fully touch and where the previous oceanic domains between Gondwana and Laurasia were only shortened, but never completely closed during the whole Palaeozoic. Thus, a subduction situation might have survived in this southeastern Variscan realm from the early to the Late Palaeozoic, while the actual collision, i.e. the docking of the megacontinents, affected predominantly the western sector of the central European Variscan orogen (see also Ziegler 1986). In this context it appears even possible that the rate of northward subduction of the Palaeotethys was strongly reinforced by the mega-continent-collision, in that the (north)westward moving Gondwana plate could have simultaneously triggered a clockwise rotation of Laurasia. Such a combined late Variscan subduction-collision scenario does, of course, not preclude Late Palaeozoic arc-continent collisions or micro-terrane accretions in the southeastern Variscan realm (see model of Frisch and Neubauer 1989). Such processes of crustal thickening could have been, for example, also responsible for the regional burial anatexis, which stood at the beginning of the plutonic evolution history of the Zentralgneise.

Acknowledgements
The authors would like to thank reviewers B. Bonin and F. Bussy for stimulating comments. W. Arming is thanked for his help in compiling the map of the western Tauern window. The investigations were supported by the Austrian "Fonds zur Förderung der wissenschaftlichen Forschung" (Grant P7353).

References

Adamia S, Lordkipanidze M, Kvantaliani I, Khutella Z (1990) Geology of the Caucasus. Field Meet of the Working group 5.1 Paleogeography and tectonics of the Paleozoic fold areas and their platform frame. Geotraverse 10 (Caucasus) Tbilisi

Alber J (1976) Seriengliederung, Metamorphose und Tektonik des Hocharngebietes (Rauristal, Salzburg). Dissertation, Univ Wien

Angel F (1954) Petrochemie der Hochalm-Ankogel-Gesteine. Jahrb Geol Bundesanst 97: 1–16

Angel F, Staber R (1952) Gesteinswelt und Bau der Hochalm-Ankogel-Gruppe. Wagner, Innsbruck

Arthaud F, Matte P (1977) Late Paleozoic strike-slip faulting in southern Europe and northern Africa: result of a right-lateral shear-zone between the Appalachians and the Urals. Geol Soc Am Bull 88: 1305–1320

Becke F (1908) Bericht über die Aufnahmen am Nord- und Ostrand des Hochalmmassivs. Sitzber Akad Wiss Math Naturwiss Kl Abt I 1908: 1–34

Benedict PC (1952) Zur Tektonik der südlichen Granatspitzgruppe (Hohe Tauern). Dissertation, Univ Zürich

Besang C, Harre W, Karl F, Kreuzer H, Lenz H, Müller R, Wendt I (1968) Radiometrische Altersbestimmungen (Rb/Sr und K/Ar) an Gesteinen des Venediger-Gebietes (Hohe Tauern, Österreich). Geol Jahrb 86: 835–844

Bonin B (1988) From orogenic to anorogenic environments: evidence from associated magmatic episodes. Schweiz Mineral Petrogr Mitt 68: 301–311

Chappell BW, White AJR (1974) Two constrasting granite types. Pac Geol 8: 173–174

Christa E (1931) Das Gebiet des Oberen Zemmgrundes in den Zillertaler Alpen. Jahrb Geol Bundesanst 81: 533–635

Cliff RA (1977) Rb-Sr measurements on granite-gneisses from the Granatspitzkern, Hohe Tauern, Austria. Verh Geol Bundesanst 1977: 101–104

Cliff RA (1981) Pre-Alpine history of the Pennine zone in the Tauern window, Austria: U-Pb and Rb-Sr geochronology. Contrib Mineral Petrol 77: 262–266

Cliff RA, Norris RJ, Oxburgh ER, Wright RC (1971) Structural, metamorphic and geochronological studies in the Reisseck and southern Ankogel Group, Eastern Alps. Jahrb Geol Bundesanst 114: 121–272

Cochérie A, Rossi P, Le Bel L (1984) The Variscan calc-alkalic plutonism of western Corsica: mineralogy and major and trace element geochemistry. Phys Earth Planet Int 35: 145–178

Cornelius HP (1930) Vorläufiger Bericht über geologische Aufnahmen im Stubach- und Kaprunertale (Glocknergruppe). Verh Geol Bundesanst 1930: 117–121

Cornelius HP (1942a) Geologisches über die Granatspitzgruppe. Z Dtsch Alpenver 1942: 61–68

Cornelius HP (1942b) Neue Aufnahmsergebnisse aus dem Matreier Tauerntal. Ber Reichsst Wien Bodenforsch 1942: 1–3

Cornelius HP, Clar E (1939) Geologie des Großglocknergebietes, Teil I. Abh Zweigst Wien Bodenforsch 25: 1–305

Exner C (1957) Erläuterungen zur geologischen Karte von Gastein. Geol Bundesanst, Wien

Exner C (1964) Erläuterungen zur Geologischen Karte der Sonnblickgruppe. Geol Bundesanst, Wien

Exner C (1983) Erläuterungen zur Geologischen Karte der Hafnergruppe. Mitt Ges Geol Bergbaustud Österr 29: 41–74

Feitzinger G (1989) Lagerstättenkundliche Untersuchungen an gangförmigen Gold-Silber Vererzungen der Sonnblickgruppe (Hohe Tauern, Kärtnen). Dissertation, Univ Salzburg

Ferre E (1989) Le plutonisme magnesio-potassique carbonifere de Corse occidentale (France). These, Univ Toulouse

Finger F, Steyrer HP (1988) Granite-types in the Hohe Tauern (Eastern Alps, Austria) – some aspects on their correlation to Variscian plate tectonic processes. Geodin Acta 2: 75–87

Finger F, Steyrer HP (1990) I Type granitoids as indicators of a late Paleozoic convergent ocean-continent margin along the southern flank of the central European Variscan orogen. Geology 18: 1207–1210

Frank W (1969) Geologie der Glocknergruppe. Wiss Alpenver Hefte 21: 95–11

Frank W, Höck V, Miller Ch (1987 a) Metamorphic and tectonic history of the Central Tauern window. In: Flügel HW, Faupl P (eds) Geodynmaics of the Eastern Alps. Deuticke, Wien pp 34–54

Frank W, Miller Ch, Pestal G (1987 b) Geologische Karte der Republik Österreich, Blatt 152 (Matrei in Osttirol) 1 : 50000. Geol Bundesanst, Wien

Frasl G (1953) Die beiden Sulzbachzungen. Jahrb Geol Bundesanst 96: 143–192

Frasl G (1958) Zur Seriengliederung der Schieferhülle in den Mittleren Hohen Tauern. Jahrb Geol Bundesanst 101: 323–472

Frasl G, Frank W (1966) Einführung in die Geologie und Petrographie des Penninikums im Tauernfenster. Aufschluß SH 15: 30–58

Frasl G, Holub B, Kruhl J, Marschallinger R (1987) Exkursion am 2. 10. in das Zentralgneis- und Migmatitgebiet im Maltatal (Östliche Hohe Tauern, Kärnten). Eskursionsführer zur Tauerntagung Mittersill/Salzburg, University of Salzburg, 9 pp

Frisch W (1977) Der alpidische Internbau der Venedigerdecke im westlichen Tauernfenster (Ostalpen). Neues Jahrb Geol Paläontol Monatsh 11: 675–696

Frisch W (1980) Tectonics of the Western Tauern window. Mitt Österr Geol Ges 71/72: 65–71

Frisch W, Neubauer F (1989) Pre-Alpine terranes and tectonic zoning in the eastern Alps. In: Dallmeyer RD (ed) Terranes in the Circum-Atlantic Paleozoic orogens. Geol Soc Am Spec Pap 230: 91–100

Fuchs G (1958) Beitrag zur Kenntnis der Geologie des Gebietes Granatspitze – Großvenediger, Hohe Tauern. Jahrb Geol Bundesanst 101: 201–248

Gundlach H, Karl F, Müller G (1967) Vergleichende geochemische Untersuchungen an ost- und südalpinen Graniten, Granodioriten und Tonaliten. Contrib Mineral Petrol 16: 285–299

Hawkesworth CJ (1976) Rb/Sr geochronology in the Eastern Alps. Contrib Mineral Petrol 54: 225–244

Jäger E, Karl F, Schmidegg O (1969) Rb-Sr Altersbestimmungen an Biotit-Muskowit Granitgneisen (Typus Augen- und Flasergneise) aus dem nördlichen Großvenedigerbereich (Hohe Tauern). Tschermaks Mineral Petrogr Mitt 13: 251–272

Karl F (1959) Vergleichende petrographische Studien an den Tonalitgraniten der Hohen Tauern und den Tonalit-Graniten einiger periadriatischer Intrusivmassive. Jahrb Geol Bundesanst 102: 1–192

Karl F (1961) Berich 1960 über Aufnahmen auf Blatt Krimml (151). Verh Geol Bundesanst 1961: A 37–A 39

Karl F, Schmidegg O (1979) Geologische Spezialkarte Blatt Krimml (151) 1 : 50000. Geol Bundesanst, Wien

Kupka E (1953) Zur geologischen Stellung des Ahornkernes in den westlichen Hohen Tauern. Skizzen zum Antlitz der Erde (Kober-Festschrift), Mitt Österr Geol Ges Wien, pp 255–272

Lambert JSR (1964) Isotopic age determinations on gneisses from the tauern Window. Verh Geol Bundesanst 1961 (1): 16–27

Lameyre J, Bowden P (1982) Plutonic rock type series: discrimination of various granitoid series and related rocks. J Volc and Geotherm Res 14: 169–186

Lammerer B (1986) Das Autochthon im westlichen Tauernfenster. Jahrb Geol Bundesanst 129: 51–67

Marschallinger R, Holub B (1990 a) Die Zentralgneise im Hochalm-Ankogel Massiv (östliches Tauernfenster, Österreich). Teil II. Zirkontypologische und geochemische Charakteristik. Mitt Österr Geol Ges 82: 19–48

Marschallinger R, Holub B (1990 b) Zentralgneise im Bereich des östlichen Tauernfensters. Mitt Österr Mineral Ges 135: 54–56

Matte P (1986) Tectonics and plate tectonic model for the Variscan belt of Europe. Tectonophysics 126: 329–374

Mielke P, Winkler HGF (1979) Eine bessere Berechnung der Mesonorm für granitische Gesteine. Neues Jahrb Miner Monatsh 1979 (10): 471–480

Morteani G (1971) Gliederung und Metamorphose der Serien zwischen Stillupptal und Schlegeistal (Zillertaler Alpen, Nordtirol). Verh Geol Bundesanst 1971 (2): 287–314

Neubauer F, Frisch W, Schmerold R, Schlöser H (1989) Metamorphosed and dismembered ophiolite suites in the basement units of the Eastern Alps. Tectonophysics 164: 49–62

Neugebauer J (1988) The Variscan plate tectonic evolution: an improved "Iapetus model". Schweiz Mineral Petrogr Mitt 68: 313–333

Pearce JA, Harris NBW, Tindle AG (1984) Trace element discrimination diagrams for the tectonic interpretation of granitic rocks. Petrol 25 (4): 956–983

Prosser E (1975) Petrographisch-Geochemische Untersuchungen im Zillertaler Zentralgneiskern. Ein Beitrag zum Zentralgneisproblem der westlichen Hohen Tauern. Dissertation, Univ München

Quadt A von (1985) Geochronologische, geochemische und isotopenchemische Untersuchungen an Gesteinen der Habachformation der Scheelitlagerstätte und des angrenzenden Altkristallins im Felbertal (Land Salzburg). Dissertation, Univ Zürich

Raab D (1983) Genese der Amphibolitfolgen der alten Schieferhülle des Tauernfensters (Ostalpen). Diplomarbeit, Univ Tübingen

Raase P (1972) Petrographie und Geologie des oberen Zillergrundes und Sundergrundes (Zillertaler Alpen, Österreich). Dissertation, Univ Kiel

Raith M (1971) Seriengliederung und Metamorphose im östlichen Zillertaler Hauptkamm (Tirol, Österreich). Verh Geol Bundesanst 1971 (1): 163–207

Raith M, Raase P, Kreuzer H, Müller P (1978) The age of the alpidic metamorphism in the Western Tauern window, Austrian Alps, according to radiometric dating. In: Closs H, Roeder D, Schmidt K (eds) Alps, Apennines, Hellenides. Schweizerbart, Stuttgart, pp 140–148

Raumer JF von (1988) Moldanubian evolution in the basement of the Alps. The external massifs of the Western Alps. In:

Kukal Z (ed) Abstracts Conference on the Bohemian massif, Geological survey. Prag, p 74
Reicherter K (1989) Geologie, Geochemie und Tektonik im Westlichen Tauernfenster (Äußeres Obersulzbachtal, Pinzgau/Österreich). Diplomarbeit, Univ Tübingen
Satir M (1974) Rb-Sr Altersbestimmungen an Glimmern der westlichen Hohen Tauern: Interpretation und geologische Bedeutung. Schweiz Mineral Petrogr Mitt 54: 213–228
Satir M, Morteani G (1982) Petrological study and radiometric age determination of the migmatites in the Penninic rocks of the Zillertaler Alpen (Tyrol, Austria). Tschermaks Mineral Petrogr Mitt 30: 59–75
Schermaier A (1991) Geologisch-petrographische Untersuchungen zur präalpidischen Entwicklung am Ostrand des Venedigermassivs (Hohe Tauern). Jahrb Geol Bundesanst 134: 345–367
Schmidegg O (1961) Geologische Übersicht der Venedigergruppe. Verh Geol Bundesanst 1961: 34–55
Scotese CR (1984) Paleozoic paleomagnetism and the assembly of Pangea. In: Van der Voo R, Scotese CR, Bonhommet N (eds) Plate reconstruction from Paleozoic paleomagnetism. Am Geophys Union Geodyn Ser 12: 1–10
Streckeisen AL (1967) Classification and nomenclature of igneous rocks Neues Jahrb Miner Abh 107: 140–240
Thiele O (1974) Tektonische Gliederung der Tauernschieferhülle zwischen Krimml und Mayrhofen. Jahrb Geol Bundesanst 117: 55–74
Vavra G (1989) Die Entwicklung des penninischen Grundgebirges im östlichen und zentralen Tauernfenster der Ostalpen – Geochemie, Zirkonmorphologie, U/Pb Radiometrie. Tübinger Geowiss Arb Reihe A, 6, 173 pp
Vavra G, Frisch W (1989) Pre-variscan back-arc and island-arc magmatism in the Tauern window (Eastern alps). Tectonophysics 169: 271–280
Wakita H, Rey R, Schmitt RA (1971) Abundances of the 14 rare-earth elements and 12 other trace-elements in Apollo 12 samples: five igneous and one breccia rocks and four soils. Proc 2nd Lumar Sci Conf, pp 1319–1329
White AJR, Chappell BW (1983) Granitoid types and their distribution in the Lachlan Fold Belt, southeastern Australia. In: Roddick JA (ed) Circum Pacific plutonic terranes. Geol Soc Am Mem 159: 21–34
Ziegler P (1986) Geodynamic model for the Paleozoic crustal consolidation of western and central Europe. Tectonophysics 126: 303–328

Chapter 3
Austro-Alpine Realm

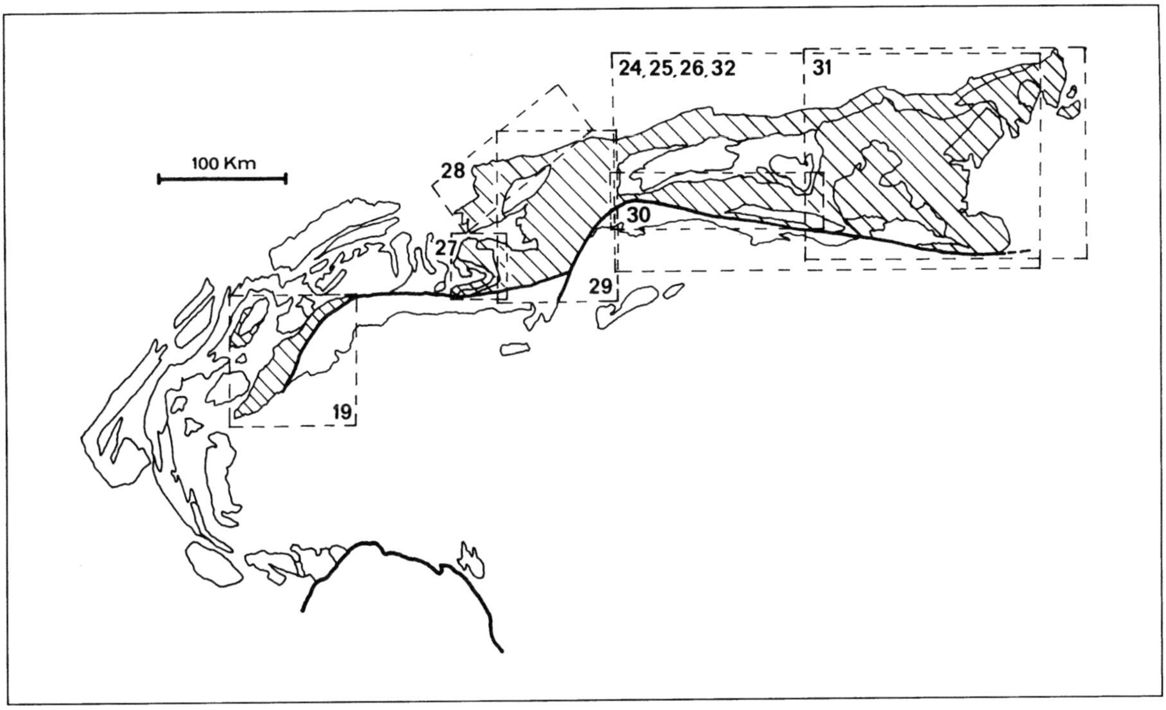

24 The Classic Fossiliferous Palaeozoic units of the Eastern and Southern Alps.
H. P. Schönlaub, H. Heinisch

25 The Austro-Alpine Quartzphyllites and related Palaeozoic formations.
F. Neubauer, FP. Sassi

26 Palaeozoic volcanism of the Eastern Alps and its palaeotectonic significance.
J. Loeschke, H. Heinisch

27 The pre-Alpine basement of the Lower Austro-Alpine nappes in the Bernina Massif (Grisons, Switzerland; Valtellina, Italy).
P. Spillmann, H. S. Büchi

28 Evolution of the Silvretta Nappe.
M. Maggetti, M. Flisch

29 Evolution of the Ötztal-Stubai, Scarl-Campo and Ulten basement units.
G. Hoinkes, M. Thöni

30 Austro-Alpine basement complex to the south of the Tauern Window.
B. Schulz, G. Nollau, H. Heinisch, G. Godizart

31 The Austro-Alpine metamorphic basement East of the Tauern Window.
F. Neubauer, W. Frisch

32 Late- and post-Variscan Sediments of the Eastern and Southern Alps.
K. Krainer

The Classic Fossiliferous Palaeozoic Units of the Eastern and Southern Alps

H. P. SCHÖNLAUB and H. HEINISCH

Abstract

In this work we review the present knowledge about stratigraphy, development of facies and tectonic evolution of the Variscan sequences of the Eastern and Southern Alps. In the Eastern Alps outcrops of fossiliferous rocks of Lower Palaeozoic age are irregularly distributed. They form a mosaic-like pattern of dismembered units incorporated into the Alpine nappe system (see Fig. 1 of Schönlaub, p. 65 this Vol.). Such areas include the Gurktal Nappe of Middle Carinthia and parts of Styria, the surroundings of Graz, a small area in southern Burgenland and the Greywacke Zone of Styria, Salzburg and Tyrol. South of the Periadriatic Line Variscan sequences are represented in the Carnic and Karawanken Alps where they form the basement of the Southern Alps. As regards the regions occupied by quartzphyllitic rocks of presumably Palaeozoic age the reader is referred to the contribution by Neubauer and Sassi (this Vol.).

Based on a comprehensive set of data a distinct geological history on either side of the Periadriatic Line is inferred. Main differences concern the distribution of fossils, the development of facies, rates of subsidence, supply area, amount of volcanism and the spatial and temporal relationship of climate sensitive rocks from north and south of the Periadriatic Line (Schönlaub this Vol.).

1 Introduction

Until about 1955 dating of sedimentary rocks of Palaeozoic age was mainly based upon macrofossils. Since the introduction of microfossil research methods in the mid-1950s, in particular conodont biostratigraphy, knowledge of sedimentary sequences increased considerably. Thus, together with sedimentological, microfacies, geochemical and structural data a very detailed subdivision of the geological record of the Eastern Alps has been established. Based on this multi-lined framework the Palaeozoic history of the Alps can be better inferred than ever before and hence, seems well constrained in today's geology of the Alps.

The occurrences of fossiliferous Palaeozoic outcrops represent different tectonic units (see Fig. 1 of Schönlaub, pp. 65 this Vol.). South of the Gailtal Fault they form the Variscan basement of the Southern Alps; to the north they belong to a huge thrust sheet named Upper Austro-Alpine Nappe. As far as their original palaeolatitudinal settings are concerned analysis of faunas and climate sensitive rock data has revealed fundamental differences between both major occurrences (Schönlaub this Vol.). In addition, the intra-Alpine facies development varies to a certain degree. For example, the Palaeozoic record from Middle Carinthia is lithologically more closely related to certain areas occupied by quartzphyllites than to any other region; the Greywacke Zone of Styria shows more similarities with the Carnic Alps than to the nearby Palaeozoic of Graz; this development reflects its own distinct setting suggesting an intermediate position between the Southern and Central Alps. Finally, in the Palaeozoic sequences of the Alps the participation of volcanic rocks varies considerably. They have been assigned to different geotectonic settings that characterized the Ordovician, Silurian and Devonian Periods (Loeschke and Heinisch this Vol.).

1.1 The Carnic and Karawanken Alps

The Carnic Alps of Southern Austria and Northern Italy represent one of the very few places in the world in which an almost continuous fossiliferous sequence of Palaeozoic age has been preserved. They extend in a West-East direction over 140 km from Sillian to Arnoldstein. In the following western Karawanken Alps the Variscan sequence is almost completely covered by Triassic rocks. To the east Lower Palaeozoic rocks are excellently exposed in the Seeberg area of the eastern Karawanken Alps south of Klagenfurt. In contrast to the Carnic Alps in this region Lower

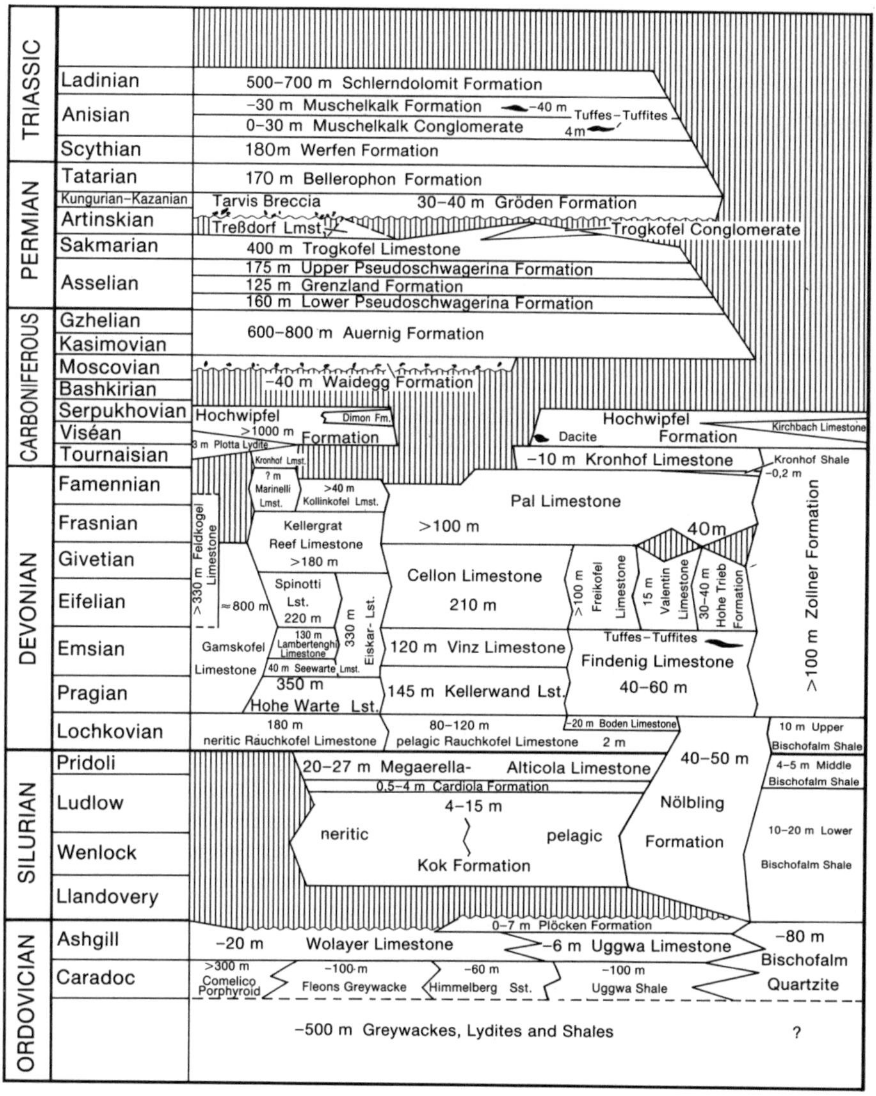

Fig. 1. Biostratigraphic scheme of the Palaeozoic sequence of the Carnic Alps. With only minor modifications this subdivision can also be applied in the Karawanken Alps. (Schönlaub 1985 b, amended by Kreutzer 1991)

Palaeozoic rocks are distributed on either side of the Periadriatic Line (Gailtal Fault). They were subdivided into a northern and a southern domain, respectively. The latter extends beyond the state border to northern Slovenia.

1.1.1 Review of Stratigraphy

Figure 1 summarizes the stratigraphy and facies relationship of various rocks of the Carnic Alps. With minor modifications this scheme is also valid for the Karawanken Alps (Schönlaub 1980; Moshammer 1989). Traditionally the sequence is subdivided into the Variscan basement rocks and its post-Variscan cover. The oldest fossiliferous rocks are Caradocian in age (Upper Ordovician) and comprise thick acid volcanics named Comelico Porphyroid and volcaniclastics of the Fleons Formation which laterally and vertically grade into the Uggwa Shale and the Himmelberg Sandstone. According to Dallmeyer and Neubauer (in press) detrital muscovites from the sandstones are characterized by apparent ages ($^{40}Ar/^{39}Ar$) of ca. 600 to 620 Ma and may thus be derived from a source area affected by Late Precambrian (Cadomian) metamorphism. They are succeeded by bioclastic limestones, i.e. the massive Wolayer Limestone and the corresponding quiet-water Uggwa Limestone, respectively. The global regression during the Hirnantian Stage (Late Ashgillian) is documented by arenaceous limestones of the Plöcken Formation. It resulted in channelling, erosion and local non-deposition. Thus basal Silurian strata generally disconformably overlie the Late Ordovician sequence.

Ordovician fossil groups include rich collections of bryozoans, brachiopods, trilobites, pelmatozoans and hyolithes occurring with varying abundances in the Uggwa Shale, and abundant conodonts in the limestones (Schönlaub, see summary in this Vol.).

In the Carnic Alps the Silurian transgression began at the very base of the Llandovery, i. e. in the graptolite zone of *Akidograptus acuminatus.* Its forerunner from the latest Ordovician, *Gl. persculptus,* was reported from the western Karawanken Alps. Due to the unconformity separating the Ordovician from the Silurian a certain pile of sediments is locally missing, which correspond to several conodont zones in the Llandovery and Wenlock in both the Carnic and Karawanken Alps. At some places even basal Lochkovian strata may disconformably rest upon Upper Ordovician limestones.

Silurian lithofacies is split up into four major facies reflecting different depths of deposition and hydraulic conditions. A shallow marine environment represents the Plöcken facies characterized in succeeding order by the pelagic Kok Formation, the Cardiola Fm. and the Megaerella-Alticola Limestones. The typical section is the 60 m thick Cellonetta profile well known for its merits for the Silurian conodont zonation established by Walliser (1964).

The Wolayer facies represents an even shallower environment. It is characterized by fossiliferous limestones with abundant orthoconic nautiloids, trilobites, small brachiopods and crinoids. Due to a period of non-deposition at the base this facies is represented by only 10 to 15 m thick limestones. The main occurrences are in the Lake Wolayer region of the central Carnic Alps.

The Findenig facies represents an intermediate facies between the shallow water and the starving basinal environment. It comprises interbedded black graptolite shales, marls and limestone beds. At its base a quartzose sandstone may locally occur.

The stagnant water graptolite facies is named the Bischofalm facies. It is represented by 60 to 80 m thick black siliceous shales, black cherty beds ("lydite") and clayish shales which contain abundant graptolites. Their distribution has been clearly outlined by the thorough work of Jaeger in the past 25 years (see Jaeger 1975; Flügel et al. 1977; Jaeger and Schönlaub 1980; Schönlaub 1985 a).

The four Silurian lithofacies reflect different rates of subsidence. During the Llandovery to the beginning of the Ludlow, sedimentation suggests a steadily subsiding basin and a transgressional regime. This tendency decreased and perhaps stopped during the Pridoli to form balanced conditions with uniform limestones being widely deposited. Simultaneously, in the Bischofalm facies black graptolite shales were replaced by greenish and greyish shales ("Middle Bischofalm Shale").

At the base of the Devonian in the Bischofalm facies the deep-water graptolite environment was restored until the end of the Lochkovian Stage. The succeeding strata of the Zollner Formation also represent a deep-water off-shore setting that lasted until the end of the Devonian or Early Carboniferous.

In comparison with the Late Ordovician and the Silurian, subsidence and mobility of the sea bottom significantly increased in the Devonian. This is documented in a Lower Devonian transgressional sequence including the Rauchkofel Limestone which is up to 180 m thick and corresponds to some 20 m of pelagic limestones ("Boden Lst."). During the Pragian and Emsian Stages the differences even increased. Within short distances of less than 10 km (Kreutzer 1991) a strongly varying facies pattern developed indicating a progressive but not uniform deepening of the basin. It was filled with thick reef and near-reef organodetritic limestones including different intertidal lagoonal deposits of more than 1000 m thickness in the Carnic Alps and some 300 m in the Karawanken Alps. They are time equivalent to some 100 m of pelagic cephalopod limestones and the pelitic Zollner Formation.

In the Carnic and Karawanken Alps reef growth started in the Lower Emsian. Main reef builders were stromatoporoids, tabulate corals and calcareous algae like *Renalcis.* For the Karawanken Alps Rantitsch (1990) concluded an arrangement of reefs resembling present-days atolls as opposed to the Carnic Alps with its barriere-type reefs (Kreutzer 1990). Depending on adequate subsidence, the location of the reef belt shifted spatially and temporarily during the Devonian. In contrast to the Carnic Alps with its 150 m thick reefs of Givetian age the Karawanken Alps show no good records of the Middle Devonian. In both areas the reef development ended in the Frasnian when the former shallow sea subsided and the reefs drowned and were partly eroded (Bandel 1972; Tessensohn 1974, 1983; Pohler 1982; Kreutzer 1990). Subsequently, with few exceptions, e.g. the Kollinkofel Lst., uniform pelagic goniatite and clymeniid limestones were deposited lasting from the Frasnian/Famennian boundary to the Late Tournaisian Stage. They were named Pal and Kronhof Lst., respectively. Generally, these wackestones contain abundant cephalopods, trilobites, radiolarians, foraminifera, ostracods, conodonts and even fish teeth.

The nature of the transition from these limestones to the following clastics of the Hochwipfel Formation raised a long-lasting controversy about the significance of tectonic events in the Lower Carboniferous. It has now been settled after recognizing a wide variety of distinct palaeokarst features in the Karawanken Alps (Tessensohn 1974) and in the Carnic Alps (Schönlaub 1990 a; Schönlaub et al. 1991) including an extensive palaeorelief with related collapse breccias, fissures, stratabound ore deposits and a silcrete regolith at the surface ("Plotta Lydite"), and caves with cave sediments, formation of speleothems and palaeokarst-related cements in the subsurface. The palaeokarst was caused by a drop in sea-level during the Late Tournaisian. Rise of sea-level and/or collapse of the carbonate basin promoted the transgression of the Hochwipfel Formation which presumably started as early as at the Tournaisian/Visean boundary.

On account of its characteristic lithology and sedimentology Tessensohn (1971, 1983), Spalletta et al. (1980), van Amerom et al. (1984), Spalletta and Venturini (1988) and others interpreted the 600 to more than 1000 m thick Hochwipfel Formation as a flysch sequence. In modern terminology the Kulm sediments indicate a Variscan active plate margin in a collisional regime following extensional tectonics during the Devonian Period. The main lithology comprises arenaceous to pelitic turbidites with intercalations of several tens of metres thick pebbly mudstones, disorganized debris flows and chert and limestone breccias in its lower part. They may represent submarine canyon fillings or inner fans. Widespread, although less abundantly, are up to 10 m thick massive sandstone beds. Upwards, and locally also laterally, the flysch grades into volcaniclastites and volcanics of the Dimon Formation.

With the exception of trace fossils the palaeontological content of the flysch series is very poor. According to van Amerom et al. (1984) and van Amerom and Schönlaub (1992) plant remains are fairly common suggesting a Middle Visean to Namurian age for the formation of parts of the flysch. Other stratigraphic data are derived from the underlying limestone beds and a few scattered limestone intercalations, i. e. the Kirchbach Limestone which provided index conodonts of the Visean/Namurian boundary (Flügel and Schönlaub 1990). Moreover, limestone clasts within the debrites are of great interest. They comprise a broad spectrum of shallow water carbonate shelf types with stratigraphically important fossils like the coral *Hexaphyllia mirabilis*, the algae *Pseudodonezella tenuissima,* the foraminifera *Howchinia bradyana* and early fusulinids. Apparently, these clasts, together with the turbidites, were supplied from a source area located originally to the north of the present Southern Alps.

In the Southern Alps the Variscan orogeny reached its climax between the Late Namurian and the Late Westphalian Stages. This time corresponds to the interval from the Early Bashkirian to the Middle or Late Moscovian Stages. According to Kahler (1983) the oldest post-Variscan transgressive sediments are Late Middle Carboniferous in age and, more precisely, correspond to the *Fusulinella bocki* zone of the Upper Miatchkovo Substage of the Moscovian Stage (Moscow Basin). In particular between Straniger Alm and Lake Zollner they rest, with a spectacular angular unconformity, upon strongly deformed basement rocks including the Hochwipfel Formation, the Silurian-Devonian Bischofalm Formation or different Devonian limestones. This basal part, named Waidegg Formation, consists mainly of basal conglomerates, disorganized pebbly siltstones and arenaceous and silty shales with thin limestone intercalations. Even metre-sized limestone boulders reworked from the basement were recognized at the base of the transgressive sequence (Fenninger et al. 1976) and named Malinfier Horizon by Italian geologists. The Bombaso Formation of the Naßfeld region, i. e. the Pramollo member, has also long been regarded as the base of the Auernig Group in this area (Venturini et al. 1982; Venturini 1990). Based on new field evidence, however, for this member a clear relationship with the Variscan Hochwipfel Formation is suggested.

South of Naßfeld the transgressive molasse-type cover comprises the 600 to 800 m thick fossiliferous Auernig Group followed by more than 1000 m thick limestones and clastics of Permian age (see Holser et al. 1991; Krainer this Vol.). They characterize a differentially subsiding carbonate platform and outer shelf setting which were affected by transgressive-regressive cycles from the Westphalian to the Artinskian Stages. This cyclicity may be explained as the response to the continental glaciation in the Southern Hemisphere (see Schönlaub this Vol.). Rocks of post-Variscan age also occur in the eastern Karawanken Alps.

In the eastern Karawanken Alps north of the Periadriatic Line rocks of Palaeozoic age have long been known. They belong to the "Diabaszug von Eisenkappel" (Fig. 2). This narrow belt extends in an W-E direction from Zell Pfarre via Schaidasattel to east of Eisenkappel and continues further east to Slovenia. In Austria this zone has a length of more than 25 km and a maximum width of 3.5 km. The 650 m thick Palaeozoic sequence comprises up to 350 m of volcanic and volcaniclastic rocks and sediments. According to Loeschke (1970–1977, 1983) the first group is dominated by basic tuffs and tuffitic

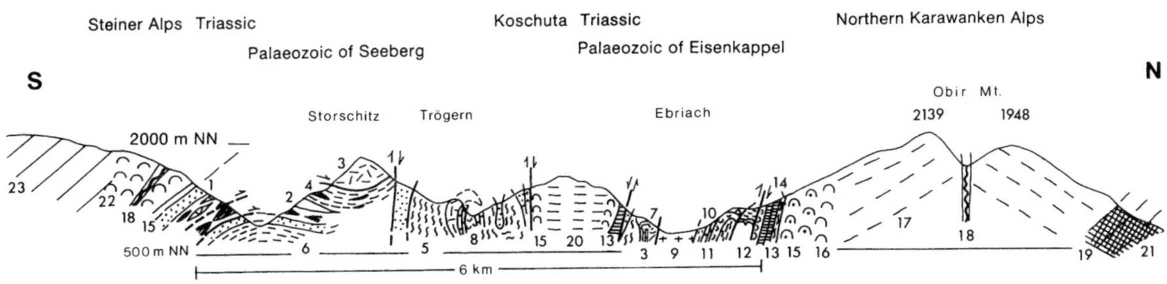

Fig. 2. N-S directed section through the Eastern Karawanken Alps. Numbers indicate *1* post-Variscan Permian and Late Carboniferous; *2* banded limestone slices; *3* Devonian limestones; *4* undated volcanics; *5* Hochwipfel Fm.; *6* Seeberg Shale; *7* Upper Ordovician and Silurian rocks; *8* volcanics of the Upper Ordovician; *9* granite of Eisenkappel; *10* pillow lava of the "Diabaszug of Eisenkappel; *11* sills; *12* Werfen Fm.; *13* Muschelkalk Fm.; *14* Partnach Fm.; *15, 16, 17* Wetterstein Lst.; *18* Raibl Fm.; *19* Rhätian to Jurassic deposits; *20* Schlern Dolomite; *21* Tertiary; *22, 23* Dachstein Lst. (Schönlaub 1979)

rocks, massive pillow lavas and basic sills of hawaiitic composition with ultrabasic layers. Sills and pillow lavas represent spilites which differentiated from alkali-olivine basalts, the original geotectonic setting of which is yet not known. Subsequent low-temperature metamorphism associated with devitrification and metasomatic replacement processes caused the spilitic mineral composition in these rocks. The sedimentary rocks are monotonous grey shales and slates with intercalations of conglomeratic greywackes, quartzitic and graphitic sandstones and thin limestone beds. The definite age of this succession is, as yet, not exactly known although some poorly preserved single cone conodonts recovered from the limestone intercalations are rather in favour of an Ordovician than any younger age (Neubauer pers. comm.).

1.1.2 Tectonic Remarks

The Palaeozoic sequence of the Carnic and Karawanken Alps represent a strongly compressed WNW-ESE-running thrust sheet complex composed of isoclinally folded anchito epimetamorphic Palaeozoic rocks. This style of deformation developed during the Variscan orogeny in the Late Namurian or Early Westphalian. It is sealed by the post-Variscan cover overlying the deformed basement with a distinct angular unconformity. Paraconformities occur at different levels within the Palaeozoic sequence, for example, at the end of the Ordovician, in the late Middle and early Upper Devonian and in the Lower Carboniferous. Presumably, they were caused by sea-level changes related to the glaciation of parts of Gondwana at the end of the Ordovician, to seismic shock events, and to a palaeokarstic event, respectively. Lowering of sea-level and/or block faulting may also have acted at the end of the Trogkofel Stage being responsible for extensive erosion and accumulation of reworked limestones, stratigraphic gaps, formation of fissures and local karstification.

For many years the complicated tectonics of the Carnic Alps was explained in terms of nine nappes produced during the Variscan orogeny. Each north-verging nappe consisted of a more or less continuous Ordovician to Devonian sequence and was separated from the next by the clastics of the Hochwipfel Formation. The extent of Alpine overprints on this pile of nappes was difficult to decide. With respect to the less deformed post-Variscan cover, however, it was concluded that the intensity of the Variscan tectonics was much stronger than the Alpine deformation. Nevertheless, the latter resulted in interferences between both and was responsible for a complex deformative pattern in the Southern Alps (Castellarin and Vai 1981).

According to Vai (1979) the horizontal shortening of the Carnic Alps during the Variscan deformation is estimated to be 75–80% of its original width. This value does not consider the assumed detachment from pre-Ordovician basement rocks.

Based on new field data, mainly from the Naßfeld area, the old concept was challenged by Venturini (1990, 1991) who proposed a new structural model. He speculated on three distinct and interacting phases that resulted in different systems of asymmetric folds and faults distributed along a N 120°–140° E direction (Fig. 3).

1. Middle or early Upper Carboniferous compressional tectonics caused a huge SSW-verging fold that affected the whole belt. Syn-kinematically a back-fold system with clear northern vergence developed on its rear side. Such smaller-scale syn- and anticlines can be recognized, for example, on Roßkofel, at Hoher Trieb, at Plöckenpaß-Kleiner Pal-Piz Timau. Perhaps even the fold structure sep-

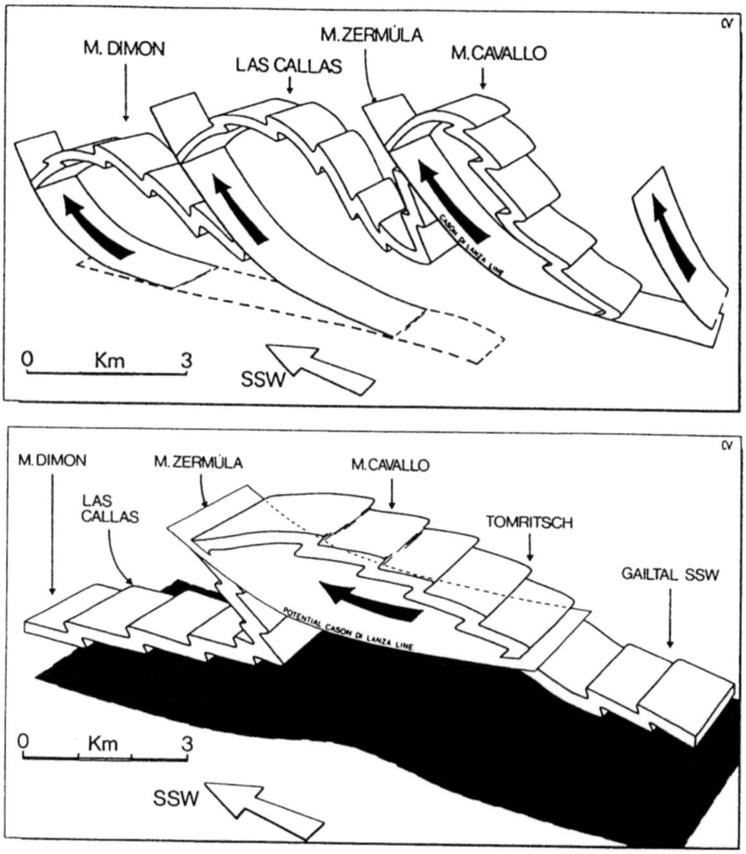

Fig. 3. Hercynian deformation of the Carnic Alps. Shown are the 1st and 3rd *(above)* deformative phases. The huge asymmetric fold affected the whole Palaeozoic belt. The 3rd phase formed thrusts with open folds which refolded the older structures of the 1st and 2nd deformative stages (Venturini 1990)

arating the Cellon subnappe from the Kellerwand-subnappe (Kreutzer 1990) can be attributed to this deformation.

2. In response to uplifting, brittle deformation occurred with development of flat fault planes along shale horizons. As a result the huge asymmetrical fold was cut into smaller tectonic slices.
3. The third phase occurred during further uplift. It produced huge open antiforms following new thrust planes and older folded structures. They were later reactivated during the Alpine compression.

The formation of sedimentary basins in the Upper Carboniferous was governed by extensional tectonics (Venturini 1990). They were related to 120° to 130° directed fault zones forming, thus, an elongated trough with an original width of not more than 15 km, shortened today to a narrow zone of some 10 km.

With regard to the Alpine deformation of the Carnic Alps the reader is referred to Eichhübl (1988) and Venturini (1990, 1991).

The tectonic framework of the eastern Karawanken Alps is characterized by the north-verging anticlinal structure of the central and southern part (Fig. 4). Its axis dips gently towards southwest. The whole area may be subdivided into two superposed allochthonous units. In addition, north of the Seeberg anticline the folded Trögern area further complicates the deformation style.

1. In the area around the Seeberg Pass, the uppermost unit is represented by the Reef unit. Near the Pass rocks of the core are well-exposed. They comprise reef and near-reef limestones, e. g. north of Plasnik (P. 1257), at Rapold, Pasterk, Storschitz and at the Grintoutz localities. Laterally this facies grades into forereef and pelagic deposits. Generally, the sequence within this unit consists of different limestones of Devonian age, followed locally by the Carboniferous Hochwipfel Formation and transgressive sediments of Late Carboniferous and Permian age. At the southern limb, geologically well-known outcrops are located such as Paulitschwand, Leßnik and Sadnikar, while on the northern side Sadonighöhe, Stanwiese, Grintoutz and Hirschfelsen have long been famous. The lateral movement of the Reef unit is estimated to be some 4.5 km.
2. The above-mentioned uppermost unit is underlain by the Bänderkalk unit ("striated banded lime-

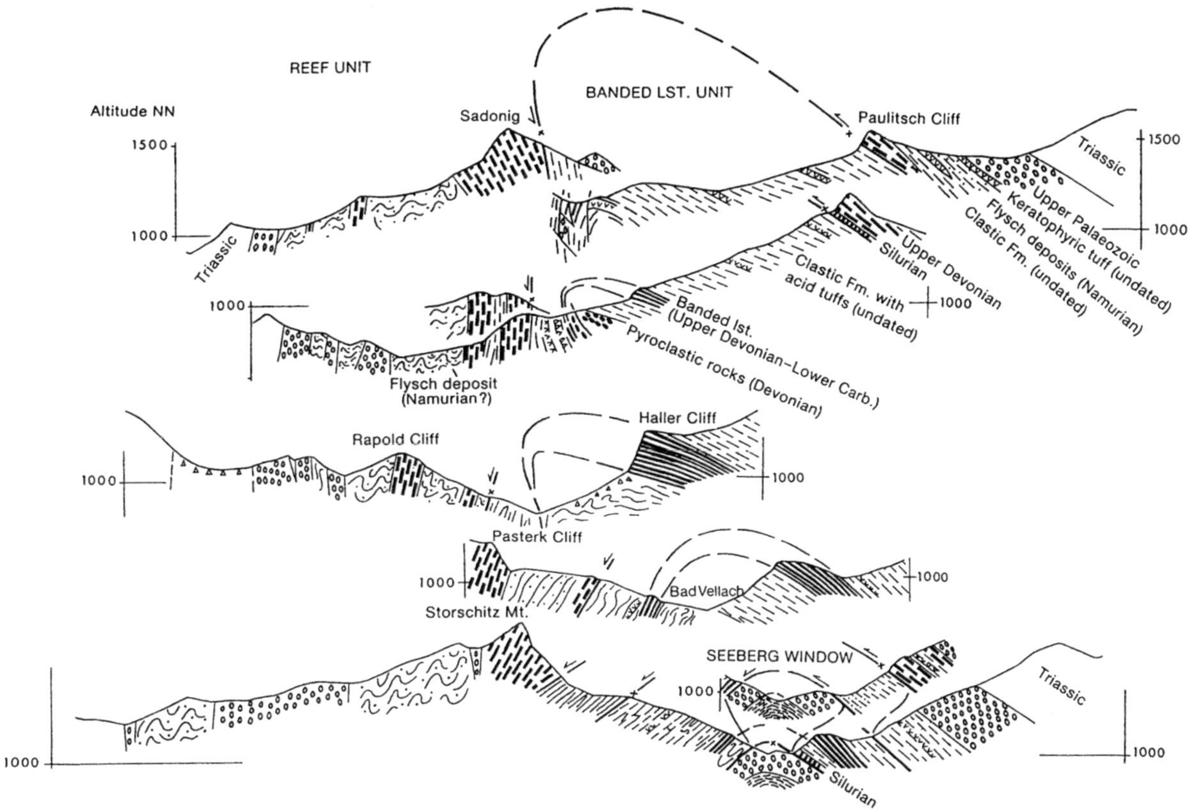

Fig. 4. N-S running sections through the Palaeozoic of the Seeberg area of the eastern Karawanken Alps. (After Rolser and Tessensohn 1974)

stone unit"). It is dominated by banded limestones as well as over and underlying clastics. Locally, at its base, nautiloid-bearing Silurian limestones and Lower Devonian tentaculite-bearing limestones occur. The amount of thrusting in this unit does not exceed 1.5 km.

3. The Basal unit is well distributed between the village of Bad Vellach and the locality "Steiner". Structurally, this unit can be regarded as a tectonic window (Fig. 5). Its sequence consists of the "Seeberg Shale" the age of which has yet not been ascertained and its transgressive cover formed by the equivalents of the Auernig Group, i. e. fusulinid-bearing limestones, shales, sandstones and quartz-rich conglomerates.

To unravel the complicated tectonic deformation of the eastern Karawanken Alps the already mentioned Late Carboniferous sediments are of critical importance as they provide clear evidence of the age of nappe-forming processes. Due to the fact that the post-Variscan molasse-type sediments are also involved in the nappe pile the main deformation in this area must be of Alpine age.

North of the anticline formed by these nappes the folded zone of Trögern occurs. It is characterized by a steep-to-vertical-dipping sequence dominated by clastic rocks of the Hochwipfel Formation. Locally also, the Devonian substratum and the post-Variscan cover ist exposed showing a mushroom and drop-like appearance due to squeezing of competent rocks between clastic layers. This zone may locally attain a width of more than 3 km.

In addition to the huge fold structures with amplitudes of several hundreds of metres small-scale folding is very common in the Seeberg area. It mainly affects those regions which are occupied by shales, i. e. the Seeberg Shale and the Hochwipfel Formation. Finally, steep faults have further subdivided the whole area into numerous small blocks. During the uplift of the whole area the Triassic cover of the Koschuta belt and the Steiner Alps detached from the underlying Late Carboniferous and Permian rocks.

The narrow belt of the "Diabaszug of Eisenkappel" from north of the Periadriatic Line is fault-bounded to the north and the south (Fig. 3). It represents a highly compressed folded and faulted north-verging zone showing several repetitions. To

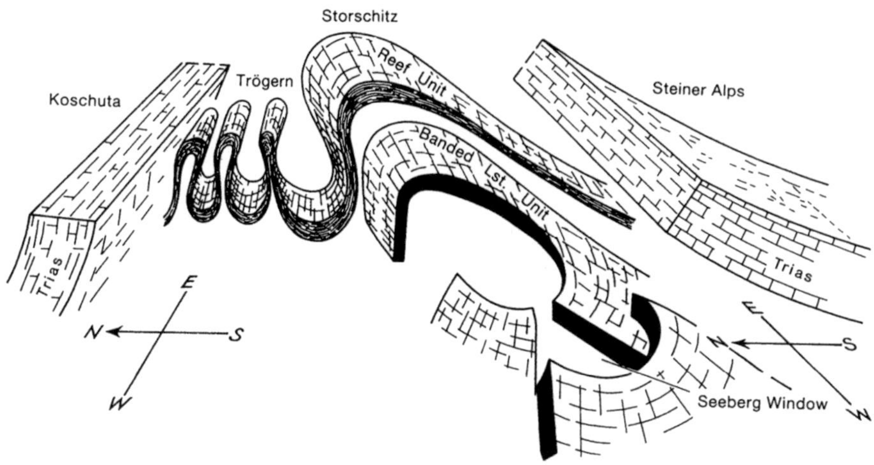

Fig. 5. Diagram showing the main tectonic units of the Seeberg area of the Eastern Karawanken Alps. (Rolser and Tessensohn 1974)

the north this belt of Palaeozoic rocks is thrust upon Late Permian and Triassic rocks. Most probably they formed the original cover of the Lower Palaeozoic volcaniclastic sequence suggesting, thus, a Variscan deformation for this Palaeozoic series. The southern boundary is formed by the northward thrust Karawanken granite. According to radiometric dating it was formed during Late Variscan times (Cliff et al. 1975). During intrusion the Diabase of Eisenkappel and its accompanying metamorphic rocks were marginally affected by contact metamorphism (Exner 1972).

2 The Carboniferous of Nötsch

The famous fossiliferous outcrops of Carboniferous age are located in the Gail Valley between Windische Höhe and the Villacher Alpe. It culminates in the peak called Badstube (1369 m) and is crossed by the Nötsch River. The name-bearing village of Nötsch, however, is situated in the Gailtal crystalline complex following to the south of the Carboniferous deposits (Fig. 6).

Since the beginning of the last century the Carboniferous of Nötsch has been famous for its abundance of fossils and thus has attracted many geologists and palaeontologists. The east-west directed exposures extend as a narrow fault-bounded wedge over a distance of 8 km, the maximum width of which is 2 km in the east. Further to the west the Carboniferous rocks are squeezed out between the above-mentioned rocks and are also covered by Quarternary deposits, respectively.

The tectonic significance of these Carboniferous rocks has raised many controversial statements in the past. In fact, the true relationship between the Carboniferous sediments and the surrounding units of the Gailtal crystalline complex and the Drauzug has long been a matter of debate and has yet not been solved satisfactorily. One of the main problems concerns the northern boundary of the Carboniferous deposits (see Fig. 6). Some authors consider it as a distinct fault zone separating the Carboniferous from the Permian and Triassic, while others assume an originally transgressive relationship between Upper Carboniferous rocks and the overlying Permian clastics. A conclusive decision about one of the two options has significant implications for the tectonic framework of the greater part of the Eastern Alps.

Based on a revised map and additional palaeontological work carried out in the last few years (Schönlaub 1985 b), knowledge of most rocks and fossils considerably increased. In the south-dipping sequence which was affected by several NNE-SSW trending distinct faults the oldest part occurs in the north and is named Erlachgraben Formation. Towards the south it is followed by the Badstub Breccia and the Nötsch Formation. Erlachgraben and Nötsch Formations display similar lithologies such as greyish blackish shales, micaceous siltstones, sandstones and quartz-rich conglomerates. Locally, fossils occur very abundantly. The disorganized Badstub Breccia is composed of mainly subrounded and rounded crystalline clasts such as amphibolites, ortho- and paragneisses, schists, micaschists, quartz, quartzites, marbles and few limestone clasts embedded in a dense green matrix of tholeiitic composition. From sedimentological evidence Schönlaub (1985 b) and subsequently Krainer and Mogessie (1991) inferred a sedimentary origin for the breccia. Previously a volcanic source was favoured for the origin of this rock. Conodonts recovered from limestone clasts indicate a formation

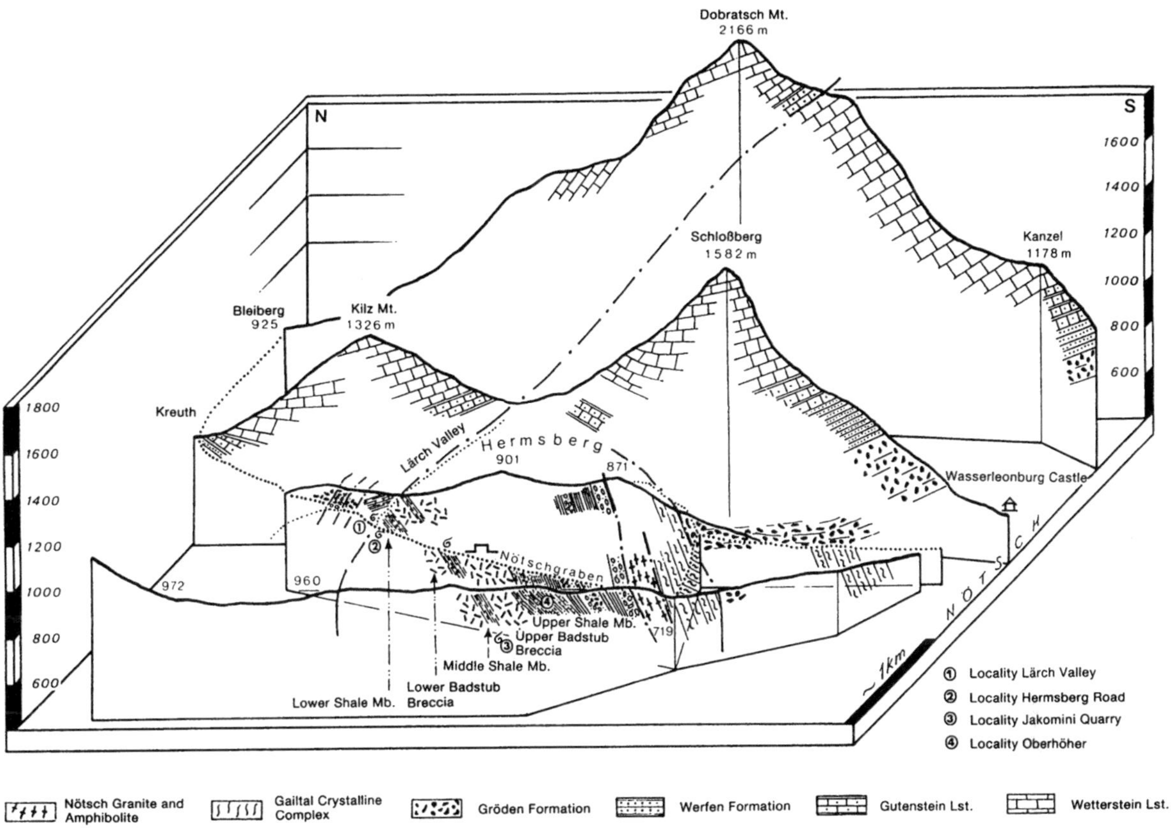

Fig. 6. Diagrammatic sketch of the eastern part of the Carboniferous of Nötsch showing the south-dipping succession along the river Nötsch bounded to the north and south by faults. To the east the Carboniferous deposits are overlain by Permian and Triassic rocks of the mountain Dobratsch (= Villacher Alpe). (Schönlaub 1985b)

after the *Paragnathodus nodosus* Zone. In terms of the presently used, chronostratigraphical subdivision of the Carboniferous, this time corresponds to the latest Visean or more probably, to the early Serpukhovian.

New and revised fossil data (see Schönlaub this Vol.) suggest an overall Late Visean to Early Westphalian age for the molasse-type Carboniferous sediments. The dominating fossil groups are brachiopods, followed by bivalves, trilobites, gastropods, corals, crinoids, bryozoans, very few cephalopods and plants; microfossils include foraminifera, ostracods and few conodonts. In addition in the clastics trace-fossils are fairly common.

Yet the basement of the transgressive Carboniferous sequence has not been found. It may either be formed by an amphibolite-grade crystalline complex or less probably, by the Gailtal quartzphyllite. Interestingly, at several places north of the village of Nötsch there is clear evidence of a transgressive relationship between the latter and overlying Permian clastics (Schönlaub 1985b). It may, thus, be concluded that the present outline of the Carboniferous basin was formed during the Alpine orogeny which affected and rejuvenated older faults and created new ones parallel to the Periadriatic Line. Extensive N-S shortening was mainly responsible for the closely neighbouring different tectonic units observed today; in addition, vertical movements promoted the preservation of Carboniferous deposits distributed today in an apparently distinct and almost exotic setting.

3 The Palaeozoic of the Gurktal Nappe

For a long time a few fossil localities only were the sole data base in this region, i. e. at Aich quarry near Althofen, at Christofberg north of Klagenfurt and in the surroundings of Feldkirchen. This changed after the introduction of research methods for microfossils, in particular conodonts. Since then many new data have been obtained:

1. A conodont-based subdivision of the Magdalensberg Group of Kahler (1953) was first established

by Strehl (1962) on the western margin of the Saualpe between Eberstein and Klein St. Paul. He recognized a 400 m thick clastic sequence with intercalations of limestones ranging from almost the base of the Silurian to the Upper Devonian. It succeeds basic volcanics and pyroclastic rocks assigned to the Upper Ordovician. Interestingly, one of the lowermost limestone lenses of the Silurian contains abundant tabulate corals and other fossil debris (crinoids, brachiopods) suggesting a bioclastic or patch-reef origin. Buchroithner (1979) confirmed this conclusion and extended the biostratigraphic information to the Upper Ordovician (Ashgillian). From the same area of Klein St. Paul, Neubauer and Pistotnik (1984) recorded basic volcanics of Lower Silurian age.

2. Clar et al. (1963) supplemented these data and provided additional biostratigraphic evidence from the area between Althofen-Mölbling and the Magdalensberg region including the western part of the Saualpe.
3. From limestone intercalations within phyllitic slates of the southeastern part of the Saualpe, Kleinschmidt and Wurm (1966) recorded Upper Silurian (Ludlowian) conodonts.
4. Neugebauer (1970) discovered spiriferid brachiopods from marbles within the "Phyllite Group" of the Saualpe region suggesting a Devonian age.
5. The Aich quarry near Althofen was further subdivided by Schönlaub (1971). It displays a 50-m-thick limestone succession ranging from the Lower Emsian to the Early Famennian. Parts of the Middle Devonian, however, are missing. The limestone sequence is overlain by shales and cherts which yielded conodonts assignable to the Late Tournaisian or the Tournaisian/Visean boundary (Neubauer and Herzog 1985).
6. At Mölbling, some 5- to 10-m-thick limestones and dolomites provided conodonts of Late Silurian age. According to Buchroithner (1979) these carbonate rocks may well represent the extended base of the nearby Aich quarry section. In addition, Upper Ordovician conodonts were found near Mölbling (Neubauer and Pistotnik 1984) suggesting an overall continuous long-ranging Ordovician to Carboniferous section in this region.
7. Additional biostratigraphic information was derived from Drasenberg near the village of Meiselding and from Schelmberg east of Guttaring (von Gosen et al. 1982; Neubauer and Herzog 1985). Based on conodonts some limestone lenses were assigned to the Lower and Upper Devonian, respectively. They occur within a siliciclastic sequence which, according to Neubauer and Herzog (1985), suggests similarities with the flysch-type Hochwipfel Formation of the Southern Alps.

In conclusion, the facies development of the Gurktal nappe system varies between a carbonate dominated and a carbonate-poor facies (Tollmann 1977; Buchroithner 1979; Ebner et al. 1990; see Fig. 7). The first represents the pelagic "Facies of Althofen" and corresponds to the Pridolian, the whole Devonian and the Lower Carboniferous. It is opposed by the "Facies of Magdalensberg" representing a more than 500-m-thick clastic-volcanic sequence with intercalations of 2- to 8-m-thick limestone horizons. It spans the interval from Middle (?) Ordovician to the Middle Devonian (Buchroithner 1979; Pistotnik 1989).

This facies variation has also been documented from north and northwest of central Carinthia, i.e. the Murau and Turrach areas (Höll 1970; Ebner et al. 1977; Buchroithner 1978, 1979; Neubauer 1979; Neubauer and Pistotnik 1984; Ebner et al. 1990; Neubauer and Sassi this Vol.). In these regions the oldest part of the sequence (?) is formed by the "Metadiabase Formation" of presumably Lower or Middle Ordovician age (according to Schnepf 1989 they may be assigned to the Silurian). These thick mafic volcanics are overlain by the more than 100-m-thick Golzeck Sandstone containing conodont-bearing Upper Ordovician limestone lenses, the 7-m-thick Golzeck Quartzporphyry and the 6-m-thick Lower Auen Dolomite, indicating, most probably, an Ashgillian age. After a sedimentary break lasting from the Lower to the Middle Silurian the following part of the sequence is represented by the 20-m-thick Middle Auen Dolomite in the Upper Silurian and the 20-m-thick Haider Marble of Lower to Middle Devonian age. The 10-m-thick Upper Auen Dolomite represents the uppermost portion of the sequence indicating a Frasnian age.

The concurrent clastic facies displays marked differences (Neubauer 1979; Neubauer and Pistotnik 1984). For example, at Prankerhöhe, distinct lithologies comprise more than 500 m of sandstones, quartz wackes and quartz arenites underlying the Emsian to Eifelian Ursch Dolomite. Also, in the surroundings of Turrach similar lithologies are distributed.

In addition, the Lower Palaeozoic sequences of the Gurktal Nappe system are characterized by volcanic activities. Volcanism occurred at different times with varying intensity and differing geochemical composition depending mainly on its palaeotectonic setting. For further details, the reader is referred to the respective chapters of this volume (Loeschke and Heinisch; Neubauer and Sassi).

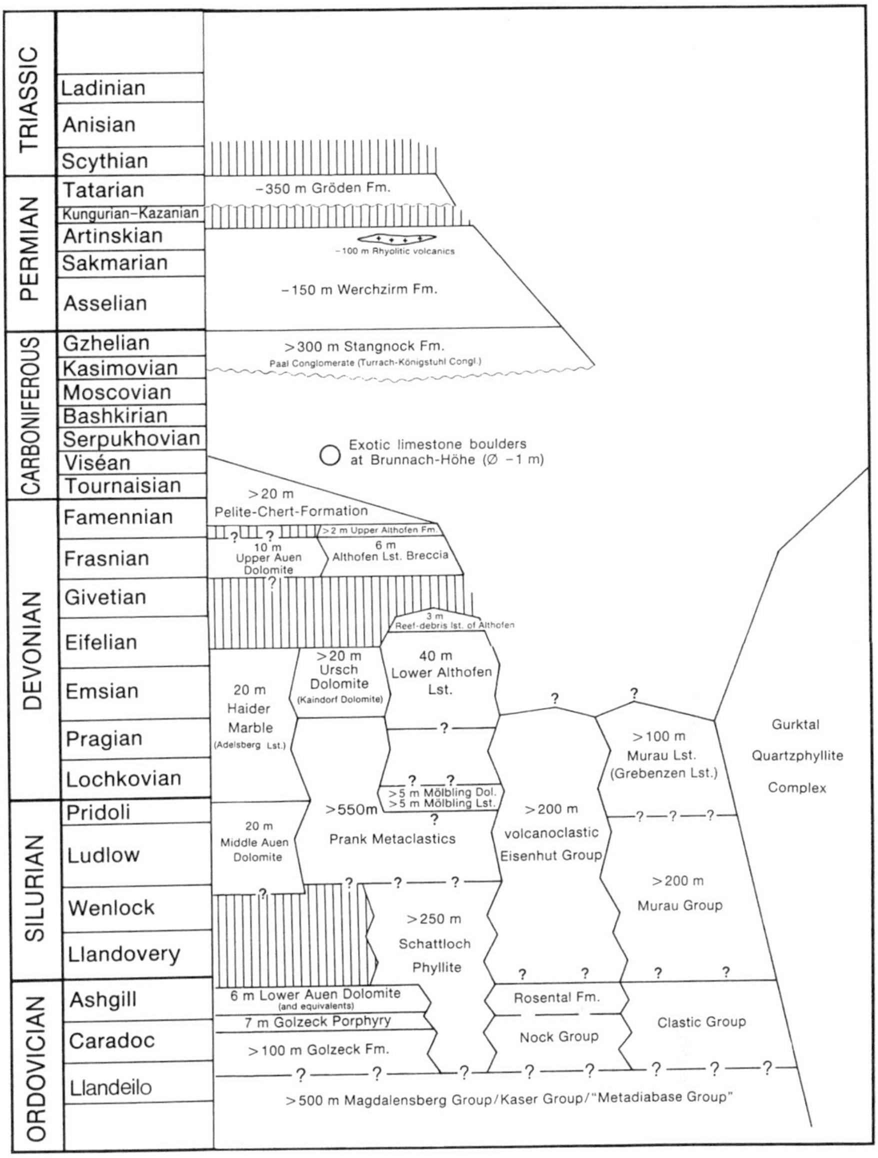

Fig. 7. Stratigraphy of the Palaeozoic sequence of the Gurktal Nappe of Middle Carinthia and the surroundings of Murau (SW Styria). Compiled from Schönlaub (1971), Buchroithner (1979), Neubauer and Pistotnik (1984), Havlicek et al. (1987), Loeschke (1989 a), Schlöser et al. (1990), Krainer (this Vol.) and unpublished data from one of the authors (Schönlaub)

4 The Palaeozoic of Graz

4.1 Introduction

The Palaeozoic sequence around Graz has long been famous for its varied lithology and its abundance of fossils of mainly Devonian age. The Palaeozoic outcrops cover an area of approx. 50 × 25 km to the east and west of the Mur river (Fig. 8). Even within the city of Graz Silurian and Devonian oucrops are widely distributed. Close to the southwestern edge the "Gosau of Kainach" transgressively overlies the Lower Palaeozoic sequence. Its eastern and northern frame is formed by crystalline rocks attributed to the Middle Austro-Alpine tectonic complex. To the south the Palaeozoic is covered by Tertiary rocks which mask the assumed continuation to the Sausal and Remschnig counterparts of southwestern Styria.

Recently, Fritz et al. (1986, 1991) subdivided the Palaeozoic nappe system into lower and upper nappes (Fig. 10). The first type is represented by the metamorphic Schöckel Nappe and the 'Angerkristallin', the latter by the anchi- to epimetamorphic Hochlantsch, Rannach, Heuberg, Laufnitzdorf and Kalkschiefer Nappes. In comparison with the former these higher nappes lack a Variscan ductile structural imprint.

Figure 8 summarizes the lithology, stratigraphic range and regional distribution of the main facies de-

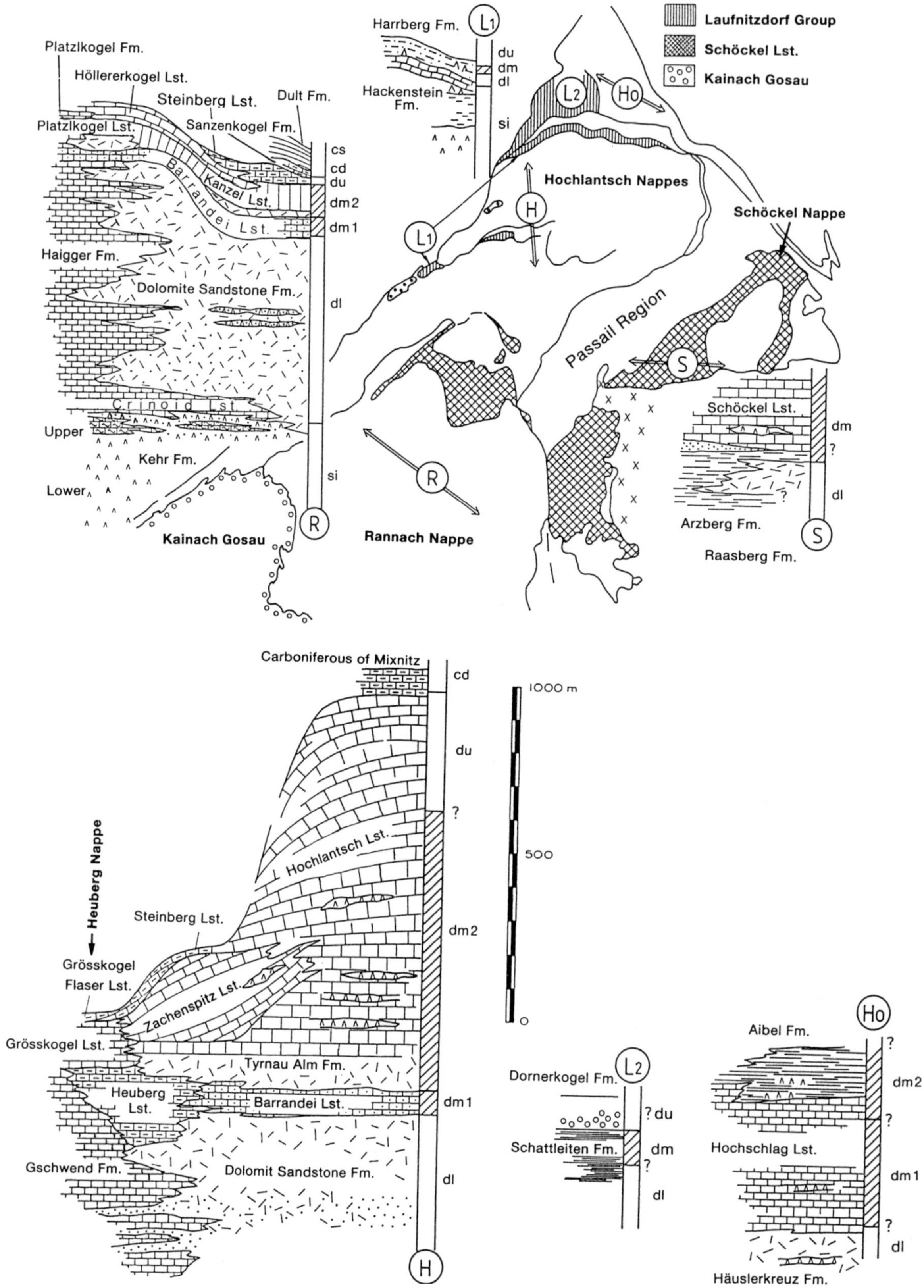
Laufnitzdorf Group
Schöckel Lst.
Kainach Gosau
Platzlkogel Fm.
Höllererkogel Lst.
Steinberg Lst.
Dult Fm.
Platzlkogel Lst.
Sanzenkogel Fm.
Barrandei Lst.
Kanzel Lst.
cs
cd
du
dm2
dm1
Haigger Fm.
Dolomite Sandstone Fm.
dl
Crinoid Lst.
Upper
Kehr Fm.
Lower
si
Kainach Gosau
R
Harrberg Fm.
L1
du
dm
dl
Hackenstein Fm.
si
L2
Ho
Hochlantsch Nappes
H
Schöckel Nappe
Passail Region
S
Rannach Nappe
Schöckel Lst.
dm
?
dl
Arzberg Fm.
Raasberg Fm.
Carboniferous of Mixnitz
cd
du
?
Heuberg Nappe
Hochlantsch Lst.
Steinberg Lst.
Grösskogel Flaser Lst.
Zachenspitz Lst.
dm2
Grösskogel Lst.
Tyrnau Alm Fm.
Heuberg Lst.
Barrandei Lst.
dm1
Gschwend Fm.
Dolomit Sandstone Fm.
dl
H
1000 m
500
0
Dornerkogel Fm.
?du
Schattleiten Fm.
dm
dl
Aibel Fm.
dm2
Hochschlag Lst.
dm1
Häuslerkreuz Fm.

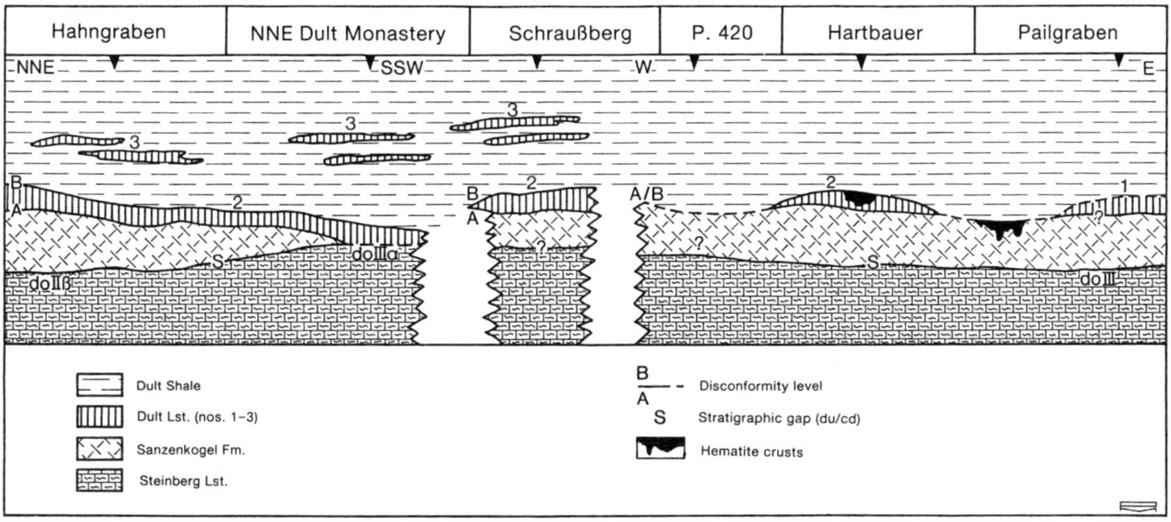

Fig. 9. Upper Devonian and Lower Carboniferous sediments of the Graz Palaeozoic. Note unconformity at the *D/C* boundary (letter *S*) and between the Sanzenkogel and the Dult Formations *(A, B)*. (Ebner 1976)

velopments of the Rannach and Hochlantsch Nappes supplemented by additional data from other units. This sketch clearly demonstrates that the individual nappes represent facies nappes.

4.2 *Review of Stratigraphy*

The Palaeozoic history of the Graz area is best displayed by the sequence of the Rannach Nappe. The oldest parts comprise basic metavolcanics of Ludlow age followed by Upper Silurian to Lower Devonian clastics and limestones. Sedimentation of different carbonates dominated from the Middle Devonian to the beginning of the early Upper Carboniferous.

New field data from the Silurian Kehr Formation indicates a sedimentation pattern controlled mainly by volcanism (Fritz and Neubauer 1988; Neubauer 1989). The eastern area is characterized by a proximal shallow water setting with lavas and coarse lapilli tuffs, while the western sections represent the distal facies displaying cinerites with intercalated lapilli-rich beds, agglomerates, shales and pelagic limestones. Beside other components the Kehr Agglomerate consists of 1–3 % quartzites, dolomites, cherts and reworked limestones. During the succeeding Lochkovian and Pragian Stages sedimentation of the 0 to 100 m thick Crinoid Beds and time-equivalent quartz-arenites followed the inherited topography. However, in the course of the Lower Devonian, the clastic input progressively ceased in favour of carbonate sedimentation, and finally in the Emsian gave way to an almost uniform formation of dolomites, i. e. the 500 to 1000 m thick Dolomit-Sandstein Formation (Fenninger and Holzer 1978). However, in its lower part a weak volcanic activity is indicated by the intercalation of pyroclastic rocks.

New field and biostratigraphic data obtained from other nappes suggest overall similar environmental conditions for the Upper Silurian portion of the respective sections of the Schöckel Nappe and the Laufnitzdorf Group. In the latter, however, pelagic nodular limestone sedimentation persisted during the Devonian (Gollner et al. 1982).

The lower 300 m thick Sandstone member of the Dolomit-Sandstein Formation comprises thick sandstone beds, dolomitic shales and bioclastic dolomites; the succeeding Dolomite member consists of lower greyish and upper blackish dolomites. In the grey dolomites Fenninger and Holzer (1978) recognized stromatolites, sheet cracks, fenestral fabrics, tepee structures, pseudomorphs of gypsum indicating a supra- and subtidal environment; the black types are rich in amphiporids suggesting a setting in a protected lagoon. As far as these types and the extent of the basal sandstone member are concerned lateral and vertical variations are quite common (see Gollner and Zier 1985).

◄

Fig. 8. Stratigraphy of the thrust system of the Palaeozoic of Graz. *Letters* of the stratigraphic columns indicate: *R* Rannach Group; L_1, L_2 Laufnitzdorf Group; *H* Hochlantsch Group; *Ho* Hochschlag Group; *S* Schöckel Group. (Flügel and Neubauer 1984)

In the Eifelian, the thick Dolomit-Sandstein Formation is succeeded by the Barrandei and the Kanzel Limestones. The first are represented by 80- to 100-m-thick greyish packstones and bioclastic limestones rich in corals, stromatoporoids, brachiopods and crinoids indicating a shallow water environment (Flügel 1975a). Locally, they are interbedded with graphitic shales and brownish marls (e.g. "Chonetenschiefer"). In the Heuberg Nappe, the Heuberg Limestone represents a lateral equivalent of the Barrandei Lst. The main difference is its increased content of clastic detritus suggesting a more near-shore setting.

The upper portion of the Barrandei Lst., represented by black *Amphipora*-bearing limestones is succeeded by the coarse-bedded and massive, up to 100-m-thick Kanzel Limestone. It comprises mudstones and wackestones with only few corals and other fossils indicating a Middle or Upper Givetian age. Presumably, they were formed in a protected and little agitated, moderately deep environment. Time equivalent deposits are Platzkogel Lst. and Größkogel Lst., respectively, which may also pass to platy limestones, thin-bedded clayish limestones and even sandstones. Most probably they represent a near-shore shallow water environment close to or shortly after the Middle/Upper Devonian boundary.

In the Hochlantsch Nappe the 140- to 500-m-thick Tyrnau Alm Formation largely corresponds to the Kanzel Lst. and its equivalents; previously it was known as "Calceola Beds". This formation comprises dolomitic rocks, sandstones, dolomitic sandstones, limestones, shales and rauhwackes in the lower part followed by a volcanic horizon and an upper limestone member. This latter contains abundant corals and stromatoporoids forming small biostromal accumulations. For details, see Gollner and Zier (1982, 1985), Zier (1982) and Gollner (1983).

The Tyrnau Alm Formation is overlain by the 350-m-thick, mostly well-bedded Zachenspitz Limestone, previously named "Quadrigeminum Lst.". They are characterized by abundant fossils such as tabulate and rugose corals, bioherms of stromatoporoids, amphiporids, crinoids and other fossil debris, suggesting a subtidal, quiet water environment with local reef growth. Locally, at the base volcaniclastic intercalations are common.

The lateral equivalent of this formation is the more than 800-m-thick Hochlantsch Limestone. The lithology comprises massive greyish to reddish limestones which have been interpreted as shallow water, back-reef deposits. They provided only few fossils like, for example, conodonts, tabulate corals and *Amphipora* indicating a Givetian to Frasnian age.

In the Hochlantsch Nappe, the uppermost beds are represented by 15-m-thick styliolinid limestones resembling the Steinberg Lst. of the Rannach Nappe. According to conodonts, a Frasnian age of these limestones has been concluded. They are disconformably succeeded by a limestone breccia forming the base of the cephalopod bearing "Carboniferous of Mixnitz". These pelagic deposits are up to 100-m-thick and correspond to the Upper Tournaisian, Visean and basal Namurian B.

In the Rannach Nappe, the Upper Devonian is represented by the greyish to yellowish or reddish and between 20- and 70-m-thick Steinberg Limestone. The transition from the underlying Kanzel Lst. is gradational starting in the *varcus* conodont Zone of the Late Givetian and ending at different time levels of the Famennian. In general, these limestones closely resemble cephalopod limestones widely distributed in the Variscan region, e.g. the Rhenish Slate Mts., the Carnic Alps or southern France. The fauna (cephalopods, trilobites, ostracods, foraminifera, conodonts) indicate a pelagic setting of moderate depths between 60 and 300 m. Corresponding units occur in the western part of the Rannach Nappe at Platzlkogel and at Höllererkogel (Höllererkogel Lst.) but also in southern slices of the Hochlantsch Nappe (Größkogel Lst.).

The 20- to 25-m-thick Sanzenkogel Formation represents the Lower Carboniferous part of the limestone succession of the Rannach Nappe (Fig. 9). This unit comprises greyish bedded limestones with intercalations of chert, phosphorite and shales in the lower part. They disconformably overly various levels of the Steinberg Lst. suggesting significant erosion and non-deposition in the Tournaisian. As a result the Lower Sanzenkogel Limestone ranging from the equivalents of the *sulcata* Zone to the Upper Tournaisian *anchoralis-latus* Zone, does not exceed 2.2 m in thickness; frequently it is even completely missing. Formation of fissures, a distinct micro- and macrorelief, internal sediment accumulations, hematite crusts, hardground and mixed conodont faunas most probably indicate local emersion and karstification at the end of the Tournaisian (Ebner 1978a, b; Fig. 9). Subsequently, limestone sedimentation continued and the Upper Sanzenkogel limestone was deposited.

In the Rannach Nappe, the more than 70-m-thick predominantly clastic Dult Formation represents the youngest rock. At its base the 10- to 15-m-thick Dult Limestone is developed which disconformably overlies the Sanzenkogel Limestone (Fig. 9). Based on conodonts, these basal parts represent shallow water limestones corresponding to the Namurian B. Due to the relief at the base these limestones have only local-

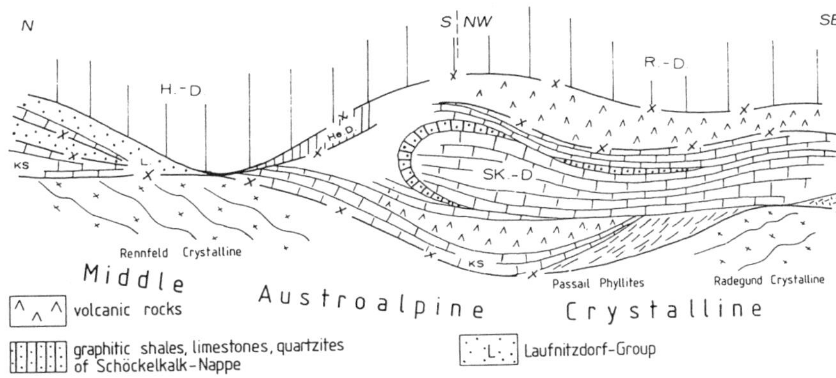

Fig. 10. Tectonic sketch of the Palaeozoic of Graz. *H.-D.* Hochlantsch Nappe; *He. D.* Heuberg Nappe; *R.-D.* Rannach Nappe; *KS* Kalkschiefer Unit; *RF* Raasberg Group (unknown age, underlying Silurian strata). (Ebner et al. 1990)

ly been deposited and preserved (Ebner 1976a, b, 1977a, b, 1978a, b, 1980a, b). Hence, these deposits too suggest local emersion which accords well with data from other parts of Central Europe (Schönlaub et al. 1991). The range of the Dult Formation is not precisely known yet; it may extend to the Westphalian as has been inferred from a few conodonts and some plants.

Around Graz the deepest tectonic unit is represented by the Hochschlag Group. However, the stratigraphy is only poorly known due to tectonic complications, metamorphic overprints and scarcity of fossils. It comprises the basal Häuslerkreuz Formation composed of arenites, marbles, dolomitic rocks and metavolcanics. A few conodonts indicate an Upper Silurian to Lower Devonian age. It is followed by the Hochschlag Formation composed of interbedded platy and shaly limestones, graphitic quartzites, black shales, dolomitic rocks and local occurrences of metavolcanics. Locally the limestones yielded Middle Devonian conodonts. The Hochschlag Fm. may be overlain by the Aibel Formation composed mainly of siltstones and shales (Thalhammer 1982) with less abundantly occurring limestones, cherts, sandstones and basic metavolcanics of unknown age. Similar rocks are widely distributed near the western margin of the Palaeozoic of Graz and north of the Hochlantsch mountain. Generally they have been described as "Kalkschiefer". According to Thalhammer and Tschelaut (1983), Flügel (1984), Tschelaut (1984) and Hubauer (1983) this lithology includes interbedded mudstones to wackestones, bioclastic limestones, sandstones, arenaceous shales, marls and intercalations of volcaniclastic rocks. Thickness of these strata varies between several 100 m and more than 800 m. Based on conodonts, the Kalkschiefer Group may range from the Upper Silurian (or Lower Devonian) to the late Middle Devonian.

4.3 Tectonic Remarks

The Palaeozoic of Graz is subdivided into a number of thrust sheets which are composed of Silurian to Carboniferous sequences of distinct facies (Figs. 8, 10). The metamorphic overprint varies from amphibolite facies at the tectonic base, i. e. the Anger crystalline complex, and greenschist facies of the Schöckel nappe *s. l.,* to nearly unmetamorphosed sequences in the Rannach and Hochlantsch thrust sheets at the tectonic top of this pile. The middle portion of the nappe pile displays a multiple imbrication of the Laufnitzdorf Nappe with Kalkschiefer nappes in the northern part of the Palaeozoic of Graz. Age of the metamorphic overprint is Cretaceous; Variscan metamorphism is restricted to the deepest thrust sheets. Ductile deformation along thrust surfaces and in the interior of sheets is related to Alpine thrusting and subsequent low-angle normal faulting, which reactivated thrust surfaces during exhumation (Fritz et al. 1991).

5 The Palaeozoic of Burgenland

In southern Burgenland, rocks of Palaeozoic age have been known since the last century (Hoffmann 1877). The Devonian age was inferred from corals and crinoids which were found in massive limestones intercalated in a predominantly shaly sequence. Such occurrences are located between the villages of Hannersdorf and Burg, near Kirchfidisch and in the surroundings of Güssing (see Pollak 1962; Schönlaub 1984, 1990b).

According to Toula (1878), the small fossil collection comprises tabulate and rugose corals, badly preserved brachiopods and crinoid stems. Based on this fauna, Toula deduced a Middle Devonian age. Recently new findings of tabulate corals, amphiporas, crinoids and a few conodonts, i.e. icriodids, confirmed the old age assignment. Lithologically these carbonates suggested a close relationship to coeval formations in the nearby Palaeozoic of Graz. Such an affinity seems well founded by subsurface data from the area in between (Ebner 1978c, 1988; Flügel 1988).

According to new stratigraphic data, interbedded shales and limestones of the Blumau Formation dominate the Silurian. The Lower Devonian Arnwiesen Group is characterized by dolomitic rocks which mainly occur in the subsurface of the Tertiary Basin of eastern Styria. Its Middle Devonian part, however, is widely exposed in Southern Burgenland. Ebner (1988) and Flügel (1988) suggested a correlation of these carbonates with the Dolomit-Sandstein Formation of Graz. Less certain are other possible equivalents, for example, the parallelization with the Barrandei Formation or the Dult Formation. Other affinities may exist with the Palaeozoic outcrops of Noth Hungary, i.e. the Szendrö-Uppony Mountains which show a very similar lithologic succession.

6 The Greywacke Zone

6.1 Introduction

The Greywacke Zone forms the Palaeozoic basement of the Northern Limestone Alps. The WSE-ENE directed belt of Palaeozoic rocks extends over a length of some 450 km and a maximum width of 23 km, from the Rhätikon of Vorarlberg to the town of Ternitz, east of the Semmering Pass. In the Vienna Basin the eastern continuation is covered by Tertiary rocks.

The Greywacke Zone and its transgressive cover, i.e. the Northern Limestone Alps, belong to the Upper Austro-Alpine nappe system. It is thrust upon crystalline complexes of the Central Alps following to the south. Generally, all rocks of the Greywacke Zone are moderately metamorphosed ranging as far as the upper greenschist grade, and display a certain degree of foliation. Consequently, most fossils have been either destroyed or are badly preserved.

In the eastern part, the Lower Palaeozoic sequence locally starts with a conglomeratic horizon named Kalwang Gneiss Conglomerate (Daurer and Schönlaub 1979). Based on its position below fossil bearing strata an age within the Ordovician can be inferred for the conglomerate. Around the town of Bruck an der Mur it rests upon a crystalline basement complex (Neubauer 1985) which by some authors has been referred to an enigmatic intra-Ordovician collision-subduction process (Frisch et al. 1984; Becker et al. 1987; Neubauer and Frisch 1988; Neubauer et al. 1989; Frisch et al. 1990).

In the provinces of Styria and Lower Austria, the Greywacke Zone is subdivided into two major tectonic units, i.e. the Veitsch Nappe below and the Noric Nappe System above (Figs. 11, 14). The first comprises shallow water limestones and molasse-type clastics of Lower to Upper Carboniferous age (see Krainer this Vol.). while the latter forms a series of strongly deformed nappes and smaller tectonic slices ranging from the Middle or Upper Ordovician to the Carboniferous. Overthrusting of the Noric and Veitsch nappes occurred during the Alpine orogeny; both nappes are separated by the "Noric Thrust Line". For the Noric Nappe System, however, a strong Variscan deformation is indicated by the transgressive "sealing" red beds of the Permian Präbichl Formation (Schönlaub 1982; Neubauer 1989).

6.2 The Eastern Greywacke Zone of Styria

6.2.1 Review of Stratigraphy

The Eastern Greywacke Zone comprises a series of nappes and tectonic slices which display a certain variation of facies development of the Lower Palaeozoic rock sequence. A pre-Caradocian age is suggested for a more than 1000 m thick unfossiliferous sequence of greenschists, arenaceous shales, a thin marble bed, metaclastics and metapyroclastic rocks underlying the Blasseneck Quartzporphyry in the southern Eisenerz Alps, north of Kalwang (Loeschke et al. 1990; Fig. 12). Near the base the Kalwang Gneiss Conglomerate forms a distinct, debris flow horizon interbedded in greenschists. Its age remains open. Near Bruck, however, a similar conglomerate contains orthogneiss boulders which crystallized at about 500 Ma (Neubauer 1985; Neubauer et al. 1987). This suggests a formation during the Ordovician.

The succeeding Blasseneck Quartzporphyry ("Blasseneck-Porphyroid") varies in thickness between a few metres less and more than 1500 m (Heinisch 1981). In the Präbichl area the quartzporphyry is underlain by some, up to 30 m thick, limestone lenses which produced a rich conodont fauna of Late Caradocian or Early Ashgillian age (Flajs and Schönlaub 1976). The volcanics represent different types of recrystallized massive ignimbrites (pumice flow deposits), unwelded tuffs and other pyroclastics. According to its chemistry, alkali-rhyolithic and rhy-

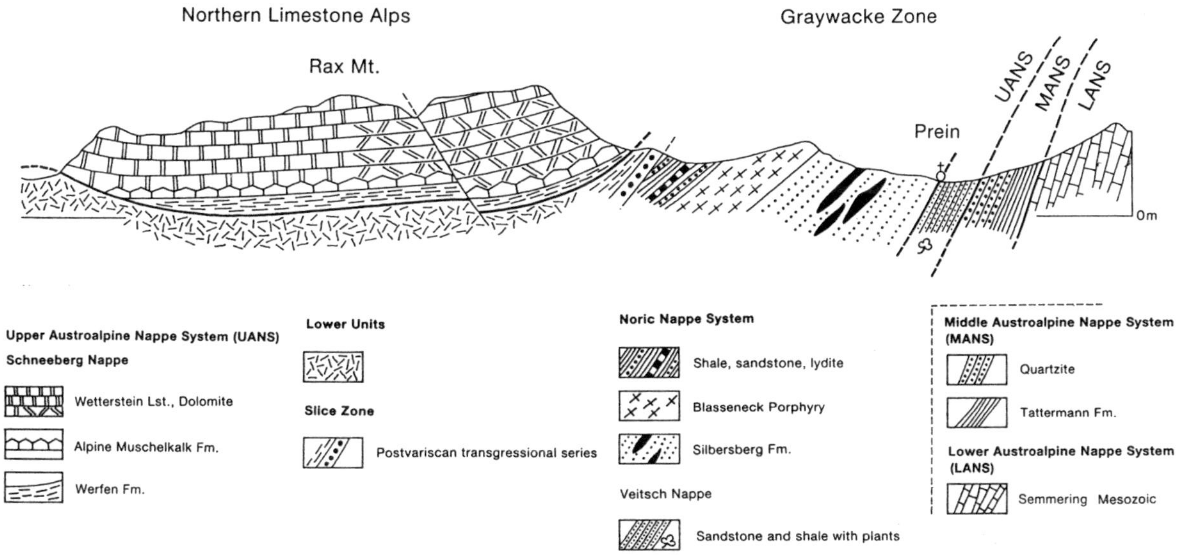

Fig. 11. NW-SE directed cross-section of the Semmering area showing the tectonic relationship between the Lower, Middle and Upper Austro-Alpine Units. Note bipartation of the Greywacke Zone into the Veitsch Nappe below and the Noric Nappe System above. The latter is unconformably overlain by the Northern Limestone Alps. (Tollmann 1963)

olithic compositions dominate over rhyodacitic, dacitic and trachyandesitic rocks. These latter types mainly occur around Eisenerz. In this northern area, the quartzporphyry is overlain by some 60 m thick quartzarenites named "Polster Quartzite" and the 13 m thick Cystoid Limestone both belonging to the Ashgillian Series (Fig. 12). In the southern part, however, the corresponding rocks may be represented by slates with intercalations of quartzarenites, greenschist and black shales comprising a thickness of a few hundred metres. Limestone intercalations in the upper part indicate a Late Llandovery or Early Wenlock age.

During the Silurian, the facies pattern varied from some 50 m thick crinoid- and nautiloid-bearing limestones in the northernmost part to black graptolitic shales in a more southern area. Upwards and also laterally they grade into interbedded limestones and shales followed by a pure limestone development during the Late Ludlow and Pridoli. Thus far, intercalations of basic volcanics of Llandovery age have only been found near the southern margin of the Greywacke Zone (Schönlaub 1976, 1982). Their geochemistry is mainly of intraplate type (Schlaegel-Blaut 1990; see also Loeschke and Heinisch this Vol.)

In the eastern Greywacke Zone limestone sedimentation continued during the Devonian. Unlike other areas of the Alps, however, splitting of facies is less pronounced during this time. The Lochkovian Stage is characterized by platy limestones which may pass laterally into variegated limestones with intercalations of shales and organodetritic limestones. Locally they range into the Pragian and Emsian Stages (Fig. 13). Also, during this time nodular dacryoconarid-bearing limestones and reddish limestones are very common.

In this part of the Greywacke Zone the majority of sections end at or close to the Lower/Middle Devonian boundary. At few places, however, some unfossiliferous limestones lying above may represent a Middle Devonian age as can be inferred from overlying rocks of Frasnian and even Famennian age. They demonstrate that limestone sedimentation may have lasted through the entire Devonian. During this time between 200 and 300 m of limestones were deposited.

The Devonian sequence is disconformably overlain by a limestone breccia and the 100- to 150-m-thick clastic Eisenerz Formation (Fig. 13). The breccia produced reworked and mixed conodonts spanning the time from the Middle Devonian to the Dinantian. Most probably the breccia was formed during a karstification event in the Visean (Schönlaub 1982). At the end of the Dinantian the palaeokarst was covered by the clastic Eisenerz Formation ranging perhaps into the lower parts of the Upper Carboniferous.

A partly modified stratigraphic framework is inferred from more eastern areas of the Styrian Greywacke Zone. According to Nievoll (1983, 1987), the lowermost part is represented by the more than 300 m thick unfossiliferous Silbersberg Formation comprising metaclastites with intercalations of quartzarenites, metaquartzconglomerates, greenschists, cherts and acid volcanics (Fig. 11). In analogy to the Eisenerz region, they are overlain by the

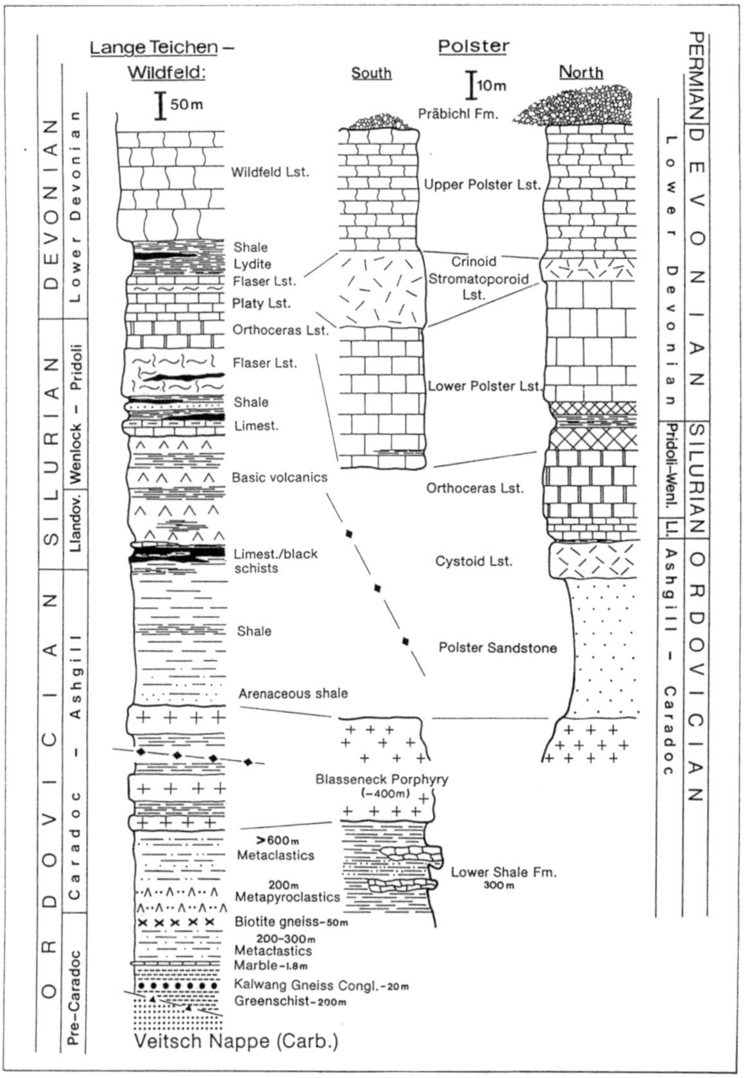

Fig. 12. Stratigraphy and facies variation of the Lower Palaeozoic sequence of the Greywacke Zone of the southern and northern Eisenerz Alps. (After Schönlaub 1979)

Blasseneck Quartzporphyry. The succession above is represented by the 300- to 1000-m-thick metapelites of the Rad Formation. Although fossils are almost lacking for the lower part of the Rad Formation (Rad unit) an Upper Ordovician and Silurian age is concluded from the position above the quartzporphyry and mainly below the occurrence of ferruginous conodont-bearing Devonian limestones. Whether or not the overlying pelites of the Stocker Unit correspond to the clastic Eisenerz Formation of the Carboniferous or represent tectonic repetitions is yet unsolved.

Based on field evidence, however, the supposed Ordovician age of the Silbersberg Fm. was questioned recently. According to Hermann et al. (1991) this formation has been interpreted as Verrucano-type sediment of Permian age overlying phyllites of probable Lower Palaeozoic age. It belongs to the newly proposed Silbersberg Nappe and thus represents an intermediate position between the Veitsch and the Noric Nappes of the eastern Greywacke Zone.

In the Enns Valley between the towns of Admont and Radstadt, the Greywacke Zone is tectonically reduced or even eliminated between crystalline rocks of the Central Alps to the south and the Limestone Alps to the north. The former continuation, however, has been confirmed by small outcrops near the village St. Martin south of Grimming (Böhm 1988) and from northeast of the town of Schladming (Schönlaub unpubl.). At the first site, black graphitic phyllites are associated with conodont-bearing marbles of early Famennian age; previously, this sequence was regarded as part of the Veitsch nappe. East of Schladming, yellowish-brownish limestones, overlying a metapelitic sequence, produced conodonts of probably Upper Silurian age.

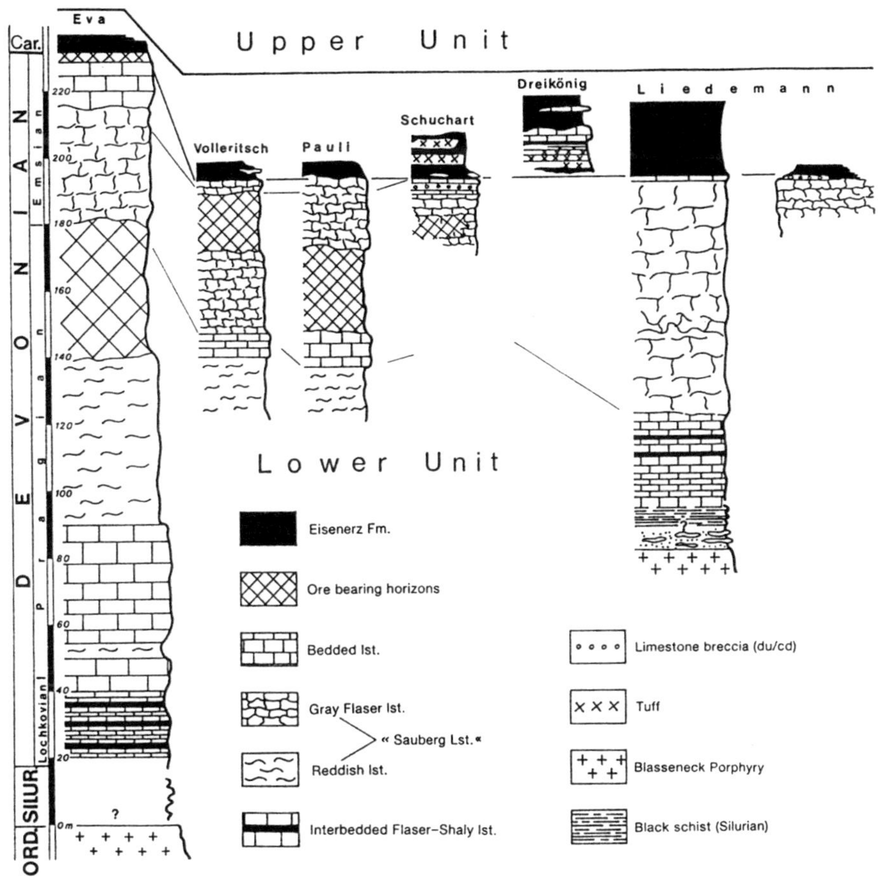

Fig. 13. Composite stratigraphic sections of the iron mine at Erzberg/Eisenerz. Note unconformity relationship between the Devonian sequence and the overlying clastic Eisenerz Formation ("Zwischenschiefer") with a limestone breccia at the base. (After Schönlaub 1979)

6.2.2 Tectonic Remarks

The main deformation of the Lower Palaeozoic sequence of the eastern Greywacke Zone occurred in the Upper Carboniferous. It resulted in southwest-facing axial surface planes accompanied locally by a foliation at very low-grade metamorphic conditions. During that time also, an extensive system of nappes, with flat-lying thrust surfaces, formed. This is documented by coarse clastic sediments of the Permian Präbichl Formation which unconformably overly different Variscan nappes, folds and faults (Flajs and Schönlaub 1976; Schönlaub 1982; Neubauer 1989).

In the Eisenerz Alps of the eastern Greywacke Zone from south to north the following Variscan nappes and tectonic slices can be distinguished (Schönlaub 1982, Fig. 14): Zeiritzkampl Nappe, Wildfeld Nappe, Reiting Nappe, Slice zone and Northern zone. Each nappe is composed of a rock sequence of mainly different stratigraphic extent and lithology. During the Alpine orogeny this pile of basement rocks of the Northern Alps was thrust upon a lower unit comprised of the Carboniferous of the Veitsch Nappe. Both major tectonic units are separated by the Noric Thrust Plane.

The Veitsch Nappe represents the lowermost unit of the Upper Austro-Alpine thrust sheet. It is thrust upon the Middle Austro-Alpine thrust system exposed in the tectonic window of Mautern (Schönlaub 1982). The lithology within the window resembles the Rannach Formation and the Lower Triassic "Scythian Quartzite" well-known from the Semmering area. In the region south of the Eisenerz Alps such clastics form the Permo-Triassic cover of the Variscan Seckau crystalline complex (Fig. 14).

6.3 The Western Greywacke Zone of Salzburg and Tyrol

6.3.1 Review of Stratigraphy

In the early 1960s in the western Greywacke Zone a team from Innsbruck University headed by Mostler started the re-examination of the stratigraphy and tectonic evolution of the Lower Palaeozoic se-

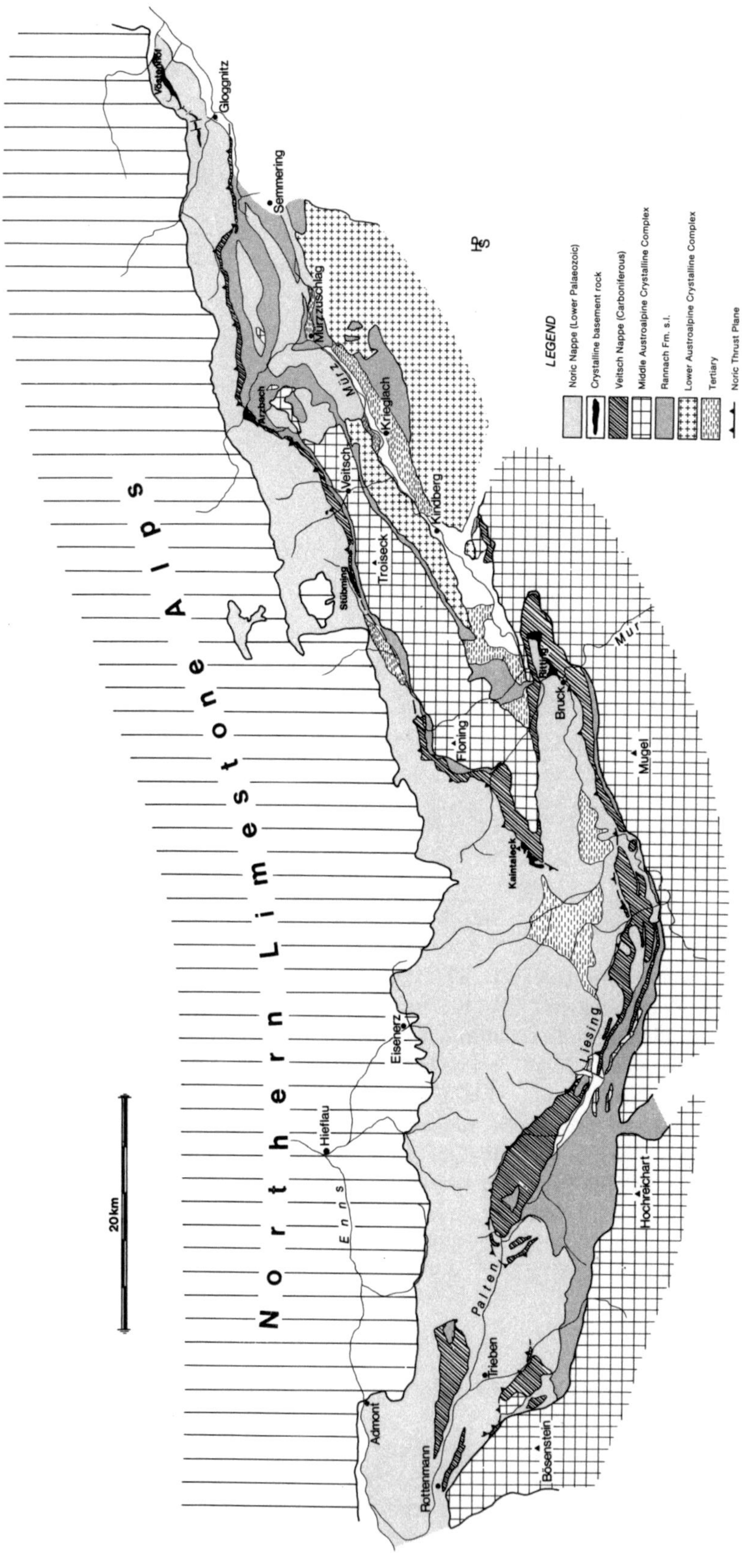

Fig. 14. Main subdivision of the eastern part of the Greywacke Zone of Styria and Lower Austria. (After Schönlaub 1979)

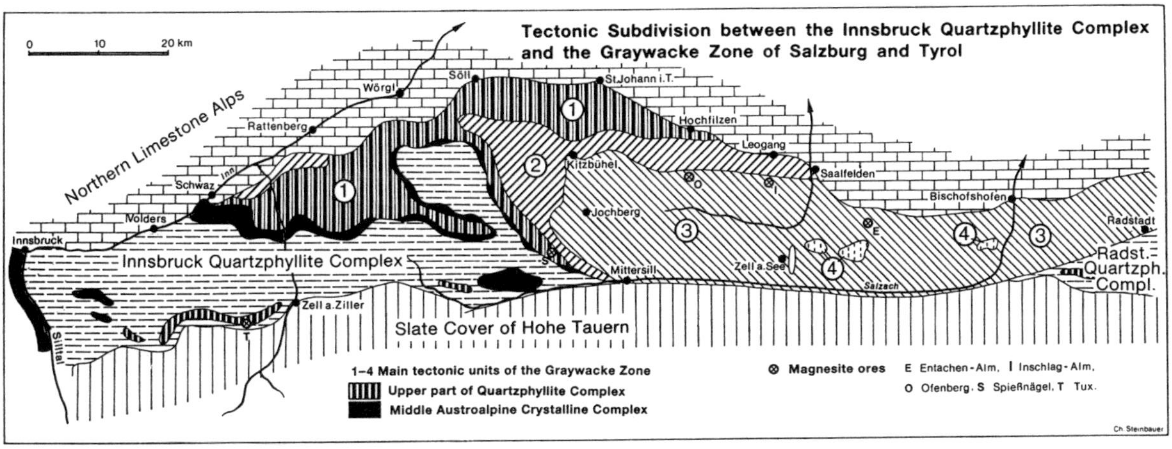

Fig. 15. Main structural subdivision of the Greywacke Zone of Salzburg and Tyrol. (After Mostler 1973)

quences of Salzburg and Tyrol. The result was a well-founded new biostratigraphic scheme based mainly on conodonts, supplemented by detailed sedimentological, petrographical and geochemical analysis (e.g. Mostler 1966, 1968, 1970, 1984; Al-Hasani and Mostler 1969; Emmanuilidis and Mostler 1970; Mavridis and Mostler 1970; Colins et al. 1980; Fig. 15). Subsequently, detailed mapping of the Kitzbühel-Saalbach area was carried out by a working group of Munich University focusing on sedimentology, stratigraphy, volcanology, petrography and structural geology (Heinisch 1986, 1988; Heinisch et al. 1987; Schlaegel 1988; Schlaegel-Blaut 1990).

Based on the comprehensive set of new and refined old data from the two groups mentioned above between Kitzbühel and Zell am See an obvious heterogenity of facies of the Palaeozoic rock record has been recognized. Within short distances two distinct facies can be distinguished (Heinisch 1988; Fig. 16). They are preserved in two nappes named Wildseeloder and Glemmtal Unit, respectively, which are separated by a polyphase composite shear zone.

The Wildseeloder Unit covers the northern part of the studied area. It is characterized by thick piles of Upper Ordovician Blasseneck Quartzporphyry and different pelagic carbonates that developed during the Silurian and the Devonian (Fig. 16).

In this unit the oldest rocks are represented by the Blasseneck Quartzporphyry. Although in this part of the Greywacke Zone fossils are missing there are good reasons to assume the same Upper Ordovician age for these volcanics as in the eastern Greywacke Zone. According to Mostler (1970, 1984) and Heinisch (1981, 1988) the up to 600 m thick rhyolitic ignimbrites extruded under subaerial to shallow marine conditions. Due to intense thrusting, rocks older than the Blasseneck Quartzporphyry are not preserved within the Wildseeloder Unit.

Conglomeratic horizons at the top of the volcanics might mask a considerable gap in sedimentation. In addition, mudstones, debris flows and sandstones partly interfinger with volcaniclastic layers. At some places the Blasseneck Quartzporphyry is overlain by limestones of Llandovery age (Mostler 1967, 1970). During the Silurian, facies differentiation was a common feature ranging from black shales with occasional occurrences of graptolites to cherts, siliceous pelagic limestones, condensed cephalopod limestones and even to dolomitic rocks. Starting in the Upper Silurian, a carbonate platform developed until the early Upper Devonian. It comprises shallow water lagoonal dolomites, a local reef development and pelagic limestone of Frasnian age (Mostler 1970; Schönlaub 1979, 1980).

The Glemmtal Unit covers the southern part of the area. It comprises several thousand metres of siliciclastic sediments named Wildschönau Group. Locally intercalations of condensed pelagic limestones with interbedded cherts and siliceous shales are found. They are the Klingler Kar Formation. Intercalations of basic magmatites ranging from a few metres to several 100 m thickness are of further importance.

The monotonous siliciclastic rocks of the Wildschönau Group consist of interbedded shales, siltstones and sandstones, with locally occurring microconglomerates, conglomerates and breccias. Relics of sedimentary features (i.e. grading, cross-stratification, plane-parallel laminations, convolute bedding, Bouma sequences) demonstrate a turbidite origin. Within short distances two intergrading facies can be distinguished (Heinisch 1986; Fig. 16):

1. The Schattberg Formation displays characteristics of proximal turbidites, e.g. coarse-grained metre-

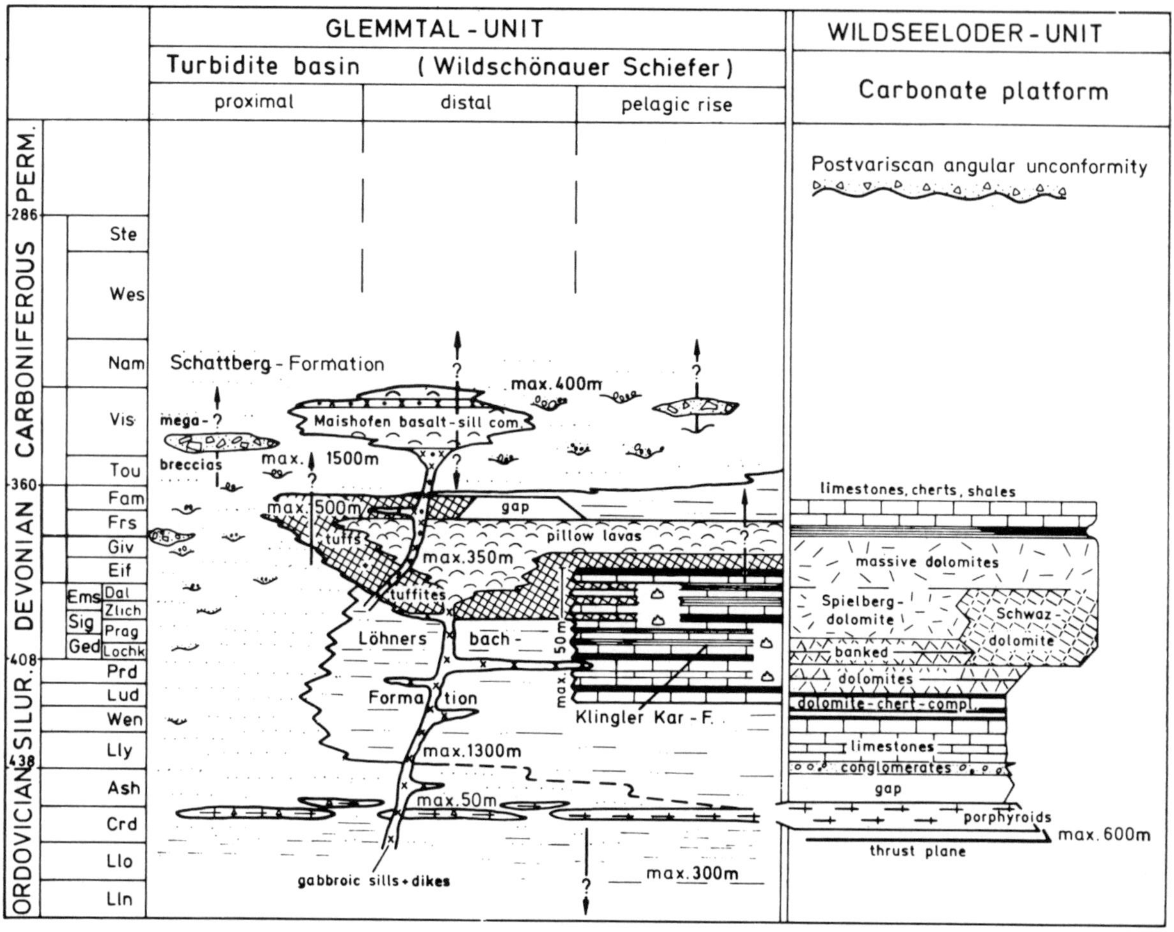

Fig. 16. Lithostratigraphic correlation and biostratigraphy of the Greywacke Zone around Zell am See and Saalbach after Heinisch 1988. Correlation within Glemmtal Unit using porphyroids and basic volcanics as key beds; biostratigraphic control based on conodonts within the Klingler Kar Formation. Correlation within the Wildseeloder Unit from Mostler (1970, modified)

thick graded sandstones, channelled microconglomerates and breccias;

2. The Löhnersbach Formation shows features indicating distal turbidites, e. g. decimetre-thick medium- to fine-grained sandstones and higher amounts of siltstones and shales.

In general, facies relations are diachronous although a certain degree of coarsening upward can be recognized. Following the basic volcanism in the Middle Devonian the Schattberg Formation was distributed over the entire Glemmtal Unit. Whether or not this formation extends into the Carboniferous is yet not known because of lack of fossils.

Olistholithic megabreccias of the Schattberg Formation comprise a varying spectrum of partly well-rounded boulders of garnet or hornblende gneisses, amphibolites, quartzites, sericitic gneisses and granitoids. According to their mineralogical composition (mostly subgreywackes), the sandstones indicate a continental source area (Heinisch 1988).

In conclusion, during the Lower Palaeozoic the Greywacke Zone of Kitzbühel and Saalbach may have acted as a marginal basin near a continent. It was supplied by turbidites connected with submarine fan and channel deposits. The adjacent eroded continent was characterized by different metamorphic rocks intruded by granitoids. This passive margin regime persisted from at least the Upper Ordovician to the Middle Devonian.

The up to 50 m thick Klingler Kar Formation represents a marine hemipelagic, deep-water setting which lasted from the Upper Silurian to the Middle Devonian. It comprises condensed cephalopod limestones, marls, greenisch and black shales, chert horizons ("lydites") and basaltic tuffs. The only fossils yet discovered are poorly preserved nautiloids (*Orthoceras* sp.) and conodonts which indicate different levels within the Upper Silurian and the Lower to Middle Devonian. In addition, they provide precise age assignments for some of the intercalated volcanic horizons (Heinisch 1986, 1988; Heinisch et al. 1987).

During a short time span, all facies zones except the carbonate platform were affected by an intense basic magmatism (Schlaegel 1988; Schlaegel-Blaut 1990). According to conodonts, basaltic volcanism apparently started in the late Lower Devonian Emsian Stage (Heinisch et al. 1987); its end has not yet been dated. The volcanism produced lavas, sills, variegated pyroclastics and tuffites described in great detail by Schlaegel-Blaut (1990). Based on trace elements and REE patterns the basaltic volcanism is of intraplate type (transitional basalts and alkali basalts) which may have formed seamounts reaching temporarily a subaerial stage (see Loeschke and Heinisch this Vol.).

In contrast to these rocks, the basalt-sill complex of Maishofen shows a different tholeiitic geochemistry. The pillow lavas and sills were formed under deeper water conditions (Schlaegel-Blaut 1990). In the past, the up to 200 m thick volcanic sequence has been interpreted as an Ordovician ocean floor basalt (Colins et al. 1980; Mostler 1984). As yet, the exact age has not been based on fossil evidence. From recently obtained field data Schlaegel-Blaut (1990) concluded an age younger than the Upper Ordovician porphyroids.

Originally, the stratigraphic base of the clastic sedimentation ("Wildschönauer Schiefer") was supposed to be in the Lower Ordovician (Mostler 1970). Recently, this idea was supported by acritarchs of Tremadocian and those of Arenigian age, respectively, recovered from metapelites near Kitzbühel (Reitz and Höll 1989, 1991). As detailed geological maps are missing in this area, it is not possible yet to link these acritarch-bearing strata with any lithostratigraphic units, as defined for the Saalbach-Zell am See area (Heinisch 1988, Fig. 16). Whether or not in this region some of the basaltic volcanics are also of Ordovician age as suggested by Mostler (1970, 1984) is presently difficult to decide due to lack of fossils and an extensive Alpine overprint.

In summary, during the Silurian and Devonian the shallow water platform facies of the Wildseeloder Unit existed contemporaneously with the turbiditic basin facies of the Glemmtal Unit in the western Greywacke Zone. The connecting link between both facies is the Blasseneck Quartzporphyry, although this rock too reflects two different settings: the Wildseeloder Unit is characterized by subaerial ignimbrites; in the Glemmtal Unit the dominating rock type is an epiclastic volcanic debris washed into the basin by sediment flows. Yet, the original palaeogeographic relationship between these two depositional areas is unknown.

6.3.2 Tectonic Remarks

Following the concept of Mostler (1973) the western Greywacke Zone is divided into four Variscan nappes which are sealed by the post-Variscan angular unconformity (Fig. 15). In the study area of Heinisch (1988) this tectonic scheme was modified to include two nappes, i. e. the Wildseeloder Unit and the Glemmtal Unit. Both nappes are characterized by distinct lithofacies. However, the intense deformation of the southern margin of the post-Variscan cover sequence (Hochfilzen Formation) strongly suggests an Alpine overprint on the Variscan nappe system of this region. Moreover the cataclastic shear zone separating the Wildseeloder and the Glemmtal nappes are further indications of later movements.

Obviously, the whole area was affected by intense, small-scale brittle Alpine block faulting as well as thrusting. Thus, the lithological boundaries are frequently sheared, in particular at the contact between rigid carbonates, porphyroids, metavolcanics and siliciclastic rock sequences.

In summary, the Glemmtal Unit forms a flat-lying sequence, with steep and distorted southern and northern flanks. Large-scale fold structures are, however, missing and small-scale folds are linked to the brittle faulting. The sequence of the Wildseeloder Unit shows vertical dipping. Tectonically unaffected stratigraphic sequences are extremely rare. Consequently, isolated stratigraphic data should be treated with great care as it is almost impossible to distinguish between Variscan and Alpine structures. Even an olistholithic origin for some of the carbonate blocks can not be excluded, in particular at the base of the Wildseeloder Unit.

The southern margin of the Greywacke Zone (Salzach Valley) is characterized by increasing Alpine metamorphism, strong ductile deformation and the existence of a complicated polyphase shear zone (Heinisch 1988).

7 Conclusions

Based on the data from the classic fossiliferous Palaeozoic regions of the Eastern und Southern Alps, there is, in the authors opinion, no unequivocal evidence for a Caledonian collision of parts of the early

Alps as proposed by Frisch et al. (1984), Frisch and Neubauer (1989) and Loeschke (1989a). In fact, neither an angular unconformity has been recognized in the geological record of the Alps nor any significant hiatus in sedimentation between Lower and Upper Ordovician sequences. The widespread Upper Ordovician acidic volcanism as documented by the Blasseneck Quartzporphyry does not necessarily need to be related to a subduction process. Rather, its partly calc-alkaline chemical composition can be interpreted as the result of anatectic melts of a thick continental crust formed during the "Pan-African tectonothermal event" (Almond 1984; Sacchi 1989).

The angular unconformity advocated by Neubauer (1985) as evidence for a Caledonian event from the base of the Palaeozoic sequence of the Eastern Greywacke Zone is undoubtedly older than Upper and even Lower Ordovician. Furthermore, the conglomerate which occurs at the base and perhaps corresponds to the Kalwang Gneiss Conglomerate has not been dated yet.

There is no true time relationship between these hypothetical events in the Alps and the Caledonian collision and closure of the Iapetus ocean in northern Europe, during the Late Silurian, followed by the Old Red sedimentation (Neugebauer 1988). Consequently, in the Alps, the term "Pan-African" should be applied when describing Cambrian and/or Ordovician accretionary events. The affinities with Gondwana seems to justify such an approach (see Schönlaub this Vol., with further references).

Also, during the Ordovician to Devonian there is definitely no proof of a well-developed ocean floor characterized by oceanic crust in the Alps. The majority of the basic rocks neither displays a relationship to an active plate margin nor to a mature oceanic ridge segment, but instead shows intraplate geochemistry. Any model dealing with the plate tectonic evolution of the early Alps must consider this fact (Flügel 1990).

Variation of facies is a widely occurring phenomenon and has been demonstrated for the Greywacke Zone and for other areas in the Alps which comprise Lower Palaeozoic strata. However, biostratigraphic data yet available are not always as sufficient as they should be for a reliable base of a highly sophisticated evolutionary model. In particular, this regards fossil data from the Early Ordovician and in greenschist grade metamorphosed strata.

Crustal extension is the most common and distinct feature during the Early Palaeozoic history of the Alps. It strongly controlled the sedimentation and the kind of volcanism from the Ordovician to the middle of the Dinantian. During the Upper Carboniferous, this changed in favour of subduction and collision r lated processes which has best been documented the Southern Alps. In contrast to this region in th Eastern Alps, the Variscan orogeny reached i paroxysm in the Late Devonian or early Lower Ca boniferous. Such a relationship is testified by fossili erous strata of early Visean age which unconformab overlie pre-Variscan and Variscan basement rocks.

References

Al-Hasani N, Mostler H (1969) Zur Geologie der Spießnäg südlich Kirchberg (Nördliche Grauwackenzone, Tirol Veröff Univ Innsbr 9, Alpenkdl Stud 5: 5–26

Almond DC (1984) The concepts of "Pan-African episode" an "Mozambique Belt" in relation to the geology of E and N Africa. Bull Fac Earth Sci Abdulaziz Univ 6: 71–87

Amerom van HWJ, Schönlaub HP (1992) Pflanzenfossilien au dem Karbon von Nötsch und der Hochwipfel Formation de Karnischen Alpen. Jahrb Geol Bundesanst 135: 195–216

Amerom van HWJ, Flajs G, Hunger G (1984) Die "Flora de Marinelli Hütte" (Mittleres Visé) aus dem Hochwipfelflysc der Karnischen Alpen (Italien). Meded Rijks Geol Dien 37-3: 1–41

Bandel K (1972) Palökologie und Paläogeographie im Devo und Unterkarbon der Zentralen Karnischen Alpen. Palaeon togr Abt A 141: 1–117

Bauer FK (1983) Perm der Vellacher Kotschna. Erläuterunge zur Geologischen Karte der Karawanken 1:25000, Ostte Geol Bundesanst, Vienna, pp 46–47

Becker LP, Frank W, Höck V, Kleinschmidt G, Neubauer F, Sas FP, Schramm JM (1987) Outlines of the pre-Alpine metamo phic events in the Austrian Alps. In: Flügel HW, Sassi FP, Gre ula P (eds) Pre-Variscan and Variscan Events in the Alpine Mediterranean mountain belts. Alfa, Bratislava, pp 69–106

Böhm F (1988) Geologie des Grimming-Westabschnittes. Mi Ges Geol Bergbaustud Österr 34/35: 151–184

Buch L von (1824) Über die Karnischen Alpen. Leonhard Miner Taschenb 18, 2 Abt: 396–437

Buchroithner MF (1978) Biostratigraphische Untersuchunge im Paläozoikum der Steiermark. Mitt Naturwiss Ver Steie mark 108: 77–94

Buchroithner MF (1979) Biostratigraphie und fazielle Unte suchungen im Paläozoikum von Mittelkärnten. Carinthia 169: 71–95

Castellarian A, Vai GB (1981) Importance of Hercynian tecton ics within the framework of the Southern Alps. J Struct Ge 3: 477–486

Clar E, Fritsch W, Meixner H, Pilger A, Schönenberg R (196 Die geologische Neuaufnahme des Saualpen-Kristallin (Kärnten), VI. Carinthia II 153/73: 23–51

Cliff RA, Holzer HF, Rex DC (1975) The age of the Eisenkap pel granite, Carinthia and the history of the Periadriati lineament. Verh Geol Bundesanst 1974: 347–350

Colins E, Hoschek G, Mostler H (1980) Geologische Entwick lung und Metamorphose im Westabschnitt der Nördliche Grauwackenzone unter besonderer Berücksichtigung de Metabasite. Mitt Österr Geol Ges 71/72: 343–378

Dallmeyer RD, Neubauer F $^{40}Ar/^{39}Ar$ age of detrital muscovite from the Carnic Alps: evidence for Cadomian basement i the Eastern Alps (in press).

Daurer A, Schönlaub HP (1979) Anmerkung zur Basis der Nördlichen Grauwackenzone. Mitt Österr Geol Ges 69: 77–87

Ebner F (1976 a) Die Schichtfolge an der Wende Unterkarbon/Oberkarbon in der Rannachfazies des Grazer Paläozoikums. Verh Geol Bundesanst 1976: 65–93

Ebner F (1979 b) Das Silur/Devon-Vorkommen von Eggenfeld – ein Beitrag zur Biostratigraphie des Grazer Paläozoikums. Mitt Abt Geol Paläontol Bergb Mus Joanneum 37: 1–33

Ebner F (1977 a) Die Gliederung des Karbons von Graz mit Conodonten. Jahrb Geol Bundesanst 120: 449–493

Ebner F (1977 b) Die Transgression der Folge der Dult (Oberkarbon, Paläozoikum von Graz). Mitt Naturwiss Ver Steiermark 107: 35–53

Ebner F (1978 a) Die sedimentäre Entwicklung des Unterkarbons in Österreich. Österr Akad Wiss Schriftenr Erdwiss Komm 3: 179–190

Ebner F (1978 b) Stratigraphie des Karbon der Rannachfazies im Paläozoikum von Graz, Österreich. Mitt Österr Geol Ges 69: 163–196

Ebner F (1978 c) Der paläozoische Untergrund in der Bohrung Waltersdorf I (S Hartberg, Oststeiermark). Mitt Österr Geol Ges 68: 5–11

Ebner F (1980 a) Conodont localities in the surroundings of Graz/Styria. In: Schönlaub HP (ed) Second European Conodont Symposiim (ECOS II), Guidebook Abstracts. Abh Geol Bundesanst 35: 101–127

Ebner F (1980 b) Steinbergkalk und Sanzenkogelschichten im Kalvarienbergzug E von Gratwein. Mitt Naturwiss Ver Steiermark 110: 47–54

Ebner F (1988) Das Paläozoikum in den RAG-Bohrungen Blumau 1, 1 a und Arnwiesen 1 (Oststeirisches Tertiärbecken). Jahrb Geol Bundesanst 131: 563–573

Ebner F, Neubauer F, Pistotnik J (1977) Vorbericht über stratigraphische Untersuchungen im Altpaläozoikum südlich und westlich von Murau. Sitzungsber Österr Akad Wiss Mat-Naturwiss Kl 1977: 21–25

Ebner F, Fenninger A, Gollner H, Holzer HL, Neubauer F, Nievoll J, Ratschbacher L, Stattegger K, Tschelaut W, Thalhammer O, Zier C (1990) Stratigraphic correlation forms of Paleozoic units in Austria. In: Sassi FP, Zanferrari A (eds) Pre-Variscan and Variscan events in the Alpine-Mediterranean belts, stratigraphic correlation forms. Rend Soc Geol Ital 12: 213–239

Eichhübl P (1988) Groß- und Kleintektonische Untersuchungen zum alpidischen und variszischen Gebirgsbau in den Östlichen Karnischen Alpen (Österreich/Italien). Diplomarbeit, Naturwiss Fak Univ Wien, 123 pp

Emmanuilidis G, Mostler H (1970) Zur Geologie des Kitzbühler Horns und seiner Umgebung mit einem Beitrag über die Barytvererzung des Spielberg-Dolomites (Nördliche Grauwackenzone, Tirol). Festband Geol Inst 300 Jahr Univ Innsbruck, pp 547–569

Exner C (1972) Geologie der Karawankenplutone östlich Eisenkappel, Kärnten. Mitt Österr Geol Ges 64: 1–108

Fenninger A, Holzer H-L (1978) Die Genese der Dolomitsandstein-Folge des Grazer Paläozoikums. Mitt Österr Geol Ges 69: 109–162

Fenninger A, Schönlaub HP, Holzer H-L, Flajs G (1976) Zu den Basisbildungen der Auernigschichten in den Karnischen Alpen (Österreich). Verh Geol Bundesanst 1976: 243–255

Flajs G, Schönlaub HP (1976) Die biostratigraphische Gliederung des Altpaläozoikums am Polster bei Eisenerz (Nördliche Grauwackenzone, Österreich). Verh Geol Bundesanst 1976: 257–303

Flügel E, Schönlaub HP (1990) Exotic limestone clasts in the Carboniferous of the Carnic Alps and Nötsch. In: Venturini C, Krainer K (eds) Field workshop on Carboniferous to Permian sequence of the Pramollo-Nassfeld Basin (Carnic Alps). Proc, pp 15–19

Flügel HW (1975 a) Erläuterungen zur Geologischen Karte des Grazer Berglandes 1:100000. Geol Bundesanst Vienna, pp 1–288

Flügel HW (1975 b) Einige Probleme des Variszikums von Neo-Europa. Geol Rundsch 64: 1–62

Flügel HW (1984) Das Grazer Paläozoikum. In: Flügel HW, Neubauer F (eds) Steiermark – Geologie der österreichischen Bundesländer in kurzgefaßten Einzeldarstellungen. Geol Bundesanst, Vienna, pp 44–54

Flügel HW (1988) Geologische Karte des prätertiären Untergrundes. Erläuterungen zu den Karten über den präteritären Untergrund des Steirischen Beckens und der Südburgenländischen Schwelle. Geol Bundesanst, pp 1–49

Flügel HW (1990) Das voralpine Basement im Alpin-Mediterranen Belt – Überblick und Problematik. Jahrb Geol Bundesanst, Vienna 133: 181–221

Flügel HW, Jaeger H, Schönlaub HP, Vai GB (1977) Carnic Alps. In: Martinsson A (ed) The Silurian-Devonian Boundary. Int Union Geol Sciences, Ottawa Ser A 5: 126–142

Flügel HW, Neubauer F (1984) Steiermark – Geologie der österreichischen Bundesländer in kurzgefaßten Einzeldarstellungen. Geol Bundesanst, Vienna, pp 1–127

Frisch W, Neubauer F (1989) Pre-Alpine terranes and tectonic zoning in the eastern Alps. Geol Soc Am Spec Pap 230: 91–100

Frisch W, Neubauer F, Satir M (1984) Concepts of the evolution of the Austroalpine basement complex (Eastern Alps) during the Caledonian-Variscan cycle. Geol Rundsch 73: 47–68

Frisch W, Neubauer F, Bröcker M, Brückmann W, Haiss N (1987) Interpretation of geochemical data from the Caledonian basement complex: In: Flügel HW, Sassi FP, Grecula P (eds) Pre-Variscan and Variscan events in the Alpine-Mediterranean mountain belts. Alfa, Bratislava, pp 209–226

Frisch W, Ménot R-P, Neubauer F, Raumer von JF (1990) Correlation and evolution of the Alpine basement. Schweiz Mineral Petrogr Mitt 70: 265–285

Fritz H, Neubauer F (1988) Geodynamic aspects of the Silurian and Early Devonian sedimentation in the Paleozoic of Graz (Eastern Alps). Schweiz Mineral Petrogr Mitt 68: 359–367

Fritz H, Gsellmann H, Neubauer F, Ratschbacher L, Agnoli F, Flügel HW, Reisinger J, Schlamberger J (1986) Strukturanalyse in einem heterogen deformierten Terrain: Das Grazer Paläozoikum (Ostalpin). Abstr Worksh Strukturgeologie, Kristallingeologie, Tektonik, Univ Tübingen, Tübingen

Fritz H, Neubauer F, Ratschbacher L (1991) Compression versus extension in the Paleozoic of Graz (Eastern Alps, Austria). Zentralbl Geol Paläontol Teil I 1991: 55–68

Gaertner HR von (1931) Geologie der Zentralkarnischen Alpen. Denkschr Österr Akad Wiss Wien Math-Naturwiss Kl 102: 113–199

Gollner H (1983) Die Geologie des Hochlantschstockes (Grazer Paläozoikum, Steiermark). Diss, Univ Graz, pp 251

Gollner H, Zier C (1982) Stratigraphic correlation forms of the Hochlantsch facies (Hochlantscheinheit, Harrbergerformation) in the Paleozoic of Graz (Austria), Geotraverse B. In: Sassi FP, Varga I (eds) IGCP Project 5, Correlation of Prevariscan and Variscan events in the Alpine-Mediterranean mountain belt. Newsletter 4, pp 38–40

Gollner H, Zier C (1985) Zur Geologie des Hochlantsch (Grazer Paläozoikum, Steiermark). Jahrb Geol Bundesanst 128: 43–73

Gollner H, Thalhammer O, Tschelaut W, Zier C (1982) Die Laufnitzdorf-Gruppe. Eine pelagische Fazies im Grazer Paläozoikum. Mitt Naturwiss Ver Steiermark 112: 63–73

Gosen W von, Haiges K-A, Thiedig F (1982) Fossilführendes Paläozoikum am Ostrand der Gurktaler Alpen (Steiermark/Kärnten-Österreich). Mitt Geol Paläontol Inst Univ Hamb 53: 151–168

Haberfelner E (1935) Die Geologie des Eisenerzer Reichenstein und des Polster. Mitt Abt Bergb Landesmus Joanneum 2: 1–32

Havlicek V, Kriz J, Serpagli E (1987) Upper Ordovician brachiopod assemblages of the Carnic Alps, Middle Carinthia and Sardinia. Boll Soc Paleontol Ital 25: 277–311

Heinisch H (1981) Zum ordovizischen "Porphyroid"-Vulkanismus der Ost- und Südalpen – Stratigraphie, Petrographie, Geochemie. Jahrb Geol Bundesanst 124: 1–109

Heinisch H (1986) Die Geologie der Nördlichen Grauwackenzone zwischen Kitzbühl und Zell am See und ihre Bedeutung für die Rekonstruktion der altpaläozoischen Geodynamik des Ostalpenraumes. Habilitationsschrift, Univ München

Heinisch H (1988) Hinweise auf die Existenz eines passiven Kontinentalrandes im Altpaläozoikum der Nördlichen Grauwackenzone – Ostalpen. Schweiz Mineral Petrogr Mitt 68: 407–418

Heinisch H, Sprenger W, Weddige K (1987) Neue Daten zur Altersstellung der Wildschönauer Schiefer und des Basaltvulkanismus im ostalpinen Paläozoikum der Kitzbühler Grauwackenzone (Österreich). Jahrb Geol Bundesanst 130: 163–173

Heritsch F (1936) Die Karnischen Alpen. Monographie einer Gebirgsgruppe der Ostalpen mit variszischen und alpidischem Bau. Geol Inst Univ Graz, Graz, pp 1–205

Hermann S, Paulus G, Handler R, Neubauer F, Nievoll JS (1991) The Silbersberg nappe in the eastern Graywacke Zone: a new nappe, its palaeogeographic and tectonic significance for the Austroalpine. Terra Abstr 3: 218–219

Hoffmann K (1877) Aufnahmebericht über das Jahr 1876. Verh Geol Reichsanst 1877: 14–23

Höll R (1970) Die Zinnober-Vorkommen im Gebiet der Turracher Höhe (Nock-Gebiet/Österreich) und das Alter der Eisenhut-Schieferserie. Neues Jahrb Geol Paläontol Monatsh 1970: 201–224

Holser WT, Schönlaub HP, Boeckelmann K, Magaritz M (1991) The Permian-Triassic of the Gartnerkofel-1 core (Carnic Alps, Austria): synthesis and conclusions. In: Holser WT, Schönlaub HP (eds) The Permian-Triassic boundary of the Carnic Alps, Austria. Abh Geol Bundesanst 45: 213–232

Hubauer N (1983) Bericht 1982 über geologische Aufnahmen auf Blatt 134 Passail. Jahrb Geol Bundesanst 126: 315

Jaeger H (1975) Die Graptolithenführung im Silur/Devon des Cellon-Profils (Karnische Alpen). Carinthia II 165/85: 111–126

Jaeger H, Schönlaub HP (1980) Silur und Devon nördlich der Gundersheimer Alm in den Karnischen Alpen (Österreich). Carinthia II 170/90: 403–444

Kahler F (1953) Der Bau der Karawanken und des Klagenfurter Beckens. Carinthia II Sh 16: 1–78

Kahler F (1983) Fusuliniden aus Karbon und Perm der Karnischen Alpen und der Karawanken. Carinthia II SH 41: 1–107

Kleinschmidt G, Wurm F (1966) Paläozoikum und epizonale Serien zwischen St. Andrä im Lavanttal und Griffen. Carinthia II 156/76: 108–140

Krainer K, Mogessie A (1991) Composition and significance of resedimented amphibolite breccias and conglomerates ("Badstubbreccia Complex") in the Carboniferous of Nötsch (Eastern Alps, Austria). Jahrb Geol Bundesanst 134: 65–81

Kreutzer LH (1990) Mikrofazies, Stratigraphie und Paläogeographie des Zentralkarnischen Hauptkammes zwischen Seewarte und Cellon. Jahrb Geol Bundesanst 133: 275–343

Kreutzer LH (1992) Palinspastische Entzerrung und Neugliederung des Devons in den Zentralkarnischen Alpen aufgrund von neuen Untersuchungen. Jahrb Geol Bundesanst 135: 261–272

Loeschke J (1970) Zur Geologie und Petrographie des Diabaszuges westlich Eisenkappel (Ebriachtal/Karawanken/Österreich). Oberrhein Geol Abh 19: 73–100

Loeschke J (1977) Kaledonischer eugeosynklinaler Vulkanismus Norwegens und der Ostalpen im Vergleich mit rezentem Vulkanismus unterschiedlicher geotektonischer Positionen: Eine Arbeitshypothese. Z Dtsch Geol Ges 128: 185–207

Loeschke J (1983) Diabaszug von Eisenkappel westlich der Vellach. In: Bauer FK, Cerny I, Exner C et al. (eds) Erläuterungen zur Geologischen Karte der Karawanken 1:25 000, Ostteil. Geol Bundesanst, Vienna, pp 24–30

Loeschke J (1989 a) Lower Palaeozoic volcanism of the Eastern Alps and its geodynamic implications. Geol Rundsch 78: 599–616

Loeschke J (1989 b) Die paläotektonische Stellung der Vulkanite der Magdalensberg-Serie (Ober-Ordovizium, Gurktaler Decke, Kärnten, Österreich). Carinthia II 179: 491–507

Loeschke J, Kretzschmar T, Langer V, Streck M (1990) Zur Geologie der Norischen Decke nordwestlich von Kalwang (Eisenerzer Alpen, Steirische Grauwackenzone/Österreich). Jahrb Geol Bundesanst 133: 345–363

Mavridis A, Mostler H (1970) Zur Geologie der Umgebung des Spielberghorns mit einem Beitrag über die Magnesitvererzung. Festb Geol Inst 300 J Univ Innsbr, pp 523–546

Metz K (1953) Die stratigraphische und tektonische Baugeschichte der steirischen Grauwackenzone. Mitt Geol Ges Wien 44: 1–84

Moshammer B (1989) Das südalpine pelagische Eisenkappeler Paläozoikum (Trögener Gruppe) der Ostkarawanken. Carinthia II 179/99: 611–640

Mostler H (1966) Zur Einstufung der "Kieselschiefer" von der Lachtal-Grundalm (Fieberbrunn, Tirol). Verh Geol Bundesanst 1966: 157–170

Mostler H (1967) Conodonten aus dem tieferen Silur der Kitzbühler Alpen (Tirol). Ann Naturhist Mus Wien 71: 295–303

Mostler H (1968) Das Silur im Westabschnitt der Nördlichen Grauwackenzone (Tirol und Salzburg). Mitt Ges Geol Bergbaustud Österr 18: 89–150

Mostler H (1970) Struktureller Wandel und Ursachen der Faziesdifferenzierung an der Ordoviz/Silur-Grenze in der Nördlichen Grauwackenzone (Österreich). Festbd Geol Inst 300 J Univ Innsbr, pp 507–522

Mostler H (1973) Alter und Genese ostalpiner Spatmagnesite unter besonderer Berücksichtigung der Magnesitlagerstätten im Westabschnitt der Nördlichen Grauwackenzone (Tirol, Salzburg). Festschrift Heissel. Veröffentl Univ Innsbruck 86: 237–266

Mostler H (1984) Erfassung basischer Massengesteine im Raum Mittersill-Zell a. See-Salzburg. Arch Lagerstättenforsch (Ostalpen) 5: 105–115

Neubauer F (1979) Die Gliederung des Altpaläozoikums südlich und westlich von Murau (Steiermark/Kärnten). Jahrb Geol Bundesanst 122: 455–511

Neubauer F (1984) Das Murauer und Gurktaler Paläozoikum. In: Flügel HW, Neubauer F (eds) Steiermark – Geologie der österreichischen Bundesländer in kurzgefaßten Einzeldarstellungen. Geol Bundesanst, Vienna, pp 55–58

Neubauer F (1985) Eine präoberordovizische Transgression in der Grauwackenzone (Ostalpen) und ihre Bedeutung für paläozoische Geodynamik und alpidischen Deckenbau. Neues Jahrb Geol Paläontol Monatsh 1985: 46–64

Neubauer F (1988) The Variscan orogeny in the Austroalpine and Southalpine domains of the Eastern Alps. Schweiz Mineral Petrogr Mitt 68: 339–349

Neubauer F (1989) Late Variscan structures of the Eastern Greywacke Zone (Eastern Alps). Neues Jahrb Paläontol Monatsh 1989: 425–432

Neubauer F, Frisch W (1988) Ordovician-Silurian geodynamic evolution of the Alps – the orogeny back-arc basin model. Schweiz Mineral Petrogr Mitt 68: 351–357

Neubauer F, Herzog U (1985) Das Karbon der Stolzalpendecke Mittelkärntens – Implikationen für die variszische Paläogeographie und Orogenese im Ostalpin. Anz Österr Akad Wiss Math-Naturwiss Kl 1985: 105–109

Neubauer F, Pistotnik J (1984) Das Altpaläozoikum und Unterkarbon des Gurktaler Deckensystems (Ostalpen) und ihre paläogeographischen Beziehungen. Geol Rundsch 73: 149–174

Neubauer F, Frisch W, Hansen BT (1987) New data on the evolution of the Austroalpine basement: a U/Pb zircon study. Terra Cognita 7: p 96

Neubauer F, Frisch W, Schmerold R, Schlöser H (1989) Metamorphosed and dismembered ophiolite suites in the basement units of the Eastern Alps. Tectonophysics 164: 49–62

Neugebauer J (1970) Altpaläozoische Schichtfolge, Deckenbau und Metamorphosenablauf im südwestlichen Saualpen-Kristallin (Ostalpen). Geotekt Forsch 35: 23–93

Neugebauer J (1988) The Variscan plate tectonic evolution: an improved "Iapetus model". Schweiz Mineral Petrogr Mitt 68: 313–333

Nievoll J (1983) Stratigraphische und strukturgeologische Untersuchungen in der Grauwackenzone bei Veitsch (Steiermark). Diss, Univ Graz, 150 pp (unpubl)

Nievoll J (1987) The Lower Paleozoic of the Noric Nappe at Veitsch (Eastern Alps, Austria). In: Flügel HW, Sassi FP, Grecula P (eds) Pre-Variscan and Variscan events in the Alpine-Mediterranean Mountain Belts, IGCP Project 5, Regional Volume, Miner Slov Monogr, pp 379–386

Pasini M (1963) Alcuni Fusulinida della serie del Monte Auernig (Alpi Carniche) e loro significato stratigrafico. Riv Ital Paleontol 69: 337–382

Pistotnik J (1989) Die Phyllitgruppe und Magdalensbergserie auf ÖK-Blatt 186 St. Veit/Glan. In: Arbeitstagung der Geol Bundesanst, Vienna 1989, Blatt 186 St. Veit an der Glan, pp 42–44

Pohler S (1982) Die Entwicklung der Flachwasserkarbonate im Mittel- und Oberdevon der Zentralen Karnischen Alpen (Seewartesockel zum S-Abfall der Hohen Warte). Diplomarbeit, Paläontol Inst Univ Bonn, 134 pp

Pollak W (1962) Untersuchungen über Schichtfolge, Bau und tektonische Stellung des österreichischen Anteils der Eisenberggruppe im südlichen Burgenland. Diss, Phil Fak Univ Wien, 108 pp

Rantitsch G (1990) Fazies und Diagenese devonischer Riffkalke des Seeberger Aufbruches (Kärnten, Österreich). Diplomarbeit, Inst Geol Paläontol Univ Graz, 120 pp

Reitz E, Höll R (1989) Unterordovizische Acritarchen aus der Nördlichen Grauwackenzone (Ostalpen). Jahrb Geol Bundesanst 132: 761–774

Reitz E, Höll R (1991) Biostratigraphischer Nachweis von Arenig in der Nördlichen Grauwackenzone (Ostalpen). Jahrb Geol Bundesanst 134: 329–344

Rolser J, Tessensohn F (1974) Alpidische Tektonik im Variszikum der Karawanken und ihre Beziehung zum Periadriatischen Lineament. Geol Jahrb A25: 23–53

Sacchi R (1989) Pan African Rejuvenation. 28th Int Geol Congr Washington, 3: 2–3 (Abstr)

Schlaegel P (1988) Geotektonische Neuinterpretation des basischen Magmatismus der Nördlichen Grauwackenzone, Ostalpen – ein Überblick. Schweiz Mineral Petrogr Mitt 68: 419–432

Schlaegel-Blaut P (1990) Der basische Magmatismus der Nördlichen Grauwackenzone, Oberostalpines Paläozoikum. Abh Geol Bundesanst 43: 1–149

Schlöser H, Kullmann J, Loeschke J (1990) Korallen-führendes Unterkarbon auf der Brunnachhöhe (Nockgebiet, Gurktaler Decke, Österreich). Carinthia II 180: 643–650

Schnepf HG (1989) Vulkanismus und Sedimentologie im Murauer Paläozoikum (Gurktaler Decke, Oberostalpin, Österreich). Arb Inst Geol-Paläontol Univ Tübingen A2: 1–183

Schönlaub HP (1971) Die Althofener Gruppe – eine neue stratigraphische Einheit im Devon Mittelkärntens (Österreich). Neues Jahrb Geol Paläontol Monatsh 1971: 288–305

Schönlaub HP (1976) Bericht 1975 über Aufnahmen in der Nördlichen Grauwackenzone auf den Blättern 131, Kalwang und 132, Trofaiach. Verh Geol Bundesanst 1976: A136–A138

Schönlaub HP (1979) Die Nördliche Grauwackenzone. In: Das Paläozoikum von Österreich. Abh Geol Bundesanst 33: 76–97

Schönlaub HP (1980) Carnic Alps. Field Trip A. In: Schönlaub HP (ed) Second European Conodont Symposium ECOS II, Guidebook Abstracts. Abh Geol Bundesanst 35: 5–57

Schönlaub HP (1982) Die Grauwackenzone in den Eisenerzer Alpen (Österreich). Jahrb Geol Bundesanst 124: 361–423

Schönlaub HP (1984) Das Paläozoikum von Sulz bei Güssing im Südburgenland. Jahrb Geol Bundesanst, Vienna, 127: 501–505

Schönlaub HP (1985 a) Das Paläozoikum der Karnischen Alpen. In: Arbeitstagung der Geologischen Bundesanstalt 1985. Geol Bundesanst, Vienna 1985, pp 34–52

Schönlaub HP (1985 b) Das Karbon von Nötsch und sein Rahmen. Jahrb Geol Bundesanst 127: 673–692

Schönlaub HP (1990 a) The collapse of the Carboniferous flysch trough in the Carnic Alps. In: Venturini C, Krainer K (eds) Field workshop on Carboniferous to Permian sequence of the Pramollo-Nassfeld Basin (Carnic Alps). Proc, Bologna, p 14

Schönlaub HP (1990 b) Das Altpaläozoikum im Südburgenland. In: Exkursionsführer Wandertagung 1990 Bernstein: 52–54. Österr Geol Ges, Vienna, pp 1–149

Schönlaub HP, Klein P, Magaritz M, Rantitsch G, Scharbert S (1991) Lower Carboniferous Paleokarst in the Carnic Alps (Austria, Italy). Facies 25: 91–118

Spalletta C, Venturini C (1988) Conglomeratic sequences in the Hochwipfel formation: a new paleogeographic hypothesis on the Hercynian flysch stage of the Carnic Alps. Jahrb Geol Bundesanst 131: 637–647

Spalletta C, Vai GB, Venturini C (1980) Il flysch ercinico nella geologia dei Monti Paularo e Dimon (Alpi Carniche). Mem Soc Geol Ital 20: 243–265

Strehl E (1962) Die geologische Neuaufnahme des Saualpen-Kristallins, Teil IV. Das Paläozoikum und sein Deckgebirge zwischen Klein St. Paul und Brückl. Carinthia II 152: 46–74

Tessensohn F (1971) Der Flysch-Trog und seine Randbereiche im Karbon der Karawanken. Neues Jahrb Geol Paläontol Abh 138: 169–220

Tessensohn F (1974) Zur Fazies paläozoischer Kalke in den Karawanken (Karawankenkalke II). Verh Geol Bundesanst 1974: 89–130

Tessensohn F (1983) Eisenkappler und Seeberger Paläozoikum. In: Bauer FK, Cerny I, Exner C et al. (eds) Erläuterungen zur Geologischen Karte der Karawanken 1:25000, Ostteil. Geol Bundesanst, Vienna, pp 32–45

Thalhammer O (1982) Das Paläozoikum nördlich der Breitenau. Diss, Univ Graz, 251 pp (unpubl)

Thalhammer O, Tschelaut W (1983) Some remarks on the northern parts of the Graz Paleozoic (Austria, geotraverse B). In: Sassi FP, Szederkenyi T (eds) Intern Geol Correlation Program Project 5. Newsletter 5, pp 187–190
Tollmann A (1963) Ostalpensynthese. Deuticke, Vienna, pp 1–256
Tollmann A (1977) Geologie von Österreich, Band 1. Deuticke, Vienna, 766 pp
Toula F (1878) Über Devon-Fossilien aus dem Eisenburger Comitate. Verh Geol Reichsanst 1878: 47–52
Tschelaut W (1984) Die Geologie des Gebietes zwischen Röthelstein und Gams bei Frohnleiten (Grazer Paläozoikum, Steiermark). Diss, Univ Graz, 155 pp (unpubl)
Vai GB (1979) Una palinspastica permiana della Catena Paleocarnica. Rend Soc Geol Ital 1: 25–27
Venturini C (1990) Geologia della Alpi Carniche Centro Orie tali. Commune Udine Ed Mus Friul St Nat 36: 1–220
Venturini C (1991) Geologia delle Alpi Carniche centro orie tali. Comune di Udine Publ 36. Ediz Mus Friulano di Sto Naturale, Udine, pp 1–220
Venturini C, Ferrari A, Spalletta C, Vai GB (1982) La disc danza ercinica, il tardorogeno e il postorogeno nella geolog del Passo di Pramollo. In: Castellarin A, Vai GB (eds) Gui alla geologia del Sudalpino centro-orientale. Guide Ge Reg Soc Geol Ital, Bologna, pp 305–319
Walliser OH (1964) Conodonten des Silurs. Abh hess La desamtes Bodenforsch 41: 1–106
Zier C (1982) Die Geologie des Gebietes zwischen Häre schützklamm und Tyrnauergraben (Grazer Paläozoiku Steiermark). Diss, Univ Graz, 364 pp

The Austro-Alpine Quartzphyllites and Related Palaeozoic Formations

F. NEUBAUER and F. P. SASSI

Abstract

The Austro-Alpine quartzphyllites and related clastic- and volcanic-rich sequences are interpreted as the infilling of an Early Ordovician to Early Carboniferous basin. The basin may have been initially formed as a back-arc basin on a Late Cadomian/Pan-African metamorphic crust during the Ordovician. A pulse of renewed rifting during Silurian and Early Devonian probably led to the opening of a nearby oceanic basin and formation of a passive margin which is mainly represented by a Middle to Late Devonian carbonate platform. Contraction and basin closure occurred during Visean to Early Westphalian. It was accompanied by: (*i*) rapid subsidence due to lithospheric loading by thrust sheets, (*ii*) formation of a foredeep with orogenic flysch deposits in front of incoming thrust sheets, and (*iii*) weak metamorphic processes.

1 Introduction

Quartzphyllites are mainly metapelitic rocks in greenschist facies which are widespread in the Austro-Alpine and Southalpine basements (Fig. 1). Although it was thought for a long time that these units consisted of a rather monotonous lithology, detailed work has shown the presence of thick mafic and subordinate acidic volcanic sequences as well as thin carbonate layers, mostly ferrugenous dolomites and also magnesites. Scarce fossil records, mostly conodonts and more recently acritarchs and other palynomorpha of Ordovician to Early Carboniferous age, gave rise to a reasonable stratigraphy of incompletely known sections (e.g. Schönlaub 1979; Scharbert and Schönlaub 1980; Sassi et al. 1987). The range of metamorphic grade covers the chlorite, biotite ad almandine zones. There are gradual transitions, on the one hand, to anchimetamorphic, sub-greenschist facies clastic-bearing, fossiliferous Palaeozoic sequences which are dominated by mafic volcanics and/or carbonates, and on the other hand, to micaschist-marble units (see contributions of Neubauer and Frisch, Loeschke and Heinisch, and Schönlaub und Heinisch, this Vol.)

Therefore, we portray in this chapter all units which are dominated by quartzphyllites, slates, and accompanying mafic and acidic metavolanics in order to pick out relationships between different areas which are largely dispersed and dismembered by Mesozoic/Tertiary tectonic displacement. Because of the complex structure, a formal stratigraphic framework is missing in most of the areas under considera-

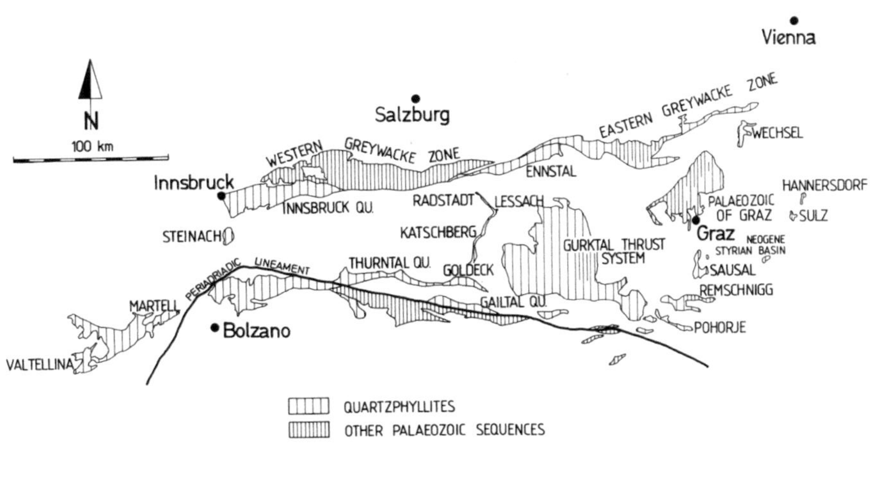

Fig. 1. Distribution of quartzphyllite-dominated units in the Austro-Alpine basement of the Eastern Alps

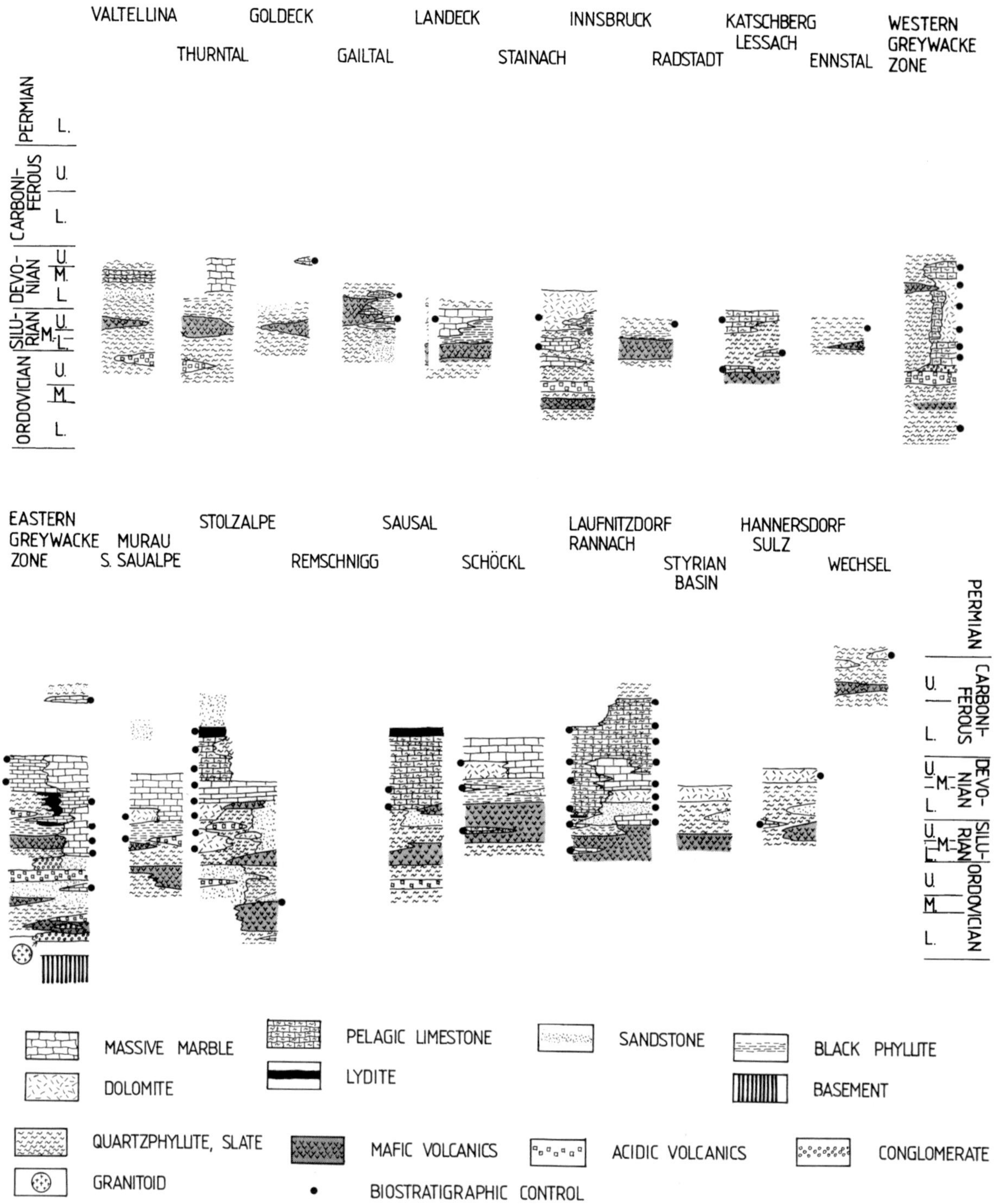

Fig. 2. Lithostratigraphic columns including evidence of protolith ages from biostratigraphic data

tion. Also, several Austro-Alpine quartzphyllite terrains, especially those east of the Tauern window, suffered a Mesozoic/Tertiary metamorphic overprint within greenschist facies between 140 and 80 Ma (Frank 1987; Frank et al. 1987) and all of them were strongly deformed and truncated at their structural base. Thicknesses are generally uncertain, with estimates of structural thicknesses ranging between 1000 and 2000 m.

We use the age table of Harland et al. (1989) for correlation of geochronological and biostratigraphical data.

2 Regional Discriptions

The Austro-Alpine quartzphyllites outcropping in different areas are poorly correlated with exceptions of some idealized sections (e.g. Schönlaub 1979; Neubauer and Pistotnik 1984; Sassi et al. 1987). In this review, we provide sections and try to establish facies variatious based on data published in the last two decades. Stratigraphic sections are given in Fig. 2. The biostratigraphic ages are presented in Table 1, the ages and the nature of volcanic episodes are listed in Table 2, and the geochronological data used for determining the age of metamorphism in Table 3.

2.1 Casanna Phyllites of the Valtellina and Martell Quartzphyllites

The Casanna phyllites occur in the Valtellina area and continue to the Martell Valley in the northeast (Fig. 1) (Andreatta 1951). Besides widespread phyllonites which are derived from medium-grade metamorphic rocks by Alpine mylonitization, greyish phyllites dominate. The boundary between quartzphyllites and higher grade mesozonal micaschists is transitional, controlled by a pre-Alpine metamorphic zonation. Garnet occurs in the higher-grade phyllites, and staurolite was also reported at the boundary towards staurolite-kyanite-almandine micaschists. A syn- to post-kinematic metamorphic overprint of Alpine age is commonly assumed: it often displays a retrograde character, but locally reached temperatures suitable for the crystallization of new staurolite.

These phyllites contain intercalations of thin quartzite, carbonatic phyllites, "prasinites" (i.e. epidote-chlorite-actinolite schists) and other low-grade metabasites, as well as a reddish to greenish, mica-bearing marble (Andreatta 1951; Hess 1953; Pozzi 1965; Schmid 1973). In the Formi and Martell Valleys, a "porphyroid" gneiss occurs within the phyllites (Dal Piaz and Martin 1980; Heinisch 1981). No detailed section is available, but the marble of the Valtellina area resembles closely the Early Devonian pelagic marble which is intercalated in quartzphyllites of the Gurktal thrust system (see later).

The metabasic rocks point to two different groups (Poli 1989): those of the Ortler area are relatively high-temperature amphibolites characterized by the presence of metamorphic diopside, whereas Valtellina "prasinites" exhibit a low-temperature assemblage that still allows the recognition of sedimentary contamination. Nevertheless, both of these groups of metabasites show low Ti, P and Nb-Ta contents, suggesting a calc alkaline affinity of the basic rocks interlayered in the phyllites.

Clastic zircons of a quartzite yielded a U-Pb zircon upper intercept age of 2050 Ma (Grauert et al. 1973), which is interpreted as an average age of zircon crystallization in the source region. The lower intercept age is approximately 500 Ma and indicates the penetrative metamorphic overprint in the source region (Grauert et al. 1973).

2.2 Thurntal, Kreuzeck and Goldeck Quartzphyllites

A sequence of south-dipping quartzphyllites occurs generally on the tectonic top of the micaschists and paragneisses south of the Tauern window (Fig. 1). These quartzphyllites form a small stripe ranging

Table 1. Stratigraphic data. Data are nearly exclusively based on conodonts except palynomorpha and brachiopods

Area	Biostratigraphic evidence[a]	Reference
Goldeck:	lD	Heinz (1987)
Gailtal:	eD	Schönlaub (1979)
Innsbruck:	O (P), lS, eD, mD	Höll and Maucher (1968) Mostler (1973)
Radstadt:	lS or eD	Schönlaub (1975)
Katschberg:	lS	Schönlaub et al. (1976) Schönlaub and Zezula (1975)
Ennstal:	S (P)	Priewalder and Schumacher (1976)
Murau nappe and Phyllite group/Saualpe:	lS, eD	Neubauer (1980 a, b) Neugebauer (1970) Niederl (1980)
Stolzalpen nappe:	mO (B), lO, eS, mS, lS, eD, mD, lD, eC	Buchroithner (1978, 1979) Havlicek et al. (1987) Höll (1970) Neubauer (1980 a, b) Neubauer and Pistotnik (1984) Neubauer and Herzog (1985) Riehl-Herwirsch (1970)
Sausal and Remschnigg	mS, lS, eD	Ebner (1975) Mioc and Ramovs (1973)
Hannersdorf – Sulz	lS, mD	Schönlaub (1984)
Basement of the Styrian Neogene basin	eD	Ebner (1978, 1988)
Eastern Grauwacken zone	lO, eS, mS, lS, eD, mD, lD, eC	Flajs and Schönlaub (1976) Nievoll (1987) Schönlaub (1982) Schönlaub et al. (1980)
Wechsel phyllites	lC-P (P)	Planderova and Pahr (1983)

[a] P, Palynomorpha; B, brachiopods. O, Ordovician; S, Silurian; D, Devonian; C, Carboniferous; e, early; m, middle; l, late.

Table 2. Evidence for timing and nature of volcanic episodes within the quartzphyllite areas

Area	Age[a]	Reference for age	Nature	Reference for nature
Martell:				
			Tholeiitic(?)	Poli (1989)
Thurntal:				
			Mafic-acidic	Heinisch and Schmidt (1984)
Kreuzeck:				
			Mafic calc-alkaline	Reimann and Stumpfl (1981)
Goldeck:				
			Mafic	Deutsch (1977) Heinz (1987)
Gailtal:				
	lS(eD?)		Mafic	Heinisch (1987)
Ennstal:				
			Mafic tholeiitic	Schlüter et al. (1984) Höll et al. (1985)
Murau nappe:				
	eS(?)	Neubauer (1980)	Mafic	
	lD-eC(?)		Mafic transitional	Neubauer (1989b)
Stolzalpe nappe:				
Magdalensberg Fm.	mO	Havlicek et al. (1987)	Mafic alkalic	Loeschke (1989a)
Nock Fm.	lO	Neubauer and Pistotnik (1984)	Calc-alkaline	Giese (1988)
Haider Fm.	lO	Neubauer (1980)	Acidic	Schnepf (1989)
"Metadiabas" Fm.	eS	Schnepf (1989)	Mafic alkalic	Schnepf (1989)
	eD-mD	Neubauer (1980b)	Mafic	
Eisenhut Fm.	eS-mS	Höll (1970)	Mafic alkalic	Giese (1988)
	lS-eD		Acidic	Giese (1988)
Sausal-Remschnigg:				
	lO(?)		Acidic	Schlamberger (1988)
	lS-eD(?)		Mafic trans.	Schlamberger (1988)
Schöckl nappe:				
Arzberg Fm.	mS	Fritz (1991)	Mafic alkalic	Weber (1990) Loeschke (1988a)
Laufnitzdorf nappe:				
	eS	Tschelaut (1984)	Mafic	
Rannach and Hochlantsch nappes:				
Kehr Fm.	mS-lS	Neubauer (1989c)	Mafic alkalic	Fritz and Neubauer (1988)
Hochlantsch	mD(?)		Mafic trans.	Loeschke (1988a)
Eastern Grauwacken zone:				
Kalwang Fm.	mO(?)		Mafic tholeiitic	Schäffer and Tarkian (1984)
Gerichtsgraben Fm.	mO(?)		Acidic	Loeschke (1989b)
	mO(?)		Mafic	
Blasseneck	lO	Flajs and Schönlaub (1976)	Acidic	Heinisch (1981)
Kragelschinken Fm.	eS-mS	Schönlaub (1982)	Mafic alkalic	Schlaegel-Blaut (1990)
Rad Fm.	lS-eD	Nievoll (1987)	Acidic	Nievoll (1987)

[a] For abbreviations, see Table 1.

from the Thurntal area in the west via the Kreuzeck mountains to Goldeck in the east (see also Schulz et al. this Vol.).

The boundary between the **Thurntal quartzphyllites** and the underlying higher-grade rocks (kyanite + staurolite ± sillimanite paragneisses and micaschists) is tectonic, as pointed out by Sassi and Zanferrari (1972) and emphasized by Heinisch and Schmidt (1976). It is questionable whether this boundary only represents a disconnected original stratigraphic boundary between a phyllitic cover and its original basement which jointly underwent the Variscan metamorphism, or if it is a major tectonic boundary between exotic terrains which did not share any common stage of their pre-Alpine metamorphic evolution. The former hypothesis seems to be supported by the post-crystalline character of this tectonic boundary (i.e. its late or post-Variscan age) and the occurrence in the Eastern Alps of situations such as those described by Neubauer (1985).

The Thurntal quartzphyllites complex (Sassi and Zanferrari 1972; Heinisch and Schmidt 1976, 1984) and the phyllitic complex of the southern Kreuzeck mountains (Lahusen 1972; Reimann and Stumpfl 1981) contain greyish quartzphyllites and fine-grained chlorite micaschists, prasinites and other

Table 3. Geochronological data

Locality	Method	Age (Ma)	Reference
Martell	U-Pb zircon upper intercept	2050	Grauert et al. (1973)
Thurntal	Rb-Sr biotite	294–258	Borsi et al. (1978)
Kreuzeck	K-Ar muscovite	301 ± 4	Brewer (1969)
	K-Ar muscovite	301–291	
	K-Ar biotite	261 ± 6	Brewer (1969)
	K-Ar biotite	261–212	
Goldeck	K-Ar muscovite	320	Brewer (1969)
	K-Ar muscovite	320–292	
	K-Ar biotite	303 ± 7	Brewer (1969)
	K-Ar biotite	303–240	
Eastern Grauwacken zone			
	Rb-Sr thin slab	246 ± 35	Jung (1981)
	Detrital and metamorphic white micas	ca. 320–180 Ma	Jung (1981)

green schists, rare marble lenses in deep and high levels of the structural section, as well as porphyroid gneisses in a middle level of the section (Fig. 2). The Thurntal phyllites sometimes bear almandine-rich garnet, indicating that temperatures pertaining to the upper greenschist facies were reached. The average *b* cell parameter of muscovite is rather low (8.992 Å; n = 57; s = 0.005; Sassi 1972; and Sassi and Spiess 1992), indicating low pressure condition of metamorphism (Guidotti and Sassi 1986).

The **Kreuzeck quarzphyllites** also display a pre-Alpine greenschist facies metamorphism, during which almandine-rich garnet (with the usual bell-shaped zoning of Fe and Mn) also crystallized within the metapelites of suitable bulk composition. Their microtextural evolution is identical to that recorded within the Thurntal phyllites, with the axial plane foliation S_2 prevailing over both the almost obliterated S_1 and the local, discontinuous S_3. The pressure character is also identical to that recorded in the Thurntal and Kreuzeck phyllites, as shown by Sassi and Spiess (1992) on the basis of the low value of the *b* cell parameter of muscovite (average value 8.993 Å; n = 60; s = 0.004) and the criteria in Guidotti and Sassi (1986). Black schists and accompanying greenstones are widespread in the Kreuzeck mountains (Lahusen 1972; Reimann and Stumpfl 1981), where they bear a stratiform mineralization. In the Thurntal area, Cu-bearing sulphides are widespread; accompanied in addition by a Sb-W mineralization at this level (Lahusen 1972; Reimann and Stumpfl 1980; Koeppel et al. this Vol.).

In the **Goldeck** mountains, Deutsch (1977) and Heinz (1987) described an approximately 1500-m-thick tectonic pile of informal stratigraphic units which are (from bottom to the top):

- Graphite-rich phyllites with metacherts and carbonates. Heinz (1987) mentioned a poor conodont fauna of Late Devonian age (based on *Palmatolepsis* sp., Schönlaub pers. comm.);
- A phyllite unit with greyish phyllites and thick calcitic and dolomitic marbles. Conodonts suggest an Early Devonian age of dolomite marbles (own unpubl. data);
- A metavolcanic unit composed of mainly mafic crystal tuffs, ash tuffs, volcanogenic breccias and minor lava flows;
- A hanging wall unit which contains phyllites, metagreywackes and quartzphyllites.

Although unresolved with respect to its age, a good correlation to the Gurktal thrust system (GTS) is obvious (Fig. 2). The graphite-rich phyllites and the carbonate-rich unit closely resemble the lithologic content of the Murau nappe within the GTS. The metavolcanic unit may be related to the mafic volcanics of the Stolzalpe nappe of the GTS (see Sect. 2.5).

Sassi and Spiess (1992) obtained a low average *b* cell dimension of muscovite from the lower graphite-rich phyllites: 8.995 Å (n = 58; s = 0.004); based on the criteria in Guidotti and Sassi (1986), they deduced from these data a clear indication of low-pressure metamorphic conditions.

K-Ar muscovite and biotite data are in the range of 320–261 Ma (Brewer 1969; also cited in Hoke 1990). The oldest biotite data refer to the age of cooling after the metamorphic climax, whereas oldest muscovite may date the minimum age of the metamorphic climax. However, both are partly rejuvenated due to increasing Cretaceous temperatures near the structural base of these quartzphyllites. Therefore, we interpret the mineral ages which have been obtained from quartzphyllites near the structural base, as mixing ages due to partial rejuvenation during Cretaceous metamorphism.

2.3 *Gailtal Quartzphyllites*

The Gailtal quartzphyllites (Fig. 1) correspond best to their term, as they contain quartzites, quartzitic phyllites, mineralized black phyllites, low-grade metabasites and dark marbles (Sassi et al. 1974; Heinisch 1987). Badly preserved conodonts from two sites testify Silurian and/or Early Devonian age of dolomitic marbles within the black phyllites (Fig. 2).

Mylonitic shear zones of Alpine age are common in this area (Unzog 1989); however, they separate rock volumes in which the pre-Alpine textural and mineral features are still preserved. Among them, the low *b* cell parameter of muscovite should be mentioned (average value 8.995, n = 75, s = 0.006; Visona 1974; Sassi and Spiess 1992): they clearly indicate the low-pressure character of this metamorphism (Guidotti and Sassi 1986).

Heinisch (1987) argued strongly for lateral facies differentiation, and for transition of quartzphyllites into medium grade metamorphic rocks. However, this transition is very rapid and the staurolite zone in the medium-grade rocks adjacent to the phyllites is thin, as reported by Heinisch (1987) and emphasized by Becker et al. (1987). Therefore, the hypothesis proposed by Sassi et al. (1974) that the Gailtal quartzphyllites represent a Palaeozoic cover and the adjacent higher grade rocks their pre-Variscan basement is valid, although it still remains a hypothesis. The low-pressure metamorphism, which altered the Palaeozoic rocks into the quartzphyllite complex, may have masked the original metamorphic gap occurring at the stratigraphic boundary, obliterating it with its rapid zonation.

The age of metamorphism is not proved by the available geochronological data, but it is constrained by a well-known angular unconformity between the metamorphic quartzphyllites and Permian sedimentary cover at the easternmost margin of the Gailtal quartzphyllites. This fact indicates a Variscan, pre-Permian age of metamorphism, comparable to that of the Thurntal and Goldeck areas.

2.4 Innsbruck, Landeck, Steinach, Radstadt, Katschberg, Lessach and Ennstal Quartzphyllites

The **Innsbruck quartzphyllite** (Fig. 1) is subdivided into three subunits. From the stratigraphic bottom to the top they are (Haditsch and Mostler 1983):

1. A basal quartzphyllite-greenschist unit, which also includes acid volcanics in a higher level (Fig. 2). Recently, Ordovician acritarchs were discovered in quartzphyllites (Reitz pers. comm.) supporting the assummed Ordovician sedimentation age of quartzphyllites which was proposed by Haditsch and Mostler (1983);
2. A carbonate-sericite phyllite unit, which is composed of marbles and quartzphyllites, low-grade metabasites and quartzites;
3. A black schist-carbonate unit, which exhibits a probable lateral facies variation between dolomite marble/magnesites and black schists/siliceous sediments in deeper stratigraphic levels.

Variation in mineralization supports the proposed subdivision into these three units (Haditsch and Mostler 1983; see also Koeppel et al. this Vol.). Only the dolomites from the hangingwall unit have yielded conodont faunas which are of late Silurian (*siluricus* zone) to Middle Devonian age (Höll and Maucher 1968; Mostler 1973).

The low-pressure character of metamorphism in the Innsbruck quartzphyllites is shown, according to the criteria in Guidotti and Sassi (1986), by the low *b* cell parameter of muscovite (average value 8.996 Å; n = 70; s = 0.004; Sassi 1972; Sassi and Spiess 1991).

The **Steinach quartzphyllite** (Fig. 1) is composed of grey-green garnet-bearing phyllites, black phyllite, chlorite schists, mafic schists, and thin lenses of ferrigenous dolomites, marbles and magnesites (Sassi and Menegazzo 1971; Frizzo and Visona 1981). The carbonates are concentrated in high structural levels.

Sassi and Menegazzo (1971) referred this metamorphism to the lower greenschist facies (the garnet has spessartine-rich composition; chlorite + muscovite is the stable phyllosilicate assemblage) and low pressure conditions (according to Sassi 1972 and Sassi and Spiess 1992, the average *b* cell parameter of muscovite is low: 8.994 Å; n = 50, s = 0.007). Sassi and Menegazzo (1971) also established that the microtextural and mineral evolution of the Steinach quartzphyllites was complete before the sedimentation of the Upper Carboniferous Nösslach conglomerate, which stratigraphically lies over the phyllites and is not affected by metamorphic crystallization.

The **Radstadt, Katschberg,** and **Lessach quartzphyllites** surround the eastern edge of the Penninic Tauern window (Fig. 1). The sequences are composed of grey quartzphyllites which bear intercalations of black phyllites, thin black quartzites, green basic schists, and rare serpentinites, as well as boudins of ferruginous dolomites and black marbles (Exner 1990). Silurian conodonts were reported from a few localities (Schönlaub 1975; Schönlaub and Zezula 1975; Schönlaub et al. 1976). A detailed section is not yet reconstructed, but some authors (e.g. Schönlaub 1979) think that the green schists form the stratigraphic base.

The **Ennstal quartzphyllite** (Fig. 1) area occurs along the northern margin of the crystalline rocks of the Niedere Tauern. The nature of the boundary to the southern adjacent medium-grade metamorphic rocks, micaschists and marbles, is uncertain. The quartzphyllite area is composed mainly of grey quartzphyllites and thick, regularly interlayered, metabasic layers with minor intercalations of black phyllites, quartzites, and rare, tens of metre-scaled marbles (Höll et al. 1985; Schlüter et al. 1984). Acritarchs suggest a predominant Silurian age for these quartzphyllites (Priewalder and Schumacher 1976). The metabasic layers exhibit a tholeiitic geochemical signature (Höll et al. 1985; Schlüter et al. 1984).

2.5 *Palaeozoic of the Gurktal Thrust System*

The Gurktal thrust system (GTS) contains two Alpine thrust sheets composed of weakly metamorphosed Palaeozoic formations, the lower Murau nappe and the Stolzalpe nappe (e.g. Neubauer and Pistotnik 1984; Fig. 3). Both nappes are separated by Permo-Mesozoic rocks along the northern margin of the GTS. To the east, the Murau nappe exhibits strong affinities to the Phyllite Group of the southern Saualpe (Kleinschmidt et al. 1975), as well as to the phyllites along the northern margin of the Karawanken, and may be followed into the Pohorje Mountains in Slovenia (Hinterlechner-Ravnik and Moine 1977). The classical Magdalensberg group of Middle Carinthia of the top of the Phyllite group is comparable to the strata of the Stolzalpe nappe at the northwestern margin of the GTS (Fig. 3).

The stratigraphy of the **Murau nappe** is uncertain because of the lack of continuous, fossil-bearing sections, and strong, partly unresolved internal structure. As a rule, the stratigraphy is consistent with that of the Schöckl unit within the Palaeozoic of Graz as described later. Green schists derived from lava flows, sills and tuffs occur at the base within a phyllite matrix (Fig. 2; Neubauer 1980 a). A phyllite-rich unit overlies the green schists. Carbonatic phyllites, black phyllites, and quartzites with minor basic green schists and orthoquartzites build up the next higher stratigraphic unit. Along the southern margin of the Gurktal nappe, widespread acidic volcanics occur (Fritsch 1961; Hajek 1963; Hurler 1972; Loeschke 1989 b). The higher stratigraphic level is made up of carbonates which exhibit lateral facies variation. Dolomites from the base of the carbonate unit in the more northern region yielded a Late Silurian to Early Devonian, respectively a Pragian to Zlichovian conodont fauna (Buchroithner 1978; Schönlaub 1979; Neubauer 1980 a, b; Niederl 1980). Calcite marbles are widespread. A single occurrence of a brachiopod supports the Late Silurian and/or Early Devonian age of the carbonates in the Saualpe (Neugebauer 1970). The marbles sometimes bear coarse-grained mafic dykes of transitional basaltic affinity (Neubauer 1989 b).

The **Stolzalpe nappe** forming tectonic klippen above the Murau nappe (Fig. 3) exhibits an internal imbrication, and probably consists of slices which are mainly differentiated in facies during Late Silurian to Early Devonian. The distinction from the units of the Murau nappe is uncertain because of often missing Permo-Mesozoic rocks along thrust surfaces and because of similar lithologies, mainly quartzphyllites, in both units.

The distinction of various stratigraphic levels is well constrained by some fossils. As a rule, thick maf-

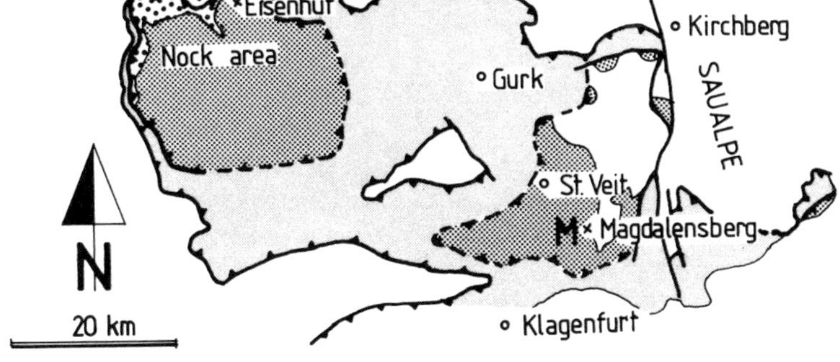

Fig. 3. Simplified map of the Gurktal thrust system (after Neubauer and Pistotnik 1984). *1* Westphalian C to Early Permian post-Variscan terrestrial molasse (conglomerates); *2* Stolzalpe nappe (with locations of type sections of facies variations which are shown in Fig. 4: *P* Pranker facies; *A* Auen facies; *M* Magdalensberg facies); *3* Ackerl complex (a metamorphic basement complex); *4* Murau nappe; *5* fault; *6* Alpidic thrust surface

ic volcanic sequences occur at the stratigraphic base (Fig. 2) and are overlain by pelite- to psammite-dominated sequences, with pelagic limestones at the top of the succession. In detail, the basal mafic-dominated volcanics occur at two distinct stratigraphic levels: a late Middle Ordovician age is proved from brachiopods for the Magdalensberg group which occurs west of the Saualpe (Havlicek et al. 1987) and conodont data (Riehl-Herwirsch 1970). The Nock Series (Giese 1988) is of Late Ordovician age (Neubauer and Pistotnik 1984). The older unit is interpreted as a basaltic-keratophyric marine island volcanic suite (Riehl-Herwirsch 1970) of alkaline geochemical characteristics (Loeschke 1989 a, b). The Nock Series exhibit calc-alkaline affinities (Giese 1988; Loeschke 1989 b).

These units contrast with an Early Silurian volcanic sequence, the Eisenhut group, which is at the northwestern edge of the GTS (Höll 1970; Neubauer and Pistotnik 1984), the Metadiabas Fm. at the northern edge of the GTS (Neubauer 1980 a; Schnepf 1989), and the Magdalensberg group west of the Saualm (Neubauer and Pistotnik 1984). However, Ordovician and Early Silurian volcanics are probably separated by a Late Ordovician unit of slates, quartzphyllites and sandstones which contain also lydites and thin carbonate layers (Neubauer 1980 a; Neubauer and Pistotnik 1984).

A slaty facies with cherts and allodapic limestones persists through the Wenlockian to the boundary of Lochkovian/Pragian (Magdalensberg facies; Fig. 4). On the other hand, thick sandstones, quartz wackes and quartz arenites occur locally, covering the same time span. This is also true for thin Late Wenlockian to Pragian dolomites, and pelagic limestones. The youngest units occurring at the Pragian/Zlichovian boundary, are reddish to greenish pelagic limestones which occur and are comparable in all units differentiated above. The absence of the conodont genus *Icriodus* in thin Early and Middle Devonian limestones and dolomites (e. g. Neubauer 1980 a) argues for a deep marine, pelagic depositional evironment.

The Zlichovian to Pragian carbonates locally include mafic volcanics (Neubauer 1980 b). These carbonates locally persist up to the Early Carboniferous and they are overlain by cherts and greywackes. The cherts yielded conodont faunas from the Tournaisean/Visean boundary (Neubauer and Herzog 1985). Rare diabase dykes and composition of sometimes volcanogenic greywackes suggest a felsic and mafic intrabasinal volcanic suite (Neubauer and Herzog 1985). Recent investigations have identified Late Visean corals in the carbonates within cover rocks stratigraphically above the quartzphyllites (Schlöser et al. 1990). Therefore, deformation and metamorphism occurred within a short intra-Visean time interval.

2.6 *The Sausal, Remschnigg and Hannersdorf Quartzphyllite Areas*

Isolated outcrops of Palaeozoic rocks surround the Neogene Styrian basin (Fig. 1) and form also part of the basement of that basin (Flügel 1988).

The **Remschnigg area** contains predominantly purple mafic ash-fall tuffs at the base, and grey slates above. Thin layers of crinoidal limestones are intercalated within the slates (Ebner 1975). Conodonts from these limestones constrain different stratigraphic lev-

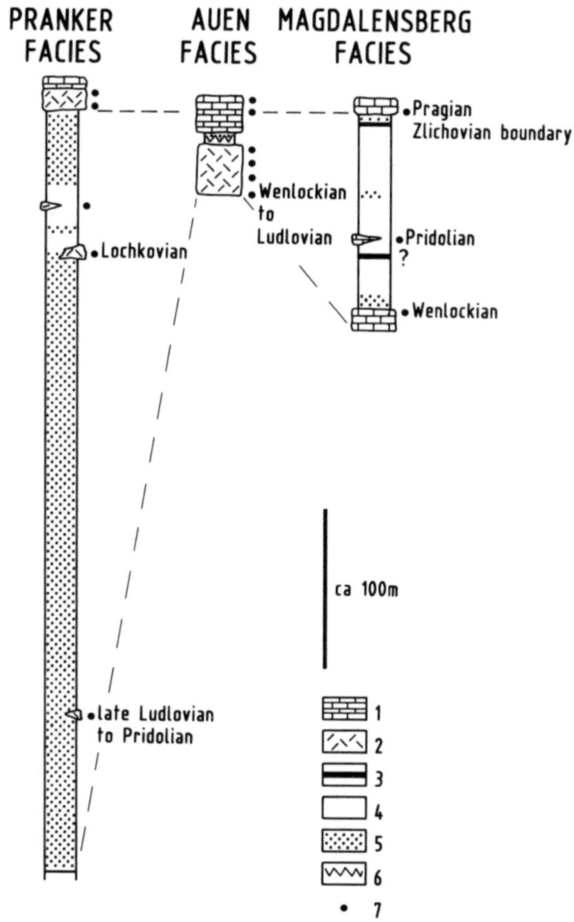

Fig. 4. Late Silurian to Early Devonian facies variation in the Stolzalpe nappe of the Gurktal thrust system (after Neubauer and Pistotnik 1984). For locations of the sections, see Fig. 3. *1* Pelagic limestones; *2* dolomite; *3* lydite; *4* phyllite, slate; *5* sandstone; *6* acidic tuff; *7* stratigraphic control by condonts

els between the Llandoverian/Wenlockian boundary and Early Devonian. Coloured pelagic limestones in the hanging wall contain Pragian/Zlichovian conodonts (Mioc and Ramovs 1973; Buggisch et al. 1975; Ebner 1975).

The **Sausal section** starts with acidic volcanic rocks which are followed by calcareous phyllites and quartzphyllites. Thin mafic tuffs, lavas as well as dykes, and sandstones form the hanging wall part of the section (Schlamberger 1988). This part of the section bears no fossils. Mica-bearing sandstones, nodular limestones, and lydites are separated from the former section. The nodular limestone yielded conodonts of Lochkovian to Pragian age (Buggish et al. 1975). The mafic lava flows and dykes are uncertain in age and show transitional basalt characteristics (Schlamberger 1988). The quartz-rich sandstones as well as metamorphic heavy minerals such as garnet were interpreted as deposition on a passive continental margin with a predominantly metamorphic source region (Schlamberger 1988).

At **Hannersdorf,** ca. 250 m of thick dolomites and subordinate limestones occur (Schönlaub 1990). Rugose and tabulate corals, crinoids and brachiopods suggest a Middle Devonian age of deposition (Toula 1878). Sandy dark slates, marls, and greenschists overlie the thick dolomite sequence. Other sections expose brownish slates, cherts, dolomites and sandy phyllites. The latter bear a probable Silurian conodont fauna (Schönlaub 1984). The tuffaceous greenschists from Hannersdorf exhibit an alkalic chemistry of within-plate basalt affinity (Gratzer 1985).

Comparable sequences have been found in wells which penetrated the **basement of the Neogene Styrian basin** (Flügel 1988). Among these, Early Devonian dolomites were proved by conodonts (Ebner 1978, 1988). Other dark slates, which also include sandstone layers, are uncertain in age (Flügel 1988).

2.7 The Palaeozoic of Graz

Besides the well-studied Silurian/Devonian carbonate-dominated sections (Flügel and Neubauer 1984), approximately one half of the Palaeozoic sequence of Graz is composed of phyllites and related lithologies which are in the greenschist facies metamorphic state and nearly completely free of fossils (Fig. 2, 5). These units occur mainly in the deeper and eastern part of the Palaeozoic section of Graz. In this area, phyllites are the main constituents of some tectonic units, especially the Anger crystalline complex which exhibits transitions to the Heilbrunner phyllite, and the Passail group (Flügel 1975; Flügel and Neubauer 1984; Gsellmann 1988). All these units consist of thick greyish phyllites at their base, which also contains light-coloured quartzite intercalations. The next formation, the

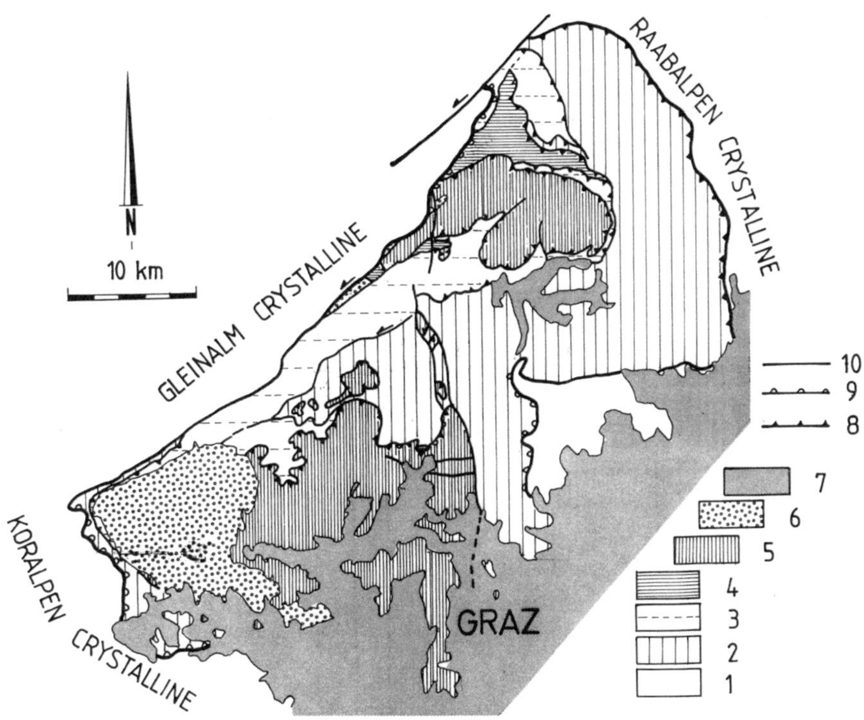

Fig. 5. Simplified tectonic map of the Palaeozoic of Graz (after Fritz et al. 1991). *1* Middle Austro-Alpine metamorphic basement; *2–5* Palaeozoic of Graz: *2* quartzphyllite-dominated nappes (Schöckl nappe *s. l.* Anger crystalline); *3* Kalkschiefer nappes; *4* Laufnitzdorf nappe; *5* Hochlantsch and Rannach nappes; *6* late Cretaceous; *7* Neogene; *8* Alpine thrust surface; *9* Alpine low angle normal fault; *10* fault

Arzberg Fm., is composed of greenschists, tuffs, blackschists grading into carbonatic phyllites (Weber 1990). This formation is the host of widespread stratabound Pb-Zn ore deposits (Weber 1990). Limestone intercalations incorporated into the Arzberg Fm. prove an Early Devonian age of deposition (Tschelaut 1985). The depositional environment is interpreted as a partly anoxic basin which is divided by intrabasinal highs from which shallow water carbonates are re-deposited in local basins. An approximately 200- to 350 m-thick calcitic marble (Schöckl marble, Hochschlag marble, Heilbrunn marble) overlies the Arzberg formation. Locally, dolomites and dolomitic quartzites (Raasberg Fm.) are interlayered within this contact. Scarce macrofossils argue for a predominant Middle Devonian age of deposition (Schönlaub 1979). A further sequence, composed mainly of slates, is the Laufnitzdorf group which contains Early to Middle Silurian mafic volcanics at the base. Slates, lydites and greywackes compose the Late Silurian to Early Devonian portion of the section, with Early to Middle Devonian pelagic limestones forming the hanging wall (Gollner et al. 1982; Tschelaut 1984; Gsellmann 1988). This section is comparable to the Magdalensberg facies of the Stolzalpe nappe of the GTS. The Laufnitzdorf group grades into the Rannach group which contains comparable alkaline mafic volcanics at the stratigraphic base (Fritz and Neubauer 1988). The higher portion of the latter section is different because of predominance of thick shallow-water carbonates during the Devonian.

The presence of a Variscan metamorphism affecting the Palaeozoic of Graz is poorly constrained by petrological and geochronological data (Becker et al. 1987). A first metamorphic event in greenschist facies is probably the result of a Variscan metamorphism (Becker et al. 1987). Only very few geochronological arguments, e.g. the presence of extraneous radiogenic argon in Alpine white mica, support a Variscan metamorphic overprint.

2.8 *Eastern Greywacke Zone*

Besides Late Ordovician to Devonian carbonate-dominated sections, the eastern Greywacke zone (Fig. 1) also contains extensive quartzphyllite regions. These well-studied quartzphyllites have special importance because of their significance for dating and interpreting other clastic-bearing Austro-Alpine sections. These quartzphyllites are related to sections with thin Devonian pelagic limestones (Schönlaub 1982). The section starts with a Cadomian crystalline basement (see Neubauer and Frisch this Vol.) which is overlain discordantly by a metaconglomerate (Neubauer 1985). The structural base is dominated by ca. 1000 m thick, partly arkosic phyllites which contain thin greenstone layers, acidic tuffs and metasandstones (Daurer and Schönlaub 1978; Schönlaub 1982; Loeschke 1989b; Loeschke et al. 1990). The lithostratigraphic upper level of these basal phyllites bear conodonts of Caradocian age (Flajs and Schönlaub 1976). Recently, Reitz and Höll (1989) described palynomorpha of Early Ordovician age within slates of a comparable section within the western Greywacke zone. The Blasseneck "porphyroid" which is embedded in a late Ordovician clastic sequence is of ignimbritic origin (Heinisch 1981; Loeschke et al. 1990). Ordovician sandstones are quartz wackes and quartz arenites which display increasing maturity during Ordovician (own unpublished data). Heavy mineral data indicate a metamorphic source region (Stattegger 1982).

The Blasseneck porphyroid is overlain locally by polymict metaconglomerates and sandstones which continue into black slates and cherts (Hermann and Neubauer unpubl. data). Laterally, thick slates occur (Rad slates) containing another acidic tuff of Late Silurian or Early Devonian age (Nievoll 1987). The slates grade into Middle and Late Devonian pelagic limestones. The age of deformation is younger than Namurian A (youngest clasts in a Carboniferous crinoidal breccia) and older than the supposed Early Permian age of the overstepping Prebichl formation. The discordance and an angular unconformity prove some (two or three) Variscan nappes as well as semiductile folding and faulting with SW-directed vergence (Schönlaub 1982; Neubauer 1989a). An internal imbrication of the Noric nappe has been recognized for a long time because of unconformable Permian overstep sequences which cut previous thrust surfaces (Flajs 1967; Schönlaub 1982). The Rb-Sr thin slab age of Carboniferous slates is 246 Ma (Jung 1981). Clastic muscovites yielded a maximum model age of approximately 320 Ma which is close to the supposed age of the Variscan metamorphism.

2.9 *Wechsel Quartzphyllites*

The Wechsel phyllites occur within the Wechsel window and other comparable windows on the northeastern margin of the Eastern Alps (Fig. 1). In the Wechsel window, the section comprises dark phyllites, black phyllites, volcanogenic greywackes and rare mafic schists and leucocratic metatuffites (Faupl

1970; Vetters 1970; Matura 1990). In other windows, the Wechsel phyllites are accompanied by the Wiesmath gneiss, an intermediate orthogneiss of igneous origin. The protolith age of the Wechsel phyllites is uncertain. Planderova and Pahr (1983) described mixed floras of sporomorpha ranging from Silurian to earliest Permian. Recent data confirmed the Late Carboniferous/Early Permian age of the Wechsel phyllites (Planderova and Pahr 1990). However, this age is unusual for quartzphyllite in the Eastern Alps and needs further confirmation.

3 Regional Correlations

Disregarding the problematic Wechsel area, the data discussed suggest some general features of nearly all the quartzphyllite-dominated areas:

1. A comparison of the lithostratigraphic sequences evidences more analogies among the many quartzphyllite areas than differences. Therefore, the assumption that each quartzphyllite area represents a part of an originally unique regional domain is reasonable, and attempts at correlating the different areas turn out to be obvious.

2. The quartzphyllite areas often contain complex lithologies which show gradual transition via thin pelagic limestones to a carbonate platform and to thick, submarine volcanogenic piles.

3. The range of biostratigraphic data suggests a general Late Ordovician to Devonian age of quartzphyllites mainly based on conodonts from carbonates. Pre-Late Orodovician sections are nearly free of carbonates. The increasing use of palynomorpha for dating may extend the common age of quartzphyllites to Early and Middle Ordovician.

4. The related mafic volcanics have an approximate Silurian age in most areas, but they have a Middle Devonian age in some areas (e.g. western Greywacke zone), or occurs at different levels from Late Ordovician to Middle Devonian in other areas.

5. The acidic volcanics are mainly of Late Ordovician age but in some areas they also occur far below the Late Ordovician level respectively above in a Late Silurian to Early Devonian sequence.

6. The sequences suggest obvious changes in facies, lithology and general trend of evolution which may be interpreted in terms of event stratigraphy. The major events are from bottom to the top of the section: (1) deposition of a thick wedge of clastic sediments; however, this base sequence only occurs in a few section; (2) the Blasseneck ignimbrite event during Late Ordovician; (3) the Silurian alkaline mafic volcanism including another widespread quartzkeratophyre; (4) the "black shale"-event including mineralization in Late Silurian and/or Early Devonian; (5) the clastic wedge of Late Silurian and Early Devonian age; (6) the pelagic nodular limestone event of earliest Zlichovian age; (7) the formation of a carbonate platform during Middle Devonian; (8) the lydite-event of Visean age which is immediately followed by (9) greywackes of probable Visean age.

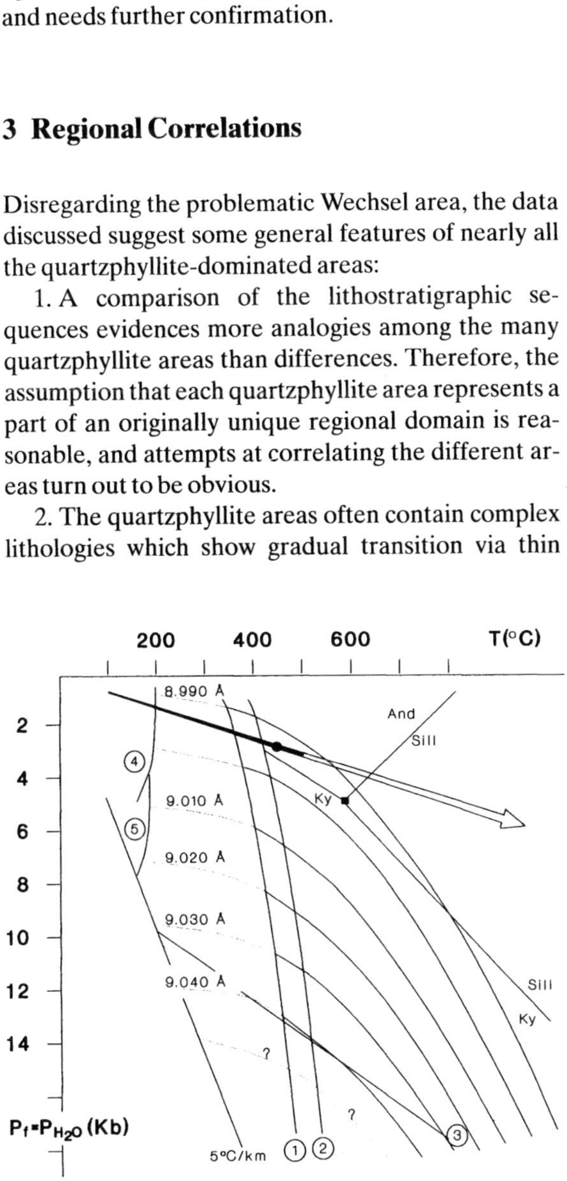

Fig. 6. Thermal gradient (*arrow:* 45 °C/km) during Variscan metamorphism, as estimated on the basis of the average value of the muscovite *b* cell parameter (8.994 Å; n = 370, s = 0.005) and the T value deduced from the mineral assemblages (about 450 °C); the *black part of the arrow* represents the temperature range which progressively affected these phyllites during Variscan heating (after Sassi and Spiess 1992)

4 Metamorphism

All these quartzphyllite sequences certainly underwent the Variscan regional metamorphism, as shown by (1) the metamorphic gap between the unmetamorphosed Late Palaeozoic cover (Permian in Gailtal, Late Visean and respectively Late Carboniferous in the Stolzalpe nappe and Steinach nappe) and the underlying phyllites, (2) the well-known angular unconformity in the eastern Greywacke zone, and (c) the radiometric geochronological data (Table 3).

Microtextural analyses performed in southwestern quartzphyllites areas indicate that this regional metamorphism had a two stage development:

1. The older stage was a syn-kinematic recrystallization during which an early folidation S_1 evolved to an axial plane foliation S_2 which, in many cases, is the main mesoscopic foliation;
2. The younger stage was a post-kinematic recrystallization during the thermal climax, and pre-dates the appearance of a local and discontinuous foliation S_3 (which is mainly post-crystalline).

The metamorphic grade covers the whole temperature range from the sub-greenschist facies (i.e. anchizone) to the upper greenschist facies (garnet zone).

The composition of potassic white micas, estimated by means of the *b* cell dimension, clearly indicates low-pressure conditions (Sassi 1972; Sassi and Scolari 1974; Guidotti and Sassi 1976, 1986), as shown in Fig. 6 based on numerous analytical data (Sassi and Spiess 1992). These *b* data jointly with the estimated temperatures indicate relatively high values of the metamorphic thermal gradient, around 45 °C/km (Sassi and Spiess 1992).

Both metamorphic grade and pressure characters as outlined are to be referred to the younger metamorphic stage. The radiometric age data in Table 3 are to be interpreted as cooling ages, considering that they were obtained from micas by means of the K-Ar method. The age of metamorphism is expected to be slightly older, and the age values (Rb-Sr whole-rock isochrons) obtained by Satir and Morteani (1979) north of the Tauern window from the Lower Austro-Alpine Steinkogel schists (322 ± 24 Ma) and Schwazer augengneisses (347 ± 30 Ma) may represent, over a regional scale, reliable age values of the two stages of the Variscan metamorphism. In this connection, it is worth mentioning here that identical age values were also obtained from the South-Alpine quartzphyllites both for the older stage (350 ± 3 Ma: Del Moro et al. 1980; 347 ± 17: Cavazzini et al. 1991) and the younger stage (317 ± 16 Ma: Del Moro et al. 1980; 326 ± 3 Ma: Del Moro et al. 1984).

The biotite age values in the range 260–180 Ma probably reflect partial rejuvenations due to the Alpine metamorphic overprint; however, considering the high sensitivity of the K/Ar isotopic system, we cannot exclude that Late Variscan and/or Meso-zoic/Tertiary deformations were responsible, in some cases, for this rejuvenation.

5 Discussion

The widespread changes in lithology suggest that all quartzphyllite areas belong to the same large basin and reflect distinct trends in basinal evolution. On the other hand, mafic volcanism occurred during different times, from Ordovician to the Early/Middle Devonian boundary; its chemical signature seems to change from tholeiitic to alkalic and again to tholeiitic (Table 2; see also Loeschke et al. this Vol.). Acidic volcanism took place at three different stratigraphic levels, during Ordovician before the Blasseneck event, the Blasseneck ignimbrite during the Late Ordovician, and afterwards approximately at the Silurian/Early Devonian boundary.

If we restore Alpidic deformation, mainly the NW-directed thrusting (Neubauer 1988; for compilation, see Ratschbacher and Frisch this Vol.), we get an idea about the relative positions of the individual quartzphyllite units (Flügel 1977; Schönlaub 1979; Neubauer 1988). This restoration is used in Fig. 7 for the interpretation of the general palaeogeography of the Austro-Alpine quartzphyllites and related Palaeozoic formations.

Sedimentation in the quartzphyllite areas started in Early Ordovician, if the new data of Reitz and Höll (1989) have a regional significance (Fig. 7a). Early sedimentation may reflect molasse-like deposits after a Late Cadomian/Pan-African orogeny (see also Ebner et al. 1987). The rifting on newly formed Cadomian continental crust may reflect the formation of a back-arc basin (Neubauer and Frisch 1988). The graded sandstones as well as the occurrence of marine fossils like conodonts show the marine character of Caradocian sedimentation, whereas the ignimbrite event reflects an uplift phase.

Strong differentiation occurred during the Silurian, with conglomerates at the one hand, and basinal slates at the other hand (Fig. 7b). Furthermore, the differentiation and the alkaline nature of the volcanism argue for a renewed rifting phase, with block faulting and fast differential tectonic subsidence. Block tilting during Late Silurian has been recently found in the Palaeozoic of Graz (Neubauer 1989c). Silurian mafic volcanics occur mainly in basinal facies.

The early rift basins were filled up during Late Silurian and Early Devonian times by clastic fan deposits as well as by probable carbonatic tempestites. The clastics suggest long transport and a mature, continental source region. The onset of pelagic limestone sedimentation indicates the transition from the rift to the drift-stage, respectively the final stage of passive continental margin formation (Fig. 7c). However,

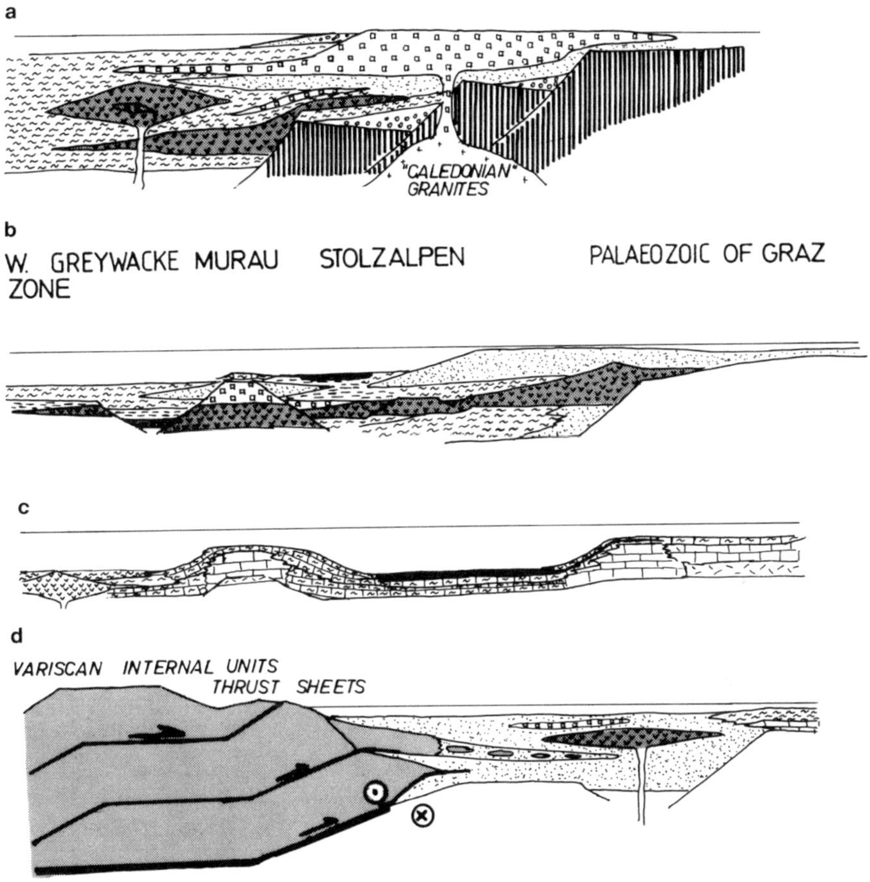

Fig. 7 a–d. Evolutionary models for sedimentation and Palaeozoic tectonism in the quartzphyllite areas. Assumed orientation of the sketch is ca. NW to SE in relation to the present-day geographic coordinates. Facies types are shown for restricted age stages. For lithostratigraphic description, see Fig. 2. **a** The Ordovician evolution is interpreted as the time of formation of a back-arc basin. This ensialic basin formed on Cadomian/Pan-African crust which has been consolidated during the Cambrian, and is infilled with clastics and bimodal volcanic suites; **b** Silurian to basal Early Devonian was the time of renewed rifting, block faulting, alkalic mafic and acidic volcanism and sedimentary exhalative-type mineralization. Afterwards the basin was filled with clastic sediments derived from the southeastern area; **c** during the late stages of Early Devonian to Early Carboniferous times the siliciclastic input lost importance except in western parts. Near-shore carbonates, lagoonal dolomites and reefs dominated the southeastern margin, and probably represent an intrabasinal high. The deeper part of the basin was filled with pelagic limestones and cherts. Volcanism occurred along the northwestern margin. **d** Contraction and basin closure by thrusting started during mid-Visean times. Subsidence may be explained by loading due to incoming crustal sheets. The basin was filled by greywackes, mafic and acidic volcanic detritus

ophiolites related to the passive margin were not identified up to now in the Eastern Alps. The widespread occurrence of mafic volcanism within quartzphyllites/slates of this time may be related to this event (Neubauer 1980b; Heinisch et al. 1987, 1988; Heinisch 1988; Schlaegel-Blaut 1990).

Late Silurian to Middle Devonian palaeogeography proposes a coastline to the southeast and a basin opening to the northwest, using restoration related to the present-day geographic framework. The siliciclastic hinterland lost its importance approximately at the Pragian/Zlichovian boundary.

Reef-related limestones may be present in the Middle Devonian marbles/limestones. However, Late Devonian to Early Carboniferous pelagic limestones are widespread. They indicate sea level rise and/or sharp subsidence. Carbonates are followed by Visean greywackes of a prograding submarine fan.

We interpret the sharp subsidence during lowermost Carboniferous to be the result of loading of this passive margin by incoming crustal thrust sheets, which are represented by other metamorphic Austro-Alpine basement units (see Neubauer and Frisch this Vol.), during continental collision (Fig. 7d). A series of nappes was formed. Major Variscan nappes are well-known in the eastern and western Greywacke zone and possibly in the Goldeck and Kreuzeck area as well as in the Gurktal thrust system. The direction of thrusting is constrained only by a very few data. These data of the eastern Greywacke zone indicate top-SW transport (Neubauer 1989a), approximately parallel to the trend of the Late Variscan palaeogeographic zonation (Neubauer 1988). This would imply an overall sinistral transpressional regime during that time. All these data suggest a deformation style of a foreland-fold-and-thrust belt typical for external parts of orogens.

The age of metamorphism, respectively of cooling after metamorphism, is Mid- to Late Carboniferous (see Table 3). Local Permian ages may reflect alteration during Permian acidic magmatism in overstepping sequences or the influence of the Alpine metamorphic overprint.

All major quartzphyllite units, except the Murau nappe and the Palaeozoic of Graz, were covered by Late Carboniferous and Permian overstep sequences (e.g. Flügel 1978).

Two final considerations concern the Variscan regional metamorphism, and specifically: (1) its relation to rather high thermal gradients, and (2) its timing.

1. The activation, development and features of this metamorphism should be consequences of the regional evolution outlined above and sketched in Fig. 7. However, the most important aspects of this metamorphism have not yet been understood. Specifically, tectonic loading and related subsidence are important factors controlling the thermal insulation of the deep rock sequences, but the heat source for this low-pressure metamorphic event is not yet known. In modelling the low pressure metamorphisms, uprising magmatic masses are a good explanation of the heat source, but in our case no geological records have been detected of magmatic masses of suitable age and volume except those that are present in other metamorphic Austro-Alpine basement units.
2. With respect to the timing of metamorphism, similar age values very close to 350 Ma and 320 Ma were obtained both in the Lower Austro-Alpine (LA) rock sequences and in the Southalpine (SA) quartzphyllites, as reported. This similarity indicates that the early regional heating related to the Variscan event and the thermal climax occurred at the same time (i.e. 350 and 320 Ma ago, respectively) in totally different structural units (LA versus SA), notwithstanding the important tectonic lines occurring in between. Therefore, the timing of the Variscan metamorphism was substantially synchronous over very large areas.

Acknowledgements
We gratefully appreciate reviews by Onno Oncken, Würzburg, and D. Robinson, Bristol, which helped us to increase the readability of the paper. Field and laboratory work of F. N. have been partly supported by the Austrian Science Foundation (grants P 5506 and P 7405). F. P. S. acknowledges MURST (40 and 60% funds) and CNR (Centro di Studio Orogeno Alpi Orientali).

References

Andreatta C (1951) Polymetamorphose and Tektonik in der Ortlergruppe. Neues Jahrb Mineral Monatsh 1951: 13–28
Becker LP, Frank W, Höck V, Kleinschmidt G, Neubauer F, Sassi FP, Schramm JM (1987) Outlines of the pre-Alpine metamorphic events in the Austrian Alps. In: Flügel HW, Sassi FP, Grecula P (eds) Pre-Variscan and Variscan events in the Alpine-Mediterranean mountain belts. Mineralia slovaca Monogr, Alfa, Bratislava pp 69–106
Borsi S, Del Moro A, Sassi FP, Zanferrari A, Zirpoli G (1978) New geopetrologic and radiometric data on the Alpine history of the Austridic continental margin south of the Tauern window (Eastern Alps). Mem Sci Geol 32: 1–19
Brewer MS (1969) Excess radiogenic argon in metamorphic micas from the Eastern Alps, Austria. Earth Planet Sci Lett 6: 321–331
Buchroithner MF (1978) Biostratigraphische Untersuchungen im Paläozoikum der Steiermark. Mitt Naturwiss Ver Steiermark 108: 77–93
Buchroithner MF (1979) Biostratigraphische und fazielle Untersuchungen im Paläozoikum von Mittelkärnten. Carinthia II 169/89: 71–95
Buggisch W, Kleinschmidt G, Lütke F (1975) Die Einstufung von altpaläozoischen Kalken im Sausal und Remschnigg (Steiermark Österreich) aufgrund von Conodonten. Neues Jahrb Geol Paläontol Monatsh 1975: 263–278
Cavazzini G, Del Moro A, Sassi FP, Zirpoli G (1991) New data on the radiometric age of the Southalpine Basement of the Eastern Alps. Convegno "Geol Basamento Italiano", Siena 21–22 Marzo 1991 (Abstr) Dep Geol Univ Siena, Siena
Dal Piaz GV, Martin S (1984) I porfiroidi nelle filladi della Valle di Forni, Falda dell'Ortles, Austroalpine Superiore. Rend Soc Geol Ital 3: 17–20
Daurer A, Schönlaub HP (1978) Anmerkungen zur Basis der Nördlichen Grauwackenzone. Mitt Österr Geol Ges 69 (1976): 77–88
Del Moro A, Sassi FP, Zirpoli G (1980) Preliminary results on the radiometric age of the Hercynian metamorphism in the South-Alpine basement of the Eastern Alps. Neues Jahrb Geol Paläontol Monatsh 12: 707–718
Del Moro A, Sassi FP, Zirpoli G (1984) Acidic gneisses from the Plan de Corones area, and chronological data on South-Alpine basement in Pusteria (Eastern Alps). Mem Sci Geol Padova 36: 403–412
Deutsch A (1977) Geologie und Petrographie der mittleren Goldeckgruppe (Kärtnen/Österreich). Jahrb Geol Bundesanst 120: 231–294
Ebner F (1975) Ein Beitrag zum Altpaläozoikum des Remschnigg (Steiermark). Verh Geol Bundesanst 1974: 281–287

Ebner F (1978) Der paläozoische Untergrund in der Bohrung Waltersdorf 1 (S Hartberg, Oststeiermark). Mitt Österr Geol Ges 68 (1975): 5–11

Ebner F (1988) Das Paläozoikum in den RAG-Bohrungen Blumau 1, 1a und Arnwiesen 1 (Oststeirisches Tertiärbecken). Jahrb Geol Bundesanst 131: 563–573

Ebner F, Neubauer F, Stattegger K (1987) The Caledonian event in the Eastern Alps: a review. In: Flügel HW, Sassi FP, Grecula P (eds) Pre-Variscan and Variscan events in the Alpine-Mediterranean mountain belts. Mineralia slovaca Monogr, Alfa, Bratislava pp 169–182

Exner Ch (1990) Erläuterungen zur Geologischen Karte des mittleren Lungaues. Mitt Ges Geol Bergbaustud Österr 36: 1–38

Faupl P (1970) Zur Geologie des NW-Abschnittes des Wechselgebietes zwischen Trattenbach (NÖ.) und Fröschnitz (Stmk.). Mitt Ges Geol Bergbaustud 19: 27–70

Flajs G (1967) Conodontenstratigraphische Untersuchungen im Raum von Eisenerz, Nördliche Grauwackenzone. Mitt Geol Ges Wien 59 (1966): 157–212

Flajs G, Schönlaub HP (1976) Die biostratigraphische Gliederung des Altpaläozoikums am Polster bei Eisenerz (Nördliche Grauwackenzone Österreich). Verh Geol Bundesanst 1976: 257–303

Flügel HW (1975) Die Geologie des Grazer Berglandes. Mitt Abt Geol Bergbau Joanneum Sdh 1: 1–288

Flügel HW (1977) Paläogeographie und Tektonik des alpinen Variszikums. Neues Jahrb Geol Paläontol Monatsh 1977: 659–674

Flügel HW (1978) Some remarks to the Postvariscic transgressions and the age of the Variscic phases in the Alps. Österr Akad Wiss Erdwiss Komm 3: 171–177

Flügel HW (1988) Geologische Karte des prätertiären Untergrundes. In: Kröll A, Flügel HW, Seiberl W, Weber F, Walach G, Zych G (eds) Erläuterungen zu den Karten über den prätertiären Untergrund des Steirischen Beckens und der Südburgenländischen Schwelle. Geol Bundesanst, Vienna, pp 1–49

Flügel HW, Neubauer F (1984) Steiermark. Erläuterungen zur Geologischen Karte der Steiermark 1:200000. Geol Bundesanst, Vienna, 127 pp

Frank W (1987) Evolution of the Austroalpine elements in the Cretaceous. In: Flügel HW, Faupl P (eds) Geodynamics of the Eastern Alps. Deuticke, Vienna, pp 379–406

Frank W, Kralik M, Scharbert S, Thöni M (1987) Geochronological data from the Eastern Alps. In: Flügel HW, Faupl P (eds) Geodynamics of the Eastern Alps. Deuticke, Vienna, pp 272–281

Fritsch W (1961) Saure Eruptivgesteine aus dem Raume nordwestlich von St. Veit an der Glan in Kärnten. Geologie 10: 67–80

Fritz H (1991) Stratigraphie, Fazies und Tektonik im nordwestlichen Grazer Paläozoikum (Ostalpen). Jahrb Geol Bundesanst 134: 227–255

Fritz H, Neubauer F (1988) Geodynamic aspects of the Silurian and Early Devonian sedimentation in the Paleozoic of Graz (Eastern Alps). Schweiz Mineral Petrogr Mitt 69: 357–365

Fritz H, Neubauer F, Ratschbacher L (1991) Extension versus compression in the Paleozoic of Graz. Zentralbl Geol Paläontol Teil 1 1991: 55–68

Frizzo P, Visona D (1981) New data regarding lithostratigraphy and metamorphism of the Steinach Nappe (Brenner Region Eastern Alps). Studi Trentini Sci Nat 58: 3–10

Giese U (1988) Lower Paleozoic volcanic evolution at the northwestern border of the Gurktal nappe, Upper Austroalpine, Eastern Alps. Schweiz Mineral Petrogr Mitt 68: 381–396

Gollner H, Thalhammer O, Tschelaut W, Zier C (1982) Die Laufnitzdorf-Gruppe – eine pelagische Fazies im Grazer Paläozoikum. Mitt Naturwiss Ver Steiermark 112: 63–74

Gosen W von, Haiges KH, Thiedig F (1982) Fossilführendes Paläozoikum am Ostrand der Gurktaler Alpen (Steiermark/Kärnten – Österreich). Mitt Geol Paläontol Inst Univ Hamb 53: 151–168

Gratzer R (1985) Vergleichende Untersuchungen an Metabasiten im Raum Hannersdorf, Burgenland. Österr Akad Wiss Math-Naturwiss Kl Abt I 194: 131–148

Grauert B, Hänny R, Soptrajanova G (1973) Age and origin of detrital zircons from the Pre-Permian basements of the Bohemian Massif and the Alps. Contrib Mineral Petrol 40: 105–130

Groschopf R (1970) Zur Petrographie und Tektonik des schwachmetamorphen Altpaläozoikums im nordöstlichen Klagenfurter Becken (Ostalpen). Mitt Ges Geol Bergbaustud 19: 215–278

Gsellmann H (1988) Zur Geologie am Nordostrand des Grazer Paläozoikums (im Bereich Hochschlag–Plankogel–Heilbrunn). PhD Thesis, Univ Graz, 102 pp

Guidotti CV, Sassi FP (1976) Muscovite as a petrogenetic indicator mineral in pelitic schists. Neues Jahrb Mineral Abh 127: 97–142

Guidotti CV, Sassi FP (1986) Classification and correlation of metamorphic facies series by means of muscovite *b* data from low-grade metapelites. Neues Jahrb Miner Abh 153: 363–380

Haditsch HG, Mostler H (1983) The succession of ore mineralization of the lower Austroalpine Innsbruck quartzphyllite. In: Schneider HJ (ed) Mineral deposits of the Alps. Springer, Berlin Heidelberg New York, pp 51–59

Hajek H (1963) Die geologischen Verhältnisse des Gebietes N Feistritz-Pulst im Glantal, Kärnten. Mitt Geol Ges Wien 55 (1962): 1–40

Harland WB, Armstrong RL, Cox AV, Craig LE, Smith AG, Smith DG (1989) A geological time scale 1989. Cambridge University Press, Cambridge, 263 pp

Havlicek V, Kriz J, Serpagli E (1987) Upper Ordovician brachiopod assemblages of the Carnic Alps, Middle Carinthia and Sardinia. Boll Soc Paleontol Ital 25 (1986): 277–311

Heinisch H (1981) Zum ordovizischen "Porphyroid"-Vulkanismus der Ost- und Südalpen – Stratigraphie, Petrographie, Geochemie. Jahrb Geol Bundeanst 124: 1–109

Heinisch H (1987) Concepts for the geological evolution of Gailtalkristallin (Kärnten – Austria). In: Flügel HW, Sassi FP, Grecula P (eds) Pre-Variscan and Variscan events in the Alpine-Mediterranean mountain belts. Mineralia slovaca Monogr, Alfa, Bratislava, pp 293–312

Heinisch H (1988) Hinweise auf die Existenz eines passiven Kontinentalrandes im Altpaläozoikum der Nördlichen Grauwackenzone – Ostalpen. Schweiz Mineral Petrogr Mitt 68: 407–418

Heinisch H, Schmidt K (1976) Zur kaledonischen Orogenese in den Ostalpen. Geol Rundsch 65: 459–482

Heinisch H, Schmidt K (1984) Zur Geologie des Thurntaler Quarzphyllites und des Altkristallins südlich des Tauernfensters (Ostalpen, Südtirol). Geol Rundsch 73: 113–129

Heinisch H, Sprenger W, Weddigge K (1987) Neue Daten zur Altersstellung der Wildschönauer Schiefer und des Basaltvulkanismus im ostalpinen Altpaläozoikum der Kitzbüheler Grauwackenzone (Österreich). Jahrb Geol Bundesanst 130: 163–173

Heinisch H, Hertogen J, Schlaegel P (1988) Ausgewählte Spurenelemente von metamorphen basischen Magmatiten aus der Nördlichen Grauwackenzone (Österreich). Jahrb Geol Bundesanst 131: 267–278

Heinz H (1987) Geologie der östlichen Goldeckgruppe (Kärnten). Jahrb Geol Bundesanst 130: 175–203
Hess W (1953) Beiträge zur Geologie der südöstlichen Engadiner Dolomiten zwischen dem oberen Münstertal und der Valle di Fraele. Eclogae Geol Helv 46: 40–142
Hinterlechner-Ravnik E, Moine B (1977) Geochemical characteristics of the metamorphic rocks of the Pohorje Mountains. Geologija 20: 107–140
Hoke E (1990) The Altkristallin of the Kreuzeck Mountains SE Tauern window Eastern Alps – basement crust in a convergent plate boundary zone. Jahrb Geol Bundesanst 133: 5–87
Höll R (1970) Die Zinnober-Vorkommen im Gebiet der Turracher Höhe (Nock-Gebiet/Österreich) und das Alter der Eisenhut-Schieferserie. Neues Jahrb Geol Paläontol Monatsh 1970: 201–224
Höll R, Maucher A (1968) Genese und Alter der Scheelit-Magnesit-Lagerstätte Tux. Sitzungsber Bayer Akad Wiss Math-Naturwiss Kl 1967: 1–11
Höll R, Maucher A (1976) The stratabound ore deposits in the Eastern Alps. Handbook of stratabound and stratiform ore deposits, Vol. 5. Elsevier, Amsterdam, pp 1–30
Höll R, Dahmen R, Seidl A, Werner W (1985) Tungsten Mineralization in the Ennstal Phyllite Series near Schladming/Eastern Alps. Monogr Ser Mineral Deposits 25: 161–176
Hurler H (1972) Geochemie und Petrographie der Metakeratophyre in der südlichen Saualpe (Kärnten). PhD Thesis, Univ Tübingen, 67 pp
Jung G (1981) Radiometrische Altersdatierung und Metamorphoseuntersuchungen der Kalkalpenbasis und der Grauwackenzone in der Radmer und am steirischen Erzberg. Frühalp Gesch Ostalpen 1: 21–27
Kleinschmidt G, Wurm F (1966) Paläozoikum und epizonale Serien zwischen St. Andrä im Lavanttal und Griffen. Carinthia II 156/66: 108–140
Kleinschmidt G, Neugebauer J, Schönenberg R (1975) Gesteinsinhalt und Stratigraphie der Phyllitgruppe in der Saualpe. Clausth Geol Abh Sonderbd 1: 11–44
Lahusen L (1972) Die schicht- und zeitgebundenen Antimonit-Scheelit-Vorkommen und Zinnobervererzungen der Kreuzeck- und Goldeckgruppe in Kärnten und Osttirol, Österreich. Mineral Deposita 7: 31–60
Loeschke J (1988a) Zur Geochemie basischer Vulkanite aus dem Grazer Paläozoikum. Mitt Naturwiss Ver Steiermark 118: 95–105
Loeschke J (1988b) Metavulkanite aus der Phyllit-Gruppe der südlichen Saualpe (Kärnten, Österreich) und ihre plattentektonische Bedeutung. Jahrb Geol Bundesanst 131: 301–304
Loeschke J (1989a) Die paläotektonische Stellung der Vulkanite der Magdalensberg-Serie (Ober-Ordovizium Gurktaler Decke, Kärnten, Österreich). Carinthia II 179/99: 491–507
Loeschke J (1989b) Lower Paleozoic volcanism of the Eastern Alps and its geodynamic implications. Geol Rundsch 78: 599–616
Loeschke J, Kretzschmar T, Langer V, Streck M (1990) Zur Geologie der Norischen Decke nordwestlich von Kalwang (Eisenerzer Alpen, Steirische Grauwackenzone/Österreich). Jahrb Geol Bundesanst 133: 345–363
Matura A (1990) Der Wechselkomplex im Wechselfenster. In: Koller F (ed) Wandertagung 1990 der Österreichischen Geologischen Gesellschaft. Österr Geol Ges, Vienna, pp 30–36
Mioc R, Ramovs A (1973) Erster Nachweis des Unterdevons im Kozjak Gebirge (Possruck) westlich von Maribor (Zentralalpen). Bull Sci 18: 135–136
Mostler H (1973) Alter und Genese ostalpiner Spatmagnesite unter besonderer Berücksichtigung der Magnesitlagerstätten im Westabschnitt der Nördlichen Grauwackenzone (Tirol, Salzburg). Veröff Univ Innsbr 86: 237–266
Neubauer FR (1979) Zum Alter von Dolomiten auf der Stolzalpe bei Murau und am Adelsberg bei Neumarkt (Stmk.) Anz Akad Wiss Math Naturwiss Kl 1979: 116–119
Neubauer FR (1980a) Die Gliederung des Altpaläozoikums südlich und westlich von Murau (Steiermark/Kärnten). Jahrb Geol Bundesanst 122: 455–311
Neubauer F (1980b) Exkursion Murauer Paläozoikum. Mitt Abt Geol Bergbau Landesmus Joanneum 41: 115–128
Neubauer F (1985) Eine präoberordovizische Transgression in der Grauwackenzone (Ostalpen) und ihre Bedeutung für paläozoische Geodynamik und alpidischen Deckenbau. Neues Jahrb Geol Paläontol Monatsh 1985: 46–64
Neubauer F (1988) Bau und Entwicklungsgeschichte des Rennfeld-Mugel- und des Gleinalmkristallins (Ostalpen). Abh Geol Bundesanst 42: 1–137
Neubauer F (1989a) Late Variscan structures of the Eastern Greywacke Zone (Eastern Alps). Neues Jahrb Geol Paläontol Monatsh 1989/7: 425–432
Neubauer F (1989b) Petrographie und Geochemie der Grüngesteine von Althofen (Altpaläozoikum der Murauer Decke, Steiermark). Jahrb Geol Bundesanst 132: 231–239
Neubauer F (1989c) Lithostratigraphie und Struktur an der Basis der Rannachdecke im zentralen Grazer Paläozoikum. Jahrb Geol Bundesanst 132: 459–474
Neubauer F, Frisch W (1988) Ordovician-Silurian evolution of the Alps – the orogeny-back arc basin model. Schweiz Mineral Petrogr Mitt 68: 351–357
Neubauer F, Herzog U (1985) Das Karbon der Stolzalpendecke Mittelkärntens: Implikationen für die variszische Orogenese im Ostalpin. Anz Österr Akad Wiss Math-Naturwiss Kl 1985/6: 106–109
Neubauer F, Pistotnik J (1984) Das Altpaläozoikum und Unterkarbon der Gurktaler Decke und ihre paläogeographischen Beziehungen. Geol Rundsch 73: 149–174
Neugebauer J (1970) Alt-paläozoische Schichtfolge, Deckenbau und Metamorphose-Ablauf im südwestlichen Saualpen-Kristallin (Ostalpen). Geotekt Forsch 35: 23–93
Niederl R (1980) Zur Geologie des Raumes Oberwölz und des Pleschaitz N-Abfalles (Murauer Paläozoikum). Mitt Abt Geol Paläontol Bergbau Landesmus Joanneum 41: 81–83
Nievoll J (1987) The Lower Paleozoic of the Noric nappe at Veitsch (Eastern Alps, Austria). In: Flügel HW, Sassi FP, Gracula P (eds) Pre-Variscan and Variscan events in the Alpine-Mediterranean mountain belts. Alfa, Bratislava, pp 379–386
Planderova E, Pahr A (1983) Biostratigraphical evaluation of weakly metamorphosed sediments of Wechsel series and their possible correlation with Harmonia Group in Male Karpaty Mountains. Mineral Slovaca 15: 385–436
Planderova E, Pahr A (1990) Biostratigraphical evaluation of weakly metamorphosed sediments of Wechsel series and their possible correlation with Harmonia group in Male Karpaty Mts. In: Minarikova D, Lobitzer H (eds) Thirty years of geological cooperation between Austria and Czechoslovakia. Ustredni ustav geologicky, Prague, pp 171–177
Poli S (1989) Pre-Hercynian magmatism in the Eastern Alps the origin of metabasites from the Austroalpine basement. Schweiz Mineral Petrogr Mitt 69: 407–421
Pozzi R (1965) Schema tettonico dell'Alta Valtellina da Livigno al Gruppo dell'Ortles. Eclogae Geol Helv 58: 21–38
Priewalder H, Schuhmacher R (1976) Petrographisch-tektonische Untersuchungen in den Ennstaler Phylliten (Niedere Tauern Steiermark) und deren Einstufung in das Silur. Verh Geol Bundesanst 1976: 59–61

Reimann C, Stumpfl EF (1981) Geochemical setting of stratabound stibnite mineralization in the Kreuzeck Mountains Austria. Trans Instn Min Metall Sect B Appl Earth Sci 90: 126–132

Reitz E, Höll R (1989) Unterordovizische Acritarchen aus der Nördlichen Grauwackenzone (Ostalpen). Jahrb Geol Bundesanst 132: 761–774

Riehl-Herwirsch G (1970) Zur Altersstellung der Magdalensbergserie, Mittelkärnten, Österreich. Mitt Ges Geol Bergbaustud 19: 195–214

Sassi FP (1972) The petrological and geological significance of the *b* values of potassic white micas in low-grade metamorphic rocks: an application to the Eastern Alps. Tschermaks Mineral Petrogr Mitt 18: 105–113

Sassi FP, Menegazzo L (1971) Contributo alla conoscenza della Falda die Steinach (Brennero). Mem Ist Geol Mineral Univ Padova 29: 1–29

Sassi FP, Scolari A (1974) The *b* value of the potassic white micas as a barometric indicator in low-grade metamorphism of pelitic schists. Contrib Mineral Petrol 45: 143–152

Sassi FP, Zanferrari A (1972) Il significato geologico del Complesso del Turntaler (Pusteria), con particolare riguardo alla successione di eventi metamorfici prealpini nel basamento austridico delle Alpi Orientali. Boll Soc Geol Ital 91: 533–557

Sassi FP, Zanferrari A, Zirpoli G, Borsi S, Del Moro A (1974) The Austrides to the south of the Tauern window and the Periadriatic lineament between Mules and Mauthen. Neues Jahrb Geol Paläontol Monatsh 7: 421–434

Sassi FP, Kalvacheva R, Zanferrari A (1984) New data on the age of deposition of the Southalpine phyllitic basement in the Eastern Alps. Neues Jahrb Geol Paläontol Monatsh 1984: 741–751

Sassi FP, Visona D, Zirpoli G (1987) Similarities among the lithostratigraphic sequences in different part of the Austrides (Eastern Alps). In: Flügel HW, Sassi FP, Grecula P (eds) Pre-Variscan and Variscan events in the Alpine-Mediterranean mountain belts. Alfa, Bratislava, pp 427–430

Sassi R, Spiess R (1992) Further data on the pre-Alpine metamorphic pressure conditions of the Austridic phyllitic complexes in the Eastern Alps. In: Carmignani L, Sassi FP (eds) Contributions to the Geology of Italy with special regard to the Palaezoic basements. IGCP No 276 Newslett 5: 297–307, Siena

Satir VM, Morteani G (1979) Kaledonische, herzynische und alpidische Ereignisse im Mittelostalpin nördlich der westlichen Hohen Tauern, abgeleitet aus petrographischen und geochronologischen Untersuchungen. Geol Rundsch 68: 1–40

Schäffer U, Tarkian M (1984) Die Genese der stratiformen Sulfidlagerstätte Kalwang (Steiermark), der Grünsteinserie und einer assoziierten silikatreichen Eisenformation. Tschermaks Mineral Petrogr Mitt 33: 169–186

Scharbert S, Schönlaub HP (1980) Das Prävariszikum und Variszikum. In: Oberhauser R (ed) Der geologische Aufbau Österreichs. Springer, Berlin Heidelberg New York Wien, pp 3–20

Schlaegel P (1988) Geotektonische Neuinterpretation des basischen Magmatismus der Nördlichen Grauwackenzone Ostalpen – ein Überblick. Schweiz Mineral Petrogr Mitt 68: 419–432

Schlaegel-Blaut P (1990) Der basische Magmatismus der Nördlichen Grauwackenzone, oberostalpines Paläozoikum. Abh Geol Bundesanst 43: 1–149

Schlamberger J (1988) Geodynamic evolution of the Sausal Paleozoic (Eastern Alps). Schweiz Mineral Petrogr Mitt 68: 369–379

Schlöser H, Kullmann J, Loeschke J (1990) Korallen-führendes Unterkarbon auf der Brunnachhöhe (Nockgebiet, Gurktaler Decke, Österreich). Carinthia II 180/100: 643–650

Schlüter J, Tarkian M, Stumpfl EF (1984) Die stratiforme Sulfidlagerstätte Walchen, Steiermark, Österreich: Geochemie und Genese. Tschermaks Mineral Petrogr Mitt 33: 287–296

Schmid SM (1973) Geologie des Umbrailgebietes. Eclogae Geol Helv 66: 101–210

Schnepf HG (1989) Vulkanismus und Sedimentologie im Murauer Paläozoikum (Gurktaler Decke/Oberostalpin/Österreich). Arb Inst Mus Geol Paläontol Univ Tübingen A 2: 1–183

Schönlaub HP (1971) Die Althofener Gruppe – eine neue stratigraphische Einheit im Devon Mittelkärntens (Österreich). Neues Jahrb Geol Paläontol Monatsh 1971: 288–305

Schönlaub HP (1975) Zum Alter der Radstädter Quarzphyllite (Unterostalpin Salzburg). Ann Naturhist Mus Wien 79: 47–55

Schönlaub HP (1979) Das Paläozoikum in Österreich. Abh Geol Bundesanst 33: 1–124

Schönlaub HP (1982) Die Grauwackenzone in den Eisenerzer Alpen (Österreich). Jahrb Geol Bundesanst 124: 361–423

Schönlaub HP (1984) Das Paläozoikum von Sulz bei Güssing im Südburgenland. Jahrb Geol Bundesanst 127: 501–505

Schönlaub HP (1985) Das Karbon von Nötsch und sein Rahmen. Jahrb Geol Bundesanst 127: 673–692

Schönlaub HP (1990) Das Altpaläozoikum im Südburgenland. Exkursionsführer Österr Geol Ges 13: 52–54

Schönlaub HP, Zezula G (1975) Silur-Conodonten aus einer Phyllonit-Zone im Muralpen-Kristallin (Lungau/Salzburg). Verh Geol Bundesanst 1975: 253–269

Schönlaub HP, Exner Ch, Nowotny A (1976) Das Altpaläozoikum des Katschberges und seiner Umgebung (Österreich). Verh Geol Bundesanst 1976: 115–146

Schönlaub HP, Flajs G, Thalmann F (1980) Conodontenstratigraphie am Steirischen Erzberg (Nördliche Grauwackenzone). Jahrb Geol Bundesanst 123: 169–229

Stattegger K (1982) Schermineraluntersuchungen in der östlichen Grauwackenzone (Steiermark/Österreich). Verh Geol Bundesanst 1982: 107–121

Strehl E (1962) Die geologische Neuaufnahme des Saualpen-Kristallins IV. Das Paläozoikum und sein Deckgebirge zwischen Klein St. Paul und Brückl. Carinthia II 152/72: 46–74

Toula F (1878) Über Devon-Fossilien aus dem Eisenburger Comitate. Verh kk Geol Reichsanst 1878: 45–52

Tschelaut W (1984) Das Paläozoikum zwischen Röthelstein und Gams bei Frohnleiten. Mitt Naturwiss Ver Steiermark 114: 133–161

Tschelaut W (1985) Über das Alter der Arzberger Schichten und der Blei-Zink-Vererzung im Grazer Paläozoikum. Jahrb Geol Bundeanst 128: 241–243

Unzog W (1989) Schertektonik im Gailtalkristallin und an seiner Begrenzung. PhD Thesis, University of Graz, 204 pp

Vetters W (1970) Zur Geologie des SW-Abschnittes des Wechselgebirges zwischen Rettenegg und Freistritzsattel (Steiermark, Österreich). Mitt Ges Geol Bergbaustud 19: 71–102

Visonà D (1974) Il significato barometrico delle muscoviti nelle filladi della Gailtal (Alpi Orientali). St Trent Sci Nat 51: 1–8

Weber L (1990) Die Blei-Zinkerzlagerstätten des Grazer Paläozoikums und ihr geologischer Rahmen. Arch Lagerstättenforsch Geol Bundesanst 12: 1–289

Palaeozoic Volcanism of the Eastern Alps and Its Palaeotectonic Significance

J. LOESCHKE and H. HEINISCH

Abstract

The rock record of Palaeozoic volcanic rocks of the Austro-Alpine realm reveals that within-plate alkalibasalts erupted during the Ordovician (Magdalensberg-Series), the Silurian (Gurktaler Alpen, Murau, eastern Greywacke zone, Saualpe), the Devonian (western Greywacke zone, Palaeozoic of Graz) and probably also in the Lower Carboniferous (Eisenkappel area, Karawanken Mountains). The eruption of these within-plate alkalibasalts and the contemporaneous continuous sedimentation indicate a long-lasting extensional process which is responsible for a general thinning of the continental crust and subsidence of marine basins. The sediments of the basins show a widespread facies differentiation depending on the source area and water depth.

In addition to these within-plate alkalibasalts a second group of volcanic rocks erupted during the Upper Ordovician which are represented by pyroclastic rocks and lavas of intermediate to acidic composition ("porphyroids" of the Greywacke zone, volcanic rocks of the Nock-Series, Gurktaler Alpen). These rocks have a calc-alkaline composition. Their plate tectonic setting is difficult to establish. Since the eruption of calc-alkaline volcanic rocks is usually connected with convergent plate boundaries a late or post-collisional volcanic event is the most likely interpretation.

1 Introduction

In the Palaeozoic parts of the Austro-Alpine realm volcanic rocks occur in many places and in various stratigraphic positions. They are found in the Greywacke zone, the Gurktaler Alpen, the Klagenfurt Basin, the Saualpe, in the Palaeozoic of Graz and in the Karawanken Mountains south of Klagenfurt (Fig. 1). They are intercalated in Ordovician to Carboniferous sediments and their geodynamic significance has been interpreted in several ways (Loeschke 1977, 1989a; Kolmer 1978; Colins et al. 1980; Heinisch 1981, 1988; Heinisch and Schmidt 1982; Frisch et al. 1984, 1987; Schlaegel 1988; Frisch and Neubauer 1989; Schlaegel-Blaut 1990).

The geotectonic models proposed for the formation of the volcanic rocks include island arc and active continental margin settings, but also mid-ocean ridge, intraplate, rift and passive continental margin settings. The models differ from each other partly in a striking way. This is due to several facts:

1. The ages of the volcanic sequences are not sufficiently well known in all localities though many biostratigraphic data have been established mainly with the help of conodonts (Schönlaub 1979; Neubauer and Pistotnik 1984; Heinisch et al. 1987).
2. Results of geochemical analyses of hydrothermally altered and metamorphic volcanic rocks can allow different interpretations, even if the "immobile elements" (Ti, Zr, Nb, Y, P, rare earth elements) are used (Pearce 1982), because similar petrogenetic processes can take place in different geotectonic environments, as for example the formation of alkali basalts on oceanic islands, in continental rifts and back-arc basins.
3. For the reconstruction of the pre-Variscan plate tectonic settings and their palaeogeographic arrangement it is necessary to estimate the direction and amount of the Variscan nappe movements.
4. In addition, the palinspastic rearrangement of the Alpine nappes must be considered, which is now again under discussion.
5. During the Alpine and probably also during the Variscan plate convergence considerable oblique, transpressive movements occurred (Laubscher and Bernoulli 1982; Neubauer 1988a) causing large-scale strike-slip displacements.
6. Variscan and Alpine metamorphism overprinted the Palaeozoic sequences.

These complexities make any interpretation of the former geotectonic setting of the Palaeozoic volcanic sequences of the Austro-Alpine realm very difficult.

The volcanic and sedimentary sequences discussed here are only those which are metamorphosed

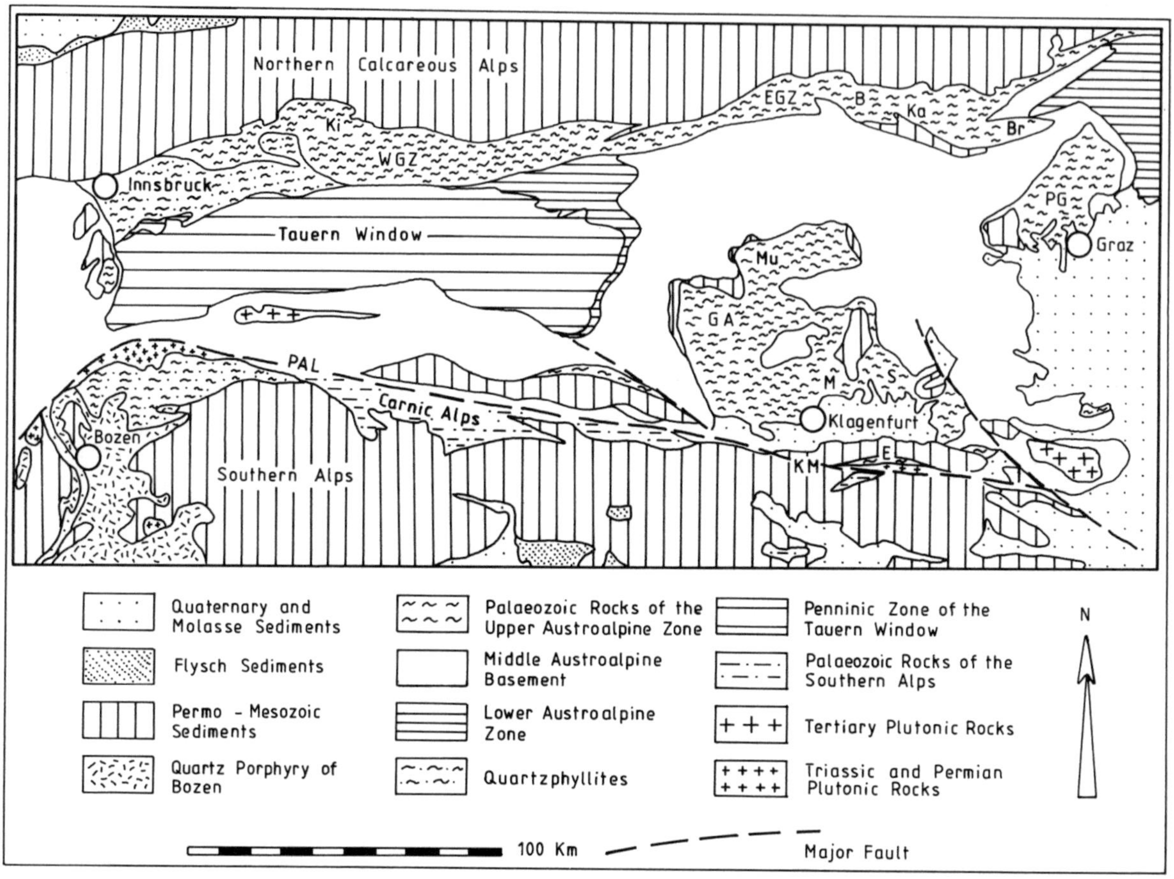

Fig. 1. Geological map of parts of the Eastern Alps showing the most important localities mentioned in the text (*B* Blasseneck; *Br* Bruck a.d. Mur; *E* Eisenkappel; *EGZ* eastern Greywacke zone; *GA* Gurktaler Alpen; *Ka* Kalwang; *Ki* Kitzbühel; *KM* Karawanken Mountains; *M* Magdalensberg Series; *Mu* Murau; *PAL* Periadriatic lineament; *PG* Palaeozoic of Graz; *WGZ* western Greywacke zone)

under very-low- or low-grade conditions, show still some primary magmatic and sedimentary structures and are, in most cases, stratigraphically dated by macro- and micro-fossils. The interpretation presented here is based on about 1000 complete geochemical major and trace element analyses and numerous detailed field work data (Heinisch 1981; Loeschke 1989a; Schlaegel-Blaut 1990 and references therein). Precision, accuracy and details of these analyses are mentioned in these references. The mineralogical composition of the volcanic rocks and the secondary geochemical alteration processes which affected them, are described in the publications cited. These questions are not treated here, as this chapter is an attempt to summarize the results of the geochemical analyses and to give a palaeotectonic interpretation. Selected representative major and trace element analyses of the rocks discussed are listed in Table 1.

2 Stratigraphy

The stratigraphic position of the investigated volcanic sequences is shown in Figs. 2 (western Greywacke zone) and 3 (rest of localities). In the western Greywacke zone (Figs. 1 and 2) the acidic volcanic rocks ("porphyroids") are assumed to be of Upper Ordovician age. This age is deduced from lithological and geochemical correlations with the eastern part of the Greywacke zone (Schönlaub 1979). At one locality they are stratigraphically overlain by dated Lower Silurian carbonates (Mostler 1967). Therefore, the age of the "porphyroids" suggested in Fig. 2 is in accordance with field evidence. The basic volcanism started in the late Lower Devonian (Zlichovian, Dalejan) which is demonstrated by conodonts for the most important sequences. The conodonts were found in thin carbonate layers within the basal parts of basic volcanic tuffs (Heinisch et al. 1987). The upper limit is uncertain. Since some se-

Table 1. Representative major and trace element analyses of various volcanic rock types treated in the text

	A[a]	B	C	D	E	F	G	H	I	J	K	L	M	N
SiO_2 (%)	49.3	53.7	47.9	49.8	66.0	70.5	46.8	53.2	49.0	47.2	46.4	66.9	46.9	47.1
TiO_2	3.0	1.6	2.6	2.6	0.59	0.33	3.3	0.84	2.1	2.6	3.3	0.37	3.9	3.5
Al_2O_3	15.6	15.5	15.1	15.3	15.7	14.1	14.6	15.9	13.0	11.0	16.2	16.5	16.1	15.3
Fe_2O_3	3.7	3.2	4.2	3.9	1.4	0.95	4.3	6.2	1.8	1.7	4.6	3.6	4.2	6.6
FeO	7.7	6.4	9.4	8.8	2.6	1.4	8.1	3.5	7.9	8.5	4.6	1.2	7.9	5.6
MnO	0.21	0.14	0.25	0.22	0.04	0.03	0.15	0.21	0.14	0.15	0.13	0.02	0.10	0.18
MgO	6.1	6.3	6.2	7.1	1.8	0.74	6.0	5.0	8.5	8.9	2.4	1.1	4.5	4.1
CaO	8.8	9.0	7.4	5.5	1.8	1.4	5.1	6.5	9.9	10.0	10.3	0.55	3.6	8.3
Na_2O	2.8	2.6	4.0	4.2	4.9	3.9	3.8	2.6	2.3	2.7	3.5	3.5	2.9	3.5
K_2O	0.11	0.08	1.7	0.76	2.1	4.3	0.14	1.3	0.66	1.0	1.7	3.8	4.6	2.2
P_2O_5	0.44	0.14	0.40	0.42	0.18	0.13	0.48	0.24	0.24	0.36	0.50	0.11	0.65	0.55
H_2O^+	2.7[b]	2.7[b]	3.5[b]	2.8[b]	2.9	1.5	4.5	3.1	2.6	2.8	1.0	1.3	3.5	2.2
CO_2	n. d.	n. d.	n. d.	n. d.	0.10	0.70	3.4	1.0	0.98	2.4	5.4	0.56	0.61	0.25
Sum	97.76[c]	98.66[c]	99.15[c]	98.6[c]	100.11	99.89	100.67	99.59	99.12	99.31	100.03	99.51	99.46	99.38
Ba (ppm)	n. d.	n. d.	n. d.	n. d.	508	549	228	453	226	324	304	783	759	n. d.
Cr	86	195	142	85	<40	<40	91	73	337	416	<62	–	<62	39
Nb	n. d.	n. d.	n. d.	n. d.	14	11	22	13	17	26	70	85	54	n. d.
Ni	14	20	24	19	<43	<43	81	–	232	207	<38	–	<38	n. d.
Rb	20	10	37	33	60	85	9	39	12	15	39	60	46	25
Sr	823	168	263	319	117	108	346	749	494	330	386	228	60	364
V	705	189	706	765	76	45	283	267	178	231	249	20	328	n. d.
Y	51	36	40	48	29	33	25	23	20	21	26	51	30	34
Zr	377	137	225	232	219	142	231	146	174	209	335	695	215	289

[a] A: Gabbroic intrusive rock, Vorderes Saalachtal, western Greywacke zone (Schlaegel-Blaut 1990, sample K 1).
B: Metabasalt lava, Vorderes Saalachtal, western Greywacke zone (Schlaegel-Blaut 1990, sample K 13).
C: Metabasalt lava, Hinteres Saalachtal, western Greywacke zone (Schlaegel-Blaut 1990, sample KP 6).
D: Gabbroic intrusive rock, Hinteres Saalachtal, western Greywacke zone (Schlaegel-Blaut 1990, sample KP 10).
E: Blasseneck "porphyroid", Leobner, eastern Greywacke zone (Kretzschmar, unpubl. Thesis, Univ. Tübingen 1989, sample BL 21).
F: Blasseneck "porphyroid", Leobner, eastern Greywacke zone (Kretzschmar, unpubl. Thesis, Univ. Tübingen 1989, sample BL 33).
G: Metabasalt lava of the Magdalensberg Series (Loeschke 1989 b, sample KS 8).
H: Greenschist of the Nock Series, Gurktaler Alpen (Giese 1988, sample UG 151).
I: Metabasalt lava of the Kaser Series, Gurktaler Alpen (Giese 1988, sample UG 142).
J: Metabasalt lava of the Eisenhutschiefer Series, Gurktaler Alpen (Giese 1988, sample UG 174).
K: Greenschist of the Saualpe (Loeschke 1988 a, sample KS 89).
L: Metarhyolite of the Saualpe (Loeschke 1988 a, sample KS 94).
M: Metabasalt lava of the Palaeozoic of Graz (Loeschke 1988 b, sample KS 57).
N: Metabasalt lava of the Karawanken Mountains (Eisenkappel area, Loeschke 1973, 1975, sample PK 7).
[b] Loss of ignition.
[c] Sum of the oxides on a volatile-free basis; n. d. = not determined.

quences are undated as yet, a certain possibility remains for the assumption of a basic Ordovician (?) volcanism as it was suggested in earlier stratigraphical correlations (Mostler 1968; Schönlaub 1979). Recently, Lower Ordovician acritarchs were found in siliciclastic sequences by Reitz and Höll (1989). Unfortunately, dating was done without mapping and the localities are far away from the investigated volcanic rocks. For the basalt-sill complex of Maishofen (Fig. 2) a post-Middle Devonian (Carboniferous?) age is proposed by Schlaegel-Blaut and Heinisch (1990). This assumption is based on detailed mapping and correlation with dated sequences. But it must be admitted, that neither conodont nor acritarch fossils have been found so far within the basalt-sill complex. The acidic volcanic rocks of the eastern Greywacke zone (Blasseneck "porphyroid"; Figs. 1 and 3, column 1) are of Upper Ordovician age (Schönlaub 1982). Below these acidic volcanic rocks thin layers of basic volcanic rocks of unknown age are found. Well-dated basic volcanic rocks occur in the Silurian (Schönlaub 1982).

In the Gurktaler Alpen (Figs. 1 and 3, columns 2 and 3) several volcanic sequences have been found (Mulfinger, unpubl. Thesis, Univ. Tübingen 1986; Giese 1988). Their stratigraphic age is only partly known, but the Eisenhutschiefer Series was dated by conodonts as Silurian and the Nock Series as Upper

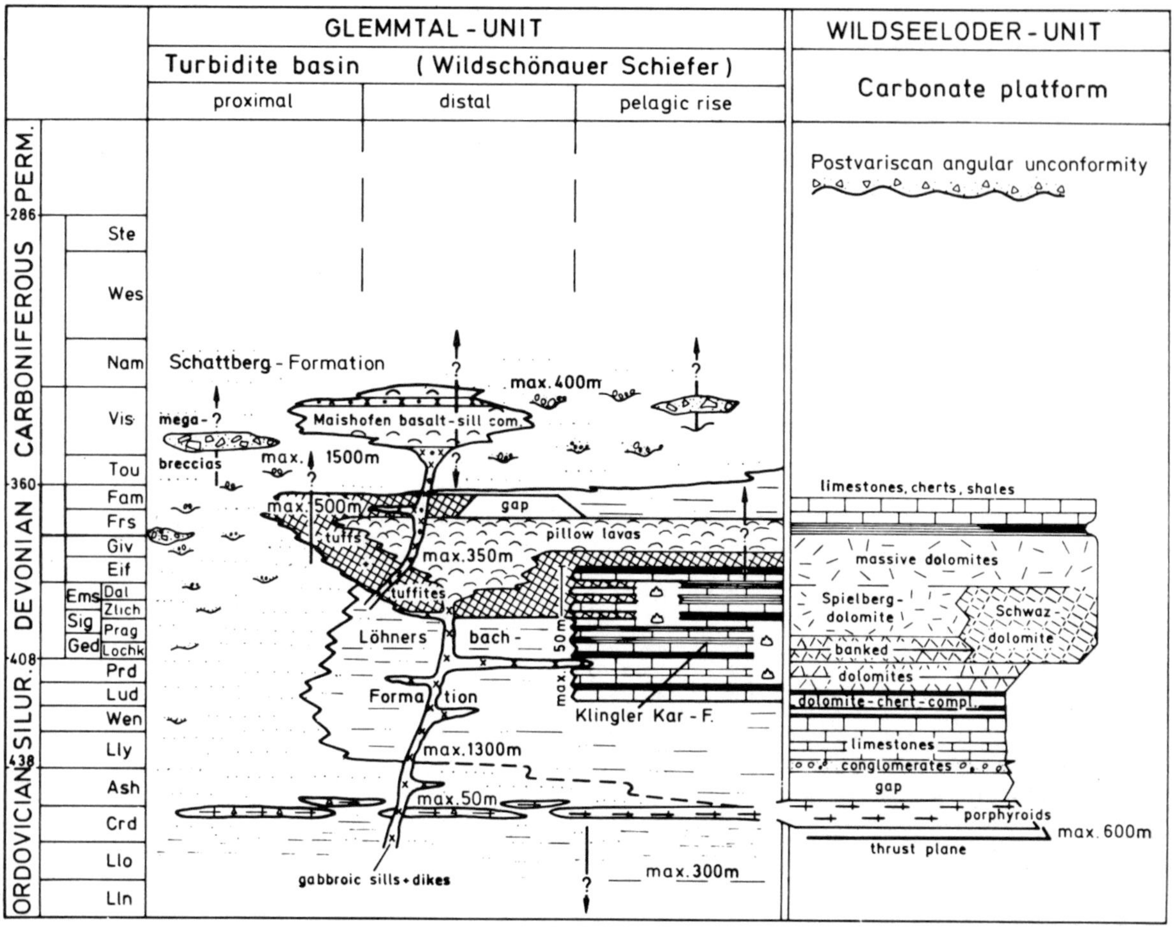

Fig. 2. Stratigraphic position and facies relations of the volcanic rocks within the western Greywacke zone (after Schönlaub 1979; Heinisch 1988). Schattberg formation and Löhnersbach formation consist of siliciclastic proximal and distal turbidite sequences, Klingler Kar formation consists of condensed, conodont-bearing carbonates, black shales, lydites, interlayered with basaltic tuffs

Ordovician (Höll 1970; Neubauer and Pistotnik 1984). The Magdalensberg Series (Figs. 1 and 3, column 4) is part of the Gurktal nappe and belongs to the Ordovician (Riehl-Herwirsch 1970, Havlicek et al. 1987). The volcanic rocks of the Saualpe (Phyllite group; Figs. 1 and 3, column 5) also belong to the Gurktal nappe, but are not well dated. The marbles which overlie the volcanic rocks are most probably of Devonian age, but the age of the underlying volcanic rocks is unknown. It could be Silurian and Upper Ordovician (Kleinschmidt et al. 1975). The volcanic rocks of the Palaeozoic of Graz are Silurian and Devonian in age (Figs. 1 and 3, column 6; Flügel and Neubauer 1984).

The Eisenkappel area in the Karawanken Mountains (Fig. 1) comprises a thick sequence of spilites (Loeschke 1973). Its age is unknown since no fossils have been found so far. The radiometric K-Ar determination of the spilites (Loeschke and Weber 1973) of 315 respectively 296 Ma represents the climax of Variscan metamorphism. The sequence is included here in the discussion, but omitted from Fig. 3.

2.1 *Western Greywacke Zone (Tyrol, Salzburg)*

In the western Greywacke zone (Tyrol, Salzburg, i. e. Kitzbühel area, see Fig. 1), according to Heinisch (1986) and Schlaegel-Blaut (1990) two different facies realms occur which are preserved in two different nappes and separated from each other by a complex shear zone. Both these realms are considered as former neighbouring palaegeographic units that were tectonically superimposed upon each other during later tectonic movements. There is no evidence for

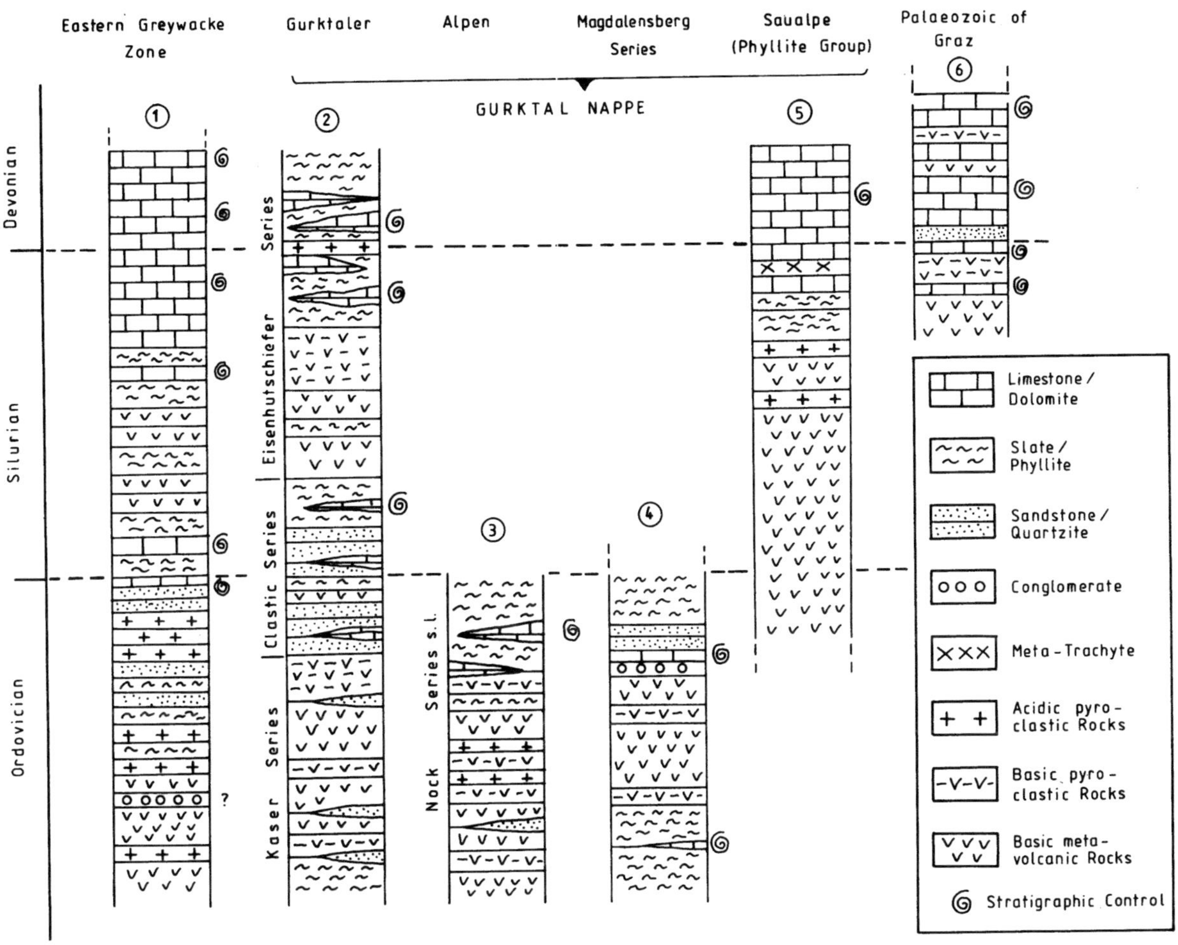

Fig. 3. Lower Palaeozoic stratigraphic sections of the Eastern Alps (compiled after Riehl-Herwirsch 1970; Kleinschmidt et al. 1975; Schönlaub 1979; Flügel and Neubauer 1984; Neubauer and Pistotnik 1984; Giese 1988; Havlicek et al. 1987; Loeschke et al. 1990)

palaeo-position, primary distance and direction of stacking of the two units (Heinisch 1988).

Within the "Wildseeloder unit" (Fig. 2) up to 600-m-thick piles of ignimbrites of rhyolitic to alkali-rhyolitic composition occur which are similar to the Blasseneck "porphyroid". Primary volcanic textures (i. e. perfect welding, shards, collapsed pumice lapilli) prove an extrusion under subaerial to shallow marine conditions (Mostler 1970; Heinisch 1981, 1988). They are overlain by platform carbonates of predominantly Devonian age (Fig. 2). The "Glemmtal unit" (Fig. 2) represents a turbiditic basin filled with thick sequences of siliciclastic rocks ("Wildschönauer Schiefer"). Only thin layers of epiclastic "porphyroids" occur which are redeposited by turbidite mechanisms. They most probably represent the distal deeper marine, contemporaneous equivalents of the ignimbrite sheets mentioned. A huge amount of basic volcanic rocks is intercalated within the metasediments which have been interpreted by Colins et al. (1980) and Mostler (1984) as ocean floor basalts. Recently, detailed investigations, combining petrographical and volcanological studies with geochemical data, including REE analyses, were carried out by Schlaegel (1988) and Schlaegel-Blaut (1990) within the entire Greywacke zone (western and eastern part). Two different types of rocks were found:

1. Intraplate basalts ("hinteres Saalachtal", indicated as Devonian in Fig. 2): The sequences consist of thin gabbroic sills, highly vesicular pillow basalts (thickness max. 350 m), various types of pyroclastic rocks (thickness max. 400 m) and volcanic detritus (tuffites). These sequences formed in a shallow marine environment (<500 m below sea level). Even angular scoria and beach boulders are found within the pyroclastic and volcano-sedimentary sequences.

   The rocks are highly enriched in immobile incompatible elements (Fig. 4). Geochemically they are

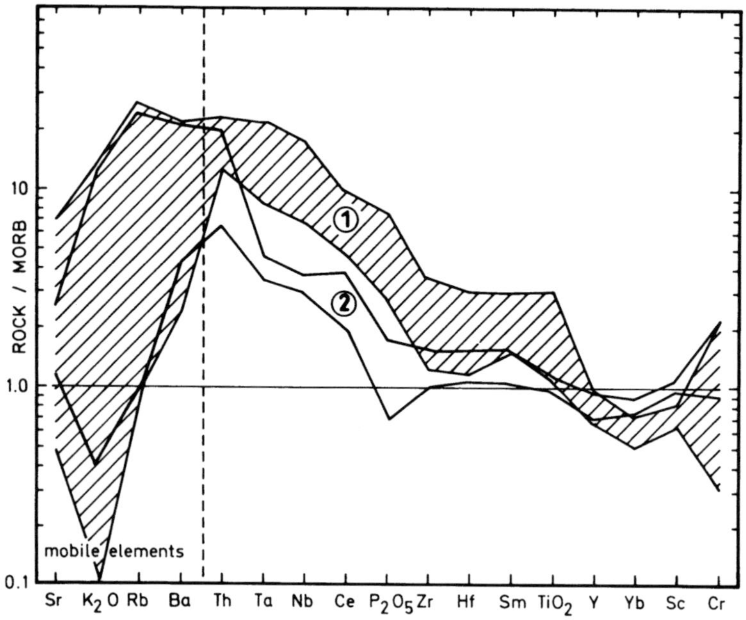

Fig. 4. Examples for the composition of metabasalts from the western Greywacke zone (Schlaegel-Blaut 1990), normalized with corresponding values of mid-ocean ridge basalts (MORB) (after Pearce 1982). The area "hinteres Saalachtal" *(1)* contains intraplate basalts, highly enriched in immobile incompatible elements, while definitely minor enriched tholeiitic basalts prevail in the basalt-sill complex of Maishofen "vorderes Saalachtal" *(2)*

transitional basalts and alkalibasalts as discussed in great detail by Schlaegel-Blaut (1990; see also Table 1, columns C and D).

Thus, an intraplate ocean island type of basalts is deduced for the Devonian magmatism of the western Greywacke zone which is consistent with volcanological and geochemical data (Heinisch 1988; Schlaegel-Blaut 1990). The scenario of growing seamounts, temporarily reaching a shoaling stage or even a stable subaerial oceanic island stage is established. The major part of the metabasalts belongs to this Devonian event (Heinisch et al. 1987).

2. Tholeiitic basalts of a marginal basin: the "basalt-sill complex of Maishofen (vorderes Saalachtal)" displays a different development. The sequence (thickness about 400 m) consists of pillow basalts and sheet-flows with few or no vesicles, many gabbroic sills, some dioritic sills and thin intercalated shale horizons (Fig. 2). Products of explosive volcanism are lacking. The volcanological investigations prove an extrusion under deeper-water conditions (>500 m below sea level).

The rocks are of tholeiitic compositon (Schlaegel-Blaut 1990). They are slightly less enriched in immobile incompatible elements compared to intraplate sequences (Fig. 4, Table 1, columns A and B). The geochemical characteristics are ambiguous. In the numerous discrimination diagrams used by Schlaegel-Blaut (1990) geochemical compositions vary between MORB-, island arc-tholeiite and intraplate basalt affinities. They are definitely not N-type mid-ocean ridge basalts due to their enrichment in immobile incompatible elements (Fig. 4); also a subduction component in the sense of Pearce et al. (1984) is missing. Thus, several possibilities remain for the plate tectonic interpretation of these rocks. The best-fitting suggestion is to correlate this independent magmatic event with extensional processes (marginal basin, oceanic plateau, sill-sediment complex connected with a continental rift zone?) adjacent to a passive continental margin (Schlaegel-Blaut 1990). According to field mapping and lateral correlation with fossiliferous rocks the age of these metabasalts is assumed to be younger, perhaps of Carboniferous age (Heinisch and Schlaegel 1989).

2.2 *Eastern Greywacke Zone*

In the eastern Greywacke zone, an area to the north of Kalwang (Fig. 1) has been investigated in detail (Loeschke et al. 1990) and the results of the geochemical analyses are shown in the Figs. 5 and 6 and in Table 1, columns E and F. The Upper Ordovician Blasseneck "porphyroid" represents massive sheets of pyroclastic rocks (and only some lavas) of rhyolitic, dacitic and andesitic composition. Its thickness reaches up to 1500 m (Heinisch 1981). Volcanic relic structures in the Blasseneck "porphyroid" show that the pyroclastic layers are pumice flow deposits. They are the result of subaerial eruptions near to a marine basin in which the pyroclastic fragments were re-

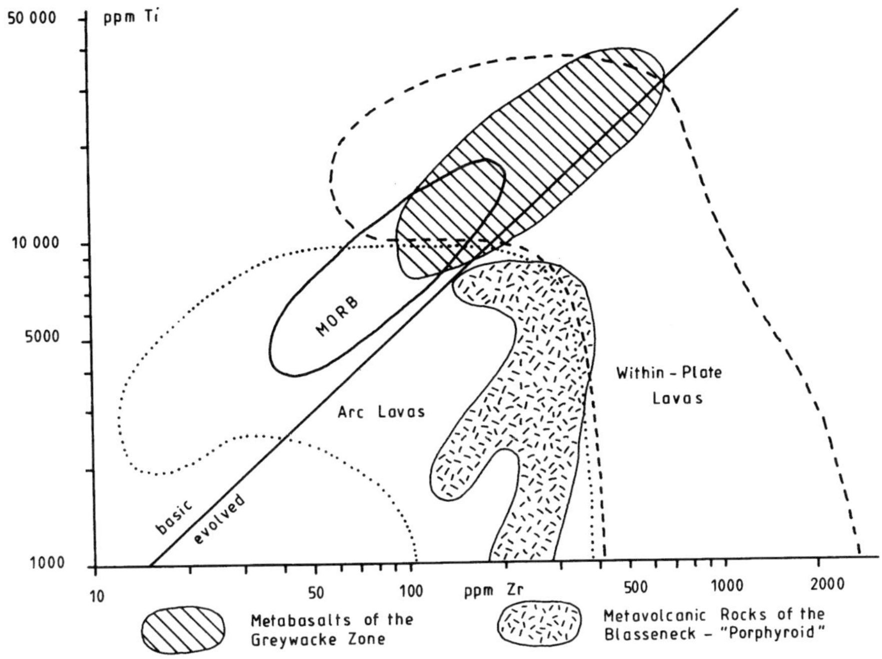

Fiig. 5. Ti-Zr diagram (Pearce 1983, unpubl. report, Open University) for metavolcanic rocks of the Blasseneck "porphyroid" and of the western Greywacke zone (compiled after Loeschke et al. 1990; Schlaegel-Blaut 1990)

worked. After the eruptions a huge caldera was probably formed (Heinisch 1981; Loeschke et al. 1990).

Figure 5 shows that the Ti and Zr contents of the Blasseneck "porphyroid" plot mainly in the field of arc lavas. In Fig. 6 the Nb and Zr contents of the Blasseneck "porphyroid" plot again in the fields of calc-alkaline and high-K calc-alkaline volcanic rocks of island arcs and active continental margins.

Acidic volcanic rocks of continental rifts and oceanic islands have compositions strikingly different from those of the Blasseneck "porphyroid" (Fig. 6). The interpretation of these data is difficult. Acidic volcanic rocks of calc-alkaline and high-K calc-alkaline composition occur mainly along island arcs and active continental margins (Ewart 1979), but they do also occur in back-arc basins as for example in the Basin and Range Province of the western USA (Best et al. 1989; see Fig. 6) during periods of extension. A typical continental rift situation similar to the Afar depression or to the continental rifts of Eastern Africa can be excluded because there per-alkaline rhyolites and trachytes occur which have not been found in the Blasseneck "porphyroid".

Metabasalts within the eastern Greywacke zone of Styria are confined to small areas, separated from each other by faults (Schlaegel-Blaut 1990). They

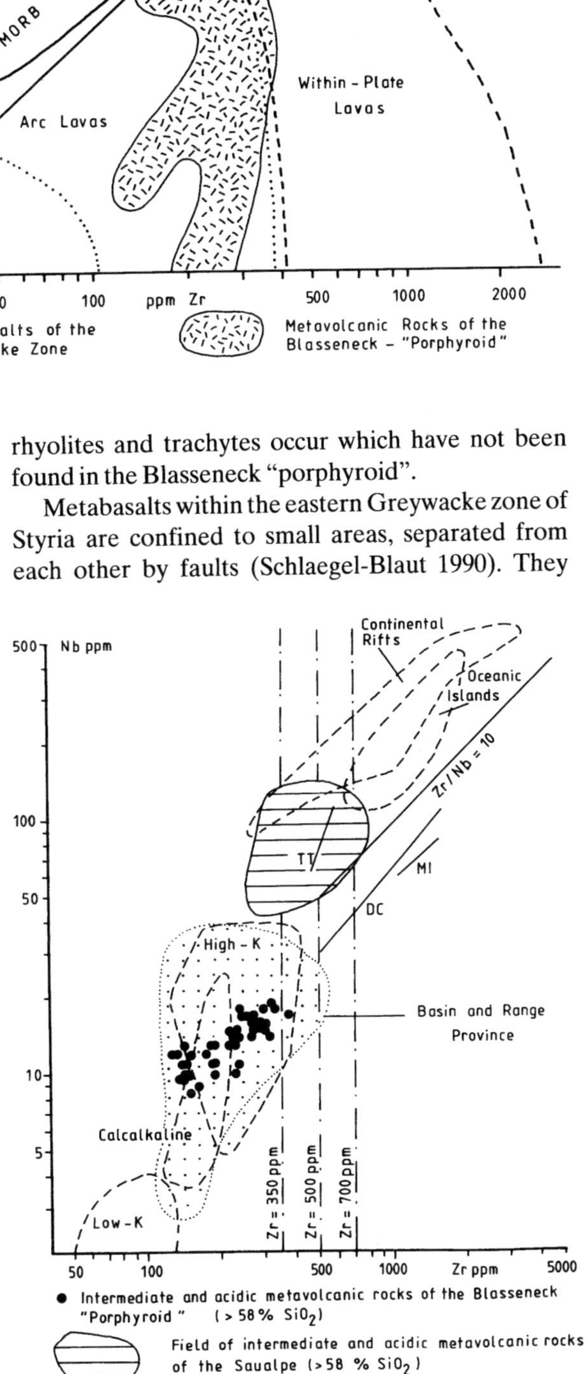

Fig. 6. Nb-Zr diagram (Leat et al. 1986) for intermediate and acidic metavolcanic rocks of the Blasseneck "porphyroid" and of the Phyllite group of the Saualpe (Austria) (after Loeschke 1988a; Loeschke et al. 1990). For comparison, evolutionary trends of the Tala Tuff (*TT*, Mexico; Mahood 1981), the Devine Canyon Tuff (*DC*, Oregon/USA; Greene 1973), Mayor Island (*MI*, New Zealand; Ewart et al. 1968) and ignimbrites of the Basin and Range Province (USA; Best et al. 1989) are also plotted on the diagram

have been investigated at Admont, Bruck a.d. Mur (Fig. 1), Kalwang-Lange Teichen, St. Lorenzen-Trieben, Niederbergalm, Pfaffendorf, Trofaiach-Gößgraben and Villmersdorf-Rottenmann. They belong mostly to the Noric nappe system (Flügel and Neubauer 1984). The dated ones of the metabasalt-sequences yielded a Silurian age (Schönlaub 1982). Gabbroic sills, vesicular pillow basalts and various volcanoclastic rocks are found, with a maximum thickness of 200 m. Metamorphism and tectonic overprinting are slightly higher than in the western Greywacke zone. Therefore greenschists in which primary magmatic textures are completely destroyed are more frequent.

The geochemical composition of these metabasalts indicates an intraplate ocean island type setting similar to that of the western Greywacke zone (not documented in this work, see Schlaegel-Blaut 1990). Only within the "Admont area" the geochemical data show different REE patterns and certain similarities to the "basalt-sill complex of Maishofen".

2.3 Gurktaler Alpen

In the Gurktaler Alpen there are two different volcanic sequences (Fig. 3, columns 2 and 3; Table 1, columns H, I and J) which have different stratigraphic ages (Giese 1988). Column 2 of Fig. 3 shows a sequence which is formed by the Kaser, Clastic and Eisenhutschiefer Series (division by Giese 1988). Primary structures such as graded bedding and cross bedding in the sandstones of the Clastic Series, the shapes of pillow lavas and cumulate layers of clinopyroxenes at the base of a thick sill in the Eisenhutschiefer Series can still be recognized and indicate in the field the stratigraphic top of the section. The Silurian age of the Eisenhutschiefer Series has been established by Höll (1970) and Neubauer and Pistotnik (1984). The Kaser Series which lies beneath the Clastic Series consists of pillow lavas, basic sills and fine-grained clastic sediments. Its age is unknown, but the field data indicate an age older than the Clastic Series. It is possibly Ordovician. The Nock Series (Fig. 3, column 3) consists of basic to intermediate calc-alkaline lavas and tuffs and of some acidic pyroclastic rocks and phyllites which are more strongly metamorphosed than the rocks of the other series. They show the crystallization of epidote, actinolite and biotite indicating greenschist facies metamorphism. The Nock Series is of Upper Ordovician age (Neubauer and Pistotnik 1984).

Figure 7 shows the Ti and Zr contents of the two sequences and their geotectonic setting. Most analyses of the Nock Series plot in the field of arc lavas, whereas the analyses of the Kaser and Eisenhutschiefer Series plot in the field of within-plate lavas. The volcanic rocks of the Nock Series are therefore probably the product of volcanic eruptions above a convergent plate boundary. In contrast to that, the volcanic rocks of the Kaser and Eisenhutschiefer Series were formed in an area of crustal extension. They are similar in composition to alkalibasalts and hawaiites. Those volcanic rocks occur on oceanic islands and in continental rifts. The Clastic Series comprises distal turbidites rich in quartz which were probably deposited along an upper Ordovician passive continental margin.

2.4 Magdalensberg Series

The Magdalensberg Series (Fig. 3, column 4) consists of pillow lavas of spilitic composition, basic sills and tuffs in which primary magmatic structures can still be recognized in many cases. According to Riehl-Herwirsch (1970), the sequence represents the remnant of a volcanic island. A conglomerate at the top of the sequence contains boulders of volcanic rocks and is a beach conglomerate. Carbonatic fossil-bearing tuffites follow in the upper part of the section and indicate a Middle Ordovician age (Havlicek et al. 1987). They are overlain by quartzites which are derived from a continental block with low relief.

Figure 8 shows that the metabasalts of the Magdalensberg Series are relatively enriched in Nb, P, Zr and Ti compared to mid-ocean ridge basalts (Loeschke 1989b). They have the characteristics of mildly alkaline within-plate basalts (Table 1, column G). Those rocks occur in continental rifts and on oceanic islands and are typical of areas of crustal extension. They show similarities to the rocks of the Kaser Series and Eisenhutschiefer Series which are also within-plate lavas. The quartzites which lie on top of the volcanic rocks of the Magdalensberg Series indicate a sedimentary environment which lay in the interior of a craton or near to a passive continental margin (Maynard et al. 1982; Dickinson et al. 1983). The geotectonic setting may have been similar to a passive continental margin with some islands which were positioned near to the shelf edge like the Canary Islands or it may have been a broad continental rift flooded by the sea like the North Sea.

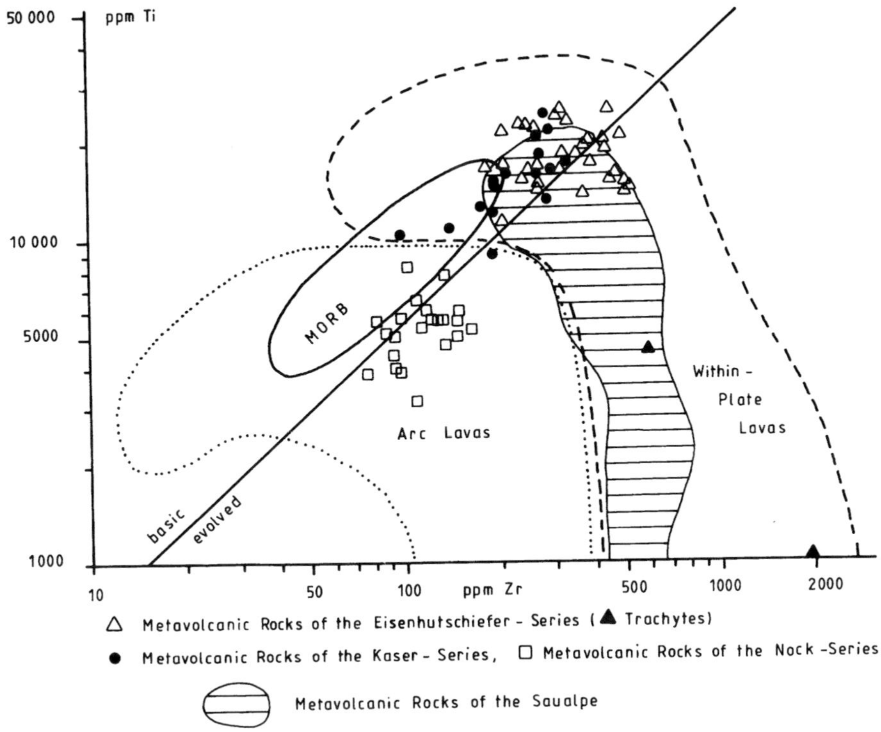

Fig. 7. Ti-Zr diagram (Pearce 1983, unpubl. report, Open University) for metavolcanic rocks of the Gurktaler Alpen (Eisenhutschiefer Series, Kaser Series, Nock Series) and the Saualpe (compiled after Giese 1988; Loeschke 1988a)

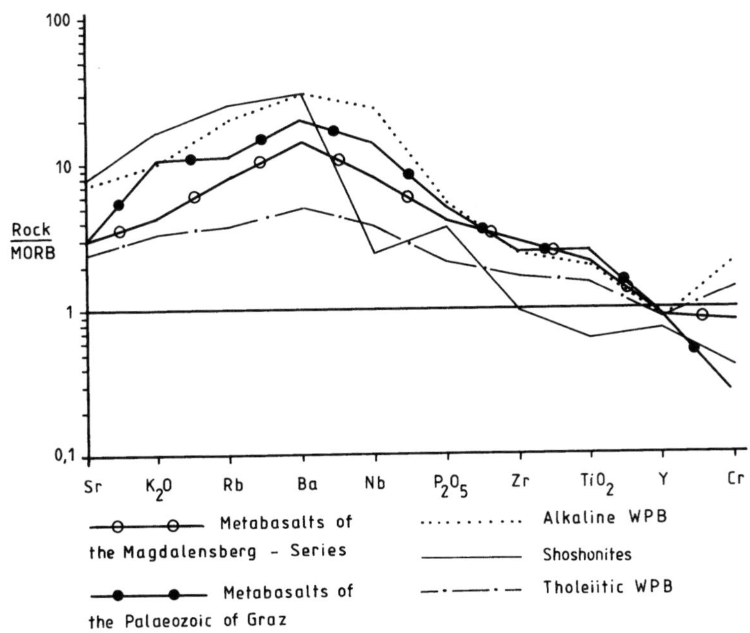

Fig. 8. Selected average major and trace element contents of metabasalts of the Magdalensberg Series (Loeschke 1989b) and the Palaeozoic of Graz (Loeschke 1988b) normalized with corresponding values of mid-ocean ridge basalts (MORB) (after Pearce 1982). For comparison, the element distribution patterns of shoshonites, alkaline within-plate basalts *(WPB)* and tholeiitic within-plate basalts (Pearce 1982) are also plotted on the diagram

2.5 *Saualpe*

The stratigraphy and petrography of the Phyllite group of the Saualpe (Fig. 3, column 5) has been established by Kleinschmidt et al. (1975). The metavolcanic rocks occur beneath Devonian marbles and various volcanic rock types can be distinguished: alkalibasalts, trachyandesites, trachytes and peralkalirhyolites (comendites, pantellerites) which are all metamorphosed under low-grade conditions (Hurler 1972; Loeschke 1988a; Table 1, columns K and L). In Fig. 7 their Ti and Zr contents plot in the field of within-plate lavas. In Fig. 6 the Nb and Zr contents of the intermediate and acidic volcanic rock types of the Saualpe occupy a field which is transitional between high-K calc-alkaline rocks of active continental margins and islands arcs and peralkaline acidic volcanic rocks of oceanic islands and continental rifts. Young volcanic rocks which also plot in this transitional field of Fig. 6 (TT, DC, MI) are mildly alkaline to peralkaline rhyolites which erupt along fracture zones on active continental margins and island arcs.

The interpretation of the geochemical data of the metavolcanic rocks of the Saualpe allows two possibilities. They are either within-plate lavas which erupted in areas of crustal extension or they are similar to lavas which erupted along fracture zones on an extinct active continental margin/island arc indicating the transition from an Ordovician convergent plate boundary to a Silurian/Devonian extensional tectonic regime.

2.6 *Palaeozoic of Graz*

In the Palaeozoic of Graz (Flügel and Neubauer 1984; Fritz and Neubauer 1988) volcanic rocks occur in Silurian strata beneath Devonian limestones. Some of them are intercalated in the Devonian limestones, especially in those of Middle Devonian age (Fig. 3, column 6). They form sills, pillow lavas (Table 1), column M) and tuffs which are partly strongly altered by secondary metasomatic alteration processes. Figure 8 shows that their Nb, P, Zr and Ti contents are relatively enriched compared to mid-ocean ridge basalts (Loeschke 1988b) and that they are similar to within-plate alkalibasalts. Since within-plate alkalibasalts are always connected with extensional tectonic regimes, the occurrence of alkalibasalts in the Palaeozoic of Graz indicates extensional processes which lasted from the Silurian to the Middle Devonian.

2.7 *Karawanken Mountains – Eisenkappel Area*

In the Eisenkappel area of the Karawanken Mountains a Palaeozoic sequence comprising shales, greywackes, sills, pillow lavas and tuffs occurs to the north of the Periadriatic lineament. The youngest rocks are greywackes which overlie the volcanic rocks (Loeschke 1970). The pillow lavas and sills are typical within-plate alkalibasalts (Loeschke 1975; Table 1, column N). Their stratigraphic age is unknown. In their geochemical composition they are very similar to the other within-plate alkalibasalts which occur in the Kaser and Eisenhutschiefer Series, in the Magdalensberg Series and in the Palaeozoic of Graz.

Since their age is unknown, to facts are important for an interpretation of the possible age and the former geotectonic setting:

1. On top of the volcanic rocks of the Eisenkappel area greywackes are deposited which are petrographically similar to the greywackes of the Hochwipfelflysch.
2. Alkalibasalts similar to the alkalibasalts of the Eisenkappel area can occur in continental rifts, on oceanic islands, along passive continental margins, in back-arc basins such as the Sea of Japan and in broad continental depressions which are floored by a thinned continental crust and covered by a shallow sea as, for example, in the North Sea.

Both facts could be interpreted in the way that the sequence of Eisenkappel indicates a period of crustal extension which caused a thinning of the continental crust prior to the deposition of flysch deposits which were equivalent to the Hochwipfelflysch. This interpretation would place the Palaeozoic sequence of Eisenkappel into the Lower Carboniferous. The geotectonic consequences of this interpretation are discussed in the next chapter.

3 Conclusions and Discussion

Figure 9 shows the geotectonic evolution of the Austro-Alpine realm (north of the Periadriatic lineament) from the Upper Ordovician to the Middle Devonian. In terms of terranes (Frisch and Neubauer 1989) Fig. 9 shows the evolution of the Noric composite terrane.

In the Austro-Alpine realm two different geotectonic settings can be recognized during the **Upper Ordovician** (Fig. 9A):

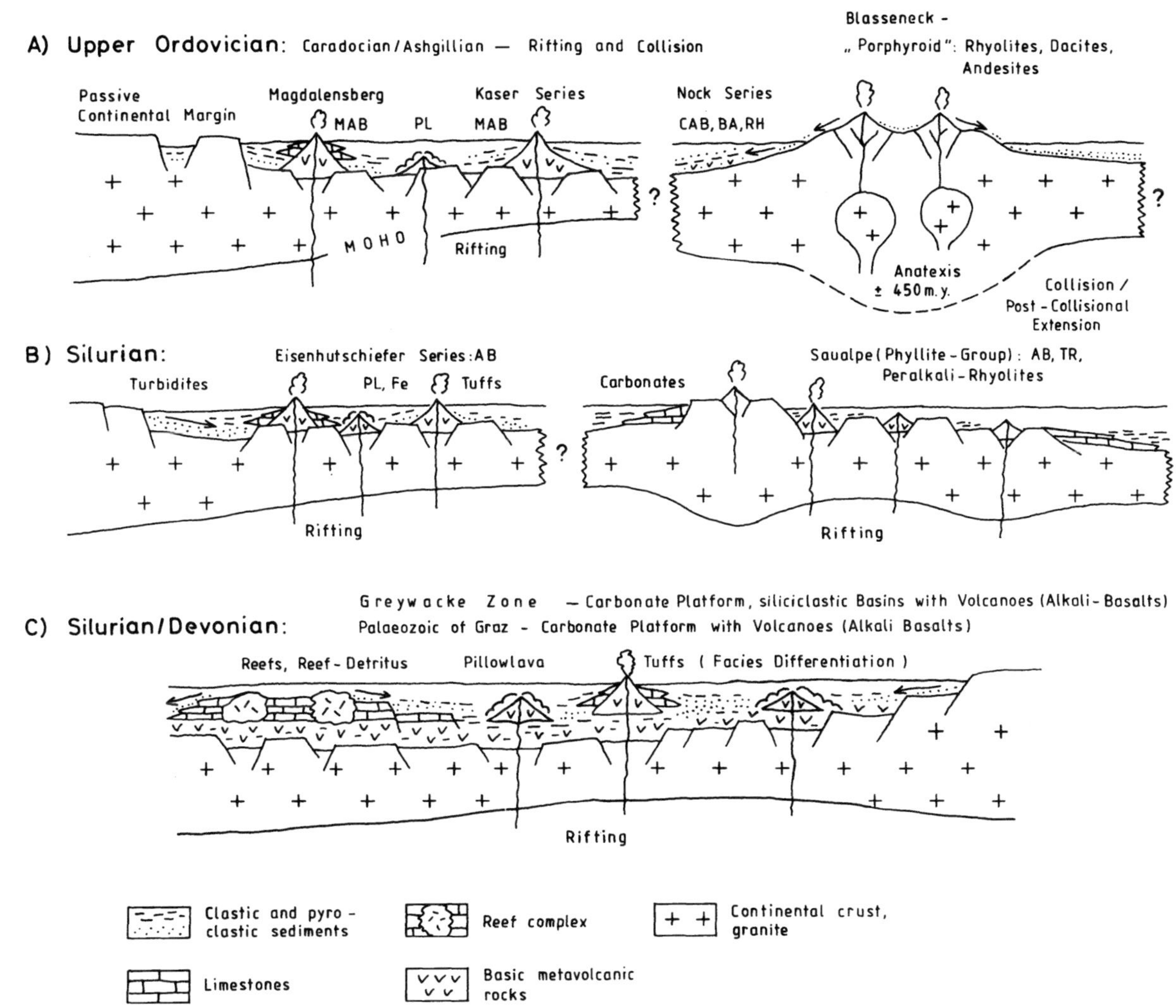

Fig. 9. Hypothetical sections showing the relations between sedimentation, volcanism and crustal structure during Upper Ordovician to Middle Devonian time in the Austro-Alpine realm. *AB* Alkaline basalts; *BA* basaltic andesites; *CAB* calc-alkaline basalts; *FE* Submarine hydrothermal iron ore deposits; *MAB* mildly alkaline basalts; *PL* pillow lava; *RH* rhyolites; *TR* trachytes (after Heinisch 1988; Loeschke 1989 a)

1. The Magdalensberg and Kaser Series represent a rift situation on thinned continental crust.
2. In contrast to that, the Blasseneck "porphyroid" and the Nock Series indicate a situation which must differ from a rift situation because they comprise great amounts of calc-alkaline volcanic rocks and hardly any alkaline rocks. The original geotectonic setting of the Blasseneck "porphyroid" is difficult to establish. The most probable solution is a situation which is similar to a late- or post-collisional volcanic event after the "Caledonian" orogeny *sensu* Frisch et al. (1987) or "Pan-African" orogeny *sensu* Heinisch (1988). It is supposed that the pyroclastic rocks of the Blasseneck "porphyroid" were related in some way to granitic and granodioritic intrusions which were formed by anatectic processes in the deeper continental crust (Heinisch and Schmidt 1982). Radiometric age determinations of orthogneisses and granites of the Eastern Alps often yield ages of about 450 Ma indicating an Upper Ordovician anatectic event (Scharbert and Schönlaub 1980). The pyroclastic rocks of the Blasseneck "porphyroid" and the Nock Series are thought to be the cogenetic volcanic products of this event. This interpretation requires either a former palaeogeographic neighbourhood between the Austro-Alpine metamorphic basement and the Blasseneck "porphyroid" or even an angular unconformity beneath the Blasseneck "porphyroid". The possible existence of a Lower Palaeozoic, "Caledonian" or "Pan-African" angular unconformity within the Alps is discussed contrarily

(Heinisch and Schmidt 1976; Frisch et al. 1984; Neubauer 1985; Heinisch 1988). This fundamental question is unsolved until now due to missing stratigraphic data and clear field evidence. The Ordovician magmatism may have been similar in style to the processes which occurred during the Permian of the Southern Alps with the eruption of the voluminous quartz porphyry of Bozen (D'Amico 1979). This eruption occurred after the Variscan collision and before the rifting processes which lead to the opening of the Mesozoic Tethys. Another useful example may be the Lower Tertiary ignimbrites and the Upper Tertiary alkaline basalts of the Basin and Range Province of western North America. The volcanic rocks there are first subduction-related and then rift-related changing their compostion from calc-alkaline to alkaline (Dickinson 1981; Nealy and Sheridan 1989). A similar development can be observed in parts of the Eastern Alps: the Upper Ordovician Blasseneck "porphyroid" and Nock Series are calc-alkaline in composition. From the beginning of the Silurian onwards alkaline volcanic rocks erupted everywhere in the Austro-Alpine realm (Fig. 9 B) which indicates the onset of widespread extensional processes. Therefore the time from the Upper Ordovician to the Lower Silurian was a time where the arrangement of the plate configuration might have changed.

The **Silurian** situation shows rifting processes indicated by the occurrence of alkaline volcanic rocks both in the Gurktaler Alpen (Eisenhutschiefer Series) and in the Saualpe (Fig. 9 B). The difference between both areas is that the rock record of the Saualpe contains much more acidic volcanic rocks than the Eisenhutschiefer Series and may therefore represent a slightly different geotectonic situation (Loeschke 1989 a). The Palaeozoic rocks of Murau (Schnepf 1989) indicate a similar environment as the rocks of the Eisenhutschiefer Series.

The **Silurian/Devonian** situation of the Greywacke zone and the Palaeozoic of Graz is characterized by crustal extension (Fig. 9 C). The sequences of the Greywacke zone are the best-preserved and thickest basic rocks of the Austro-Alpine Palaeozoic. The basic volcanism of the Greywacke zone is not linked to Caledonian events, as suggested in some earlier published plate tectonic models (Höll and Maucher 1976; Loeschke 1977; Colins et al. 1980).

There is no evidence for a completely developed oceanic crust, neither of Ordovician age as assumed by Mostler (1984) nor of Devonian or Carboniferous age. The overwhelming majority of the basic rocks displays neither a relationship to an active plate margin nor to a mature ocean ridge segment. Most of the volcanic sequences generated on volcanic seamounts or island volcanoes with intraplate geochemistry and within a shallow marine enviroment on thinned continental crust – in the eastern Greywacke zone during the Silurian and in the western Greywacke zone during the Devonian. This is in good accordance with sedimentological data (Heinisch 1986, 1988) which prove a nearby passive continental margin, yielding huge amounts of siliciclastic, terrigenous detritus to the continuously subsiding Palaeozoic basins.

Only within small areas (Maishofen), minor enriched, tholeiitic basalt sequences are found which extruded within a deeper water environment. No final interpretation can be given for this slightly differing situation, as affinities to an oceanic plateau as well as to sill-sediment complexes connected with a continental rift zone do exist (Schlaegel-Blaut 1990). Even the link with Variscan plate convergence cannot be excluded, having induced small-scale back-arc spreading centres in a marginal basin.

Thus, with the exception of the Blasseneck "porphyroid" and the Nock Series, a long-lasting extensional process can be seen in the Eastern Alps which was accompanied by volcanic events from the Upper Ordovician to the Middle Devonian. This slow but continuous, general thinning of the continental crust led to subsidence and deposition of thick sedimentary and volcanic sequences. It is important to point out, that ophiolite sequences indicating a mature oceanic ridge stage, are not documented within the considered realms. The peak of crustal thinning and subsidence was probably reached in Lower Carboniferous (pre-Upper Visean) time shortly before the beginning of the deposition of flysch greywackes in the Eisenkappel area which are petrographically similar to the Hochwipfelflysch (Flügel 1990). The region which was affected by these rifting processes is assumed to have been the northern margin of Gondwana (Heinisch 1986; Neugebauer 1988; Frisch et al. 1990).

The **Karawanken Mountains** belong partly to the South-Alpine realm (south of the Periadriatic lineament) and partly to the Austro-Alpine realm (north of the Periadriatic lineament) (see Fig. 1). Since the volcanic rocks of the Eisenkappel area which are situated directly north of the Periadriatic lineament are included here in the discussion, the situation of the South-Alpine and Austro-Alpine part of the Karawanken Mountains must be considered (Loeschke 1970, 1973; Tessensohn 1971, 1974; Mooshammer 1989; Kullmann pers. comm.).

The Middle Devonian of the Karawanken Mountains was characterized by a facies differentiation including reefs and deeper basins which were filled with reef detritus and some clastic sediments. During the time between the Upper Devonian and the Lower Carboniferous the facies differentiation increased because continuous sedimentation, but also erosion and redeposition of parts of the older limestones can be observed (Tessensohn 1983; Ebner 1990). This is interpreted here as the beginning of the culmination of a rifting process which lead to uplift and erosion of rift shoulders and subsidence of basins in which pelagic limestones and radiolarian chert were deposited. It is suggested that the pillow lavas and sills of the Eisenkappel area occurred at a time when the rifting process reached its peak (Lower Carboniferous, pre-Upper Visean). The thinned continental crust was intruded by alkalibasaltic magma in the centre of the rift. The basin had the character of a broad intracontinental rift floored by thinned continental crust and accompanied by margins similar to passive continental margins. It represented the western continuation of the Palaeotethys to the east (Neugebauer 1988; Frisch and Neubauer 1989).

During the Upper Visean to Lower Namurian the plate tectonic constellation changed from a divergent to a convergent plate boundary in the Karawanken Mountains. The clastic input of the Hochwipfelflysch comprises turbidites, mudflows and olistostromes. The huge blocks of Devonian limestones in the Seeberg area (south of the Periadriatic lineament) are thought to be enormous olistolites which slid down from an active continental margin to the north (and a passive continental margin to the south). A similar constellation at Namurian time is suggested by Neubauer (1988b). The alkalibasalts of the Eisenkappel area are thought to represent caught-up remnants of seamounts which were incorporated into an imbricate stack during northward underthrusting in Namurian times. The collision between the northern margin of Gondwana to the south and the Middle-Austro-Alpine crystalline basement to the north was completed during the Stephanian. It is suggested that the Periadriatic lineament is the site of two Palaeozoic deep-reaching fault systems, i. e. an Upper Devonian to Lower Carboniferous rift system and an Uper Visean to Namurian thrust zone which have been reactivated as a right-lateral horizontal shear zone during the Alpine orogeny.

Acknowledgements
We thank Prof. Dr. W. Frisch, Prof. Dr. J. Kullmann (Tübingen) and Dr. P. Schlaegel-Blaut (München) for stimulating discussions. The critical, accurate and detailed review was very helpful. The Deutsche Forschungsgemeinschaft supported the investigations with financial assistance which is gratefully acknowledged.

References

Best MG, Christiansen EH, Blank RH (1989) Oligocene caldera complex and calc-alkaline tuffs and lavas of the Indian Peak volcanic field, Nevada. Geol Soc Am Bull 101: 1076–1090

Colins E, Hoschek G, Mostler H (1980) Geologische Entwicklung und Metamorphose im Westabschnitt der Nördlichen Grauwackenzone unter besonderer Berücksichtigung der Metabasite. Mitt Österr Geol Ges 71/72: 343–378

D'Amico C (1979) General picture of Hercynian magmatism in the Alps, Calabria-Peloritani and Sardinia-Corsica. International Geological Correlation Programme Proj 5, Newslett 1, Inst Miner Petrol Univ Padua, pp 33–68

Dickinson WR (1981) Plate tectonic evolution of the southern Cordillera. In: Dickinson WR, Payne WB (eds) Relations of tectonics to ore deposits in the southern Cordillera. Ariz Geol Soc Digest 14: 113–135

Dickinson WR, Beard IS, Brakenridge GR, Erjavec JL, Ferguson RC, Inman KF, Knepp RA, Lindberg FA, Ryberg PT (1983) Provenance of North American Phanerozoic sandstones in relation to tectonic setting. Geol Soc Am Bull 94: 222–235

Ebner F (1990) Circummediterranean Carboniferous Preflysch sedimentation. In: Venturini C, Krainer K (eds) Field Work Shop on the Carboniferous to Permian sequence of the Pramollo – Nassfeld Basin (Carnic Alps). Preprint, Udine, pp 20–32

Ewart A (1979) A review of the mineralogy and chemistry of Tertiary-Recent dacitic, latitic, rhyolitic, and related volcanic rocks. In: Baker F (ed) Trondhjemites, dacites and related rocks. Elsevier, Amsterdam, pp 291–306

Ewart A, Taylor SR, Capp A (1968) Geochemistry of the pantellerites of Mayor Island, New Zealand. Contrib Mineral Petrol 17: 11–140

Flügel H (1990) Das voralpine Basement im Alpin-Mediterranen Belt – Überblick und Problematik. Jahrb Geol Bundesanst 133: 181–221

Flügel H, Neubauer F (1984) Steiermark – Erläuterungen zur geologischen Karte der Steiermark 1:200000. Geol Bundesanst, Wien

Frisch W, Neubauer F (1989) Pre-Alpine terranes and tectonic zoning in the Eastern Alps. Geol Soc Am Spec Pap 230: 91–100

Frisch W, Neubauer F, Satir M (1984) Concepts of the evolution of the Austroalpine basement complex (Eastern Alps) during the Caledonian-Variscan cycle. Geol Rundsch 73: 47–68

Frisch W, Neubauer F, Bröcker M, Brückmann W, Haiss N (1987) Interpretation of geochemical data from the Caledonian basement within the Austroalpine basement complex. In: Flügel H, Sassi FP, Grecula P (eds) Pre-Variscan and Variscan events in the Alpine-Mediterranean mountain belt. Alfa, Bratislava, pp 209–226

Frisch W, Menot RP, Neubauer F, von Raumer J (1990) Correlation and evolution of the Alpine basement. Schweiz Mineral Petrogr Mitt 70: 265–285

Fritz H, Neubauer F (1988) Geodynamic aspects of the Silurian and Early Devonian sedimentation in the Paleozoic of Graz (Eastern Alps). Schweiz Mineral Petrogr Mitt 68: 359–367

Giese U (1988) Altpaläozoischer Vulkanismus am NW-Rand der Gurktaler Decke – Geologie, Petrologie, Geochemie. PhD Thesis, Univ Tübingen, 227 pp

Greene RC (1973) Petrology of the welded tuff of Devine Canyon, southeastern Oregon. US Geol Surv Prof Pap 797: 1–26

Havlicek V, Kriz J, Serpagli E (1987) Upper Ordovician brachiopod assemblages of the Carnic Alps, Middle Carinthia and Sardinia. Boll Soc Paleontol Ital 25: 277–311

Heinisch H (1981) Zum ordovizischen "Porphyroid"-Vulkanismus der Ost- and Südalpen; Stratigraphie, Petrographie, Geochemie. Jahrb Geol Bundesanst 124: 1–109

Heinisch H (1986) Die Geologie der Nördlichen Grauwackenzone zwischen Kitzbühel und Zell am See und ihre Bedeutung für die Rekonstruktion der altpaläozoischen Geodynamik des Ostalpenraumes. Habilitationsschrift, Univ München

Heinisch H (1988) Hinweise auf die Existenz eines passiven Kontinentalrandes im Altpaläozoikum der Nördlichen Grauwackenzone – Ostalpen. Schweiz Mineral Petrogr Mitt 68/3: 407–418

Heinisch H, Schlaegel P (1989) Bericht 1988 über geologische Aufnahmen in der Nördlichen Grauwackenzone auf Blatt 123 Zell am See. Jahrb Geol Bundesanst 132/2: 568–569

Heinisch H, Schmidt K (1976) Zur Kaledonischen Orogenese in den Ostalpen. Geol Rundsch 65: 459–482

Heinisch H, Schmidt K (1982) Zur Genese der Augengneise im Altkristallin der Ostalpen. Neues Jahrb Geol Paläontol Monatsh 1982: 211–239

Heinisch H, Sprenger W, Weddige K (1987) Neue Daten zur Altersstellung der Wildschönauer Schiefer und des Basaltvulkanismus im ostalpinen Paläozoikum der Kitzbüheler Grauwackenzone (Österreich). Jahrb Geol Bundesanst 130/2: 163–173

Höll R (1970) Die Zinnober-Vorkommen im Gebiet der Turracher Höhe (Nock-Gebiet/Österreich) und das Alter der Eisenhutschieferserie. Neues Jahrb Geol Paläontol Monatsh 1970: 201–224

Höll R, Maucher A (1976) The strata-bound ore deposits in the Eastern Alps. In: Wolf KH (ed) Handbook of strata-bound and stratiform ore deposits. Elsevier, Amsterdam, pp 1–36

Hurler H (1972) Geochemie und Petrographie der Metakeratophyre in der südlichen Saualpe. PhD Thesis, Univ Tübingen, 53 pp

Kleinschmidt G, Neugebauer J, Schönenberg R (1975) Gesteinsinhalt und Stratigraphie der Phyllitgruppe in der Saualpe. In: Pilger A, Schönenberg R, Weißenbach N (eds) Geologie der Saualpe. Clausth Geol Abh Sonderb 1: 11–44

Kolmer H (1978) Die Verteilung von Ti, Sr, Y und Zr in spilitischen Gesteinen der Steiermark. Mitt Naturwiss Ver Steiermark 108: 31–43

Laubscher H, Bernoulli D (1982) History and deformation of the Alps. In: Hsü KJ (ed) Mountain building processes. Academic Press, London, pp 169–180

Leat PT, Jackson SE, Thorpe RS, Stillman CJ (1986) Geochemistry of bimodal basalt – subalkaline/peralkaline rhyolite provinces within the southern British Caledonides. J Geol Soc Lond 143: 259–273

Loeschke J (1970) Zur Geologie und Petrographie des Diabaszuges westlich Eisenkappel (Ebriachtal/Karawanken/Österreich). Oberrhein Geol Abh 19: 73–100

Loeschke J (1973) Zur Petrogenese paläozoischer Spilite aus den Ostalpen. Neues Jahrb Mineral Abh 119: 20–56

Loeschke J (1975) Spurenelement-Daten von paläozoischen Spiliten aus den Ostalpen und ihre Bedeutung für geotektonische Interpretationen. Geol Rundsch 64: 62–74

Loeschke J (1977) Kaledonischer eugeosynklinaler Vulkanismus Norwegens und der Ostalpen im Vergleich mit rezentem Vulkanismus unterschiedlicher geotektonischer Positionen: eine Arbeitshypothese. Z Dtsch Geol Ges 128: 185–207

Loeschke J (1988a) Metavulkanite aus der Phyllit-Gruppe der südlichen Saualpe (Kärnten/Österreich) und ihre plattentektonische Bedeutung. Jahrb Geol Bundesanst 131: 301–314

Loeschke J (1988b) Zur Geochemie basischer Vulkanite aus dem Grazer Paläozoikum. Mitt Naturwiss Ver Steiermark 118: 95–105

Loeschke J (1989a) Lower Palaeozoic volcanism of the Eastern Alps and its geodynamic implications. Geol Rundsch 78: 599–616

Loeschke J (1989b) Die paläotektonische Stellung der Vulkanite der Magdalensberg-Serie (Ober-Ordovizium, Gurktaler Decke, Kärnten, Österreich). Carinthia II 179/99: 491–507

Loeschke J, Weber K (1973) Geochemie und Metamorphose paläozoischer Tuffe und Tonschiefer aus den Karawanken (Österreich). Neues Jahrb Geol Paläontol Abh 142: 115–138

Loeschke J, Kretzschmar T, Langer V, Streck M (1990) Zur Geologie der Norischen Decke nordwestlich Kalwang (Eisenerzer Alpen, Steirische Grauwackenzone/Österreich). Jahrb Geol Bundesanst 133/2: 345–363

Mahood GA (1981) Chemical evolution of a Pleistocene rhyolitic center: Sierra La Primavera, Jalisco, Mexico. Contrib Mineral Petrol 77: 129–149

Maynard JB, Valloni R, Ho-Shing Y (1982) Composition of modern deep-sea sands from arc-related basins. In: Legget JK (ed) Trench-forearc geology: sedimentation and tectonics on modern and ancient active plate margins. Blackwell, Oxford, pp 551–561

Mooshammer B (1989) Das südalpine pelagische Eisenkappeler Paläozoikum (Trögener Gruppe) der Ostkarawanken. Carinthia II 179/99: 611–640

Mostler H (1967) Conodonten aus dem tieferen Silur der Kitzbüheler Alpen (Tirol). Ann Naturhist Mus Wien 71: 295–303

Mostler H (1968) Das Silur im Westabschnitt der Nördlichen Grauwackenzone (Tirol und Salzburg). Mitt Ges Geol Bergbaustud Wien 18: 89–150

Mostler H (1970) Struktureller Wandel und Ursachen der Faziesdifferenzierung an der Ordoviz/Silur-Grenze in der Nördlichen Grauwackenzone (Österreich). Festbd Geol Inst 300 J Univ Innsbr, pp 507–522

Mostler H (1984) Erfassung basischer Massengesteine im Raum Mittersill – Zell a. See – Salzburg. Arch Lagerstättenforsch (Ostalpen) 5: 105–115

Nealy LD, Sheridan MF (1989) Post-Laramide volcanic rocks of Arizona and northern Sonora, Mexico, and their inclusions. In: Jenney JP, Reynolds SJ (eds) Geological evolution of Arizona. Ariz Geol Soc Digest 17: 609–647

Neubauer F (1985) Eine präoberordovizische Transgression in der Grauwackenzone (Ostalpen) und ihre Bedeutung für paläozoische Geodynamik und alpidischen Deckenbau. Neues Jahrb Geol Paläontol Monatsh 1985: 46–64

Neubauer F (1988a) Bau und Entwicklungsgeschichte des Rennfeld-Mugel- und des Gleinalm-Kristallins (Ostalpen). Abh Geol Bundesanst 42: 137 pp

Neubauer F (1988b) The Variscan orogeny in the Austroalpine and Southalpine domains of the Eastern Alps. Schweiz Mineral Petrogr Mitt 68: 339–349

Neubauer F, Pistotnik J (1984) Das Altpaläozoikum und Unterkarbon des Gurktaler Deckensystems (Ostalpen) und ihre paläogeographischen Beziehungen. Geol Rundsch 73: 149–174

Neugebauer J (1988) The Variscan plate tectonic evolution: an improved "Iapetus -model". Schweiz Mineral Petrogr Mitt 68: 313–333

Pearce J (1982) Trace element characteristics of lavas from destructive plate boundaries. In: Thorpe RS (ed) Andesites. Wiley, Chichester, pp 525–548

Pearce J, Lippard S, Roberts S (1984) Characteristics and tectonic significance of supra-subduction zone ophiolites. Geol Soc Spec Publ 16: 77–94

Reitz E, Höll R (1989) Unterordovizische Acritarchen aus der Nördlichen Grauwackenzone (Ostalpen). Jahrb Geol Bundesanst 132/4: 761–774

Riehl-Herwirsch G (1970) Zur Altersstellung der Magdalensbergserie – Mittelkärnten, Österreich. Mitt Geol Bergbaustud 19: 195–214

Scharbert S, Schönlaub HP (1980) Das Prävariszikum und Variszikum. In: Oberhauser R (ed) Der geologische Aufbau Österreichs. Springer, Berlin Heidelberg New York Wien, pp 3–20

Schlaegel P (1988) Geotektonische Neuinterpretation des basischen Magmatismus der Nördlichen Grauwackenzone, Ostalpen – ein Überblick. Schweiz Mineral Petrogr Mitt 68: 419–432

Schlaegel-Blaut P (1990) Der basische Magmatismus der Nördlichen Grauwackenzone. Oberostalpines Paläozoikum, Österreich. Abh Geol Bundesanst 43: 149 pp

Schlaegel-Blaut P, Heinisch H (1990) The Devonian basic intraplate volcanism of the northern Greywacke zone, Eastern Alps and its relation to crustal extension. International Geological Correlation Programme Proj 276 Newslett 2, Mém Géol Lausanne 8

Schnepf HG (1989) Vulkanismus und Sedimentologie im Murauer Paläozoikum (Gurktaler Decke/Oberostalpin/Österreich). Arb Inst Geol Paläontol Univ Tübingen A 2: 1–183

Schönlaub HP (1979) Das Paläozoikum in Österreich. Abh Geol Bundesanst 33: 3–124

Schönlaub HP (1982) Die Grauwackenzone in den Eisenerzer Alpen (Österreich). Jahrb Geol Bundesanst 124: 361–423

Tessensohn F (1971) Der Flysch-Trog und seine Randbereiche im Karbon der Karawanken. Neues Jahrb Geol Paläontol Abh 138: 169–220

Tessensohn F (1974) Zur Fazies paläozoischer Kalke in den Karawanken (Karawankenkalke II). Verh Geol Bundesanst 1974: 89–130

Tessensohn F (1983) Eisenkappeler und Seeberger Paläozoikum. In: Geologische Bundesanstalt (ed) Erläuterungen zur Geologischen Karte der Karawanken 1:25 000, Ostteil. Geol Bundesanst, Wien, pp 32–45

The Pre-Alpine Basement of the Lower Austro-Alpine Nappes in the Bernina Massif (Grisons, Switzerland; Valtellina, Italy)

P. SPILLMANN and H. J. BÜCHI

Abstract

Pre-Alpine basement rocks in the Bernina Massif occur in the cores of large recumbent anticlines which define Eo-Alpine nappe structures. Polyphase Alpine overprinting led to a strong recrystallization of the lower parts of the nappe pile. Basement rocks with a pre-Alpine tectonic and metamorphic overprinting were the country rocks of late Variscan intrusive rocks, which show exclusively Alpine overprints. Namurian intrusives (326 Ma) display a transition from calc-alkaline to alkaline igneous activity. Uplift and erosion led to pre-Permian exhumation of calc-alkaline and alkaline intrusive rocks. In the lowermost Permian (285 Ma), rhyolitic extrusive activity was followed by continental clastic sedimentation marking the onset of the Alpine sedimentation cycle. Incipient rifting is indicated by basic igneous activity which preceded the opening of the oceanic Piemonte basin and the formation of a passive continental margin in Jurassic time.

1 Introduction

1.1 Alpine Structures

The structure of the lower Austro-Alpine nappe pile in the Bernina Massif (Grison, Switzerland; Valtellina, Italy), located at the westernmost edge of the Austro-Alpine nappes of the Eastern Alps (Fig. 1), is outlined by large recumbent isoclinal folds, facing west. The cores of the anticlines are formed of pre-Mesozoic basement units. Mesozoic metasediments are squeezed between the basement rocks where they define narrow synclines separating the nappes (Fig. 2).

The lower unit, the Margna nappe, is separated from the upper unit, the Bernina nappe *s. l.*, by a continuous and well-defined syncline composed of Mesozoic sediments.

The internal structure of these units is defined by thin slices of tectonically reduced sequences of Mesozoic metasediments which subdivide the Margna nappe into Fora and Maloja subnappes (Liniger and Guntli 1988), whereas the Bernina nappe *s. l.* is composed of Sella, Corvatsch, Bernina *s. str.* and Stretta nappes (Staub 1914, 1946). West of Poschiavo Valley at Sassalbo mountain an analogous syncline links the

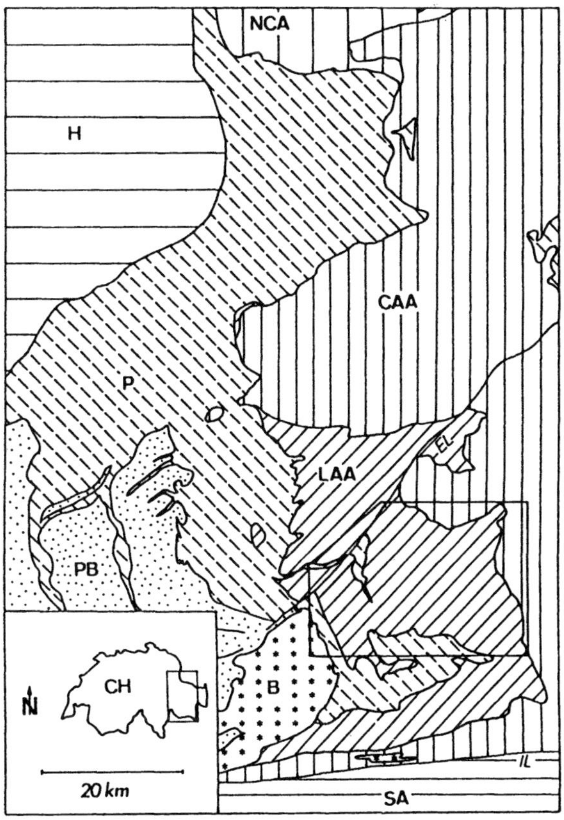

Fig. 1. Tectonic sketch map of the westernmost edge of the Austro-Alpine nappes of the Eastern Alps with location of the area mapped in Fig. 2. *NCA* Northern Calcareous Alps; *CAA* Central Austro-Alpine; *LAA* lower Austro-Alpine; P Penninic Mesozoic to Palaeogene sediments and ophiolites; *PB* Penninic basement rocks; H Helvetic massifs and nappes; *SA* Southern Alps; *B* Tertiary intrusives; *EL* Engadine line; *IL* Insubric line

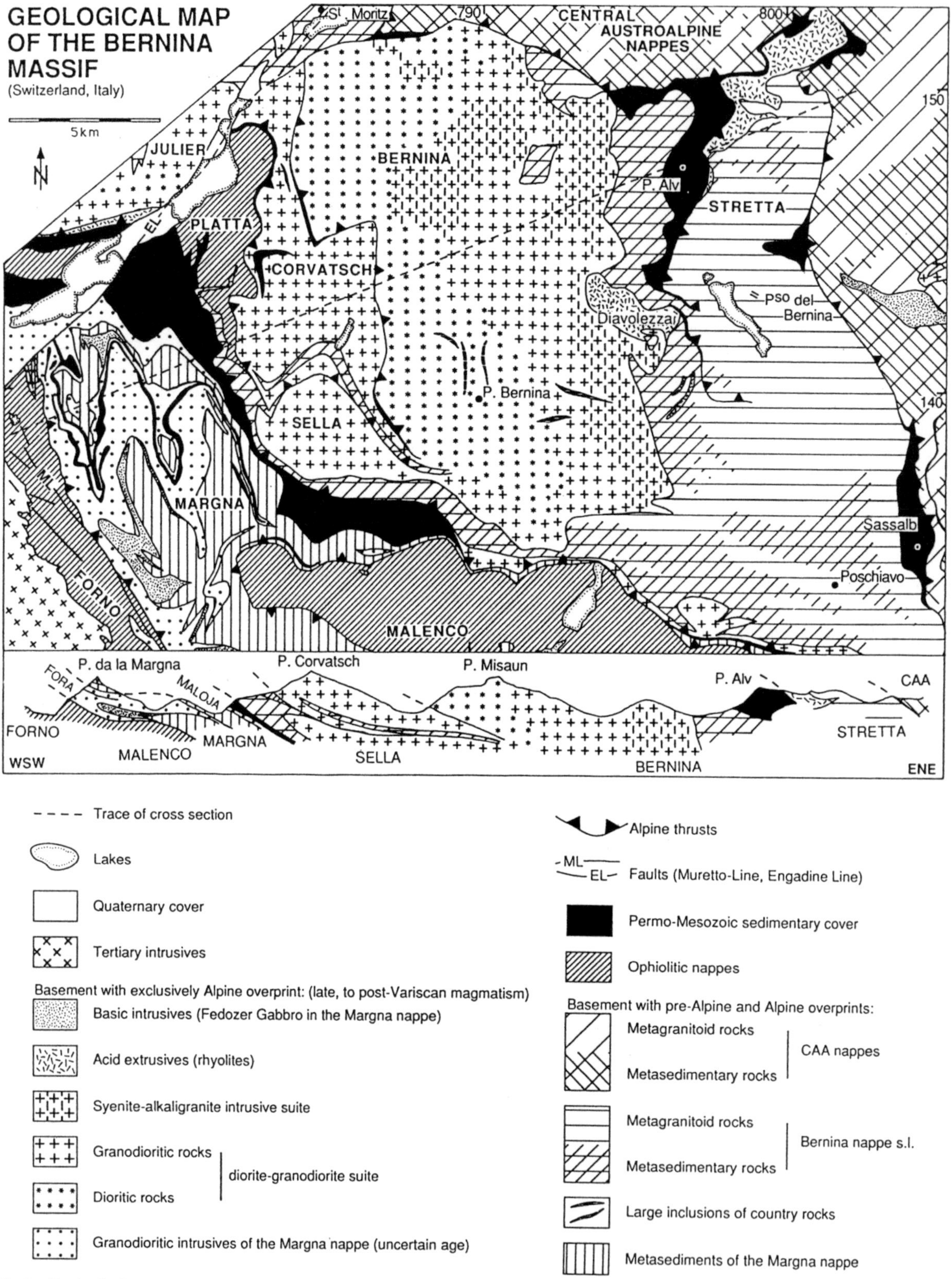

Fig. 2. Geological map of the Bernina Massif. (After Staub 1946; mappings carried out for diploma theses at ETH Zürich and own mapping)

Bernina nappe *s.l.* to the Central Austro-Alpine nappes (Spitz and Dyhrenfurth 1913; Zehnder 1975).

A palinspastic reconstruction of the nappes suggests a continuous sheet of continental crust in pre-Alpine times. A complication is presented by the imbrication of the nappe pile with the underlying ophiolitic units. Parts of the ophiolitic Platta nappe occur between lower Austro-Alpine nappes. This can be explained by Eo-Alpine tectonics or by a complex palaeogeographic configuration.

Tectonic and metamorphic overprints during Eo-Alpine and later orogenic phases led to an intense recrystallization of the Margna nappe and most parts of Sella and Corvatsch nappes. The rocks of the Bernina nappe *s. str.* and the Stretta nappe underwent almost no Alpine tectonic overprinting and only moderate Alpine recrystallization. The Alpine metamorphic grade ranged from middle- to upper-greenschist facies. Therefore most of the pre-Alpine features in the basement are lost for direct observation and interpretation. (All rocks in the area described underwent Alpine metamorphism. As a consequence, all rock names should be preceded by the prefix "meta", but for rocks with weak Alpine overprinting the prefix "meta" will be omitted.)

The nappe pile was later backfolded and overturned to its present vertical position along the Tonale line in the south (Cornelius 1912, 1915; Zyndel 1912). Important lateral and vertical movements occurred along the Engadine line to the northwest (Gyr 1967; Trümpy 1977; Mützenberg 1986; Liniger and Guntli 1988). The Tertiary intrusives of the Bregaglia Masino Massif in the west unconformably cut the nappe structures (Cornelius 1913; Staub 1918).

1.2 *Lithologies in the Basement* (Fig. 2)

Basement lithologies can be classified from their orogenic histories. Mesozoic sediments and ophiolitic units show exclusively Alpine overprinting and are not discussed here. Basement rocks showing no evidence of pre-Alpine overprinting are interpreted as products of late Variscan magmatism. The metamorphic rocks displaying pre-Alpine (Variscan or older) and Alpine overprintings were the country rocks of the late Variscan intrusives. They can be subdivided into metamorphic rocks of igneous or sedimentary origin (after Staub 1946).

The relationship between late Variscan intrusives and their country rocks is clear in the Bernina nappe *s. str.* (Staub 1916; Rageth 1984; Büchi 1987). A large plutonic body, mainly composed of diorites in the west and granodiorites to granites in the east, comprises the highest peaks of the Bernina massif. These intrusives conspicuously display intrusive contacts with pre-Alpine metamorphic gneisses and schists at their eastern limit. The western border of the Bernina nappe *s. str.* is a tectonic contact. The mainly dioritic rocks are thrust onto granodioritic rocks of the Corvatsch nappe, which tectonically overlies the Sella nappe. These thrusts are of Alpine age and consist of thin slivers of Mesozoic sediments, which parallel the predominant schistosity and the mylonite zones observed in part of these nappes. As a consequence of this strong Alpine overprint the western and southern contacts of the intrusives with their country rocks are tectonized and exhibit almost no intrusive features. Primary magmatic textures are only observable in the cores of larger intrusive bodies.

The Margna nappe was nearly completely recrystallized during Alpine metamorphisms. Its basement is composed of a banded metasedimentary sequence, as well as of gabbroic to granitic meta-igneous rocks.

2 Pre-Alpine Evolution of the Margna Basement

2.1 *Metasedimentary Rocks*

Metasedimentary rocks in the basement of the Margna nappe have been mapped as "Fedozserie" by Staub (1946). They are a polymetamorphic sedimentary sequence, including banded metapelitic schists, with marble and calc-silicate horizons and amphibolitic layers. The polymetamorphic nature of the sequence is evidenced by two Alpine regional-metamorphic overprints of upper- and middle-greenschist grade and relicts of mineral assemblages, indicating a pre-Alpine high-grade metamorphism (Guntli and Liniger 1989; Spillmann 1989). The petrography of pre-Alpine mineral assemblages is listed below.

1. Metapelitic schists: the widespread garnet-bearing micaschists include coarse porphyroclastic garnets in addition to fine-grained euhedral garnets of Alpine formation. The clasts are clearly pre-kinematic relative to the first Alpine schistosity and are therefore products of a pre-Alpine metamorphism. Columnar aggregates (2–8 cm), composed of fine-grained muscovite, paragonite, quartz and ilmenite, are interpreted as reaction products of Alpine retrogressive metamorphism of andalusite or staurolite (Schumacher 1975; Guntli und Liniger 1989).

2. Amphibolites: the Alpine metamorphic grade of the metabasic rocks is epidote-amphibolite facies. The retrograde character of this paragenesis is observable in some occurrences of garnet-amphibolites which have reacted to a mineral assemblage including actinolite, albite, clinozoisite and chlorite.
3. Marbles and calc-silicates: coarse-grained calcitic marbles of the "Fedozserie" contain diopside as a widespread constituent. Nodule-shaped antigorite and chlorite crystals in the marbles are interpreted as breakdown products of a paragenesis which included olivine and spinel. Fine-grained dolomitic marbles in the Mesozoic cover have a lower grade mineral assemblage of dolomite, phlogopite, and quartz.

Besides the existence of relict mineral assemblages, which demonstrate a high-grade pre-Alpine metamorphic overprint, there is some evidence for a penetrative pre-Alpine tectonic overprint. Interference patterns of two isoclinal folding phases, the younger of which generated the Alpine schistosity, were observed at some outcrops in the metapelitic schists. Only one isoclinal folding phase is detectable in the metasediments of the Mesozoic cover, which places the first generation of isoclinal folds in the basement to a pre-Alpine tectonic event.

2.2 *Granodioritic to Granitic Meta-Igneous Rocks*

Metagranitoid rocks make up about 50 % of the basement in the western outcrops of the Margna nappe. They are isoclinally folded in metasedimentary rocks of the "Fedozserie" and exhibit a purely Alpine recrystallization. Magmatic textures are overprinted to form leucocratic augengneisses or mylonites. Geochemically, these metagranitoids of granodioritic to granitic composition (Guntli 1987) are similar to late Variscan intrusive rocks of the Bernina nappe *s. l.* Greenschists of dioritic composition appear to represent the intermediate member of this intrusive suite (Sidler 1990).

The existence of a pre-Alpine metamorphic and tectonic overprint cannot be excluded a priori for the metagranitoids. Thus, the age of the protolith of these gneisses is uncertain (Fig. 2). Inclusions of garnet-amphibolites may be the record of pre-Alpine regional or contact metamorphisms.

2.3 *Gabbroic Meta-Igneous Rocks*

Metagabbroic rocks occur in the internal parts of the basement forming the cores of the recumbent anticlines which define the subnappe structure of the Margna nappe. Staub (1917a, 1946) introduced the term "Fedozer Gabbro" for these rocks. Metagabbros with preserved magmatic textures occur northwest of Monte Braccia in the Malenco Valley (south of the area mapped in Fig. 2; Cornelius 1915). In all other outcrops Alpine overprinting has led to the formation of banded or flaser-type Ab-Chl-Act greenschists. A detailed geologic and petrographic description of the "Fedozer Gabbro" is given in Gautschi (1980).

The variation in the magmatic mineral assemblage: plag(an 50)-cpx-opx-(Spl)-(Ol) leads to modal gabbro, gabbronorite and Ol-gabbronorite (after Streckeisen 1976). The transitions among these compositional variations are gradual and define a magmatic layering on outcrop scale. The equigranular, fine- to medium-grained rocks exhibit a weakly defined foliation in hand specimens.

Corona-forming reactions between olivine and plagioclase, and between ilmenite and plagioclase, indicate post-intrusive subsolidus re-equilibration. Mineral chemistry of opx/cpx pairs defines granulite-facies conditions (800–825 °C; >6 kb) (Gautschi 1980) for these reactions. The formation of these corona structures can be explained by isobaric cooling of the plutonic body after its intrusion into deeper continental crust.

Small bodies of metagabbros and metamonzodiorites occur in the Margna basement north of the Tonale line (Tunesi et al. 1991). The magmatic relict minerals: cpx-plag(an 80)-krs-zrn-ap-mag are interpreted as cumulates.

Neither the mineral assemblages nor the tectonic textures of the metagabbroic rocks are suggestive of the pre-Alpine tectono-metamorphic overprint described for the metasedimentary sequence in the Margna basement. An acute angular unconformity occurs between the layering of the metasedimentary sequence and the metagabbro contact.

This leads to the conclusion that the "Fedozer Gabbro" is the product of late Variscan or even Mesozoic basic plutonism (see also Sect. 5).

3 Pre-Alpine Evolution of the Bernina Basement *s. l.*

3.1 Pre-intrusive Basement Rocks

Gneisses and micaschists are widespread in the eastern part of the Bernina nappe *s. l.* They comprise the whole Stretta nappe, the eastern part of the Bernina nappe *s. str.* and continue as highly Alpine-tectonized phyllites in the Sella nappe (Fig. 2).

Based on the mapping of Staub (1946), metamorphic rocks of igneous and of sedimentary origin can be distinguished (Fig. 2); however, distinguishing between these groups can sometimes be difficult. Some alkalifeldspar-rich augengneisses can be interpreted as meta-arkosic rocks by geochemical arguments (Rageth 1982). Other augengneisses show tectonized magmatic contacts to banded metasediments. The metasedimentary rocks in the Bernina nappe *s. l.* differ from the ones in the Margna nappe by the absence of marble horizons and by a more psammitic character. Marbles in banded metapelitic rocks are recorded in the Sella nappe, which is closely linked to the Margna nappe (Hermann 1991; Müntener 1991). A clearly pre-intrusive, penetrative schistosity is preserved in regions of weak or absent Alpine overprinting. Pre-Alpine mineral assemblages including biotite, oligoclase and garnet indicate a pre-Alpine metamorphism of upper greenschist to lower amphibolite facies.

3.2 Late Variscan Igneous Rocks

The description of late Variscan igneous rocks has been mainly concentrated on observations on the Bernina nappe *s. str.* (Rageth 1982, 1984; Büchi 1987). Intrusive rocks in the underlying Corvatsch and Sella nappes (Staub 1915; Müller 1982; Münger 1982; Benning 1990; Hermann 1991) are analogous, but many magmatic features have been lost due to Alpine overprinting. Nevertheless, petrographic observations at minor deformed outcrops, and geochemical data confirm a strong similarity of the igneous rocks in the entire Bernina nappe *s. l.*

Based on petrographic and geochemical arguments four magmatic units have been established:

a. A calc-alkaline diorite-granodiorite suite;
b. an alkaline syenite-alkaligranite suite;
c. an alkaline subvolcanic to volcanic rhyolitic body;
d. basic dykes.

The main characteristics of each unit are as follows:
a. The calc-alkaline diorite-granodiorite suite makes up the main mass of intrusive rocks of the Bernina nappe *s. l.* Both members of the suite have intrusive contacts with Variscan metamorphic and tectonized country rocks. The granodiorites, bearing rounded dioritic inclusions, extend as apophyses into the diorites and are therefore the younger member of the suite (Rageth 1984; Büchi 1987). Minor outcrops of serpentinites near the contact zone to the country rocks are interpreted as altered ultramafic cumulates of pyroxenitic composition (Büchi 1987).

The dioritic member comprises biotite- and hornblende-bearing, coarse-grained gabbros and diorites. They were intruded in an earlier stage of fine-grained gabbros and diorites, which contain, in addition, minor amounts of pyroxenes of pigeonitic composition. The amphibole is magnesia-hornblende, some with cores of cummingtonitic composition. Accessory minerals are apatite and zircon.

The granodioritic member includes tonalitic to granodioritic, coarse-grained rocks of a characteristic blue colour with biotite and magnesia-hornblende as mafic constituents. Apatite, zircon, and orthite appear as primary magmatic accessories.

The presence of biotite and cummingtonite in rocks of gabbroic composition indicates elevated H_2O partial pressure in the melt (Wood and Carmichael 1973; Wones and Gilbert 1982).

The diorite-granodiorite suite is crosscut by aplitic dykes. No aplites have been found in the younger syenite-alkaligranite suite (Büchi 1987). Consequently, the aplitic dykes seem to be related to the diorite-granodiorite suite, which is confirmed by geochemical data as well.
b. The syenitic rocks are the oldest member of the syenite-alkaligranite suite, and occur as minor outcrops at the border of larger alkaligranitic intrusive bodies. The coarse-grained syenites, bearing biotite and hornblende as mafic minerals, grade into a coarse-grained alkaligranite, in which biotite and hornblende occur together with fluorite, zircon and orthite as accessory minerals. Rapakivi structures have been observed at some outcrops. Fine-grained leucogranites, which are almost free of mafic minerals, occur as plugs and veins in the alkaligranites. The gradual contacts between syenites, alkaligranites and leucogranites suggest a cogenetic succession, representing an alkaline intrusive suite.

The syenite-alkaligranite suite was emplaced after the diorite-granodiorite suite. Dioritic inclusions in the alkaligranites have an angular habit, whereas the older granodiorites contain dioritic xenoliths with rounded shapes (Rageth 1984). A syenitic dyke, cross-cutting granodiorites occurs near Pontresina (Büchi 1987). The contacts between alkaligranites and granodiorites are poorly defined. Near their contacts, granodiorites exhibit a slight contact-parallel foliation defined by preferred orientation of mafic constituents (Büchi 1987). The alkaline intrusive suite post-dates the calc-alkaline suite but intruded before the granodioritic member was completely solidified. Assuming emplacement at a shallow crustal level (see Sect. 5), there must be a relatively short time span between the two intrusive phases.

c. Rhyolitic rocks are present as dykes, cross-cutting the two formerly described intrusive suites as well as their pre-Alpine metamorphic country rocks. In the area of Diavolezza, larger masses of rhyolitic extrusives cover alkaligranites and their country rocks unconformably (Staub 1946; Rageth 1984). East of the Bernina Valley, in the core of the anticline forming the Stretta nappe, rhyolites crop out between pre-Alpine metamorphic basement rocks and clastic sediments of Permian to Lower Triassic age, containing clasts of rhyolitic and other basement rocks.

In the area of Diavolezza (Rageth 1984), the stratigraphic contacts of the Permo-Mesozoic sedimentary cover are disturbed by Alpine imbrication.

The rhyolitic rocks are composed of 70 to 90 % of a very fine-grained groundmass in which drop-like quartz, euhedral alkalifeldspar and minor amounts of plagioclase crystals represent phenocrysts in a formerly glassy or microfelsitic matrix. The anhedral quartz grains, up to 3 mm in diameter, have a typical volcanic habit with corrosion bays due to magmatic resorption.

Rageth (1984) described two types of rhyolites in the Diavolezza area: a poorly structured Piz Trovat type and a more stratified Sass Queder type. The fabric of the rhyolites is generally isotropic, but some interlayers exhibit a laminar flow layering covered by a more nodular type.

Flow layering and nodule formation are the only recognizable volcanic features in outcrops. It is assumed that the extrusive events were characterized by extrusion of coherent effusive flows and domes, as well as by explosive pyroclastic flows (ignimbrites). Leucocratic, silica-rich layers (with up to 85 wt. % SiO_2) are interpreted as products of late magmatic silicification due to fumarolic activity.

The rhyolites are considered to be products of an independent magmatic event. Rhyolitic extrusive rocks lie directly over the alkaligranites, which are cross-cut by rhyolitic dykes. A period of uplift and erosion must have taken place between the emplacement of the intrusives and the extrusion of the rhyolites.

d. As products of the youngest magmatic activity, basic dykes cross-cut all the previously described basement rocks. The subvertical, mostly north-south striking dykes, have sharply defined planar contacts with the country rock (Eikenberg 1984; Rageth 1984; Büchi 1987). A 2–4 cm wide chilled margin borders the dykes which have an average width of 2 m. Due to strong Alpine recrystallization, primary magmatic mineral assemblages and textures are almost unrecognizable. Fine-grained alteration products like chlorite, white mica, epidote, and stilpnomelane represent a former ophitic magmatic texture. Columnar amphiboles and scarcer biotite, clinopyroxene, and zoned plagioclase are recognized as phenocrysts.

4 Geochemical Data

Whole rock analyses have been made on late Variscan igneous rocks of the Bernina Massif. Because of complete recrystallization in highly tectonized parts of the nappes, and strong alteration of magmatic minerals even in weakly deformed rocks of the Bernina nappe *s. str.*, these data must be interpreted carefully. Alteration of calcic plagioclase led to the formation of albite, epidote, and white-mica. Most of the amphiboles are strongly altered to chlorite or are overgrown by actinolite. Titanium released during reequilibration of biotite formed sagenitic rutile. The mobility of elements involved in these reactions could have a major effect on the whole rock chemistry.

Nevertheless, the petrographic distinction of two intrusive suites in the Bernina nappe *s. str.* is confirmed by geochemical data (Rageth 1984; Büchi 1987). The AFM plot in Fig. 3 illustrates, a calc-alkaline trend for the diorite-granodiorite intrusive suite and a more alkaline trend for the syenite-alkaligranite intrusive suite. The rhyolites plot within this second group. Acid alkaline rocks were found exclusively in the eastern part of the Bernina nappe *s. l.* Granodioritic rocks with dioritic inclusions make up the main mass of the underlying Corvatsch and Sella nappes (Staub 1915; Müller 1982; Münger 1982; Benning 1990; Hermann 1991). A highly stretched intrusive body (Musella-granite) at the base of the Sella nappe represents, with the exception of aplitic dykes, the most acid member of the calc-alkaline intrusive suite (Hermann 1991).

Strongly deformed meta-igneous rocks of dioritic to granitic composition in the Margna basement follow the calc-alkaline trend of the diorite-granodiorite suite of the Bernina nappe *s. l.* (Guntli 1987; Benning 1990; Sidler 1990; Müntener 1991). Metagabbroic

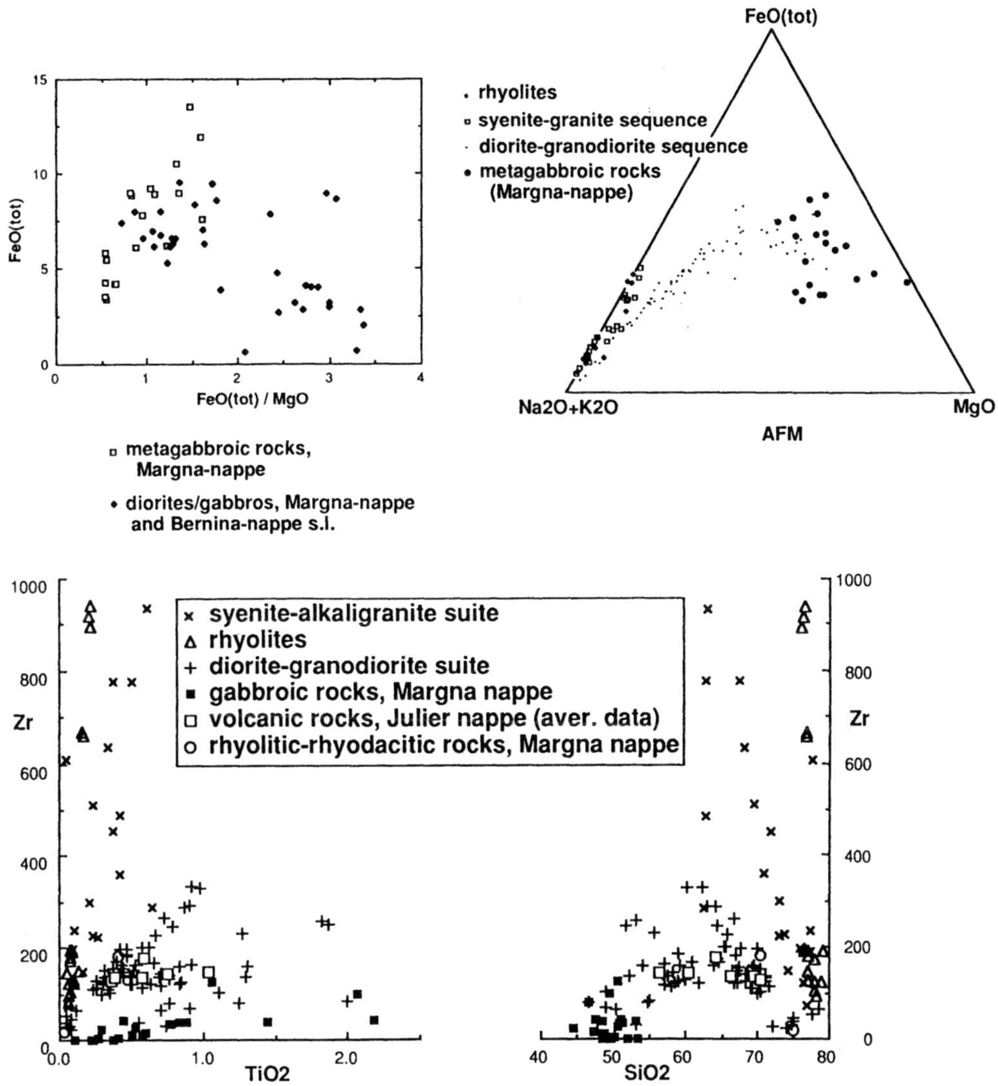

Fig. 3. Graphic representation of geochemical data (references in the text)

rocks (Gautschi 1980; Liniger 1987) in the Margna nappe can be interpreted as the most basic member of this suite. On the other hand, field observations provide no evidence for a genetic link between the metagabbros and the more acid meta-igneous rocks in the Margna basement. The metagabbroic rocks seem to follow a different fractionation trend which is illustrated by the FeO(total) vs FeO(total)/MgO plot in Fig. 3 which shows a positive slope for the metagabbros, and a negative slope for the calc-alkaline rocks. The lack of amphibole or biotite in the metagabbros, as well as their low K-content suggest a transitional or tholeiitic affinity. A tholeiitic evolution was proposed by Gautschi (1980); however, a ferro-gabbroic member could not be recognized. An alkaline affinity was proposed by Tunesi et al. (1991) for the kaersutite-bearing gabbroic rocks in the southern parts of the Margna nappe.

Therefore a subdivision into two magmatic suites can be made for the meta-igneous rocks in the Margna nappe: an alkaline to tholeiitic gabbroic body in the internal part of the Margna basement and a calc-alkaline suite with a strong geochemical affinity to the intrusives of the Bernina nappe *s. l.*

Zr plotted against SiO_2 and TiO_2 (Fig. 3) delineates the two intrusive suites in the Bernina nappe *s. l.* The concentrations of Zr and other LIL-elements like La, Ce or Nd increase in alkaline rocks relative to rocks of the calc-alkaline suite. The rhyolites with high Zr-concentrations are dykes which are not directly linked to the extrusive rhyolites in the Diavolezza area (Münger 1982; Büchi 1987). Additionally, averaged analyses of metavolcanic rocks cropping out in the Julier area are plotted (Mercolli 1989). Their andesitic-basaltic to rhyolitic compositions follow the calc-alkaline trend of the Bernina intrusives. The rhyolitic rocks of the Bernina nappe *s. str.* show no geochemical similarity with these meta-volcanics. Two analyses of rhyolitic and rhyodacitic metavolcanics of the Margna basement are available (Sidler 1990). The rhyodacitic sample plots within the calc-alkaline trend and may belong to the volcanic suite described in the Julier area.

Basic dykes, representing the youngest magmatic activity in the area, have been analyzed (Rageth 1982; Eikenberg 1984; Büchi 1987; Hermann 1991). The total alkali to silica ratio of these dykes suggests that they are alkaline basalts, trachybasalts and trachyandesites (after Miyashiro 1978; Cox et al. 1979). Some of the dykes are of basaltic-andesitic composition. Two analyses, with basanitic compositions, are available from metavolcanic rocks within middle Triassic sediments (Rageth 1982; Eikenberg 1984). Following the discrimination method of Pearce and Cann (1973), the relative concentrations of Ti, Zr, and Y in basanites and basalts indicate a within-plate tectonic setting for this youngest magmatic activity.

5 Age Problems, Discussion

The products of late Variscan igneous activity occur as basement rocks with exclusively Alpine tectonic and metamorphic overprintings. These rocks intruded pre-Alpine metamorphic gneisses and schists. U/Pb-isotope geochemistry on zircon crystals from intrusive rocks of the Bernina nappe *s. str.* defines a discordia line with an upper intercept at 326 ± 13 Ma (von Quadt et al. in prep.), interpreted as the intrusive age, and a lower intercept about 70 Ma indicating an Eo-Alpine overprint. The discordia line was constructed using isotopic ratios derived from zircon crystals from both the calc-alkaline and the alkaline intrusive suite. The alkaline suite is younger, but as pointed out earlier, the time span between the emplacements of the suites seems to be small.

A discordant U/Pb-Zircon age of 285 ± 5 Ma (von Quadt et al. in prep.) was determined for the extrusive rhyolites in the Diavolezza area. The time gap between intrusive and extrusive activity is necessary to explain the situation in the Diavolezza area where alkaligranites are overlain by extrusive rhyolites. The depth of emplacement of the intrusive rocks was estimated to be lower than 3 kb from the Al-content in magmatic amphiboles (Büchi 1987). A late Variscan erosion rate of 0.2–0.6 mm/a can be estimated using the age data. With the exception of this late Variscan uplift there is no evidence for late Variscan tectonic activity affecting igneous rocks. A pre-Triassic tectonic phase was proposed in the Julier area where an inverted succession of calc-alkaline metavolcanic rocks have been intruded by a syntectonic granite and covered by a normal succession of Triassic sediments (Mercolli 1989). This syntectonic development of late Variscan calc-alkaline magmatism, as reported for instance in the Aar Massif (Oberhänsli et al. 1988), was not recognized in the Bernina Massif.

The plutonic rocks of the Bernina nappe *s. str.* grade from calc-alkaline to alkaline. This transition from calc-alkaline to alkaline igneous activity in pre-Permian time supports the idea that late Variscan calc-alkaline igneous activity was not directly related to an active subduction zone, but rather to magma generation in an upper hybridized mantle. Calc-alkaline magmas are produced as long as hydrous melts can be generated from a once-subducted slab of oceanic lithosphere after the collision of continental lithosphere (Bonin 1988).

The extrusion of rhyolites in the lowermost Permian, after a phase of uplift and erosion, initiated a new tectonic setting leading to cratonization of this segment of continental crust.

The Alpine sedimentation cycle started by the deposition of massive to layered conglomerates and sandstones (Schüpbach 1976; Naef 1987). The mostly terrestrial sediments contain rhyolitic and other basement rocks as coarse detrital components. No bio- or chronostratigraphic data are available to date the onset of these apparently strongly heterochronous deposits. The Tethyan transgression is marked by the transition to finer clastic marine deposits and finally to carbonatic Middle and Upper Triassic platform sediments (Naef 1987).

Permo-Triassic deposits are weakly developed in the lower Austro-Alpine nappes. Their thickness increases irregularly towards the Central Austro-Alpine nappes in the east (Naef 1987). This implies the formation of areas with differing subsidence rates. The segment of continental crust now forming the lower Austro-Alpine nappes, was a structural high until Middle Jurassic time (Staub 1917b; Trümpy 1975; Naef 1987; Froitzheim and Eberli 1990). This configuration marked the beginning of a distensive stress regime and led to the formation of intracontinental basins. The onset of this new stress field triggered the extrusion of the rhyolites in the Bernina Massif and in other parts of the inactive Variscan orogenic belt (cf. the "270 Ma critical period" of Bonin 1988).

The alkaline basic dykes, cross-cutting the rhyolites, and therefore emplaced later than 285 Ma have no direct surface equivalent. The basanitic sills in Middle Triassic sediments may be genetically linked to them. No dykes have been found cross-cutting Upper Triassic carbonates. Therefore, these dykes can be attributed to a Permian to Lower Triassic alkaline magmatic activity which is also described in other parts of the European, Apulian and African continental crust (cf. the comprehensive study of Bonin 1988).

An erosional gap seems to be present between rhyolites and the base of Permo-Triassic clastic sediments (Furrer 1985; Naef 1987). The lack of volcanic surface equivalents, associated with the basic dykes, can be explained by complete disaggregation of the basic volcanics during erosion, which are consequently not present in the psephitic fraction of the rhyolite-bearing conglomerates.

As observed already, the metagabbroic rocks of the Margna nappe are not cogenetic with the calc-alkaline intrusive suite, but belong to a late Variscan or Mesozoic intrusive event. A genetic link with the alkaline dykes might be postulated for the gabbroic rocks of alkaline affinity, described in the southern part of the Margna nappe (Tunesi et al. 1991), but hardly for the main gabbroic body, which is clearly subalkaline (Gautschi 1980). A Mesozoic intrusive age can neither be proved nor excluded. Late Variscan intrusive rocks of the calc-alkaline and the alkaline suite are in stratigraphic contact with Mesozoic sediments or even with Early Permian extrusives, thus indicating a pre-Permian exhumation. On the other hand, gabbroic rocks crop out exclusively in internal parts of the Margna basement. Therefore, a Permo-Mesozoic age of the "Fedozer Gabbro" cannot be excluded. A minimum age is given by the onset of Eo-Alpine deformations in the Upper Cretaceous.

Crustal thinning processes, establishing a passive continental margin, are recorded in the Mesozoic sedimentary cover by the onset of turbiditic megacycles caused by fragmentation of the Triassic carbonate platform by intense block faulting. In the Bernina nappe this onset has been dated as Toarcian (Eberli 1988). Synsedimentary Jurassic detachment faults have been traced into basement rocks in the lower Austro-Alpine nappes north of the Engadine line (Froitzheim and Eberli 1990). These low-angle faults are not recorded within the Bernina basement, but Jurassic fault scarp breccias have been found in the Mesozoic cover at Piz Alv (Spitz and Dyhrenfurth 1913; Schüpbach 1970) between Stretta and Bernina nappe *s. str.*, and also in highly deformed metasediments between the Sella and Margna nappes (Spillmann 1989).

Hypothetically, the intrusion of gabbroic rocks in the Margna basement may have been caused by upwelling of mantle lithosphere, evoking the break-up of the overlying continental crust and later formation of ophiolitic crust in the Piemonte oceanic basin. An interpretation, also proposed for the Lower Triassic layered gabbroic suite of the Dent Blanche nappe (Dal Piaz et al. 1977), situated in a comparable tectonic position in the Western Alps (Staub 1917a).

6 Conclusion

The Bernina Massif is built up of lower Austro-Alpine nappes and upper Penninic ophiolitic nappes. The lower Austro-Alpine nappes are palinspastically linked to the Central Austro-Alpine nappes, representing continental crust of the Apulian Promontory. The basement lithologies of the lower Austro-Alpine nappes differ from the Central Austro-Alpine basement by the predominance of late Variscan intrusive rocks. The basement of the Southern Alps exhibit a completely different Alpine tectonic evolution. Consequently, the palaeogeographic relationships to this domain are not well understood. The palinspastic connection to the Penninic and Helvetic basement is made more difficult because the lower Austro-Alpine nappes border to the west as well as to the north a zone where palinspastic continuity is disturbed due to subduction, lateral shifting or even rotation of crustal blocks (Trümpy 1975).

Consequently, this study contributes to the knowledge of the pre-Alpine evolution of a segment of Austro-Alpine continental crust.

In summary, the geological history of the Bernina Massif, deduced from the data presented is sketched as follows:

1. pre-Namurian: single or polyphase tectonic and metamorphic overprinting causes the formation of high grade metamorphic gneisses and schists. Uplift of metamorphic rocks into shallower crustal levels.
2. Namurian and Upper Carboniferous: emplacement of calc-alkaline intrusive rocks followed by an alkaline syenite-alkaligranite intrusive suite. Uplift and erosion leading to exhumation of the intrusive rocks.
3. Permian and Triassic: extrusion of alkaline rhyolitic rocks. Erosion and deposition of clastic sediments into continental troughs followed by marine carbonatic platform sedimentation. Transition

from alkaline to tholeiitic basic igneous activity indicating an incipient rifting phase.
4. Jurassic and Lower Cretaceous: progressive fragmentation of the continental crust. Formation of a passive continental margin, bordering the oceanic Piemonte Basin.
5. Upper Cretaceous and Cenozoic: thrusting of the Austro-Alpine nappes over the Penninic ophiolitic units. Polyphase Alpine deformation and metamorphism.

Acknowledgements

The authors would like to thank V. Trommsdorff, M. Liniger, and N. Froitzheim for their critical reviews of the manuscript. Special thanks are extended to J. Connolly who made the English somewhat more English.

References

Benning L (1990) Metamorphose und Struktur im Malenco-Serpentinit und in der Margna- und Selladecke, zwischen Pass d'Ur und Pizzo Scalino. Val Malenco (Provincia di Sondrio), Teil 2. Diploma Thesis, ETH Zürich

Bonin B (1988) From orogenic to anorogenic environments: evidence from associated magmatic episodes. Schweiz Mineral Petrogr Mitt 68: 301–311

Büchi HJ (1987) Geologie und Petrographie der Bernina IX. Das Gebiet zwischen Pontresina und Morteratschgletscher. Diploma Thesis, Univ Zürich

Cornelius HP (1912) Über die Rhätische Decke im Oberengadin und den südlich benachbarten Gebieten. Centralbl Mineral Geol Paläontol 20: 632–638

Cornelius HP (1913) Geologische Beobachtungen im Gebiete des Fornogletschers (Engadin). Centralbl Mineral Geol Paläontol 8: 246–252

Cornelius HP (1915) Zur Kenntnis der Wurzelregion im unteren Veltlin. Neues Jahrb Mineral Geol Paläontol 40: 253–363

Cox KG, Bell JD, Pankhurst RJ (1979) The interpretation of igneous rocks. Allen and Unwin, London, 450 pp

Dal Piaz GV, De Vecchi Gp, Hunziker J (1977) The Austroalpine layered gabbros of the Matterhorn and Mt. Collon-Dents de Bertol. Schweiz Mineral Petrogr Mitt 57: 59–88

Eberli GP (1988) The evolution of the southern continental margin of the Jurassic Tethys Ocean as recorded in the Allgäu Formation of the Austroalpine nappes of Graubünden (Switzerland). Eclogae Geol Helv 88: 175–214

Eikenberg J (1984) Geologie und Petrographie der Bernina VIII. Diploma Thesis, ETH Zürich

Froitzheim N, Eberli GP (1990) Extensional detachment faulting in the evolution of a Tethys passive continental margin, Eastern Alps, Switzerland. Geol Soc Am Bull 102: 1297–1308

Furrer H (ed) (1985) Field workshop on Triassic and Jurassic sediments in the Eastern Alps of Switzerland. Mitt Geol Inst ETH/Univ Zürich NR 248

Gautschi A (1980) Metamorphose und Geochemie der basischen Gesteine des Bergeller Ostrandes. Dissertation, ETH Zürich

Guntli P (1987) Geologie und Petrographie der Margna südlich des Silsersees, Oberengadin. Diploma Thesis, ETH Zürich

Guntli P, Liniger M (1989) Metamorphose in der Margna-Decke im Bereich Piz da la Margna und Piz Fedoz (Oberengadin). Schweiz Mineral Petrog Mitt 69: 289–301

Gyr T (1967) Geologische und petrographische Untersuchungen am Ostrande des Bergeller Massifs. Dissertation ETH Zürich

Hermann J (1991) Geologische Untersuchungen südlich der Berninagruppe, Teil 1. Diploma Thesis, ETH Zürich

Liniger M (1987) Geologie und Petrographie der Margna südlich des Silsersees, Oberengadin. Diploma Thesis, ETH Zürich

Liniger M, Guntli P (1988) Bau und Geschichte des zentralen Teils der Margna-Decke. Schweiz Mineral Petrogr Mitt 68: 41–54

Mercolli I (1989) The volcanic suite of the Julier area (Grisons). Part 1: Volcanic and tectonic evolution. Schweiz Mineral Petrogr Mitt 69: 423–433

Miyashiro A (1978) Nature of alkalic volcanic rock series. Contrib Mineral Petrol 66: 91–104

Müller D (1982) Geologie und Petrographie der Bernina I. Kristallistion, Deformation und Geochemie des südlichen Corvatschgranits im Oberengadin, GR. Diploma Thesis, Univ Zürich

Münger R (1982) Geologie und Petrographie der Bernina II. Kristallisation, Deformation und Geochemie im Nordteil der Corvatschdecke. Diploma Thesis, ETH Zürich

Müntener O (1991) Geologische Untersuchungen südlich der Berninagruppe, Teil 2. Diploma Thesis, ETH Zürich

Mützenberg St (1986) Ergebnisse geologischer Studien südlich Maloja. Schweiz Mineral Petrogr Mitt 66: 466–472

Naef MH (1987) Ein Beitrag zur Stratigraphie der Trias-Serien im Unterostalpin Graubündens. Dissertation ETH Zürich

Oberhänsli R, Schenker F, Mercolli I (1988) Indications of Variscan nappe tectonics in the Aar Massif. Schweiz Mineral Petrogr Mitt 68: 509–520

Pearce JA, Cann JR (1973) Tectonic setting of basic volcanic rocks determined using trace element analyses. Earth Planet Sci Lett 19: 290–300

Rageth R (1982) Geologie und Petrographie der Bernina. Teil III. Diavolezza. Diploma Thesis, ETH Zürich

Rageth R (1984) Intrusiva und Extrusiva der Berninadecke zwischen Morteratsch und Berninapass (Graubünden). Schweiz Mineral Petrogr Mitt 64: 83–109

Schüpbach MA (1970) Der Sedimentzug Piz Alv – Val da Fain (Unterostalpine Decken, Berninapass) Vierteljahrsschr Naturforsch Ges Zür 115: 231–237

Schüpbach MA (1976) Tektonik im Gebiet des Berninapasses und der Val Chamuera (Engadin). Eclogae Geol Helv 69: 63–73

Schumacher T (1975) Geologisch-petrographische Untersuchungen des Malencoserpentinits und der Margnagesteine bei Chiareggio. Diploma Thesis, Univ ETH Zürich

Sidler D (1990) Strukturen und Metamorphose in der Margna- und Sella-Decke und im Malenco-Serpentinit zwischen Pass d'Ur und Pizzo Scalino (Prov di Sondrio, Italien), Teil 1. Diploma Thesis, Univ Zürich

Spillmann P (1989) Struktur und Metamorphose der Margna-Decke im obersten Valmalenco. Schweiz Mineral Petrogr Mitt 69: 147–150

Spitz A, Dyhrenfurth G (1913) Die Triaszonen am Berninapass (Piz Alv) und im östlichen Puschlav (Sassalbo). Kurze Mitteilung. Verh k k Geol Reichsanst 16: 403–415

Staub R (1914) Zur Tektonik des Berninagebirges. Vierteljahrsschr Naturforsch Ges Zür 58: 359–370

Staub R (1915) Petrographische Studien im westlichen Berninagebirge. Vierteljahrsschr Naturforsch Ges Zür 60: 55–336

Staub R (1916) Tektonische Studien im östlichen Berninagebirge. Vierteljahrsschr Naturforsch Ges Zür 61: 324–404

Staub R (1917a) Das Aequivalent der Dentblanchedecke in Bünden. Vierteljahrsschr Naturforsch Ges Zür 62: 349–370

Staub R (1917b) Über Faziesverteilung und Orogenese in den südöstlichen Schweizeralpen. Beitr Geol Karte Schweiz NF 46/1

Staub R (1918) Geologische Beobachtungen am Bergellermassiv. Vierteljahrsschr Naturforsch Zür 63: 1–18

Staub R (1946) Geologische Karte der Berninagruppe, 1: 50000. Spez.-Karte 118. Schweiz Geol Komm, Basel

Streckeisen A (1976) To each plutonic rock its proper name. Earth Sci Rev 12: 1–33

Trümpy R (1975) Penninic-Austroalpine boundary in the Swiss Alps: a presumed former continental margin and its problems. Am J Sci 275: 209–238

Trümpy R (1977) The Engadine line: a sinistral wrench fault of the Central Alps. Mem Geol Soc China 2: 1–12

Tunesi A, Mariconti M, Meciani L (1991) Rifting and Eo-Alpine imprints in the Margna nappe, Central Alps. Terra Abstr 3/1: 218

von Quadt A, Büchi HJ, Grünenfelder M U-Pb Zircon ages from different igneous rocks of the Bernina nappe system (Grisons, Switzerland) (in prep.)

Wones DR, Gilbert MC (1982) Amphiboles in the igneous environment. In: Veblen DR, Ribbe PH (eds) Reviews in mineralogy, Mineral Soc Am, Washington DC, pp 355–390

Wood BJ, Carmichael ISE (1973) P_{total}, P_{H2O} and the occurrence of cummingtonite in volcanic rocks. Contrib Mineral Petrol 40: 149–158

Zehnder K (1975) Zur Geologie des Sassalbo (Val Poschiavo, Graubünden). Vierteljahrsschr Naturforsch Ges Zür 120: 189–194

Zyndel F (1912) Über den Gebirgsbau Mittelbündens. Beitr Geol Karte Schweiz NF 41: 1–40

Evolution of the Silvretta Nappe

M. MAGGETTI and M. FLISCH

Abstract

The Silvretta basement nappe had a complex Precambrian and Palaeozoic history. During the Proterozoic, sedimentation of mostly greywackes occurred in a back-arc situation, probably contemporaneously with voluminous intrusions/extrusions of MOR basaltic magmas. A first metamorphic overprint in amphibolite facies conditions preceded a high pressure event ("eclogitic" peak conditions 550–650 °C, min. 14–16 kb). In Late Proterozoic time, intrusions of basic to intermediate magmas (older orthogneisses) occurred. A Rb-Sr isochron of 895 + 130/−140 Ma is interpreted as crystallization age. The geochemical composition indicates, as for the eclogitic event, a collisional regime for the generation of these magmas. A second HP event occurred before the intrusion/extrusion of acid magmas (younger orthogneisses). Their Rb-Sr isochron of 451 ± 2 Ma is interpreted as an Ordovician crystallization age. This magmatism occurred probably during an Early Palaeozoic continental rifting causing their generation and emplacement. The Variscan deformation and metamorphism was a polyphasic amphibolite facies event. The metamorphic peak (600–650 °C, 5.5–7.5 kb) was dated with Rb-Sr small scale isochron at 370 ± 17 Ma. An uplift caused the crystallization of andalusite in the rocks and in veins. The following compressional event (assumed 450–550 °C, ~4 kb) led to the formation of kilometre-long folds. Rapid rise of the Silvretta crystalline complex and extensional tectonics in the Pangea resulted in the intrusion of Late Carboniferous (310–280 Ma?) rift tholeiites, which are preserved as dykes. Continuous extensional tectonics resulted in the formation of horst and graben structures, filled with Permian acid volcanic detritic material and ignimbritic flows. The Silvretta nappe then began to subside and terrigeneous material as well as dolomites were deposited in Triassic-Jurassic time. The Alpine metamorphism (110–90 Ma) reached anchizonal facies conditions in the west and greenschist facies conditions in the east of the Silvretta. Then the Silvretta was uplifted differentially and detached from the basement. Pseudotachylites (~75 Ma old) formed at this time and document fossil hypocentres of earthquakes. The overthrust of the Silvretta nappe onto the Pennine foreland occurred about 60–55 Ma ago.

1 Introduction

The Upper Austro-Alpine Silvretta nappe comprises approx. 1600 km^2 of crystalline rocks which, in the west, are covered by autochthonous Permian volcanic and Triassic sedimentary rocks (Fig. 1). The crystalline rocks in the Silvretta are about one third metabasites (mainly eclogites, amphibolites, metagabbros, metadiorites), one third metagranitoids (augengneisses and/or fine-grained leucocratic gneisses) and one third paragneisses (biotite-plagioclase gneisses, micaschists, feldspar knotted gneisses). Migmatites, marbles, quartzites and calc-silicate rocks are present in subordinate amounts.

Kilometre-long folds ("Schlingen") are the characteristic tectonic element. They are found primarily in the southwest of the Silvretta nappe (Fig. 2) and fold the main foliation (S2) without forming a new one. Maggetti et al. (1990a) report on the history of the scientific exploration of the Silvretta and give an exhaustive literature. For the problem of the tectonic position of the phyllite gneiss zone and the quartzphyllite zone see Amann (1985) and Rockenschaub (1990). The Silvretta nappe is the result of several complex sedimentary, magmatic and tectono-metamorphic events (Table 1), which is discussed in greater detail later.

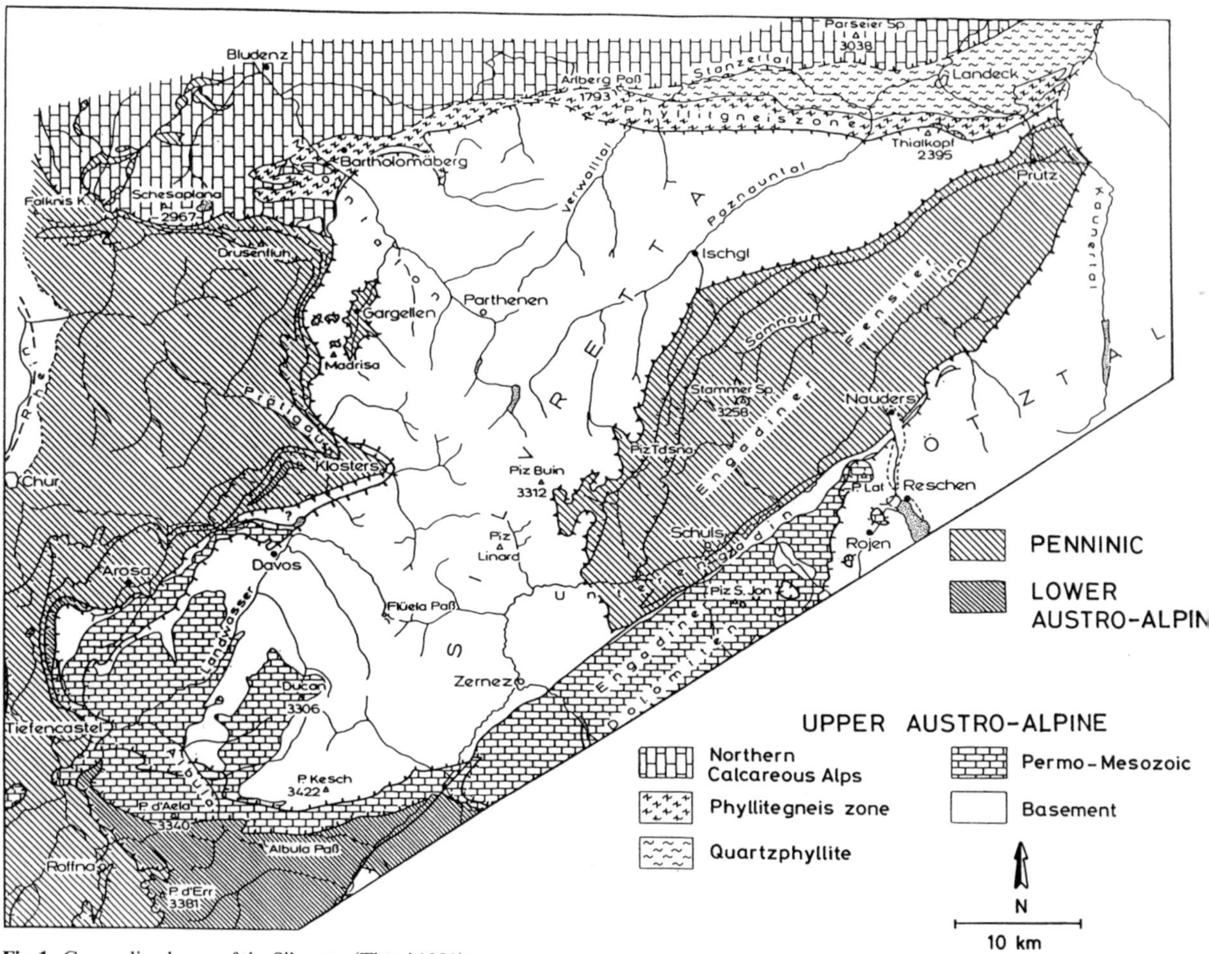

Fig. 1. Generalized map of the Silvretta (Thöni 1981)

2 Synopsis of the Evolution of the Silvretta Nappe

2.1 *Event 1: Precambrian Sedimentation and Attendant Magmatism*

The oldest rocks are, without doubt, the protoliths of the paraseries (principally biotite-plagioclase gneisses and alumosilicate-micaschists) and the eclogites, respectively amphibolites. This series was named the gneiss-amphibolite association by Frisch et al. (1984) and considered to be pre-Ordivician.

Earlier work interprets the *paragneisses* as an extensive series of primordial clastic rocks (greywackes to shales) with sporadic intercalations of carbonates (now marbles) and marls (now calc-silicate rocks). Alternatively, Bröcker (1985), Frisch and Neubauer (1984) and Frisch et al. (1987) advocate a volcanic, or volcano-detrital origin of the biotite-plagioclase gneisses, for which geochemical evidence indicates rhyodacitic to dacitic magma sources. The calc-alkaline chemical characteristics of these rocks and their association with tholeiitic amphibolites is considered by Frisch et al. (1984) to indicate an island arc or active continental margin environment. Grauert (1969) suggests a Cambrian sedimentation age, but Pasteels (1964) and Grauert and Arnold (1968) indicate the detrital zircons origin from a Precambrian parent rock (U-Pb age of the zircons: at least 1500 Ma). According to Frisch et al. (1984, 1987), the sedimentation age could even be Proterozoic. In the light of the newest dates (see event 3), a Precambrian sedimentation age appears likely.

At the contacts of the more extensive *amphibolite* and paragneiss areas, the two rock types are closely interbedded. This sequence is interpreted as a primary structure formed either contemporaneously by sedimentation or insignificantly younger intrusions/extrusions of basic magmas. Escher (1921), Streckeisen (1928) and Spaenhauer (1932) interpret the large amphibolite areas as orthogenetic deriva-

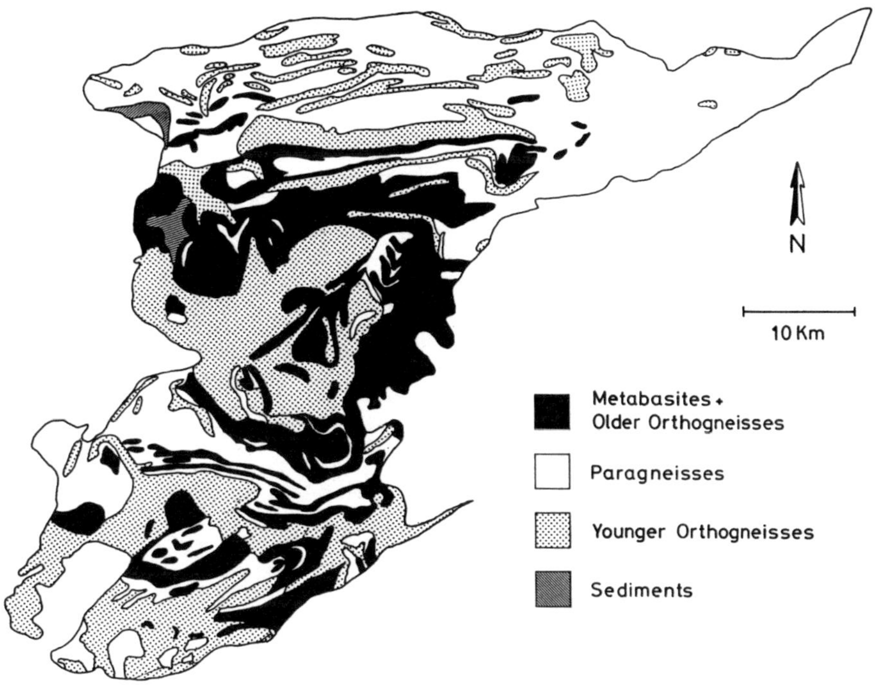

Fig. 2. Simplified geological map of the Silvretta basement rocks (after Spicher 1972). The metabasites comprise eclogites and amphibolites as well as basic and intermediate "older orthogneisses". The sediments belong to the upper Austro-Alpine or the Penninic realm. The post-Permian sedimentary cover of the Silvretta nappe is not shown

tives. On the other hand, Streckeisen et al. (1966) postulated a paragenetic origin of the amphibolites with carbonate inclusions and of the amphibolites closely interbedded with paragneisses. The last are interpreted, following Blumenthal (1926), as sedimentary tuffaceous deposits. Detailed geochemical studies by Frisch and Neubauer (1984), Maggetti and Galetti (1984), Frisch et al. (1987), Maggetti et al. (1987) and Maggetti and Galetti (1988) document the orthogenetic, tholeiitic nature and provide no evidence of a sedimentary origin. If trace and minor element discriminant diagrams for recent basalts can be successfully applied to these Precambrian rocks, then the protoliths could be MORBs (Fig. 3). The association of detrital sediments with such MORBs poses problems, but would be compatible with an incipient rifting of a continent.

Another problem is the formation of the banded amphibolites. According to Grubenmann (1904) the light bands (1–20 cm width) are injections of aplitic material. Spaenhauer (1932) interprets the light and dark bands as a sedimentary interbedding or as metamorphic differentiation strata. According to Bröcker (1985) and Frisch et al. (1987), the light strata are andesitic to dacitic and the dark strata have tholeiitic compositions. The banded amphibolites are therefore the product of a metamorphic recrystallization of an earlier, bimodal volcanic sequence of andesitic-dacitic and basaltic tuffs. The corresponding geotectonic model, just as for the paragneisses, would be an active continental margin or an island arc. Recent analyses by Bollin (1989), however, revealed granitic chemistry for some light bands, and so Grubenmann's (1904) interpretation remains possible. It has not yet been determined whether these aplites should be assigned to the "older" or the "younger" orthogneisses.

2.2 Event 2: Precambrian Deformation and Metamorphism

It was clear to earlier investigators that prior to the Variscan metamorphism, an even earlier metamorphic event affected the Silvretta nappe (e.g. Grauert 1969; Hoernes 1971). Relicts of this first metamorphism are inclusions found in some varieties of the "older orthogneisses" (see event 3). These relicts are amphibolites in the metagabbros of the Val Sarsura (Thierrin 1982, 1983); amphibolites, retrograded eclogites, calcsilicate rocks and paragneisses in the meta-quartzdiorites of the Val Gonda (Wenk 1934; Grauert 1969), nodular schists ("Fleckengesteine") in the Mönchalp-type metagranites (Streckeisen 1928; Streckeisen et al. 1966; von der Crone 1989a,b); as well as older generations of garnets and alumosilicates in metapelites (Hoernes 1971; Flisch 1987). These rocks are present in the gneissic terranes and found as angular-edged fragments in the rocks that preserve magmatic textures. Consequently, meta-

Table 1. Evolution scheme of the Silvretta

Event	Age	Sedimentation	Magmatism	Tectonism	Metamorphism
1	Precambrian	Clastic (greywackes, pelites)	Tholeiitic MORBs and/or tholeiitic intrusives and/or andesitic/basaltic pyroclastites		
2	Precambrian $\geqslant$900 Ma			D_{1a}(polyphasic) $\rightarrow S_{1a}$	Epidote-amphibolite facies, 4–5 kb, 560–620 °C Migmatites?
				D_{1b}(polyphasic) $\rightarrow S_{1b}$	HP I(eclogites) 14–16 kb min, 550–650 °C Amphibolite facies (symplectites) 12–13 kb, 550–650 °C for the incipient decompression
3	Precambrian 900 Ma		Gabbroic, dioritic granodioritic. granitic intrusives (Older Orthogneisses)		
4	Precambrian(?)				HP I
5	Ordovician 451 ± 2 Ma		Intermediate to acid extrusives or intrusives (Younger Orthogneisses)		
6	Variscan 370–350 Ma			$D_2 \rightarrow S_2$	Amphibolite facies 5.5–7.5 kb, 600–680 °C Post-D_2 andalusite sillimanite in metapelites Post-D_2 andalusite quartz-rods 2–3 kb, 550–650 °C
	340–310 Ma			D_3 "Schlingen" $\rightarrow S_3$ (locally)	~ 4 kb, 450–550 °C
				D_4 Shear bands	P=?, ~275 ± 50 °C
7	Permo-Carboniferous 310–280 Ma		Tholeiitic dykes (diabases)	Rapid uplift + strong erosion + rifting	
	~280 Ma	Clastic (greywackes, conglomerates, sands)	Rhyolitic pyroclastites + ignimbrites		
	Triassic-Jurassic 245–?	Carbonates (dolomites)			
8	Alpine 110–90 Ma and later			3 km overburden D_5	Anchizone-epizone 2–4 kb, 225–500 °C
	~75 Ma			Silvretta detachment pseudotachylites	
	approximately 60–55 Ma			Silvretta over the Penninic elements of the Engadine window	
	60–35 Ma			Upwelling of the Engadine window	
	35–3/2 Ma			Homogeneous uplift	P ~1 kb, 100 °C
	< 3–2 Ma			Tilting to the east	

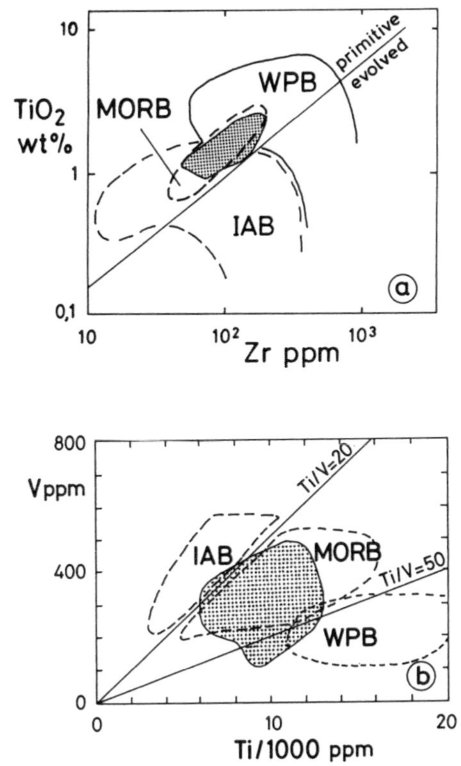

Fig. 3. **a** Geotectonic position of the protoliths of the eclogites and amphibolites from the Silvretta. TiO_2-Zr diagram (Pearce and Cann 1973). The 106 "filtered" (Pearce 1983) analyses of amphibolites and eclogites *(shaded area)* fall within the MORB field. *IAB* Island arc basalts; *MORB* mid-ocean ridge basalts; *WPB* within-plate basalts. **b** V-Ti/1000 plot (Shervais 1982; Hodder 1985). 40 eclogite and 63 amphibolite analyses *(shaded area)* representing former basaltic liquids (Pearce 1983) fall mainly within the MORB field. (Chemical analyses from Maggetti and Galetti 1984, 1988; Bröcker 1985; Michael 1985a; Maggetti et al. 1987; Hayoz 1988; Bollin 1989 and unpubl. data)

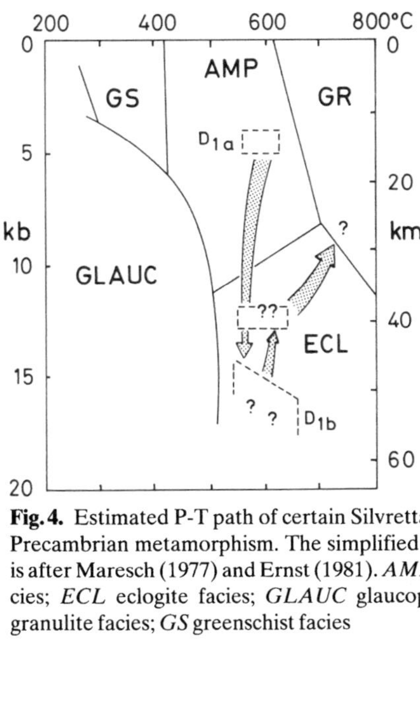

Fig. 4. Estimated P-T path of certain Silvretta rocks during the Precambrian metamorphism. The simplified petrogenetic grid is after Maresch (1977) and Ernst (1981). *AMP* Amphibolite facies; *ECL* eclogite facies; *GLAUC* glaucophane facies; *GR* granulite facies; *GS* greenschist facies

morphism must be older than the intrusions. The fine-grained marginal areas of the quartzdioritic hornblende-gneisses and the sharp contacts described by Grauert (1969) further supports this hypothesis.

The inclusions are clearly foliated (S1), even in the "older orthogneisses" with magmatic texture and may still show fold structures (von der Crone 1989a,b). This indicates that no contact-metamorphic overprint occurred. The paragneiss inclusions are not suitable for T-P estimates, but indicate at least amphibolite facies conditions based on their mineralogy. According to von der Crone (1989a,b), the large staurolites and andalusites in the nodular schists grew along the first foliation (S1a) at about 560–620 °C and 4–5 kb (Fig. 4). The subsequent breakdown of both porphyroblast types to fine-grained garnet, biotite, quartz and kyanite suggests conditions of over 700 °C and more than 8.5 kb.

In the eclogites of the Silvretta nappe, too, mineral relicts of earlier amphibolite-facies metamorphism have been preserved in the cores of the garnets (Maggetti and Galetti 1988). This first tectono-metamorphic event was followed by an eclogitic episode, which has been studied in detail by Thierrin (1982, 1983), Michael (1985a,b) and Maggetti and Galetti (1988). The polyphase structural overprint (D1b), that is the recrystallization, climaxed at temperatures of 550–650 °C (garnet-omphacite geothermometer) and minimum pressures of 14–16 kb (Jd-content of omphacites). Decompression lead to the multistage decomposition of the omphacites into a symplectitic aggregate of plagioclase + clinopyroxene or plagioclase + amphibole. For the first stage of these transformations Maggetti and Galetti (1988) assume 550–650 °C and calculate minimum pressures of 12–13 kb (Jd-content of clinopyroxene).

One of the paragneiss types is the feldspar knotted gneiss (Hammer 1914; Escher 1921; Streckeisen 1928; Spaenhauer 1932; Frapolli 1975; Krähenbühl 1984; Flisch 1987). In it the plagioclase porphyroblasts contain large quantities of vermicular quartz, characteristic atoll garnet, rutile and biotite. According to Streckeisen (1928) and Spaenhauer (1932) the growth of these plagioclases may be attributed to contact-metamorphic, pneumatolytic processes that occurred during the intrusion of the magmas now represented by the "younger orthogneisses". Frapolli (1975), on the other hand, dates them to the time of the formation of the Variscan andalusites (post-D2, pre-D3, see event 5). Amann (1985) found no spatial relationship between these gneisses and the younger orthogneisses. According to this author, they could have been formed during the Variscan main deformation. Krähenbühl (1984) and Flisch (1987) assume a pre-S2 age (see event 6), at least for a large number of the plagioclase-blasts, which means they would also be Precambrian. We correlate the atoll-garnet and the rutile with the high-pressure event.

The age of the first amphibolite facies (D1a) overprint is not well constrained. Earlier authors held it to be "Caledonian". A minimum age has, however, now been given by the Rb-Sr ages of the "older orthogneisses" (approx. 900 Ma), resulting in a Precambrian, probably Proterozoic age.

Dating of the high pressure event (D1b) is also problematic. Hoernes (1971) gave a "Caledonian" (Ordovician) age. In this present work it is assigned to the Precambrian metamorphism. The reasons are: (1) the presence of rocks that are interpreted as retrograded eclogites (symplectitic garnet-amphibolites) in the massive Mönchalp-type metagranites (Streckeisen et al. 1966; von der Crone 1989a) and (2) the assumption of similar intrusion ages of this granite and the "older orthogneisses". At present, no absolute dating is available to support this argument. Field observations (von der Crone 1989a) lead to the conclusion that basic magmas (= metagabbros of the "older orthogneisses") co-existed with acid magmas (= metagranites of the Mönchalp type), suggesting that this granite be indirectly assigned to the "older orthogneisses" group.

The question posed by the high-pressure overprint of the "older orthogneisses" (see event 4) is whether there are really two high-pressure events occurring at two entirely different times (one before and one after the intrusion of the "older orthogneisses" magmas) or whether all of these high pressure relicts formed during a single high-pressure event?

2.3 *Event 3: Intrusion of the Magmas of the "Older Orthogneisses"*

Grauert (1969) was the first to assume a pre-Ordorician intrusion age for a group of intermediary to granitic gneisses; this was on the basis of the isotopic composition and comparison with similar rocks in the Oetztal and the Gotthard massif. Later research has shown that this group contains metamorphic ultramafic and gabbroid rocks (Spaenhauer 1932; Bröcker 1985; Hayoz 1988; von der Crone 1989a; Maggetti et al. 1990a,b); dioritic to tonalitic rocks (Spaenhauer 1932; Wenk 1934; Thierrin 1982, 1983; Bröcker 1985; Hayoz 1988; Bollin 1989; Maggetti et al. 1990a,b); and granodioritic to granitic rocks (Streckeisen 1928; Wenk 1934; Grauert 1969; von der Crone 1989a). The latter also includes augengneisses of the Mönchalp-type and the "Rotbüel granite" (Flisch 1981, 1987), which are of importance in understanding the evolution of the Silvretta crystalline rocks.

Reference has already been made to the fact that the ultramafic, gabbroid and granitic types commonly still possess relict magmatic, massive structures. The primary magmatic paragenesis is, however, only partially preserved in these structures. In the cumulative harzburgitic ultramafites, it is the association of olivine + orthopyroxene + clinopyroxene + ilmenite + apatite + biotite. The plagioclase is completely saussuritized. The massive parts of Mönchalp granite consist of large potash feldspars, bluish quartz, biotite and grey-green "pinite" after cordierite. The plagioclases have been almost completely recrystallized to a finer grain size.

Rb-Sr total rock analysis of 14 "older orthogneisses" from gabbroic to tonalitic composition give an isochron of 895 (+130/−140) Ma (Fig. 5, Table 2) which is interpreted as crystallization age. A Precambrian, Proterozoic intrusion age is therefore plausible.

There has been a great deal of speculation about the genesis of these magmas. According to Streckeisen (1928), the protoliths of the Mönchalp granitic magmas were formed by anatexis of paragneisses. Wenk (1934), on the other hand, places them in the Flüela granitic group (= event 5) and attributes their specific character to the assimilation of metasediments (in situ contamination). Grauert (1969) interprets them as diatexites, Flisch (1987) as anatexites (S-type granites). Models have also been presented concerning the genesis of the compositionally intermediate orthogneisses of dioritic (now hornblende-gneisses) and tonalitic (now biotite-gneisses) signature. For example, assimilation of gneisses (Spaenhauer 1932) and/or amphibolitic material (Grauert 1969) by granitic magmas. But extensive assimilation, especially of amphibolites, is unrealistic for granitic magmas which are rarely overheated! Flisch (1987) pointed out the I-type character of these protoliths, and this is confirmed by Maggetti et al. (1990b). Based on the REE-distribution pattern, Maggetti et al. (1990b) suggested that the intermediate magmas were formed by a partial anatexis of garnet-bearing metabasites. The strong fractionation of the light REE is consistent with the REE patterns of intermediate partial melts of garnet-containing, basic sources as presented by Kay and Kay (1986) and Rudnick and Taylor (1986). The gabbros and metagabbros of the "older orthogneisses" have flat REE distributions, and their positive Eu-anomaly is probably the result of plagioclase cumulation. Maggetti et al.

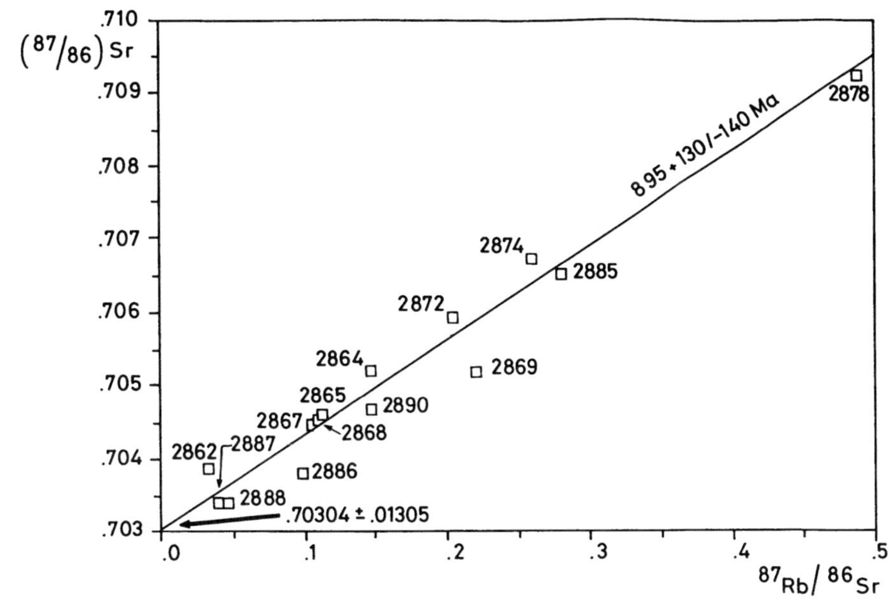

Fig. 5. Rb-Sr total rock isochron of the "older orthogneisses" (metagabbros, metadiorites and metatonalites)

Table 2. Rb-Sr total rock analyses from 30-kg samples of the older orthogneisses (Fig. 5)

Sample	Type coordinates	Rb total (ppm)	Sr total (ppm)	$^{87}Sr/^{86}Sr$ ($\pm 2\sigma$)	$^{87}Rb/^{86}Sr$ ($\pm 2\sigma$)
	(Meta-)Gabbro				
KAW 2862	799675/176200	3.8167	348.42	0.703864 ± 0.000015	0.03168 ± 0.00063
KAW 2865	799675/176200	14.261	366.13	0.704557 ± 0.000014	0.11266 ± 0.00225
KAW 2867	799675/176200	11.746	321.41	0.704477 ± 0.000012	0.10570 ± 0.00211
KAW 2868	799675/176200	13.253	356.22	0.704509 ± 0.000012	0.10761 ± 0.00215
KAW 2886	802375/186680	14.117	413.70	0.703819 ± 0.000016	0.09869 ± 0.00197
KAW 2887	802375/186680	8.3897	558.05	0.703398 ± 0.000016	0.04348 ± 0.00087
KAW 2888	802375/186680	6.9459	520.80	0.703415 ± 0.000016	0.03857 ± 0.00077
	Metadiorite				
KAW 2890	802625/185780	15.094	297.43	0.704688 ± 0.000014	0.14679 ± 0.00293
	Metatonalite				
KAW 2864	799675/176200	18.083	355.81	0.705216 ± 0.000013	0.14701 ± 0.00294
KAW 2869	798225/176050	24.153	315.12	0.705114 ± 0.000012	0.22170 ± 0.00443
KAW 2872	797580/175710	22.464	321.09	0.705931 ± 0.000010	0.20238 ± 0.00405
KAW 2874	797540/175660	26.742	300.54	0.706729 ± 0.000014	0.25741 ± 0.00515
KAW 2878	797580/175710	38.771	231.65	0.709248 ± 0.000010	0.48431 ± 0.00968
KAW 2885	802360/176500	28.859	299.13	0.706535 ± 0.000013	0.27910 ± 0.00558

(1990b) suggest that these tholeiitic, MORB-like melts originate from a depleted mantle.

A compilation of analyses from the literature demonstrate the I-type characteristics of the intermediate magmas. However, the Mönchalp-type augengneisses overlaps the I- and S-type field (Fig. 6a). In the geotectonic discrimination diagram after Pearce et al. (1984), most analyses of the intermediate and acid rocks plot in the VAG domain (Fig. 6b). The overlapping with other domains is explained with magmatic fractionation effects (Maggetti et al. 1990b).

Only the gabbroic and granitic types possess relict plutonic structures. The dioritic hornblende-gneisses, which are the dominant field occurrence are always foliated, and the schistose tonalitic biotite-gneisses, which are less abundant, could have volcanic protoliths. However, it appears to us that their association with true basic and acid plutonites is an important argument for the plutonic nature of the protoliths also of the intermediate types.

The intrusion level of this gabbroic to granitic plutonite association still awaits clarification. The sharp contacts, the angular contours of the inclusions in the

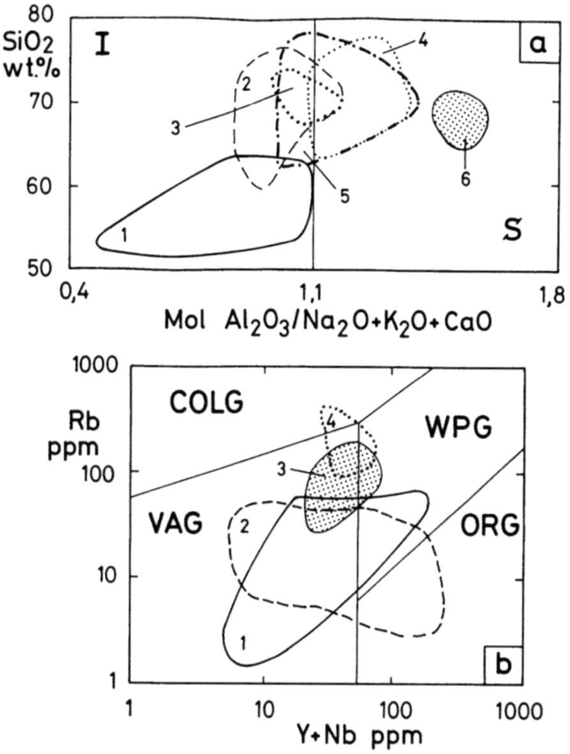

Fig. 6. a Classification diagram (Chappell and White 1974) for 160 granitoid analyses from the Silvretta. *1* 51 Metadiorites; *2* 27 metatonalites; *3* 7 Tschuggen-type augengneisses; *4* 53 Flüela-type-, Güstizia-type- and Macun-type augengneisses; *5* 10 Mönchalp-type augengneisses; *6* 12 Rotbüel-type augengneisses. **b** Geotectonic discrimination diagram (Pearce et al. 1984). *COLG* Collision granite; *ORG* ocean ridge granite; *VAG* volcanic arc granite; *WPG* within-plate granite. *1* 51 Metadiorites; *2* 27 metatonalites; *3* 18 Mönchalp- and Rotbüel-type augengneisses; *4* 32 Flüela associations. (Chemical analyses from Spitz and Dyhrenfurth 1914; Escher 1921; Streckeisen 1928; Krähenbühl 1984; Bröcker 1985; Flisch 1987; Müller 1989; von der Crone 1989a; Maggetti et al. 1990b and unpubl. data)

massive parts, and the fine-grained chilled margins indeed indicate a rather shallow level. Contact-metamorphic relicts in the adjoining rock have not been observed so far.

2.4 Event 4: Second HP Event

A high-pressure overprint of the "older orthogneisses" is documented by the presence of coronitic gabbros (Spaenhauer 1932; Thierrin 1982, 1983; von der Crone 1989a), rutile and zoisite relicts in the metadiorites and metatonalites (Maggetti 1986), and garnet coronas around biotite in the Mönchalp-type metagranites (Maggetti 1986; von der Crone 1989a). Fuchs et al. (1986) postulated 10–12 kb and approximately 1000 °C for an HP event documented in metagabbros in association with peridotites and pyroxenites. It is not established if this event belongs to the eclogitic one (= event 2) or to this second HP event. The time and geotectonic interpretation of this event is not clear. Since these relicts have not been recognized till now in the "younger orthogneisses", this episode probably occurred prior to the intrusion of the Ordovician magmas and can be assigned to the Pan-African tectonic event at least.

2.5 Event 5: Intrusion of the Magmas of the "Younger Orthogneisses"

The "younger orthogneisses" group comprises coarse- to medium-grained 2-mica augengneisses (Flüela-type, Frauenkirch-type, Tschuggen-type, Radönt-type, Urezzas-type), medium- to fine-grained muscovite gneisses (Güstizia-type, Macun-type), meta-aplites and meta-pegmatites as well as undeformed pegmatites. Relict magmatic textures are found only in the pegmatites (Grubenmann 1904; Hayoz 1988; von der Crone 1989a) and inclusions are extremely rare. According to Amann (1985), the pegmatites could be very young, about 290 Ma. According to Flisch (1987), the protoliths of the augengneisses seem to have been intruded before those of the medium- to fine-grained gneisses.

In the field, transitions from the Frauenkirch-type to the Flüela-type can be frequently observed. The Frauenkirch-type is interpreted as a more highly differentiated derivative of the former. Geochemically, the Flüela granitic types are S-granitoids (Flisch 1987; see Fig. 6a). The Tschuggen-type, however, covers both fields! As shown in Fig. 6b, the geotectonic position of the protoliths is ambiguous. Their intrusion level is not known, but the contacts are never discordant.

There have been intensive isotopic and geochemical studies of the "younger orthogneisses" in the Silvretta (Jäger et al. 1961; Jäger 1962; Pasteels 1964; Grauert and Arnold 1968; Grauert 1969; Thöni 1980, 1981; Flisch 1986, 1987). According to Flisch the Rb-Sr total rock isochron gives an Ordovician age of 451 ± 2 Ma (Fig. 7 and Table 3). This is interpreted as the crystallization age. The total rock isochron of the Frauenkirch-type augengneisses (approx. 365 Ma) is interpreted by Grauert (1969) as the intrusion age, but by Flisch (1987) as rejuvenated recrystallization age during the Variscan main deformation. Ordovician ages of orthogneisses are common in the Austro-

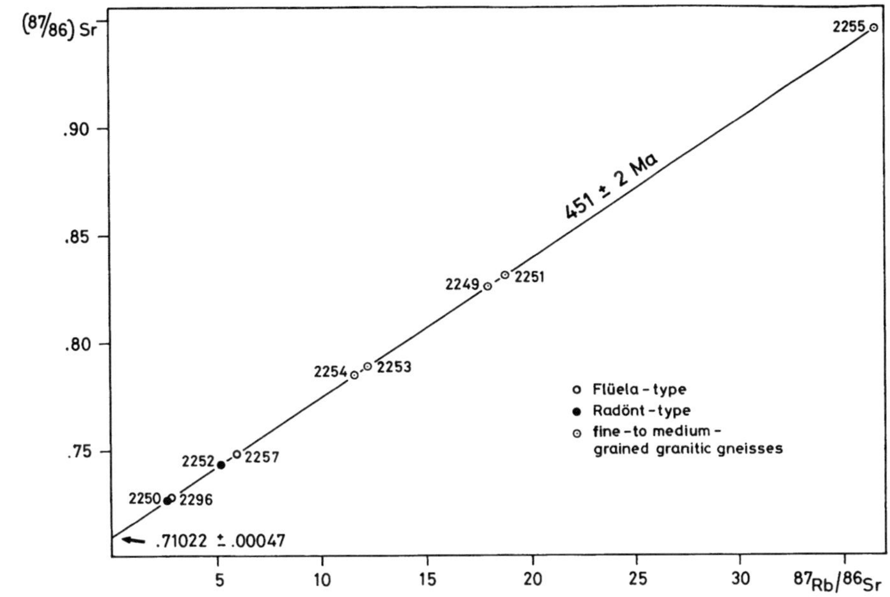

Fig. 7. Rb-Sr total rock isochron of the "younger orthogneisses" (Flisch 1987)

Table 3. Rb-Sr total rock analyses from 30 kg samples of the younger orthogneisses (Fig. 7)

Sample	Type coordinates	Rb total (ppm)	Sr total (ppm)	$^{87}Sr/^{86}Sr$ ($\pm 1\sigma$)	$^{87}Rb/^{86}Sr$ ($\pm 1\sigma$)
	Flüela-type				
KAW 2296	793 210/189 190	136.7	138.10	0.72840 ± 0.00005	2.849 ± 0.015
KAW 2257	791 400/182 265	193.8	94.61	0.74822 ± 0.00005	5.951 ± 0.030
	Radönt-type				
KAW 2250	792 590/178 750	116.3	130.3	0.72707 ± 0.00003	2.589 ± 0.013
KAW 2252	793 100/177 975	188.6	105.0	0.74343 ± 0.00002	5.215 ± 0.026
	Fine- to medium-grained granite gneisses				
KAW 2249	793 045/178 960	210.1	34.23	0.82595 ± 0.00003	17.96 ± 0.09
KAW 2251	793 235/178 220	246.5	38.60	0.83236 ± 0.00003	18.70 ± 0.09
KAW 2253	793 550/178 375	225.7	53.46	0.78838 ± 0.00004	12.31 ± 0.06
KAW 2254	793 580/178 980	252.3	63.16	0.78505 ± 0.00005	11.64 ± 0.06
KAW 2255	793 330/178 850	255.0	20.63	0.94475 ± 0.00003	36.60 ± 0.18

Alpine domain. There are widely differing opinions regarding the geotectonic interpretation of this "Caledonian Event" (e.g. Sassi et al. 1987; Heinisch 1988). If the interpretation of an Ordovician intrusion age is correct, we must ask ourselves why such large masses of palingenic granite magmas intrude if there was no prior tectonic event. The lack of relictic magmatic texture, the frequent thin interlaying of these types and the absence of true exogenic inclusions in these enormous masses of augengneisses, when compared with the aspects of the small outcrops of the Mönchalp-type orthogneisses, is puzzling. Such a behaviour can be explained if a subaeral origin of those rocks is assumed. Heinisch and Schmidt (1982) interprete the augengneisses of similar tectonic position in the Eastern Alps as ignimbritic metavolcanics of rhyolitic-rhyodacitic composition, genetically related to the Upper Ordovician "porphyroids" found in less metamorphosed areas of the Eastern Alps (e.g. Heinisch 1981). Weber (1984) discussed a model of formation of this magmatism by an Early Palaeozoic continental rifting. If we favour such a genetic relationship, then some portion of the Proterozoic metasediments could also belong to a Palaeozoic sedimentary cover. This could be true for some protoliths of the micaschists and the marbles.

2.6 Event 6: Variscan Deformation and Metamorphism

Post-Ordovician deformation and recrystallization history is polyphase and plurifacial and has been discussed in detail by a number of authors. Recently, Frapolli (1975) and Flisch (1987) have worked to clarify the complex structures. Practically all the rocks of the Silvretta nappe are characterized by a marked penetrative foliation. The corresponding fold-axes F2 can be observed in the field (Fig. 8). During D2, for example, the assemblage:

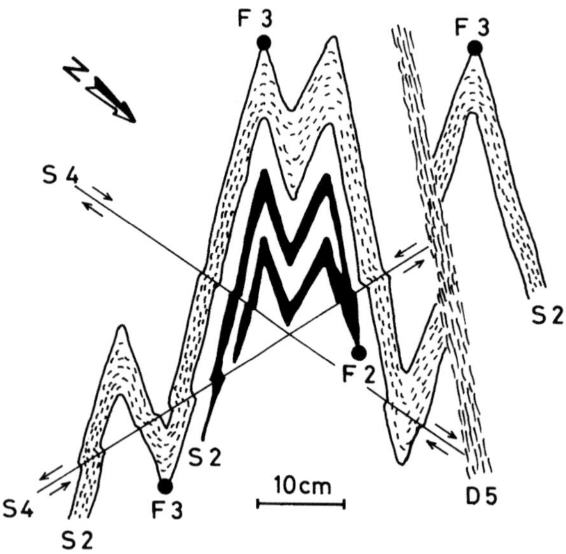

Fig. 8. Sketch of the Variscan (F2, F3, S4) and Alpine (D5) structural evolution of the Silvretta nappe in the Radönt area (after Flisch 1987). *D* Deformation; *F* fold axis; *S* schistosity

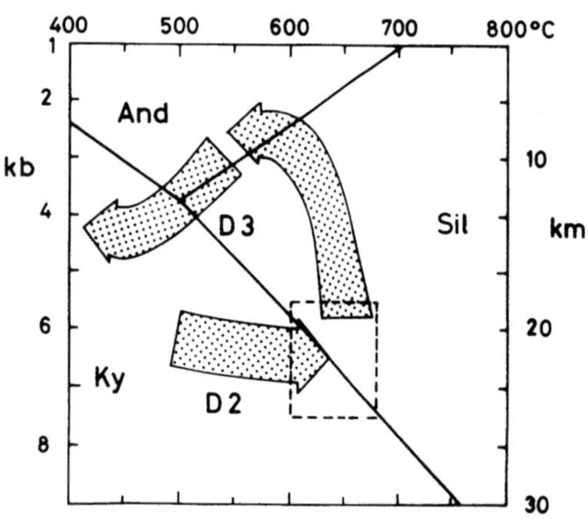

Fig. 9. Estimated P-T path of the Silvretta basement for the Variscan evolution (after Flisch 1987). *Rectangle* P-T conditions of the metamorphic peak

garnet + staurolite + kyanite + plagioclase + tourmaline + white mica + biotite + quartz

was stable in the metapelites (Hoernes 1971; Frapolli 1975; Amann 1985; Flisch 1987).

According to mineralogical and chemical studies by Flisch (1987), the climax of the metamorphism, which was reached syn- to post-D2, corresponds to the following T-P-conditions: 600–680 °C and 5.5–7.5 kb (Fig. 9). Amann (1985) advocates 500–650 °C and 4–6.5 kb. The temperature of 680 °C was only exceeded for a short time, as testified by the presence of zoned garnets in the alumosilicate gneisses and the absence of Variscan anatexites.

Rb-Sr small-scale isochrons on an Ordovician banded gneiss (= meta-aplitic layers in dark augengneiss of the Tschuggen type) give an age of 370 ±17 Ma for the Variscan main deformation D2 (Fig. 10 and Table 4).

Surprisingly, the main foliation of the Proterozoic eclogites (S1b) is in conformity with the Variscan s2! Did the Variscan main deformation occur in the same direction as the eclogitic or did the Variscan amphibolite-facies recrystallization, at least in the amphibolites, take place with no specific movements? The latter would then have affected only the protoliths of the "younger gneisses". Or are some amphibolites not only older than the eclogites but perhaps of the same age, i.e. have some amphibolites been formed at the same time as the eclogites from an identical protolith, but under more hydrous conditions? That would explain the identical schistosity. All authors agree, however, that the majority of the amphibolites were formed during event 6.

After D2 the Silvretta nappe rose quickly, and in the metapelites staurolite broke down to sillimanite/andalusite + biotite. Kyanite was also replaced by andalusite and fibrolithic sillimanite after biotite is widespread (Hoernes 1971; Frapolli 1975; Flisch 1987). The andalusite occurs also as large porphyroblasts and idiomorphic prisms in post-kinematic, extensional rods and veins of quartz (Gramann 1899; Streckeisen 1928; Spaenhauer 1933; Flisch 1987). The latter author estimates, for this episode, temperatures of 550–600 °C and pressures of 2–3 kb.

A further compression phase (D3) with assumed conditions of 450–550 °C and 4 kb (Flisch 1987) led to the formation of kilometre-long folds ("Schlingen") and metre-centimetre size parasitic folds. The latter are widespread. Only in few places does a new foliation S3 develop (Flisch 1987). According to Grauert (1981), this deformation occurred between 375–300 Ma. Flisch (1987) places this same event in the range between 340–310 Ma.

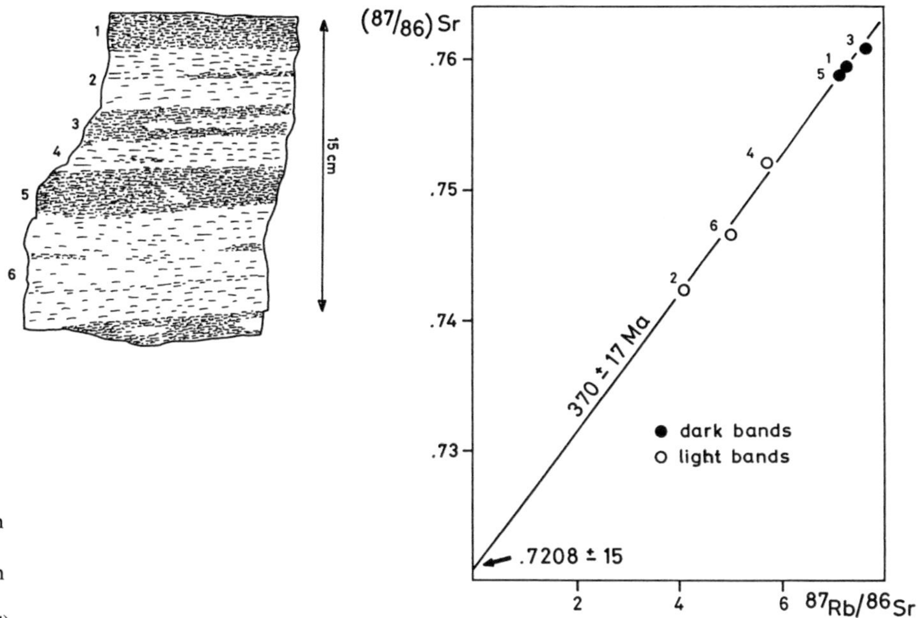

Fig. 10. Dating of the main Variscan deformation D2. Rb-Sr small-scale isochron of sample KAW 2356, a banded gneiss (Flisch 1987)

Table 4. Total rock analyses of the banded gneiss Tschuggen type from the quarry Tschuggen (Fig. 10)

Sample	Band	Rb total (ppm)	Sr total (ppm)	$^{87}Sr/^{86}Sr$ ($\pm 1\sigma$)	$^{87}Rb/^{86}Sr$ ($\pm 1\sigma$)
KAW 2356/1	Dark	163.3	65.12	0.75951 ± 0.00004	7.291 ± 0.036
KAW 2356/2	Light	128.7	91.60	0.74236 ± 0.00005	4.080 ± 0.020
KAW 2356/3	Dark	181.4	68.91	0.76084 ± 0.00003	7.658 ± 0.038
KAW 2356/4	Light	154.9	79.13	0.75217 ± 0.00002	5.689 ± 0.028
KAW 2356/5	Dark	168.7	68.45	0.75883 ± 0.00004	7.167 ± 0.036
KAW 2356/6	Light	163.7	94.58	0.74667 ± 0.00006	5.026 ± 0.025

The third Variscan deformation D4 is documented by combined shear bands, by a minute, centimetre-sized folding or by the distortion of large rock blocks. Temperatures reached for the actually exposed rocks, based on a partially quartz recrystallization (Voll 1976), are at least 275 ± 50 °C (Flisch 1987).

Hoernes (1971), Bertle (1972), Pirkl (1980) and Amann (1985) postulate a Late Variscan regional diaphthoresis which reached the sericite-chlorite subfacies of the greenschist facies with temperatures at a minimum of 400 °C.

2.7 Event 7: Permo-Carboniferous Rift Magmatism and Subsequent Triassic-Jurassic Sedimentation

The various Variscan compression phases were followed by the rapid rising of the Silvretta crystalline complex. As a result of extension in Pangea, basic magma rose and crystallized within the Silvretta nappe as narrow feeder-dykes (diabases: Hammer 1930; Bearth 1932). Bearth mapped 176 dykes in the Swiss part of the Silvretta nappe alone! The first results of recent chemical studies of these dyke clusters are available (Hellermann 1990). The 235 analyzed samples show continental, strongly differentiated tholeiitic magma compositions. They evolved by olivine + clinopyroxene + plagioclase fractionation at depth, and were extensively contaminated by crustal material, as shown by LILE and Sr isotopic composition. The frequent occurrences of chilled margins shows that the magmas intruded a largely cooled and uplifted basement. The present level of erosion would not have been deeply buried at that time. Since the basic dykes do not penetrate the sedimentary, autochthonous cover, they should be pre-Permian, but post-D3 (310–280 Ma?).

The present surface was exposed during the Permian, at least in the western part of the Silvretta nappe. Continuous extensional tectonics resulted in the formation of horst and graben structures, related to an active continental margin (Mercolli and Oberhänsli 1988). Voluminous deposits consisting of volcanic detritic material and ignimbrite flows have survived from this time (Spitz and Dyhrenfurth 1914; Eugster 1934; Dössegger 1974; Furrer 1985; Graf 1987; Zaugg 1987; Mercolli and Oberhänsli 1988). This acid magmatism has high-Al rhyolite chemistry.

The Silvretta nappe then began to subside and formed the base of a platform on which terrigenous materials, followed mainly by dolomites, were deposited. Sporadic volcanic episodes are documented by thin tuffaceous layers or discordant veins. Sedimentation also continued in the Jurassic, but the youngest remaining Silvretta sediment is Rhät (approx. 210 Ma). Furrer (1985) gives a good synopsis of the Permian volcanic activity and the Triassic sedimentation.

2.8 Event 8: Alpine Deformation and Metamorphism

Amann (1985), Flisch (1986) and Hurford et al. (1989) have treated the Alpine history in detail. The climax of Alpine metamorphism was reached about 110–90 Ma ago. Conditions extend from anchizonal in the west (Dunoyer de Segonzac and Bernoulli 1976; Giger 1985) to greenschist facies in the east and south (e.g. formation of stilpnomelane in the metagranites, Grauert 1969). This zoning of the Alpine metamorphism is documented both by age data (Flisch 1986) and by the mineral paragenesis (Thöni 1981; Amann 1985) (Fig. 11). The Silvretta crystalline rocks mostly underwent brittle deformation during Alpine events and formed cataclasites (D5). The eastern region of the Silvretta nappe is thought to have been 12–13 km deep at that time, while the western region was at a depth of about 7 km (Amann 1985: ~300 °C and 1–2 kb in the western, ~400 °C and 2–3 kb in the eastern parts of the Silvretta). Since the autochthonous sediments in the western part had a maximum of 4 km thickness (Furrer 1985), the Eo-Alpine metamorphism must have been caused by the overload of a higher tectonic unit (in the west), about 3 km thick.

The climax of the Eo-Alpine metamorphism was followed by a differential uplift and the detachement of the Silvretta nappe. At this time pseudotachylites were formed (Hammer 1914, 1930; Bearth 1933; Masch 1970, 1973; Tollmann 1977; Thöni 1981; Schmutz 1986, 1989). These extremely fine-grained rocks, which are found only in a closely defined area

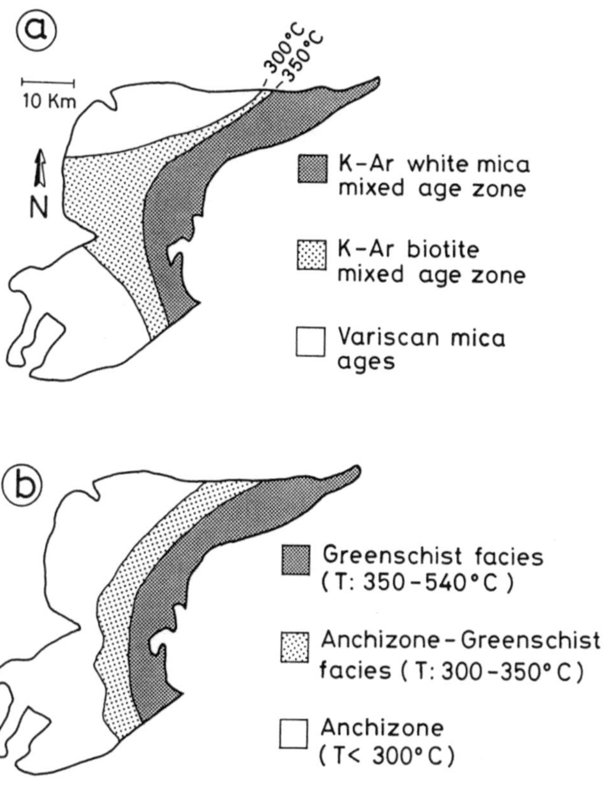

Fig. 11a, b. Alpine metamorphic zonation as deduced from **a** the age determination (Hurford et al. 1989; Maggetti et al. 1990) and **b** the mineral distribution (Thöni 1981)

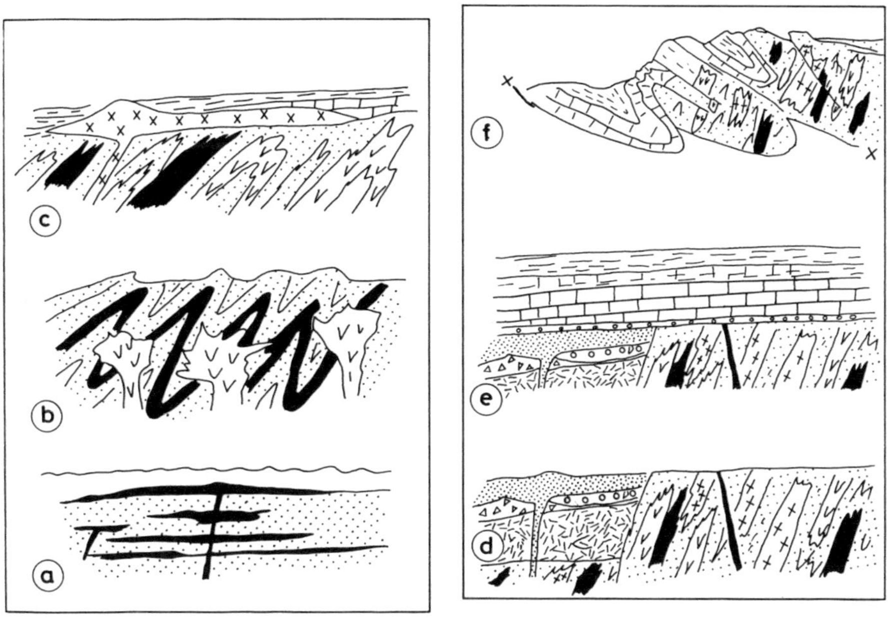

Fig. 12a–f. Schematic cartoon of the Silvretta evolution (see text for further explanation)

at the border of the Penninic Lower Engadine window and which occur in all rock types, are interpreted by Masch (1970, 1973) and Schmutz (1986, 1989) as rapidly cooled, frictional cataclasites (i. e. melts). The pseudotachylites document a fossil hypocentre of earthquakes (Laubscher 1983) which shook the massif in the course of the shearing off of the Silvretta basement (Masch 1970: 0.8–1.5 kb, 900–950 °C; Thöni 1988: 3–4 kb, 300–600 °C). According to Thöni (1988), they are about 75 Ma old.

The overthrust of the Silvretta nappe onto the Penninic foreland, and the underplating of the latter, occurred about 60–55 Ma ago (youngest sediments in the Lower Engadine window: Early Palaeocene to Late Eocene; Rudolph 1982; Oberhauser 1983).

In the time span 60–35 Ma the Lower Engadine window vaulted. Between 35 and 3–2 Ma the Silvretta nappe was uplifted homogeneously at the rate of approx. 0.1 mm/year (Hurford et al. 1989). The last event is the Chur uplift, which began 3–2 Ma ago along with a tilting of the Silvretta terrain about 2.2 ° to the east.

3 Conclusion

In Fig. 12, a schematic "geopoetic" cartoon of the evolution of the Silvretta is given. Proterozoic time is characterized by clastic sedimentation and probably contemporaneous intrusion/extrusion of MOR basaltic magmas (Fig. 12a). Subduction processes lead to the deformation and metamorphism in amphibolite facies, followed by a high-pressure event, ending with the formation of eclogites; at the end of this episode, collision-related Proterozoic magmas (older orthogneisses) intruded (Fig. 12b). A new HP event overprinted these magmatic rocks and deformed (?) them; Ordovician acid magmatism is documented in the younger orthogneisses which could have protoliths of plutonic or volcanic character; the latter hypothesis is preferred here (Fig. 12c). The cartoon shows that, for this episode, carbonatic as well as pelitic sedimentation could have occurred with this rift volcanism.

During the Variscan orogeny, the basement was intensily deformed and metamorphosed in amphibolite facies; Late Carboniferous tholeiitic dykes scattered through the basement. Permian rifting lead to the formation of horst and graben structures, filled with acid volcanoclastic material and ignimbrite flows (Fig. 12d). In Triassic to Jurassic time the Silvretta was a shallow epicontinental platform (Fig. 12e). Thereafter compression began leading to a maximum greenschist facies Alpine metamorphism, followed by the detachment and overthrust of the Silvretta onto the Penninic domain. Whereas the basement reacted in a brittle way, the Permo-Triassic cover was folded (Fig. 12 f).

Acknowledgements
We thank A. Benghezal and R. Bollin for the typing of the manuscript, J. Charrière for the drawings and the Swiss National Foundation for the financial sup-

port (grants 2.920-0.83, 2.856-0.85, 2.858-0.88, 20-26481.89 and 20-27956.89). Comments by L. Benciolini, R. Bollin, B. Hellermann Furrer, F. Neubauer, J. von Raumer and J. C. Schumacher greatly improved the manuscript. Thanks are due to both co-editors and Springer-Verlag for the perfect editing.

References

Amann A (1985) Zur Metamorphose des nördlichen Silvrettakristallins. Thesis, Univ Innsbruck

Bearth P (1932) Die Diabasgänge der Silvretta. Schweiz Mineral Petrogr Mitt 12: 147–178

Bearth P (1933) Über Gangmylonite der Silvretta. Schweiz Mineral Petrogr Mitt 13: 347–355

Bertle H (1972) Zur Geologie des Fensters von Gargellen (Vorarlberg) und seines kristallinen Rahmens. Mitt Ges Geol Bergbaustud 22: 1–59

Blumenthal M (1926) Zur Tektonik des Westrandes der Silvretta zwischen oberem Prätigau und oberen Montafon. Jahresber Naturforsch Ges Graubünden 64: 51–83

Bollin R (1989) Geologie des Val Urezzas. Diploma Thesis, Univ Freiburg, Switzerland

Bröcker M (1985) Petrographische und geochemische Untersuchungen im Val Tuoi (Silvretta, Schweiz). Diploma Thesis, Univ Tübingen

Chappell BW, White AJR (1974) Two contrasting granite types. Pac Geol 8: 173–174

von der Crone M (1989a) Geologie und Petrographie im Val Barlasch. (Graubünden). Diploma Thesis, Univ Freiburg, Switzerland

von der Crone M (1989b) Fleckengesteine der Silvretta: Relikte einer vorvariszischen (panafrikanischen?) Metamorphose. Schweiz Mineral Petrogr Mitt 69: 141–144

Dössegger R (1974) Verrucano und Bundsandstein in den Unterengadiner Dolomiten. Mitt Geol Inst ETH und Univ Zürich, NF 235

Dunoyer De Segonzac G, Bernoulli D (1976) Diagenèse et métamorphisme des argiles dans le Rhétien Sud-alpin et Austro-alpin (Lombardie et Grisons). Bull Soc Geol Fr 7 (XVIII): 1283–1293

Ernst WG (1981) Petrogenesis of eclogites and peridotites from the Western and Ligurian Alps. Am Mineral 66: 443–472

Escher F (1921) Petrographische Untersuchungen in den Bergen zwischen Davos und Piz Kesch. Jahresber Naturforsch Ges Graubünden LX: 5–55

Eugster H (1934) Bergün-Ducanpass-Sertig- oder Monstein. Geol Führer Schweiz, XIII 90: 1026–1033

Flisch M (1981) Geologie und Petrographie der Rotbüelgruppe. Lic Thesis, Univ Berne

Flisch M (1986) Die Hebungsgeschichte der oberostalpinen Silvretta-Decke seit der mittleren Kreide. Bull Ver Schweiz Petrol Geol Ing 53 (123): 23–49

Flisch M (1987) Teil 1: Geologische, petrographische und isotopengeologische Untersuchungen an Gesteinen des Silvrettakristallins. Teil 2: Die Hebungsgeschichte der oberostalpinen Silvretta-Decke seit der mittleren Kreide. Thesis, University of Berne

Frapolli G (1975) Petrografia della regione del passo della Flüela. Schweiz Mineral Petrogr Mitt 55: 307–364

Frisch W, Neubauer F (1984) Gneis-Amphibolitfolgen und kaledonische Orogenese im ostalpinen Kristallin. Fortschr Mineral 62: 67–68

Frisch W, Neubauer F, Satir M (1984) Concepts of the evolution of the Austroalpine basement complex (Eastern Alps) during the Caledonian Variscan cycle. Geol Rundsch 73 (1): 47–68

Frisch W, Neubauer F, Bröcker M, Brückmann W, Haiss N (1987) Interpretation of geochemical data from the Caledonian basement within the Austroalpine basement complex. In: Flügel HW, Sassi FP, Grecula P (eds) Pre-Variscan and Variscan events in the Alpine-Mediterranean mountains belts. Mineralia slovaca Monogr. Alfa, Bratislava, pp 209–236

Fuchs G, Kurat G, Ntaflos T (1986) Ein Peridotit – Vorkommen im Silvretta – Kristallin südlich von Galtür. Jahrb Geol Bundesanst 129: 283–290

Furrer H (1985) Field workshop on Triassic and Jurassic sediments in the Eastern Alps of Switzerland. Guide Book. Mitt Geol Inst ETH und Univ Zürich NF 248

Giger M (1985) Petrographisch-geologische Untersuchungen der Davoser Dorfberg-Decke und ihrer benachbarten tektonischen Einheiten im Gebiet Weissfluhjoch-Schafläger-Dorfberg (Davos, Kanton Graubünden). Lic Thesis, Univ Bern

Graf K (1987) Geologische Untersuchungen der Sedimente und Vulkanite der Landwasserkette zwischen Valbellahorn und Wiesen (Graubünden). Lic Thesis, Univ Berne

Gramann A (1899) Über die Andalusitvorkommnisse im rätischen Flüela- und Scalettagebiet und die Färbung der alpinen Andalusite. Vierteljahrsschr Naturforsch Ges Zür XLIV: 7–57

Grauert B (1969) Die Entwicklungsgeschichte des Silvrettakristallins aufgrund radiometrischer Altersbestimmungen. Thesis, Univ Berne

Grauert B (1981) Das Alter der Schlingentektonik im Silvretta-Ötztalkristallin aufgrund radiometrischer Altersbestimmungen. Fortschr Mineral 59 (1): 54–56

Grauert B, Arnold A (1968) Deutung diskordanter Zirkonalter der Silvrettadecke und des Gotthardmassivs (Schweizer Alpen). Contrib Mineral Petrogr 20: 34–56

Grubenmann U (1904) Pneumatolyse und Pegmatite. Mit einem Anhang über den Turmalinpegmatit vom Piz Cotschen im Unterengadin. Vierteljahrsschr Naturforsch Ges Zür IL: 376–391

Hammer W (1914) Das Gebiet der Bündner Schiefer im tirolischen Oberinntal. Jahrb K K Geol Reichsanst 64: 555–562

Hammer W (1930) Über Pseudotachylite in den Ostalpen. Jahrb Geol Bundesanst 80: 571–585

Hayoz P (1988) Geologie der Region Ardez (Unterengadin). Diploma Thesis, Univ Freiburg, Switzerland

Heinisch H (1981) Zum ordovizischen "Porphyroid"-Vulkanismus der Ost- und Südalpen, Stratigraphie, Petrographie, Geochemie. Jahrb Geol Bundesanst 124 (1): 1–109

Heinisch H (1988) Hinweise auf die Existenz eines passiven Kontinentalrandes im Altpaläozoikum der Nördlichen Grauwackenzone-Ostalpen. Schweiz Mineral Petrogr Mitt 68: 407–418

Heinisch H, Schmidt K (1982) Zur Genese der Augengneise im Altkristallin der Ostalpen. Neues Jahrb Geol Paläontol Monatsh 4: 211–239

Helfer M (1990) Geologie des Val Tuoi. Diploma Thesis, Univ Freiburg, Switzerland

Hellermann E B (1990) Geochemistry of Diabase Dikes of the Silvretta (Switzerland). Schweiz Mineral Petrogr Mitt 70: 109–114

Hodder APW (1985) Depth of origin of basalts inferred from Ti/V ratios and a comparison with the K_2O-depth relationship for island-arc volcanics. Chem Geol 48: 3–16
Hoernes S (1971) Petrographische Untersuchungen an Paragneisen des polymetamorphen Silvrettakristallins. Tschermaks Mineral Petrogr Mitt 15: 56–70
Hurford A, Flisch M, Jaeger E (1989) Unravelling the thermotectonic evolution of the Alps – a contribution from fission track analyses and mica dating. Geol Soc Spec Publ 45: 369–398
Jäger E (1962) Rb-Sr age determination on micas and total rocks from the Alps. J Geophys Res 67 (13): 5293–5306
Jäger E, Geiss J, Niggli E, Streckeisen A, Wenk E, Wüthrich H (1961) Rb-Sr-Alter an Gesteinsglimmern der Schweizer Alpen. Schweiz Mineral Petrogr Mitt 41 (2): 255–272
Kay RW, Kay M (1986) Petrology and geochemistry of the lower continental crust: an overview. In: Dawson JB, Carswell DA, Hall I, Wedepohl KH (eds) The nature of the lower continental crust. Geol Soc Spec Publ 24: 147–159
Krähenbühl R (1984) Petrographisch-geologische Untersuchungen in der Silvretta-Masse vom Flüela-Weisshorn zu Gorihorn und Rosställispitz (Kanton Graubünden). Lic Thesis, University of Berne
Laubscher HP (1983) Detachment, shear and compression in the Central Alps. Geol Soc Am Mem 158: 191–211
Maggetti M (1986) Petrographische Kriterien zur Altersstellung des Hochdruckereignisses in der Silvrettamasse. Fortschr Mineral 64 (1): 103
Maggetti M, Galetti G (1984) Chemie und geotektonische Position von Metabasiten aus dem Südosten der Silvretta (Schweiz). Schweiz Mineral Petrogr Mitt 64: 423–450
Maggetti M, Galetti G (1988) Evolution of the Silvretta eclogites: metamorphic and magmatic events. Schweiz Mineral Petrogr Mitt 68: 467–484
Maggetti M, Galetti G, Stosch H-G (1987) Eclogites from the Silvretta Nappe (Switzerland): geochemical constraints on the nature and geotectonic setting of their protoliths. Chem Geol 64: 319–334
Maggetti M, Flisch M, Bollin R (1990a) Bericht über die Exkursion der Schweiz. Mineralogischen und Petrographischen Gesellschaft ins Silvretta-Kristallin und in den Westrand des Unterengadiner Fensters (11.–14. September 1989). Schweiz Mineral Petrogr Mitt 70: 121–157
Maggetti M, Galetti G, Stosch H-G (1990b) Geochemische Argumente zur Genese der „Älteren Orthogneise" der Silvretta. Schweiz Mineral Petrogr Mitt 70: 103–107
Maresch WG (1977) Experimental studies on glaucophane: an analysis of present knowledge. Tectonophysics 43: 109–125
Masch L (1970) Die Pseudotachylite der Silvretta – eine Untersuchung ihrer Deformation und Aufschmelzung. Thesis, University of München
Masch L (1973) Untersuchung der Aufschmelzung und Deformation der Pseudotachylite der Silvretta (Österreich, Schweiz). Neues Jahrb Mineral Monatsh 11: 486–509
Mercolli I, Oberhänsli R (1988) Variscan tectonic evolution in the Central Alps: a working hypothesis. Schweiz Mineral Petrogr Mitt 68: 491–500
Michael G (1985a) Geologie und Petrographie im Macun (Unterengadin). Diploma Thesis, Univ Freiburg, Switzerland (unpubl)
Michael G (1985b) Géologie et Pétrographie de Macun (Silvretta). Schweiz Mineral Petrogr Mitt 65 (2/3): 369–372
Müller B (1989) Petrographisch – geologische Untersuchungen in der Silvretta-Decke im Gebiet Ober Silvretta bis Chremerchöpf (Kanton Graubünden). Zirkonmorphologische Untersuchungen an Orthogneisen des Silvrettakristallins. Lic Thesis, Univ Berne
Oberhauser R (1983) Mikrofossilfunde im Nordwestteil des Unterengadiner Fensters sowie im Vespalaflysch des Rätikon. Jahrb Geol Bundesanst 26 (1): 71–93
Pasteels P (1964) Mesures d'âges sur les zircons de quelques roches des Alpes. Schweiz Mineral Petrogr Mitt 44: 519–541
Pearce JA (1983) A "users" guide to basalt discrimination diagrams. Report, The Open University, Milton Keynes, 37 pp (unpubl)
Pearce JA, Cann J (1973) Tectonic setting of basic volcanic rocks determined using trace element analysis. Earth Planet Sci Lett 19: 290–300
Pearce JA, Harris NBW, Tindle AG (1984) Trace element discrimination diagrams for the tectonic interpretation of granitic rocks. J Petrol 25 (4): 956–983
Pirkl HR (1980) Geologische Beschreibung der Silvretta-, Samnaun- und Verwallgruppe. In: Oberhauser R (cd) Der geologische Aufbau Oesterreichs. Springer Wien, New York, pp 332–347
Rockenschaub MJ (1990) Die tektonische Stellung der Landecker Quarzphyllit- und Phyllitgneiszone. Jahrb Geol Bundesanst 133 (4): 619–633
Rudnick RL, Taylor SR (1986) Geochmical constraints on the origin of Archaean tonalitic-trondhjemitic rocks and implication for the lower crustal composition. In: Dawson JB, Carswell DA, Hall J, Wedepohl KH (eds) The Nature of the Lower Continental Crust. Geol Soc Spec Publ 24: 179–191
Rudolph J (1982) Tieferes Tertiär im oberen Fimbertal. Neues Jahrb Geol Paläontol Monatsh 3: 181–183
Sassi FP, Zanferrari G, Zirpoli G (1987) The Caledonian event in the Eastern Alps: a review. In: Flügel HW, Sassi FP, Grecula P (eds) Pre-Variscan and Variscan events in the alpine-mediterranean mountain belts. Alfa, Bratislava, pp 431–434
Schmutz HU (1986) Die Pseudotachylite der Silvretta am NW-Rand des Unterengadiner Fensters, eine strukturelle und petrographische Beschreibung. Schweiz Nationalfonds, Bull Abt 2, Info 2, Heft 4: 28
Schmutz HU (1989) Die Pseudotachylite der Silvretta am NW-Rand des Unterengadiner Fensters, eine strukturelle und petrographische Beschreibung. Schweiz Nationalfonds, Bull Abt 2, Info 2, Heft 10: 25
Shervais JW (1982) Ti-V plots and the petrogenesis of modern and ophiolithic lavas. Earth Planet Sci Lett 59: 101–118
Spaenhauer F (1932) Petrographie und Geologie der Grialetsch-Vadret-Sarsure-Gruppe. Schweiz Mineral Petrogr Mitt 12: 27–146
Spaenhauer F (1933) Die Andalusit- und Disthenvorkommen der Silvretta. Schweiz Mineral Petrogr Mitt 13: 323–346
Spicher A (1972) Geologische Karte der Schweiz 1:500000. Schweiz Geol Komm, Bern
Spitz A, Dyhrenfurth G (1914) Monographie der Engadiner Dolomiten zwischen Schuls, Scanfs und dem Stilfserjoch. Beitr Geol. Karte Schweiz, NF, XLIV. Lief
Streckeisen A (1928) Geologie und Petrographie der Flüelagruppe. Schweiz Mineral Petrogr Mitt 8: 87–239
Streckeisen A, Grauert B, Peters Tj (1966) Bericht über die Exkursion der Schweiz. Mineralogischen und Pegrographischen Gesellschaft ins Silvretta-Kristallin und in den Totalp-Serpentin. Schweiz Mineral Petrogr Mitt 46 (2): 704–722
Thierrin J (1982) Géologie et pétrographie du Val Sarsura, Grisons. Diploma Thesis, Univ Freiburg, Switzerland
Thierrin J (1983) Les éclogites et le complexe gabbroïque du Val Sarsura (Silvretta). Schweiz Mineral Petrogr Mitt 63: 479–496

Thöni M (1980) Distribution of pre-Alpine and Alpine metamorphism of the southern Ötztal Mass and the Scarl Unit based on K/Ar determinations. Mitt Österr Geol Ges 71/72: 139–165

Thöni M (1981) Degree and evolution of the alpine metamorphism in the Austroalpine unit W of the Hohe Tauern in the light of K/Ar age determination on micas. Jahrb Geol Bundesanst 124 (1): 111–174

Thöni M (1988) Rb-Sr isotopic resetting in mylonites and pseudotachylites: implications for the detachment and thrusting of the Austroalpine basement nappes in the Eastern Alps. Jahrb Geol Bundesanst 131 (1): 169–201

Tollmann A (1977) Geologie von Österreich, Bd I. Die Zentralalpen. Deuticke, Wien

Voll G (1976) Recrystallization of quartz, biotite and feldspars from Erstfeld to the Leventina nappe, Swiss Alps, and its geological significance. Schweiz Mineral Petrogr Mitt 56: 641–647

Weber K (1984) Variscan events: early Palaeozoic continental rift metamorphism and Late Palaeozoic crustal shortening. In: Hutton D M W, Sanderson D J (eds) Variscan tectonics of the North Atlantic region. Geol Soc Spec Publ 14: 3–22

Wenk E (1934) Beiträge zur Petrographie und Geologie des Silvrettakristallins. Schweiz Mineral Petrogr Mitt 14: 196–278

Zaugg F (1987) Geologische Untersuchung der Sedimente und Vulkanite der Landwasserkette zwischen Sandhubel und Wiesen (Graubünden). Lic Thesis, Univ Berne

Evolution of the Ötztal-Stubai, Scarl-Campo and Ulten Basement Units

G. Hoinkes and M. Thöni

Abstract

The Austro-Alpine basement, between the Tauern window to the E and the Western Alps to the W, treated in this contribution may be divided into three subunits due to differing lithology and metamorphic history. These three subunits, the Ötztal-Stubai-, the Scarl-Campo- and the Ulten basement, are divided by two major tectonic lines, the Schlinig line (Ötztal-Stubai/Scarl-Campo) and the Peio line (Scarl-Campo/Ulten). Although each of the three units consists of quartzo-feldspatic and pelitic metasediments, acid to intermediate metamagmatites, in addition to metabasites and metacarbonates, the dominant rock types of the Ötztal-Stubai-, the Campo- and the Ulten-unit are metasediments. The Scarl unit however is mainly composed of acid metamagmatites. Ages of the acid magmatic protoliths are most probably Variscan in the Scarl unit and Caledonian in the Ötztal-Stubai unit.

Additional characteristic features of the three subunits are: pre-Variscan eclogites in the Ötztal-Stubai-, Late-Variscan pegmatites in the Campo and Variscan metaperidotites in the Ulten-units. In terms of pre-Alpine metamorphic evolution the Ötztal-Stubai- and Scarl-Campo-units suffered from common, medium to high grade, metamorphic conditions (with relics of high-pressure conditions), whereas the Ulten unit shows indications of very high grade (granulite facies) conditions.

In Alpine times the Ötztal-Stubai and Scarl-Campo basement units were both overprinted with increasing intensity from northwest to southeast, whereas the Ulten zone shows almost no Alpine metamorphism.

1 Regional Geologic Outline

The Austro-Alpine basement units exposed between the Tauern window and the Western-Alps constitute one of the most important reference areas for unravelling the pre-Alpine evolution history in the Alps. This is mainly due to the fact that over large parts in this area the rocks escaped, almost completely Alpine deformation and metamorphism, hence pre-Alpine mineral parageneses are frequently well preserved. Detailed geological mapping has been carried out since the first decades of this century by Hammer (1912, 1923), Sander (1929), Spitz and Dyhrenfurth (1914), and Andreatta (1951 b) among others.

Regionally, the area may be subdivided into different subunits, like the Silvretta, the Ötztal-Stubai, the Scarl-Campo and the Ulten complex (Fig. 1). These subunits may be mainly a result of fragmentation and dismembering of a greater Austro-Alpine basement unit during Alpine off-shearing, nappe stacking and tectonic transport, although pre-Alpine structural elements may well have been reactivated during these processes. Basically, these different subunits are thought to have maintained a rather original position between themselves. A major shear zone of regional importance is, however, developed along the Schlinig thrust (Schmid and Haas 1989). The overall nappe character of the different basement subunits is also evident from Fig. 1: They presently overlie tectonically, series of the Mesozoic Piemontais-Penninic ocean and/or of the distal European continental margin, as exposed in the Tauern window and in the Engadine window.

Lithologically as well as with respect to the metamorphic history and age, four groups of rocks are discerned:

a. "Altkristallin" s. l.;
b. Phyllites;
c. Permo-Mesozoic (meta-) sediments;
d. Dykes.

While not dealing with the Silvretta basement here (see the contribution by Flisch and Maggetti this Vol.), the remaining area is divided into three different subunits:

1. The Ötztal-Stubai complex;
2. The Scarl-Campo complex;
3. The Ulten zone.

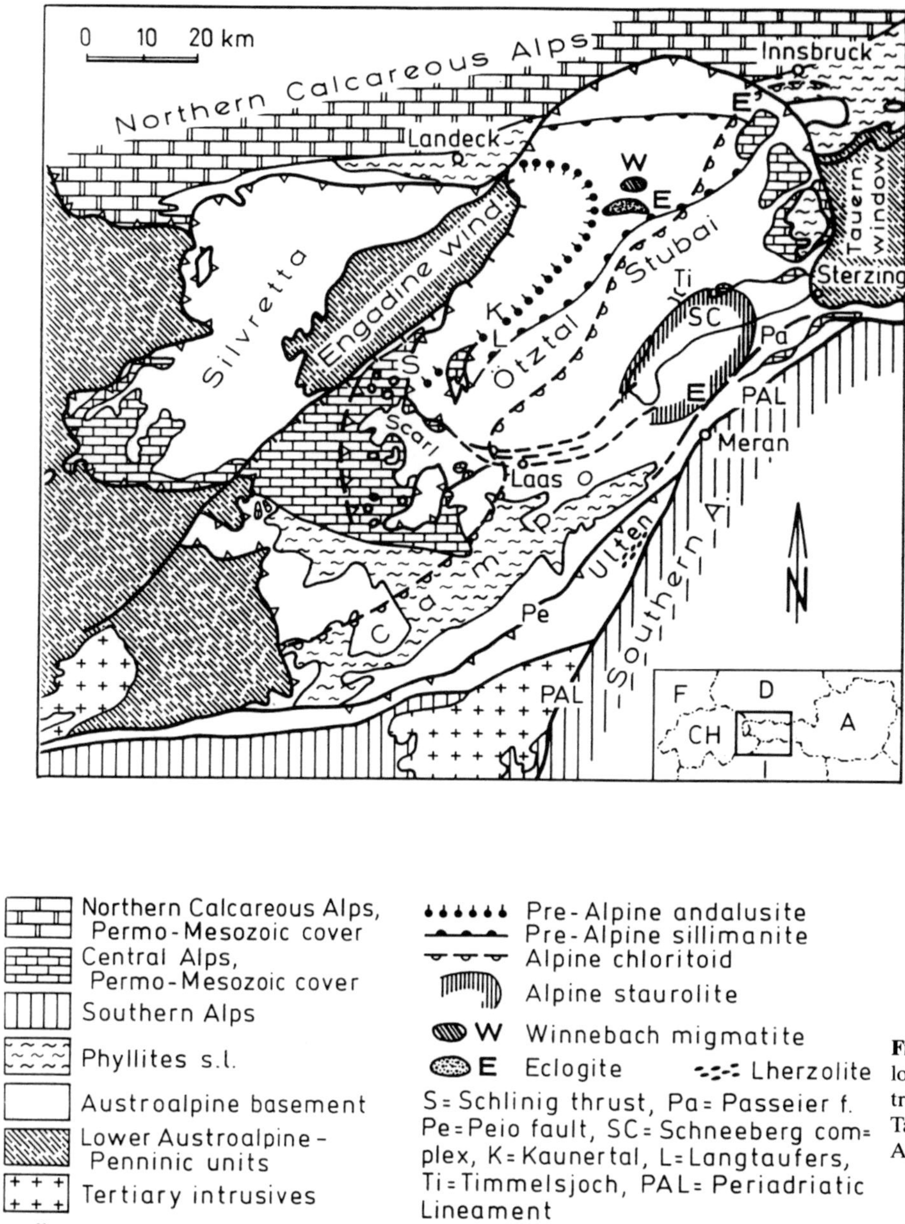

Fig. 1. Geologic and petrologic sketch map of the Austro-Alpine units between Tauern window and Western Alps

2 Ötztal-Stubai Complex

The Ötztal-Stubai complex is the tectonically bounded thrust mass immediately to the west of the Penninic Tauern window. The base of this thrust mass is only exposed in the southwest where a major thrust plane (Schlinig thrust) is recognized between the Ötztal-Stubai complex and the Sesvenna unit below. This thrust plane marks the southern border of the Ötztal-Stubai complex in the Vinschgau (Val Venosta). In the area of Meran, the tectonic style changes from local shearing to large-scale folding due to increasing temperatures during this tectonic event (Schmid and Haas 1989). Therefore, exact borders of the Ötztal-Stubai basement, in terms of a unit with a common pre-Alpine evolution, cannot be drawn in the area between Meran and Sterzing. A possible line marking the pre-Alpine border of the Ötztal-Stubai complex in this area lies between the Schneeberg synclines and the Laas series (Fig. 1). The southeastern border of the Ötztal-Stubai complex as a unit with a common Alpine history is represented by the tectonic line Passeier – Sterzing. This is a main suture dividing pre-Alpine crystalline basement to the east from the basement with strong Alpine metamorphic evolution to the west (Ötztal-Stubai complex).

2.1 Lithology and Petrography

The Ötztal-Stubai complex consists mainly of amphibolite facies-grade metasedimentary quartzofeldspatic rocks and subordinated metapelites. Intercalated are numerous acid to intermediate metamagmatites, some amphibolites, eclogites and metacarbonates (especially in the Schneeberg complex and the Laas series). A significant tectonic and lithologic feature is the Schneeberg complex representing a complicated E-W trending synclinorium of mainly metapelites, metacarbonates, metamarls and amphibolites which are deeply folded into the quartzofeldspatic basement rocks (Fig. 1; Schmidegg 1933; Helbig and Schmidt 1978). A mean crustal residence age of about 760 Ma for the Schneeberg metasediments is given by a Sm/Nd-model age of coarse grained garnet-micaschists from the Timmelsjoch area (Thöni unpubl. data). This age is interpreted as an upper age limit for the sedimentation of the Schneeberg metasediments and as a mean crustal residence time for the protolith material. The Schneeberg synform is accompanied to the south by another lithologically distinct metacarbonate-rich sequence, the Laas series (Sander 1929). Schistosity and fold axes are clearly east-west trending in the northern part of the Ötztal-Stubai crystalline block but the tectonic style changes to steep axes and huge open folds of kilometre-scale ("Schlingen") in the southern part (Fig. 1). On the meso- to microscopic scale different structural elements related to at least four distinct deformation phases (D1–D4) were recognized across the Ötztal-Stubai basement by van Gool et al. (1987). D2, characterized by intense foliation defining the easily recognizable microlayering on a mesoscopic to microscopic scale and isoclinal folding, is regarded as the most important one and is probably related to the dominant mineral-forming event. D1, on the other hand, is rarely observerd as relict microstructures only. D3-type structures are represented by chevron-type open folds affecting the whole Ötztal-Stubai block. The significance and age of D3 is still under discussion: its age is probably Late Variscan in the Alpine low grade parts (Thöni 1988) and most probably Cretaceous in the Alpine higher grade parts to the south. D4 is represented by local shear zones and open folding and is of Alpine age (Thöni 1986, 1988). A relict Mesozoic (meta-)sedimentary cover is exhibited in the eastern- (and western) most parts of Ötztal-Stubai block. The stratigraphic sequence generally comprises Scythian to Neocomian sediments in the area between Sterzing and Innsbruck (Fig. 1). A post-Variscan but pre-Alpine magmatic event is shown by basaltic to andesitic dykes with tholeiitic geochemical signature cutting the Ötztal-Stubai crystalline basement but not the Mesozoic cover (Purtscheller and Rammlmair 1982).

Orthogneisses occur as concordant layers or discordant stocks with sharp contacts to the paragneiss basement. They vary in composition from metagranites over -granodiorites, -tonalites to -quartzdiorites and even -diorites. Detailed geochemical and petrologic results are still missing but investigations are in progress. Radiometric age determinations on mainly muscovite-bearing S-type metagranites using U/Pb-dating on zircons and Rb/Sr-dating on whole-rocks (Miller et al. 1967; Schmidt et al. 1967; Harre et al. 1968; Satir 1975; Borsi et al. 1980; Thöni 1983; see Thöni 1986 for compilation) result in an age-range of approxy 470 to 420 Ma for single kilogram-size whole rock samples. This age interval may either mark the time of crystallization or (and) of a pre-Variscan metamorphic event. The only significantly younger Rb/Sr whole-rock ages were measured on two orthogneiss samples from the westernmost Ötztal unit by Grauert (1981). These data were taken as evidence for an Early Variscan (ca. 360 Ma) age for the "Schlingentektonik" of this special area.

The only pre-Caledonian whole-rock Rb/Sr-"ages" until now were obtained from orthogneisses of the northernmost Ötztal-Stubai basement (Eichhorn unpubl). Unfortunately, the data show a large scatter and due to the poor fit of the data points no reliable age calculation is possible; hence the Sr-isotopic initial ratio is poorly constrained as well, but points to an I-type rather than S-type origin for these outcrops. For comparison Flisch and Maggetti (this Vol.) report pre-Caledonian ages on I-type granitoids ("older orthogneisses") from the nearby Silvretta basement. A unique feature in the Ötztal-Stubai basement is the Winnebach-migmatite in the northern part of the Ötztal-Stubai block (Fig. 1). It represents an area of about 25 km^2 where in situ partial melting of the quartzofeldspatic gneisses occurred. Recent determinations of Rb/Sr-muscovite ages from the neosome part of the migmatite vary between 293 and 461 Ma. These data are interpreted as strong evidence for a Caledonian (or older) anatexis (Chowanetz 1990), but a Variscan age of the anatexis (Hoinkes et al. 1972) cannot be ruled out totally on the basis of these data. Söllner and Hansen (1987) deduced a "Pan-African" age of migmatization, mainly on the basis of U-Pb data on zircons. Due to the very complex zoning of the zircon grains, this interpretation is, however, not very conclusive.

Metapelites occur as intercalations in the paragneisses and are distributed over the whole area. They are significantly concentrated, however, in the northeastern Ötztal-Stubai block (Sellrain) and in the Schneeberg synclines. The only detailed investigations of metapelites of the pre-Alpine basement until now (Purtscheller 1969; Veltman 1986) results in three distinct mineral zones; a northern and southern kyanite zone, a central sillimanite zone and, within the sillimanite zone, a westernmost region where all three polymorphs kyanite, sillimanite and andalusite occur in the thin section scale. However, reaction textures of sillimanite formation from andalusite point to a polymetamorphic pre-Alpine history. Staurolite is present over the whole area but retrogression to chloritoid (obviously due to Eo-Alpine overprint) occurs in the central region (Fig. 1). The significant paragenesis for the upper temperature limit is muscovite and quartz. Metapelites of the Schneeberg synclines are characterized by single metamorphic textures derived from continuous garnet zoning and the stable co-existence of paragonite and quartz. Staurolite occurs in tiny amounts and is stabilized by significant Zn-contents (Hoinkes 1981).

Metabasic rocks occur as tens of metres to metre-wide intercalations with paragneisses in the whole Ötztal-Stubai basement and the Schneeberg synclines. A massive occurrence, however, is mapped in the central Ötztal. These rocks are the key to deciphering the pre-Alpine metamorphic history and will be discussed in more detail in the Section 2.2 dealing with metamorphic evolution.

Concerning the lithology, amphibolites contain relics of eclogites and sometimes also the pre-eclogitic magmatic precursor rocks identified by gabbroic textures. Closely connected with these metabasites are lenses of meta-ultrabasites (peridotites) and occasionally metacarbonates of clearly metasedimentary origin due to $\delta^{18}O$-values of +18.2 to +22 (Hoernes pers. comm.).

Worth mentioning is a pre-Alpine mylonite zone bordering the central amphibolite-eclogite mass and which is marked by lenses of the matacarbonates mentioned (Hoernes and Hoffer 1973; Ehlers and Hoinkes 1987; Fügenschuh unpubl.). These are indications of a possible major pre-Alpine fault separating the Ötztal-Stubai complex into two blocks. This is in accordance with the interpretation of the central Ötztal amphibolite mass as a relict dismembered ophiolite of the Speik complex defined by Frisch and Neubauer (1989). Very recently, new findings of relict eclogites were also made in the northernmost part of the Ötztal-Stubai complex by Eichhorn (unpubl.) as part of an I-type acid-basic suite with a pre-Caledonian memory (see earlier). The metabasites of the central and northern Ötztal area represent metamorphic tholeiitic magmas as derived by geochemical discrimination methods whereas amphibolites from the Schneeberg complex have metasedimentary geochemical signatures (Mogessie et al. 1985). Geochemical evidence for continental basaltic magmatism was proposed by Poli (1989) for amphibolites from the basement south of the Schneeberg complex.

2.2 Metamorphic Evolution

At least three metamorphic events can be distinguished from textural criteria within the central amphibolite mass: the oldest metamorphic overprint is represented by eclogite facies conditions which, in turn, were overprinted by amphibolite facies and finally greenschist facies conditions. The final metamorphic event is clearly of Cretaceous age and increasing conditions could be proved from northwest to southeast ending up in an Cretaceous eclogite facies (550–600 °C at ⩾11 kb) in the Passeier area (Hoinkes et al. 1991). Continuous decrease in Cretaceous conditions to the northwest is best recorded in metapelitic rocks and three main "isograds" could be mapped: (1) disappearance of Cretaceous staurolite and appearance of chloritoid approximately with the northern border of the Schneeberg synclines; (2) disappearance of Cretaceous chloritoid and (3) appearance of stilpnomelane in the northwestern region of the Ötztal mass. This zonation is also confirmed by radiometric age dating on micas which gives Cretaceous ages (75–100 Ma) in the southernmost Passeier area, mixed ages in the Cretaceous chloritoid zone and Hercynian ages of 320 Ma in the stilpnomelane zone (Miller et al. 1967; Schmidt et al. 1967; Harre et al. 1968; Thöni 1981). The Schneeberg complex is situated at the Cretaceous greenschist- to amphibolite facies transition.

Detailed investigations on micaschists and metacarbonates of the Schneeberg synclines compared to the adjacent Ötztal-Stubai crystalline area, including the Laas series, prove a single metamorphic evolution of Cretaceous age of the Schneeberg complex and a polymetamorphic evolution of surrounding crystalline basement. This implies a pre-Alpine and post-Variscan tectonic emplacement of the Schneeberg rocks in close vicinity to the lithologically comparable Laas series. Much less information exists about the P- T- t conditions of the pre-Alpine metamorphic evolution (Table 1).

Table 1. Evolution scheme of some Austro-Alpine units W of the Tauern window

	Lithology	Presumed stratigraphic and/or magmatic age	P-T conditions of metamorphic events Pre-Variscan	Variscan	Alpine
Ötztal-Stubai	Paragneisses-Micaschists	Pre-cambrian to Pre-"Caledonian"	Unknown	~650 °C/7 kb	Continuous increase to 600 °C/10 kb in the S
	Metacarbonates				
	Amphibolites-eclogites		~690 °C/~14 kb	650 °C 7 kb	550 °C/≥12 kb
	Orthogneisses	Partly "Caledonian"	Unknown	Medium grade	Very low to medium grade
	Migmatites	Precambrian to "Caledonian"	≤700 °C/?	Medium grade	Low grade
	Pegmatites	Late Variscan			Low grade
	Schneeberg metasediments	Palaeozoic		≤400 °C/?	580 °C/9–10 kb
	Post-Variscan metasediments	Permo-Mesozoic			N 450 °C/? S 530 °C/?
	Magmatic dykes	Post-Variscan			300–550 °C
Scarl-Campo	Micaschists				
	Metacarbonates		Unknown	Medium grade	
	Orthogneisses		Unknown	Medium grade	Low to very low grade
	Pegmatites	Late Variscan			
	Phyllites			Low grade	
	Post-Variscan metasediments	Permo-Mesozoic			
	Magmatic dykes	Post-Variscan			
Ultental	Paragneisses-micaschists		Unknown	750 °C/15 kb	
	Orthogneisses		Unknown	Medium grade	Very low to low grade
	Ultramafics	Variscan		700–800 °C/16–23 kb	
	Migmatites			Medium grade	

The dominant pre-Alpine metamorphism is clearly of Variscan age. The oldest Rb/Sr-ages were measured on coarse-grained muscovites from orthogneisses of the Variscan zone. They range between 349 and 369 Ma (Thöni 1986) and are probably climax ages of the Variscan metamorphism.

Recently, still older Rb-Sr ages of close to 440 Ma were found on muscovite cores from weakly-deformed orthogneisses of the Kaunertal area (unpubl. data; see Fig. 1). The overall range of K-Ar and Rb-Sr ages on micas from medium-grained, foliated rocks of the Variscan zone is the following:

K-Ar	biotite:	272–316 Ma
Rb-Sr	biotite:	272–301 Ma
K-Ar	white mica:	297–317 Ma
Rb-Sr	white mica:	292–327 Ma

Applying a common blocking temperature concept (Purdy and Jäger 1976) the Variscan temperature evolution must have been from about 500 °C at about 320 Ma to 300 °C at about 270 Ma. Temperatures of the climax conditions are not known exactly but must have been below 700 °C due to the stable co-existence of muscovite and quartz. Veltman (1986) derived mean temperatues of <700 °C (and a huge range from 600–750 °C) for the sillimanite zone and >600 °C (and a much smaller range from 570–650 °C) for the northern kyanite zone. Applying the Gibbs method (Spear 1988) to two garnets of the northern kyanite and the sillimanite zone respectively P-T conditions of 8 kb and 500 °C for the cores and about 6 kb and 570 °C (kyanite zone) or 650 °C (sillimanite zone) for the rims are deduced. Present petrologic work on metapelites in the westernmost areas (Langtaufers, Kaunertal) gave preliminary results of 650 °C and 7 kb using garnet rim composition, for the pre-Alpine ("Variscan") event. Garnet cores equilibrated with plagioclase inclusions at pressures between 10–11 kb. In this area all three aluminium silicate polymorphs kyanite, fibrolite and andalusite co-exist in some samples (Gregnanin and Piccirillo 1972). Thorough textural and microchemical investigations must be applied to these parageneses to decipher the time relationships of these three polymorphs. It is not known exactly whether the Variscan metamorphism was a single stage event or polyphase (Mogessie and Purtscheller 1986). The role of Caledonian metamorphism for the formation of the present mineral assemblages is unknown too. The pre-Alpine mineral-zonation with sillimanite in the central and kyanite in the northern and southern parts of the Ötztal-Stubai complex (Purtscheller 1969) need not be the consequence of a single metamorphic event. It is, however, obvious from radiometric age dating that Variscan metamorphic temperatures at least have been in the range of 550–600 °C but could have been higher (but below 700 °C) too. A Variscan metamorphic zonation seems therefore possible but is not conclusive. A pre-Variscan metamorphic event has been suggested for a long time and is indicated by eclogites as precursors of amphibolites in the central and northern Ötztal-Stubai basement. An unpublished age of 497 Ma for the Caledonian high-pressure event was derived by Söllner and Gebauer (unpubl.) from zircons separated from eclogites of the central amphibolite mass. The recently obtained strong radiometric indications for a Caledonian anatexis accompanied by the formation of fibrolite in the Winnebach migmatite supports this view of a dominant Caledonian metamorphism in the Ötztal-Stubai basement. Unfortunately, there are too few petrologic data as yet to answer the question about the portion of pre-Variscan metamorphic minerals in the Ötztal-Stubai basement rocks. However, moderate eclogite-facies conditions of 690 °C and 14 kb were derived by Miller (1970, 1990 pers. comm.) for the Caledonian eclogite stage. Indications of high-pressure conditions at an early stage of metamorphic evolution were also observed by Mogessie et al. (1986) from chemical zonation patterns of amphiboles from the central Ötztal amphibolite mass.

3 Scarl-Campo Complex

The regional extension of this composite basement complex is tectonically fairly well defined by the Engadine line to the NW, the Lower Austro-Alpine/Penninic units to the W, the Peio line to the S and the Schlinig thrust to the N. Towards the NE, however, no clear tectonic separation of the eastern Campo crystalline area and the southern Ötztal basement (S of the Schneeberg complex, Fig. 1) is possible. On the contrary, lithology and Alpine metamorphic zonation argue against complete tectonic disruption of the two units. Following the interpretation by Schmid and Haas (1989), the Ötztal-Stubai basement has overridden the Scarl-Campo units in a westward direction, along the Schlinig thrust, by at least some 50 km. This regionally important tectonic element is present as a discrete thrust plane in the western area, but towards the east, it develops continuously into a complicated shear zone of some hundreds of metres in thickness and the deformation is finally accommodated in the area NW of Meran by intense plastic flow, large-scale folding and backthrusting processes. Over large areas, the Scarl-Campo units can, therefore, be described as a tectonically lower element, with respect to the higher tectonic position of the Ötztal-Stubai complex.

3.1 Lithology and Petrography

The Scarl basement is mainly composed of leucocrate muscovite-granite gneisses, with biotite-granite gneisses being comparatively rare. Amphibolites and paragneisses make up only a small portion of this unit. As a whole, the Scarl basement forms a cupola-like structure, with its highest updoming in the area of Val Mustair-Scarl-Piz Sesvenna. The lithologic analogy of the Scarl crystalline with some orthogneisses of the nearby Silvretta basement further to the west is evident. Generally, the Scarl-Campo and Silvretta subunits are regarded as a coherent basement complex prior to the Cretaceous-Tertiary disruption, which was basically active as normal faulting along the Engadine lineament.

The greater part of the Scarl basement is covered by Permo-Mesozoic sediments, known as the Engadine dolomites (Spitz and Dyhrenfurth 1914; Dössegger and Müller 1976; Dösegger et al. 1982) which are in continuous but partly tectonic contact with the Ortler-Fraele zone to the S and W. Marine sedimentation lasted from the Lower Permian into mid-Cretaceous (Albian to probably Turonian) times (Caron et al. 1982; Mader 1987). The Scarl sediments display typical stockwork tectonics: the higher parts ("Oberbau", Schmid and Haas 1989) are successively detached and thrust towards the NW. This process was probably induced by the overriding Ötztal nappe. The Campo crystalline area forms the stratigraphic basis of the Ortler sediments. It is composed of medium-grade micaschists to quartzitic schists and paragneisses, with intercalations of pegmatites and some granitic gneisses and amphibolites. Among the granitic rocks, the Martell granite S of Laas is the best-studied example. It is an aplitic to pegmatitic body of ca. 3 × 9 km in dimension, intimately intruding the country rocks. Of special interest is an E-W trending zone S of the Vinschgau valley, rich in white calcite-marbles which are extensively mined near Laas/Vinschgau (Hammer 1912). From analogy to widespread massive Devonian carbonate series in low metamorphic Austro-Alpine units, a Devonian age of the sedimentary protoliths of the Laas series is assumed (Frank et al. 1987). Interestingly, a very similar rock sequence, known in the literature also as Laas series (Sander 1929) accompanies the Schneeberg rocks (Fig. 1) to the S and may be interpreted as an eastern continuation of the Laas series at the type locality in the northern Campo crystalline, which was, however, offset or distorted due to shearing along the Schlinig thrust zone, W of Meran.

Over large areas, but mainly in the central part, the medium-grade Campo crystalline is overlain by phyllitic rocks of weaker metamorphic grade (type quartzphyllites), forming an ENE-WSW trending synclinorium. Because of polyphase deformation and metamorphism the primary contacts between the phyllites and the medium-grade schists are obscured, but following Andreatta (1951 a) continuous gradations and intercalations of and between the two rock types may be observed. In the area of Ortler/Cevedale and, further to the W, these phyllites are unconformably overlain by the Permo-Mesozoic sequences of the Ortler-Fraele zone.

The Campo crystalline is also rich in dykes of variable but frequently andesitic composition (Gatto et al. 1976; Dal Piaz and Martin 1980; Purtscheller and Mogessie 1988). In the Ortler area, a post-Triassic age of dykes is partly proven from field observations (Andreatta 1951 a).

3.2 Metamorphic Evolution

Intense foliation is characteristic of large parts of the Scarl crystalline. The microstructures point to recrystallization of the magmatic parageneses at elevated metamorphic temperatures. Direct information for the effective P-T conditions during this metamorphism are not available, but local occurrence of kyanite in some biotite-bearing gneisses, as well as of amphibolites, may point to recrystallization at medium-grade metamorphic conditions. A comparison with the nearby Silvretta basement is, therefore, plausible.

The Scarl orthogneisses yielded a Rb-Sr whole-rock age of 336 ± 7 Ma (Thöni 1981). The related Sr initial ratio of 0.717 ± 0.002 points to isotopic homogenization during metamorphism rather than to a primary magmatic derivation of the orthogneiss material. This is in accordance with the intense structural overprint of the metagranitoids. The age could reflect the time when the last peak-metamorphic conditions were reached during the Variscan evolution. It is, however, fairly close to the 350 Ma age measured on some (comparable) highly differentiated aplitic-pegmatitic muscovite-granitegneisses from the adjoining Silvretta (Grauert 1969), and interpreted there as of primary magmatic age. Rb-Sr and K-Ar ages on coarse-grained white mica from the Scarl orthogneisses range between 295 and 327 Ma, tracing Variscan cooling. K-Ar and Rb-Sr ages on biotites range between 223 and 296 Ma.

Most of the Scarl orthogneisses show variable imprints of a retrogressive low-temperature metamorphism, leading to intense recrystallization of quartz, neoformation of phengitic white mica and partly of stilpnomelane. Based on radiometric data, this low-grade overprint is related to deformation and fluid infiltration, released most probably from the overlying Permo-Mesozoic sediments during Cretaceous times. Within the clastic to pelitic portions of the Scarl sediments (mainly Verrucano), the Cretaceous metamorphism reaches lowest greenschist facies conditions. Metamorphic ages on fine-grained micas from the Ortler-Scarl area are in the range of 100–70 Ma, with a probable temperature peak close to 90 Ma (Thöni 1981).

The metamorphic evolution of the Campo crystalline is rather poorly constrained. Within the higher-grade basement, centimetre-sized porphyroblasts of garnet and staurolite point to amphibolite-grade metamorphism. The Laas marbles contain plagioclase, mica, quartz and dolomite and also diopside and tremolite. They exhibit intense deformation by small- to large-scale folding and shearing. The typical paragenesis of the phyllites is quartz, muscovite, chlorite and accessories; variable amounts of biotite and garnet are observed. The overall metamorphic zonation is that of a normal sequence, with the intensity of the metamorphic grade dying out towards the top of the rock pile. The age of this amphibolite- to greenschist-facies-grade (phyllites) metamorphism is not known, but from a regional comparison and based on a few K-Ar and Rb-Sr data on micas (Thöni 1981) it is regarded as Variscan.

The Martell "granite" is an inhomogeneous leucogranite composed of quartz, plagioclase, K-feldspar, white mica and containing minor amounts of garnet, tourmaline and biotite. A well-defined Rb-Sr whole-rock isochron of 271 ± 3 Ma is interpreted as age of intrusion (Bockemühl 1988). The complicated and strongly discordant contacts give further evidence that the age of emplacement is post-metamorphic, with respect to the main structural and metamorphic events in the encasing rocks. Geochemical and Sr isotopic data support a derivation of the melts from upper crustal precursors. Other occurrences of pegmatites from the surrounding area (e.g. Matsch pegmatites, southern Ötztal basement) compare well with this type of "granite" and support the widespread Late Variscan magmatic activity in the Austro-Alpine basement.

As in the Scarl unit, the (presumably) Variscan parageneses show clear imprints of a retrogressive overprint that seems to become more intense towards the SE in the Campo complex. This overprinting event is petrographically observed by the chloritization and/or sericitization of garnet, biotite, staurolite and feldspar and by the formation of a second finer-grained mineral generation like chlorite, white mica, chloritoid and probably also garnet (Bockemühl 1988). In some places in the vicinity of the Alpine higher-grade area of the southern Ötztal basement a second generation of tiny staurolites has also been observed (Schmid and Haas 1989). From the regional distribution pattern of K-Ar and Rb-Sr ages in micas it is deduced that this retrogression is of Cretaceous age. Because of the partly almost complete resetting of these isotopic systems, Alpine metamorphic temperatures are estimated to have reached middle greenschist facies conditions. Regionally, this Alpine chloritoid zone (Purtscheller 1978; Thöni 1981; Fig. 1) stretches without interruption from near Innsbruck towards the SW, reaching the Lower Austro-Alpine/Penninic units in the Bernina area. It is only slightly offset by the syn- to late-metamorphic Schlinig thrust zone near Laas/Vinschgau (Fig. 1).

4 The Ulten-Zone

The Ulten zone to the southwest of Meran should be considered as a separate unit within the Eastern Alpine crystalline basement west of the Tauern window because of its distinct lithology and an obviously higher grade metamorphic overprint. In the Val di Rabbi – Val di Peio area the rocks show partly steep large-scale folding, similar to the "Schlingenbau" in the Ötztal basement. The Ulten zone is part of the Tonale zone and bounded by the Peio line to the north (which follows the Ulten valley resp. Val Ultimo) and the Judicarie line to the south. Both tectonic lines conjugate in the area of Meran and hence terminate the Ulten zone there. A possible continuation, to the east of Meran, may be represented by the crystalline basement rocks between the Passeier line and the Periadriatic lineament. This is not part of the Ötztal-Stubai block to the northwest because of lack of Cretaceous metamorphism (a similarity with the crystalline rocks of the Ulten zone). The main **lithology** of the Ulten zone comprises strongly foliated quartzofeldspatic gneisses, coarse garnet-kyanite micaschists, migmatites, barren pegmatites and lens-shaped ultramafic bodies of primary spinel-lherzolites showing transitions to fine-grained metamorphic garnet lherzolites (Hammer 1902; Andreatta 1935; Obata and Morten 1987). LREE-enriched trace element patterns of the spinel lherzolites are explained by mantle metasomatism, whereas the very pronounced increase in K, Sr and Li, observed in all garnet peridotites, indicates an additional lower crustal metasomatic influence (Morten and Obata 1990).

Metamorphic evolution in this area is fairly well known for the ultramafics from the petrologic work of Herzberg et al. (1977) and Obata and Morten (1987). The latter authors argument in favour of a mantle origin of the spinel lherzolites at 20 to 25 kb and 1200 °C and a deep-seated crustal metamorphic evolution to garnet lherzolites at about 20 kb and 700 to 800 °C. The question is whether this metamorphic adjustment occurred in situ, together with the present crustal host rocks, or in deeper crustal regions before tectonic emplacement to the present position. Detailed structural investigations of the contact zones between the ultramafic lenses and the granulite host rocks have still not been completed and tectonic contacts are not obvious from the outcrops. Preliminary petrologic investigations of the quartzo-feldspatic country rocks show mineral assemblages which are indicative of high pressure-, high temperature formation (alkali feldspar + kyanite, almandine + rutile). However, the highest pressures and temperatures deduced by geothermobarometric methods until now are about 14 kb and 750 °C. On the other hand, the existence of well-preserved microcline and the paragenesis muscovite and quartz in some samples are indicative for strong retrograde effects after the P-T peak of eclogite facies metamorphism. In these retrograde samples geothermobarometric calculations indicate a pressure of about 8 kb and a temperature of 600 °C. Preservation of the granulite facies assemblages in some areas and retrogression to amphibolite facies assemblages in others may be due to differing amounts of fluid available during overprinting. Time of eclogite facies metamorphism was determined by Gebauer and Grünenfelder (1978) by means of concordant U/Pb-zircon data from three ultramafic lenses with 332 to 326 Ma. Highly variable K-Ar and Rb-Sr-mica ages from acid gneisses of this area range between about 300 and 100 Ma (Thöni 1981). These data point to very weak (lowest greenschist facies) grade of Alpine metamorphism in the Ulten crystalline basement.

Conclusions

Although the petrological results are still too few to give a complete picture of metamorphic evolution of the three subunits of the Austro-Alpine basement between the Tauern window and the Western Alps, the following differences in terms of lithology and metamorphic evolution must be pointed out.

The Ötztal-Stubai subunit of quartzo-feldspatic metasediments and acid to basic metamagmatites is the only one which has a memory of pre-Variscan magmatic and high-pressure metamorphic events. During Alpine orogenesis the Ötztal-Stubai unit overrode the Scarl-Campo-unit of clearly distinct lithology. It consists mainly of acid granite gneisses of Variscan age (Scarl subunit) and metapelitic schists and marbles with numerous intrusions of pegmatites (Campo subunit) of late Variscan age. Both units had a common Variscan and Alpine metamorphic history of medium- to high- and medium- to low-grade respectively. The Ulten basement unit is a lithologically distinct block and is characterized by numerous lenticular ultramafic bodies within high-grade quartzo-feldspatic metasedimentary country rocks indicating the beginning of eclogite facies conditions. This high-grade metamorphism is thought to have effected both the ultramafics as well as the country rocks during Variscan orogeny. Indications for pre-Variscan events are not known from the Ulten basement unit now. The Ulten unit must be considered as having been a completely different basement block during Variscan times, as compared to the closely related Ötztal-Stubai and Scarl-Campo units. It is separated from the Scarl-Campo basement to the north by a major tectonic line (Peio line) which most probably also borders the Austro-Alpine basement which suffered from Alpine orogeny to the south, as does the Passeier-Jaufen line further to the east.

References

Andreatta C (1935) La formazione gneissico-kinzigitica e le oliviniti di Val d'Ultimo (Alto Adige). Mem Mus Storia Nat Ven Trid III (1): 1–160

Andreatta C (1948) La tettonica "a vortici" nei monti della Val di Sole. Rend Accad Naz Lincei ser VIII (8): 60–67

Andreatta C (1951 a) Il metamorfismo delle formazioni del gruppo dell'Ortler. Ric Sci 21 (2): 190–196

Andreatta C (1951 b) Carta Geologica delle Tre Venezie, Fo 9, Monte Cevedale, Firenze

Bockemühl C (1988) Der Marteller Granit (Südtirol, Italien); Petrographie, Geochemie, Altersbestimmungen. Inauguraldissertation, Univ Basel, 144 pp

Borsi S, Del Moro A, Sassi FP, Zirpoli G (1980) New petrographic and radiometric data on the Oetztal and Stubai orthogneisses (Eastern Alps). Neues Jahrb Mineral Monatsh 1980: 1–68

Caron M, Dössegger R, Steiger R, Trümpy R (1982) Das Alter der jüngsten Sedimente der Ortler-Decke (Oberostalpin) in der Val Trupchun (Schweizerischer Nationalpark, Graubünden). Eclogae Geol Helv 75 (1): 159–169

Chowanetz E (1990) Der Winnebachmigmatit (Ötztal/Tirol) – Argumente für eine altpaläozoische Anatexis. Diplomarbeit, Univ Wien, 73 pp

Dal Piaz GV, Martin S (1980) I porfiroidi nelle filladi della Valle dei Forni, Falda dell'Ortles, Austroalpine superiore. Rend Soc Geol Ital 3: 17–20

Dössegger R, Müller HW (1976) Die Sedimentserien der Engadiner Dolomiten und ihre lithostratigraphische Gliederung. Eclogae Geol Helv 69 (1): 229–238

Dössegger R, Furrer H, Müller HW (1982) Die Sedimentserien der Engadiner Dolomiten und ihre lithostratigraphische Gliederung, Teil 2. Eclogae Geol Helv 75: 303–330

Ehlers K, Hoinkes G (1987) Titanian chondrodite and clinohumite in marbles from the Ötztal crystalline basement. Mineral Petrol 36: 13–25

Frank W, Hoinkes G, Purtscheller F, Thöni M (1987) The Austroalpine unit west of the Hohe Tauern: the Ötztal-Stubai complex as an example for the Eoalpine metamorphic evolution. In: Flügel HW, Faupl P (eds) Geodynamics of Eastern Alps. Deuticke, Vienna, pp 272–281

Frisch W, Neubauer F (1989) Pre-Alpine terranes and tectonic zoning in the Eastern Alps. Geol Soc Am Spec Pap 230: 91–100

Gatto GO, Gregnanin A, Molin GM, Piccirillo EM, Scolari A (1976) Le manifestazioni "andesitiche" polifasiche dell'Alto Adige occidentale nel quadro geodinamico alpino. Studi Trent Sci Nat, N S 53 (5 A): 21–47

Gebauer D, Grünenfelder M (1978) U-Pb zircon dating of alpine-type garnet peridodites, example Val d'Ultimo (eastern Alps, northern Italy). US Geol Surv Open-File Rep 78–101: 135–137

van Gool JAM, Kemme MMJ, Schreurs GMMF (1987) Structural Investigations along an E-W cross-section in the Southern Ötztal Alps. In: Flügel HW, Faupl P (eds) Geodynamics of the Eastern Alps. Deuticke, Wien, pp 214–225

Grauert B (1969) Die Entwicklungsgeschichte des Silvretta-Kristallins auf Grund radiometrischer Altersbestimmung. Dissertation Phil Naturwiss Fak Univ Bern, 166 pp (Photodruck, München)

Grauert B (1981) Das Alter der Schlingentektonik im Silvretta-Ötztalkristallin aufgrund radiometrischer Altersbestimmungen. Fortschr Mineral 59 (1): 54–56

Gregnanin A, Piccirillo EM (1972) Litostratigrafia, tettonica e petrologia negli scisti austridici di alta e bassa pressione dell'area Passiria-Venosta (Alto Adige). Mem Ist Mineral Petrogr Univ Padova 28: 1–55

Hammer W (1902) Die krystallinen Alpen des Ultenthales, I. Das Gebirge südlich der Faltschauer. Jahrb kk Geol Reichsanst 52: 105–134

Hammer W (1912) Erläuterungen zur Geologischen Karte SW-Gruppe Nr. 66, Glurns und Ortler. Geol Reichsanst, Wien, 72 pp

Hammer W (1923) Erläuterungen zur Geologischen Spezial-Karte der Republik Österreich, Blatt Nauders (5245). Geol Bundesanst, Wien

Harre W, Kreuzer H, Lenz H, Müller P, Wendt I, Schmidt K (1968) Rb/Sr- and K/Ar-Altersbestimmungen an Gesteinen des Ötztalkristallins (Ostalpen). Geol Jahrb 86: 797–826

Helbig P, Schmidt K (1978) Zur Tektonik und Petrogenese am W-Ende des Schneeberger Zuges (Ostalpen). Jahrb Geol Bundesanst 121 (2): 177–217

Herzberg C, Riccio L, Chiesa S, Fornoni A, Gatto GO, Gregnanin A, Piccirillo EM, Scolari A (1977) Petrogenetic evolution of a spinel-garnet-lherzolite in the Austridic crystalline basement from Val Clapa (Alto Adige, northeastern Italy). Mem Ist Geol Minerol Univ Padova 30: 1–29

Hoernes S, Hoffer E (1973) Der Amphibolitzug des mittleren Ötztales (Tirol). Veröff Tirol Landesmus Ferdinandeum 53: 159–180

Hoinkes G (1981) Mineralreaktionen und Metamorphosebedingungen in Metapeliten des westlichen Schneebergerzuges und des angrenzenden Altkristallins (Ötztaler Alpen). Tschermaks Mineral Petrogr Mitt 28: 31–54

Hoinkes G, Purtscheller F, Schantl J (1972) Zur Petrographie und Genese des Winnebachgranites (Ötztaler Alpen, Tirol). Tschermaks Mineral Petrogr Mitt 18: 292–311

Hoinkes G, Kostner A, Thöni M (1991) Petrologic constraints for Eo-Alpine eclogite facies metamorphism in the Austro-Alpine Ötztal basement. Mineral Petrol 43: 237–254

Mader P (1987) Die Jura- und Kreideablagerungen im Lischana-Gebiet (Oberostalpine Scharl-Decke, Unterengadin). Eclogae Geol Helv 80: 633–653

Miller Ch (1970) Petrology of some eclogites and metagabbros of the Oetztal Alps, Tirol, Austria. Contrib Mineral Petrol 28: 42–56

Miller DS, Jäger E, Schmidt K (1967) Rb/Sr-Altersbestimmungen an Biotiten der Raibler Schichten des Brennermesozoikums und am Muskovitgranitgneis von Vent (Ötztaler Alpen). Eclogae Geol Helv 60: 537–541

Mogessie A, Purtscheller F (1986) Polymetamorphism of the Oetztal-Stubai basement complex based on amphibolite petrology. Jahrb Geol Bundesanst 129 (1): 69–91

Mogessie A, Purtscheller F, Tessadri R (1985) Geochemistry of amphibolites from the Oetztal-Stubai complex (northern Tyrol/Austria). Chem Geol 51: 103–113

Mogessi A, Purtscheller F, Tessadri R (1986) High alumina calcic amphiboles (alumino pargasite-magnesio sadanagaite) from metabasites and metacarbonates of central Oetztal, Eastern Alps (northern Tyrol, Austria). Neues Jahrb Miner Abh 154 (1): 21–39

Morten L, Obata M (1990) Rare earth abundances in the eastern Alpine peridotites, Nonsberg area, northern Italy. Eur J Mineral 2: 643–653

Obata M, Morten L (1987) Transformation of spinel lherzolite to garnet lherzolite in ultramafic lenses of Austridic crystalline complex, northern Italy. J Petrol 28: 599–623

Poli S (1989) Pre-Hercynian magmatism in the Eastern Alps: the origin of metabasites from the Austroalpine basement. Schweiz Mineral Petrogr Mitt 69: 407–421

Purdy JW, Jäger E (1976) K-Ar ages of rock-forming minerals from the Central Alps. Mem Ist Geol Mineral Univ Padova 30: 1–31

Purtscheller F (1969) Petrographische Untersuchungen an Alumosilikatgneisen des Ötztaler-Stubaier Altkristallins. Tschermaks Mineral Petrogr Mitt 13: 35–54

Purtscheller F (1978) Ötztaler und Stubaier Alpen. Sammlung geologischer Führer 53. Gebr. Bornträger, Berlin, Stuttgart, pp 1–128

Purtscheller F, Mogessie A (1988) Dikes from Ortler, Sarntal Alps and Brixen granite: mineralogy, chemical composition and petrogenesis. Mineral Petrol 38: 17–35

Purtscheller F, Rammlmair D (1982) Alpine metamorphism of diabase dikes in the Ötztal-Stubai metamorphic complex. Tschermaks Mineral Petrol Mitt 29: 205–221

Sander B (1929) Erläuterungen zur geologischen Karte Meran-Brixen. Schlernschr 16: 1–111

Satir M (1975) Die Entwicklungsgeschichte der westlichen Hohen Tauern und der südlichen Ötztalmasse auf Grund radiometrischer Altersbestimmungen. Mem Ist Geol Mineral Univ Padova 30: 1–84

Schmid S, Haas R (1989) Transition from near-surface thrusting to intrabasement decollement, Schlinig Thrust, Eastern Alps. Tectonics 8 (4): 697–718

Schmidegg O (1933) Neue Ergebnisse in den südlichen Ötztaler Alpen. Verh Geol Bundesanst 1933: 83–95

Schmidt K, Jäger E, Grünenfelder M, Grögler N (1967) Rb-Sr- und U-Pb-Altersbestimmungen an Proben des Ötztalkristallins und des Schneeberger Zuges. Eclogae Geol Helv 60: 529–536

Servizio Geologica d'Italia (1970) Carta Geologica d'Italia 1:100000, Foglio 4 (Merano), Salerno

Söllner F, Hansen BT (1987) "Pan-afrikanisches" und "kaledonisches" Ereignis im Ötztal-Kristallin der Ostalpen: Rb-Sr- und U-Pb-Altersbestimmungen an Migmatiten und Metamorphiten. Jahrb Geol Bundesanst 130 (4): 529–569

Spear FS (1988) The Gibbs method and Duhem's theorem: the quantitative relationships among P, T, chemical potential, phase composition and reaction progress in igneous and metamorphic systems. Contrib Mineral Petrol 99: 249–256

Spitz A, Dyhrenfurth G (1914) Monographie der Engadiner Dolomiten zwischen Schuls, Scanfs und dem Stilfserjoch. Beitr Geol Kart Schweiz, NF 74: 1–235

Thöni M (1981) Degree and evolution of the Alpine metamorphism in the Austroalpine unit W of the Hohe Tauern in the light of K/Ar and Rb/Sr age determinations on micas. Jahrb Geol Bundesanst 124: 111–174

Thöni M, with contrib by Hoinkes G (1983) The thermal climax of the early Alpine metamorphism in the Austroalpine thrust sheet. Mem Sci Geol 36: 211–238

Thöni M (1986) The Rb-Sr thin slab isochron method – an unreliable method for dating geologic events in polymetamorphic terrains? Mem Sci Geol 38: 283–352

Thöni M (1988) Rb-Sr isotopic resetting in mylonites and pseudotachylites: implications for the detachment and thrusting of the Austroalpine basement nappes in the Eastern Alps. Jahrb Geol Bundesanst 131 (1): 169–201

Veltman Ch (1986) Zur Polymetamorphose pelitischer Gesteine im Ötztal-Stubaier Altkristallin. Ph D Thesis, Univ Innsbruck

Austro-Alpine Basement Complex to the South of the Tauern Window

B. SCHULZ, G. NOLLAU, H. HEINISCH and G. GODIZART

Abstract

The pre-Upper Ordovician units of the Altkristallin and the Ordovician to Upper Devonian sedimentary and volcanic rocks of the phyllitic/fossiliferous Palaeozoic sequences show similar and parallel post-Upper Ordovician deformational structures but different P-T evolutions. In the Altkristallin, early prograde high pressure – low temperature metamorphism followed by a high pressure eclogitic stage and later decompression, evolved synchronous to a foliation-forming shearing deformation. Temperature-dominated amphibolite-facies overprinting with local formation of first melts post-dates this deformation. Greenschist-facies low pressure metamorphism, locally reaching the amphibolite facies during early shearing, is observed in the phyllitic sequences. A pre-Alpine age of medium-grade metamorphism of fossiliferous Palaeozoic sequences is not definitely proved. Folding of the foliation and later formation of shear bands are related to a common late Variscan uplift/cooling history and overprinted the foliation-parallel lithological contacts of the units. A Variscan process of progressive crustal stacking which first reworked the Altkristallin to the NW and then affected the phyllitic/fossiliferous Palaeozoic sequences to the SE can be inferred. An increasing degree of Alpine greenschist- to amphibolite-facies ductile overprinting from south to north is observed in the northern parts of the basement. In the southern parts, pre-Alpine structures and mineral assemblages remained well preserved during late Alpine brittle deformations.

1 Geological Setting in the Alpine Framework

The Austro-Alpine basement complex (ABC) to the south of the Tauern Window, in the western part the area of "classical" research ("Die Zone der alten Gneise" of Sander 1912, 1920, 1925, 1929; Bianchi 1934; Dal Piaz 1934), is subdivided and surrounded by several late Alpine shear zones and faults (Fig. 1 a, b). To the south, the Pustertal line (P) and the Gailtal line (G), which are parts of the Periadriatic lineament (PL), mark a significant border to the South Alpine realm. NW-SE trending dextral riedel shears to the south of the Tauern window are related to late Alpine dextral strike-slip movements along the PL. One of these faults, the Mölltal line (ML), marks the eastern border of the discussed part of the Austro-Alpine basement complex (ABC). Other similarly oriented faults geographically subdivide the crystalline basement into the Merano-Mules-Anterselva- and Defer-egger Alps to the west and into the Schobergruppe, Kreuzeckgruppe and Goldeckgruppe to the east (Fig. 1). The border of the Drauzug Permo-Mesozoic (D) and the internal structure of the Gailtal metamorphic basement (GMB) are, as well, controlled by late Alpine polyphase brittle transpressive faulting.

Several W-E trending late Alpine shear zones and faults further subdivide the crystalline basement. The Defereggen-Antholz-Vals line (DAV) is assembled by a northern mylonitic and a southern cataclastic domain and separates, in the western part of the basement, a northern block, with Alpine metamorphism and ductile deformation, from a southern block, lacking pervasive Alpine ductile overprinting (Borsi et al. 1978 a; Hofmann et al. 1983; Stöckhert 1985; Kleinschrodt 1987; Schulz 1989). The regional distribution pattern of late Alpine radiometric cooling ages, ranging from 15 to 28 Ma in the northern block and contrasting late Variscan mica ages in the southern block (Borsi et al. 1978 a), points to a considerable differential vertical movement component during sinistral strike-slip displacement along the shear zone (Borsi et al. 1978 a; Kleinschrodt 1987).

No discrete Alpine tectonic line separating Alpine and pre-Alpine parts of the basement has been reported from the Schobergruppe. Alpine structural and metamorphic overprinting in the northern part of the region continuously increases to the north of the line Hochschoberhütte – Lienzerhütte (Behrmann 1990). In the Kreuzeckgruppe, the

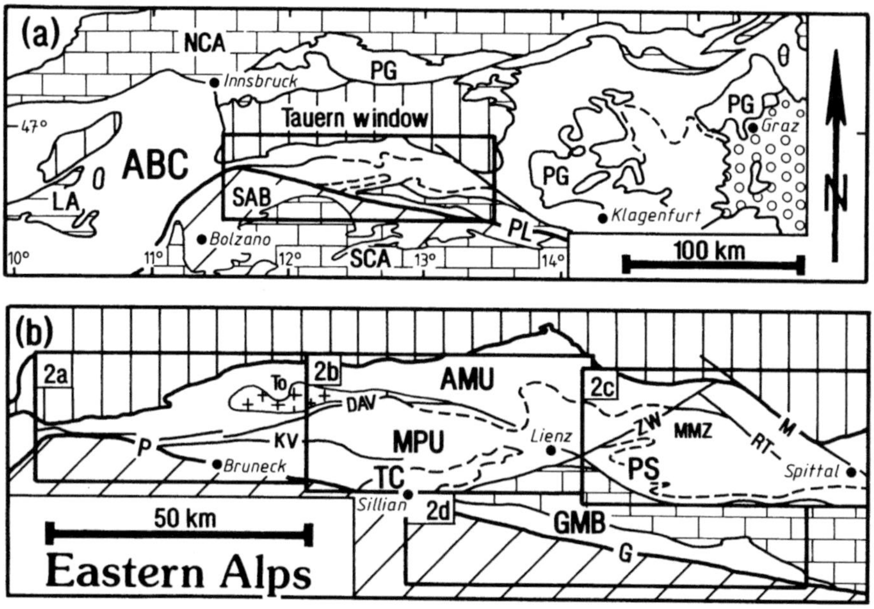

Fig. 1. a General geological setting of the Austro-Alpine basement complex to the south of the Tauern window. *ABC* Austro-Alpine basement complex; *D* Permo-Mesozoic of Drauzug; *LA* Lower Austro-Alpine; *NCA* northern Calcareous Alps; *PG* Palaeozoic sequences of Grauwackenzones, Gurktal, Graz; *PL* Periadriatic lineament; *SAB* South-Alpine basement; *SCA* southern Calcareous Alps. **b** Geological units in the Austro-Alpine basement complex to the south of the Tauern window; *insets* refer to Fig. 2.
Pre-Alpine units: *AMU* metapsammopelite-amphibolite-marble unit; *MPU* metapsammopelitic unit; *GMB* Gailtal metamorphic basement: *PS* phyllitic sequences of Kreuzeck- and Goldeckgruppe; *TC* Thurntaler complex (part of the phyllitic sequences); *To* late Alpine Rieserferner tonalite.
Late Alpine tectonic lines: *DAV* Defereggen-Antholz-Vals line; *G* Gailtal line (part of the Periadriatic lineament; *KV* Kalkstein-Vallarga line; *M* Mölltal line; *MMZ* main mylonitic zone in the Kreuzeckgruppe; *PL* Pustertal line (part of the Periadriatic lineament); *RT* Ragga-Teuchl fault zone; *ZW* Zwischenbergen-Wöllatratten line.

steeply southward-dipping main mylonite zone (MMZ) divides the Alpine overprinted northern Polinik unit from a southern Strieden unit where Alpine metamorphism did not exceed lower greenschist facies conditions. A NW-directed movement of the Strieden unit over the Polinik unit between 90 and 40 Ma has been reported (Hoke 1990).

Only brittle Alpine deformation is observed along the other important minor lines like the Kalkstein-Vallarga line (KV) (Borsi et al. 1978a; Guhl and Troll 1977, 1987), the Zwischenbergen-Wöllatratten line (ZW), the Ragga-Teuchl fault zone (RT) (Hoke 1990) and the northern and southern border faults of the Drauzug Permo-Mesozoic (Heritsch and Paulitsch 1958; van Bemmelen and Meulenkamp 1965; Heinisch et al. 1984; Heinisch and Sprenger 1988). Relics of a post-Variscan sedimentary cover of the crystalline basement are also found immediate to the south of the DAV and along KV, PL and ZW lines (Fig. 2; Table 1). The Permo-Triassic sediments of Mauls (Schindelmayr 1968; Frank et al. 1977; Hammerschmidt 1982), Staller Alm (Senarclens-Grancy 1964; Schulz 1989), Kalkstein (Guhl and Troll 1977, 1987; Niedermayr et al. 1984), Gödnachgraben (Exner 1962) and the Drauzug Permo-Mesozoic (Bechstädt 1978) were preserved from erosion due to their tectonic position in former graben or pull-apart systems which were closed during late Alpine times (van Bemmelen and Meulenkamp 1965; Sprenger and Heinisch 1992).

An increasing degree of early and late Alpine ductile structural and metamorphic reworking from southern to northern parts of the northern block, northern Schobergruppe and the Polinik unit is observed (Contini and Sassi 1980; Stöckhert 1982, 1985; Behrmann 1990; Hoke 1990). Early Alpine foliations and folds (Stöckhert 1984; Behrmann 1990; Hoke 1990) were refolded during late Alpine times and a structural continuity between the Austro-Alpine unit and the underlying Mesozoic rocks of the Matreier zone/Upper Schieferhülle (MZ/USH) is obvious (Barnick 1965; Nollau 1969; Baggio et al. 1971; Hammerschmidt 1977; Stöckhert 1985; Behrmann and Wallis 1987). In the western part of the basement pressures about 8 kb at 400 to 450 °C (Stöckhert 1984) are related to an early Alpine underthrusting of the northern part of the Austro-Alpine continental margin beneath the Adriatic Plate; 550 °C at 4 kb

(Schulz 1990b) were reached in course of a subsequent syndeformational uplift. According to K-Ar ages from small muscovites of the Alpine main foliation in late Variscan pegmatites, this early Alpine event appears to be older than 100 Ma (Prochaska 1981b; Stöckhert 1984). Amphibolite-facies early Alpine metamorphism with crystallization of staurolite and kyanite under conditions of 6.25 ± 1.25 kb at 620 ± 70 °C was reported from the Polinik unit in the Kreuzeckgruppe (Hoke 1987, 1990). In the northern regions of the eastern basement, K-Ar ages of cooling after early Alpine burial metamorphism range around 80 Ma (Oxburgh et al. 1966; Brewer 1969, 1970; Lambert 1970; Hawkesworth 1976; Waters 1976; Deutsch 1988; data compiled by Hoke 1990).

Illite crystallinity in the Permo-Mesozoic sequences in the southern parts of the basement indicates an epizonal Alpine metamorphism there (Niedermayr et al. 1984; Guhl and Troll 1987; Schulz 1989). Accordingly, in the neighbouring crystalline rocks, the Alpine metamorphic conditions rarely exceeded 300 °C and 2.5 kb (Stöckhert 1982), and in consequence late Variscan mica ages or Alpine-Variscan "mixing ages" occur (Borsi et al. 1978a; Brewer 1969; Hoke 1990). Alpine deformation in the southern parts of the basement is restricted to faulting, kink folding and local large-scale ductile-brittle folding (Sassi et al. 1974a; Guhl and Troll 1977, 1987; Schulz 1989, 1991).

The granodioritic-tonalitic plutonic bodies of the Rensen (Nollau 1974; Borsi et al. 1978b), Rieserferner-Zinsnock (Borsi et al. 1979; Bellieni 1980; Bellieni et al. 1976, 1981, 1982; Gratzer 1984) and Lienz (Borsi et al. 1978a) intruded around 30 Ma and led to local contact metamorphism (Prochaska 1981a; Mager 1985; Spaeth and Kreutzer 1989). The DAV cuts the contact metamorphic zone and southern parts of the Rieserferner pluton. Small bodies of tonalites are found within the PL (Bögel 1975; Zanferrari 1976; von Gizycki and Schmidt 1978; Heinisch and Sprenger 1988) and sometimes show slight ductile deformation. Pre-, syn- and post-intrusive dykes associated with the plutons are widespread in the basement. Porphyritic dykes in the eastern parts of the basement (30–40 Ma, Deutsch 1988) cut the MMZ but terminate at the brittle ZW and RT fault zones (Hoke 1990).

Rb-Sr ages of 28 – 15 Ma from biotites (Borsi et al. 1978a) refer to a final late Alpine uplift and cooling of the western northern block after the early Alpine metamorphism. To the east, late Alpine ages are restricted to the Penninic unit of the Tauern window. Generally, the Late Alpine deformation in the ABC can be related to movements and rotations of the Adriatic Plate along the PL. An earlier sinistral transpressional regime with NE-SW directed compression, possibly induced by an Oligocene short-staged N-NE-directed push of the Adriatic plate, was responsible for the sinistral strike-slip movement along the DAV (Kleinschrodt 1987). A later dextral transpressional regime with NW-SE directed compression is related to the dextral movement of the Adriatic plate along the PL and may explain late Alpine faults, kink folds and large-scale folds in the southern parts of the basement. In the Gailtal metamorphic basement (GMB) the intrusion of Oligocene tonalites and their subsequent ductile deformation is followed by four phases of late Alpine deformations involving two acts of pseudotachylite formation which are adjoined to, probably dextral, strike-slip faulting (Heinisch and Sprenger 1988; Sprenger and Heinisch 1992).

Apatite fission-track ages in the basement are continuously younger from the south to the north (20.5–8 Ma) and display a homogeneous post-Oligocene uplift postdating the significant vertical displacements along the DAV, MMZ and RT lines. Uplift rates of 0.2–0.3 mm/a in the western part (Grundmann and Morteani 1985) and of 0.1–0.16 mm/a in the eastern part of the basement from 20 Ma were reported (Stauffenberg 1987).

2 Pre-Alpine Tectono-Stratigraphic Units and Lithologies

The Austro-Alpine basement complex to the south of the Tauern window can be subdivided into four pre-Alpine lithologic associations of metamorphic rocks (Figs. 1, 2). A northern metapsammopelite-amphibolite-marble unit (AMU), mainly overprinted by the Alpine events, underlies a southern monotonous metapsammopelitic unit MPU. These two units are called "Liegend"- and "Hangend-Komplex" in the Schobergruppe and "northern zone" (Polinik unit)/"southern zone" (Strieden unit) in the central Kreuzeckgruppe. From the eastern parts of the Kreuzeckgruppe and the Goldeckgruppe such a lithological subdivision has not yet been reported. Both lithological units assemble the classical "Altkristallin" of the basement complex. The phyllitic sequences of the Thurntaler complex (TC), of the Kreuzeckgruppe, and the Goldeckgruppe occur on the southern border of the "Altkristallin". A forth regional-scale lithological unit is the Gailtal metamorphic basement (GMB) which is made up of four different rock associations labelled I–IV. Associations I–III may be assigned to the MPU. Association IV consists of fossiliferous early-Palaeozoic (Ordovician

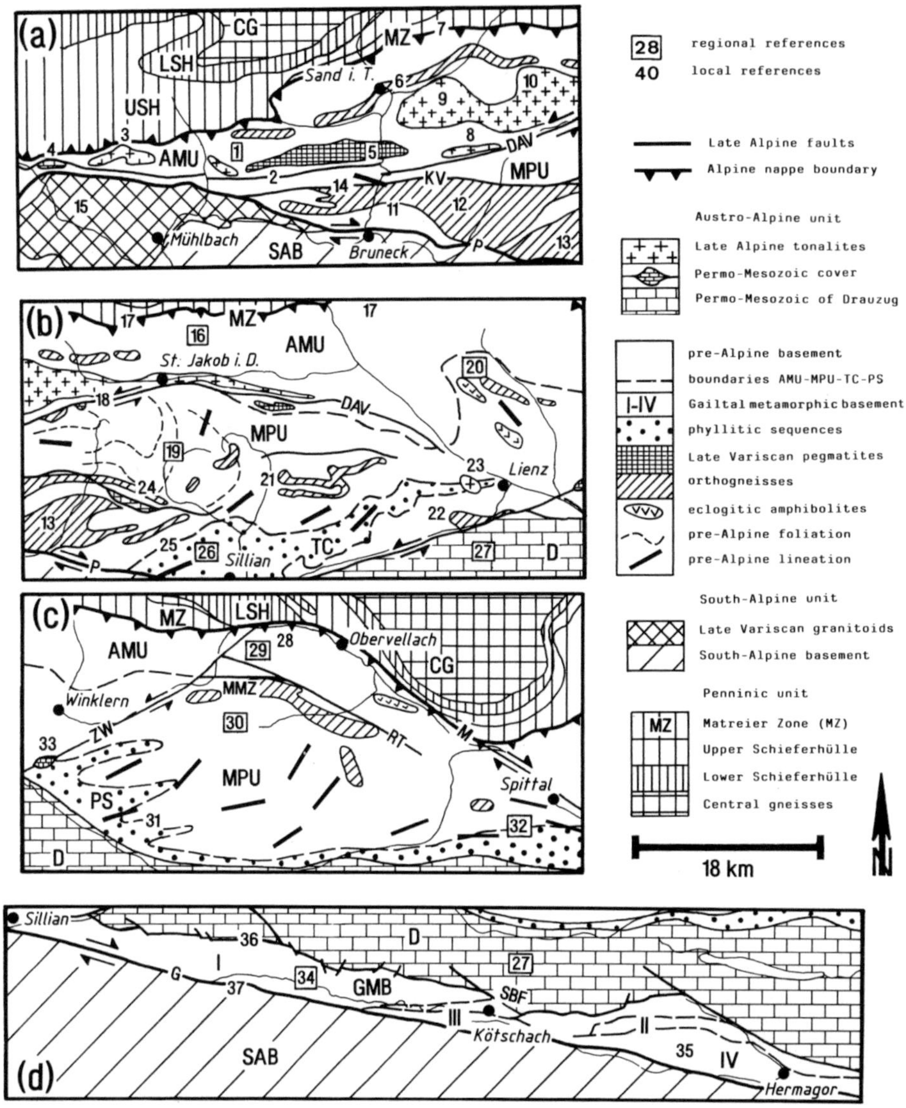

Fig. 2. Geological setting and structures in the Austro-Alpine basement complex to the south of the Tauern window (see Fig. 1 b for overall view). *Numbers* indicate local geological features and references listed in Table 1.

to Lower Devonian) metasediments. Acid orthogneisses and ortho-augengneisses occur in the AMU, MPU and in associations I–III of the GMB. These rocks are not found in the phyllitic sequences and in the association IV of the GMB. This allows us to subdivide the basement complex into the two major lithotectonic groups "Altkristallin" and "phyllitic sequences/fossiliferous Palaezoic sequences" both with partly different pre-Alpine histories.

2.1 Metapsammopelite-Amphibolite-Marble Unit (AMU)

The AMU accompanies the southern rim of the Tauern window (Figs. 1 b, 2, 3) and similar sequences of metamorphic rocks are reported from the western, central and eastern parts of the basement (Angel 1930; Angel and Krajicek 1939; Exner 1955, 1956; Oxburgh 1966; Oxburgh et al. 1966; Troll et al. 1976; Deutsch 1977; Hofmann et al. 1983; Behrmann and Wallis 1987; Heinz 1987; Behrmann 1990; Hoke 1990). In the northern and lowermost parts of the unit, chloritic garnet-muscovite-micaschists are widespread. These uniform rocks, called "Cima-Dura-phyllites" in the western part of the basement complex (Dal Piaz 1934; Baggio et al. 1971; Hammerschmidt 1977) were completely overprinted by Alpine events, as is obvious from radiometric and structural data, (Borsi et al. 1978a; Stöckhert 1985; Behrmann and Wallis 1987; Deutsch 1988; Hoke 1990).

Biotite micaschists and biotite gneisses, partly with garnet, staurolite, kyanite-sillimanite-andalusite and cordierite (Bellieni and Visona 1981) prevail in

Table 1. List of regional descriptive terms and related references from the Austro-Alpine basement complex to the south of the Tauern window. Numbers 1–15 refer to Fig. 2a; numbers 16–27 refer to Fig. 2b; numbers 28–33 refer to Fig. 2c and numbers 34–37 are found in Fig. 2d.

a

1 Western part of the metapsammopelite-amphibolite-marble unit (AMU): Sander (1929); Bianchi (1934); Dal Piaz (1934); Baggio et al. (1971); Borsi et al. (1978a); Contini and Sassi (1980); Stöckhert (1982, 1985); Hofmann et al. (1983); Höhler (1984); Grundmann and Morteani (1985); Kleinschrodt (1987)

2 Western part of the Defereggen-Antholz-Vals line (DAV): Borsi et al. (1978a), Stöckhert (1982); Kleinschrodt (1987)

3 Rensen pluton and Austro-Alpine – Penninic transition: Nollau (1969, 1974); Borsi et al. (1978b); von Gizycki and Schmidt (1978)

4 Trias of Mauls: Schindelmayr (1968); Frank et al. (1977); Hammerschmidt (1982)

5 Aplitic-pegmatitic rocks of Uttenheim (Villa Ottone): Borsi et al. (1980); Stöckhert (1984, 1987)

6 Ortho-augengneiss of Sand i. T. (Campo Tures): Hammerschmidt (1981); Mager (1981)

7 Matreier Zone (MZ): Baggio et al. (1971); Hammerschmidt (1977)

8 Metabasites of Croda Nera and southern contact of late Alpine Rieserferner pluton: Bellieni and Visonà (1981); Frizzo (1981); Mager (1985)

9 Rieserferner and Zinsnock (Vedrette di Ries and Cima Vila) plutons: Bellieni et al. (1976, 1981, 1982); Borsi et al. (1979); Bellieni (1980); Gratzer (1984); Mager (1985)

10 Northern contact zone of the Rieserferner pluton: Prochaska (1981 a, b)

11 Western part of the metapsammopelitic unit (MPU): Dal Piaz (1934); Sassi and Zirpoli (1973); Sassi et al. (1974a); Borsi et al. (1973, 1978a); Contini and Sassi (1980); Stöckhert (1982, 1985); Kleinschrodt (1987)

12 Orthogneiss of Antholz (Anterselva): Borsi et al. (1973, 1978a); Satir (1975); Peccerillo et al. (1979); Mager (1981)

13 Orthogneiss of Gsies (Casies): Borsi et al. (1973, 1978a) Peccerillo et al. (1979); Guhl and Troll (1987)

14 Kalkstein-Vallarga line (KV): Sassi et al. (1974a); Borsi et al. (1978a); Stöckhert (1982); Kleinschrodt (1987)

15 Late Variscan granitoid of Brixen (Bressanone): Borsi et al. (1978a)

b

16 Central part of the metapsammopelite-amphibolite-marble unit (AMU): Senarclens-Grancy (1964, 1972); Troll and Hölzl (1974); Troll et al. (1976); Troll (1978); Schulz (1990b)

17 Transition Austro-Alpine – Matreier zone: Barnick (1965); Behrmann and Wallis (1987); Wallis (1988); Behrmann (1990)

18 Eastern part of the Defereggen-Antholz-Vals line (DAV): Senarclens-Grancy (1964); Lunzer (1985); Steininger (1985); Schulz (1989); Spaeth and Kreutzer (1990)

19 Central part of the metapsammopelitic unit (MPU): Schmidegg (1936, 1937); Senarclens-Grancy (1964); Borsi et al. (1973, 1978a); Sassi and Zirpoli (1973); Sassi et al. (1974a, b); Guhl and Troll (1977, 1987); Troll (1978); Contini and Sassi (1980); Ntaflos (1981); Schulz (1988a, b, c, 1990a); Behrmann (1990)

20 Eclogitic amphibolites and orthogneisses of the Schobergruppe: Clar (1927a, b); Exner and Wanderer (1962); Richter (1973); Troll and Hölzl (1974); Troll et al. (1976, 1980); Brack (1977); Troll (1978); Behrmann (1990); Godizart and Schulz (1992); Schulz (1993)

21 Orthogneisses to the east of Winkeltal: Cliff (1980)

22 Orthogneiss of Leisach: Senarclens-Grancy (1964); Spaeth and Kreutzer (1989)

23 Tonalitic pluton of Pölland/Lienz: Borsi et al. (1978a); Spaeth and Kreutzer (1989)

24 Permotrias of Kalkstein: Schmidegg (1937); Guhl and Troll (1977, 1987); Niedermayr et al. (1984)

25 Transition between metapsammopelitic unit (MPU) and Thurntaler complex (TC) in the Markinkele area: Sassi and Zanferrari (1972); Krol (1974); Heinisch and Schmidt (1976, 1984)

26 Phyllitic sequences of the Thurntaler complex (TC): Sassi (1972); Sassi and Zanferrari (1972); Krol (1974); Schönlaub (1979); Heinisch and Schmidt (1976, 1982, 1984); Heinisch (1981); Spaeth and Kreutzer (1989, 1990); Schulz (1991)

27 Permo-Mesozoic of Drauzug: van Bemmelen and Meulenkamp (1965); Mostler (1972); Bechstädt (1978); Niedermayr et al. (1984)

c

28 Eastern part of the metapsammopelite-amphibolite-marble unit (AMU): Angel (1930); Oxburgh (1966); Oxburgh et al. (1966); Brewer (1969, 1970); Lambert (1970); Waters (1976); Hoke (1990)

29 Polinik unit (AMU) in the Kreuzeckgruppe: Hoke (1987, 1990)

30 Eastern part of the metapsammopelitic unit (MPU), Strieden unit in the Kreuzeckgruppe: Angel and Krajicek (1939); Exner (1955, 1956); Holzer (1958); Brewer (1969, 1970); Cliff and Rex (1976); Hawkesworth (1976); Hoke (1987, 1990)

31 Phyllitic sequences and mineralizations of the Kreuzeckgruppe (PS): Friedrich (1963); Lahusen (1972); Reimann (1980); Reimann and Stumpfl (1981)

32 Altkristallin and phyllitic sequences of the Goldeckgruppe: Angel and Krajicek (1939); Schwinner (1943); Brewer (1970); Hawkesworth (1976); Deutsch (1977, 1988); Heinz (1987)

33 Mesozoic of Gödnachgraben: Exner (1962)

d

34 Gailtal metamorphic basement (GMB): von Klebelsberg (1935); Heritsch and Paulitsch (1958); Visonà (1974); Heinisch et al. (1984); Heinisch (1987); Heinisch and Sprenger (1988)

35 Palaeozoic sequences (unit IV) of the Gailtal metamorphic basement (GMB): Schönlaub (1979); Heinisch (1987)

36 Southern border fault of Permo-Mesozoic of Drauzug: van Bemmelen and Meulenkamp (1965); Sprenger and Heinisch (1992)

37 Gailtal line (G) (part of the Periadriatic lineament): Sassi and Zanferrari (1971); Sassi et al. (1974a); Bögel (1975); Heinisch and Sprenger (1988); Sprenger and Heinisch (1992)

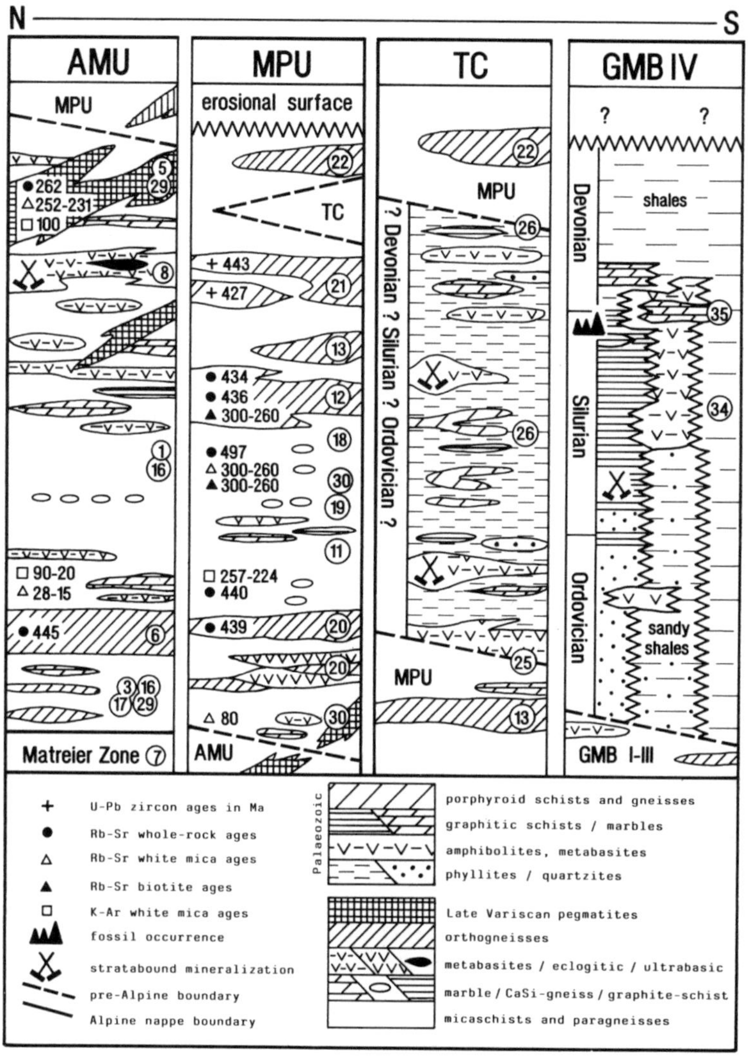

Fig. 3. Schematic lithological succession of pre-Alpine metamorphic rocks in metapsammopelite-amphibolite-marble unit *(AMU);* metapsammopelitic unit *(MPU);* phyllitic sequences of Thurntaler complex *(TC)* and Palaeozoic sequences *(unit IV)* of the Gailtal metamorphic basement *(GMB)* with radiometric and biostratigraphic data (in million years, Ma) *Bold encircled numbers* indicate references listed in Table 1. (For geological profiles through the basement, see Senarclens-Grancy 1964; Troll et al. 1976, 1980; Borsi et al. 1978a; Behrmann 1990; Hoke 1990; Schulz 1991)

the southern part of the AMU. The metapsammopelites with calcsilicatic, graphitic and quartzitic layers are interlayered by several W-E trending ortho-augengneisses, metabasites, marbles and acid aplitic-pegmatitic rocks.

Höhler (1984) demonstrated the regional inhomogeneity of Alpine reworking of pre-Alpine tremolite-bearing marbles and calcsilicate rocks by grain size and microstructural analysis. Coarse-grained parts of the marble complexes remain as relics of pre-Alpine recrystallization, smaller grains were generated during Alpine overprinting.

Microcline-bearing ortho-augengneisses are found in layers only a few metres thick to the north, and in large bodies extending tens of kilometres to the south of the garnet-muscovite-micaschists. Morphological features of zircons suggest an intrusive origin of the augengneisses of Campo Tures (Sand in Taufers) (Mager 1981). A Rb-Sr whole-rock isochron of 445 ± 24 Ma ($Sr_{init.}$ 0.7078 ± 0.0013) Hammerschmidt 1981) dates the intrusion of the acid plutonic rocks. This age corresponds to Upper-Ordovician isochrons from other orthogneisses of the basement complex (discussed later).

Amphibolites in the western part of the AMU are metamorphosed basic magmatites. Pre-Alpine high-grade assemblages (brown amphibole + bytownite + salite + Fe-Ti-oxide) and greenschist facies assemblages of early Alpine overprinting can be distinguished in these rocks. For both types, annealed fabrics indicate post-deformational crystallization. The structures were overprinted by a late Alpine deformation which induced numerous retrograde replacement reactions. Erratic distribution patterns displayed by most major and large ion lithophile elements reflect redistribution of these elements by metamorphic processes. In contrast, consistent variation trends in the high field-strength elements (HFSE) and rare earth elements (REE) suggest no significant change of these elements during metamorphism. Based on these elements, two major metabasite groups with distinct chemical characteristics along various mafic and ultramafic cumulate compositions were recognized: group 1 metabasites are low in incompatible elements and show affinities to MORB with respect to HFSE absolute abundances and element ratios like Zr/Y, Zr/Nb and Ti/V. REE patterns are LREE-depleted to flat with $La_N/Yb_N =$ 0.5–1.1. Group 2 metabasites with transitional to alkaline compositions are similar to OIB in all chemical aspects. Compared to group 1 metabasites, they are significantly enriched in the HFSE and LREE and depleted in the HREE. The REE patterns are fractionated with La_N/Yb_n 14.6–19.6 (Figs. 4a–d, g, h). Both groups cannot be related to each other by simple magmatic fractionation processes or different degrees of partial melting. A generation of group 1 metabasites from a LREE-depleted mantle source in a divergent plate boundary setting and a generation of the group 2 metabasites from an enriched mantle source in a within-plate setting, therefore is a satisfactory explanation for the chemical signatures of these rocks (Godizart 1989).

To the south of the Rieserferner (Vedrette de Ries) Massif, a 200-m-thick sequence of foliated and folded metabasites (metabasites of Schwarze Wand, Croda Nera) outcrops (Fig. 2a). Amphibolites (meta-alkalibasalts according to Godizart 1989) and epidote-amphibolites with intercalated clinopyroxene-garnet gneisses, clinopyroxene-calcite layers and a hornblendite alternate with pegmatoid gneisses and garnet-biotite gneisses (Bianchi 1934; Bellieni and Visona 1981; Mager 1985). The morphologies of zircons in the metabasites point to basic volcanic tuffs (Mager 1985) which are associated with stratiform scheelite mineralizations (Frizzo 1981).

One group of amphibolites in the Polinik unit of the Kreuzeckgruppe contains garnet enclosed in symplectites after clinopyroxene with relict omphacite. These rocks have been derived from eclogites with the original mineral assemblage omphacite-garnet-rutile-quartz, preserved in the cores of amphibole pods (Hoke 1990).

Aplitic and pegmatitic rocks are interlayered with the biotite gneisses and follow the southern border of the AMU in the western, central and eastern parts. A pegmatite-rich zone with garnet-sillimanite-biotite bearing metapelitic host rocks is found immediately to the south of the DAV (Fig. 2b) (Senarclens-Grancy 1964; Schulz 1988a). Pegmatite gneisses in the Polinik unit yielded a minimum age of 254 Ma (Hoke 1990). In the northern and lowermost part of the adjacent Strieden unit, sillimanite- and andalusite-bearing metapelites, with evidence for partial melting, occur. Interlayered pegmatites with kalifeldspar-plagioclase-quartz-sillimanite and absent primary muscovite in the region produced a Rb-Sr whole-rock age of 266 ± 6 Ma (Cliff pers. comm. in Hoke 1990) and postdate the formation of post-deformational sillimanite and andalusite in the host rocks. These pegmatite-bearing zones seem to form the uppermost part of the AMU and the lowermost part of the overlying MPU unit respectively. At Uttenheim (Villa Ottone, Ahrn Valley, Southern Tyrol), muscovite pegmatites bearing tourmaline and garnet, and kilometre-scale muscovite-granite bodies outcrop. No parental granite has been found yet. A Rb-Sr whole-rock isochron of 262 ± 5 Ma (Sr_{init} 0.7157 ± 0.0004) and Rb-Sr ages of 252-231 Ma from large muscovites in the rocks (Borsi et al. 1973, 1980) are interpreted to date the intrusion. This points to a late Variscan magmatism. According to Stöckhert (1987), these pegmatitic rocks can be derived from small-scale lit-par-lit intrusions of water-rich acid melts which were produced by partial anatexis of the host rocks in lower parts of the crust. Early-Alpine deformation then led to a recrystallization of white micas and a penetrative foliation of the pegmatites which were refolded at late Alpine time (Stöckhert 1984, 1987; Kleinschrodt 1987).

2.2 *Metapsammopelitic Unit (MPU)*

To the east, the metapsammopelitic unit MPU concordantly overlies the AMU along a pre-Alpine foliation-parallel contact (Figs. 1b, 2, 3). This contact, referred as "Schuppenzone" by Clar (1927b) separates the Liegendkomplex (AMU) from the Hangendkomplex (MPU) in the Hochschober area of the Schobergruppe (Troll and Hölzl 1974; Troll et al. 1976; Behrmann 1990). To the west, both units are divided by the DAV. Foliated eclogitic amphibolites of the Prijakt and Schleinitz area of the Schobergruppe (Richter 1973; Troll and Hölzl 1974; Troll et al. 1976, 1980) and in the Teuchlgraben area of the Kreuzeckgruppe (Angel 1930; Exner and Wanderer 1962) interlayer the lower part of a monotonous sequence of paragneisses and micaschists.

The eclogitic amphibolites of the Prijakt area (Schobergruppe) show an overall tholeiitic composition with a complex trace element geochemistry. The large ion lithophile elements have been completely

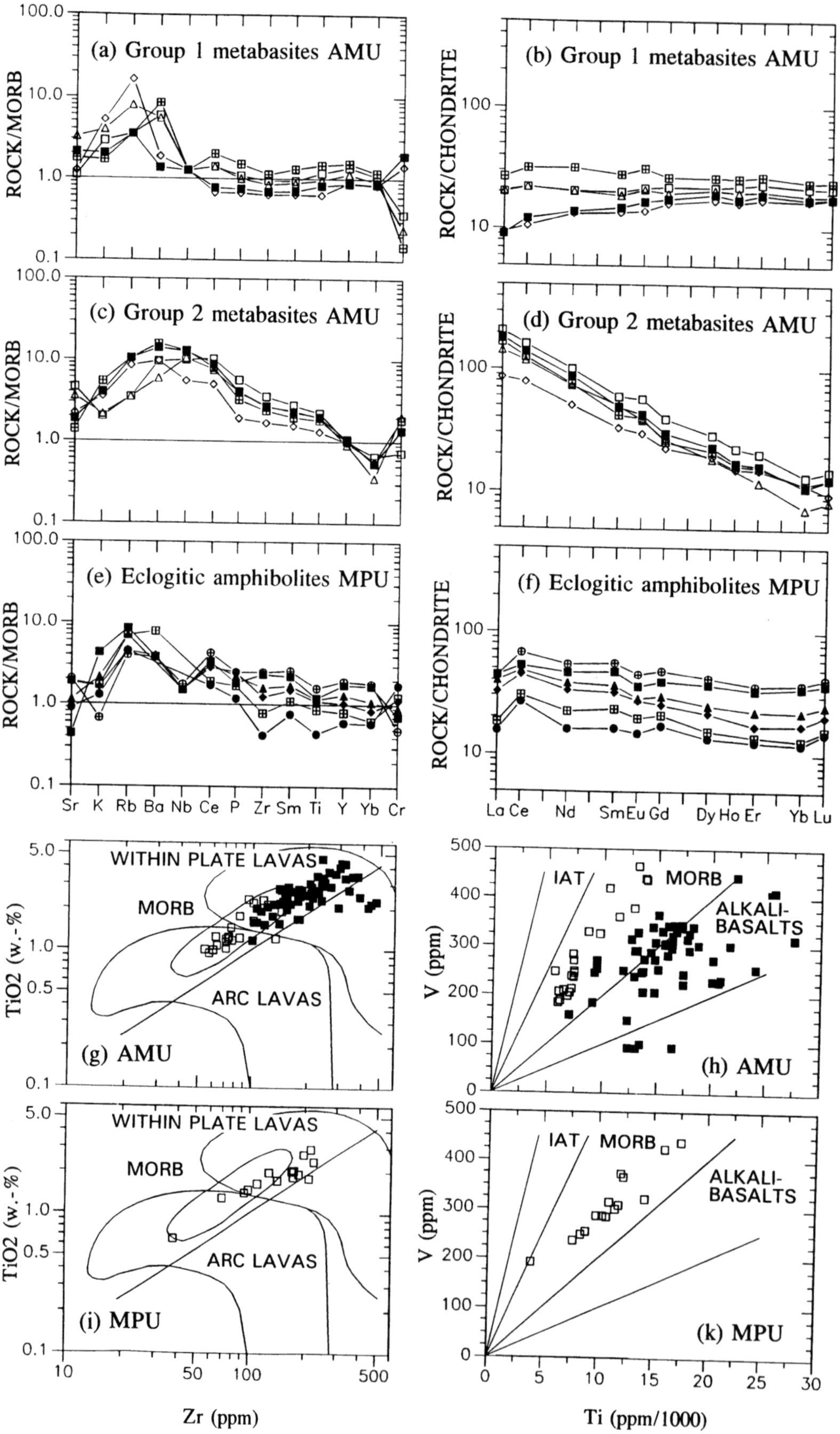
(a) Group 1 metabasites AMU
ROCK/MORB
100.0
10.0
1.0
0.1
(b) Group 1 metabasites AMU
ROCK/CHONDRITE
100
10
(c) Group 2 metabasites AMU
ROCK/MORB
(d) Group 2 metabasites AMU
ROCK/CHONDRITE
(e) Eclogitic amphibolites MPU
ROCK/MORB
Sr K Rb Ba Nb Ce P Zr Sm Ti Y Yb Cr
(f) Eclogitic amphibolites MPU
ROCK/CHONDRITE
La Ce Nd Sm Eu Gd Dy Ho Er Yb Lu
WITHIN PLATE LAVAS
MORB
ARC LAVAS
(g) AMU
TiO2 (w.-%)
5.0
1.0
0.5
0.1
IAT
MORB
ALKALI-BASALTS
(h) AMU
V (ppm)
500
400
300
200
100
0
WITHIN PLATE LAVAS
MORB
ARC LAVAS
(i) MPU
TiO2 (w.-%)
10
50
100
500
Zr (ppm)
IAT
MORB
ALKALI-BASALTS
(k) MPU
V (ppm)
0
5
10
15
20
25
30
Ti (ppm/1000)

redistributed. LREE and P were also affected by limited mobilization. High field-strength element absolute abundances and element ratios like Ti/V are in the range of present-day MORB (Figs. 4e, f, i, k). However, provided that high La/Ta ratios and a small negative Nb-Ta anomaly are primary characteristics and not due to possible metamorphic redistribution, they could indicate involvement of a subduction-related component or crustal contamination in magma genesis.

REE patterns are slightly enriched in the LREE with Ce_N/Yb_N ranging from 1.4 to 3.2. The observed compositional variations can be attributed to fractional crystallization processes, as is indicated by the wide range in mg-values, a strong iron enrichment trend, good negative correlations between compatible and incompatible elements, and nearly parallel patterns in MORB- and chondrite-normalized variation diagrams (Godizart and Schulz 1992).

A several kilometre-thick monotonous metapsammopelitic sequence of banded quartzitic paragneisses and micaschists bearing garnet, staurolite, kyanite-sillimanite-andalusite constitutes the major part of the MPU (Schmidegg 1936, 1937; Exner 1955, 1956; Senarclens-Grancy 1964; Schulz 1988a). Rarely interlayered thin amphibolites show no textural evidence of former eclogite assemblages. Thin layers of marbles, graphitic gneisses and quartz-feldspar gneisses are rarely found. Sometimes calcsilicate-gneisses with zoned contacts to their host rocks form cigar- and tongue-shaped bodies extending several metres parallel to a stretching lineation (Schulz 1988c). Borsi et al. (1973) obtained a Rb-Sr whole-rock isochron of 497 ± 38 Ma from paragneisses, which seems to date a first metamorphism in the MPU. However, the date was controversially interpreted with Variscan isochrons by Satir (1975) and (Sassi et al. (1985). Rb-Sr ages from micas in para- and orthogneisses to the south of the DAV vary from 260 to 310 Ma (Borsi et al. 1973, 1978a; Satir 1975), K-Ar mica ages from orthogneisses are 224–297 Ma (Cliff 1980). In the southern Kreuzeckgruppe, K-Ar mica and whole-rock ages range between 224 and 257 Ma (Cliff and Rex 1976) respectively 175–320 Ma (Brewer 1969, 1970; see Hoke 1990 for compilation of data). A late Variscan cooling after a pre-Alpine metamorphism is indicated by the data.

Microcline-bearing orthogneisses of the Wangenitzsee area (Schobergruppe) which are associated with eclogitic amphibolites gave a whole-rock isochron of 439 ± 20 Ma (Sr_{init} 0.7107 ± 0.0016) (Troll et al. 1976; Brack 1977). A similar whole-rock Rb-Sr isochron of 434 ± 4 Ma (Sr_{init} 0.7080 ± 0.0034) (Borsi et al. 1973) and of 436 ± 17 Ma (Satir 1975) has been obtained from the muscovite-bearing orthogneisses of Antholz (Anterselva). U-Pb isotopic data from zircons in tonalitic and leucogranitic orthogneisses in the eastern part of the Deferegger Alps yielded evidence for primary zircon formation at 443 and 427 Ma (Cliff 1980). These radiometric data are interpreted to date the intrusion of granitoid bodies at Upper Ordovician times. The basement which had been intruded by the granitoids must be of pre-Upper Ordovician age.

Geochemical investigations on the orthogneisses of Antholz, Gsies (Anterselva, Casies), Trotten, Döllach (Gail Valley) and of the orthogneisses of Campo Tures (AMU unit) by Peccerillo et al. (1979) confirmed strong genetic links between the granitic, granodioritic and tonalitic plutonic bodies, and acid metavolcanic rocks found in the phyllitic sequences. No association relationships between acid plutonites and metamorphic basic and ultrabasic rocks occur. The Casies orthogneisses have a fractionated LREE-pattern, poorly fractionated HREE and a small, but significant, negative Eu-anomaly. More acidic samples show a higher HREE fractionation. The granitoids with a calc-alkaline affinity display the character of continental plutonic provinces and were generated by a regional crustal anatexis (Peccerillo et al. 1979).

2.3 Phyllitic Sequences (Thurntaler Complex, Kreuzeckgruppe, Goldeckgruppe)

Quartzphyllites and micaschists of the Thurntaler complex (TC) are interlayered by amphibolites, epidote-amphibolites, chlorite micaschists and rare marbles in the lower and upper parts of the sequence (Figs. 1b, 2b, 3). Porphyroid gneisses are common in the intermediate part of the pile (Krol 1974; Heinisch and Schmidt 1984; Spaeth and Kreutzer 1989). At present, no radiometric or biochronological ages are available from the sequence. Sassi et al. (1974a), Heinisch and Schmidt (1976, 1984) and Schönlaub (1979) presumed an Ordovician to Early Devonian age of deposition of the acid, basic volcanic and sedi-

Fig. 4a–k. Geochemistry of metabasites. **a, c, e** MORB-normalized variation diagrams. **b, d, f** REE patterns. **g, i** Ti-Zr diagram for group 1 metabasites *(open squares)*, group 2 metabasites *(solid squares)* of the AMU and for eclogitic amphibolites from the Prijakt area, Schobergruppe (MPU, see **e, f**). **h, k** Ti-V discrimination for group 1 metabasites *(open squares)*, group 2 metabasites *(solid squares)* of the AMU, and for eclogitic amphibolites (Prijakt area, Schobergruppe, MPU)

mentary rocks by lithological comparison with less metamorphosed Early Palaeozoic sequences of the Eastern Alps.

The metabasites are volcanogenic tholeiitic rocks and occasionally bear stratiform CuS- and Sb-W-mineralizations (Krol 1974; Heinisch and Schmidt 1984). Porphyroid schists and gneisses seem to be metatuffites. They have alkali-rhyolitic, rhyolitic and dacitic compositions and probably were deposited in a marine environment (Heinisch and Schmidt 1976, 1984). Geochemistry and structures of these rocks are similar to dated Upper Ordovician porphyroids elsewhere in the Austro-Alpine basement and they are considered to be the volcanic equivalents of the Upper Ordovician acid plutonism in the adjacent MPU and AMU units (Peccerillo et al. 1979; Schönlaub 1979; Heinisch 1981; Heinisch and Schmidt 1982).

A 2000 m-thick sequence of phyllites and micaschists with intercalated amphibolites, epidote-amphibolites, metacherts, graphite-phyllites, marbles and dolomite-marbles with magnesite and siderite, is found on the southern rim of the MPU unit in the Kreuzeckgruppe (Fig. 2c). Amphibolites are widespread. In the western part, their thickness does not exceed 30 m; in the central part up to 500 m thickness is observed. The rocks are metamorphic equivalents of ocean-floor basalts (Reimann 1980). Stratabound lense-shaped and cross-cutting stibnite mineralizations and numerous massive sulphide deposits occur within the sequence (Lahusen 1972). The mineralizations are associated with basic and acid metavolcanic rocks, the latter show rhyolitic, rhyodacitic and andesitic compositions. A conceptual model of mineralization formation includes submarine basic to acid volcanicity and corresponding exhalative activity during coeval clastic sedimentation which probably occurred in a shallow back-arc basin (Reimann and Stumpfl 1981).

Three different lithological zones, a northern Altkristallin zone, a lithological border zone ("Grenzzone") and a southern phyllitic zone with micaschists, metabasites and marbles, are found in the Goldeckgruppe (Angel and Krajicek 1939; Deutsch 1977; Heinz 1987). The lowermost northern part of the Altkristallin underwent early Alpine greenschist facies reworking with blastesis of idiomorphic garnets. In the southern part of the Altkristallin and in the phyllitic sequences, only weak Alpine overprinting is observed. A continuous pre-Alpine lithological and structural transition between the Altkristallin with garnet-micaschists, partly with staurolite (Deutsch 1977), amphibolites, acid orthogneisses, and the southern phyllitic sequences is observed (Deutsch 1977, 1988; Heinz 1987). A Palaeozoic (Upper Devonian) age of deposition of these phyllitic sequences is ensured by conodonts in dolomite-marbles (Schönlaub, in Heinz 1987).

2.4 *Gailtal Metamorphic Basement (GMB)*

A small zone of pre-Alpine basement appears to the south of the Permo-Mesozoic of Drauzug (Figs. 1b, 2d, 3) and follows the Periadriatic lineament. Primary and overprinted angular unconformities between the basement and the Permo-Mesozoic cover are found. From N to S four main units (I–IV) with different lithologies and tectono-metamorphic evolution were distinguished.

Zone I ("Granatglimmerschiefer-Oligoklasblastengneis-Zone") is mainly made up of monotonous metapsammopelitic series with garnet-micaschists, quartz-micaschists, quartzites and minor intercalations of amphibolites. Layers of augengneisses concentrate at the southern border. To the W near Obertilliach and Kartitsch, porphyroblastic staurolite-oligoclase-gneisses and amphibolites occur. A small granitic intrusive body was found at Pfannegg. Zone II ("Phyllonitzone") is characterized by highly strained phyllonites, resembling phyllites in the field. Evidently these rocks and incorporated graphite-phyllites, greenschists and augengneisses were diaphthorized in a shear zone. Zone III ("Staurolith-Granat-Glimmerschieferzone") is again constituted by monotonous metapsammopelitic series bearing garnet and staurolite and local minor intercalations of amphibolites and augengneisses. Mineral assemblages show a higher grade of metamorphism and different relative time relations to deformation compared to zone I.

Zone IV ("Gailtal-Paläozoikum, mittelgradig metamorph") displays a quite different various lithology. Within diaphthoritic micaschists and quartz-micaschists frequently occur intercalations of greenschists, amphibolites, banded quartzites, black graphitic shales and marbles. Locally late stage post-kinematic garnet and staurolite document a medium-grade metamorphic overprint. Following this part of zone IV to the east, a Palaeozoic, probably Silurian age is evident from conodonts (Schönlaub 1979). A tentative lithostratigraphic model of the Ordovician to Devonian facies evolution in this zone IV was established by Heinisch (1987; Fig. 3). Summarizing, for zone IV alone, the parent rocks are evidently Palaeozoic. The other three zones are made up of monotonous siliciclastic series of unknown, Paleozoic or Proterozoic age with only a few basaltic intercala-

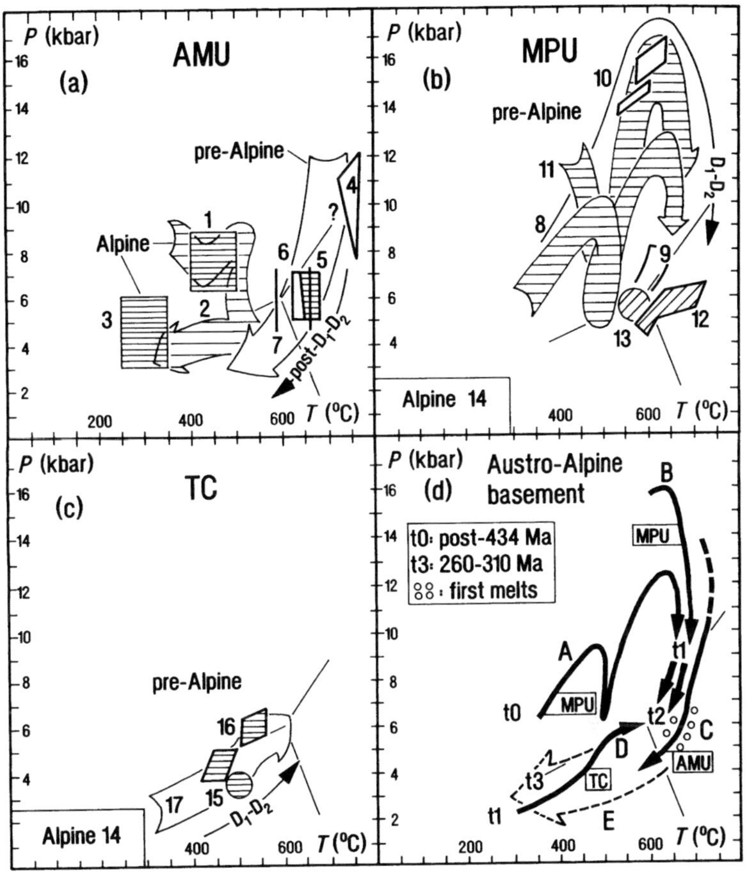

Fig. 5 a–d. P-T data from the Austro-Alpine basement complex to the south of the Tauern window. *Bold numbers* indicate references. **a** Metapsammopelite-amphibolite-marble unit *(AMU): 1* Early Alpine conditions (Stöckhert 1985); *2* Early Alpine P-T path from garnet-micaschists (Schulz 1990 b); *3* Late Alpine conditions (Stöckhert 1985); *4* pre-Alpine metamorphism of Croda Nera metabasites (Bellieni and Visonà 1981); *5* conditions of partial anatexis deduced from aplitic-pegmatitic rocks of Uttenheim (Villa Ottone) (Stöckhert 1985, 1987); *6* geothermobarometry from clinopyroxene-garnet-bearing metabasites (Croda Nera) and associated metapelites (Mager 1985); *7* P-T data from brown amphiboles co-existing with bytownite in metabasites (Godizart 1989). **b** Metapsammopelitic unit *(MPU): 8* P-T-deformation path from micaschists, southern Defereggger Alps (Schulz 1990 a); *9* P-T data from amphibole end member equilibria in amphibolite (Schulz 1990 a); *10* P-T data from clinopyroxene-garnet equilibria in an eclogitic amphibolite from the Prijakt area, Schobergruppe (Schulz 1993); *11* P-T-deformation path from micaschists near the eclogitic amphibolites, Prijakt area, Schobergruppe (Schulz 1993); *12* P-T conditions deduced from micaschists which display formation of first melts, Strieden unit of the Kreuzeckgruppe (Hoke 1990); *13* data from garnet-biotite-plagioclase in contact geothermobarometry from a micaschist (Stöckhert 1985); *14* Alpine P-T conditions in the southern part of the MPU, from fluid inclusions (Stöckhert 1985). **c** Phyllitic sequences of the Thurntaler complex *(TC): 15* P-T data from muscovite-b_0 barometry (Sassi 1972); *16* P-T data from garnet-biotite-oligoclase-muscovite assemblage in a garnet-phyllite (Schulz 1991); *17* P-T path from amphibole end member equilibria in an amphibolite (Schulz 1991). **d** Pre-Alpine P-T-t evolution in the Austro-Alpine basement complex to the south of the Tauern window, compiled from **a–c.** *t0* Whole-rock isochron (434 Ma) from orthogneiss (Borsi et al. 1973); *t1* relative time marker, Tmax in the MPU, start of metamorphism in the TC; *t2* relative time marker, Tmax in the TC, retrograde stage in the MPU; *t3* cooling ages (310–260 Ma) of micas in the MPU (Borsi et al. 1978 a). Whole-rock isochron (262 Ma) and Rb-Sr ages of white micas (252–231 Ma) from aplitic-pegmatitic rocks of Uttenheim (Villa Ottone) (Borsi et al. 1973, 1980). Common uplift/cooling of TC and MPU during t2–t3. In lower parts of the MPU and in parts of the AMU formation of first anatectic melts occurs during the uplift. Stages of pre-Alpine (Variscan) metamorphic evolution in the Austro-Alpine basement complex: *A* early prograde high pressure – low temperature stage; *B* high-pressure eclogitic stage; *C* amphibolite-facies stage, local occurrence of first melts; *D* prograde metamorphism (greenschist to amphibolite facies at low to intermediate pressures) of Palaeozoic phyllitic sequences; *E* common retrogressive metamorphism in all lithological units of the basement

tions and augengneisses. The latter could be metamorphic equivalents of acid volcanics, but rather more probably, the augengneisses are the highly sheared equivalents of granitic intrusions (Heinisch and Schmidt 1982). Thus, lithological similarities of zones I–III with the AMU and MPU to the north clearly exist.

3 Pre-Alpine Structures and Metamorphism

3.1 Metapsammopelite-Amphibolite-Marble Unit (AMU)

Relics of pre-Alpine mineral assemblages and structures are found in southern parts of the AMU where metre-scale domains remained unaffected by Alpine overprinting. A foliation S2 (S1 is represented by layers of pure quartz), sometimes slightly folded (F3) is observed (Stöckhert 1985, 1987). Bellieni and Visona (1981) identified a range of pre-Alpine metamorphic conditions of 700–750 °C at 7–11 kb (Fig. 5 a) by using clinopyroxene-garnet-quartz assemblages in the metabasites of Croda Nera. An Ordovician age of this HP-granulite metamorphism was assumed by the authors. Kyanite-sillimanite-staurolite-garnet- followed by staurolite-sillimanite/andalusite-cordierite-garnet assemblages in adjacent metapelites were attributed to a later temperature-dominated Variscan metamorphism (Bellieni and Visona 1981). Mager (1985) obtained 620–650 °C at 5–7 kb using clinopyroxene-garnet geothermometry on the Croda-Nera metabasites, and brown Mg-hornblendes co-existing with bytownite in amphibolites of the AMU may indicate 620–650 °C at pressures around 5–7 kb (Godizart 1989) during a Variscan metamorphism. According to Stöckhert (1985, 1987), 650 ± 50 °C at 6 ± 1 kb were reached during the late Variscan formation of first melts. The medium pressure – high temperature metamorphism post-dates the generation of the pre-Alpine foliation S2 as is evident from a metablastic post-deformational blastesis of the related minerals (Stöckhert 1985; Godizart 1989). Additionally, aplitic and pegmatitic rocks which were generated by a partial anatexis in the deeper parts of the crust, cross-cut the pre-Alpine foliation S2 and folds F3 (Stöckhert 1987; Hoke 1990).

3.2 Metapsammopelitic Unit (MPU)

Detailed information about the pre-Alpine history has been obtained from the central part of the MPU to the south of the DAV. The structures of foliated and folded oligoclase-bearing xenoliths inside the Upper Ordovician granitoids are discordant to the penetrative surrounding foliation of the orthogneisses. This implies an intrusion of the granitoids into already metamorphosed and deformed country rocks (Borsi et al. 1973). U-Pb zircon ages of 443 Ma from a strongly foliated tonalite gneiss and of 427 Ma from a locally unfoliated leucocratic granite gneiss were interpreted to bracket the age of an early foliation-forming deformation D1 (Cliff 1980). However, for several centimetres beside the unfoliated area, Cliff (1980) described a strong foliation, giving evidence of a post-426 Ma foliation-forming deformation of the plutonites and their host rocks. Furthermore, the local absence of penetrative foliation in the orthogneiss may be caused by deformation partitioning during shearing, or a modal paucity of micas. Thus, the data from the orthogneisses provide only indirect information to the conclusively post-427 Ma age of syn- to post-deformational amphibolite facies mineral assemblages in concordantly foliated metapelitic host rocks. It has been additionally stated by Cliff (1980) that a subsequent Variscan metamorphism, evident from mica ages (Cliff and Rex 1976) has caused no lead loss in the zircons.

In paragneisses and micaschists, a dominant foliation S2 (pure quartz layers are considered to represent an older S1) is axial-planar to isoclinal and sheath-like folds F2, all generated by a deformation D1–D2. A mineral lineation L2 in quartzitic paragneisses and calcsilicate-gneisses runs parallel and subparallel to the long axes of sheath folds and calcsilicate-gneiss bodies (Schulz 1988 a, c). Asymmetric augen structures and linear (L-) as well as linear-planar (L-S-) fabrics are found in the Upper Ordocivian orthogneisses. Microstructures associated with the linear fabrics indicate a shearing direction parallel to the lineation during deformation D1–D2 (Schulz 1988 b, c, 1990 a). The dominant linear as well as planar structures in orthogneisses (L–S) and their metapelitic host rocks (S2–L2) are parallel and it is obvious that these structures were simultaneously generated during D1–D2 in both rock types. This deformation, therefore, must be younger than the Upper Ordovician intrusion of the granitoids and of Variscan age (Borsi et al. 1973; Stöckhert 1985; Schulz 1988 a).

F2 folds and S2 were refolded and overprinted by open to tight folds and crenulations F3, Cr3, which developed during deformation D3. The hinges of F3 folds mostly run parallel to L2. F3 folds with steeply plunging hinges in some regions (Fig. 2b) are minor structures of large-scale syn- and antiforms, called "Schlingen" structures (Schmidegg 1936; Borsi et al. 1978a). Directions of lineations L2 and/or L3 (Fig. 2b, c) were tentatively compiled from Holzer (1958); Deutsch (1977), Borsi et al. (1978a, Heinz (1987) and Schulz (1988a, 1991).

Amphibolite-facies mineral assemblages involving staurolite, kyanite-sillimanite-andalusite, garnet, plagioclase in micaschists and green Ca-amphibole and andesine in amphibolites are widespread in the MPU (Sassi et al. 1974b), and crystallized synchronous and/or posterior to the formation of the main foliation S2 and F3 folds (Troll et al. 1976; Stöckhert 1985; Guhl and Troll 1987; Schulz 1988a, 1990a; Behrmann 1990; Hoke 1990). Due to its close relation to the formation-forming deformation D1–D2 (Di of Behrmann 1990; Dn of Hoke 1990), this amphibolite facies metamorphism is conclusively of post-Upper Ordovician or of Variscan age respectively.

Furthermore, detailed mineralogical-microstructural studies in micaschists (Ntaflos 1981; Schulz 1990a, 1993) yielded a mineralogical evolution with early pre/syn-S2 staurolite-out (garnet + kyanite + albite/oligoclase), intermediate syn-S2 staurolite-in (garnet + staurolite + kyanite + oligoclase) assemblages both preserved in microlithons surrounded by S2, and late syn/post-S2 staurolite-in (staurolite + kyanite-sillimanite-andalusite + andesine ± garnet) assemblages (Schulz 1990a). Syncrystalline-rotated garnets of the early and intermediate assemblages display complex and strong mineral zonations with spessartine-rich cores to pyrope-rich rims and significant variations of the Ca- (grossularite) contents (Ntaflos 1981; Hoke 1990; Schulz 1990a, 1993). By involving coexistent biotites and anorthite-poor plagioclases, P-T-d (deformation) paths, with marked isothermal decompression in their prograde sections, were estimated from the garnet zonations. The principal shapes of these P-T-d paths from samples lying several kilometres and tens of kilometres apart are similar and display a complex prograde pre/syn-S2 metamorphic evolution (Fig. 5b). High pressures and high temperatures (650 °C/12 kb and later 680 °C/8 kb) were reached in the MPU of the southern Defereggen Alps (Schulz 1990a).

Zoned garnets (pyrope 15–20%, grossularite 27–25%) and preferentially oriented clinopyroxenes (jadeite 35–42%) co-existing with albite (anorthite 7–9%) and rutile give evidence of an early high-pressure metamorphism (Troll et al. 1976; Behrmann 1990) at 550–650 °C/14–16 kb of eclogitic amphibolites in the Prijakt area of the Schobergruppe (Schulz 1993). In adjacent micaschists, similar P–T conditions were recorded by successive syn-S2 garnet + kyanite and garnet + staurolite + kyanite assemblages which coexisted with albite and/or zoisite (Schulz 1993). A subsequent amphibolite-facies overprinting of the early high pressure metamorphism is obvious from pervasive crystallization of Ca-amphiboles in the eclogitic amphibolites and of post-S2 staurolite + kyanite-sillimanite-andalusite + oligoclase/andesine + garnet assemblages in micaschists. Stöckhert (1985) estimated 580 °C/6.5 kb from a micaschist in the western part of the MPU, Hoke (1990) reported a post-deformational high temperature metamorphism at 650 ± 100 °C/5.5 ± 1 kb from the Strieden unit in the basal part of the MPU. There, staurolite, sillimanite and andalusite overgrew foliation and folds, local partial melting occurs and late Variscan pegmatites which crystallized at 665 °C/3 kb cross-cut the earlier structures (Fig. 5b).

A further, retrograde stage of metamorphism in the MPU is characterized by retrogressive replacement of garnet by biotite and chlorite and a marginal to complete sericitization of the alumosilicates and staurolite (Sassi et al. 1974b; Stöckhert 1985; Schulz 1988a).

3.3 Phyllitic Sequences (Thurntaler Complex, Goldeckgruppe)

Mesoscopic isoclinal folds F2 and an axial-planar foliation S2 (layers of pure quartz are considered to represent older S1) were generated in the rocks of the Thurntaler complex during a deformation sequence D1–D2 (Schulz 1991). Directions of S2 and a corresponding mineral lineation L are parallel to the orientations of S2 and L2–L3 in the adjacent MPU (Schmidegg 1937; Heinisch and Schmidt 1984; Spaeth and Kreutzer 1989, 1990; Schulz 1991). As is similarly observed in the adjacent MPU, microstructures in garnet-phyllites, amphibolites and porphyroid gneisses indicate a lineation-parallel direction of shearing during deformation D1–D2 (Schulz 1991). Open to tight mesoscopic folds F3 overprinted the earlier structures. Locally late mylonitic shear zones S5 (Heinisch and Schmidt 1984; Spaeth and Kreutzer 1989) with top-to-NW and top-to-WSW sense of shear affected the TC and the lithological contacts between MPU and TC. A pre-Alpine age of the mylonitic shear zones seems possible due to pervasive recrystallization of quartz in these zones (Spaeth and Kreutzer 1989) as it has not been observed in deformed Permotriassic sandstones of Kalkstein (Guhl and Troll 1987) and in the southern part of the TC (Schulz 1991).

Sassi (1972) and Sassi and Zanferrari (1972) estimated 500 °C at 3.5 kb from quartzphyllites by using b_0-values of white micas. Syncrystalline-rotated spessartine-rich first-generation garnets in garnet-phyllites grew during D1–D2. Post-S2 second-generation porphyroblasts coexisted with oligoclase (An 27), biotite, muscovite, chlorite and recorded P–T conditions of 510–540 °C/6.25 kb (Fig. 5c). Strongly zoned amphiboles with actinolitic cores and green hornblende rims from amphibolites with the assemblage Ca-amphibole + epidote + plagioclase (albite and oligoclase) + chlorite + titanite + quartz + calcite were used to estimate a temperature-dominated prograde syn/post-D1–D2 evolution from 300 °C/2.5 kb to 600 °C/6 kb by amphibole end member equilibria (Schulz 1991).

The retrograde P–T evolution in the TC is characterized by retrogressive replacement reactions within greenschist facies conditions during deformations D3 (folding F3), D4 (shear planes S4) and D5 (shear zones S5) (Heinisch and Schmidt 1984; Schulz 1991).

In the Goldeckgruppe, a continuous lithological/structural transition between the northern Altkristallin and the southern phyllitic sequences is observed. Due to the continuous transition respectively zoning to lower and upper greenschist facies conditions reached in the phyllitic sequences, the formation of staurolite in micaschists of the Altkristallin was assigned to a Variscan amphibolite facies metamorphism. A subsequent diaphthorism as well was attributed to the Variscan era (Deutsch 1977) because the relic transgressive Permo-Triassic cover shows no corresponding metamorphic phenomena. Heinz (1987) reported similar structural and metamorphic relationships and an equivalent temporal assignment of the deformational and metamorphic events to the Variscan orogeny from the eastern Goldeckgruppe phyllitic sequences.

3.4 *Gailtal Metamorphic Basement (GMB)*

Though the Gailtal metamorphic basement is not very suitable for studying pre-Alpine events due to the strong brittle Alpine overprint, at least four main phases of deformation and metamorphism can be distinguished. D1/M1 and D2/M2 events are probably of Variscan age.

Within zone I the D1/M1 event is syn- to post-kinematic with rotational deformation of staurolite and garnet. There is proof of metamorphic zoning (medium-grade in the W, almandine low-grade in the E) within this unit. In zone III, the corresponding D1–M1 event is characterized by post-kinematic growth of garnet, staurolite and sillimanite. Large parts of zone IV show a low-grade D1–M1 event of definitely Variscan age. A small domain with post-kinematic garnet and staurolite was found in this zone. Thus, a Variscan medium-grade metamorphism for a small area of fossiliferous Palaeozoic sequences seems to be possible, but not proved due to lack of radiometric data (Heinisch 1987; Heinisch and Sprenger 1988). Structural features of the D1 event, preserved as relics are a penetrative foliation, metamorphic layering and tight to isoclinal folds. Therefore it is concluded that deformation was polyphase during the D1 event.

The D2/M2 event commonly affected all units distinguished under low-grade conditions. Shear bands with grain size reduction and pressure solution cross-cutting foliation planes and intensely folded quartz rods are the corresponding structural features. Together with numerous internal shear zones, the zone II phyllonites were formed, probably as a major ductile thrust plane or ductile wrench fault, coupling basement rocks with different earlier tectono-metamorphic histories. A Variscan age of this phyllonitisation is derived from the occurrence of phyllonite boulders within the basal Post-Variscan conglomerates (Mostler 1972). Visona (1974) studied b_0-values of muscovites in phyllonites and concluded a Variscan low-pressure metamorphism, as reported by Sassi and Zanferrari (1972) from other phyllitic sequences of the Eastern Alps.

A D3/M3 event of Alpine age under low-grade to very low-grade metamorphic conditions was deduced, reaching roughly an estimated 400 °C/3.5 kb at the southern border of the GMB and 270 °C/1.5–2 kb in the Post-Variscan sediments (Niedermayr et al. 1984). Oligocene tonalites show early ductile deformation.

The tectono-metamorphic history ends in an intense late Alpine brittle deformation D4 (subdivided into four phases K1 to K4), partly under very low-grade conditions M4 and related to dextral movements along the Gailtal-line. High bulk strain values within the Gailtal megashear led to a rotation of the Variscan foliation into a straight E-W-directed vertical position. Hinge lines of relic tight to isoclinal folds and the stretching lineations dip moderately to the E or W. It seems possible that a pre-Alpine regional-scale metamorphic zoning is cross-cut and offset by the younger lateral movements, forming the observed overall lenticular pattern of the units I–IV.

4 Pre-Alpine Tectono-Metamorphic Evolution

To sum up, the Austro-Alpine basement complex to the south of the Tauern window is made up of two significantly different groups of lithological associations. In the first group, the "Altkristallin" with AMU, MPU and GMB I–III units, intrusions uf Upper Ordovician granitoids are observed and a lithological subdivision, into a monotonous metapsammopelitic part (MPU, GMB I–III) and a varied sequence (AMU) with more frequent metabasites and marbles, is possible. The second group, the phyllitic sequences of the TC, Kreuzeckgruppe, Goldeckgruppe and GMB IV, contains probably the volcanoclastic equivalents of the Upper Ordovician acid magmatism as well as fossiliferous Palaeozoic (Ordovician to Upper Devonian) sediments.

AMU, MPU and GMB I–III represent a pre-Upper Ordovician continental crust, positioned to the NW according to palaeogeographic models of the pre-Alpine arrangement of Palaeozoic units. The phyllitic sequences are assigned to Palaeozoic sedimentation and volcanism in a continental margin environment to the SE (Schönlaub 1979; Heinisch 1987).

Similar structures and parallel orientations of foliation, lineation, and fold-hinge lines are identified in the lithological contact zones of both associations. Early deformation D1–D2 (Di, Dn) with a shearing direction parallel to a stretching lineation led to isoclinal folds and a main foliation (S2) which were later refolded (D3) and overprinted by late shear zones, phyllonitization and diaphthorism. Different pre-Alpine stages of metamorphism occur in both lithological associations. In the MPU, an early high pressure – low temperature stage, a high pressure eclogitic stage, followed by a temperature-dominated pervasive amphibolite facies overprinting with formation of first melts and a final retrogressive stage can be distinguished (Sassi et al. 1974b; Troll et al. 1976; Behrmann 1990; Hoke 1990; Schulz 1990a). In the AMU, an early high pressure granulitic stage (Bellieni and Visona 1981) and a post-deformational high temperature metamorphism with formation of first anatectic melts is observed (Stöckhert 1985, 1987).

Greenschist-facies low-pressure metamorphism is mainly found in the phyllitic sequences. Additionally, there is evidence that prograde syn-D1-D2 temperature-dominated metamorphism reached amphibolite facies conditions with low to intermediate pressures in some parts (Schulz 1991). Not definitely proved is a pre-Alpine medium-grade metamorphism of fossiliferous Palaeozoic sequences.

The intrusion of the Upper Ordovician granitoids into the Altkristallin and the deposition of their, probably, volcano-clastic equivalents in the Palaeozoic phyllitic sequences provides a well-documented and correlatable time marker within the basement complex to the south of the Tauern window. Thus, it is convenient to distinguish a pre-Upper Ordovician and a post-Upper Ordovician history. The granitoids, generated by crustal anatexis (Peccerillo et al. 1979), represent the final stage of a pre-Upper Ordovician metamorphic, tectonic and geochemical evolution in AMU, MPU, GMB I–III. Structures are preserved inside rare, unconformably interlayered, xenoliths in the orthogneisses and little Rb–Sr whole-rock data, interpreted to be related to metamorphism, exist (Borsi et al. 1973; 1978a). As palaeomagnetic data suggest a Gondwana affinity of the basement complex, this pre-Upper-Ordovician history may be characterized by the term "Pan-African" (Heinisch 1986).

Due to the lack of radiometric data from Croda Nera metabasites, eclogitic amphibolites and amphibolites in the Altkristallin, it is not clear whether these basic magmatic rocks were part of the pre-Upper Ordovician crust or intruded after the granitoids.

A conclusively Palaeozoic basaltic magmatism is obvious from the metabasites in the phyllitic sequences, but available geochemical and geochronological data are still incomplete from these rocks.

Despite the clear two-stage subdivision of the pre-Alpine history in the basement complex by the Upper Ordovician acid magmatism, the temporal assignment of tectono-metamorphic events within the Altkristallin and the structural relationship between Altkristallin and phyllitic sequences still give rise to controversial discussion.

In many papers, a "Caledonian" age of the amphibolite facies stage of metamorphism in the MPU as well as a direct relation between this metamorphism and the Upper Ordovician acid magmatism by crustal anatexis has been strongly pleaded (e.g. Borsi et al. 1978a; Sassi et al. 1985) and adopted (Troll et al. 1980; Behrmann 1990). Following these early interpretations, mainly based on radiometric data, the post-Upper Ordovician (Variscan) foliation-forming deformation of the granitoids should have taken place under low-pressure greenschist facies conditions with a high (>30 °C/km) geothermal gradient. Cooling after this metamorphism then led to late Variscan mica ages in the southern parts of the Altkristallin (Borsi et al. 1978a). Apart from arguments produced by the xenoliths and a questionable interpretation of Cliff's (1980) data, this "Caledonian" model is mainly based on the generally concordant tectonic structures of the Altkristallin and the phyllitic sequences. The latter display prevailing conclusively Variscan low-pressure greenschist facies metamorphism, contrasting the dominant amphibolite facies metamorphism in the Altkristallin. Sassi and Zanferrari (1972) proposed a post-"Caledonian" angular unconformity and transgressive deposition of the Thurntaler rocks upon the Altkristallin which previously should have sufferred a "Caledonian" amphibolite facies metamorphism. Both units then should together have been affected by a Variscan low-pressure greenschist facies metamorphism with high geothermal gradients and a foliation-forming deformation.

Several new observations, from the area concerned, contradict this early model of a "Caledonian" main metamorphism and a subsequent Variscan overprinting. The supposed "basal conglomerates" of the Markinkele area (No. 25, Fig. 2c) turned out to be mylonites with isolated quartz rods in a fine-grained basaltic matrix. These mylonites were generated by shearing postdating and overprinting the lithological contact between MPU and TC (Heinisch and Schmidt 1984; Schulz 1991). Similarly, masked angular unconformities were presumed to be in the GMB (von Klebelsberg 1935; Sassi et al. 1974a). But, there, the contact between partly fossiliferous Palaeozoic and the Altkristallin is obscured by later tectonism as well (Heinisch 1987). At present, no evident proof exists of a post-Upper Ordovician "Caledonian" angular unconformity in the ABC to the south of the Tauern window. Furthermore, it has been stated by Cliff (1980), Bellieni and Visona (1981), Stöckhert (1982, 1985), Schulz (1988a), Hoke (1990) that amphibolite facies mineral assemblages in the MPU and AMU are partly younger than the main foliation S2 of the metapsammopelites. S2 is parallel to a corresponding foliation of the Upper Ordovician granitoids, therefore, this metamorphism post-dates the acid intrusions.

Additionally, the construction of P-T-t-deformation paths from the rocks, by combining and relating the mineral chemistry evolution within mineral assemblages to microstructures of progressive deformation, provided further arguments supporting a mainly Variscan tectono-metamorphism of the basement. It has been shown in the MPU that a "polymetamorphism" of the rocks from an early HP-LT stage to a HP-eclogitic stage and then to amphibolite facies conditions evolved synchronous and later than the S2 foliation-forming shearing D1–D2 and within a single P–T loop (Schulz 1990a, 1993). A post-Upper Ordovician and possibly early Variscan age of this multistage metamorphism is conclusive from the deformation of the granitoids by D1–D2. Similarly, the pre-Alpine P–T data from the AMU may tentatively be connected to a single path which shows mainly decompression and meets P–T conditions of partial melting during a final stage of uplift (Fig. 5a).

The parallel structures in the Altkristallin and the phyllitic sequences may be explained by a common deformation history with a syndeformational-synmetamorphic integration of both associations. However, the different "type" of metamorphism, underlined by different P–T–d paths, points to a tectonic integration at different stages of each metamorphic evolution (Fig. 5d). The close relation between prograde P–T evolution and shearing deformation D1–D2 in the MPU indicates an early Variscan process of crustal thickening by tectonic stacking in the course of a continental collision (Schulz 1990a). During an early stage of Variscan stacking, the "Pan-African" crust was reworked by burial and subsequent uplift. The Palaeozoic sediments of the adjacent continental margin were involved in the progressing crustal stacking in a later stage and suffered prograde temperature-dominated metamorphism. A subsequent common late Variscan retrograde evolution (cooling and uplift) in both lithological associations then probably evolved in a dextral transpressional system of NW-SE directed compression with deformations D3, D4, D5.

That way, the pre-Alpine history of the Austro-Alpine basement complex to the south of the Tauern window can be briefly summarized and categorized into a poorly known pre-Upper Ordovician evolution of "Pan-African" basement, terminated by pervasive Upper Ordovician acid magmatism, and a complete reworking of this basement and an adjacent continental margin by deformation and metamorphism of a Variscan orogeny.

Acknowledgements
We commemorate G. Troll who was not able to participate in the preparation of this manuscript. Our studies on the basement rocks to the south of the Tauern window were mainly financed by grants from the Deutsche Forschungsgemeinschaft (DFG) and supported further by several single projects. Mapping in the Gailtal metamorphic basement was granted by the Geologische Bundesanstalt Wien.

References

Angel F (1930) Gesteine der Kreuzeckgruppe (Kärnten). Mitt Naturwiss Ver Steiermark 67: 7–35

Angel F, Krajicek E (1939) Gesteine und Bau der Goldeckgruppe. Carinthia II 49: 26–57

Baggio P, Dalla Porta A, Zirpoli G (1971) Il complesso austroalpino di Cima Dura nelle zona di Monte Spico (Alto Adige Orientale). Studi Trent Sci Nat 49: 3–31

Barnick H (1965) Tektonische Untersuchungen in der westlichen Matreier Zone (Gebiet der Neuen Reichenberger Hütte, Ost-Tirol). Neues Jahrb Geol Paläontol Monatsh 1965: 575–592

Bechstädt T (1978) Faziesanalyse permischer und triadischer Sedimente des Drauzuges als Hinweis auf eine großräumige Lateralverschiebung innerhalb des Ostalpins. Jahrb Geol Bundesanst 121: 1–12

Behrmann JH (1990) Zur Kinematik der Kontinentkollision in den Ostalpen. Geotekt Forsch 76: 1–180

Behrmann JH, Wallis SR (1987) Hangendverschuppung des Tauernfenster-Südrandes bei Kals (Osttirol) als Zeuge von eo-alpinem Underplating. Jahrb Geol Bundesanst 130–133–138

Bellieni G (1980) The Cima di Vila (Zinsnock) massif: geochemical features and comparisons with the Vedrette di Ries (Rieserferner) pluton (Eastern Alps – Italy). Neues Jahrb Miner Abh 138: 244–258

Bellieni G, Visonà D (1981) Metamorphic evolution of the Austroalpine schists outcropping between the intrusive masses of Vedrette di Ries and Cima di Vila, Eastern Alps, Italy. Neues Jahrb Geol Paläontol Monatsh 1981: 586–602

Bellieni G, Comin-Chiaramonti P, Visonà D (1976) Contributo alla conoscenca del plutone delle Vedrette di Ries (Rieserferner) (Alpi Orientali). Boll Soc Geol Ital 95: 351–370

Bellieni G, Peccerillo A, Poli G (1981) The Vedrette di Ries (Rieserferner) plutonic complex: petrological and geochemical data bearing on its genesis. Contrib Mineral Petrol 78: 145–156

Bellieni G, Peccerillo A, Poli G (1982) REE distribution in the Cima di Vila (Zinsnock) granodioritic complex and its petrogenetic significance. Neues Jahrb Miner Abh 145: 50–65

Bemmelen RW van, Meulenkamp JE (1965) Beiträge zur Geologie des Drauzuges (Kärnten, Österreich). Die Lienzer Dolomiten und ihre geodynamische Bedeutung für die Ostalpen. Jahrb Geol Bundesanst 108: 213–268

Bianchi A (1934) Studi petrografici sull'Alto Adige Orientale e regione limitrofe. Mem Ist Geol Univ Padova 10: 1–243

Bögel H (1975) Zur Literatur über die "Periadriatische Naht". Verh Geol Bundesanst 1975: 163–199

Borsi S, Del Moro A, Sassi FP, Zirpoli G (1973) Metamorphic evolution of the Austridic rocks to the south of the Tauern Window (Eastern Alps): radiometric and geopetrologic data. Mem Soc Geol Ital 12: 549–571

Borsi S, Del Moro A, Sassi FP, Zanferrari A, Zirpoli G (1978a) New geopetrologic and radiometric data on the Alpine history of the Austridic continental margin south of the Tauern window. Mem Ist Geol Minerol Univ Padova 32: 1–19

Borsi S, Del Moro A, Sassi FP, Zirpoli G (1978b) On the age of the Periadriatic Rensen massif (Eastern Alps). Neues Jahrb Geol Paläontol Monatsh 1978: 267–272

Borsi S, Del Moro A, Sassi FP, Zirpoli G (1979) On the age of the Vedrette de Ries (Riesenferner) massif and its geodynamic significance. Geol Rundsch 68: 41–60

Borsi S, Del Moro A, Sassi FP, Visonà D, Zirpoli G (1980) On the existence of Hercynian aplites and pegmatites in the lower Aurina Valley (Ahrntal, Austrides, Eastern Alps). Neues Jahrb Miner Monatsh 1980: 501–514

Brack W (1977) Geochronologische Untersuchungen an Gesteinen des Altkristallins in der Schobergruppe, Österreich. Dissertation, Fakultät für Geowissenschaften der Ludwig-Maximilians-Universität, München

Brewer MS (1969) Excess radiogenic argon in metamorphic micas from the Eastern Alps, Austria. Earth Planet Sci Lett 6: 321–331

Brewer MS (1970) K-Ar age studies in the Eastern Alps – the Oberostalpindecke of Kärnten. Thesis, Oxford University

Clar E (1927a) Ein Beitrag zur Geologie der Schobergruppe bei Lienz in Tirol. Mitt Naturwiss Ver Steiermark 63: 72–90

Clar E (1927b) Ein interesssantes Profil aus den südlichen Vorlagen der Schobergruppe. Verh Geol Bundesanst 1927: 229–231

Cliff RA (1980) U-Pb isotopic evidence from zircons for lower Palaeozoic tectonic activity in the Austroalpine nappe, the Eastern Alps. Contrib Mineral Petrol 71: 283–288

Cliff RA, Rex DC (1976) Mineral ages from the Austroalpine nappe in the Deferegger Mountains, East Tyrol, Austria. Annu Rep Res Inst Afr Geol Univ Leeds 20: 78–79

Contini R, Sassi FP (1980) Su alcuni effecti metamorfici Alpini nel basemento Austridico in Pusteria (Alpi Orientali). Mem Ist Geol Mineral Univ Padova 34: 187–194

Dal Piaz G (1934) Studi geologici sull'Alto Adige e regione limitrofe. Mem Ist Geol Univ Padova 10: 1–238

Deutsch A (1977) Geologie und Petrographie der mittleren Goldeckgruppe (Kärnten/Österreich). Jahrb Geol Bundesanst 120: 231–294

Deutsch A (1984) Young Alpine dykes south of the Tauern window (Austria): A K-Ar and Sr isotope study. Contrib Mineral Petrol 85: 45–57

Deutsch A (1988) Die frühalpidische Metamorphose in der Goldeck-Gruppe (Kärnten) – Nachweis anhand von Rb-Sr-Altersbestimmungen und Gefügebeobachtungen. Jahrb Geol Bundesanst 131: 553–562

Exner C (1955) Aufnahmen (1954) in den östlichen Hohen Tauern und in südlichen Randgebieten (Blätter 154, 155, 180, 191). Verh Geol Bundesanst 1955: 17–22

Exner C (1956) Geologische Beobachtungen (1955) in der Kreuzeck-, Sadnig-, Rieserferner- und Reißeckgruppe. Verh Geol Bundesanst 1956: 24–27

Exner C (1962) Die Perm-Trias-Mulde des Gödnachgrabens an der Störungslinie von Zwischenbergen (Kreuzeckgruppe, östlich Lienz). Verh Geol Bundesanst 1956: 24–27

Exner C, Wanderer E (1962) Zur Kenntnis des Eklogit-Amphibolits im Debanttal (Schobergruppe, Osttirol). Karinthin 45/46: 228–234

Frank W, Alber J, Thöni M (1977) Jungalpine K/Ar Alter von Hellglimmern aus dem Permotriaszug von Mauls-Penser Joch (Südtirol). Anz math-naturwiss Kl Österr Akad Wiss Jg 1977 7: 6 pp

Friedrich OM (1963) Die Lagerstätten der Kreuzeckgruppe: Monographie Kärntner Lagerstätten, 3. Teil. Arch Lagerstättenforsch Ostalpen 1

Frizzo P (1981) Manifestazioni a scheelite nelle metamorfiti austroalpine a sud delle Vedrette di Ries (Alto Adige Orientale). Boll Soc Geol Ital 100: 489–498

Gizycki P von, Schmidt K (1978) Zur Genese der Plutone im SW des Tauernfensters (Ostalpen). Neues Jahrb Geol Paläontol Monatsh 1978: 657–673

Godizart G (1989) Gefüge, retrograde Metamorphose und Geochemie der Amphibolite im ostalpinen Altkristallin südlich des westlichen Tauernfensters (Südtirol, Italien). Dissertation, Naturwissenschaftliche Fakultät der Universität Erlangen-Nürnberg

Godizart G, Schulz B (1992) Geochemistry of eclogitic amphibolites from the Schobergruppe, Austroalpine basement to the south of the Tauern window, Eastern Alps. Schweiz Mineral Petrogr Mitt (in prep)

Gratzer R (1984) Ein Beitrag zur Petrologie der Rieserferner Intrusion in Ost- und Südtirol. Mitt Ges Geol Bergbaustud Österr 30/31: 319–342
Grundmann G, Morteani G (1985) The young uplift and thermal history of the Central Eastern Alps (Austria/Italy), evidence from apatite fission track ages. Jahrb Geol Bundesanst 128: 197–216
Guhl M, Troll G (1977) Mehrphasige Faltengefüge in Altkristallin und Permotrias von Kalkstein in Osttirol, Österreich. Verh Geol Bundesanst 1977: 45–52
Guhl M, Troll G (1987) Die Permotrias von Kalkstein im Altkristallin der südlichen Deferegger Alpen (Österreich). Jahrb Geol Bundesanst 130: 37–60
Hammerschmidt K (1977) Die obere Schieferhülle, die Matreier Zone und die Cima-Dura-Serie im oberen Buinlandtal (Ahrntal). Schlern 47: 97–118
Hammerschmidt K (1981) Isotopengeologische Untersuchungen am Augengneis vom Typ Campo Tures bei Rain in Taufers, Südtirol. Mem Ist Geol Mineral Univ Padova 34: 273–300
Hammerschmidt K (1982) K/Ar and $^{40}Ar/^{39}Ar$ age resolution from illites of the Trias of Mauls; Mesozoic cover of the Austroalpine basement, Eastern Alps (South Tyrol). Schweiz Mineral Petrogr Mitt 62: 113–133
Hawkesworth CJ (1976) Geochronology in the Eastern Alps. Contrib Mineral Petrol 54: 255–244
Heinisch H (1981) Zum ordovizischen "Porphyroid"-Vulkanismus der Ost- und Südalpen, Stratigraphie, Petrographie, Geochemie. Jahrb Geol Bundesanst 124: 1–109
Heinisch H (1986) Die Geologie der Nördlichen Grauwackenzone zwischen Kitzbühel und Zell am See und ihre Bedeutung für die Rekonstruktion der altpaläozoischen Geodynamik des Ostalpenraumes. Habilitationsschrift, Fakultät für Geowissenschaften der Ludwig-Maximilians-Universität München
Heinisch H (1987) Concepts for the geological evolution of the Gailtalkristallin (Kärnten – Austria). In: Sassi FP, Grecula P (eds) Pre-Variscan and Variscan events in the Alpine-Mediterranean mountain belts. Alfa, Bratislava, pp 293–312
Heinisch H, Schmidt K (1976) Zur kaledonischen Orogenese in den Ostalpen. Geol Rundsch 65: 459–482
Heinisch H, Schmidt K (1982) Zur Genese der Augengneise im Altkristallin der Ostalpen. Neues Jahrb Geol Paläontol Monatsh 1982: 211–239
Heinisch H, Schmidt K (1984) Zur Geologie des Thurntaler Quarzphyllits und des Altkristallins südlich des Tauernfensters (Ostalpen, Südtirol). Geol Rundsch 73: 113–129
Heinisch H, Sprenger W (1988) Mehrphasige Deformation und Pseudotachylitbildung im Gailtalkristallin und am Periadriatischen Lineament zwischen Sillian und Kötschach-Mauthen (Osttirol/Kärnten, Österreich). Erlanger Geol Abh 116: 41–52
Heinisch H, Schmidt K, Schuh H (1984) Zur geologischen Geschichte des Gailtal-Kristallins im unteren Lesachtal westlich von Kötschach-Mauthen (Kärnten, Österreich). Jahrb Geol Bundesanst 126: 477–486
Heinz H (1987) Geologie der östlichen Goldeckgruppe (Kärnten). Jahrb Geol Bundesanst 130: 175–203
Heritsch H, Paulitsch P (1958) Erläuterungen zur Karte des Kristallins zwischen Birnbaum und Pressegger See, Gailtal. Jahrb Geol Bundesanst 101: 191–200
Höhler S (1984) Gefügeprägung der Marmore und Kalksilikatbestände im Altkristallin zwischen Eisack- und Ahrntal in Südtirol. Dissertation, Naturwissenschaftliche Fakultät der Universität Erlangen-Nürnberg
Hofmann KH, Kleinschrodt R, Lippert R, Mager D, Stöckhert B (1983) Geologische Karte des Altkristallins südlich des Tauernfensters zwischen Pfunderer Tal und Tauferer Tal (Südtirol). Schlern 57: 572–590
Hoke L (1987) Geology of part of the Altkristallin (Eastern Alps). Thesis, Cambridge University
Hoke L (1990) The Altkristallin of the Kreuzeck Mountains, SE Tauern Window, Eastern Alps – basement crust in a convergent plate boundary zone. Jahrb Geol Bundesanst 133: 5–87
Holzer H (1958) Zur photogeologischen Karte der Kreuzeckgruppe. Geologische Luftbildinterpretation II. Jahrb Geol Bundesanst 102: 187–191
Klebelsberg R von (1935) Geologie von Tirol. Borntraeger, Berlin
Kleinschrodt R (1987) Quarzkorngefügeanalyse im Altkristallin südlich des westlichen Tauernfensters (Südtirol/Italien). Erlanger Geol Abh 114: 1–82
Krol W (1974) Geologisch-lagerstättenkundliche Untersuchungen im Gebiet nördlich von Sillian (Osttirol/Österreich). Dissertation, Naturwissenschaftliche Fakultät der Ludwig-Maximilians-Universität München
Lahusen L (1972) Schicht- und zeitgebundene Antimonit-Scheelit-Vorkommen und Zinnober-Vererzungen in Kärnten und Osttirol/Österreich. Mineral Deposita 7: 31–60
Lambert RSJ (1970) A potassium-argon study of the margin of the Tauernfenster at Döllach, Austria. Eclogae Geol Helv 63: 197–205
Lunzer H (1985) Bericht (1984) über geologische Aufnahmen im Altkristallin auf Blatt 178 Hopfgarten im Defereggental. Jahrb Geol Bundesanst 128: 310–312
Mager D (1981) Vergleichende morphologische Untersuchungen an Zirkonen des altkristallinen Augengneisses von Sand in Taufers (Südtirol) und einiger benachbarter Gesteine. Neues Jahrb Miner Monatsh 1981: 385–397
Mager D (1985) Geologische und petrographische Untersuchungen am Südrand des Rieserferner-Plutons (Südtirol) unter Berücksichtigung des Intrusionsmechanismus. Dissertation, Naturwissenschaftliche Fakultät der Universität Erlangen-Nürnberg
Mostler H (1972) Die permoskythische Transgressionsserie der Gailtaler Alpen. Verh Geol Bundesanst 1972: 143–149
Niedermayr G, Mullis J, Niedermayr E, Schramm JM (1984) Zur Anchimetamorphose permo-skythischer Sedimentgesteine im westlichen Drauzug, Kärnten – Osttirol (Österreich). Geol Rundsch 73: 207–221
Nollau G (1969) Kleintektonische Strukturen am SW-Rand des Tauernfensters und ihre Einbeziehung in großtektonische Konzepte. Geol Rundsch 58: 755–788
Nollau G (1974) Petrographische und tektonische Untersuchungen am Periadriatischen Rensengranit in Südtirol. Erlanger Geol Abh 98: 1–92
Ntaflos T (1981) Polymetamorphose im Altkristallin der Deferegger Alpen südlich der Rieserfernerintrusion. Fortsch Mineral Bh 59(1): 143–144
Oxburgh ER (1966) Faltungsphasen im Altkristallin am SE-Rand des Tauernfensters. Verh Geol Bundesanst 1966: 33–47
Oxburgh ER, Lambert RS, Baadsgard H, Simons JG (1966) Potassium-Argon age studies across the southeast margin of the Tauern window in the Eastern Alps. Verh Geol Bundesanst 1966: 17–33
Peccerillo A, Poli G, Sassi FP, Zirpoli G, Mezzacasa G (1979) New data on the Upper Ordovician acid plutonism in the Eastern Alps. Neues Jahrb Miner Abh 137: 162–183
Prochaska W (1981 a) Die Rieserfernerintrusion und ihre Kontaktaureole. Fortschr Mineral Bh 59(2): 118–125
Prochaska W (1981 b) Einige Ganggesteine der Rieserfernerintrusion mit neuen radiometrischen Altersdaten. Mitt Ges Geol Bergbaustud Österr 27: 161–171

Reimann C (1980) Lithogeochemie und schichtgebundene Vererzung, Kreuzeckgruppe, Kärnten, Österreich. Dissertation, Montanuniversität Leoben

Reimann C, Stumpfl EF (1981) Geochemical setting of stratabound stibnite mineralization in the Kreuzeck Mountains, Austria. Trans Instn Min Metall Sect B 90: 126–132

Richter W (1973) Vergleichende Untersuchungen an ostalpinen Eklogiten. Tschermaks Mineral Petrogr Mitt 19: 1–34

Sander B (1912) Über einige Gesteinsgruppen des Tauernwestendes. Jahrb k k Geol Reichsanst 62: 219–288

Sander B (1920) Geologische Studien am Westende der Hohen Tauern. Jahrb Geol Staatsanst Wien 1920: 273–296

Sander B (1925) Note illustrative della Carta geologica delle Tre Venezie, foglio Bressanone 1:100000. Uffidrogr d R magistr all Sez geol 56, Padova

Sander B (1929) Erläuterungen zur Geologischen Karte des Brixner und Meraner Gebietes. Schlernschriften 16: 1–111

Sassi FP (1972) The petrological and geological significance of the b_0 values of potassic white micas in low-grade metamorphic rocks. An application to the Eastern Alps. Tschermaks Mineral Petrogr Mitt 18: 105–113

Sassi FP, Zanferrari A (1971) Osservazioni sulla granodiorite affiorante lungo la linea della Gail presso Hollbruck (Tirolo Orientale, Austria). Studi Trent Sci Nat A 48: 271–281

Sassi FP, Zanferrari A (1972) Il significato geologico del Complesso de Thurntaler (Pusteria), con particolare riguardo alla successione di eventi metamorfici prealpini nel basamento austridico dell'Alpi Orientali. Boll Soc Geol Ital 91: 533–557

Sassi FP, Zirpoli G (1973) Sulla distribuzione di sillimanite nel basamento austroalpino a sud della Finestra dei Tauri (Alpi Orientali). Boll Soc Geol Ital 92: 831–839

Sassi FP, Zirpoli G (1980) Quadro delle conoscence sulla storia pre-alpine delle Alpi Orientali. Rend Soc Ital Mineral Petrol 36: 45–48

Sassi FP, Zanferrari A, Zirpoli G, Borsi S, Del Moro A (1974a) The Austrides to the south of the Tauern window and the Periadriatic lineament between Mules and Mauthen. Neues Jahrb Geol Paläontol Monatsh 1974: 421–435

Sassi FP, Zanferrari A, Zirpoli G (1974b) Aspetti dinamici dell'evento "caledoniano" nell'Austroalpino a sud della Finestra dei Tauri. Mem Mus Tridentino Sci Nat 20: 3–32

Sassi FP, Cavazzini G, Visonà D, Del Moro A (1985) Radiometric geochronology in the Eastern Alps: results and problems. Rend Soc Ital Mineral Petrol 40: 187–224

Satir M (1975) Die Entwicklungsgeschichte der Westlichen Hohen Tauern und der Südlichen Ötztalmasse aufgrund radiometrischer Altersbestimmungen. Mem Ist Geol Mineral Univ Padova 30: 1–82

Schindelmayr WE (1968) Geologische Untersuchungen in der Umgebung von Mauls und Stilfes in Südtirol. Dissertation, Naturwissenschaftliche Fakultät der Ludwig-Maximilians-Universität München

Schmidegg O (1936) Steilachsige Tektonik und Schlingenbau auf der Südseite der Tiroler Zentralalpen. Jahrb Geol Bundesanst 86: 115–149

Schmidegg O (1937) Der Triaszug von Kalkstein im Schlingengebiet der Villgrater Berge (Osttirol). Jahrb Geol Bundesanst 87: 111–132

Schönlaub HP (1979) Das Paläozoikum in Österreich. Abh Geol Bundesanst 33: 1–124

Schulz B (1988a) Deformation und Metamorphose im ostalpinen Altkristallin südlich des Tauernfensters (südliche Deferegger Alpen, Österreich). Schweiz Mineral Petrogr Mitt 68: 397–406

Schulz B (1988b) Mikrogefüge in Paragneisen des ostalpinen Altkristallins südlich der Defereggen-Antholz-Vals-Linie (Osttirol/Österreich). Mitt Österr Geol Ges 81: 245–253

Schulz B (1988c) Quarz- und Mikrogefüge zonierter Kalksilikatgneis-Körper im ostalpinen Altkristallin (südliche Deferegger Alpen, Österreich). Erlanger Geol Abh 116: 117–122

Schulz B (1989) Jungalpidische Gefügeentwicklung entlang der Defereggen-Antholz-Vals-Linie (Osttirol, Österreich). Jahrb Geol Bundesanst 132: 775–789

Schulz B (1990a) Prograde-retrograde P-T-t-deformation path of Austroalpine micaschists during Variscan continental collision (Eastern Alps). J Metamorphic Geol 8: 629–642

Schulz B (1990b) Tectonic significance of an early-Alpine P-T-deformation path from Austroalpine micaschists to the south of the Tauern window, Eastern Alps. Schweiz Mineral Petrogr Mitt 70: 403–417

Schulz B (1991) Deformation und Metamorphose im Thurntaler Komplex (Ostalpen). Jahrb Geol Bundesanst 134: 369–391

Schulz B (1993) Mineral chemistry, geothermobarometry and pre-Alpine high-pressure metamorphism of eclogitic amphibolites and mica schists from the Schobergruppe, Austroalpine basement, Eastern Alps. Mineral Mag 57: 189–202

Schwinner R (1943) Paläozoikum in der nordwestlichen Goldeckgruppe (ober Sachsenburg, Kärnten). Ber Reichsamt Bodenforsch Wien 1943: 147–156

Senarclens-Grancy W (1932) Beiträge zur Geologie der Deferegger Berge und der westlichen Schobergruppe in Osttirol. Centralbl Mineral Abt B 1932: 481–490

Senarclens-Grancy W (1964) Zur Grundgebirgs- und Quartärgeologie der Deferegger Alpen und ihrer Umgebung. Z Dtsch Geol Ges 116: 502–511

Senarclens-Grancy W (1972) Geologische Karte der westlichen Deferegger Alpen 1:25000. Geol Bundesanst, Wien

Spaeth G, Kreutzer S (1989) Bericht 1988 über geologische Aufnahmen im Thurntaler Quarzphyllit und Altkristallin auf Blatt 179 Lienz. Jahrb Geol Bundesanst 132: 595–597

Spaeth G, Kreutzer S (1990) Bericht 1989 über geologische Aufnahmen im Altkristallin und Thurntaler Quarzphyllit auf Blatt 179 Lienz. Jahrb Geol Bundesanst 133: 482–484

Sprenger W, Heinisch H (1992) Late Oligocene to Recent brittle transpressive deformation along the Periadriatic Lineament in the Lesach Valley (Eastern Alps). Ann Tect 6/2: 134–149

Stauffenberg H (1987) Apatite fission-track evidence for postmetamorphic uplift and cooling history of the eastern Tauern window and the surrounding Austroalpine (central Eastern Alps, Austria). Jahrb Geol Bundesanst 130: 571–586

Steininger H (1985) Bericht (1984) über geologische Aufnahmen im Altkristallin auf Blatt 178 Hopfgarten im Defereggental. Jahrb Geol Bundesanst 128: 312–313

Stöckhert B (1982) Deformation und retrograde Metamorphose im Altkristallin südlich des westlichen Tauernfensters (Südtirol). Dissertation, Naturwissenschaftliche Fakultät der Universität Erlangen-Nürnberg

Stöckhert B (1984) K-Ar determinations on muscovites and phengites from deformed pegmatites and the minimum age of the Old Alpine deformation in the Austridic basement south of the western Tauern Window (Ahrn Valley, southern Tyrol, Eastern Alps, Neues Jahrb Miner Abh 150: 103–120

Stöckhert B (1985) Pre-alpine history of the Austridic basement to the south of the western Tauern window (southern Tyrol, Italy) – Caledonian versus Hercynian event. Neues Jahrb Geol Paläontol Monatsh 1985: 618–642

Stöckhert B (1987) Das Uttenheimer Pegmatit-Feld (Ostalpines Altkristallin, Südtirol). Genese und alpine Überprägung. Erlanger Geol Abh 114: 83–106

Troll G (1978) The "Altkristallin" of eastern Tyrol between Tauern window and Periadriatic lineament. In: Closs H, Roeder D, Schmidt K (eds) Alps, Apennines, Hellenides Inter-Union Commission on Geodynamics Scientific Rep No 38. Schweizerbart, Stuttgart, pp 149–154

Troll G, Hölzl E (1974) Zum Gesteinsaufbau des Altkristallins der zentralen Schobergruppe, Osttirol. Jahrb Geol Bundesanst 117: 1–16

Troll G, Forst R, Söllner F, Brack W, Köhler H, Müller-Sohnius D (1976) Über Bau, Alter und Metamorphose des Altkristallins der Schobergruppe, Osttirol. Geol Rundsch 65: 483–511

Troll G, Baumgartner S, Daiminger W (1980) Zur Geologie der südwestlichen Schobergruppe (Osttirol, Österreich). Mitt Ges Geol Bergbaustud Österr 26: 277–295

Visona D (1974) Il significato barometrico delle muscoviti nelle filladi dello Gailtal (Alpi Orientali). Stud Rent Sci Nat 51: 94–104

Vohryzka K (1967) Bericht über geologische Aufnahmen in der Schobergruppe, Osttirol im Jahre 1966. Verh Geol Bundesanst 1967: A52–A53

Vohryzka K (1968) Bericht über geologische Aufnahmen 1967 in der Schobergruppe, Osttirol, auf Blatt 179/2 und 179/4 der österreichischen Karte 1:25000. Verh Geol Bundesanst 1967: A69–A70

Wallis SR (1988) The structural and kinematic development of the Austroalpine-Penninic boundary, SE Tauern, Eastern Alps. Thesis, Oxford University

Waters DJ (1976) Structural, metamorphic and geochronological studies in the south-east Tauern. Ph D Thesis, Oxford University

Zanferrari A (1976) On the occurrence of a Permo-Scythian syncline outcropping in the middle Lesachtal along the Gail-tal Line (Carinthia, Austria). Neues Jahrb Geol Paläontol Monatsh 1976: 109–117

The Austro-Alpine Metamorphic Basement East of the Tauern Window

F. NEUBAUER and W. FRISCH

Abstract

The Austro-Alpine metamorphic basement east of the Tauern window contains various tectono-stratigraphic units which are highly diverse with respect to protolith and metamorphic ages and also geodynamic environment of formation. Early Precambrian elements are represented in certain units, by detrital minerals or possible magmatic rocks. The continuous history started with Late Proterozoic to Cambrian island arc and ophiolite sequences, which were accreted during Late Cambrian and Ordovician tectono-thermal events. These are overlain by fossiliferous Silurian-Devonian shelf sequences. The shelf sequence is contrasted by the Silurian to Devonian active continental margin with partly S-type granite magmatism. Final collision of all units occurred during the Early Carboniferous, involving deep-crustal thrusting, metamorphism, collisional-type magmatism, and subsequent uplift. Sedimentary overstep sequences were deposited at different times from the Early Carboniferous to the Permian, and indicate the progradation of the deformation front within the internal Variscan belt.

1 Introduction

In contrast to the Austro-Alpine basement west of the Tauern window, the eastern part is highly diverse. Its metamorphic grade ranges from nearly unmetamorphosed, fossiliferous Palaeozoic sequences through low-grade rocks in the "quartzphyllite units" to medium- and high-grade metamorphic units in the "Altkristallin". The weakly metamorphosed units are treated in several chapters of this Volume (Schönlaub and Heinisch; Loeschke and Heinisch; Neubauer and Sassi). In this Section we review fossil-free lithotectonic units of the Altkristallin which underwent pre-Alpine and/or Alpine, medium to high grade metamorphism (Fig. 1).

The litho-tectonic units occur at different tectonic levels within the Alpine nappe edifice (see also Ratschbacher and Frisch this Vol.). Some of the major Alpine units are separated by Permo-Mesozoic rocks mainly along their northern margins. The following major Alpine basement nappes are distinguished (Tollmann 1977):

1. Upper Austro-Alpine nappes which include the crystalline Kaintaleck basement slices and the Ackerl complex.
2. The Middle Austro-Alpine nappes which consist of distinct lithotectonic basement units, in particular the Muriden complexes, including the "Core", the Speik and the Micaschist-Marble complexes, and the Koriden gneiss complex, with its associated micaschist complexes, in particular the Plankogel complex.
3. The Lower Austro-Alpine units which include the higher "Semmering system" with the "Grobgneis" or "Raabalpen" unit (corresponding to the Tatric unit at the northeastern margin of the Eastern Alps) and the lower "Wechsel system" with the Waldbach and Wechsel gneiss complexes.

Cretaceous metamorphism affected nearly all these Altkristallin rocks. It reached greenschist facies in most parts, and up to amphibolite facies and possibly eclogite facies in the central and southern part of the Middle and Lower Austro-Alpine units (Schimana 1986; Frank 1987; Frank et al. 1987; Deutsch 1988; Moine et al. 1989; Miller 1990; Peindl et al. 1990). Therefore pre-Alpine P-T paths have not yet been established in detail.

2 Lithostratigraphic Units

We describe the different lithotectonic basement units from the base to the top of the Alpine nappe edifice. A summary of significant features of distinct lithotectonic units is provided in Table 1. Geochronological data are presented in Table 2. Representative chemical analyses from various basement complexes are compiled for comparison in Table 3.

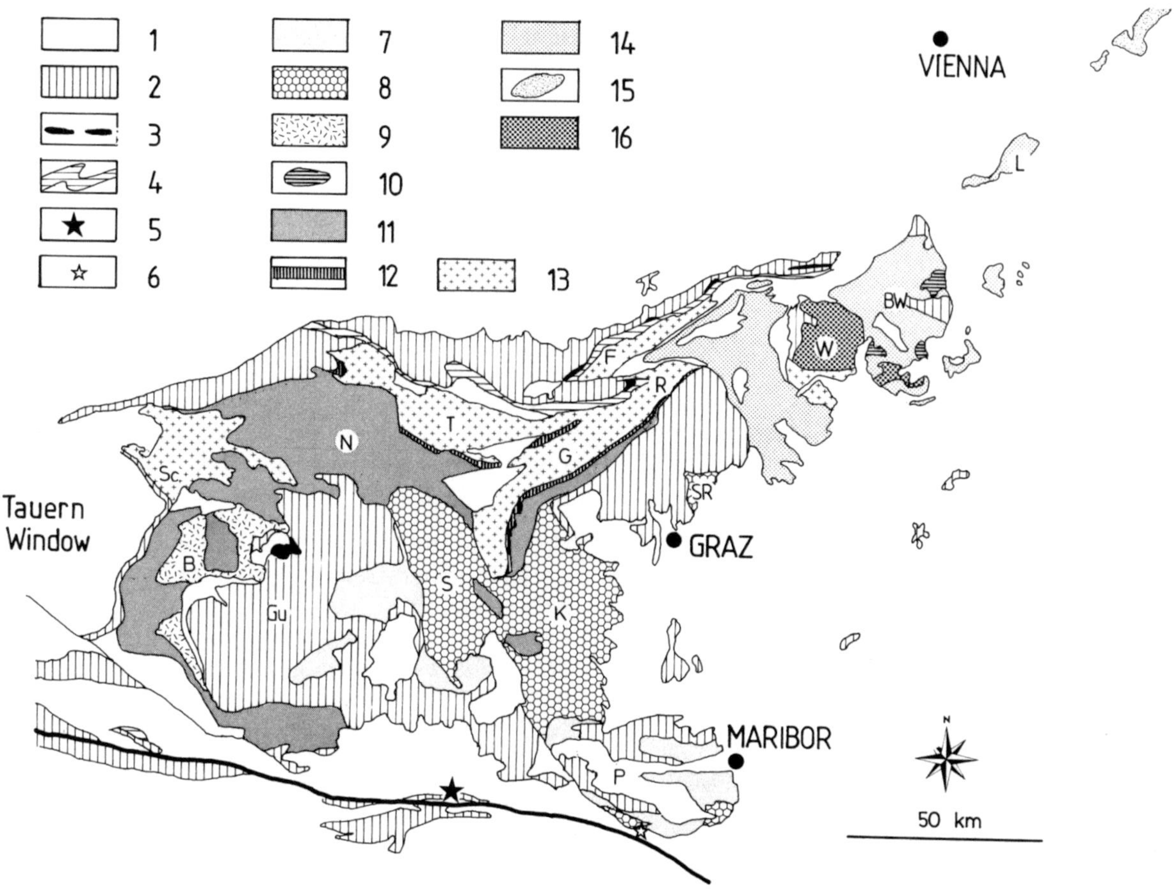

Fig. 1. Simplified map of Austro-Alpine basement units east of the Tauern window. Alpine structures not shown. **1** Permian to Caenozoic formations; **2** fossiliferous Palaeozoic formations and quartzphyllites; **3** Kaintaleck slices and Ackerl complex; **4** Carboniferous and Permian of the Veitsch nappe; **5** Eisenkappel crystalline complex; **6** Pohorje garnet-peridotite-granulite complex; **7** Plankogel complex and related micaschist complexes; **8** Koriden gneiss complex; **9** Bundschuh complex; **10** Sieggrabener complex; **11** Micaschist-Marble complex; **12** Speik complex; **13** Core complex; **14** "Grobgneis" complex; **15** Tatric unit; **16** Wechsel gneiss complex. Geographic markers: **B** Bundschuh; **BW** Bucklige Welt; **F** Troiseck-Floning mountains; **G** Gleinalm; **Gu** Gurktaler Alpen; **K** Koralpe; **L** Leithagebirge; **N** Niedere Tauern; **P** Pohorje mountains; **R** Rennfeld; **S** Saualpe; **Sc** Schladminger Tauern; **SR** St. Radegund; **T** Seckauer Tauern; **W** Wechsel

2.1 Tatric Unit

The crystalline complex forming the transitional zone of Eastern Alps to Carpathians belongs to the Tatric unit (Fig. 1). Its Permo-Mesozoic cover correlates better with the Western Carpathians than with the Alps. It is composed of paragneiss, staurolite-bearing micaschist, quartzitic biotite phyllite and greenschist with carbonate layers (Wessely 1961; Tollmann and Spendlingwimmer 1978). A penetrative Alpine metamorphic overprint is not observed. Both paragneiss and micaschist are intruded by a granitoid pluton which is made up of two-mica granite, granodiorite and leucogranite and is comparable to the adjacent Bratislava pluton of the Little Carpathians to the northeast. The Tatric unit phyllite is correlated by Tollmann and Spendlingwimmer (1978) with the Lamac formation of the Little Carpathians. In the Little Carpathians, the metamorphism is probably Devonian in age, and is constrained by a Rb-Sr whole-rock isochron of 380 ± 20 Ma (Cambel and Kral 1989). The Bratislava pluton gave a Rb-Sr whole-rock age of 347 ± 4 Ma (Cambel and Kral 1989).

2.2 Wechsel Gneiss Complex

The Wechsel gneiss complex is exposed within the Wechsel window as well as within some smaller isolated windows at the margin of the Pannonian basin (Tollmann 1978; Fig. 2). A detailed correlation between different windows does not yet exist.

Table 1. Comparative table of significant characteristics of Austro-Alpine lithotectonic basement units east of the Tauern window.

Lithotectonic unit	Nature of protolith	Age of protolith	Nature of orogenic magmatism	Age of orogenic magmatism	Age of pre-Alpine metamorphism
Tatric unit	Metamorphic rise of unknown origin	Pre-Devonian	Granite	Early Carboniferous	Devonian
Wechsel complex	Continental magmatic arc	Pre-Carboniferous	–	–	Devonian (?) to Early Carboniferous
Waldbach c.	Magmatic arc (?)	Pre-Carboniferous	Porphyric granite and pegmatite	?	?
Grobgneis unit	Magmatic rise, epi-continental	Pre-Carboniferous	I- and S-type granitoids	Early Carboniferous	Pre-Carboniferous to Carboniferous
Core complex	Magmatic arc	Late Proterozoic to Early Paleozoic	S- and I-type magmatism	Early Carboniferous	Early Carboniferous
Speik complex	Ophiolite	Pre-Silurian	Porphyric granite	Early Carboniferous	Early Carboniferous
Micaschist-Marble complex	Shelf	Late Ordovician to Devonian	Pegmatite, granite	Early Carboniferous	Early Carboniferous
Koriden gneiss c.	Continental margin accretion wedge	Late Proterozoic (?) Devonian (?)	Tholeiitic gabbro, Pegmatite	Permian	Pre-Permian and Permian
Sieggrabener c.	Subduction complex	Pre-Carboniferous	–	–	?
Plankogel complex	Ophiolitic accretion wedge	Pre-Carboniferous	Pegmatite	Permian	Pre-Permian and Permian
Bundschuh-Einach c.	Clastic wedge	Pre-Devonian	S-type granitoids	Devonian	Late Silurian to Devonian
Ritting complex	Ophiolite	Pre-Mid Cambrian	–	–	Cambrian
Frauenberg complex	Metamorphic rise	Precambrian	–	–	Cambrian
Prieselbauer c.	Metamorphic rise	Pre-Devonian	Aplite	Devonian	Devonian
Ackerl complex	Metamorphic rise	Pre-Carboniferous	–	–	Pre-Permian
Eisenkappel complex	Metamorphic rise	Pre-Late Carboniferous	Granite to gabbro	Permian	Pre-Permian
Granulite unit/ Pohorje mountains	Lower crust	Pre-Permian	–	–	Pre-Permian

The Wechsel gneiss complex consists of a thick package of uniform albite porphyroblast paragneiss ("Monotonous Wechsel gneiss unit") and a structurally overlying unit which includes other lithologies such as quartzite, epidote amphibolite and greenschist, sheet-like acidic orthogneiss and black schist and black paragneiss as well as the albite porphyroblast gneiss ("Variegated Wechsel gneiss unit"; Fig. 3) (Faupl 1970, 1972; Vetters 1970; Neubauer 1990, 1993). The footwall boundary of the Wechsel gneiss is not exposed. The hanging wall is formed by a series of quartzphyllitic rocks, the Wechsel phyllites, and Permo-Mesozoic cover rocks. The Variegated Wechsel gneiss unit dominates in the smaller windows. Albite porphyroblast gneiss, black quartzites and gneisses and the thick leucocratic Wiesmath orthogneiss are exposed (Fuchs 1962; Pahr 1980).

The thick package of albite porphyroblast paragneiss exhibits petrographic as well as chemical features of chemically altered greywackes and shales (Neubauer 1990). Analyses of the major, minor, trace and rare earth elements shows that the greenschists and epidote amphibolites have a mafic subalkaline chemistry comparable to that of a modern calc-alkaline supra-subduction zone environment (Fig. 4). The orthogneisses were strongly altered during metamorphism but they also display chemical features comparable with those of supra-subduction zone environments. The protolith ages of all these rocks are uncertain.

The Wechsel gneiss complex is interpreted as the product of an active continental margin (Neubauer 1990). The arguments for this interpretation are the chemistry of the igneous rocks, the presence of polycyclic, metasedimentary zircons in meta-quartzarenites, which indicate a nearby mature continental source region for detritus, and the occurrence of massive Fe-Cu sulphides at the hanging wall boundary of the Wechsel gneiss complex (Tufar 1968).

Metamorphism of the Wechsel gneiss complex is polyphase. In the southern part there is evidence of early amphibolite facies metamorphism overprinted

Table 2. Compilation of significant geochronological data of Austro-Alpine basement rocks east of the Tauern window[a]

Rock	Method[b]	Age (error)	Rb-Sr initial ratio (error)	Reference
Tatric unit				
Metamorphic rocks	Rb-Sr WR	380 ± 20	0.7101 ± 0.0004	Cambel and Kral (1989)
Bratislava granite	Rb-Sr WR	347 ± 4	0.7036 ± 0.0013	Cambel and Kral (1989)
Wechsel gneiss complex, Fertörakos complex:				
Pegmatite	Rb-Sr mu	351 ± 9		Kovach and Svingor (1981)
Raabalpen crystalline complex:				
Metatonalite	Rb-Sr WR	343 ± 20	0.7057 ± 0.0003	Peindl et al. (1990)
Porphyric granite gneiss	Rb-Sr WR	338 ± 12	0.7071 ± 0.0006	Scharbert (1990)
Two-mica granite gneiss	Rb-Sr WR	326 ± 11	0.7068 ± 0.0019	Peindl et al. (1990)
Two-mica granite gneiss	Rb-Sr WR	243 ± 17	0.7234 ± 0.0017	Scharbert (1990)
Muriden, core complex:				
Plagioclase gneiss	Rb-Sr WR	518 ± 44	0.7044 ± 0.0012	Frank et al. (1976)
Orthogneiss	Rb-Sr WR	432 ± 16	0.71158 ± 0.00235	Scharbert (1981)
Garnet amphibolite	U-Pb z. l.	ca. 435		Neubauer et al. (1993)
Augen gneiss	U-Pb z. l.	ca. 425		Neubauer et al. (1993)
Metatonalite ("Metablastic amphibolite")	U-Pb z. l.	356		Neubauer et al. (1993)
Two-mica metagranite	Rb-Sr WR	354 ± 16	0.7047 ± 0.0007	Scharbert (1981)
Trondhjemite gneiss	U-Pb z. c.	353 ± 2		Neubauer et al. (in prep.)
Augengneiss	Rb-Sr WR	331 ± 25	0.7058 ± 0.0029	Frank et al. (1983)
	Rb-Sr mu	258 ± 1		Frank et al. (1983)
Metagranite	Rb-Sr mu	331 ± 7		Scharbert (1981)
Pegmatite	Rb-Sr mu	329 ± 12		Scharbert (1981)
Pegmatite	K-Ar mu	347 ± 7		Hejl (1984)
Pegmatite	K-Ar mu	340 ± 18		Hejl (1984)
Biotite amphibolite	K-Ar bio	276 ± 15		Hejl (1984)
Pegmatite	K-Ar mu	265 ± 14		Hejl (1984)
Plagioclase gneiss	K-Ar mu	282 ± 15		Hejl (1984)
Muriden, Micaschist-Marble complex:				
Villach orthogneiss	Rb-Sr WR	445 ± 44	0.7122 ± 0.0014	Frimmel (1988)
Wolfsberg granite	Rb-Sr WR	258 ± 11	0.7046 ± 0.0028	Morauf (1980)
Pegmatite	Rb-Sr mu	347 ± 6		Frank et al. (1983)
Pegmatite	Rb-Sr mu	277 ± 1		Frank et al. (1983)
Pegmatite	Rb-Sr mu	256 ± 30		Jäger and Metz (1971)
Koriden				
Eclogite	Sm-Nd WR-gt	693 ± 39		Manby and Thiedig (1988)
Micaschist	U-Pb zi. u.	1200 ± 60		Paquette and Gebauer (1989)
"Plattengneis"	U-Pb zi. u.	2015 ± 49		Paquette and Gebauer (1989)
	l. i.	297 ± 3		
	Rb-Sr t. s. i.	256 ± 35	0.7158 ± 0.0012	Frank et al. (1981)
	Rb-Sr t. s. i.	249 ± 6		Frank et al. (1983)
Metagabbro	Sm-Nd m. i.	275 ± 18		Thöni and Jagoutz (1991)
Pegmatite	Rb-Sr WR	ca. 280		Scharbert, cited in Göd (1989)
	Rb-Sr mu	277 ± 5		Frank et al. (1983)
	Rb-Sr mu	265 ± 25		Morauf (1981)
	Rb-Sr mu	263 ± 25		Morauf (1981)
	Rb-Sr mu	263 ± 25		Frank et al. (1983)
	Rb-Sr mu	258 ± 16		Morauf (1981)
	Rb-Sr mu	249 ± 16		Morauf (1981)
	Rb-Sr mu	247 ± 15		Morauf (1981)
	Rb-Sr mu	246 ± 16		Morauf (1981)
	Rb-Sr mu	246 ± 32		Morauf (1981)
	Rb-Sr mu	244 ± 26		Morauf (1981)
Pegmatite (St. Radegund)	Rb-Sr WR	313 ± 18	0.7147 ± 0.0010	Neubauer and Stattegger (unpubl. data)
	Rb-Sr mu	308 ± 16		Neubauer and Stattegger (1981)
	Rb-Sr mu	303 ± 65		Neubauer and Stattegger (1981)
Bundschuh area				
Bundschuh orthogneiss	Rb-Sr WR	372 ± 29	0.738 ± 0.002	Hawkesworth (1976)
	Rb-Sr mu	352 ± 4		Frimmel (1986 a, b)

Table 2. Continued

Rock	Method[b]	Age (error)	Rb-Sr initial ratio (error)	Reference
"Kaintaleck slices"				
Hornblende-garnet gneiss	U-Pb z. u.	2,530		Neubauer et al. (1989)
	U-Pb z. l.	516		
Paragneiss	U-Pb z. l.	391 ± 2		Neubauer et al. (in prep.)
Aplitic orthogneiss	U-Pb z. u.	365 ± 20		Neubauer et al. (in prep.)
Orthogneiss boulder	U-Pb z. u.	502 ± 6		Neubauer et al. (1987)
Orthogneiss boulder	U-Pb z. u.	500 ± 20		Neubauer et al. (1987)
Eisenkappel				
Karawanken granite	K-Ar hb	252 ± 9		Cliff et al. (1974)
	Rb-Sr bio	232 ± 9		Scharbert (1975)

[a] Older data are recalculated with a decay constant of 1.42×10^{-11} a.
[b] WR, whole-rock age, z. l., zircon lower intercept age; z. u., zircon upper intercept age; m. i., mineral isochron; t. s. i., thin slab isochron; mu, muscovite; gt, garnet; hb, amphibole; bio, biotite

under upper greenschist facies conditions with retrogressive growth of albite porphyroblasts. Preliminary Rb-Sr data (ca. 335 ± 7 Ma) of white mica postdating the albite porphyroblast growth suggest a pre-Late Carboniferous age of the second metamorphic event (Müller et al. unpublished data). This event was accompanied by polyphase folding, suggesting N-S compression with regard to the present geographic framework.

The hanging wall boundary of the Wechsel gneiss complex with the Wechsel phyllites was previously interpreted as a possibly primary contact (see discussion in Schönlaub 1979). Recent investigations show, however, that this boundary is a zone of ductile W to SW-directed thrusting or, more likely, a zone of ductile low-angle normal-faulting which indicates the final Variscan emplacement of a Wechsel phyllite nappe after the metamorphic peak of the Wechsel gneiss complex (Neubauer and Pischinger unpubl. data).

The Fertörakos complex in westernmost Hungary is interpreted as part of the Wechsel gneiss unit by most authors (e. g. Kovach and Svingor 1981; Lelkes-Felvari et al. 1984). In this complex amphibolites dominate, with some phyllitic intercalations and marble. A Rb-Sr age of pegmatitic white mica was determined as 354 Ma (Kovach and Svingor 1981).

2.3 Waldbach Crystalline Complex

The Waldbach crystalline complex overlies the southern margin of the Wechsel gneiss complex along a low-angle normal-fault (Peindl et al. 1990). It consists of phyllitic micaschist at the structural base, intermediate orthogneiss (hornblende orthogneiss), various amphibolites, thin garnet micaschist, ore-bearing black schist and quartzite, with stratiform sulphides on top of the amphibolites (Tufar 1963; Huska 1970; Faupl 1972; Flügel and Neubauer 1984). Detailed investigations have not yet been carried out. The hornblende gneiss and the amphibolites may represent an island arc magmatic sequence of unknown, pre-Alpine age. Pre-Alpine metamorphism reached amphibolite facies conditions and locally partial melting occurred (Faupl 1972). Discordant pegmatites postdate both the metamorphic peak and the associated ductile deformation.

The Waldbach crystalline complex is assumed to be related to the Core complex of the Middle Austro-Alpine basement nappe system because of their similar lithologies.

2.4 The "Grobgneis" Complex

The "Grobgneis" complex is exposed to the northeast and west of the Wechsel window (Fig. 2). Other terms for this lithotectonic unit are the Raabalpen crystalline complex west of the Wechsel window or the Eselsberg complex east of the Wechsel window (Tollmann 1978; Flügel and Neubauer 1984).

The "Grobgneis" complex is composed of migmatitic paragneiss (Strallegg gneiss), micaschist (Tommer schist), and phyllonites (Mürztal and Birkfeld quartzphyllites) with some minor intercalations of talc schists, clinopyroxene-bearing amphibo-

Table 3. Representative chemical analyses of Austro-Alpine amphibolites

Sample No.	1 R146	2 RF115	3 RF143	4 RF118	5 RF145	6 556	7 23	8 SK1	9 GE2	10 34	11 333b	12 GR-D	13 GR115
SiO_2	49.88	48.30	47.87	53.68	47.12	48.56	48.88	42.40	48.74	48.91	49.28	44.65	48.05
TiO_2	1.18	1.62	0.64	1.12	0.15	1.30	3.44	0.42	2.54	1.35	3.18	1.30	2.22
Al_2O_3	15.09	15.12	19.98	16.85	19.75	13.28	14.58	28.78	14.55	14.77	14.22	16.15	14.22
Fe_2O_3	2.02	2.99	1.03	1.29	0.45	2.10	1.48	0.80	1.27	1.26	2.07	11.04*	12.57
FeO	6.76	7.44	5.45	5.38	6.50	11.44	10.05	6.00	9.30	9.24	10.96	n. d.	n. d.
MnO	0.18	0.21	0.12	0.14	0.15	0.25	0.16	0.09	0.18	0.22	0.31	0.17	0.17
MgO	6.80	6.96	9.68	5.11	9.99	6.67	6.54	4.59	8.13	10.19	5.76	5.57	5.83
CaO	9.27	11.63	9.65	7.93	12.55	9.91	7.19	15.10	12.10	10.90	8.61	11.93	11.31
Na_2O	3.71	2.81	3.00	3.54	1.21	3.00	3.94	1.10	2.88	0.88	3.07	3.14	3.11
K_2O	0.10	0.31	0.47	2.08	0.13	0.20	–	0.06	0.09	0.15	0.21	0.42	0.52
P_2O_5	0.19	0.21	0.15	0.62	0.19	0.12	0.50	0.05	0.18	0.06	0.44	0.13	0.22
L.O.I.	2.78	1.74	1.71	2.23	1.64	0.48	1.00	1.07	1.23	1.18	0.46	4.69	1.24
Cr	180	200	143	116	135	72	372	248	206	282	54	111	175
Nb	<2	n. d.	n. d.	12	n. d.	n. d.	68	8	5	<5	45	<7	18
Ni	13	101	219	<5	81	54	123	59	73	149	29	74	125
Rb	6	2	16	73	3	<3	11	7	<4	10	7	12	11
Sr	366	183	255	766	232	112	554	271	182	98	347	291	368
Y	18	26	18	34	14	28	24	17	38	24	25	22	16
Zr	100	91	54	281	15	81	307	35	130	68	211	84	158

[a] Major and minor element oxides in wt%, trace elements in ppm. n. d., Not determined; *, Fe_2O_3 as total Fe. 1: Epidote amphibolite from the Wechsel gneiss complex (Neubauer 1993). 2: Plagioclase amphibolite from the "Core complex" with supposed calc-alkaline basaltic origin (Neubauer 1988a). 3: Plagioclase amphibolite from the "Core complex" with supposed gabbroic origin (Neubauer 1988a). 4: "Metablastic amphibolite" (metatonalite) from the "Core complex" (Neubauer 1988a). 5: Metagabbro of the Speik complex (Neubauer 1988a). 6: Garnet amphibolite of the Speik complex (Schlöser 1987). 7: Alkalic garnet amphibolite of the Micaschist-Marble complex (Schlöser 1987). 8: Low-Ti eclogite of the Koriden complex (Miller et al. 1988). 9: High-Ti eclogite of the Koriden complex (Miller et al. 1988). 10: Garnet amphibolite of the Plankogel complex (Schmerold 1988). 11: Biotite amphibolite of the Plankogel complex (Schmerold 1988). 12: Garnet-zoisite amphibolite of the Ritting complex (unpubl. analysis). 13: Amphibolite of the Frauenberg complex (unpubl. analysis).

lites, quartzites, and tourmalinites (Cornelius 1952; Moreau 1981; Flügel and Neubauer 1984; Peindl 1990). The age of the sequence is constrained by the intrusion of various Carboniferous granites, but is otherwise uncertain. Geochemical patterns of rocks surrounding the talc deposit of Rabenwald suggest an evaporitic environment (Moreau 1981).

Granitic gneisses are widespread. Field and detailed Rb-Sr geochronological work (see Table 2) revealed complex intrusion relationships. The widespread "Grobgneis" is a coarse-grained, porphyric, sheet-like granite gneiss with rare lenses of tonalite gneiss. The Rb-Sr whole-rock age of the Grobgneis is 338 ± 12 Ma (Sr_o = 0.7071 ± 0.0006). A small body of fine-grained, porphyric metatonalite crops out in the southern area (Peindl 1990; Peindl et al. 1990). Its Rb-Sr age is 343 ± 20 Ma with an Sr_o value of 0.7057 ± 0.0003. Rare earth element patterns (Kiesl et al. 1983) favour a collisional origin of the "Grobgneis". A suite of isolated muscovite-biotite granite bodies is scattered throughout the southeastern Raabalpen crystalline complex. They exhibit subsolvus granite mineralogy, and often have well-preserved discordant contact aureoles with the country rocks. These two-mica granites are more differentiated with respect to the Grobgneis. A Rb-Sr errorchron of these rocks yielded an age of 326 ± 11 Ma and a Sr_o value of 0.7068 ± 0.0019 (Peindl 1990; Peindl et al. 1990).

Another suite of isolated garnet-bearing muscovite-biotite granite gneisses is exposed to the west of the latter (Fig. 2). These rocks, however, yielded a Rb-Sr errorchron of 243 ± 12 Ma and a Sr_o value of ca. 0.7234 (Scharbert 1990).

Metagabbro and diorite with well-preserved magmatic fabric occur along the margins of the Grobgneis, e. g. the gabbros of Birkfeld and Landsee (Fig. 2). Further metagabbro occurrences are found within amphibolites of the Kulm (Purkert 1927) at the southwestern margin of the "Grobgneis complex" (Moyschewitz pers. comm.). These metagabbros may be of Permian age because they postdate the pre-Alpine metamorphism (see later).

Although the Alpine metamorphic overprint reached greenschist to amphibolite facies grade, the pre-Alpine mineral assemblages are well preserved in some areas (Koller and Wieseneder 1981). A recent estimate of the Alpine metamorphism in the southwestern area is 500–550 °C and 8–9 kb (Moine

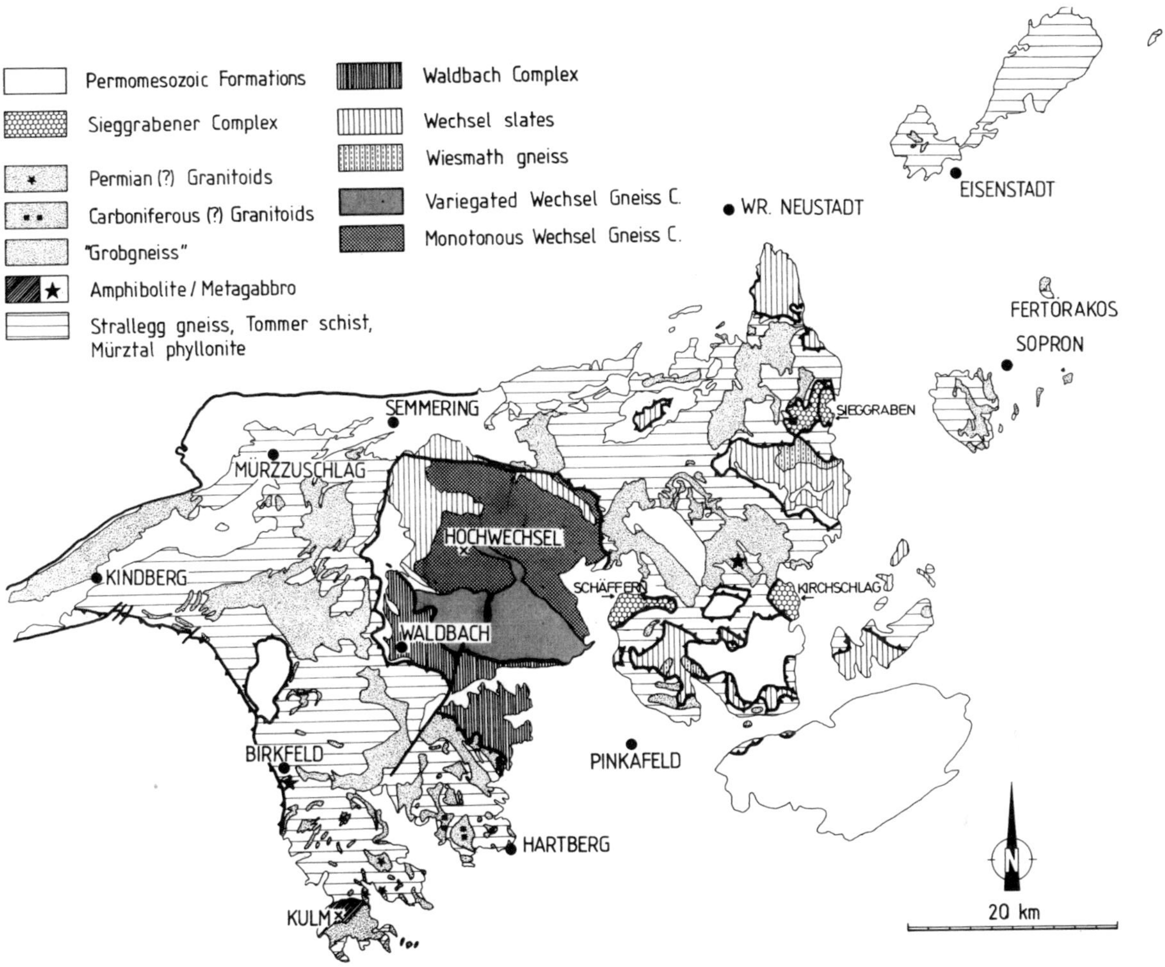

Fig. 2. Geological map of the Lower and Middle Austro-Alpine basement nappes at the eastern margin of the Eastern Alps

et al. 1989). Different stages of pre-Alpine metamorphism can be recognized because of their relationship to Carboniferous and Permian intrusions (Peindl 1990; Peindl et al. 1990). A first stage predated Carboniferous intrusions and caused partial melting of metapelites. Peak metamorphic conditions were reached after the intrusion of the Early Carboniferous muscovite-biotite granites but before the emplacement of Permian granites. They resulted in progressive decomposition of muscovite to potash feldspar and sillimanite in some Carboniferous granites (Peindl 1990) and may be described as localized granulite facies.

2.5 Muriden

The term Muriden was created for basement units which differ from the Koriden and Raabiden (Raabalpen complex) (e.g. Kober 1938). The Muriden complex contains three major lithotectonic units (Figs. 1, 5):

1. The Micaschist-Marble complex as the uppermost unit;
2. The Speik complex;
3. The "Core complex" which appears in structural domes or cores as the lowermost unit.

2.5.1 "Core Complex"

The "Core complex", best studied in the Gleinalm and Rennfeld-Mugel areas, is also exposed in the Troiseck-Floning, the Seckauer Tauern and Bösenstein, and the Schladminger Tauern (Fig. 1).

In each locality, the Core complex consists of the following lithological units (see for example, Fig. 5):

1. A thick pile of strongly foliated biotite plagioclase paragneisses form the country rocks of all the other magmatic lithological units.

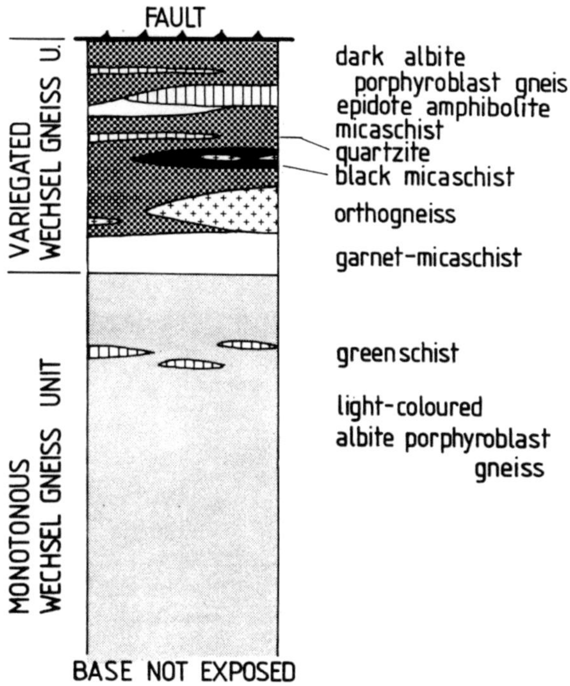

Fig. 3. Log of the Wechsel gneiss complex. (After Neubauer 1993)

2. Plagioclase orthogneisses appear in large cupolas mantled by biotite plagioclase paragneiss.
3. Large masses of various amphibolites are enclosed within plagioclase paragneiss and orthogneiss (Becker 1981; Schedl 1981; Frisch et al. 1987; Neubauer 1988a; Haiss 1991). The most common amphibolite type is a banded amphibolite which includes orthogneiss layers in the scale of centimetres to metres (Frank et al. 1976; Frisch et al. 1987). This amphibolite strongly resembles the leptinite-amphibolite complexes of Variscan basement massifs in Western Europe (e.g. Santallier et al. 1988). The banded amphibolite is in close spatial relationship to the plagioclase orthogneiss (Fig. 5).
4. A layered metatonalite complex ("metablastic amphibolites") is accompanied by ultramafic rocks, metagabbros and rare acidic rocks (Becker and Schumacher 1973; Frisch et al. 1987; Teich 1987a, b; Neubauer 1988a).
5. Major plutons of granitic, granodioritic and tonalitic rocks occur in all domes. A sheet-like augen gneiss forms the hanging wall boundary of the "Core complex".

The biotite plagioclase paragneiss contains thin marbles, calc-silicate rocks and garnet micaschists, which include manganese quartzites, as marker horizons. The gneiss itself exhibits a locally well-preserved, sedimentary layering which may reflect graded bedding of greywackes (Neubauer 1988a). Both the geochemistry and the zircon habitus suggest that the paragneiss may be derived from volcanogenic greywackes (Frisch et al. 1987; Neubauer 1988a; Haiss 1991). U-Pb zircon data suggest Late Proterozoic to Early Palaeozoic origin of the clastic zircons (Neubauer et al. 1993). We interpret the pla-

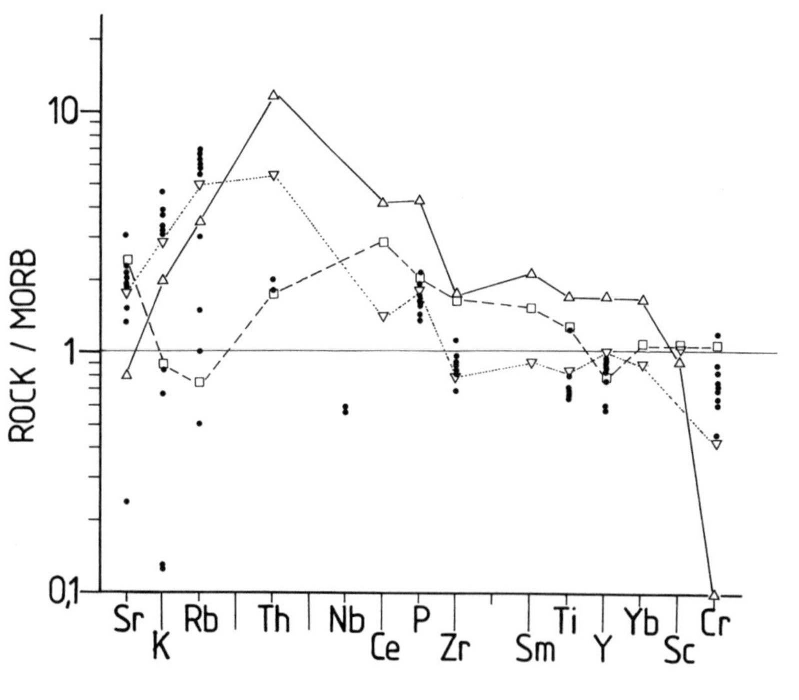

Fig. 4. Spiderdiagram of epidote amphibolites and greenschists from the Wechsel gneiss complex (Neubauer 1990)

gioclase paragneiss sequence as greywacke prism on the margin of a magmatic arc complex which has been interrupted by sedimentation on an abyssal plane with pelites and siliceous, manganese-bearing rocks.

The plagioclase orthogneisses form a major pluton in the Gleinalm area (Fig. 5) with a Rb-Sr whole-rock errorchron of 518 ± 50 Ma (Frank et al. 1976). Haiss (1991) found a U-Pb minimum age of approximately 500 Ma based on nearly concordant zircons. These rocks, in association with the banded amphibolites, were interpreted as a bimodal volcanic sequence (Frank et al. 1976). However, the dome-like shape of the orthogneisses cross-cutting other lithologies (Fig. 5), the rock distribution, the occurrence of amphibolite xenoliths and mappable, discordant contacts between plagioclase orthogneiss and country rocks argue for a plutonic origin (Neubauer 1988a, 1989b). The layering is the result of mylonitization and ductile shearing of both rock types after the intrusion of the orthogneiss protoliths into the amphibolites (Neubauer 1989b).

The amphibolites display a wide range of petrographic and chemical compositions. Because of intense ductile deformation, relics of magmatic mineralogy are preserved in only a few bodies. Well-preserved gabbros are mainly related to the suite of the "metablastic amphibolite", e.g. the Utschgraben metagabbro (Schatzmayer et al. 1990). These bodies contain relics of orthopyroxene and clinopyroxene (Hermann 1972). In contrast, Haiss (1991) has described amphibolites which are derived from subalkaline basalts with an intraplate component. Other major amphibolite bodies show a more calc-alkaline composition according to their major, minor, trace and REE chemistry (Fig. 6). Cumulate gabbroic rocks may form a major portion of the amphibolites as suggested by rare relict minerals, strong positive Eu anomalies, low Ti content and other chemical features (Fig. 7). Gabbro-derived amphibolites dominate volumetrically within both the plagioclase paragneiss and the orthogneiss. All these data indicate different origins and variable ages of the amphibolites. Protolith ages of these amphibolites are un-

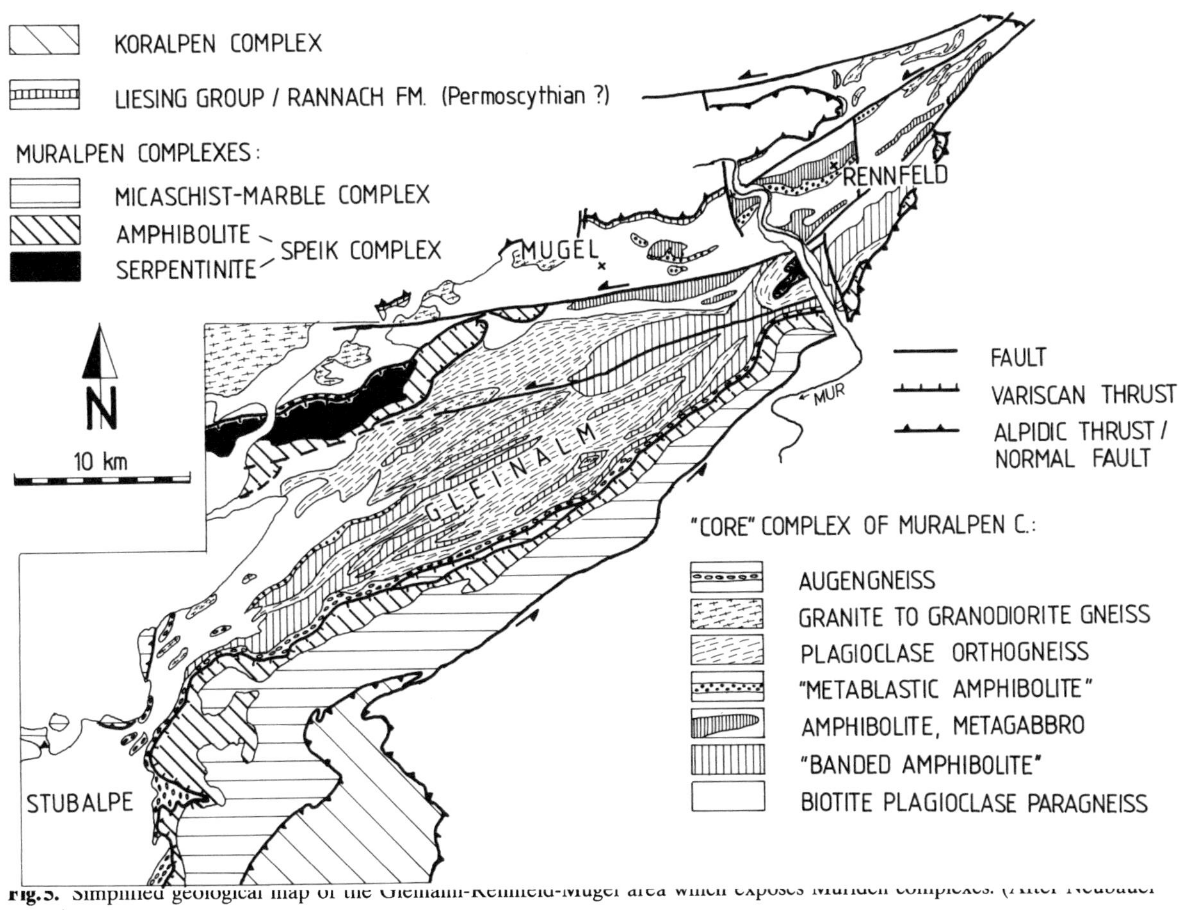

Fig. 5. Simplified geological map of the Gleinalm-Rennfeld-Mugel area which exposes Muralpen complexes. (After Neubauer 1988a)

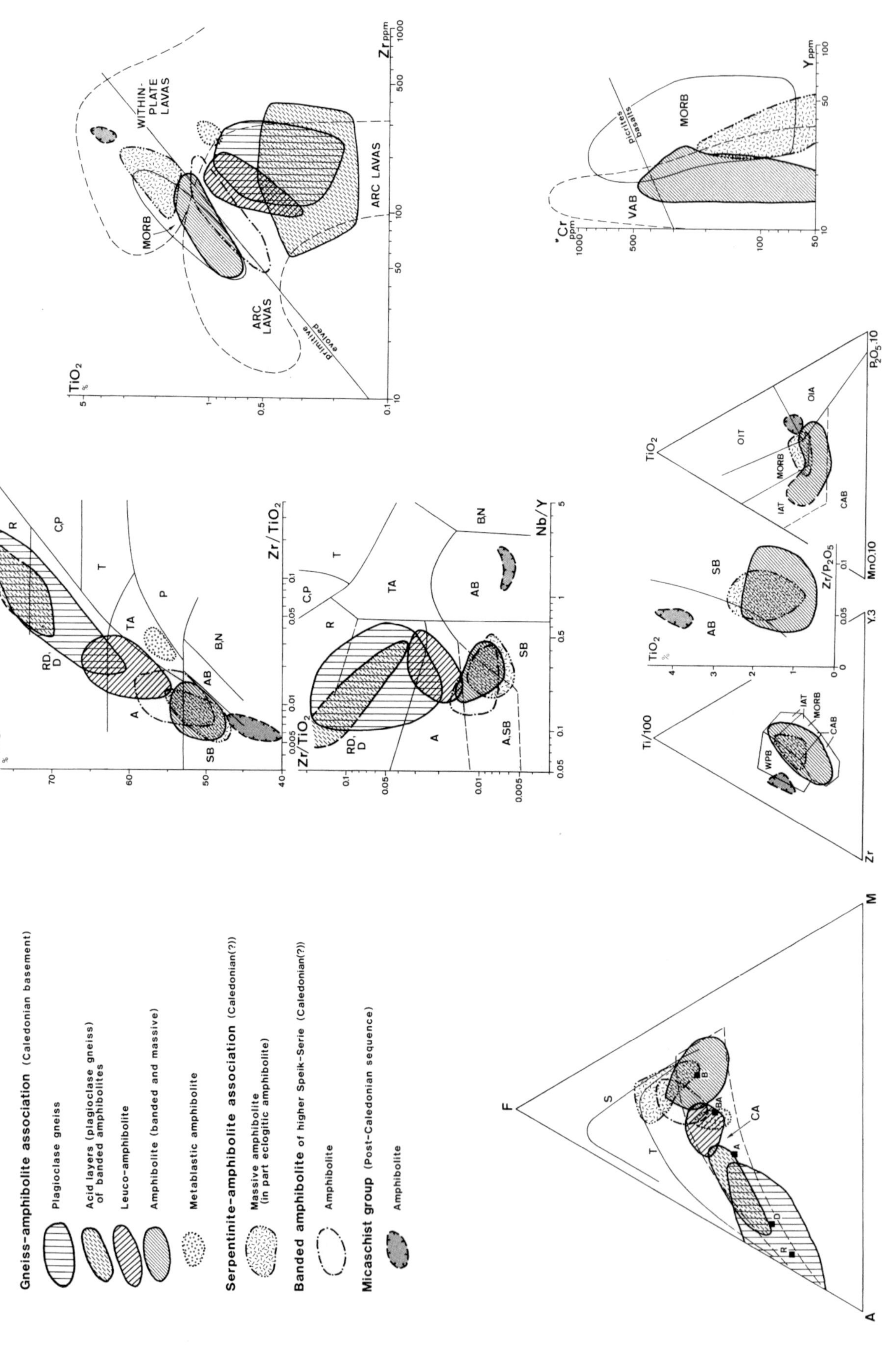
Gneiss-amphibolite association (Caledonian basement)
Plagioclase gneiss
Acid layers (plagioclase gneiss) of banded amphibolites
Leuco-amphibolite
Amphibolite (banded and massive)
Metablastic amphibolite
Serpentinite-amphibolite association (Caledonian(?))
Massive amphibolite (in part eclogitic amphibolite)
Banded amphibolite of higher Speik-Serie (Caledonian(?))
Amphibolite
Micaschist group (Post-Caledonian sequence)
Amphibolite
F
A
M
S
T
CA
SiO2 %
Zr/TiO2
Nb/Y
R
C,P
T
TA
P
RD, D
A
AB
B,N
SB
A,SB
TiO2 %
Zr ppm
MORB
WITHIN-PLATE LAVAS
ARC LAVAS
primitive
evolved
Ti/100
Zr
Y.3
WPB
IAT
CAB
Zr/P2O5
MnO.10
P2O5.10
OIT
OIA
Cr ppm
Y ppm
VAB
picrites
basalts

certain. An unusual, highly differentiated garnet amphibolite yielded a U-Pb zircon age of ca. 425 Ma (lower intercept age; Neubauer et al. 1993).

Some major amphibolite bodies exhibit migmatitic fabrics near contacts to country rocks (Neubauer 1988 a). Trondhjemite gneisses form leucosomes, hornblende felses the palaeosomes. The association is interpreted as result of partial melting of amphibolites (Neubauer 1988 a, 1993). Nearly concordant U-Pb zircons at 353 ± 2 Ma indicate a Variscan age of trondhjemite formation, and, therefore, of high-grade metamorphism (Neubauer et al. 1993).

The plutonic suite of the "metablastic amphibolite", a metatonalite, is widespread throughout the Core complex. It outcrops over a distance of 70 km mostly at the boundary between biotite plagioclase paragneiss and the banded amphibolite. The most prominent rock type of this suite is a medium-grained, massive plagioclase amphibolite or hornblende gneiss which includes lenses of metapyroxenite, metawehrlite, hornblendite, metagabbros and augengneisses which are derived from porphyric biotite granite. This augengneiss differs from the sheet-like augen gneisses at the hanging wall boundary of the Core complex. The plutonic suite is probably derived from more than one magma. The average metatonalite shows differentiation to more hornblende-poor types and to mafic and ultramafic cumulates. Our own, unpublished, Sr isotopic data indicate contamination with radiogenic Sr of some metagabbros which are enclosed within the metatonalite. The trace element signature of the augengneiss is different from, and the REE contents are lower than, those of the metatonalite. The lower intercept U-Pb zircon age of the metatonalite is 356 ± 5 Ma (Neubauer et al. 1993) which is interpreted as the age of Variscan metamorphism. The upper intercepts of the metatonalite near 3 Ga and of the augengneiss at 2.25 Ga indicate contamination by old, Archaean to Proterozoic zircons derived from continental crust during melt formation and magma rise. One of the augengneiss lenses yielded a lower intercept U-Pb zircon age of ca. 425 Ma. A preliminary Rb-Sr errorchron with 390 ± 49 Ma contradicts the U-Pb zircon age of the metatonalite. A reliable interpretation fitting all evidence is the following one: age of tonalite intrusion is approximately 360 Ma, age of metamorphic overprint approximately 360–350 Ma.

Most granitic, granodioritic and rare tonalitic and dioritic orthogneisses postdate formation of other lithologies. An orthogneiss from the Seckauer Tauern with unknown contact with the country rocks yielded a Rb-Sr age of 432 ± 16 Ma (Scharbert 1981). A nearly concordant U-Pb zircon age (ca. 500 Ma) from a tonalite gneiss is interpreted as the minimum formation age (Haiss 1991). Other granitoids are fine-grained granite and granodiorite gneisses which form discordant plutons and sheet-like augengneisses (Teich 1979; Neubauer 1988 a). A characteristic two-mica augengneiss forms the hanging wall boundary of the Core complex in the Gleinalm, Seckauer Tauern and Schladminger Tauern (Becker 1981). The Gleinalm augengneiss yielded a Rb-Sr errorchron of 330 ± 30 Ma (Frank et al. 1983). The widespread granitoids in the Gleinalm and Rennfeld area are mostly mica-poor. Various flaser granite gneisses and two-mica granites like the Zinken granite occur in the Seckauer Tauern (Metz 1976 a; Scharbert 1981). The Zinken metagranite yielded a Rb-Sr age of 354 ± 16 Ma (Scharbert 1981) indicating an Early Carboniferous age of intrusion. The source of early Variscan granitic rocks is uncertain, because detailed investigations have been not yet been made. The presence of white mica in some granitoids argues for an S-type origin.

We interpret the "Core complex" as a Late Proterozoic to Early Palaeozoic root of a magmatic arc which was flanked by sedimentary basins. The protoliths of the biotite plagioclase paragneiss may have formed in an arc-related basin which was later buried and intruded by mafic and acidic magmas. A major step of evolution is indicated by the most frequent age group in the range of 460–425 Ma which is exclusively related to magmatic rocks. Granites intruded during this period for the first time (orthogneiss in the Seck-

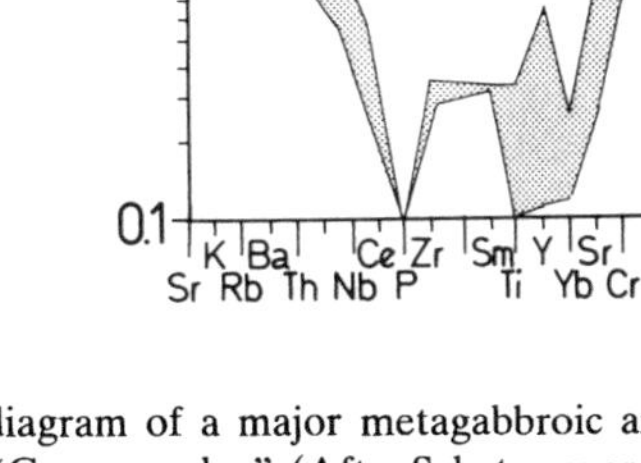

Fig. 6. Geochemical discrimination of magmatogenic amphibolites and gneisses of the Muriden units. (After Frisch et al. 1987)

Fig. 7. Spiderdiagram of a major metagabbroic amphibolite body from the "Core complex". (After Schatzmayer et al. 1990)

auer Tauern). Crust-derived radiogenic Sr isotopic composition as well as Archaean and Early Proterozoic memories in magmatic zircon indicate the presence of old continental crust underlying the "Core complex" at this time. Therefore, the subduction process ceased during this time by subduction of or collision with a plate composed of old continental crust which provided the Precambrian isotopic signature of the calc-alkaline melts. We do not know whether there is a continuous transition of plutonism to the Carboniferous granitoids. The Carboniferous granitoids are related to final collision of the Core complex with other basement units of the Eastern Alps, e.g. the Speik complex.

2.5.2 Speik Complex

The Speik complex (Fig. 5, 8) consists of a package of several hundred metres thick garnet amphibolites and banded amphibolites (metabasaltic rocks), garnet-zoisite and plagioclase amphibolites (metagabbroic rocks), thin augengneiss or serpentinite lenses up to 20 km in length, and rare metasedimentary rocks (El Ageed et al. 1980; Haditsch et al. 1981; Frisch et al. 1987; Neubauer 1988a; Neubauer et al. 1989b). The complex is exposed along the southern Gleinalm, at Kraubath, Traföß and Hochgrößen. An eclogite occurrence is also reported from the Hochgrössen (Wieseneder 1969; Richter 1973). All these rocks are intensely foliated.

Petrographic and chemical data provide evidence that the Speik complex contains all members of an ophiolite. Refractory ultramafic rocks are mainly exposed at Kraubath and Hochgrössen (El Ageed et al. 1980). Ultramafic and mafic cumulates are common within the Traföss body (Neubauer 1988a). The chemical patterns of garnet amphibolites of the southern Gleinalm are close to those of back-arc basin ocean floor basalts (Neubauer et al. 1989b). Locally occurring sulphide-rich quartzites and impure marbles are interpreted as the oceanic sedimentary layer (Neubauer 1988a).

The Speik complex has not yet been dated by geochronological methods. Its age is presumed to predate the Micaschist-Marble complex in the hanging wall for which a late Ordovician to Devonian age is assumed based on lithostratigraphic correlations (see later).

2.5.3 Micaschist-Marble Complex

The Micaschist-Marble complex is widespread throughout the Middle Austro-Alpine basement (Fig. 1). It occurs along the southern margin of the Gleinalm, within the Niedere Tauern, and is again widespread southeast of the Tauern window and in the Kreuzeck Mts. south of the Tauern window (Becker 1981; von Gosen 1989a; Hoke 1990; Fig. 1). Feldspar-rich and quartz-rich micaschists dominate in the structurally lower part. The middle portion contains garnet micaschists, pale-coloured quartzites, thin amphibolites and black schists (Becker and Schumacher 1973; Metz 1976b; Becker 1981; Neubauer 1988a). Siliceous and pure marbles, which are also intercalated with biotite amphibolites, some garnet micaschist and sheet-like pegmatites, dominate the upper portion of the sequence (Fig. 9). The transition of micaschist to carbonates is characterized by calc-silicate rocks, and sometimes by scheelite-bearing tourmalinites and dolomitic marbles (Becker 1977, 1981; Neubauer 1988a; Raith 1988).

Geochronological data of protolith ages are not available. The sequence has been correlated with the fossil-bearing Late Ordovician to Early Carboniferous sequences of the Eastern Alps: the micaschists are presumed to be of Late Ordovician to Early Devonian age, the marbles of mainly Middle Devonian age (Becker 1977, 1981; Neubauer 1988a).

The field relationship of the augengneiss in the lower Muriden complex and pegmatites in the Micaschist-Marble complex suggest cogenetic, Early Carboniferous formation (Neubauer 1989a). The Li- and Be-mineralization of some pegmatites (e.g. Koller et al. 1983) classify them as highly fractionated melts derived from a major granite.

2.5.4 Pre-Alpine Tectono-Thermal Evolution of the Muriden

The Core complex, the Speik complex and the Micaschist-Marble complex have a common Variscan thermal history. The "Caledonian thermal event" is evidenced only in the "Core complex" by U-Pb zircon lower-intercept data of the augengneiss within the metatonalite suite, and the garnet amphibolite between 450 and 425 Ma (Table 2). In addition, scattered paragneiss zircons are close to the discordia of this intercept (Neubauer et al. 1993). A 432 Ma age is found in an orthogneiss of the Seckauer Tauern (Scharbert 1981). All these data indicate magma crystallization during this time. Evidence of defor-

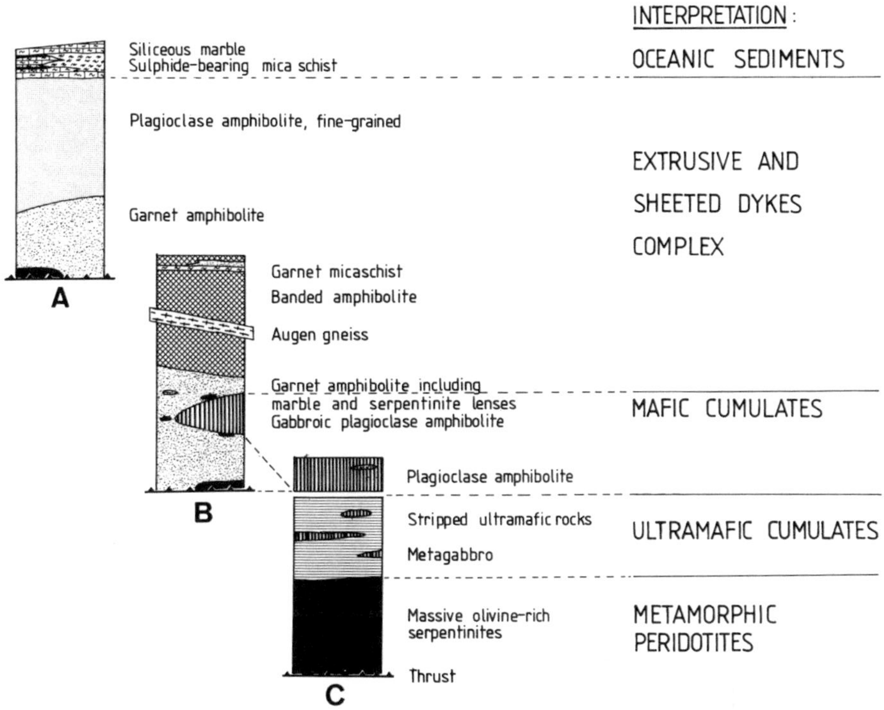

Fig. 8 A–C. Lithostratigraphy of the Speik complex. (After Neubauer et al. 1989a)

mation and metamorphism during this time span is lacking.

The superposition of both Speik and Micaschist-Marble complexes on the Core complex most probably predates peak thermal conditions of pre-Alpine metamorphism. Thrusting is accompanied by intrusion of granites, the protoliths of the augengneiss on top of the Core complex and within the Speik complex, into the zone of thrusting (Neubauer 1988b, 1989a, b). Ductile deformation of hot granites and amphibolitic country rocks is attributed to large shear-strain which caused the mylonitic fabric and the layering of gneisses and most banded amphibolites (which derived most probably from veined amphibolites). The fact that the augengneiss horizon climbs southwestwards from the top of the Core complex through the Speik complex into the Micaschist-Marble complex indicates top-to-the-southwest shear. Because of the Early Carboniferous age of the augengneiss, a Variscan age is presumed for thrusting (Neubauer 1989a, b). The Speik complex is therefore considered to be a large Variscan ophiolite nappe (Fig. 10) by Neubauer (1988a) because of the Variscan age of granite emplacement into the zone of thrusting and the Variscan age of penetrative fabrics. The age of the Speik complex is assumed to be pre-Silurian. The metamorphic conditions are assumed to have reached the stability field of staurolite in the Early Carboniferous.

2.6 *Koriden Complex ("Gneiss Group")*

The Koriden Gneiss complex is exposed in the Koralpe (Beck-Mannagetta 1970; Frank et al. 1983), Saualpe (Weissenbach 1975) and Pohorje mountains (Hinterlechner-Ravnik and Moine 1977) and consists of a thick sequence of kyanite-bearing paragneiss, paragneiss with kyanite paramorphs after andalusite, enclosing up to kilometre-sized lenses of eclogite and amphibolitic eclogite (Fig. 1). Intercalations of rare relics of metagabbro, marbles, manganese quartzites, calc-silicate rocks, and widespread pegmatites are present (Weissenbach 1975; Hinterlechner-Ravnik and Moine 1977; Raith 1988).

The age and evolution of the Gneiss group is the subject of lively discussion. The paragneiss is interpreted as metamorphosed greywacke and shale (Pacher and Riepl 1978; Heritsch 1980). The eclogites are derived from two different protoliths: kyanite-bearing, low-Ti eclogites from gabbros (clinopyroxene-bearing cumulates), kyanite-free, high-Ti eclogites from N- and E-type MOR basaltic rocks (see Table 3; Manby and Thiedig 1988; Miller et al. 1988; Miller 1990). The nature of the protolith and the metamorphic ages are uncertain. Manby and Thiedig (1988) and Manby et al. (1989) argued for Late Precambrian eclogite metamorphism based on a Sm-Nd garnet-whole rock age of 693 Ma. Thöni and Jagoutz (1991) favour a Permian age of one of the gabbroic

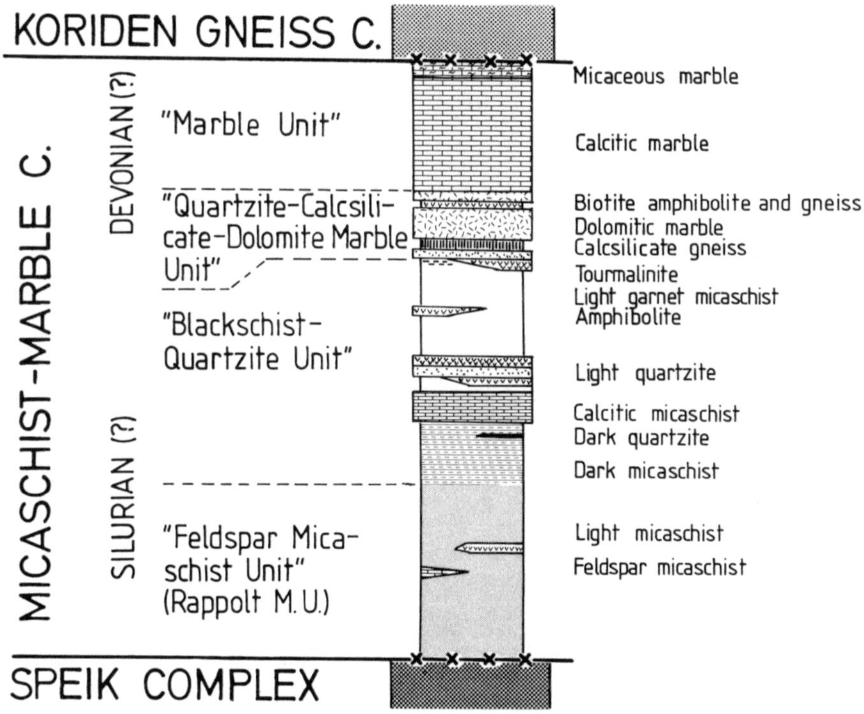

Fig. 9. Lithostratigraphy of the Micaschist-Marble complex, Gleinalm area. (After Neubauer 1988a)

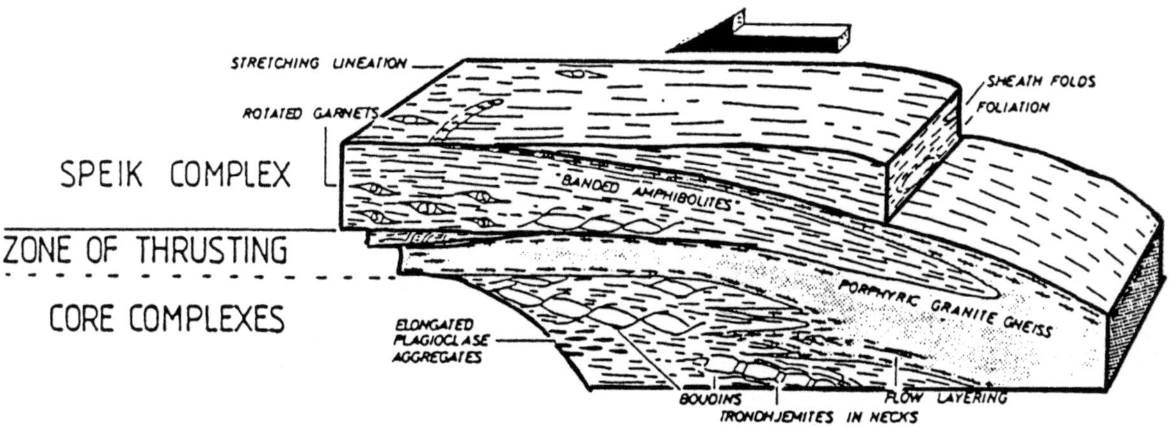

Fig. 10. Model for Variscan, melt-enhanced emplacement of the Speik ophiolite nappe. (After Neubauer 1989b)

protoliths and a Cretaceous age of eclogite metamorphism. A pegmatite within the eclogite amphibolite, however, yielded a Permian Rb-Sr age (Scharbert, cited in Göd 1989). Coarse-grained pegmatitic muscovite from pegmatite in paragneissic country rock yielded Rb-Sr ages in the range of 267–244 Ma (Morauf 1981; Frank et al. 1983). These ages are considered to be slightly rejuvenated due to Alpine metamorphic overprint under amphibolite facies conditions (Rittmann 1984; Frank 1987; Frank et al. 1987; Krohe 1987). Rb-Sr thin slab isochrons of the mylonitic "Plattengneis" also gave intra-Permian ages (Frank et al. 1981, 1983) somewhat younger than U-Pb zircon ages (Paquette and Gebauer 1989). Most workers agree that a Variscan metamorphic complex with low-pressure characteristics (andalusite stability) was intruded by numerous pegmatites in late Variscan time (Frank et al. 1983; Becker et al. 1987; Göd 1989). The crucial relationship of pegmatites to eclogites suggest a Permian or post-Permian age of eclogitization (for further discussion, see Neubauer 1991).

East of the Wechsel window, the "Grobgneis" complex is overlain by several klippen of the *"Sieg-*

grabener complex" (Kümel 1935) (for location, see Figs. 1, 2). Their superposition on the "Grobgneis" complex is Cretaceous in age. The Sieggrabener complex is composed of biotite-garnet paragneiss, kyanite gneiss, eclogite amphibolite, calc-silicate rocks, marble, and a large, kilometre-scale serpentinized peridotite body (Kümel 1935; Richter 1973). Most authors correlate the sequence with the gneiss complex in the Koralpe and Saualpe. The data base is poor. The chemistry of the eclogites is that of N-type MOR basaltic rocks (Kiesl and Weinke, cited in Koller 1990). Garnet-clinopyroxene thermometry indicates a temperature of approximately 630 °C, and phengite and pyroxene composition a pressure of ca. 10 kb (Kiesl and Weinke in Koller 1990).

The serpentinite body has in general a normative harzburgitic composition. It contains spinel pyroxenite and rodingite (Evren 1972; Koller and Richter 1980).

2.7 *Plankogel Complex and Accompanying Micaschists*

A package of micaschist units overlies the gneiss group along the southern margin of the Saualpe and Koralpe (Weissenbach 1975; Kleinschmidt and Ritter 1976) and in the Pohorje Mountains (Hinterlechner-Ravnik and Moine 1977). The most prominent units are the Plankogel complex at the bottom and the Kräuping complex above it which in turn is structurally overlain by phyllitic micaschists.

The Plankogel complex consists of a staurolite-garnet micaschist matrix in which up to kilometre-sized lenses of marbles, amphibolites, manganese quartzites and ultramafic rocks are embedded (Kleinschmidt 1975; Weissenbach 1975). Schmerold (1988) subdivided the Plankogel complex into two subunits (Figs. 11, 12): (1) The "manganese quartzite unit" contains plagioclase and biotite micaschists, amphibolites with an alkali basaltic chemistry, silicate-rich marbles and manganese-bearing quartzites. This unit is interpreted in terms of an oceanic island environment. (2) The "serpentinite-bearing unit" is characterized by coarse-grained micaschists, serpentinites, amphibolites with an ocean-floor tholeiitic chemistry and pure marbles. The normative mineral contents of the serpentinites argue for harzburgite, lherzolite, olivine orthopyroxenite and olivine websterite protoliths (Schmerold 1988). The Plankogel complex is interpreted as an ophiolitic melange zone marking a pre-Alpine ophiolitic suture zone (Frisch et al. 1984; Neubauer et al. 1989b). Both protolith age and the age of metamorphism are uncertain.

The Kräuping complex consists mainly of light-coloured micaschists, thin quartzitic layers and thick bodies of amphibolites (Weissenbach 1975). The amphibolites have a predominantly alkali-basaltic chemistry (Schmerold 1988).

2.8 *Bundschuh-Einach Complex*

In the area between the Tauern window and the Gurktal thrust system, the Micaschist-Marble complex is overlain by monotonous paragneiss (Einach-Priedröf complex) which contains lenses of Bundschuh orthogneiss (Fig. 1). The Einach-Priedröf paragneiss may be correlated with the gneisses of the Koriden because of their lithological similarities. The protolith's nature and age are unknown. The Bundschuh orthogneiss, a granitic to granodioritic gneiss, differs strikingly from other Austro-Alpine orthogneisses in its high content of muscovite, its high Sr_0 value of 0.738 and its Devonian Rb-Sr model age (Hawkesworth 1976; Frimmel 1986 a, b; 1988).

2.9 *"Kaintaleck Slices" and Ackerl Complex*

The "Kaintaleck slices" occur within the eastern Greywacke zone along 70 km of a thrust surface (Hauser 1938; Cornelius 1941). The thrust surface separates the lower Veitsch nappe with mainly Carboniferous formations (Ratschbacher 1987) from the upper Noric nappe of Ordovician to Early Carboniferous rocks. The structural thickness of the Kaintaleck slices does not exceed two hundred metres. In some areas these slices are intensely imbricated with footwall and hanging wall rocks (Neubauer et al. 1987). Lithology, geochronological data, geological relationships and structural position suggest that the Kaintaleck slices consist of various lithotectonic units (Fig. 13).

1. The Ritting complex occurs near Bruck/Mur, at the Kaintaleck and west of Vöstenhof. It is composed of garnet-zoisite amphibolite, serpentinite, micaschist and thin marble layers. The chemistry of the amphibolites corresponds to tholeiitic basalts (Neubauer et al. 1989 b). They contain a disseminated Cu mineralization. Chromite occurs in the serpentinites (Neubauer and Thalhammer 1989). The protolith age of the Ritting complex is uncertain. An amphibole concentrate yielded a disturbed total gas $^{40}Ar/^{39}Ar$

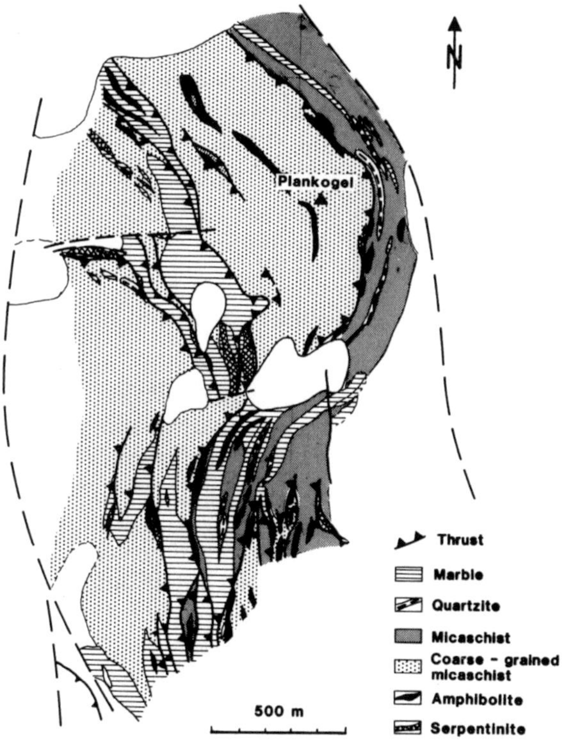

Fig. 11. Map showing the internal structure (ophiolitic melange) of the Plankogel complex. (After Neubauer et al. 1989 a)

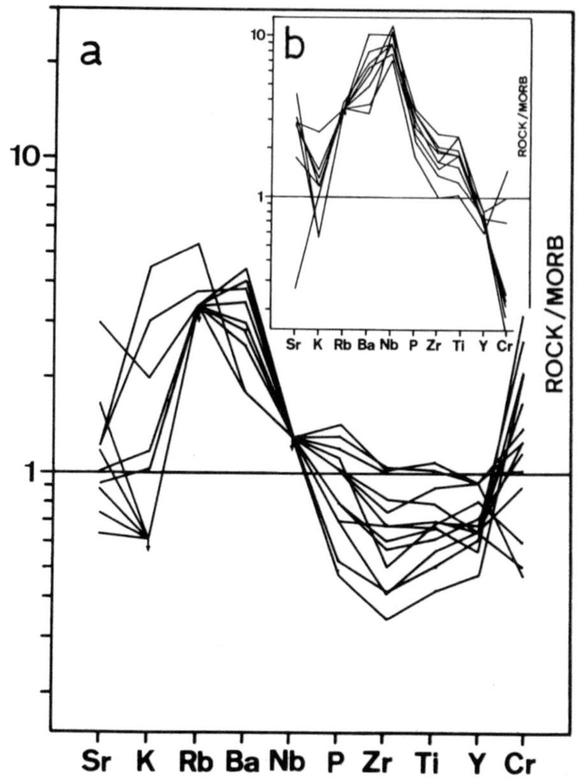

Fig. 12. Spidergram of Plankogel complex amphibolites displaying differentiation into two units with **a** tholeictic and **b** alkali basaltic trends. (After Neubauer et al. 1989 a)

age which reflects a probable Cadomian age of metamorphism (Dallmeyer and Neubauer unpubl. data).

2. The Frauenberg complex is mainly composed of paragneiss containing plagioclase amphibolite and marble lenses. The petrographic characteristics contrast with those of the Ritting complex. The amphibolites exhibit a mildly alkaline trend. A hornblende-garnet gneiss, which is included in the plagioclase amphibolite, contains numerous rounded zircon crystals with smooth surfaces which are interpreted as the result of high grade metamorphism. Pressure and temperature conditions are uncertain, but a high pressure event is indicated by the high pyrope content of garnet (ca. 40 mol% pyrope). The zircons yielded a U-Pb upper (intercept age of ca. 2.53 Ga and a lower intercept age of 516 Ma: Neubauer et al. 1989 b). Because of the highly discordant pattern the upper intercept age is considered to be the protolith age, assuming an igneous source of the hornblende-garnet gneiss, and the lower intercept the late Cadomian age of metamorphism (Neubauer et al. 1989 b). Brown, metamict zircons yielded a different upper intercept age of approximately 2.8 Ga. These zircons are interpreted as respresenting crustal contamination of the igneous source.

3. The Prieselbauer complex is composed of migmatitic augen paragneiss and micaschists which include minor occurrences of amphibolite as well as concordant and discordant aplites. The paragneiss yielded a lower intercept U-Pb zircon age of ca. 391 Ma which is interpreted as the approximate age of metamorphism. A discordant aplitic vein yielded an upper intercept zircon age of 363 Ma which reflects the zircon crystallization within the protolith (Neubauer et al. 1987; in prep.). The Prieselbauer complex is interpreted as the basement of the Carboniferous sediments of the Veitsch nappe (Neubauer et al. 1987).

The Ritting and Frauenberg complexes are transgressively covered by the Kalwang gneiss conglomerate which forms the stratigraphic base of the fossiliferous sequence of the eastern Greywacke zone (Daurer and Schönlaub 1978; Neubauer 1985). Apart from various basement fragments such as amphibolites and serpentinites, trondhjemitic orthogneiss components dominate in the conglomerate. The orthogneiss is interpreted to derive from a source in a supra-subduction zone environment because of its geochemical patterns especially those of the REEs.

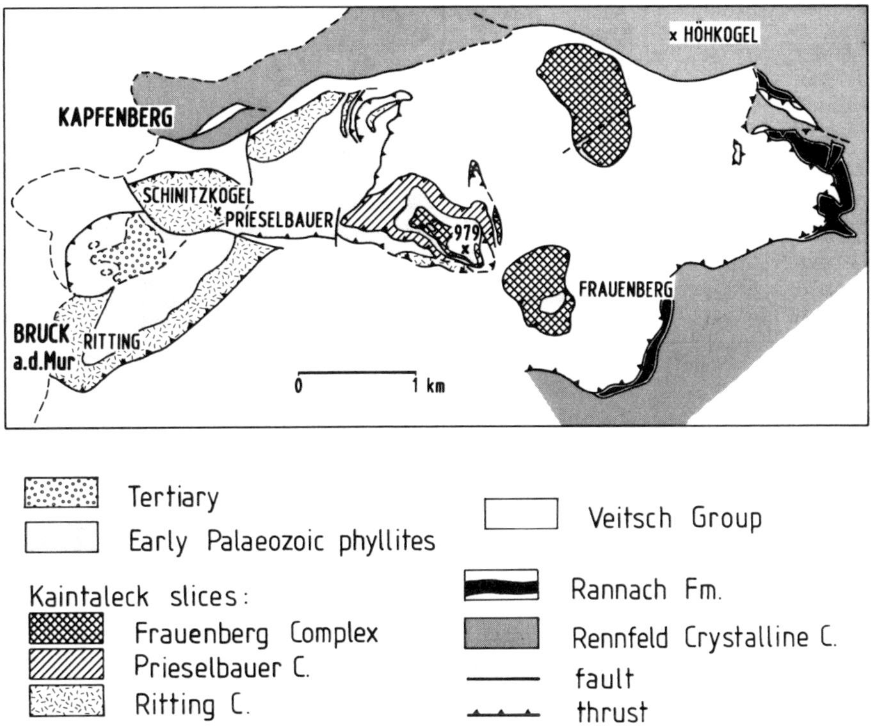

Fig. 13. Map showing internal structure of the Kaintaleck slices near Bruck/Mur. (After Neubauer et al. 1987)

The U-Pb zircon upper intercept ages of two boulders from different localities are approximately 500 Ma (Neubauer et al. 1987). The predominance of the orthogneiss clasts suggest the presence of a major pluton in the hinterland of the conglomerates. This orthogneiss is not known from the Kaintaleck slices.

The Ackerl complex at the northwestern margin of the Upper Austro-Alpine Gurktal thrust system (Fig. 1) is another medium-grade metamorphic complex and is in a similar tectonic position to the Kaintaleck slices. The Ackerl complex is composed of staurolite-bearing biotite plagioclase gneiss, partly phyllonitized micaschists, rare amphibolites, acidic orthogneisses and quartzites (Neubauer 1980). The Ackerl complex seemingly escaped a major Alpine metamorphic overprint. However, geochronological data constraining protolith ages and age of metamorphism are not available.

2.10 *Eisenkappel and Eastern Gailtal Crystalline*

A small tectonic segment of crystalline basement rocks occur immediately north of the Periadriatic lineament in the eastern Karawanken (Exner 1972; Fanninger 1978). It consists of a metamorphic sequence and the Karawanken granite pluton. The metamorphic sequence comprises biotite plagioclase paragneiss as the main lithology, with graphite-rich quartzites and gneisses, amphibolites and microcline orthogneiss (Exner 1972; von Gosen 1989 a, b). Protolith ages and metamorphic history are unknown.

The Karawanken granite pluton is composed mainly of granite with subordinate granodiorite, diorite and gabbro as well as granodiorite dykes, pegmatite, aplite and lamprophyre dykes. The age of the pluton is poorly constrained. A K-Ar amphibole age (252 ± 9 Ma; Cliff et al. 1974) and an Rb-Sr biotite age (232 ± Ma; Scharbert 1975) reflect Late Permian/Early Triassic cooling after intrusion.

Granite, diorite and augengneiss which is comparable to metagranitoids of the eastern Karawanken occur as tectonic slices along shear zones in the eastern Gailtal crystalline (Exner 1985).

2.11 *Garnet Peridotites and Granulites of the Pohorje Mts.*

Although first described long ago, no attention has been paid to the granulites and associated garnet peridotites on the southern slopes of the Pohorje mountains in Slovenia. This unit is situated immediately to the north of the Periadriatic lineament in a comparable tectonic position to other Austro-Alpine granulitic and high-pressure ultramafic rocks (Frisch et al. 1990; see also contribution of Hoinkes and Thöni, this Vol.). A new description of the Pohorje mountains was provided by Hinterlechner-Ravnik (1982, 1988) and Hinterlechner-Ravnik and Moine (1977). The metaperidotite-bearing sequence is composed of serpentinized dunite, harzburgite and garnet peridotite enclosed within eclogitic rocks. Kyanite-bearing eclogite typically associated with the garnet-peridotites suggests a relationship with the eclogite-bearing Koriden complex (Hinterlechner-Ravnik 1982, 1988). The garnet peridotite suffered polyphase metamorphism: a metamorphic event with pressures higher than 12 to 15 kb is postulated, which was overprinted under amphibolite and greenschist facies conditions (Hinterlechner-Ravnik 1988).

Protoliths and metamorphic ages are unknown. The garnet peridotite and granulite have a comparable position to the garnet peridotites and granulites in the Ulten Valley for which Variscan metamorphism is constrained by U-Pb zircon data (see Hoinkes and Thöni this Vol.).

3 Discussion

The medium- to high-grade metamorphic basement units east of the Tauern window exhibit a high degree of diversity with respect to their lithological composition, the age of granitoid intrusions and metamorphism. The lithotectonic complexes have different positions within the Alpine nappe pile. During the early Alpine orogeny lithotectonic units accumulated in certain levels. This is espescially true for the Kaintaleck slices.

Protolith ages for the metasedimentary sequences are largely a matter of speculation. Several major groups of units are distinguished with regard to their tectonothermal histories according to their geochronological ages (Tables 1, 2):

1. Rock units which were mainly affected by Early Palaeozoic magmatism and metamorphism (550–500 Ma) such as the Ritting and Frauenberg complexes, which escaped Variscan metamorphic overprint;
2. Rock units with Ordovician to Silurian, "Caledonian" ages (450–425 Ma) which were later affected by Early Carboniferous metamorphism and intrusions;
3. Rock units which underwent Devonian metamorphism and magmatism (400–360 Ma), e.g. the Prieselbauer slice, the Bundschuh area and possibly the Tatric unit;
4. Rocks with Early Carboniferous ages of metamorphism; these units are the most widespread;
5. Units, such as the Koriden gneiss group, for which the data base is uncertain or the lithological association is different from adjacent units, and whose history is not constrained by age determinations.

The geodynamic settings of most of these units appear to be different.

One set of units is regarded as the product of a Late Precambrian/Early Palaeozoic subduction system. This system includes units derived from an ensialic magmatic arc such as the Core complex, and from ocean floor (Ritting and Speik ophiolite complexes) related to it. Some portions of this system suffered Cambrian metamorphism and accompanying deformation with subsequent uplift. Parts of the newly consolidated crust acted as the basement to an Ordovician to Early Carboniferous shelf and rise sequence of a new sedimentary cycle. This is constrained by data from some Kaintaleck slices which show basement-cover relationship. The Micaschist-Marble complex is correlated by its lithostratigraphy to the fossiliferous Late Ordovician to Carboniferous sequences. We interpret all units which contain this Late Ordovician to Carboniferous cover sequence including its highly metamorphic equivalents as members of the Noric composite terrane (Frisch and Neubauer 1989).

The Tatric unit, the Bundschuh-Einach complex and the Prieselbauer complex are from a completely different setting. Evidence of Devonian metamorphism and magmatism are not compatible with the evolution of a Devonian shelf and slope sequence in the Noric composite terrane. These units are therefore interpreted as parts of another terrane, the Pannonian terrane (Frisch and Neubauer 1989), which formed a continental collisional belt in the Devonian as part of the West-European Ligerian Cordillera.

The evolution of the other lithotectonic units is badly constrained. The Plankogel complex is interpreted to be the Variscan ophiolitic suture which separates the foreland sequences from Variscan metamorphic internides, especially from those of the Pannonian terrane (Frisch and Neubauer 1989).

Geochronological data indicate an Early Carboniferous climax of metamorphism and magmatism for the Murides, the Wechsel gneiss complex and the Grobgneis complex. S- and I-type granitoids of Early Carboniferous age are widespread in the Grobgneis and Core complexes. These data suggest that the lower lithotectonic units within the Alpine nappe pile were involved in Early Carboniferous plate collision which amalgamated different Austro-Alpine basement units. Burial, deep-crustal thrusting under the influence of melts and uplift, shortly after the metamorphic peak, are the evolutionary steps recorded in this zone. The collisional event is also indicated by Viséan flysch sedimentation in the Austro-Alpine/Southalpine foredeep (e.g. Neubauer and Sassi this Vol.).

Only little evidence is available for the Late Carboniferous evolution. Permian radiometric data obtained from pegmatitic muscovites, granite and gabbro intrusions, and Rb-Sr thin slab errorchrons are widespread. The meaning of these data is not clear. A reliable explanation is a major Late Carboniferous to Permian rift event which resulted in intrusion of gabbros and formation of tholeiitic basaltic melts, high temperature/low pressure metamorphism in a regime of crustal thinning. The rift led finally to formation of the Tethys. However, most mineral data were interpreted by previous authors as the result of partial resetting of Carboniferous ages by Cretaceous metamorphism. Well-documented magmatic activity, which is also preserved in acidic tuffs in Permian overstep sequences (see Krainer this Vol.), indicates high heat flow within the Variscan crust, and the mineral ages may therefore reflect postmetamorphic cooling caused by crustal extension and uplift.

Acknowledgements.
P. O'Shea corrected the English text of an earlier version of this contribution. We appreciate helpful reviews by G. Nollau and E. Stein. The work was supported by both the Austrian Research Fund (grant P 7405-GEO) and German Research Foundation (Grant Fr 610/1)

References

Becker LP (1977) Zum geologischen und tektonischen Aufbau des Stubalpenzuges (Stmk.) mit einem Vergleich seiner Position zur NE-Saualpe (Kärnten). Carinthia II 167 (87): 113–125

Becker LP (1981) Zur Gliederung des Obersteirischen Altkristallins (Muriden). Verh Geol Bundesanst 1981: 3–17

Becker LP, Schumacher R (1973) Metamorphose und Tektonik in dem Gebiet zwischen Stub- und Gleinalpe, Stmk. Mitt Geol Ges Wien 65 (1972): 1–32

Becker LP, Frank W, Höck V, Kleinschmidt G, Neubauer F, Sassi FP, Schramm JM (1987) Outlines of the pre-Alpine metamorphic events in the Austrian Alps. In: Flügel HW, Sassi FP, Grecula P (eds) Pre-Variscan and Variscan events in the Alpine-Mediterranean mountain belts. Alfa, Bratislava, pp 69–106

Beck-Mannagetta P (1970) Über den geologischen Aufbau der Koralpe. Verh Geol Bundesanst 1970: 491–496

Cambel B, Kral J (1989) Isotopic geochronology of the western Carpathian crystalline complex: the present state. Geol Carpath 40: 387–410

Cliff RA, Holzer HF, Rex D (1974) The age of the Eisenkappel granite, Carinthia and the history of the Periadriatic lineament. Verh Geol Bundesanst 1974: 347–350

Cornelius HP (1941) Das Vorkommen altkristalliner Gesteine im Ostabschnitt der nordalpinen Grauwackenzone. Mitt Reichsst Bodenforsch Zweigst Wien 2: 19–52

Cornelius HP (1952) Die Geologie des Mürztalgebietes. Jahrb Geol Bundesanst Sonderb 2: 1–94

Daurer A, Schönlaub HP (1978) Anmerkungen zur Basis der Nördlichen Grauwackenzone. Mitt Österr Geol Ges 69 (1976): 77–88

Deutsch A (1988) Die frühalpidische Metamorphose in der Goldeck-Gruppe (Kärnten) – Nachweis anhand von Rb-Sr-Altersbestimmungen und Gefügebeobachtungen. Jahrb Geol Bundesanst 131: 553–562

El Ageed A, Saager R, Stumpfl EF (1980) Pre-Alpine ultramafic rocks in the eastern Central Alps, Styria, Austria. Proc Int Ophiolite Symp Nicosia Cyprus, pp 601–606

Evren I (1972) Die Serpentingesteine von Bernstein und Steinbach (Burgenland). Tschermaks Mineral Petrogr Mitt 17: 101–122

Exner Ch (1972) Geologie der Karawankenplutone östlich Eisenkappel, Kärnten. Mitt Geol Ges Wien 64 (1971): 1–108

Exner Ch (1985) Petrographie und Tektonik des Granitzuges von Nötsch (Kärnten). Jahrb Geol Bundesanst 127: 557–570

Fanninger E (1978) Plutonic emplacement in the eastern Karawanken Alps. Geologija 21: 81–87

Faupl P (1970) Zur Geologie des NW-Abschnittes des Wechselgebietes zwischen Trattenbach (NÖ.) und Fröschnitz (Stmk.). Mitt Geol Bergbaustud 19: 27–70

Faupl P (1972) Zur Geologie und Petrographie des südlichen Wechselgebietes. Mitt Geol Ges Wien 63 (1970): 22–51

Flügel HW, Neubauer F (1984) Erläuterungen zur Geologischen Karte der Steiermark 1:200000. Geol Bundesanst, Vienna, 127 pp

Frank W (1987) Evolution of the Austroalpine elements in the Cretaceous. In: Flügel HW, Faupl P (eds) Geodynamics of the Eastern Alps. Deuticke, Vienna, pp 379–406

Frank W, Klein P, Nowy W, Scharbert S (1976) Die Datierung geologischer Ereignisse im Altkristallin der Gleinalpe (Steiermark) mit der Rb/Sr-Methode. Tschermaks Mineral Petrogr Mitt 23: 191–203

Frank W, Frey I, Jung G, Roetzel R, Thöni M (1981) Wie intensiv war die altalpine Metamorphose und Strukturprägung im SE-Teil des ostalpinen Kristallin? Jahresber Hochschulschwerpunkt S 15 (1): 13–20

Frank W, Esterlus M, Frey I, Jung G, Krohe A, Weber J (1983) Die Entwicklungsgeschichte von Stub- und Koralpenkristallin und die Beziehung zum Grazer Paläozoikum. Jahresber Hochschulschwerpunkt S 15 (3): 263–293

Frank W, Kralik M, Scharbert S, Thöni M (1987) Geochronological data from the Eastern Alps. In: Flügel HW, Faupl P (eds) Geodynamics of the Eastern Alps. Deuticke, Vienna, pp 272–281

Frimmel H (1986a) Petrographie, Gefügemerkmale und geochronologische Daten von Kristallingeröllen aus dem Oberkarbon der Gurktaler Decke im Vergleich zum benachbarten Altkristallin. Mitt Ges Geol Bergbaustud Österr 32: 39–65

Frimmel H (1986b) Isotopengeologische Hinweise für die paläogeographische Nachbarschaft von Gurktaler Decke (Oberostalpin) und dem Altkristallin östlich der Hohen Tauern (Österreich). Schweiz Mineral Petrogr Mitt 66: 193–208

Frimmel H (1988) Metagranitoide am Westrand der Gurktaler Decke (Oberostalpin) – Genese und paläotektonische Interpretation. Jahrb Geol Bundesanst 131: 575–592

Frisch W, Neubauer F (1989) Pre-Alpine terranes and tectonic zoning in the Eastern Alps. Geol Soc Am Spec Publ 232: 99–100

Frisch W, Neubauer F, Satir M (1984) Concepts of the evolution of the Austroalpine basement complex (Eastern Alps) during the Caledonian Variscan cycle. Geol Rundsch 73: 47–68

Frisch W, Neubauer F, Bröcker M, Brückmann W, Haiss N (1987) Interpretation of geochemical data from the Caledonian basement within the Austroalpine basement complex. In: Flügel HW, Sassi FP, Grecula P (eds) Pre-Variscan and Variscan events in the Alpine-Mediterranean mountain belts. Alfa, Bratislava, pp 209–226

Frisch W, Menot RP, Neubauer F, Raumer J von (1990) Evolution of the Alpine basement. Schweiz Mineral Petrogr Mitt 70: 265–285

Fuchs G (1962) Neue tektonische Untersuchungen im Rosaliengebirge. Jahrb Geol Bundesanst 105: 19–37

Göd R (1989) The spodumene deposit at "Weinebene" Koralpe Austria. Mineral Deposita 24: 270–278

Haditsch JG, Petersen-Krauss D, Yamac Y (1981) Beiträge für eine geologisch-lagerstättenkundliche Beurteilung hinsichtlich einer hydrometallurgischen Verwertung der Kraubather Ultramafititmasse. Mitt Abt Geol Paläontol Bergbau Landesmus Joanneum 42: 23–78

Haiss N (1991) Untersuchungen zur Genese von Plagioklasgneisen im Basiskristallin der Ostalpen (Gleinalm-, Ötztal- und Silvrettakristallin. PhD Thesis, Univ Tübingen, 142 pp

Hauser (1938) Die geologischen und petrographischen Verhältnisse im Gebiete der Kaintaleckschollen. Jahrb Geol Bundesanst 88: 217–259

Hawkesworth CJ (1976) Rb/Sr geochronology in the Eastern Alps. Contrib Mineral Petrol 54: 225–244

Hejl E (1984) Geochronologische und petrologische Beiträge zur Gesteinsmetamorphose der Schladminger Tauern. Mitt Ges Geol Bergbaustud Österr 30/31: 289–318

Heritsch H (1980) Einführung zu Problemen der Petrologie der Koralpe. Mitt Abt Geol Landesmus Joanneum 41: 9–44

Hermann U (1972) Die Umwandlung eines gabbroiden Gesteines aus dem Utschgraben bei Bruck an der Mur, Steiermark. Mitt Naturwiss Ver Steiermark 102: 15–23

Hinterlechner-Ravnik A (1982) Pohorski eklogit. Geologija 25: 251–288

Hinterlechner-Ravnik A (1988) Garnet peridotite from the Pohorje Mountains. Geologija 30: 149–181

Hinterlechner-Ravnik A, Moine B (1977) Geochemical characteristics of the metamorphic rocks of the Pohorje Mountains. Geologija 20: 107–140

Hoke L (1990) The Altkristallin of the Kreuzeck Mountains, SE Tauern Window, Eastern Alps – Basement Crust in a Convergent Plate Boundary Zone. Jahrb Geol Bundesanst 133: 5–87

Huska G (1970) Zur Geologie der Umgebung von Waldbach, südwestliches Wechselgebiet, Steiermark. Verh Geol Bundesanst 1970: 61–65

Jäger E, Metz K (1971) Das Alter der Pegmatite des Raumes Bretstein – Pusterwald (Wölzer Tauern, Steiermark). Schweiz Mineral Petrogr Mitt 51: 411–414

Kiesl W, Wieseneder H, Kluger F (1983) Untersuchungen des Vorkommens der Seltenen Erden und von Thorium in Gesteinen des unterostalpinen Kristallins des Semmering-Wechselfensters. Sitzungsber Österr Akad Wiss math-naturwiss Kl I (1983): 1–20

Kleinschmidt (1975) Die „Plankogelserie" in der südlichen Koralpe unter besonderer Berücksichtigung von Manganquarziten. Verh Geol Bundesanst 1974: 351–362

Kleinschmidt G, Ritter U (1976) Geologisch-petrographischer Aufbau des Koralpenkristallins südlich von Soboth/Steiermark. Carinthia II 166/86: 57–91

Kober L (1938) Der geologische Aufbau Österreichs. Springer, Berlin Heidelberg New York Vienna, 204 pp

Koller F (1990) Wandertagung 1990 der Österreichischen Geologischen Gesellschaft in Bernstein. Österr Geol Ges, Vienna, 149 pp

Koller F, Richter W (1980) Rodingites in ultramafic rocks of the Eastern Alps. Ofioliti 5: 73–78

Koller F, Wieseneder H (1981) Gesteinsserien und Metamorphose der Rechnitzer Serie im Burgenland und des Unterostalpins der Oststeiermark. Fortschr Mineral Beih 39 (2): 167–178

Koller F, Götzinger MA, Neumayer R, Niedermayer G (1983) Beiträge zur Mineralogie und Geochemie der Pegmatite des St. Radegunder Kristallins und der Gleinalpe. Arch Lagerstättenforsch Geol Bundesanst 3: 47–65

Kovach A, Svingor E (1981) On the age of metamorphism of Fertörakos metamorphic complex NW Hungary. Verh Geol Bundesanst 1981: 73–81

Krohe (1987) Kinematics of Cretaceous nappe tectonics in the Austroalpine basement of the Koralpe region (eastern Austria). Tectonophysics 136: 171–196

Kümel F (1935) Die Sieggrabener Deckscholle im Rosaliengebirge (N.Ö. und Bgld.). Mineral Petrogr Mitt 47: 141–184

Lelkes-Felvari G, Sassi FP, Visona D (1984) Pre-Alpine and Alpine developments of the Austridic basements in the Sopron area (Eastern Alps, Hungary). Rend Soc Ital Mineral Petrol 39: 593–612

Manby GM, Thiedig F (1988) Petroplogy of eclogites from the Saualpe, Austria. Schweiz Mineral Petrogr Mitt 68: 441–466

Manby GM, Thiedig F, Millar I (1989) Textural chemical and isotopic constraints on the age of the Saualpe eclogites (Carinthia/Austria). In: Sassi FP, Bourrouilh R (eds) IGCP 5 Newslett 7 (1987): 195–202

Metz K (1976a) Der geologische Bau der Seckauer und Rottenmanner Tauern. Jahrb Geol Bundesanst 119: 151–205

Metz K (1976b) Der geologische Bau der Wölzer Tauern. Mitt Naturwiss Ver Steiermark 106: 51–75

Miller Ch (1990) Petrology of the type locality eclogites from the Koralpe and Saualpe (Eastern Alps), Austria. Schweiz Mineral Petrogr Mitt 70: 287–300

Miller C, Stosch HG, Hoernes S (1988) Geochemistry and origin of eclogites from the type locality Koralpe and Saualpe Eastern Alps, Austria. Chem Geol 67: 103–118

Moine B, Fortune JP, Moreau P, Viguer F (1989) Comparative mineralogy, geochemistry and conditions of formation of two metasomatic talc and chlorite deposits: Trimouns (Pyrenees France) and Rabenwald (Eastern Alps Austria). Econ Geol 4: 1398–1416

Morauf W (1980) Die permische Differentiation und die alpidische Metamorphose des Granitgneises von Wolfsberg, Koralpe, SE-Ostalpen, mit Rb-Sr- und K-Ar-Isotopenbestimmungen. Tschermaks Mineral Petrogr Mitt 27: 169–185

Morauf W (1981) Rb-Sr- und K-Ar-Isotopenalter an Pegmatiten aus Kor- und Saualpe, SE-Ostalpen, Österreich. Tschermaks Mineral Petrogr Mitt 28: 113–129

Moreau P (1981) Le massif du Rabenwald (Autriche) et ses mineralisations (talc, chlorite, disthéne, leusophyllite). These Doctorale, 3eme cycle Besancon University Franche-Comté, 327 pp

Neubauer (1980) Zur tektonischen Stellung des Ackerlkristallins (Nordrand der Gurktaler Decke). Mitt Österr Geol Ges 73: 39–53

Neubauer F (1985) Eine präoberordovizische Transgression in der Grauwackenzone (Ostalpen) und ihre Bedeutung für paläozoische Geodynamik und alpidischen Deckenbau. Neues Jahrb Geol Paläontol Monatsh 1985: 46–64

Neubauer F (1988a) Bau und Entwicklungsgeschichte des Rennfeld- Mugel- und des Gleinalm-Kristallins (Ostalpen). Abh Geol Bundesanst 42: 1–127

Neubauer F (1988b) The Variscan orogeny in the Austroalpine and Southalpine domains of the Eastern Alps. Schweiz Mineral Petrogr Mitt 68: 339–349

Neubauer F (1989a) A plutonic source for pegmatites in the Austroalpine basement of the Gleinalm region (Eastern Alps). Neues Jahrb Geol Paläontol Monatsh 1989: 615–624

Neubauer F (1989b) The leptinite-amphibolite complexes – a key for correlation of mid-European Variscides? Abstr IGCP Proj 233, Athens/GA, July 21–22, 1989, pp 79–82

Neubauer F (1990) Das südliche Wechselkristallin (Ostalpen): Relikt eines aktiven Kontinentalrandes? Kurzfassungen TSK III, Graz, 19.–21. April 1990, pp 143–146

Neubauer F (1991) Kinematic indicators in the Koralm and Saualm eclogites, Eastern Alps. Zentralbl Geol Paläontol Teil 1 1991: 139–155

Neubauer F (1993) The tectonic environment of the Wechsel gneiss complex (Eastern Alps): a pre-Carboniferous active continental margin? Geol Rdsch (submitted)

Neubauer F On the genesis of trondhjemite by partial melting of amphibolite: evidence from the Rennfeld region, Eastern Alps. Mineral Petrol (submitted)

Neubauer F, Stattegger K (1981) Untersuchungen zur tektonischen und metamorphen Geschichte des Radegunder Kristallins. Jahresber Hochschulschwerpunkt S 15 (2): 122–128

Neubauer F, Thalhammer O (1989) Deformation, metamorphism and mineralization of the "Panafrican" basement in the eastern Graywacke zone, eastern Alps. Terra Abstr 1: 45

Neubauer F, Frisch W, Hansen BT (1987) Time relations between Eoalpine metamorphism and thrusting: evidence from the crystalline basement of the eastern Greywacke zone. In: Flügel HW, Faupl P (eds) Geodynamics of the Eastern Alps. Deuticke, Vienna, pp 263–271

Neubauer F, Frisch W, Schmerold R, Schlöser H (1989a) Metamorphosed and dismembered ophiolite suites in the basement units of the Eastern Alps Tectonophysics 164: 49–62

Neubauer F, Frisch W, Hansen BT (1989b) A Late Archean rock in the Eastern Alps? Terra Abstr 1 (1): 5

Neubauer F, Frisch W, Hansen BT (1993) The pre-Alpine tectonothermal evolution of the Rennfeld region, Eastern Alps: a U-Pb zircon study. Lithos (submitted)

Neubauer F, Frisch W, Hansen BT Contrasting pre-Variscan basement complexes in the eastern Grauwacken zone (Eastern Alps): evidence from U-Pb zircon data. (in prep)

Pacher F, Riepl F (1978) Über die chemische Zusammensetzung von Gneisen und Glimmerschiefern der Koralpe, Steiermark. Mitt Naturwiss Ver Steiermark 108: 45–54

Pahr A (1980) Das Rosalien- und Leithagebirge sowie die Hainburger Berge In: Oberhauser R (ed) Der geologische Aufbau Österreichs. Springer, Berlin Heidelberg New York Wien, pp 326–331

Paquette JL, Gebauer D (1989) U-Pb zircon and Sm-Nd isotopic study on eclogitized meta-basic and meta-acidic rocks of the Koralpe and Saualpe, Eastern Alps, Austria. Terra Abstr 1: 18

Peindl P (1990) Variszische und alpidische Entwicklungsgeschichte des südöstlichen Raabalpenkristallins (Steiermark). PhD Thesis, Univ Graz, 252 pp

Peindl P, Neubauer F, Moyschewitz G, Reindl H, Wallbrecher E (1990) Die geologische Entwicklung des südlichen Raabalpen- und Wechselkristallins. Excursion guide TSK III „Raabalpen- und Wechselkristallin" 24 April 1990. Geol Inst Univ Graz, pp 3–21

Purkert R (1927) Geologie des Kulm bei Weiz. Mitt Naturwiss Ver Steiermark 63: 91–110

Raith JG (1988) Tourmaline rocks associated with stratabound scheelite mineralization in the Austroalpine crystalline complex, Austria. Mineral Petrol 39: 265–288

Ratschbacher L (1987) Stratigraphy, tectonics and paleogeography of the Veitsch nappe (Graywacke zone, Eastern Alps: a rearrangement. In: Flügel HW, Sassi FP, Grecula P (eds) Pre-Variscan and Variscan events in Alpine-Mediterranean mountain belts. Mineralia slovaca Monogr., Alfa, Bratislava, pp 407–414

Richter W (1973) Vergleichende Untersuchungen an ostalpinen Eklogiten. Tschermaks Mineral Petrogr Mitt 19: 1–50

Rittmann KL (1984) Argon in Hornblende, Biotit und Muskowit bei der geologischen Abkühlung: $^{40}Ar/^{39}Ar$ Untersuchungen. Diss, Naturwiss Math Fak Univ Heidelberg, 231 pp

Santallier D, Briand B, Menot RP, Piboule M (1988) Les complexes leptyno-amphiboliques (C.L.A.): revue critique et suggestions pour un meilleur emploi de ce terme. Bull Soc Geol Fr 1988 (8): 3–12

Scharbert S (1975) Radiometrische Altersdaten von Intrusivgesteinen im Raum Eisenkappel (Karawanken, Kärnten). Verh Geol Bundesanst 1975: 301–304

Scharbert S (1981) Untersuchungen zum Alter des Seckauer Kristallins. Mitt Ges Geol Bergbaustud Österr 27: 173–188

Scharbert S (1990) Rb-Sr-Daten aus dem Raabalpenkristallin. Excursion guide TSK III Exkursion „Raabalpen- und Wechselkristallin" 24 April 1990. Geol Inst Univ Graz, pp 22–25

Schatzmayer M, Neubauer F, Frisch W (1990) Die geodynamische Signifikanz des Utschgraben-Metagabbro, Muralpenkristallin, Steiermark. Abstracts TSK III, 19–21 April 1990, Geol Inst Univ Graz, pp 195–198

Schedl A (1981) Geologische, geochemische und lagerstättenkundliche Untersuchungen im ostalpinen Kristallin der Schladminger Tauern. PhD Thesis, Univ Vienna, 209 pp

Schimana R (1986) Neue Ergebnisse zur Entwicklungsgeschichte des Kristallins um Radenthein (Kärnten, Österreich). Mitt Ges Geol Bergbaustud 33: 221–232

Schlöser H (1987) Die Entwicklung des mittelostalpinen Altkristallins im Neuhofgraben, Steiermark – Petrographie, Geochemie und Metamorphose. Diploma Thesis, Univ Tübingen, 140 pp

Schmerold R (1988) Die Plankogelserie im ostalpinen Kristallin von Kor- und Saualpe (Kärnten Steiermark – Österreich) als ophiolithische Sutur. PhD Thesis, Univ Tübingen, 161 pp

Schönlaub HP (1979) Das Paläozoikum in Österreich. Abh Geol Bundesanst 33: 1–124

Teich T (1979) Die Genese des Augengneiszuges in der zentralen und südlichen Stubalpe mit einer Zusammenfassung über den Augengneiszug der Hochalpe-Gleinalpe-Stubalpe, Steiermark-Kärnten. Mitt Naturwiss Ver Steiermark 109: 39–54

Teich T (1987a) Die „metablastischen" Amphibolit-Typen im Rachaugraben der Gleinalpe, Steiermark. Mitt Naturwiss Ver Steiermark 117: 35–38

Teich T (1987b) „Fremdgesteinseinschlüsse" im „metablastischen" Amphibolitzug der Gleinalpe, Steiermark. Mitt Naturwiss Ver Steiermark 117: 39–48

Thöni M, Jagoutz E (1991) Eoalpine eclogites in the Eastern Alps: new Sm-Nd, Rb-Sr and Pb-Pb isotopic data. Terra abstracts 3: p 89

Tollmann A (1977) Geologie von Österreich, Band I. Die Zentralalpen. Deuticke, Vienna, 766 pp

Tollmann A (1978) Eine Serie neuer tektonischer Fenster des Wechselsystems am Ostrand der Zentralalpen. Mitt Österr Geol Ges 68 (1975): 129–142

Tollmann E, Spendlingwimmer R (1978) Crinoiden im Anis (Mitteltrias) der Tatriden der Hainburger Berge (Niederösterreich). Mitt Österr Geol Ges 68 (1975): 59–77

Tufar W (1963) Die Erzlagerstätten des Wechselgebietes. Joanneum Mineral Mitteilungsbl 1963: 1–60

Tufar W (1968) Die Kupferlagerstätte von Trattenbach (Niederösterreich). Tschermaks Mineral Petrogr Mitt 12: 140–181

Vetters W (1970) Zur Geologie des SW-Abschnittes des Wechselgebietes zwischen Rettenegg und Feitritztal (Steiermark, Österreich). Mitt Ges Geol Bergbaustud 19: 71–102

von Gosen W (1989a) Gefügeentwicklungen, Metamorphosen und Bewegungen der ostalpinen Baueinheiten zwischen Nockgebiet und Karawanken (Österreich). Geotekt Forsch 72: 1–247

von Gosen W (1989b) Fabric development and the evolution of the Periadriatic lineament in southeast Austria. Geol Mag 126: 55–71

Weissenbach N (1975) Gesteinsinhalt und Seriengliederung des Hochkristallins in der Saualpe. Clausth Geol Abh Sonderb 1: 61–114

Wessely G (1961) Geologie der Hainburger Berge. Jahrb Geol Bundesanst 104: 273–349

Wieseneder H (1969) Der Eklogitamphibolit von Hochgrössen, Steiermark. Joanneum Mineral Mitteilungsbl 1969: 156–164

Late- and Post-Variscan Sediments of the Eastern and Southern Alps

K. Krainer

Abstract

In the Southern and Eastern Alps late- to post-Variscan sedimentation processes started during the Late Carboniferous (Late Moscovian/Cantabrian). The sediments of the Carboniferous sequence of Nötsch and Veitsch, which are also briefly discussed, are interpreted as "synorogenic sediments".

The late to post-Variscan Molasse sediments of the Southern and Eastern Alps were deposited in NE-SW and E-W to ESE-WNW oriented intramontane basins which probably formed as a result of crustal thinning and transtensional and transpressional tectonics along megashear zones due to the eastward drift to the Eurasian plate and westward drift of the African plate.

Tectonic processes and climate, including climatic changes, were the most important controlling factors of sedimentation processes. Based on major tectonic and climatic events, as well as on fossil assemblages, a close correlation of the late to post-Variscan sequences of the Southern and Eastern Alps is possible.

The late to post-Variscan sequence of the Southern and Eastern Alps is divided into two evolutionary cycles, which are separated by a major intra-Permian tectonic event.

The lower cycle (Late Carboniferous/Early Permian) is characterized by the formation of intramontane basins which were filled with sediments of different environments and with volcanic rocks. Transgressive-regressive clastic-carbonate cycles, within the shallow marine sequence of the Carnic Alps (Southern Alps), are related to eustatic sea-level fluctuations caused by the Gondwana glaciation. The climatic shift from humid to semiarid conditions near the Carboniferous/Permian boundary caused a significant change in sedimentary processes in the continental sequences of the Eastern Alps.

During the upper cycle (?Middle – Late Permian to Early Triassic) sedimentation patterns were more uniform and similar sequences developed in the Southern and Eastern Alps. At this time, sedimentation was characterized by the transgression of the Tethys Sea from SE to NW and interfingering of shallow marine sediments with continental deposits. Sedimentation processes were also influenced by a climatic shift near the Permian/Triassic boundary and sedimentary cycles of the Scythian are probably caused by different spreading rates of mid-oceanic ridges. The question where to draw the boundary between the "late to post-Variscan" Molasse sediments and the following Alpidic sedimentation cycle still remains open.

1 Introduction

Late to post-Variscan (Late Carboniferous to Early Triassic) Molasse sedimentation in the Eastern and Southern Alps was strongly controlled by tectonic processes of the Late Variscan orogenic cycle, and climate, respectively climatic changes. As there is no general agreement, when late- to post-Variscan sedimentation processes started and how they are divided from synorogenic sediments, a brief definition of the term late- to post-Variscan sedimentation will be given.

The Variscan orogenic belt in Europe was formed during Late Palaeozoic (Devonian/Carboniferous) convergence of Laurasia and Gondwana (Ziegler 1982). During this time a number of basins and small oceans of varying widths and types were formed and closed (Vai 1980; Behr et al. 1984; Vai and Cocozza 1986 and others). For the South-Alpine domain Vai and Cocozza (1986) proposed, that the large sialic belt between Laurasia and Gondwana was repeatedly transected by mostly dextral strike-slip fault systems during Middle Cambrian to Late Carboniferous times. These mega-transform systems and associated strike-slip movements led to the closure of the lapetus ocean and then to the opening of the Palaeotethys and penecontemporaneous closing of the Rheic ocean.

According to the model presented by Frisch and Neubauer (1989), the Rheic ocean, represented for

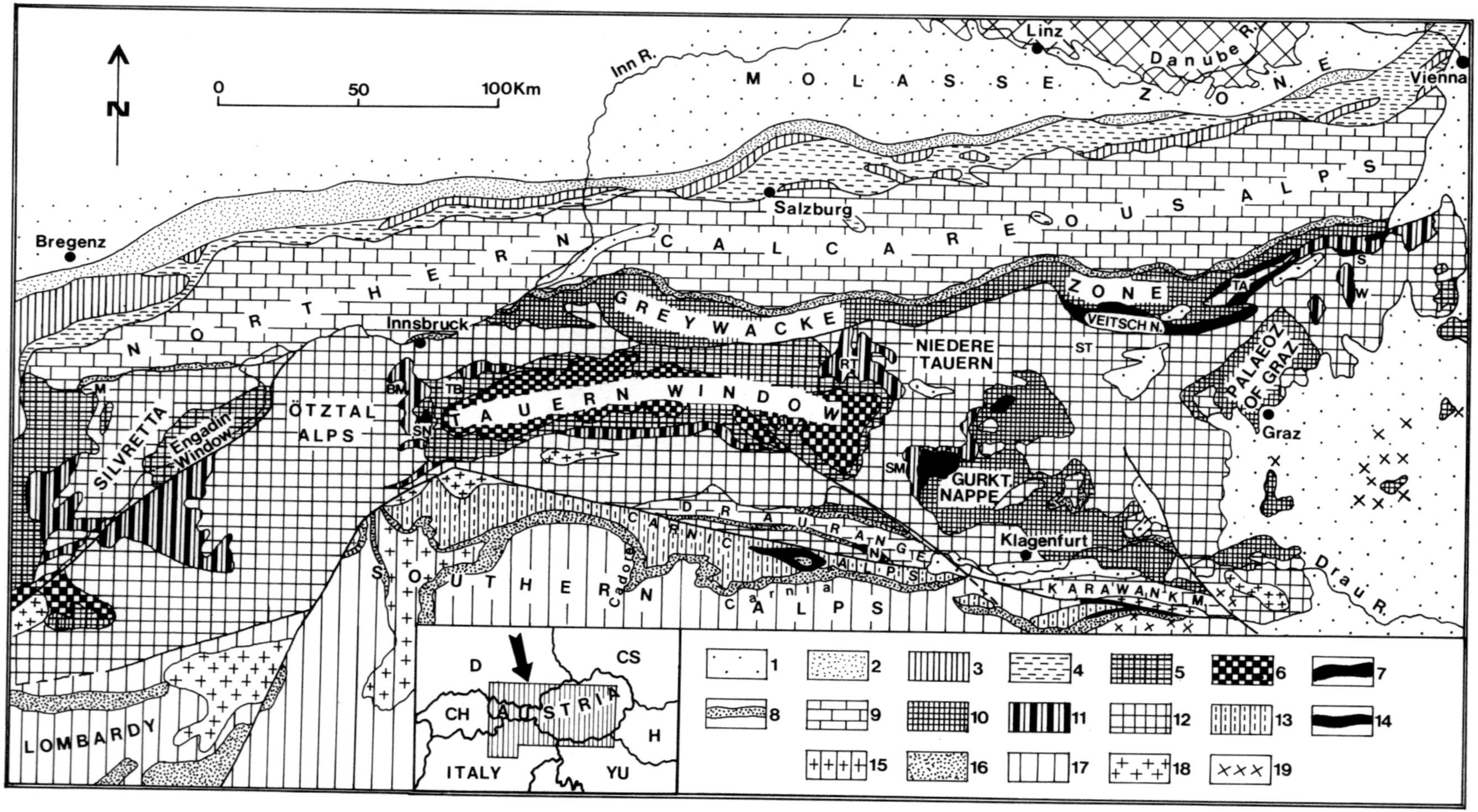

Fig. 1. Geologic sketch map of the Eastern and Southern Alps (after Flügel and Faupl 1987) with the most significant outcrops of late to post-Variscan sediments. **1** Molasse zone; **2** Sub-Alpine Molasse; **3** Helvetic and Klippen zone; **4** Rhenodanubian Flysch; **5** metasedimentary rocks of the Penninic zone; **6** crystalline basement of the Penninic Zone; **7** Carboniferous sediments of the Eastern Alps (Upper Austro-Alpine); **8** Permian/Early Triassic (Scythian) sediments of the Eastern Alps (Upper Austro-Alpine); **9** Mesozoic sediments (Upper Austro-Alpine); **10** Palaeozoic rocks (Ord.-Early Carb.) of the Eastern Alps; **11** Permo-Mesozoic in Central-Alpine facies (Lower and Middle Austro-Alpine); **12** metamorphic basement ("Altkristallin"); **13** Palaeozoic rocks (Ord.-Early Carb.) of the Southern Alps; **14** Late Carboniferous sediments of the Southern Alps; **15** Bolzano volcanic complex; **16** Permian/Early Triassic (Scythian) sediments of the Southern Alps; **17** Mesozoic sediments of the Southern Alps; **18** Peri-Adriatic intrusive masses; **19** Neogene volcanics. **M** Montafon; **BM** Brenner Mesozoic; **SN** Steinacher nappe; **TB** Tarntaler Berge; **RT** Radstätter Tauern; **SM** Stangalm Mesozoic; **N** Nötsch; **ST** Seckauer Tauern; **TA** Thörl/Aflenz; **S** Semmering; **W** Wechsel.

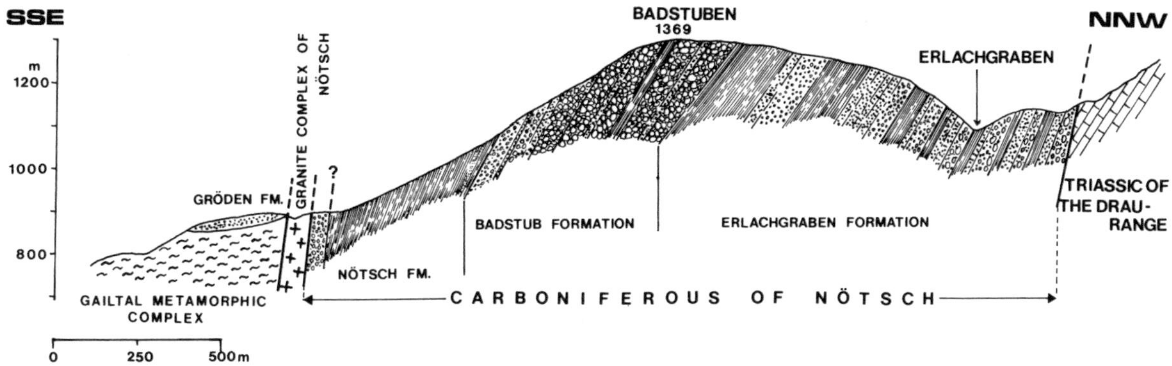

Fig. 2. Schematic cross-section through the Carboniferous sequence of Nötsch. (After Krainer and Mogessie 1991)

example by the Plankogel and Koriden terranes, was closed during Devonian-Early Carboniferous by the northwestward movement of a microplate termed Noric terrane, which was part of Africa (Gondwana). As a consequence, synorogenic basins formed by extensional and compressional strike-slip movements between the microplate(s) and Gondwana (Vai and Cocozza 1986). This synorogenic basin ("foredeep") was filled with Visean to Namurian/Westphalian shallow- to deep-marine sediments of the Veitsch nappe, Nötsch area and South-Alpine Hochwipfel and Dimon formations. During the compressional orogenic phase ("Carnic Phase" of Vai 1975; Castellarin and Vai 1981) in the Westphalian, the sediments of the synorogenic basin in the Carnic Alps, were tectonically shortened by strong folding and formation of a south verging "Schuppenbau" ("Palaeocarnic Chain"), the western part of the Carnic Alps was slightly metamorphosed.

The Visean to Namurian/Westphalian sediments of the Nötsch area and Veitsch nappe, which were formed in a synorogenic basin complex, too, must also have been affected by these Late Carboniferous orogenic movements. In the Carnic Alps, the folded Variscan Complex (Palaeocarnic Chain) is overlain by Late Carboniferous-Permian Molasse sediments with a classical major unconformity (Fenninger et al. 1976; Vai 1976; Vai et al. 1980). This major unconformity, although not known from the Carboniferous of Nötsch and Veitsch due to strong Alpine tectonic overprinting, separates the synorogenic sedimentary sequences of Visean to Early Westphalian age from the late to post-Variscan Molasse sediments. Late to post-Variscan Molasse sedimentation, therefore, started not before the uppermost Moscovian, in most of the Austro-Alpine units during the lowermost Permian.

Within the late to post-Variscan sequence of all tectonic units two evolutionary cycles, separated by a major unconformity, can clearly be recognized.

Late Palaeozoic (Late Carboniferous/Lower Permian) sediments of the lower cycle accumulated in intramontane Molasse basins which were formed by fracture tectonics as a result of crustal thinning and mega-shear zones due to the eastward drift of the Eurasian plate and westward drift of the African plate (Arthaud and Matte 1977). As a consequence, approximately E-W or ESE-WNW-trending basins formed which were filled with thick continental and shallow marine shelf sediments and volcanic rocks.

The climate was humid throughout the Late Carboniferous and became increasingly arid during the Permian. During Late Permian-Early Triassic (upper cycle) tectonic activity diminished and sedimentation patterns were more uniform over wide areas. Whereas marine sedimentation in the Carnic Alps and southern Karawanken Mountains started as early as Late Carboniferous and in the Dolomites in the Late Permian, marine sedimentation did not start before the Early Triassic within the Austro-Alpine units (Eastern Alps) in most cases. Based on major tectonic and climatic events, it was possible to correlate these sedimentary sequences of the Eastern and Southern Alps, although they were formed in different depositional environments.

These major events are the climatic shift near the Carboniferous/Permian boundary, eustatic sea level fluctuations during the Late Carboniferous/Early Permian caused by the Gondwana glaciation, a major tectonic event (hiatus) within the Permian ("Saalian movements") at the base of the Gröden formation in the Southern and Eastern Alps (boundary between the lower and upper cycle), a climatic shift at the Permian/Triassic boundary, and sea level changes during the Scythian, probably caused by different spreading rates.

Intense investigations of the late to post-Variscan sediments of the Eastern and Southern Alps during the last years greatly enlarged our knowledge of their evolution and depositional environment.

In the following sections an attempt will be made to give a brief and up-to-date summary of the most important data concerning stratigraphy, fossil content and depositional environment of the late to post-Variscan sedimentary sequences of the Eastern and Southern Alps. The most significant outcrops of these sediments are shown in Fig. 1. Before starting the discussion of the late to post-Variscan sedimentary sequences of the Eastern and Southern Alps, a brief summary of the synorogenic Carboniferous sediments of Nötsch and Veitsch will be presented.

2 Carboniferous of Nötsch

The tectonically isolated Carboniferous of Nötsch, extending about 9 km in an east-west direction and 3 km in a north-south direction, is exposed at the southern margin of the Drau Range (Fig. 1). The Carboniferous of Nötsch is fault-bounded to the north, bordering the Permian-Triassic sediments of the Drau Range, and to the south, bounded by the Granite Complex of Nötsch and Gailtal Metamorphic Complex (Fig. 2). The eastern margin is poorly exposed and the Carboniferous of Nötsch seems to be overthrust by Permian-Triassic sediments of the Dobratsch Massif (see also Kodsi and Flügel 1970; Schönlaub 1985).

From the base (north) to the top (south) the Carboniferous sequence of Nötsch is divided into Erlachgraben formation, Badstub formation and Nötsch formation.

The several hundred metre thick Erlachgraben formation, in its lower part, is composed of polymict conglomerates, immature sandstones (lithic arenites and lithic wackes), siltstones and shales. The sediments were formed by debris flows and different types of turbidity currents, and show fining upward sequences, which are up to some tens of metres thick. Trace fossils of the *Nereites* association (Tessensohn 1972) point to a deep marine environment, the sediments were probably deposited on the lower slope. The material is derived from a metamorphic hinterland composed of gneisses, different kinds of schists and phyllites.

The "Pölland Folge" of Kodsi and Flügel (1970), which is considered as part of the Nötsch Formation by Schönlaub (1985), does not show any differences compared to the Erlachgraben formation and therefore cannot be separated.

The upper part of the Erlachgraben formation is composed of fossiliferous siltstones and shales with intercalated sandstones and pebbly mudstones, which originated from turbidity currents. Intercalated are also biomicrites and algal limestones. The trace fossil *Dictyodora liebeana* and intercalated turbidite layers indicate sedimentation in a deeper marine environment, most likely on the upper part of the slope (Krainer 1992).

The Erlachgraben formation grades upward into the Badstub formation ("Badstubbreccie"), an approximately 400 m thick sequence of amphibolite breccias, conglomerates, sandstones and fossiliferous siltstones. Although this sequence has been interpreted differently by various authors, – as diabase (Frech 1894), volcanic breccia (Kieslinger 1956), diabase breccia (Schönlaub 1973), tectonic breccia (Angel 1932), metamorphic tholeiitic basalt (Teich 1982) and sedimentary breccia (Felser 1936; Schönlaub 1985), from structural and textural features and from the fossil content in the upper part of the sequence, a sedimentary origin is proved (Krainer and Mogessie 1991).

The Badstub formation is interpreted as a sedimentary sequence of submarine resedimented breccias, conglomerates, sandstones and siltstones formed by sediment gravity flows on a proximal fan or slope along an active fault zone. Due to the lack of exposure, the lateral facies distribution and transport direction is not known. The high content of amphibolite-derived material, which is highly susceptible to weathering processes, up to about 80 % of the conglomerates and breccias are composed of amphibolite clasts, and the coarse grain size, indicate very rapid sedimentation caused by tectonic movements at an active fault zone. Geochemical studies showed that the amphibolite clasts are derived from metamorphosed tholeiitic ocean floor basalts (Krainer and Mogessie 1991).

The Nötsch formation is a several hundred metre thick sequence of shales and siltstones, in the lower part rich in fossils and with intercalated sandstones and conglomerates of turbidite origin. The facies is similar to the upper part of the Erlachgraben formation. In the lower part, the sediments contain a rich fauna of brachiopods, echinoderms, bivalves, bryozoans, trilobites, gastropods and others (see summary in Schönlaub 1985; for trilobites Hahn and Hahn 1987).

The coarse clastic sediments in the southernmost part of the Carboniferous of Nötsch are quite similar to the Erlachgraben formation and are probably tectonically placed into this position.

Age of the Carboniferous of Nötsch: the rich fauna from the upper part of the Erlachgraben formation is probably of Late Visean-Namurian age according to

Schönlaub (1985). Conodonts from limestone clasts within amphibolite breccias of the Badstub formation indicate Late Visean age (Schönlaub 1985) and the rich trilobite fauna from the lower part of the Nötsch formation points to Late Visean age (Hahn and Hahn 1987). Therefore, the Erlachgraben formation is of Visean age, the Badstub formation of Late Visean age and the Nötsch formation of Late Visean-Namurian age.

3 Carboniferous of the Veitsch Nappe

In the western part of the Veitsch nappe, Ratschbacher (1984, 1987) subdivided the approximately 650 m thick Carboniferous sequence into Steilbachgraben Fm. (Early Visean) at the base, the Triebenstein Fm. (Visean-Namurian) and the Sunk Fm. (?Westphalian) at the top (Fig. 3). According to Ratschbacher (1984, 1987) there is no detectable stratigraphic break in this sequence.

The Steilbachgraben Fm. (maximum thickness 230 m) is predominantly composed of fine grained, laminated clastic sediments, interbedded with sandstone layers rich in detrital feldspars (up to 50%). Dolomite- and magnesite-lenses are intercalated. The sediments of this sequence are interpreted as having been formed in a clastic, littoral and hypersaline shelf lagoon environment, associated with the bioherms of the Triebenstein Fm. Intercalated are thin volcanic layers (metatuffs), which are interpreted as derived from tholeiitic intraplate basalts due to their chemical composition (Prohaska and Ebner 1989).

The Triebenstein Fm. is a sequence of up to about 300 m thick, pure, massive carbonates with lensoid geometry and some preserved fossils (corals, crinoids), interpreted as a subtidal carbonate facies with scattered bioherms.

The Sunk Fm., 50–150 m thick, comprises a sequence of clastic and carbonate sediments with a coarsening upward trend, composed of several second-order cycles. The sequence starts with carbonates at the base and grades upward into shales, siltstones, sandstones, and finally conglomerates with intercalated graphite-layers (sometimes containing some plant fossils). The sediments of this coarsening-upward cycle are interpreted as deposits of a regressive shoreline. Conglomerates probably represent channel-fill deposits and fossiliferous limestones, shales and siltstones are believed to represent distributary bay deposits (for details, see Ratschbacher 1984, 1987).

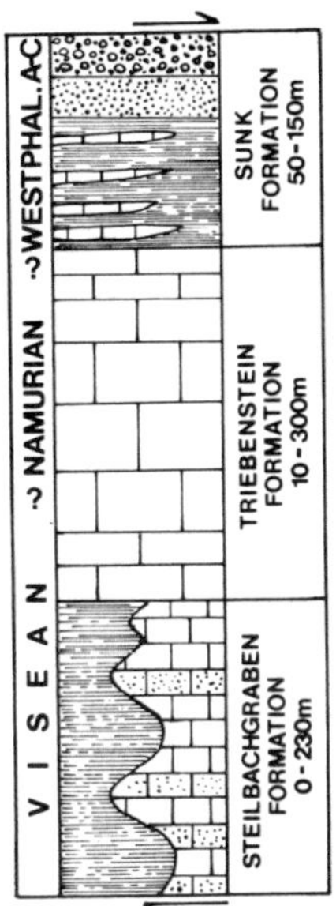

Fig. 3. Schematic columnar section through the Carboniferous sequence of the Veitsch nappe. (After Ratschbacher 1984, 1987)

A correlation with the Carboniferous sequence of the Veitsch area is not possible. In the Veitsch area the sequence is composed of fine- and coarse-grained clastic sediments and intercalated carbonate rocks (magnesite and dolomite). From dolomites biosiltites with abundant fossil fragments (brachiopods, gastropods, echinoderms, corals, bryozoans, trilobites and ostracods) and crinoid grainstones are reported. The microfauna of the dolomites indicates deposition on a shallow shelf with water depths ranging from 10–50 m, and normal salinity (Krainer 1992). The trilobite fauna is of Early Visean age (Hahn and Hahn 1977), corals indicate Late Visean age (Briegleb 1971; Felser 1977).

From several localities plant fossils are known from the upper part of the Carboniferous sequence of the Veitsch nappe (e.g. Klamm/Semmering, Wurmalpe near Leoben, Kaisersberg). The poorly preserved flora contains only about 15 species (see Jongmans 1938; van Amerom and Boersma 1974; Tenchov 1980). Of stratigraphic significance (according to Wagner 1984) are *Alethopteris lonchitica, Alethopteris decurrens, Neuropteris heterophyllum* and *Sphenophyllum cuneifolium,* indicating Westphalian A–C. As the last Variscan orogenic movements are dated Westphalian C/D in the Carnic Alps ("Carnic Phase"), the whole Carboniferous sequence of the Veitsch nappe is composed of "synorogenic" sediments. In the Veitsch area, late to post-Variscan sedimentation most probably started during the lowermost Permian (Präbichl Fm.).

	Penninic Unit		Lower Austroalpine Units					
	Tauern Window		Radstätter Tauern		Tarntaler Berge			
	Hochstegen Facies	Brennkogel Facies	Hochfeind Facies	Pleisling Facies	Hippold Facies	Reckner Facies	Semmering	Wechsel
Scythian	"Permoskyth-quartzite" (0–30 m)	"Wustkogelserie": Alpine Röt, Semmering-quarzit (40 m)	Lantschfeld-quarzit (40 m)	Alpine Röt (20 m) Lantschfeld-quarzit (150 m)	"Permoskyth-quarzit" (150 m)	Alpine Röt (5 m) Lantschfeld-quarzit (40 m)	Alpine Röt (10 m) Semmering-quarzit (200 m)	Alpine Röt (10 m) Semmering-quarzit (100 m)
Permian		"Wustkogelserie": Alpine Verrucano (150 m)	Alpine Verrucano (30 m)	Alpine Verrucano (130 m)		Alpine Verrucano (5 m)	Alpine Verrucano (150 m)	"Arkose-schist breccia-porphyroid-series" (10–150 m)

Fig. 4. Simplified Permian/Early Triassic (Scythian) stratigraphic nomenclature of the Penninic and Lower Austro-Alpine units

	Thörl/Aflenz	Seckauer Tauern	Stangalm Mesozoic	Brenner Mesozoic
Scythian	Alpine Röt (25 m)	Semmering quarzit (Plattelquarzit)	"Permoskyth-quarzit" (15–20 m) (probably of Late Scythian Age)	Late Scythian Fan-delta – shelf clastics (15–20 m)
Permian	Alpine Verrucano (100 m)	Alpine Verrucano (Rannachserie)		

Fig. 5. Simplified Permian/Early Triassic (Scythian) stratigraphic nomenclature of some Middle Austro-Alpine units

4 Eastern Alps

In the Eastern Alps, late- to post-Variscan sediments are found in all major tectonic units (Penninic unit, Lower Austro-Alpine, Middle Austro-Alpine = Central Austro-Alpine, and Upper Austro-Alpine unit). In the Penninic (Upper Schieferhülle), Lower and Middle Austro-Alpine units the late- to post-Variscan sequence is very similar and composed of coarse- to fine-grained clastic sediments, in the lower (? Permian) part with intercalated rhyolithic volcanic rocks and reworked volcanic clasts.

All late- to post-Variscan sequences of the units mentioned have been deformed and overprinted by Alpine metamorphism to phengite- and sericite-schists, arkose-schists and arkose-gneisses, quartzites and porphyroids.

The sequence, probably completely corresponding to the upper cycle (? Middle Permian-Early Triassic) is usually less than 100 m, rarely reaching 300 m in thickness. Based on lithologic correlations and comparison with the late- to post-Variscan sequence of the Upper Austro-Alpine units, the sequence is generally divided into a coarse-grained (? Middle-Late) Permian unit called "Alpine Verrucano" and a finer-grained unit composed of various types of quartzites and schists ("Skythquarzite"), correlated to the Scythian and overlain by evaporitic sediments (Alpine Röt, Reichenhaller Rauhwacke) in most cases (Figs. 4, 5).

In the Lower Austro-Alpine of the Semmering area, the Late- to post-Variscan sequence consists of the Alpine Verrucano (sericite schists and sericite-quartz-schists) with intercalated rhyolithic volcanics, some tens of metres thick, overlain by the Semmering quartzite (laminated and cross-bedded quartzites) and Alpine Röt (up to 20 m-thick shales and quartzites with thin intercalations of evaporitic sediments).

In the Radstätter Tauern, the sequence is very similar: 130 m Alpine Verrucano (sericite-schists, sericite-quartzites, conglomeratic quartzites; Permian), 150 m "Lantschfeldquarzit" (quartzites; Scythian) and 20 m Alpine Röt (shales and quartzites; Late Scythian) in the "Pleislingfazies", and 30 m Alpine Verrucano (Permian), 40 m Lantschfeldquarzit (Scythian) in the "Hochfeindfazies" (Tollmann 1964, 1977).

The late- to post-Variscan sequence of the Stangalpe area (Middle Austro-Alpine *sensu* Tollmann; Fig. 1) is only about 20 m thick and rests on the metamorphic basement (mica schists and gneisses). The sequence shows a fining-upward trend with quartzites and conglomeratic interlayers, sometimes displaying trough crossbedding at the base. The metasandstones consist of quartz-rich lithic arenites with feldspar-contents ranging from 4–24%. The sequence grades upwards into sandy dolomites of Anisian age. The basal part is probably of fluvial origin whereas the upper part of the sequence was formed in a shallow marine environment (see Krainer 1984).

A very similar late to post-Variscan sequence occurs at the base of the "Brenner Mesozoics" in the western part of the Central Alps (Tyrol, Fig. 1). The approximately 20 m thick sequence is built up of quartz-rich ortho- and paraconglomerates, and trough cross-bedded, laminated and massive sandstones in the lower part. Towards the top, carbonate content increases and the sequence grades into dolomitic sediments of Anisian age ("Basisdolomit"). The lower part is interpreted as fan delta, the upper part as shallow marine shelf deposits (Krois and Stingl 1989). Based on an event-stratigraphic correlation with late to post-Variscan sediments of the Southern and Northern Alps, the sequence of both, the Stangalpe area and Brenner Mesozoics are presumed to be of Late Scythian (Campill) age.

In the Niedere Tauern, a very thick sequence (up to several kilometres) of Permian Alpine Verrucano and Scythian Semmeringquarzit (conglomerates, quartzites, sericite-quartzite-schists, sericite-phyllites), termed "Rannachserie", overlies the metamorphic basement (Tollmann 1977).

Age and depositional environment of all these metasediments are not exactly known due to severe deformation and metamorphic overprinting. Further information is given in Frasl and Frank (1966), Frank (1972) (Penninic unit); Tollmann and Faupl (1972), Tollmann (1964) (Lower Austro-Alpine unit); Claasen et al. (1987), Krainer (1984), Krois and Stingl (1989), Schünemann et al. (1982) (Middle Austro-Alpine unit), for summaries see also Tollmann (1964, 1972, 1977), Oberhauser (1980).

In the following section, therefore, the more detailed discussion will be limited to the late to post-Variscan sequence of the Upper Austro-Alpine units.

4.1 Upper Austro-Alpine Units

Late to post-Variscan sedimentation within the Upper Austro-Alpine unit of Carboniferous (Stephanian) age are known only from a few places (at the NW margin of the Gurktal Nappe, the Steinach Nappe) and bulk sedimentation did not start before the earliest Permian (see Fig. 6).

Major progress in stratigraphy and depositional history has been achieved by extensive investigations in recent years (for example Niedermayr 1975, 1985; Angerer et al. 1976; Haditsch et al. 1978; Krainer 1982, 1985, 1987a, b, 1989a, c, d, 1990a, b; Niedermayr and Scheriau-Niedermayr 1982; Stingl 1982, 1983, 1984, 1987, 1989;

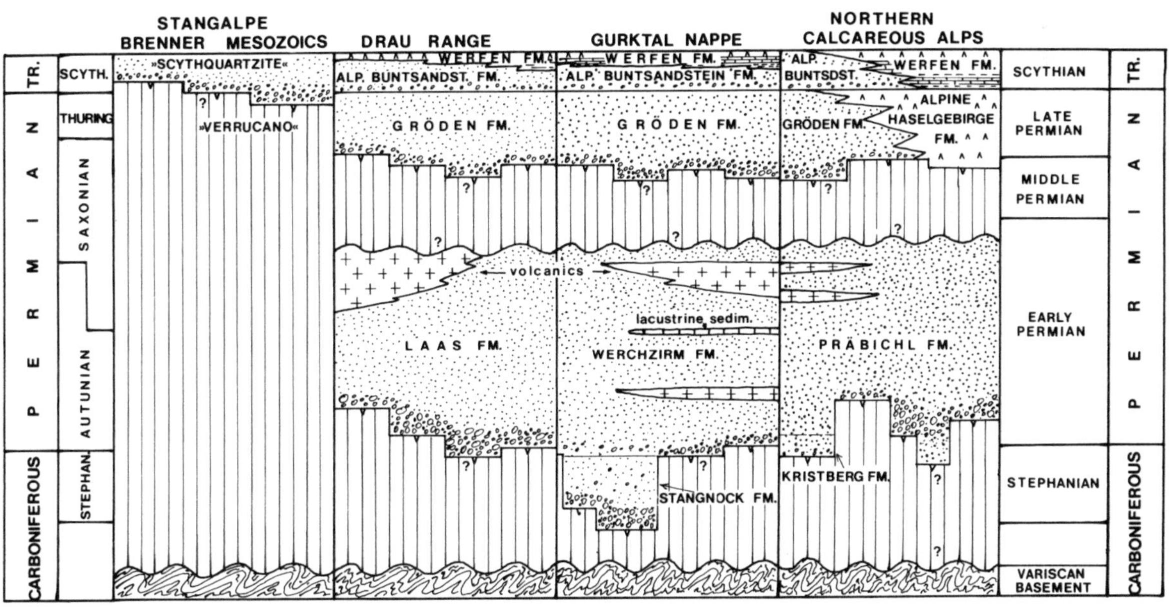

Fig. 6. Schematic stratigraphic framework of late to post-Variscan (Late Carboniferous to Early Triassic) sediments in some Austro-Alpine units (Central-Alpine and Upper Austro-Alpine). *Legend*, see Fig. 11

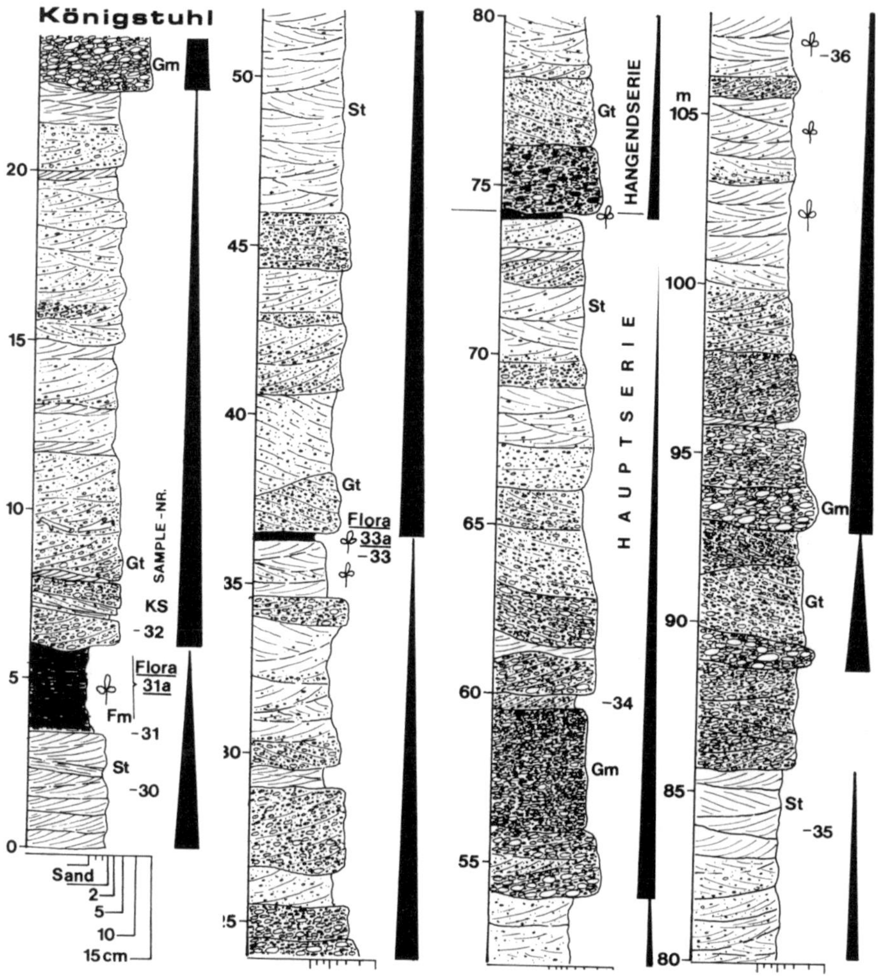

Fig. 7. Detailed log through the main part of the Stangnock formation (Hauptserie) at the northern side of the Königstuhl Mountain. Several fining-upward mega-sequences are developed, formed by massive and trough cross-bedded conglomerates (lithofacies *Gm* and *Gt*), trough cross-bedded sandstones (lithofacies *St*) and dark siltstones and shales containing abundant plant fossils (lithofacies *Fm*, overbank deposits) at the top of some of the mega-sequences. (After Krainer 1989c, Fig. 13)

Mostler and Rossner 1984; Poscher 1985; Krainer and Stingl 1986a, b; Krainer and Spötl 1989; Sylvester 1989a, b).

4.1.1 *Late Carboniferous – Early Permian (Lower Cycle)*

The best – exposed example of Late Carboniferous sediments, which has been intensively studied over the last years, is the **Stangnock formation** at the NW margin of the Gurktal nappe (Krainer 1989a, c and references therein; Fritz et al. 1990). The Stangnock formation comprises a sequence of more than 400 m thick intermontane Molasse sediments and can be divided into three units:

1. Basal sequence consists of polymict conglomerates rich in gneiss clasts, and intercalated immature, coarse-grained sandstones (feldspathic lithic arenites). These sediments are interpreted as deposits of a proximal fluvial system.

2. The main part of the Stangnock formation (Hauptserie) is built up by a few, indistinctly developed megasequences. At the base, these megasequences are characterized by a sharp, erosive boundary, starting with sediments of a gravelly braided river system, grading upwards into a gravel-sandstone facies, frequently showing features of a meandering river system. At the top of these sequences usually dark shales with thicknesses up to a few metres occur, which contain abundant, well-preserved plant fossils. The shales, interpreted as overbank fines deposited on flood plains and in oxbow lakes, sometimes are overlain by thin anthracite seams (Fig. 7).

Conglomerates are very rich in quartz clasts (> 90%), sandstones are classified as moderately sorted, subangular lithic arenites and sublitharenites, subordinate lithic wackes, all containing high amounts of polycrystalline quartz.

3. The top sequence (Hangendserie) shows similar features, slight differences exist concerning the com-

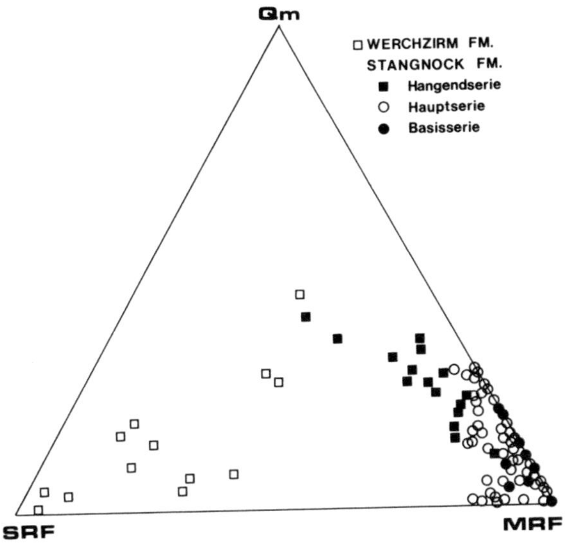

Fig. 8. Sandstone compositions of the Stangnock formation (Stephanian) and Werchzirm formation (Autunian) in the ternary diagram *Qm* (monocrystalline quartz); *S*RF sedimentary rock fragments; *MRF* metamorphic rock fragments. The upper part of the Stangnock formation (Hangendserie) is characterized by larger amounts of Qm (partly volcanic quartz) and metasedimentary rock fragments compared to the basal and main part (Basisserie, Hauptserie) of the Stangnock formation. More significant is the difference to the sandstones of the Werchzirm formation which contain high contents of sedimentary rock fragments. (After Krainer 1989a, c)

position of the sediments: sandstones contain volcanic rock fragments, esp. volcanic quartz (Fig. 8), referring to first volcanic activity during the Late Carboniferous in this area.

On account of current directions which show a significant eastward trend, it is concluded that the intramontane basin developed in an approximately east-west direction.

The sharp erosive base of the megasequences within the main sequence and top sequence is referred to as synsedimentary fracture tectonics (block faulting).

Plant fossils of the Stangnock formation (72 different taxa) have been studied from more than 20 localities and different horizons within the sequence (summary in Fritz et al. 1990). The following megafloral zones (*sensu* Wagner 1984) can be distinguished from base to top: *Odontopteris cantabrica* Zone, *Lobatopteris lamuriana* Zone, *Alethopteris zeilleri* Zone in the lower part of the Stangnock formation (dominance of *Linopteris neuropteroides* and Lycophyta), *Sphenophyllum angustifolium* Zone in the middle part, characterized by Pecopterids, Alethopterids, *Neuropteris cordata* and *Neuropteris ovata,* and *Callipteris conferta* Zone in the uppermost part of the Stangnock formation (see Fig. 13).

Based on petrographical and geochronological investigations of the gneiss clasts, which he compared with the gneisses of the metamorphic basement ("Middle Austro-Alpine" *sensu* Tollmann 1977), Frimmel (1986a, b) showed that the gneiss clasts of the Upper Austro-Alpine Stangnock formation most likely are derived from the underlying "Middle Austro-Alpine" metamorphic rocks (Bundschuh-orthogneiss, see also discussion in Frank 1987).

The Stangnock formation is overlain by Early Permian red beds (Werchzirm formation, Fig. 8).

Within the small Upper Austro-Alpine Steinach nappe in western Austria (Tyrol) Late Carboniferous (Stephanian) sediments overlie the presumably Early Palaeozoic Steinach quartzphyllite, a diaphthoritic overprinted metamorphic complex (Frizzo and Visona 1981).

The poorly exposed Stephanian sequence is composed of various channel-, bar- and overbank sediments and thin anthracite seams which have been mined. The composition is very similar to the sediments of the Stangnock formation: conglomerates are again very rich in quartz clasts. Sandstones are classified as lithic arenites, sometimes as lithic wackes (Krainer 1990a, see also Schmidegg 1949; Karl 1956). According to the depositional history and composition the Late Carboniferous (Stephanian) sequence of the Steinach nappe is very similar to the Stangnock formation, especially to the conglomerate-sandstone facies of the main sequence.

The Late Carboniferous sediments of the Steinach nappe are also interpreted as intramontane Molasse sediments, which probably have been formed in the same basin system as the Stangnock formation.

Shales contain plant fossils, first reported by Pichler (1859). The first taphoflora lists were published by Stur (1871) and Stache (1872). The first comprehensive paper of the plant fossils was published by Kerner (1897), the last by Jongmans (1938).

The flora contains about 30 different taxa. The occurrence of *Neuropteris scheuchzeri, Sphenophyllum oblongifolium, Sphenophyllum emarginatum, Pecopteris arborescencs* and *Pecopteris hemitelioides* indicates that the flora is of Cantabrian/Barruelian age (*Odontopteris cantabrica* Zone and *Lobatopteris lamuriana* Zone according to Wagner 1984). Overlying Permian (and younger) sediments are eroded.

The Permian sequence of the Upper Austro-Alpine unit, which rarely contains fossils, is divided into two lithostratigraphic units, separated by a major hiatus ("Saalian movements"): the Early Permian lithostratigraphic unit of the lower cycle, and the Late Permian lithostratigraphic unit of the upper cycle.

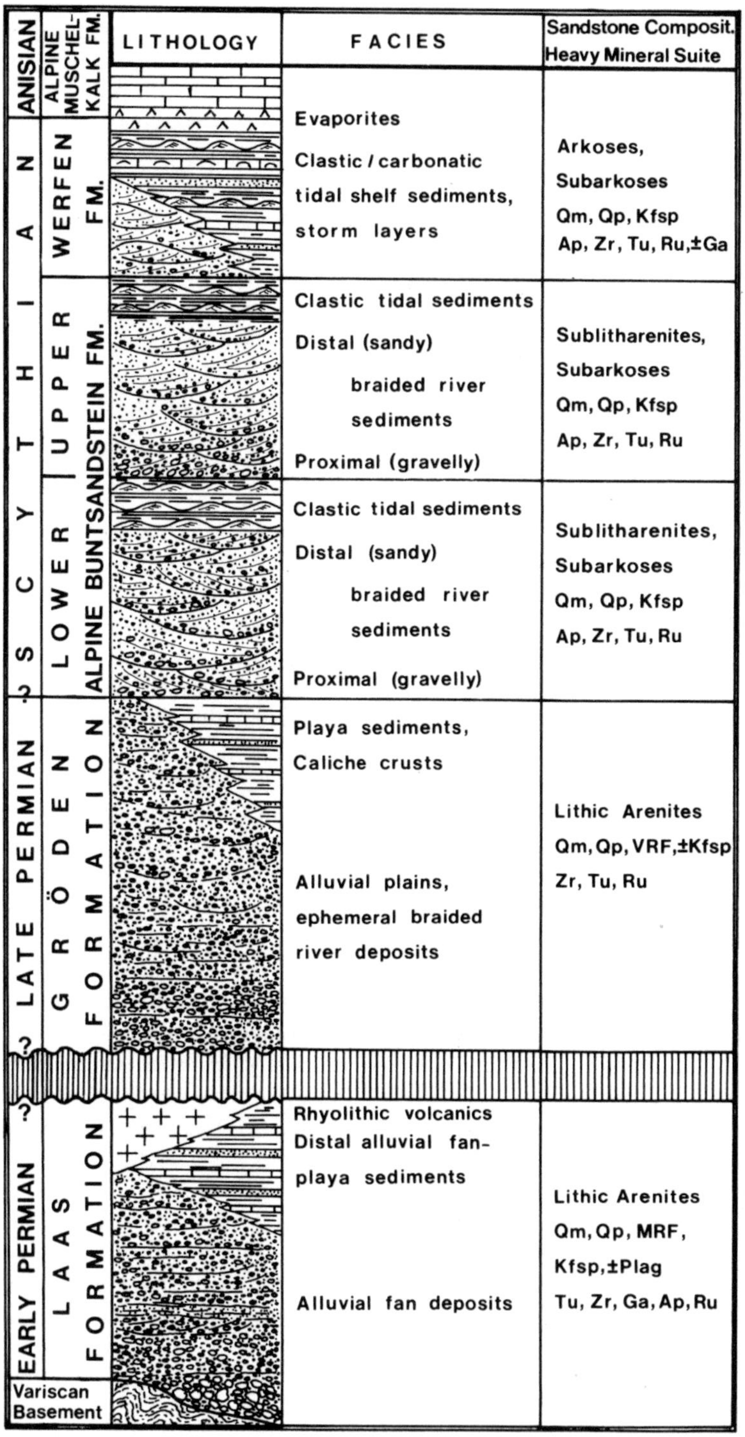

Fig. 9. Generalized lithostratigraphic column of Permian and Early Triassic (Scythian) sediments of the Drau Range with depositional environment, sandstone composition. (*Qm* Monocrystalline, *Qp* polycrystalline quartz; *Kfsp* kalifeldspar; *Plag* plagioclase; *VRF* volcanic rock fragments; *MRF* metamorphic rock fragments) and heavy mineral assemblages. *Ap* apatite; *Ga* garnet; *Ru* rutile; *Tu* tourmaline; *Zr* zircon

Proximal to distal alluvial fan sediments (red coloured breccias, conglomerates, immature sandstones), grading into fine-grained sandflat-playa complexes with caliche crusts and rare thin algal layers at some places, characterize the Early Permian throughout all tectonic units (**Basalbreccia** and **Präbichl formation** in the Northern Calcareous Alps, **Laas formation** in the Drau Range, **Werchzirm formation** in the Gurktal nappe, maximum thickness about 150 m; see Niedermayr and Schierau-Niedermayr 1982; Stingl 1983; Krainer and Stingl 1986a, b; Krainer 1987b, 1989b, 1990b; Sylvester 1989a, b). In most cases, these sediments overlie the Variscan basement, which is formed of crystalline rocks (schists, gneisses etc., Drau Range) or different, weakly metamorphosed Variscan sediments (Figs. 6, 9). At the NW margin of the Gurktal nappe, Early Permian red beds overlie Late Carboniferous sediments (Stangnock formation). The Early Permian sediments are rich in clasts derived from the local basement, especially weakly metamorphosed old Palaeozoic rocks (polymict conglomerates, lithic arenites and wackes, see Fig. 9).

In some places plant fossils are known from the base of this Early Permian sequence (van Amerom et al. 1982; Fritz and Boersma 1987 a, b, 1988; Fritz et al. 1990). The occurrence of *Callipteris conferta, Ernestiodendron filiciformae, Walchia* sp. and *Sphenophyllum angustifolium* in the basal Werchzirm, Laas and Kristberg Fms. is of stratigraphic significance, pointing to lowermost Permian age (*Callipteris conferta* Zone).

In the Drau Range, Gurktal nappe and the westernmost part of the Northern Calcareous Alps, rhyolithic volcanics (ignimbrites and pyroclastic flows) with thicknesses up to about 100 m are widespread on top of this Early Permian sequence (Figs. 6, 9). From these volcanics no radiometric age determinations exist, but based on palynological data from lacustrine sediments within the equivalent Bolzano Volcanic Complex of the Southern Alps, the Early Permian sequence, including the rhyolithic volcanics, probably reaches up to Late Artinskian-Kungurian (Hartkopf-Fröder and Krainer 1990).

4.1.2 Late Permian – Early Triassic (Upper Cycle)

With a hiatus caused by block faulting ("Saalian movements"), Late Permian siliciclastic sediments (conglomerates, sandstones, shales of red bed type) of ephemeral braided rivers and playas termed **Gröden formation** overlie Early Permian sediments and mark a sudden and significant change in sedimentation (Figs. 6, 9). This hiatus marks the boundary between the lower and upper cycle and is equivalent to that between the Bolzano Volcanic Complex and Gröden formation or between the Trogkofel limestone and Tarvis breccia in the Southern Alps. The sediments of the Gröden formation contain large amounts of reworked volcanic fragments derived from the volcanic rocks on top of the Early Permian sequence, and are lacking stratigraphically significant fossils (Niedermayr and Scheriau-Niedermayr 1982; Stingl 1983; Krainer 1987 b, 1989 d, 1990 b; Sylvester 1989 a, b). In the Drau Range and Gurktal nappe the maximum thickness is about 350 m, in the Northern Calcareous Alps about 800 m.

In the central and eastern realm of the Northern Calcareous Alps a few hundred metres of halite-bearing marine evaporites were deposited in an approximately E-W-trending Late Permian rift, now represented by a strongly deformed decollement horizon at the base of the Northern Calcareous Alps (Spötl 1987, 1988 a, b, 1989). Stratigraphical and palaeoenvironmental data (S-isotopes of sulphate minerals, bromide content in halites, pollen and spores as well as rare marine bivalves) unequivocally support a marine origin for this Late Permian evaporites ("Alpine Haselgebirge") (Spötl 1989).

The boundary between the Late Permian Gröden formation and Scythian Alpine Buntsandstein formation is documented by an abrupt change in the depositional environment and composition of the sediments, caused by a climatic change to slightly more humid conditions. This boundary presumably corresponds to the Permian/Triassic boundary (Brandner et al. 1986; Krainer 1987 a).

Based on transgressional and regressional events, the Scythian sequence of the Drau Range and Gurktal nappe can be subdivided into three fining upward megasequences: **Lower and Upper Alpine Buntsandstein formation** and **Werfen formation** (Krainer 1985, 1987 a; Figs. 9, 10).

Lower and Upper Alpine Buntsandstein formations are built up by quartz- rich conglomerates of a gravelly braided river system, grading upward into small-scale fining-upward cycles of a sandy braided river system. Some sections contain clastic shallow-marine sediments developed on top. Compared to the Permian sandstones (uncemented, poorly sorted angular lithic arenites rich in volcanic fragments), Scythian sandstones are better sorted and rounded, well-cemented by authigenic overgrowths of quartz and feldspar and carbonate cements in most cases, and contain more quartz and detrital feldspars. The heavy mineral composition also differs significantly. The regressive event at the base of the Upper Buntsandstein formation can be correlated with the regressive "Campill Event" in the Southern Alps based on palynomorphs.

The dominantly siliciclastic sediments of the **Werfen formation**, containing thin intercalated fossiliferous limestone beds, were deposited on a shallow-marine, tidally influenced epicontinental shelf. Micro- and macrofauna, especially *Meandrospira pusilla* (HO) (foraminifer), *Natiria costata* Münster, *Costatoria costata* (Zenker), *Tirolites* and *Dinarites* indicate a Late Scythian age. The Werfen Fm. of the Drau Range and Gurktal nappe corresponds to the Val Badia and Cencenighe member of the Southern Alps. The regressive event on top of the Werfen Fm., corre-

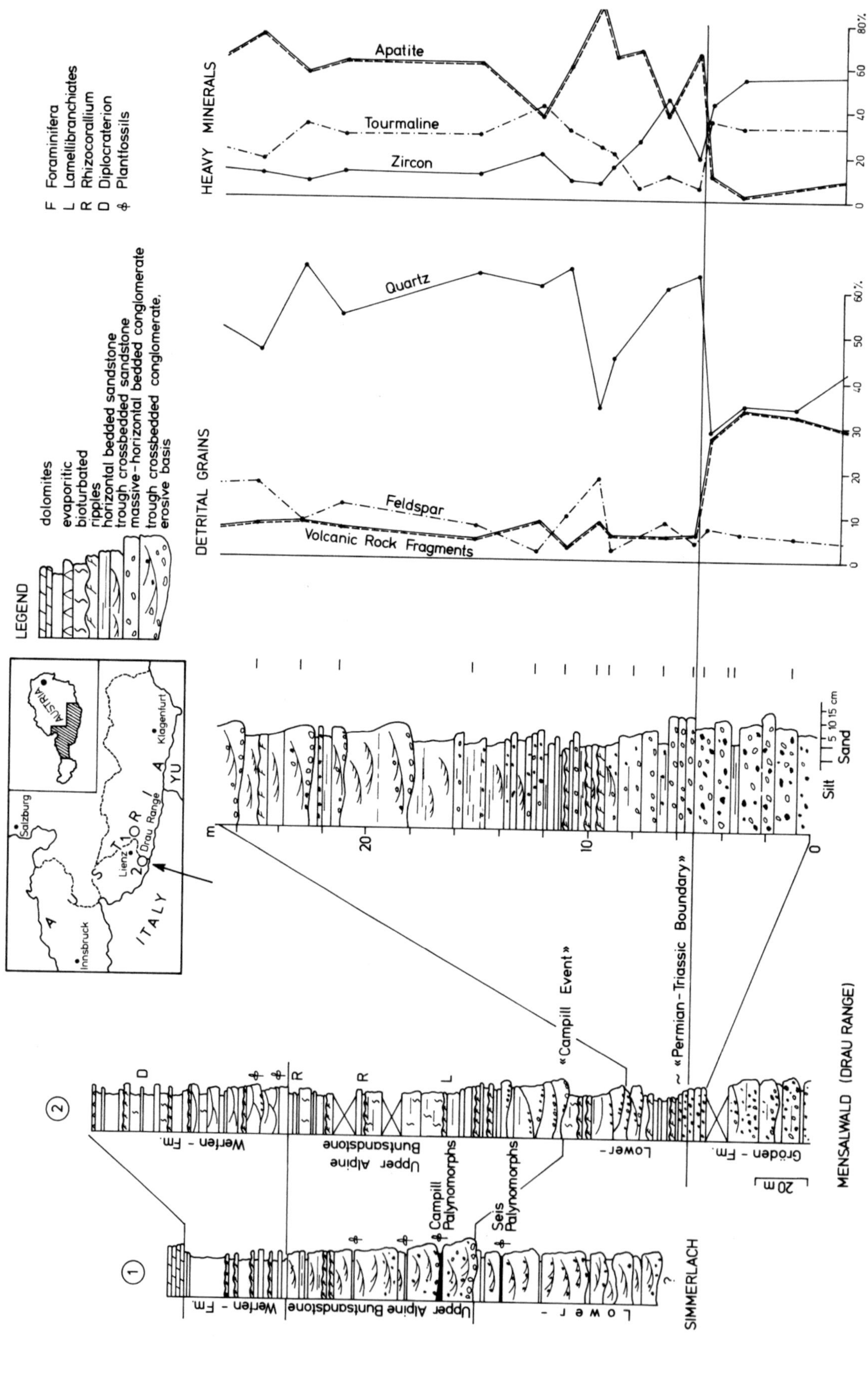

LEGEND
dolomites
evaporitic
bioturbated
ripples
horizontal bedded sandstone
trough crossbedded sandstone
massive-horizontal bedded conglomerate
trough crossbedded conglomerate, erosive basis
F Foraminifera
L Lamellibranchiates
R Rhizocorallium
D Diplocraterion
Plantfossils
HEAVY MINERALS
Apatite
Tourmaline
Zircon
DETRITAL GRAINS
Quartz
Feldspar
Volcanic Rock Fragments
Silt
Sand
5 10 15 cm
«Campill Event»
«Permian-Triassic Boundary»
Werfen-Fm.
Upper Alpine Buntsandstone
Lower-
Gröden-Fm.
Campill Palynomorphs
Seis Palynomorphs
SIMMERLACH
MENSALWALD (DRAU RANGE)
20 m
Innsbruck
Salzburg
Lienz
Drau Range
Klagenfurt
ITALY
YU
AUSTRIA

sponding to the San Lucano member of the Werfen Fm. of the Southern Alps, is marked by an evaporitic horizon (Krainer 1985, 1987 a; Niedermayr 1985).

In the Northern Calcareous Alps, a similar evolution took place during the Scythian. In the eastern and middle part, the whole Scythian is represented by the "Werfen beds" which interfinger with fluvial to marginal marine sediments of the Alpine Buntsandstein to the west (Stingl 1984, 1987, 1989; Poscher 1985). Microfacies analysis of the upper Werfen Fm. in the central part of the Northern Calcareous Alps indicates clastic and carbonate sedimentation in shallow, subtidal environment with oolite-shoals. Low-diversity of microfauna (foraminifers, gastropods, ostracods, ophiurs, crinoid fragments, conodonts and fish remains) indicates normal salinity. Interbedded are thin tempestite layers which are interpreted as being formed during storm events (Mostler and Rossner 1984). In the westernmost part of the Northern Calcareous Alps, the whole Scythian is represented by the Alpine Buntsandstein (Krainer 1982).

On top of the Werfen formation, an evaporitic horizon is developed (Reichenhaller Rauhwacke), expressing a regressive event at the Scythian/Anisian boundary. This sedimentation pattern within the Northern Calcareous Alps was caused by the transgression of the "Werfen sea" from SE to NW, furthermore similar transgressional and regressional events permit a similar subdivision of the Scythian sequence like in the Drau Range or Gurktal nappe.

The illite-crystallinity index of Permian and Scythian sediments of the Drau range, Gurktal nappe and Northern Calcareous Alps marks Alpine anchizonal metamorphism (Schramm 1977, 1982; Schramm et al. 1983; Niedermayr et al. 1984; Kralik et al. 1987).

5 Southern Alps

In the Southern Alps, Late Carboniferous-Permian sediments lie on the Variscan basement with a classical unconformity. This basement was folded and, in the western part, slightly metamorphosed during Westphalian times (Fig. 11). Within the late to post-Variscan sedimentary sequence, two cycles can be recognized, which are separated by a main unconformity (see Massari 1986).

Fig. 10. Detailed logs through Late Permian/Early Triassic (Scythian) sequences in the western part of the Drau Range indicating changes in sandstone composition (detrital grains, heavy minerals) at the Permian/Triassic boundary

The first cycle is characterized by Late Carboniferous-Early Permian sediments and volcanic rocks which were accumulated in intramontane basins formed by block- and wrench-faulting. In the Carnic Alps and southern Karawanken Mountains, the sequence of the first cycle is represented by cyclic deltaic and shallow marine clastic and carbonate sediments, the Bombaso formation, Auernig, Rattendorf, and Trogkofel groups, ranging from Late Moscovian to Late Artinskian.

In the Dolomites, the first cycle is formed by the local development of thin alluvial fan sediments (Ponte Gardena/Waidbruck conglomerate). It is followed by a thick volcanic sequence of latiandesitic to rhyolitic tuffs, ignimbrites and lavas with intercalated fluvial and lacustrine sediments (Bolzano Volcanic Complex) of Early Permian age. In the Lombardian Alps, south of the Adamello Massif, different intramontane basins formed, which were filled with thick sequences of volcanic rocks and clastic sediments (Collio formation).

With a hiatus (?)Middle – Late Permian sediments of the second cycle, which are more widely distributed but not restricted to discrete basins, unconformably overlie Early Permian or even older rocks. In the southern Karawanken Mountains, Carnic Alps and Dolomites, the second cycle starts with continental to shallow marine clastic sediments of the Gröden formation which repeatedly interfinger with and grade upwards into the shallow marine evaporitic and carbonate sediments of the Bellerophon formation. The Bellerophon transgression prograded slowly westwards over a very low-gradient and flat landscape. Sabkha sediments formed on this very shallow and wide, evaporitic transition zone. During prograding transgressional events, fossiliferous limestones of open-lagoon shallow water environments were deposited. West of the Dolomites, the second cycle is represented by continental red beds of the "Verrucano Lombardo" and by shallow marine sediments of the Servino formation (see also Fig. 11).

5.1 Carnic Alps/Southern Karawanken Mountains

After the Variscan orogenic phase which was active during Westphalian times, block- and wrench-faulting resulted in the formation of discrete basins mainly oriented WNW-ESE (for example the "Pramollo Basin") and NE-SW (Venturini 1982, 1990a, b). These basins were filled with deltaic to shallow-marine sediments of the Late Carboniferous Bombaso formation, Auernig group and Early Permian Rattendorf group (Fig. 12).

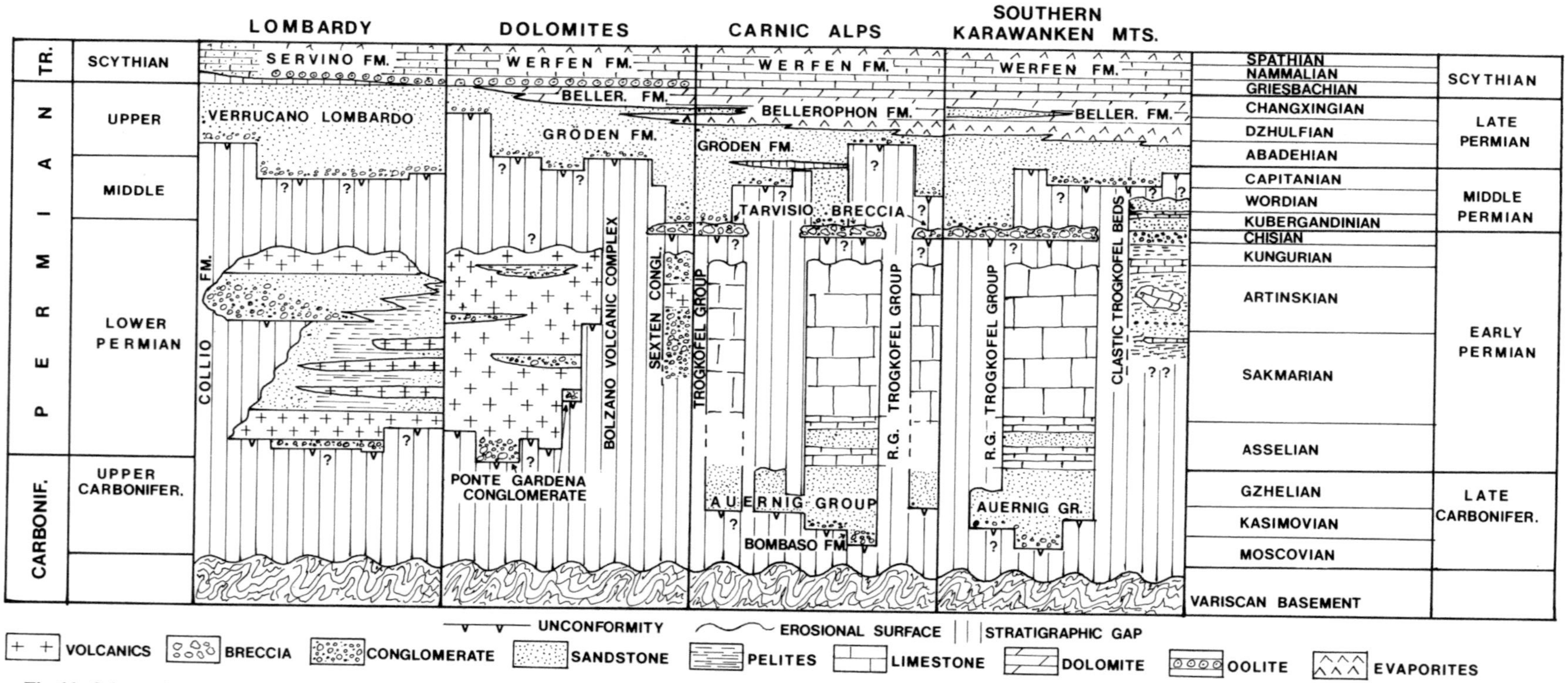

Fig. 11. Schematic stratigraphic framework of late to post-Variscan (Late Carboniferous to Early Triassic) sediments of the Southern Alps (after Italian Research Group 1986, p. 37 and Massari et al. 1988, verified and supplemented). *RG* Rattendorf group

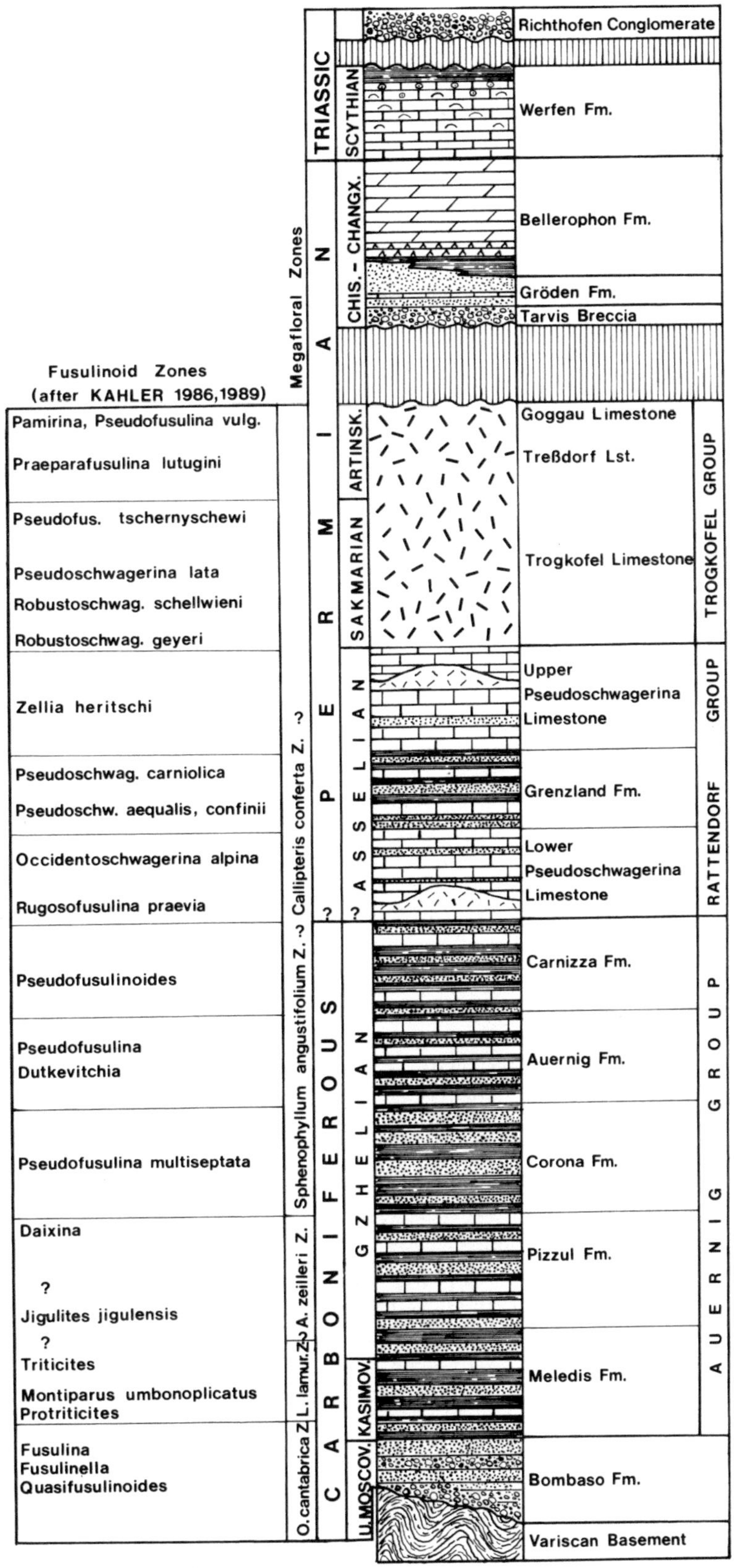

Fig. 12. Generalized stratigraphic column through the late to post-Variscan (Late Carboniferous to Early Triassic) sequence of the Carnic Alps in the Trogkofel – Gartnerkofel area

Fusulinid Zones (after KAHLER 1986)		CARNIC ALPS (shallow marine)		Megaflora Zones (after WAGNER 1984)	EASTERN ALPS (Gurktal Nappe, continental)	N-SPAIN
Zellia heritschi	Upper ASSELIAN	RATTENDORF GROUP	Upper Pseudoschwagerina Limestone	Callipteris conferta Zone	Werchzirm Formation	AUTUNIAN
Pseudoschwagerina carniolica Pseudoschw. aequalis	Middle		Grenzland Formation			
Occidentoschwagerina alpina Rugosofusulina praevia	Lower		Lower Pseudoschwagerina Limestone			
Pseudofusulinoides	GZHELIAN	AUERNIG GROUP	Carnizza Fm.	Sphenophyllum angustifolium Zone	STANGNOCK FORMATION Upper Part	STEPHANIAN C
Pseudofusulina Dutkevitchia			Auernig Fm.			
Pseudofusulina multiseptata			Corona Fm.			
Daixina ? Jigulites jigulensis ?			Pizzul Fm.	Alethopteris zeilleri Zone	Middle Part	STEPHANIAN B
Triticites rossicus Montiparus umbonoplic Protriticites	KASIMOVIAN		Meledis Fm.	? ? Lobatopteris lamuriana Zone	Lower Part	BURRUELIAN
Fusulina Fusulinella Quasifusulinoides	LATE MOSCOVIAN (MJATCHKOVIAN)	Bombaso Formation		Odontopteris cantabrica Zone		CANTABRIAN

Fig. 13. Correlation of the Late Carboniferous/Early Permian shallow marine sedimentary sequence of the Carnic Alps (Southern Alps) with the continental sedimentary sequence of the Gurktal nappe (Eastern Alps)

5.1.1 Late Carboniferous – Early Permian (Lower Cycle)

The folded Variscan basement is unconformably overlain by the **Bombaso formation,** which is formed by immature clastic wedges fed by the Palaeocarnic Chain (Fenninger et al. 1976; Venturini 1990a, b). These sediments are composed of radiolarian chert, arenitic and volcanic clasts (Pramollo member) or Silurian to Early Dinantian limestone clasts (Malinfier horizon). The overlying **Auernig group** with a maximum thickness of approximately 1200 m is divided into **Meledis-, Pizzul-, Corona-, Auernig- and Carnizza formations** according to Selli (1963) (Untere kalkarme Schichtgruppe, Untere kalkreiche Schichtgruppe, Mittlere kalkarme Schichtgruppe, Obere kalkreiche Schichtgruppe, Obere kalkarme Schichtgruppe of Heritsch et al. 1933). Based on fusulinids (see Kahler 1983, 1985, 1986, 1989), the Bombaso formation is of uppermost Moscovian (Miatchkovian) age, the Meledis formation of Early Kasimovian age, the Pizzul formation of Late Kasimovian – Early Gzhelian age, and Corona-, Auernig-, and Carnizza formations all correspond to Gzhelian age. Other stratigraphically significant fossils are trilobites (Hahn and Hahn 1987; Hahn et al. 1989) and brachiopods (Gauri 1965).

The main lithofacies-types are quartz-rich conglomerates (deltaic-beach environment), trough-crossbedded sandstones, hummocky-crossbedded sandstones (shoreface), bioturbated, sometimes fossiliferous siltstones and shales, and fossiliferous limestones rich in calcareous algae *(Anthracoporella spectabilis, Archaeolithophyllum missouriense, Epimastopora, Eugonophyllum)* sometimes forming very flat algal mud mounds, and fusulinids. Other fossils within the limestones are small foraminifers like *Ammovertella, Bradyina, Calcitornella, Endothyra, Tetrataxis, Textularia, Tuberitina* amongst others, sphinctozoans (Kügel 1987), bryozoans (Kodsi 1967), ostracods, echinoderms and the microproblematicum *Tubiphytes.* In the Meledis formation, small Auloporid mounds occur (Flügel and Krainer 1991). Algal mounds are known from the Pizzul- and

Auernig formations (Boeckelmann 1985; Krainer 1990c).

These lithofacies types form prominent clastic-carbonate transgressive and regressive cycles ("cyclothemes") with thicknesses of 10–40 m, similar to the Pennsylvanian cyclothemes of the Western Interior (USA), especially in the upper part of the Auernig group (Corona, Auernig and Carnizza formations). In general, sea-level lowstands are marked by coarse clastic sediments (conglomerates), whereas sea-level highstands are marked by limestones.

Cycle duration lies in the order of 100000 years and the cycle formation is related to eustatic sea-level changes caused by the Gondwana glaciation (see Massari and Venturini 1990 for detailed information).

Well preserved plant fossils (more than 100 taxa) are known from about 30 localities and different horizons within the Auernig group of the Carnic Alps. The plant fossil assemblage is very similar to that of the Late Carboniferous Stangnock formation, and the same plant fossil assemblages (*Odontopteris cantabrica* Zone, *Lobatopteris lamuriana* Zone, *Alethopteris zeilleri* Zone, *Sphenophyllum angustifolium* Zone and *Callipteris conferta* Zone in the sense of Wagner 1984) can be recognized from the base to the top allowing good correlation with the continental Stangnock formation (Fig. 13; see Fritz et al. 1990 for detailed information).

The Permian of the Carnic Alps and southern Karawanken Mountains is a thick sequence of different, in most cases shallow, marine carbonate and clastic sediments, divided into Rattendorf group, Trogkofel group (lower cycle), Tarvis breccia, Gröden formation and Bellerophon formation (upper cycle) from base to top (Figs. 11, 12).

The **Rattendorf group** shows a similar facies assemblage in comparison to the Auernig group. It is composed of the Lower Pseudoschwagerina limestone (approx. 160 m), Grenzland formation (approx. 125 m) and Upper Pseudoschwagerina limestone (approx. 70 m). The sediments are of shallow marine environments, sedimentation patterns changed from near-coast and inner shelf environments to outer shelf environments.

The **Lower Pseudoschwagerina limestone** (LPL) is a sequence of thin clastic sediments (sandstones, siltstones, seldom conglomerates) and well-bedded to massive shallow marine limestones formed in a near-coastal and inner shelf environment. Algal mud mounds and reef mounds formed by phylloid algae (Flügel 1977) are also developed. The sediments have been studied by Flügel (1974), Homann (1969), Buggisch et al. (1976). Micritic limestones are rich in calcareous algae, especially *Anthracoporella, Atractyliopsis, Epimastopora, Mizzia, Anchicodium, Eugonophyllum, Archaeolithophyllum, Komia, Ungdarella* and *Gymnocodium* (Homann 1969), contain also foraminifers, gastropods and echinoderms. Based on fusulinids *(Occidentoschwagerina alpina)*, the LPL is of Early Asselian age (Kahler 1985). Within the LPL four transgressive and regressive cycles, similar to those of the Auernig group, have been recognized (Homann 1969), caused by glacio-eustatic sea-level changes of the Gondwana glaciation.

The **Grenzland formation** is a clastic sequence of quartz-rich conglomerates, crossbedded sandstones, siltstones and red shales, intercalated are thin limestone horizons, from which the following microfacies types have been described by Buttersack and Boeckelmann (1984) and Boeckelmann (1985): oncoid wackestones, oncoid grainstones rich in phylloid algae, wackestones rich in terrigenous material, and bioclastic wackestones rich in sessile foraminifers. The sediments were deposited in a shallow-marine near-coast environment. Based on fusulinids *(Pseudoschwagerina aequalis)* the Grenzland formation is of Middle Asselian age according to Kahler (1985). From one locality Fritz and Boersma (1984) and Boersma and Fritz (1990) described a small plant fossil assemblage (16 taxa) which contains *Callipteris* cf. *conferta* and *Sphenophyllum* cf. *angustifolium*.

The **Upper Pseudoschwagerina limestone** (UPL) is characterized by dark-grey, thin bedded limestones with shale partings and a few thin clastic intercalations (sandstones and siltstones). The sediments are interpreted as open-marine platform carbonates by Flügel (1971, 1977; Flügel et al. 1971) and are characterized by more diverse biota and microfacies types in comparison to the Lower Pseudoschwagerina limestone and Grenzland formation. The limestones are rich in fossils, particularly in calcareous algae (Homann 1972), small foraminifers (Flügel 1971), fusulinids, corals, bryozoans, brachiopods, gastropods, pelecypods and echinoderm fragments. Microfacies types have been described by Flügel (1968) and Buttersack and Boeckelmann (1984). Cycles indicate repeated shifting from very shallow water to off-shore environments in an open-marine shelf lagoon with normal water circulation according to Flügel (1977). The Upper Pseudoschwagerina limestone is of Upper Asselian age based on the occurrence of the fusulinid *Zellia heritschi* (Kahler 1985).

The Trogkofel group

Sediments of the Rattendorf group are overlain by approximately 400 m thick bedded and massive limestones (Trogkofel limestone, Treßdorf limestone, Goggau limestone), from which Flügel (1980a, b,

1981) described 15 microfacies types. The sediments all formed in shallow, restricted and open marine shelf-lagoons with only minor bathymetrical differences. Shelf edge carbonates are represented by different types of mud mounds (*Tubiphytes/Archaeolithoporella* build-ups) composed of sediment-binding organisms (encrusting foraminifers, phylloid algae, *Tubiphytes, Archaeolithoporella* and byrozoans). Calcareous algae (46 species) have been studied by Flügel and Flügel-Kahler (1980) and the fusulinid fauna (about 70 species) by Kahler and Kahler (1980).

Based on fusulinids, the lower part of the Trogkofel limestone is dated as Tastubian (Early Sakmarian) *(Pseudofusulina moelleri),* the Treßdorf limestone as Early Artinskian *(Praeparafusulina lutugini)* and the Goggau limestone as Late Artinskian *(Pseudofusulina vulgaris* – group and *Pamirina darvasica)* (see Flügel 1980a, b; Kahler and Kahler 1980; Kahler 1986; Fig. 12).

In the southern Karawanken Mountains, the Trogkofel group seems to be represented by the "Clastic Trogkofel beds" (Ljubljana beds), first recognized by Ramovs (1963) in Slovenia (see also Ramovs 1968, 1972, 1974, 1983; Kochansky-Devide et al. 1973; Buser 1974; Kullmann and Ramovs 1980). The very thick sequence (500–2000 m) is composed of red and grey, sometimes fossiliferous shales, siltstones, sandstones and quartz conglomerates and of intercalated thin layers of dark limestones containing conodonts. Within this sequence, there seem to be large blocks (olistoliths) of shallow-water limestones (Trogkofel limestone). The sequence, which is not well studied with regard to sedimentological parameters and stratigraphic range (reworked clasts of Trogkofel limestone, described by Kochansky-Devide et al. 1973, indicate an age younger than that of the Trogkofel limestone), seems to be of deep marine origin. Similar sediments are described from a deep borehole in the northern Adriatic Sea (Amands bis 1 well) by Rozza and Sartorio (1990).

5.1..2 Late Permian – Early Triassic (Upper Cycle)

With a hiatus due to severe block faulting ("Saalian movements") the **Tarvis breccia** overlies the Trogkofel formation (Fig. 12). The thickness of this breccia which is interpreted as a scarpfoot fan deposit and proximal debris flows, ranges from a few metres up to about 200 m. Limestone clasts (maximum diameter about 25 cm) derived from the Trogkofel formation are the most frequent clast-type. Subordinate are quartz-, rhyolithic volcanic-, sandstone-, radiolarian chert- and phyllite clasts. The matrix is formed by red and grey shales and marls. Interbedded red siltstones and claystones with calcareous concretions are interpreted as being formed in an evaporitic depositional environment of a coastal sabkha by Kober (1984).

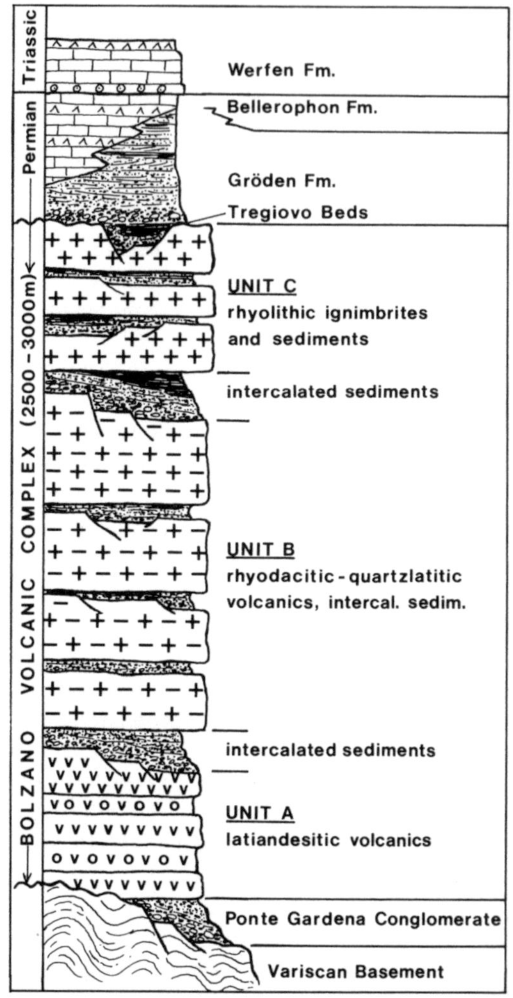

Fig. 14. Idealized stratigraphic column of the Permian/Early Triassic (Scythian) sequence in the western Dolomites. (After Brandner and Mostler 1982)

Freshwater limestone-cements and pedogenic structures indicate a regressive event (subaerial exposure) on top of the Trogkofel formation (Buggisch and Flügel 1980). The Tarvis breccia probably corresponds to the *Misellina* Zone (Cisjanskian) in age (Buggisch and Flügel 1980; Kahler 1980) and is referred to the "Saalian orogenetic phase" by many authors.

The boundary between the Tarvis breccia and the conglomerates of the overlying Gröden formation is marked by the first appearance of rhyolithic volcanic clasts.

The **Gröden formation** with widely varying thicknesses (0–800 m) rests upon rocks of different age:

Tarvis breccia, Auernig group, Hochwipfel formation or even Devonian reef limestones in the Tarvisio area, indicating strong block faulting during this period. In the Carnic Alps and southern Karawanken Mountains, the Gröden formation is formed by carbonatic shales and siltstones with intercalated thin dolomite layers and sandstones.

The sediments are of fluvial (alluvial fan, alluvial plain), playa and shallow marine origin (see Buggisch et al. 1976; Buggisch 1978; Ori and Venturini 1981; Farabegoli et al. 1986). In the Julian Alps near Bled marine sedimentation is characterized by the occurrence of spong-algal-reefs.

In the upper part sediments of the Gröden formation interfinger with the **Bellerophon formation,** which has been studied in the Carnic Alps (Reppwand) by Buggisch (1974). The sequence, approximately 200 m thick, is composed of evaporitic sediments at the base, followed by bituminous dolomites, well-bedded intraclast grainstones with algae, foraminifers, echinoderm fragments, ostracods (open-marine shelf sediments) and bioclastic mudstones formed in a restricted shelf environment (Buggisch 1974). Low-energy biomicritic limestones containing a highly diverse fauna and flora of the uppermost Bellerophon formation are overlain by shallow marine, sub-, inter- and supratidal sediments of the Werfen formation (Scythian), showing similar facies associations and transgressive-regressive trends compared to the Dolomites, although the characteristic basal Tesero horizon is not developed (Buggisch and Noe 1986; Carulli et al. 1986; Boeckelmann 1988, 1990; Schönlaub 1990).

5.2 Dolomites

In the Dolomites (Alto Adige-Trentino) late to post-Variscan sedimentation started during lowermost Permian and is controlled by strong block-faulting, formation of horst-graben topography and volcanic activity. The sequence shows two evolutionary cycles which are separated by a major unconformity: (a) a lower cycle of Early Permian age (Ponte Gardena/Waidbruck conglomerate and Bolzano Volcanic Complex) and (b) an upper cycle of (?) Middle to Late Permian age (Gröden formation and Bellerophon formation).

5.2.1 Early Permian (Lower Cycle)

The low-grade metamorphic Variscan basement (Brixen quartzphyllite) is overlain locally by coarse-grained red beds of alluvial fan environment **(Ponte Gardena/Waidbruck conglomerate)** deposited in fault-bounded basins under semiarid climatic conditions (Fig. 14). The sediments (conglomerates and sandstones) with a maximum thickness of 200 m are composed of fragments derived from the underlying basement and of volcanic rocks of the Bolzano volcanic complex (Dal Cin 1966, 1972; Wopfner 1984; Krainer 1989b). Varying thickness of the sequence results from synsedimentary block-faulting. At many places, the Variscan basement is directly overlain by the Bolzano volcanic complex, or Late to post-Variscan sedimentation may even have started with sediments of the Gröden formation.

The **Bolzano Volcanic Complex** covers an area of approximately 4000 km^2, the central part is up to about 3000 m thick (Fig. 14). It is a sequence of latiandesitic to rhyolithic volcanic rocks (lavas, ignimbrites, tuffs) with intercalated fluvial and lacustrine sediments. The sequence is divided into three units according to Brandner and Mostler (1982): unit A at the base, composed of latiandesites and dacites (300–800 m); unit B composed of quartzlatites and rhyodacites (1000 m), and unit C at the top, composed of rhyolithes (mostly ignimbrites, 1000 m) (see also D' Amico and Segatta 1979; Bargossi et al. 1981; D' Amico 1986; Di Battistini et al. 1988, 1989). Within the Bolzano Volcanic Complex, fluvial and lacustrine sediments with thicknesses up to about 200 m are developed at several horizons. These sediments, which mark interruptions of volcanic activity, were deposited in small basins formed by block faulting according to volcano-tectonics. The clastic sediments (conglomerates, sandstones and siltstones) are deposits of proximal to distal alluvial fan environments and are composed of detritus derived from volcanic rocks of the Bolzano volcanic complex. At the top of these sequences, lacustrine sediments composed of black siltstones and shales, thin carbonate layers (algal layers) and thin silica layers occur sparsely. Within such lacustrine sediments, from the upper part of the Bolzano Volcanic Complex, NW of Bolzano, between units B and C, a rich palynomorph assemblage was found. The palynomorphs are very well preserved within thin silica layers, which were precipitated from the alkaline lake water, and are of stratigraphic significance (upper Early Permian to lower Middle Permian, Artinsk-Kungur; Hartkopf-Fröder and Krainer 1990). Radiometric age determinations on biotite from rhyodacitic and rhyolithic ignimbrites showed ages around 270 Ma (D' Amico et al. 1980; D' Amico 1986). Age-equivalent sediments of the Carnic Alps and southern Karawanken Mountains are lacking volcanic material.

In the Sexten Dolomites, the Sexten (Sesto) conglomerate, an equivalent of the Ponte Gardena conglomerate at the base of the Bolzano Volcanic Complex, is overlain by the Tarvis breccia (see Buggisch and Flügel 1980; Flügel and Kraus 1986).

5.2.2 *Late Permian – Early Triassic (Upper Cycle)*

With an unconformity, (?) Middle-Late Permian sediments of the upper cycle overlie the Bolzano Volcanic Complex or the Variscan basement directly. The sediments are similar to those of the Carnia and southern Karawanken Mountains and can be divided into two lithostratigraphic units: Gröden formation (Val Gardena sandstone) and Bellerophon formation (Figs. 11, 14).

The **Gröden formation** shows rapid variations in thickness (0–500 m) due to strongsynsedimentary tectonics. The clastic sequence is subdivided into three terrigenous bodies separated from each other by carbonate and evaporitic sediments of the Bellerophon formation (interfingering of Gröden formation and Bellerophon formation) due to tectonically controlled transgressive and regressive events of second order (Farabegoli and Viel 1982 b).

Several facies assemblages were recognized, ranging from alluvial fan environment, braided stream and meandering stream deposits to sandy lobes of terminal fan settings and coastal sabkha sediments. Pedogenic calcretes, local presence of pedogenic gypsum, thin coal beds with rich floras and tetrapod footprints point to semiarid climatic conditions. Sandstones are classified as feldspathic litharenites and lithic arkoses (Dal Cin 1965, 1972; Bosellini and Dal Cin 1968; Buggisch et al. 1976; Buggisch 1978; Ori and Venturini 1981; Wopfner 1984; Conti et al. 1986; Ori 1986; Massari et al. 1988; see esp. Italian Research Group 1986).

The fossil flora is characterized by Pteridophytes (*Lepidodendron, Pecopteris* etc.) and Gymnosperms (*Voltzia, Walchia, Ortiseia*) (Leonardi 1968).

Tetrapod footprints have been recognized from different horizons within the Gröden formation. The famous and well studied Bletterbach/Butterloch section in particular contains one of the richest tetrapod footprint assemblages of the European Permian (see for example Leonardi et al. 1975; Conti et al. 1977, 1980, 1986; Ceoloni et al. 1986, 1988). *Pachypes dolomiticus, Rhynchosauroides pallinii, Chirotherium* sp., *Dicynodontipus geinitzi, Laoporus* sp., *Tridactylichnium leonardii, Synaptichnium* sp., *Chelichnus* cf. *duncani* and *Parodoxichnium* are of stratigraphic importance.

From the middle part of the Bletterbach/Butterloch section, a marine fauna dominated by nautiloids (*Lopingoceras, Pleuronautilus, Mojsvaroceras, Tainoceras* etc.) is known (Mutschlechner 1933; Conti et al. 1977).

Bivalves and eurytopic foraminifers (*Earlandia, Glomospira* etc.) occur within dolomitic layers and in the uppermost part of the Gröden formation.

Based on palynomorphs, the basal sediments of this section are of ?Latest Capitanian – Earliest Abadehian age (Conti et al. 1986; Massari et al. 1988; see also Klaus 1963).

The **Bellerophon formation** up to about 500 m thick (in the Cadore area), interfingers at the base with clastic sediments of the Gröden formation and can be subdivided into large-scale second-order cycles with thicknesses of 50–200 m. It is assumed that during the Late Permian (Bellerophon formation) the climate was more arid than before (Assereto et al. 1973).

In the western Dolomites the lower cycle starts with coastal sabkha sediments (dolomites, marls, mudstones, nodular chicken-wire and laminated gypsum), grading upward into open-lagoon fossiliferous dolomitic limestones (Fiamazza facies). This cycle is overlain by fine-grained red bed sediments, which mark a regressive event and grade upward into dolomites of a restricted lagoon environment, and finally into shallow shelf, dark coloured fossiliferous limestones (Badiota facies) (Bosellini and Hardie 1973).

In Cadore and SW Carnia, probably separated from the western area by a structural high, the Bellerophon formation is thicker, due to higher subsidence rates. The lower part comprises a thick sequence of laminated gypsum, overlain by transgressive limestone-dolomite cycles. A regression is marked by vuggy dolomites and the uppermost part is characterized by shallow-marine fossiliferous limestones with a high diversity fauna (Badiota facies).

The distribution of fossils within the Bellerophon formation is highly controlled by the depositional environment. Restricted lagoon carbonate and evaporitic sediments are rare in fossils (*Bakevellidae* and a few eurytopic microfossils like *Aeolisaccus, Glomospira* and *Cyclogyra*).

Open-lagoon sediments (Badiota facies) contain a rich macro- and microfauna: *Bellerophon, Aviculopecten, Bakevillia, Tirolonautilus, Peripetoceras,* foraminifers (*Geinitzia, Globivalvulina, Hemigordius, Staffelia* etc.), algae (*Mizzia, Atractyliopsis, Gymnocodium*) and echinoderm remains (Neri and Posenato 1985; Pasini 1985; Italian Research Group 1986).

The uppermost part of the Bellerophon formation in the Dolomites, Cadore and Carnia is composed of

0.5–2 m thick dark marls and limestones with a characteristic brachiopod fauna (*Comeliciana, Janiceps, Araxathyris, Ombonia, Orthothetina*) and fusulinids (*Palaeofusulina* and others) pointing to Dorashamian age according to Neri and Posenato (1985).

Ombonia and some Permian microfossils are also found within the lowermost part of the Tesero horizon suggesting that the Permian/Triassic boundary lies within the lower part of this basal oolitic horizon of the Werfen formation (Pasini 1985; Broglio-Loriga et al. 1986 a, b; Italian Research Group 1986).

From west to east, the Scythian **Werfen formation** (Servino formation of the Lombardian region) overlies the Verrucano Lombardo, Val Gardena sandstone, sabkha facies and shallow shelf limestone facies of the Bellerophon formation. The Werfen formation is represented by a sequence of carbonatic and clastic sediments reaching 150 m in Lombardy and about 700 m in Cadore. The sequence is subdivided into nine lithostratigraphic units (horizons and members). The sediments were deposited in a shallow shelf environment ranging from subtidal (below fair-weather wave base, frequently affected by storm waves), tidal flat and supratidal mudflat environment (Broglio-Loriga et al. 1983, 1986 a). The basal oolitic unit (Tesero horizon) is only developed west of the Cadore-Comelico area and is lacking in the east (Carnia, southern Karawanken Mountains).

The Werfen formation is subdivided into transgressive-regressive cycles based on periodically intercalated supratidal sediments and clastic influx, which are probably controlled by eustatic sea-level changes and tectonics (Assereto et al. 1973; Farabegoli and Viel 1982 a; Broglio-Loriga et al. 1983; Brandner 1984). Similar transgressive-regressive events have been recognized in the Scythian sequence of the Upper Austro-Alpine units (Brandner et al. 1984; Krainer 1985, 1987 a).

The reader is also referred to the results presented by the Italian Research Group (1986) and to Neri and Pasini (1985), Perri and Andraghetti (1987) Brandner (1988), Broglio-Loriga et al. (1986 b), Buggisch and Noe (1986), Oddone and Vannucci (1986) and Vissher and Brugman (1986) for a detailed discussion of the Permo-Triassic -boundary and Werfen formation.

5.3 *Western Southern Alps (Lombardy)*

In the western part of the Southern Alps, late- to post-Variscan (? Late Carboniferous-Permian) sediments with intercalated volcanic rocks, which were deposited in tectonically formed graben-like basins (Collio basin, Tione basin, Boario basin, Orobic basin) overlie the metamorphic Variscan basement (Cassinis 1985, 1986). The sequence is again subdivided into two cycles (Fig. 11).

The lower cycle (?Late Carnoniferous-Lower Permian) is represented by a thick sequence of rhyolithic volcanic rocks, fluvial and lacustrine sediments **(Collio formation),** divided into members A–G according to Cassinis (1966) and Ori et al. (1986). Ceoloni et al. (1986, 1987, 1988) described Permian tetrapod footprints from members C and F of the Collio formation in Val Trompia. The tetrapod footprint assemblage is similar to that of the Oberhof and Rotteröd formation of the German Rotliegendes. Tetrapod footprints and plant fossils from the upper part of member C have been dated as Upper Autunian by Haubold and Katzung (1975) and in Middle-Upper Saxonian by Remy and Remy (1978).

The second cycle (?Middle-Late Permian) is characterized by clastic red beds of fluvial origin (Verrucano Lombardo), time-equivalent to the Gröden formation (Val Gardena sandstone) and Bellerophon formation in the Dolomites (Cassinis 1966; Cassinis et al. 1986; Ori et al. 1986). The Verrucano Lombardo comprises a thick fining-upward sequence with braided stream deposits at the base, grading into meandering stream deposits and finally into sheet-flow deposits according to Cassinis et al. (1986). The Verrucano Lombardo is overlain by the **Servino formation,** which corresponds to the Werfen formation of the Dolomites (see Cassinis et al. 1986).

6 Discussion and Conclusion

It is still a matter of discussion, at what time "late to post-Variscan Molasse sedimentation" started and where to draw the boundary of the sedimentary sequence of the following Alpine cycle.

In the Southern Alps (Carnic Alps, Karawanken Mountains) the first late to post-Variscan Molasse sediments formed during the Late Moscovian (Bombaso formation). With a classical unconformity the Molasse sediments overlie the folded Variscan basement.

In the Eastern Alps the situation is still unclear. The Carboniferous sequence of Nötsch, ranging from Visean to Namurian (?Early Westphalian), is interpreted as Molasse by Schönlaub (1985), the Carboniferous sequence of the Veitsch nappe is also interpreted as Molasse by Ebner (1990) and Neubauer (pers. comm.). Both, the Carboniferous sequence of Nötsch and Veitsch, show a tectonic lower and upper boundary, the basement is unknown.

On the NW-margin of the Gurktal nappe system and Steinach Nappe the late to post-Variscan Molasse sediments unconformably overlie the folded and slightly metamorphosed Variscan basement. As in the Carnic Alps, the oldest Molasse sediments are dated as Cantabrian by plant fossils (*Odontopteris cantabrica* Zone).

The Carboniferous sequence of Nötsch and Veitsch most likely formed in the same basin (or basin complex) as the Hochwipfel formation ("Variscan Flysch") of the Southern Alps (Carnic Alps, Karawanken Mountains) (see, for example, Neubauer 1988; Fig. 8). The deepwater sediments of the Hochwipfel formation were strongly affected by orogenic movements during the Late Westphalian ("Carnic Phase"), resulting in severe thrusting and folding and weak metamorphism in the western part of the Carnic Alps.

The Carboniferous of Nötsch and Veitsch, which were deposited in a synorogenic basin during the same time span as the Hochwipfel formation, have probably also been affected by tectonic movements, although this cannot be seen in outcrop today, and consequently are termed "synorogenic sediments". In the Eastern Alps, therefore, "late to post-Variscan Molasse" sedimentation started as early as during the Cantabrian, when NE-SW and E-W to ESE-WNW oriented intramontane basins were formed. The same directions have been recognized in many of the late to post-Variscan intramontane basins of Middle Europe (e.g. Falke 1974; Lorenz and Nicholls 1976; Schwab 1977; Lützner 1981, 1987). The intramontane basins formed as a result of crustal thinning and extension due to widespread melting processes in the lower crust and granitic intrusions accompanied by volcanic processes in the upper crust.

According to Arthaud and Matte (1977), the intramontane basins were formed as a result of transtensional and transpressional tectonic processes along megashear zones, due to the eastward drift of the Eurasian plate and the westward drift of the African plate. Tectonic processes and climate, including climatic changes, were the most significant controlling factors of late to post-Variscan sedimentation processes.

Based on a major tectonic event within the Permian ("Saalian movements"), the late to post-Variscan sequence of the Southern and Eastern Alps is divided into two evolutionary cycles. The sediments of the lower cycle (Late Carboniferous/Early Permian) were deposited in discrete intramontane basins, which were filled with sediments of different environments and volcanic rocks.

In the eastern part of the Southern Alps (Karawanken Mountains, Carnic Alps) clastic and carbonate shallow marine sediments (Bombaso formation, Auernig, Rattendorf, and Trogkofel groups) were deposited. Within the upper part of the Auernig group and the Rattendorf group clastic – carbonate cycles are well developed, caused by eustatic sea-level fluctuations due to the Gondwana glaciation.

In the western part of the Southern Alps (Dolomites, Lombardy), the lower cycle is represented by Early Permian continental red beds (e.g. Ponte Gardena/Waidbruck conglomerate) and up to about 3000 m thick latiandesitic to rhyolithic volcanic rocks with intercalated alluvial fan and lacustrine sediments ("Bolzano Volcanic Complex").

In the Eastern Alps, Late Carboniferous/Early Permian sediments of the lower cycle are all of continental origin. Late Carboniferous sediments (e.g. Stangnock formation), containing abundant plant fossils, were formed under humid climatic conditions, whereas during the Early Permian, red beds (e.g. Werchzirm formation) with only a few plant fossils at the base, were deposited, pointing to increasingly arid conditions. The climatic shift from humid to more arid conditions was near the Carboniferous/Permian boundary (within the *Callipteris conferta* Zone). Within the Early Permian, red bed sequences of the Eastern Alps rhyolithic volcanic rocks also occur.

Correlation of the continental Late Carboniferous/Early Permian (Cantabrian-Autunian) sequence of the Eastern Alps, with the shallow marine sequence of the Carnic Alps, is possible based on megaflora zones. Both sequences are characterized by the same plant fossil assemblages, ranging from the *Odontopteris cantabrica* Zone (Cantabrian) at the base to the *Callipteris conferta* Zone (Autunian).

A major tectonic event ("Saalian movements"), characterized by strong block faulting, separates the lower from the upper cycle. The boundary is marked by a stratigraphic gap and a significant change in sedimentation, although there is no detectable angular unconformity. The sediments of the upper cycle conformably overlie the sequence of the lower cycle.

Strong synsedimentary block faulting is expressed by the huge amount of reworked material from the lower cycle within the sediments of the upper cycle. In parts of the Carnic Alps, for example, the sediments of the lower cycle have been completely reworked due to strong uplifting of discrete crustal blocks, and the basal sediments of the upper cycle (Tarvis breccia) rest upon limestones of Devonian age.

In the Eastern Alps, Late Permian sediments of the upper cycle (Gröden formation) contain high percentages of reworked volcanic rocks of the lower cycle, indicating again strong block faulting and rejuvenation of the relief. In some parts of the Eastern Alps (Middle Austro-Alpine unit), late- to post-Variscan sedimentation did not start until Late Permian/Early Triassic.

During the Late Permian/Early Triassic, sedimentation was influenced by the transgression of the NW-Tethys Sea from SE to NW and interfingering of shallow marine sediments with continental deposits. In the western part of the Southern Alps (Dolomites), the first marine sediments are of Late Permian age, in the Eastern Alps in most cases the first marine sediments occur within the Early Triassic (Alpine Buntsandstein and Werfen formation).

A significant change in sedimentation is recognized at the Permian/Triassic boundary, caused by a climatic shift. Sedimentary cycles within the Scythian are related to eustatic sea-level fluctuations, probably caused by different spreading rates of mid-ocean ridges (Brandner 1984; Krainer 1987a). According to Wopfner (1984), the Permian Molasse sediments of the Southern Alps between Val Camonica and the Karawanken Mountains were formed in intramontane Molasse basins of a complex rift system, which he relates to rifting processes initiating the Alpidic sedimentation phase.

Sediments of the upper cycle (Gröden formation) have been termed "Verrucano" by Cassinis et al. (1979). According to their new definition, "Verrucano" stands for a "tectofacies" of the South European region (including the Alps), and Cassinis et al. opposed the inclusion of the "Verrucano" in the "late- to post-Variscan Molasse sediments". According to these authors, the Verrucano sediments were deposited during an early stage of continental rift processes introducing the Alpidic cycle. Following this conception, the hiatus between the lower and upper cycle separates the late- to post-Variscan sediments from the sedimentary sequence of the following Alpidic cycle.

As it is still unclear if the major tectonic event within the Permian, which originally was defined as a Variscan orogenic phase ("Saalian phase") by Stille (1924), or the rifting processes during the Middle Triassic mark the beginning of the Alpidic cycle, the sedimentary sequence of the upper cycle (Late Permian/Early Triassic) has been included in the discussion of the "late to post-Variscan sediments" of the Southern and Eastern Alps.

Acknowledgements
The author wishes to thank F. Neubauer (Graz), K. Stattegger (Kiel) and an unknown reviewer for reading through an early draft of the manuscript and for their critical comments and helpful discussions.

References

Amerom HWJ van, Boersma M (1974) Vorläufige Untersuchungsergebnisse an älteren und neu aufgesammelten jungpaläozoischen Pflanzenfunden der Ostalpen (Österreich). Carinthia 164 (84): 9–15

Amerom HWJ van, Angerer H, Mostler H (1982) Über eine Autuno-Stephanische Flora aus den Kristbergschichten im Montafon, Vorarlberg (Österreich). Jahrb Geol Bundesanst 124: 283–323

Angel F (1932) Diabase und deren Abkömmlinge in den österreichischen Ostalpen. Mitt Naturwiss Ver Steiermark 69: 5–24

Angerer H, Haditsch JG, Leichtfried W, Mostler H (1976) Disseminierte Kupfererze im Perm des Montafon (Vorarlberg). Geol Paläont Mitt Innsbr 6: 1–57

Arthaud F, Matte P (1977) Late Paleozoic strike-slip faulting in southern Europe and northern Africa: result of a right lateral shear zone between the Appalachians and the Urals. Geol Soc Am Bull 88: 1305–1320

Assereto R, Bosellini A, Fantini-Sestini N, Sweet WC (1973) The Permian-Triassic boundary in the Southern Alps (Italy). In: Logan A, Hills LV (eds) The Permian and Triassic systems and their mutual boundary. Can Soc Petrol Geol Mem 2: 176–199

Bargossi GM, D'Amico CD, Di Battistini G (1981) Le K-Andesiti inferoiri Atesine tra Nogare e Serso (Trento). Rend Soc Ital Mineral Petrol 37: 719–738

Behr H-J, Engel W, Franke W, Giese P, Weber K (1984) The Variscan Belt in central Europe: main structures, geodynamic implications, open questions. Tectonophysics 109: 15–40

Boeckelmann K (1985) Mikrofazies der Auernig-Schichten und Grenzland-Bänke westlich des Rudnig-Sattels (Karbon-Perm; Karnische Alpen). Facies 13: 155–174

Boeckelmann K (1988) Die Werfener Schichten in den Karnischen Alpen und westlichen Karawanken (Österreich und Italien). PhD Thesis, RWTH Aachen, 213 pp

Boeckelmann K (1990) The uppermost Permian and the Permian-Triassic boundary in the Gartnerkofel 1-Core. In: Venturini C, Krainer K (eds) Field workshop on Carboniferous to Permian sequence of the Pramollo-Nassfeld basin (Carnic Alps). Proc (preprint) Arti Grafiche Friulani, Udine, pp 74–79

Boersma M, Fritz A (1990) Die Paläofloren Kärntens: Ober-Karbon/Unter-Perm. In: Fritz A, Boersma M, Krainer K (eds) Steinkohlenzeitliche Pflanzenfossilien aus Kärnten. Carinthia II, Sonderh 49: 133–172

Bosellini A, Dal Cin R (1968) Sedimentologia delle Arenarie di Val Gardena. Sezione di S. Martino in Badia (Bolzano). Boll Soc Geol Ital 87: 401–423

Bosellini A, Hardie LA (1973) Depositional theme of a marginal marine evaporite. Sedimentology 20: 5–27

Brandner R (1984) Meeresspiegelschwankungen und Tektonik in der Trias der NW-Tethys. Jahrb Geol Bundesanst 126: 435–475

Brandner R (1988) The Permian-Triassic boundary in the Dolomites (Southern Alps, Italy), San Antonio section. Ber Geol Bundesanst 15 Rare events in geology, IGCP 199: 49–56

Brandner R, Mostler H (1982) Der geologische Aufbau des Schlerngebietes und seiner weiteren Umgebung. In: Mostler H (ed) Exkursionsführer zur Jahrestagung der Österreichischen Geologischen Gesellschaft. Seis am Schlern, Österreichische Geol Ges. Innsbruck, pp 1–42

Brandner R, Donofrio DA, Krainer K, Mostler H, Resch W, Stingl V (1984) Correlation of transgressional and regressional events in the Lower Triassic of the Northern and Southern Alps (Buntsandstein-, Servino-, Werfen Formation). 5th Eur Regional Meet Sedimentol (IAS), Marseille, April 1984, Poster Presentation. 2pp (Abstr)

Brandner R, Donofrio DA, Krainer K, Mostler H, Nazarow W, Resch W, Stingl V, Weissert H (1986) Events at the Permian-Triassic boundary in the Southern and Northern Alps. In: Field Conference on Permian and Permian-Triassic boundary in the South-Alpine segment of the western Tethys, and additional reports. Brescia, pp 15–16 (Abstr)

Briegleb D (1971) Geologie der Magnesitlagerstätte am Sattlerkogel in der Veitsch (Steiermark). Berg Hüttenmänn Monatsh 116: 359–375

Broglio Loriga C, Masetti D, Neri C (1983) La Formazione di Werfen (Scitico) delle Dolomiti occidentali: sedimentologia e biostratigrafia. Riv Ital Paleontol Stratigr 88: 501–598

Broglio Loriga C, Neri C, Posenato C (1986a) The Lower Triassic of the Dolomites and Cadore. In: Field Conference on Permian and Permian-Triassic boundary in the South-Alpine segment of the Western Tethys, Field Guidebook. Società Geologica Italiana, Commerciale Pavese, Pavia, pp. 29–34.

Broglio Loriga C, Neri C, Pasini M, Posenato R (1986b) Marine fossil assemblages from Upper Permian to lowermost Triassic in the western Dolomites (Italy). Mem Soc Geol Ital 34: 5–44

Buggisch W (1974) Die Bellerophonschichten der Reppwand (Gartnerkofel) (Oberperm, Karnische Alpen). Untersuchungen zur Fazies und Geochemie. Carinthia II 164 (84): 17–26

Buggisch W (1978) Die Grödener Schichten (Perm, Südalpen). Sedimentologische und geochemische Untersuchungen zur Unterscheidung mariner und kontinentaler Sedimente. Geol Rundsch 67: 149–180

Buggisch W, Flügel E (1980) Die Trogkofel-Schichten der Karnischen Alpen. Verbreitung, geologische Situation und Geländebefund. In: Flügel E (ed) Die Trogkofel-Stufe im Unterperm der Karnischen Alpen. Carinthia II, Sonderh 36: 13–50

Buggisch W, Noe S (1986) Upper Permian and Permian-Triassic boundary of the Carnia (Bellerophon Formation, Tesero horizon, northern Italy). Mem Soc Geol Ital 34: 91–106

Buggisch W, Flügel E, Leitz F, Tietz G–F (1976) Die fazielle und paläogeographische Entwicklung im Perm der Karnischen Alpen und in den Randgebieten. Geol Rundsch 65: 649–690

Buser S (1974) Neue Feststellungen im Perm der westlichen Karawanken. Carinthia II 164 (84): 27–37

Buttersack E, Boeckelmann K (1984) Palaeoenvironmental evolution during the Upper Carboniferous and the Permian in the Schulter – Trogkofel area (Carnic Alps, northern Italy). Jahrb Geol Bundesanst 126: 349–358

Carulli GB, Radrizzani CP, Ponton M (1986) The Permian-Triassic boundary in the Paularo area (Carnia). Mem Soc Geol Ital 34: 107–120

Cassinis G (1966) La Formazione di Collio nell'area tipo dell'alta Val Trompia (Permiano inferiore bresciano). Riv Ital Paleontol Stratigr 72: 507–588

Cassinis G (1985) Il Permiano nel Gruppo dell'Adamello alla luce delle ricerche sui coevi terreni delle aree contermini. Mem Soc Geol Ital 26 (1983): 119–132

Cassinis G (1986) The western continental area of the Permian Southalpine Tethys. In: Field Conference on Permian and Permian-Triassic boundary in the South-Alpine segment of the western Tethys, Field Guidebook. Società Geologica Italiana, Commerciale Pavese, Pavia, pp 35–37

Cassinis G, Giobbi E, Neri C, Ori GG (1986) The continental Permian south of the Adamello Massif. In: Field Conference on Permian and Permian-Triassic boundary in the South-Alpine segment of the western Tethys, Field Guidebook. Società Geologica Italiana, Commerciale Pavese, Pavia, pp 135–148

Cassinis G, Elter G, Rau A, Tongiorgi M (1979) Verrucano: a tectofacies of the Alpine – Mediterranean southern Europe. Mem Soc Geol Ital 20: 135–149

Castellarin A, Vai GB (1981) Importance of Hercynian tectonic within the framework of the Southern Alps. J Struct Geol 3: 477–486

Ceoloni P, Conti MA, Mariotti N, Nicosia U (1986) New Permian tetrapod footprints from Southern Alps. Mem Soc Geol Ital 34: 45–65

Ceoloni P, Conti MA, Mariotti N, Mietto P, Nicosia U (1987) Tetrapod footprints from Collio Formation. Mem Soc Geol Ital 39: 213–233

Ceoloni P, Conti MA, Mariotti N, Mietto P, Nicosia U (1988) Tetrapod footprint Faunas from southern and central Europe. Z Geol Wiss 16: 895–906

Claassen Th, Gosen W von, Sylvester H, Thiedig F (1987) Die Permotrias und ihr Grundgebirge zwischen Faaker See und Turiawald südöstlich von Villach (Kärnten/Österreich). Jahrb Geol Bundesanst 130: 391–413

Conti MA, Leonardi G, Mariotti N, Nicosia U (1977) Tetrapod footprints of the "Val Gardena Sandstone" (north Italy). Their paleontological, stratigraphic and paleoenvironmental meaning. Paleontol Ital 70: 1–91

Conti MA, Leonardi G, Mariotti N, Nicosia U (1980) A review of a group of Permian tetrapod footprints from northern Italy. Mem Sci Geol 34: 143–152

Conti MA, Fontana D, Mariotti N, Massari F, Neri C, Nicosia U, Pasini M, Pittau D (1986) The Bletterbach-Butterloch section (Val Gardena Sandstone and Bellerophon Formation). In: Field Conference on Permian and Permian-Triassic boundary in the South-Alpine segment of the western Tethys, Excursion Guidebook. Società Geologica Italiana, Commerciale Pavese, Pavia, pp 91–110.

Dal Cin R (1965) Ricerche morfoscopiche sulle Arenarie di Val Gardena della Regione Dolomitica. Ann Univ Ferrara Sez 9: 163–172

Dal Cin R (1966) Caratteristiche morfometriche e granulometriche di alcuni affioramenti di Conglomerato basale (Verrucano alpino) delle Dolomiti. Atti Symp Verrucano, Pisa. Soc Tosc Sc Nat, pp 275–290

Dal Cin R (1972) I conglomerati tardo-paleozoici post-ercinici delle Dolomiti. Verh Geol Bundesanst 1972: 47–74

D'Amico C (1986) Volcanic sequence in Trentino-Alto Adige. In: Field Conference on Permian and Permian-Triassic boundary in the South-Alpine segment of the western Tethys, Excursion Guidebook. Società Geologica Italiana, Commerciale Pavese, Pavia, pp 16–22

D'Amico C, Segatta G (1979) Le lava afiriche riolitiche-trachiriolitiche del Lagorai sud-occidentale (Trentino) nel sistema vulcanico atesino del Permiano, gruppo inferiore. Mineral Petrogr Acta 23: 239–255

D'Amico C, Del Moro A, Freddo A, Pardini G (1980) Studio radiometrico delle ignimbriti riolitiche atesine, gruppo superiore. Rend Soc Ital Mineral Petrol 36: 703–716

Di Battistini G, Bargossi GM, Spotti G, Toscani L (1988) Andesites of the Late Hercynian volcanic sequence in Trento-Alto Adige (northern Italy). Rend Soc Ital Mineral Petrol 43: 1087–1100

Di Battistini G, Gallo F, Giammetti F, Vernia L (1989) Permian andesites from Val d'Isarco and Val di Funes volcanic sequence (Bolzano, northern Italy). Mineral Petrogr Acta 32: 123–137

Ebner F (1990) Carboniferous sediments in Austria, Hungary and Slovakia: correlation, paleogeography and Implications for Variscan Orogeny. IGCP Project No. 276: Paleozoic in the Tethys, Meeting 1990, Sept. 21–30, Algeria pp 10–13 (Abstr)

Falke H (1974) Die Unterschiede in den Sedimentationsvorgängen zwischen dem Autunien und Saxonien von Mittel- und Westeuropa. Geol Rundsch 63: 819–849

Farabegoli E, Viel G (1982a) Litostratigrafia della Fm. di Werfen (Trias inferiore) delle Dolomiti occidentali. Ind Miner 6: 3–14

Farabegoli E, Viel G (1982b) Il Permo-Scitico delle Alpi Meridionali. Relazione finale Contr CREST 092.79.7 MPPI. Samim Pertulosa, 48 pp

Farabegoli, E, Flügel E, Levanti D, Noe S, Pasini M, Perri MC, Venturini C (1986) The Tarvisio Breccia and Val Gardena Sandstone: Lithostratigraphic and Paleoenvironments. The Permian/Triassic Boundary – Carnic and Comelico. In: Field Conference on Permian and Permian-Triassic Boundary in the South-Alpine Segment of the Western Tethys, Excursion Guidebook. Società Geologica Italiana, Commerciale Pavese, Pavia, pp 55–72

Felser KO (1936) Die Badstub-Breccie der Karbonscholle von Nötsch im Gailtal (Kärnten). Zentralbl Mineral Geol Paläontol B 8: 305–308

Felser KO (1977) Die stratigraphische Stellung der Magnesitvorkommen in der östlichen Grauwackenzone (Steiermark, Österreich). Berg Hüttenmänn Monatsh 122: 17–23

Fenninger A, Schönlaub HP, Holzer HL, Flajs G (1976) Zu den Basisbildungen der Auernigschichten in der Karnischen Alpen (Österreich). Verh Geol Bundesanst 1976: 243–255

Flügel E (1968) Bericht über fazielle und stratigraphische Untersuchungen im Perm der Karnischen Alpen. Carinthia II 158 (78): 38–65

Flügel E (1971) Palökologische Interpretation des Zottachkopf-Profiles mit Hilfe von Kleinforaminiferen (Oberer Pseudoschwagerinen-Kalk, unteres Perm: Karnische Alpen). Carinthia II Sonderh 28: 61–96

Flügel E (1974) Fazies-Interpretation der unterpermischen Sedimente in den Karnischen Alpen. Carinthia II 164 (84): 43–62

Flügel E (1977) Environmental models for Upper Paleozoic benthic calcareous algal communities. In: Flügel E (ed) Fossil algae. Springer, Berlin Heidelberg New York, pp 314–343

Flügel E (1980a) Die Trogkofel-Stufe im Unterperm der Karnischen Alpen. Carinthia II Sonderh 36: 1–260

Flügel E (1980b) Die Mikrofazies der Kalke in den Trogkofel-Schichten der Karnischen Alpen. In: Flügel E (ed) Die Trogkofel-Stufe im Unterperm der Karnischen Alpen. Carinthia II Sonderh 36: 51–99

Flügel E (1981) Lower Permian Tubiphytes/Archaeolithoporella buildups in the Southern Alps (Austria and Italy). In: Toomey DF (ed) European fossil reef models. SEPM Spec Publ 30: 143–160

Flügel E, Flügel-Kahler E (1980) Algen aus den Kalken der Trogkofel-Schichten der Karnischen Alpen. In: Flügel E (ed) Die Trogkofel-Stufe im Unterperm der Karnischen Alpen Carinthia II Sonderh 36: 113–182

Flügel E, Homann W, Tietz G-F (1971) Litho- und Biofazies eines Detailprofils in den Oberen Pseudoschwagerinen-Schichten (Unter-Perm) der Karnischen Alpen. Verh Geol Bundesanst 1971: 10–42

Flügel E, Krainer K (1991) Allogenic and autogenic controls of reef mound formation: Late Carboniferous auloporid coral buildups from the Carnic Alps, Italy. Neues Jahrb Geol Paläontol Monatsh 185: 39–62

Flügel E. Kraus S (1986) The Lower Permian Sexten breccia (Sexten Dolomites) and the Tarvis breccia (Carnic Alps): microfacies, depositional environment and paleotectonic implications. Mem Soc Geol Ital 34: 67–90

Flügel HW, Faupl P (eds) (1987) Geodynamics of the Eastern Alps. F. Deuticke, Vienna, 418 pp

Frank W (1972) Permoskyth im Penninikum der Hohen Tauern. Verh Geol Bundesanst 1972: 151–153

Frank W (1987) Evolution of the Austroalpine elements in the Cretaceous. In: Flügel HW, Faupl P (eds) Geodynamics of the Eastern Alps. Deuticke, Vienna, pp 379–406

Frasl G, Frank W (1966) Einführung in die Geologie und Petrographie des Penninikums im Tauernfenster. Aufschluß Sonderh 15: 30–58

Frech F (1894) Die Karnischen Alpen. Niemeyer, Halle, 514 pp

Frimmel H (1986a) Petrographie, Gefügemerkmale und geochronologische Daten von Kristallingeröllen aus dem Oberkarbon der Gurktaler Decke im Vergleich zum benachbarten Altkristallin. Mitt Ges Geol Bergbaustud Österr 32: 39–65

Frimmel H (1986b) Isotopengeologische Hinweise für die paläogeographische Nachbarschaft von Gurktaler Decke (Oberostalpin) und dem Altkristallin östlich der Hohen Tauern (Österreich). Schweiz Mineral Petrogr Mitt 66: 193–208

Frisch W, Neubauer F (1989) Pre-Alpine terranes and tectonic zoning in the Eastern Alps. Geol Soc Am Spec Pap 230: 91–100

Fritz A, Boersma M (1984) Fundberichte über Pflanzenfossilien aus Kärnten 1984, Beitrag 8: Grenzlandbänke (Rudnigsattel, Unterperm), Karnische Alpen. Carinthia II 174 (94): 59–69

Fritz A, Boersma M (1987a) Fundberichte über Pflanzenfossilien aus Kärnten 1987, Beitrag 15: Wunderstätten (Unterperm), St. Pauler Berge. Carinthia II 177 (97): 381–394

Fritz A, Boersma M (1987b) Fundberichte über Pflanzenfossilien aus Kärnten 1987, Beitrag 16: Kötschach, Gailtaler Alpen (Unterperm). Carinthia II 177 (97): 395–407

Fritz A, Boersma M (1988) Fundberichte über Pflanzenfossilien aus Kärnten 1988, Beitrag 17: Ulrichsberg (Unterperm), Kärnten. Carinthia II 178 (98): 437–446

Fritz A, Boersma M, Krainer K (1990) Steinkohlenzeitliche Pflanzenfossilien aus Kärnten. Carinthia II Sonderh 49: 1–189

Frizzo P, Visona D (1981) New data regarding lithostratigraphy and metamorphism of the Steinach Nappe (Brenner Region, Eastern Alps). Studi Trent Sci Nat 58: 3–10

Gauri KL (1965) Uralian stratigraphy, trilobites and brachiopods of the western Carnic Alps (Austria). Jahrb Geol Bundesanst Sonderb 11: 1–94

Haditsch JG, Leichtfried W, Mostler H (1978) Intraskythische, exogen (mechanisch)-sedimentäre Cu-Vererzung im Montafon (Vorarlberg). Geol Paläontol Mitt Innsbr 8: 183–207

Hahn G, Hahn R (1977) Trilobiten aus dem Unter-Karbon der Veitsch (Steiermark, Österreich). Neues Jahrb Geol Paläontol Monatsh 1977 (3): 137–143

Hahn G, Hahn R (1987) Trilobiten aus dem Karbon von Nötsch und aus den Karnischen Alpen Österreichs. Jahrb Geol Bundesanst 129: 567–619

Hahn G, Hahn R, Schneider G (1989) Neue Trilobitenfunde aus der Waidegg Formation (hohes Oberkarbon) der Karnischen Alpen (Österreich). Jahrb Geol Bundesanst 132: 645–664

Hartkopf-Fröder C, Krainer K (1990) Fluviatile und lakustrine Sedimente innerhalb der permischen Bozener Quarzporphyrabfolge (Südtirol/Italien): Ihre biostratigraphische und palökologische Bedeutung. Sediment '90, Bonn (Abstr), Geol Inst Univ Bonn, 3 pp

Haubold H, Katzung G (1975) Die Position der Autun/Saxon – Grenze (Unteres Perm) in Europa und Nordamerika. Schriftenr Geol Wiss 3: 87–138
Heritsch F, Kahler F, Metz K (1933) Die Schichtfolge von Oberkarbon und Unterperm. In: Heritsch F (ed) Stratigraphie von Oberkarbon und Unterperm in den Karnischen Alpen. Mitt Geol Ges Wien 26: 162–189
Homann W (1969) Fazielle Gliederung der Unteren Pseudoschwagerinenkalke (Unter-Perm) der Karnischen Alpen. Neues Jahrb Geol Paläontol Monatsh 1969: 265–280
Homann W (1972) Unter- und tief-mittelpermische Kalkalgen aus den Rattendorfer Schichten, dem Trogkofel-Kalk und dem Treßdorfer Kalk der Karnischen Alpen (Österreich). Senckenb Lethaea 53: 135–313
Italian Research Group – IGCP Proj No 203 (1986) Field Conference on Permian and Permian-Triassic boundary in the South-Alpine segment of the western Tethys. Excursion Field Guidebook. Società Geologica Italiana, Commerciale Pavese, Pavia, 158 pp
Jongmans WJ (1938) Paläobotanische Untersuchungen im österreichischen Karbon. Berg Hüttenmänn Monatsh 86: 97–104
Kahler F (1980) Zur Definition der Saalischen Phase im marinen Bereich der Südalpen. In: Flügel E (ed) Die Trogkofel-Stufe im Unterperm der Karnischen Alpen Carinthia II Sonderh 36: 259–260
Kahler F (1983) Fusuliniden aus Karbon und Perm der Karnischen Alpen und der Karawanken. Carinthia II Sonderh 41: 1–107
Kahler F (1985) Oberkarbon und Unterperm der Karnischen Alpen. Ihre Biostratigraphie mit Hilfe der Fusuliniden. Carinthia II Sonderh 42: 1–93
Kahler F (1986) Ein Normalprofil der Fusuliniden-Stratigraphie im Oberkarbon und Unterperm der Karnischen Alpen. Carinthia II 176 (96): 1–17
Kahler F (1989) Die Fusuliniden. In: Ebner F, Kahler F (eds) Catalogus Fossilium Austriae, Heft II/b/1: Foraminifera Palaeozoica. Österr Akademie der Wissenschaften, Wien, pp 87–295
Kahler F, Kahler G (1980) Fusuliniden aus den Kalken der Trogkofel-Schichten der Karnischen Alpen. In: Flügel E (ed) Die Trogkofel-Stufe im Unterperm der Karnischen Alpen Carinthia II Sonderh 36: 183–254
Karl F (1956) Die Komponenten des oberkarbonen Nößlach-Konglomerates (Tirol). Mitt Geol Ges Wien 48: 71–87
Kerner F (1897) Die Carbonflora des Steinacherjoches. Jahrb k k Geol Reichsanst 47: 365–386
Kieslinger F (1956) Die nutzbaren Gesteine Kärntens. Carinthia II Sonderh 17: 1–384
Klaus W (1963) Sporen aus dem Südalpinen Perm. Jahrb Geol Bundesanst 106: 229–361
Kober V (1984) Zur Genese der Tarviser Breccie in den Karawanken, NW-Jugoslawien. Geol Inst Univ Köln Sonderveröff 56: 1–155
Kochansky-Devide V, Buser S, Cajhen J, Ramovs A (1973) Detailliertes Profil durch die Trogkofel-Schichten am Bache Kosutnik in den Karawanken (German summary). Razprave Slov akad znam umetn IV razr 16: 169–200
Kodsi GM (1967) Die Fauna der Bank s des Auernig (Oberkarbon; Karnische Alpen, Österreich). Carinthia II 157 (77): 59–81
Kodsi GM, Flügel HW (1970) Lithofazies und Gliederung des Karbons von Nötsch. Carinthia II 160 (80): 7–17
Krainer K (1982) Zur Sedimentologie und Vererzung der "Hangendquarzite" im Stanzer Tal/Arlberggebiet (Tirol). Geol Paläontol Mitt Innsbr 12: 81–94
Krainer K (1984) Sedimentologische Untersuchungen an permischen und untertriadischen Sedimenten des Stangalm-Mesozoikums (Kärnten/Österreich). Jahrb Geol Bundesanst 127: 159–179
Krainer K (1985) Zur Sedimentologie des Alpinen Buntsandsteins und der Werfener Schichten (Skyth) Kärntens. Geol Paläontol Mitt Innsbr 14: 21–81
Krainer K (1987 a) Zusammensetzung und fazielle Entwicklung des Alpinen Buntsandsteins und der Werfener Schichten im westlichen Drauzug (Kärnten/Osttirol). Jahrb Geol Bundesanst 130: 61–91
Krainer K (1987 b) Das Perm der Gurktaler Decke: eine sedimentologische Analyse. Carinthia II 177 (97): 49–92
Krainer K (1989 a) Molassesedimentation im Oberkarbon der Ostalpen am Beispiel der Stangnock-Formation am NW-Rand der Gurktaler Decke (Österreich). Zentral Geol Paläontol Teil I (1988): 807–820
Krainer K (1989 b) Composition and evolution of Lower Permian molasse sediments (Ponte Gardena Conglomerate) at the base of the Bolzano Volcanic Complex, Southern Alps (N Italy). Neues Jahrb Geol Paläontol Monatsh 1989: 400–424
Krainer K (1989 c) Die fazielle Entwicklung der Oberkarbonsedimente (Stangnock-Formation) am NW-Rand der Gurktaler Decke. Carinthia II 179 (99): 563–601
Krainer K (1989 d) Zum gegenwärtigen Stand der Permoskythforschung im Drauzug. Carinthia II 179 (99): 371–382
Krainer K (1990 a) Ein Beitrag zum Oberkarbon der Steinacher Decke ("Karbon des Nößlacher Joches". Tirol). Mitt Ges Geol Bergbaustud Österr 36: 87–99
Krainer K (1990 b) Fazielle und sedimentpetrographische Untersuchungen im Perm des Drauzuges. Mitt Österr Geol Ges 82 (1989): 49–78
Krainer K (1990 c) The limestone facies of the A4 and A5 Formations (Auernig Group). In: Venturini C (ed) Field workshop on Carboniferous to Permian sequence of the Pramollo-Nassfeld basin (Carnic Alps), Guidebook. Arti Grafiche Friulani, Udine, pp 76–80
Krainer K (1992) Fazies, Sedimentationsprozesse und Paläogeographie im Karbon der Ost- und Südalpen. Jahrb Geol Bundesanst 135: 99–193
Krainer K, Mogessie A (1991) Composition and significance of resedimented amphibolite breccias and conglomerates (Badstub Formation) in the Carboniferous of Nötsch (Eastern Alps, Austria). Jahrb Geol Bundesanst 134: 65–81
Krainer K, Spötl Ch (1989) Detrital and authigenic feldspars in Permian and Early Triassic sandstones, Eastern Alps (Austria). Sediment Geol 62: 59–77
Krainer K, Stingl V (1986 a) Alluviale Schuttfächersedimente im Ostalpinen Perm am Beispiel der Präbichlschichten an der Typuslokalität bei Eisenerz/Steiermark (Österreich). Mitt Österr Geol Ges 78 (1985): 231–249
Krainer K, Stingl V (1986 b) Perm, Unter- und Mitteltrias im Bereich von Wörgl bis Saalfelden (Exkursion E am 3. und 4. April 1986). Jahresber Mitt Oberrhein Geol Ver NF 68: 93–103
Kralik M, Krumm H, Schramm JM (1987) Low grade and very low grade metamorphism in the northern Calcareous Alps and in the Greywacke Zone: illite-crystallinity data and isotopic ages. In: Flügel HW, Faupl P (eds) Geodynamics of the Eastern Alps Deuticke, Vienna, pp 164–178
Krois P, Stingl V (1989) Die stratigraphische Neueinstufung der basalen klastischen Gesteine im Brennermesozoikum (Stubaier Alpen/Tirol). Zentralbl Geol Paläontol Teil I 1989 (9/19): 1455–1466

Kügel HW (1987) Sphinctozoen aus den Auernigschichten des Naßfeldes (Oberkarbon, Karnische Alpen, Österreich). Facies 16: 143–156

Kullmann J, Ramovs A (1980) Cephalopoden aus dem Oberkarbon (Gzhelium) und Unterperm der Karawanken. Geol Palaeontol 14: 195–208

Leonardi P (1968) Le Dolomiti. Geologia dei monti tra Isarco e Piave, vol 2. Manfrini, Trento, 1019 pp

Leonardi P, Conti MA, Leonardi G, Mariotti N, Nicosia U (1975) *Pachypes dolomiticus* n. gen. n sp.: pareiasaur footprint from the "Val Gardena Sandstone" (Middle Permian) in the western Dolomites (N Italy). Atti Accad Naz Lincei Rend Cl Sci Fis Mat Nat 8 (57): 221–232

Lorenz V, Nicholls IA (1976) The Permocarboniferous basin and range province of Europe. An application of plate tectonics. In: Falke H (ed) The continental Permian in central, west and south Europe Reidel, Dordrecht, pp 313–339

Lützner H (1981) Sedimentatin der variszischen Molasse im Thüringer Wald. Schriftenr Geol Wiss 17: 1–217

Lützner H (1987) Sedimentary and volcanic Rotliegendes of the Saale depression (Excursion Guidebook – Symposium on Rotliegendes in Central Europe). Acad Sci GDR, Central Inst Phys Earth, 197 pp

Massari F (1986) Some thoughts on the Permo-Triassic evolution of the South-Alpine area (Italy). Mem Soc Geol Ital 34: 179–188

Massari F, Venturini C (1990) The significance of the Auernig Group cyclicity. In: Venturini C (ed) Field workshop on Carboniferous to Permian sequence of the Pramollo-Nassfeld basin (Carnic Alps), Guidebook. Arti Grafiche Friulani, Udine, pp 81–86

Massari F, Conti MA, Fontana D, Helmold K, Mariotti N, Neri C, Nicosia U, Ori GG, Pasini M, Pittau P (1988) The Val Gardena Sandstone and Bellerophon Formation in the Bletterbach Gorge (Alto Adige, Italy): biostratigraphy and sedimentology. Mem Sci Geol 40: 229–273

Mostler H, Rossner R (1984) Mikrofazies und Palökologie der höheren Werfener Schichten (Untertrias) der Nördlichen Kalkalpen. Facies 10: 87–144

Mutschlechner G (1933) Cephalopodenfauna im Grödener Sandstein (Vorbericht). Verh Geol Bundesanst 76: 136

Neri C, Pasini M (1985) A mixed fauna at the Permian-Triassic boundary, Tesero section, western Dolomites (Italy). Boll Soc Paleontol Ital 23: 113–117

Neri C, Posenato R (1985) New biostratigraphical data on uppermost Werfen Formation of western Dolomites (Trento, Italy). Geol Paläontol Mitt Innsbr 14: 83–107

Neubauer F (1988) The Variscan orogeny in the Austroalpine and Southalpine domains of the Eastern Alps. Schweiz Mineral Petrogr Mitt 68: 339–349

Niedermayr G (1975) Gedanken zur lithofaziellen Gliederung der postvariszischen Transgressions-Serie der westlichen Gailtaler Alpen, Österreich. Mitt Geol Ges Wien 66/67: 105–126

Niedermayr G (1985) Fluvial braidplain passing into an intertidal belt at the margin of the Tethys Sea in the Alpine Buntsandstein of the Drauzug in Carinthia and eastern Tyrol (Austria). In: Mader D (ed) Aspects of fluvial sedimentation in the Lower Triassic Buntsandstein of Europe. Lecture Notes in Earth Sciences 4. Springer, Berlin Heidelberg New York, pp 487–496

Niedermayr G, Scheriau-Niedermayr E (1982) Zur Nomenklatur, Seriengliederung und Lithofazies der permo-skythischen Basisschichten des westlichen Drauzuges. Verh Geol Bundesanst 1982: 33–51

Niedermayr G, Mullis J, Niedermayr E, Schramm JW (1984) Zur Anchimetamorphose permo-skythischer Sedimentgesteine im westlichen Drauzug, Kärnten – Osttirol (Österreich). Geol Rundsch 73: 207–221

Oberhauser R (ed) (1980) Der Geologische Aufbau Österreichs. Springer, Berlin Heidelberg New York Wien

Oddone M, Vannucci R (1986) PGE and REE geochemistry at the B-W boundary in the Carnic and Dolomite Alps (Italy). Mem Soc Geol Ital 34: 129–139

Ori GG (1986) The nature of Permian rivers in Southern Alps. Mem Soc Geol Ital 34: 155–160

Ori GG, Venturini C (1981) Facies continentali permiane nelle Alpi Carniche. Rend Soc Geol Ital 3: 21–24

Ori GG, Dalla S, Cassinis G (1986) Depositional history of the Permian continental sequence in the Val Trompia – Passo Croce Domini area (Brescian Alps, Italy). Mem Soc Geol Ital 34: 141–154

Pasini M (1985) Biostratigrafia con i Foraminiferi del limite Fm. a Bellerophon – Fm. di Werfen fra Recoaro e la Val Badia (Alpi Meridionali). Riv Ital Paleontol Stratigr 90: 463–480

Perri MC, Andraghetti M (1987) Permian-Triassic boundary and Early Triassic conodonts from the Southern Alps. Riv Ital Paleontol Stratigr 93: 291–328

Pichler A (1859) Aus dem Inn- und Wippthale (Beiträge zur Geognosie Tirols). Z Ferdinandeum Tirol Vorarlb 3. Folge, H 8: 137–232

Poscher G (1985) Klastisch-terrestrisches Permoskyth am Südrand der Nördlichen Kalkalpen – eine Faziesanalyse. Master's Thesis, Univ Innsbruck, 123 pp

Prohaska W, Ebner F (1989) Geochemische Untersuchungen an Metavulkaniten der Veitscher Decke/Grauwackenzone. Sitzungsber Österr Akad Wiss Math Naturwiss Kl Abt I 197: 191–205

Ramovs A (1963) Biostratigraphie der Trogkofel-Stufe in Jugoslawien. Neues Jahrb Geol Paläontol Monatsh 1963: 382–388

Ramovs A (1968) Biostratigraphie der klastischen Entwicklung der Trogkofelstufe in den Karawanken und Nachbargebieten. Neues Jahrb Geol Paläontol Abh 131: 72–77

Ramovs A (1972) Mittelpermische Klastite und deren marine Altersäquivalente in Slowenien, NW Jugoslawien. Verh Geol Bundesanst 1972: 35–45

Ramovs A (1974) Fortschritte in der Untersuchung des Unter- und Mittel-Perms in Slowenien, NW Jugoslawien. Carinthia II 164 (94): 99–103

Ramovs A (1983) Unterperm-Conodonten aus den Karawanken/Slowenien, NW Jugoslawien. Neues Jahrb Geol Paläontol Abh 164: 414–427

Ratschbacher L (1984) Beitrag zur Neugliederung der Veitscher Decke (Grauwackenzone) in ihrem Westabschnitt (Obersteiermark, Österreich). Jahrb Geol Bundesanst 127: 423–453

Ratschbacher L (1987) Stratigraphy, tectonics, and paleogeography of the Veitsch nappe (Greywacke zone, Eastern Alps, Austria): a rearrangement. In: Flügel HW, Sassi FP, Grecula P (eds) Pre-Variscan and Variscan events in the Alpine-Mediterranean mountain belts Alfa Bratislava, pp 407–414

Remy W, Remy R (1978) Die Flora des Perms im Trompia-Tal und die Grenze Saxon/Thuring in den Alpen. Argum Paleobot 5: 57–90

Rozza R, Sartorio D (1990) Evidence of Permian sediments in north Adriatic Sea: the Amanda 1 BIS Well. In: Venturini C, Krainer K (eds) Field Workshop on Carboniferous to Permian Sequence of the Pramollo-Nassfeld Basin (Carnic Alps), Proc (pre print). Arti Grafiche Friulani, Udine, p 69

Schmidegg O (1949) Der geologische Bau der Steinacher Decke mit dem Anthrazitkohlenflöz am Nößlachjoch (Brenner-Gebiet). Veröff Mus Ferdinandeum 26/29: 1–19

Schönlaub HP (1973) Zur Kenntnis des Nord-Süd-Profiles im Nötschgraben westliche Villach. Verh Geol Bundesanst 1973: 359–365

Schönlaub HP (1985) Das Karbon von Nötsch und sein Rahmen. Jahrb Geol Bundesanst 127: 673–692

Schönlaub H P (1990) The Permian/Triassic boundary in the Nassfeld region. In: Venturini C, Krainer K (eds) Field workshop on Carboniferous to Permian sequence of the Pramollo-Nassfeld Basin (Carnic Alps), Proc (preprint) Arti Grafiche Friulani, Udine, pp 72–73

Schramm JM (1977) Über die Verbreitung epi- und anchimetamorpher Sedimentgesteine in der Grauwackenzone und in den Nördlichen Kalkalpen (Österreich). Geol Paläontol Mitt Innsbr 7: 3–20

Schramm JM (1982) Überlegungen zur Metamorphose des klastischen Permoskyth der Nördlichen Kalkalpen vom Alpenostrand bis zum Rätikon (Österreich). Verh Geol Bundesanst 1982: 73–83

Schramm JM, Gosen W von, Seeger M, Thiedig F (1983) Zur Metamorphose variszischer und postvariszischer Feinklastika in Mittel- und Ostkärnten (Österreich). Mitt Geol Paläontol Inst Univ Hamb 53: 169–179

Schünemann M, Gosen W von, Thiedig F (1982) Die Viktringer Permotrias und ihre Beziehung zur Gurktaler Decke (Kärnten – Österreich). Mitt Geol Paläontol Inst Univ Hamb 53: 191–206

Schwab M (1977) Zur paläotektonischen Entwicklung des Halleschen Permosilesgebietes (Nordöstlicher Saaletrog). Hall Jahrb Geowiss 1: 69–84

Selli R (1963) Schema geologico delle Alpi Carniche e Giulie occidentali. G Geol 2, (30): 1–136

Spötl Ch (1987) Eine klastisch-evaporitische Oberperm-Entwicklung im Hallstätter Salzberg (Salzkammergut, Österreich). Mitt Österr Geol Ges 80 (1987): 115–142

Spötl Ch (1988a) Schwefelisotopendatierungen und fazielle Entwicklung permoskythischer Anhydrite in den Salzbergbauen von Dürrnberg/Hallein und Hallstatt (Österreich). Mitt Ges Geol Bergbaustud Österr 34/35: 209–229

Spötl Ch (1988b) Zur Altersstellung permoskythischer Gipse im Raum des östlichen Karwendelgebirges (Tirol). Geol Paläontol Mitt Innsbr 14: 197–212

Spötl Ch (1989) The Alpine Haselgebirge Formation, northern Calcareous Alps (Austria): Permo-Scythian evaporites in an alpine thrust system. Sediment Geol 65: 113–125

Stache G (1872) Über die Steinkohlenformation der Centralalpen. Verh k k Geol Reichsanst 1872: 78–81

Stille H (1924) Grundfragen der vergleichenden Tektonik. Borntraeger, Berlin, 443 pp

Stingl V (1982) Sedimentologie und Vererzung des Alpinen Verrucano im Stanzertal (Tirol). Geol Paläontol Mitt Innsbr 12: 71–80

Stingl V (1983) Ein Beitrag zur Fazies der Prebichlschichten zwischen St. Johann i. T. und Leogang (Tirol/Salzburg). Geol Paläontol Mitt Innsbr 12: 207–233

Stingl V (1984) Alpiner Buntsandstein und Werfener Schichten bei Leogang (Salzburg). Geol Paläontol Mitt Innsbr 14: 1–19

Stingl V (1987) Die fazielle Entwicklung des Alpinen Buntsandsteins (Skyth) im Westabschnitt der Nördlichen Kalkalpen (Tirol/Salzburg, Österreich). Geol Rundsch 76: 647–664

Stingl V (1989) Marginal marine sedimentation in the basal Alpine Buntsandstein (Scythian) in the western part of the northern Calcareous Alps (Tyrol/Salzburg, Austria). Palaeogeogr Palaeoclimatol Palaeoecol 72: 249–262

Stur D (1871) Geologie der Steiermark. Verl Geogn Mont Ver Stmk, Graz, 654 pp

Sylvester H (1989a) Vorbericht über sedimentologisch-sedimentpetrographische Untersuchungen in den Perm/Skyth-Siliziklastika der ostalpinen Decken (Kärnten/Österreich). Carinthia II 179 (99): 401–424

Sylvester H (1989b) Vergleich von Perm/Skyth-Profilen des Ober-, Mittel- und Unterostalpins. Jahrb Geol Bundesanst 132: 791–821

Teich T (1982) Zum Chemismus der Badstubbrekzie im Unterkarbon von Nötsch in Kärnten. Carinthia II 172 (92): 91–96

Tenchov Y (1980) Die paläozoische Megaflora von Österreich. Eine Übersicht. Verh Geol Bundesanst 1980: 161–174

Tessensohn F (1972) Einige neue Beobachtungen im Karbon von Nötsch, Kärnten. Carinthia II 162 (82): 143–147

Tollmann A (1964) Das Permoskyth in den Ostalpen sowie Alter und Stellung des "Haselgebirges". Neues Jahrb Paläontol Monatsh 5: 270–299

Tollmann A (1972) Alter und Stellung des Alpinen Verrucano in den Ostalpen. Verh Geol Bundesanst 1972: 83–95

Tollmann A (1977) Geologie von Österreich, Bd. I: Die Zentralalpen. Deuticke, Wien

Tollmann A, Faupl P (1972) Alpiner Verrucano im Semmering- und Wechselgebiet. Verh Geol Bundesanst 1972: 107–118

Vai GB (1975) Hercynian basin evolution of the Southern Alps. In: Coy Squyres (ed) Geology of Italy, vol 1, the Earth Sciences Society of the Libyen Arab Republic, Tripoli, pp 293–298

Vai GB (1976) Stratigrafia e paleogeografia ercinica delle Alpi. Mem Soc Geol Ital 13: 7–37

Vai GB (1980) Tracing the Hercynian structural zones across "Neo-Europa": an introduction. Mem Soc Geol Ital 20: 39–45

Vai GB, Cocozza T (1986) Tentative schematic zonation of the Hercynian chain in Italy. Bull Soc Geol Fr 8, II, 1: 95–114

Vai GB, Francavilla F, Ferrari A, Contarini MT (1980) La sezione del Monte Carnizza (Carbonifero superiore, Alpi Carniche). Mem Soc Geol Ital 20: 267–276

Venturini C (1982) Il Bacino tardoercinico di Pramollo (Alpi Carniche): un evoluzione regolata dalla tettonica sinsedimentaria. Mem Soc Geol Ital 24: 23–42

Venturini C (ed) (1990a) Field workshop on Carboniferous to Permian Sequence of the Pramollo-Nassfeld Basin (Carnic Alps), Guidebook Arti Grafiche Friulani, Udine, pp 1–159

Venturini C (1990b) Geologia delle Alpi Carniche centro orientali. Edizioni del Museo Friulano di Storia Naturale, Udine Publ 36: 1–222

Vissher H, Brugman WA (1986) The Permian-Triassic boundary in the Southern Alps: a palynological approach. Mem Soc Geol Ital 34: 121–128

Wagner RH (1984) Megafloral zones of the Carboniferous. IX Congr Int Strat Geol Carb, C R 2: 109–134

Wopfner H (1984) Permian deposits of the Southern Alps as product of initial alpidic taphrogenesis. Geol Rundsch 73: 259–277

Ziegler PA (1982) Geological atlas of western and central Europe. Elsevier, Amsterdam, 130 pp

Chapter 4
Southern Alps

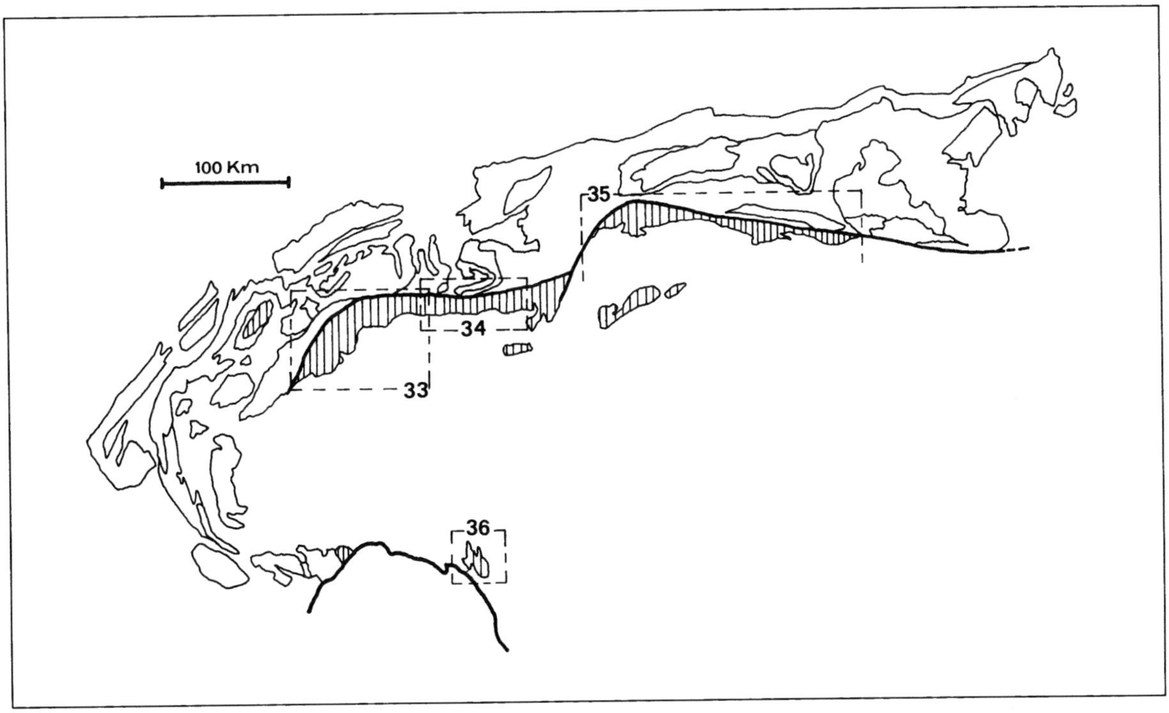

33 Ivrea-zone and adjacent Southern Alpine basement.
S. M. Schmid

34 Pre-Alpine Structural and Metamorphic histories in the Orobic Southern Alps, Italy.
G. B. Siletto, M. I. Spalla, A. Tunesi, J. M. Lardeaux, A. Colombo

35 The South-Alpine metamorphic basement in the Eastern Alps.
F. P. Sassi, R. Spiess

36 The Pre-Alpine Basement in the Alpi Apuane (Northern Apennines, Italy).
P. Conti, A. di Pisa, M. Gattiglio, M. Meccheri

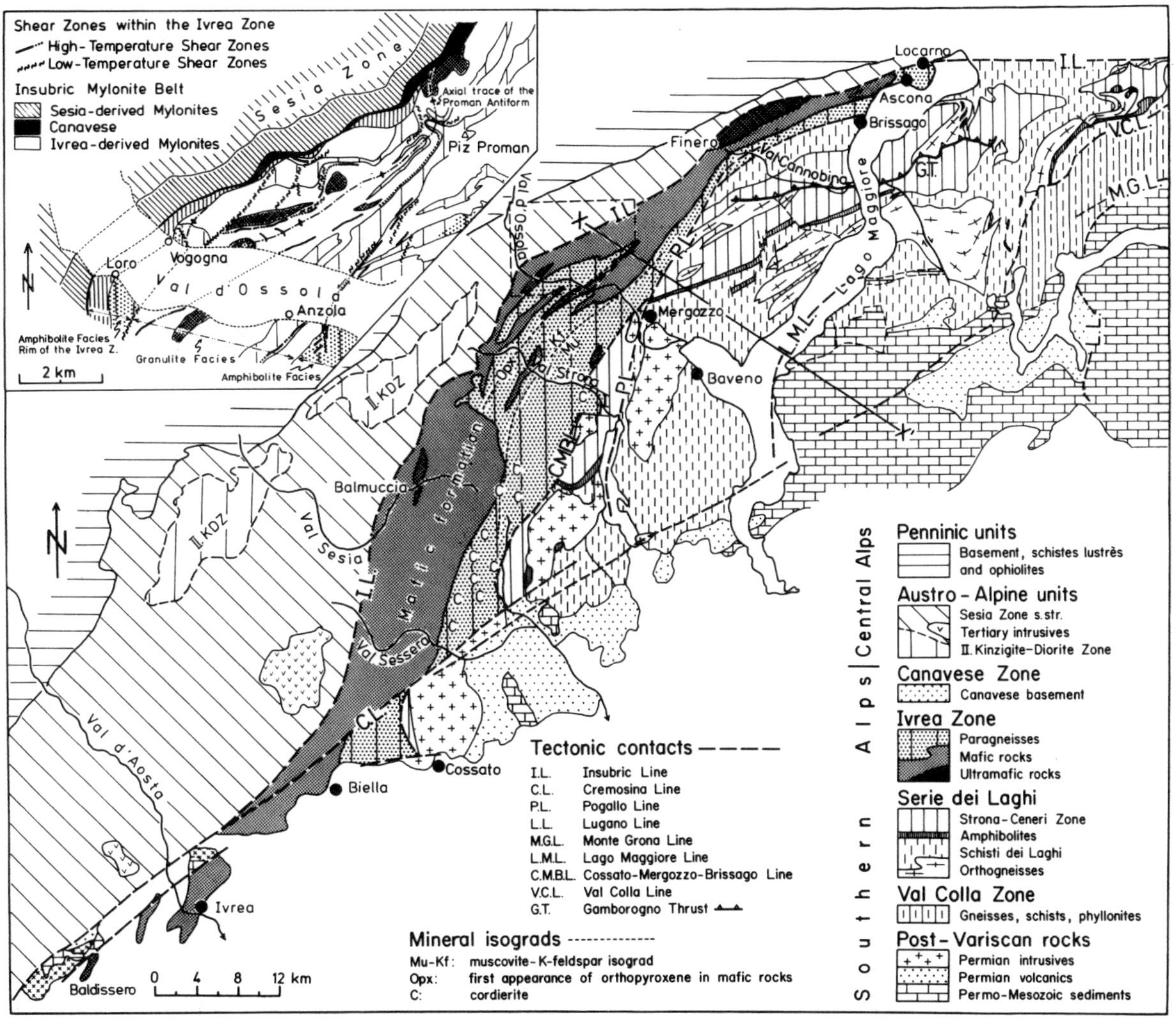

Fig. 1. Map of Ivrea zone and adjacent Southern Alpine basement (after Boriani et al. 1990a,b; Reinhard 1964; Zingg et al. 1990). *X-X'* Location of profile in Fig. 2A; **inset** detailed geological map of the northwestern rim of the Ivrea zone in the Val d'Ossola. (After Schmid 1967; Brodie and Rutter 1987; Schmid et al. 1987)

still a matter of debate: simple Alpine tilting as depicted in Fig. 2 (Handy 1987; Zingg et al. 1990), local tilting associated with ramp anticlines of Alpine age (Schumacher 1990) or preservation of Variscan and Late Palaeozoic structures with Alpine movements of only minor importance (Vai and Cocozza 1986; Boriani et al. 1990a,b) have been proposed.

Towards the Po plain the Serie dei Laghi (and the rest of the Southern Alpine basement further to the E) are unconformably overlain by Upper Carboniferous to Lower Permian volcanoclastic and volcanic series, Permo-Carboniferous terrigenous sediments, and, eventually the Mesozoic series. The latter characterize the present-day area of the Southern Alps as a passive continental margin setting at the WNW edge of the Apulia (or Adria) microcontinent (Weissert and Bernoulli 1985) which formed during Late Triassic to Early Jurassic. Needless to say that stretching caused by passive continental margin formation is expected to leave a thermal and deformational imprint on the basement underlying the sediments and recently this has been directly documented in an area further to the E (Bertotti 1990, 1991), where basement-cover relationships are better preserved than in our area of interest. Another area of Permo-Mesozoic cover rocks (Canavese sediments) belongs to the very distal part of this passive continental margin and follows the Insubric line. Since these sediments are intensely mylonitized, forming part of the Insubric mylonite belt (Schmid et al. 1987), their exact position with respect to the adjecent Ivrea zone is obscured. In the area of the Canavese *s. str.*, i.e. near the town of Ivrea where stratigraphic basement-cover contacts are better preserved, Biino and Com-

Ivrea Zone and Adjacent Southern Alpine Basement

S. M. Schmid

Abstract

The available literature is critically assessed and existing controversies are emphasized. Based on a synthesis of published ideas the extended discussion proposes the following evolutionary scheme (compare sketches in Fig. 3): (1) an impressive volume of Proterozoic metasediments in the Ivrea zone and Serie dei Laghi has first been juxtaposed during accretionary wedge formation and underplating above a subducted Ivrea oceanic crust during a Late Proterozoic to Early Palaeozoic event ("Cadomian" or Caledonian). (2) A high geotherm was established by Ordovician times (460–480 Ma) as evidenced by anatexis in paragneisses of the Ivrea zone and granitic intrusion associated with granitization of adjacent metasediments in the Serie dei Laghi. This time marks the onset of a long-lasting history of deep burial under elevated temperatures for both Ivrea zone and Serie dei Laghi. (3) Variscan orogeny at around 320 Ma led to crustal thickening and decoupling between lower crust (Ivrea zone) and intermediate crustal levels (Serie dei Laghi). Large parts of the Serie dei Laghi were exhumed during Variscan orogeny, thereby preserving pressure-dominated Variscan mineral assemblages while the Ivrea zone remained in a lower crustal environment resulting in widespread re-equilibration of mineral assembleges to moderate pressures. (4) An important Late Palaeozoic magmatic event of Permian age is unrelated to Variscan orogeny since it post-dates re-establishment of normal crustal thickness. Synchronous magmatic activity is documented at all crustal levels from the still granulite facies metamorphic crust-mantle transition to the sedimentary cover and is best explained in the context of intracontinental wrench tectonics. (5) Crustal thinning associated with passive continental margin formation in the Early Mesozoic leads to considerable uplift of the Ivrea zone and cooling to temperatures below 300 °C. (6) Late Oligocene to Early Miocene deformation is responsible for the final exposure of a vertical section through attenuated continental crust.

1 Introduction

Recently a number of review-type articles have been published about the area concerned. For details and extensive references the reader is referred to these articles (i.e. Zingg 1983, 1990; Boriani et al. 1990a,b; Zingg et al. 1990). This contribution merely offers a brief overview introducing the reader who is unfamiliar with the area. The discussion will focus on selected points of interest with special emphasis on controversies and still open questions.

2 Geological Setting

It is best to start with the present setting of the Ivrea zone and adjacent basement units (Serie dei Laghi in the sense of Boriani et al. 1990a) within the framework of the Alpine orogeny since earlier tectonic settings are largely overprinted and, consequently, much more difficult to reconstruct.

The area concerned (Fig. 1) is located at the W end of a basement high within the Southern Alps (i. e. that part of the Alpine chain located to the SE and S of the Alpine Insubric line). This basement high closely follows the trend of the Insubric line: further to the E it runs parallel to the trend of the E-W-striking Tonale line, a segment of the Insubric line, E of Locarno (Orobic basement), in order to swing into a NNE-SSW-strike further to the W (Ivrea zone and Serie dei Laghi), again roughly parallel to the Canavese line (western segment of the Insubric line). Thus, there seems to be an intimate relationship between the Insubric line and this Southern Alpine basement high, both in terms of its location and structural trend. It is widely accepted that the section through the Ivrea zone (lower crust) and Serie dei Laghi (upper crust) offers a unique cross section through the Southern Alpine crust (Fig. 2). This implies that later tectonic movements are responsible for exposing such a crustal section at the earth's surface. But age, kinematics and geometry of these tectonic movements are

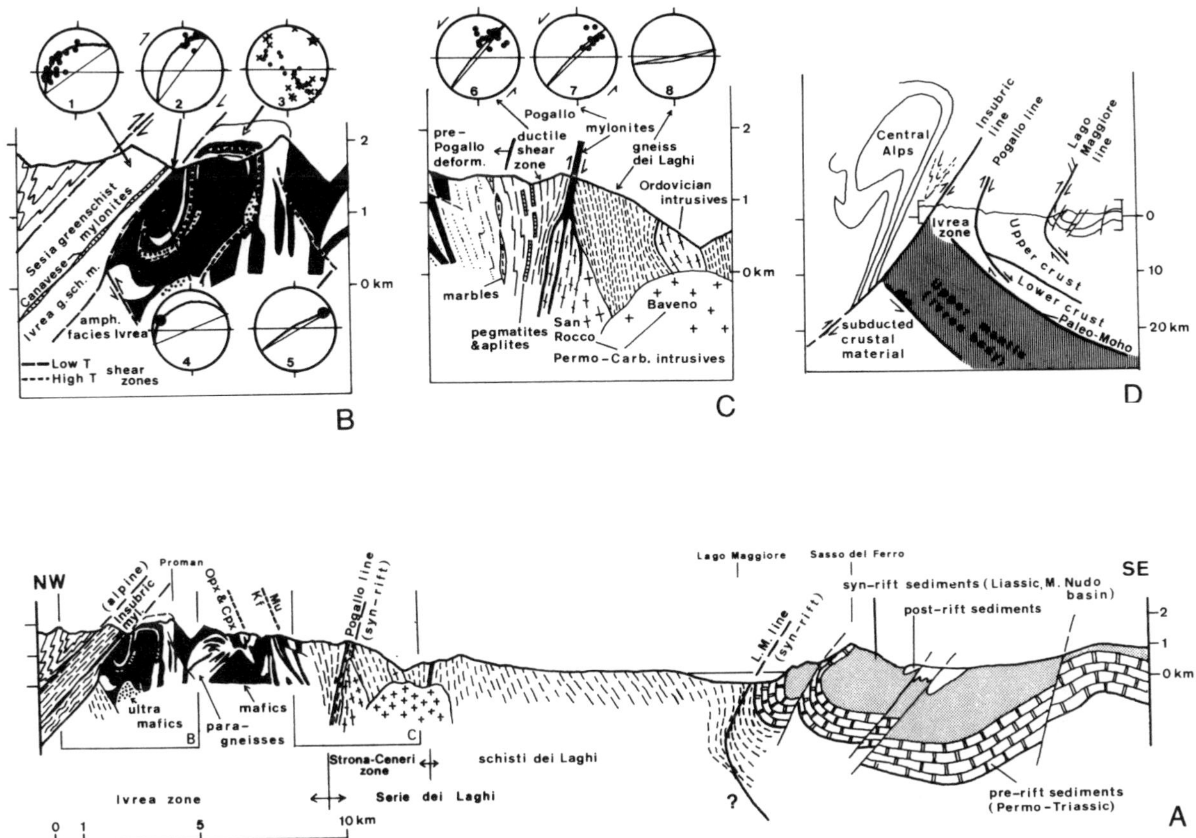

eations *(dots)* of the Insubric mylonites (*projections 1* and *2* from Schmid et al. 1987); measurements from the Proman antiform (*projection 3: dots* foliations poles; *crosses* slickenside lineations; *star* fold axis orientation from Zingg et al. 1990), average orientation of foliation and stretching lineation within the high-temperature shear zones (*projections 4 and 5,* from Brodie and Rutter 1987); **C** details of the area around the Pogallo ductile fault zone (*projection 6* indicates average foliation and lineation orientation within the Pogallo ductile fault zone; *projection 7* refers to the mylonites at the Pogallo line: *projection 8* give the average orientation of foliation in the Strona-Ceneri zone; *projections 6–8* from Handy 1987); **D** sketch illustrating the position of the profile depicted in Fig. 3A within a larger context

pagnioni (1989) report evidence that the Canavese sediments were deposited onto a low-grade Variscan basement, the "Canavese basement", which is in tectonic contact with the Ivrea zone.

Geophysically, the Ivrea zone is considered to represent the surface expression of a much larger high-density body (the Ivrea body) leading to a huge NNE-SSW-striking anomaly at the inner arc of the Western Alps. The original suggestion of a single ESE-dipping slice of mantle material (Berckhemer 1968), underplated by crustal material of lower density has recently been modified in the sense that this mantle slice associated with the Ivrea zone is only the most internal of several wedges of mantle and lower crustal material (Rey et al. 1990). However, mantle material directly underlying the Ivrea zone, together with the ESE-dipping subsurface continuation of the Ivrea lower crustal material is still considered to be the main cause of the observed gravity anomaly (see Figs. 8, 9, 11 in Rey et al. 1990). If these model sections, satisfying the gravity and magnetic anomalies in an admittedly non-unique way, are taken at face value they would argue for simple large-scale tilting of a crustal section into a near-vertical attitude.

Because the Ivrea zone and the Serie dei Laghi have been affected by tectonism, both related to Alpine orogeny and passive continental margin formation, a reconstruction of the pre-Alpine geological setting (Variscan and/or Caledonian) is very difficult. This difficulty is agravated by the fact that the Ivrea zone represents a lower crustal level during Palaeozoic orogeny. In analogy to the findings of Eisbacher et al. (1989) in the Schwarzwald-Vosges area, lower crustal levels are expected to ignore basically the main convergent tectonic boundaries due to detachment of the upper crust during convergence and due to late Hercynian overprint associated with the rise of granitic plutons and the formation of a Late Palaeozoic Neo-Moho. Yet another difficulty lies in the, still largely debated, metamorphic evolution during the Palaeozoic leading to difficulties in the timing of the major deformational events associated with this metamorphic evolution.

In spite of these difficulties, the general observation of a decrease in metamorphic grade from W to E within the basement directly below the Permo-Carboniferous transgression was used by Vai and Cocozza (1986) to argue that the western part of the Southern Alpine basement (our area of interest with its granulite to amphibolite grade metamorphism) represents the more internal parts of the Variscan orogen, with the externides being situated in the easternmost Southern Alpine basement (Comelico and Carnian Alps with a weak sub-greenschist facies metamorphic overprint onto Ordovician to Carboniferous foreland sediments). It has to be emphasized, however, that the Ivrea zone, and consequently the granulite facies part of the Southern Alpine basement is nowhere in direct contact with the Permo-Carboniferous transgression (as is the case for parts of the amphibolite grade Serie dei Laghi). Thus, this metamorphic zonation does not simply reflect a gradient at the Earth's surface before deposition of the Permo-Carboniferous sediments. Consequently, a large portion of the exhumation process, particularly in our area of interest, may well postdate the Variscan orogeny and has to be the product of passive continental margin formation and/or Alpine tectonics. Furthermore, the observation of low-grade metamorphism within the Canavese basement in the westernmost parts of the Southern Alpine basement mentioned already (Biino and Compagnoni 1989) argues against a simple metamorphic gradient below the Permo-Carboniferous transgression from W to E. Additional difficulties arise from the observation of pre-Permo-Carboniferous greenschist retrogradation of higher grade metamorphism in the basement of the Orobic Alps (Milano et al. 1988) and the Serie dei Laghi (Borghi 1987). Evidence for an even younger (Permo-Mesozoic) metamorphic overprint affecting parts of the Orobic basement is given by Siletto et al. (this Vol).

3 Lithological Units and Contacts Between These Units

It is customary to subdivide the basement rocks of the western part of the Southern Alpine basement into major lithological units, often refered to as "zones". Contacts between these zones may be gradual or marked by major faults.

3.1 The Ivrea Zone

The Ivrea-Verbano zone (Schmid 1967) was originally referred to as formazione diorito-kinzigitica (Novarese 1929). The term is still used for the same lithological unit now situated N of the Insubric line ("seconda zona diorito-kinzigitica"). The Ivrea zone is separated by the greenschist facies Insubric mylonite belt from the Central Alps.

This mylonite zone not only consists of retrogressed protoliths of the Ivrea zone and the Central Alps (Sesia zone) but also comprises an intermediate zone, the Canavese zone. Since the contact between the Canavese and Ivrea zones is heavily tectonized within the Insubric mylonite belt, the original relationship between the Ivrea and Canavese zones remains unclear. Generally pre-Alpine metamorphic grade increases within the Ivrea zone towards the Insubric line from amphibolite to granulite grade. However, heavily tectonized slivers of amphibolite grade basement rocks (inset in Fig. 1), reminiscent of the SE rim of the Ivrea zone (Schmid et al. 1987) or distinctly different from the Ivrea zone (granite and diorite, intrusive into low-grade metamorphics mapped as Canavese basement in Fig. 1, Biino and Compagnoni 1989) locally reappear within or adjacent to the Insubric mylonite zone. This Canavese basement is associated with the Canavese Permo-Mesozoic cover. These Canavese cover rocks only underwent low grade metamorphism of Alpine age (Zingg and Hunziker 1990). Thus, highly attenuated remnants of the South Alpine upper crustal levels reappear in the immediate vicinity of the Insubric line. It is difficult to decide as to how much of this attenuation had already occurred during crustal thinning during the Early Mesozoic and how much is due to Alpine movements along the Insubric line. Traditionally, the Canavese zone was interpreted as representing an independent palaeogeographic and tectonic unit (Argand 1910) forming the root of Austro-Alpine nappes. However, the close affinities of the Permo-Mesozoic Canavese cover with the distal part of the passive continental margin in the Southern Alps argues for the hypothesis that the Canavese zone simply represents a remnant of upper crustal levels originally overlying the Ivrea zone. Zingg et al. (1990) suggested that the overall structure of the Ivrea zone may be viewed as that of a huge asymmetric antiform of Late Alpine age (whose core is only locally preserved, Proman antiform in the Val d'Ossola, Fig. 2) with a highly attenuated NW limb (Canavese zone) and a much wider SE limb (the main part of the Ivrea zone) with its well-known transition from granulite facies metamorphism near the core to amphibolite facies conditions further to the SE.

The Ivrea zone is predominantly made up of paragneisses (referred to as kinzigites in the amphibolite grade part and stronalites in the granulite facies part) and mafic rocks. Thin marble horizons and pegmatitic dykes concentrate along the SE margin of the Ivrea zone and, more sporadically, near the Insubric line. Peridotites occur as isolated lenses predominantly within the NW part of the Ivrea zone, in the vicinity of the Insubric line, with the exception of one peridotite body at the southern border of the Ivrea zone (Boriani and Peyronel Pagliani 1968), associated with eclogitic amphibolites.

The ultramafic rocks include peridotites *s. l.* and pyroxenites. Spinel peridotites and lherzolites are predominant in the Baldissero and Balmuccia bodies (Lensch 1971; Shervais 1979; Rivalenti et al. 1981). The Finero body contains phlogopite and hornblende peridotite resulting from crustal contamination (Hunziker and Zingg 1982). The layered group adjacent to the Balmuccia body is made up of spinel-bearing dunite and harzburgite interlayered with gabbro and norite (Rivalenti et al. 1981). In the Balmuccia region Sm-Nd data suggest an event of crust formation about 600 Ma ago (Voshage et al. 1987). However, younger ages (Ordovician and/or Late Palaeozoic) for deep-seated intrusions and partial melting within the layered complex are very likely (see discussion in Voshage et al. 1987). Evidence for a Late Palaeozoic event (about 270 Ma) is particularly compelling (mafic and ultramafic dykes in the Balmuccia peridotite and phlogopite metasomatism associated with crustal contamination in the Finero complex, Voshage et al. 1987, 1988). In a later publication Voshage et al. (1990) no longer believe that there is any basis for the above mentioned 600 Ma event published in earlier publications (Voshage et al. 1987, 1988). Instead they propose magmatic underplating at 270 Ma to be responsible for the formation of the entire Mafic complex in Val Sesia.

Three types of mafic rocks experienced regional metamorphism and deformation of Palaeozoic age (predating the 270 Ma magmatic event, but more precise dating is still a matter for debate): (1) mafic rocks of oceanic origin, alternating with paragneiss (Sills and Tarney 1984), (2) large bodies of the "Anzola gabbro" type, and (3) parts of the banded mafic rocks within the granulite facies part of the Ivrea zone such as those within the layered complex mentioned already. It has to be emphasized, however, that a fourth type preserved magmatic features: the diorite rim of the Mafic formation in Val Sesia with orthopyroxene of magmatic origin (Zingg 1980). This rim yields ages of 286 Ma (Pin 1986) and Bürgi and Klötzli (1990) obtain similar ages (274 Ma) for anatectic paragneisses in the contact aureole of the Mafic formation. These particular anatectic melts related to the contact aureole of the Mafic formation define an important time-marker in that they (and consequently also the intrusion of the Mafic formation) postdate deformation associated with regional metamorphism.

Amongst the paragneisses, sillimanite bearing metapelites (kinzigites and stronalites) are predominant. Within the eastern part of the Ivrea zone, the mineral isograd zonation runs parallel to the predominant strike (NW-SE to NNW-SSE) in order to run discordantly across compositional banding in the vicinity of the Mafic formation (particularly well documented by the muscovite-K-feldspar isograd, Zingg 1980). Widespread occurrences of cordierite and andalusite within the metapelites adjacent to the Mafic formation suggest final equilibration at lower pressures as compared with the NE part of the Ivrea zone (Zingg 1980). This suggests that a former zonation into granulite and amphibolite regional metamorphism running parallel to the major structures in the Ivrea zone was later overprinted in the SW part of the Ivrea zone. This overprint is most probably associated with the thermal event due to the intrusion of at least parts of the Mafic formation at around 270 Ma.

The marble horizons concentrating along the SE rim of the Ivrea zone are in immediate neighbourhood to amphibolite horizons, interbedded with paragneisses whose chemistry indicates mid-oceanic basalt origin (Sills and Tarney 1984). The accretionary wedge scenario proposed by Sills and Tarney (1984) is very attractive also for explaining: (1) the lack of a granitic basement underlying metapelites and marbles, and (2) the occurrence of high-P relics such as eclogitic amphibolites (Boriani and Peyronel Pagliani 1968; Buletti 1983; Borghi 1987, 1988; Boriani et al. 1990b) and kelyphitic peridotites (Lensch and Rost 1972) found both within Ivrea zone and Serie dei Laghi. Concerning the age of accretionary wedge formation, it can only be said that it post-dates the Early Proterozoic age of the upper intercept of the zircon discordia (Köppel 1974) obtained for the paragneisses and pre-dates Ordovician granites found in the Serie dei Laghi (Boriani et al. 1990a).

Structurally, the Ivrea zone is characterized by the steep attitude of compositional banding and main foliation. E-plunging stretching lineations are of remarkably constant orientation. The structural section published by Schmid (1967) is still the only complete section available through the Ivrea zone (his profile has been incorporated into the profile depicted in Fig. 2). At least two phases of folding are contemporaneous with Palaeozoic amphibolite to granulite grade metamorphism (Kruhl and Voll 1976; Steck and

Tièche 1976). A prominent antiform next to the Insubric line (Proman antiform) formed under greenschist facies conditions and is of Alpine age (Schmid et al. 1987). Handy and Zingg (1991) subdivide the tectonometamorphic evolution into three episodes: Episode 1 includes large-scale magmatism and regional metamorphism, responsible for the general compositional and metamorphic zonation presently observed and characterized by annealed microstructures in all lithologies. Crustal attenuation during the Late Palaeozoic (transtension) and Early Mesozoic rifting lead to episode 2 deformation, resulting in localized shear under variable metamorphic conditions indicated by the synkinematic overprint of previously annealed microstructures. Episode 3 structures are related to polyphase Alpine transpressional tectonics.

Pressure and temperature estimates scatter enormously. For the granulite facies part pressure estimates vary between 6.5 and 9 kb (Sills 1984) while temperature estimates are between 750 and 950 °C (see also references and discussions in Zingg 1983 and Zingg et al. 1990). For the amphibolite facies part pressures around 5–6 kb and temperatures between 550 and 750 °C are reported. Sills (1984) reports a substantial pressure gradient across strike within the granulite facies part (around 0.5 kb/km) suggesting tectonic thinning after equilibration.

The large scatter in P-T estimates is not surprising given the long-lasting tectono-metamorphic evolution of the Ivrea zone. These estimates are often taken to indicate the P-T conditions of the main event of regional metamorphism during episode 1 of Handy and Zingg (1991). However, as already emphasized by Handy and Zingg (1991) this episode 1 is certainly not a single event, given the ongoing controversy over its age, discussed later. Reequilibrations are also expected during the Late Palaeozoic magmatic event (around 270 Ma) associated with initial exhumation in a transtensional environment (episode 2 of Handy and Zingg 1991).

3.2 The Serie dei Laghi

Terminology concerning the basement between Ivrea zone and the main part of the Permo-Triassic cover of the Southern Alps is confusing. Italian workers collectively refer to this basement as Serie dei Laghi (Boriani et al. 1990b) while Swiss authors (Schmid 1967; Zingg 1983) use the term Strona-Ceneri zone (here referred to as Strona-Ceneri zone *s. l.*). According to Boriani et al. (1990b), the Serie dei Laghi can be further subdivided into the Schisti dei Laghi and the Strona-Ceneri zone in his sense (here referred to as Strona-Ceneri zone *s. str.*, identical to the older term "gneiss dei Laghi").

The Serie dei Laghi predominantly consists of amphibolite grade metasediments, intruded by metagranite bodies of Ordovician age (466 Ma, Boriani et al. 1982/83), and plutonic rocks belonging to the Late Palaeozoic (285–275 Ma) magmatic cycle (a suite of small stocks and dyke swarms of predominantly basaltic composition, the "appinites" of Boriani et al. 1990a, and epiplutonic Permian granite intrusions, known as "Graniti dei Laghi" or "Baveno-M. Orfano suite"). The subdivision of pre-Ordovician metasediments into Schisti dei Laghi and Strona-Ceneri zone *s. str.* is based on several arguments (Boriani et al. 1990b):

1. The metasediments within the Schisti dei Laghi are predominantly pelitic. They correspond to the "schiefrige Biotit-Plagioklasgneise" and "Giumello-Gneise" of Reinhard (1964). Meta-sediments of the Strona-Ceneri zone *s. str.* are predominantly arenaceous, sometimes with abundant clay matrix. According to Boriani et al. (1990b), older metasandstones (Cenerigneiss of Reinhard 1964, containing small boudins of hornfels and calc-silicates), are followed by "gneiss minuti" ("Hornfelsgneis" of Reinhard 1964), interpreted as clay-rich metasiltites and finer-grained metasandstones. The metasediments within the Strona-Ceneri zone *s. str.* still preserve sedimentary structures in spite of later granitization interpreted to be coeval with the Ordovician intrusions.
2. In many places, a continuous amphibolite horizon defines an important marker horizon between the two subunits (Fig. 1). Boriani et al. (1990b) interpret this horizon in terms of a stratigraphic intercalation of metasedimentary origin (coarse-grained basaltic tuffites). However, amphibolites are not restricted to this main horizon but are also found within the Schisti dei Laghi of the Gamborogno area where they contain eclogite facies relics (Borghi 1988). Furthermore, this marker horizon contains peridotite lenses near the eastern termination of the Serie dei Laghi adjacent to the Val Colla line (Reinhard 1964).
3. The Schisti dei Laghi are interpreted to be polymetamorphic, still recording a pre-Ordovician phase of high-grade metamorphism, whereas the Strona-Ceneri zone *s. str.* is supposedly still unmetamorphic at the time of intrusion of the Ordovician granites.

The subdivision based on criteria 1 and 2 is probably justified. The evidence of a single (Variscan) regional metamorphic event in the Strona-Ceneri zone *s. str.*, however, is not very convincing and based on: (1) the interpretation of Al-silicate nodules within metasediments to have formed during contact metamorphism (Bigioggero and Boriani 1975); (2) the view that granitization of the Strona-Ceneri country rocks, due to the Ordovician intrusion, apparently argues for the absence of metamorphism within these country rocks (Boriani et al. 1990b); (3) the assumption that the amphibolites mentioned form a stratigraphic marker horizon, and (4) structural arguments indicating that the dominant D1 schistosity of post-Ordovician age transposes an earlier schistosity in the Serie dei Laghi whereas it transposes sedimentary bedding in the Strona-Ceneri zone *s. str.* Should this interpretation be correct, it would have severe consequences for unravelling Variscan tectonics since the Schisti dei Laghi would represent an older basement complex with respect to the cover of the Strona-Ceneri zone *s. str.* (Borghi 1987).

Structurally, compositional layering and main foliation of the Serie dei Laghi remain in a subvertical orientation, W of Lago Maggiore. The general strike (WSW-ENE), however, is distinctly discordant to that of the Ivrea zone over a considerable distance along the Pogallo fault discussed later. Along the Val d'Ossola section, depicted in Fig. 2, the Strona Ceneri zone *s. str.* is directly juxtaposed with the Ivrea zone, followed by the Schisti dei Laghi further to the S. Hence, the "younging" direction would be towards the N, if the Schisti dei Laghi are, indeed, older. This would imply that the lower crustal section across the Ivrea zone was overlain by an inverted mid- and upper crustal section represented by the Serie dei Laghi after removing later tilting of the entire section. Further complications with the concept of the Serie dei Laghi representing an older basement, arise from the fact that the Serie dei Laghi reappear in direct contact with the Ivrea zone, in an area further to the NE (Val Cannobina) according to Boriani et al. (1990b). This area is characterized by later folding of the main foliation around subvertical E plunging axes ("Schlingen") continuing into the Gamborogno area east of Lago Maggiore (Reinhard 1964; Borghi 1989). This folding around subvertical axes is interpreted to be the result of Late Palaeozoic strike-slip tectonics by Boriani at al. (1990a). However, if verticalization of the main schistosity is attributed to Alpine tilting (Schmid et al. 1987) the attitude of the fold axes would have been subhorizontal at the time of their formation. The stretching lineations within the Serie dei Laghi exhibit considerable scatter (Borghi 1989) probably due to this same large-scale refolding event.

The Ordovician granites provide a time marker not available in the Ivrea zone. The age of sedimentation of the Schisti dei Laghi is most probably Proterozoic or early Palaeozoic, given high-P relics pre-dating the Ordovician granites. The age of regional metamorphism coeval with the formation of the main foliation is constrained to be post-Ordovician in the Serie dei Laghi according to Boriani et al. (1990b). Relics of a pre-Ordovician phase of high-pressure metamorphism are found within the Schisti dei Laghi in the Gamborogno area where Borghi (1988) estimates pressures between 15 and 20 kb and temperatures > 700 °C. Widespread occurrences of kyanite document relatively high pressures for the main post-Ordovician metamorphism as well. During this Variscan event there appears to be little spatial variation in metamorphic grade except near the contact with the Ivrea zone where sillimanite appears. Sillimanite grows post-kinematically indicating that the thermal peak (600–650 °C) followed the baric peak (6–8 kb; P-T estimates according to Borghi 1988 and Boriani et al. 1990b). Borghi (1987, 1988) reports a second greenschist metamorphic and deformational overprint related to exhumation but pre-dating Late Palaeozoic magmatism (Boriani et al. 1990b). This observation supports interpretations based on the cooling age concept indicating that the Serie dei Laghi represent a higher structural level, that cooled earlier than the Ivrea zone, which still occupies a deep crustal level during the Late Palaeozoic (anatexis related to the Late Palaeozoic intrusion of parts of the main Mafic formation). This is important since Boriani et al. (1990a) cast serious doubts on the cooling age concept and its tectonic consequences, discussed later.

The subvertically oriented contact between Serie dei Laghi and Ivrea zone is of tectonic nature along the Pogallo line which comprises the Pogallo ductile fault zone and a narrow fault zone associated with greenschist facies mylonites and cataclasites (Handy 1987). The kinematics of movement are those of oblique sinistral strike-slip in present-day coordinates. The synkinematic metamorphic gradient (increasing metamorphic grade across strike towards the Ivrea zone and along strike towards the NE) led Handy (1987) to confirm a hypothesis first published by Hodges and Fountain (1984), namely that of a tilted normal fault of Mesozoic age. The asymmetry of synkinematic metamorphic conditions precludes a strike-slip scenario and is typical for normal faults (see Mancktelow 1990, for a particularly well-documented Alpine example). The age is inferred from applying the cooling age concept and will be discussed later.

Boriani et al. (1990a) do not share this view and interpret the Pogallo fault as a strike-slip zone formed during the late Variscan but before intrusion of the appinites. This dating is at odds with the observation of tectonic overprint of Permian intrusions by tectonic activity along the Pogallo fault (Handy 1987). This alternative view results in an earlier age for verticalization of the Ivrea zone and Serie dei Laghi (Variscan as opposed to Alpine) and is incompatible with evidence from the cooling ages (Handy 1987), Alpine deformation in the Ivrea zone (Schmid et al. 1987) and palaeomagnetic evidence for post-Oligocene 60 ° rotation of the Ivrea zone (Schmid et al. 1989).

According to Boriani et al. (1990a) another fault zone is believed to be closely associated with the Pogallo fault, in terms of kinematics and time of formation: the "Cossato-Mergozzo-Brisago line" (CMB line). According to Boriani et al. (1990a), this "line" marks the transition between the Ivrea zone and Serie dei Laghi where the Pogallo fault breaks away from this lithological contact (SW of Val d'Ossola) and where the Pogallo fault no longer appears as a structural discontinuity (Val Cannobina) due to higher metamorphic conditions during fault movement (Handy 1987). The principal argument for the interpretation of both CMB line and Pogallo line in terms of a Late Variscan strike-slip zone is that the appinite suite is spatially and temporally associated with the CMB line, the appinites having intruded into transtensional segments of this vertically oriented fault zone. In view of the undisputable fact that the Pogallo fault deforms Permian intrusives (Handy 1987), the timing of the CMB and Pogallo lines as part of a late Variscan fault zone becomes very unlikely.

In the absence of modern structural studies along the CMB line it cannot yet be decided with certainty, when, and if at all, such a high-grade fault zone was active. The occurrence of (Early Palaeozoic?) high pressure relics at and to both sides of this "line", together with difficulties in delineating a clear lithological and/or structural boundary away from the Pogallo line (Val Cannobina, SW part of Val d'Ossola) argues for a common geological history of both the Ivrea zone and Serie dei Laghi for the final stages of crustal evolution, before the onset of fault movement along the Pogallo line. However, considerable detachment must have occurred between Ivrea zone and Serie dei Laghi during Variscan orogeny as discussed in the following discussion section.

Towards the E, the Serie dei Laghi are delimited by another fault zone, the Val Colla line, from the Val Colla zone. This fault zone is characterized by greenschist facies overprint of presumably Variscan age (Reinhard 1964) and its southern part was reactivated during Alpine orogeny according to Schumacher (1990). No major jump in Variscan metamorphic grade is observed across the Val Colla line. The kinematics of movement along the Cremosina line are still unknown. The map pattern (Fig. 1) suggests dextral transtensive movements during Alpine orogeny.

4 Discussion

4.1 *Age and Geotectonic Setting of Pre-Late Palaeozoic Deformation and Metamorphism*

A pre-Variscan event related to high-pressure metamorphism (only sporadically documented in Ivrea zone and Serie dei Laghi) and associated deformation (documented only in parts of the Serie dei Laghi) is interpreted to be responsible for pile up and juxtaposition of the various metasedimentary units which now occupy an impressive volume of rocks within Ivrea zone and Serie dei Laghi. At least parts of the mafic lithologies were incorporated during this event as well (mafics of oceanic origin, amphibolites with high-P relics). The age of intrusion of the Anzola-type gabbro and the mafics within the layered complex remains unknown. The pre-Ordovician age of this event is documented in the Serie dei Laghi. Accretionary wedge formation and underplating of a hypothetical continent above a subducted Ivrea oceanic crust is the most likely scenario for explaining the lack of an older granitic basement for the Ivrea zone and Serie dei Laghi metasediments (Fig. 3A). Supposing such a scenario to be correct any attempt at reconstructing a simple stratigraphic sequence for these metasediments would be bound to fail. The age of accretionary wedge formation can only be further constrained if the Sm-Nd data of around 600 Ma (Voshage et al. 1987, 1988, however, cf. another view in Voshage et al. 1990) are taken to date the earliest event of crust (more precisely oceanic crust accepting an accretionary wedge scenario) formation near the crust-mantle boundary in the Ivrea zone. In such a case Late Proterozoic to Early Palaeozoic ("Cadomian" or Caledonian) age is most likely. Radiometric dating of an Ordovician tectono-metamorphic event in anatectic paragneisses (478 Ma Rb-Sr whole rock isochron of Hunziker and Zingg 1980, however, cf. Pin 1990 for a reinterpretation) and contemporaneous intrusion of granites in the Serie dei Laghi indicate restauration of a high geotherm by Ordovician times (Fig. 3B). The age of intrusion of the Anzola-

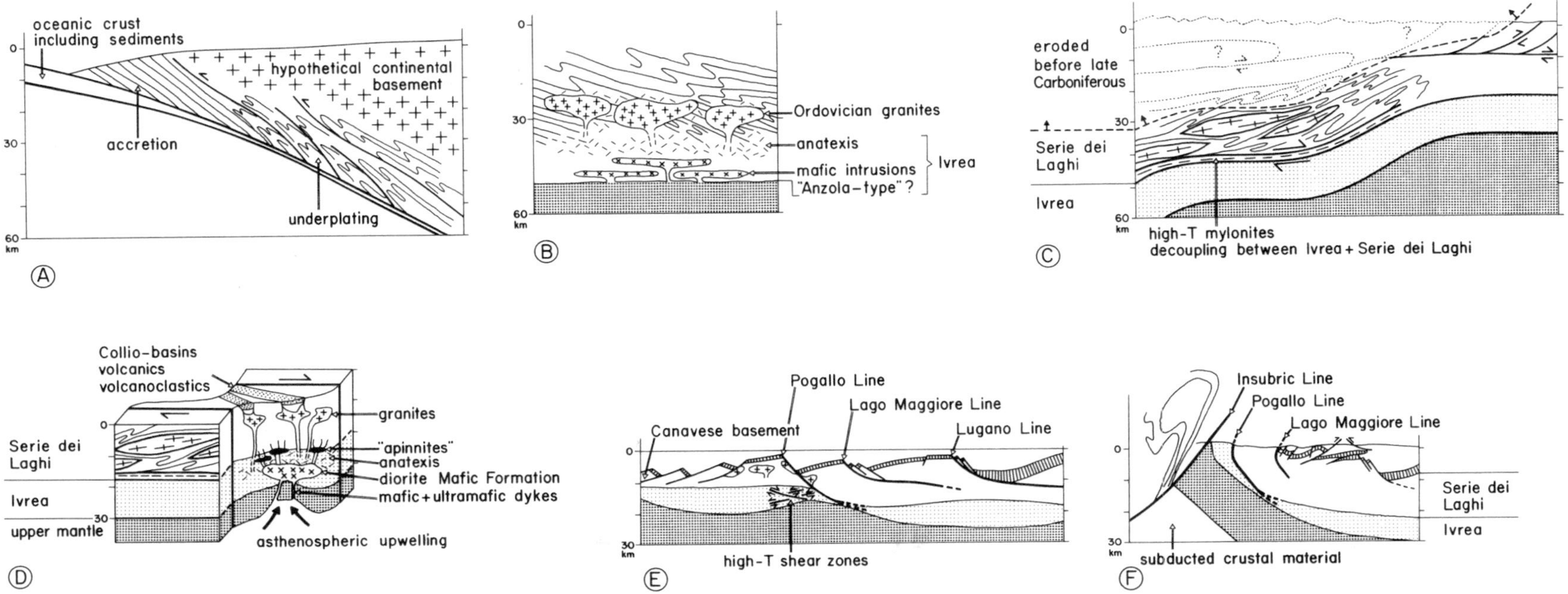

Fig. 3A–F. Largely speculative sketches illustrating the evolutionary scheme discussed in the text. All sketches have the same scaling (vertically and horizontally) and are very approximately viewed towards the N (W on the *left-hand side*). **A** Late Proterozoic or Early Palaeozoic; **B** Ordovician; **C** Variscan orogeny; **D** Late Palaeozoic (very Late Carboniferous – Early Permian) wrench tectonics and associated magmatic activity; **E** Early Mesozoic crustal attenuation; **F** Alpine (mainly Late Oligocene to Early Miocene) orogenic movements

type gabbros is unknown but tentatively assumed to be contemporaneous with this first event of anatexis in the paragneisses (Fig. 3B).

Much of the controversy about the "age of regional metamorphism" in the Ivrea zone is redundant in the sense that a long-lasting history of polyphase deformation, dehydration, anatexis and melt mobilization produced different effects on the various isotopic systems used for dating (see discussion in Zingg et al. 1990). For example, the youngest ages of around 270 Ma related to anatexis found within and adjacent to the Mafic formation (Pin 1986; Bürgi and Klötzli 1990) do not date the "age of metamorphism" but most probably a magmatic event at the end of a long-lasting history of deep burial (episode 1 of Handy and Zingg 1991) of the Ivrea zone. This was immediately followed by or perhaps contemporaneous with initial stages of uplift indicated by the cooling ages (for a compilation of cooling age data, see Zingg et al. 1990 and Handy and Zingg 1991). The metamorphic zonation probably does not have an "age" either since it is expected to be the final product of this long-lasting history. Individual deformational episode 1 events (Handy and Zingg 1991) associated with this high-grade metamorphism cannot be tied to individual metamorphic events, since the microstructures are all strongly annealed.

In case of the Ivrea zone, a Variscan tectonometamorphic event can only be indirectly inferred from the pervasive polyphase deformational overprint of the various lithological units previously juxtaposed in the Early Palaeozoic accretionary wedge setting. Its age is poorly constrained from radiometric age dating (see discussion in Zingg et al. 1990). But the deformation related to this event certainly pre-dates the 270 Ma thermal overprint in the contact zone of the Mafic formation.

Given the Ordovician granites as a time-marker (466 Ma) the distinction between pre-Ordovician (pre-intrusion) and Variscan (post-intrusion) structures is easier for the Serie dei Laghi. The inferred Variscan age for the main metamorphic and deformational overprint, as seen today, is primarily based on structural observations indicating that mineral growth is related to the main schistosity post-dating the Ordovician granite intrusion (Borghi 1987, 1988). Interpretation of radiometric age determinations, however, is more problematic. Zircon U-Pb and Rb-Sr total rock ages scatter between 550 and 450 Ma, while mineral ages, including hornblende K-Ar and muscovite Rb-Sr data (scattering between 390 and 260 Ma) indicate that temperatures of at least 500 °C persisted during the Variscan event (see compilation of age data in Zingg et al. 1990). Boriani et al. (1982/83) suggest that the muscovite Rb-Sr ages around 325 Ma date this Variscan event more precisely. Some of the monazite U-Pb ages (Köppel 1974; Köppel and Grünenfelder 1978/79) also indicate a Variscan event between 320 and 285 Ma. On the other hand, Zingg (1983) casts doubts on the interpretation of these ages as formation ages. The cooling curve for the southern part of the Serie dei Laghi is reasonably well established for the region S of Brissago (see Fig. 4b in Handy and Zingg 1991) and suggests rapid cooling immediately post-dating the 325 Ma muscovite Rb-Sr age. Since rapid exhumation between 325 and about 310 Ma (biotite ages closing at around 300 °C) is probably related to Variscan orogeny (see Handy and Zingg 1991 and the following discussion) the 325–310 Ma cooling period for the southern part of the Serie dei Laghi probably dates the final stages of Variscan orogeny.

In spite of these difficulties in interpreting radiometric data we regard the structural observations of Borghi (1987, 1988) as convincing evidence for a Variscan age of the main tectono-metamorphic event in the Serie dei Laghi. The relatively high pressures (6–8 kb) indicated for this Variscan event characterized by the widespread occurrence of kyanite are in contrast to those of the amphibolite facies part of the Ivrea zone (5–6 kb) where sillimanite is the predominant phase. Two interpretations are possible: (1) the Ivrea zone and Serie dei Laghi do not represent a crustal section during Variscan orogeny at all and are juxtaposed in late Variscan times by movements along the CMB line (Boriani et al. 1990a); or (2) an earlier baric peak was completely overprinted by subsequent temperature dominated metamorphism and/or Late Palaeozoic magmatism in the Ivrea zone where high temperature conditions persisted much longer. In fact, kyanite has also been reported from the Ivrea zone in the Val Sesia, Biella and Val Grande region (Bertolani 1959; Capedri 1971; Boriani and Sacchi 1973; Handy 1986).

Arguments against late Variscan activity along the CMB line have already been discussed. Additionally, the following argument favours the second interpretation: both sillimanite and kyanite are found within the Serie dei Laghi adjacent to the Ivrea zone in Val Cannobina (Boriani et al. 1977), sillimanite growing later and post-kinematically. Borghi (1988) reports a substantial decrease in pressure at constant temperature associated with this post-kinematic metamorphic stage, also found in the Gamborogno area. Thus, it seems that a later thermal overprint, albeit weaker, is also present in the Serie dei Laghi. However, this thermal overprint was closely followed by uplift (margarite coronas around kyanite reported by

Borghi (1988) and did not reach completion within the Serie dei Laghi.

Two additional features of the Variscan tectonometamorphic event in the Serie dei Laghi contrast with findings in the Ivrea zone: (1) the absence of a metamorphic zonation except near the contact with the Ivrea zone; and (2) the presence of greenschist facies deformational overprint pre-dating Late Palaeozoic magmatism. This suggests decoupling between the lower crust (Ivrea zone) and intermediate crustal levels (Serie dei Laghi) during the Variscan orogeny (Fig. 3C) rather than late Variscan juxtaposition along the CMB line. This decoupling could be held responsible for major differences in structural style between the Ivrea zone and Serie dei Laghi (see also Handy and Zingg 1991): large-scale Pre-Alpine folding leading to variable orientations of stretching lineations ("Schlingen" folds) is restricted to the Serie dei Laghi. Such large-scale folding and/or faulting, associated with crustal thickening and subsequent exhumation during the pre-Late Palaeozoic greenschist facies event of Borghi (1988), could be held responsible for the disruption of metamorphic gradients and early exhumation of large parts of the Serie dei Laghi. Due to the considerable lack of detailed structural investigations in the Serie dei Laghi, no more can be said with respect to the Variscan orogeny and its geotectonic setting.

It has to be emphasized, however, that the exhumation history of the SE part of the Serie dei Laghi definitely pre-dates that of the Ivrea zone and thus cannot be related to Late Palaeozoic or Mesozoic events discussed in the next section. In the SE part of the Serie dei Laghi, exhumation occurred during the closing stages of Variscan orogeny (325–310 Ma) and was accompanied by very substantial amounts of erosion (indicated in Fig. 3C) as documented by the Permo-Carboniferous transgression onto amphibolite grade basement. No direct signs of exhumation and cooling pre-dating Late Palaeozoic and Mesozoic events have been found in the Ivrea zone so far. The Ivrea lower crust is completely detached from higher crustal levels and remains in a regime of elevated temperatures. However, since the considerable amounts of erosion in the Serie dei Laghi suggest isostatic uplift of overthickened intermediate and upper crustal levels, isothermal uplift of the lower crust is to be expected. The pressure estimates of Sills (1984) for the lowermost granulite facies part (< 9 kb) indeed suggest that normal crustal thickness of around 30 km was reinstalled before final equilibration of the geobarometers. It is this isostatic and isothermal uplift which could lead to temperature-dominated overprint in the Ivrea zone, related to the re-installation of normal crustal thickness and strong annealing of microstructures.

4.2 Age and Geotectonic Setting of Magmatism and Uplift During the Late Palaeozoic and Early Mesozoic

Diametrically opposed views have been proposed with regard to the Late Palaeozoic and Early Mesozoic evolution. One is that expressed by Boriani et al. (1990a,b) who emphasize the importance of transcurrent movements during the Late Palaeozoic (Permian) period. They invoke a subduction zone with oblique incidence along the CMB line connected with uplift of the Ivrea zone and emplacement of dioritic stocks and dykes in transtensional sectors, immediately followed by an Early Permian magmatic arc represented by the Graniti dei Laghi and related volcanics. Stille and Buletti (1987) even propose SE-directed subduction of Ivrea oceanic crust during Permian time, a scenario similar to that proposed in the previous Section for Early Palaeozoic times. They suggest that final closure of the Proto-Tethys did not occur before Permian times. These views are not only incompatible with the regional metamorphic evolution during the closing stages of Variscan orogeny discussed so far, but also with the large-scale scenario proposed by Ziegler (1986) indicating Westphalian-Early Stephanian (about 300 Ma) consolidation of the Variscan fold belt in Europe.

Another view assumes that no crustal thickening occurred since Early Permian times and that the present-day thickness of Ivrea zone and Serie dei Laghi in the Val d'Ossola profile allows for an estimate of the total crustal thickness of about 30 km for the Southern Alpine basement (sketched in Fig. 3D) before the onset of Late Palaeozoic events (Handy 1987). Thereby, the effects of Variscan orogeny and crustal thickening are not taken into consideration, i. e. it is tacitly inferred that a normal crustal thickness (30 km) was re-installed after the Variscan orogeny and before deposition of the Permo-Carboniferous cover about 300 Ma ago. These assumptions are reasonable in the light of the discussions of the previous chapter and the findings of Eisbacher et al. (1989) for the Schwarzwald-Vosges area. It is important to recall that estimates on the amount of crustal thinning during Late Palaeozoic and Early Mesozoic crustal attenuation (Handy 1987) rely on these assumptions, which are not unanimously accepted.

Deformations related to episode 2 deformation, in

the sense of Handy and Zingg (1991), clearly post-date annealed microstructures for the high-temperature shear zones observed in many parts of the Ivrea zone, some of them forming under similar or even higher metamorphic conditions than those prevailing during episode 1. Brodie and Rutter (1987) inferred some 2 km of crustal thinning produced by an anastomosing network of mylonite zones and small-scale shear zones within the lowermost 5 km of the Ivrea zone adjacent to and within ultramafic bodies. They infer E-W-directed extension by restoring the stretching lineations into their pre-Alpine orientation. Handy (1987) estimated a total amount of crustal thinning of around 8–10 km produced by normal faulting along the contact between Ivrea zone and Serie dei Laghi at and in the vicinity of the Pogallo fault, again indicating E-W extension when restoring the Ivrea zone into a subhorizontal attitude. There, although extension initiated under peak metamorphic conditions, most of the movement occurred under retrograde conditions. Assuming an original custal thickness of 30 km, the total amount of stretching results in a stretching factor of $\beta = 1.5$–1.7.

Difficulties arise during attempts to date these extensional features more exactly within the large 270–180 Ma time interval indicated by the cooling ages in the Ivrea zone. Direct observational links between stretching and Late Palaeozoic magmatic activity have not been established yet. Brodie et al. (1989) date initiation of crustal thinning in the lowermost Ivrea crust at around 280 Ma, interpreting Ar-Ar data obtained from extensional shear zones to be cooling ages. This age closely corresponds to the age of intrusion of parts of the Mafic formation in Val Sesia. Thus the manifestations of Late Palaeozoic magmatic activity are contemporaneous with the initial stages of E-W-directed extension according to these authors.

It follows from this discussion that the initial phases of crustal attenuation may be Late Palaeozoic. Handy and Zingg (1991) proposed that these initial stretching stages reflect a tectonic regime which is distinctly different from passive continental margin formation in the Mesozoic. Based on the geometry of presently ENE-WSW-trending Collio-Verrucano basins they propose sinistral transtension of the Southern Alpine crust, coeval with Permian magmatism in all crustal levels, and compatible with the observed E-W stretching lineations in the high-temperature mylonites of Brodie and Rutter (1987). Handy and Zingg (1991) suggest that such local sinistral transtension would have to be conjugate with large-scale dextral shear between Africa and Europe (Arthaud and Matte 1977; Ziegler 1986).

Concerning the Pogallo fault, Handy (1987) estimates that about 3 km of crust was thinned at the Ivrea zone – Serie dei Laghi – contact sometime between Early Permian and Middle Triassic times, based on an interpretation of cooling ages adjacent to the Pogallo line. However, an additional 5–7 km have to be kinematically linked to a new tectonic scenario: the opening of the NNE–SSW–trending asymmetrical rifts formed during the Late Triassic to Early Jurassic (Bernoulli 1964; Bertotti 1990, 1991). The idea, that earlier crustal attenuation related to wrench tectonics was actually related to movements along the Pogallo fault zone, as seen today, is not compelling. The structural discordance across the greenschist facies mylonites and cataclasites at the southern margin of the Pogallo fault zone certainly formed during the Early Jurassic when temperature in the Ivrea zone fell to below 300 °C.

Late Palaeozoic magmatism is coeval at all crustal levels (Fig. 3D). The new data and observations of Bürgi and Klötzli (1990) lend additional support to the suggestion first proposed by Fountain (1986) that intrusion of the Mafic formation caused partial melting in deep crustal metasediments which in turn led to Permian plutonic and volcanic activity in higher crustal levels. Additionally, crustal contamination within phlogopite peridotites and intrusion of mafic and ultramafic dykes into the Balmuccia peridotite (Voshage et al. 1987, 1988) are contemporaneous with this event. This points to a deep seated origin of magmatic activity. Whole crustal failure due to Late Palaeozoic wrench tectonics (Ziegler 1986) could have allowed rapid magma ascent (Handy and Zingg 1991). Potentially, the Ivrea zone offers an unique opportunity to study the deep crustal manifestations of this, often underestimated, major tectonic and magmatic event. Since this event post-dates exhumation of the Serie dei Laghi and restauration of a normal crustal thickness after the Variscan orogeny, the term "Late Variscan" is misleading. In fact, Arthaud and Matte (1977) suggest that this wrench faulting event is independent of the previous evolution and forms a peculiar class of tectonic movement affecting the entire rigid crust.

Details on contemporaneous strain partitioning between lower crustal extension by anastomosing conjugate shear zones, localized shear along the Pogallo fault and faulting directly associated with basin formation (Fig. 3E) remain problematic, due to dating problems with regard to the high-temperature shear zones in the lower crust. However, we do not regard the arguments for Late Palaeozoic initiation of shear zone formation proposed by Brodie et al. (1989) as conclusive. According to Brodie and Rutter

(1987), the geometric characteristics of these conjugate shear zones suggest substantial shortening perpendicular to layering (i. e. in a vertical direction after retro-deformation of Alpine tilting). This type of strain is hardly compatible with wrench tectonics. Consequently, we favour an Early Mesozoic age for their formation (Fig. 3E). From large-scale considerations and from a sedimentological point of view (Bertotti 1990, 1991) it is more appropriate to separate clearly Late Palaeozoic wrench tectonics from the onset of crustal attenuatiuon during the Middle Triassic. In fact, a Middle Triassic magmatic event related to the initiation of rifting has recently been reported by Stähle at al. (1990) from the Ivrea zone.

4.3 Emplacement Tectonics During Alpine Orogeny

The cooling ages indicate that temperatures fell below 300 °C before the Middle Jurassic for the entire Southern Alpine basement section exposed today. Therefore, emplacement tectonics during the Alpine orogeny was only associated with moderate uplift and the temperatures prevailing during Alpine deformation did not allow for a penetrative structural overprint. Consequently, the Ivrea zone acted as a relatively rigid indenter during post-collisional transpressive movements, at a retraining bend of the Insubric line (Merle et al. 1989; Schmid et al. 1989).

Nevertheless, the pre-Alpine main foliation, including high-grade mylonite horizons related to Late Palaeozoic and Early Mesozoic exhumation were affected by flexural slip folding producing major antiforms immediately S of the Insubric line. Schmid et al. (1987) regard these antiforms as synchronous with backthrusting along the Insubric line during the Late Oligocene. The tilted cross-section through the Southern Alpine crust in the Val d'Ossola geometrically corresponds to the SE limb of the Proman antiform with granulite facies Ivrea zone in its core (Figs. 2 and 3F). Its NW limb is constituted by strongly mylonitized remnants of higher crustal levels (Canavese and amphibolite facies Ivrea zone). Extreme attenuation of the NW limb and lack of lateral continuity between individual antiforms along strike is due to later overprint by strike-slip movements along the Insubric line. Conflicting views exist on the metamorphic conditions associated with another antiform at Finero (greenschist facies conditions, according to Kruhl and Voll 1976 versus amphibolite facies conditions, reported by Steck and Tièche 1976). The eastern plunge of yet another antiform (undoubtedly formed under greenschist facies conditions) of inferred Alpine age near Ascona (Schmid unpubl. results) is reponsible for the eastern termination of the Ivrea zone which is interpreted to be at much greater depth (around 20 km) along the NFP 20 southern traverse (Bernoulli et al. 1990). On the other hand, tilting due to large-scale folding seems to persist towards the SW and into the ECORS profile (Rey et al. 1990). The basement-cover contact, under the cover of the Po plain, is gently SE dipping along the ECORS profile (Roure et al. 1990), indicating a hinge zone between the verticalized Ivrea cross-section and this gently dipping basement-cover contact. During the Late Oligocene to Early Miocene, i.e. contemporaneous with verticalization due to folding, this basement-cover contact was shortened by minor thrusting (Roure et al. 1990).

East of Lago Maggiore, the Alpine deformation is more complex and appears to be characterized by a network of thrusts (e.g. the Gamborogno thrust mapped in Fig. 1) and transfer zones forming in response to an interference between E-W and N-S compression during several phases of Alpine overprint (Schumacher 1990). Thus, the basement high, E of Lago Maggiore, cannot be uniquely related to tilting of a cross section through the Southern Alpine basement. Even further to the E (Grignia section, Laubscher 1985) it may be entirely related to ramp antiforms forming during S-vergent thrusting.

Only a limited amount of fission track data are presently available for radiometrically dating Alpine emplacement tectonics (Hurford 1986; Hurford et al. 1989; Bürgi and Klötzli 1990). According to the apatite fission track data tilting of the Ivrea zone and vertical components of relative movement along the Insubric line are older than 12 Ma. The zircon data exhibit a large scatter, possibly due to rejuvenation near the Insubric line. However, the age of 64 Ma reported by Bürgi and Klötzli points to a pre-Oligocene phase of uplift, predating verticalization during the Late Oligocene. Earlier events are also indicated by K-Ar mineral ages found within the Canavese sediments along the SW part of the Insubric line (Zingg and Hunziker 1990), a segment of the Insubric line where pre-Oligocene mylonites are preserved (Schmid et al. 1989).

Late Cretaceeous and/or Palaeocene orogeny affected a large part of the Southern Alpine basement, presently exposed in the Central Alps (Sesia zone, including the "seconda zona diorito kinzigitica", Dal Piaz this Vol). Schmid et al. (1987) speculated that the density inversion at the base of the Ivrea geophysical body is due to underplating by a SE-dipping continuation of the Sesia zone during this early Alpine event.

Brack (1981) documented pre-Late Eocene deformation near the Adamello intrusion, a compressional phase held responsible for much of the thrusting E of Lago Maggiore by Schumacher (1990). Zingg and Hunziker (1990) report Cretaceous mica ages in the Canavese sediments at the SW end of the Insubric line. No structures related to this early phase of Alpine tectonics have been documented so far in the Ivrea zone.

Other phases of Alpine overprint are postulated for the cover of the Southern Alps (Late Eocene to Early Miocene and Late Miocene thrusting, Bernoulli et al. 1989). However, their effect is not documented west of Lago Maggiore where large-scale tilting of the SE limb of the Proman antiform during the Late Oligocene and Early Miocene is held responsible for exposure of this unique cross section through the entire crust (Fig. 3F).

4.4 Do Ivrea Zone and Serie dei Laghi Expose a Typical Cross-Section Through Continental Crust?

It was argued that the major effect of Alpine orogeny was large-scale tilting in the absence of a penetrative overprint and/or major thrusting along the Val d'Ossola section. Supposing this interpretation to be correct (for opposing views see Boriani et al. 1990a,b) a cross-section through pre-Alpine crust is indeed exposed. However, does this cross-section expose a typical section through continental crust in general, a Palaeozoic orogen, or, alternatively thinned continental crust in a passive continental margin environment?

It is hoped that the previous discussions made it clear that this section is the result of a long-lasting complex evolution. It began with accretion of thick packages of metasediments and mafic intrusions above a subducting oceanic lithosphere and below an old continental basement, not preserved within the area of interest, followed by a long-lasting period of consolidation in a lower crustal environment and partial exhumation due to Late Palaeozoic and Early Mesozoic extension tectonics. In this sense, the presently exposed section is a composite of an evolution which is particular for this zone. Therefore, it can certainly not provide a universally valid "type section" of continental crust. Very likely the search after a "type section" is a fruitless enterprise, anyway, in the sense thast it would reflect a highly immobilistic view of crustal evolution.

In regard to the Variscan orogeny, the potential advantage of having an exposed crustal section is counteracted by the disadvantage that higher crustal levels are not preserved, due to substantial amounts of erosion, before deposition of the Permo-Carboniferous onto amphibolite grade rocks. Consequently, only deeper crustal levels of this orogen are exposed. A comparison of P-T conditions between the Ivrea zone and Serie dei Laghi suggests that P-T-conditions did not equilibrate at the same time within these units which were decoupled during orogeny. Hence, the present metamorphic zonation does not reflect the P-T conditions at a particular instant of time during Palaeozoic orogeny. This is most dramatically exemplified by the contemporaneity between granulite facies conditions in the Ivrea zone and transgression of the Permo-Carboniferous cover onto amphibolite grade Serie dei Laghi. The most important consequences of the interpreations regarding Variscan orogeny are: (1) decoupling between lower und intermediate crust during Variscan orogeny; and (2) re-equilibration to normal crustal thickness during the late stages of Variscan orogeny.

The most direct insight offered in this cross section relates to the Late Palaeozoic magmatic activity (Fig. 3D) and to extensional tectonics during the Early Mesozoic (Fig. 3E) which affected a piece of continental crust which re-equilibrated to normal thickness before the onset of both these events. The section offers a unique opportunity to study contemporaneous magmatic activity associated with Late Palaeozoic wrench tectonics at all crustal levels, an opportunity which largely remains unexploited by magmatic petrologists so far. Due to dating problems, it cannot be excluded that this Late Palaeozoic magmatic activity is contemporaneous with the initiation of crustal thinning in the Ivrea zone and along the NE part of the contact between Ivrea zone and Serie dei Laghi.

Attempts to reconstruct the geometry and kinematics of crustal attenuation during the Early Mesozoic have been made (Brodie and Rutter 1987; Handy 1987; Schmid et al. 1987; Bertotti 1990, 1991; see Fig. 3E) and suggest that each crustal level had its own particular style for acommodating extension: listric or domino-type faults in upper crustal levels, localized faulting (Pogallo fault) at mid-crustal levels, and extension of the lower crust by a network of conjugate shear zones. However, these reconstructions assume contemporaneous stretching during the Early Mesozoic at all crustal levels.

Seismologists early recognized the potential offered by this Southern Alpine cross-section for interpreting seismic properties of the crust, in particular of the highly reflective lower crust (e.g. Fountain 1976). Recently, this cross-section has been sampled for laboratory studies on seismic wave velocities (e. g. Burke and Fountain 1990). These studies confirm the view that Ivrea zone and Serie dei Laghi represent an excellent example of attenuated continental crust.

The reconstruction of the evolutionary scheme as depicted in Fig. 3 is admittedly speculative, in particular regarding the stages pre-dating Mesozoic extension. Nevertheless, such a reconstruction was attempted with the main purpose of stimulating further research directed towards solving fascinating problem-oriented questions. The final stages of the evolution (Fig. 3E,F) and, in particular, the Alpine age of tilting of a crustal section are much better constrained by the available data. Concluding, we firmly believe that Ivrea zone and Serie dei Laghi offer a unique opportunity to study a section through attenuated continental crust and to further investigate deep crustal manifestations of pre-Alpine orogeny.

Acknowledgements
J. Hunziker and M. Handy improved an earlier draft of the manuscript. A very critical review of A. Boriani and E. Giobbi was very valuable for further improvements. A. Zingg is thanked for first introducing me to this interesting area and for many discussions over the years.

References

Argand E (1910) Sur la racine de la nappe rhétique. Beitr Geol Karte Schweiz NF 24: 17–19

Arthaud F, Matte Ph (1977) Late Paleozoic strike-slip faulting in southern Europe and northern Africa: result of a right-lateral shear zone between the Appalachians and the Urals. Geol Soc Am Bull 88: 1305–1320

Berckhemer H (1968) Topographie des "Ivrea-Körpers", abgeleitet aus seismischen und gravimetrischen Daten. German Research Group for Explosion Seismology. Schweiz Mineral Petrogr Mitt 48: 235–246

Bernoulli D (1964) Zur Geologie des Monte Generoso (Lombardische Alpen. Beitr Geol Karte Schweiz NF118: 135 pp

Bernoulli D, Bertotti G, Zingg A (1989) Northward thrusting of the Gonfolite Lombarda ("South-Alpine Molasse") onto the Mesozoic sequence of the Lombardian Alps: implications for the deformation history of the Southern Alps. Eclogae Geol Helv 82: 841–856

Bernoulli D, Heitzmann P, Zingg A (1990) Central and Southern Alps in southern Switzerland. Tectonic evolution and first results of reflection seismics. In: Roure F, Heitzmann P, Polino R (eds) Deep structure of the Alps. Soc Géol Fr Mém 156: 289–302

Bertolani M (1959) La formazione basica "Ivrea-Verbano" e la sua posizione nel quadro geologico-petrografico della Basse Valsesia e del Biellese. Period Mineral 28: 151–209

Bertotti G (1990) The deep structure of the Monte Generoso basin: an extensional basin in the south-Alpine Mesozoic passive continental margin. In: Roure F, Heitzmann P, Polino R (eds) Deep structure of the Alps. Soc Géol Fr Mém 156: 303–308

Bertotti G (1991) Early Mesozoic extension and Alpine shortening in the western Southern Alps: the geology of the area between Lugano and Menaggio (Lombardy, northern Italy). Mem Sci Geol Padova 43: 17–123

Bigioggero B, Boriani A (1975) I noduli a silicati di Al negli gneiss minuti della Strona-Ceneri. Boll Soc Geol Ital 94: 2073–2084

Biino G, Compagnoni R (1989) The Canavese zone between the Serra d'Ivrea and the Dora Baltea River (Western Alps). Eclogae Geol Helv 82: 413–427

Borghi A (1987) Osservationi strutturali ed implicazioni geologiche nel settore N-E della "Serie dei Laghi" (Alpi Meridionali). Rend Soc Geol Ital 10: 75–78

Borghi A (1988) Evoluzione metamorfica del settore nord-est della Serie dei Laghi (Alpi Meriodionali – Canton Ticino). Rend Soc Geol Ital 11: 165–170

Borghi A (1989) L'evoluzione metamorphico-strutturale del settore nord-orientale della Serie dei Laghi (Alpi Meridionali). Dissertatione finale, Università di Torino, 187 pp

Boriani A, Sacchi R (1973) Geology of the junction between the Ivrea-Verbano and Strona-Ceneri zones. Mem Ist Geol Mineral Univ Padova 28: 1–35

Boriani A, Peyronel Pagliani G (1968) Rapporti fra le putoniti erciniche e le metamorfiti del "Massiccio dei Laghi" nella zona del M. Cerano (bassa Val d'Ossola). Rend Soc Ital Mineral Petrol 24: 111–142

Boriani A, Bigioggero B, Origoni Giobbi (1977) Metamorphism, tectonic evolution and tentative stratigraphy of the "Serie dei Laghi" – geological map of the Verbania area (northern Italy). Mem Sci Geol 32: 1–25

Boriani A, Origoni Giobbi E, Del Moro A (1982/83) Composition, level of intrusion and age of the Serie dei Laghi orthogneisses (northern Italy-Ticino, Switzerland). Rend Soc Ital Mineral Petrol 38: 191–205

Boriani A, Burlini L, Sacchi R (1990a) The Cossato-Mergozzo-Brissago line and the Pogallo line (Southern Alps, northern Italy) and their relationships with the Late Hercynian magmatic and metamorphic events. Tectonophysics 182: 91–102

Boriani A, Giobbi Origoni E, Borghi A, Caironi V (1990b) The evolution of the "Serie dei Laghi" (Strona-Ceneri and Schisti dei Laghi): the upper component of the Ivrea-Verbano crustal section; Southern Alps, North Italy and Ticino, Switzerland. Tectonophysics 182: 103–118

Brack P (1981) Structures in the southwestern border of the Adamello intrusion (Alpi bresciane, Italy). Schweiz Mineral Petrogr Mitt 61: 37–50

Brodie KH, Rutter EH (1987) Deep crustal extensional faulting in the Ivrea zone of northern Italy. Tectonophysics 140: 193–212

Brodie KH, Rex D, Rutter EH (1989) On the age of deep crustal extensional faulting in the Ivrea zone, northern Italy. In: Coward MP, Dietrich D, Park RG (eds) Alpine tectonics. Geol Soc Lond Spec Publ 45: 203–210

Buletti M (1983) Zur Geochemie und Entstehungsgeschichte der Granat-Amphibolite des Gamborognogebietes, Ticino, Südalpen. Schweiz Mineral Petrogr Mitt 63: 233–247

Bürgi A, Klötzli U (1990) New data on the evolutionary history of the Ivrea zone (northern Italy). Bull Swiss Assoc Petrol Geol Eng 56 (130): 49–70

Burke MM, Fountain DM (1990) Seismic properties of rocks from an exposure of extended continental crust – new laboratory measurements from the Ivrea Zone. Tectonophysics 182: 119–146

Capedri S (1971) Sulle rocce della formazione basica Ivrea-Verbano. 2. Petrogratia delle granuliti e rocce derivate affioranti nella Val Mastallone (Vercelli) e loro evoluzione petrogenetica. Mem Soc Geol Ital 10: 277–312

Cassinis G, Mattavelli I, Morelli GI (1978) Studio petrografico e mineralogico delle formazione di Collio nel Permiano inferiore nell'alta Val Trompia (Prealpi Bresciane). Mem Soc Geol Univ Padova 32: 1–13

Eisbacher GH, Lüschen E, Wickert F (1989) Crustal-scale thrusting and extension in the Hercynian Schwarzwald and Vosges, central Europe. Tectonics 8: 1–21

Fountain DM (1976) The Ivrea-Verbano and Strona-Ceneri zones, northern Italy: a cross-section of the continental crust – new evidence from seismic velocities. Tectonophysics 33: 145–166

Fountain DM (1986) Implications of deep crustal evolution for seismic reflection seismology. In: Barazangi M, Brown L (eds) Reflection seismology; the continental crust. Am Geophys Union Geodyn Ser 14: 1–7

Handy MR (1986) The structure and rheological evolution of the Pogallo Fault zone, a deep crustal dislocation in the Southern Alps of northwestern Italy (Prov. Novara). Inauguraldissertation, Phil-Naturwiss Fak Univ Basel, 327 pp

Handy MR (1987) The structure, age and kinematics of the Pogallo Fault zone; Southern Alps, northwestern Italy. Eclogae Geol Helv 80: 593–632

Handy MR, Zingg A (1991) The tectonic and rheological evolution of an attenuated cross-section of the continental crust: Ivrea crustal section, Southern Alps, northwestern Italy and southern Switzerland. Geol Soc Am Bull 103: 236–253

Hodges KV, Fountain DM (1984) Pogallo line, South Alps, northern Italy: an intermediate crustal level, low angle normal fault? Geology 12: 151–155

Hunziker JC, Zingg A (1980) Lower Paleozoic amphibolite to granulite facies metamorphism in the Ivrea-Zone (Southern Alps-northern Italy). Schweiz Mineral Petrogr Mitt 60: 181–213

Hunziker JC, Zingg A (1982) Zur Genese der ultrabasischen Gesteine der Ivrea-Zone. Schweiz Mineral Petrogr Mitt 62: 483–486

Hurford AJ (1986) Cooling and uplift patterns in the Lepontine Alps (south-central Switzerland) an an age of vertical movement on the Insubric fault line. Contrib Mineral Petrol 92: 413–427

Hurford AJ, Flisch M, Jäger E (1989) Unraveling the thermotectonic evolution of the Alps: a contribution from fission track analysis and mica dating. In: Coward MP, Dietrich D, Park RG (eds) Alpine tectonics. Geol Soc Spec Publ 45: 369–398

Köppel V (1974) Isotopic U-Pb ages on monazites and zircons from the crust-mantle transition and adjacent units of the Ivrea and Ceneri zones (Southern Alps, Italy). Contrib Mineral Petrol 43: 55–70

Köppel V, Grünenfelder M (1978/79) Monazite and zircon U-Pb ages from the Ivrea and Ceneri zones. Abstr, 2nd Symp Ivrea-Verbano, Varallo. Mem Sci Geol 33: 257

Kruhl JH, Voll G (1976) Fabrics and metamorphism from the Monte Rosa root zone into the Ivrea zone near Finero, southern margin of the Alps. Schweiz Mineral Petrogr Mitt 56: 627–633

Laubscher HP (1985) Large-scale, thin-skinned thrusting in the southern Alps: kinematic models. Geol Soc. Am Bull 96: 710–718

Lensch G (1971) Die Ultramafitite der Zone von Ivrea. Ann Univ Sarav 9: 1–146

Lensch G, Rost F (1972) Keliphitperidotite in der mittleren Ivreazone zwischen Val d'Ossola und Val Strona. Ein Beitrag zur Herkunftstiefe der Ultramafitite der Ivreazone. Schweiz Mineral Petrogr Mitt 52: 237–250

Mancktelow NS (1990) The Simplon Fault zone. Beitr Geol Karte Schweiz NF163: 74 pp

Merle O, Cobbold PR, Schmid SM (1989) Tertiary kinematics in the Lepontine dome. In: Coward MP, Dietrich D, Park RG (eds) Alpine tectonics Geol Soc Lond Spec Publ 45: 113–134

Milano FP, Pennacchioni G, Spalla IM (1988) Alpine and pre-Alpine tectonics in the central Orobic Alps (Southern Alps). Eclogae Geol Helv 81: 273–294

Novarese V (1929) La zona del Canavese e le formazioni adiacenti. Mem Descr Carta Geol Ital 22: 65–212

Pin C (1986) Datation U-Pb sur zircons a 285 M.a. du complexe gabbro-dioritique du Val Sesia – Val Mastallone etage tardihercynien du métamorphisme granulitique de la zone Ivrea-Verbano (Italie). C R Acad Sci Paris 303/II: 827–830

Pin C (1990) Evolution of the lower crust in the Ivrea zone: a model based on isotopic and geochemical data. In: Vielzeuf D, Vidal Ph (eds) Granulites and crustal evolution. Kluwer, Pordrecht, pp 87–110

Reinhard M (1964) Über das Grundgebirge des Sottoceneri im Süd-Tessin und die darin auftretenden Ganggesteine. Beitr Geol Karte Schweiz NF117: 89 pp

Rey D, Quarta T, Mouge P, Miletto M, Lanza R, Galdeano A, Carozzo MT, Armando E, Bayer R (1990) Gravity and aeromagnetic maps of the western Alps: contribution to the knowledge on the deep structures along the Ecors-Crop seismic profile. In: Roure F, Heitzmann P, Polino R (eds) Deep structure of the Alps. Soc Géol Fr Mém 156: 107–122

Rivalenti G, Garuti G, Rossi A, Siena F, Sinigoi S (1981) Existence of different peridotite types and a layered igneous complex in the Ivrea zone of the Western Alps. J Petrol 22: 127–153

Roure F, Polino R, Nicolich R (1990) Early Neogene deformation beneath the Po plain: constraints on post-collisional Alpine evolution. In: Roure F, Heitzmann P, Polino R (eds) Deep structure of the Alps. Soc Géol Fr Mém 156: 309–321

Schmid R (1967) Zur Petrographie und Struktur der Zone Ivrea-Verbano zwischen Valle d'Ossola und Val Grande (Prov. Novara, Italien). Schweiz Mineral Petrogr Mitt 47: 935–1117

Schmid SM, Zingg A, Handy M (1987) The kinematics of movements along the Insubric line and the emplacement of the Ivrea zone. Tectonophysics 135: 47–66

Schmid SM, Aebli HR, Heller F, Zingg A (1989) The role of the Periadriatic line in the tectonic evolution of the Alps. In: Coward MP, Dietrich D, Park RG (eds) Alpine tectonics. Geol Soc Spec Publ 45: 153–171

Schumacher ME (1990) Alpine basement thrusts in the eastern Seengebirge, Southern Alps (Italy/Switzerland). Eclogae Geol Helv 83: 645–663

Shervais JW (1979) Thermal emplacement model for the Alpine Lherzolite massif at Balmuccia, Italy. J Petrol 20: 795–820

Sills JD (1984) Granulite facies metamorphism in the Ivrea zone, N.W. Italy. Schweiz Mineral Petrogr Mitt 64: 169–191

Sills JD, Tarney J (1984) Petrogenesis and tectonic significance of amphibolites interlayered with metasedimentary gneisses in the Ivrea zone, Southern Alps, northwest Italy. Tectonophysics 107: 187–206

Stähle V, Frenzel G, Kober B, Michard A, Puchelt H, Scheider W (1990) Zircon syenite pegmatites in the Finero peridotite (Ivrea zone): evidence for a syenite from a mantle source. Earth Planet Sci Lett 101: 196–205

Steck A, Tièche J-C (1976) Carte géologique de l'antiforme de Finero avec des observations sur les phases de déformation et de recristallisation. Bull Suisse Minéral Pétrogr 56: 501–512

Stille P, Buletti M (1987) Nd-Sr isotopic characteristics of the Lugano volcanic rocks and constraints on the continental crust formation in the South Alpine domain (N-Italy-Switzerland). Contrib Mineral Petrol 96: 140–150

Vai GB, Cocozza T (1986) Tentative schematic zonation of the Hercynian chain in Italy. Bull Soc Géol Fr 8 (2): 95–114

Voshage H, Hunziker JC, Hofmann AW, Zingg A (1987) A Nd and Sr isotopic study of the Ivrea zone, southern Alps, N-Italy. Contrib Mineral Petrol 97: 31–42

Voshage H, Sinigoi S, Mazzucchelli M, Demarchi G, Rivalenti G, Hofmann AW (1988) Isotopic constraints on the origin of ultramafic and mafic dykes in the Balmuccia peridotite (Ivrea Zone). Contrib Mineral Petrol 100: 261–267

Voshage H, Hofmann AW, Mazzucchelli M, Rivalenti G, Sinigoi S, Raczek I, Demarchi G (1990) Isotopic evidence from the Ivrea zone for a hybrid lower crust formed by magmatic underplating. Nature 347: 731–736

Weissert HJ, Bernoulli D (1985) A transform margin in the Mesozoic Tethys: evidence from the Swiss Alps. Geol Rundsch 74: 665–679

Ziegler PA (1986) Geodynamic model for the Paleozoic crustal consolidation of western and central Europe. Tectonophysics 126: 303–328

Zingg A (1980) Regional metamorphism in the Ivrea zone (Southern Alps, N-Italy): field and microscopic investigations. Schweiz Mineral Petrogr Mitt 60: 153–179

Zingg A (1983) The Ivrea and Strona-Ceneri zones (Southern Alps, Ticino and N-Italy) – a review. Schweiz Mineral Petrogr Mitt 63: 361–392

Zingg A (1990) The Ivrea crustal cross-section (northern Italy and southern Switzerland). In: Salisbury MH, Fountain DM (eds) Exposed cross-sections of the continental crust. Kluwer, Dordrecht, pp 1–19

Zingg A, Hunziker JC (1990) The age of movements along the Insubric Line west of Locarno (northern Italy and southern Switzerland). Eclogae Geol Helv 83: 629–644

Zingg A, Handy MR, Hunziker JC, Schmid SM (1990) Tectonometamorphic history of the Ivrea zone and its relationships to the crustal evolution of the Southern Alps. Tectonophysics 182: 169–192

Pre-Alpine Structural and Metamorphic Histories in the Orobic Southern Alps, Italy

G. B. SILETTO, M. I. SPALLA, A. TUNESI, J. M. LARDEAUX and A. COLOMBO

Abstract

Petrological and structural data on the Orobic South-Alpine basement are reviewed in order to reconstruct its thermomechanical history. P-T-t paths are discussed for Val Vedello – Passo S. Marco and Lario areas, i.e. for sectors traditionally attributed to different metamorphic grades (greenschist and amphibolite facies respectively). Structural investigations evidenced that, in these zones, the first recognizable phase of deformation (D_1) developed in all the Orobic basement under amphibolite facies conditions during the Variscan period (ca. 330 Ma). Although some metagranitoids, emplaced in the Orobic basement, are considered to be Ordovician, no relics testify to the real pre-Variscan basement during the intrusion of the granites. After the D_1 phase in the central-eastern areas (Val Vedello – Passo S. Marco) during D_2 phase, a greenschist retrogression is observed. In the westernmost area (Lario basement) the D_2 phase is characterized by an increase in temperature under low pressure regime. During the D_2 phase

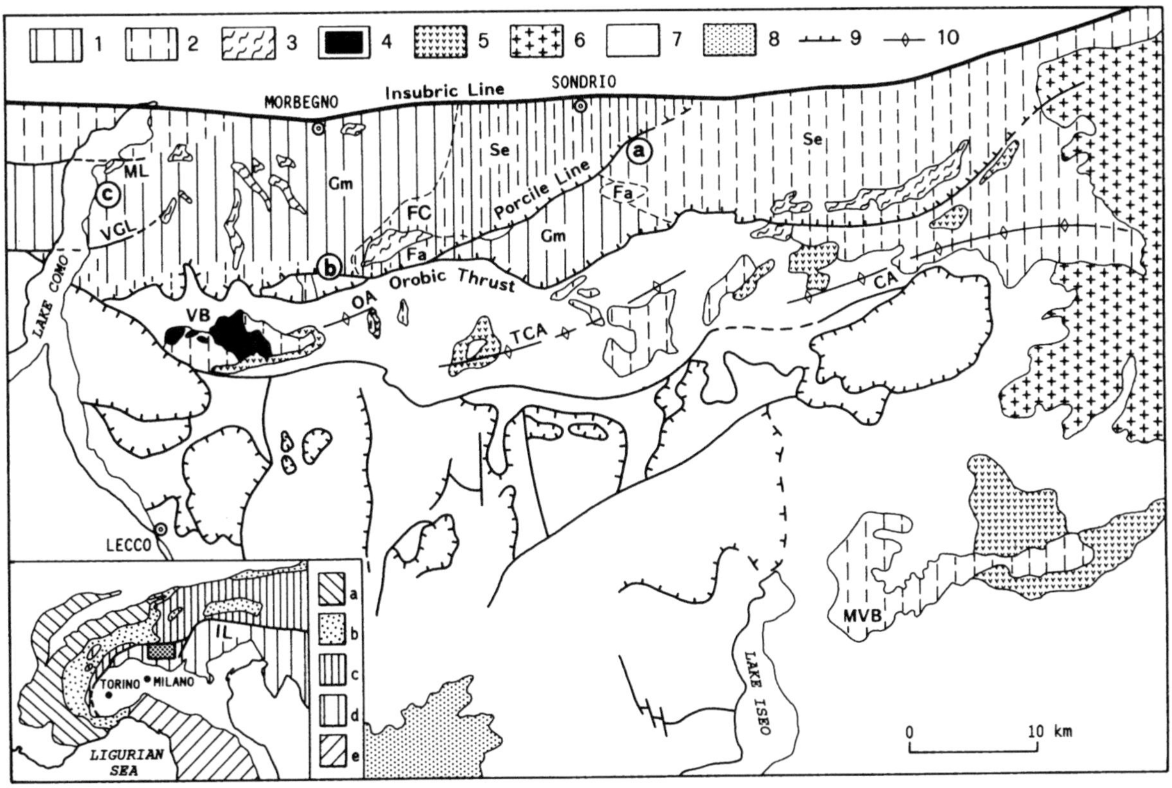

Fig. 1. Sketch map of the metamorphic zonation proposed for the Variscan South-Alpine chain (from the Metamorphic Map of the Alps, Niggli 1973 and Castellarin 1981 redrawn for the Orobic Alps). *1* Medium-grade metamorphic rocks; *2* low-grade metamorphic rocks; *3* Ordovician orthogneisses; *4* late-Variscan plutons; *5* Permian volcanites; *6* Adamello pluton; *7* Permo-Cenozoic sedimentary cover; *8* Po plain. *9* thrusts; *10* anticlines; *CA* Cedegolo Anticline; *Fa* Filladi di Ambria; *FC* Fioraro complex; *Gm* Gneiss di Morbegno; *MVB* Massiccio delle Tre Valli Bresciane; *OA* Orobic Anticline; *Se* Scisti di Edolo; *TCA* Trabuchello-Cabianca anticline; *VB* Val Biandino pluton; *VGL* Val Grande line; *a* Val Vedello area; *b* Passo S. Marco area; *c* Lario area

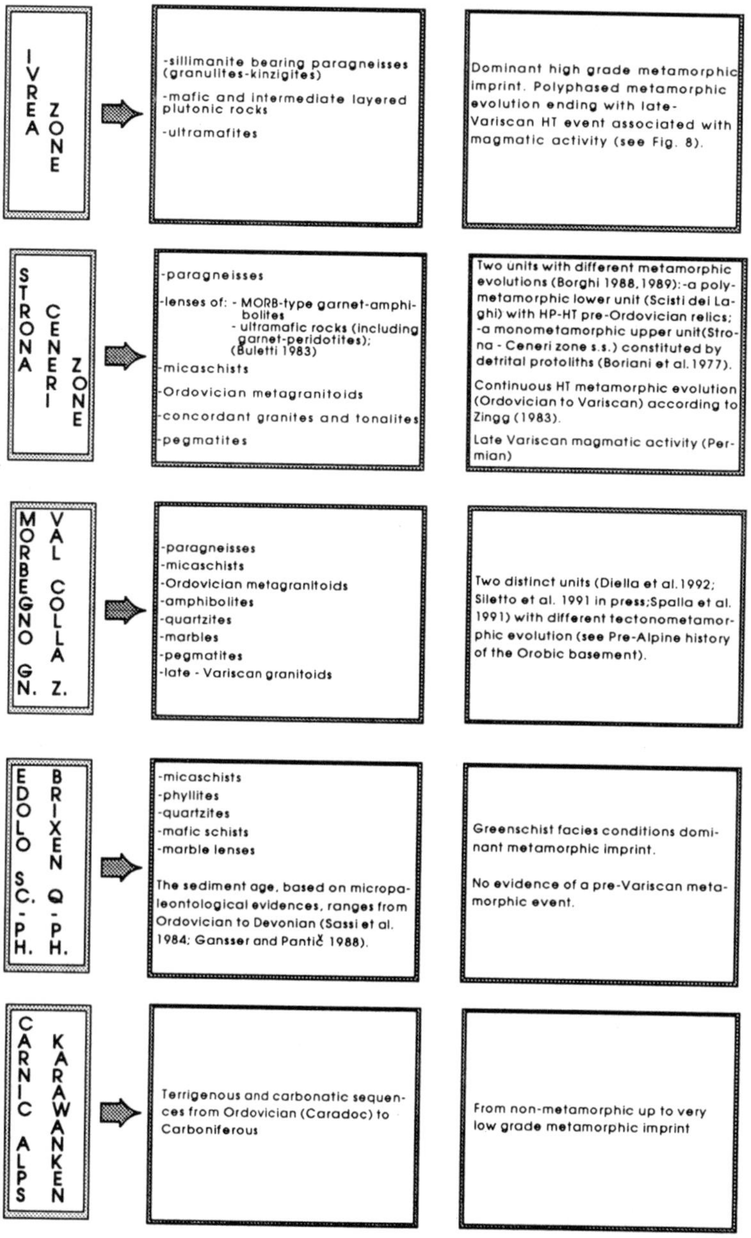

Fig. 2. Lithologies and main metamorphic imprints affecting the different tectonic units of the South-Alpine basement (after Frisch et al. 1990, with modifications). If the metamorphic evolution of the Strona Ceneri zone and the Edolo schists or the Carnic sequences are compared, it is evident that the tectono-metamorphic setting of the South-Alpine crust has been heterogeneous since the Ordovician intrusions. Probably, the different tectono-metamorphic units already existed in the South-Alpine domain before the Variscan orogeny

of deformation, considering the available geochronological data, a late Variscan post-thickening uplift is suggested for the Val Vedello-Passo S. Marco areas, and a Permo-Mesozoic uplift related to an extensional tectonic regime for the Lario area. These contrasted evolutions result from the tectonic piling up of various crustal segments of the South-Alpine basement, each representative of different tectonic histories. The polyphased character of the two tectono-metamorphic histories appears to be in contradiction with the metamorphic zoneography traditionally accepted for the South-Alpine domain.

1 Introduction

The South-Alpine tectonic domain is considered to be a portion of the Adria plate traditionally separated from the Alpine domain by the Insubric line (Fig. 1).

The central part of the South-Alpine domain, from Lake Como to the Adamello massif, is known as the "Orobic Southern Alps". Its present setting is due to Alpine convergence which is responsible for the formation of a complex system of south-verging folds and stacked thrust sheets. The northernmost sheets consist, almost exclusively, of basement rocks, whereas in the south a thick cover, formed of clastic to carbonatic Permo-Cenozoic sediments, occurs. Some tectonic windows (e.g. Mezzoldo, Valmoresca and Massiccio delle Tre Valli Bresciane) and geophysical data (Cassano et al. 1986; CroP 1989; Roeder 1989) demonstrate that the basement is also involved in the southernmost part of the Orobic thrust system.

In the basement, the main metamorphic imprint is traditionally considered to be Variscan with increasing metamorphic conditions from east (near Adamello: greenschist facies conditions; Fig. 1) to west (Lario region near Lake Como: amphibolite facies conditions; Fig. 1). In this scheme the Ivrea granulites represent the higher grade (Fig. 2) zone of the regional metamorphic zoneography proposed for the South-Alpine basement (Vai and Cocozza 1986 and references therein).

On the contrary, recent structural and petrological investigations strongly indicate a complex deformational history (Milano et al. 1988; Siletto 1990, 1991) and a retrograde metamorphic evolution (Diella et al. 1992).

2 Geological Setting

After the first studies, performed at the end of the last century (Curioni 1844; Taramelli 1880, 1890; Benecke 1884; Melzi 1891 a, b; Stella 1894; Repossi 1904, 1914) describing the sedimentary cover and the mineralogy of the basement, the present understanding of the evolution of the Orobic basement is the result of the re-mapping (1:100000 scale) of the Orobic Alps (sheets Numbers 7–18 Pizzo Bernina-Sondrio, and Number 19 Tirano; Beltrami et al. 1971; Bonsignore et al. 1971) carried out in the 1970s. In these works, the authors demonstrated that the Orobic basement is mainly constituted of paragneisses, phyllites, micaschists, mostly derived from psammitic to pelitic sequences, quartzites and minor marbles. Magmatic rocks are represented by both acid and basic products, but are subordinated in amount with respect to the metasedimentary rocks (Fig. 2). All these rocks are affected by pre-Alpine metamorphic events.

Previous authors established (mainly for the metasedimentary rocks) lithostratigraphic units and preserved the historical names proposed for some lithologies at the beginning of this century. Beltrami et al. (1971) and Bonsignore et al. (1971) recognized in the eastern area (near the Adamello pluton) a pelitic sequence with micaschists, phyllites, rare quartzites and marbles (Scisti di Edolo, Salomon 1901; Filladi di Ambria, Dozy 1933) mainly characterized by the widespread occurrence of white mica and chlorite. In the central region (south of Morbegno) biotite, garnet, kyanite and staurolite-bearing paragneisses (Gneiss di Morbegno, Fig. 1) were interpreted as a thick sequence of prevailing metapsammitic rocks by Cornelius (1916).

The presence of chloritoid, instead of staurolite, in the tectonic window of the Massiccio delle Tre Valli Bresciane (Fig. 1), seems to be related to differences in the bulk-chemistry of some pelitic layers as suggested by Giobbi Origoni and Gregnanin (1983).

More to the west, metapelitic rocks with a widespread growth of sillimanite and local occurrence of andalusite, amphibolites and pegmatites ("Dervio-Olgiasca Zone", El Tahlawi 1965) crop out in the northern Lake Como region (Lario area).

In all these units, metagranitoids occur as small bodies; most of them are presumed to be Late Variscan in age (e.g. Bonsignore et al. 1971), but geochronological data are lacking. The largest acidic stock emplaced in the Orobic basement is the Fioraro complex (Fig. 1). It preserves magmatic textures in the core but a schistosity is developed at its margin and in the apophyses. The latter are deformed from at least the second phase of deformation recognized in the surrounding schists, and is associated with the retrogradation into greenschist facies conditions (Siletto 1991). This phase of deformation is clearly Variscan in age, because it does not affect the Permo-Triassic sedimentary cover; as a consequence the Fioraro complex can not be interpreted as belonging to late Variscan magmatism. Its geochemical features are consistent with a pre-Variscan, probably Ordovician magmatic event (Colombo et al. 1991).

The chemical composition is granitic and subordinately granodioritic (silica between 62.5% and 75%) and indicates a peraluminous character. This body shows low Ba and high Rb, Th, Zr, Nb, Y, La, Ce contents (Fig. 3), which are typical features of post-collisional granites (low Rb/Zr ratio) according to Pearce et al. (1984) and Harris et al. (1986).

The Fioraro rocks show an ambiguous behaviour because of the similarity with VAG (volcanic arc granites) or WPG (within-plate granites); this feature is also characteristic of many post-collisional granites. The REE pattern is quite homogeneous, moderately fractionated (La/Yb ratio between 8 and 21 with a medium value around 13.5) and with a strong Eu anomaly (Eu/Eu* between 0.25 and 0.50).

All the geochemical characteristics are consistent with those of the Ordovician metagranitoids of the Serie dei Laghi in the western South-Alpine basement (Boriani et al. 1982–83) and for example of the Massif Central – France (Downes and Duthou 1988). The Fioraro complex shows a more alkaline signature

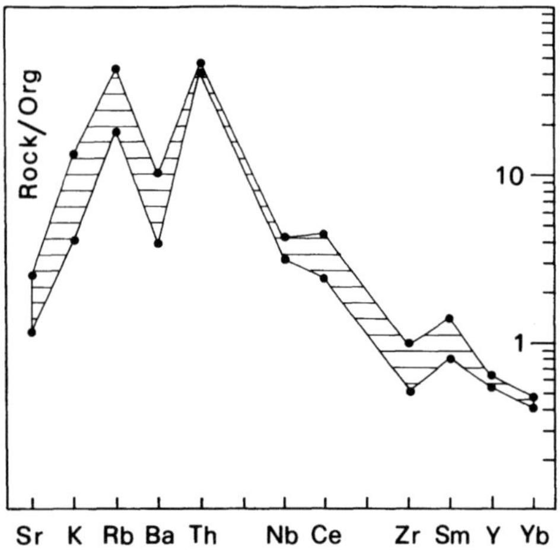

Fig. 3. ORG-normalized patterns of the Fioraro metagranitoids *(dashed area)* here redrawn after Colombo et al. (1991); normalizing values after Pearce et al. (1984)

(high Nb, Y, Zr) than the Serie dei Laghi metagranitoids and it appears, therefore, to represent a more evolved magmatic episode.

Lenses of K-feldspar augengneisses are interlayered in the Orobic basement and appear quite different in composition with respect to metagranitoids; they present a syncollisional character (Colombo and Tunesi, unpubl. data). Near the tectonic contact between basement and sedimentary cover leucocratic magmatic rocks frequently occur: the Gneiss Chiari del Corno Stella (Stella 1894; Reinhard 1953, 1964). These gneisses bear quartz, chessboard albite, muscovite, K-feldspar and rare biotite. Boriani and Colombo (1979) suggested a derivation from felsic volcanics and proposed a pre-Variscan age for these rocks. Metabasic rocks are mostly represented by lenses of amphibolites with plagioclase and garnet.

Basic dykes are present in all the Orobic Alps in successive generations. The older group, mostly represented by alkaline rocks (e.g. Diabasi di Edolo), pre-dates the Alpine structures and is intruded in the basement or within extensional shear zones affecting both the basement and the Permian cover; it is related to the opening of Permian basins (Bigioggero et al. 1981, and references therein; Cadel 1986; Cadel et al. 1987, 1990). The younger group is related to the Oligocene post-collisional magmatic activity and linked to tensional phases post-dating the nappe piling. These dykes cross-cut the Alpine structures (Cassinis et al. 1986) and range in composition from basalts to andesites with a calc-alkaline affinity (Beccaluva et al. 1983).

3 Alpine History

The first tectonic studies of Porro (1897, 1899, 1903, 1911, 1931) already indicated that the basement was thrust upon the cover along the roughly E-W trending Orobic line (Fig. 1). The existence of a complex system of thrusts dissecting basement and cover was demonstrated in the 1930s (e.g. Wennekers 1932) when researchers from the Leiden University mapped nearly all the Orobic Alps (see De Sitter and De Sitter-Koomans 1949 for a complete synthesis), but it is only in the 1980s that kinematic models have been proposed (Laubscher 1985, 1990; Roeder 1989). Following current interpretations, the deformational structures of Alpine age are confined to a discrete, though complex, system of thrust faults dissecting the crystalline basement and the sedimentary cover (e.g. the more recent deformational episodes recognized in the eastern Orobic basement: D_3-D_4 by Milano et al. 1988; Siletto 1990, 1991). Cataclasites and ultracataclasites, unaffected by mineralogical recrystallization, are associated with these faults. Pseudotachylites are locally found, indicating high strain rates at a relatively shallow depth. There is no evidence of pervasive metamorphism that could be attributed to the Alpine tectonics, although in some places stilpnomelane (Crespi et al. 1981, 1982) seems to be associated with the regional distribution of the Alpine lines (e.g. the Insubric line).

In the sedimentary cover, the location of the Alpine thrusts was strongly influenced by major décollements, guided by the ancestral Permo-Mesozoic palaeogeographic framework and lithostratigraphy (Castellarin and Vai 1982; Gaetani and Jadoul 1987).

The regional distribution of the Alpine tectonic lines seems to be related to a roughly N-S compression (Castellarin 1979; Gaetani and Jadoul 1979), although there are some indications of a dextral component of movement. Actually, the orientation of the Orobic anticlines (Orobic, Trabuchello-Cabianca and Cedegolo anticlines) and of the Porcile line (Fig. 1), in addition to some kinematic indicators along the Orobic line, are compatible with a phase of dextral movement (top to the ESE; Forcella and Jadoul 1990; Siletto 1990).

The limited age data for the Alpine structures in the Orobic Alps, allow us to situate the compressive tectonics within a time span ranging from Cretaceous (Doglioni and Bosellini 1987) up to ca. 10 Ma. Indeed, the first evidence of compression can be found on stratigraphic basis in the Cenomanian-Turonian deposits (Bersezio and Fornaciari 1987, 1988), and the Adamello intrusives (ca. 40–30 Ma; Del Moro et al. 1983) truncate the pre-existing folds and thrusts between basement and Permo-Mesozoic cover (Brack 1981, 1983, 1984; Cassinis et al. 1986). Basic dykes cross-cutting thrust planes in the sedimentary cover have been recently dated (Zanchi et al. 1990) to 50 Ma (K/Ar, whole-rock). The last evidence of compressive tectonics is of Burdigalian-Tortonian age, as demonstrated by the the Gonfolite Lombarda back thrust (Chiasso-Stabio back thrust, Bernoulli et al. 1989) and the subsurface tectonics in the Po plain (Pieri and Groppi 1981; Pieri 1983).

Therefore these data suggest that the fold and thrust system in the Southern Alps was still active in Neo-Alpine time, but started during Eo-Alpine tectonic phases.

4 Pre-Alpine History of the Orobic Basement

Several authors proposed, in the South-Alpine basement, a broad-scale metamorphic zoneography, characterized by an increase in metamorphic grade from north east to south west (Mottana and Schiavinato 1973; Boriani et al. 1974 with references; Boriani and Mottana 1978; Vai et al. 1981; Boriani and Sacchi 1983; Vai and Cocozza 1986). However, discrepancies are focused in the regional zonational reconstruction by some authors (Boriani and Mottana 1978; Boriani and Sacchi 1983), e.g. the Dervio-Olgiasca zone, which was interpreted as a prograde sequence under local low pressure regime, from chlorite (El Tahlawi 1965) or from biotite (Bocchio et al. 1980) in the south, to sillimanite zone in the north.

Bocchio et al. (1981) and Mottana et al. (1985) first identified, on a metamorphic and chronological basis, two areas with different metamorphic and chronological evolutions: an "intermediate pressure unit", with a Variscan metamorphic imprint and a "low pressure unit", i.e. the Dervio-Olgiasca zone, overprinted by a low-pressure – high-temperature metamorphism, Late Permian to Early Jurassic in age. In order to explain this post-Variscan high-temperature overprint in the Dervio-Olgiasca zone, Mottana et al. (1985, p. 362) suggested a "localized temperature increase from below, i.e. a heat flow related to factors such as crustal thinning or mantle upwelling".

Considering the works of Milano et al. (1988) and Diella et al. (1992) and new data from Siletto (1990, 1991) the structural and metamorphic evolution of metapelites can be discussed in three zones of the Orobic basement (Fig. 1): (a) the Val Vedello area; (b) the Passo S. Marco area; and (c) Lario area (including the Dervio-Olgiasca zone, El Tahlawi 1965). The first two domains (a, b) are traditionally regarded as being metamorphosed under greenschist facies conditions, and the latter one (c) under amphibolite facies conditions.

4.1 Val Vedello and Passo S. Marco Areas

In these areas structural and microstructural-petrological investigations allow us to recognize two generations of ductile structures (hereafter called D_1 and D_2) associated with pre-Alpine polyphase metamorphic re-equilibrations, followed by the Alpine deformation phases (D_3 and D_4), not coeval with mineral growth and related to Alpine thrusts (Fig. 4).

According to Milano et al. (1988) and Siletto (1991) the two earlier sets of folds (D_1 and D_2) are isoclinal with well-developed axial plane foliations (S_1 and S_2). The earlier deformative stage (D_1) reveals amphibolite facies grade with the mineral parageneses:

a. Grt+red-brown Bt + Ms ± Pl + Qtz + Rt
b. Grt+red-brown Bt + Ky + Ilm + St + Pl ± Ms + Qtz

followed, during the D_2 deformation, by greenschist facies grade mineral assemblage (Fig. 5a):

MsII+green BtII+Chl+Ab±Ep±Cal+Qtz

D_1 and D_2 structures have a pre-Alpine age since they do not affect the Permo-Triassic non-metamorphic cover: these structures are transected by Alpine cataclastic thrust planes affecting the cover.

Thermobarometrical determinations (Siletto 1991; Spalla et al. 1991; Diella et al. 1992), performed utilizing independent well-calibrated thermobarometers, indicate that the amphibolite facies metamorphic imprint is characterized by P = 7–9 kb and T = 570–610 °C, whereas the retrograde greenschist facies event is associated with P < 4 kb and T < 500 °C (Fig. 6; see discussion in Diella et al. 1992).

4.2 Lario Basement Area

In the northern Lake Como region three generations of ductile structures were recognized (hereafter called D_1, D_2 and D_3).

D_1 structures consist mainly of relict small-scale rootless folds associated with a crenulation cleavage (S_1). D_2 structures are represented by asymmetric tight to isoclinal folds associated with a mylonitic foliation. Although the D_2 overprint is intense in this re-

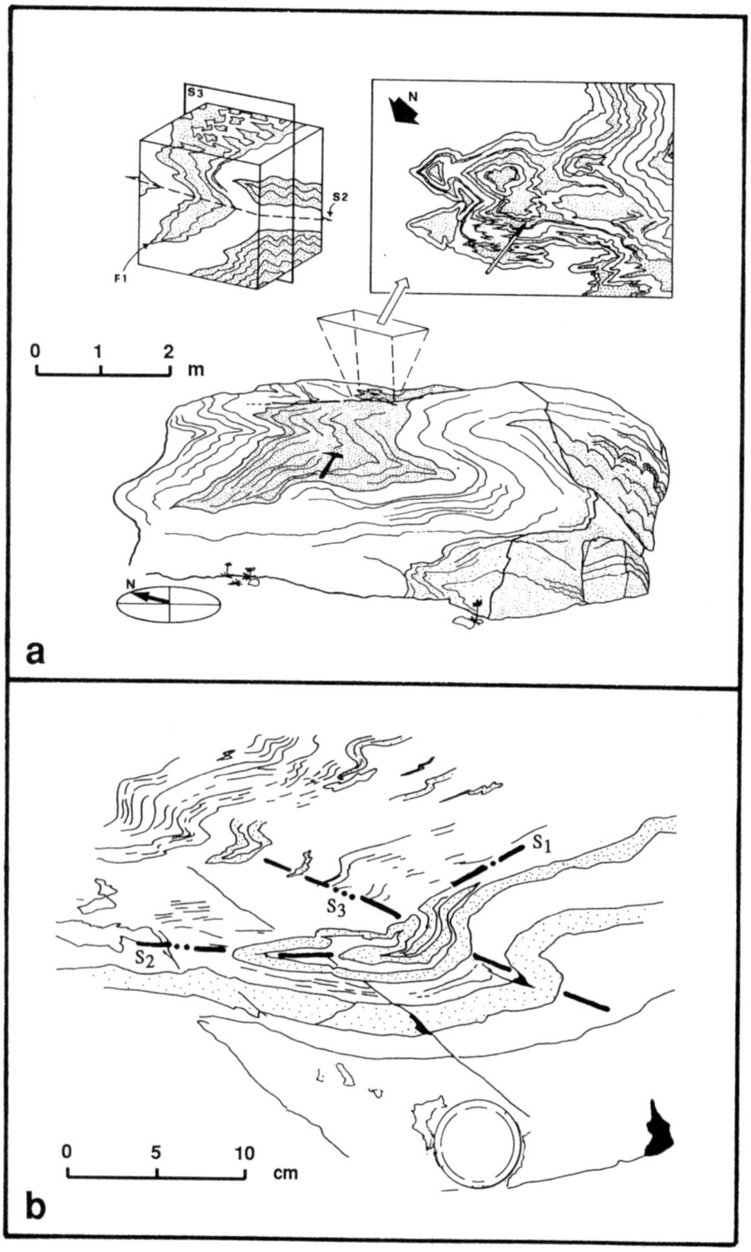

Fig. 4. a Fold interference patterns in Val Vedello; and **b** Passo S. Marco areas. Types of interference patterns. **a** Type 3 of Ramsay (1967) formed by F_2 and F_1 folds and type 2 pattern formed by F_3 and F_2 or F_1 folds; **b** type 3 interference pattern between F_3, F_2 and F_1 folds

gion, it is possible to define a mineralogical association that predates the development of S_2 foliation:

Grt+Bt+Qtz+St+Ky±Pl±Kfs(?)+Ms±Rt+Ilm+Tour

During D_2 deformation, a decompression, accompanied by an increase in temperature, is documented by the breakdown of garnet, which is mainly replaced by sillimanite and red-brown Ti-rich biotite. This interpretation is also supported by textural evidences of staurolite and white mica break-down and by the observation that S_2 foliation is laterally replaced by domains, up to centimetre-size, characterized by granoblastic quartz, plagioclase, minor K-feldspar, and red-brown biotite with lobate margins. A successive decrease in temperature is marked by andalusite growth (Fig. 5b).

Sillimanite and Ti-rich biotite underline clearly a stretching lineation (trending NW-SE) and shear and foliation planes (both plunging roughly towards the south) developed during D_2 non-coaxial deformation. Pegmatite bodies, at low angle with S_2 foliation, are generally localized within shear bands and present undeformed cores, whereas "s-c" structures are developed at their margins. Where these structures are observed, on sections perpendicular to the foliation plane and parallel to the stretching lineation, their kinematics indicates extension (Fig. 7). Thus, all these features support a pegmatite emplacement during D_2 extensional tectonics.

A third phase of deformation (D_3) is often associated with mylonitic bands developed under greenschist facies conditions (Fig. 5 b) and with rare metre-scale chevron folds that seldom display an axial plane foliation. Fold axes (b_3) are subhorizontal and east-west trending: D_3 axial surfaces plunge towards northwest and southeast. It must be underlined that D_2 and D_3 structures are reworked and rotated during the Alpine compression.

According to Diella et al. (1992) thermobarometrical calculations give P values of 7–9 kb and a T range of 550–630 °C for the first metamorphic stage. During the D_2 deformation P values of 4–5.5 kb and T of 650–750 °C were obtained. The final stage of the Lario basement uplift is marked by the destabilization of high temperature assemblages into Chl+MsII, yielding temperatures below 500 °C (Fig. 6).

In summary, in both the Val Vedello and Passo S. Marco areas, two earlier pre-Alpine synmetamorphic phases of deformation (D_1 and D_2) were identified; they developed under amphibolite and greenschist conditions respectively. In the Lario area, on the contrary, the pervasive D_2 deformation phase is characterized by increasing in temperature during decompression. Here, D_2 almost completely obliterates the earlier structures, evident only as preserved relics in D_2 microlithons. In addition, the geometrical features of the D_2 structures (extensional crenulation cleavage and large-scale shear bands, Fig. 7) suggest that the main active mechanism during D_2 deformation was prevailing progressive simple shearing (Bell

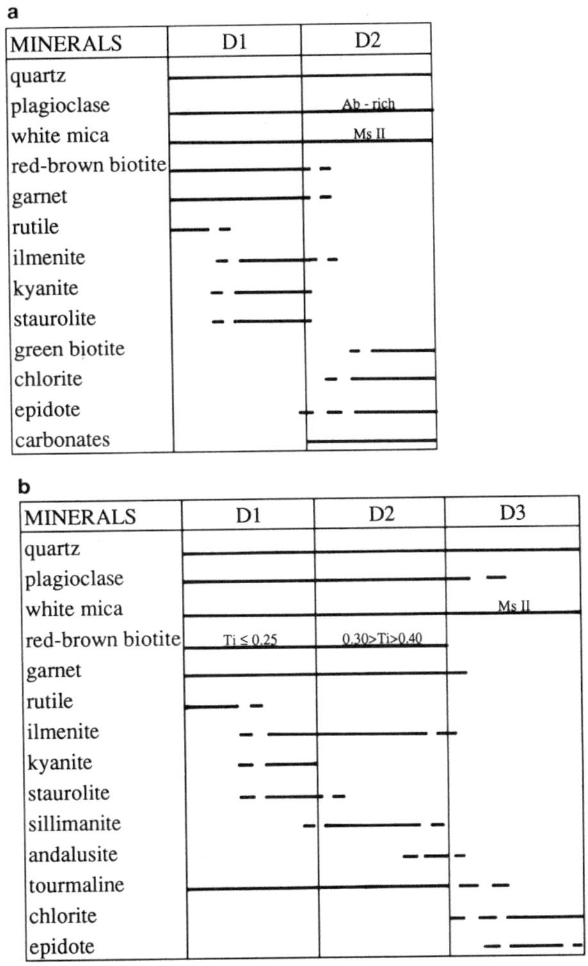

Fig. 5 a, b. Synoptic scheme of the mineral growth versus deformation relationships for **a** Val Vedello and Passo S. Marco areas; and **b** Lario area. Detailed microstructural observations can be found in Milano et al. (1988) and Siletto (1991) for **a** and in Diella et al. (1992) for **b**

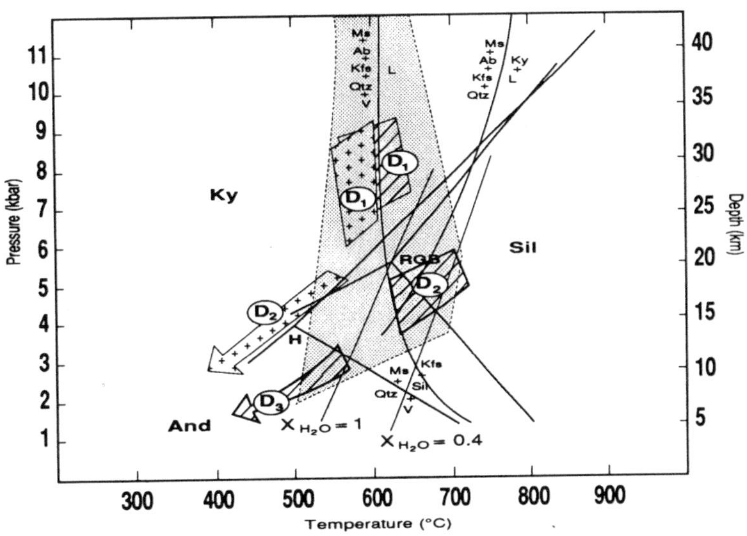

Fig. 6. P-T-relative time of deformation paths showing the estimated metamorphic conditions of Val Vedello – Passo S. Marco basement *(crosses)* and Lario basement *(oblique lines)* (redrawn from Diella et al. (1992). The stability field of staurolite + quartz is *dotted.* Al-silicates triple point: *H* Holdaway (1971); *RGB* Richardson et al. (1969). Initial melting of pelite (wet and dry) after Thompson and Tracy (1979). *Ms + Qtz = Kfs + Sill + V* for KFASH after Vielzeuf (1984)

Fig. 7. Shear *(c)* and foliation *(s)* planes, developed during D_2 deformation, affecting both pegmatite boundaries and the country rocks (Dervio-Olgiasca zone, Lario basement)

1981, 1985) in an extensional tectonic regime (Siletto et al. 1990).

The later phases of deformation (D_3, D_4) in the Val Vedello and Passo S. Marco areas are non metamorphic, and can be referred to as Alpine thrust tectonics. On the other hand, the third phase of deformation (D_3) in the Lario area is associated with a greenschist retrogression. Although there is no evidence to define its age, the similarity between the metamorphic overprint related to D_3 and the greenschist metamorphism associated with the Val Grande-Val Larga line (Bertotti, 1990) allow us to correlate them. In addition the D_3 chevron folds (lacking axial plane foliation) are spatially related to the D_3 mylonitic bands. The D_3 mylonitic foliation become prevalent approaching the Val Grande-Val Larga line (Fig. 1). Bertotti (1990) interpreted the Val Grande-Val Larga line as the deeper prolongation of the Lugano-Grona Liassic listric normal fault, later deformed during Alpine compressive tectonics. Accepting this hypothesis, we can consider the D_3 generation of structures as developed during Liassic extensional tectonics.

In conclusion, the sectors result in quite different styles of deformation, dominant deformation mechanism and metamorphic evolutions. The thermobarometrical results (Diella et al. 1992) show that the shapes of these P-T paths are also different. Considering the Val Vedello – Passo S. Marco basement, a cooling during the uplift history can be demonstrated. In the Lario basement, on the other hand, the decompression history is associated with a temperature increase giving rise to an abnormally high thermal gradient at the end of the tectonic evolution. Thus the Orobic Southern Alps must be regarded as a composite domain in which crustal slices with contrasted thermomechanical evolutions are now juxtaposed.

5 Age of Metamorphisms: Radiometric Data

For a discussion of the chronological evolution of the Orobic Alps, it is useful to consider the available geochronological data obtained in the central Southern Alps (a few scattered data from Hanson et al. 1966, Mottana et al. 1985 with references therein) but also the ages proposed for the whole South-Alpine domain (see Table 1 and Fig. 8 for complete references).

The oldest ages are obtained from detrital zircons from garnet-staurolite schists (2050 Ma, upper intercept, U/Pb; Grauert et al. 1973). Palynomorphs of Lower Palaeozoic age have been observed in the Scisti di Edolo of the easternmost Orobic basement (Gansser and Pantič 1988).

The sedimentary Permo-Cenozoic cover, unconformably overlying the metamorphic basement, has traditionally suggested a Variscan age for the main metamorphic imprint; however, some age determinations suggest that an older thermal event may also be recorded. From metasediments of the Ivrea zone, Hunziker and Zingg (1980) reported an isotopic homogenization (peak of metamorphism) of Caledonian age (478 ± 20 Ma, Rb/Sr isochron). Similar ages are obtained for the metasediments around Cima d'Asta (446 ± 13 Ma, Rb/Sr isochron; Del Moro

Table 1. Geochronological data from the Orobic South-Alpine basement

Rock type	Sample No	Location	Material	Method	Age (Ma)	Reference
Pegmatite	P-4	D.O.Z.	Ms	K/Ar	207±10	Hanson et al. (1966)
Pegmatite	P-10	D.O.Z.	Ms	K/Ar	215±11	Hanson et al. (1966)
Pegmatite	P-15	D.O.Z.	Ms	K/Ar	218±11	Hanson et al. (1966)
Pegmatite	P-27	D.O.Z.	Ms	K/Ar	198±10	Hanson et al. (1966)
Pegmatite	P-30	D.O.Z.	Ms	K/Ar	219±11	Hanson et al. (1966)
Pegmatite	P-31	D.O.Z.	Bt	K/Ar	239±12	Hanson et al. (1966)
Pegmatite	P-4	D.O.Z.	Ms	Rb/Sr	286	Hanson et al. (1966)
			Ms		250	
Pegmatite	P-15	D.O.Z.	Ms	Rb/Sr	223	Hanson et al. (1966)
Micaschist	DAT-2	Corenno Plinio	Ms	K/Ar	180±3	Bocchio et al. (1981)
			Bt		218±5	
			WR		247±5	
Bt-Sil schist	DAT-3	Laghetto di Piona	Ms	K/Ar	190±6	Bocchio et al. (1981)
			Bt		218±6	
			WR		264±8	
Bt-Sil schist	DAT-4	Laghetto di Piona	Ms	K/Ar	221±5	Bocchio et al. (1981)
			Bt		223±5	
			WR		224±5	
Pegmatite	DAT-5	Laghetto di Piona	Ms	K/Ar	228±6	Bocchio et al. (1981)
Fine-grained gn.	DAT-8	Tremenico	Ms	K/Ar	238±7	Mottana et al. (1985)
			Bt		269±9	
			WR		259±7	
Fine-grained gn.	DAT-9	Bondo	Ms	K/Ar	231±5	Mottana et al. (1985)
			Bt		252±6	
			WR		402±9	
Amphibolite	DAT-12	Musso	Bt	K/Ar	254±8	Mottana et al. (1985)
			WR		201±5	
Quartzdiorite	DAT-15	Val Biandino	Bt	K/Ar	271±8	De Capitani et al. (1988)
			Hbl		331±10	
			WR		214±6	
Quartzdiorite	DAT-15a	Val Biandino	Bt	K/Ar	274±8	De Capitani et al. (1988)
			WR		229±7	
Quartzdiorite	DAT-15b	Val Biandino	Bt	K/Ar	291±9	De Capitani et al. (1988)
			WR		238±7	
Quartzdiorite	DAT-15c	Val Biandino	Bt	K/Ar	282±8	De Capitani et al. (1988)
			Hbl		348±13	
Quartzdiorite	DAT-15d	Val Biandino	Bt	K/Ar	275±8	De Capitani et al. (1988)
Granite	DAT-15e	Val Biandino	Bt	K/Ar	246±7	De Capitani et al. (1988)
Granite	DAT-15f	Val Biandino	Bt	K/Ar	301±9	De Capitani et al. (1988)
Gabbrodiorite	DAT-18	Zucco di Cornisella	Bt	K/Ar	307±9	De Capitani et al. (1988)
			WR		242±7	
Tonalite	DAT-19	Cima delle Forcelle	Bt	K/Ar	294±9	De Capitani et al. (1988)
			WR		275±8	
Granite	DAT-20	Zucco di Cornisella	Bt	K/Ar	290±8	De Capitani et al. (1988)
Paragneiss	DAT-1	Bellano	Ms	K/Ar	330±10	Bocchio et al. (1981)
			Bt		331±10	
			WR		313±9	
Paragneiss	DAT-6	Morbegno	Ms	K/Ar	330±8	Bocchio et al. (1981)
			Bt		276±8	
			WR		210±5	
Fine-grained gn.	DAT-7	Val Tartano	Ms	K/Ar	306±7	Bocchio et al. (1981)
			Bt		291±6	
			WR		315±4	
Paragneiss	DAT-10	Bema	Ms	K/Ar	242±7	Mottana et al. (1985)
			Bt		292±9	
			WR		228±7	
Fine-grained gn.	DAT-11	Fiesso	Ms	K/Ar	170±6	Mottana et al. (1985)
			WR		134±4	
Fine-grained gn.	DAT-13	Cosio	Ms	K/Ar	186±6	Mottana et al. (1985)
			Bt		229±7	
			WR		43±2	
Schist		Val Trompia	Ms	Rb/Sr	364±15	Del Moro in Riklin (1983)

D.O.Z.: Dervio Olgiasca zone

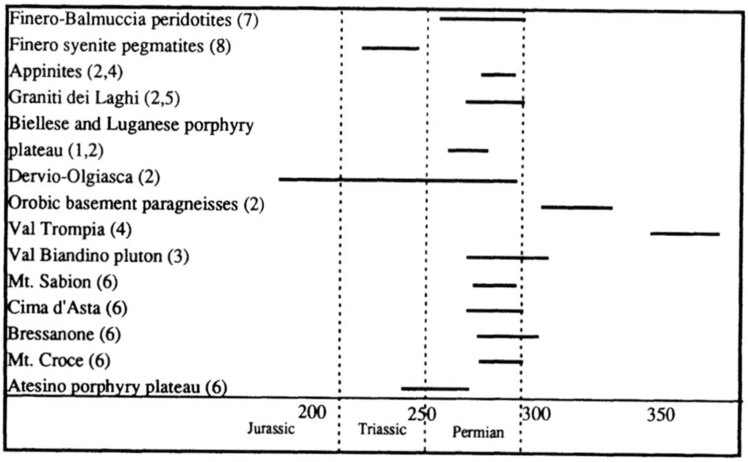

Fig. 8. Published radiometric age determinations for the Late Variscan magmatic activity and Variscan metamorphism. References: *1* Stille and Buletti (1987); Bakos et al. (1990); *2* Boriani et al. (1985 with references); *3* De Capitani et al. (1988); *4* Del Moro in Riklin (1983); *5* Pinarelli et al. (1988); *6* Sassi et al. (1985 with references); *7* Voshage et al. (1987, 1988); *8* Stähle et al. (1990)

1989) and for the Scisti di Edolo (459 ± 66 Ma, Rb/Sr isochron; Del Moro pers. comm. in Boriani et al. 1985). The orthogneisses of the Serie dei Laghi gave an age of 446 ± 5 (Rb/Sr isochron; Boriani et al. 1982–83), interpreted as the age of the intrusion.

In addition to these data testifying an eo-Variscan event in the Orobic Alps, the older ages (Mottana et al. 1985 with references therein) are obtained from schists in areas with metamorphic evolution similar to the Val Vedello-Passo S. Marco areas. They range from 330 to 300 Ma (K/Ar, Rb/Sr; Table 1). On the other hand, the ages obtained from samples (schists and pegmatites) of the Dervio-Olgiasca zone with high-T metamorphic imprint range from 270 to 180 Ma (K/Ar, Rb/Sr; Table 1; Hanson et al. 1966; Mottana et al. 1985 with references). These ages (180–270 Ma), widespread in all the South-Alpine basement, are too young to be attributed to the Variscan metamorphism, and were generally explained as being due to radiogenic loss during some geologically unidentified Permian thermal event (related to extension by Ferrara and Innocenti 1974). However, in the last decade geochronological studies on the igneous rocks emplaced at different crustal levels in the Alpine chain revealed Permian ages (late-Variscan) in most of the cases (Fig. 8).

The geological meaning of these radiometric ages and the magmatic activity are still under discussion and they might be considered to mark the end of the Variscan orogeny or the beginning of rifting phases.

6 Discussion and Conclusions

The pre-Alpine tectono-metamorphic evolution of the Orobic Alps shows a retrograde character. Indeed, in Lario and in Val Vedello-Passo S. Marco basements the reconstructed evolutions, even if contrasted, imply the uplift of deep-seated crustal slices.

For interpreting the geodynamic context of these uplifts it is necessary to compare the calculated P-T shapes with the thermal models calculated for active zones (Fig. 9; England and Thompson 1984; Ruppel et al. 1988). If we consider the normal heat supply reached after crustal thickening (maximally relaxed geotherm; England and Thompson 1984), it appears that the thermal history of the Val Vedello – Passo S. Marco basement can be compatible with an uplift controlled by isostatic/erosion processes after crustal thickening (Fig. 9). Such interpretation is in accordance with the geochronological data which indicate the presence of a Variscan tectono-metamorphic

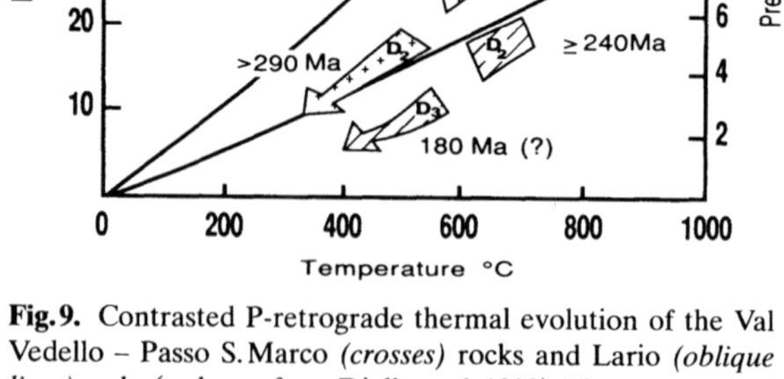

Fig. 9. Contrasted P-retrograde thermal evolution of the Val Vedello – Passo S. Marco *(crosses)* rocks and Lario *(oblique lines)* rocks (redrawn from Diella et al. 1992). The unperturbed geotherm *(Ve)* and the maximal relaxed geotherm *(V∞)* are from England and Thompson (1984)

event. On the contrary, the palaeothermal gradients calculated for the Lario basement are much higher and cannot reflect a post-collisional uplift history. Such a pattern, implying an abnormal heat supply, could be better explained in an extensional regime (see discussion in Brodie and Rutter 1987; Ruppel et al. 1988; Brodie et al. 1989).

If we consider the available chronological constraints, they range from 270–260 and 200–180 Ma, thus during Permo-Jurassic times. It is ascertained that such a period corresponds to widespread magmatic activity in all crustal levels in the South-Alpine domain, coeval with the emplacement of ultramafic bodies (Ivrea, Finero, Balmuccia). The volume of rocks emplaced is much larger than any other magmatic event recorded within the Alpine chain (e.g. Oligocene magmatism). In the Orobic basement small plutons (e.g. Val Biandino, De Capitani et al. 1988 with references), dykes and volcanic rocks (Origoni Giobbi et al. 1979; Crisci et al. 1984) referring to Permo-Triassic times occur. It is important to note that most of this magmatism is characterized by both acid and mafic products with an alkaline to calc-alkaline affinity. The genesis of acidic magmas was earlier interpreted as due to the melting of continental crustal component, and only recently has been linked to mantle-derived products, possibly related to an extensional regime (Ferrara and Innocenti 1974; Crisci et al. 1984).

Moreover, the recognition on structural and petrological grounds of an extensional episode in the Orobic basement is in accordance with the opening of the Permo-Triassic basins much-studied in this domain of the Alpine belt (Bernoulli 1964; Winterer and Bosellini 1981; Wopfner 1984; Cadel 1986; Cadel et al. 1987; Gaetani and Jadoul 1987; Handy 1987; Schmidt et al. 1987, 1989; Bertotti 1990). Thus, we regard the Lario basement as a crustal slice uplifted by tectonic unroofing during a major Permo-Jurassic crustal thinning episode, as proposed for some units of the Sesia-Lanzo zone (Lardeaux and Spalla 1991), on the base of pre-Alpine metamorphic relics still recognizable after the Alpine orogenesis.

In conclusion, the Orobic Southern Alps are composed of two main crustal slices, the Val Vedello – Passo S. Marco basement reflecting the post-collisional history of the Variscan chain, the Lario basement preserving the tectonic and metamorphic imprint of the Permo-Jurassic extensional tectonics. During the Tertiary Alpine convergence, portions of the Val Vedello – Passo S. Marco and Lario units, previously decoupled during rifting, were re-stacked as superficial crustal peels along non-metamorphic thrusts.

Acknowledgements

This work was supported by the research grant from Ministero Università Ricerca Scientifica Tecnologica (60 %) and CNR Centro di Studio per la Stratigrafia e Petrografia delle Alpi Centrali-Milano. We are grateful to V. Diella, A. Ferrario, G. Gosso and M. Scambelluri for useful discussions and comments on the manuscript. J. F. von Raumer, H. Fritz and F. Neubauer critically reviewed the paper and improved it with useful suggestions. M. I. Spalla gratefully acknowledges the financial support of the Fondation Scientifique de Lyon et du Sud-Est and of the C. N. R. for the post-doctoral scholarships which allowed her study period at the E. N. S., Lyon.

References

Bakos F, Del Moro A, Visonà D (1990) The Hercynian volcano-plutonic association of Ganna (Lake Lugano, central Southern Alps, Italy). Eur J Mineral 2: 373–383

Beccaluva L, Bigioggero B, Chiesa S, Colombo A, Fanti G, Gatto GO, Gregnanin A, Montrasio A, Piccirillo EM, Tunesi A (1983) Post-collisional orogenic dyke magmatism in the Alps. Mem Soc Geol Ital 26: 341–359

Bell TH (1981) Foliation developement: the contribution, geometry and significance of progressive bulk inhomogeneous shortening. Tectonophysics 75: 273–296

Bell TH (1985) Deformation partitioning and porphyroblast rotation of metamorphic rocks: a radical reinterpretation. J Metamorphic Geol 3: 109–118

Beltrami G, Bianchi A, Bonsignore G, Callegari E, Casati P, Crespi R, Dieni I, Gnaccolini M, Liborio G, Montrasio A, Mottana A, Ragni U, Schiavinato G, Zanettin B (1971) Note illustrative della Carta Geologica d'Italia alla scala 1 : 100 000 Foglio 19: Tirano. Nuova Tecnica Grafica, Roma for Servizio Geologico d'Italia

Benecke EW (1884) Erläuterungen zu einer geol. Karte des Grignagebirges. Neues Jahrb Mineral Geol Paleontol 3: 171–251

Bernoulli D (1964) Zur Geologie des Monte Generoso (Lombardische alpen). Beitr Geol Karte Schweiz N, Bern 118: 1–134

Bernoulli D, Bertotti G, Zingg A (1989) Northward thrusting of the Gonfolite Lombarda ("South-Alpine Molasse") onto the Mesozoic sequence of the Lombardian Alps: implications for the deformation history of the Southern Alps. Eclogae Geol Helv 82(3): 841–856

Bersezio R, Fornaciari M (1987) Cretaceous sequences in the Lombardy Basin: stratigraphic outline between the lakes of Lecco and Iseo. Mem Soc Geol Ital 40: 187–197

Bersezio R, Fornaciari M (1988) Geometria e caratteri stratigrafici della Sequenza Cenomaniana nel Bacino Lombardo (Alpi Meridionali). Riv Ital Paleontol Stratigr 94: 425–454

Bertotti G (1990) The deep structure of the Monte Generoso Basin: an extensional basin in the South-Alpine Mesozoic passive continental margin. In: Roure F, Heitzman P, Polino R (eds) Deep structure of the Alps. Mem Soc Géol Fr NS 156: 303–308

Bigioggero B, Colombo A, Lanza R, Sacchi R (1981) Dati geochimici e paleomagnetici preliminari su filoni nel Sudalpino della zona del Biellese. Rend Soc Geol Ital 4: 271–274

Bocchio R, Crespi R, Liborio G, Mottana A (1980) Variazioni composizionali delle miche chiare nel metamorfismo progrado degli scisti sudalpini dell'alto lago di Como. Mem Sci Geol 34: 153–176

Bocchio R, De Capitani L, Liborio G, Mottana A, Nicoletti M, Petrucciani C (1981) K-Ar radiometric age determinations of the South-Alpine metamorphic complex, western Orobic Alps (Italy). Neues Jahrb Mineral Monatsh 7: 289–307

Bonsignore G, Casati P, Crespi R, Fagnani G, Liborio G, Montrasio A, Mottana A, Ragni U, Schiavinato G, Venzo S (1971) Note illustrative della Carta Geologica d' Italia alla scala 1 : 100 000 Fogli 7 e 18: Pizzo Bernina e Sondrio. Nuova Tecnica Grafica, Roma for Servizio Geologico d' Italia, p 130

Borghi A (1988) L' evoluzione metamorfico-strutturale del settore nord-orientale della Serie dei Laghi (Alpi Meridionali). Thesis, Univ Genova, Genova

Borghi A (1989) Evoluzione metamorfica del settore nord-est della Serie dei Laghi (Alpi Meridionali – Canton Ticino). Rend Soc Geol Ital 11: 165–170

Boriani A, Colombo A (1979) Gli Gneiss Chiari tra la Val Sesia e il Lago di Como. Rend Soc Ital Mineral Petrol 35(1): 299–312

Boriani A, Mottana A (1978) Southern Alps (from Torino to the Adamello Massif). In: Zwart HJ (ed) Metamorphic map of Europe. 1 : 2 500 000 Explanatory text. Unesco, Paris, pp 225–227

Boriani A, Sacchi R (1983) The western South Alpine basement and its tectonic setting. In: Second EGT workshop. The southern segment, 7–9 Feb 1985, Venice. European Science Foundation Strasbourg, pp 87–92

Boriani A, Dal Piaz Gv, Hunziker JC, von Raumer J, Sassi FP (1974) Caratteri, distribuzione ed età del metamorfismo prealpino nelle Alpi. Mem Soc Geol Ital 13: 165–225

Boriani A, Bigioggero B, Giobbi Mancini E (1977) Metamorphism, tectonic evolution and tentative stratigraphy of the "Serie dei Laghi". Geological map of the Verbania area (northern Italy). Mem Soc Geol Ital 32: 26

Boriani A, Origoni Giobbi E, Del Moro A (1982–83) Composition, level of intrusion and age of the "Serie dei Laghi" orthogneisses (northern Italy – Ticino, Switzerland). Rend Soc Ital Mineral Petrol 38(1): 191–205

Boriani A, Colombo A, Macera P (1985) Radiometric geochronology of Central Alps. Rend Soc Ital Mineral Petrol 40: 139–186

Brack P (1981) Structures in the southwestern border of the Adamello intrusion (Alpi Bresciane, Italy) Schweiz Mineral Petrogr Mitt 61: 37–50

Brack P (1983) Multiple intrusion examples from the Adamello batholith (Italy) and their signifacance on the mechanisms of intrusion. Mem Soc Geol Ital 26: 145–157

Brack P (1984) Geologie der Intrusiva und Rahmengesteine des Südwest Adamello (Nord Italien). Thesis 7612, ETH Zürich

Brodie KH, Rutter EH (1987) Deep crustal extensional faulting in the Ivrea zone of northern Italy. Tectonophysics 140: 193–212

Brodie KH, Rex D, Rutter EH (1989) On the age of deep crustal extensional faulting in the Ivrea zone, northern Italy. In: Coward MP, Dietrich D, Park RG (eds) Alpine tectonics. Blackwell, Oxford, Geol Soc Spec Publ 45: 203–210

Buletti M (1983) Zur Geochemie und Entstehungsgeschichte der Grant-Amphibolite des Gambarognogebiets, Ticino. Schweiz Mineral Petrogr Mitt 63: 233–247

Cadel G (1986) Geology and uranium mineralization of the Collio Basin (central Southern Alps, Italy). Uranium 2: 215–240

Cadel G, Fuchs Y, Meneghel L (1987) Uranium mineralization associated with the evolution of a Permo-Carboniferous volcanic field. Examples from Novazza and Val Vedello (northern Italy). Uranium 3: 407–421

Cadel G, Cosi M, Pennacchioni G, Spalla MI (1990) Structural map of the central Orobic Alps (Val Vedello-Novazza uranium province. Southern Alps, Italy). Mem Soc Geol Ital 45: 145–147

Cassano E, Anelli L, Fichera R, Cappelli V (1986) Pianura Padana. Interpretazione integrata di dati geofisici e geologici. In: 73° Congr Soc Geol Ital – symp, 29 Sept–4 Oct 1986, Roma, AGIP, S. Donato Milanese, p 27

Cassinis G, Dal Piaz GV, Eusebio A, Gosso G, Martinotti G, Massari F, Milano PF, Pennacchioni G, Perello M, Pessina CM, Roman E, Spalla MI, Tosetto S, Zerbato M (1986) Report on a structural and sedimentological analysis in the uranium province of the Orobic Alps, Italy. Uranium 2: 241–260

Castellarin A (1979) Il problema dei raccorciamenti crostali nel Sudalpino. Rend Soc Geol Ital 1: 21–23

Castellarin A (ed) (1981) Carta tettonica delle Alpi Meridionali alla scala 1 : 200 000. Pubbl 441 del Progetto Finalizzato Geodinamica CNR, p 220

Castellarin A, Vai GB (1982) Guida alla geologia del Sudalpino centro-orientale. Guide Geologiche Regionali. Soc Geol Ital, Bologna, p 386

Colombo A, Siletto GB, Tunesi A (1991) Pre-Variscan magmatism in central Southern Alps: the Fioraro complex. Terra Abstr 3(1): 210

Cornelius HP (1916) Zur Kenntnis der Wurzelregion im unteren Veltlin. Neues Jahrb Mineral Geol Paleontol 40: 253–263

Grespi R, Liborio G, Mottana A (1981) Metamorfismo tardoalpino di grado bassissimo nel basamento a sud della Linea Insubrica. Rend Soc Ital Mineral Petrol 37: 813–824

Crespi R, Liborio G, Mottana A (1982) On a widespread occurrence of stilpnomelane to the south of the Insubric line, Central Alps, Italy. Neues Jahrb Mineral Monatsh 6: 265–271

Crisci CM, Ferrara G, Mazzuoli R, Rossi PM (1984) Geochemical and geochronological data on Triassic volcanism of the Southern Alps of Lombardy (Italy): genetic implications. Geol Rundsch 73: 279–292

CroP (1989) Rapporto preliminare dei profili CroP "Alpi Centrali". Esagrafica Roma for Consiglio Nazionale delle Ricerche. Progetto strategico CROsta Profonda, p 105

Curioni G (1844) Stato geologico della Lombardia. In: Cattaneo C (ed) Notizie naturali e civili della Lombardia. Bernardoni Milano, pp 27–88

De Capitani L, Delitala MC, Liborio G, Mottana A, Nicoletti M, Petrucciani C (1988) K-Ar dating of the Val Biandino plutonic complex (Orobic Alps, Italy) Mem Sci Geol 40: 285–294

Del Moro A (1989) Il metodo Rb-Sr: cronometro dei processi geologici. In: Tunesi A, Boriani A (eds) Lezioni e seminari di Geocronologia e geochimica isotopica. Ric Sci Ed Per, Università Milano, 64: 43–64

Del Moro A, Ferrara G, Tonarini S, Callegari E (1983) Rb-Sr systematics on rocks from the Adamello batholith (Southern Alps). Mem Soc Geol Ital 26: 261–284

De Sitter LU, De Sitter-Koomans CM (1949) Geology of the Bergamasc Alps, Lombardia, Italy. Leidse Geol Meded 14: 1–257

Diella V, Spalla MI, Tunesi A (1992) Contrasted thermomechanical evolutions in the Southalpine metamorphic basement of the Orobic Alps (Central Alps, Italy). J Metamorphic Geol 10: 203–219

Doglioni C, Bosellini A (1987) Eoalpine and mesoalpine tectonics in the Southern Alps. Geol Rundsch 76: 735–754
Downes H, Duthou J-L (1988) Isotopic and trace-elements arguments for the lower-crustal origin of Hercynian granitoids and Pre-Hercynian orthogneisses, Massif Central (France). Chem Geol 68: 291–308
Dozy JJ (1933) Die Geologie der Catena Orobica zwischen Corno Stella und Pizzo del Diavolo di Tenda. Leidse Geol Meded 6: 133–230
El Tahlawi MR (1965) Geologie und Petrographie des Nordöstlichen Comerseegebietes (Provinz Como, Italien). Thesis 27, ETH Zürich
England PC, Thompson AB (1984) Pressure-temperature-time paths of regional metamorphism, I. Heat transfer during the evolution of region of thickened continental crust. J Petrol 25 (4): 894–928
Ferrara G, Innocenti F (1974) Radiometric age evidences of a Triassic thermal event in the Southern Alps. Geol Rundsch 63: 572–581
Forcella F, Jadoul F (1990) Deformazioni post-Adamello nelle Alpi Orobie. Stato dell'arte ed ipotesi di lavoro. Studi Geol Camerti vol spec, pp 139–141
Frisch W, Menot P, Neubauer F, von Raumer JF (1990) Correlation and evolution of the Alpine basement. Schweiz Mineral Petrogr Mitt 70: 265–285
Gaetani M, Jadoul F (1979) The structure of the Bergamasc Alps. Rend Atti Accad Naz Lincei VIII, LXVI: 5
Gaetani M, Jadoul F (1987) Controllo ancestrale sui principali lineamenti strutturali delle prealpi lombarde centrali. Rend Soc Geol Ital 10: 21–24
Gansser A, Pantič N (1988) Prealpine events along the eastern Insubric line (Tonale line, northern Italy). Eclogae Geol Helv 81 (3): 567–577
Giobbi Origoni E, Gregnanin A (1983) The crystalline basement of the "Massiccio delle Tre Valli Bresciane": new petrographic and geochemical data. Mem Soc Geol Ital 26: 133–144
Grauert B, Hanny R, Soptrajanova G (1973) Age and origin of detrital zircon from Pre-Permian basements of the Bohemian Massif and in the Alps. Contrib Mineral Petrol 40: 105–130
Handy MR (1987) The structure, age and kinematics of the Pogallo Fault zone; Southern Alps, northwestern Italy. Eclogae Geol Helv 80: 593–632
Hanson GN, El Tahlawi MR, Weber W (1966) K-Ar and Rb-Sr ages of pegmatites in the south-central Alps. Earth Planet Sci Lett 1: 407–413
Harris NBW, Pearce JA, Tindle AG (1986) Geochemical characteristics of collision zone magmatism. In: Coward MP, Ries AC (eds) Collision tectonics. Blackwell, Oxford, Geol Soc Spec Publ 19: 67–81
Holdaway MJ (1971) Stability of andalusite and the aluminum silicate phase diagram. Am J Sci 271: 97–131
Hunziker JC, Zingg A (1980) Lower Paleozoic amphibolite to granulite facies metamorphism in the Ivrea zone (Southern Alps, N Italy). Schweiz Mineral Petrogr Mitt 60: 181–213
Lardeaux JM, Spalla MI (1991) From granulites to eclogites in the Sesia Zone (Italian Western Alps): a record of the opening and closure of the Piedmont Ocean. In: Compagnoni R, Piccardo GB (eds) High pressure metamorphism in the Western Alps. J Metamorphic Geol 9: 35–59
Laubscher HP (1985) Large-scale thin-skinned thrusting in Southern Alps: kinematic models. Geol Soc Am Bull 96: 710–718
Laubscher HP (1990) The problem of deep structure of the Southern Alps: 3-D material balance considerations and regional consequences. Tectonophysics 176: 103–121
Melzi G (1891 a) Ricerche microscopiche sulle rocce del versante valtellinese della Catena Orobica occidentale. G Mineral Crist Pet 2: 1–34
Melzi G (1891 b) Sunto di alcune osservazioni stratigrafiche e petrografiche sul versante valtellinese delle Prealpi Orobie occidentali. Rend Reg Ist Lomb Sci Lett 24: 705–716
Milano F, Pennacchioni G, Spalla MI (1988) Alpine and pre-Alpine tectonics in the central Orobic Alps (Southern Alps). Eclogae Geol Helv 81: 273–293
Mottana A, Schiavinato G (1973) Metamorfismo regionale e di contatto nel settore nord-occidentale del massiccio dell'Adamello. Mem Ist Geol Minerol Univ Padova 29: 1–71
Mottana A, Nicoletti M, Petrucciani C, Liborio G, De Capitani L, Bocchio R (1985) Pre-alpine and alpine evolution of the Southalpine basement of the Orobic Alps. Geol Rundsch 74: 353–366
Niggli E (ed) (1973) Metamorphic map of the Alps 1 : 1 000 000. Subcom Cart Metam Belts of the World, Unesco, Paris
Origoni Giobbi E, Peyronel Pagliani G, Zanchini Camerini RM (1979) Contributo alla conoscenza chimico-petrografica delle vulcaniti Permiane affioranti tra la Val Trompia e le Valli Giudicarie. Rend Soc Ital Mineral Petrol 35 (1): 277–298
Pearce JA, Harris NBW, Tindle AG (1984) Trace element discrimination diagrams for the tectonic interpretation of granitic rocks. J Petrol 25: 956–983
Pieri M (1983) Three seismic profiles through the Po plain. In: Bally AW (ed) Seismic expression of structural styles, a picture and work atlas. Am Assoc Petrol Geol 15 (3): 3.4.1
Pieri M, Groppi G (1981) Subsurface geological structure of the Po plain, Italy. CNR, Progetto finalizzato Geodinamica, 414: 1–13
Pinarelli L, Del Moro A, Boriani A (1988) Rb-Sr geochronology of Lower Permian plutonism in Massiccio dei Laghi, Southern Alps (NW Italy). Rend Soc Ital Mineral Petrol 42 (2): 411–428
Porro C (1897) Cenni preliminari ad un rilevamento geologico dalla Vasassina al Monte Venerocolo. Rend Reg Ist Lomb Sci Lett 32: 408–419
Porro C (1899) Cenni preliminari ad un rilievo geologico nelle Alpi Orobie (Valsassina e Pizzo dei Tre Signori). Rend Reg Ist Lomb Sci Lett 30: 619–633
Porro C (1903) Alpi Bergamasche. Note illustrative della Carta Geologica e delle Sezioni. Tip degli Operai. Milano, p 30
Porro C (1911) Note geologiche sulle Alpi Bergamasche e Bresciane. Rend Reg Ist Lomb Sci Lett Ser II 44: 863–883
Porro C (1931) Rapporti fra gli scisti cristallini a Sud-Est di Gerola e le formazioni permiane. Catena Orobica. Rend Reg Ist Lomb Sci Lett 64: 711–787
Ramsay JG (1967) Folding and fracturing of rocks. McGraw-Hill, New York, p 568
Reinhard M (1953) Über das Grundgebirge des Sottoceneri im südlichen Tessin. Eclogae Geol Helv 46: 214–222
Reinhard M (1964) Über das Grundgebirge des Sottoceneri im Süd-Tessin und die darin auftretenden Ganggesteine. Beitr Geol Karte Schweiz NF 117: 1–89
Repossi E (1904) Osservazioni geologiche e petrografiche sui dintorni di Musso (lago di Como). Atti Soc Ital Sci Nat 43 (3): 261–303
Repossi E (1914) I filoni pegmatitici di Olgiasca. Rinvenimento in essi di minerali di uranio. Atti Soc Ital Sci Nat 52: 487–513
Richardson SW, Gilbert MC, Bell PM (1969) Experimental determination of the kyanite-andalusite and andalusite- sillimanite equilibria; the aluminum silicate triple point. Am J Sci 267: 259–272
Riklin KA (1983) Kontaktmetamorphose Permischer Sandsteine im Adamello Massif. Thesis, ETH Zürich

Roeder D (1989) South-Alpine thrusting and trans-Alpine convergence. In: Coward MP, Dietrich D, Park RG (eds) Alpine tectonics. Blackwell, Oxford, Geol Soc Spec Publ 45: 211–227
Ruppel C, Royden L, Hodges KV (1988) Thermal modelling of extensional tectonics: application to pressure-temperature-time histories of metamorphic rocks. Tectonics 75: 947–957
Salomon W (1901) Über neue geologische Aufnahmen der östlichen Hälfte der Adamellogruppe. Sitzber K Preuss Akad Wiss 1: 170–185
Sassi FP, Kalvacheva R, Zanferrari A (1984) New data on the age of deposition of the Southalpine phyllitic basement in the Eastern Alps. Neues Jahrb Geol Paleontol Monats 1984: 741–751
Sassi FP, Cavazzini G, Visonà D, Del Moro A (1985) Radiometric geochronology in the Eastern Alps: results and problems. Rend Soc Ital Mineral Petrol 40: 187–224
Schmid SM, Zingg A, Handy M (1987) The kinematics of movement along the Insubric line and the emplacement of the Ivrea zone. Tectonophysics 35: 47–66
Schmid SM, Aebli HR, Heller F, Zingg A (1989) The role of the Periadriatic line in the tectonic evolution of the Alps. In: Coward MP, Dietrich D, Park RG (eds) Alpine tectonics. Blackwell, Oxford, Geol Soc Spec Publ 45: 153–171
Siletto GB (1990) Polyphase tectonics in the orobic basement of Passo S. Marco (Upper Val Brembana, Southern Alps, Italy). Mem Soc Geol Ital 45: 101–105
Siletto GB (1991) Cronologia relativa dei sovrascorrimenti in aree selezionate del basamento orobico. Thesis, Univ Milano, Milano
Siletto GB, Spalla MI, Tunesi A, Nardo M, Soldo L (1990) Structural analysis in the Lario basement (Central Southern Alps, Italy). Mem Soc Geol Ital 45: 93–100
Spalla MI, Diella V, Siletto GB, Tunesi A, Gosso G, Lardeaux JM (1991) Contrasted pressure-retrograde paths in the orobic basement: a record of the Late-Variscan uplift and Permo-Jurassic rifting in the Southern Alps, Central Alps, Italy. Terra Abstr 3 (1): 217
Stähle V, Frenzel G, Kober B, Michard A, Puchelt H, Schneider W (1990) Zircon syenite pegmatites in the Finero peridotite (Ivrea Zone): evidence for a syenite from a mantle source. Earth Planet Sci Lett 101: 196–205
Stella A (1894) Contributo alla conoscenza delle formazioni pretriassiche nel versante meridionale delle Alpi Centrali. Boll Reg Comit Geol Ital 25 (1): 83–114
Stille P, Buletti M (1987) Nd-Sr Isotopic characteristics of the Lugano rocks and constraints on the continental crust formation in the South Alpine domain (N-Italy-Switzerland). Contrib Mineral Petrol 96: 140–150
Taramelli T (1880) Il Canton Ticino meridionale e i paesi finitimi. Spiegazione del foglio 24 Dufour colorito geologicamente da Spreafico, Negri, Stoppani. Mater Carta Geol Svizz ser I, 17: 1–230
Taramelli T (1890) Spiegazione della carta geologica della Lombardia. Tip. Fusi. Pavia, p 58
Thompson AB, Tracy RJ (1979) Model systems for anatexis of pelitic rocks. II. Facies series melting and reactions in the system CaO-$KAlO_2$-$NaAlO_2$-Al_2O_3-SiO_2-H_2O. Contrib Mineral Petrol 70: 429–438
Vai GB, Cocozza T (1986) Tentative schematic zonation of the Hercynian chain in Italy. Bull Soc Géol Fr 8 (2): 95–114
Vai GB, Boriani A, Rivalenti G, Sassi FP (1981) Catena Ercinica e Paleozoico nelle Alpi Meridionali. In: Cento Anni di Geologia Italiana, Jubilee volume. Pitagora, Bologna. Soc Geol Ital (1984), pp 133–154
Vielzeuf D (1984) Relations de phases dans le facies granulite et implications géodynamiques. Les réactions minéralogiques témoins de leur rétromorphose. Thesis, Clermont-Ferrand II University, Clermont-Ferrand
Voshage H, Hunziker JC, Hofmann AW, Zingg A (1987) A Nd and Sr isotopic study of Ivrea zone, Southern Alps, N-Italy. Contrib Mineral Petrol 97: 31–42
Voshage H, Sinigoi S, Mazzucchelli M, Demarchi G, Rivalenti G, Hofmann AW (1988) Isotopic constraints on the origin of ultramafic and mafic dykes in the Balmuccia peridotite (Ivrea Zone). Contrib Mineral Petrol 100: 261–267
Wennekers JHL (1932) The structure of the Bergamo Alps compared with that of the Highlands of Scotland. Leidse Geol Meded 4: 83–93
Winterer EL, Bosellini A (1981) Subsidence and sedimentation on Jurassic passive continental margin, Southern Alps, Italy. AAPG Bull 65: 394–421
Wopfner H (1984) Permian deposits of the Southern Alps as product of initial alpidic taphrogenesis. Geol Rundsch 73 (1): 259–277
Zanchi A, Chiesa S, Gillot PY (1990) Tectonic evolution of the Southern Alps in the Orobic structural chain/and geochronological indications for pre-Tertiary compressive tectonics. Mem. Soc. Geol. Ital. 45: 77–82
Zingg A (1983) The Ivrea and Strona-Ceneri Zones (Southern Alps, Ticino and N-Italy) – a review. Schweiz Mineral Petrogr Mitt 63: 361–392

The South-Alpine Metamorphic Basement in the Eastern Alps

F. P. Sassi and R. Spiess

Abstract

A model of lithostratigraphic sequence is presented, ranging from Cambrian to Devonian; its chronological frame is, however, mainly hypothetical due to the lack of fossils. The interposition of a volcano-sedimentary complex (VSC) makes the distinction between an upper (UPC) and a lower complex (LPC) possible, within this rather monotonous pelitic-psammitic sequence. The LPC is assumed to be pre-Caradocian: it includes the level in which acritarch data demonstrated the occurrence of Cambrian metasediments in the Alps for the first time. The VSC, assumed to be Late Ordovician due to interregional correlations, records an important volcanic activity, mainly acidic and locally basic in composition. The metamorphism developed under low pressure, greenschist facies conditions during two main stages, one synkinematic (S_2 foliation) and one post-kinematic, the latter recording the thermal climax. Its Variscan age is demonstrated by isotopic data. The value of 350 Ma has been related to the early regional heating, 320 Ma to the thermal climax, and 316–300 Ma to the cooling. The whole set of the Variscan age values in the Eastern Alps indicates that Variscan metamorphism was synchronous over large areas. The regional trend of the main foliation (S_2) and crenulation axis is reported. E-W strikes and verticality are typical along the Periadriatic lineament.

An extensional regime during Ordovician could explain the widespread Upper Ordovician volcanism in the Southern Alps; it also represents a suitable frame to understand both the whole "Upper Ordovician granite-rhyolite association" and the "Caledonian" metamorphism occurring in the Austro-Alpine, for which Pan-African pertinence has been proposed.

1 Introduction

In the Eastern Alps, the South-Alpine crystalline basement (SCB) consists of low-grade metamorphic rocks injected by Permo-Carboniferous granitoid bodies. It outcrops in three areas (Fig. 1):

1. Sarentino (Sarntal)-Bressanone (Brixen)-Pusteria (Pustertal)-Comelico area, in which the low-grade metamorphic rocks are intruded by the Ivigna (Ifinger)-Bressanone (Brixen) granitoid body and other minor intrusions;
2. Valsugana-Cima d'Asta-Agordo-Cereda area, in which the low grade metamorphic rocks are crosscut by the Cima d'Asta granitoid pluton and other minor intrusions;
3. Recoaro area, which only consists of low grade metamorphic rocks.

An easily observable unconformity separates the SCB from the overlying Upper Palaeozoic clastic sediments ["Verrucano" conglomerates and Val Gardena (Grödener) sandstones].

The SCB certainly escaped the Alpine metamorphism, as radiometric data and the lack of metamorphic effects in the overlying Upper Palaeozoic sediments demonstrate. Therefore, its petrogenetic history is only pre-Alpine.

The main deformational history (micro-, meso- and larger-scale folding, probably including nappe emplacement) also is pre-Alpine; however, gentle folds and brittle deformations of Alpine age occur, too.

The eastern boundary between the SCB (Comelico area) and the easterly adjacent, anchimetamorphic, Palaeozoic sequences of the Carnic Alps and Southern Karawanken is probably transitional, related to the gradual vanishing of the Variscan metamorphic grade. However, the exact location of this boundary is not yet known (Arkai et al. 1991).

As mentioned, two main rock associations make up the SCB:

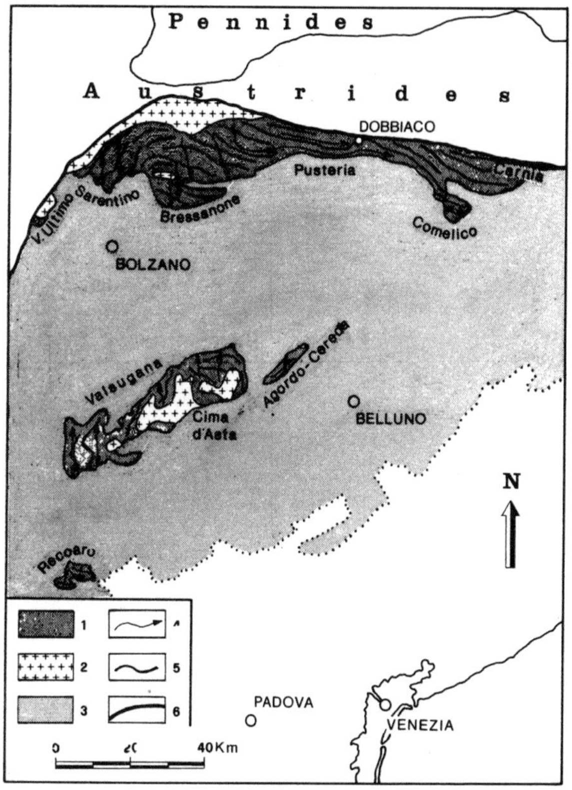

Fig. 1. The outcropping areas of the South-Alpine crystalline basement in the Eastern Alps, and general trends of the main foliation (S_2) and crenulation axis. *1* Metamorphic basement; *2* Permo Carboniferous granitoids; *3* post-Variscan cover; *4* crenulation axis; *5* main foliation (S_2); *6* Periadriatic lineament

1. The low-grade metamorphic sequence, which makes up the bulk of the SCB; SCB mostly consists of silicate metasediments, with minor interlayered metavolcanics and local carbonate-bearing intercalations;
2. Non-metamorphic, late Variscan, igneous rocks, making up variously sized plutonic bodies which cross-cut the low-grade metamorphic sequence; they are mostly acidic in composition, but minor intermediate and basic types also occur.

The present chapter only deals with the metamorphic rocks. The late Variscan intrusives are considered in a specific chapter by Bonin et al. (this Vol.).

2 The Metamorphic Lithostratigraphic Sequence

When identification of the lithostratigraphic sequence in the SCB is attempted, great difficulties are encountered (for many reasons: the large distances separating the different SCB areas from each other, the monotonous character of the rock types making up the bulk of the SCB, and the discontinuity of their exposures. However, the main difficulties are the lack of biochronologic data and the complexity of the structural situation, which in turn is difficult to interpret for the already mentioned reasons.

Aware of these difficulties, Sassi and Zirpoli (1965, 1968, 1983) followed the only possible approach, i.e. focussing their attention on rock levels which represent potential reference horizons. In this context, they concentrated upon the acidic metavolcanic-metavolcanoclastic horizon (known as the "porphyroid" plateau) occurring in several SCB areas, the stratabound polymetallic concentrations and other lithologic peculiarities. On these grounds, they proposed and gradually improved a general model of lithostratigraphic sequence for the whole SCB.

The conceptual basis of this model is the assumption that the volcanism recorded by these "porphyroids" was of regional importance in monitoring the SCB geodynamic history (Sassi and Zirpoli 1965, 1968): as a consequence, the "porphyroids" were admitted to have played a role as key-horizon for interregional correlations within the SCB, and the name of "pre-Hercynian porphyric plateau" (here, PPP for short) was proposed. The PPP was consequently used as a lithostratigraphic marker in order to distinguish and underlying phyllitic complex from an overlying one; and the SCB was therefore divided into three lithostratigraphic complexes (Sassi et al. 1974).

The validity of this three-fold classification and related chronological hypotheses was confirmed by further research work. Consequently, Sassi and Zirpoli (1989) defined their model better on the basis of all available data, outlining the lithology, the recorded magmatic activity, the metamorphic features and the geochronological data or constraints for each lithostratigraphic complex. The lithostratigraphic complexes in this model are, as follows (from bottom to top).

a. *Lower Pelitic-Psammitic Complex* (LPC). Its assumed sedimentation age is pre-Caradocian. It includes the level in which Cambrian acritarchs have been found (Sassi et al. 1984a, b; Kalvacheva et al. 1986), demonstrating the occurrence of Cambrian metasediments in the Alps for the first time.

The LPC mainly consists of quartz-phyllites in which synkinematic metamorphic differentiation enhanced and transposed the primary sedimentary banding. The only record of magmatic activity in it is represented by gneisses of dacitic composition outcropping in the Plan de Corones (Kronplatz) area (Del Moro et al. 1984).

Two main metamorphic episodes have been recognized in the LPC, both displaying complex relationships between crystallization and deformation. Metamorphic conditions cover the whole temperature range of the greenschist facies, and display regional zoneography (see Sect. 4).

The metamorphic effects are Variscan in age, as some radiometric data indicate; however, a problematic age value of 463 Ma was also obtained from the Plan de Corones area (see Sect. 5).

b. *Volcano-Sedimentary Complex* (VSC). Its assumed pre-metamorphic age is Late Ordovician. This chronological hypothesis is based on long-distance correlation (Sassi et al. 1979) between the rock sequence in which the Comelico "porphyroids" occur and that described and dated by Flajs and Schönlaub (1976) in the northern Grauwacken Zone.

The rock types occurring in the VSC make up a metapelitic-psammitic sequence in which many, lithologically varied, intercalations characteristically occur: acidic metavolcanics and metavolcanoclastics of rhyolitic to rhyodacitic composition, now represented by the "porphyroids"; basic metavolcanics and metavolcanoclastics, now represented by epidote-amphibolites (Visona' and Zanferrari 1985) and epidote-chlorite-actinolite schists; strata-bound Fe, Cu (Zn, Pb), mineralizations ("Kieslager" type), a review of which was given by Frizzo (1983); ilmenite-rich phyllites; a discountinuous horizon of carbonate schists and carbonate-bearing phyllites; carbonaceous black phyllites and quartzites. These "porphyroids" make up the "pre-Hercynian porphyric plateau" (Sassi and Zirpoli 1968), later renamed the "Upper Ordovician Rhyolitic Plateau" and included in the Austro-Alpine "Upper Ordovician granite-rhyolite association" occurring in the Eastern Alps (Bellieni and Sassi 1981). The chemistry of the "porphyroids" indicates calc-alkaline affinity, and is consistent with a genetic model through crustal anatexis (Sassi and Zirpoli 1979; Bellieni and Sassi 1981).

Therefore, two compositionally different types of volcanism are recorded within the VSC: one acidic and one basic. If they both belong to the same volcanic cycle, a definitely two-modal character should be assigned to it, with a well-expressed Daly gap.

The metamorphic features of the VSC do not differ from those recorded in the LPC: the only differences in microtexture and mineralogy are encountered in some peculiar rock levels (e.g. carbonate schists, metarhyolites), due to bulk compositional constraints and consequent peculiar behaviour during deformation and recrystallization.

c. *Upper Pelitic-Psammitic Complex* (UPC for short). Its assumed age is Silurian-Devonian, although no direct evidence of this age is available.

The rock types making up the UPC are quartz-phyllites, identical in grain size, mineral composition and mineral banding to the rock types making up the LPC. Specifically, the UPC metamorphic and microtextural features, as well as the crystallization-deformation relationships, are identical to those recorded in the LPC (Mazzoli and Sassi 1988). Consequently, UPC and LPC rocks can only be distinguished from each other with objective certainty after structural reconstruction of the rock sequences and recognition of their position respectively below or above the VSC.

This outlined lithostratigraphic model is a workable tool for understanding and describing the situations occurring in all SCB areas. Zanferrari (1985; see also Poli and Zanferrari 1989) proposed a more detailed lithostratigraphic model, assigning the rank of "formation" to what we cautiously only consider as "complexes", and giving a regional meaning to rock intercalations which may be only local. In particular, this author refers all metabasites occurring in the SCB to the "Gudon (Gufidaun) formation", assigning them to a Lower-?Middle Silurian age. Such a detailed approach seems to us rather premature, considering the poor database available, and taking into account the following regional situation occurring within the Austro-Alpine phyllitic complexes of the Eastern Alps: the basic volcanism there is not only Silurian, but also Middle Devonian in some areas and, in others, it occurs at different levels, from Late Ordovician to Middle Devonian (Neubauer and Sassi, this Vol.). However, the Zanferrari model, although requiring further confirmation on a regional scale and further chronological support, is worthy of being used as a working tool.

3 Regional Structural Trends

The SCB quartz-phyllites are commonly crenulated and display a well-developed S_2 foliation. Figure 1 schematically shows the regional pattern of the S_2 surfaces and crenulation axes; it is based on data by Agterberg (1961), Sassi and Zirpoli (1968), Blackburn et al. (1968) and new data.

In Pusteria (Pustertal), S_2 is subvertical and strikes E-W, whereas the crenulation axis steeply plunges towards the E. Eastwards to Comelico and Carnia, the crenulation axis tends to become subhorizontal, slightly plunging towards the SE or NW. S_2 planes are subvertical and their strike ranges between W and NW. In the Bressanone (Brixen) area S_2 dips S-SW, and the amount of dip increases from W to E and from S to N.

The Bressanone area is the transition region between the EW subvertical attitude of S_2 dominating in Pusteria, and the subhorizontal attitude of S_2 prevailing in the westernmost (Sarentino, Sarntal) and southernmost areas (Valsugana, Agordo-Cereda, Recoaro). The northeastern part of the Bressanone area is also the "fitting region" between the NS crenulation axes prevailing within the western and southern areas, and the EW crenulation axes prevailing in Pusteria, Comelico and Carnia.

It should be stressed that: (1) EW strikes and verticality are typical in SCB quartz-phyllites along the Periadriatic lineament; (2) S_2 surfaces in the central Valsugana area roughly conform to the shape of the Cima d'Asta granitoid pluton.

4 Metamorphic Features

Disregarding details and local pecularities, the *metamorphic history* of the SCB can be described in terms of two main stages of crystallization: one synkinematic and one post-kinematic.

The synkinematic stage took place under low greenschist facies conditions. It consists of two different synkinematic crystallization phases (Cr_1 and Cr_2 respectively), which developed under increasing temperatures, each having produced a foliation, respectively S_1 and S_2. S_1 is only recognizable as oriented inclusion trails within some porphyroblasts and rootless microfolds within some lepidoblastic layers. S_2 is commonly the main foliation both on meso- and microscopic scales: it is an axial plane foliation whose orientation also depends on the granoblasts/lepidoblasts ratio within the given layer.

The post-kinematic crystallization stage (Cr_3) records the thermal climax in these rocks. It occurred under greenschist facies conditions, producing mineral assemblages characterized respectively by the stability of chlorite or biotite or almandine, spatially arranged according to the regional zonation. It is important to point out that the Cr_3 mineral assemblages crystallized after a deformation stage during which S_2 was variously folded, and flattening effects around the existing porphyroblasts were produced.

A later, mainly post-kinematic crystallization stage is also clearly recognizable in some areas, where albite microporphyroblasts occur. They clearly include the mineral grains crystallized during Cr_3 (e.g. post-kinematic chloritoid and mica flakes), as well as oriented inclusion trails geometrically consistent with the external folded pattern. The albite porphyroblasts are sometimes zoned with oligoclase rims, and locally display a snowball texture.

The crystallization of the albite porphyroblasts is commonly older than the local appearance of a discontinuous S_3 foliation. This planar anisotropy is genetically related to a late deformation stage not accompanied by any regionally significant crystallization. However, retrograde metamorphic effects occur locally, and low-temperature minerals crystallized within the pressure shadows.

Figure 2 roughly sketches this outlined evolution as recorded in the Recoaro quartz-phyllites, i.e. in an area where Cr_3 is only represented by zoned albite microporphyroblasts and S_2 was only gently folded. Figure 3 represents the T-t evolution reconstructed in the Sarentino quartz-phyllites, where Cr_2 had a peculiar development and late, retrograde overprints are well recorded (Cr_4, Cr_5).

The traditional interpretation of the inclusion trails was adopted by the authors in both the above cases. However, recent literature (e.g. Bell and Rubenach 1983; Bell 1985) is inclined to reject porphyroblast rotation and hold that it is the surrounding foliation that rotates. Furthermore, it is now commonly admitted that porphyroblasts grow within the shortening domains, and inclusion trails in them record the geometrical pattern of the crenulation cleavage as it was in that given development stage. Therefore, the porphyroblasts of the SCB quartz-phyllites may be re-interpreted according to the new ideas, admitting that they crystallized over an evolving crenulation cleavage under prograde conditions, Cr_3 representing the thermal climax.

The *pressure character* of the Variscan metamorphism as recorded in the South-Alpine quartz-phyllites has been estimated by means of the average B cell parameter of muscovite (Sassi 1972; Guidotti and Sassi 1986). The pressure value turns out to be relatively low, and consequently the metamorphic thermal gradient relatively high, close to 40 °C/km.

Only a few results are available concerning the *P-T-t path* recorded in the SCB. A semi-quantitative path has been reconstructed by Cardin et al. (1985), and a more detailed path has been published by Hammerschmidt and Stöckhert (1987).

As concerns *regional metamorphic zonation,* Sassi and Zirpoli (1968) and Sassi et al. (1974) pointed out an increase in metamorphic grade from the east (Comelico area: chlorite zone) to the west (Sarentino, Sarntal Alps: almandine zone), and from the south

(Recoaro area: mainly chlorite zone) to the north (Sarentino Alps). However, such a pattern only is a rough general indication, which cannot be directly related to the palaeogeothermic situation: in fact, if it is used to estimate the metamorphic thermal gradient, unbelievably low values are obtained (due to the great thickness of the assumed mineral zones), in sharp contrast with the relatively high value which has been estimated on petrographic bases. This regional pattern outlined may only be referred to the regional structural situation, as a combined result of folding of the isothermal surfaces from one hand, and different depths of erosion on the other (Fig. 4). In this context, i.e. if we admit that the Variscan metamorphic isogrades were folded due to postmetamorphic Variscan (and Alpine?) tectogenesis, it is also possible to explain the true metamorphic zonations which may be observed locally, and are geometrically consistent with the relatively high values of the metamorphic thermal gradient.

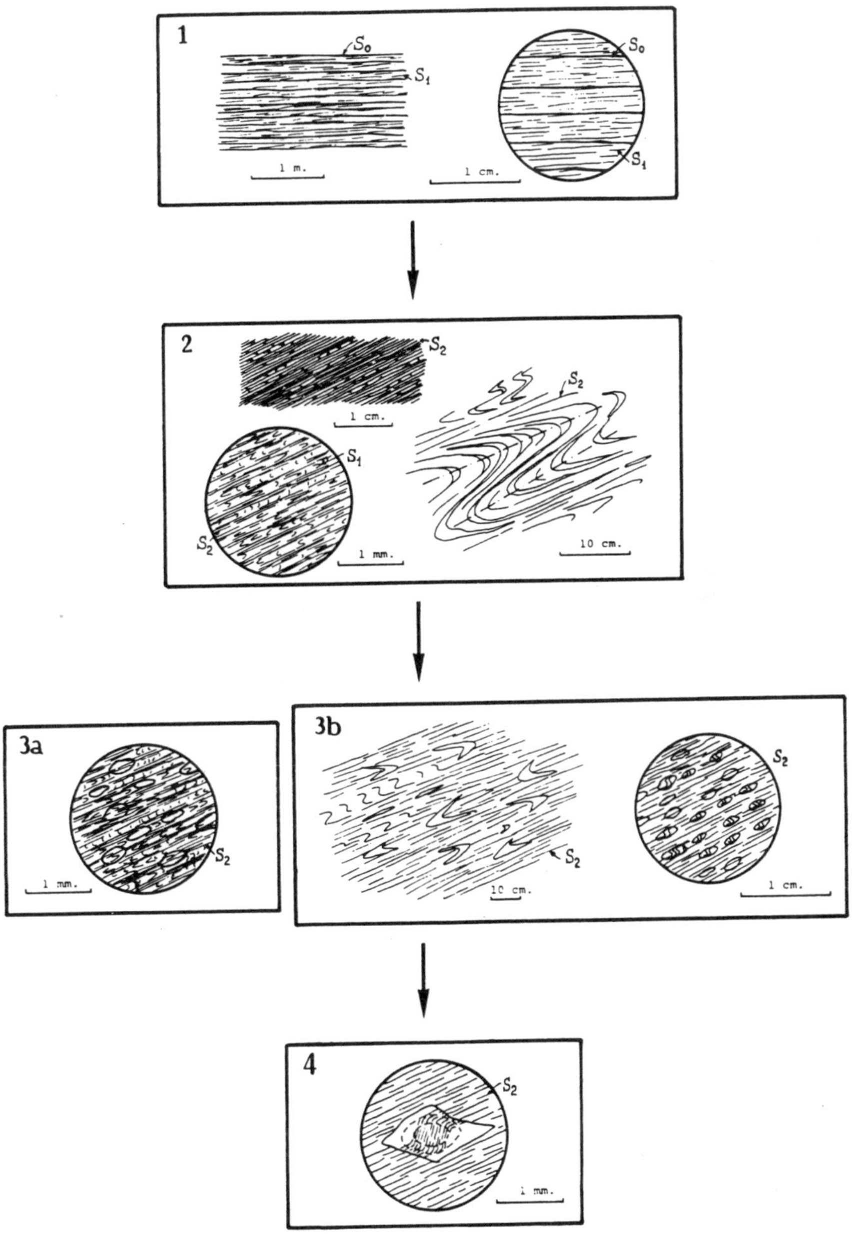

Fig. 2. Microstructural evolution of the quartz-phyllites in the Recoaro area. The stages *1* to *4* are in chronological succession; *3a* and *3b* refer to different scales (Blackburn et al. 1968)

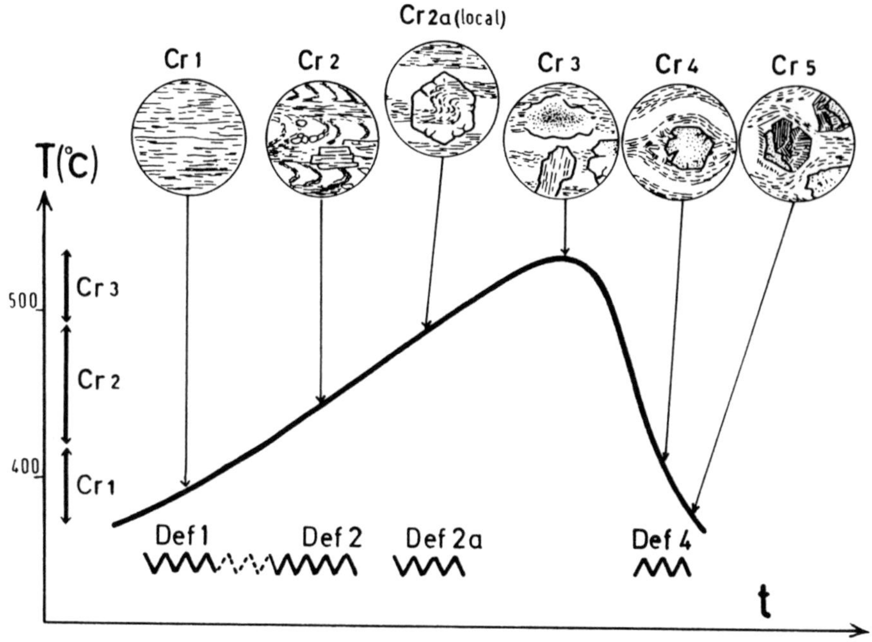

Fig. 3. Temperature-time and microstructural evolution of the quartz-phyllites in the Sarentino (Sarntal) area (Cardin et al. 1985)

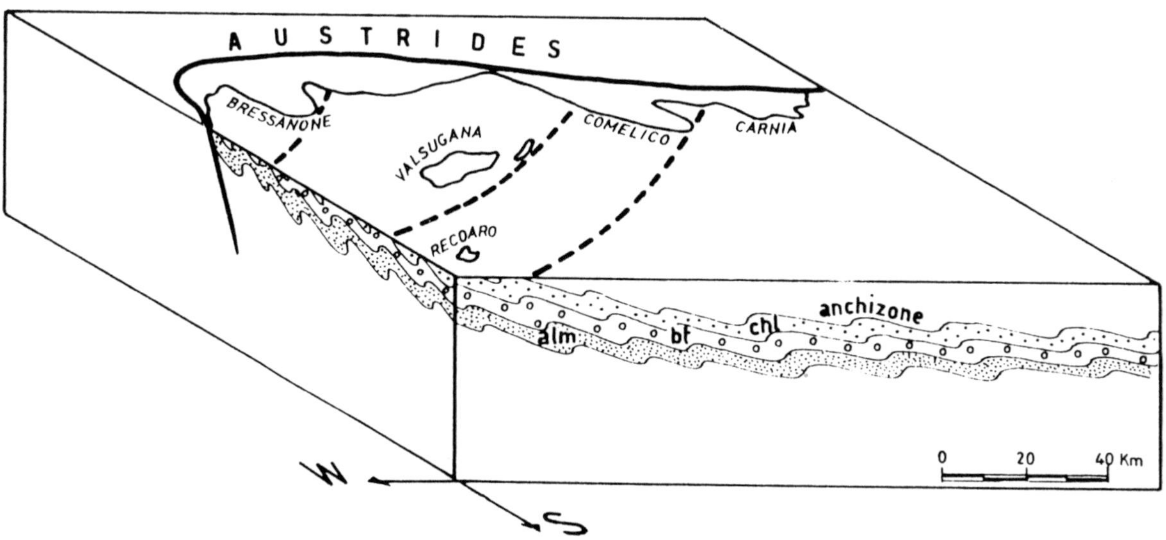

Fig. 4. Schematic pattern of the regional metamorphic zonation (Mazzoli and Sassi 1988)

5 Timing of Metamorphism

As mentioned, the pre-Alpine age of this metamorphism has long been known, because pebbles of quartz-phyllites occur within the non-metamorphic Permian conglomerates; and, before radiometric data were not available, the hypothesis of a Variscan age prevailed over the Precambrian hypothesis. On the other hand, the two-stage development of this pre-Alpine metamorphism was pointed out over 20 years ago (e. g. Sassi and Zirpoli 1968) by means of microtextural analyses.

Radiometric data (Table 1) confirmed both the hypothetized Variscan age and the two-stage development of this metamorphism. They assigned an age value close to 350 Ma (corresponding to the Viséan according to the timescale proposed by Odin 1982) to the older stage, and close to 320 Ma (corresponding to the Namurian-Viséan boundary) to the younger one (Rb-Sr mineral and whole-rock isochrons: Del Moro et al. 1980, 1984; Cavazzini et al. 1991). The value of 350 Ma is to be interpreted as related to the early regional heating, that of 320 Ma to the thermal climax, while a muscovite K-Ar age of

Table 1. Geochronological data (WR = whole rock; age values in Ma)

Locality	Rock type	Method		Age value	Reference[c]
Funes (Villnoss)	Metarhyolites	Rb-Sr	WR isochron	350 ± 3	(1)
"	Phyllite	Rb-Sr	WR-Ms-Grt	354 ± 10	"
"	Metavolcanoclastics	Rb-Sr	WR isochron	317 ± 16	"
"	Phyllite	Rb-Sr	WR-Bt	314 ± 5	"
"	"	Rb-Sr	Ms	314 ± 5	"
"	"	Rb-Sr	Ms-Pl (+Qtz)	312 ± 6	"
Plan de Corones (Kronplatz)	Phyllite	Rb-Sr	WR-Ms-Ab	463 ± 36	(2)
"	Acidic gneiss	"	WR-Ab-Kfs-Bt	326 ± 3	"
"	"	"	WR-Ab-Kfs-Bt	321 ± 1	"
"	"	"	WR-Ab-Kfs	309 ± 24	"
"	"	"	WR-Ab-Kfs	302 ± 4	"
"	"	"	WR-Ab-Kfs	293 ± 6	"
Valdaora (Olang)	Phyllite	K-Ar	Ms	316 ± 8[a]	(3)
"	"	^{40}Ar-^{39}Ar	Ms	319 ± 5.5[b]	"
Agordo	Phyllite	Rb-Sr	WR isochron	347 ± 17	(4)

[a] Range of several measurements.
[b] Plateau age.
[c] References: (1) Del Moro et al. (1980); (2) Del Moro et al. (1984); (3) Hammerschmidt and Stöckhert (1987); (4) Cavazzini et al. (1991).

316 ± 8 Ma from Valdaora phyllites (Olang, Pustertal) has been interpreted as cooling age by Hammerschmidt and Stöckhert (1987), consistently with the Rb-Sr mineral ages close to 300 Ma obtained by Del Moro et al. (1984).

An intriguing age value of 463 ± 18 Ma (albite-muscovite Rb-Sr internal isochron) has been obtained from a quartz-phyllite outcropping at Plan de Corones (Kronplatz) near Brunico (Bruneck, Pustertal) (Del Moro et al. 1984). It may represent the crystallization age of the analyzed minerals, indicating that parts of the SCB were also affected by a "Caledonian" metamorphism. This hypothesis implies the existence of Cambrian and/or Precambrian protoliths, a requirement which agrees very well with the above-mentioned acritarch data from the Agordo-Cereda phyllites (Sassi et al. 1984a, b). However, further radiometric data are necessary before we can admit with objective certainty that a "Caledonian" metamorphism is recorded in some parts of the SCB.

6 Final Considerations

The features and geological evolution of the SCB, although known in their main lines, still require better definition under many, not negligible aspects. Specifically:

1. Lithostratigraphy must be better known.
2. The sedimentation age of the different rock complexes is almost unknown (only one fossil finding, although important, has been reported).
3. The age of metamorphism should be better determined on a regional scale. The available radiometric data only refer to a relatively small part of the northernmost phyllitic area. Although they sufficiently define the Variscan age of metamorphism, they are still problematic as concerns a possible Ordovician metamorphic event. Furthermore, as regards the Variscan metamorphism, its two-stage development established by means of petrographic criteria should be better tested radiometrically.
4. The age values of 350 and 320 Ma obtained for the two Variscan stages of metamorphism (see Sect. 5) are identical to those obtained far to the north from Lower Austro-Alpine rocks (Satir and Morteani 1978). Therefore, within the limits of the analytical errors, the timing of the early regional heating (350 Ma) and thermal climax (320 Ma) turns out to be identical on a very large regional scale, notwithstanding the important tectonic lines occurring in between. The Variscan metamorphism seems to have been substantially synchronous over very large areas.
5. Metamorphic features should be better defined in terms of P-T-t paths and regional zoneography (which is only broadly known). The low-pressure character of this Variscan metamorphism is well established, but a suitable model for explaining it is not yet available. From the geophysical point of view, a geologically plausible heat source operating on a regional scale should be recognized. From the geological point of view, a tectonic cover should be recognized, sufficiently thick to have produced, during the Namurian (i. e. 320 Ma ago), a pressure of about 3 kb in Silurian and perhaps Devonian rocks.

6. The eastern boundary towards the non-metamorphic (partly anchimetamorphic) rock sequences of Carnia and Karawanken should be defined with respect to both its location and its meaning. This boundary is probably only a metamorphic boundary, related to the eastwards vanishing of the Variscan metamorphic grade.
7. The meaning of the volcanic activity recorded in the SCB should also be more deeply interpreted. Bellieni and Sassi (1981) assumed an extensional context to explain the Upper Ordovician rhyolitic plateau. Such a hypothesis should be carefully explored, taking into account that a rifting regime during the Ordovician may also represent a suitable frame to explain both the whole "Upper Ordovician granite-rhyolite association" and the "Caledonian" metamorphism occurring in the Austro-Alpine basement of the Eastern Alps (Sassi and Schmidt 1982; Ebner et al. 1987; Frisch et al. 1987; Sassi et al. 1987; and references quoted therein), for which Pan-African pertinence has been proposed (Sassi 1989).

Acknowledgements
This report was written within the ambit of the research activities of IGCP Project No. 276, Italian CNR (Centro di Studio per i Problemi dell'Orogeno delle Alpi Orientali, Padova) and MURST (40 % FPS).

References

Agterberg FP (1961) Tectonics of the crystalline basement of the dolomites in north Italy. Geol Ultraiectina Utrecht 8: 1–232
Arkai P, Sassi FP, Zirpoli G (1991) On the location of the boundary between the low and very low-grade Southalpine basement in the Eastern Alps. Terra Abstr 3: 100
Bell TH (1985) Deformation partitioning and porphyroblast rotation in metamorphic rocks: a radical reinterpretation. J Metamorphic Geol 3: 109–118
Bell TH, Rubenach MJ (1983) Sequential porphyroblast growth and crenulation cleavage development during progressive deformation. Tectonophysics 92: 171–194
Bellieni G, Sassi FP (1981) New chemical data and a review on the South-Alpine "Pre-Hercynian Rhyolitic plateau" in the Eastern Alps. In: Karamata S, Sassi FP (eds) IGCP Proj 5 Newslett 3: 22–27
Blackburn CE, Sassi FP, Zulian T (1968) Il basamento cristallino di Recoaro: assetto strutturale ed evoluzione tettonico-metamorfica. Mem Accad Patav SS LL AA, Cl Sci Mat Nat 81: 1–19
Cardin A, Pisani F, Sassi FP, Visonà D, Zirpoli G (1985) Le metapeliti e le anfiboliti del basamento sudalpino delle Alpi Sarentine (Alpi Orientali). Mem Sci Geol Padova 37: 379–406
Cavazzini Gc, Del Moro A, Sassi FP, Zirpoli G (1991) New data on the radiometric age of the Southalpine basement of the Eastern Alps. Geologia del Basamento Italiano. Convegno in memoria di Tommaso Cocozza Siena 21–22 marzo 1991, Abstr vol (Siena)
Del Moro A, Sassi FP, Zirpoli G (1980) Preliminary results on the radiometric age of the Hercynian metamorphism in the South-Alpine basement of the Eastern Alps. Neues Jahrb Geol Paläontol Monatsh 12: 707–718
Del Moro A, Sassi FP, Zirpoli G (1984) Acidic gneisses from Plan de Corones area, and chronological data on South-Alpine basement in Pusteria (Eastern Alps). Mem Ist Geol Minerol Univ Padova 36: 403–412
Ebner F, Neubauer F, Statteger K (1987) The Cadedonian event in the Eastern Alps: a review. In: Flügel HW, Sassi FP, Grecula P (eds) Pre-Variscan and Variscan events in the Alpine-Mediterranean mountain belts. Mineralia slovaca IGCP 5 Monogr, Alfa, Bratislava, pp 169–182
Flajs G, Schönlaub HP (1976) Die biostratigraphische Gliederung des Altpaläozoikums am Polster bei Eisenerz. Verh Geol Bundesanst 2: 257–303
Frisch W, Neubauer F, Brocker M, Bruckmann W, Haiss H (1987) Interpretation of geochemical data from the Caledonian basement within the Austroalpine basement complex. In: Flügel HW, Sassi FP, Grecula P (eds) Pre-Variscan and Variscan events in the Alpine-Mediterranean mountain belts. Mineralia slovaca IGCP 5 Monogr, Alfa Bratislava, pp 209–226
Frizzo P (1983) Pre-metamorphic Paleozoic mineralizations in the Austro-Alpine and South-Alpine basement of the Eastern Alps. Relationships between paleogeodynamics, stratigraphy and metallogenesis. In: Sassi FP, Szederkenyi T (eds) IGCP Proj 5 Newslett 5: 41–48
Guidotti CV, Sassi FP (1986) Classification and correlation of metamorphic facies series by means of muscovite *b* data from low-grade metapelites. Neues Jahrb Miner Abh 153: 363–380
Hammerschmidt K, Stöckhert B (1987) A K-Ar and $^{40}Ar/^{39}Ar$ study on white micas from Brixen quartzphyllite, Southern Alps. Evidence for argon loss at low temperatures. Contrib Mineral Petrol 95: 393–406
Kalvacheva R, Sassi FP, Zanferrari A (1986) Acritarch evidence for the Cambrian age of phyllites in the Agordo area (South-Alpine basement of Eastern Alps, Italy). Rev Palaebot Palynot 48: 311–336
Mazzoli C, Sassi R (1988) Caratteri del metamorfismo ercinico nella fillade sudalpina a ovest di Bressanone. Mem Sci Geol Padova 40: 295–314
Odin GS (1982) The Phanerozoic time scale revised. Episodes 1982 (3): 3–9
Poli ME, Zanferrari A (1989) Caratteri stratigrafici e strutturali del basamento cristallino di Agordo (Belluno). Rend Soc Geol Ital 13 (in press)
Sassi FP (1972) The petrological and geological significance of the b_0 values of potassic white micas in low-grade metamorphic rocks. An application to the Eastern Alps. Tschermaks Mineral Petrogr Mitt 18: 105–113
Sassi FP (1989) Correlation of pre-Hercynian and Hercynian events in the Alpine-Mediterranean Belt: final results of IGCP Proj 5. 28th Int Geol Congr Washington (Abstr) 3: 27 (Abstr vol, Washington)
Sassi FP, Schmidt K (1982) The "Caledonian" event in the Eastern Alps: Guidelines for its geodynamic interpretation. In: Sassi FP, Varga I (eds) IGCP Proj 5 Newslett 4: 78–81
Sassi FP, Zirpoli G (1965) Contributo alla conoscenza degli scisti cristallini del Comelico (Cadore). Mem Accad Pat SS LL AA 78: 35–72
Sassi FP, Zirpoli G (1968) Il basamento cristallino di Recoaro. Studio petrografico. Mem Soc Geol Ital 7: 227–245
Sassi FP, Zirpoli G (1979) Pre-Variscan acidic magmatism in the

Eastern Alps. In: Sassi FP, Visonà D (eds) IGCP Proj 5 Newslett 1: 111–121
Sassi FP, Zirpoli G (1983) A contribution to the lithostratigraphic classification of the Southalpine basement in the Eastern Alps. IGCP Proj 5, Internal Rep
Sassi FP, Zirpoli G (1989) The lithostratigraphic sequence in the South-Alpine basement of the Eastern Alps. In: Sassi FP, Zanferrari A (eds) Pre-Variscan and Variscan events int he Alpine-Mediterranean belts, stratigraphic correlation forms. Rend Soc Geol Ital 12: 397–402
Sassi FP, Zanferrari A, Zirpoli G (1974) Some considerations on the Southalpine basement of the Eastern Alps. Neues Jahrb Paläontol Monatsh 10: 609–624
Sassi FP, Zanferrari A, Zirpoli G (1979) Nuovi dati sulla stratigrafia e i caratteri chimico-petrografici dei porfiroidi del Comelico (Alpi Orientali). Mem Soc Geol Ital 20: 425–434
Sassi FP, Kalvacheva R, Zanferrari A (1984 a) New data on the age of deposition of the South-Alpine phyllitic basement in the Eastern Alps. Neues Jahrb Geol Paläontol Monatsh 12: 741–751
Sassi FP, Zanferrari A, Kalvacheva R (1984 b) Acritarch dating of phyllites in the Agordo area (South-Alpine basement of the Eastern Alps, Italy). 27th Int Geol Congr Moscow (Abstr) 1: 172–173 (Abstr vol, Moscow)
Sassi FP, Zanferrari G, Zirpoli G (1987) The Caledonian event in the Eastern Alps: a review. In: Flügel HW, Sassi FP, Grecula P (eds) Pre-Variscan and Variscan events in the Alpine-Mediterranean mountain belts. Mineralia slovaca IGCP 5 Monogr, Alfa, Bratislava, pp 431–434
Satir M, Morteani G (1978) Kaledonische, herzynische und alpidische Ereignisse im Mittelostalpin nördlich der westlichen Hohen Tauern, abgeleitet aus petrographischen und geologischen Untersuchungen. Geol Rundsch 68: 1–40
Visonà D, Zanferrari A (1985) Rifting ercinico nel basamento sudalpino delle Alpi orientali. In: Cocozza T, Ricci CA (eds) Evoluzione stratigrafica, tettonica, metamorfica, magmatica del Paleozoico italiano, Siena, 13–14 dicembre 1985, p 111 (Abstr vol, Siena)
Zanferrari A (1985) Stratigrafia e tettonica del basamento sudalpino delle Alpi Orientali. In: Cocozza T, Ricci CA (eds) Evoluzione stratigrafica, tettonica, metamorfica magmatica del Paleozoico italiano, Siena, 13–14 dicembre 1985, pp 112–113 (Abstract vol, Siena)

The Pre-Alpine Basement in the Alpi Apuane (Northern Apennines, Italy)

P. Conti, A. Di Pisa, M. Gattiglio and M. Meccheri

Abstract

The Alpi Apuane is a well-known tectonic window in the Northern Apennines (Tuscany), in which a pre-Alpine lithostratigraphic sequence crops out widely and which appears to have been deeply affected, together with the Alpine cycle cover, by the Tertiary tectono-metamorphic evolution of the Apenninic chain.

The authors illustrate the main lithological and structural features of these formations and discuss the relationships between the Apuan Variscan basement and the regional Variscan system of southern Europe.

1 Introduction

The Northern Apennines are a nappe belt that formed during the Tertiary tectono-metamorphic evolution of the Italo-Dinaric palaeomargin. The palaeogeographic domains involved in this evolution are, from inland to foreland: Ligurian domain, Subligurian domain, Tuscan domain (internal and external), Umbrian domain and Marchean domain.

Several tectonic units originating from those domains are overthrust from the internal unit onto the more and more external ones. At present, the most complete exposure of this nappe belt is visible in the tectonic window of the Alpi Apuane where, from top to bottom, the following units are stacked: Ligurian units (Mt. Gottero unit and Helmintoidic Flysch unit), Canetolo unit, Tuscan nappe, Massa unit and "Autochthonous" Auct.[1] unit. The units belonging to the Umbrian domain crop out only beyond the main crest of the Tuscan-Emilian Apennines, but very likely their more internal sector is buried under the Apuan window.

The Massa unit and the "Autochthonous" Auct. unit (hereafter MU and AU, respectively) represent the external Tuscan domain and form the "Apuan metamorphic complex" (hereafter AMC). The AMC, unlike the overlying tectonic units, is characterized by a greenschist facies metamorphism of Tertiary age that accompanied a polyphase, ductile and pervasive deformation. This deformation is represented by an ensialic shear zone and by later uplift of this sector of the Apennines chain.

This tectono-metamorphic history affected both the Tuscan stratigraphic succession and the underlying Palaeozoic basement (Fig. 1); this fact and the Tertiary uplift of the Alpi Apuane provide us with the widest and best-exposed outcrops of Palaeozoic rocks in the Northern Apennines.

2 The Alpine Structural Frame of the Alpi Apuane

The Apuan structure is due to the already mentioned Tertiary deformation which involved two tectono-metamorphic events. They were connected (Carmignani et al. 1978; Carmignani and Giglia 1984 and references therein) to first, tectonic tangential shortening, linked to continental collision (the tectogenetic event), and second to a dome-like uplift related to the beginning of a generalized extensional regime that affected progressively lower and lower structural levels (the orogenetic event: Carmignani et al 1987; Carmignani and Kligfield 1990).

2.1 *The Tectogenetic Event (D1)*

The tectogenetic event (D1) is manifest as an Upper Oligocene thick, ensialic shear zone and results in a

[1] Auct. (Auctorum): Latin word which indicates a geological concept introduced by previous authors and at present partly or completely obsolete. In this case (and in other papers during the 1980s) wie use an "obsolete" geological concept (that is, "autochthonous") because of its large diffusion in the Italian literature concerning the Alpi Apuane.

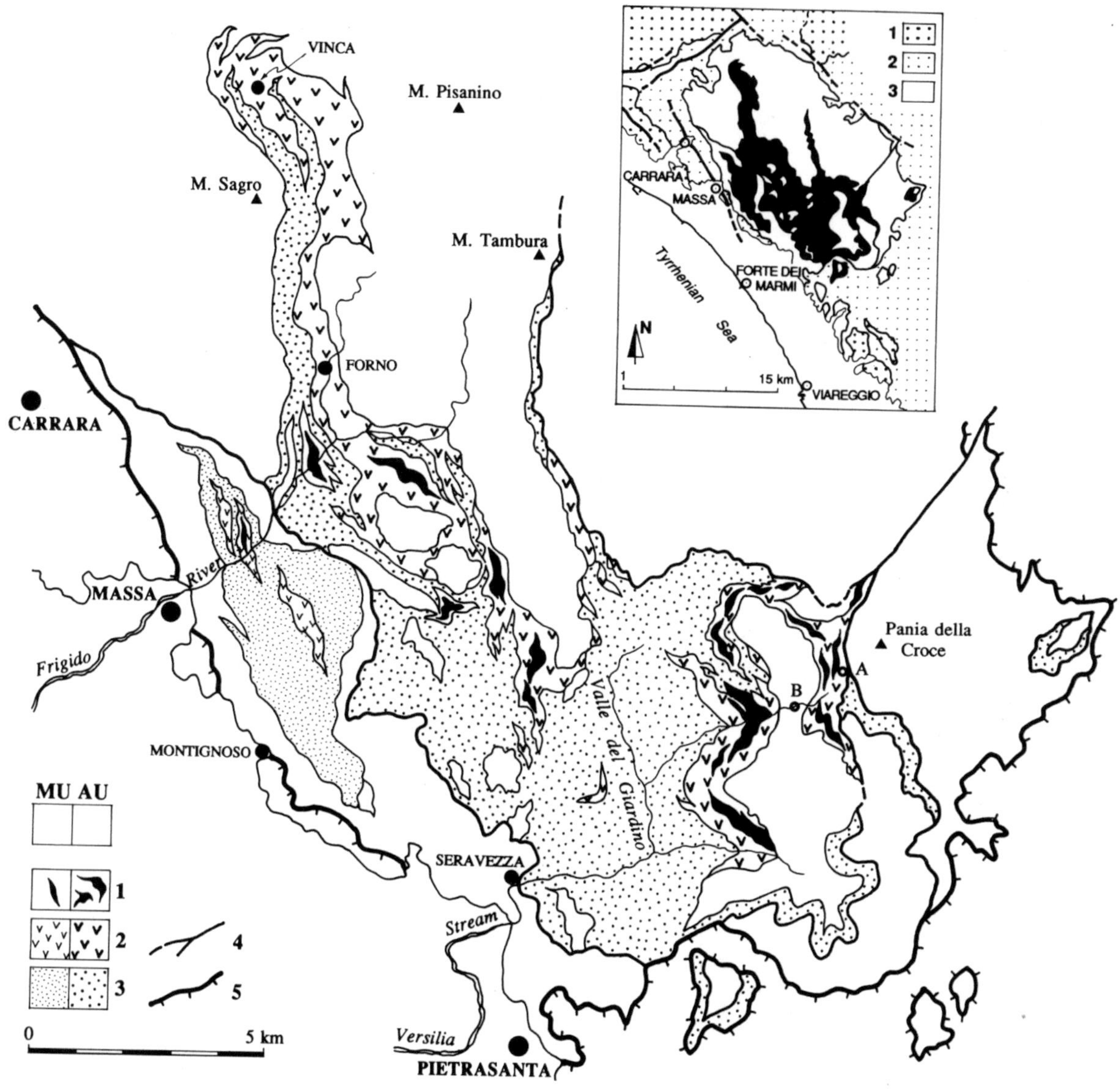

Fig. 1. Geological map of the Alpi Apuane Palaeozoic basement. *MU* Massa unit; *AU* "Autochthonous" Auct. unit; *box Symbols:* Alpine stratigraphic covers and recent sediments; *1* carbonate-phyllitic metasediments of Silurian-?Devonian age; 2 Pf+PfS+MQZ; *3* LP; *4* D1 and D2 ductile tectonic contacts and/or faults (north of Pania della Croce) inside the AMC; *5* tectonic boundary between the AMC and the overlying Tuscan nappe. On the map: *circle A* Foce di Mosceta; *circle B* Canale delle Volte. Regional tectonic sketch map, **(inset)** *1* Ligurian units and Canetolo unit; *2* Tuscan nappe; *3* AMC, the *black field* represents the Palaeozoic basement

complex structural setting, whose main structures are mentioned here:

1. Northeast-vergent mega-anticlines and mega-synclines with a similar geometry, and many minor folds ranging in size from microscopic to kilometric;
2. A pervasive axial plane foliation (S1) accompanies these folds, which more or less completely transposes the former sedimentary bedding and forms small angles with the shear zone boundaries;
3. Ductile shear horizons, from a microscopic size up to several metres thick, develop parallel to the S1 and are characterized by extreme elongation (on average SW/NE trending) and even strong obliteration of almost any stratigraphic feature. These shear horizons can reach regional importance as tectonic surfaces separating metamorphic subunits: the overturned flank of Mt. Tambura anticline (Fig. 1) is an example of this structure.

The L1 extension lineation everywhere indicates southwest-northeast tectonic transport. In contrast, the orientation of the fold axes is highly variable: in the western sector of the AU (i.e. in the uppermost structural levels of this unit) they trend mainly northwest, while in the central and eastern areas (i.e. towards deeper and deeper structural levels) they progressively become parallel to the L1 lineation, owing to a passive rotation of the fold axes on their own axial planes (i.e. S1 foliation: Carmignani et al. 1978; Günther and Wallbrecher 1979), according to the models of progressive simple shear deformation outlined by J.G. Ramsay (1967) and D.M. Ramsay (1979), Sanderson (1974), Escher and Watterson (1974), Rhodes and Gayer (1977), Williams (1978).

The regional attitude of all structural elements related to this event demonstrates that the tectonic movements had an Apenninic sense of motion (Carmignani and Giglia 1983), that is, from the internal (southwest) to the external (northeast) palaeogeographic domains.

The MU constitutes a tectonic wedge located between the AU and the anchimetamorphic Tuscan nappe. This structural position facilitated the development of many ductile shear zones which, together with S1, constitute the main structural features of this unit. As a consequence, its Triassic succession is preserved only in few and restricted areas. In spite of this, the MU may be considered a first-order anticline with a wide Palaeozoic core that is internally sheared and overthrust on the southwestern boundary of AU.

It must be noted that during this tangential episode the overlapping of all tectonic units (the nappe building) reaches its complete configuration and provides the lithologic charge under which the synkinematic metamorphism occurred.

2.2 The Orogenetic Event (D2)

This event, which seems to follow the first one without any appreciable break in time, is connected to the general uplift of the Apuan region. It affected both the AMC and the overlying units and led to a northwest-southeast trending dome structure that appears to have been controlled by the D1 antiformal stack (Carmignani and Kligfield 1990).

In the metamorphic units, this event deformed the S1 foliation into a complicated antiform with a thinned hinge and thickened limbs (Carmignani and Giglia 1979). In this antiform, many open folds developed with amplitudes up to a kilometre or more; these folds are asymmetric with vergences toward the flanks of the main structure, and they are accompanied by subhorizontal to slightly inclined axial plane schistosity (S2).

3 The Palaeozoic Basement of the Alpi Apuane

The Palaeozoic basement of the AMC is represented by a ?Late Cambrian–?Devonian lithostratigraphic succession, which has been recognized mainly on the basis of:

1. Rare palaeontological traces in some rocks (nautiloids: Meneghini 1880; conodonts: Vai 1972; Bagnoli and Tongiorgi 1980);
2. Constant and consistent geometric relationships between the various lithologic associations;
3. Comparison (Di Pisa et al. 1988) of these associations with the better known and dated formations of the Palaeozoic successions of southeastern Sardinia (Carmignani et al. 1982b, 1986; Carosi and Gattiglio 1989).

The recognized informal stratigraphic units are shown in a restored succession (Fig. 2) and are described next from oldest to youngest.

"Filladi Inferiori" (Barberi and Giglia 1966) (Lower Phyllite: hereafter LP; ?Late Cambrian-?Early Ordovician). The LP is made up of three main lithotypes: light to dark grey quartzite, dark grey and/or grey-greenish phyllitic quartzite and phyllite.

The most common and widespread lithologic combination is constituted by 10 cm or more thick layers of light quartzite, that alternate with bands of grey or greenish phyllite. Locally where the quartzite component dominates, lenticular quartzitic bodies form with lengths up to several tens of metres and thickness up to 10–15 m. These layers show rare but clear relics of sedimentary structures (gradation, bedding and cross-bedding). In other places, the main lithology is a homogeneous, dark grey or grey-green phyllite, with minor and small lenses of graphitic, blackish metapelite.

The top of LP is marked locally by the presence of a discontinuous level of matrix-supported paraconglomerate that contains quartzitic pebbles and grains of magmatic Qtz and Ab.[2] Where absent, this psephitic layer may be replaced by a few metre thick, whitish, homogeneous quartzite that shows a high textural and mineralogical maturity. Both the para-

---

[2] Mineral symbols after Kretz (1983).

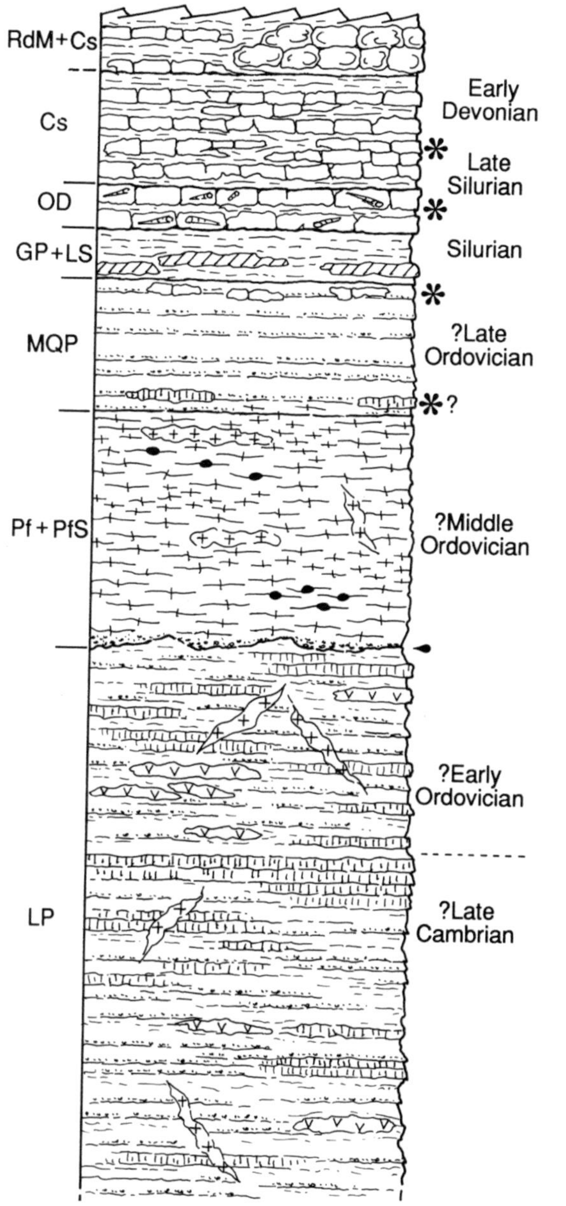

Fig. 2. The lithostratigrafic succession of the Alpi Apuane Palaeozoic basement. *1* Nodular metalimestone; *2* calcschist; *3* dolostone; *4* phyllite; *5* lydite; *6* metasandstone; *7* crinoid ossicle-bearing metalimestone; *8* carbonate metasiltite; *9* Pf with large K-feldspar phenocrysts; *10* Pf with small K-feldspar phenocrysts; *11* PfS; *12* subalkaline metabasite; *13* alkaline metabasite; *14* quartzite; *15* metaconglomerate; *16* inferred "Caledonian" unconformity; *17* fossils

conclomerate and the quartzite could outline an unconformity; unfortunately, the Alpine overprint (see later) obliterated the typical signs of such a structure, apart from the already described clastic metasediments. We will discuss this situation in the last Section.

In the field, the S1 Alpine foliation is the main regional schistosity everywhere, and it generally coincides with the described layering. It either develops as a spaced strain-slip cleavage in the quartzites or as a flow cleavage in the more phyllitic levels. The most common mineral association is Qtz + Ms + Chl + Ab, with Ap + -Ep + -Rt + -Zrn + -Tur + -Fe-oxides and hydroxides as accessories. An older foliation is recognizable in the "lithons" between the S1 surfaces where it often forms uprooted, intrafolial fold hinges. This planar anisotropy corresponds to a metamorphic layering and is usually marked by thin granoblastic Qtz-Ab-rich layers that pervasively alternate with micaceous films.

All of these features allow us to refer to the LP as primary pelitic-arenitic sediments that were deposited in a basin environment. Until now, no palaeontologic data have been available, nevertheless, the pre-Alpine foliation testifies to a tectono-metamorphic event that pre-dates the Alpine stratigraphic cycle.

Lenses of metabasic rocks (MBP, see later), consisting of greenish and grey-greenish basaltic metavolcanics and metagreywakes, form lenticular bodies inside the LP. They show a blastoporphyritic texture with subidiomorphic relic Pl porphyroclasts and often homogeneously distributed, abundant and stick-shaped, leucoxene-rimmed Ilm, in a Pl + -Chl + -Ep + -Fe-carbonate matrix.

"Porfiroidi e Scisti Porfirici" (Barberi and Giglia 1966) (porphyroid and porphyritic schist: hereafter Pf and PfS respectively; ?Middle Ordovician). These rocks represent the metavolcanic complex of the Apuan basement and are made up of two main varieties: porphyroid with small and large K-feldspar phenocrysts and porphyritic schist. On the whole, these rocks are homogeneous and massive and give abrupt and sharp features to the morphology.

The porphyroid containing small K-feldspars (Pf) is the more frequent type. It shows an augen structure of 3–4 mm Qtz and K-feldspar grains in a Qzt + Ms + Chl matrix. Qtz phenocrysts have resorbed boundaries (Barberi and Giglia 1966) and are clearly magmatic. The porphyroid with large K-feldspar porphyroclasts (up to 7–8 cm) is also present, but not as common as the previous type.

According to Bonatti (1938), Barberi and Giglia (1966) and Tucci (1980), these rocks derived from primary, rhyolitic-rhyodacitic, volcanic-volcanoclastic rocks; Puxeddu et al. (1984) related them to original acidic-intermediate volcanics (Na-K rhyolite).

The PfS represents primitive, arkosic to orthoquartzitic terrigenous sediments derived from erosion of the acidic volcanics under subaerial conditions (Barberi and Giglia 1966; Gianelli and Puxeddu 1980; Tucci 1980; Puxeddu et al. 1984). As already reported by Tucci (1980), a pre-Alpine foliation is recognizable in Pf and PfS, although it is not as well displayed as in the LP (at the scale of the whole massif).

Minor basaltic to intermediate sub-alkaline bodies (blastophitic metabasites: MBO, see later) are also associated with this metavolcanic complex, that show original subvolcanic relationships with the underlying LP.

Metasandstone, quartzite and phyllite (MQP). This formation consists of a variety of lithologies (metasandstone, metagreywacke, phyllitic quartzite and phyllite) that were probably derived from the erosion of the pre-existing acidic to intermediate volcanics. In many cases, the actual metavolcanics (that is, Pf + PfS) show a gradual transition to the basal levels (containing minor arkosic metasandstone) of this ensemble.

The most characteristic lithotype of this suite is a metasandstone that consists of a compact, dark grey to grey-greenish Qtz-arenite and Qtz-wacke with a widespread blastopsammitic texture. It typically contains large amounts of magmatic Qtz grains that often form equigranular, mature, grain-supported horizons. Less commonly, this metasandstone also contains feldspar grains.

The micaceous component in these metasandstones, although generally scarce, locally creates more or less phyllitic varieties. The complicated tectono-metamorphic history has caused irregular distribution and re-organization of these rocks, however, sometimes it is possible to observe clear remnants of sedimentary structures, such as several metre thick sequences of levels that show internal sorting.

Rare carbonate levels are locally intercalated in the lower portion of MQP, near the boundary with the Pf + PfS: these consist of major brownish metasiltite with variable carbonate components, that are typically riddled and enriched with Fe oxides and hydroxides. These discontinuous, thin (few metres thick) horizons could correspond to the "schists truè" of the sequences of southeastern Sardinia, but up to now they have not provided any palaeontological content.

Other carbonate rocks (previously reported by Barberi and Giglia 1966) alternate with the dominant clastic metasediments and consist of dolomitic grey levels (up to 20 cm thick) with rare crinoid ossicles. Where the calcareous matrix occurs among metasiltites and phyllites, it is always variously enriched in Fe-oxides and hydroxides and Tur.

The characteristics of MQP suggest that sedimentary processes were active over a coastal plane that with time developed towards an epicontinental shallow sea environment. In these rocks a pre-S1 metamorphic foliation is easily recognizable.

Graphitic phyllite and lydian stone (hereafter GP and LS, respectively). The GP has a homogeneous, prevalently micaceous (mainly white micas) composition; it contains widespread and abundant Gr and minor Qtz, and shows a very fine lepidoblastic texture.

Within the GP there are rare layers of LS, that consist of a few centimetre thick beds of microcrystalline, granoblastic quartzite alternating with lepidoblastic films of GP. The quartzite beds often show the pre-S1 metamorphic layering, which constitutes the only evidence of a pre-Alpine deformation in these rocks.

Due to its composition, the GP has been severely thinned and strongly deformed during the Variscan and Alpine tectono-metamorphic evolutions. Nevertheless, reliable original relationships with both the MQP suite and the overlying Orthoceras-bearing metasediments are occasionally well preserved.

The GP has been referred to as a primary, graptolite-bearing, prevalently siltitic sediment (Gortani 1933; Vai 1972) which, together with carbonate sediments, formed a wide and uniform sedimentary "cover" over the entire Euro-African Silurian sea. The only palaeontologic report is from Gortani (1933), who found indeterminable relics of graptolites in the Vinca and Foce di Mosceta areas (see Fig. 1).

Orthoceras-bearing dolostone (hereafter OD). This is a well-known rock in the Apuan basement and consists of granoblastic-xenoblastic re-crystallized Dol and minor Cal, with scattered Qtz, Ms and xenoblastic Ab. In the field it is grey with a typical brown-reddish alteration patina. The dolostone appears massive, sometimes coarsely bedded, but locally the presence of thin graphitic-phyllitic levels provides a foliation.

These carbonate metasediments have yelded crinoid ossicles and several specimens of orthoceratides (Meneghini 1880, 1881; Zaccagna 1932; Gortani 1933); on the base of these fossils, many authors assigned those rocks to an undetermined Palaeozoic age. Mainly on the base of facies correlations with analogous and palaeontologically dated horizons in the Alps, and because of the poor preservation of the orthoceratides, Barberi and Giglia (1966) assigned a Triassic age to their "Dolomie scistose". Despite this, a pre-Variscan age of the Alpi Apuane basement has been accepted as a result of work by Vai (1972) and later Bagnoli and Tongiorgi (1980), who reported conodonts of Lower-Middle Ludlow (Frigido River) and to Upper Ludlow (Canale delle Volte: see Fig. 1).

Calcschist (hereafter Cs). This rock, first examined by Bagnoli and Tongiorgi (1980), consists of 10 cm thick layers of light grey and whitish carbonate that alternate with thinner (maximum 1–2 cm) intercalations of greenish and bright phyllites and carbonate phyllites. Both varieties are close to parallel with the S1 Alpine foliation.

The carbonate horizons, which are dominantly composed of Cal and minor Dol, microcrystalline Qtz, Ms and Chl, often contain abundant crinoid ossicles and show a granoblastic to lepidoblastic texture. The Chl content within phyllites and calcareous phyllites usually gives a typical green colour.

Red nodular dolomitic metalimestone (hereafter RdM). Rare and discontinuous outcrops of this rock have been found in the Seravezza inland and north of Pania della Croce. In both cases the RdM occupies the same stratigraphic position above the OD/Cs sequence.

The RdM consists of dark reddish-purple, calcareous and dolomitic metasediments that lack regular stratification and exhibit a sort of nodular, possibly original structure. Anastomosing layers and/or irregular films of phyllosilicates (mainly Ms and Chl) surround the nodular carbonate bodies emphasizing this particular structure on weathered surfaces.

The RdM has not yet provided palaeontological data but Bagnoli and Tongiorgi (1980) proposed a Devonian age on the base of striking analogies between these rocks and carbonate formations in the identical stratigraphic position in pre-Variscan sequences of southwestern Europe and Sardinia.

4 Geochemical Data of the Metabasic Rocks of the Alpi Apuane Variscan Basement

Although the metabasic rocks in the Apuan basement are quite rare, they show some interesting features that help us to infer a more complete geodynamic setting for the stratigraphic and tectonic evolution which has been described already and is discussed later in the text.

We must remember that the basic rocks we are dealing with have suffered two tectono-metamorphic cycles (Variscan and Alpine) and that such a complex history may have strongly influenced the absolute concentration of many of the elements usually detected in routine chemical analysis (expecially K, Na, Ca, Rb, Sr, Ba).

The original petrography is never preserved. Albitized Pl, Ap and leucoxene-rimmed Ilm phenocrysts represent the few relic mineral phases in some samples; by contrast, some well preserved magmatic textures are often present, now referable to blastoporphyritic and blastophitic types.

The mafic phases are completely metamorphosed into mineral aggregates that include neoblastic Chl ± Ep ± Cal±opaques and are fairly well distributed throughout the rock. Metamorphic condition with P(fluid) = P(tot) is marked by the widespread formation of these OH-bearing silicates and carbonates. Microtextural evidence shows that Cal formed in pre-Alpine times and that CO_2 was present in the metamorphic fluid phase since Variscan times. This condition seems, at least in these rocks, not to have been repeated during the Alpine metamorphism.

Any discrimination among the Apuan basement metabasites is only possible utilizing the HFSE (high field strength elements) which are considered immobile during metamorphism and alteration processes (assuming that the above mentioned two metamorphic cycles did not produce significant modification at least in their original ratios).

In the Variscan basement of the AMC two distinct suites (MBP: blastoporphyritic metabasites; MBO: blastophitic metabasites) of metabasic rocks occur. Their characters were previously discussed by Conti et al. (1988) who have shown a generally transitional chemical composition for both types, but a contrasting petrogenetic affinity. The composition of these metabasites is here reproposed with a more complete set of their chemical data (Table 1).

The MBP rocks that always appear effusive (see end of LP section) with a blastoporphyritic texture show a general basaltic composition and an alkaline petrogenetic affinity. The latter is pointed out by their high TiO_2 contents and Nb/Y ratios (see Table 1 A), as well as by their REE chondrite normalized patterns (Fig. 3 A). These patterns are characterized by high LREE/HREE fractionation, no Eu anomalies and similar $(La/Sm)_n$ = (2.4–2.2) and $(Gd/Yb)_n$ = (2.2–1.9) ratios, in spite of the difference of their total abundances (ΣREE = 168 and 105 ppm).

Similar characters are also displayed by other metabasic rocks, which crop out all over the Variscan chain in Europe (e. g. in Sardinia and in Maures Massif) and are generally linked to continental extensional or transpressive regimes of different ages from Tremadoc (Maures) to Carboniferous through Caradoc (Sardinia) Ricci and Sabatini 1978; Seyler and Boucarut 1979; Memmi et al. 1983; Seyler 1983; Di Pisa et al. 1991).

The MBO rocks appear as subvolcanic bodies (see end of Pf+PfS Sect.) with a blastophytic texture,

Table 1. Representative chemical analyses of metabasites of the Apuan Palaeozoic basement

A MBP metabasites

	R3-60	VG-2	T4-51	T4-78	T4-74	T4-75
SiO_2	45.34	45.77	45.28	49.41	46.25	46.71
TiO_2	3.72	1.92	1.87	2.12	1.75	1.87
Al_2O_3	13.56	16.13	15.21	17.79	15.71	17.39
Fe_2O_3	17.03	11.30	11.98	11.98	12.18	12.41
MnO	0.16	0.11	0.15	0.10	0.19	0.12
MgO	4.66	6.23	6.00	9.14	7.33	9.94
CaO	5.89	5.55	7.50	0.57	5.36	2.29
Na_2O	4.14	4.50	3.34	3.49	3.86	3.34
K_2O	0.40	0.04	0.60	0.30	0.23	0.23
P_2O_5	0.55	0.40	0.45	0.34	0.40	0.39
LOI	4.32	7.84	9.28	4.76	6.74	5.31
SUM	99.77	99.79	99.21	100.00	100.00	100.00
Ba (ppm)	106	6	219	138	83	106
Rb	16	<5	21	21	14	14
Sr	221	380	348	91	147	154
Nb	38	26	24	32	38	39
Y	38	28	31	30	37	35
Zr	263	220	213	288	335	328
Ni	51	142	113	137	105	104
Cr	58	141	121	168	91	99
V	394	129	139	195	166	18
Nb/Y	1.00	0.93	0.77	1.07	1.03	1.11
La	31.77	18.67	18.37	18	23	21
Ce	71.00	44.83	42.46	41	53	47
Nd	33.67	20.64	20.92	–	–	–
Sm	8.38	5.38	5.85	–	–	–
Eu	2.49	1.67	1.73	–	–	–
Gd	7.46	4.83	5.58	–	–	–
Dy	6.55	4.51	5.05	–	–	–
Er	3.21	2.29	2.58	–	–	–
Yb	2.75	2.06	2.33	–	–	–
Lu	0.59	0.35	0.39	–	–	–
ΣREE	167.87	105.23	105.26	–	–	–
(La/Yb)n	7.8	6.1	5.3	–	–	–
(La/Sm)n	2.4	2.2	2.0	–	–	–
(Gd/Yb)n	2.2	1.9	1.9	–	–	–

B MBO metabasites

	VG-1b	T4-76	T4-93	R3-34	R3-33	T7-35
SiO_2	47.17	56.07	56.30	56.49	57.07	57.78
TiO_2	1.37	0.96	0.84	1.24	1.08	0.90
Al_2O_3	16.51	16.62	15.90	16.83	17.95	16.26
Fe_2O_3	8.44	8.54	6.77	9.67	8.17	7.89
MnO	0.14	0.13	0.09	0.11	0.12	0.11
MgO	5.84	3.89	2.89	4.99	5.43	4.89
CaO	6.33	3.12	4.03	1.43	0.80	2.26
Na_2O	3.64	5.98	4.90	5.62	4.72	4.68
K_2O	1.11	0.31	1.37	0.16	0.93	0.72
P_2O_5	0.35	0.53	0.42	0.63	0.41	0.45
LOI	8.21	3.85	6.50	2.83	3.37	4.05
SUM	99.11	100.00	100.01	100.00	100.01	99.99
Ba (ppm)	220	75	260	64	272	128
Rb	33	18	58	13	57	30
Sr	284	149	143	189	166	131
Nb	10	20	17	16	17	18
Y	28	36	31	36	32	32
Zr	168	409	340	366	348	358
Ni	65	19	18	22	18	15
Cr	166	27	22	24	23	26
V	162	115	103	137	103	110
Nb/Y	0.36	0.55	0.55	0.44	0.53	0.56
La	17.45	33.39	28.80	29.69	22.05	27.55
Ce	41.55	71.17	63.14	68.08	49.44	62.11
Nd	20.40	33.83	28.17	32.92	23.40	28.80
Sm	4.98	7.37	6.30	7.39	5.56	6.50
Eu	1.30	1.62	1.52	1.72	1.55	1.54
Gd	4.40	6.08	5.27	6.44	4.91	5.40
Dy	4.34	5.69	4.91	5.73	4.65	4.93
Er	2.44	3.18	2.72	3.20	2.65	2.79
Yb	2.26	3.14	2.80	3.12	2.55	2.80
Lu	0.30	0.51	0.50	0.57	0.48	0.46
ΣREE	99.42	165.98	144.13	158.86	117.24	142.88
(La/Yb)n	5.2	7.2	6.9	6.4	5.8	6.6
(La/Sm)n	2.2	2.8	2.9	2.5	2.5	2.7
(Ga/Yb)n	1.6	1.6	1.5	1.6	1.5	1.6

Major and trace elements were determined by XRF after Franzini et al. (1975) and Leoni and Saitta (1976). The same method was employed for La and Ce determinations of samples T4-78, T4-74 and T4-75. REE were determined by plasma emission spectroscopy at C. R. P. G. of Nancy. The samples R3-60, VG-2, T4-51 and VG-1b were completely analyzed at C. R. P. G.

range from basaltic to intermediate compositions and exhibit a subalkaline affinity (see Table 1B). They show TiO_2 contents and Nb/Y ratios lower than MBP and a general negative correlation of Ti and V vs. any independent fractionation index (e.g. FeO tot/MgO, SiO_2 or Zr): this fact suggests fractionation processes under relatively high fO_2 (Miyashiro and Shido 1975; Shervais 1982).

The REE patterns (Fig. 3B) differ from the previous ones essentially by either the presence of a slight negative Eu anomaly, in both analyzed samples, or a less pronounced HREE fractionation $(Gd/Yb)_n$ = 1.5–1.6. Such a REE distribution has been encountered in many andesitic rocks and indicates, together with the other previously discussed chemical characters, a calc-alkaline affinity for the MBO samples.

The available geochemical data here reported for the Apuan basement metabasites, also supported by geological field observations, lead to hypothesize the occurrence of two magmatic suites representing two different geodynamic settings.

5 The Pre-Alpine Structure and the Alpine Tectonic Overprint

As already pointed out, the traces of a pre-Alpine tectono-metamorphic event are usually detectable in

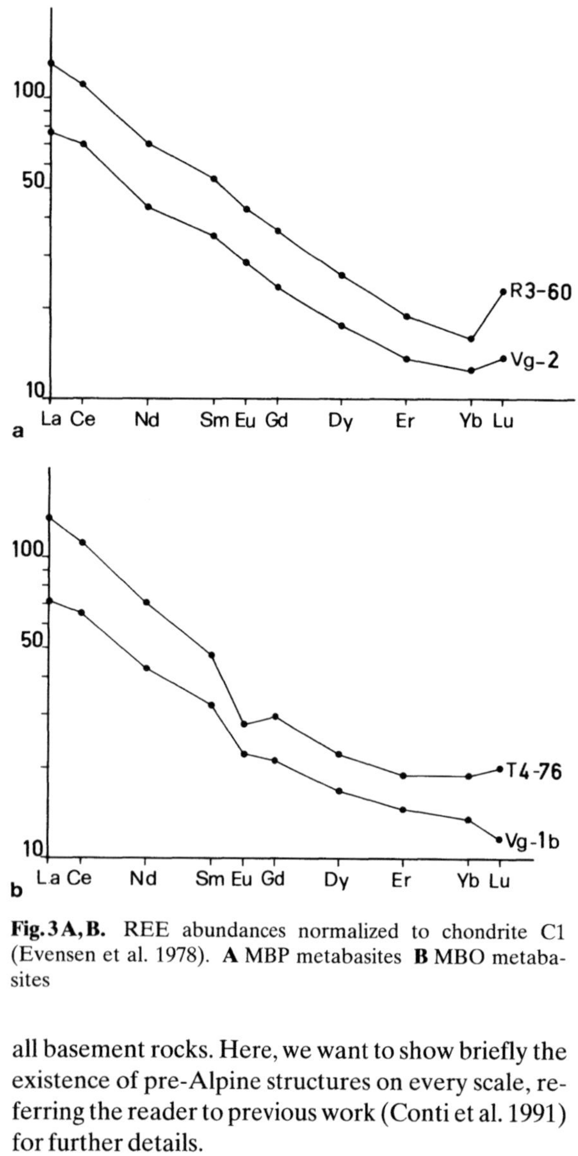

Fig. 3 A, B. REE abundances normalized to chondrite C1 (Evensen et al. 1978). **A** MBP metabasites **B** MBO metabasites

all basement rocks. Here, we want to show briefly the existence of pre-Alpine structures on every scale, referring the reader to previous work (Conti et al. 1991) for further details.

On a regional scale, the predominantly dolomitic Upper Triassic Vinca and "Grezzoni" formations represent the beginning of the Alpine sedimentary cycle of the AU. These units lie uncomformably on the Palaeozoic sequences, which are both upright and overturned with respect to the stratigraphic layering of the first Alpine sediments. This indicates that the latter were deposited on an already deformed and eroded basement.

In spite of the severe and pervasive deformation that involved stretching during the D1 Alpine event (Carmignani et al. 1978, 1980), some features of the old structures, such as folds and probably also thrusts, are still recognizable. In fact, the distribution of the Silurian-?Devonian rocks in relation to S1, reveals the existence of ancient synclines that developed at different geometric levels within the pre-Alpine building. Taking into account the Alpine kinematics, we can tentatively estimate that the pre-Alpine trend of these folds would have been almost E-W (± 30°).

The regional interference pattern between Alpine and pre-Alpine structures may be referred to as the "oblique type 1" as described by Ramsay (1967).

On a mesoscopic scale it is directly possible, in some outcrops, to map the pre-Alpine/Alpine folds interference geometry. Both anticlines and synclines with well-preserved limbs can be distinguished on a metre to hundreds of metres scale. Also, in this case the "oblique type 1" of Ramsay is recognizable.

On a smaller scale, all rocks, especially LP, show a pre-existing foliation inside "lithons" between the S1 Alpine-spaced surfaces. This geometric relationship within the Silurian OD has a particular geological meaning. The interference patterns all correspond to types 3 and 4 of Bell and Rubenach (1983).

6 Palaeotectonic Evolution of the Apuan Basement

The severe overprint of Alpine tectonics on the Palaeozoic sequence, that was already deformed and metamorphosed during the Variscan history, makes it rather difficult to reconstruct the palaeotectonic evolution of the Apuan basement. Most of the sedimentary and palaeontological record was nearly completely destroyed by the two orogenies and, as a consequence, the lithostratigraphic succession has been restored only on the basis of seldom preserved fossils (which indicate a Late Silurian age) and reciprocal relationships between the mapped lithologies.

The reconstruction of the stratigraphic sequence, shown in Fig. 2, has been performed through comparison with other Palaeozoic sequences in the Mediterranean area, particularly those cropping out along the "Zona a Falde" (nappe zone) of the Sardinian sector of the Variscan chain (Carmignani et al. 1977; Di Pisa et al. 1988; Gattiglio 1988).

The geometric relationships between the various lithotypes, and clear, although few data on the stratigraphic polarity suggest that the LP is the oldest formation of the Apuan Palaeozoic sequence. The lithologic and sedimentary features of LP indicate that it was a primary pelitic-arenitic sequence that may be related to the basal, silico-clastic deposits of the most important tectonic units of Variscan Sardinia.

In Sardinia these sediments provided a Late Cambrian to Early Ordovician (Arenig) age (Barca et al. 1982 a, b; Tongiorgi et al. 1984; Albani et al. 1985) and have been interpreted as turbiditic deposits in a submarine fan or basin floor environment (Barca and Marini 1980; Barca et al. 1982).

The same age can be attributed to the alkaline metabasites which stratigraphically lie within the LP, as already pointed out by Tongiorgi and Bagnoli (1981 a, b) and Puxeddu et al. (1984). In fact, alkaline volcanism of this age is reported from several Variscan massifs of South Europe, for instance from the "Serie du Collobrières" of the Maures Massif (Seyler and Boucarut 1979), where such rocks have been dated at 498 ± 12 Ma (U/Pb on Zrn: Lancelot et al. in Seyler 1983) corresponding to the Cambrian-Ordovician boundary (Harland et al. 1985).

Following Seyler's proposal for the Maures Massif, we assume an extensional regime within continental crust during the LP deposition. This regime would also be suggested by presence of the alkaline metabasites, and continued at least up to the Early Ordovician (Tremadoc). It could have been linked to development of a passive continental margin, rather than to a continental rifting: no sedimentary evidence is available for the latter situation. On the contrary, compressional events would have occurred during Early-Middle Ordovician. These events are well known in Sardinia where the uppermost portion of the Cambrian-Ordovician terrigenous sediments represents the progressive infilling of the basin (Tongiorgi et al. 1984), during a regression which ultimately led to its emersion.

In the Alpi Apuane, this emersion would be represented by rare metaconglomerate horizons, containing sandstone, quartzite and minor volcanic pebbles, which are interbedded along the boundary between LP and Pf + PfS (Gattiglio et al. 1989). In Sardinia the top of the Cambrian-Ordovician terrigenous sediments shows erosional channels and depressions filled by conglomerates. The latter correspond to an important unconformity ("Discordanza Sarda" in southwest Sardinia: Stille 1939; "Discordanza Sarrabese" in southeast Sardinia: Calvino 1961) related to "Caledonian tectonic movements" ("Sardische Phase" by Stille 1939).

These movements, which developed in a time span between Arenig (488–478 Ma) and Caradoc (458–448 Ma) Barca et al. 1987), were not accompanied by important tectono-metamorphic phenomena (Barca et al. 1985) and generated only gentle folds with a subvertical cleavage in southwest Sardinia (Arthaud 1963, 1970; Poll and Zwart 1964; Poll 1966; Carmignani et al. 1982 a), and angular unconformities in the other Variscan sequences of the island.

In the Alpi Apuane, we believe the metaconglomerates at the top of LP have the same significance, that is, they represent a "Caledonian event", even though the two subsequent orogenies (Variscan and Alpine) have, until now, prevented us from recognizing any "Caledonian" structures (but, it is possible that the "Caledonian" tectonics would not have been important, as in Sardinia).

It is therefore likely that the geodynamic setting changed from extensional, during Late Cambrian-Tremadoc, to compressional, during Arenig-Middle Ordovician.

The widespread and abundant, acidic to minor intermediate volcanics (Pf + PfS) represent the most significant marker of such a change. They consist of original ignimbrite covers and lava flows (and joined erosional products) and represent a coeval lithologic association in any Variscan sequence of South Europe (Austrides, Southern Alps, Calabria, Sardinia, etc.).

In the Sardinia Variscan sequences, continental volcanics unconformably overlie ("Sarrabese-Sarda" Unconformity) the Late Cambrian-Arenig metasandstone and are, in turn, covered by fossiliferous metasediments of Caradoc-Ashgill age (Calvino 1961; Naud 1979; Carmignani et al. 1986, 1989; Carosi et al. 1987). These volcanics, composed mainly of porphyritic acidic lavas and rhyolitic-rhyodacitic ignimbrites with minor andesitic lavas and basalts, represent a well-defined subalkaline suite (Memmi et al. 1983) resulting from magmatism which occurred during the evolution of an active continental margin.

These volcanics can be correlated with the Apuan units Pf and PfS, thereby assigning them to an age ranging from Early to Late Ordovician (Di Pisa et al. 1988). At present, radiometric ages of these rocks are not available. The geochemical data that exist on the rare basic rocks of intermediate composition indicate a subalkaline character: thus, also the subalkaline basites that are present as dykes within LP (Conti et al. 1989) might belong to the same magmatic cycle.[3]

The Pf + PfS suite of the Apuan basement can also be correlated with analogous rocks in the sequences of the Austro-Alpine nappes (Flajs and Schönlaub 1976) and South-Alpine Palaeozoic, where they are present either as wide volcanic flows ("Piattaforma

[3] These dyke-like products were previously reported also from inside Pf and related to a generic, Late to post-Ordovician magmatism (Conti et al. 1989). More careful field analyses have since raised some uncertainty as to whether the relationship between the two lithologies is actually intrusive. Moreover, recent findings of rare basic rocks of intermediate composition in association with Pf, lead us to think that the subalkaline lithologies must relate, more likely, to the Middle Ordovician cycle. We cannot, however, exclude the possibility of younger magmatic cycles.

porfirica pre-ercinica": Sassi and Zirpoli 1965, 1968; Sassi et al. 1978. "Porphyroid-platte": Mostler 1970. Interpreted as Late Ordovician products: Dal Piaz et al. 1975; Flajs and Schönlaub 1976; Sassi et al. 1980), or as intrusive bodies within the prevolcanic basement (450–420 Ma old orthogneiss: Sassi and Zirpoli 1980; Borsi et al. 1980; Del Moro et al. 1984).

Both Austro-Alpine and Southalpine granitoids and porphyroids of Ordovician age (Ordovician "evento granitico-riolitico": Schmidt and Söllner 1982) were derived from crustal anatexis and show a calc-alkaline character (Peccerillo et al. 1979; Sassi and Zirpoli 1980; Sassi et al. 1980). This magmatism has been related (Sassi and Zirpoli 1980) to a "Caledonian" tectono-metamorphic episode (Sassi and Zanferrari 1972; Borsi et al. 1973; Sassi et al. 1974, 1978, 1980; Purtscheller and Sassi 1975; Peccerillo et al. 1979; Söllner and Schmidt 1981).

We consider that the Apuan basement subalkaline magmatism and the Austro-Alpine, Southalpine and Sardinian calc-alkaline magmatism developed during the same tectonic processes (the "Caledonian" event), in a geodynamic setting typical of an active continental margin.

During and after this subalkaline magmatism, the volcanoes were partly eroded and variously covered by the products of this erosion. These products show an evolution indicating progressive transition from subaerial to submarine sedimentation (metasandstone, carbonate metasiltite and minor, crinoid ossicle-rich metalimestone: Gattiglio et al. 1989), although only a general transgression is inferrable because, at present, no palaeontological data are available.

In the Sardinia sequences, typical transgressive features are also represented by the deposits between the subalkaline volcanics and the Silurian fossiliferous sediments: these deposits correspond to the "Trasgressione Caradociana" Auct. and contain populous fossil associations of Caradoc (458–448 Ma) and Ashgill (448–438 Ma) age (Maccagno 1965; Laufeld 1973; Naud 1979).

The MQP sequence of the Apuan basement may also have been deposited in the Late Ordovician, and represents transgression on the previous volcanics. This phenomenon is common in the South Europe Variscan sequences (for instance, in the "Catena Paleocarnica": Spalletta et al. 1982) and is probably due to:

1. The melting of the Sahara inland ice;
2. The initiation of extension throughout the crustal sector that was previously deformed and thickened by the "Caledonian" event.

This extensional period (indicated by Spalletta et al. 1982, als "Stadio di piattaforma strutturale soggetta a rifting abortivo") caused important lateral variations within the MQP deposits of the Alpi Apuane. In Sardinia, extension is also marked by rare, basic alkaline lavas (within-plate basalts: WPB, Beccaluva et al. 1981; Memmi et al. 1983) that are interlayered with the Late Ordovician sediments.

The subsequent palaeotectonic history of the Apuan Variscan cycle is analogous to that in the other Variscan massifs in South Europe: slow crustal extension led to a Silurian epicontinental shallow sea with neritic sedimentation of Gr-bearing pelite and minor lydite (GP+LS), and deposition of fossiliferous dolomite (OD). Everywhere in the Mediterranean province, during this period the basin was relatively stable, with reduced sedimentation (Spalletta et al. 1982). Gradually, this environment changed to a more open sea in which the first carbonate platforms began to develop.

Although the subsequent evolution is difficult to detect in the Apuan basement because of the scarcity of the youngest sediments, the few RdM outcrops reveal (according to the reconstruction of Bourrouilh 1980, based on nodular limestone genesis) a new rifting episode during the ?Devonian which caused the drowning of the carbonate platforms. This extension is more completely documented in the Carnia Alps (Vai and Spalletta 1982) and in Sardinia; in the latter area, the nodular calcareous cover is widespread (Bourrouilh 1980) and metagabbroic bodies with a continental alkaline character (Memmi et al. 1983) form intrusions inside the blackish Silurian schist (northwest sector of the island).

The onset of the Variscan deformation caused cessation of the pelagic carbonate deposition in the Apuan sequence, in which synorogenic foredeep sediments have not been found.

A tentative location of the Apuan basement in the circum-Mediterranean Variscan chain must take into account either the absence of flysch-like, pre-/synorogenic deposits, or the pre-Alpine metamorphism (up to green schist facies) and deformation. Both factors indicate a primary position of Apuan Palaeozoic between the Variscan axial zone (with higher deformation and metamorphism) and the foreland domains (with thick and widespread synorogenic deposits). Such a position might be recognized as the "Zona a Falde" (nappe zone) of the Sardinian Variscan chain.

The palaeotectonic frame of the Alpi Apuane Variscan basement, discussed in this Section, is outlined mainly on the basis of its correlation with the wide Sardinia Palaeozoic sector, whose geodynamic and palaeogeographic evolution has recently been described by Carmignani et al. (1991).

References

Albani R, Di Milia A, Minzoni N, Tongiorgi M (1985) Nuovi dati palinologici e considerazioni geologiche sull'età delle Arenarie di Solanas (Cambro-Ordoviciano, Sardegna centrale). Soc Tosc Sci Nat Atti Mem Ser A 91: 1–20

Arthaud F (1963) Un exemple de tectoniques superposées dans le Paléozoique de l'Iglesiente (Sardaigne). Soc Geol Fr C R Comm 9: 303–304

Arthaud F (1970) Etude tectonique et microtectonique comparée de deux domains Hercyniens: les nappes de la Montagne Noire (France) et l'anticlinorium de l'Iglesiente (Sardaigne). Publ 1 Sér Géol Struct USTLA, Montpellier

Bagnoli G, Tongiorgi M (1980) New fossiliferous Silurian (M. Corchia) and Devonian (Monticiano) layers in the Tuscan Paleozoic. Soc Geol Ital Mem 20: 301–313

Barberi F, Giglia G (1966) La serie scistosa basale dell'autoctono delle Alpi Apuane. Soc Geol Ital Boll 84: 41–92

Barca S, Marini C (1980) Lineamenti litostratigrafici e sedimentologici del "Postgotlandiano" del Fluminese (Sardegna SW). Soc Geol Ital Mem 20: 203–216

Barca S, Del Rio M, Pittau Demelia P (1982 a) Acritarcs in the "Arenarie di San Vito" of southeast Sardinia: stratigraphical and geological implications. Soc Geol Ital Boll 100: 369–375

Barca S, Cocozza T, Del Rio M, Pittau Demelia P (1982 b) Discovery of Lower Ordovician acritarcs in the "Postgotlandiano" sequence of southwestern Sardinia (Italy): age and tectonic implications. Soc Geol Ital Boll 100: 377–392

Barca S, Carmignani L, Cocozza T, Franceschelli M, Ghezzo C, Memmi I, Minzoni N, Pertusati PC, Ricci CA (1985) The Caledonian events in Sardinia. In: Gee DG, Sturt BA (eds) The Caledonide orogen – Scandinavia and related area. Wiley and Sons, New York, pp 1195–1199

Barca S, Cocozza T, Del Rio M, Pillola GL, Pittau Demelia P (1987) Datation de l'Ordovicien inférieur par Dictyonema flabelliforme et Acritarches dans la partie supérieure de la formation "Cambrienne" de Cabitza (SW de la Sardaigne, Italie): consequences geodinamiques. C R Acad Sci Paris 305: 1109–1113

Beccaluva L, Leone F, Maccioni L, Macciotta G (1981) Petrology and tectonic setting of the Paleozoic basic rocks from Iglesiente-Sulcis (Sardinia, Italy). Neues Jahrb Miner Abh 140: 184–201

Bell TH, Rubenach MJ (1983) Sequential porphyroblast growth and crenulation cleavage development during progressive deformation. Tectonophysics 92: 171–194

Bonatti S (1938) Studio petrografico delle Alpi Apuane. Carta Geol Ital Mem Descr 26

Borsi S, Del Moro A, Sassi FP, Zirpoli G (1973) Metamorphic evolution of the Austridic rocks of the Tauern window (Eastern Alps). Radiometric and geopetrological data. Soc Geol Ital Mem 12: 549–571

Borsi S, Del Moro A, Sassi FP, Zirpoli G (1980) New petrologic and radiometric data on the Ooetzal and Stubai orthogneisses (Eastern Alps). Neues Jahrb Miner Monatsh 2: 75–87

Bourrouihl R (1980) Orthoceratitico-rosso et Goniatitico-rosso: facies marqueurs de la naissance et de l'evolution de paleomarges au Paleozoique. In: Farinacci A, Elmi S (eds) Rosso Ammonitico Symposium Proceedings. Edizioni Tectoscienza, Roma, pp 39–55

Calvino F (1961) Lineamenti strutturali del Sarrabus-Gerrei (Sardegna sud-orientale). Serv Geol Ital Boll 81: 489–556

Carmignani L, Giglia G (1979) Large scale reverse "drag folds" in the late Alpine building of the Apuane Alps (Northern Apennines). Soc Tosc Sci Nat Atti Mem Ser A 86: 109–125

Carmignani L, Giglia G (1983) Il problema della doppia vergenza sulle Alpi Apuane e la struttura di M. Corchia. Soc Geol Ital Boll 26: 515–525

Carmignani L, Giglia G (1984) "Autoctono Apuano" e Falda Toscana: sintesi dei dati e interpretazioni più recenti. In: Cento anni di geologia italiana. Vol Giub I° Centenario Soc Geol Ital, pp 199–214

Carmignani L, Kligfield R (1990) Crustal extension in the Northern Apennines: the transition from compression to extension in the Alpi Apuane Core Complex. Tectonics 9: 1275–1303

Carmignani L, Rau A, Squarci P, Tongiorgi M, Vai GB (1977) Le successioni Paleozoico-Triassiche della Sardegna centrale e dell'"Autoctono" apuano: analogie e possibili correlazioni. In: Vai GB (ed) Escursione in Sardegna 1977: risultati e commenti. Gruppo Lav Paleoz Boll 2: 11–14

Carmignani L, Giglia G, Kligfield R (1978) Structural evolution of the Apuane Alps: an example of continental margin deformation in the Northern Apennines, Italy. J Geol 86: 487–504

Carmignani L, Giglia G, Kligfield R (1980) Nuovi dati sulla zona di taglio ensialica delle Alpi Apuane. Soc Geol Ital Mem 21: 93–100

Carmignani L, Cocozza T, Gandin A, Pertusati PC (1982 a) Lineamenti della geologia dell'Iglesiente-Sulcis. In: Soc Geol Ital (ed) – Guide Geologiche Regionali – Guida alla geologia del Paleozoico sardo. pp 55–77

Carmignani L, Cocozza T, Ghezzo C, Pertusati PC, Ricci CA (1982 b) Guida alla geologia del Paleozoico Sardo. Guide Geologiche Regionali Soc Geol Ital Mem 24

Carmignani L, Cocozza T, Ghezzo C, Pertusati PC, Ricci CA (1986) Guide-book to the excursion on the Paleozoic basement of Sardinia. IGCP 5 Newsletter Spec Iss

Carmignani L, Gattiglio M, Kalin O, Meccheri M, Di Pisa A, Franceschelli M, Leoni L, Memmi I, Conti P, Vietti N (1987) Guida all'escursione sul Complesso Metamorfico delle Alpi Apuane. Summer School "Geologia e Petrologia dei Basamenti Cristallini" CNR/Univ Siena

Carmignani L, Cherchi A, Ricci CA (1989) Basement structure and Mesozoic-Cenozoic evolution of Sardinia. In: The lithosphere in Italy. Advances in Earth Sciences Research. Accad Naz Lincei 80: 63–92

Carmignani L, Barca S, Cappelli B, Di Pisa A, Gattiglio M, Oggiano G, Pertusati PC (1991) A tentative geodynamic model for the Hercynian basement of Sardinia. In: Cappelli B, Liotta D (eds) Geologia del basamento italiano Univ Siena 21–22/3/1991, pp 23–26

Carosi R, Gattiglio M (1989) Geotraverse B in Sardinia (SCF – Stratigraphic Correlation Form): new data on the Sardinian Paleozoic basement. In: Sassi FP, Zanferrari A (eds) Pre-Variscan and Variscan events in the Alpine-Mediterranean mountain belts: stratigraphic correlation forms. Soc Geol Ital Rend 12: 143–172

Carosi R, Elter FM, Gattiglio M, Oggiano G, Pertusati PC (1987) Some specification on pre-Caradocian volcanism in central Sardinia. IGCP 5 Newsletter 7: 27–29

Conti P, Gattiglio M, Meccheri M (1991) The overprint of the Alpine tectono-metamorphic evolution on the Hercynian orogen: an example from the Apuane Alps (Northern Apennines, Italy). Tectonophysics 191: 335–346

Conti P, Di Pisa A, Gattiglio M, Meccheri M, Vietti N (1988) Nuovi dati sulle metabasiti della Valle del Giardino nel basamento paleozoico apuano (Appennino settentrionale). Soc Tosc Sci Nat Atti Mem Ser A 95: 89–103

Dal Piaz GV, von Raumer J, Sassi FP, Zanettin B, Zanferrari A (1975) Geological outline of the Italian Alps. In: Squyres C (ed) Geology of Italy. Earth Sci Soc Libyan Arab Rep, pp 299–375

Del Moro A, Sassi FP, Zirpoli G (1984) Acidic gneisses from Plan de Corones area, and chronological data on South-Alpine basement in Pusteria (Eastern Alps). Sci Geol Mem Padova 36: 403–412

Di Pisa A, Franceschelli M, Leoni L, Meccheri M (1985) Regional variations of the metamorphic temperature across the Tuscanid 1 Unit and its implications on the Alpine metamorphism (Apuan Alps, northern Tuscany). Neues Jahrb Miner Abh 151: 197–211

Di Pisa A, Gattiglio M, Meccheri M (1988) Le unità metamorfiche di basso grado della Calabria (Unità di Bocchigliero e Paleozoico di Stilo-Bivongi): considerazioni strutturali e analogie con l'Ercinico sardo e toscano. Soc Geol Ital Rend 11: 29–32

Di Pisa A, Gattiglio M, Oggiano G (1991) Some specifications on the pre-Hercynian basic magmatic activity in the Nappe Zone of Sardinia: evidence of two alkaline magmatic cycles. In: Cappelli B, Liotta D (eds) Geologia del basamento italiano. Univ Siena 21–22/3/1991, pp 148–149

Escher A, Watterson J (1974) Stretching fabrics, folds and crustal shortening. Tectonophysics 22: 223–231

Evensen NM, Hamilton PJ, O'Nions RK (1978) Rare earth abundances in chondritic meteorites. Geochim Cosmochim Acta 42: 1199–1212

Flajs G, Schönlaub HP (1976) Die biostratigraphische Gliederung des Altpaläozoikums am Polster bei Eisenerz (Nördliche Grauwackenzone, Österreich). Verh Geol Bundesanst Wien, pp 257–303

Franzini M, Leoni L, Saitta M (1975) Revisione di una metodologia analitica per fluorescenza X, basata sulla correzione completa degli effetti di matrice. Soc Ital Mineral Petrol Rend 31: 365–378

Gattiglio M (1988) Confronti tra il basamento ercinico sardo e il basamento paleozoico delle Alpi Apuane. Tesi di Dottorato, University of Pisa

Gattiglio M, Meccheri M (1987) Preliminary considerations on the lithostratigraphic succession of the Apuane Alps Paleozoic basement (northern Tuscany, Italy). In: Sassi FP, Bourrouilh R (eds) IGCP 5 Newslett 7: 57–59

Gattiglio M, Meccheri M, Tongiorgi M (1989) Geotraverse B in central Italy (SCF – stratigraphic correlation form:) new data on the Tuscan Paleozoic basement. In: Sassi FP, Zanferrari A (eds) Pre-Variscan and Variscan events in the Alpine-Mediterranean mountain belts: stratigraphic correlation forms. Soc Geol Ital Rend 12: 247–257

Gianelli G, Puxeddu M (1980) An attempt at classifying the Tuscan Paleozoic: geochemical data. Soc Geol Ital Mem 20: 435–446

Gortani M (1933) Il Paleozoico delle Alpi Apuane. Rend Reg Accad Sci Ist Bologna Cl Sci Fis 37: 53–57

Günther K, Wallbrecher E (1979) Fabric patterns in the tectonic units of the Apuan Alps. In: Closs H, Roeder DH, Schmidt K (eds) Alps, Apennines, Hellenides. Schweizerbart, Stuttgart, pp 267–271

Harland WB, Cox AV, Llewellin PG, Pickton CAG, Smith AG, Walters R (1985) Geological time scale. University Press, Cambridge, pp 131

Kretz R (1983) Symbols for rock-forming minerals. Am Mineral 68: 277–279

Laufeld S (1973) Ordovician Chitinozoa from Portixeddu, Sardinia. Soc Paleontol Ital Boll 12: 3–7

Leoni L, Saitta M (1976) X-ray fluorescence analysis of 29 trace elements in rock and mineral standards. Sol Ital Mineral Petrol Rend 32: 497–510

Maccagno AM (1965) Contributo alla conoscenza della fauna della Sardegna. Gli Echinodermi di Portixeddu. Nota 1. Accad Sci Fis Mat Atti Napoli 5: 149–195

Memmi I, Barca S, Carmignani L, Cocozza T, Elter FM, Franceschelli M, Gattiglio M, Ghezzo C, Minzoni N, Naud G, Pertusati PC, Ricci CA (1983) Further geochemical data on the pre-Hercynian igneous activities of Sardinia and on their geodynamic significance. IGCP 5 Newsletter 5: 87–91

Meneghini G (1880) Nuovi fossili delle Alpi Apuane. Soc Tosc Sci Nat Atti Proc Verb II 14/11/1880 4

Meneghini G (1881) Ulteriori osservazioni sui Trilobiti di Sardegna e sui fossili paleozoici delle Alpi Apuane. Soc Tosc Sci Nat Atti Proc Verb II 8/5/1881 2

Miyashiro A, Shido F (1975) Tholeiitic and calc-alkalic series in relation to the behaviour of titanium, vanadium, chromium and nickel. Am J Sci 275: 265–277

Mostler H (1970) Struktureller Wandel und Ursachen der Faziesdifferenzierung an der Ordoviz/Silur-Grenze in der Nördlichen Grauwackenzone (Österreich). Festb Geol Inst 300-Jahrfeier Univ, Innsbr, pp 507–522

Naud G (1979) Les shales de Rio Canoni, formation-repère fossilifère de l'Ordovicien supérieur de la Sardaigne sud-orientale. Consequences stratigraphiques et structurales. Soc Geol Fr Bull XXI: 155–159

Peccerillo A, Poli G, Sassi FP, Zirpoli G, Mezzacasa G (1979) New data on the Upper Ordovician acid plutonism in the Eastern Alps. Neues Jahrb Miner Abh 137: 162–183

Poll JJK (1966) The geology of the Rosas-Terreseo area, Sulcis, South Sardinia. Leidse Geol Meded 35: 117–208

Poll JJK, Zwart HJ (1964) On the tectonics of the Sulcis area, South Sardinia. Leidse Geol Meded 43: 144–146

Purtscheller F, Sassi FP (1975) Some thought on the pre-Alpine metamorphic history of the Austridic basement of the Eastern Alps. Tschermaks Mineral Petrogr Mitt 22: 175–199

Puxeddu M, Saupè F, Dechomets R, Gianelli G, Moine B (1984) Geochemistry and stratigraphic correlations – application to the investigation of geothermal and mineral resources of Tuscany, Italy. Chem Geol 43: 77–113

Ramsay DM (1979) Analysis of rotation of folds during progressive deformation. Geol Soc Am Bull 90: 732–738

Ramsay JG (1967) Folding and fracturing rocks. McGraw-Hill, New York

Ricci CA, Sabatini G (1978) Petrogenetic affinity and geodynamic significance of metabasic rocks from Sardinia, Corsica and Provence. Neues Jahrb Miner Monatsh 1: 23–38

Rhodes S, Gayer RA (1977) Non-cylindrical folds, linear structures in the X direction and mylonite structures developed during translation of the Caledonian Kalak Nappe Complex of Finnmark. Geol Mag 114: 329–408

Sanderson DJ (1974) The development of fold axes oblique to the regional trend. Tectonophysics 16: 55–70

Sassi FP, Zanferrari A (1972) Il significato geologico del complesso del Turntale (Pusteria), con particolare riguardo alla successione di eventi metamorfici prealpini nal basamento austridico delle Alpi Orientali. Soc Geol Ital Boll 91: 533–557

Sassi FP, Zirpoli G (1965) Contributo alla conoscenza degli scisti cristallini del Comelico (Cadore). Mem Accad Patav SS LL AA 78: 35–72

Sassi FP, Zirpoli G (1968) Il basamento cristallino di Recoaro: studio petrografico. Soc Geol Ital Mem 7: 227–245

Sassi FP, Zirpoli G (1980) Stato delle conoscenze sulla storia pre-ercinica delle Alpi Orientali. Soc Geol Ital Mem 20: 11–20

Sassi FP, Zanferrari A, Zirpoli G, Borsi S, Del Moro A (1974) The Austrides to the south of the Tauern window and the Periadriatic lineament between Mules and Mauthen. Neues Jahrb Geol Paleontol Monatsh 7: 421–434

Sassi FP, Schönlaub HP, Zanferrari A (1978) The pre-Variscan history of the Eastern Alps. Österr Akad Wiss 3: 261–277

Sassi FP, Zanferrari A, Zirpoli G (1980) Nuovi dati sulla stratigrafia e i caratteri chimico-petrografici dei porfiroidi del Comelico (Alpi orientali). Soc Geol Ital Mem 20: 425–434

Schmidt K, Söllner F (1982) Proposal for the geodynamic interpretation of the "Caledonian event" in central and southern Europe, with special reference to the Eastern Alps. IGCP 5 Newslett 4: 82–93

Seyler M (1983) Geologie de la partie centrale del Maures. Consequences sur l'evolution geodinamique du massif. Geol Fr 1: 81–96

Seyler M, Boucarut M (1979) Existence d'un paleovolcanisme alcalin et transitionel dans le massif hercynien del Maures (Var). Etude geochimique del amphibolites et resultats preliminaires concernant la chimie des leptynites. Soc Geol Fr Bull 21: 11–20

Shervais JW (1982) Ti-V plots and the petrogenesis of modern and ophiolitic lavas. Earth Planet Sci Lett 59: 101–118

Söllner F, Schmidt K (1981) Rb/Sr- und U/Pb-Datierungen am Winnebach-Migmatit (Ötztaler Alpen, Österreich) (Abstr). Fortschr Mineral 59: 188 pp

Spalletta C, Vai GB, Venturini C (1982) La Catena Paleocarnica. In Guide Geologiche Regionali Guida alla – Geologia del Sudalpino centro-orientale. Soc Geol Ital, pp 281–292

Stille H (1939) Bemerkungen betreffend die "sardische" Faltung und den Ausdruck "ophiolithisch". Z Dtsch Geol Ges 91: 771–773

Tongiorgi M, Bagnoli G (1981 a) Geotraverse B in central Italy: SCF (stratigrafic correlation forms). IGCP 5 Newslett 3: 152–155

Tongiorgi M, Bagnoli G (1981 b) Stratigraphie du socle paléozoique de la bordure continentale de l'Apennin septentrional (Italie centrale). Soc Geol Fr Bull 23: 319–323

Tongiorgi M, Albani R, Di Milia A (1984) The Solanas sandstones of central Sardinia: new paleontological data (acritarcs) and an attempt of geological interpretation (a "post-Sardinian" molasse?). Soc Geol Fr Bull 26: 665–680

Tucci P (1980) Le metamorfiti della "serie scistosa basale" del Monte Corchia (Alpi Apuane). Per Mineral 49: 57–148

Vai GB (1972) Evidence of Silurian in the Apuane Alps (Tuscany, Italy). G Geol 38: 349–372

Vai GB, Spalletta C (1982) Devonian Palaeo-Tethyan carbonate platform-basin system of the Southern Alps: evolutionary trend and comparative analysis. Terra Cognita 5: 109–115

Williams GD (1978) Rotation of contemporary folds into the X direction during overthrust processes in Laksefjord, Finnmark. Tectonophysics 48: 29–40

Zaccagna D (1932) Descrizione geologica delle Alpi Apuane. Carta Geol Ital Mem Descr 25: 440 pp

Part IV

Conclusions

Late Precambrian and Palaeozoic Evolution of the Alpine Basement – An Overview

J. F. von Raumer and F. Neubauer

Abstract

Situated between Gondwana and Laurussia, the pre-Mesozoic Alpine basement registered all the main orogenic periods since the Precambrian. During this long evolution, rifting, asymmetric drift, strike-slip tectonics, and rotation led to the stepwise transformation of Gondwana-type pieces (microcontinents detached from Gondwana) and to their final collision in the Variscan orogenic belt from Late Devonian to Carboniferous time.

Geochronological data and large-scale comparision of lithologies provide evidence for the existence of a major Late Proterozoic/Early Cambrian oceanic suture which is preserved in the basement units of the actual Penninic domain and adjacent areas, separating a Moldanubian domain (parts of the Helvetic area) from the more "African" bloc (Southern Alps, Austro-Alpine) where Pan-African/Cadomian elements are better preserved.

Basically, most Austro-Alpine sequences formed a Silurian-Devonian passive margin which was separated from the active continental margin of the "Ligerian Cordillera", the latter including the Helvetic and some Austro-Alpine units. The two domains collided during the Early Carboniferous as part of the south-directed extra-Alpine Variscides.

1 Introduction

Pre-Mesozoic basement is incorporated in all Alpine mega-units of the Alps and covers almost 50 percent of the Alpine domain. The actual pattern displays contrasting tectonostratigraphic units which may have been formed in distant places and brought together during the Alpine or older orogenies. Frisch et al. (1990) gave a summary of the differences and similarities of pre-Mesozoic evolution of the distinct Alpine tectonometamorphic units (Fig. 1):

1. The Helvetic zone, which is correlated with the European foreland of the Alps and is part of the Alpine lower plate;
2. The Penninic zone, which is also a continuation of the European foreland but separated from it by crustal thinning and the formation of microplates;
3. The Austro-Alpine and South-Alpine zones, which represent the upper plate of the Alpine edifice but were separated from each other by the Periadriatic lineament which was active during several extended periods before and during the Tertiary.

As these three Alpine tectonostratigraphic units were accreted during the Mesozoic-Tertiary orogenic cycle, they may display a distinct pre-Alpine history and have to be treated separately. The Alpine evolution throughout the Alps and the palaeotectonic reconstruction at the beginning of the Mesozoic are the subject of chapters by de Graciansky (this Vol.), Pfiffner (this Vol.), and Ratschbacher and Frisch (this Vol.). The aim of this chapter is to give a summary and comprehensive review of the pre-Mesozoic basement evolution and to discuss correlations and differences between various units, as well as their respective correlation with their Variscan framework (see also Neubauer and von Raumer, this Vol.). Polyphase deformation and metamorphism, including Alpine events, veil the earlier history and the nature of the protoliths. After more than 200 years of research on the pre-Mesozoic basement referred to in von Raumer and Neubauer (this Vol.), the wealth of stratigraphical, petrological, geochemical, and geochronological data demonstrates the complex histories of the tectonostratigraphic units. These data allow models to be proposed for both the geotectonic environments in which the protoliths were formed and the pre-Mesozoic evolution of the units. In comparing the geological evolution across the Helvetic, Penninic, and Austro-Alpine – South-Alpine basement zones, certain similarities and general lines of evolution become evident, but there are also striking differences.

Knowledge derived from the observations of many working groups made within the Alpine do-

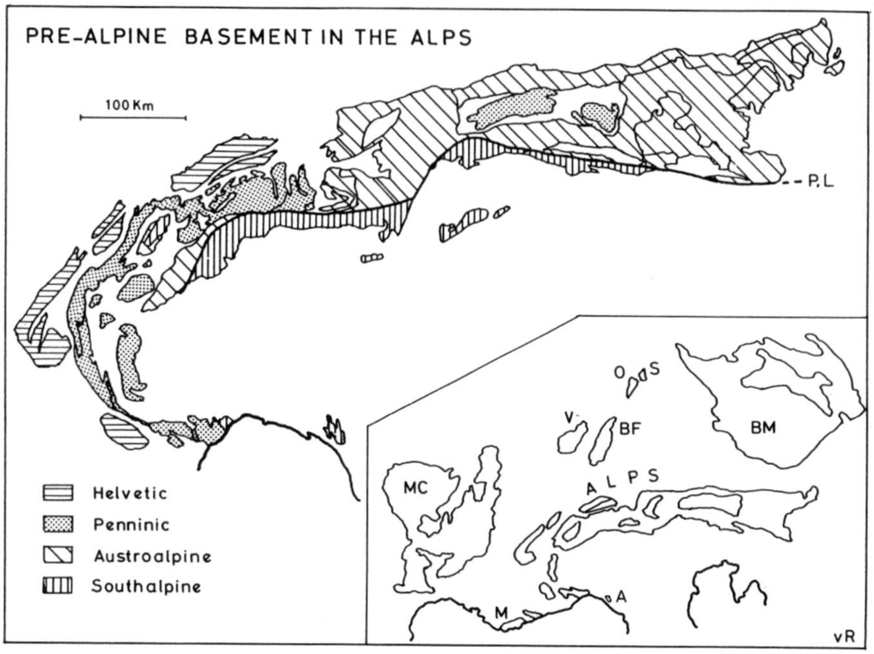

Fig. 1 Sketch map of pre-Alpine basement units in the Alps and their affiliation to the Alpine tectonostratigraphic zones. *P. L.* Periadriatic lineament. *Inset* The pre-Alpine basement areas in the European framework. *A* Apuane Alps; *BF* Black Forest; *BM* Bohemian Massif; *M* Massif des Maures; *MC* Massif Central; *O* Odenwald; *S* Spessart; *V* Vosges

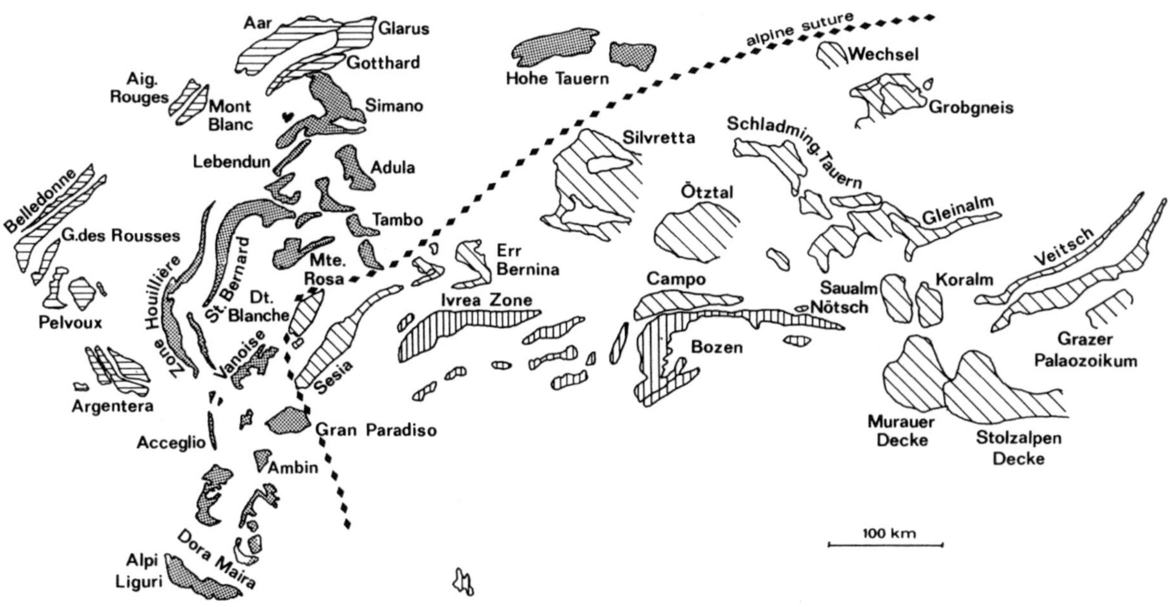

Fig. 2 Pre-Mesozoic basement units in the Alps in their approximate pre-Mesozoic distribution (after Neubauer 1988; De Graciansky this Vol.; Pfiffner this Vol.; Ratschbacher and Frisch this Vol.) with indication of the Alpine suture and names of the different units. *Hatched areas,* see Fig. 1

main allows verification of interpretations and assumptions based on observations made in the Variscan framework and projected into the Alps. Nonetheless, the authors are aware of the immense uncertainties which still exist in the reconstruction of pre-Mesozoic spatial relationships between basement zones (Figs. 1, 2).

As during the long geological evolution the palaeotectonic situation of a given basement zone changed, in the following sections comparable geological time periods will be chosen as those used by Harris and Fettes (1988) for the Caledonian-Appalachian orogen to present the palaeotectonic evolution of the different basement zones through time (Fig. 3 A–F).

2 Evolution of the Alpine Basement Through Time and Space

Frisch and Neubauer (in Keppie and Dallmeyer 1989) presented a sketch map of the Palaeozoic terranes and tectonostratigraphic units but with a late Alpine distribution of basement units. In contrast to the actual distribution of basement units (Fig. 1), Fig. 2 is based on the palinspastic reconstructions of de Graciansky (this Vol.), Pfiffner (this Vol.), and Ratschbacher and Frisch (this Vol.), and illustrates the tectonostratigraphic basement units in a configuration corresponding to their approximate pre-Mesozoic distribution. In addition, based on the configuration of Fig. 2, we present maps to show the different geodynamic settings for given time periods (Fig. 3 A–F).

2.1 Precambrian – Early Palaeozoic – Observations and Data

It is extremely difficult to develop models for the early evolution of Alpine basement units and their spatial relationships and, in consequence, Precambrian events are a subject of controversy, whereas the geological history from the Ordovician to the Permian, is commonly well constrained by data. Direct or indirect proof for the Precambrian history comes from very different methods (Neubauer 1991; Gebauer this Vol.; Köppel et al. this Vol.; Schönlaub this Vol.).

Field observations reveal the existence of an older basement:

1. A local unconformity of crystalline basement covered by low-grade Ordovician metasediments has been described by Neubauer (1985) from the Kaintaleck slices in the Austro-Alpine basement (see Neubauer and Frisch this Vol.).
2. Within Late Ordovician to Early Devonian detrital sediments (Greywacke zone), heavy minerals and magmatic and metamorphic clasts in conglomerates reveal the presence of a continental hinterland of pre-Late Ordovician age (Ebner et al. 1987; Heinisch 1988).
3. Neubauer (1991) discusses the Late Proterozoic to Early Palaeozoic tectonothermal evolution in the Eastern Alps.
4. From observations in the Silvretta area (Austro-Alpine basement), Maggetti et al. (1990) deduce a whole sequence of Precambrian metamorphic events, attributed to events of at least Pan-African age.
5. Very different criteria of comparative lithostratigraphy across the different basement units of the Alpine domain (Frisch et al. 1990) have led to the assumption that Precambrian lithologies should have existed in all tectonostratigraphic units.

The *fossil record* (scarce sporomorphs and acritarchs) gives evidence of Late Riphean, Lower Vendian and Cambrian metasediments from the Habach Series (Tauern window; Reitz and Höll 1988, 1989; Gilg et al. 1989; Höll and Reitz 1991), from the external massifs (Giorgi et al. 1979), and from the Southern Alps (Sassi et al. 1984) (see Schönlaub this Vol.). Such local occurrences of fossils are not related to continuous Late Ordovician to Carboniferous sedimentary sequences and suggest the existence of Late Riphean to Lower Vendian sedimentary basins.

Different *radiometric data* give evidence for the existence of Precambrian basement units in large parts of the pre-Mesozoic basement in the Alps or in the source region of sedimentary and magmatic rocks:

1. Gebauer et al. (1988) discuss a Late Riphaean age of eclogite protoliths (U-Pb zircon, SHRIMP) from the Gotthard Massif. Garnet-peridotite protoliths from Alpe Arami underwent metamorphism around 600 Ma (U-Pb zircon, upper intercept; Gebauer et al. 1991). U-Pb zircon ages (Schaltegger 1991) reveal the existence of Late Proterozoic detrital zircons in the northern Aar Massif.
2. Stille and Tatsumoto (1985), Schenk-Wenger and Stille (1989) and Biino et al (1993) give ages around 1000 Ma for metabasic rocks (Sm-Nd rock isochron) from basement relics of the Penninic and Helvetic domains, and suggest an Upper Proterozoic ophiolite series being preserved in these tectonostratigraphic units. Amphibolites from the Tauern window in the same tectonostratigraphic domain yield U-Pb zircon (upper intercept) and Sm/Nd (mineral isochron of zircon size fractions) age of ca. 650 Ma (von Quadt 1992).
3. Preliminary isotopic determinations from the Silvretta suggest a protolith formation around 900 Ma for basic, intermediate and acidic magmatic rocks (Rb-Sr total; Sm-Nd total; Maggetti and Flisch this Vol.).
4. Rb/Sr-derived sedimentation ages older than 600 Ma are discussed for Austro-Alpine paragneisses (Silvretta; Grauert 1966) and Ivrea metapelites (Hunziker and Zingg 1980).
5. Detrital muscovites from Late Ordovician sandstones in the Carnic Alps yield $^{40}Ar/^{39}Ar$ plateau ages of about 620–640 Ma interpreted as the age of post-metamorphic, early Cadomian cooling of the

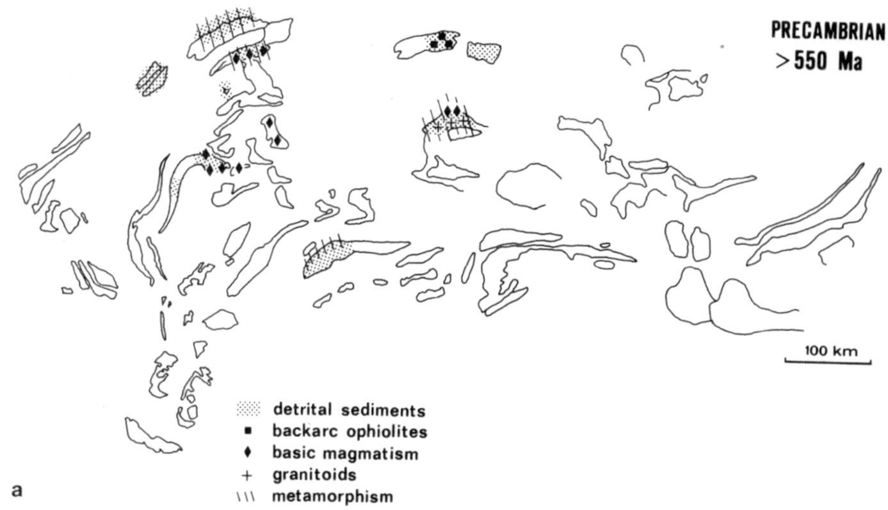

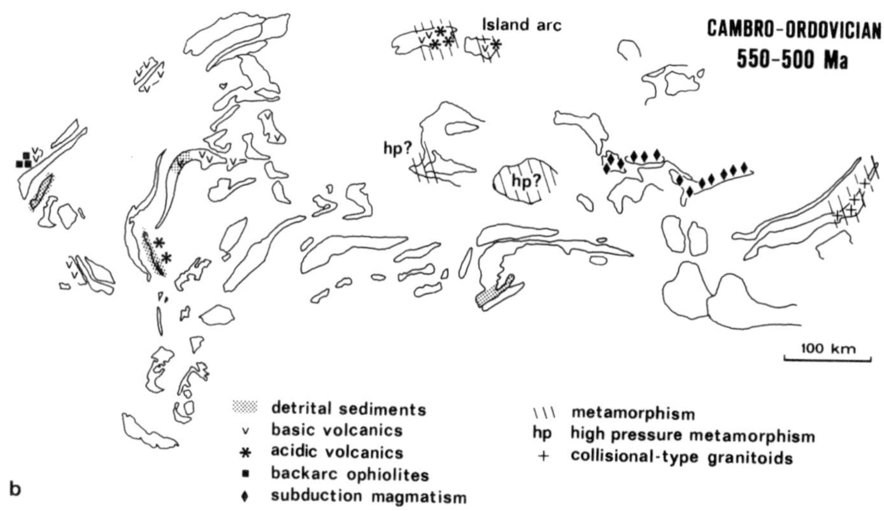

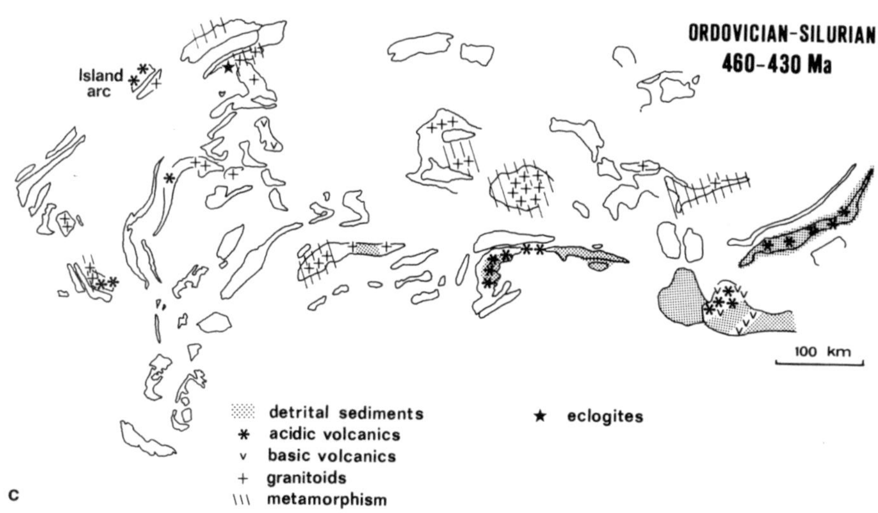

Fig. 3a–f. Pre-Mesozoic basement in the Alps (approximate late Variscan configuration, see Fig. 2) and distribution of tectonostratigraphic events for distinct time periods (explanations in the text). **a** Precambrian: presence of Precambrian relics in all Alpine lithostratigraphic units; **b** Cambrian: period of Palaeozoic rifting in the NW, of subduction in the SW; **c** Ordovician: widespread formation of acidic volcanics and granitoids, formation of eclogites in the Gotthard area, Helvetic domain;

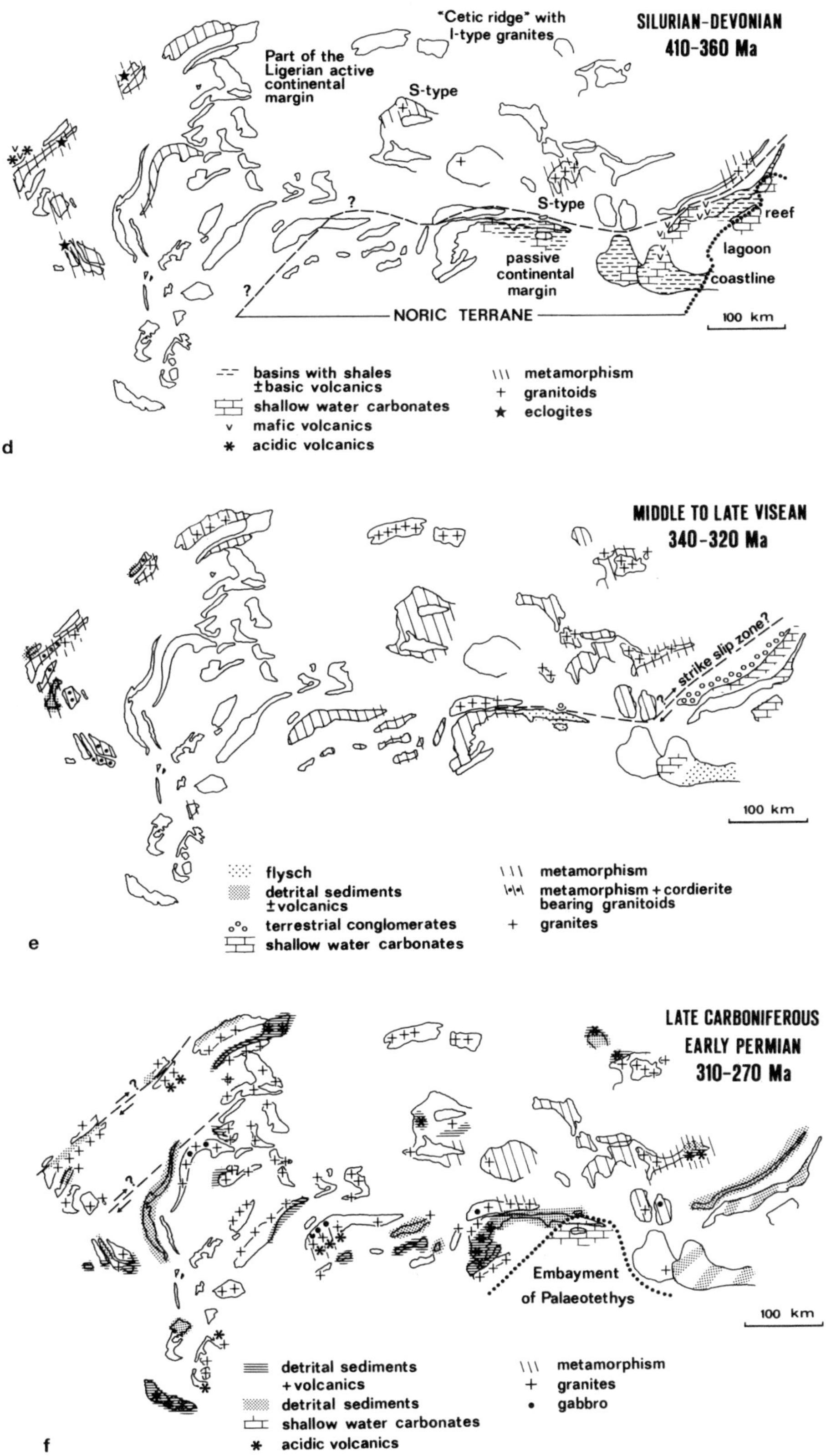

Fig. 3d–f. **d** Silurian-Devonian: mainly the Helvetic domain is part of the Ligerian active continental margin, whereas the SE is formed by the Noric terrane, a passive continental margin covered by sediments. Parts of the Penninic domain have been the place of the pre-Variscan suture with formation of oceanic crust; **e** Visean: nappe stacking and final collision; **f** Late Carboniferous – Early Permian: in most parts post-orogenic collapse and exhumation

source region (Dallmeyer and Neubauer 1991). The authors postulate two age provinces of detrital minerals, one with ca. 500 Ma zircons (metamorphism and magmatites yield comparable ages), and another with 620 Ma muscovites, reflecting probably two different basement terranes in the metamorphic hinterland.

6. Upper intercept ages (2–2.3 Ga; U/Pb zircon) of detrital zircons from Palaeozoic sediments suggest the existence of an Early Proterozoic hinterland (Grauert and Arnold 1968; Pidgeon et al. 1970; Köppel and Grünenfelder 1971; Köppel 1974; Grauert et al. 1973; Söllner and Hansen 1987).
7. Inherited lead isotopic components (3.2–0.8 Ga; U/Pb) of zircons in Palaeozoic and Tertiary magmatic rocks and their metamorphic derivatives give evidence for old crust (Cliff 1981; Söllner and Hansen 1987; Gebauer et al. 1988; Neubauer et al. 1989; Paquette et al. 1989; Gebauer this Vol.) in all Alpine lithostratigraphic units.
8. Lead isotopic data from stratiform ore deposits indicate the existence of a 2–3 Ga old lead component (Köppel 1983). The corresponding lead isotope pattern suggests the existence of two main domains (Köppel 1983): the Southern Alps and Austro-Alpine domain, with a "dominant component from an old lower crust", and the Penninic domain and more northern parts of the Alps, where "dominant components of such older crust are lacking". The decreasing influence of an eroded old crust could indicate that sediments deposited in the Penninic and more northern domains further from the hinterland (see also Köppel et al. this Vol.).

2.2 *Precambrian Evolution*

Precambrian crustal elements ("Gondwana-type" elements; Ziegler 1984) were present in all four tectonostratigraphic units of the Alps. The high-grade metamorphism recording crustal thickening and collision in the Silvretta area (Maggeti et al. 1990), the regional anatexis in the Ötztal area (Söllner and Hansen 1987) and the Precambrian cooling ages (Dallmeyer and Neubauer 1991) indicate that the oldest parts of the Austro-Alpine domain were affected by events of Pan-African – Cadomian age. For the Silvretta area even earlier phases (Pan-African I? or older) cannot be excluded (Maggetti et al. this Vol.), confirming the close relationship to the evolution of Gondwana. This holds also for preserved relics from the Penninic and Helvetic domains, where recent observations (Abrecht et al. 1991; Schaltegger 1991; Gebauer this Vol.; Biino 1992) strongly suggest a larger presence of the Pan-African event and a former attachment to the Gondwana continent.

Late Precambrian sediments may have been deposited on a platform, an environment best documented over large areas (Jell 1974; Piper 1985), or on oceanic crust in an accretionary wedge environment. The palaeogeodynamic interpretation given by Zwart and Dornsiepen (1978) and Cogne and Wright (1980) for the Late Proterozoic or Cadomian structural plan in Central Europe, in consequence, should be enlarged to include also the basement of the Alpine domain situated between Gondwana and Laurasia. Fracture zones of Late Proterozoic or Early Cambrian age influenced the formation of sedimentary troughs in Germany (Zeman 1980; Freyer and Suhr 1987), in the Bohemian Massif (Hirschmann 1988), where they were the place of Late Proterozoic rifting and subduction. In the African shield Buggisch and Flügel (1988); Rahman et al. (1990) and Harms (oral comm.), and in the North American shield Stanley and Ratcliffe (1985) and Skehan (1988) observe a regular pattern of late Proterozoic rift zones. Taking in account such observations and regarding the Precambrian evolution in the Alpine domain, we may, in consequence, envisage large areas with deposits of a continental-margin environment or the existence of intracontinental sedimentary troughs guided by major fracture zones. In fact, the most important lithologic association comprises metagreywackes, metapelites, a few quartzites and very few carbonates, and locally metavolcanics. This association is found in the Helvetic domain, in the basement of the Eastern Alps, in parts of the Penninic zone, and in the Southern Alps, implying the existence of such sedimentological environments.

It is evident that, besides such more continental domains, the time span between the Late Proterozoic (about 700 Ma) and the Cambrian is marked by high crustal mobility (Fig. 3 A, B). Rifting and formation of oceanic crust may have occurred in the Helvetic and Penninic zones as indicated by the emplacement of the Gotthard eclogite protoliths (Biino 1992). Amphibolites (former tholeiitic MOR basalts) occur in different zones and indicate the beginning of rifting over a large domain. Such rift zones could represent reactivated older structural elements. Not all of the metabasites must have originated in oceanic domains, as part of them could have formed on transform faults.

The Penninic basement of the Eastern Alps is part of an approximately coeval, Late Proterozoic, possibly intra-oceanic island arc system. An origin from intra-Cambrian arc systems is also postulated for the Austro-Alpine "gneiss-amphibolite complexes" which, possibly, were related to arc ophiolites. Such ophiolites are known from the Ötztal, the Speik complex, and the Ritting complex. The ages of these ophiolites are not known exactly but are constrained by minimum ages for the first metamorphic events. These minimum ages are about 500 Ma for the Ötztal region (eclogite metamorphism, nearly concordant zircon ages, Söllner and Hansen 1987), and about 520 Ma for the Kaintaleck complex (U-Pb zircon, lower intercept; metamorphic peak conditions; Neubauer and Frisch this Vol.). Based on $^{40}Ar/^{39}Ar$ mineral age data Handler et al. (in prep.) argued that major portions of the Kaintaleck complex were rejuvenated by a subsequent tectonothermal overprint between 470 and 400 Ma. Both ages indicate cessation of subduction and back-arc spreading followed by crustal consolidation of restricted crustal segments.

The coeval rifting and back-arc spreading as well as subduction magmatism and eclogite metamorphism may suggest a Pacific-type margin along which these controversial processes occur at the same time. Interestingly, Murphy and Nance (1989) came to a similar conclusion for the Avalon superterrane, which extended from North America to Armorica. In this way, the Alpine basement may have been part of the Pan-African – Cadomian – Baikalides belt between Gondwana and Laurasia. According to Frisch and Neubauer (1989), the Noric terrane of the Austro-Alpine basement together with the South-Alpine basement was temporarily attached to Gondwana.

2.3 *Cambro-Ordovician Break-Up and Subduction*

The irregular pattern of continental fractures and distribution of ophiolites observed for the early history of Iapetus (Skehan 1988) holds also for the margin of the Gondwana continent, to which the oldest parts of the Southern Alps and the Austro-Alpine basement were probably attached. Plutonic and volcanic activity around 500 Ma (Sm-Nd; Pin and Carme 1987) is discussed by Neubauer et al. (1987), Ménot et al. (1988), Guillot et al. (1991), and Santini (1991) and suggests crustal extension with magmatic activity and formation of sedimentary troughs. In contrast to the Cambrian-Early Ordovician continental break-up, observed in the basement of the Helvetic and Penninic domain, subduction and back-arc complexes apparently continued in some Penninic units (Tauern window), in the Austro-Alpine domain and, locally, also in the Southern Alps. The Belledonne back-arc magmatic association and the main oceanic suture in the Penninic domain and related areas of Gotthard and the Eastern Alps certainly represent different palaeogeographic units. The large Austro-Alpine-Penninic mobile zone (Fig. 4), at least since this time period, but probably earlier, became an important structural element separating most parts of the Helvetic basement from the more southern parts of the Gondwana block.

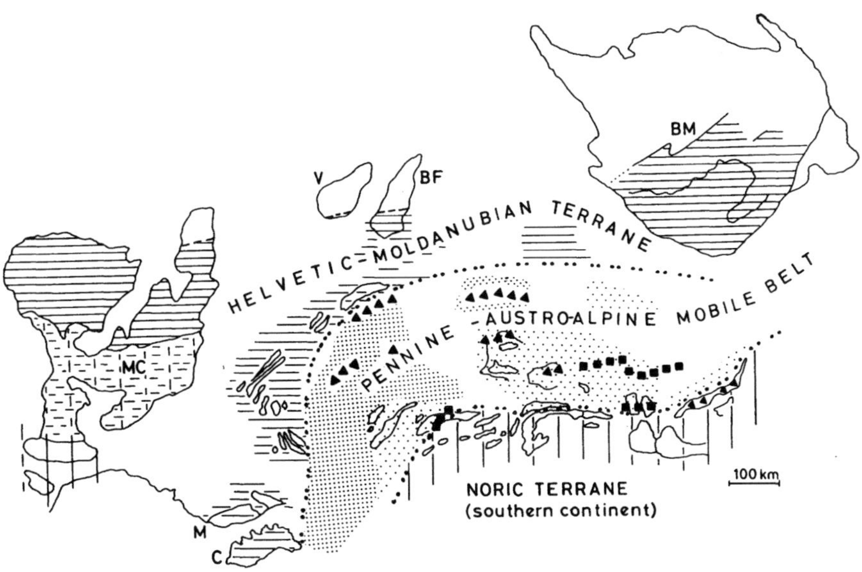

Fig. 4. The situation of the Alpine basement (pre-Mesozoic reconstruction, see Fig. 2) in the Variscan European framework. The Helvetic continental margin in the NW and the Noric terrane (Fig. 3d) in the SE are separated by a large mobile zone, mainly Penninic *(finely stippled)* and Austro-Alpine *(widely stippled)* basement units where, during different time periods since the Late Proterozoic, and at different places, formation of oceanic crust occurred. *Triangles* Proterozoic ophiolites; *squares* oceanic crust of uncertain age. *BF* Black Forest; *BM* Bohemian Massif; *C* Corsica; *M* Massif des Maures; *MC* Massif Central; *V* Vosges

2.4 *Ordovician Geotectonic Setting*

Ordovician palaeomagnetic data yield high latitudes for central and southern Europe and thus show a position close to Gondwana (Torsvik et al. 1990), and the sedimentary and faunal records (Schönlaub this Vol.) suggest a situation in temperate latitudes. There is no indication that basement zones of the Alps can be correlated with Fennoscandia in northern Europe.

Distinct tectonostratigraphic units underwent a different evolution during this time period (Fig. 3c). Parts of the Helvetic units underwent HP metamorphism with formation of eclogites (Gebauer this Vol.), and/or granulite facies (discussion: Abrecht et al. 1991), but the corresponding data might be the result of the Ordovician thermal event and the following collision. Granitoids indicate an island arc situation and, although incertain, a Laurasia-directed subduction is discussed. Metamorphism and anatexis around 450 Ma are documented from the northern part of the Aar Massif (U-Pb zircon data, Schaltegger 1991). The gneiss-amphibolite association, as exemplified by the Ötztal region, underwent final decompression after eclogite metamorphism at 500 Ma and local anatexis (ca. 460 Ma, Söllner and Hansen 1987) or else a collisional type of magmatism (Thöni 1986) by intrusion of large volumes of porphyric granites (ca. 450–440 Ma). A subduction type of metamorphism and calc-alkaline granites followed by post-orogenic granitoids is observed in the Southern Alps (Boriani et al. 1991). In general, granites (480–420 Ma) cover a large region comprising the Helvetic external massifs, areas of the Penninic zone in the Central Alps, the Southern Alps, and large volumes within the Austro-Alpine gneiss-amphibolite association. In most cases, these granites are now sheet-like augengneisses derived mainly from porphyritic granites. First geochemical data indicate an intraplate to arc situation for the granitoid rocks from the external massifs (Chiaradia 1992), and the chemical evolution of basic volcanics (Ménot and Paquette this Vol.) provides evidence also for an initial stage of rifting for large parts of the western external massifs. Based on scarce geochemical data from the Austro-Alpine domain (Thöni 1986), the granites are considered collisional, post-dating high pressure and anatexis. This evolution path has to be reconciled with the evolution of the Austro-Alpine and South-Alpine quartzphyllite and other sedimentary sequences, which resulted from Ordovician subsidence (Loeschke and Heinisch this Vol.; Neubauer and Sassi this Vol.). New biostratigraphic data (Reitz and Höll 1991; Heinisch pers. comm.; Neubauer unpubl. data) suggest sedimentation as early as Early Ordovician, most probably in a back-arc basin (Neubauer and Frisch 1988; Loeschke 1989). Superficial subsidence and sedimentation, uplift of the metamorphic belt and intrusion of granites may have occurred in the same belt, indicating a back-arc environment like the basin and range province of the US Cordillera. Schönlaub (this Vol.) discusses formation of granitoids as anatectic melts from a thick continental crust formed during the Pan-African event. Bonin et al. (this Vol.) attribute the general formation of calc-alkaline granites to a long-lived subduction regime, where a complete orogenic cycle was not achieved. In contrast, the Southern Alps are interpreted as a distinct terrane before the main collision. The observations lead to very contrasting plate tectonic interpretations for the different domains, including collisional and extensional regimes. The latter could be either the expression of opening of the Tethys I ocean (Cogné and Lefort 1985), or the products of transform rifting (Vai and Cocozza 1986), indicating detachment from the Gondwana continent (Ziegler 1988).

The regional comparative study of Late Palaeozoic granitoids (Bonin 1988, et al. this Vol.) shows that evolution of granites changes rapidly through time and space. In consequence, much more good data are needed to overcome the difficulties and to provide insight into the obviously contradictory interpretations of the Ordovician granitoids and the corresponding "thermal event". They might represent an evolution going from early convergence to collision and followed by post-collisional collapse.

2.5 *Silurian-Devonian – The Ligerian-Variscan Orogeny*

For this period very different tectonic zones can be recognized across the pre-Mesozoic basement. In the SE (Upper Austro-Alpine units, Fig. 3d) shelf sediments were deposited on a passive margin. All evidence suggests crustal extension with formation of basins and highs, alkaline volcanism, and mature detrital sediments. Towards the NW, the sediments indicate deeper water and the proximity of an oceanic basin separating this margin from other Alpine basement units (Heinisch 1988; Neubauer and Sassi this Vol.). In the middle Austro-Alpine (E of the Tauern window, Fig. 3d) true S-type granites are known from the Bundschuh area.

The evolution of the Penninic basement during the late Ordovician to Carboniferous period remains rather obscure. Vavra (1989) proposed that the island arc in the Tauern window attained maturity with the production of large volumes of rhyodacitic melts which were transformed into antectic granitoids in the Carboniferous.

During this period, at the latest, the external massifs are considered to have formed part of the internal zone of the Variscides, as they were the site of high-pressure metamorphism (eclogites, U/Pb, zircon fractions; Paquette et al. 1989) and which might be the consequence of early collision (von Raumer et al. this Vol.). The Helvetic basement is characterized by Devonian thrusting and crustal thickening, the pressure-temperature path indicating a collisional type of evolution. As a consequence of uplift and erosion and/or unroofing, large volumes of anatectic melts formed at the thermal peak after the Late Devonian. This late evolution may be interpreted as the time of post-orogenic collapse after the main crustal thickening (England 1987).

Consistent histories in the extra-Alpine Variscides lead to the conclusion (von Raumer 1988) that the external massifs were a link between the Ligerian (southern Brittany) and Moldanubian zones (Bohemian Massif). If compared with the Atlantic province, the compressional evolution of the external massifs could represent the continuation of structures called the Ligerian cordillera (Autran and Cogné 1980; Cogné and Lefort 1985), the early Variscan precursor of the main Variscan orogenic belt. The eclogites are interpreted to represent an advanced stage of underthrusting in a continental crust (Bouchardon et al. 1989). Depending on the main direction of shortening and the symmetry of approaching blocks (Fig. 3 d), different sectors of the orogen, during this early time period, might have seen different styles of tectonics, changing between tangential tectonics in one part and strike-slip in the other part. Thermal modelling on eclogites reveals different P-T-t paths from east to west for the western, adjacent areas of the Alpine domain (Mercier et al. 1991).

Such reconstructions necessarily have to be revised in the light of the following tectonic evolution when large domains changed site through regional strike-slip. First discussions give evidence of such events, combining the less metamorphic units with polymetamorphic basement units in the external massifs (Guillot 1991).

Contemporaneous crustal thinning is reported, further to the west, in the Belledonne area (external massifs, Ménot 1987), where the Devonian Rioupéroux-Livet plutonic and volcanic complex indicated continental thinning and rifting (Ménot and Paquette this Vol.), accompanied by the intrusion of trondhjemitic melts. In this way, the external massifs, after the main phase of compression, underwent an extensional regime, observed also in the extra-Alpine Variscan domains (Downes and Wilson 1991).

Some minor units of the Austro-Alpine basement, like the Bundschuh area, reveal a history comparable to that of the Helvetic basement, characterized by Devonian deformation, high-temperature metamorphism, and abundant granitoid intrusions. It is therefore considered also as part of the internal zone of the Variscides (Neubauer 1988; Frisch and Neubauer 1989) which, together with the Ligerian Zone and the Helvetic basement, formed an active continental margin referred to as the Ligerian Cordillera (Matte 1986).

2.6 Late Devonian – Early Carboniferous – Time of Collision

The Austro-Alpine basement, a Devonian passive margin, and the Helvetic basement, a Devonian compressive zone, accreted together during the Carboniferous collision.

The South-Alpine basement and the Austro-Alpine Noric passive margin were part of the southern (southeastern) branch of the Variscan orogen and a continuation of the Montagne Noire and the Cantabrian zone in southwestern Europe. It experienced distension and subsidence until overthrusting started in the Early Carboniferous (about mid-Visean). The former oceanic domain between the southeastern domains and the Bundschuh region then closed. Flysch sediments were deposited on the lower plate (Fig. 3e), and thrusting prograded from internal to external parts. Although the thermal climax was reached in the Early Carboniferous (northern parts of the middle Austro-Alpine; Fig. 3e), granite intrusions are much less abundant than in the internal domains, probably due to the lower plate position during Variscan collision. In contrast, in the Helvetic basement, granite intrusions are widespread. Earlier granites, in the Aar Massif, were deformed by south-vergent nappe tectonics (Oberhänsli et al. 1988) and followed by younger, undeformed granites. In the Belledonne area, the south-directed nappe tectonics has been dated by Ménot et al. (1987, K-Ar on amphiboles; 324 ± 12 Ma, and dextral strike-slip accompanied by anatexis yielded an age of 317 Ma (U/Pb monazite, Bussy and von Raumer, 1993). Krohe and Eisbacher (1988) reported such evolution from the Black Forest, and Feybesse et al. (1988) from the eastern parts of the French Central Massif. The Pannonian terrane of the Eastern Alps suffered a comparable evolution and should be included, in consequence, in the same tectonic zone. Such nappe tectonics could be the product of wrench tectonics as a local response to the general strike-slip and transpressional regime which, since the Early Carboniferous, dominated southern Europe (Arthaud and Matte 1977; Neugebauer 1988, 1989).

2.7 *Late Variscan Post-Collisional Evolution*

From the Carboniferous on, the entire domain, in all Alpine basement units, suffered the multiple transformations of post-collisional evolution accompanied by uplift and erosion.

The most striking regional feature in response to the general strike-slip tectonics appears in the palinspastic maps (de Graciansky this Vol.; Pfiffner this Vol.; Ratschbacher and Frisch this Vol.), major longitudinal zones, where either granitoids, sediments and/or volcanics appear (Fig. 3F). Schulz and von Raumer (1992), for the Aiguilles Rouges, discuss syndeformational uplift during late Variscan strike-slip. Strike-slip favoured the intrusion of granitoids, and major volcanic activity, as in the Western and Ligurian Alps, might have been at the sites of pull-apart. The general late Variscan evolution of the Alpine basement has to be seen in an extensional environment (see also Dal Piaz this Vol.) triggering the emplacement of very different magmatic rocks and giving way to sedimentary basins at the surface. Graf (1992) compared the evolution with that of the basin-and-range-province during the Cretaceous-Cenozoic times.

Granitoids are well known from all parts of the "Alpine" basement. In the Aar Massif, they evolve from calc-alkalic rocks in the Late Carboniferous to intraplate granites with an increasing alkalic component in the Permian (Mercolli and Oberhänsli 1988). Following Bonin et al. (this Vol.), most basement areas of the Helvetic and Penninic domains display a distensive regime giving way to the formation of granitic plutonic bodies and locally some volcanic suites. In contrast, in the Eastern and Southern Alps late Variscan granitoids seem to document a plate margin convergence at the southern rim of the Variscan Belt.

Sedimentary sequences in the basement of the Austro-Alpine domain with continuous deposits since Early Carboniferous have been deformed during Late Carboniferous (Namurian B/C – Westphalian B/C) forming a foreland thrust and fold belt. Late Carboniferous molasse-type sediments were deposited in elongated transtensional basins like the Zone Houillère separating the Penninic and Helvetic basement domains (Escher 1988) or the sedimentary trough of Late Carboniferous to Permian age of northern Switzerland (Matter 1987). In Austria, Krainer and Mogessie (1991) found convincing sedimentological evidence for the existence of a major strike-slip zone at Nötsch, the Nötsch-Veitsch-Ochtina zone separating the weakly metamorphosed parts of the Austro-Alpine basement from the highly metamorphosed parts. Such sedimentary troughs follow the same trend and indicate a Carboniferous extensional or transtensional regime. The transition from terrestrial to thick marine molasse deposits of the Tethys is exposed in the eastern part of the Southern Alps, where the Late Carboniferous Auernig formation exhibits a far-reaching faunal relationship which can be followed to the east along the Tethyan margin.

The position of the Penninic basement relative to the Helvetic and Austro-Alpine basement zones during the Variscan orogeny is not well constrained. Major late Variscan strike-slip faulting may have severely changed the palaeogeographic positions. Restoration of the Alpine history implies that in post-Variscan times the Penninic basement was directly attached to the Helvetic basement and that the Austro-Alpine – South Alpine zone was situated further to the east (Mercolli and Oberhänsli 1988; Ratschbacher et al. 1989). Intense syn- to late-orogenic magmatic activity in the Penninic basement underlines the spatial neighbourhood of the Helvetic and Penninic basement domains during the Carboniferous and Permian (Mercolli and Oberhänsli 1988; Oberhänsli et al. 1991).

3 Conclusion and Discussion

The preceding short overview strengthens the insight that the pre-Mesozoic Alpine basement represents a complex collage (Keppie 1985) of composite terranes, the domain of the Eastern Alps (Frisch and Neubauer 1989) serving as the best example of complications. These composite terranes, the Alpine tectonostratigraphic units of Fig. 1, were welded together during Eo-Alpine and Tertiary events.

Recently, attempts have been made to reconstruct the zonal evolution and accretion between Gondwana and Laurentia (Lefort 1989). Such reconstructions across the Atlantic Ocean as well as the palaeogeographic reconstitutions during the Palaeozoic (Puchkov 1988, Ziegler 1988) now receive additional support from the observations made in the Alpine domain. The pre-Alpine basement in the Alps, situated between Laurasia and Gondwana, has been the site of distension and crustal shortening at different moments and, based on the actual interpretation of isotopic data, all main orogenic time periods since the Precambrian can be recognized.

The preceding cartoon-like review is based on many assumptions and on the interpretation of isotopic data: it suggests a stepwise collision and closure of the suture between Gondwana and Laurasia. The scenario would change considerably if the Sm/Nd and U/Pb isotopic data from eclogites received a new interpretation through advanced methods. However, the general line of evolution will remain valid. The pre-Mesozoic basement of the Alps is composed of pieces of the North-Gondwana margin (Puchkov 1988, Paris and Robardet 1990, Oczlon 1992, von Raumer and Neubauer 1992) that drifted towards Laurasia. Enlarged through crustal growth, after the Devonian the final collision welded together the two mega continents.

It is understood that this evolution has to be consistent with the adjacent Variscan framework outside the Alps, the Alpine basement occupying only a small sector in the pre-Permian Variscan mountain chains. It is evident that it will be easier to see parallels in the *post-collisional* structures as older events are mostly hidden by the high-grade metamorphic evolution. Parallels to the east Mediterranean mountain chains are discussed by Neubauer and von Raumer (this Vol.).

The Helvetic basement in the Central and Western Alps represents the northernmost basement zones in the Alpine basement and preserves the main zone of Variscan collision. Boreholes show that the basement of the external massifs and the northern foreland hidden under sediments is continuous, and that the Black Forest area is only separated by a trough of Carboniferous-Permian detrital sediments (Matter 1987). Following von Raumer et al. (this Vol.), the basement areas of the Black Forest and the external massifs belong to the same crustal type having suffered comparable geologic histories, characteristic for the Moldanubian zone of Kossmat (1927) and comprising the Bohemian Massif, Black Forest, Vosges, and continuing into the French Central Massif.

In the western part, the N-S cross-section through the Massif Central has been interpreted as a collision orogen (Matte 1986). Large-scale gneiss nappes (Ledru et al. 1989) contain many occurrences of eclogites, and similar gneiss-eclogite units appear in the Black Forest and Vosges and in the Alpine external massifs (von Raumer and Ménot 1989). Recent observations (von Raumer et al. this Vol.) from the external massifs reveal the existence of high-grade anatectic rock series with lenses of eclogites, granulites, and serpentinites (former lherzolites). Such occurrences recall comparable structures from the Monts de Lyonnais (Gardien et al. 1990) and from the Gföhl and Drosendorf units of the Bohemian Massif (Matte et al. 1990). We are thus tempted to place the external domain of the Alpine basement zones near a kind of major suture (Fig. 5C), which might correspond to the limit of one of the crustal wedges produced during continental collision. A comparable situation can be deduced for the most northern

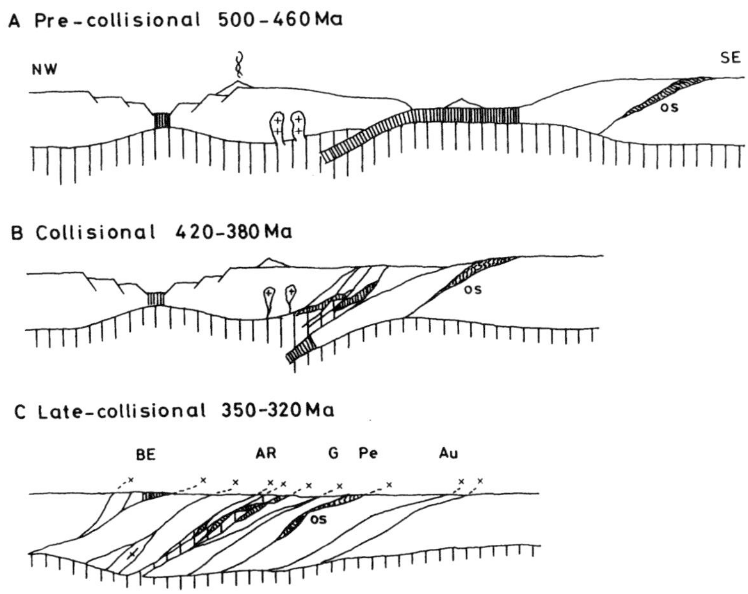

Fig. 5. Palaeotectonic cross-section through the Variscan orogen in the western Central Alps. **A** Pre-collisional (500–460 Ma); **B** Collisional (420–380 Ma); **C** Late-collisional time period (350–320 Ma). *AR* Aiguilles Rouges area; *Au* Austro-Alpine basement domains, undifferentiated; *BE* Belledonne area (Chamrousse ophiolite); *G* Gotthard area; *Pe* Penninic basement areas, undifferentiated: *os* older suture

Austro-Alpine elements which should have occupied a place near the suture of the Moravian basement and the Moldanubian crustal wedges.

The *pre-collisional* evolution in this zone is difficult to unravel, but it is supposed that relics of a cryptic suture are preserved. A candidate for a major pre-Variscan, complex suture having separated the passive Gondwana margin from the active Laurasian margin is probably represented by the basement units in the Alpine Penninic domain and adjacent areas (e.g. Gotthard Massif, Helvetic domain; parts of the lower Austro-Alpine units; see also Figs. 3c, 5A), called the Penninic-Austro-Alpine mobile belt in Fig. 4. It is likely to represent a special type of palaeotectonic environment, which might have been influenced by a large zone of thinned crust already since the Proterozoic and comparable to those Precambrian rift-zones observed from the African shield (Rahman et al. 1990). During different time periods, in this large zone (Fig. 3, 4), and at different places, rifting and formation of oceanic crust occurred, representing a telescoped mega-suture between Laurasia and Gondwana. It is evident that the difficulties in identifying protoliths in this zone prevent an exact localization of this suture which, in its global extension, has been discussed by Pin (1990) and Neubauer (1991). Pieces of lower crust, obviously, have been transported over larger distances during collisional, tangential tectonics (Figs. 3d, 5B) and, during the late collisional period (Fig. 5C), when the lower metamorphic, external domains were thrusted over the high metamorphic areas. It would be interesting to apply different geotectonic models, e.g. the extension by simple shear (Stampfli and Marthaler 1990), but data are not sufficient to fulfill all the necessary criteria.

It is evident that late Variscan strike-slip and tangential tectonics overprinted Pre- and early Variscan structures. Taking into account several hundreds of kilometers of tectonic transport for each period of opening, strike-slip and convergence during the Alpine events (Winkler 1988) and applying comparable rates of transport for the time since the Lower Carboniferous, the palinspastic reconstruction for the Lower Carboniferous might change considerably, and resulting in a kind of late Variscan, large-scale "buckling" pattern, the latter triggering the emplacement of magmatic rocks. The question arises regarding the structural pattern, pre-dating the late Variscan overprint. The distribution of monometamorphic rock units and pieces of polymetamorphic basement, after restoration, should give insight into older major structures. Besides the Carboniferous flysch (Franke 1989), lower metamorphic rocks should appear south of the Variscan mountain chain. Lower metamorphic units are observed west of the external massifs and also in the western part of the Penninic domain, where they have locally a Cambrian age (Guillot et al. 1991). The polymetamorphic Palaeozoic basement in the Alps, consequently, could have been limited in the west by a more N-S trending, major structure, a rift zone or a transform fault being the place of tangential tectonics during collision. Differences in tectonic evolution with changing style of tectonics on the western and northern limits could have their origin in an asymmetric collision (by indenter; see Neubauer and von Raumer this Vol.). The existence of such a major fault could have an influence on the continuation of the major ensialic or oceanic suture discussed already. It might be possible that the ensialic rift zone of the Central Massif could have, through a major dislocation, its corresponding continuation in the collision zone, preserved in the basement of the external and Penninic domain of the Alps. Detailed comparative work will be necessary to acquire better insight into the major regional correlation across the Alpine border of the Western Alps.

Placing all the events into the general crustal evolution of the Variscan fold belt, the cross-section through the pre-Mesozoic basement in the Alps includes a major suture between Gondwana and Laurasia, a highly condensed (through Alpine events) cross-section through the Variscan fold belt, where early subduction was followed by obduction and collision, and where the late Variscan evolution through post-orogenic collapse, strike-slip and exhumation prepared the main tectonic lines for the Alpine evolution.

Acknowledgements

A great number of ideas discussed in this review were stimulated by the results published by the many working groups in the Alps. We want to express our gratitude to all authors for their very stimulating and constructive collaboration. We would like to mention in particular the great effort of P.C. de Graciansky (Paris), O.A. Pfiffner (Bern), W. Frisch and L. Ratschbacher (Tübingen), as their palinspastic reconstructions for the pre-Mesozoic period were the palaeotectonic basis of our review. We are indebted for the critical review and very helpful suggestions of G. Biino (Bern), M. Brown (College Park, Maryland), W. Franke (Giessen), J.C. Hunziker (Lausanne), G. Martinotti (Torino), R.P. Ménot (Saint Etienne), John Rodgers (New Haven, Connecticut), and E. Stein (Darmstadt). Thanks also to Massimo Chiaradia (Fribourg) who, patiently, prepared the final drawings.

References

Abrecht J, Biino GG, Mercolli I, Stille P (1991) Pre-Alpine evolution of the Helvetic domain, Central Switzerland. Terra Abstr 3 (1): 479

Arthaud F, Matte P (1977) Late Paleozoic strike-slip faulting in southern Europe and northern Africa: result of a right-lateral shear zone between Appalachians and the Urals. Bull Geol Soc Am 88: 1305–1320

Biino G (1992) The pre-Alpine evolution of the Aar-, Tavetscher and Gotthard massifs. Thesis, Bern 1992 (Unpubl.)

Boriani A, Caironi V, del Moro A, Giobbi Origoni E, Pezzotta F, Pinarelli L (1991) Ordovician and Permian orogenic plutonism and the growth of the South Alpine crust, NW Italy and Ticino, Switzerland. Terra Abstr 3: 35

Bouchardon JL, Santallier D, Briand B, Ménot RP, Piboule M (1989) Eclogites in the French Palaeozoic Orogen: geodynamic significance. Tectonophysics 169: 317–332.

Buggisch W, Flügel E (1988) The Precambrian/Cambrian boundary in the Anti-Atlas (Morocco) Discussion and new results. In: Jacobshagen VH (ed) The Atlas system of Morocco. Springer, Berlin Heidelberg New York, pp 81–90

Bussy F. & von Raumer J.F. 1993: U-Pb dating of Palaeozoic events in the Mont-Blanc crystalline massif, Western Alps. Terra Nova 5, Abstracts I: 382–383

Chiaradia M. 1992: The scheelite skarn of Salanfe (Valais, Switzerland). Thesis Fribourg 225 pp (unpubl.)

Cliff RA (1981) Pre-alpine history of the Penninic zone in the Tauern window, Austria: U-Pb and Rb–Sr geochronology. Contrib Mineral Petrol 77: 262–266

Cogné J, Lefort JP (1985) The Ligerian orogeny: a Proto-Variscan event, related to the Siluro-Devonian evolution of the Tethys I ocean. In: Gee DG, Sturt B (eds) The Caledonian orogen – Scandinavia and related areas. Wiley, Chichester, pp 1185–1193

Cogné J, Wright AE (1980) L'orogène cadomien. Int Geol Congr 26. Sess, Paris C6, pp 29–54

Dallmeyer RD, Neubauer F (1991) $^{40}Ar/^{39}Ar$ age of detrital muscovite, Carnic Alps: Evidence for Cadomian basement in the Eastern Alps. Terra Abstr 3: 12–13

Downes H, Wilson M (1991) A Review of age and origin of oceanic crustal fragments in the Variscan belt. Terra Abstr 3 (1): 202

Ebner F, Neubauer F, Stattegger K (1987) The Caledonian event in the Eastern Alps; a review. In: Flügel HW, Sassi FP, Grecula P (eds) Pre-Variscan and Variscan events in the Alpine Mediterranean mountain belts. Alfa, Bratislava, pp 169–182

England PC (1987) Diffuse continental deformation; length scales, rates and metamorphic evolution. Philos Trans R Soc Lond A 321: 3–22

Escher A (1988) Structure de la nappe du Grand Saint-Bernard entre le Val de Bagnes et les Mischabel. Rapp Geol Serv Hydrol Geol Natl 7: 1–26

Feybesse JL, Lardeaux JM, Johan V, Tegvey M, Dufour E, Lemière B, Delfour J (1988) La série de la Brévenne (Massif central français): une unité dévonienne charriée sur le complexe métamorphique des Monts du Lyonnais à la fin de la collision varisque. CR Acad Sci 307 II: 991–996

Franke W (1989) Tectonostratigraphic units in the Variscan belt of Central Europe. Geol Soc Am Spec Pap 230: 67–90

Freyer G, Suhr P (1987) Über Archaeocyathiden-Funde und den lithologischen Aufbau des Unterkambriums im Gebiet von Torgau. Z Geol Wiss 15 (6): 605–680

Frisch W, Neubauer F (1989) Pre-Alpine terranes and tectonic zoning in the Eastern Alps. Geol Soc Am Spec Pap 230: 91–99

Frisch W, Neubauer F, Ménot RP, von Raumer JF (1990) Correlation and evolution of the Alpine basement. Schweiz Mineral Petrogr Mitt 70: 265–286

Gardien V, Tegyey M, Lardeaux M, Misseri M, Dufour E (1990) Crust-mantle relationships in the French Variscan chain: the example of the southern Monts du Lyonnais unit (eastern French Massif Central). J Metamorphic Geol 8: 477–492

Gebauer D, Quadt A, Compston W, Williams I, Grünenfelder M (1988) Archaische Komponenten in retrograden Eklogiten des Gotthardmassivs. Schweiz Mineral Petogr Mitt 68: 485–490

Gilg HA, Höll R, Kupferschmid MP, Reitz E, Stärk H, Weber-Diefenbach K (1989) Die Basisschieferfolge in der Habachformation im Felber- und Amertal (Tauernfenster, Salzburg). Mitt Österr Geol Ges 81: 65–91

Giorgi L, Giraud P, Vachard D (1979) Sur la présence de microorganismes d'âge cambrien dans les schistes cristallins du versant occidental du massif cristallin externe des Grandes Rousses (Alpes occidentales). C R Acad Sci Paris 288-D: 1079–1082

Graf K (1992) A geodynamic comparision of late Variscan volcanism in the Alpine fold belt and Tertiary to Cenozoic volcanism of the Cordilleran fold belt in the Western United States. Unpublished Thesis, Bern 1992.

Grauert B (1966) Rb-Sr age determinations on orthogneisses of the Silvretta (Switzerland). Earth Planet Sci Lett I: 139–142

Grauert B, Arnold A (1968) Deutung diskordanter Zirkonalter der Silvrettadecke und des Gotthardmassivs. Contrib Mineral Petrol 20: 34–56

Grauert B, Hänny R, Sobtrajanova B (1973) Age and origin of detrital zircons from the pre-Permian basements of the Bohemian Massif and the Alps. Contrib Mineral Petrol 40: 195–230

Guillot F, Liégois JP, Fabre J (1991) Des granophyres du Cambrien terminal dans le Mont Pourri (Vanoise, zone briançonnais): première datation U-Pb sur zircon d'un socle des zones internes des Alpes françaises. C R Acad Sci Paris 313 II: 239–244

Guillot PL (1991) Place et significance de la "série satinée" de Belledonne dans l'édifice de napps du bâtit varisque français. C R Acad Sci 312 II: 1581–1585

Handler R, Dallmeyer RD, Neubauer F, Hermann S, Frank W: $^{40}Ar/^{39}Ar$ and Rb-Sr mineral age data from the Kaintaleck Nappe, Austroalpine basement, eastern Alps: significance for closure and resetting of isotopic systems. J Geol Soc London (in prep)

Harris AL, Fettes DJ (eds) (1988) The Caledonian-Appalachian orogen. Geol Soc Spec Publ 38, Oxford, pp 1–643

Heinisch H (1988) Hinweise auf die Existenz eines passiven Kontinentalrandes im Altpaläozoikum der nördlichen Grauwackenzone – Ostalpen. Schweiz Mineral Petrogr Mitt 68: 407–418

Hirschmann G (1987) Some remarks on the position of the Rhenisch Massif between the Variscides and the Caledonides. In: Vogel A, Miller H, Greiling R (eds) The Rhenisch Massif – structure, evolution, mineral deposits and present geodynamics. Vieweg, Braunschweig, pp 103–112

Hirschmann G (1988) Connections between Moldanubicum, Bohemicum and Saxothuringicum from a lithostratigraphic viewpoint. In: Proceedings of the 1st Int Conference on the Bohemian Massif, Prague, 26.9.–3.10.1988: 112–120

Höll R, Reitz E (1991) Palynological evidence for an Upper Proterozoic age in the older schist cover and for lower cretaceous age in the younger schist cover of the Tauern window, Eastern Alps. Terra Abstr 3 (1): 337

Hunziker JC, Zingg A (1980) Lower Paleozoic Amphibolite to Granulite Facies metamorphism in the Ivrea Zone (Southern Alps, Northern Italy). Schweiz Mineral Petrogr Mitt 60: 181–213

Jell PA (1974) Faunal provinces and possible planetary reconstruction of the middle Cambrian. J Geol 82: 319–350

Keppie JD (1985) The Appalachian collage. In: Gee DG, Sturt BA (eds) The Caledonide orogen – Scandinavia and related areas. Wiley, Chichester, pp 1217–1226

Keppie JD, Dallmeyer RD (1989) Tectonic map of pre-Mesozoic terranes in circum-Atlantic Phanerozoic orogenes. IGCP Proj 23, Terranes in the Circum-Atlantic Palaeozoic Orogens. Halifax

Köppel V (1974) Isotopic U-Pb ages of monazites and zircons from the crust-mantle transition and adjacent units of the Ivrea and Ceneri Zones (Southern Alps, Italy). Contrib Mineral Petrol 43: 55–70

Köppel V (1983) Summary of lead isotope data from ore deposits of the Eastern and Southern Alps. Some metallogenetic and geotectonic implications. In: Schneider HJ (ed) Springer, Berlin Heidelberg New York, pp 162–168

Köppel V, Grünenfelder M (1971) A study of inherited and newly formed zircons from paragneisses and granitised sediments of the Strona-Ceneri-Zone (Southern Alps). Schweiz Mineral Petrogr Mitt 51: 385–409

Kossmat F (1927) Gliederung des varistischen Gebirgsbaues. Abh Sächs Geol Landesamt 1: 1–59

Krainer K, Mogessie A (1991) Composition and significance of resedimented amphibolite breccias and conglomerates (Badstub formation) in the Carboniferous of Nötsch (Eastern Alps, Austria). Jahrb Geol Bundesanst 134: 65–81

Krohe A, Eisbacher GH (1988) Oblique crustal detachment in the Variscan Schwarzwald, southwestern Germany. Geol Rundsch 77: 25–43

Ledru P, Lardeaux JM, Santallier D, Autran A, Quenardel JM, Floc'h JP, Lerouge G, Maillet N, Marchand J, Ploquin A (1989) Ou sont les nappes dans le Massif Central Français? Bull Soc Géol Fr (8) 5–3: 605–618

Lefort JP (1989) Basement correlation across the North Atlantic. Springer Berlin Heidelberg New York, 148 pp

Loeschke J (1989) Lower Palaeozoic volcanism of the Eastern Alps and its geodynamic implications. Geol Rundsch 78: 599–616

Lorenz V, Nicholls IA (1984) Plate and intraplate processes of Hercynian Europe during the Late Paleozoic. Tectonophysics 107: 25–56

Maggetti M, Flisch M, Bollin R (1990) Bericht über die Exkursion der Schweiz. Mineralogischen und Petrographischen Gesellschaft ins Silvretta Kristallin und in den Westrand des Unterengadiner Fensters. Schweiz Mineral Petrogr Mitt 70: 121–157

Matte Ph (1986) Tectonic and plate tectonics model for the Variscan belt of Europe. Tectonophysics 126: 329–374

Matte Ph, Maluski H, Rajlich P, Franke W (1990) Terrane boundaries in the Bohemian massif: result of large-scale Variscan shearing. Tectonophysics 177: 151–170

Matter A (1987) Faziesanalyse und Ablagerungsmilieu des Permokarbons im Nordschweizer Trog. Eclogae Geol Helv 80: 345–368

Ménot RP (1987) Magmatismes paléozoiques et structuration carbonifere du Massif de Belledonne (Alpes Françaises). These, Univ Lyon, Mém et Doc Centre Armoricain Etudes Structurales des Socles, Rennes, 1988 (21): 1–465

Ménot RP, Bonhomme MG, Vivier G (1987) Structuration tectono-métamorphique carbonifère dans le massif de Belledonne (Alpes occidentales fraçaises). Apport de la géochronologie K/Ar des amphiboles. Schweiz Mineral Petrogr Mitt 67: 273–284

Ménot RP, Peucat JJ, Scarenzi D, Piboule M (1988) 496 My age of plagiogranites in the Chamrousse ophiolite complex (external crystalline massifs in the French Alps): evidence of a Lower Paleozoic oceanization. Earth Planet Sci Lett 88: 82–92

Mercier L, Lardeaux JM, Davy P (1991) On the tectonic significance of retrograde P-T-t paths in eclogites of the French Massif Central. Tectonics 10: 131–140

Mercolli I, Oberhänsli R (1988) Variscan tectonic evolution in the Central Alps. Schweiz Mineral Petrogr Mitt 68: 491–500

Murphy JB, Nance RD (1989) Model for the evolution of the Avalonian-Cadomian belt. Geology 17: 735–738

Neubauer F (1985) Eine präoberordovizische Transgression in der Grauwackenzone (Ostalpen) und ihre Bedeutung für paläozoische Geodynamik und alpidischen Deckenbau. Neues Jahrb Geol Paläontol Monatsh 1985: 46–64.

Neubauer F (1988) Bau und Entwicklungsgeschichte des Rennfeld-Mugel und des Gleinalm-Kristallins (Ostalpen). Abh Geol Bundesanst 42: 1–137

Neubauer F (1991) Late Proterozoic and Early Palaeozoic tectonothermal evolution of the Eastern Alps. In: Dallmeyer RD, Lecorché JP (eds) The West African orogens and circum-Atlantic correlatives. Springer, Berlin Heidelberg New York, pp 307–314

Neubauer F, Frisch W (1988) Ordovician-Silurian geodynamic evolution of the Alps – the orogeny back-arc basin model. Schweiz Mineral Petrogr Mitt 68: 351–357

Neubauer F, Frisch W, Hansen BT (1987) Time relations between eo-Alpine metamorphism and thrusting. Evidence from the crystalline basement of the Greywacke Zone. In: Faupl P, Flügel HW (eds) Geodynamics of the Eastern Alps. Deuticke, Wien, pp 132–139

Neubauer F, Frisch W, Schmerold R, Schlöser H (1989) Metamorphosed and dismembered ophiolite suites in the basement of the Eastern Alps. Tectonophysics 164: 49–62

Neugebauer J (1988) The Variscan plate tectonic evolution: an improved "Iapetus model". Schweiz Mineral Petrogr Mitt 68: 313–333

Neugebauer J (1989) The Iapetus model: a plate tectonic concept for the Variscan belt of Europe. Tectonophysics 169: 229–256

Oberhänsli R, Schenker F, Mercolli I (1988) Indications of Variscan nappe tectonics in the Aar Massif. Schweiz Mineral Petrogr Mitt 68: 509–520

Oberhänsli R, Krähenbühl U, Stille P (1991) Contrasting REE characteristics in meta-lamprophyres from Variscan massifs of the Central Swiss Alps and REE patterns in lamprophyres from Variscan terranes of Western Europe. Schweiz Mineral Petrogr Mitt 71: 53–62

Oczlon M (1992) Gondwana and Laurussia before and during the Variscan orogeny in Europe and related areas. Heidelberger Geow Abhandl 53: 1–56

Paquette JL (1987) Comportement des systèmes isotopiques U-Pb et Sm-Nd dans le métamorphisme éclogitique. Chaîne hercynienne et chaîne alpine. Mem Doc Centre Arm Struct Socle 14: 1–190

Paquette JL, Ménot RP, Peucat JJ (1989) REE, Sm-Nd and U-Pb zircon study of eclogites from the Alpine external massifs (Western Alps): evidence for crustal contamination. Earth Planet Sci Lett 96: 181–198

Paris F, Robardet M (1990) Early Palaeozoic palaeobiogeography of the Variscan regions. Tectonophysics 177: 193–213

Pidgeon RT, Köppel V, Grünenfelder M (1970) U-Pb isotopic relations in zircon suites from a para- and orthogneiss from the Ceneri zone, southern Switzerland. Contrib Mineral Petrol 26: 1–11

Pin C (1990) Variscan oceans: ages, origins and geodynamic implications inferred from geochemical and radiometric data. Tectonophysics 169: 333–340

Pin C, Carme Chr (1987) A Sm-Nd isotopic study of 500 Ma old oceanic crust in the Variscan belt of Western Europe: the Chamrousse ophiolite complex, Western Alps (France). Contrib Mineral Petrol 96: 406–413

Piper JDA (1985) Continental movements and breakup in Late Precambrian-Cambrian times: prelude to Caledonian orogenesis. In: Gee DG, Sturt Ba (eds) The Caledonian orogen – Scandinavia and related areas. Wiley, Chichester, pp 19–34

Puchkov V.N. (1988) Correlation and geodynamic features of pre-Alpine tectonic movements throughout and around the Alpine orogen. Studia Geologica Polonica XCI: 78–92

von Quadt A (1992) U-Pb zircon and Sm-Nd geochronology of mafic and ultramafic rocks from the central part of the Tauern Window (Eastern Alps). Contrib Mineral Petrol 110: 57–67

Rahman EM Abdel, Harms U, Schandelmeier H, Franz G, Darbyshire DPF, Horn P, Müller-Sonius D (1990) A new ophiolite occurrence in the Sudan – constraints on late Proerozoic tectonism. Terra Nova 2: 363–376

Ratschbacher L, Frisch W, Neubauer F, Schmidt SM, Neugebauer J (1989) Extension in compressional orogenic belts: the eastern Alps. Geology 17: 404–407

von Raumer JF (1988) Caledonian-Variscan structures in the Alps: an introduction. Schweiz Mineral Petrogr Mitt 68: 291–299

von Raumer JF Ménot RP (1989) Evolution paleozoique du socle ouest-alpin et place des Massifs Cristallins Externes dans l'orogene Varisque. CR Acad Sci Paris 309-II: 397–402

von Raumer JF, Neubauer F (1992) The Palaeozoic evolution of the Alps. Terra Nova 4, suppl abstracts 2: 56–57

Reitz E, Höll R (1988) Jungproterozoische Mikrofossilien aus der Habachformation in den mittleren Hohen Tauern und dem nordostbayerischen Grundgebirge. Jahrb Geol Bundesanst 131: 329–340

Reitz E, Höll R (1989) Unterordovizische Akritarchen aus der Nördlichen Grauwackenzone (Ostalpen). Jahrb Geol Bundesanst 132: 761–774

Reitz E, Höll R (1991) Acritarch dating of lower Ordovician age in the northern Graywacke zone and Innsbruck quartzphyllite group, Eastern Alps. Terra Abstr 3 (1991): 338

Santini L (1991) Géochimie et géochronologie dans les paragenèses de haute pression de la Nappe d'Adula (Alpes centrales, Suisse). Thesis, Lausanne, 85 pp (unpubl.)

Sassi FP, Kalvacheva R, Zanferrari A (1984) New data on the age of deposition of the Southalpine phyllitic basement in the Eastern Alps. Neues Jahrb Geol Paläontol Monatsh 1984: 741–751

Schaltegger U (1991) Die polymetamorphe Geschichte des Aarmassivs – neue U-Pb-Resultate. Schweiz Min Petr Ges Jahrestagung Chur 1991 Abstracts: 20

Schenk-Wenger K, Stille P (1989) Geochemical and isotope evidence for an extensive Proterozoic ophiolite-suite in the Central Alps (Switzerland). Eur J Mineral 1 (1): 160

Schmidt K (1977) Der altpaläozoische Magmatismus und seine Stellung in der tektonischen Geschichte Mittel- und Südwesteuropas. Z Dtsch Geol Ges 128: 121–141

Schulz B, von Raumer JF (1992) Syndeformational uplift of Variscan high pressure rocks (Val de Bérard, Aiguilles Rouges, Western Alps). Ztschr Dt Geol Ges Vol 144 (in press)

Skehan JW (1988) Evolution of the Iapetus Ocean and its borders in pre-Arenig times: a synthesis. In: Harris AL, Fettes DJ (eds) The Caldeonian-Appalachian orogen. Geol Soc Spec Publ 38: 185–229

Söllner F, Hansen BT (1987) "Pan-afrikanisches" und "kaledonisches" Ereignis im Ötztalkristallin der Ostalpen. Jahrb Geol Bundesanst 130 (4): 529–569

Stampfli G, Marthaler M (1990) Divergent and convergent margins in the north-western Alps – confrontation to actualistic models. Geodyn Acta (Paris) 4 (3): 159–184

Stanley RS, Ratcliffe NM (1985) Tectonic synthesis of the Taconian orogeny in western New England. Bull Geol Soc Am 96: 1227–1250

Stille P, Tatsumoto M (1985) Precambrian tholeiitic-dacitic rock-suites and Cambrian ultramafic rocks in the Penninic nappe system of the Alps: evidence from Sm-Nd isotopes and rare earth elements. Contrib Mineral Petrol 89: 184–192

Thöni M (1986) The Rb-Sr thin slab isochron method – an unreliable geochronologic method for dating geologic events in polymetamorphic terrains? Mem Sci Geol Padova XXXVI-II: 283–352

Vai GB, Cocozza T (1986) Tentative schematic zonation of the Hercynian chain in Italy. Bull Soc Géol Fr (8) II 1: 95–114

Vavra G (1989) Die Entwicklung des penninischen Grundgebirges im östlichen und zentralen Tauernfenster der Ostalpen. Geochemie, Zirkonmorphologie, U/Pb-Radiometrie. Dissertation Univ Tübingen, 170 pp

Winkler W (1988) Mid- to early Late Cretaceous flysch and melange formations in the western part of the Eastern Alps. Jb Geol BA 131: 341–389

Zeman J (1980) Die Krustenentwicklung im präkambrisch und paläozoisch mobilen Europa außerhalb der Osteuropäischen Tafel. Z Geol Wiss 8: 393–404

Ziegler PA (1980) Northwestern Europe: subsidence patterns of post-Variscan basins. Int Geol Congr 26. Sess, Paris C6: 249–280

Ziegler PA (1984) Caledonian and Hercynian crustal consolidation of western and central Europe – a working hypothesis. Geol Mijnbouw 63: 93–108

Ziegler PA (1985) Late Caledonian framework of western and central Europe. In: Gee DG, Sturt Ba (eds) The Caledonian orogen – Scandinavia and related areas. Wiley, Chichester, pp 3–18

Ziegler PA (1988) Laurussia – the old red continent. In: McMillan NJ, Embry AF, Glass DJ (eds) Devonian of the World, vol. I. Can Soc Petr Geol, pp 15–48

Zwart HJ, Dornsiepen UF (1978) The tectonic framework of central and western Europe. Geol Mijnbouw 57: 627–654

The Alpine Basement – Linkage Between Variscides and East-Mediterranean Mountain Belts

F. Neubauer and J. v. Raumer

Abstract

The basement of eastern Mediterranean Alpine mountain belts and the extra-Alpine Variscides forms a collage of various tectonostratigraphic units which have been accreted and consolidated during the Variscan orogenic cycle. In consequence, most basement units of the entire Alpine-Mediterranean mountain belts, part of the southern, south-directed segment of the central to west European Variscan collisional belt, have an extension in extra-Alpine Variscides. The evolution of this segment of Variscides reflects processes of an active continental margin since late Precambrian time with both subduction and formation of back-arc basins, and accretion of suspected terranes along the northern Gondwanian margin. Final transpressional collision between crustal pieces of distinct Early Palaeozoic evolution occurred during Early to Late Carboniferous and is followed by molasse-like red bed sedimentation. Subsequent transgression of the Tethyan Sea along the southern margin of Variscides in Late Carboniferous to Permian was associated with dextral transtensive megashearing between Gondwana and Laurussia.

1 Introduction

The Alpine basement resulted from Carboniferous plate collision and accompanying orogenic diastrophism which affected various crustal pieces between Laurussia and Gondwana (for recent reviews, see Frisch et al. 1990; von Raumer and Neubauer this Vol.). The Variscan orogenic belt is well known from basement massifs around the Alps in western and central Europe (Lorenz and Nicholls 1984; Matte 1986; Ziegler 1986; Franke 1989 a, b; Keppie and Dallmeyer 1989; Martinez-Catalan 1990) and in other Mediterranean mountain belts like the Carpathians, Dinarides, Apennines (Fig. 1; Flügel 1990; contributions in Flügel et al. 1987; Sassi and Zanferrari 1990). Basement rocks are also known in back-arc and foreland basins within the arcuate Alpine mountain belts, mainly from wells which penetrated, e. g., the bottom of the Adriatic sea and of the Pannonian basin (Vai and Cocozza 1986; Arkai and Lelkes-Felvari 1987). The correlation of all these various Alpine basement units with the extra-Alpine Variscides and with the basement of Alpine-Mediterranean mountain belts is the topic of this contribution (Fig. 1). An approach to correlate the intra-Alpine basement has been recently provided by Flügel (1990), which is a very useful summary of suggested correlations of fossil-bearing sequences.

Major problems restricting the correlation are rejuvenation by Mesozoic and Cenozoic deformation and metamorphism, disperson of continuous basement zones by large-scale Mesozoic and Tertiary strike-slip faulting and escaping microplates during the Mesozoic and Caenozoic transpressive collision of Africa and Europe. Furthermore, the database, especially that of reliable geochronological data, e. g. U-Pb zircon and Rb-Sr data, is poor in all metamorphic areas under consideration. For all these reasons, any reconstruction of pre-Mesozoic events in the Alpine-Mediterranean mountain belt includes many uncertainties at present.

However, our reconstructions are based on recent data compilations of extra-Alpine Variscides (Matte 1986; Ziegler 1986; Franke 1989 a, b; Martinez-Catalan 1990; Pin 1990) and of Alpine Mediterranean mountain belts (Flügel 1990; contributions in Flügel et al. 1987, and Sassi and Zanferrari 1990) and on recent contributions which are not included in these compilations.

Within the arcuate Alpine-Mediterranean mountain belts, the Alps form the linkage between the Apennines, Carpathians and Dinarides (Fig. 1). Basically, tectonic units of the Alpine-Mediterranean mountain belts have been transported both onto the adjacent continents, stable Europe and Africa, and onto the Apulian microplate in the interior of the Alpine-Mediterranean belt during Tertiary plate collision (Burchfiel 1980; Dercourt et al. 1986; Dewey et

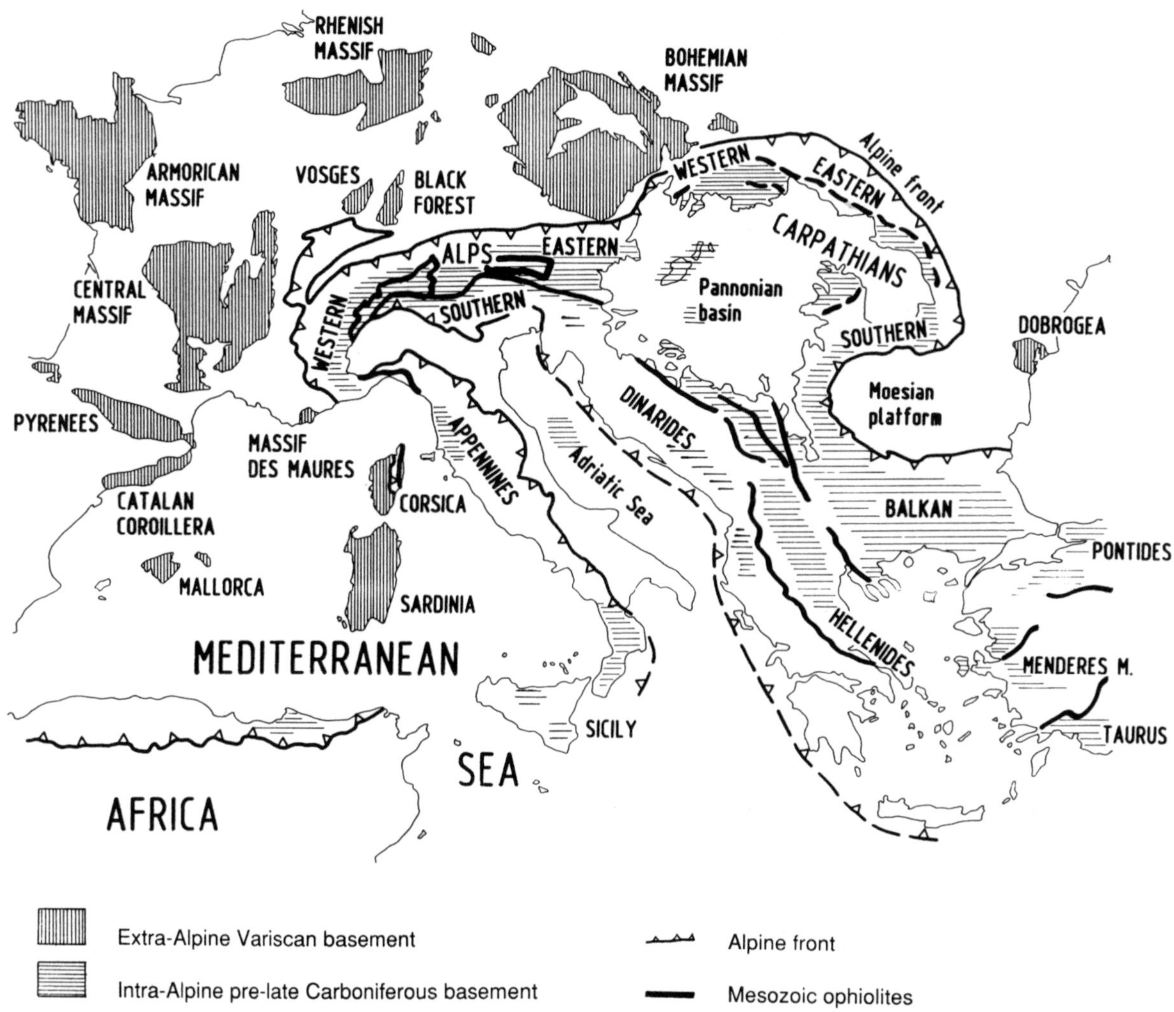

Fig. 1. Pre-Permian exposures in the Variscides of Western Europe (Iberian Massif excluded) and the Alpine-Mediterranean mountain belts

al. 1989; Durand-Delga and Fontboté 1980; Le Pichon et al. 1988; Robertson and Dixon 1984; Trümpy 1989). The following major tectonic units have been distinguished from the interior to the external Alpine-Mediterranean mountain belts:

Apennines, Southern Alps, and Dinarides are thrust onto the Adriatic realm which is a stable platform in the interior of the Mediterranean mountain belts. One major ophiolitic suture (Fig. 2) is the Ligurian/Piemontais ophiolite which separates the Austro-Alpine from the Helvetic/Penninic basement (Pfiffner this Vol.). In the western Carpathians, the ophiolite-like assemblage of the Cimmerian Meliata unit may represent a further suture within the Austro-Alpine domain. Its southern extension are the Mures ophiolite and the Vardar zone, respectively. The latter separates the Serbomacedonian-Rhodope-Balkan basement massifs from the Dinarides to the west. The Vardar zone is commonly traced to the Pontides in northern Turkey (Sengör et al. 1988). Large-scale Tertiary lineaments with preferred strike-slip motion, e.g. the Periadriatic lineament and its possible extension within the Pannonian basin, displaced basement blocks.

From these Mesozoic and Tertiary sutures and lineaments mentioned, we may separate the following basement blocks which may have a distinct history being different from those juxtaposed by Mesozoic/Tertiary suturing (Fig. 2):

1. The Apulian microplate includes the stable platform of the Adriatic sea, and detached basement slices inside suture zones in mountain belts surrounding the Adriatic Sea.
2. Further basement units are outside of the inner, Cimmerian Meliata suture, e.g. the Veporic and

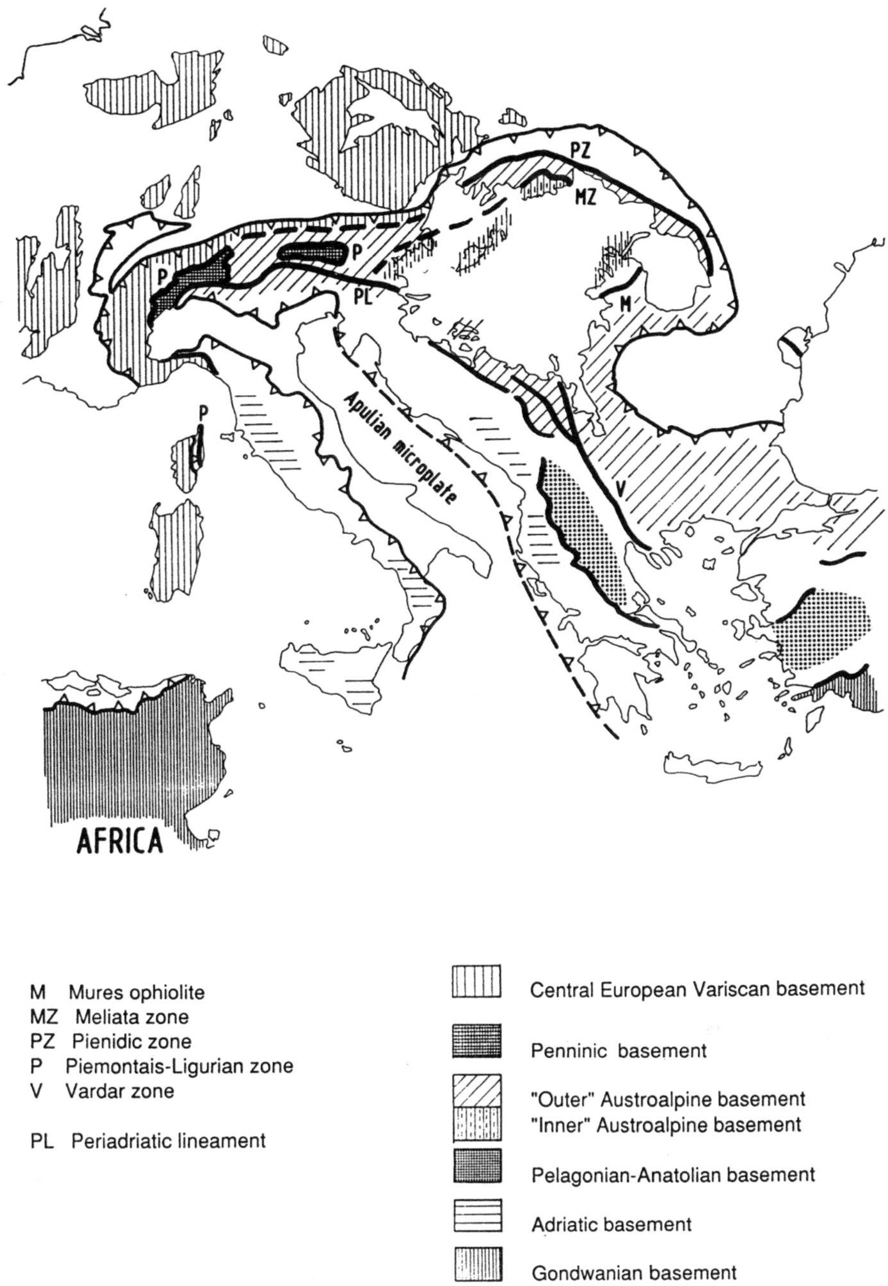

Fig. 2. Mesozoic and Cenozoic ophiolitic sutures in the Alpine Mediterranean mountain belts. These sutures separate basement units with distinct evolution

Tatric units of western Carpathians ("Outer Austro-Alpine basement units").

3. Other basement units are located inside of the suggested Cimmerian Meliata suture ("Inner Austro-Alpine basement units").
4. The Penninic basement is related to the European foreland from which it is, in part, separated by the Valais basin (Pfiffner this Vol.; de Graciansky this Vol.; Trümpy 1989).
5. External Alpine basement units correlate to the European foreland without any intervening Mesozoic ophiolite basin. The Helvetic unit of the Alps belongs to this group, and probably the Danubic unit in the southern Carpathians.

Basement zones with distinct features and geodynamic significance can cross from one Mesozoic/Tertiary microplate, as outlined, to another. From the location of Mesozoic/Tertiary suture zones a puzzle pattern results for pre-Mesozoic basement units which leaves many unresolved detailed relationships.

2 Ophiolites in Extra- and Intra-Alpine Variscides

Although some authors explain the European Variscides mainly as an intracontinental fold belt after the Devonian (e.g. Neugebauer 1988, 1990), many scattered dismembered and metamorphosed pre-Mesozoic ophiolites are part of the extra- and intra-Alpine Variscides (Belov et al. 1986; Neubauer et al. 1989; Bouchardon et al. 1989; Pin 1990). These ophiolites suggest separation of continental pieces by oceanic crust, and indicate major mobility during Palaeozoic evolution, as is also inferred from palaeomagnetic data (e.g. Perroud et al. 1984). Known pre-Mesozoic ophiolite complexes in the basement are listed in Table 1 and shown on Fig. 3.

For most ophiolites, the ages of protolith formation are uncertain (Table 1). However, geochronological data and inferred ages suggest that oceanic separation was restricted to certain times (Fig. 3, Table 1):

1. Late Proterozoic to Early Cambrian (e.g. northern Balkans: Haydoutov 1989; Alps: von Raumer and Neubauer this Vol.);
2. Ophiolites which apparently cluster around 500 Ma (Chamrousse: von Raumer et al. this Vol.; Ötztal: Söllner and Hansen 1987);
3. Ophiolites of supposed and confirmed Devonian age, respectively (e.g. Rakovec Group in Carpathians: Grecula 1987, 1990; Grecula and Hovorka 1987);
4. Ophiolites of early Dinantian age (e.g. Polish Sudetes: Pin 1990).

These groups give evidence of four major periods of extension, although most of these ophiolites bear a back-arc basin signature (for further discussion, see Bouchardon et al. 1989; Neubauer et al. 1989; Pin 1990) and therefore suggest a relationship to subduction zones.

3 Correlation of the Alpine Basement with East Mediterranean Mountain Belts

A series of maps contains the basic features of distinct time stages (Figs. 4, 5, 7–10, 12–14). We renounced the restoration of Alpine displacement of basement units in the Carpathians/Hellenides/Dinarides because all such models are under considerable discussion (e.g. Robertson and Dixon 1984; Dewey et al. 1989; Trümpy 1989; Stämpfli et al. 1991).

With respect to the metamorphic overprint, independent of the age of metamorphism, the Alpine-Mediterranean basement may be subdivided into an outer rim of highly metamorphic basement with poor control of chronostratigraphy, and into an inner zone of fossil-bearing, very low- to low-grade rocks in Alpine mountain belts which surround the Adriatic and the western Mediterranean Sea (Vai and Cocozza 1986; Cocozza et al. 1987; Flügel 1990). Moreover, Flügel (1990) distinguished three zones, based on Devonian and Carboniferous stratigraphy (Figs. 4, 6):

1. The Mediterranean Crystalline zone (MCZ);
2. The Noric-Bosnian zone (NBZ);
3. The Betic-Serbian zone (BSZ).

In the following, we use this division, introduce further subdivisons and add the Pelagonian-Anatolian zone (PAZ).

1. The **Mediterranean Crystalline zone** (MCZ) comprises mainly metamorphic rocks and huge masses of pre-Mesozoic granites. This zone is a result of predominantly Variscan metamorphism and magmatism although a strong Mesozoic/Tertiary metamorphic overprint is common. Pieces of pre-Variscan metamorphic crust are identified within large portions of this zone. The MCZ comprises the Helvetic, Penninic and Middle/Lower Austro-Alpine basement of Western and Eastern Alps, the outer portion of internal Carpathians, i.e. the Veporic and Tatric units, the Zemplinicum and Marmarosh Massif of Eastern Carpathians (Vozarova 1989; Rudakov 1990), the basement units of Southern Carpathians, especially the Danubian, Getic, and Supragetic basement units (Kräutner 1989; Nastaseanu and Kräutner 1990). Further south, the Serbomacedonian and Rhodope Massifs belong to this unit, followed by the Karakaya microcontinent in Turkey (Sengör et al. 1988). The MCZ is a composite of various plutono-metamorphic complexes of distinct origin and time of formation. Nearly all underwent Variscan metamorphism during Carboniferous post-dating sometimes such of Silurian/Devonian or also older metamorphic events. Therefore, we suggest that the MCZ basically include a mid-Palaeozoic active continental margin system.

The MCZ is separated from the NBZ in the Eastern Alps and Carpathians by the **Noetsch-Veitsch-Ochtina zone** (Flügel 1990; North Gemeric-Veitsch zone after Neubauer and Vozarova 1990). This zone forms a widely traceable Alpine thrust sheet. It contains Early Carboniferous siliciclastic shallow water deposits which are followed by reef limestones and Late Carboniferous terrestrial conglomerates. The relationship indicates an overall molasse-type setting which started earlier than in all other basement units. The primary basement is uncertain. The presence of

Fig. 3. Distribution of Pre-Mesozoic ophiolites in extra- and intra-Alpine Variscides (further information, see Table 1)

Table 1. Pre-Mesozoic ophiolites of extra-Alpine Variscides and Mediterranean mountain belts

Name	Protolith age	Age of tectonic emplacement	Reference
A Basement in the Alpine-Mediterranean belt			
Stubach complex	ca. 650–500 Ma	Carboniferous	Neubauer et al. (1989)
Ritting complex	>520 Ma	520 Ma	Neubauer et al. (1989)
Chamrousse complex	ca. 500 Ma	Early Carboniferous	von Raumer et al. (this Vol.)
Speik complex	pre-Silurian (?)	Early Carboniferous	Neubauer et al. (1989)
Rakovec Group	Devonian (?)	Devonian (?)	Grecula (1987)
Plankogel complex	Devonian (?)	Carboniferous (?)	Neubauer et al. (1989)
Balkan	Late Proterozoic	Early Palaeozoic	Haydoutov (1989)
Volvi complex	pre-Permian	?	Robertson and Dixon (1984)
B Selected ophiolite-like complexes in extra-Alpine, central southern Variscides			
Letovice Fm.	pre-Carboniferous	Carboniferous	Schulmann et al. (1991)
Marianske Lasne c.	pre-Carboniferous	Carboniferous	Franke (1989 a), Pin (1990)
Baie d'Audierne	Late Proterozoic	Carboniferous	Bouchardon et al. (1989)
Camptoceaux complex	Precambrian or Ordovician/Silurian	Silurian	Bouchardon et al. (1989)
Leptinite-amphibolite and related complexes	Ordovician	Silurian	Bouchardon et al. (1989) Pin (1990)
Gießen complex	Middle Devonian	Early Carboniferous	Franke (1989 a)
Polish Sudetes	Early Carboniferous	Early Carboniferous	Pin (1990)
Jesenik/Sobotin	Devonian (?)	Carboniferous	Pin (1990)

Devonian crystalline basement slices in close relationship with the Noetsch-Veitsch-Ochtina zone in the Eastern Alps argues for a pre-Carboniferous metamorphic unit pre-dating age of deposition (Frisch and Neubauer 1989; Neubauer and Frisch this Vol.). Comparable, undated metamorphic units, the Rakovec and Klatov groups, are found in Western Carpathians (Vozarova and Vozar 1987, 1988; Neubauer and Vozarova 1990). Therefore, we interpret the Noetsch-Veitsch-Ochtina zone as (southern) extension of the MCZ.

The North Gemeric-Veitsch zone is interpreted by Neubauer (1988), Vozarova and Vozar (1987, 1988), and Neubauer and Vozarova (1990) as an overstepping successor basin amalgamating two major Variscan microplates.

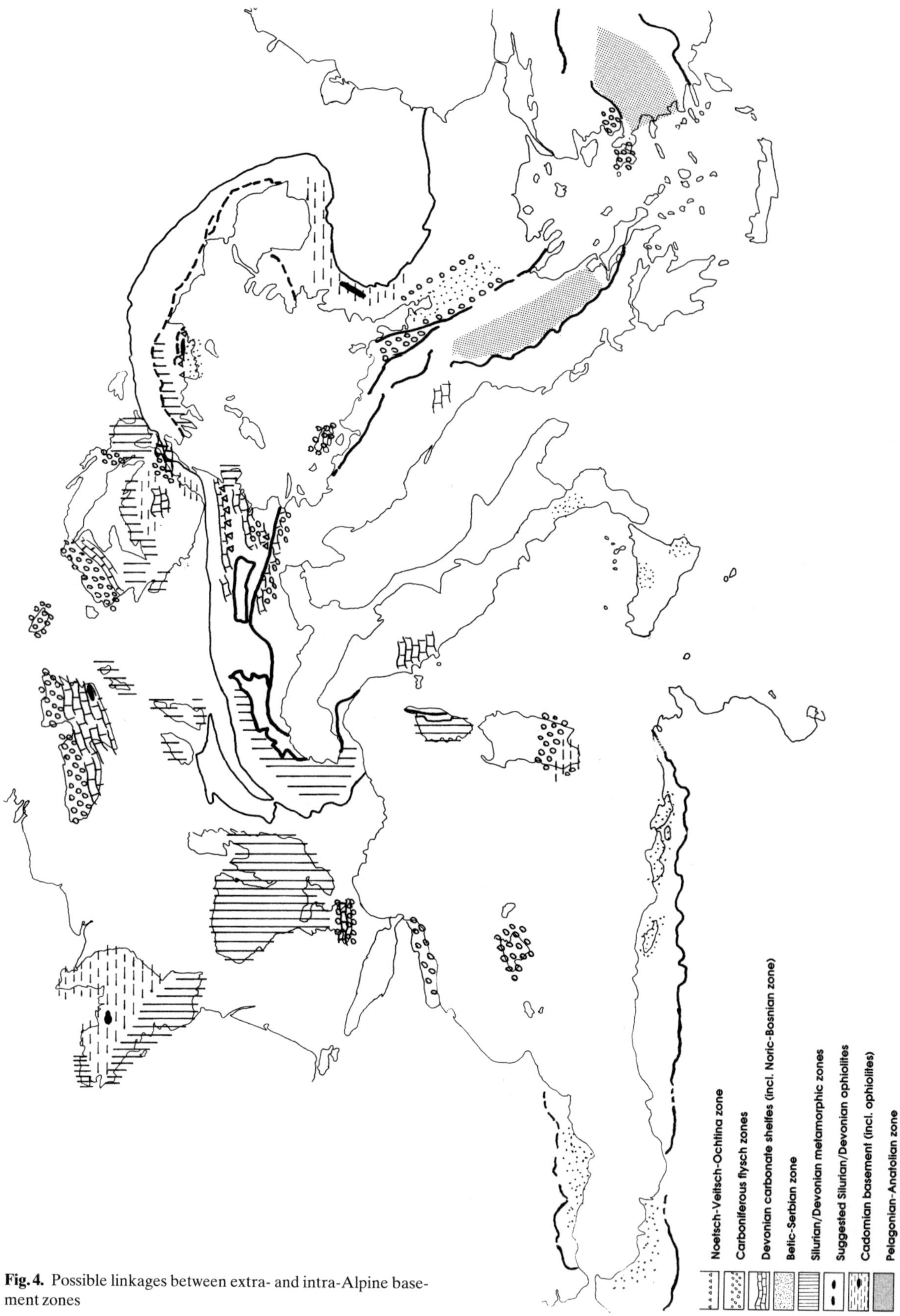

Fig. 4. Possible linkages between extra- and intra-Alpine basement zones

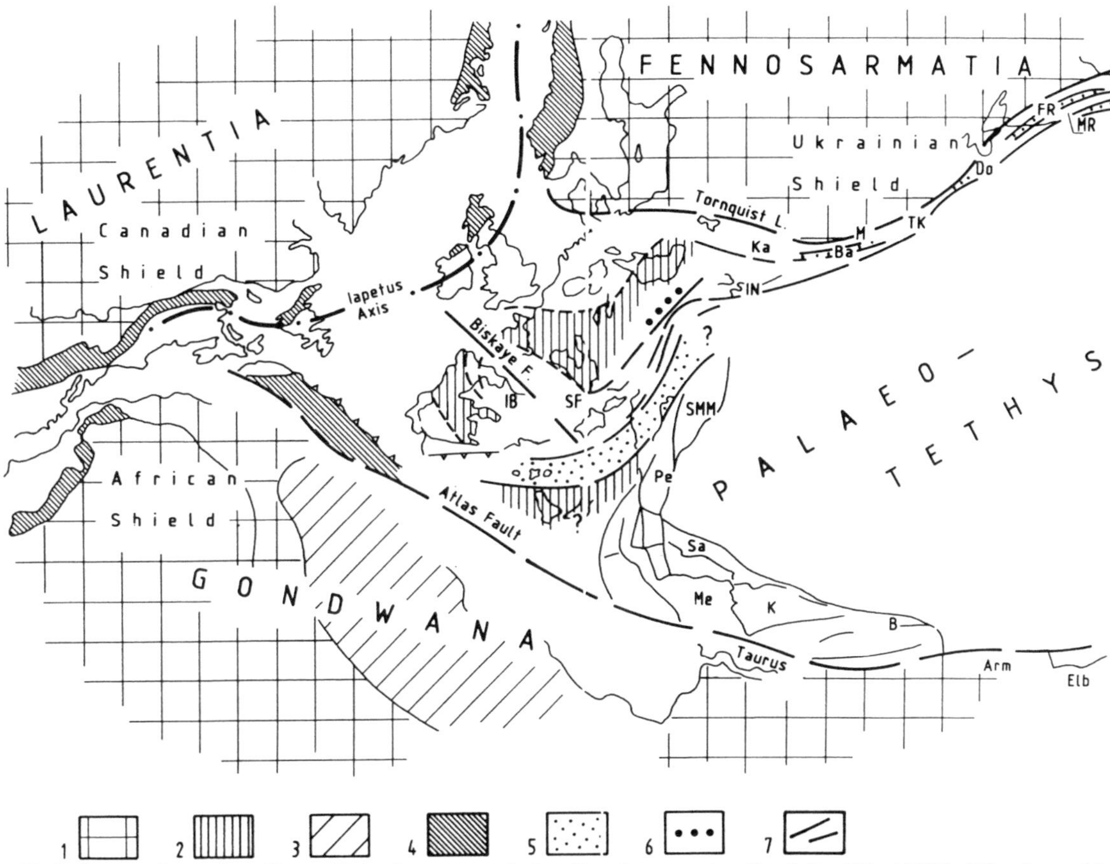

Fig. 5. Proposed zonation of intra-Alpine basement units in a Permian reconstruction (after Flügel 1990). *1* Gondwana (*Arm* Armenia; *Elb* Elburs); *2* metamorphic zones; *3* marine Carboniferous on Gondwana; *4* circum-Atlantic pre-Mesozoic zones; *5* Betic-Serbian zone; *6* Nötsch-Ochtina zone (Carboniferous); *7* Mediterranean zones (*IB* Iberia; *SF* southern France; Fennosarmatian Pacific-type margin with: *IN* Istanbul nappe; *Ka* Eastern/Southern Carpathians; *B* Balkanides; *Do* northern Dobrogea; *M* Moesian platform; *Tk* trans-Caucasus; Caucasus: *Fr* Forerange zone; *Mr* Mainrange zone; *SV* Svaneti zone, Pelagonian-Anatolian plate: *SMM* Serbo-Macedonian Massif (?); *Pe* Pelagonian Massif; *Sa* Sakarya continent; *Me* Menderes-Cyclades Massif; *K* Kirsehir Massif; *B* Bitlis Massif)

2. The **Noric-Bosnian zone** (NBZ) is defined by predominantly Devonian shallow water carbonates (Flügel 1990) including pelagic limestones which may have been deposited outside of a reef belt. The carbonates are overlain by flysch which was apparently propagating against the Adriatic Sea during Carboniferous time (Vai and Cocozza 1986; Cocozza et al. 1987; Flügel 1990).

According to pre-Devonian stratigraphy, the Noric-Bosnian zone may be subdivided into three subzones (see also Fig. 6):

a. In the first subzone, there is nearly continuous sedimentation since Cambrian times. It bears Cambrian archaeocyatha limestones. The only occurrences within the Alpine belts, of the area under consideration, are those of western Macedonia (Ramovs et al. 1990), although detailed data are missing from this region, and of the Taurides of southern Turkey (Demirtasli 1990). Archaeocyatha limestones have a close relationship to Sardinia, Montagne Noire, Cantabrian Mountains and northern Armorica in western Europe (see also Fig. 8). Similar sequences are present in the subsurface below the Aquitanian basin (Paris and Robardet 1990; Robardet et al. 1990). In the Armorican Massif, these Cambrian limestones are deposited on a Cadomian basement. The Carnic Alps and Karawanken of the Southern Alps do not include prelate Ordovician rocks, basically the Cambrian limestones. The basement of all these areas is unknown or questionable. A nearby Cadomian basement recently has been suggested by a 600 Ma age of detrital muscovites in Late Ordovician sandstones of the Southern Alps (Dallmeyer and Neubauer 1991).

b. In the next group the Cambrian archaeocyatha limestone is lacking. Sedimentation started mainly in the Early Ordovician. The low-grade metamorphic units of the Eastern Alps which have little pieces of a Cambrian, 550–500 Ma old basement at the base

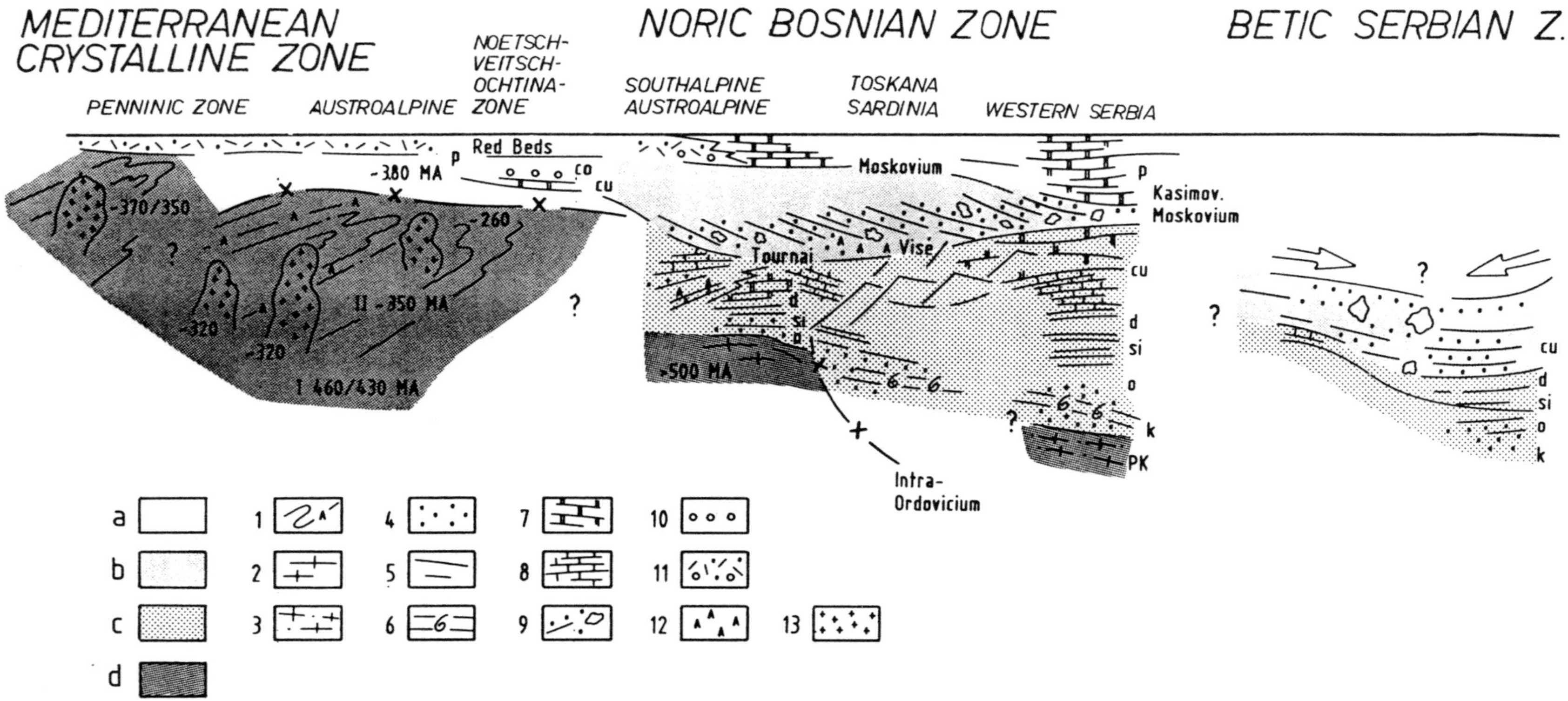

Fig. 6. Schematic reconstruction (section) of the NW-Mediterranean Late Palaeozoic shelf and its basement (after Flügel 1990). *a* Post-Variscan overstep sequences; *b* Variscan flysch sequences; *c* Early Palaeozoic to Late Devonian/Tournaisean passive continental margin deposits; *d* metamorphic basement zones. In detail: *1* Variscan metamorphic zones; *2* Proterozoic metamorphic basement; *3* pre-mid-Ordovician metamorphic basement; *4* clastic sediments; *5* pelitic sediments; *6* archaeocyatha limestone; *7* shallow water limestones; *8* pelagic limestones; *9* turbidites; *10* molasse and deltaic deposits; *11* red beds (with fluvial deposits); *12* volcanics; *13* plutonic rocks

belong to that zone (see also, von Raumer and Neubauer this Vol.). Therefore, sedimention started much later than in the first zone which includes Cambrian limestones. However, primary relationships between both units remain unclear.

Both units, those with and those without Cambrian limestones, contain Late Ordovician to Devonian marine shallow water carbonate deposits which grade into pelagic limestones (Spassov 1987; Ebner et al. 1990; Grecula et al. 1990; Kräutner 1990; Tenchov 1990). The widespread occurrence of coeval formations of similar depositional setting, like Late Silurian orthoceras limestones and graptolite shales, the Early Devonian pelagic limestones, Middle Devonian reef-like deposits, indicates deposition within the same basin. The Middle Devonian sequences argue for a wide oceanic basin with contourites, and pelagic sediments along margins due to oceanic currents (Oczlon 1990).

c. The **Betic-Serbian zone** (BSZ) lacks thick Devonian carbonate sequences; it mainly contains Cambrian to Devonian sandstones, slates, and flysch with thick olistolithic sequences from Tournaisean up to the Moscovian, respectively (Fig. 6). The Betic-Serbian zone comprises sequences of the lower thrust sheet of Chios island in Greece, the Drina and Prakac Palaeozoic units of Dinarides, the Palaezoic southern Gemeric units of Western Carpathians (all located in eastern Mediterranean mountain belts; see Flügel 1990; Grecula et al. 1990; Krstic and Maslarevic 1990), the Betic cordillera, Menorca, Rif and Kabylie of northern Africa and the Calabrian arc in the western Mediterranean (Bourrouilh et al. 1980; Buchroithner et al. 1980; Chalouan 1987; Herbig and Stattegger 1988). In the western Mediterranean, this zone is identical with the Alcabeca microplate which escaped to the west during Tertiary time and caused deformation in Alpine belts around the western Mediterranean Sea (Durand-Delga and Fontboté 1980). The Carboniferous flysch deposits of the western Mediterranean contain detritus of Silurian to Carboniferous carbonatic shallow water deposits (Buchroithner et al. 1980; Herbig and Stattegger 1989).

4. We propose a further zone with mainly metamorphic rocks, the **Pelagonian-Anatolian zone** (PAZ), corresponding to the Pelagonian-Anatolian terrane of Flügel (1990), which occurs in the interior of the former (MCZ) in the external Hellenides/Dinarides, and of western Turkey (Kröner and Sengör 1990), respectively. It is surrounded by Mesozoic ophiolites (Fig. 2). The PAZ comprises the Pelagonian-Thessalian Massif, the Afyon zone, the Sakarya Massif, and the Menderes zone (Jacobshagen 1986; Papanikolaou and Demirtasli 1987; Papanikolaou 1989). The Pelagonian-Thessalian Massif contains mainly thick micaschist in its lower part, and marbles of unknown age in the hanging wall part (Papanikolaou 1989). They are intruded by Late Carboniferous granites (Jacobshagen 1986). Whatever the time of formation, this zone forms part of the internal, plutono-metamorphic zone of the Variscan orogen. The original relationships of SBZ and NBZ are uncertain.

From this zonation we conclude that SBZ and the subzone of NBZ with Cambrian limestones are possibly related to an epicontinental sea or oceanic seaway which opened during the Cambrian. This seaway remained open up to the Early Carboniferous. On the other hand, stratigraphy calls for a Precambrian basement underlying this zone. In contrast, major portions of the NBZ apparently have an Early Palaeozoic metamorphic basement, and the transition from rifting to drifting occurred during Early and Middle Devonian. These relationships suggests that we have two distinct, passive, continental margins which are apparently unrelated together by time of formation.

4 Correlation of the Alpine Basement with Central and Western Europe

In this section, we discuss possible relations of the Alpine basement with the Variscan basement to the north, west and southwest of the Alps, which is exposed in the Bohemian Massif, Black Forest, Vosges, from the French Massif Central and the Massif des Maures and from Sardinia and Corsica (Figs. 1, 4).

The **Bohemian Massif** is composed of four major tectonostratigraphic units, the Moravo-Silesian zone to the east, Drosendorf and Gföhl units, both part of the Moldanubian zone, and the Tepla-Barrandian zone to the north (Fuchs and Matura 1976; Franke 1989 b; Matte et al. 1990). The southeastern Bohemian Massif may be explained by the Variscan dextral transpressive closure of an intra-Moldanubian oceanic realm exposed in the Letovice formation of the Svratka dome in Bohemia (e.g. Schulmann 1990; Schulmann et al. 1991; Dallmeyer et al. 1992; Fritz and Neubauer 1993) which resulted from pushing by a palaeo-Alpine block. The Moravo-Silesian zone contains a Cadomian basement (Breemen et al. 1982; Finger et al. 1989) which links West European Cadomia with intra-Alpine Cadomian crustal elements present in the Austro-Alpine (von Raumer and Neubauer this Vol.) as well as with coeval granites (Grünenfelder et al. 1983), back-arc ophiolite and the

Cambrian island arc sequences of this region of the Balkan/Southern Carpathians, respectively (Haydoutov 1989). Moreover, the Moldanubian Drosendorf unit also contains Cadomian protoliths (Gebauer and Grünenfelder 1982; Frank et al. 1990) suggesting a wide distribution of Cadomian units.

Other Moldanubian units, mainly granulites apparently yielded intra-Ordovician protolith and metamorphic ages (Arnold and Scharbert 1973; Frank et al. 1990).

Recent investigations argue for a Variscan age of granulite metamorphism and for protoliths of ultramafic rocks and their uprise and final emplacement together with granulite nappes (Carswell and Jamtveit 1990; Carswell 1991). Granulites and associated ultramafic rocks with similar geochronologic ages occur in the Ulten zone of the Eastern Alps which is part of a zone traceable from the Ivrea zone in Southern Alps via the Tonale-Ulten zone to the southeasternmost Eastern Alps (Frisch et al. 1990). Furthermore, comparable units with ultramafic rocks occur in the internal Carpathians (Hovorka 1985) and Dinarides (Pamic 1989a). On the other hand, the Ivrea zone (Voshage et al. 1990) is compared with coeval granulites of Calabria which is joined via granulite-like rocks in northern Sardinia to the Ivrea zone (e.g. Schenk 1981; Vai and Cocozza 1986). However, the granulite-serpentinite association does not argue for a primary relationship and zonation but rather for a similar coeval deep-crustal tectonometamorphic process and subsequent crustal-scale thrusting which exposed these rocks in high structural levels.

The Moldanubian zone of the Bohemian Massif extends to the **Black Forest, Vosges** and northern **Massif Central** to the west. Boreholes display that the basement of the Alpine external massifs and the northern foreland hidden under Mesozoic and Caenozoic sediments is continuous. The Black Forest is separated only by a trough of Carboniferous-Permian detrital sediments from Alpine external massifs (Matter 1989). The comparative discussion leads to the conclusion (von Raumer and Menot 1989; von Raumer et al. this Vol.) that the basement areas of the Black Forest and the external massifs belong to the same crustal type having suffered very similar geologic histories that are characteristic for the Moldanubian zone of Kossmat (1927) comprising Vosges, Black Forest and the Moldanubian tectonic units of the Bohemian Massif.

The northern Massif Central, Vosges and Black Forest are separated into metamorphic subunits with separate tectonothermal histories by Late Devonian – Carboniferous basinal sedimentary sequences (e.g. Wickert 1988; Eisbacher et al. 1989) which are closely related to western Aiguilles Rouges (von Raumer et al. this Vol.). These sediments, mainly greywackes and shales, which are interlayered with calc-alkaline volcanics, are interpreted as an expression of subduction and dextral transpressive shortening during Early Carboniferous (Eisbacher et al. 1989; von Raumer et al. this Vol.).

The highly metamorphic units of all these massifs surrounding the Alps are composed of monotonous paragneisses which contain leptinite-amphibolite complexes (LACs), sometimes with eclogites and high-pressure granulites, all those rocks occurring in basal portions of crustal-scale thrust sheets (Matte 1986). The protolith age of LACs is in a few cases proved to be Early Ordovician (for a review, see Santallier et al. 1988; Bouchardon et al. 1989; Pin 1990) mostly interpreted as rift zones which sometimes prograded into continental separation because of the ophiolites associated with it.

Comparing the different occurrences of eclogites, Bouchardon et al. (1989) came to the conclusion that only those of southern Brittany were of true oceanic domains, whereas in the more eastern parts of the French Massif Central, basic rocks and eclogites represent one or several narrow tensional zones within thinned continental crust. Besides a few occurrences of large ophiolite bodies many highly metamorphosed and dismembered ophiolites are observed. Major portions of early Palaeozoic oceanic crust were possibly destroyed by subduction or obduction (Pin 1990).

Eclogites associated with LACs indicate deep burial and overriding by crustal contraction during the Late Ordovician, Silurian and Early Devonian assigned to the eastern continuation of the Ligerian Cordillera of the southern Armorican Massif. This evolutionary step is closely comparable to the evolution of the Helvetic basement in the external massifs.

In this western part, the N-S cross-section through the Massif Central has been interpreted as a collisional orogen (Matte 1986). Large-scale gneiss nappes (Ledru et al. 1989) contain eclogite lenses. Similar gneiss-eclogite units appear in the Black Forest, Vosges, but also in the Alpine external massifs (von Raumer and Menot 1989). Similar lithologies, namely monotonous paragneisses and associated keliphytic amphibolites/leptinites at the base of high-grade metamorphic thrust sheets are widespread in Western Carpathians (M. Janak pers. comm.). In this region, some Late Devonian mineral cooling ages (Cambel and Kral 1989) suggest intra-Devonian metamorphism as in the Massif Central.

Another common feature of these extra-Alpine basement massifs is the widespread Early Carboniferous granitoid magmatism which is common for the whole Moldanubian region from the Bohemian Massif to Sardinia. Huge masses of mainly coarse-grained granite has been emplaced in all these basement massifs. Recently, Finger and Steyrer (1990) interpreted the magmatism of the Bohemian Massif/Alps cross-section as the result of N-directed subduction of the inboard side of a Palaeo-Tethyan ocean after major Variscan plate movements. On the other hand, the magmatism is approximately coeval with continent-continent collision and shearing which occurred in all basement massifs (Black Forest: Eisbacher et al. 1989; Massif Central: Matte 1986). This relationship to the time of deformation suggests rather, collisional origin of granite magmatism than subduction although the granites often bear I-type geochemical signatures.

All these comparable lithologies, mainly the huge masses of Early and Late Carboniferous granites, argue for a close relationship between the extra-Alpine Moldanubian zone, the Corso-Sardinian block (Menot and Orsini 1990) and the Helvetic, Penninic and Lower Austro-Alpine basement of the Alps (Mercolli and Oberhänsli 1988; Finger and Steyrer 1990). Coeval granites are present in nearly all Alpine basement units, both MCZ and PAZ, (Cambel and Kral 1989; Sengör et al. 1988). Furthermore, such granitoids are common in the Southern Carpathians (Kräutner, 1987), the Dinarides (e.g. Pamic 1989b), the Rhodope/Serbomacedonian Massifs (Katerinopoulos and Marcopoulos 1987) and the Thessalian/Pelagonian Massifs (Jacobshagen 1986). On the other hand, scarce remnants of another, older granite belt are preserved along northern margin of Eastern Alps which separates the Bohemian Massif from the Eastern Alps. Frasl and Finger (1988) introduced the "Cetic Massif" which was probably derived from the southern margin of the European continental plate during Alpine plate collision. It consists of a quartzdiorite-tonalite-granodiorite I-type association which is preserved in almost unmetamorphosed blocks along thrust surfaces on the Alpine front (Frasl and Finger 1988). Preliminary Rb-Sr whole-rock and mineral ages indicate a pre-Carboniferous, probably Late Devonian age of formation and cooling of the "Cetic Massif" (Thöni pers. comm.).

The most obvious linkage between Alps and the French Massif Central is the apparent extension of the Visean flysch sequences from its southern edge in the Montagne Noire via Sardinia (Vai and Cocozza 1986; Carosi and Gattiglio 1990) into the Southern Alps (e.g. Franke and Engel 1988; Franke 1989b). It continues into the highest Austro-Alpine nappe complex north of the Periadriatic lineament (Neubauer and Herzog 1985), into possible occurrences in the subsurface of the Pannonian basin (Flügel 1990; Kovacs and Lelkes-Felvari 1990), widespread Visean flysch occurrences of the Dinarides and Hellenides (Ramovs et al. 1987, 1990; Stattegger 1989; Flügel 1990) and Pontides (Demirtasli 1990). All these flysch sequences may be part of the same basin, basically of the Noric-Bosnian zone and of the Betic-Serbian zone according to Flügel (1990), and also of the Catalonian Range (Julivert and Duran 1990; Julivert et al. 1990) in the western Mediterranean. This Carboniferous flysch belt forms the southern, external limit of Variscan crust near Gondwana. We interpret the formation of this zone with rapid subsidence of the Gondwanian lithosphere by loading by incoming thrust sheets of the MCZ and its extension in highly metamorphic thrust sheets of the Massif Central (Matte 1986; Ledru et al. 1989).

5 Evolution of Mid-European, Extra- and Intra-Alpine Variscides

The French Massif Central contains units similar to the Alpine basement and may serve, therefore, as the base for projection of these zones into the easterly adjacent Alpine belt. This zonation includes the Variscan metamorphic crust on the northern part of the section, and the Carboniferous foredeep with flysch which may serve roughly as the southern edge. The data discussed already clearly indicate differences in the evolution path of the Alpine-Mediterranean basement units. The most obvious differences are related to the evolution of (1) Late Cambrian to mid-Ordovician; (2) to Silurian and Devonian; and (3) to Early to Mid-Carboniferous.

Although the history of Archaean and Early Proterozoic rocks is well-known by inherited zircons (e.g. Gebauer et al. 1988, 1989) the more precisely known history started with the Late Proterozoic (Fig. 7). Cadomian/Pan-African events of around 670 to 600 Ma are recorded in different parts of the region under consideration. The more pronounced evidence includes granitoid intrusions in southern Carpathians, Bohemian Massif, and metamorphism, recorded by Austro-Alpine migmatites and South-Alpine detrital sediments. The formation of a magmatic arc and back-arc basin crust in the Penninic zone of the Eastern Alps and of ophiolites in the Balkan – Southern Carpathians is apparently coeval. These data indicate the operation of subduction systems and collisional-type orogeny happened in certain areas of extra- and intra-Alpine Variscides. Otherwise, the Late Proterozoic mobile zone suggest a close relationship to the

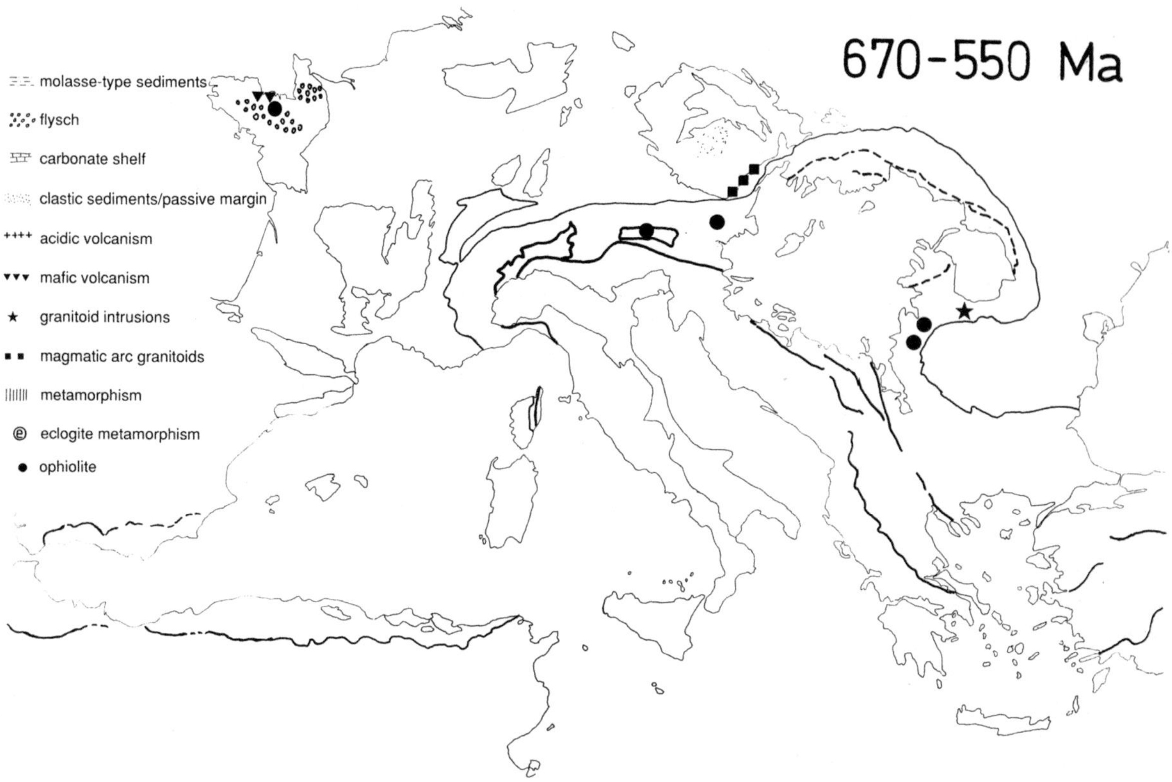

Fig. 7. Late Proterozoic to Early Cambrian geodynamic settings in extra- and intra-Alpine Variscides

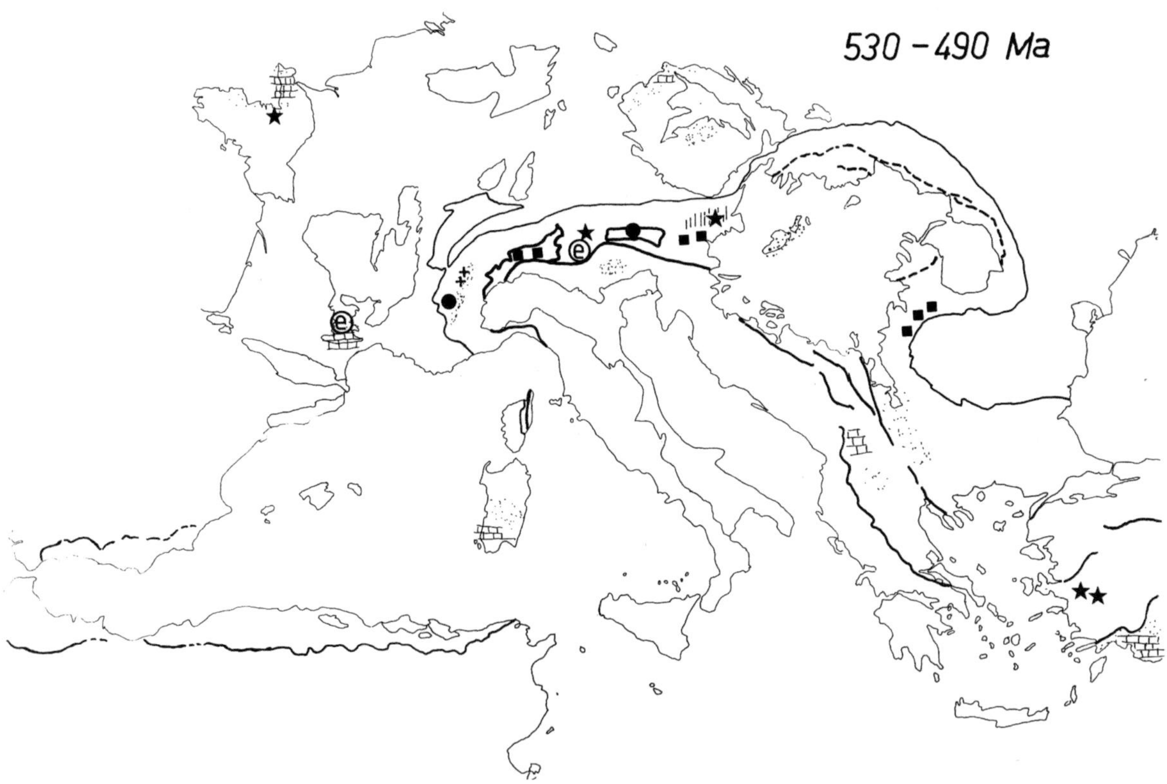

Fig. 8. Late Cambrian to Early Ordovician geodynamic settings in extra- and intra-Alpine Variscides. For explanations to symbols, see Fig. 7

Cadomian domain in Western Europe in which similar accretion history can be observed between 670 and 500 Ma (Strachan et al. 1991).

The Mid-Cambrian to Mid-Ordovician evolution step (Fig. 8) includes both sedimentation in epicontinental/passive continental margin settings (both SBZ and NBZ) and further operation of subduction systems. The epicontinental sequences contain archaeocyatha limestones and other shallow water deposits often overstepping Cadomian crust which has been stabilized before. Most authors agree that this epicontinental sea was formed along the northern margin of Africa (Bourrouilh et al. 1980).

Other zones of the MCZ, especially the Penninic, Helvetic, and Balkan units belong to a subduction system with subduction-related magmatic complexes and back-arc ophiolites during the mid-Cambrian to Ordovician period (Fig. 8). Basically, some of these complexes display a continuous evolution from Late Proterozoic complexes. However, the subduction system is not compatible with the epicontinental sedimentary sequences which include Cambrian archaeocyatha limestones and represent, therefore, another terrane. Major portions of this subduction system may have been accreted at about the Cambrian/Ordovician boundary giving rise to the intra-Ordovician tectonothermal events. Early to mid-Ordovician high-pressure events have been reported from the Helvetic and western Austro-Alpine units (Gebauer et al. 1988; Söllner and Hansen 1987; von Raumer et al. this Vol.). Otherwise, oceanic crust, such as the Chamrousse ophiolite, has been formed at the Cambrian/Ordovician boundary which possibly remained open up to the Carboniferous (von Raumer et al. this Vol.).

However, rift processes proceeded again during the Middle and Late Ordovician forming new sedimentary basins with alkaline rift magmatism either by formation and widening of back-arc basins or by intracontinental rifting (Fig. 9). Basically, this zone separates the Armorican shelf from the MCZ. Passive continental margins have been formed on both sides of a presumed sea. Remnants of the northern margin are preserved in the leptynite-amphibolite complexes of the Moldanubian region (Massif Central up to Bohemian Massif, especially to eastern Sudetes). The role of large Late Ordovician porphyric granite bodies and widespread coeval acidic volcanism is not clear (Söllner et al. 1982). Geochemical data indicate a collisional-type setting rather than subduction. Another explanation argues for the formation of magmas by back-arc extension (Neubauer and Frisch 1988; Loeschke 1989).

This northern margin is converted to compression with subduction including eclogite metamorphism

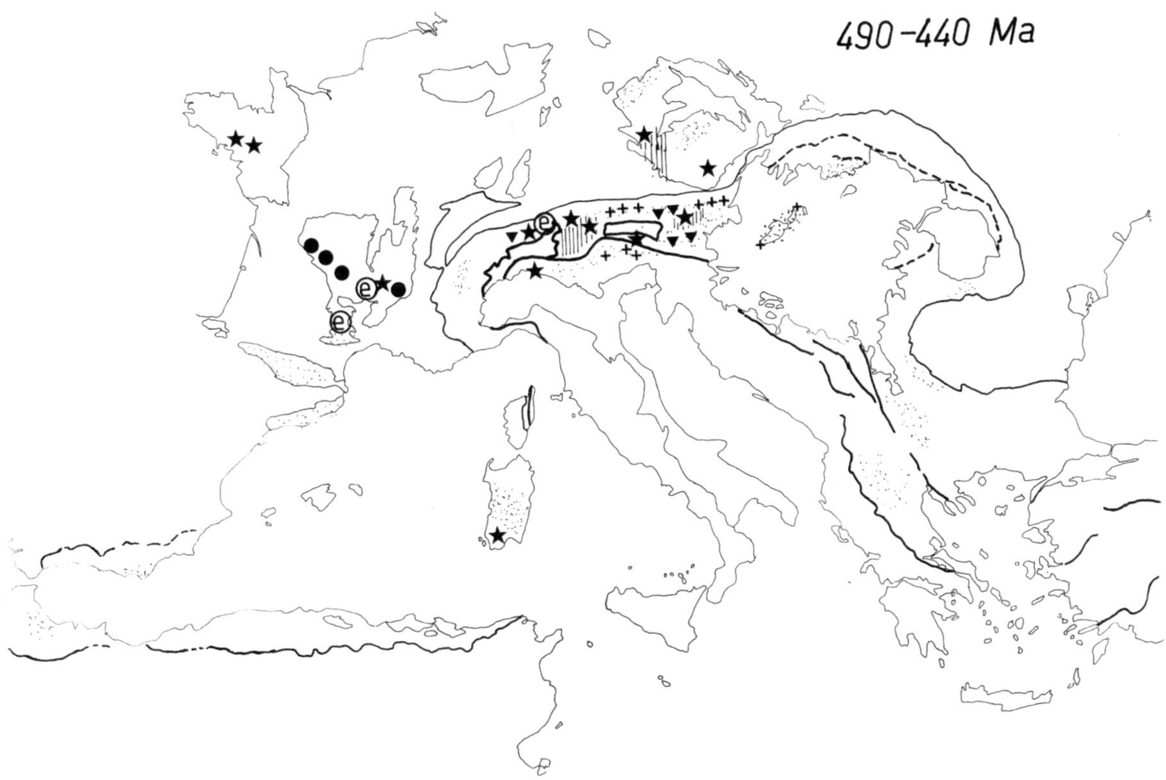

Fig. 9. Middle to Late Ordovician geodynamic settings in extra- and intra-Alpine Variscides. For explanation to symbols, see Fig. 7

during Late Ordovician, Silurian and Early Devonian (Figs. 10a; 11); this is mainly evidenced by geochronological data but not by the record of sediments. This zone is an extension of the Ligerian Cordillera of the southern Armorican Massif and the French Massif Central of the West European Variscides. Most authors agree with northwards directed subduction and associated crustal stacking (Matte 1986; Ledru et al. 1989) suggesting an open ocean to the south of this zone during the Silurian and Devonian.

In contrast, we recognize continuous subsidence of the intra-Alpine Noric-Bosnian zone and the Serbian-Betic zone with Silurian alkaline rift magmatism, deepening and facies variation of basins and formation of a carbonate platform during the Devonian (Fig. 10b). The carbonate platform includes barrier-type reefs (Vai and Spalletta 1982), pelagic swells as well as indications of far-reaching faunal exchange and possible bottom currents in a wide ocean (Oczlon 1990). Therefore, we suggest a Silurian to Devonian separation of this passive margin from the Ligerian Cordillera and its Alpine extension by an intervening oceanic realm which might have been situated on the southern margin of that oceanic seaway (see also von Raumer and Neubauer this Vol.). Ophiolites of an appropriate age have not been proved up to now. The Rakovec group of the Carpathians might be a member of that oceanic crust.

Early Palaeozoic faunal and floral relationships remain unclear and controversial. Early Devonian faunas of the Southern Alps display some relationship to the Uralo-Tienshan faunas (Rau and Tongiorgi 1981; Schönlaub this Vol.). On the other hand, faunas and floras of the Austro-Alpine zone, especially those of the Palaeozoic of Graz, are related to the Rhenohercynian belt along the northern margin of Variscides (e.g. Hubmann 1990). These arguments may indicate that both domains belong to different parts of the same passive continental margin.

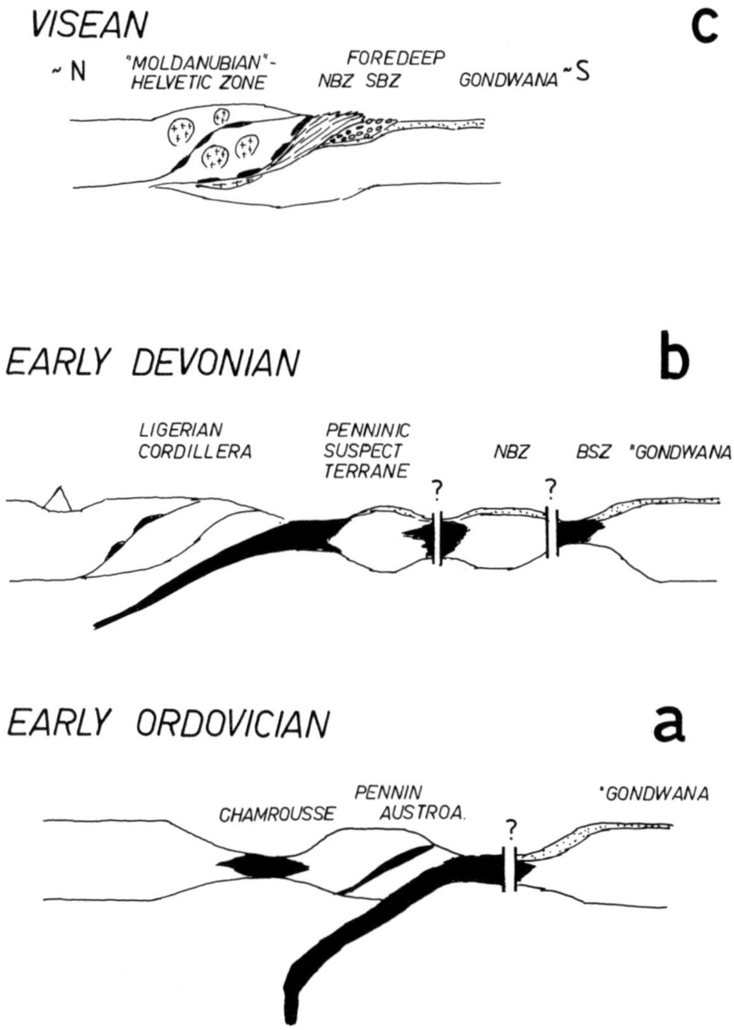

Fig. 10. Schematic sketches (sections) of some important evolution stages of intra- and extra-Alpine Variscides. **a** (Late Cambrian to) Early Ordovician stage: rifting, back-arc spreading; **b** (end of) Early Devonian: subduction systems, and passive margin systems both separated by a south European oceanic realm; **c** Visean: plate collision, formation of a foredeep and thrusting. The location of *SBZ* and *NBZ* in relation to "Gondwana" in **a** and **b** is uncertain

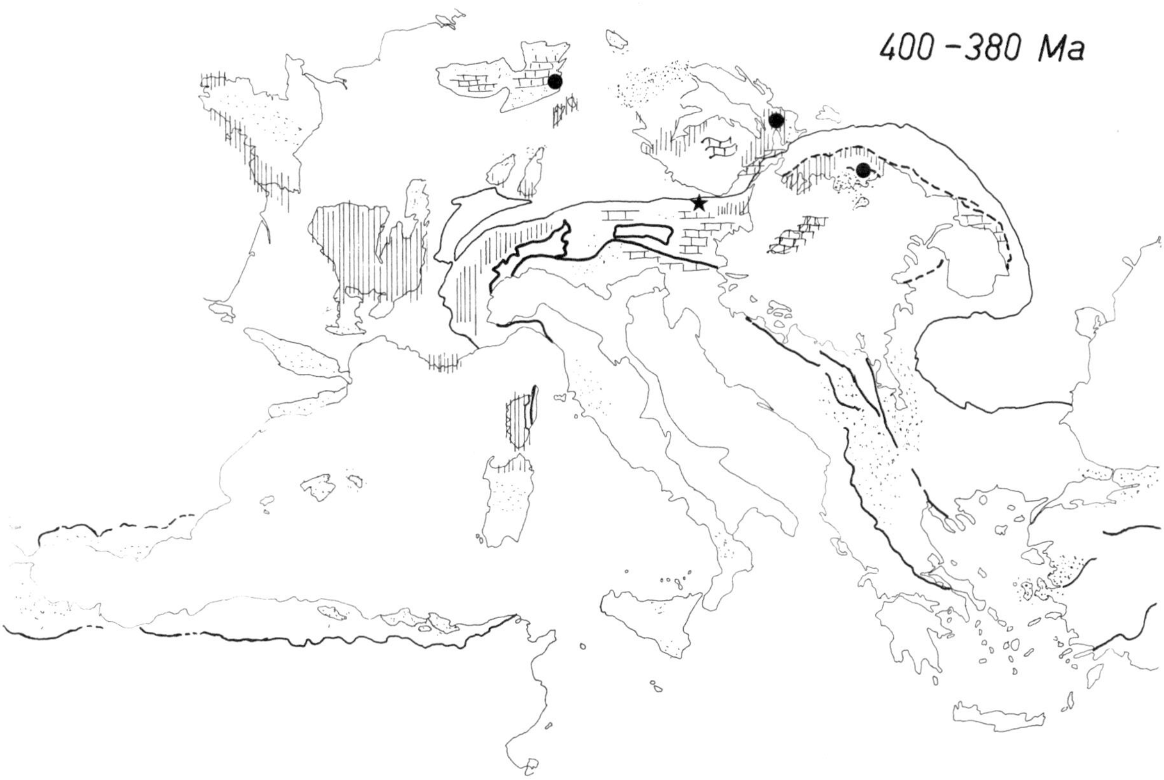

Fig. 11. Early to Middle Devonian geodynamic settings in extra- and intra-Alpine Variscides. For explanation to symbols, see Fig. 7

Extension of the passive continental margin is converted to contraction during Early Carboniferous (Figs. 10c, 12). The most prominent feature is the rapid subsidence of the NBZ carbonate platform during Visean time and the onset of flysch sedimentation. The flysch belt indicates a foredeep which has been formed by loading of the former passive continental margin by another continental plate and sedimentation in front of incoming thrust sheets. In this way, the flysch belt is a key for locating the Variscan suture within the Mediterranean mountain belts. The flysch belt may be followed from the Cantabrian mountains via the Montagne Noire into the Eastern and Southern Alps and further through the inner (eastern) Dinarides up to Chios island and to northern Turkey, respectively. The other extension to the western Mediterranean is via Betic Cordillera, Rif to the Kabylie along the northern margin of the African craton. The flysch belt is always situated inside of, and to the south of the MCZ and the Moldanubian zone, respectively. This relationship indicates that orogenic polarity has generally been towards south or southeast (Fig. 6). The time of onset of flysch sedimentation is variable. In the Serbian-Betic zone, flysch sedimentation started at the Devonian/Carboniferous boundary and persisted up to the Moscovian stage, whereas the onset of the Noric-Bosnian flysch propagated from basal Visean to Moscovian. Carboniferous shallow water carbonates like reef and lagoonal limestones occur as clasts within olistostromes of the Betic flysch (Buchroithner et al. 1980). The origin of these clasts is uncertain. They are found nowhere else in nearby Carboniferous sequences. These clasts are presumed to be derived from a southern continent (Herbig pers. comm.). Metamorphic clasts might be derived from the MCZ. Ophiolitic detritus suggests the presence of obducted ophiolites in the source region of greywackes (Stattegger 1989).

Final closure and thrusting within the flysch belt occurred during Late Carboniferous in the western Mediterranean up to the Alps and probably in parts of Dinarides. On the other hand, in Dinarides and Hellenides a clear cut is not always obvious and flysch sedimentation may have persisted into Permian (Papanikolaou 1989). Some other Alpine-Mediterranean basement units lack the Carboniferous flysch and any convincing evidence for tectonothermal activities. These areas include the Taurides of southern Turkey which are part of the Variscan Turkey-Iran magmatic gap (Sengör et al. 1988).

The Moldanubian zone of the Variscides is connected with the MCZ which is also a zone of

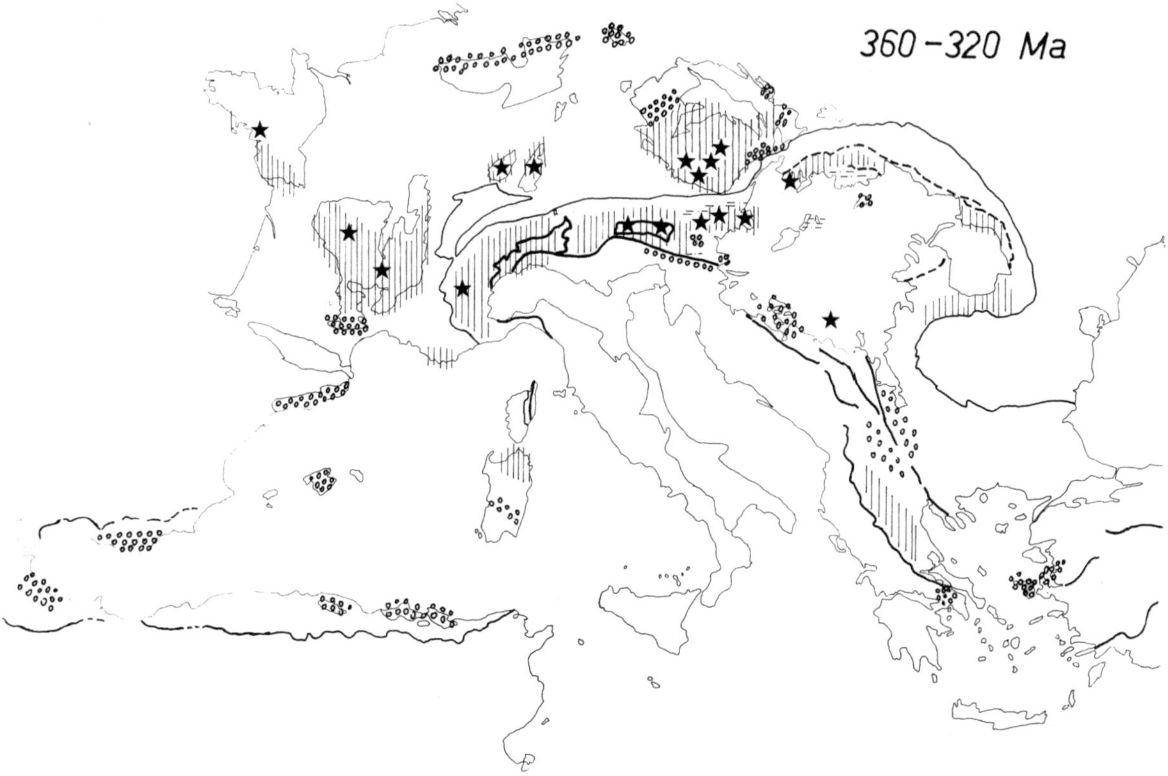

Fig. 12. Dinantian geodynamic settings in extra- and intra-Alpine Variscides. For explanation to symbols, see Fig. 7

Carboniferous metamorphism, deformation and granitoid intrusions, sometimes superimposed on portions of the Silurian/Devonian Ligerian Cordillera, sometimes on units not metamorphosed previously. Carboniferous metamorphism has been of a low-pressure type suggesting uprising crust and heating by large plutons. The Early Carboniferous plutons display an interesting feature with more S-type plutons in the external northern parts, and more calc-alkaline I-type plutons within the Alps. Whatever the reason may be, this pattern indicates variable sources, especially the presence of juvenile crust, possibly the Cadomian to Early Palaeozoic ophiolites and magmatic arc sequences in the Alpine region.

A major part of the seaway remained open (Dercourt et al. 1986). In the eastern Mediterranean, onlap of Tethyan sediments can be observed on top of deformed Variscan crust (Fig. 13). Towards the north, Variscan metamorphic crust is overlain by Late Carboniferous terrestrial coal-bearing molasse-like beds and Permian red beds (Nastaseanu 1987; Tenchov 1987; Vozarova and Vozar 1987, 1988; Paproth 1989) which often include calc-alkaline acidic and intermediate volcanics (Harkovska 1987; Stan 1987; Bonin 1988; Bonin et al. this Vol.). These volcanic sequences often are related to major plutons which are interpreted in terms of a basin-and-range-type model (Bonin et al. this Vol.).

6 Discussion

From the data discussed previously we conclude that the extra- and intra-Alpine Variscides are composed of tectonostratigraphic units which suffered distinct geodynamic evolution histories throughout Silurian and Devonian times. These major units are:

1. The Gondwanan margin of Africa and its detached pieces in Taurides, Menderes Massif, southern Sardinia and the Montagne Noire. These units may be part of a major microplate, Armorica, north of Africa. They possibly formed an old Cambrian passive continental margin which lasted through Ordovician to the Devonian and which was consumed in major portions during the Carboniferous.

The SBZ with mainly clastic Early Palaeozoic sequences may represent an outboard extension of the Gondwanan shelf with continental slope and rise deposits.

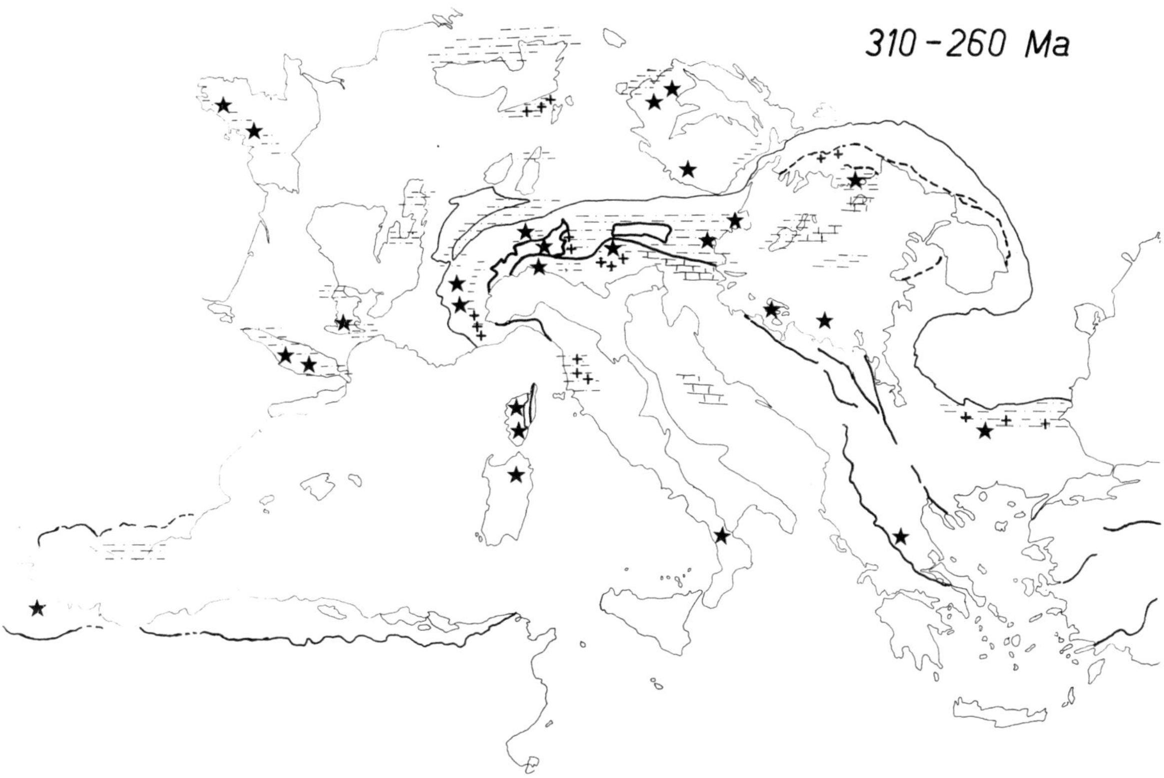

Fig. 13. Late Carboniferous to Early Permian geodynamic settings in extra- and intra-Alpine Variscides. For explanation to symbols, see Fig. 7

2. Major portions of the NBZ do not include deposits of Cambrian age. It was deposited on a crust which was most probably formed and stabilized during the Cambrian and Ordovician. In particular Early to mid-Ordovician metamorphic crust was formed, most probably during an orogenic event. Some arguments advocate this interpretation: (a) the presence of eclogites of Late Cambrian to mid-Ordovician metamorphic ages which are included in mid-Ordovician anatectic regions; (b) the widespread distribution of highly differentiated Late Ordovician granites which apparently have a collision-type chemical signature.

The apparent lack of orogenic flysch which is often advocated to reject the concept of Ordovician orogeny could have two sources: (a) restricted knowledge; there are thick metamorphic clastic sequences, especially monotonous paragneisses, of pre-Late Ordovician age all over the MCZ which often have characteristics of turbidities; (b) evidence for intra-Ordovician orogeny comes from internal, metamorphic areas of the supposed Ordovician orogen which does not include flysch deposits.

This NBZ shows evidence for Silurian to Devonian extension with widespread volcanic sequences during the Silurian and a carbonate platform formed during the Devonian. Although there is strong evidence for Devonian continental separation and drift there exist only a few appropriate ophiolites. The Plankogel complex of Eastern Alps and the Rakovec group are two possible Devonian ophiolites which could be related to this separation.

Although no reasonable palaeomagnetic data are available from this zone, a drift towords the equator is apparent from the lithological data. Ordovician sequences lack carbonates except Ashgillian crinoid limestones. Influence from Gondwanan (?) glaciation is advocated by Schönlaub (this Vol.). Limestone deposition increased from Silurian onwards with reefs and reef-related carbonates during Middle Devonian suggesting a near-equator position of this zone. Faunal, floral, and lithologic data argue for a different relationship for South-Alpine and Austro-Alpine Devonian. The Austro-Alpine basement contains many mature Early Devonian sandstones, and some red sandstones in Eifelian sequences which suggest relationship to the Old Red Continent. A similar relationship is supposed from reef floras and faunas (e.g. *Zdimir* sp., Udotaceae) which are closely related to those of the Rhenohercynian zone (Hubmann

1990). In contrast, the South-Alpine floras and faunas are related rather to the Ural and Tienshan although sandstone deposits are completely missing there.

Both, the SBZ and NBZ, and their extension in the West European Variscides form a large Carboniferous foredeep with subsidence due to downwarping lithosphere which was filled up mainly with greywackes. The axis of this flysch is prograding through time.

3. The Penninic basement units, composed of Late Proterozoic to Early Palaeozoic island arc sequences, of both the Western and Eastern Alps are different from the extra-Alpine Variscides. Therefore, we regard this zone as a suspect terrane which was accreted to Variscides during the Carboniferous orogeny. The next possible extension of these island arc belts is the Ossa-Morena zone of the southern Iberian Massif (Ribeiro et al. 1990) and the Avalon-Cadomian superterrane on a global scale (Murphy and Nance 1989).

4. We summarize all sequences with Late Ordovician/Silurian/Devonian metamorphism into one major zone although pieces of this zone have been widely dispersed by later processes. The essential reason for this is that this setting contrasts the evolution of the Gondwanan margin and of its northern extension. Basic features of this zone are high-pressure metamorphism followed by low-pressure metamorphism, early onset of plutonism during the Devonian, and basically south-directed emplacement of nappes. We relate these features to long-lasting Late Ordovician/Silurian/Devonian subduction and subsequent collision with the Gondwanan margin.

The contrasting Silurian/Devonian evolution histories of the Gondwanan margin and of the Ligerian Cordillera imply a major oceanic basin between both units. This ocean had been formed before the Silurian and matches, therefore, characteristics of several ophiolites and rift complexes which are close to this margin (Chamrousse, portions of the LACs, Speik complex etc.).

Final continent-continent collision occurred during the Visean. The collision is shown by (a) the formation of foredeep with flysch sedimentation which forms a large and broad belt along southern margin of the Variscides (Fig. 12) (b) deep-crustal thrusting in the Massif Central, Black Forest, and within the MCZ, (c) a broad belt of granitoids which forms the entire internal belt and the MCZ, (d) the narrow Visean zones which include subduction-related volcanic sequences (Vogesen/ Badenweil-Lenzkirch zone). The age of metamorphism in internal zones matches the time of burial of the external part of the NBZ/SBZ.

Collision occurred with a large transpressive component of displacement along the now, east-west trending Variscan belt. Besides the well-known Late Carboniferous to Early Permian E-W trending zones in Western Europe which are interpreted as part of an indentation system of Africa into the Iberian Armorican arc (e. g. Matte 1986), a further indentation is necessary to explain all structures around the present Alpine chain (Fig. 14). Basic arguments for this interpretation are:

1. The northward motion of the entire Bohemian Massif which is due to a straight southern boundary during earliest Permian (Handler et al. 1991) as well as of the basement block between Bohemian Massif and Black Forest (Weber 1984).

2. The dextral motion along the E-W trending Baden-

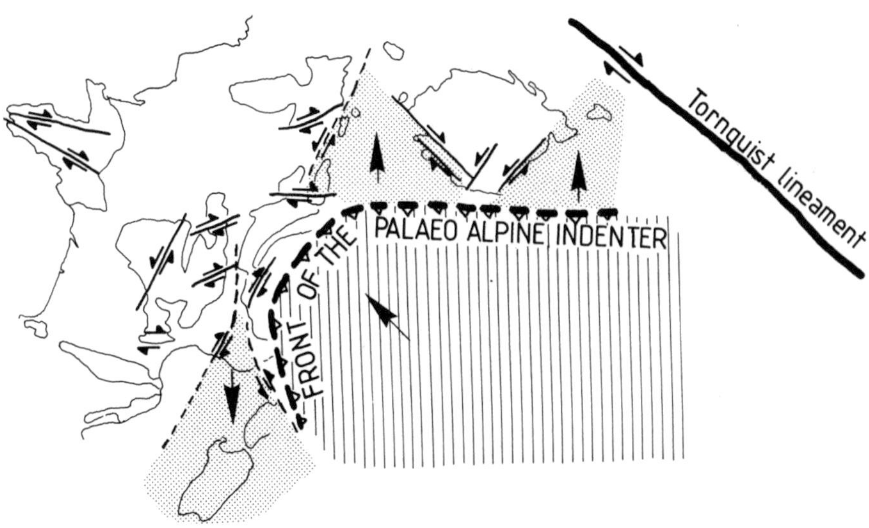

Fig. 14. Suggested model for Late Carboniferous to Early Permian wrench tectonics

weil-Lenzkirch zone during Late Carboniferous (Eisbacher et al. 1989).

3. The sinistral displacement along the N-S trending Sillon Houilliere within the Massif Central (Ledru et al. 1989).

4. A further, hypothetical sinistral wrench fault is necessary between the Massif Central and the Alpine external massif, Massif de Maures and Corsica to explain the southward displacement of the Corso-Sardinian/Massif des Maures block. Because of the predominance of Devonian metamorphism the latter resembles rather the northern, internal part of the Massif Central than its southern margin. Joye (1990; unpubl.) described dextral displacement along NE-SW trending shear zones within the external massifs separating the escaping block against the Alps to the east.

The operation of all these fault systems was accompanied by sedimentation in narrow intramontane basins which might have been formed basically along releasing bends of arcuate transcurrent faults (e.g. along Sillon Houilliere). The deposition of Permian sediments followed similar trends (Paproth 1989). Continental red bed deposits of central Europe are accompanied by a broad shelf along the southern margin of the Variscides which includes southern Hellenides, Sicily-Tunesia, Dinarides and Southern Alps. This shelf displays subsidence of Variscan metamorphic and deformed crust and may be interpreted as result of the formation of the Tethys on the far side of Africa during dextral megashearing between Laurussia and Gondwana (Matte 1986).

In summary, we can distinguish a major megacycle in evolution of extra- and intra-Alpine Variscides. The megacycle started with step-wise stabilization of a Cadomian belt which is preserved in major portions of extra- and intra-Alpine Variscides. Possible linkages are to Pan-African (Neubauer 1990) as well as to Baikalian events (Khain 1985). The megacycle started with subsidence and sedimentation following the early Cadomian orogenic event and resulted in two major zones: the first one includes early Palaeozoic passive continental settings with the "Gondwanan" margin, NBZ and SBZ, the second one the long-lasting Early to mid-Palaeozoic Pacific-type active continental margin settings of the South Armorican zone, Massif Central and MCZ. Such a setting would explain the fast and regular change of short-lived rifts to contractional zones in the later. Strike-slip motion along this margin would explain the incomplete record of several zones, geodynamic complexities, and the presence of suspect terranes.

Acknowledgements

The paper benefited considerably from discussions of the first author with coworkers within the IGCP No. 5 and 233 projects. Especially, we appreciate valuable suggestions and reviews, respectively, by H.W. Flügel, J. Friebe, H. Fritz (Graz), F. Ebner (Leoben), S. Kovacs (Budapest) and B. Krstic (Beograd). The paper is partially based on data achieved by work within a project supported by the Austrian Research Fund (project P7405–GEO).

References

Arkai P, Lelkes-Felvari G (1987) Very low- and low-grade metamorphic terrains in Hungary. In: Flügel HW, Sassi FP, Grecula P (eds) Pre-Variscan and Variscan events in the Alpine-Mediterranean mountain belts. Mineralia slovaca Monogr. Alfa, Bratislava, pp 51–68

Arnold A, Scharbert HG (1973) Rb-Sr-Altersbestimmungen an Granuliten der südlichen Böhmischen Masse in Österreich. Schweiz Mineral Petrogr Mitt 53: 61–78

Belov AA, Gatinsky YG, Mossakovsky AA (1986) A precise on pre-Alpine tectonic history of Tethyan Paleooceans. Tectonophysics 127: 197–211

Belov AA, Sl'Usar' BS, Lartchenkov EP (1987) Pre-Mesozoic formations of the north-west Black Sea region – stratigraphic correlation forms and tectonics. In: Sassi FP, Bourrouilh R (eds) IGCP 5 Newslett 7: 119–128

Bonin B (1988) From orogenic to anorogenic environments: evidence from associated magmatic episodes. Schweiz Mineral Petrogr Mitt 68: 301–311

Bouchardon JL, Santallier D, Briand B, Menot RP, Piboule M (1989) Eclogites in the French Palaeozoic orogen: geodynamic significance. Tectonophysics 169: 317–332

Bourrouilh R, Cocozza T, Demange M, Durand-Delga M, Gueirard S, Guitard G, Julivert M, Nartinez FJ, Massa D, Mitrouse R, Orsini JB (1980) Essai sur l'evolution paleogeographique, structurale et metamorphique du Paleozoique du Sud de la France et de l'Ouest de la Mediterranee. In: Cogné J, Slanski M (eds) Geology of Europe from Precambrian to post-Hercynian sedimentary basins. 26th IGC, Coll C6; BRGM, Villeneuve d'Asqu, pp 159–188

Breemen Ov, Aftalion M, Bowes RD, Dudek A, Misar Z, Povondra P, Vrana S (1982) Geochronological studies of the Bohemian Massif, Czechoslokaia and their significance in the evolution of central Europe. Trans R Soc Edinb Earth Sci 73: 89–108

Buchroithner MF, Flügel E, Flügel HW, Stattegger K (1980) Die Devongerölle des paläozoischen Flysch von Menorca und ihre stratigraphische Bedeutung. Neues Jahrb Geol Paläontol Abh 159: 172–224

Burchfiel BC (1980) Eastern Alpine system and the Carpathian orocline as an example of collision tectonics. Tectonophysics 63: 31–62

Cambel B, Kral J (1989) Isotopic geochronology of the western Carpathian crystalline complex the present state. Geol Carpath 40: 387–410

Carswell DA (1991) Variscan high P-T metamorphism and uplift history in the Moldanubian zone of the Bohemian Massif in lower Austria. Eur J Mineral 3: 323–342

Carswell DA, Jamtveit B (1990) Variscan Sm-Nd ages for the high-pressure metamorphism in the Moldanubian zone of the Bohemian Massif, lower Austria. Neues Jahrb Miner Abh 162: 69–78

Carosi R, Gattiglio M (1990) Stratigraphic correlation forms concerning the Sardinian Paleozoic basement (Italy). Rend Soc Geol Ital 12 (1989): 143–172

Chalouan A (1987) Paleozoic nappes of the Ghomarides (Internal Rif Mts., Morocco): review of stratigraphy, paleogeography, and Variscan structurations. In: Flügel HW, Sassi FP, Grecula P (eds) Pre-Variscan and Variscan events in the Alpine-Mediterranean mountain belts. Mineralia slovaca Monogr. Alfa, Bratislava, pp 107–134

Cocozza T, Decandia FA, Lazzarotto A, Pasini M, Vai GB (1987) The marine Carboniferous sequence in southern Tuscany: its bearing for Hercynian paleogeography and tectofacies. In: Flügel HW, Sassi FP, Grecula P (eds) Pre-Variscan and Variscan events in the Alpine-Mediterranean mountain belts. Mineralia slovaca Monogr. Alfa, Bratislava, pp 135–144

Dallmeyer RD, Neubauer F (1991) $^{40}Ar/^{39}Ar$ age of detrital muscovite, Carnic Alps: evidence for Cadomian basement in the Eastern Alps. Terra Abstr 3: 12–13

Dallmeyer RD, Neubauer F, Hoeck V (1992) $^{40}Ar/^{39}Ar$ age control on the chronology of Late Paleozoic tectonothermal activity in the southeastern Bohemian Massif, Austria (Moldanubian and Moravian zones). Tectonophysics 210

Demirtasli E (1990) Stratigraphic correlation form of Turkey. Rend Soc Geol Ital 12 (1989): 183–211

Dercourt J, Zonenshain L, Ricou L, Knipper A, Grandjaquet C, Sbortsshikov I, Geyssant J, Leprevier C, Pechersky D, Boulin J, Sibouet J, Savostin L, Sorokhtin O, Westphal M, Bazhenov M, Lauer J, Bijuduval B (1986) Geological evolution of the Tethys belt from the Atlantic to the Pamirs since the Lias. Tectonophysics 123: 241–315

Dewey JF, Helman Ml, Turco E, Hutton DHW, Knott SD (1989) Kinematics of the western Mediterranean. Geol Soc Spec Publ 45: 265–283

Durand-Delga M, Fontbote JM (1980) Le cadre structurale de la Mediterranee occidentale. Mem Bur Rech Geol Min 15: 67–85

Dvorak J (1973) Synsedimentary tectonics of the Palaeozoic of the Drahany uplands (Sudeticum, Moravia, Czechoslovakia). Tectonophysics 17: 359–391

Ebner F, Fenninger A, Gollner H, Holzer HL, Neubauer F, Nievoll J, Ratschbacher L, Stattegger K, Tschelaut W, Thalhammer O, Zier C (1990) Stratigraphic correlation forms of Paleozoic units in Austria. Rend Soc Geol Ital 12 (1989): 213–239

Eisbacher GH, Lüschen E, Wickert F (1989) Crustal-scale thrusting and extension in the Hercynian Schwarzwald and Vosges, central Europe. Tectonics 8: 1–21

Finger F, Steyrer HP (1990) I-type granitoids as indicators of a Late Paleozoic convergent ocean-continent margin along the southern flank of the central European Variscan orogen. Geology 18: 1207–1210

Finger F, Frasl G, Höck V, Steyrer HP (1989) The granitoids of the Moravian zone of northeast Austria: products of a Cadomian active continental margin. Precambrian Res 45: 235–245

Flügel HW (1990) Das voralpine Basement im Alpin-Mediterranen Belt – Überblick und Problematik. Jahrb Geol Bundesanst 133: 181–221

Flügel HW, Sassi FP, Grecula P (eds) (1987) Pre-Variscan and Variscan events in the Alpine-Mediterranean mountain belts. Mineralia slovaca Monogr. Alfa, Bratislava

Frank W, Scharbert S, Thöni M, Popp F, Hammer S (1990) Isotopengeologische Neuergebnisse zur Entwicklungsgeschichte der Böhmischen Masse. Proterozoische Gesteinsserien und variszische Hauptorogenese. Österr Beitr Meteorol Geophys 3: 185–228

Franke W (1989 a) Variscan plate tectonics in central Europe – current ideas and open questions. Tectonophysics 169: 221–228

Franke W (1989 b) Tectonostratigraphic units in the Variscan belt of central Europe. Geol Soc Am Spec Pap 230: 67–90

Franke W, Engel W (1988) Tectonic settings of synorogenic sedimentation in the Variscan Belt of Europe. In: Bessly BM, Kelling G (eds) Sedimentation in a synorogenic basin complex. Blackie, Glasgow, pp 8–17

Frasl G, Finger F (1988) The "Cetic Massif" below the Eastern Alps – characterized by its granitoids. Schweiz Mineral Petrogr Mitt 68: 433–439

Frisch W, Neubauer F (1989) Pre-Alpine terranes and tectonic zoning in the Eastern Alps. Geol Soc Am Spec Pap 230: 91–100

Frisch W, Menot RP, Neubauer F, Raumer J von (1990) Correlation and evolution of the Alpine basement. Schweiz Mineral Petrogr Mitt 70: 265–285

Fritz H, Neubauer F Displacement partitioning on a transpressive plate margin: the southeastern Bohemian Massif. Geology (submitted)

Fuchs G, Matura A (1976) Die Geologie des Kristallins der südlichen Böhmischen Masse. Jahrb Geol Bundesanst 119: 1–43

Gebauer D, Grünenfelder M (1982) Geological development of the Hercynian belt of Europe based on age and origin of high-grade and high-pressure mafic and ultramafic rocks. Abstr 5th Int Conf Geochronology Isotope Geol, Kyoto/Japan, pp 111–112

Gebauer D, Quadt A von, Compston W, Williams I, Grünenfelder M (1988) Archaean zircons in a retrograded, Caledonian eclogite of the Gotthard Massif (Central Alps, Switzerland). Schweiz Mineral Petrogr Mitt 68: 485–490

Gebauer D, Williams IS, Compston W, Grünenfelder M (1989) The development of the central European continental crust since the Early Archean based on conventional and ion-microprobe dating of up to 3.84 b.y. old detrital zircons. Tectonophysics 157: 81–96

Grecula P (1987) Variscan nappes in the tectonic framework of the Gemeric unit, Western Carpathians. In: Flügel HW, Sassi FP, Grecula P (eds) Pre-Variscan and Variscan events in the Alpine-Mediterranean mountain belts. Mineralia slovaca Monogr. Alfa, Bratislava, pp 237–250

Grecula P (1990) Paleozoic lithostratigraphic units of the Gemericum, West Carpathians. Rend Soc Geol Ital 12 (1989): 265–269

Grecula P, Hovorka D (1987) Early Paleozoic volcanism of the Western Carpathians. In: Flügel HW, Sassi FP, Grecula P (eds) Pre-Variscan and Variscan events in the Alpine-Mediterranean mountain belts. Mineralia slovaca Monogr. Alfa, Bratislava, pp 251–270

Grecula P, Vozarova A, Vozar J (1990) Characteristics of lithotectonic profiles in geotraverse C, West Carpathians (Czechoslovakia). Rend Soc Geol Ital 12 (1989): 275–282

Gründenfelder M, Popescu G, Soriou M, Arsenescu S, Berza T (1983) K-Ar and U-Pb dating of metamorphic formations and the associated igneous bodies of the central South Carpathians. Anuarul Institutului de Geologie si Geofizica 61: 37–46

Handler R, Brandmayr M, Wallbrecher E (1991) The Rodl shear zone in the southern Bohemian Massif. Zentralbl Geol Paläontol Teil 1 (1991): 69–86

Harkovska AV (1987) Comparative analysis of the Variscan and Late Alpine orogenic volcanism in Bulgaria. In: Flügel HW, Sassi FP, Grecula P (eds) Pre-Variscan and Variscan events in the Alpine-Mediterranean mountain belts. Mineralia slovaca Monogr. Alfa, Bratislava, pp 271–278

Haydoutov I (1989) Precambrian ophiolites, Cambrian island arc and Variscan suture in the South Carpathian-Balkan region. Geology 17: 905–908

Herbig HG, Stattegger K (1989) Late Palaeozoic heavy minerals and clast modes from the Betic Cordillera (southern Spain): transition from a passive to an active continental margin. Sediment Geol 63: 93–108

Hovorka D (1985) Ultramafic rocks of the Western Carpathians, Czechoslovakia. Geologicky ustav Dionyza Stura, Bratislava, pp 1–258

Hubmann B (1990) Udotaceen (Grünalgen) aus dem Grazer Paläozoikum/Österreich (Barrandeikalke, Eifelium). Facies 22: 147–158

Jacobshagen V (1986) Geologie von Griechenland. Borntraeger, Berlin, 363 pp

Joye JB (1990) L'evolution pression – temperature – deformation dans le massif des Aiguilles Rouges, massif externe alpin. These, Universite Fribourg, 135 pp

Julivert M, Duran H (1990) Paleozoic stratigraphy of the central and northern part of the Catalonian Coastal Ranges (NE Spain). Acta Geol Hisp 25: 3–12

Julivert M, Duran H, Cardellach E (1990) Stratigraphic correlation form of the Paleozoic sequence in the Catalonian Coastal Ranges (north-eastern Mediterranean). Rend Soc Geol Ital 12 (1989): 289–294

Katerinopoulos A, Marcopoulos T (1987) Acid magmatism in the Pelagonian and Rhodope belts (Macedonia – Greece). In: Flügel HW, Sassi FP, Grecula P (eds) Pre-Variscan and Variscan events in the Alpine-Mediterranean mountain belts. Mineralia slovaca Monogr. Alfa, Bratislava, pp 323–328

Keppie JD, Dallmeyer RD (1989) Tectonic map of pre-Mesozoic terranes in circum-Atlantic Phanerozoic orogens. IGCP Proj 233 Terranes. Dept. of Mines and Energy, Halifax

Khain VE (1985) Geology of the USSR Part 1: Old cratons and Paleozoic fold belts. Bornträger, Berlin, 272 pp

Kossmat (1927) Gliederung des varistischen Gebirgsbaues. Abh Sächs Geol Landesamt 1: 1–39

Kovacs S, Lelkes-Felvari G (1990) Stratigraphic correlation forms of Paleozoic units in Hungary. Rend Soc Geol Ital 12 (1989): 295–303

Kräutner HG (1987) The metamorphic Paleozoic of the Romanian Carpathians. In: Flügel HW, Sassi FP, Grecula P (eds) Pre-Variscan and Variscan events in the Alpine-Mediterranean mountain belts. Mineralia slovaca Monogr. Alfa, Bratislava, pp 329–350

Kräutner HG (1989) Similar trends of prealpine geological evolution in the Carpathians and in the Variscan central Europe. Geol Soc Greece Spec Publ 1: 43–58

Kräutner HG (1990) Stratigraphic correlation forms in the east Carpathians. Rend Soc Geol Ital 12 (1989): 305–312

Kröner A, Sengör AMC (1990) Archean and Proterozoic ancestry in Late Precambrian to Early Paleozoic crustal elements of southern Turkey as revealed by single-zircon dating. Geology 18: 1186–1190

Kristic B, Maslarevic L (1990) Depositional environment of the marine Paleozoic in the Hercynid Kucaj zone, eastern Serbia. Bull T. CII Acad Serb Sci Arts Sci Nat 32: 29–37

Ledru P, Lardeaux JM, Santallier D, Autran A, Quenardel JM, Floc'h JP, Lerouge G, Mailett N, Marchand J, Ploquin A (1989) Ou sont les nappes dans le Massif central francais? Bull Soc Geol Fr 1989 (8) V: 605–618

Le Pichon X, Bergerat F, Roulet MJ (1988) Plate kinematics and tectonics leading to the Alpine belt formation; a new analysis. Geol Soc Am Spec Pap 218: 111–131

Liew TC, Hofmann AW (1988) Precambrian crustal components, plutonic associations, plate environment of the Hercynian Fold Belt of central Europe: Indications from a Nd and Sr isotopic study. Contrib Mineral Petrol 98: 129–138

Loeschke J (1989) Lower Palaeozoic volcanism of the Eastern Alps and its geodynamic implications. Geol Rundsch 78: 599–616

Lorenz V, Nicholls IA (1984) Plate and intraplate processes of the Hercynian Europe during the Late Palaeozoic. Tectonophysics 107: 25–56

Martinez-Catalan JR (1990) A non-cylindrical model for the northwest Iberian allochthonous terranes and their equivalents in the Hercynian belt of Western Europe. Tectonophysics 179: 253–272

Matte Ph (1986) Tectonics and plate tectonics model for the Variscan belt of Europe. Tectonophysics 126: 329–374

Matte P, Maluski H, Rajlich P, Franke W (1990) Terrane boundaries in the Bohemian Massif: result of large-scale Variscan shearing. Tectonophysics 177: 151–170

Matter A (1989) Faziesanalyse und Ablagerungsmilieus des Permokarbons im Nordschweizer Trog. Eclogae Geol Helv 80: 345–367

Menot RP, Orsini JP (1990) Evolution du socle antestephanien de Corse: evenements magmatiques et metamorphiques. Schweiz Mineral Petrogr Mitt 70: 35–53

Mercolli I, Oberhänsli R (1988) Variscan tectonic evolution in the Central Alps: a working hypothesis. Schweiz Mineral Petrogr Mitt 68: 491–500

Murphy JB, Nance RD (1989) Model for the evolution of the Avalonian-Cadomian belt. Geology 17: 735–738

Nastaseanu S (1987) Upper Paleozoic molasse deposits in the Romanian South Carpathians. In: Flügel HW, Sassi FP, Grecula P (eds) Pre-Variscan and Variscan events in the Alpine-Mediterranean mountain belts. Mineralia slovaca Monogr. Alfa, Bratislava, pp 371–378

Nastaseanu S, Kräutner HG (1990) Geotraverse D in the South Carpathians: stratigraphic correlation forms. Rend Soc Geol Ital 12 (1989): 339–348

Neubauer F (1988) The Variscan orogeny in the Austroalpine and Southalpine domains of the Eastern Alps. Schweiz Mineral Petrogr Mitt 68: 339–349

Neubauer F (1990) Late Proterozoic and Early Paleozoic tectonothermal evolution of the Eastern Alps. In: Dallmeyer RD, Lecorche JP (eds) The West African orogens and Circum-Atlantic correlatives. Springer, Berlin Heidelberg New York, pp 307–314

Neubauer F, Frisch W (1988) Ordovician-Silurian geodynamic evolution of the Alps – the orogeny-back arc basin model. Schweiz Mineral Petrogr Mitt 68: 351–357

Neubauer F, Herzog U (1985) Das Karbon der Stolzalpendecke Mittelkärntens – Implikationen für die variszische Orogenese im Ostalpin. Anz Österr Akad Wiss Math-Naturwiss Kl 1985: 105–109

Neubauer F, Vozarova A (1990) The Noetsch-Veitsch-North Gemeric zone of Alps and Carpathians: correlation, paleogeography and significance for Variscan orogeny. In: Minarikova D, Lobitzer H (eds) Thirty years of geological cooperation between Austria and Czechoslovakia. Ustredni ustav geologicky, Prague, pp 167–171

Neubauer F, Frisch W, Schmerold R, Schlöser H (1989) Metamorphosed and dismembered ophiolite suites in the basement of the Eastern Alps. Tectonophysics 164: 49–62
Neugebauer J (1988) The Variscan plate tectonic evolution: an improved "Iapetus model". Schweiz Mineral Petrogr Mitt 68: 313–333
Neugebauer J (1990) The Iapetus model: a plate tectonic concept for the Variscan belt of Europe. Tectonophysics 169: 229–256
Oczlon MS (1990) Ocean currents and unconformities: the North Gondwana Middle Devonian. Geology 18: 509–512
Pamic JJ (1989 a) Ultramafics from metamorphic complexes of the Slavonian Mountains (northern Croatia, Yugoslavia). In: Sassi FP, Bourrouilh R (eds) IGCP 5 Newslett 7: 221–227
Pamic J (1989 b) Hercynian and Alpine granitic-metamorphic complexes of the adjoining area of the Dinarides and Pannonian basin in Yugoslavia as related to geodynamics. Geol Carpath 40: 259–280
Papanikolaou DJ (1989) Are the Medial Crystalline Massifs of the eastern Mediterranean drifted Gondwanian fragments? Geol Soc Greece Spec Publ 1: 63–90
Papanikolaou DJ, Demirtasli E (1987) Geological correlation between the Alpine segments of the Hellenides-Balkanides and Taurides-Pontides. In: Flügel HW, Sassi FP, Grecula P (eds) Pre-Variscan and Variscan events in the Alpine-Mediterranean mountain belts. Mineralia slovaca Monogr. Alfa, Bratislava, pp 387–396
Paproth E (1989) Die paläogeographische Entwicklung Mittel-Europas im Karbon. Geol Jahrb Hessen 117: 53–68
Paris F, Robardet M (1990) Early Palaeozoic palaeobiography of the Variscan regions. Tectonophysics 177: 193–213
Perroud H, van der Voo R, Bonhommet N (1984) Paleozoic evolution of the Armorica plate on the basis of paleomagnetic data. Geology 12: 579–582
Pin C (1990) Variscan oceans: ages, origins and geodynamic implications inferred from geochemical and radiometric data. Tectonophysics 177: 215–227
Prost AE, Becq-Giraudon JF (1989) Evidence for mid-Permian compressive tectonics in Western Europe supported by a comparison with the Alleghanian geodynamic evolution. Tectonophysics 169: 333–340
Rajlich P (1990) Strain and tectonic styles related to Variscan transpression and transtension in the Moravo-Silesian Culmian basin, Bohemian Massif, Czechoslovakia. Tectonophysics 174: 351–367
Ramovs A, Sremac J, Kulenovic E (1987) Results of biostratigraphical investigations in western Yugoslav Paleozoic realized in IGCP Project No 5. In: Flügel HW, Sassi FP, Grecula P (eds) Pre-Variscan and Variscan events in the Alpine-Mediterranean mountain belts. Mineralia slovaca Monogr. Alfa, Bratislava, pp 397–406
Ramovs A, Hinterlechner-Ravnik A, Kalenic M, Karamata S, Kochansky-Devide, Krstic B, Kulenovic E, Mirkovic M, Petkovsky P, Sremac J, Temkova V (1990) Stratigraphic correlation forms of the Yugoslav Paleozoic. Rend Soc Geol Ital 12 (1989): 359–383
Rau A, Tongiorgi M (1981) Some problems regarding the Paleozoic paleogeography in Mediterranean Western Europe. J Geol 89: 663–673
Ribeiro A, Quesada C, Dallmeyer RD (1990) Geodynamic evolution of the Iberian Massif. In: Dallmeyer RD, Martinez-Garcia E (eds) Pre-Mesozoic geology of Iberia. Springer, Berlin Heidelberg New York, pp 399–409
Robardet M, Paris F, Reacheboeuf PR (1990) Palaeogeographic evolution of southwestern Europe during Early Palaeozoic times. In: McKerrow WS, Scotese CR (eds) Palaeozoic palaeogeography and biogeography. Geol Soc Mem 12: 411–419
Robertson AHF, Dixon JE (1984) Introduction aspects of the geological evolution of the eastern Mediterranean. In: Dixon JE, Robertson AHF (eds) The geological evolution of the eastern Mediterranean. Geol Soc Spec Publ 17: 1–74
Rudakov SG (1987) Classification of pre-Mesozoic zones of the Carpathian-Balkan region according to the type of evolution during the Late Proterozoic and Paleozoic. In: Flügel HW, Sassi FP, Grecula P (eds) Pre-Variscan and Variscan events in the Alpine-Mediterranean mountain belts. Mineralia slovaca Monogr. Alfa, Bratislava, pp 415–426
Rudakov SG (1990) Pre-Mesozoic history of the Marmarosh Massif, East Carpathians (Soviet part). Rend Soc Geol Ital 12 (1989): 385–389
Santallier D, Briand B, Menot RP, Piboule M (1988) Les complexes leptynoamphiboliques (C. L. A.): revue critique et suggestions pour un meilleur emploi de ce terme. Bull Soc Geol Fr 1988 (8) IV/1: 3–12
Sassi FP, Zanferrari A (eds) (1990) Pre-Variscan and Variscan events in the Alpine-Mediterranean belts. Stratigraphic correlation forms. Rend Soc Geol Ital 12: 95–436
Scharbert S, Vesela M (1990) Rb-Sr systematics of intrusive rocks from the Moldanubicum around Jihlava. In: Minarikova D, Lobitzer H (eds) Thirty years of geological cooperation between Austria and Czechoslovakia. Ustredni ustav geologicky, Prague, pp 262–272
Schenk V (1981) Synchronous uplift of the lower crust of the Ivrea Zone and of southern Calabria and its possible consequences for the Hercynian orogeny in southern Europe. Earth Planet Sci Lett 56: 305–320
Schulmann K (1990) Fabric and kinematic study of the Bites orthogneiss (southwestern Moravia): result of large-scale northeastward shearing parallel to the Moldanubian/Moravian boundary. Tectonophysics 177: 229–244
Schulmann K, Ledru P, Autran A, Melka R, Lardeaux JM, Urban M, Lobkowicz M (1991) Evolution of nappes in the eastern margin of the Bohemian Massif: a kinematic interpretation. Geol Rundsch 80: 73–92
Sengör AMC, Satir M, Akkök R 1984) Timing of tectonic events in the Menderes Massif, western Turkey: implications for tectonic evolution for pan-African basement in Turkey. Tectonics 3: 693–707
Sengör AMC, Altmer D, Cin A, Ustaömer T, Hsü KJ (1988) Origin and assembly of the Tethyside orogenic collage at the expense of Gondwana Land. In: Audley-Charles MG, Hallam A (eds) Gondwana and Tethys. Geol Soc Spec Publ 37: 119–181
Söllner F, Hansen BT (1987) "Pan-afrikanisches" und "kaledonisches" Ereignis im Ötztalkristallin der Ostalpen. Jahrb Geol Bundesanst 130: 529–569
Söllner F, Schmidt K, Baumann A, Hansen BT (1982) Zur Altersstellung des Winnebach-Migmatits im Ötztal (Ostalpen). Verh Geol Bundesanst 1982, pp 95–106
Spalletta C, Vai GB (1990) Stratigraphic correlation forms of the Calabrian-Peloritan arc (southern Italy). Rend Soc Geol Ital 12 (1989): 411–416
Spassov Ch (1987) The Devonian system in Bulgaria. In: Flügel HW, Sassi FP, Grecula P (eds) Pre-Variscan and Variscan Events in the Alpine-Mediterranean mountain belts. Mineralia slovaca Monogr. Alfa, Bratislava, pp 435–444
Stämpfli G, Marcoux J, Bund A (1991) Tethyan margins in space and time. Palaeogeography Palaeochimatology Palaeoecology 87: 373–409
Stan N (1987) Upper Carboniferous and Permian volcanics from Romania. In: Flügel HW, Sassi FP, Grecula P (eds) Pre-

Variscan and Variscan events in the Alpine-Mediterranean mountain belts. Mineralia slovaca Monogr. Alfa, Bratislava, pp 445–456

Stattegger K (1989) Heavy mineral-paths in the Mediterranean Carboniferous and Variscan plate tectonics. IGCP 276, Newslett 7: 252–255

Strachan RA, Brown M, Dallmeyer RD, D'Lemos RS (1991) Tectonothermal history of Cadomian terrane accretion in the north Armorican Massif. Terra Abstr (Suppl) 4: 29–30

Superceanu CI (1985) Die Banater Erzprovinz. Geol Rundsch 74: 417–439

Tenchov YG (1987) Coincidence of the Variscan and Alpine chains in Bulgaria as attested by the distribution of clastic sediments. In: Flügel HW, Sassi FP, Grecula P (eds) Pre-Variscan and Variscan events in the Alpine-Mediterranean mountain belts. Mineralia slovaca Monogr. Alfa, Bratislava, pp 457–467

Tenchov YG (1990) Stratigraphic correlation forms of the Paleozoic in Bulgaria. Rend Soc Geol Ital 12 (1989): 423–433

Tollmann A (1982) Großräumiger variszischer Deckenbau im Moldanubikum und neue Gedanken zum Variszikum Europas. Geotekt Forsch 64: 1–91

Trümpy R (1989) A possible Jurassic-Cretaceous transform system in the Alps and the Carpathians. Geol Soc Am Spec Pap 218: 93–109

Vai GB, Cocozza T (1986) Tentative schematic zonation of the Hercynian chain in Italy. Bull Soc Geol Fr 1986 (8) II/1: 95–114

Vai GB, Spalletta C (1982) Devonian Paleo-Tethyan carbonate platform – basin system of the Southern Alps: evolutionary trend and comparative analysis. Terra Cognita 2: 109–115

von Raumer JF, Menot RP (1989) Evolution paleozoique du socle ouest-alpin et place des Massifs Cristallins Externes dans l'orogene Varisque. C R Acad Sci Paris 309 (II): 397–402

Voshage H, Hofmann AW, Mazzucchelli M, Rivalenti G, Sinigoi S, Raczek I, Demarchi G (1990) Isotopic evidence from the Ivrea Zone for a hybrid lower crust formed by magmatic underplating. Nature 347: 731–736

Vozarova A (1989) Petrology of crystalline rocks of Zemplinicum (West Carpathians). Geol Soc Greece Spec Publ 1: 97–104

Vozarova A, Vozar J (1987) West Carpathians Late Paleozoic and its paleotectonic development. In: Flügel HW, Sassi FP, Grecula P (eds) Pre-Variscan and Variscan events in the Alpine-Mediterranean mountain belts, Mineralia slovaca Monogr. Alfa, Bratislava, pp 469–487

Vozarova A, Vozar J (1988) Late Paleozoic in West Carpathians. Geologicky ustav Dionyza Stura, Bratislava, 314 pp

Weber K (1984) Variscan events: early Palaeozoic continental rift metamorphism and Late Palaeozoic crustal shortening. In: Hutton DHW, Sanderson DJ (eds) Variscan tectonics of the North Atlantic region. Geol Soc Spec Pap 14: 3–22

Wickert F (1988) Paleozoic evolution of continental crust in the Beaujolais-Lyonnais area, northeastern part of the Massif Central, France. Geol Rundsch 77: 467–482

Ziegler PA (1986) Geodynamic model for the Palaeozoic crustal consolidation of western and central Europe. Tectonophysics 126: 303–328

Subject Index

G

H

MIX
Papier aus verantwortungsvollen Quellen
Paper from responsible sources
FSC® C105338

If you have any concerns about our products,
you can contact us on
ProductSafety@springernature.com

In case Publisher is established outside the EU,
the EU authorized representative is:
Springer Nature Customer Service Center GmbH
Europaplatz 3, 69115 Heidelberg, Germany

Printed by Libri Plureos GmbH
in Hamburg, Germany